PEARSON ALWAYS LEARNING

Applied Calculus with Linear Programming

For Business, Economics, Life Sciences, and Social Sciences

Business Calculus I & II, 1425 & 1476

Taken From:

Calculus for Business, Economics, Life Sciences, and Social Sciences, Twelfth Edition
by Raymond A. Barnett, Michael R. Ziegler and Karl E. Byleen

Applied Calculus for Business, Economics, Life Sciences, and Social Sciences, Fifth Edition
by Raymond A. Barnett, Michael R. Ziegler and Karl E. Byleen

Additional Calculus Topics to Accompany *Calculus*, 10e
and *College Mathematics*, 10e, Tenth Edition
by Raymond A. Barnett, Michael R. Ziegler and Karl E. Byleen

College Mathematics for Business, Economics, Life Sciences, and Social Sciences, Tenth Edition
by Raymond Barnett, Michael Ziegler and Karl Byleen

Additional Calculus Topics to Accompany *Calculus*, 12e
and *College Mathematics*, 12e, Twelfth Edition
by Raymond Barnett, Michael Ziegler and Karl Byleen

Cover Art: Courtesy of Photodisc/Getty Images and Brown X Pictures.

Taken from:

Calculus for Business, Economics, Life Sciences, and Social Sciences, Twelfth Edition
by Raymond A. Barnett, Michael R. Ziegler and Karl E. Byleen

Published by Prentice Hall
Upper Saddle River, New Jersey 07458

Applied Calculus for Business, Economics, Life Sciences, and Social Sciences, Fifth Edition
by Raymond A. Barnett, R. Ziegler and Karl E. Byleen

Published by Prentice Hall

Additional Calculus Topics to Accompany *Calculus,* Tenth Edition and *College Mathematics,* Tenth Edition
by Raymond A. Barnett, Michael R. Ziegler, and Karl E. Byleen

Published by Prentice Hall

College Mathematics for Business, Economics, Life Sciences, and Social Sciences, Tenth Edition
By Raymond A. Barnett, Michael R. Ziegler, and Karl E. Byleen

Published by Prentice Hall

Additional Calculus Topics to Accompany *Calculus*, Twelfth Edition and *College Mathematics*, Twelfth Edition
by Raymond Barnett, Michael Ziegler and Karl Byleen

Published by Prentice Hall

Pearson Learning Solutions, 501 Boylston Street, Suite 900, Boston, MA 02116
A Pearson Education Company
www.pearsoned.com

Printed in the United States of America

3 4 5 6 7 8 9 10 V0CR 16 15 14 13 12

000200010270777762

CG

ISBN 10: 1-256-29818-2
ISBN 13: 978-1-256-29818-2

CONTENTS

1) *Calculus for Business, Economics, Life Sciences, and Social Sciences*, Twelfth Edition by Raymond A. Barnett, Michael R. Ziegler and Karl E. Byleen

2) *Applied Calculus for Business, Economics, Life Sciences, and Social Sciences*, Fifth Edition by Raymond A. Barnett, Michael R. Ziegler and Karl E. Byleen

3) *Additional Calculus Topics* to Accompany *Calculus*, 10e and *College Mathematics*, 10e, Tenth Edition by Raymond A. Barnett, Michael R. Ziegler and Karl E. Byleen

4) *College Mathematics for Business, Economics, Life Sciences and Social Sciences*, Tenth Edition by Raymond Barnett, Michael Ziegler and Karl Byleen

5) *Additional Calculus Topics* to Accompany *Calculus*, 12e and *College Mathematics*, 12e, Twelfth Edition by Raymond Barnett, Michael Ziegler and Karl Byleen

Limits and the Derivative

Introduction

How do algebra and calculus differ? The two words *static* and *dynamic* probably come as close as any to expressing the difference between the two disciplines. In algebra, we solve equations for a particular value of a variable—a static notion. In calculus, we are interested in how a change in one variable affects another variable—a dynamic notion.

Isaac Newton (1642–1727) of England and Gottfried Wilhelm von Leibniz (1646–1716) of Germany developed calculus independently to solve problems concerning motion. Today calculus is used not just in the physical sciences, but also in business, economics, life sciences, and social sciences—any discipline that seeks to understand dynamic phenomena.

In Chapter 1 we introduce the *derivative*, one of the two key concepts of calculus. The second, the *integral*, is the subject of Chapter 4. Both key concepts depend on the notion of *limit*, which is explained in Sections 1-1 and 1-2. We consider many applications of limits and derivatives. See, for example, Problems 69 and 70 in Section 1-3 on the concentration of a drug in the bloodstream.

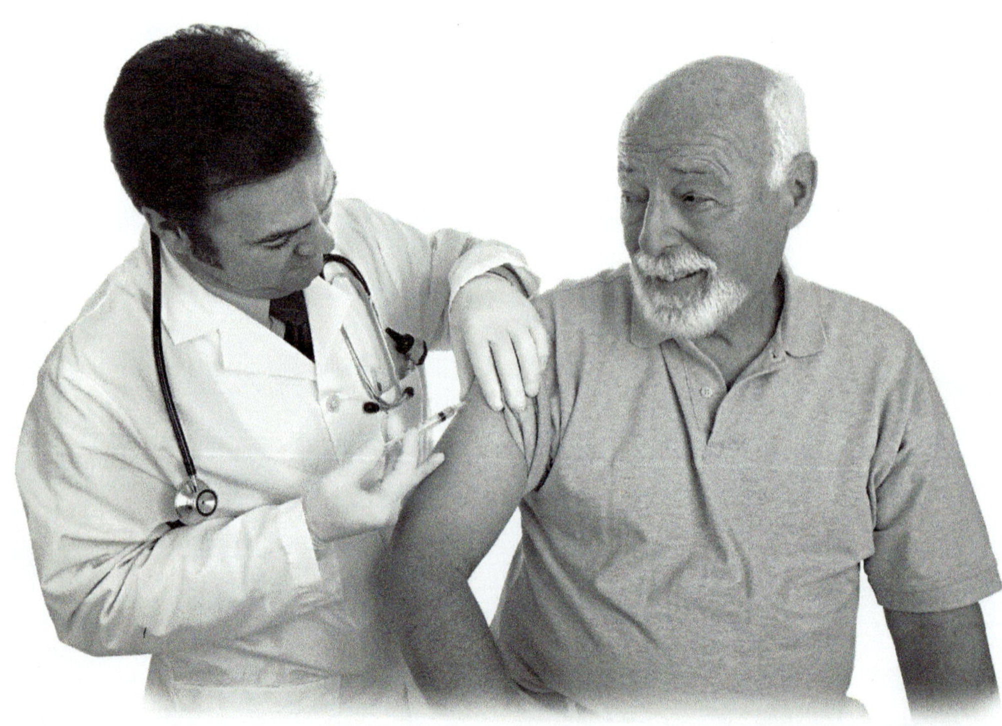

1-1 Introduction to Limits

Basic to the study of calculus is the concept of a *limit*. This concept helps us to describe, in a precise way, the behavior of $f(x)$ when x is close, but not equal, to a particular value c. In this section, we develop an intuitive and informal approach to evaluating limits.

Functions and Graphs: Brief Review

The graph of the function $y = f(x) = x + 2$ is the graph of the set of all ordered pairs $(x, f(x))$. For example, if $x = 2$, then $f(2) = 4$ and $(2, f(2)) = (2, 4)$ is a point on the graph of f. Figure 1 shows $(-1, f(-1))$, $(1, f(1))$, and $(2, f(2))$ plotted on the graph of f. Notice that the domain values -1, 1, and 2 are associated with the x axis and the range values $f(-1) = 1$, $f(1) = 3$, and $f(2) = 4$ are associated with the y axis.

Given x, it is sometimes useful to read $f(x)$ directly from the graph of f. Example 1 reviews this process.

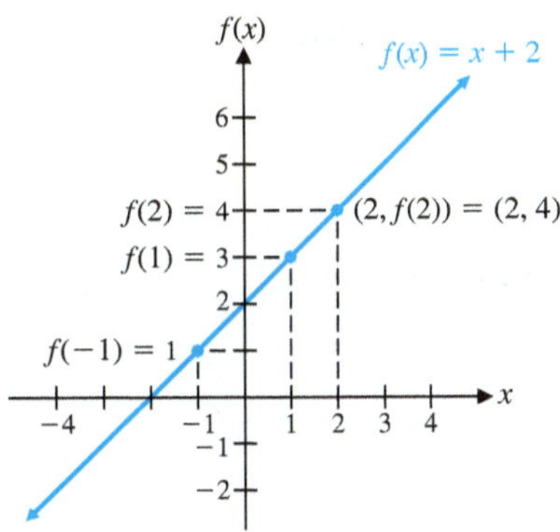

Figure 1

EXAMPLE 1 **Finding Values of a Function from Its Graph** Complete the following table, using the given graph of the function g.

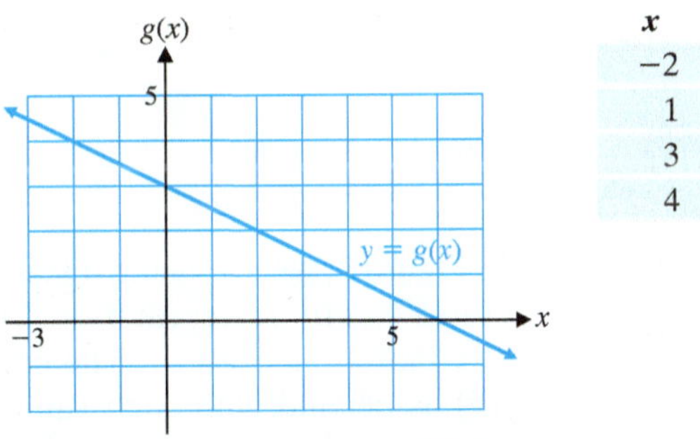

x	$g(x)$
-2	
1	
3	
4	

SOLUTION To determine $g(x)$, proceed vertically from the x value on the x axis to the graph of g and then horizontally to the corresponding y value $g(x)$ on the y axis (as indicated by the dashed lines).

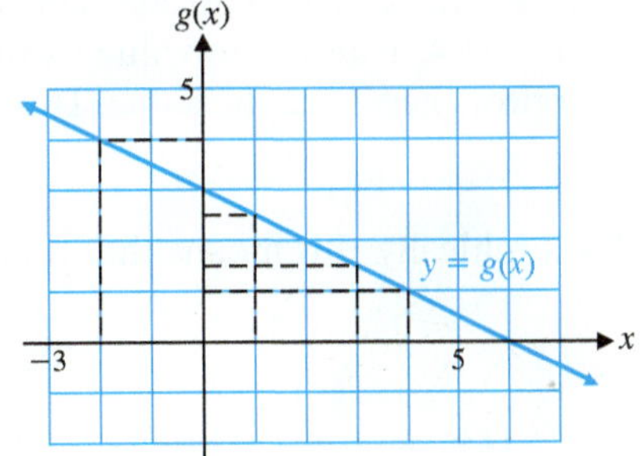

x	$g(x)$
-2	4.0
1	2.5
3	1.5
4	1.0

Matched Problem 1 Complete the following table, using the given graph of the function h.

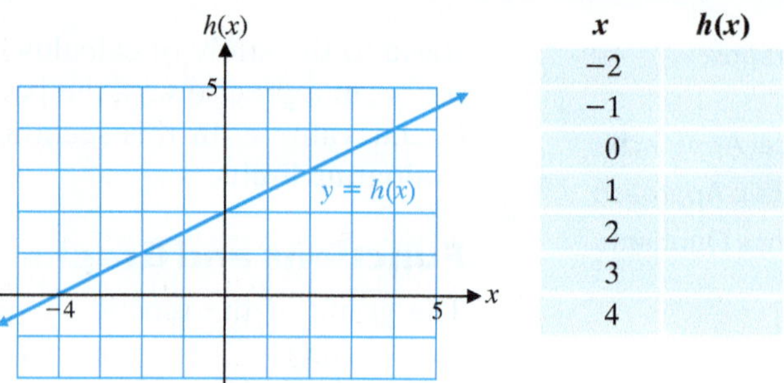

x	$h(x)$
−2	
−1	
0	
1	
2	
3	
4	

Limits: A Graphical Approach

We introduce the important concept of a *limit* through an example, which leads to an intuitive definition of the concept.

EXAMPLE 2 **Analyzing a Limit** Let $f(x) = x + 2$. Discuss the behavior of the values of $f(x)$ when x is close to 2.

SOLUTION We begin by drawing a graph of f that includes the domain value $x = 2$ (Fig. 2).

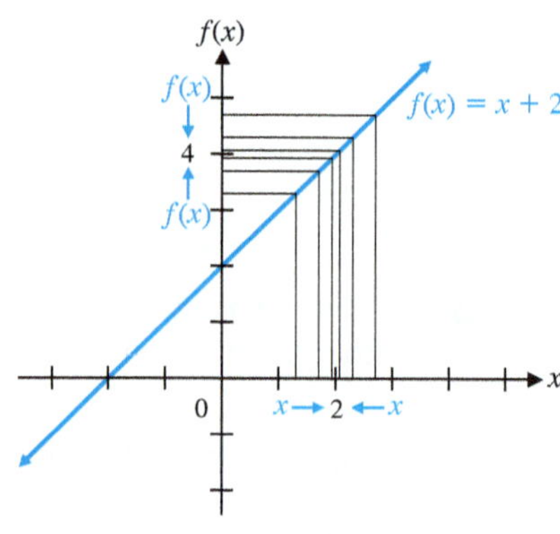

Figure 2

In Figure 2, we are using a static drawing to describe a dynamic process. This requires careful interpretation. The thin vertical lines in Figure 2 represent values of x that are close to 2. The corresponding horizontal lines identify the value of $f(x)$ associated with each value of x. [Example 1 dealt with the relationship between x and $f(x)$ on a graph.] The graph in Figure 2 indicates that as the values of x get closer and closer to 2 on either side of 2, the corresponding values of $f(x)$ get closer and closer to 4. Symbolically, we write

$$\lim_{x \to 2} f(x) = 4$$

This equation is read as "The limit of $f(x)$ as x approaches 2 is 4." Note that $f(2) = 4$. That is, the value of the function at 2 and the limit of the function as x approaches 2 are the same. This relationship can be expressed as

$$\lim_{x \to 2} f(x) = f(2) = 4$$

Graphically, this means that there is no hole or break in the graph of f at $x = 2$.

Matched Problem 2 Let $f(x) = x + 1$. Discuss the behavior of the values of $f(x)$ when x is close to 1.

We now present an informal definition of the important concept of a limit. A precise definition is not needed for our discussion, but one is given in a footnote.*

DEFINITION Limit

We write

$$\lim_{x \to c} f(x) = L \quad \text{or} \quad f(x) \to L \quad \text{as} \quad x \to c$$

if the functional value $f(x)$ is close to the single real number L whenever x is close, but not equal, to c (on either side of c).

Note: The existence of a limit at c has nothing to do with the value of the function at c. In fact, c may not even be in the domain of f. However, the function must be defined on both sides of c.

The next example involves the **absolute value function**:

$$f(x) = |x| = \begin{cases} -x & \text{if } x < 0 \\ x & \text{if } x \geq 0 \end{cases}$$

$f(-2) = |-2| = -(-2) = 2$

$f(3) = |3| = 3$

The graph of f is shown in Figure 3.

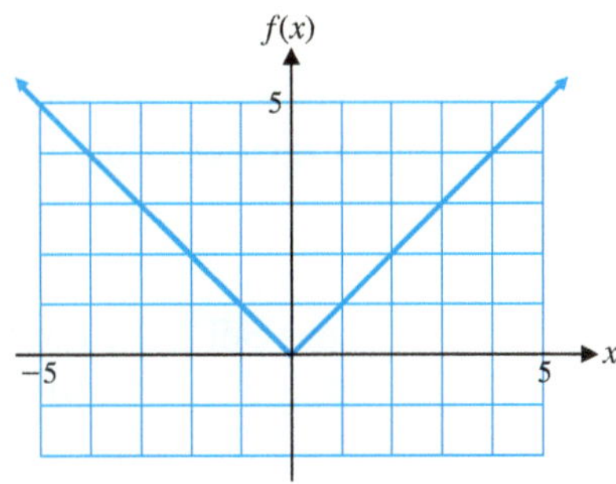

Figure 3 $f(x) = |x|$

EXAMPLE 3 **Analyzing a Limit** Let $h(x) = |x|/x$. Explore the behavior of $h(x)$ for x near, but not equal, to 0. Find $\lim_{x \to 0} h(x)$ if it exists.

SOLUTION The function h is defined for all real numbers except 0. For example,

$$h(-2) = \frac{|-2|}{-2} = \frac{2}{-2} = -1$$

$$h(0) = \frac{|0|}{0} = \frac{0}{0} \qquad \text{Not defined}$$

$$h(2) = \frac{|2|}{2} = \frac{2}{2} = 1$$

In general, $h(x)$ is -1 for all negative x and 1 for all positive x. Figure 4 illustrates the behavior of $h(x)$ for x near 0. Note that the absence of a solid dot on the vertical axis indicates that h is not defined when $x = 0$.

When x is near 0 (on either side of 0), is $h(x)$ near one specific number? The answer is "No," because $h(x)$ is -1 for $x < 0$ and 1 for $x > 0$. Consequently, we say that

$$\lim_{x \to 0} \frac{|x|}{x} \text{ does not exist}$$

Figure 4

Neither $h(x)$ nor the limit of $h(x)$ exists at $x = 0$. However, the limit from the left and the limit from the right both exist at 0, but they are not equal.

*To make the informal definition of *limit* precise, we must make the word *close* more precise. This is done as follows: We write $\lim_{x \to c} f(x) = L$ if, for each $e > 0$, there exists a $d > 0$ such that $|f(x) - L| < e$ whenever $0 < |x - c| < d$. This definition is used to establish particular limits and to prove many useful properties of limits that will be helpful in finding particular limits.

Matched Problem 3 Graph

$$h(x) = \frac{x - 2}{|x - 2|}$$

and find $\lim_{x \to 2} h(x)$ if it exists.

In Example 3, we see that the values of the function $h(x)$ approach two different numbers, depending on the direction of approach, and it is natural to refer to these values as "the limit from the left" and "the limit from the right." These experiences suggest that the notion of **one-sided limits** will be very useful in discussing basic limit concepts.

DEFINITION One-Sided Limits

We write

$$\lim_{x \to c^-} f(x) = K$$ $x \to c^-$ is read "x approaches c from the left" and means $x \to c$ and $x < c$.

and call K the **limit from the left** or the **left-hand limit** if $f(x)$ is close to K whenever x is close to, but to the left of, c on the real number line. We write

$$\lim_{x \to c^+} f(x) = L$$ $x \to c^+$ is read "x approaches c from the right" and means $x \to c$ and $x > c$.

and call L the **limit from the right** or the **right-hand limit** if $f(x)$ is close to L whenever x is close to, but to the right of, c on the real number line.

If no direction is specified in a limit statement, we will always assume that the limit is **two-sided** or **unrestricted**. Theorem 1 states an important relationship between one-sided limits and unrestricted limits.

THEOREM 1 On the Existence of a Limit

For a (two-sided) limit to exist, the limit from the left and the limit from the right must exist and be equal. That is,

$$\lim_{x \to c} f(x) = L \text{ if and only if } \lim_{x \to c^-} f(x) = \lim_{x \to c^+} f(x) = L$$

In Example 3,

$$\lim_{x \to 0^-} \frac{|x|}{x} = -1 \qquad \text{and} \qquad \lim_{x \to 0^+} \frac{|x|}{x} = 1$$

Since the left- and right-hand limits are *not* the same,

$$\lim_{x \to 0} \frac{|x|}{x} \text{ does not exist}$$

EXAMPLE 4 **Analyzing Limits Graphically** Given the graph of the function f in Figure 5, discuss the behavior of $f(x)$ for x near (A) -1, (B) 1, and (C) 2.

SOLUTION (A) Since we have only a graph to work with, we use vertical and horizontal lines to relate the values of x and the corresponding values of $f(x)$. For any x near -1 on either side of -1, we see that the corresponding value of $f(x)$, determined by a horizontal line, is close to 1.

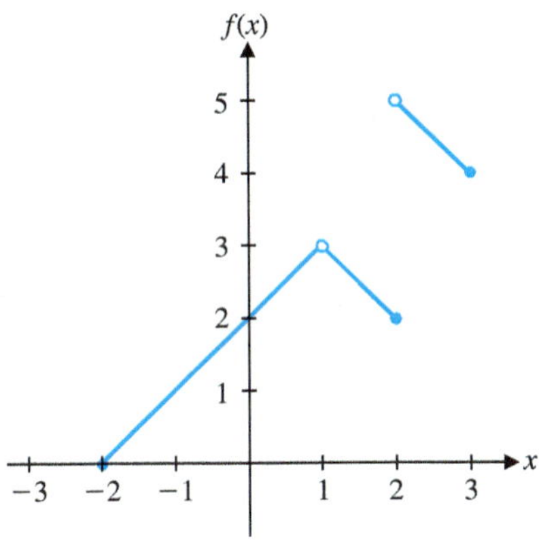

Figure 5

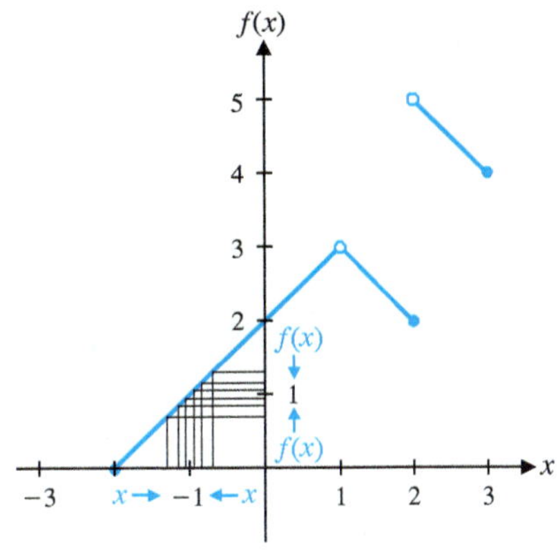

$$\lim_{x \to -1^-} f(x) = 1$$
$$\lim_{x \to -1^+} f(x) = 1$$
$$\lim_{x \to -1} f(x) = 1$$
$$f(-1) = 1$$

(B) Again, for any x near, but not equal to, 1, the vertical and horizontal lines indicate that the corresponding value of $f(x)$ is close to 3. The open dot at $(1, 3)$, together with the absence of a solid dot anywhere on the vertical line through $x = 1$, indicates that $f(1)$ is not defined.

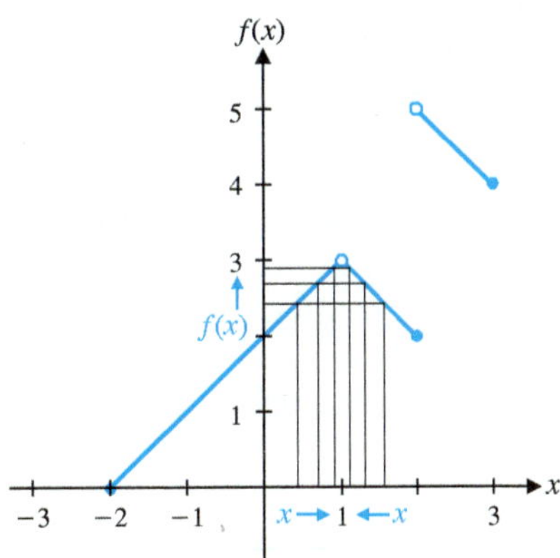

$$\lim_{x \to 1^-} f(x) = 3$$
$$\lim_{x \to 1^+} f(x) = 3$$
$$\lim_{x \to 1} f(x) = 3$$
$f(1)$ not defined

(C) The abrupt break in the graph at $x = 2$ indicates that the behavior of the graph near $x = 2$ is more complicated than in the two preceding cases. If x is close to 2 on the left side of 2, the corresponding horizontal line intersects the y axis at a point close to 2. If x is close to 2 on the right side of 2, the corresponding horizontal line intersects the y axis at a point close to 5. This is a case where the one-sided limits are different.

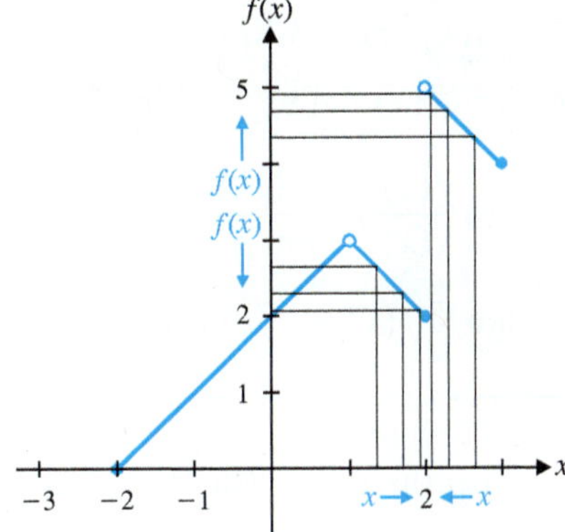

$$\lim_{x \to 2^-} f(x) = 2$$
$$\lim_{x \to 2^+} f(x) = 5$$
$\lim_{x \to 2} f(x)$ does not exist
$$f(2) = 2$$

Matched Problem 4 Given the graph of the function f shown in Figure 6, discuss the following, as we did in Example 4:

(A) Behavior of $f(x)$ for x near 0

(B) Behavior of $f(x)$ for x near 1

(C) Behavior of $f(x)$ for x near 3

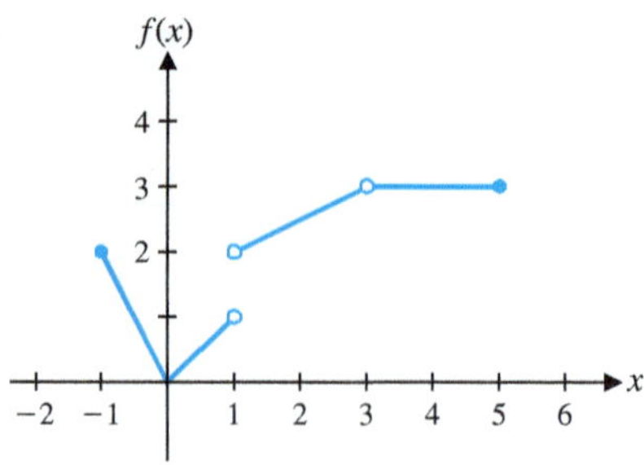

Figure 6

CONCEPTUAL INSIGHT

In Example 4B, note that $\lim_{x \to 1} f(x)$ exists even though f is not defined at $x = 1$ and the graph has a hole at $x = 1$. In general, the value of a function at $x = c$ has no effect on the limit of the function as x approaches c.

Limits: An Algebraic Approach

Graphs are very useful tools for investigating limits, especially if something unusual happens at the point in question. However, many of the limits encountered in calculus are routine and can be evaluated quickly with a little algebraic simplification, some intuition, and basic properties of limits. The following list of properties of limits forms the basis for this approach:

THEOREM 2 Properties of Limits

Let f and g be two functions, and assume that

$$\lim_{x \to c} f(x) = L \qquad \lim_{x \to c} g(x) = M$$

where L and M are real numbers (both limits exist). Then

1. $\lim_{x \to c} k = k$ for any constant k
2. $\lim_{x \to c} x = c$
3. $\lim_{x \to c}[f(x) + g(x)] = \lim_{x \to c} f(x) + \lim_{x \to c} g(x) = L + M$
4. $\lim_{x \to c}[f(x) - g(x)] = \lim_{x \to c} f(x) - \lim_{x \to c} g(x) = L - M$
5. $\lim_{x \to c} kf(x) = k \lim_{x \to c} f(x) = kL$ for any constant k
6. $\lim_{x \to c}[f(x) \cdot g(x)] = [\lim_{x \to c} f(x)][\lim_{x \to c} g(x)] = LM$
7. $\lim_{x \to c} \dfrac{f(x)}{g(x)} = \dfrac{\lim_{x \to c} f(x)}{\lim_{x \to c} g(x)} = \dfrac{L}{M}$ if $M \neq 0$
8. $\lim_{x \to c} \sqrt[n]{f(x)} = \sqrt[n]{\lim_{x \to c} f(x)} = \sqrt[n]{L}$ $L > 0$ for n even

Each property in Theorem 2 is also valid if $x \to c$ is replaced everywhere by $x \to c^-$ or replaced everywhere by $x \to c^+$.

EXPLORE & DISCUSS 1

The properties listed in Theorem 2 can be paraphrased in brief verbal statements. For example, property 3 simply states that *the limit of a sum is equal to the sum of the limits.* Write brief verbal statements for the remaining properties in Theorem 2.

EXAMPLE 5 **Using Limit Properties** Find $\lim_{x \to 3}(x^2 - 4x)$.

SOLUTION

$$\lim_{x \to 3}(x^2 - 4x) = \lim_{x \to 3} x^2 - \lim_{x \to 3} 4x \qquad \text{Property 4}$$

$$= \left(\lim_{x \to 3} x\right) \cdot \left(\lim_{x \to 3} x\right) - 4 \lim_{x \to 3} x \qquad \text{Properties 5 and 6}$$

$$= 3 \cdot 3 - 4 \cdot 3 = -3 \qquad \text{Property 2}$$

With a little practice, you will soon be able to omit the steps in the dashed boxes and simply write

$$\lim_{x \to 3}(x^2 - 4x) = 3 \cdot 3 - 4 \cdot 3 = -3$$

Matched Problem 5 Find $\lim_{x \to -2}(x^2 + 5x)$.

What happens if we try to evaluate a limit like the one in Example 5, but with x approaching an unspecified number, such as c? Proceeding as we did in Example 5, we have

$$\lim_{x \to c}(x^2 - 4x) = c \cdot c - 4 \cdot c = c^2 - 4c$$

If we let $f(x) = x^2 - 4x$, we have

$$\lim_{x \to c} f(x) = \lim_{x \to c}(x^2 - 4x) = c^2 - 4c = f(c)$$

That is, this limit can be evaluated simply by evaluating the function f at c. It would certainly simplify the process of evaluating limits if we could identify the functions for which

$$\lim_{x \to c} f(x) = f(c) \qquad (1)$$

since we could use this fact to evaluate the limit. It turns out that there are many functions that satisfy equation (1). We postpone a detailed discussion of these functions until the next section. For now, we note that if

$$f(x) = a_n x^n + a_{n-1}x^{n-1} + \cdots + a_0$$

is a polynomial function, then, by the properties in Theorem 1,

$$\begin{aligned}\lim_{x \to c} f(x) &= \lim_{x \to c}(a_n x^n + a_{n-1}x^{n-1} + \cdots + a_0)\\ &= a_n c^n + a_{n-1}c^{n-1} + \cdots + a_0 = f(c)\end{aligned}$$

and if

$$r(x) = \frac{n(x)}{d(x)}$$

is a rational function, where $n(x)$ and $d(x)$ are polynomials with $d(c) \neq 0$, then by property 7 and the fact that polynomials $n(x)$ and $d(x)$ satisfy equation (1),

$$\lim_{x \to c} r(x) = \lim_{x \to c} \frac{n(x)}{d(x)} = \frac{\lim_{x \to c} n(x)}{\lim_{x \to c} d(x)} = \frac{n(c)}{d(c)} = r(c)$$

These results are summarized in Theorem 3.

THEOREM 3 Limits of Polynomial and Rational Functions

1. $\lim_{x \to c} f(x) = f(c)$ for f any polynomial function.
2. $\lim_{x \to c} r(x) = r(c)$ for r any rational function with a nonzero denominator at $x = c$.

EXAMPLE 6 **Evaluating Limits** Find each limit.

(A) $\lim_{x \to 2}(x^3 - 5x - 1)$ (B) $\lim_{x \to -1} \sqrt{2x^2 + 3}$ (C) $\lim_{x \to 4} \frac{2x}{3x + 1}$

SOLUTION (A) $\lim_{x \to 2}(x^3 - 5x - 1) = 2^3 - 5 \cdot 2 - 1 = -3$ Theorem 3

(B) $\lim_{x \to -1} \sqrt{2x^2 + 3} = \sqrt{\lim_{x \to -1}(2x^2 + 3)}$ Property 8

$= \sqrt{2(-1)^2 + 3}$ Theorem 3

$= \sqrt{5}$

(C) $\lim_{x \to 4} \frac{2x}{3x + 1} = \frac{2 \cdot 4}{3 \cdot 4 + 1}$ Theorem 3

$= \frac{8}{13}$

Matched Problem 6 Find each limit.

(A) $\lim_{x \to -1}(x^4 - 2x + 3)$ (B) $\lim_{x \to 2} \sqrt{3x^2 - 6}$ (C) $\lim_{x \to -2} \frac{x^2}{x^2 + 1}$

EXAMPLE 7 **Evaluating Limits** Let

$$f(x) = \begin{cases} x^2 + 1 & \text{if } x < 2 \\ x - 1 & \text{if } x > 2 \end{cases}$$

Find each limit.

(A) $\lim_{x \to 2^-} f(x)$ (B) $\lim_{x \to 2^+} f(x)$ (C) $\lim_{x \to 2} f(x)$ (D) $f(2)$

SOLUTION (A) $\lim_{x \to 2^-} f(x) = \lim_{x \to 2^-}(x^2 + 1)$ If $x < 2$, $f(x) = x^2 + 1$.

$= 2^2 + 1 = 5$

(B) $\lim_{x \to 2^+} f(x) = \lim_{x \to 2^+}(x - 1)$ If $x > 2$, $f(x) = x - 1$.

$= 2 - 1 = 1$

(C) Since the one-sided limits are not equal, $\lim_{x \to 2} f(x)$ does not exist.

(D) Because the definition of f does not assign a value to f for $x = 2$, only for $x < 2$ and $x > 2$, $f(2)$ does not exist.

Matched Problem 7 Let

$$f(x) = \begin{cases} 2x + 3 & \text{if } x < 5 \\ -x + 12 & \text{if } x > 5 \end{cases}$$

Find each limit.

(A) $\lim_{x \to 5^-} f(x)$ (B) $\lim_{x \to 5^+} f(x)$ (C) $\lim_{x \to 5} f(x)$ (D) $f(5)$

It is important to note that there are restrictions on some of the limit properties. In particular, if

$$\lim_{x \to c} f(x) = 0 \quad \text{and} \quad \lim_{x \to c} g(x) = 0, \quad \text{then finding } \lim_{x \to c} \frac{f(x)}{g(x)}$$

may present some difficulties, since limit property 7 (the limit of a quotient) does not apply when $\lim_{x \to c} g(x) = 0$. The next example illustrates some techniques that can be useful in this situation.

EXAMPLE 8 **Evaluating Limits** Find each limit.

(A) $\lim_{x \to 2} \frac{x^2 - 4}{x - 2}$ (B) $\lim_{x \to -1} \frac{x|x + 1|}{x + 1}$

SOLUTION (A) Algebraic simplification is often useful when the numerator and denominator are both approaching 0.

$$\lim_{x \to 2} \frac{x^2 - 4}{x - 2} = \lim_{x \to 2} \frac{(x - 2)(x + 2)}{x - 2} = \lim_{x \to 2}(x + 2) = 4$$

(B) One-sided limits are helpful for limits involving the absolute value function.

$$\lim_{x \to -1^+} \frac{x|x + 1|}{x + 1} = \lim_{x \to -1^+}(x) = -1 \quad \text{If } x > -1, \text{ then } \frac{|x + 1|}{x + 1} = 1.$$

$$\lim_{x \to -1^-} \frac{x|x + 1|}{x + 1} = \lim_{x \to -1^-}(-x) = 1 \quad \text{If } x < -1, \text{ then } \frac{|x + 1|}{x + 1} = -1.$$

Since the limit from the left and the limit from the right are not the same, we conclude that

$$\lim_{x \to -1} \frac{x|x + 1|}{x + 1} \quad \text{does not exist}$$

Matched Problem 8 Find each limit.

(A) $\lim_{x \to -3} \frac{x^2 + 4x + 3}{x + 3}$ (B) $\lim_{x \to 4} \frac{x^2 - 16}{|x - 4|}$

CONCEPTUAL INSIGHT

In the solution to Example 8A we used the following algebraic identity:

$$\frac{x^2 - 4}{x - 2} = \frac{(x - 2)(x + 2)}{x - 2} = x + 2, \qquad x \neq 2$$

The restriction $x \neq 2$ is necessary here because the first two expressions are not defined at $x = 2$. Why didn't we include this restriction in the solution? When x approaches 2 in a limit problem, it is assumed that x is close, but not equal, to 2. It is important that you understand that both of the following statements are valid:

$$\lim_{x \to 2} \frac{x^2 - 4}{x - 2} = \lim_{x \to 2}(x + 2) \quad \text{and} \quad \frac{x^2 - 4}{x - 2} = x + 2, \; x \neq 2$$

Limits like those in Example 8 occur so frequently in calculus that they are given a special name.

DEFINITION Indeterminate Form

If $\lim_{x \to c} f(x) = 0$ and $\lim_{x \to c} g(x) = 0$, then $\lim_{x \to c} \frac{f(x)}{g(x)}$ is said to be **indeterminate**, or, more specifically, a **0/0 indeterminate form.**

The term *indeterminate* is used because the limit of an indeterminate form may or may not exist (see Example 8A and 8B).

CAUTION The expression 0/0 does not represent a real number and should never be used as the value of a limit. If a limit is a 0/0 indeterminate form, further investigation is always required to determine whether the limit exists and to find its value if it does exist.

If the denominator of a quotient approaches 0 and the numerator approaches a nonzero number, then the limit of the quotient is not an indeterminate form. In fact, a limit of this form never exists.

THEOREM 4 Limit of a Quotient

If $\lim_{x \to c} f(x) = L, L \neq 0$, and $\lim_{x \to c} g(x) = 0$,

then

$$\lim_{x \to c} \frac{f(x)}{g(x)} \quad \text{does not exist}$$

EXPLORE & DISCUSS 2

Use algebraic and/or graphical techniques to analyze each of the following indeterminate forms.

(A) $\lim_{x \to 1} \frac{x - 1}{x^2 - 1}$ (B) $\lim_{x \to 1} \frac{(x - 1)^2}{x^2 - 1}$ (C) $\lim_{x \to 1} \frac{x^2 - 1}{(x - 1)^2}$

Limits of Difference Quotients

Let the function f be defined in an open interval containing the number a. One of the most important limits in calculus is the limit of the **difference quotient**,

$$\lim_{h \to 0} \frac{f(a + h) - f(a)}{h} \tag{2}$$

If

$$\lim_{h \to 0} [f(a + h) - f(a)] = 0$$

as it often does, then limit (2) is an indeterminate form.

EXAMPLE 9 **Limit of a Difference Quotient** Find the following limit for $f(x) = 4x - 5$:

$$\lim_{h \to 0} \frac{f(3 + h) - f(3)}{h}$$

SOLUTION

$$\begin{aligned} \lim_{h \to 0} \frac{f(3 + h) - f(3)}{h} &= \lim_{h \to 0} \frac{[4(\mathbf{3} + \boldsymbol{h}) - 5] - [4(\mathbf{3}) - 5]}{h} \\ &= \lim_{h \to 0} \frac{12 + 4h - 5 - 12 + 5}{h} \\ &= \lim_{h \to 0} \frac{4h}{h} = \lim_{h \to 0} 4 = 4 \end{aligned}$$

Since this is a 0/0 indeterminate form and property 7 in Theorem 2 does not apply, we proceed with algebraic simplification.

Matched Problem 9 Find the following limit for $f(x) = 7 - 2x$: $\lim_{h\to 0} \frac{f(4+h) - f(4)}{h}$.

EXAMPLE 10 **Limit of a Difference Quotient** Find the following limit for $f(x) = |x + 5|$:

$$\lim_{h\to 0} \frac{f(-5+h) - f(-5)}{h}$$

SOLUTION

$$\lim_{h\to 0} \frac{f(-5+h) - f(-5)}{h} = \lim_{h\to 0} \frac{|(-5+h) + 5| - |-5 + 5|}{h}$$

$$= \lim_{h\to 0} \frac{|h|}{h} \text{ does not exist}$$

Since this is a 0/0 indeterminate form and property 7 in Theorem 2 does not apply, we proceed with algebraic simplification.

Matched Problem 10 Find the following limit for $f(x) = |x - 1|$: $\lim_{h\to 0} \frac{f(1+h) - f(1)}{h}$.

EXAMPLE 11 **Limit of a Difference Quotient** Find the following limit for $f(x) = \sqrt{x}$:

$$\lim_{h\to 0} \frac{f(2+h) - f(2)}{h}$$

SOLUTION

$$\lim_{h\to 0} \frac{f(2+h) - f(2)}{h} = \lim_{h\to 0} \frac{\sqrt{2+h} - \sqrt{2}}{h}$$

$$= \lim_{h\to 0} \frac{\sqrt{2+h} - \sqrt{2}}{h} \cdot \frac{\sqrt{2+h} + \sqrt{2}}{\sqrt{2+h} + \sqrt{2}}$$

$$= \lim_{h\to 0} \frac{2 + h - 2}{h(\sqrt{2+h} + \sqrt{2})}$$

$$= \lim_{h\to 0} \frac{1}{\sqrt{2+h} + \sqrt{2}}$$

$$= \frac{1}{\sqrt{2} + \sqrt{2}} = \frac{1}{2\sqrt{2}}$$

This is a 0/0 indeterminate form, so property 7 in Theorem 2 does not apply. Rationalizing the numerator will help.

$(A - B)(A + B) = A^2 - B^2$

Matched Problem 11 Find the following limit for $f(x) = \sqrt{x}$: $\lim_{h\to 0} \frac{f(3+h) - f(3)}{h}$.

Exercises 1-1

A

In Problems 1–8, use the graph of the function f shown to estimate the indicated limits and function values.

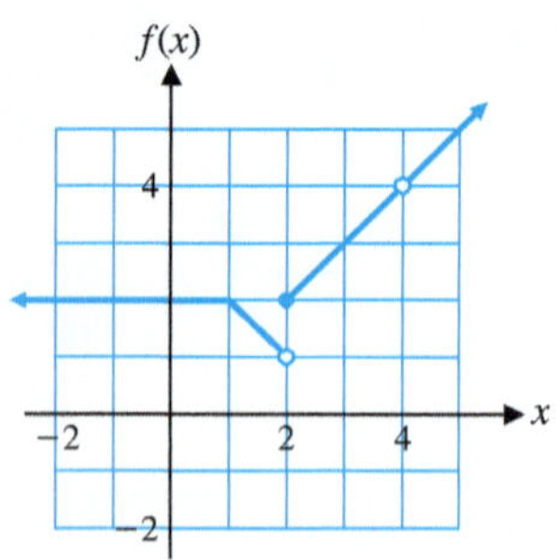

Figure for 1–8

1. $f(-0.5)$ **2.** $f(-1.5)$

3. $f(1.75)$ **4.** $f(1.25)$

5. (A) $\lim_{x\to 0^-} f(x)$ (B) $\lim_{x\to 0^+} f(x)$
(C) $\lim_{x\to 0} f(x)$ (D) $f(0)$

6. (A) $\lim_{x\to 1^-} f(x)$ (B) $\lim_{x\to 1^+} f(x)$
(C) $\lim_{x\to 1} f(x)$ (D) $f(1)$

7. (A) $\lim_{x\to 2^-} f(x)$ (B) $\lim_{x\to 2^+} f(x)$
(C) $\lim_{x\to 2} f(x)$ (D) $f(2)$
(E) Is it possible to redefine $f(2)$ so that $\lim_{x\to 2} f(x) = f(2)$? Explain.

8. (A) $\lim_{x\to 4^-} f(x)$ (B) $\lim_{x\to 4^+} f(x)$
(C) $\lim_{x\to 4} f(x)$ (D) $f(4)$
(E) Is it possible to define $f(4)$ so that $\lim_{x\to 4} f(x) = f(4)$? Explain.

In Problems 9–16, use the graph of the function g shown to estimate the indicated limits and function values.

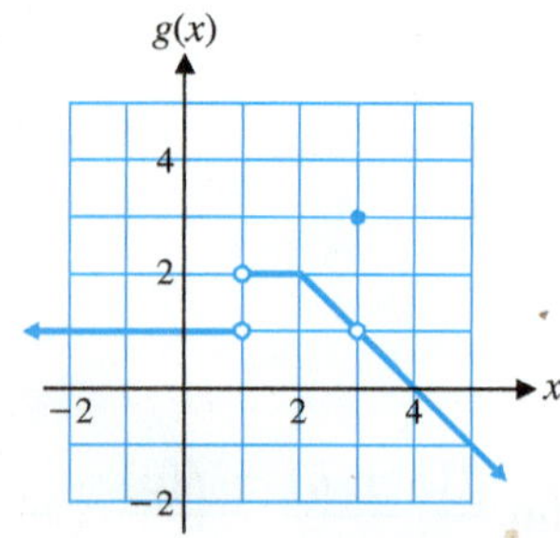

Figure for 9–16

9. $g(1.9)$ **10.** $g(0.1)$

11. $g(3.5)$ **12.** $g(2.5)$

13. (A) $\lim_{x\to 1^-} g(x)$ (B) $\lim_{x\to 1^+} g(x)$
(C) $\lim_{x\to 1} g(x)$ (D) $g(1)$
(E) Is it possible to define $g(1)$ so that $\lim_{x\to 1} g(x) = g(1)$? Explain.

14. (A) $\lim_{x\to 2^-} g(x)$ (B) $\lim_{x\to 2^+} g(x)$
(C) $\lim_{x\to 2} g(x)$ (D) $g(2)$

15. (A) $\lim_{x\to 3^-} g(x)$ (B) $\lim_{x\to 3^+} g(x)$
(C) $\lim_{x\to 3} g(x)$ (D) $g(3)$
(E) Is it possible to redefine $g(3)$ so that $\lim_{x\to 3} g(x) = g(3)$? Explain.

16. (A) $\lim_{x\to 4^-} g(x)$ (B) $\lim_{x\to 4^+} g(x)$
(C) $\lim_{x\to 4} g(x)$ (D) $g(4)$

In Problems 17–20, use the graph of the function f shown to estimate the indicated limits and function values.

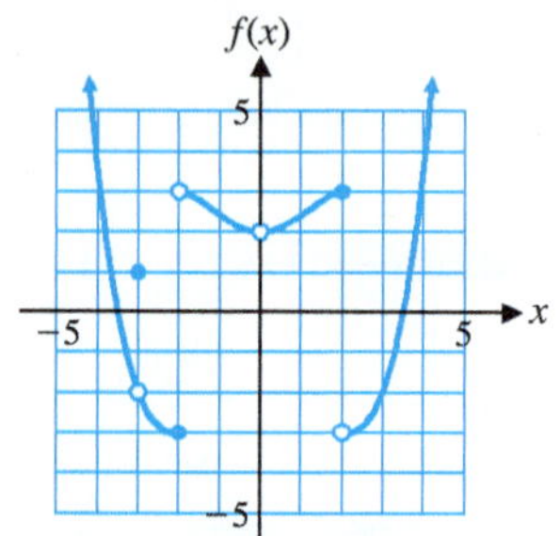

Figure for 17–20

17. (A) $\lim_{x\to -3^+} f(x)$ (B) $\lim_{x\to -3^-} f(x)$
(C) $\lim_{x\to -3} f(x)$ (D) $f(-3)$
(E) Is it possible to redefine $f(-3)$ so that $\lim_{x\to -3} f(x) = f(-3)$? Explain.

18. (A) $\lim_{x\to -2^+} f(x)$ (B) $\lim_{x\to -2^-} f(x)$
(C) $\lim_{x\to -2} f(x)$ (D) $f(-2)$
(E) Is it possible to define $f(-2)$ so that $\lim_{x\to -2} f(x) = f(-2)$? Explain.

19. (A) $\lim_{x\to 0^+} f(x)$ (B) $\lim_{x\to 0^-} f(x)$
(C) $\lim_{x\to 0} f(x)$ (D) $f(0)$
(E) Is it possible to redefine $f(0)$ so that $\lim_{x\to 0} f(x) = f(0)$? Explain.

20. (A) $\lim_{x\to 2^+} f(x)$ (B) $\lim_{x\to 2^-} f(x)$
(C) $\lim_{x\to 2} f(x)$ (D) $f(2)$
(E) Is it possible to redefine $f(2)$ so that $\lim_{x\to 2} f(x) = f(2)$? Explain.

In Problems 21–30, find each limit if it exists.

21. $\lim_{x\to 3} 4x$ **22.** $\lim_{x\to -2} 3x$

23. $\lim_{x\to -4} (x + 5)$ **24.** $\lim_{x\to 5} (x - 3)$

25. $\lim_{x\to 2} x(x - 4)$ **26.** $\lim_{x\to -1} x(x + 3)$

27. $\lim_{x\to -3} \frac{x}{x + 5}$ **28.** $\lim_{x\to 4} \frac{x - 2}{x}$

29. $\lim_{x\to 1} \sqrt{5x + 4}$ **30.** $\lim_{x\to 0} \sqrt{16 - 7x}$

Given that $\lim_{x\to 1} f(x) = -5$ *and* $\lim_{x\to 1} g(x) = 4$, *find the indicated limits in Problems 31–38.*

31. $\lim_{x\to 1}(-3)f(x)$

32. $\lim_{x\to 1} 2g(x)$

33. $\lim_{x\to 1}[2f(x) + g(x)]$

34. $\lim_{x\to 1}[g(x) - 3f(x)]$

35. $\lim_{x\to 1}\dfrac{2 - f(x)}{x + g(x)}$

36. $\lim_{x\to 1}\dfrac{3 - f(x)}{1 - 4g(x)}$

37. $\lim_{x\to 1}\sqrt{g(x) - f(x)}$

38. $\lim_{x\to 1}\sqrt[3]{2x + 2f(x)}$

In Problems 39–42, sketch a possible graph of a function that satisfies the given conditions.

39. $f(0) = 1;\ \lim_{x\to 0^-} f(x) = 3;\ \lim_{x\to 0^+} f(x) = 1$

40. $f(1) = -2;\ \lim_{x\to 1^-} f(x) = 2;\ \lim_{x\to 1^+} f(x) = -2$

41. $f(-2) = 2;\ \lim_{x\to -2^-} f(x) = 1;\ \lim_{x\to -2^+} f(x) = 1$

42. $f(0) = -1;\ \lim_{x\to 0^-} f(x) = 2;\ \lim_{x\to 0^+} f(x) = 2$

B

In Problems 43–58, find each indicated quantity if it exists.

43. Let $f(x) = \begin{cases} 1 - x^2 & \text{if } x \le 0 \\ 1 + x^2 & \text{if } x > 0 \end{cases}$. Find

(A) $\lim_{x\to 0^+} f(x)$ (B) $\lim_{x\to 0^-} f(x)$
(C) $\lim_{x\to 0} f(x)$ (D) $f(0)$

44. Let $f(x) = \begin{cases} 2 + x & \text{if } x \le 0 \\ 2 - x & \text{if } x > 0 \end{cases}$. Find

(A) $\lim_{x\to 0^+} f(x)$ (B) $\lim_{x\to 0^-} f(x)$
(C) $\lim_{x\to 0} f(x)$ (D) $f(0)$

45. Let $f(x) = \begin{cases} x^2 & \text{if } x < 1 \\ 2x & \text{if } x > 1 \end{cases}$. Find

(A) $\lim_{x\to 1^+} f(x)$ (B) $\lim_{x\to 1^-} f(x)$
(C) $\lim_{x\to 1} f(x)$ (D) $f(1)$

46. Let $f(x) = \begin{cases} x + 3 & \text{if } x < -2 \\ \sqrt{x + 2} & \text{if } x > -2 \end{cases}$. Find

(A) $\lim_{x\to -2^+} f(x)$ (B) $\lim_{x\to -2^-} f(x)$
(C) $\lim_{x\to -2} f(x)$ (D) $f(-2)$

47. Let $f(x) = \begin{cases} \dfrac{x^2 - 9}{x + 3} & \text{if } x < 0 \\ \dfrac{x^2 - 9}{x - 3} & \text{if } x > 0 \end{cases}$. Find

(A) $\lim_{x\to -3} f(x)$ (B) $\lim_{x\to 0} f(x)$
(C) $\lim_{x\to 3} f(x)$

48. Let $f(x) = \begin{cases} \dfrac{x}{x + 3} & \text{if } x < 0 \\ \dfrac{x}{x - 3} & \text{if } x > 0 \end{cases}$. Find

(A) $\lim_{x\to -3} f(x)$ (B) $\lim_{x\to 0} f(x)$
(C) $\lim_{x\to 3} f(x)$

49. Let $f(x) = \dfrac{|x - 1|}{x - 1}$. Find

(A) $\lim_{x\to 1^+} f(x)$ (B) $\lim_{x\to 1^-} f(x)$
(C) $\lim_{x\to 1} f(x)$ (D) $f(1)$

50. Let $f(x) = \dfrac{x - 3}{|x - 3|}$. Find

(A) $\lim_{x\to 3^+} f(x)$ (B) $\lim_{x\to 3^-} f(x)$
(C) $\lim_{x\to 3} f(x)$ (D) $f(3)$

51. Let $f(x) = \dfrac{x - 2}{x^2 - 2x}$. Find

(A) $\lim_{x\to 0} f(x)$ (B) $\lim_{x\to 2} f(x)$
(C) $\lim_{x\to 4} f(x)$

52. Let $f(x) = \dfrac{x + 3}{x^2 + 3x}$. Find

(A) $\lim_{x\to -3} f(x)$ (B) $\lim_{x\to 0} f(x)$
(C) $\lim_{x\to 3} f(x)$

53. Let $f(x) = \dfrac{x^2 - x - 6}{x + 2}$. Find

(A) $\lim_{x\to -2} f(x)$ (B) $\lim_{x\to 0} f(x)$
(C) $\lim_{x\to 3} f(x)$

54. Let $f(x) = \dfrac{x^2 + x - 6}{x + 3}$. Find

(A) $\lim_{x\to -3} f(x)$ (B) $\lim_{x\to 0} f(x)$
(C) $\lim_{x\to 2} f(x)$

55. Let $f(x) = \dfrac{(x + 2)^2}{x^2 - 4}$. Find

(A) $\lim_{x\to -2} f(x)$ (B) $\lim_{x\to 0} f(x)$
(C) $\lim_{x\to 2} f(x)$

56. Let $f(x) = \dfrac{x^2 - 1}{(x + 1)^2}$. Find

(A) $\lim_{x\to -1} f(x)$ (B) $\lim_{x\to 0} f(x)$
(C) $\lim_{x\to 1} f(x)$

57. Let $f(x) = \dfrac{2x^2 - 3x - 2}{x^2 + x - 6}$. Find

(A) $\lim_{x\to 2} f(x)$ (B) $\lim_{x\to 0} f(x)$
(C) $\lim_{x\to 1} f(x)$

58. Let $f(x) = \dfrac{3x^2 + 2x - 1}{x^2 + 3x + 2}$. Find

(A) $\lim_{x\to -3} f(x)$ (B) $\lim_{x\to -1} f(x)$
(C) $\lim_{x\to 2} f(x)$

In Problems 59–64, discuss the validity of each statement. If the statement is always true, explain why. If not, give a counterexample.

59. If $\lim_{x\to 1} f(x) = 0$ and $\lim_{x\to 1} g(x) = 0$, then $\lim_{x\to 1}\dfrac{f(x)}{g(x)} = 0$.

60. If $\lim_{x\to 1} f(x) = 1$ and $\lim_{x\to 1} g(x) = 1$, then $\lim_{x\to 1}\dfrac{f(x)}{g(x)} = 1$.

61. If f is a polynomial, then, as x approaches 0, the right-hand limit exists and is equal to the left-hand limit.

62. If f is a rational function, then, as x approaches 0, the right-hand limit exists and is equal to the left-hand limit.

63. If f is a function such that $\lim_{x\to 0} f(x)$ exists, then $f(0)$ exists.

64. If f is a function such that $f(0)$ exists, then $\lim_{x\to 0} f(x)$ exists.

Compute the following limit for each function in Problems 65–68.

$$\lim_{h\to 0}\frac{f(2+h)-f(2)}{h}$$

65. $f(x) = 3x + 1$

66. $f(x) = 5x - 1$

67. $f(x) = x^2 + 1$

68. $f(x) = x^2 - 2$

C

69. Let f be defined by

$$f(x) = \begin{cases} 1 + mx & \text{if } x \le 1 \\ 4 - mx & \text{if } x > 1 \end{cases}$$

where m is a constant.

(A) Graph f for $m = 1$, and find

$$\lim_{x\to 1^-} f(x) \quad \text{and} \quad \lim_{x\to 1^+} f(x)$$

(B) Graph f for $m = 2$, and find

$$\lim_{x\to 1^-} f(x) \quad \text{and} \quad \lim_{x\to 1^+} f(x)$$

(C) Find m so that

$$\lim_{x\to 1^-} f(x) = \lim_{x\to 1^+} f(x)$$

and graph f for this value of m.

(D) Write a brief verbal description of each graph. How does the graph in part (C) differ from the graphs in parts (A) and (B)?

70. Let f be defined by

$$f(x) = \begin{cases} -3m + 0.5x & \text{if } x \le 2 \\ 3m - x & \text{if } x > 2 \end{cases}$$

where m is a constant.

(A) Graph f for $m = 0$, and find

$$\lim_{x\to 2^-} f(x) \quad \text{and} \quad \lim_{x\to 2^+} f(x)$$

(B) Graph f for $m = 1$, and find

$$\lim_{x\to 2^-} f(x) \quad \text{and} \quad \lim_{x\to 2^+} f(x)$$

(C) Find m so that

$$\lim_{x\to 2^-} f(x) = \lim_{x\to 2^+} f(x)$$

and graph f for this value of m.

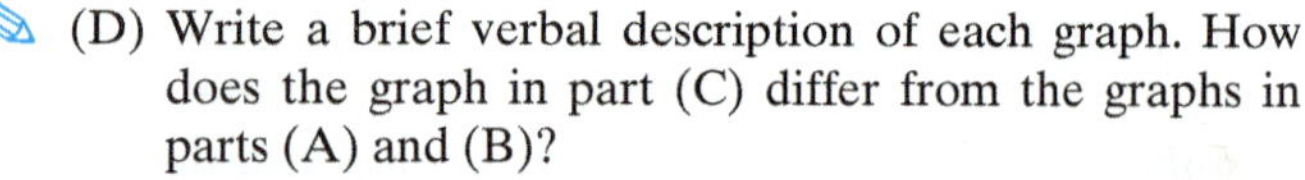

(D) Write a brief verbal description of each graph. How does the graph in part (C) differ from the graphs in parts (A) and (B)?

Find each limit in Problems 71–74, where a is a real constant.

71. $\lim_{h\to 0}\dfrac{(a+h)^2 - a^2}{h}$

72. $\lim_{h\to 0}\dfrac{[3(a+h) - 2] - (3a - 2)}{h}$

73. $\lim_{h\to 0}\dfrac{\sqrt{a+h} - \sqrt{a}}{h}$, $a > 0$

74. $\lim_{h\to 0}\dfrac{\dfrac{1}{a+h} - \dfrac{1}{a}}{h}$, $a \ne 0$

Applications

75. Telephone rates. A long-distance telephone service charges $0.99 for the first 20 minutes or less of a call and $0.07 per minute for each additional minute or fraction thereof.

(A) Write a piecewise definition of the charge $F(x)$ for a long-distance call lasting x minutes.

(B) Graph $F(x)$ for $0 < x \le 40$.

(C) Find $\lim_{x\to 20^-} F(x)$, $\lim_{x\to 20^+} F(x)$, and $\lim_{x\to 20} F(x)$, whichever exist.

76. Telephone rates. A second long-distance telephone service charges $0.09 per minute or fraction thereof for calls lasting 10 minutes or more and $0.18 per minute or fraction thereof for calls lasting less than 10 minutes.

(A) Write a piecewise definition of the charge $G(x)$ for a long-distance call lasting x minutes.

(B) Graph $G(x)$ for $0 < x \le 40$.

(C) Find $\lim_{x\to 10^-} G(x)$, $\lim_{x\to 10^+} G(x)$, and $\lim_{x\to 10} G(x)$, whichever exist.

77. Telephone rates. Refer to Problems 75 and 76. Write a brief verbal comparison of the two services described for calls lasting 20 minutes or less.

78. Telephone rates. Refer to Problems 75 and 76. Write a brief verbal comparison of the two services described for calls lasting more than 20 minutes.

A company sells custom embroidered apparel and promotional products. Table 1 shows the volume discounts offered by the company, where x is the volume of a purchase in dollars. Problems 79 and 80 deal with two different interpretations of this discount method.

Table 1

Volume Discount (Excluding Tax)

Volume ($x)	Discount Amount
$300 ≤ x < $1,000	3%
$1,000 ≤ x < $3,000	5%
$3,000 ≤ x < $5,000	7%
$5,000 ≤ x	10%

79. Volume discount. Assume that the volume discounts in Table 1 apply to the entire purchase. That is, if the volume x satisfies $300 ≤ x < $1,000, then the entire purchase is discounted 3%. If the volume x satisfies $1,000 ≤ x < $3,000, the entire purchase is discounted 5%, and so on.

(A) If x is the volume of a purchase before the discount is applied, then write a piecewise definition for the discounted price $D(x)$ of this purchase.

(B) Use one-sided limits to investigate the limit of $D(x)$ as x approaches $1,000. As x approaches $3,000.

80. Volume discount. Assume that the volume discounts in Table 1 apply only to that portion of the volume in each interval. That is, the discounted price for a $4,000 purchase would be computed as follows:

$$300 + 0.97(700) + 0.95(2{,}000) + 0.93(1{,}000) = 3{,}809$$

(A) If x is the volume of a purchase before the discount is applied, then write a piecewise definition for the discounted price $P(x)$ of this purchase.

(B) Use one-sided limits to investigate the limit of $P(x)$ as x approaches \$1,000. As x approaches \$3,000.

(C) Compare this discount method with the one in Problem 79. Does one always produce a lower price than the other? Discuss.

81. **Pollution.** A state charges polluters an annual fee of \$20 per ton for each ton of pollutant emitted into the atmosphere, up to a maximum of 4,000 tons. No fees are charged for emissions beyond the 4,000-ton limit. Write a piecewise definition of the fees $F(x)$ charged for the emission of x tons of pollutant in a year. What is the limit of $F(x)$ as x approaches 4,000 tons? As x approaches 8,000 tons?

82. **Pollution.** Refer to Problem 81. The fee per ton of pollution is given by $A(x) = F(x)/x$. Write a piecewise definition of $A(x)$. What is the limit of $A(x)$ as x approaches 4,000 tons? As x approaches 8,000 tons?

83. **Voter turnout.** Statisticians often use piecewise-defined functions to predict outcomes of elections. For the following functions f and g, find the limit of each function as x approaches 5 and as x approaches 10.

$$f(x) = \begin{cases} 0 & \text{if } x \le 5 \\ 0.8 - 0.08x & \text{if } 5 < x < 10 \\ 0 & \text{if } 10 \le x \end{cases}$$

$$g(x) = \begin{cases} 0 & \text{if } x \le 5 \\ 0.8x - 0.04x^2 - 3 & \text{if } 5 < x < 10 \\ 1 & \text{if } 10 \le x \end{cases}$$

Answers to Matched Problems

1.

x	−2	−1	0	1	2	3	4
$h(x)$	1.0	1.5	2.0	2.5	3.0	3.5	4.0

2. $\lim_{x\to 1} f(x) = 2$

3.

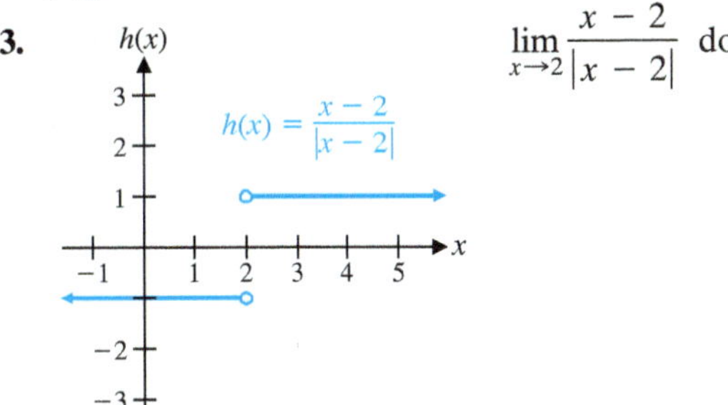

$\lim_{x\to 2} \frac{x-2}{|x-2|}$ does not exist

4. (A) $\lim_{x\to 0^-} f(x) = 0$
$\lim_{x\to 0^+} f(x) = 0$
$\lim_{x\to 0} f(x) = 0$
$f(0) = 0$

(B) $\lim_{x\to 1^-} f(x) = 1$
$\lim_{x\to 1^+} f(x) = 2$
$\lim_{x\to 1} f(x)$ does not exist
$f(1)$ not defined

(C) $\lim_{x\to 3^-} f(x) = 3$
$\lim_{x\to 3^+} f(x) = 3$
$\lim_{x\to 3} f(x) = 3$
$f(3)$ not defined

5. −6

6. (A) 6 (B) $\sqrt{6}$ (C) $\frac{4}{5}$

7. (A) 13 (B) 7 (C) Does not exist (D) Not defined

8. (A) −2 (B) Does not exist

9. −2 **10.** Does not exist **11.** $1/(2\sqrt{3})$

1-2 Infinite Limits and Limits at Infinity

- Infinite Limits
- Locating Vertical Asymptotes
- Limits at Infinity
- Finding Horizontal Asymptotes

In this section, we consider two new types of limits: infinite limits and limits at infinity. Infinite limits and vertical asymptotes are used to describe the behavior of functions that are unbounded near $x = a$. Limits at infinity and horizontal asymptotes are used to describe the behavior of functions as x assumes arbitrarily large positive values or arbitrarily large negative values. Although we will include graphs to illustrate basic concepts, we postpone a discussion of graphing techniques until Chapter 3.

Infinite Limits

The graph of $f(x) = \frac{1}{x-1}$ (Fig. 1) indicates that

$$\lim_{x\to 1^+} \frac{1}{x-1}$$

does not exist. There does not exist a real number L that the values of $f(x)$ approach as x approaches 1 from the right. Instead, as x approaches 1 from the right, the values of $f(x)$ are positive and become larger and larger; that is, $f(x)$ increases without bound (Table 1). We express this behavior symbolically as

$$\lim_{x\to 1^+} \frac{1}{x-1} = \infty \quad \text{or} \quad f(x) = \frac{1}{x-1} \to \infty \quad \text{as} \quad x \to 1^+ \tag{1}$$

Since ∞ is a not a real number, *the limit in (1) does not exist.* We are using the symbol ∞ to describe the manner in which the limit fails to exist, and we call this situation an **infinite limit**. If x approaches 1 from the left, the values of $f(x)$ are negative and become larger and larger in absolute value; that is, $f(x)$decreases through negative values without bound (Table 2). We express this behavior symbolically as

$$\lim_{x \to 1^-} \frac{1}{x-1} = -\infty \quad \text{or} \quad f(x) = \frac{1}{x-1} \to -\infty \quad \text{as} \quad x \to 1^- \tag{2}$$

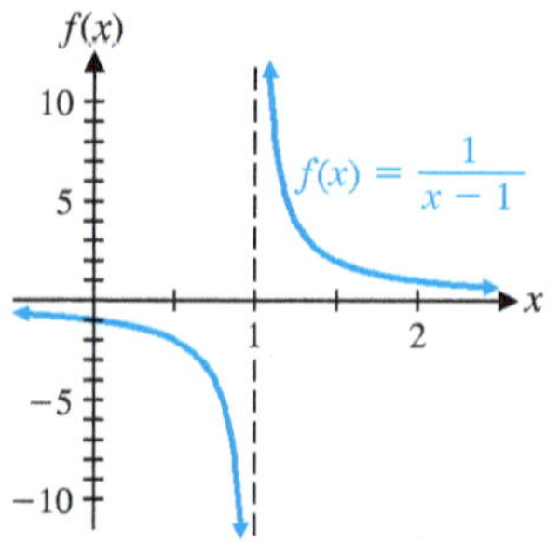

Figure 1

Table 1

x	$f(x) = \frac{1}{x-1}$
1.1	10
1.01	100
1.001	1,000
1.0001	10,000
1.00001	100,000
1.000001	1,000,000

Table 2

x	$f(x) = \frac{1}{x-1}$
0.9	−10
0.99	−100
0.999	−1,000
0.9999	−10,000
0.99999	−100,000
0.999999	−1,000,000

The one-sided limits in (1) and (2) describe the behavior of the graph as $x \to 1$ (Fig. 1). Does the two-sided limit of $f(x)$ as $x = 1$ exist? No, because neither of the one-sided limits exists. Also, there is no reasonable way to use the symbol ∞ to describe the behavior of $f(x)$ as $x \to 1$ on both sides of 1. We say that

$$\lim_{x \to 1} \frac{1}{x-1} \text{ does not exist}$$

EXPLORE & DISCUSS 1

Let $g(x) = \dfrac{1}{(x-1)^2}$

Construct tables for $g(x)$ as $x \to 1^+$ and as $x \to 1^-$. Use these tables and infinite limits to discuss the behavior of $g(x)$ near $x = 1$.

We used the dashed vertical line $x = 1$ in Figure 1 to illustrate the infinite limits as x approaches 1 from the right and from the left. We call this line a *vertical asymptote.*

DEFINITION Infinite Limits and Vertical Asymptotes

The vertical line $x = a$ is a **vertical asymptote** for the graph of $y = f(x)$ if

$$f(x) \to \infty \quad \text{or} \quad f(x) \to -\infty \quad \text{as} \quad x \to a^+ \quad \text{or} \quad x \to a^-$$

[That is, if $f(x)$ either increases or decreases without bound as x approaches a from the right or from the left].

Locating Vertical Asymptotes

How do we locate vertical asymptotes? If f is a polynomial function, then $\lim_{x \to a} f(x)$ is equal to the real number $f(a)$ [Theorem 3, Section 1-1]. So *a polynomial function has no vertical asymptotes*. Similarly (again by Theorem 3, Section 1-1), *a vertical asymptote of a rational function can occur only at a zero of its denominator.* Theorem 1 provides a simple procedure for locating the vertical asymptotes of a rational function.

THEOREM 1 Locating Vertical Asymptotes of Rational Functions

If $f(x) = n(x)/d(x)$ is a rational function, $d(c) = 0$ and $n(c) \neq 0$, then the line $x = c$ is a vertical asymptote of the graph of f.

If $f(x) = n(x)/d(x)$ and both $n(c) = 0$ and $d(c) = 0$, then the limit of $f(x)$ as x approaches c involves an indeterminate form and Theorem 1 does not apply:

$$\lim_{x \to c} f(x) = \lim_{x \to c} \frac{n(x)}{d(x)} \quad \frac{0}{0} \text{ indeterminate form}$$

Algebraic simplification is often useful in this situation.

EXAMPLE 1 **Locating Vertical Asymptotes** Let $f(x) = \dfrac{x^2 + x - 2}{x^2 - 1}$

Describe the behavior of f at each zero of the denominator. Use ∞ and $-\infty$ when appropriate. Identify all vertical asymptotes.

SOLUTION Let $n(x) = x^2 + x - 2$ and $d(x) = x^2 - 1$. Factoring the denominator, we see that

$$d(x) = x^2 - 1 = (x - 1)(x + 1)$$

has two zeros: $x = -1$ and $x = 1$.

First, we consider $x = -1$. Since $d(-1) = 0$ and $n(-1) = -2 \neq 0$, Theorem 1 tells us that the line $x = -1$ is a vertical asymptote. So at least one of the one-sided limits at $x = -1$ must be either ∞ or $-\infty$. Examining tables of values of f for x near -1 or a graph on a graphing calculator will show which is the case. From Tables 3 and 4, we see that

$$\lim_{x \to -1^-} \frac{x^2 + x - 2}{x^2 - 1} = -\infty \quad \text{and} \quad \lim_{x \to -1^+} \frac{x^2 + x - 2}{x^2 - 1} = \infty$$

Table 3

x	$f(x) = \dfrac{x^2 + x - 2}{x^2 - 1}$
−1.1	−9
−1.01	−99
−1.001	−999
−1.0001	−9,999
−1.00001	−99,999

Table 4

x	$f(x) = \dfrac{x^2 + x - 2}{x^2 - 1}$
−0.9	11
−0.99	101
−0.999	1,001
−0.9999	10,001
−0.99999	100,001

Now we consider the other zero of $d(x)$, $x = 1$. This time $n(1) = 0$ and Theorem 1 does not apply. We use algebraic simplification to investigate the behavior of the function at $x = 1$:

$$\begin{aligned} \lim_{x \to 1} f(x) &= \lim_{x \to 1} \frac{x^2 + x - 2}{x^2 - 1} && \frac{0}{0} \text{ indeterminate form} \\ &= \lim_{x \to 1} \frac{(x - 1)(x + 2)}{(x - 1)(x + 1)} \\ &= \lim_{x \to 1} \frac{x + 2}{x + 1} && \text{Reduced to lowest terms (see Appendix A-4)} \\ &= \frac{3}{2} \end{aligned}$$

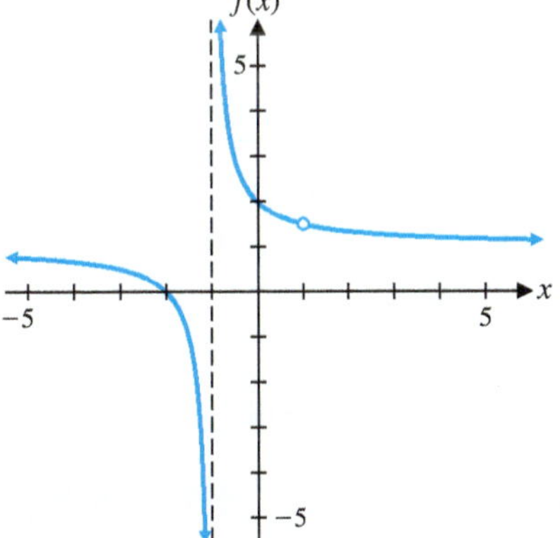

Figure 2 $f(x) = \dfrac{x^2 + x - 2}{x^2 - 1}$

Since the limit exists as x approaches 1, f does not have a vertical asymptote at $x = 1$. The graph of f (Fig. 2) shows the behavior at the vertical asymptote $x = -1$ and also at $x = 1$.

Matched Problem 1 Let $f(x) = \dfrac{x - 3}{x^2 - 4x + 3}$.

Describe the behavior of f at each zero of the denominator. Use ∞ and $-\infty$ when appropriate. Identify all vertical asymptotes.

EXAMPLE 2 **Locating Vertical Asymptotes** Let $f(x) = \frac{x^2 + 20}{5(x - 2)^2}$

Describe the behavior of f at each zero of the denominator. Use ∞ and $-\infty$ when appropriate. Identify all vertical asymptotes.

SOLUTION Let $n(x) = x^2 + 20$ and $d(x) = 5(x - 2)^2$. The only zero of $d(x)$ is $x = 2$. Since $n(2) = 24 \neq 0$, f has a vertical asymptote at $x = 2$ (Theorem 1). Tables 5 and 6 show that $f(x) \to \infty$ as $x \to 2$ from either side, and we have

$$\lim_{x \to 2^+} \frac{x^2 + 20}{5(x - 2)^2} = \infty \quad \text{and} \quad \lim_{x \to 2^-} \frac{x^2 + 20}{5(x - 2)^2} = \infty$$

Table 5

x	$f(x) = \frac{x^2 + 20}{5(x - 2)^2}$
2.1	488.2
2.01	48,080.02
2.001	4,800,800.2

Table 6

x	$f(x) = \frac{x^2 + 20}{5(x - 2)^2}$
1.9	472.2
1.99	47,920.02
1.999	4,799,200.2

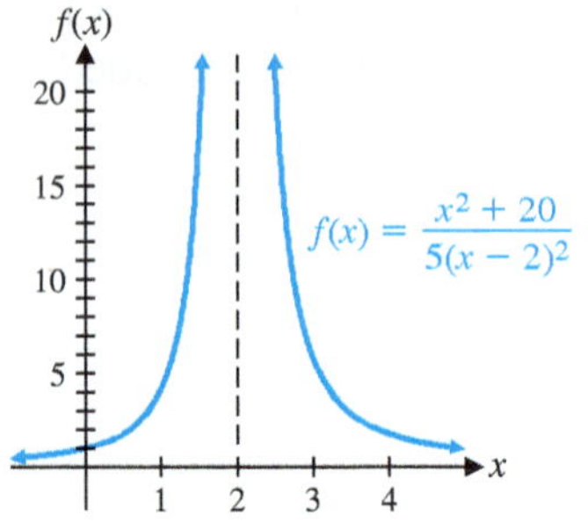

The denominator d has no other zeros, so f does not have any other vertical asymptotes. The graph of f (Fig. 3) shows the behavior at the vertical asymptote $x = 2$. Because the left- and right-hand limits are both infinite, we write

$$\lim_{x \to 2} \frac{x^2 + 20}{5(x - 2)^2} = \infty$$

Figure 3

Matched Problem 2 Let $f(x) = \frac{x - 1}{(x + 3)^2}$.

Describe the behavior of f at each zero of the denominator. Use ∞ and $-\infty$ when appropriate. Identify all vertical asymptotes.

CONCEPTUAL INSIGHT

When is it correct to say that a limit does not exist, and when is it correct to use $\pm\infty$? It depends on the situation. Table 7 lists the infinite limits that we discussed in Examples 1 and 2.

Table 7

Right-Hand Limit	Left-Hand Limit	Two-Sided Limit
$\lim_{x \to -1^+} \frac{x^2 + x - 2}{x^2 - 1} = \infty$	$\lim_{x \to -1^-} \frac{x^2 + x - 2}{x^2 - 1} = -\infty$	$\lim_{x \to -1} \frac{x^2 + x - 2}{x^2 - 1}$ does not exist
$\lim_{x \to -2^+} \frac{x^2 + 20}{5(x - 2)^2} = \infty$	$\lim_{x \to 2^-} \frac{x^2 + 20}{5(x - 2)^2} = \infty$	$\lim_{x \to 2} \frac{x^2 + 20}{5(x - 2)^2} = \infty$

The instructions in Examples 1 and 2 said that we should use infinite limits to describe the behavior at vertical asymptotes. If we had been asked to *evaluate* the limits, with no mention of ∞ or asymptotes, then the correct answer would be that **all of these limits do not exist**. Remember, ∞ is a symbol used to describe the behavior of functions at vertical asymptotes.

Limits at Infinity

The symbol ∞ can also be used to indicate that an independent variable is increasing or decreasing without bound. We write $x \to \infty$ to indicate that x is increasing without bound through positive values and $x \to -\infty$ to indicate that x is decreasing without bound through negative values. We begin by considering power functions of the form x^p and $1/x^p$ where p is a positive real number.

If p is a positive real number, then x^p increases as x increases. There is no upper bound on the values of x^p. We indicate this behavior by writing

$$\lim_{x\to\infty} x^p = \infty \quad \text{or} \quad x^p \to \infty \quad \text{as} \quad x \to \infty$$

Since the reciprocals of very large numbers are very small numbers, it follows that $1/x^p$ approaches 0 as x increases without bound. We indicate this behavior by writing

$$\lim_{x\to\infty} \frac{1}{x^p} = 0 \quad \text{or} \quad \frac{1}{x^p} \to 0 \quad \text{as} \quad x \to \infty$$

Figure 4 illustrates the preceding behavior for $f(x) = x^2$ and $g(x) = 1/x^2$, and we write

$$\lim_{x\to\infty} f(x) = \infty \quad \text{and} \quad \lim_{x\to\infty} g(x) = 0$$

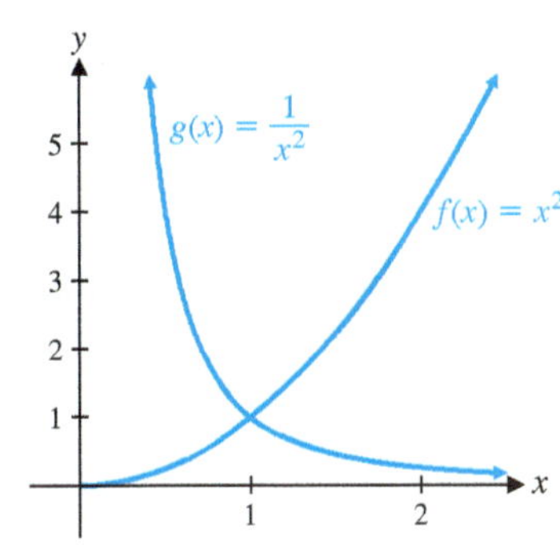

Figure 4

Limits of power forms as x decreases without bound behave in a similar manner, with two important differences. First, if x is negative, then x^p is not defined for all values of p. For example, $x^{1/2} = \sqrt{x}$ is not defined for negative values of x. Second, if x^p is defined, then it may approach ∞ or $-\infty$, depending on the value of p. For example,

$$\lim_{x\to-\infty} x^2 = \infty \quad \text{but} \quad \lim_{x\to-\infty} x^3 = -\infty$$

For the function g in Figure 4, the line $y = 0$ (the x axis) is called a *horizontal asymptote*. In general, a line $y = b$ is a **horizontal asymptote** of the graph of $y = f(x)$ if $f(x)$ approaches b as either x increases without bound or x decreases without bound. Symbolically, $y = b$ is a horizontal asymptote if either

$$\lim_{x\to-\infty} f(x) = b \quad \text{or} \quad \lim_{x\to\infty} f(x) = b$$

In the first case, the graph of f will be close to the horizontal line $y = b$ for large (in absolute value) negative x. In the second case, the graph will be close to the horizontal line $y = b$ for large positive x. Figure 5 shows the graph of a function with two horizontal asymptotes: $y = 1$ and $y = -1$.

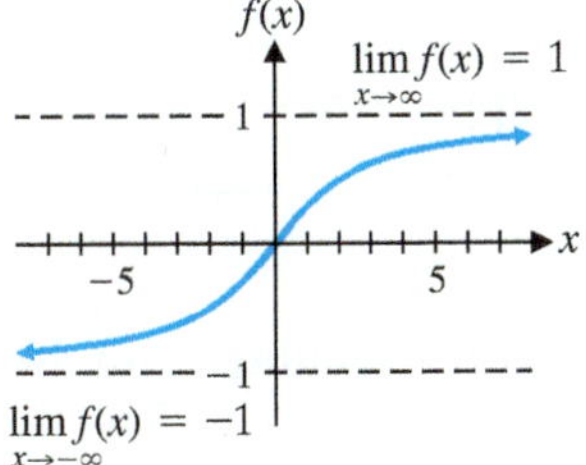

Figure 5

Theorem 2 summarizes the various possibilities for limits of power functions as x increases or decreases without bound.

THEOREM 2 Limits of Power Functions at Infinity

If p is a positive real number and k is any real number except 0, then

1. $\lim_{x\to-\infty}\frac{k}{x^p} = 0$
2. $\lim_{x\to\infty}\frac{k}{x^p} = 0$
3. $\lim_{x\to-\infty} kx^p = \pm\infty$
4. $\lim_{x\to\infty} kx^p = \pm\infty$

provided that x^p is a real number for negative values of x. The limits in 3 and 4 will be either $-\infty$ or ∞, depending on k and p.

How can we use Theorem 2 to evaluate limits at infinity? It turns out that the limit properties listed in Theorem 2, Section 1-1, are also valid if we replace the statement $x \to c$ with $x \to \infty$ or $x \to -\infty$.

EXAMPLE 3 **Limit of a Polynomial Function at Infinity** Let $p(x) = 2x^3 - x^2 - 7x + 3$. Find the limit of $p(x)$ as x approaches ∞ and as x approaches $-\infty$.

SOLUTION Since limits of power functions of the form $1/x^p$ approach 0 as x approaches ∞ or $-\infty$, it is convenient to work with these reciprocal forms whenever possible. If we factor out the term involving the highest power of x, then we can write $p(x)$ as

$$p(x) = 2x^3\left(1 - \frac{1}{2x} - \frac{7}{2x^2} + \frac{3}{2x^3}\right)$$

Using Theorem 2 in Section 1-3 and Theorem 2 in Section 1-1, we write

$$\lim_{x\to\infty}\left(1 - \frac{1}{2x} - \frac{7}{2x^2} + \frac{3}{2x^3}\right) = 1 - 0 - 0 + 0 = 1$$

For large values of x,

$$\left(1 - \frac{1}{2x} - \frac{7}{2x^2} + \frac{3}{2x^3}\right) \approx 1$$

and

$$p(x) = 2x^3\left(1 - \frac{1}{2x} - \frac{7}{2x^2} + \frac{3}{2x^3}\right) \approx 2x^3$$

Since $2x^3 \to \infty$ as $x \to \infty$, it follows that

$$\lim_{x\to\infty} p(x) = \lim_{x\to\infty} 2x^3 = \infty$$

Similarly, $2x^3 \to -\infty$ as $x \to -\infty$ implies that

$$\lim_{x\to-\infty} p(x) = \lim_{x\to-\infty} 2x^3 = -\infty$$

So the behavior of $p(x)$ for large values is the same as the behavior of the highest-degree term, $2x^3$.

Matched Problem 3 Let $p(x) = -4x^4 + 2x^3 + 3x$. Find the limit of $p(x)$ as x approaches ∞ and as x approaches $-\infty$.

The term with highest degree in a polynomial is called the **leading term**. In the solution to Example 3, the limits at infinity of $p(x) = 2x^3 - x^2 - 7x + 3$ were the same as the limits of the leading term $2x^3$. Theorem 3 states that this is true for any polynomial of degree greater than or equal to 1.

THEOREM 3 Limits of Polynomial Functions at Infinity

If

$$p(x) = a_n x^n + a_{n-1} x^{n-1} + \cdots + a_1 x + a_0,\ a_n \neq 0, n \geq 1$$

then

$$\lim_{x \to \infty} p(x) = \lim_{x \to \infty} a_n x^n = \pm\infty$$

and

$$\lim_{x \to -\infty} p(x) = \lim_{x \to -\infty} a_n x^n = \pm\infty$$

Each limit will be either $-\infty$ or ∞, depending on a_n and n.

A polynomial of degree 0 is a constant function $p(x) = a_0$, and its limit as x approaches ∞ or $-\infty$ is the number a_0. For any polynomial of degree 1 or greater, Theorem 3 states that the limit as x approaches ∞ or $-\infty$ cannot be equal to a number. This means that **polynomials of degree 1 or greater never have horizontal asymptotes**.

A description of the behavior of the limits of any function f at infinity is called the **end behavior** of f. $\text{Lim}_{x \to \infty} f(x)$ determines the right end behavior, and $\lim_{x \to -\infty} f(x)$ determines the left end behavior. According to Theorem 3, the right and left end behavior of a nonconstant polynomial $p(x)$ is always an infinite limit.

EXAMPLE 4 **End Behavior of a Polynomial** Describe the end behavior of each polynomial.

(A) $p(x) = 3x^3 - 500x^2$ (B) $p(x) = 3x^3 - 500x^4$

SOLUTION (A) According to Theorem 3, as x increases without bound to the right, the right end behavior is

$$\lim_{x \to \infty} (3x^3 - 500x^2) = \lim_{x \to \infty} 3x^3 = \infty$$

and as x increases without bound to the left, the end behavior is

$$\lim_{x \to -\infty} (3x^3 - 500x^2) = \lim_{x \to -\infty} 3x^3 = -\infty$$

(B) As x increases without bound to the right, the end behavior is

$$\lim_{x \to \infty} (3x^3 - 500x^4) = \lim_{x \to \infty} (-500x^4) = -\infty$$

and as x increases without bound to the left, the end behavior is

$$\lim_{x \to -\infty} (3x^3 - 500x^4) = \lim_{x \to -\infty} (-500x^4) = -\infty$$

Matched Problem 4 Describe the end behavior of each polynomial.

(A) $p(x) = 300x^2 - 4x^5$ (B) $p(x) = 300x^6 - 4x^5$

Finding Horizontal Asymptotes

Since a rational function is the ratio of two polynomials, it is not surprising that reciprocals of powers of x can be used to analyze limits of rational functions at infinity. For example, consider the rational function

$$f(x) = \frac{3x^2 - 5x + 9}{2x^2 + 7}$$

Factoring the highest-degree term out of the numerator and the denominator, we write

$$f(x) = \frac{3x^2}{2x^2} \cdot \frac{1 - \frac{5}{3x} + \frac{3}{x^2}}{1 + \frac{7}{2x^2}}$$

$$\lim_{x\to\infty} f(x) = \lim_{x\to\infty} \frac{3x^2}{2x^2} \cdot \lim_{x\to\infty} \frac{1 - \frac{5}{3x} + \frac{3}{x^2}}{1 + \frac{7}{2x^2}} = \frac{3}{2} \cdot \frac{1 - 0 + 0}{1 + 0} = \frac{3}{2}$$

The behavior of this rational function as x approaches infinity is determined by the ratio of the highest-degree term in the numerator ($3x^2$) to the highest-degree term in the denominator ($2x^2$). Theorem 4 generalizes this result to any rational function and lists the three possible outcomes.

THEOREM 4 Limits of Rational Functions at Infinity and Horizontal Asymptotes of Rational Functions

(A) If $f(x) = \dfrac{a_m x^m + a_{m-1}x^{m-1} + \cdots + a_1 x + a_0}{b_n x^n + b_{n-1}x^{n-1} + \cdots + b_1 x + b_0}$, $a_m \neq 0, b_n \neq 0$

then $\lim_{x\to\infty} f(x) = \lim_{x\to\infty} \dfrac{a_m x^m}{b_n x^n}$ and $\lim_{x\to-\infty} f(x) = \lim_{x\to-\infty} \dfrac{a_m x^m}{b_n x^n}$

(B) There are three possible cases for these limits:

1. If $m < n$, then $\lim_{x\to\infty} f(x) = \lim_{x\to-\infty} f(x) = 0$, and the line $y = 0$ (the x axis) is a horizontal asymptote of $f(x)$.
2. If $m = n$, then $\lim_{x\to\infty} f(x) = \lim_{x\to-\infty} f(x) = \dfrac{a_m}{b_n}$, and the line $y = \dfrac{a_m}{b_n}$ is a horizontal asymptote of $f(x)$.
3. If $m > n$, then each limit will be ∞ or $-\infty$, depending on m, n, a_m, and b_n, and $f(x)$ does not have a horizontal asymptote.

Notice that in cases 1 and 2 of Theorem 4, the limit is the same if x approaches ∞ or $-\infty$. So, **a rational function can have at most one horizontal asymptote** (see Fig. 6).

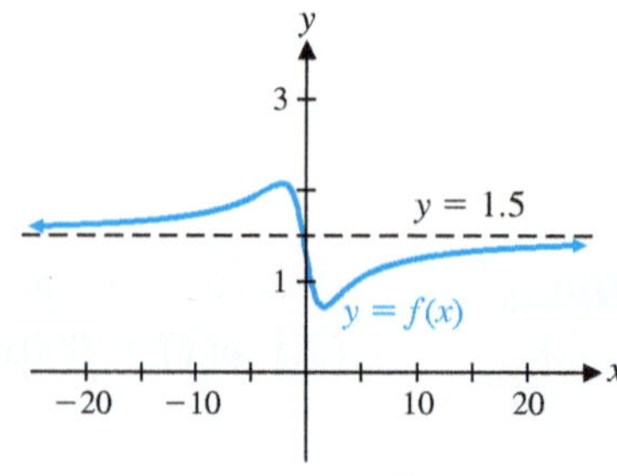

Figure 6 $f(x) = \dfrac{3x^2 - 5x + 9}{2x^2 + 7}$

CONCEPTUAL INSIGHT

The graph of f in Figure 6 dispels the misconception that the graph of a function cannot cross a horizontal asymptote. Horizontal asymptotes give us information about the graph of a function only as $x \to \infty$ and $x \to -\infty$, not at any specific value of x.

EXAMPLE 5 **Finding Horizontal Asymptotes** Find all horizontal asymptotes, if any, of each function.

(A) $f(x) = \dfrac{5x^3 - 2x^2 + 1}{4x^3 + 2x - 7}$ (B) $f(x) = \dfrac{3x^4 - x^2 + 1}{8x^6 - 10}$

(C) $f(x) = \dfrac{2x^5 - x^3 - 1}{6x^3 + 2x^2 - 7}$

SOLUTION We will make use of part A of Theorem 4.

(A) $\lim_{x\to\infty} f(x) = \lim_{x\to\infty} \dfrac{5x^3 - 2x^2 + 1}{4x^3 + 2x - 7} = \lim_{x\to\infty} \dfrac{5x^3}{4x^3} = \dfrac{5}{4}$

The line $y = 5/4$ is a horizontal asymptote of $f(x)$.

(B) $\lim_{x\to\infty} f(x) = \lim_{x\to\infty} \dfrac{3x^4 - x^2 + 1}{8x^6 - 10} = \lim_{x\to\infty} \dfrac{3x^4}{8x^6} = \lim_{x\to\infty} \dfrac{3}{8x^2} = 0$

The line $y = 0$ (the x axis) is a horizontal asymptote of $f(x)$.

(C) $\lim_{x\to\infty} f(x) = \lim_{x\to\infty} \dfrac{2x^5 - x^3 - 1}{6x^3 + 2x^2 - 7} = \lim_{x\to\infty} \dfrac{2x^5}{6x^3} = \lim_{x\to\infty} \dfrac{x^2}{3} = \infty$

The function $f(x)$ has no horizontal asymptotes.

Matched Problem 5 Find all horizontal asymptotes, if any, of each function.

(A) $f(x) = \dfrac{4x^3 - 5x + 8}{2x^4 - 7}$ (B) $f(x) = \dfrac{5x^6 + 3x}{2x^5 - x - 5}$

(C) $f(x) = \dfrac{2x^3 - x + 7}{4x^3 + 3x^2 - 100}$

An accurate sketch of the graph of a rational function requires knowledge of both vertical and horizontal asymptotes. As we mentioned earlier, we are postponing a detailed discussion of graphing techniques until Section 3-4.

EXAMPLE 6 **Finding the Asymptotes of a Rational Function** Find all the asymptotes of the function

$$f(x) = \frac{2x^2 - 5}{x^2 + 4x + 4}$$

SOLUTION Let $n(x) = 2x^2 - 5$ and $d(x) = x^2 + 4x + 4 = (x + 2)^2$. Since $d(x) = 0$ only at $x = -2$ and $n(-2) = 3$, the only vertical asymptote of f is the line $x = -2$ (Theorem 1). Since

$$\lim_{x\to\infty} f(x) = \lim_{x\to\infty} \frac{2x^2 - 5}{x^2 + 4x + 4} = \lim_{x\to\infty} \frac{2x^2}{x^2} = 2$$

the horizontal asymptote is the line $y = 2$ (Theorem 3).

Matched Problem 6 Find all the asymptotes of the function $f(x) = \dfrac{x^2 - 9}{x^2 - 4}$.

Exercises 1-2

A

Problems 1–8 refer to the following graph of $y = f(x)$.

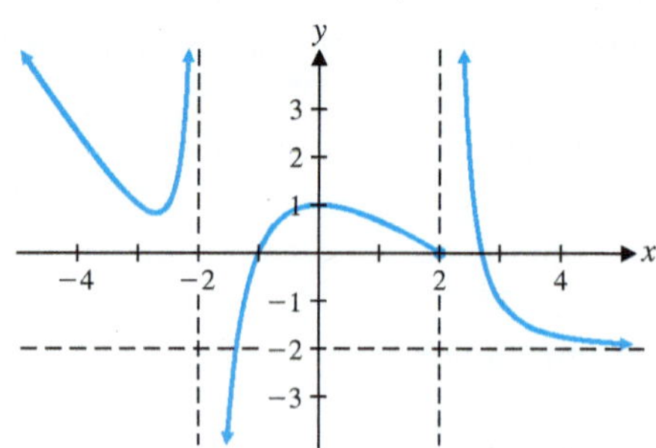

Figure for 1–8

1. $\lim_{x\to\infty} f(x) = ?$
2. $\lim_{x\to-\infty} f(x) = ?$
3. $\lim_{x\to-2^+} f(x) = ?$
4. $\lim_{x\to-2^-} f(x) = ?$
5. $\lim_{x\to-2} f(x) = ?$
6. $\lim_{x\to2^+} f(x) = ?$
7. $\lim_{x\to2^-} f(x) = ?$
8. $\lim_{x\to2} f(x) = ?$

In Problems 9–16, find each limit. Use $-\infty$ and ∞ when appropriate.

9. $f(x) = \dfrac{x}{x-5}$
   (A) $\lim_{x\to5^-} f(x)$ (B) $\lim_{x\to5^+} f(x)$ (C) $\lim_{x\to5} f(x)$

10. $f(x) = \dfrac{x^2}{x+3}$
    (A) $\lim_{x\to-3^-} f(x)$ (B) $\lim_{x\to-3^+} f(x)$ (C) $\lim_{x\to-3} f(x)$

11. $f(x) = \dfrac{2x-4}{(x-4)^2}$
    (A) $\lim_{x\to4^-} f(x)$ (B) $\lim_{x\to4^+} f(x)$ (C) $\lim_{x\to4} f(x)$

12. $f(x) = \dfrac{2x+2}{(x+2)^2}$
    (A) $\lim_{x\to-2^-} f(x)$ (B) $\lim_{x\to-2^+} f(x)$ (C) $\lim_{x\to-2} f(x)$

13. $f(x) = \dfrac{x^2+x-2}{x-1}$
    (A) $\lim_{x\to1^-} f(x)$ (B) $\lim_{x\to1^+} f(x)$ (C) $\lim_{x\to1} f(x)$

14. $f(x) = \dfrac{x^2+x+2}{x-1}$
    (A) $\lim_{x\to1^-} f(x)$ (B) $\lim_{x\to1^+} f(x)$ (C) $\lim_{x\to1} f(x)$

15. $f(x) = \dfrac{x^2-3x+2}{x+2}$
    (A) $\lim_{x\to-2^-} f(x)$ (B) $\lim_{x\to-2^+} f(x)$ (C) $\lim_{x\to-2} f(x)$

16. $f(x) = \dfrac{x^2+x-2}{x+2}$
    (A) $\lim_{x\to-2^-} f(x)$ (B) $\lim_{x\to-2^+} f(x)$ (C) $\lim_{x\to-2} f(x)$

In Problems 17–20, find the limit of each polynomial $p(x)$ (A) as x approaches ∞ (B) as x approaches $-\infty$

17. $p(x) = 4x^5 - 3x^4 + 1$
18. $p(x) = 4x^3 - 3x^4 + x^2$
19. $p(x) = 2x^5 - 2x^6 - 11$
20. $p(x) = 2x^4 - 2x^3 + 9x$

B

In Problems 21–30, use $-\infty$ or ∞ where appropriate to describe the behavior at each zero of the denominator and identify all vertical asymptotes.

21. $f(x) = \dfrac{1}{x+3}$
22. $g(x) = \dfrac{x}{4-x}$
23. $h(x) = \dfrac{x^2+4}{x^2-4}$
24. $k(x) = \dfrac{x^2-9}{x^2+9}$
25. $F(x) = \dfrac{x^2-4}{x^2+4}$
26. $G(x) = \dfrac{x^2+9}{9-x^2}$
27. $H(x) = \dfrac{x^2-2x-3}{x^2-4x+3}$
28. $K(x) = \dfrac{x^2+2x-3}{x^2-4x+3}$
29. $T(x) = \dfrac{8x-16}{x^4-8x^3+16x^2}$
30. $S(x) = \dfrac{6x+9}{x^4+6x^3+9x^2}$

In Problems 31–38, find each function value and limit. Use $-\infty$ or ∞ where appropriate.

31. $f(x) = \dfrac{4x+7}{5x-9}$
    (A) $f(10)$ (B) $f(100)$ (C) $\lim_{x\to\infty} f(x)$

32. $f(x) = \dfrac{2-3x^3}{7+4x^3}$
    (A) $f(5)$ (B) $f(10)$ (C) $\lim_{x\to\infty} f(x)$

33. $f(x) = \dfrac{5x^2+11}{7x-2}$
    (A) $f(20)$ (B) $f(50)$ (C) $\lim_{x\to\infty} f(x)$

34. $f(x) = \dfrac{5x+11}{7x^3-2}$
    (A) $f(8)$ (B) $f(16)$ (C) $\lim_{x\to-\infty} f(x)$

35. $f(x) = \dfrac{7x^4-14x^2}{6x^5+3}$
    (A) $f(-6)$ (B) $f(-12)$ (C) $\lim_{x\to-\infty} f(x)$

36. $f(x) = \dfrac{4x^7-8x}{6x^4+9x^2}$
    (A) $f(-3)$ (B) $f(-6)$ (C) $\lim_{x\to-\infty} f(x)$

37. $f(x) = \dfrac{10-7x^3}{4+x^3}$
    (A) $f(-10)$ (B) $f(-20)$ (C) $\lim_{x\to-\infty} f(x)$

38. $f(x) = \dfrac{3+x}{5+4x}$
    (A) $f(-50)$ (B) $f(-100)$ (C) $\lim_{x\to-\infty} f(x)$

In Problems 39–52, find all horizontal and vertical asymptotes.

39. $f(x) = \frac{2x}{x + 2}$

40. $f(x) = \frac{3x + 2}{x - 4}$

41. $f(x) = \frac{x^2 + 1}{x^2 - 1}$

42. $f(x) = \frac{x^2 - 1}{x^2 + 2}$

43. $f(x) = \frac{x^3}{x^2 + 6}$

44. $f(x) = \frac{x}{x^2 - 4}$

45. $f(x) = \frac{x}{x^2 + 4}$

46. $f(x) = \frac{x^2 + 9}{x}$

47. $f(x) = \frac{x^2}{x - 3}$

48. $f(x) = \frac{x + 5}{x^2}$

49. $f(x) = \frac{2x^2 + 3x - 2}{x^2 - x - 2}$

50. $f(x) = \frac{2x^2 + 7x + 12}{2x^2 + 5x - 12}$

51. $f(x) = \frac{2x^2 - 5x + 2}{x^2 - x - 2}$

52. $f(x) = \frac{x^2 - x - 12}{2x^2 + 5x - 12}$

In Problems 53–58, discuss the validity of each statement. If the statement is always true, explain why. If not, give a counterexample.

53. A rational function has at least one vertical asymptote.

54. A rational function has at most one vertical asymptote.

55. A rational function has at least one horizontal asymptote.

56. A rational function has at most one horizontal asymptote.

57. A polynomial function of degree ≥ 1 has neither horizontal nor vertical asymptotes.

58. The graph of a rational function cannot cross a horizontal asymptote.

C

59. Theorem 3 states that

$$\lim_{x \to \infty} (a_n x^n + a_{n-1} x^{n-1} + \cdots + a_0) = \pm\infty.$$

What conditions must n and a_n satisfy for the limit to be ∞? For the limit to be $-\infty$?

60. Theorem 3 also states that

$$\lim_{x \to -\infty} (a_n x^n + a_{n-1} x^{n-1} + \cdots + a_0) = \pm\infty.$$

What conditions must n and a_n satisfy for the limit to be ∞? For the limit to be $-\infty$?

Describe the end behavior of each function in Problems 61–68.

61. $f(x) = 2x^4 - 5x + 11$

62. $f(x) = 3x - 7x^5 - 2$

63. $f(x) = 7x^2 + 9x^3 + 5x$

64. $f(x) = 4x^3 - 5x^2 - 6x^6$

65. $f(x) = \frac{x^2 - 5x - 7}{x + 11}$

66. $f(x) = \frac{5x^4 + 7x^7 - 10}{x^2 + 6x^4 + 3}$

67. $f(x) = \frac{5x^5 + 7x^4 - 10}{-x^3 + 6x^2 + 3}$

68. $f(x) = \frac{-3x^6 + 4x^4 + 2}{2x^2 - 2x - 3}$

Applications

69. Average cost. A company manufacturing snowboards has fixed costs of \$200 per day and total costs of \$3,800 per day for a daily output of 20 boards.

(A) Assuming that the total cost per day $C(x)$ is linearly related to the total output per day x, write an equation for the cost function.

(B) The average cost per board for an output of x boards is given by $\overline{C}(x) = C(x)/x$. Find the average cost function.

(C) Sketch a graph of the average cost function, including any asymptotes, for $1 \leq x \leq 30$.

(D) What does the average cost per board tend to as production increases?

70. Average cost. A company manufacturing surfboards has fixed costs of \$300 per day and total costs of \$5,100 per day for a daily output of 20 boards.

(A) Assuming that the total cost per day $C(x)$ is linearly related to the total output per day x, write an equation for the cost function.

(B) The average cost per board for an output of x boards is given by $\overline{C}(x) = C(x)/x$. Find the average cost function.

(C) Sketch a graph of the average cost function, including any asymptotes, for $1 \leq x \leq 30$.

(D) What does the average cost per board tend to as production increases?

71. Energy costs. Most appliance manufacturers produce conventional and energy-efficient models. The energy-efficient models are more expensive to make but cheaper to operate. The costs of purchasing and operating a 23-cubic-foot refrigerator of each type are given in Table 8. These costs do not include maintenance charges or changes in electricity prices.

Table 8 **23-ft³ Refrigerators**

	Energy-Efficient Model	Conventional Model
Initial cost	\$950	\$900
Total volume	23 ft³	23 ft³
Annual cost of electricity	\$56	\$66

(A) Express the total cost $C_e(x)$ and the average cost $\overline{C}_e(x) = C_e(x)/x$ of purchasing and operating an energy-efficient model for x years.

(B) Express the total cost $C_c(x)$ and the average cost $\overline{C}_c(x) = C_c(x)/x$ of purchasing and operating a conventional model for x years.

(C) Are the total costs for an energy-efficient model and for a conventional model ever the same? If so, when?

(D) Are the average costs for an energy-efficient model and for a conventional model ever the same? If so, when?

(E) Find the limit of each average cost function as $x \to \infty$ and discuss the implications of the results.

72. Energy costs. Most appliance manufacturers produce conventional and energy-efficient models. The energy-efficient models are more expensive to make but cheaper to operate. The costs of purchasing and operating a 36,000-Btu central air conditioner of each type are given in Table 9. These costs do not include maintenance charges or changes in electricity prices.

Table 9 36,000 Btu Central Air Conditioner

	Energy-Efficient Model	Conventional Model
Initial cost	$4,000	$2,700
Total capacity	36,000 Btu	36,000 Btu
Annual cost of electricity	$932	$1,332

(A) Express the total cost $C_e(x)$ and the average cost $\overline{C}_e(x) = C_e(x)/x$ of purchasing and operating an energy-efficient model for x years.

(B) Express the total cost $C_c(x)$ and the average cost $\overline{C}_c(x) = C_c(x)/x$ of purchasing and operating a conventional model for x years.

(C) Are the total costs for an energy-efficient model and for a conventional model ever the same? If so, when?

(D) Are the average costs for an energy-efficient model and for a conventional model ever the same? If so, when?

(E) Find the limit of each average cost function as $x \to \infty$ and discuss the implications of the results.

73. Drug concentration. A drug is administered to a patient through an injection. The drug concentration (in milligrams/milliliter) in the bloodstream t hours after the injection is given by $C(t) = \frac{5t^2(t + 50)}{t^3 + 100}$. Find and interpret $\lim_{t\to\infty} C(t)$.

74. Drug concentration. A drug is administered to a patient through an IV drip. The drug concentration (in milligrams/milliliter) in the bloodstream t hours after the drip was started is given by $C(t) = \frac{5t(t + 50)}{t^3 + 100}$. Find and interpret $\lim_{t\to\infty} C(t)$.

75. Pollution. In Silicon Valley, a number of computer-related manufacturing firms were contaminating underground water supplies with toxic chemicals stored in leaking underground containers. A water quality control agency ordered the companies to take immediate corrective action and contribute to a monetary pool for the testing and cleanup of the underground contamination. Suppose that the monetary pool (in millions of dollars) for the testing and cleanup is given by

$$P(x) = \frac{2x}{1 - x} \qquad 0 \le x < 1$$

where x is the percentage (expressed as a decimal) of the total contaminant removed.

(A) How much must be in the pool to remove 90% of the contaminant?

(B) How much must be in the pool to remove 95% of the contaminant?

(C) Find $\lim_{x\to 1^-} P(x)$ and discuss the implications of this limit.

76. Employee training. A company producing computer components has established that, on average, a new employee can assemble $N(t)$ components per day after t days of on-the-job training, as given by

$$N(t) = \frac{100t}{t + 9} \qquad t \ge 0$$

(A) How many components per day can a new employee assemble after 6 days of on-the-job training?

(B) How many days of on-the-job training will a new employee need to reach the level of 70 components per day?

(C) Find $\lim_{t\to\infty} N(t)$ and discuss the implications of this limit.

77. Biochemistry. In 1913, biochemists Leonor Michaelis and Maude Menten proposed the rational function model (see figure)

$$v(s) = \frac{V_{max}\, s}{K_M + s}$$

for the velocity of the enzymatic reaction v, where s is the substrate concentration. The constants V_{max} and K_M are determined from experimental data.

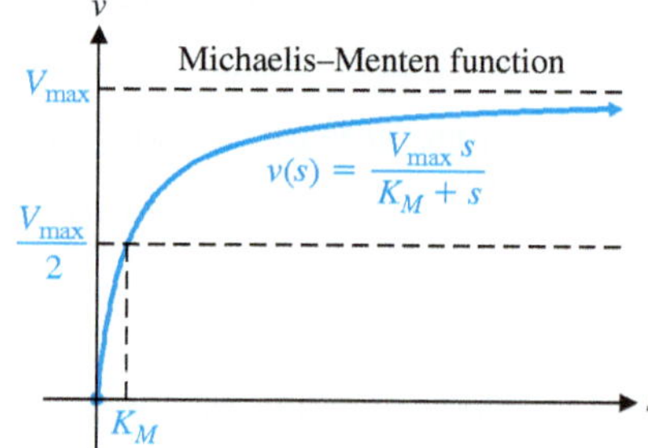

Figure for 77

(A) Show that $\lim_{s\to\infty} v(s) = V_{max}$.

(B) Show that $v(K_M) = \frac{V_{max}}{2}$.

(C) Table 10* lists data for an enzyme treated with the substrate saccharose.

Plot the preceding points on graph paper and estimate V_{max} to the nearest integer. To estimate K_M, add the

Table 10

s	v
5.2	0.866
10.4	1.466
20.8	2.114
41.6	2.666
83.3	3.236
167	3.636
333	3.636

*Michaelis and Menten (1913) *Biochem. Z.* 49, 333–369.

horizontal line $v = \frac{V_{max}}{2}$ to your graph, connect successive points on the graph with straight-line segments, and estimate the value of s (to the nearest multiple of 10) that satisfies $v(s) = \frac{V_{max}}{2}$.

(D) Use the constants V_{max} and K_M from part C to form a Michaelis–Menten function for the data in Table 10.

(E) Use the function from part D to estimate the velocity of the enzyme reaction when the saccharose is 15 and to estimate the saccharose when the velocity is 3.

78. Biochemistry. Table 11* lists data for the enzyme invertase treated with the substrate sucrose. We want to model these data with a Michaelis–Menten function.

Table 11

s	v
2.92	18.2
5.84	26.5
8.76	31.1
11.7	33
14.6	34.9
17.5	37.2
23.4	37.1

(A) Plot the points in Table 11 on graph paper and estimate V_{max} to the nearest integer. To estimate K_M, add the horizontal line $v = \frac{V_{max}}{2}$ to your graph, connect successive points on the graph with straight-line segments, and estimate the value of s (to the nearest integer) that satisfies $v(s) = \frac{V_{max}}{2}$.

(B) Use the constants V_{max} and K_M from part (A) to form a Michaelis–Menten function for the data in Table 11.

(C) Use the function from part (B) to estimate the velocity of the enzyme reaction when the sucrose is 9 and to estimate the sucrose when the velocity is 32.

79. Physics. The coefficient of thermal expansion (CTE) is a measure of the expansion of an object subjected to extreme temperatures. To model this coefficient, we use a Michaelis–Menten function of the form

$$C(T) = \frac{C_{max}\,T}{M + T} \qquad \text{(Problem 77)}$$

where C = CTE, T is temperature in K (degrees Kelvin), and C_{max} and M are constants. Table 12[†] lists the coefficients of thermal expansion for nickel and for copper at various temperatures.

Table 12 Coefficients of Thermal Expansion

T (K)	Nickel	Copper
100	6.6	10.3
200	11.3	15.2
293	13.4	16.5
500	15.3	18.3
800	16.8	20.3
1,100	17.8	23.7

(A) Plot the points in columns 1 and 2 of Table 12 on graph paper and estimate C_{max} to the nearest integer. To estimate M, add the horizontal line CTE $= \frac{C_{max}}{2}$ to your graph, connect successive points on the graph with straight-line segments, and estimate the value of T (to the nearest multiple of fifty) that satisfies $C(T) = \frac{C_{max}}{2}$.

(B) Use the constants $\frac{C_{max}}{2}$ and M from part (A) to form a Michaelis–Menten function for the CTE of nickel.

(C) Use the function from part (B) to estimate the CTE of nickel at 600 K and to estimate the temperature when the CTE of nickel is 12.

80. Physics. Repeat Problem 79 for the CTE of copper (column 3 of Table 12).

Answers to Matched Problems

1. Vertical asymptote: $x = 1$; $\lim_{x\to 1^+} f(x) = \infty$, $\lim_{x\to 1^-} f(x) = -\infty$
$\lim_{x\to 3} f(x) = 1/2$ so f does not have a vertical asymptote at $x = 3$
2. Vertical asymptote: $x = -3$; $\lim_{x\to -3^+} f(x) = \lim_{x\to -3^-} f(x) = -\infty$
3. $\lim_{x\to\infty} p(x) = \lim_{x\to -\infty} p(x) = -\infty$
4. (A) $\lim_{x\to\infty} p(x) = -\infty$, $\lim_{x\to -\infty} p(x) = \infty$
(B) $\lim_{x\to\infty} p(x) = \infty$, $\lim_{x\to -\infty} p(x) = \infty$
5. (A) $y = 0$ (B) No horizontal asymptotes
(C) $y = 1/2$
6. Vertical asymptotes: $x = -2$, $x = 2$; horizontal asymptote: $y = 1$

*Institute of Chemistry, Macedonia.

[†]National Physical Laboratory

1-3 Continuity

Theorem 3 in Section 1-1 states that if f is a polynomial function or a rational function with a nonzero denominator at $x = c$, then

$$\lim_{x \to c} f(x) = f(c) \tag{1}$$

Functions that satisfy equation (1) are said to be *continuous* at $x = c$. A firm understanding of continuous functions is essential for sketching and analyzing graphs. We will also see that continuity properties provide a simple and efficient method for solving inequalities—a tool that we will use extensively in later sections.

Continuity

Compare the graphs shown in Figure 1. Notice that two of the graphs are broken; that is, they cannot be drawn without lifting a pen off the paper. Informally, a function is *continuous over an interval* if its graph over the interval can be drawn without removing a pen from the paper. A function whose graph is broken (disconnected) at $x = c$ is said to be *discontinuous* at $x = c$. Function f (Fig. 1A) is continuous for all x. Function g (Fig. 1B) is discontinuous at $x = 2$ but is continuous over any interval that does not include 2. Function h (Fig. 1C) is discontinuous at $x = 0$ but is continuous over any interval that does not include 0.

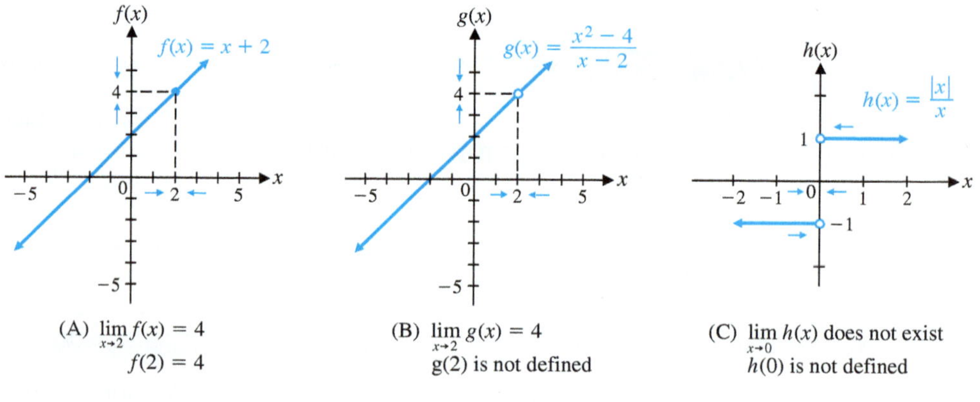

(A) $\lim_{x \to 2} f(x) = 4$, $f(2) = 4$

(B) $\lim_{x \to 2} g(x) = 4$, $g(2)$ is not defined

(C) $\lim_{x \to 0} h(x)$ does not exist, $h(0)$ is not defined

Figure 1

Most graphs of natural phenomena are continuous, whereas many graphs in business and economics applications have discontinuities. Figure 2A illustrates temperature variation over a 24-hour period—a continuous phenomenon. Figure 2B illustrates warehouse inventory over a 1-week period—a discontinuous phenomenon.

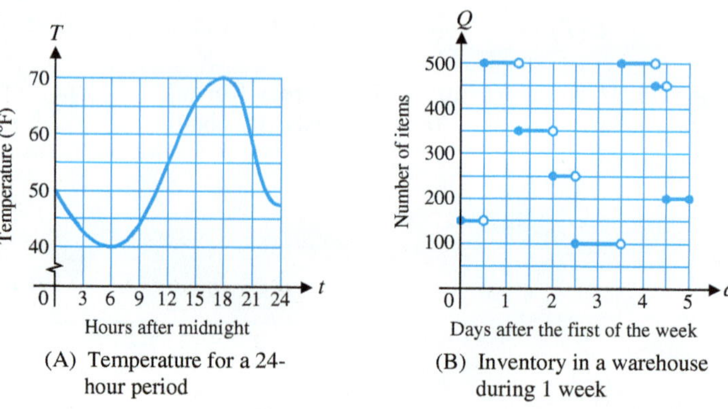

(A) Temperature for a 24-hour period

(B) Inventory in a warehouse during 1 week

Figure 2

Explore & Discuss 1

(A) Write a brief verbal description of the temperature variation illustrated in Figure 2A, including estimates of the high and low temperatures during the period shown and the times at which they occurred.

(B) Write a brief verbal description of the changes in inventory illustrated in Figure 2B, including estimates of the changes in inventory and the times at which those changes occurred.

The preceding discussion leads to the following formal definition of continuity:

DEFINITION Continuity

A function f is **continuous at the point $x = c$ if**

1. $\lim_{x \to c} f(x)$ exists **2.** $f(c)$ exists **3.** $\lim_{x \to c} f(x) = f(c)$

A function is **continuous on the open interval (a, b)** if it is continuous at each point on the interval.

If one or more of the three conditions in the definition fails, then the function is **discontinuous** at $x = c$.

EXAMPLE 1 **Continuity of a Function Defined by a Graph** Use the definition of continuity to discuss the continuity of the function whose graph is shown in Figure 3.

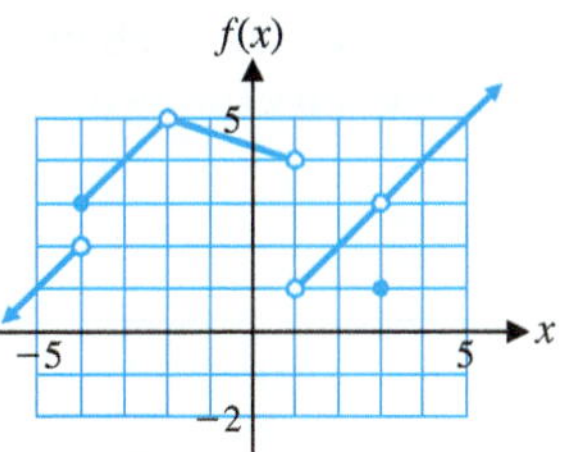

Figure 3

SOLUTION We begin by identifying the points of discontinuity. Examining the graph, we see breaks or holes at $x = -4, -2, 1$, and 3. Now we must determine which conditions in the definition of continuity are not satisfied at each of these points. In each case, we find the value of the function and the limit of the function at the point in question.

Discontinuity at $x = -4$:

$$\lim_{x \to -4^-} f(x) = 2$$

$$\lim_{x \to -4^+} f(x) = 3$$

Since the one-sided limits are different, the limit does not exist (Section 1-1).

$$\lim_{x \to -4} f(x) \text{ does not exist}$$

$$f(-4) = 3$$

So, f is not continuous at $x = -4$ because condition 1 is not satisfied.

Discontinuity at $x = -2$:

$$\lim_{x \to -2^-} f(x) = 5$$

$$\lim_{x \to -2^+} f(x) = 5$$

$$\lim_{x \to -2} f(x) = 5$$

$$f(-2) \text{ does not exist}$$

The hole at $(-2, 5)$ indicates that 5 is not the value of f at -2. Since there is no solid dot elsewhere on the vertical line $x = -2$, $f(-2)$ is not defined.

So f is not continuous at $x = -2$ because condition 2 is not satisfied.

Discontinuity at $x = 1$:

$$\lim_{x \to 1^-} f(x) = 4$$
$$\lim_{x \to 1^+} f(x) = 1$$
$$\lim_{x \to 1} f(x) \text{ does not exist}$$
$$f(1) \text{ does not exist}$$

This time, f is not continuous at $x = 1$ because neither of conditions 1 and 2 is satisfied.

Discontinuity at $x = 3$:

$$\lim_{x \to 3^-} f(x) = 3$$ The solid dot at (3, 1) indicates that $f(3) = 1$.
$$\lim_{x \to 3^+} f(x) = 3$$
$$\lim_{x \to 3} f(x) = 3$$
$$f(3) = 1$$

Conditions 1 and 2 are satisfied, but f is not continuous at $x = 3$ because condition 3 is not satisfied.

Having identified and discussed all points of discontinuity, we can now conclude that f is continuous except at $x = -4, -2, 1$, and 3.

CONCEPTUAL INSIGHT

Rather than list the points where a function is discontinuous, sometimes it is useful to state the intervals on which the function is continuous. Using the set operation **union,** denoted by ∪, we can express the set of points where the function in Example 1 is continuous as follows:

$$(-\infty, -4) \cup (-4, -2) \cup (-2, 1) \cup (1, 3) \cup (3, \infty)$$

Matched Problem 1 Use the definition of continuity to discuss the continuity of the function whose graph is shown in Figure 4.

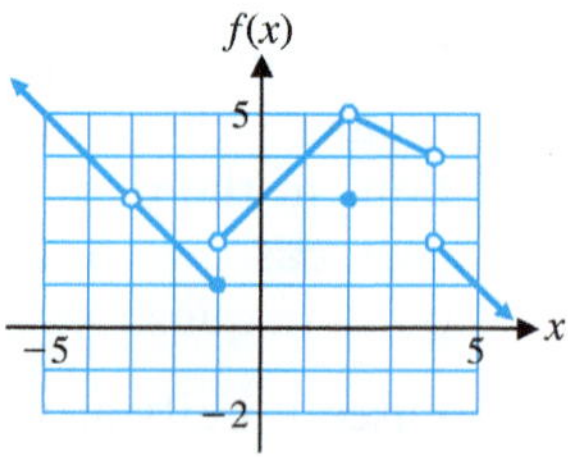

Figure 4

For functions defined by equations, it is important to be able to locate points of discontinuity by examining the equation.

EXAMPLE 2 **Continuity of Functions Defined by Equations** Using the definition of continuity, discuss the continuity of each function at the indicated point(s).

(A) $f(x) = x + 2$ at $x = 2$

(B) $g(x) = \dfrac{x^2 - 4}{x - 2}$ at $x = 2$

(C) $h(x) = \dfrac{|x|}{x}$ at $x = 0$ and at $x = 1$

SOLUTION (A) f is continuous at $x = 2$, since

$$\lim_{x\to 2} f(x) = 4 = f(2) \quad \text{See Figure 1A.}$$

(B) g is not continuous at $x = 2$, since $g(2) = 0/0$ is not defined (see Fig. 1B).

(C) h is not continuous at $x = 0$, since $h(0) = |0|/0$ is not defined; also, $\lim_{x\to 0} h(x)$ does not exist.

h is continuous at $x = 1$, since

$$\lim_{x\to 1} \frac{|x|}{x} = 1 = h(1) \quad \text{See Figure 1C.}$$

Matched Problem 2 Using the definition of continuity, discuss the continuity of each function at the indicated point(s).

(A) $f(x) = x + 1$ at $x = 1$ (B) $g(x) = \frac{x^2 - 1}{x - 1}$ at $x = 1$

(C) $h(x) = \frac{x - 2}{|x - 2|}$ at $x = 2$ and at $x = 0$

We can also talk about one-sided continuity, just as we talked about one-sided limits. For example, a function is said to be **continuous on the right** at $x = c$ if $\lim_{x\to c^+} f(x) = f(c)$ and **continuous on the left** at $x = c$ if $\lim_{x\to c^-} f(x) = f(c)$. A function is **continuous on the closed interval [a, b]** if it is continuous on the open interval (a, b) and is continuous both on the right at a and on the left at b.

Figure 5A illustrates a function that is continuous on the closed interval $[-1, 1]$. Figure 5B illustrates a function that is continuous on the half-closed interval $[0, \infty)$.

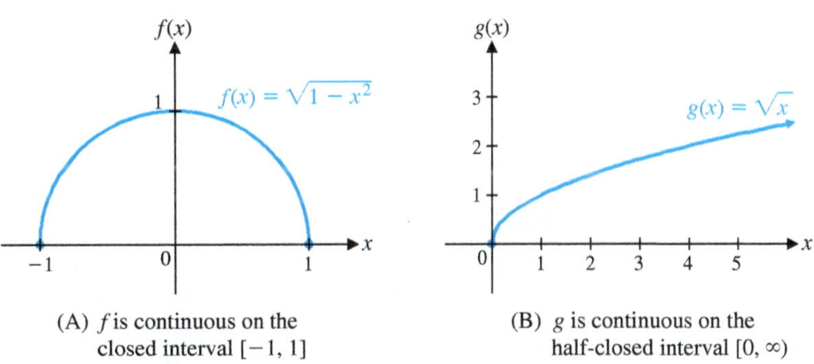

Figure 5 Continuity on closed and half-closed intervals

Continuity Properties

Functions have some useful **general continuity properties:**

If two functions are continuous on the same interval, then their sum, difference, product, and quotient are continuous on the same interval except for values of x that make a denominator 0.

These properties, along with Theorem 1, enable us to determine intervals of continuity for some important classes of functions without having to look at their graphs or use the three conditions in the definition.

THEOREM 1 Continuity Properties of Some Specific Functions

(A) A constant function $f(x) = k$, where k is a constant, is continuous for all x.

$f(x) = 7$ is continuous for all x.

(B) For n a positive integer, $f(x) = x^n$ is continuous for all x.

$f(x) = x^5$ is continuous for all x.

(C) A polynomial function is continuous for all x.

$2x^3 - 3x^2 + x - 5$ is continuous for all x.

(D) A rational function is continuous for all x except those values that make a denominator 0.

$\frac{x^2 + 1}{x - 1}$ is continuous for all x except $x = 1$, a value that makes the denominator 0.

(E) For n an odd positive integer greater than 1, $\sqrt[n]{f(x)}$ is continuous wherever $f(x)$ is continuous.

$\sqrt[3]{x^2}$ is continuous for all x.

(F) For n an even positive integer, $\sqrt[n]{f(x)}$ is continuous wherever $f(x)$ is continuous and nonnegative.

$\sqrt[4]{x}$ is continuous on the interval $[0, \infty)$.

Parts C and D of Theorem 1 are the same as Theorem 3 in Section 1-1. They are repeated here to emphasize their importance.

EXAMPLE 3 **Using Continuity Properties** Using Theorem 1 and the general properties of continuity, determine where each function is continuous.

(A) $f(x) = x^2 - 2x + 1$

(B) $f(x) = \frac{x}{(x + 2)(x - 3)}$

(C) $f(x) = \sqrt[3]{x^2 - 4}$

(D) $f(x) = \sqrt{x - 2}$

SOLUTION

(A) Since f is a polynomial function, f is continuous for all x.

(B) Since f is a rational function, f is continuous for all x except -2 and 3 (values that make the denominator 0).

(C) The polynomial function $x^2 - 4$ is continuous for all x. Since $n = 3$ is odd, f is continuous for all x.

(D) The polynomial function $x - 2$ is continuous for all x and nonnegative for $x \geq 2$. Since $n = 2$ is even, f is continuous for $x \geq 2$, or on the interval $[2, \infty)$.

Matched Problem 3 Using Theorem 1 and the general properties of continuity, determine where each function is continuous.

(A) $f(x) = x^4 + 2x^2 + 1$

(B) $f(x) = \frac{x^2}{(x + 1)(x - 4)}$

(C) $f(x) = \sqrt{x - 4}$

(D) $f(x) = \sqrt[3]{x^3 + 1}$

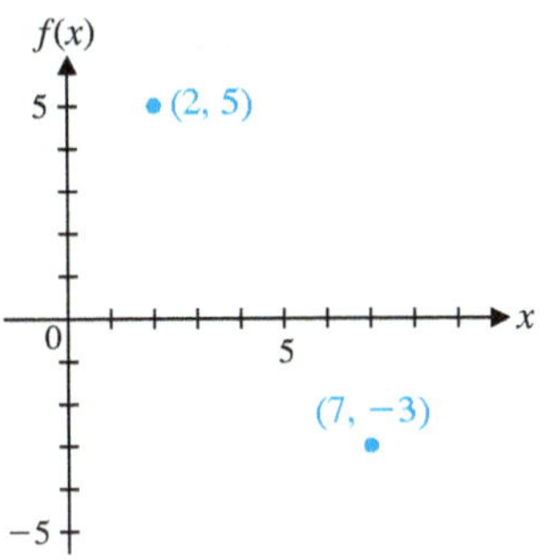

Figure 6

Solving Inequalities Using Continuity Properties

One of the basic tools for analyzing graphs in calculus is a special line graph called a *sign chart.* We will make extensive use of this type of chart in later sections. In the discussion that follows, we use continuity properties to develop a simple and efficient procedure for constructing sign charts.

Suppose that a function f is continuous over the interval $(1, 8)$ and $f(x) \neq 0$ for any x in $(1, 8)$. Suppose also that $f(2) = 5$, a positive number. Is it possible for $f(x)$ to be negative for any x in the interval $(1, 8)$? The answer is "no." If $f(7)$ were -3, for example, as shown in Figure 6, then how would it be possible to join the points $(2, 5)$ and $(7, -3)$ with the graph of a continuous function without crossing the x axis between 1 and 8 at least once? [Crossing the x axis would violate our assumption that $f(x) \neq 0$ for any x in $(1, 8)$.] We conclude that $f(x)$ must be positive for all x in $(1, 8)$. If $f(2)$ were negative, then, using the same type of reasoning, $f(x)$ would have to be negative over the entire interval $(1, 8)$.

In general, **if f is continuous and $f(x) \neq 0$ on the interval (a, b), then $f(x)$ cannot change sign on (a, b).** This is the essence of Theorem 2.

> **THEOREM 2 Sign Properties on an Interval (a, b)**
> If f is continuous on (a, b) and $f(x) \neq 0$ for all x in (a, b), then either $f(x) > 0$ for all x in (a, b) or $f(x) < 0$ for all x in (a, b).

Theorem 2 provides the basis for an effective method of solving many types of inequalities. Example 4 illustrates the process.

EXAMPLE 4 **Solving an Inequality** Solve $\dfrac{x + 1}{x - 2} > 0$.

SOLUTION We start by using the left side of the inequality to form the function f.

$$f(x) = \frac{x + 1}{x - 2}$$

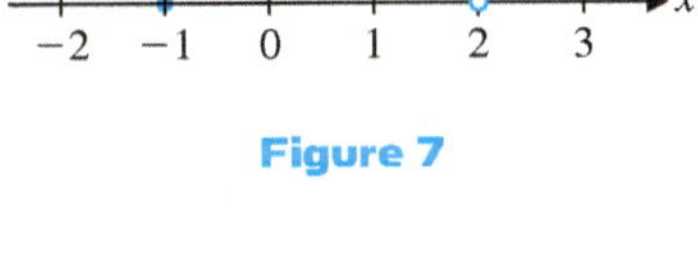

Figure 7

The rational function f is discontinuous at $x = 2$, and $f(x) = 0$ for $x = -1$ (a fraction is 0 when the numerator is 0 and the denominator is not 0). We plot $x = 2$ and $x = -1$, which we call *partition numbers,* on a real-number line (Fig. 7). (Note that the dot at 2 is open because the function is not defined at $x = 2$.) The partition numbers 2 and -1 determine three open intervals: $(-\infty, -1)$, $(-1, 2)$, and $(2, \infty)$. The function f is continuous and nonzero on each of these intervals. From Theorem 2, we know that $f(x)$ does not change sign on any of these intervals. We can find the sign of $f(x)$ on each of the intervals by selecting a **test number** in each interval and evaluating $f(x)$ at that number. Since any number in each subinterval will do, we choose test numbers that are easy to evaluate: -2, 0, and 3. The table in the margin shows the results.

Test Numbers

x	$f(x)$
-2	$\frac{1}{4}\ (+)$
0	$-\frac{1}{2}\ (-)$
3	$4\ (+)$

The sign of $f(x)$ at each test number is the same as the sign of $f(x)$ over the interval containing that test number. Using this information, we construct a **sign chart** for $f(x)$ as shown in Figure 8.

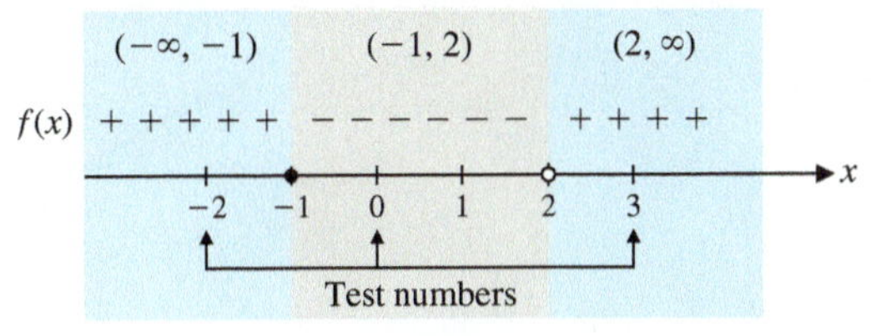

Figure 8

From the sign chart, we can easily write the solution of the given nonlinear inequality:

$$f(x) > 0 \quad \text{for} \quad \begin{array}{ll} x < -1 \text{ or } x > 2 & \text{Inequality notation} \\ (-\infty, -1) \cup (2, \infty) & \text{Interval notation} \end{array}$$

Matched Problem 4 Solve $\dfrac{x^2 - 1}{x - 3} < 0$.

Most of the inequalities we encounter will involve strict inequalities ($>$ or $<$). If it is necessary to solve inequalities of the form $\geq$ or $\leq$, we simply include the endpoint x of any interval if f is defined at x and $f(x)$ satisfies the given inequality. For example, from the sign chart in Figure 8, the solution of the inequality

$$\frac{x+1}{x-2} \geq 0 \quad \text{is} \quad \begin{array}{ll} x \leq -1 \text{ or } x > 2 & \text{Inequality notation} \\ (-\infty, -1] \cup (2, \infty) & \text{Interval notation} \end{array}$$

In general, given a function f, a **partition number** is a value of x such that f is discontinuous at x or $f(x) = 0$. **Partition numbers determine open intervals in which $f(x)$ does not change sign.** By using a test number from each interval, we can construct a sign chart for $f(x)$ on the real-number line. It is then easy to solve the inequality $f(x) < 0$ or $f(x) > 0$.

We summarize the procedure for constructing sign charts in the following box:

PROCEDURE Constructing Sign Charts

Given a function f,

Step 1 Find all partition numbers:

(A) Find all numbers such that f is discontinuous. (Rational functions are discontinuous for values of x that make a denominator 0.)

(B) Find all numbers such that $f(x) = 0$. (For a rational function, this occurs where the numerator is 0 and the denominator is not 0.)

Step 2 Plot the numbers found in step 1 on a real-number line, dividing the number line into intervals.

Step 3 Select a test number in each open interval determined in step 2 and evaluate $f(x)$ at each test number to determine whether $f(x)$ is positive (+) or negative (−) in each interval.

Step 4 Construct a sign chart, using the real-number line in step 2. This will show the sign of $f(x)$ on each open interval.

Exercises 1-3

A

In Problems 1–6, sketch a possible graph of a function that satisfies the given conditions at $x = 1$ and discuss the continuity of f at $x = 1$.

1. $f(1) = 2$ and $\lim_{x\to 1} f(x) = 2$
2. $f(1) = -2$ and $\lim_{x\to 1} f(x) = 2$
3. $f(1) = 2$ and $\lim_{x\to 1} f(x) = -2$
4. $f(1) = -2$ and $\lim_{x\to 1} f(x) = -2$
5. $f(1) = -2$, $\lim_{x\to 1^-} f(x) = 2$, and $\lim_{x\to 1^+} f(x) = -2$
6. $f(1) = 2$, $\lim_{x\to 1^-} f(x) = 2$, and $\lim_{x\to 1^+} f(x) = -2$

Problems 7–14 refer to the function f shown in the figure. Use the graph to estimate the indicated function values and limits.

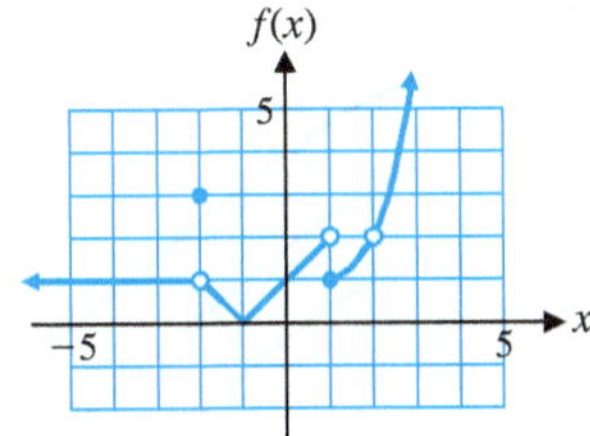

Figure for 7–14

7. $f(0.9)$
8. $f(0.1)$
9. $f(-1.9)$
10. $f(-0.9)$
11. (A) $\lim_{x\to 1^-} f(x)$ (B) $\lim_{x\to 1^+} f(x)$
(C) $\lim_{x\to 1} f(x)$ (D) $f(1)$
(E) Is f continuous at $x = 1$? Explain.
12. (A) $\lim_{x\to 2^-} f(x)$ (B) $\lim_{x\to 2^+} f(x)$
(C) $\lim_{x\to 2} f(x)$ (D) $f(2)$
(E) Is f continuous at $x = 2$? Explain.
13. (A) $\lim_{x\to -2^-} f(x)$ (B) $\lim_{x\to -2^+} f(x)$
(C) $\lim_{x\to -2} f(x)$ (D) $f(-2)$
(E) Is f continuous at $x = -2$? Explain.
14. (A) $\lim_{x\to -1^-} f(x)$ (B) $\lim_{x\to -1^+} f(x)$
(C) $\lim_{x\to -1} f(x)$ (D) $f(-1)$
(E) Is f continuous at $x = -1$? Explain.

Problems 15–22 refer to the function g shown in the figure. Use the graph to estimate the indicated function values and limits.

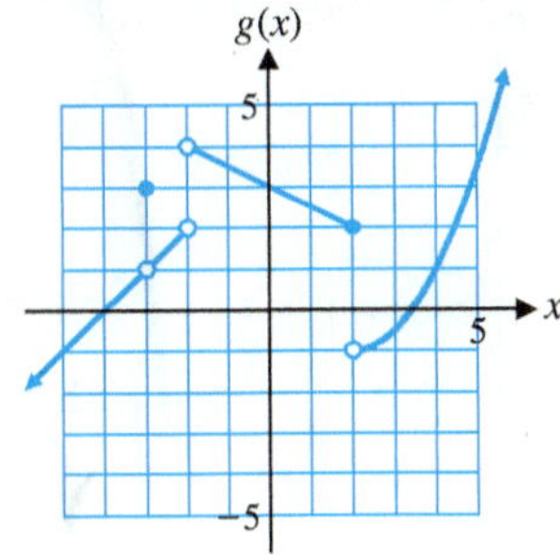

Figure for 15–22

15. $g(-3.1)$
16. $g(-2.1)$
17. $g(1.9)$
18. $g(-1.9)$
19. (A) $\lim_{x\to -3^-} g(x)$ (B) $\lim_{x\to -3^+} g(x)$
(C) $\lim_{x\to -3} g(x)$ (D) $g(-3)$
(E) Is g continuous at $x = -3$? Explain.
20. (A) $\lim_{x\to -2^-} g(x)$ (B) $\lim_{x\to -2^+} g(x)$
(C) $\lim_{x\to -2} g(x)$ (D) $g(-2)$
(E) Is g continuous at $x = -2$? Explain.
21. (A) $\lim_{x\to 2^-} g(x)$ (B) $\lim_{x\to 2^+} g(x)$
(C) $\lim_{x\to 2} g(x)$ (D) $g(2)$
(E) Is g continuous at $x = 2$? Explain.
22. (A) $\lim_{x\to 4^-} g(x)$ (B) $\lim_{x\to 4^+} g(x)$
(C) $\lim_{x\to 4} g(x)$ (D) $g(4)$
(E) Is g continuous at $x = 4$? Explain.

Use Theorem 1 to determine where each function in Problems 23–32 is continuous.

23. $f(x) = 3x - 4$
24. $h(x) = 4 - 2x$
25. $g(x) = \dfrac{3x}{x + 2}$
26. $k(x) = \dfrac{2x}{x - 4}$
27. $m(x) = \dfrac{x + 1}{(x - 1)(x + 4)}$
28. $n(x) = \dfrac{x - 2}{(x - 3)(x + 1)}$
29. $F(x) = \dfrac{2x}{x^2 + 9}$
30. $G(x) = \dfrac{1 - x^2}{x^2 + 1}$
31. $M(x) = \dfrac{x - 1}{4x^2 - 9}$
32. $N(x) = \dfrac{x^2 + 4}{4 - 25x^2}$

B

33. Given the function

$$f(x) = \begin{cases} 2 & \text{if } x \text{ is an integer} \\ 1 & \text{if } x \text{ is not an integer} \end{cases}$$

(A) Graph f.
(B) $\lim_{x\to 2} f(x) = ?$
(C) $f(2) = ?$
(D) Is f continuous at $x = 2$?
(E) Where is f discontinuous?

34. Given the function

$$g(x) = \begin{cases} -1 & \text{if } x \text{ is an even integer} \\ 1 & \text{if } x \text{ is not an even integer} \end{cases}$$

(A) Graph g.
(B) $\lim_{x\to 1} g(x) = ?$
(C) $g(1) = ?$
(D) Is g continuous at $x = 1$?
(E) Where is g discontinuous?

In Problems 35–42, use a sign chart to solve each inequality. Express answers in inequality and interval notation.

35. $x^2 - x - 12 < 0$

36. $x^2 - 2x - 8 < 0$

37. $x^2 + 21 > 10x$

38. $x^2 + 7x > -10$

39. $x^3 < 4x$

40. $x^4 - 9x^2 > 0$

41. $\dfrac{x^2 + 5x}{x - 3} > 0$

42. $\dfrac{x - 4}{x^2 + 2x} < 0$

43. Use the graph of f to determine where

(A) $f(x) > 0$ (B) $f(x) < 0$

Express answers in interval notation.

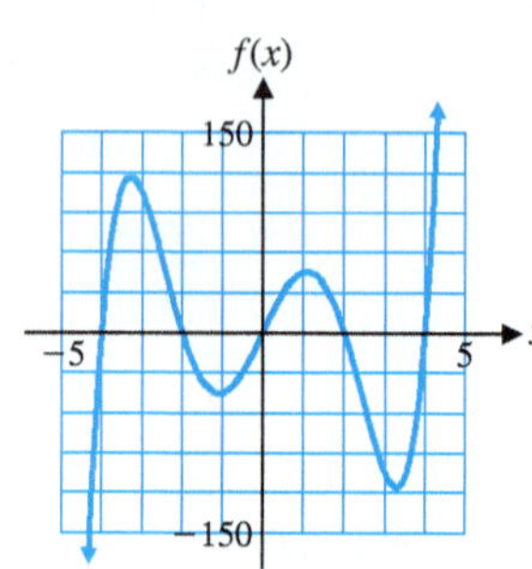

Figure for 43

44. Use the graph of g to determine where

(A) $g(x) > 0$ (B) $g(x) < 0$

Express answers in interval notation.

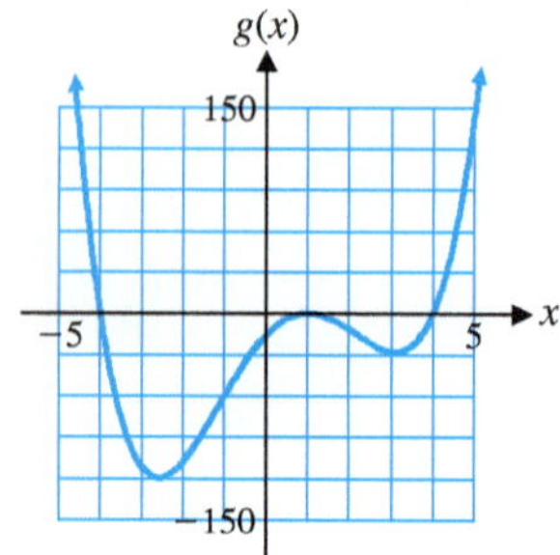

Figure for 44

In Problems 45–48, use a graphing calculator to approximate the partition numbers of each function f(x) to four decimal places. Then solve the following inequalities:

(A) $f(x) > 0$ *(B)* $f(x) < 0$

Express answers in interval notation.

45. $f(x) = x^4 - 6x^2 + 3x + 5$

46. $f(x) = x^4 - 4x^2 - 2x + 2$

47. $f(x) = \dfrac{3 + 6x - x^3}{x^2 - 1}$

48. $f(x) = \dfrac{x^3 - 5x + 1}{x^2 - 1}$

Use Theorem 1 to determine where each function in Problems 49–56 is continuous. Express the answer in interval notation.

49. $\sqrt{x - 6}$

50. $\sqrt{7 - x}$

51. $\sqrt[3]{5 - x}$

52. $\sqrt[3]{x - 8}$

53. $\sqrt{x^2 - 9}$

54. $\sqrt{4 - x^2}$

55. $\sqrt{x^2 + 1}$

56. $\sqrt[3]{x^2 + 2}$

In Problems 57–62, graph f, locate all points of discontinuity, and discuss the behavior of f at these points.

57. $f(x) = \begin{cases} 1 + x & \text{if } x < 1 \\ 5 - x & \text{if } x \ge 1 \end{cases}$

58. $f(x) = \begin{cases} x^2 & \text{if } x \le 1 \\ 2x & \text{if } x > 1 \end{cases}$

59. $f(x) = \begin{cases} 1 + x & \text{if } x \le 2 \\ 5 - x & \text{if } x > 2 \end{cases}$

60. $f(x) = \begin{cases} x^2 & \text{if } x \le 2 \\ 2x & \text{if } x > 2 \end{cases}$

61. $f(x) = \begin{cases} -x & \text{if } x < 0 \\ 1 & \text{if } x = 0 \\ x & \text{if } x > 0 \end{cases}$

62. $f(x) = \begin{cases} 1 & \text{if } x < 0 \\ 0 & \text{if } x = 0 \\ 1 + x & \text{if } x > 0 \end{cases}$

C

Problems 63 and 64 refer to the ***greatest integer function****, which is denoted by* $[x]$ *and is defined as*

$$[x] = \textit{greatest integer} \le x$$

For example,

$$[-3.6] = \textit{greatest integer} \le -3.6 = -4$$
$$[2] = \textit{greatest integer} \le 2 = 2$$
$$[2.5] = \textit{greatest integer} \le 2.5 = 2$$

The graph of $f(x) = [x]$ *is shown. There, we can see that*

$$[x] = -2 \quad \textit{for} \quad -2 \le x < -1$$
$$[x] = -1 \quad \textit{for} \quad -1 \le x < 0$$
$$[x] = 0 \quad \textit{for} \quad 0 \le x < 1$$
$$[x] = 1 \quad \textit{for} \quad 1 \le x < 2$$
$$[x] = 2 \quad \textit{for} \quad 2 \le x < 3$$

and so on.

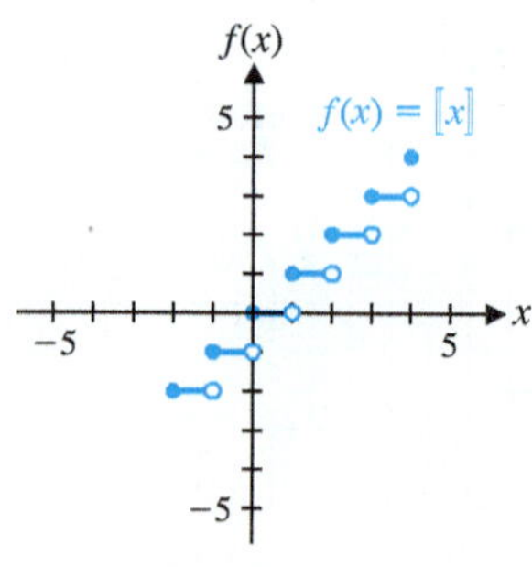

Figure for 63 and 64

63. (A) Is f continuous from the right at $x = 0$?
(B) Is f continuous from the left at $x = 0$?
(C) Is f continuous on the open interval $(0, 1)$?
(D) Is f continuous on the closed interval $[0, 1]$?
(E) Is f continuous on the half-closed interval $[0, 1)$?

64. (A) Is f continuous from the right at $x = 2$?
(B) Is f continuous from the left at $x = 2$?
(C) Is f continuous on the open interval $(1, 2)$?
(D) Is f continuous on the closed interval $[1, 2]$?
(E) Is f continuous on the half-closed interval $[1, 2)$?

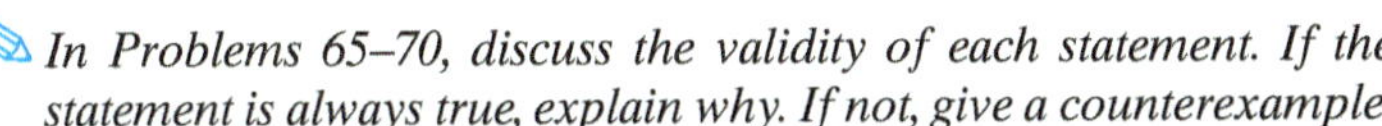

In Problems 65–70, discuss the validity of each statement. If the statement is always true, explain why. If not, give a counterexample.

65. A polynomial function is continuous for all real numbers.

66. A rational function is continuous for all but finitely many real numbers.

67. If f is a function that is continuous at $x = 0$ and $x = 2$, then f is continuous at $x = 1$.

68. If f is a function that is continuous on the open interval $(0, 2)$, then f is continuous at $x = 1$.

69. If f is a function that has no partition numbers in the interval (a, b), then f is continuous on (a, b).

70. The greatest integer function (see Problem 61) is a rational function.

In Problems 71–74, sketch a possible graph of a function f that is continuous for all real numbers and satisfies the given conditions. Find the x intercepts of f.

71. $f(x) < 0$ on $(-\infty, -5)$ and $(2, \infty)$; $f(x) > 0$ on $(-5, 2)$

72. $f(x) > 0$ on $(-\infty, -4)$ and $(3, \infty)$; $f(x) < 0$ on $(-4, 3)$

73. $f(x) < 0$ on $(-\infty, -6)$ and $(-1, 4)$; $f(x) > 0$ on $(-6, -1)$ and $(4, \infty)$

74. $f(x) > 0$ on $(-\infty, -3)$ and $(2, 7)$; $f(x) < 0$ on $(-3, 2)$ and $(7, \infty)$

75. The function $f(x) = 2/(1 - x)$ satisfies $f(0) = 2$ and $f(2) = -2$. Is f equal to 0 anywhere on the interval $(-1, 3)$? Does this contradict Theorem 2? Explain.

76. The function $f(x) = 6/(x - 4)$ satisfies $f(2) = -3$ and $f(7) = 2$. Is f equal to 0 anywhere on the interval $(0, 9)$? Does this contradict Theorem 2? Explain.

Applications

77. **Postal rates.** First-class postage in 2009 was \$0.44 for the first ounce (or any fraction thereof) and \$0.17 for each additional ounce (or fraction thereof) up to a maximum weight of 3.5 ounces.
(A) Write a piecewise definition of the first-class postage $P(x)$ for a letter weighing x ounces.
(B) Graph $P(x)$ for $0 < x \le 3.5$.
(C) Is $P(x)$ continuous at $x = 2.5$? At $x = 3$? Explain.

78. **Telephone rates.** A long-distance telephone service charges \$0.07 for the first minute (or any fraction thereof) and \$0.05 for each additional minute (or fraction thereof).
(A) Write a piecewise definition of the charge $R(x)$ for a long-distance call lasting x minutes.
(B) Graph $R(x)$ for $0 < x \le 6$.
(C) Is $R(x)$ continuous at $x = 3.5$? At $x = 3$? Explain.

79. **Postal rates.** Discuss the differences between the function $Q(x) = 0.44 + 0.17[x]$ and the function $P(x)$ defined in Problem 77.

80. **Telephone rates.** Discuss the differences between the function $S(x) = 0.07 + 0.05[x]$ and the function $R(x)$ defined in Problem 78.

81. **Natural-gas rates.** Table 1 shows the rates for natural gas charged by the Middle Tennessee Natural Gas Utility District during summer months. The customer charge is a fixed monthly charge, independent of the amount of gas used per month.

Table 1 Summer (May–September)

Base charge	\$5.00
First 50 therms	0.63 per therm
Over 50 therms	0.45 per therm

(A) Write a piecewise definition of the monthly charge $S(x)$ for a customer who uses x therms* in a summer month.
(B) Graph $S(x)$.
(C) Is $S(x)$ continuous at $x = 50$? Explain.

82. **Natural-gas rates.** Table 2 shows the rates for natural gas charged by the Middle Tennessee Natural Gas Utility District during winter months. The customer charge is a fixed monthly charge, independent of the amount of gas used per month.

Table 2 Winter (October– April)

Base charge	\$5.00
First 5 therms	0.69 per therm
Next 45 therms	0.65 per therm
Over 50 therms	0.63 per therm

*A British thermal unit (Btu) is the amount of heat required to raise the temperature of 1 pound of water 1 degree Fahrenheit, and a therm is 100,000 Btu.

(A) Write a piecewise definition of the monthly charge $S(x)$ for a customer who uses x therms in a winter month.

(B) Graph $S(x)$.

(C) Is $S(x)$ continuous at $x = 5$? At $x = 50$? Explain.

83. Income. A personal-computer salesperson receives a base salary of \$1,000 per month and a commission of 5% of all sales over \$10,000 during the month. If the monthly sales are \$20,000 or more, then the salesperson is given an additional \$500 bonus. Let $E(s)$ represent the person's earnings per month as a function of the monthly sales s.

(A) Graph $E(s)$ for $0 \le s \le 30{,}000$.

(B) Find $\lim_{s \to 10{,}000} E(s)$ and $E(10{,}000)$.

(C) Find $\lim_{s \to 20{,}000} E(s)$ and $E(20{,}000)$.

(D) Is E continuous at $s = 10{,}000$? At $s = 20{,}000$?

84. Equipment rental. An office equipment rental and leasing company rents copiers for \$10 per day (and any fraction thereof) or for \$50 per 7-day week. Let $C(x)$ be the cost of renting a copier for x days.

(A) Graph $C(x)$ for $0 \le x \le 10$.

(B) Find $\lim_{x \to 4.5} C(x)$ and $C(4.5)$.

(C) Find $\lim_{x \to 8} C(x)$ and $C(8)$.

(D) Is C continuous at $x = 4.5$? At $x = 8$?

85. Animal supply. A medical laboratory raises its own rabbits. The number of rabbits $N(t)$ available at any time t depends on the number of births and deaths. When a birth or death occurs, the function N generally has a discontinuity, as shown in the figure.

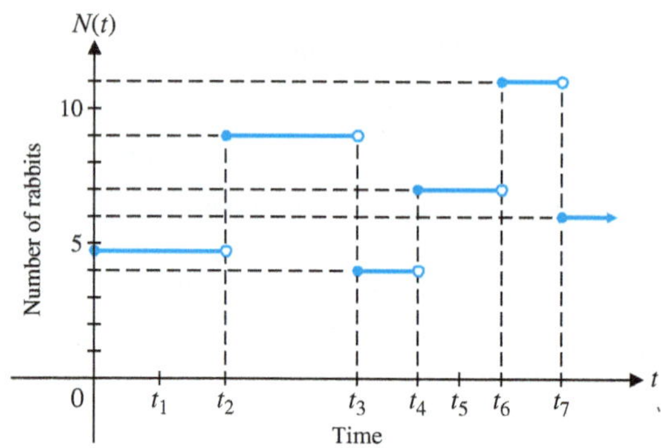

Figure for 85

(A) Where is the function N discontinuous?

(B) $\lim_{t \to t_5} N(t) = ?$; $N(t_5) = ?$

(C) $\lim_{t \to t_3} N(t) = ?$; $N(t_3) = ?$

86. Learning. The graph shown represents the history of a person learning the material on limits and continuity in this book. At time t_2, the student's mind goes blank during a quiz. At time t_4, the instructor explains a concept particularly well, then suddenly a big jump in understanding takes place.

(A) Where is the function p discontinuous?

(B) $\lim_{t \to t_1} p(t) = ?$; $p(t_1) = ?$

(C) $\lim_{t \to t_2} p(t) = ?$; $p(t_2) = ?$

(D) $\lim_{t \to t_4} p(t) = ?$; $p(t_4) = ?$

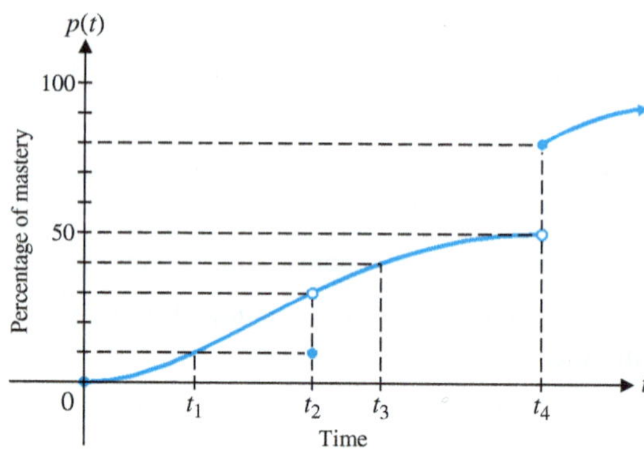

Figure for 86

Answers to Matched Problems

1. f is not continuous at $x = -3, -1, 2$, and 4.

$x = -3$: $\lim_{x \to -3} f(x) = 3$, but $f(-3)$ does not exist

$x = -1$: $f(-1) = 1$, but $\lim_{x \to -1} f(x)$ does not exist

$x = 2$: $\lim_{x \to 2} f(x) = 5$, but $f(2) = 3$

$x = 4$: $\lim_{x \to 4} f(x)$ does not exist, and $f(4)$ does not exist

2. (A) f is continuous at $x = 1$, since $\lim_{x \to 1} f(x) = 2 = f(1)$.

(B) g is not continuous at $x = 1$, since $g(1)$ is not defined.

(C) h is not continuous at $x = 2$ for two reasons: $h(2)$ does not exist and $\lim_{x \to 2} h(x)$ does not exist.
h is continuous at $x = 0$, since $\lim_{x \to 0} h(x) = -1 = h(0)$.

3. (A) Since f is a polynomial function, f is continuous for all x.

(B) Since f is a rational function, f is continuous for all x except -1 and 4 (values that make the denominator 0).

(C) The polynomial function $x - 4$ is continuous for all x and nonnegative for $x \ge 4$. Since $n = 2$ is even, f is continuous for $x \ge 4$, or on the interval $[4, \infty)$.

(D) The polynomial function $x^3 + 1$ is continuous for all x. Since $n = 3$ is odd, f is continuous for all x.

4. $-\infty < x < -1$ or $1 < x < 3$; $(-\infty, -1) \cup (1, 3)$

1-4 The Derivative

We will now make use of the limit concepts developed in Sections 1-1, 1-2, and 1-3 to solve the two important problems illustrated in Figure 1. The solution of each of these apparently unrelated problems involves a common concept called the *derivative.*

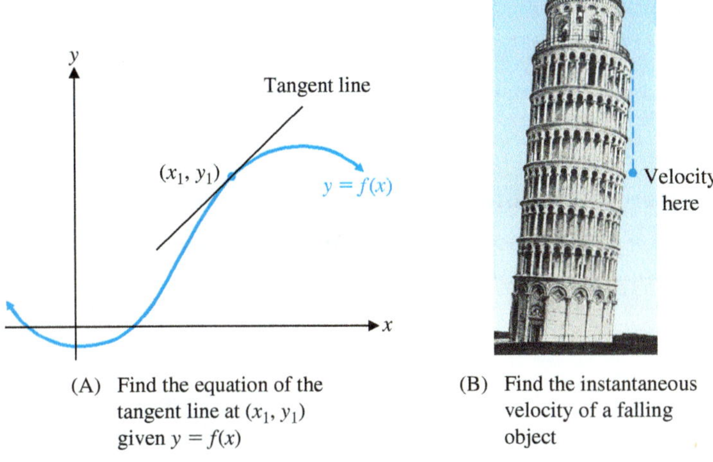

(A) Find the equation of the tangent line at (x_1, y_1) given $y = f(x)$

(B) Find the instantaneous velocity of a falling object

Figure 1 Two basic problems of calculus

Rate of Change

Let us start by considering a simple example.

EXAMPLE 1 **Revenue Analysis** The revenue (in dollars) from the sale of x plastic planter boxes is given by

$$R(x) = 20x - 0.02x^2 \qquad 0 \le x \le 1{,}000$$

and is graphed in Figure 2.

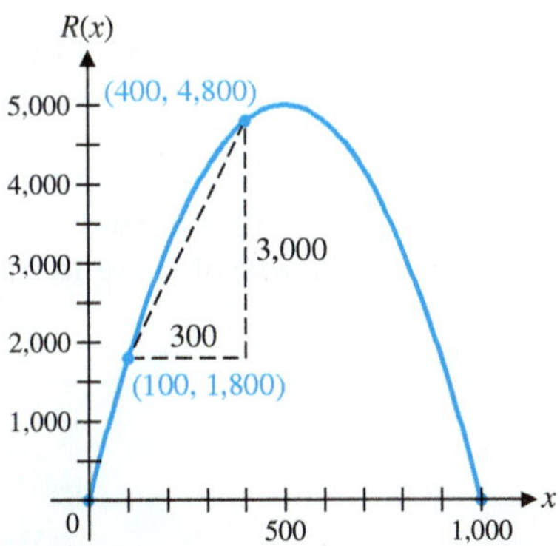

Figure 2 $R(x) = 20x - 0.02x^2$

(A) What is the change in revenue if production is changed from 100 planters to 400 planters?

(B) What is the average change in revenue for this change in production?

SOLUTION (A) The change in revenue is given by

$$R(400) - R(100) = 20(400) - 0.02(400)^2 - [20(100) - 0.02(100)^2]$$
$$= 4{,}800 - 1{,}800 = \$3{,}000$$

Increasing production from 100 planters to 400 planters will increase revenue by $3,000.

(B) To find the average change in revenue, we divide the change in revenue by the change in production:

$$\frac{R(400) - R(100)}{400 - 100} = \frac{3{,}000}{300} = \$10$$

The average change in revenue is $10 per planter when production is increased from 100 to 400 planters.

Matched Problem 1 Refer to the revenue function in Example 1.

(A) What is the change in revenue if production is changed from 600 planters to 800 planters?

(B) What is the average change in revenue for this change in production?

In general, if we are given a function $y = f(x)$ and if x is changed from a to $a + h$, then y will change from $f(a)$ to $f(a + h)$. The *average rate of change is the ratio of the change in y to the change in x.*

DEFINITION Average Rate of Change

For $y = f(x)$, the **average rate of change from $x = a$ to $x = a + h$** is

$$\frac{f(a + h) - f(a)}{(a + h) - a} = \frac{f(a + h) - f(a)}{h} \qquad h \neq 0 \qquad (1)$$

As we noted in Section 1-1, equation (1) is called the **difference quotient**. The preceding discussion shows that the difference quotient can be interpreted as an average rate of change. The next example illustrates another interpretation of this quotient: the velocity of a moving object.

EXAMPLE 2 **Velocity** A small steel ball dropped from a tower will fall a distance of y feet in x seconds, as given approximately by the formula

$$y = f(x) = 16x^2$$

Figure 3 shows the position of the ball on a coordinate line (positive direction down) at the end of 0, 1, 2, and 3 seconds.

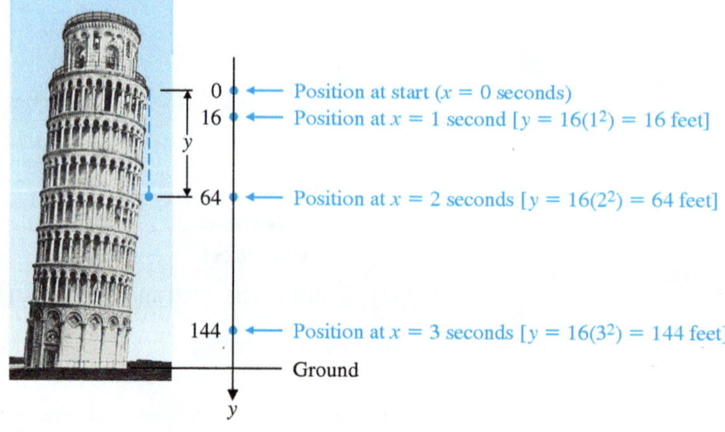

Figure 3 Note: Positive y direction is down.

(A) Find the average velocity from $x = 2$ seconds to $x = 3$ seconds.

(B) Find and simplify the average velocity from $x = 2$ seconds to $x = 2 + h$ seconds, $h \neq 0$.

(C) Find the limit of the expression from part B as $h \to 0$ if that limit exists.

(D) Discuss possible interpretations of the limit from part C.

SOLUTION (A) Recall the formula $d = rt$, which can be written in the form

$$r = \frac{d}{t} = \frac{\text{Distance covered}}{\text{Elapsed time}} = \text{Average velocity}$$

For example, if a person drives from San Francisco to Los Angeles (a distance of about 420 miles) in 7 hours, then the average velocity is

$$r = \frac{d}{t} = \frac{420}{7} = 60 \text{ miles per hour}$$

Sometimes the person will be traveling faster and sometimes slower, but the average velocity is 60 miles per hour. In our present problem, the average velocity of the steel ball from $x = 2$ seconds to $x = 3$ seconds is

$$\begin{aligned}\text{Average velocity} &= \frac{\text{Distance covered}}{\text{Elapsed time}} \\ &= \frac{f(3) - f(2)}{3 - 2} \\ &= \frac{16(3)^2 - 16(2)^2}{1} = 80 \text{ feet per second}\end{aligned}$$

We see that if $y = f(x)$ is the position of the falling ball, then the average velocity is simply the average rate of change of $f(x)$ with respect to time x, and we have another interpretation of the difference quotient (1).

(B) Proceeding as in part A, we have

$$\begin{aligned}\text{Average velocity} &= \frac{\text{Distance covered}}{\text{Elapsed time}} \\ &= \frac{f(2 + h) - f(2)}{h} && \text{Difference quotient} \\ &= \frac{16(2 + h)^2 - 16(2)^2}{h} && \text{Simplify this 0/0 indeterminate form.} \\ &= \frac{64 + 64h + 16h^2 - 64}{h} \\ &= \frac{h(64 + 16h)}{h} = 64 + 16h \qquad h \neq 0\end{aligned}$$

Notice that if $h = 1$, the average velocity is 80 feet per second, which is the result in part A.

(C) The limit of the average velocity expression from part B as $h \to 0$ is

$$\begin{aligned}\lim_{h \to 0} \frac{f(2 + h) - f(2)}{h} &= \lim_{h \to 0} (64 + 16h) \\ &= 64 \text{ feet per second}\end{aligned}$$

(D) The average velocity over smaller and smaller time intervals approaches 64 feet per second. This limit can be interpreted as the velocity of the ball at the *instant* that the ball has been falling for exactly 2 seconds. Therefore, 64 feet per second is referred to as the **instantaneous velocity** at $x = 2$ seconds, and we have solved one of the basic problems of calculus (see Fig. 1B).

Matched Problem 2 For the falling steel ball in Example 2, find

(A) The average velocity from $x = 1$ second to $x = 2$ seconds

(B) The average velocity (in simplified form) from $x = 1$ second to $x = 1 + h$ seconds, $h \neq 0$

(C) The instantaneous velocity at $x = 1$ second

The ideas introduced in Example 2 are not confined to average velocity, but can be applied to the average rate of change of any function.

DEFINITION Instantaneous Rate of Change

For $y = f(x)$, the **instantaneous rate of change at $x = a$** is

$$\lim_{h \to 0} \frac{f(a + h) - f(a)}{h} \qquad (2)$$

if the limit exists.

The adjective *instantaneous* is often omitted with the understanding that the phrase **rate of change** always refers to the instantaneous rate of change and not the average rate of change. Similarly, **velocity** always refers to the instantaneous rate of change of distance with respect to time.

Slope of the Tangent Line

So far, our interpretations of the difference quotient have been numerical in nature. Now we want to consider a geometric interpretation. A line through two points on the graph of a function is called a **secant line.** If $(a, f(a))$ and $(a + h, f(a + h))$ are two points on the graph of $y = f(x)$, then we can use the slope formula to find the slope of the secant line through these points (Fig. 4).

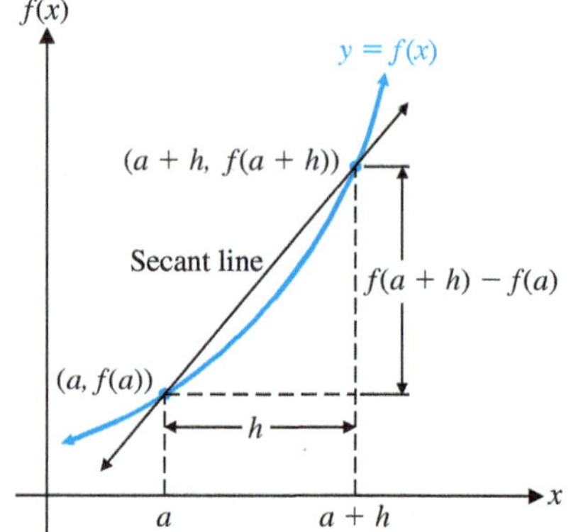

Figure 4 Secant line

$$\textbf{Slope of secant line} = \frac{f(a + h) - f(a)}{(a + h) - a}$$

$$= \frac{f(a + h) - f(a)}{h} \qquad \text{Difference quotient}$$

The difference quotient can be interpreted as both the average rate of change and the slope of the secant line.

CONCEPTUAL INSIGHT

If (x_1, y_1) and (x_2, y_2) are two points in the plane with $x_1 \neq x_2$ and L is the line passing through these two points, then

Slope of L	Point-slope form for L
$m = \dfrac{y_2 - y_1}{x_2 - x_1}$	$y - y_1 = m(x - x_1)$

These formulas will be used extensively in the remainder of this chapter.

EXAMPLE 3 **Slope of a Secant Line** Given $f(x) = x^2$,

(A) Find the slope of the secant line for $a = 1$ and $h = 2$ and 1, respectively. Graph $y = f(x)$ and the two secant lines.

(B) Find and simplify the slope of the secant line for $a = 1$ and h any nonzero number.

(C) Find the limit of the expression in part B.

(D) Discuss possible interpretations of the limit in part C.

SOLUTION (A) For $a = 1$ and $h = 2$, the secant line goes through $(1, f(1)) = (1, 1)$ and $(3, f(3)) = (3, 9)$, and its slope is

$$\frac{f(1+2) - f(1)}{2} = \frac{3^2 - 1^2}{2} = 4$$

For $a = 1$ and $h = 1$, the secant line goes through $(1, f(1)) = (1, 1)$ and $(2, f(2)) = (2, 4)$, and its slope is

$$\frac{f(1+1) - f(1)}{1} = \frac{2^2 - 1^2}{1} = 3$$

Figure 5 Secant lines

The graphs of $y = f(x)$ and the two secant lines are shown in Figure 5.

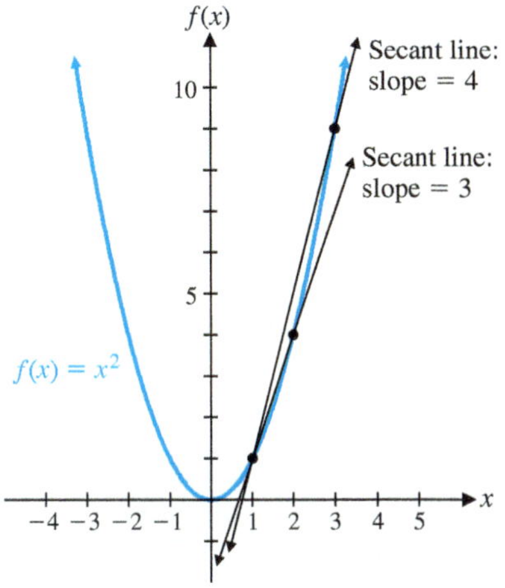

(B) For $a = 1$ and h any nonzero number, the secant line goes through $(1, f(1)) = (1, 1)$ and $(1 + h, f(1 + h)) = (1 + h, (1 + h)^2)$, and its slope is

$$\begin{aligned}\frac{f(1+h) - f(1)}{h} &= \frac{(1+h)^2 - 1^2}{h} && \text{Square the binomial.}\\ &= \frac{1 + 2h + h^2 - 1}{h} && \text{Combine like terms and factor the numerator.}\\ &= \frac{h(2+h)}{h} && \text{Cancel.}\\ &= 2 + h \qquad h \neq 0\end{aligned}$$

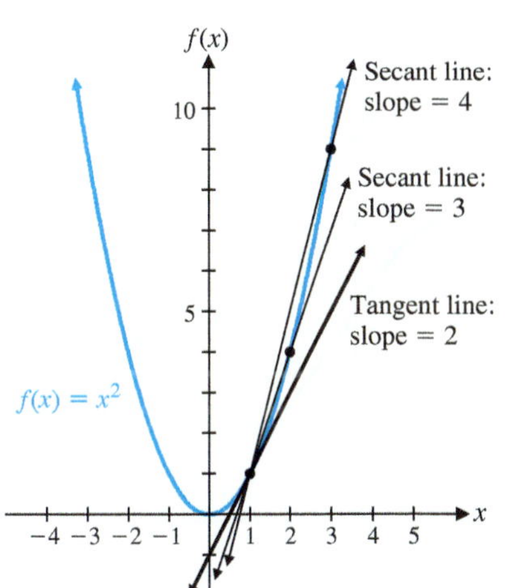

Figure 6 Tangent line

(C) The limit of the secant line slope from part B is

$$\lim_{h \to 0} \frac{f(1+h) - f(1)}{h} = \lim_{h \to 0} (2 + h) = 2$$

(D) In part C, we saw that the limit of the slopes of the secant lines through the point $(1, f(1))$ is 2. If we graph the line through $(1, f(1))$ with slope 2 (Fig. 6), then this line is the limit of the secant lines. The slope obtained from the limit of slopes of secant lines is called the *slope of the graph* at $x = 1$. The line through the point $(1, f(1))$ with this slope is called the *tangent line*. We have solved another basic problem of calculus (see Fig. 1A on page 41).

Matched Problem 3 Given $f(x) = x^2$,

(A) Find the slope of the secant line for $a = 2$ and $h = 2$ and 1, respectively.

(B) Find and simplify the slope of the secant line for $a = 2$ and h any nonzero number.

(C) Find the limit of the expression in part B.

(D) Find the slope of the graph and the slope of the tangent line at $a = 2$.

The ideas introduced in the preceding example are summarized next:

DEFINITION Slope of a Graph

Given $y = f(x)$, the **slope of the graph** at the point $(a, f(a))$ is given by

$$\lim_{h \to 0} \frac{f(a + h) - f(a)}{h} \tag{3}$$

provided that the limit exists. The slope of the graph is also the **slope of the tangent line** at the point $(a, f(a))$.

CONCEPTUAL INSIGHT

If the function f is continuous at a, then

$$\lim_{h \to 0} f(a + h) = f(a)$$

and limit (3) will be a 0/0 indeterminate form. As we saw in Examples 2 and 3, evaluating this type of limit typically involves algebraic simplification.

From plane geometry, we know that a line tangent to a circle is a line that passes through one and only one point of the circle (Fig. 7A). Although this definition cannot be extended to graphs of functions in general, the visual relationship between graphs of functions and their tangent lines (Fig. 7B) is similar to the circle case. Limit (3) provides both a mathematically sound definition of a tangent line and a method for approximating the slope of the tangent line.

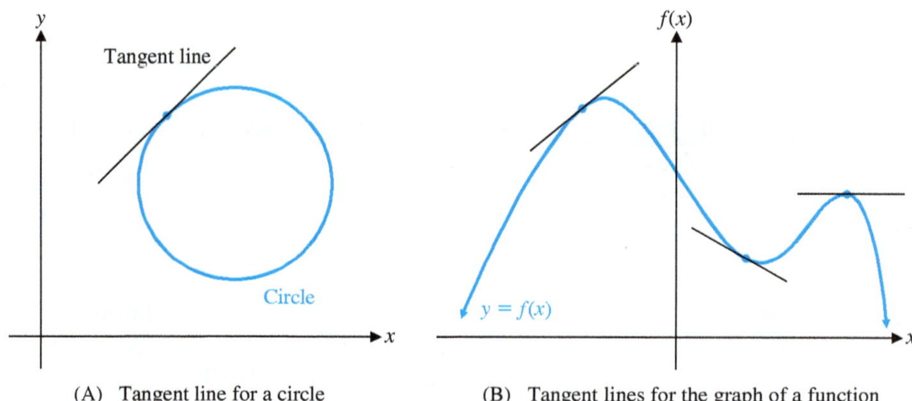

(A) Tangent line for a circle

(B) Tangent lines for the graph of a function

Figure 7

The Derivative

We have seen that the limit of a difference quotient can be interpreted as a rate of change, as a velocity, or as the slope of a tangent line. In addition, this limit provides solutions to two of the three basic problems stated at the beginning of the chapter. We are now ready to introduce some terms that refer to that limit. To follow customary practice, we use x in place of a and think of the difference quotient

$$\frac{f(x + h) - f(x)}{h}$$

as a function of h, with x held fixed as h tends to 0.

DEFINITION The Derivative

For $y = f(x)$, we define the **derivative of f at x**, denoted by $f'(x)$, to be

$$f'(x) = \lim_{h \to 0} \frac{f(x + h) - f(x)}{h} \quad \text{if the limit exists}$$

If $f'(x)$ exists for each x in the open interval (a, b), then f is said to be **differentiable** over (a, b).

(Differentiability from the left or from the right is defined by using $h \to 0^-$ or $h \to 0^+$, respectively, in place of $h \to 0$ in the preceding definition.)

The process of finding the derivative of a function is called **differentiation**. The derivative of a function is obtained by **differentiating** the function.

SUMMARY Interpretations of the Derivative

The derivative of a function f is a new function f'. The domain of f' is a subset of the domain of f. The derivative has various applications and interpretations, including the following:

1. *Slope of the tangent line.* For each x in the domain of f', $f'(x)$ is the slope of the line tangent to the graph of f at the point $(x, f(x))$.
2. *Instantaneous rate of change.* For each x in the domain of f', $f'(x)$ is the instantaneous rate of change of $y = f(x)$ with respect to x.
3. *Velocity.* If $f(x)$ is the position of a moving object at time x, then $v = f'(x)$ is the velocity of the object at that time.

Example 4 illustrates the **four-step process** that we use to find derivatives in this section. In subsequent sections, we develop rules for finding derivatives that do not involve limits. However, it is important that you master the limit process in order to fully comprehend and appreciate the various applications we will consider.

EXAMPLE 4 **Finding a Derivative** Find $f'(x)$, the derivative of f at x, for $f(x) = 4x - x^2$.

SOLUTION To find $f'(x)$, we use a four-step process.

Step 1 Find $f(x + h)$.

$$\begin{aligned} f(x + h) &= 4(x + h) - (x + h)^2 \\ &= 4x + 4h - x^2 - 2xh - h^2 \end{aligned}$$

Step 2 Find $f(x + h) - f(x)$.

$$\begin{aligned} f(x + h) - f(x) &= 4x + 4h - x^2 - 2xh - h^2 - (4x - x^2) \\ &= 4h - 2xh - h^2 \end{aligned}$$

Step 3 Find $\dfrac{f(x + h) - f(x)}{h}$.

$$\begin{aligned} \frac{f(x + h) - f(x)}{h} &= \frac{4h - 2xh - h^2}{h} = \frac{h(4 - 2x - h)}{h} \\ &= 4 - 2x - h, \quad h \neq 0 \end{aligned}$$

Step 4 Find $f'(x) = \lim_{h \to 0} \dfrac{f(x + h) - f(x)}{h}$.

$$f'(x) = \lim_{h \to 0} \frac{f(x + h) - f(x)}{h} = \lim_{h \to 0} (4 - 2x - h) = 4 - 2x$$

So if $f(x) = 4x - x^2$, then $f'(x) = 4 - 2x$. The function f' is a new function derived from the function f.

Matched Problem 4 Find $f'(x)$, the derivative of f at x, for $f(x) = 8x - 2x^2$.

The four-step process used in Example 4 is summarized as follows for easy reference:

PROCEDURE The four-step process for finding the derivative of a function f:

Step 1 Find $f(x + h)$.

Step 2 Find $f(x + h) - f(x)$.

Step 3 Find $\dfrac{f(x + h) - f(x)}{h}$.

Step 4 Find $\lim\limits_{h \to 0} \dfrac{f(x + h) - f(x)}{h}$.

EXAMPLE 5 **Finding Tangent Line Slopes** In Example 4, we started with the function $f(x) = 4x - x^2$ and found the derivative of f at x to be $f'(x) = 4 - 2x$. So the slope of a line tangent to the graph of f at any point $(x, f(x))$ on the graph is

$$m = f'(x) = 4 - 2x$$

(A) Find the slope of the graph of f at $x = 0$, $x = 2$, and $x = 3$.

(B) Graph $y = f(x) = 4x - x^2$ and use the slopes found in part (A) to make a rough sketch of the lines tangent to the graph at $x = 0$, $x = 2$, and $x = 3$.

SOLUTION (A) Using $f'(x) = 4 - 2x$, we have

$$f'(0) = 4 - 2(0) = 4 \quad \text{Slope at } x = 0$$
$$f'(2) = 4 - 2(2) = 0 \quad \text{Slope at } x = 2$$
$$f'(3) = 4 - 2(3) = -2 \quad \text{Slope at } x = 3$$

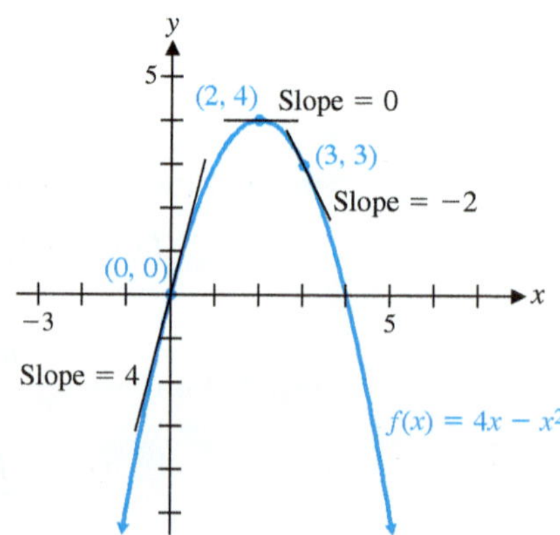

Matched Problem 5 In Matched Problem 4, we started with the function $f(x) = 8x - 2x^2$. Using the derivative found there,

(A) Find the slope of the graph of f at $x = 1$, $x = 2$, and $x = 4$.

(B) Graph $y = f(x) = 8x - 2x^2$, and use the slopes from part (A) to make a rough sketch of the lines tangent to the graph at $x = 1$, $x = 2$, and $x = 4$.

EXPLORE & DISCUSS 1

In Example 4, we found that the derivative of $f(x) = 4x - x^2$ is $f'(x) = 4 - 2x$. In Example 5, we graphed $f(x)$ and several tangent lines.

(A) Graph f and f' on the same set of axes.

(B) The graph of f' is a straight line. Is it a tangent line for the graph of f? Explain.

(C) Find the x intercept for the graph of f'. What is the slope of the line tangent to the graph of f for this value of x? Write a verbal description of the relationship between the slopes of the tangent lines of a function and the x intercepts of the derivative of the function.

EXAMPLE 6 **Finding a Derivative** Find $f'(x)$, the derivative of f at x, for $f(x) = \sqrt{x} + 2$.

SOLUTION We use the four-step process to find $f'(x)$.

Step 1 Find $f(x + h)$.

$$f(x + h) = \sqrt{x + h} + 2$$

Step 2 Find $f(x + h) - f(x)$.

$$\begin{aligned} f(x + h) - f(x) &= \sqrt{x + h} + 2 - (\sqrt{x} + 2) \quad \text{Combine like terms.} \\ &= \sqrt{x + h} - \sqrt{x} \end{aligned}$$

Step 3 Find $\dfrac{f(x + h) - f(x)}{h}$.

$$\begin{aligned} \frac{f(x + h) - f(x)}{h} &= \frac{\sqrt{x + h} - \sqrt{x}}{h} \\ &= \frac{\sqrt{x + h} - \sqrt{x}}{h} \cdot \frac{\sqrt{x + h} + \sqrt{x}}{\sqrt{x + h} + \sqrt{x}} \\ &= \frac{x + h - x}{h(\sqrt{x + h} + \sqrt{x})} \\ &= \frac{h}{h(\sqrt{x + h} + \sqrt{x})} \\ &= \frac{1}{\sqrt{x + h} + \sqrt{x}} \qquad h \neq 0 \end{aligned}$$

We rationalize the numerator (Appendix A, Section A-6) to change the form of this fraction. Combine like terms. Cancel.

Step 4 Find $f'(x) = \lim_{h \to 0} \dfrac{f(x + h) - f(x)}{h}$.

$$\begin{aligned} \lim_{h \to 0} \frac{f(x + h) - f(x)}{h} &= \lim_{h \to 0} \frac{1}{\sqrt{x + h} + \sqrt{x}} \\ &= \frac{1}{\sqrt{x} + \sqrt{x}} = \frac{1}{2\sqrt{x}} \qquad x > 0 \end{aligned}$$

So the derivative of $f(x) = \sqrt{x} + 2$ is $f'(x) = 1/(2\sqrt{x})$, a new function. The domain of f is $[0, \infty)$. Since $f'(0)$ is not defined, the domain of f' is $(0, \infty)$, a subset of the domain of f.

Matched Problem 6 Find $f'(x)$ for $f(x) = \sqrt{x + 4}$.

EXAMPLE 7 **Sales Analysis** A company's total sales (in millions of dollars) t months from now are given by $S(t) = \sqrt{t} + 2$. Find and interpret $S(25)$ and $S'(25)$. Use these results to estimate the total sales after 26 months and after 27 months.

SOLUTION The total sales function S has the same form as the function f in Example 6. Only the letters used to represent the function and the independent variable have been changed. It follows that S' and f' also have the same form:

$$S(t) = \sqrt{t} + 2 \qquad f(x) = \sqrt{x} + 2$$

$$S'(t) = \frac{1}{2\sqrt{t}} \qquad f'(x) = \frac{1}{2\sqrt{x}}$$

Evaluating S and S' at $t = 25$, we have

$$S(25) = \sqrt{25} + 2 = 7 \qquad S'(25) = \frac{1}{2\sqrt{25}} = 0.1$$

So 25 months from now, the total sales will be \$7 million and will be increasing at the rate of \$0.1 million (\$100,000) per month. If this instantaneous rate of change of sales remained constant, the sales would grow to \$7.1 million after 26 months, \$7.2 million after 27 months, and so on. Even though $S'(t)$ is not a constant function in this case, these values provide useful estimates of the total sales.

Matched Problem 7 A company's total sales (in millions of dollars) t months from now are given by $S(t) = \sqrt{t + 4}$. Find and interpret $S(12)$ and $S'(12)$. Use these results to estimate the total sales after 13 months and after 14 months. (Use the derivative found in Matched Problem 6.)

In Example 7, we can compare the estimates of total sales by using the derivative with the corresponding exact values of $S(t)$:

	Exact values	Estimated values
$S(26) = \sqrt{26} + 2 =$	$7.099\ldots$	≈ 7.1
$S(27) = \sqrt{27} + 2 =$	$7.196\ldots$	≈ 7.2

For this function, the estimated values provide very good approximations to the exact values of $S(t)$. For other functions, the approximation might not be as accurate.

Using the instantaneous rate of change of a function at a point to estimate values of the function at nearby points is an important application of the derivative.

Nonexistence of the Derivative

The existence of a derivative at $x = a$ depends on the existence of a limit at $x = a$, that is, on the existence of

$$f'(a) = \lim_{h \to 0} \frac{f(a + h) - f(a)}{h} \tag{4}$$

If the limit does not exist at $x = a$, we say that the function f is **nondifferentiable at** $\boldsymbol{x = a}$, or $\boldsymbol{f'(a)}$ **does not exist.**

EXPLORE & DISCUSS 2 Let $f(x) = |x - 1|$.

(A) Graph f.

(B) Complete the following table:

h	-0.1	-0.01	-0.001	$\to 0 \leftarrow$	0.001	0.01	0.1
$\dfrac{f(1 + h) - f(1)}{h}$	?	?	?	$\to ? \leftarrow$	?	?	?

(C) Find the following limit if it exists:

$$\lim_{h \to 0} \frac{f(1 + h) - f(1)}{h}$$

(D) Use the results of parts (A)–(C) to discuss the existence of $f'(1)$.
(E) Repeat parts (A)–(D) for $\sqrt[3]{x-1}$.

How can we recognize the points on the graph of f where $f'(a)$ does not exist? It is impossible to describe all the ways that the limit of a difference quotient can fail to exist. However, we can illustrate some common situations where $f'(a)$ fails to exist (see Fig. 8):

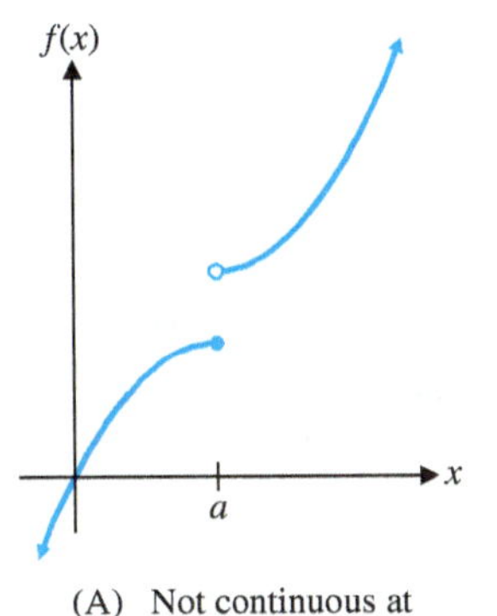

(A) Not continuous at $x = a$

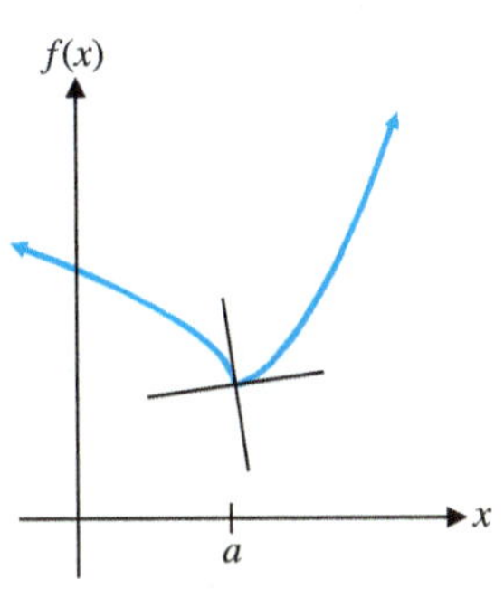

(B) Graph has sharp corner at $x = a$

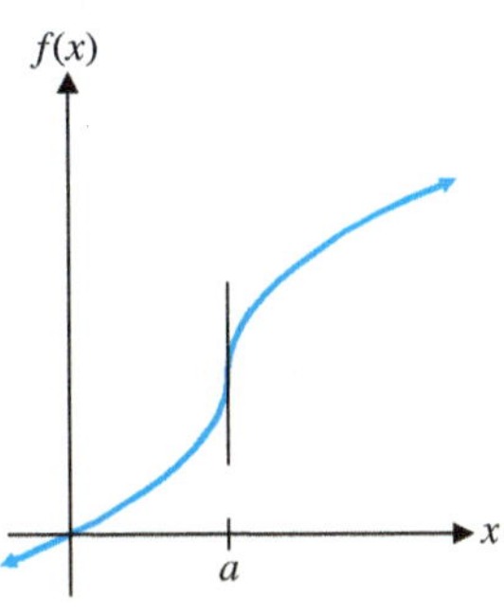

(C) Vertical tangent at $x = a$

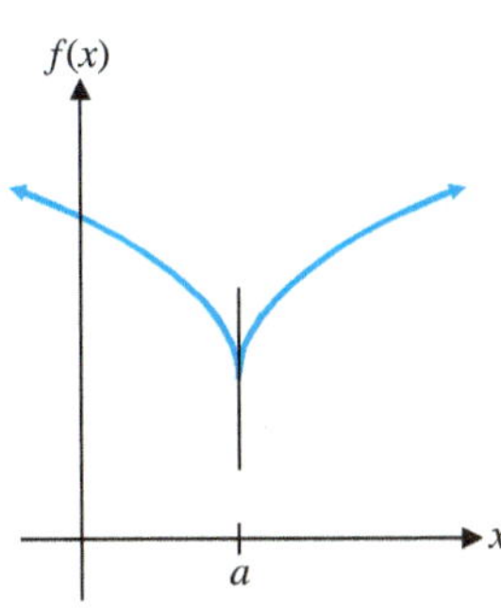

(D) Vertical tangent at $x = a$

Figure 8 The function f is nondifferentiable at $x = a$.

1. If the graph of f has a hole or a break at $x = a$, then $f'(a)$ does not exist (Fig. 8A).
2. If the graph of f has a sharp corner at $x = a$, then $f'(a)$ does not exist, and the graph has no tangent line at $x = a$ (Fig. 8B). (In Fig. 8B, the left- and right-hand derivatives exist but are not equal.)
3. If the graph of f has a vertical tangent line at $x = a$, then $f'(a)$ does not exist (Fig. 8C and D).

Exercises 1-4

A

In Problems 1 and 2, find the indicated quantity for $y = f(x) = 5 - x^2$ and interpret that quantity in terms of the following graph.

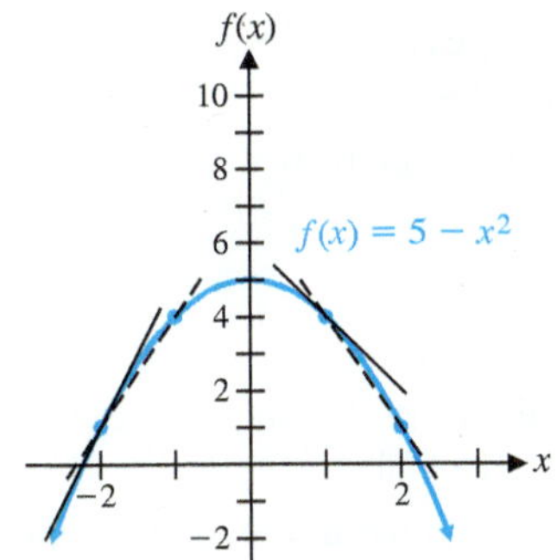

1. (A) $\dfrac{f(2) - f(1)}{2 - 1}$ (B) $\dfrac{f(1 + h) - f(1)}{h}$

(C) $\lim_{h \to 0} \dfrac{f(1 + h) - f(1)}{h}$

2. (A) $\dfrac{f(-1) - f(-2)}{-1 - (-2)}$

(B) $\dfrac{f(-2 + h) - f(-2)}{h}$

(C) $\lim_{h \to 0} \dfrac{f(-2 + h) - f(-2)}{h}$

3. Find the indicated quantities for $f(x) = 3x^2$.

(A) The average rate of change of $f(x)$ if x changes from 1 to 4.

(B) The slope of the secant line through the points $(1, f(1))$ and $(4, f(4))$ on the graph of $y = f(x)$.

(C) The slope of the secant line through the points $(1, f(1))$ and $(1 + h, f(1 + h))$, $h \neq 0$. Simplify your answer.

(D) The slope of the graph at $(1, f(1))$.

(E) The instantaneous rate of change of $y = f(x)$ with respect to x at $x = 1$.

(F) The slope of the tangent line at $(1, f(1))$.

(G) The equation of the tangent line at $(1, f(1))$.

4. Find the indicated quantities for $f(x) = 3x^2$.

(A) The average rate of change of $f(x)$ if x changes from 2 to 5.

(B) The slope of the secant line through the points $(2, f(2))$ and $(5, f(5))$ on the graph of $y = f(x)$.

(C) The slope of the secant line through the points $(2, f(2))$ and $(2 + h, f(2 + h))$, $h \neq 0$. Simplify your answer.

(D) The slope of the graph at $(2, f(2))$.

(E) The instantaneous rate of change of $y = f(x)$ with respect to x at $x = 2$.

(F) The slope of the tangent line at $(2, f(2))$.

(G) The equation of the tangent line at $(2, f(2))$.

In Problems 5–26, use the four-step process to find $f'(x)$ and then find $f'(1)$, $f'(2)$, and $f'(3)$.

5. $f(x) = -5$

6. $f(x) = 9$

7. $f(x) = 3x - 7$

8. $f(x) = 4 - 6x$

9. $f(x) = 2 - 3x^2$

10. $f(x) = 2x^2 + 8$

11. $f(x) = x^2 + 6x - 10$

12. $f(x) = x^2 + 4x + 7$

13. $f(x) = 2x^2 - 7x + 3$

14. $f(x) = 2x^2 + 5x + 1$

15. $f(x) = -x^2 + 4x - 9$

16. $f(x) = -x^2 + 9x - 2$

17. $f(x) = 2x^3 + 1$

18. $f(x) = -2x^3 + 5$

19. $f(x) = 4 + \frac{4}{x}$

20. $f(x) = \frac{6}{x} - 2$

21. $f(x) = 5 + 3\sqrt{x}$

22. $f(x) = 3 - 7\sqrt{x}$

23. $f(x) = 10\sqrt{x + 5}$

24. $f(x) = 16\sqrt{x + 9}$

25. $f(x) = \frac{3x}{x + 2}$

26. $f(x) = \frac{5x}{3 + x}$

B

Problems 27 and 28 refer to the graph of $y = f(x) = x^2 + x$ shown.

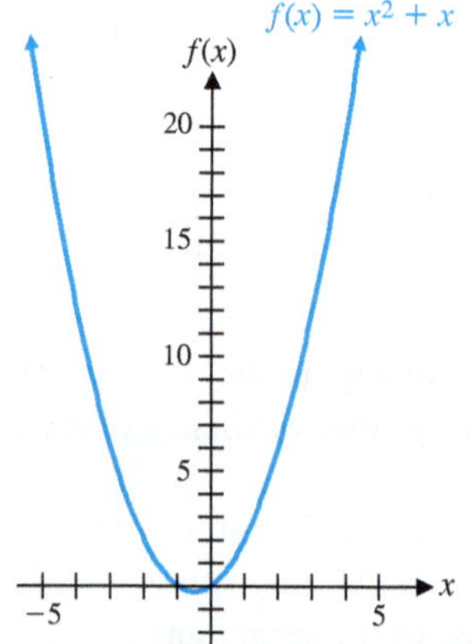

27. (A) Find the slope of the secant line joining $(1, f(1))$ and $(3, f(3))$.

(B) Find the slope of the secant line joining $(1, f(1))$ and $(1 + h, f(1 + h))$.

(C) Find the slope of the tangent line at $(1, f(1))$.

(D) Find the equation of the tangent line at $(1, f(1))$.

28. (A) Find the slope of the secant line joining $(2, f(2))$ and $(4, f(4))$.

(B) Find the slope of the secant line joining $(2, f(2))$ and $(2 + h, f(2 + h))$.

(C) Find the slope of the tangent line at $(2, f(2))$.

(D) Find the equation of the tangent line at $(2, f(2))$.

In Problems 29 and 30, suppose an object moves along the y axis so that its location is $y = f(x) = x^2 + x$ at time x (y is in meters and x is in seconds). Find

29. (A) The average velocity (the average rate of change of y with respect to x) for x changing from 1 to 3 seconds

(B) The average velocity for x changing from 1 to $1 + h$ seconds

(C) The instantaneous velocity at $x = 1$ second

30. (A) The average velocity (the average rate of change of y with respect to x) for x changing from 2 to 4 seconds

(B) The average velocity for x changing from 2 to $2 + h$ seconds

(C) The instantaneous velocity at $x = 2$ seconds

Problems 31–38 refer to the function F in the graph shown. Use the graph to determine whether $F'(x)$ exists at each indicated value of x.

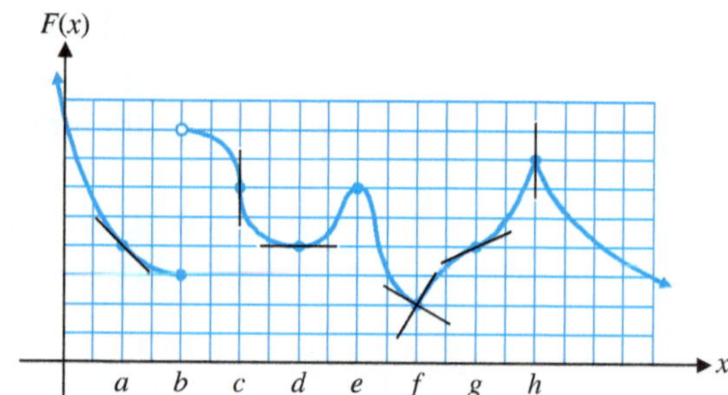

31. $x = a$

32. $x = b$

33. $x = c$

34. $x = d$

35. $x = e$

36. $x = f$

37. $x = g$

38. $x = h$

39. Given $f(x) = x^2 - 4x$,

(A) Find $f'(x)$.

(B) Find the slopes of the lines tangent to the graph of f at $x = 0, 2$, and 4.

(C) Graph f and sketch in the tangent lines at $x = 0, 2$, and 4.

40. Given $f(x) = x^2 + 2x$,

(A) Find $f'(x)$.

(B) Find the slopes of the lines tangent to the graph of f at $x = -2, -1$, and 1.

(C) Graph f and sketch in the tangent lines at $x = -2, -1$, and 1.

41. If an object moves along a line so that it is at $y = f(x) = 4x^2 - 2x$ at time x (in seconds), find the instantaneous velocity function $v = f'(x)$ and find the velocity at times $x = 1, 3$, and 5 seconds (y is measured in feet).

42. Repeat Problem 41 with $f(x) = 8x^2 - 4x$.

43. Let $f(x) = x^2$, $g(x) = x^2 - 1$, and $h(x) = x^2 + 2$.

(A) How are the graphs of these functions related? How would you expect the derivatives of these functions to be related?

(B) Use the four-step process to find the derivative of $m(x) = x^2 + C$, where C is any real constant.

44. Let $f(x) = -x^2$, $g(x) = -x^2 - 1$, and $h(x) = -x^2 + 2$.

(A) How are the graphs of these functions related? How would you expect the derivatives of these functions to be related?

(B) Use the four-step process to find the derivative of $m(x) = -x^2 + C$, where C is any real constant.

In Problems 45–50, discuss the validity of each statement. If the statement is always true, explain why. If not, give a counterexample.

45. If $f(x) = C$ is a constant function, then $f'(x) = 0$.

46. If $f(x) = mx + b$ is a linear function, then $f'(x) = m$.

47. If a function f is continuous on the interval (a, b), then f is differentiable on (a, b).

48. If a function f is differentiable on the interval (a, b), then f is continuous on (a, b).

49. The average rate of change of a function f from $x = a$ to $x = a + h$ is less than the instantaneous rate of change at $x = a + \frac{h}{2}$.

50. If the graph of f has a sharp corner at $x = a$, then f is not continuous at $x = a$.

C

In Problems 51–54, sketch the graph of f and determine where f is nondifferentiable.

51. $f(x) = \begin{cases} 2x & \text{if } x < 1 \\ 2 & \text{if } x \geq 1 \end{cases}$

52. $f(x) = \begin{cases} 2x & \text{if } x < 2 \\ 6 - x & \text{if } x \geq 2 \end{cases}$

53. $f(x) = \begin{cases} x^2 + 1 & \text{if } x < 0 \\ 1 & \text{if } x \geq 0 \end{cases}$

54. $f(x) = \begin{cases} 2 - x^2 & \text{if } x \leq 0 \\ 2 & \text{if } x > 0 \end{cases}$

In Problems 55–60, determine whether f is differentiable at x = 0 by considering

$$\lim_{h \to 0} \frac{f(0 + h) - f(0)}{h}$$

55. $f(x) = |x|$

56. $f(x) = 1 - |x|$

57. $f(x) = x^{1/3}$

58. $f(x) = x^{2/3}$

59. $f(x) = \sqrt{1 - x^2}$

60. $f(x) = \sqrt{1 + x^2}$

61. A ball dropped from a balloon falls $y = 16x^2$ feet in x seconds. If the balloon is 576 feet above the ground when the ball is dropped, when does the ball hit the ground? What is the velocity of the ball at the instant it hits the ground?

62. Repeat Problem 61 if the balloon is 1,024 feet above the ground when the ball is dropped.

Applications

63. Revenue. The revenue (in dollars) from the sale of x infant car seats is given by

$$R(x) = 60x - 0.025x^2 \qquad 0 \leq x \leq 2{,}400$$

(A) Find the average change in revenue if production is changed from 1,000 car seats to 1,050 car seats.

(B) Use the four-step process to find $R'(x)$.

(C) Find the revenue and the instantaneous rate of change of revenue at a production level of 1,000 car seats, and write a brief verbal interpretation of these results.

64. Profit. The profit (in dollars) from the sale of x infant car seats is given by

$$P(x) = 45x - 0.025x^2 - 5{,}000 \qquad 0 \leq x \leq 2{,}400$$

(A) Find the average change in profit if production is changed from 800 car seats to 850 car seats.

(B) Use the four-step process to find $P'(x)$.

(C) Find the profit and the instantaneous rate of change of profit at a production level of 800 car seats, and write a brief verbal interpretation of these results.

65. Sales analysis. A company's total sales (in millions of dollars) t months from now are given by

$$S(t) = 2\sqrt{t + 10}$$

(A) Use the four-step process to find $S'(t)$.

(B) Find $S(15)$ and $S'(15)$. Write a brief verbal interpretation of these results.

(C) Use the results in part (B) to estimate the total sales after 16 months and after 17 months.

66. Sales analysis. A company's total sales (in millions of dollars) t months from now are given by

$$S(t) = 2\sqrt{t + 6}$$

(A) Use the four-step process to find $S'(t)$.

(B) Find $S(10)$ and $S'(10)$. Write a brief verbal interpretation of these results.

(C) Use the results in part (B) to estimate the total sales after 11 months and after 12 months.

67. Mineral consumption. The U.S. consumption of tungsten (in metric tons) is given approximately by

$$p(t) = 164t^2 + 161t + 12{,}326$$

where t is time in years and $t = 0$ corresponds to 2005.

(A) Use the four-step process to find $p'(t)$.

(B) Find the annual production in 2015 and the instantaneous rate of change of production in 2015, and write a brief verbal interpretation of these results.

68. Mineral consumption. The U.S. consumption of copper (in thousands of metric tons) is given approximately by

$$p(t) = 29t^2 - 258t + 4{,}658$$

where t is time in years and $t = 0$ corresponds to 2005.

(A) Use the four-step process to find $p'(t)$.

(B) Find the annual production in 2017 and the instantaneous rate of change of production in 2017, and write a brief verbal interpretation of these results.

69. Electricity consumption. Table 1 gives the retail sales of electricity (in billions of kilowatt-hours) for the residential and commercial sectors in the United States. (*Source:* Energy Information Administration)

Table 1 Electricity Sales

Year	Residential	Commercial
2000	1,192	1,055
2002	1,265	1,104
2004	1,292	1,230
2006	1,352	1,300
2008	1,379	1,352

(A) Let x represent time (in years) with $x = 0$ corresponding to 2000, and let y represent the corresponding residential sales. Enter the appropriate data set in a graphing calculator and find a quadratic regression equation for the data.

(B) If $y = R(x)$ denotes the regression equation found in part (A), find $R(20)$ and $R'(20)$, and write a brief verbal interpretation of these results. Round answers to the nearest tenth of a billion.

70. Electricity consumption. Refer to the data in Table 1.

(A) Let x represent time (in years) with $x = 0$ corresponding to 2000, and let y represent the corresponding commercial sales. Enter the appropriate data set in a graphing calculator and find a quadratic regression equation for the data.

(B) If $y = C(x)$ denotes the regression equation found in part (A), find $C(20)$ and $C'(20)$, and write a brief verbal interpretation of these results. Round answers to the nearest tenth of a billion.

71. Air pollution. The ozone level (in parts per billion) on a summer day in a metropolitan area is given by

$$P(t) = 80 + 12t - t^2$$

where t is time in hours and $t = 0$ corresponds to 9 A.M.

(A) Use the four-step process to find $P'(t)$.

(B) Find $P(3)$ and $P'(3)$. Write a brief verbal interpretation of these results.

72. Medicine. The body temperature (in degrees Fahrenheit) of a patient t hours after taking a fever-reducing drug is given by

$$F(t) = 98 + \frac{4}{t + 1}$$

(A) Use the four-step process to find $F'(t)$.

(B) Find $F(3)$ and $F'(3)$. Write a brief verbal interpretation of these results.

Answers to Matched Problems

1. (A) $-\$1{,}600$ (B) $-\$8$ per planter
2. (A) 48 ft/s
   (B) $32 + 16h$
   (C) 32 ft/s
3. (A) 6, 5 (B) $4 + h$
   (C) 4 (D) Both are 4
4. $f'(x) = 8 - 4x$
5. (A) $f'(1) = 4, f'(2) = 0, f'(4) = -8$
   (B)

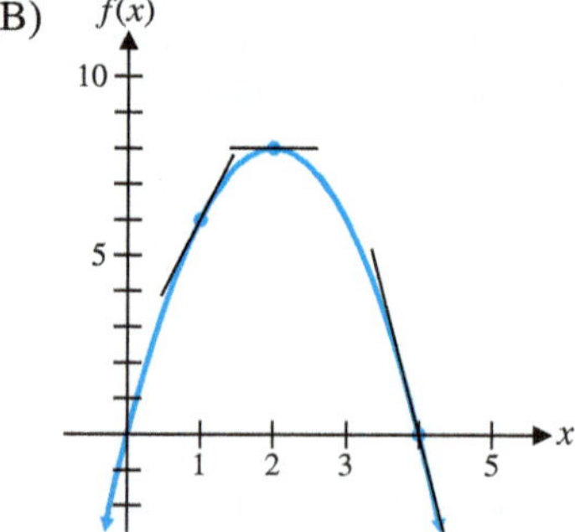

6. $f'(x) = 1/(2\sqrt{x + 4})$
7. $S(12) = 4$, $S'(12) = 0.125$; 12 months from now, the total sales will be \$4 million and will be increasing at the rate of \$0.125 million (\$125,000) per month. The estimated total sales are \$4.125 million after 13 months and \$4.25 million after 14 months.

1-5 Basic Differentiation Properties

- Constant Function Rule
- Power Rule
- Constant Multiple Property
- Sum and Difference Properties
- Applications

In Section 1-4, we defined the derivative of f at x as

$$f'(x) = \lim_{h \to 0} \frac{f(x + h) - f(x)}{h}$$

if the limit exists, and we used this definition and a four-step process to find the derivatives of several functions. Now we want to develop some rules of differentiation. These rules will enable us to find the derivative of many functions without using the four-step process.

Before exploring these rules, we list some symbols that are often used to represent derivatives.

NOTATION The Derivative

If $y = f(x)$, then

$$f'(x) \qquad y' \qquad \frac{dy}{dx}$$

all represent the derivative of f at x.

Each of these derivative symbols has its particular advantage in certain situations. All of them will become familiar to you after a little experience.

Constant Function Rule

If $f(x) = C$ is a constant function, then the four-step process can be used to show that $f'(x) = 0$. Therefore,

The derivative of any constant function is 0.

THEOREM 1 Constant Function Rule

If $y = f(x) = C$, then

$$f'(x) = 0$$

Also, $y' = 0$ and $dy/dx = 0$.

Note: When we write $C' = 0$ or $\frac{d}{dx}C = 0$, we mean that $y' = \frac{dy}{dx} = 0$ when $y = C$.

CONCEPTUAL INSIGHT

The graph of $f(x) = C$ is a horizontal line with slope 0 (Fig. 1), so we would expect that $f'(x) = 0$.

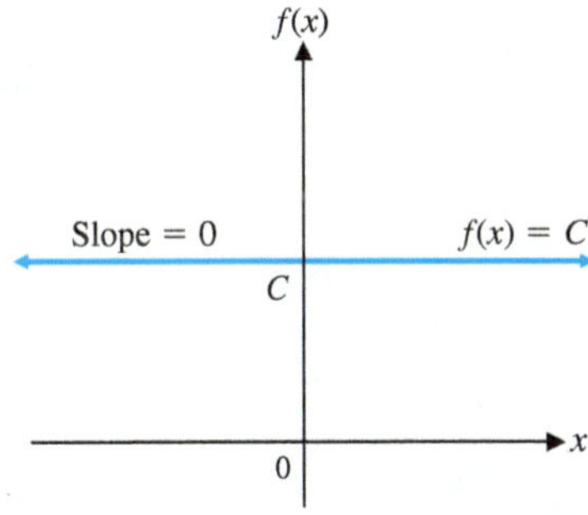

Figure 1

EXAMPLE 1 **Differentiating Constant Functions**

(A) If $f(x) = 3$, then $f'(x) = 0$.

(B) If $y = -1.4$, then $y' = 0$.

(C) If $y = \pi$, then $\frac{dy}{dx} = 0$.

(D) $\frac{d}{dx}23 = 0$

Matched Problem 1 Find

(A) $f'(x)$ for $f(x) = -24$

(B) y' for $y = 12$

(C) $\frac{dy}{dx}$ for $y = -\sqrt{7}$

(D) $\frac{d}{dx}(-\pi)$

Power Rule

A function of the form $f(x) = x^k$, where k is a real number, is called a **power function**. The following elementary functions are examples of power functions:

$$f(x) = x \qquad h(x) = x^2 \qquad m(x) = x^3$$
$$n(x) = \sqrt{x} \qquad p(x) = \sqrt[3]{x} \tag{1}$$

EXPLORE & DISCUSS 1

(A) It is clear that the functions f, h, and m in (1) are power functions. Explain why the functions n and p are also power functions.

(B) The domain of a power function depends on the power. Discuss the domain of each of the following power functions:

$$r(x) = x^4 \qquad s(x) = x^{-4} \qquad t(x) = x^{1/4}$$
$$u(x) = x^{-1/4} \qquad v(x) = x^{1/5} \qquad w(x) = x^{-1/5}$$

The definition of the derivative and the four-step process introduced in Section 1-4 can be used to find the derivatives of many power functions. For example, it can be shown that

$$\begin{aligned}
&\text{If} \quad f(x) = x^2, \quad \text{then} \quad f'(x) = 2x.\\
&\text{If} \quad f(x) = x^3, \quad \text{then} \quad f'(x) = 3x^2.\\
&\text{If} \quad f(x) = x^4, \quad \text{then} \quad f'(x) = 4x^3.\\
&\text{If} \quad f(x) = x^5, \quad \text{then} \quad f'(x) = 5x^4.
\end{aligned}$$

Notice the pattern in these derivatives. In each case, the power in f becomes the coefficient in f' and the power in f' is 1 less than the power in f. In general, for any positive integer n,

$$\text{If} \quad f(x) = x^n, \quad \text{then} \quad f'(x) = nx^{n-1}. \tag{2}$$

In fact, more advanced techniques can be used to show that (2) holds for *any* real number n. We will assume this general result for the remainder of the book.

THEOREM 2 Power Rule

If $y = f(x) = x^n$, where n is a real number, then

$$f'(x) = nx^{n-1}$$

Also, $y' = nx^{n-1}$ and $dy/dx = nx^{n-1}$.

EXAMPLE 2 **Differentiating Power Functions**

(A) If $f(x) = x^5$, then $f'(x) = 5x^{5-1} = 5x^4$.

(B) If $y = x^{25}$, then $y' = 25x^{25-1} = 25x^{24}$.

(C) If $y = t^{-3}$, then $\dfrac{dy}{dt} = -3t^{-3-1} = -3t^{-4} = -\dfrac{3}{t^4}$.

(D) $\dfrac{d}{dx}x^{5/3} = \dfrac{5}{3}x^{(5/3)-1} = \dfrac{5}{3}x^{2/3}$.

Matched Problem 2 Find

(A) $f'(x)$ for $f(x) = x^6$

(B) y' for $y = x^{30}$

(C) $\dfrac{dy}{dt}$ for $y = t^{-2}$

(D) $\dfrac{d}{dx}x^{3/2}$

In some cases, properties of exponents must be used to rewrite an expression before the power rule is applied.

EXAMPLE 3 **Differentiating Power Functions**

(A) If $f(x) = 1/x^4$, we can write $f(x) = x^{-4}$ and

$$f'(x) = -4x^{-4-1} = -4x^{-5}, \quad \text{or} \quad \frac{-4}{x^5}$$

(B) If $y = \sqrt{u}$, we can write $y = u^{1/2}$ and

$$y' = \frac{1}{2}u^{(1/2)-1} = \frac{1}{2}u^{-1/2}, \quad \text{or} \quad \frac{1}{2\sqrt{u}}$$

(C) $$\frac{d}{dx}\frac{1}{\sqrt[3]{x}} = \frac{d}{dx}x^{-1/3} = -\frac{1}{3}x^{(-1/3)-1} = -\frac{1}{3}x^{-4/3}, \text{ or } \frac{-1}{3\sqrt[3]{x^4}}$$

Matched Problem 3 Find

(A) $f'(x)$ for $f(x) = \frac{1}{x}$ (B) y' for $y = \sqrt[3]{u^2}$ (C) $\frac{d}{dx}\frac{1}{\sqrt{x}}$

Constant Multiple Property

Let $f(x) = ku(x)$, where k is a constant and u is differentiable at x. Using the four-step process, we have the following:

Step 1 $f(x + h) = ku(x + h)$

Step 2 $f(x + h) - f(x) = ku(x + h) - ku(x) = k[u(x + h) - u(x)]$

Step 3 $$\frac{f(x + h) - f(x)}{h} = \frac{k[u(x + h) - u(x)]}{h} = k\left[\frac{u(x + h) - u(x)}{h}\right]$$

Step 4 $$\begin{aligned} f'(x) &= \lim_{h\to 0}\frac{f(x + h) - f(x)}{h} \\ &= \lim_{h\to 0} k\left[\frac{u(x + h) - u(x)}{h}\right] && \lim_{x\to c} kg(x) = k\lim_{x\to c} g(x) \\ &= k\lim_{h\to 0}\left[\frac{u(x + h) - u(x)}{h}\right] && \text{Definition of } u'(x) \\ &= ku'(x) \end{aligned}$$

Therefore,

The derivative of a constant times a differentiable function is the constant times the derivative of the function.

THEOREM 3 **Constant Multiple Property**

If $y = f(x) = ku(x)$, then

$$f'(x) = ku'(x)$$

Also,

$$y' = ku' \qquad \frac{dy}{dx} = k\frac{du}{dx}$$

EXAMPLE 4 **Differentiating a Constant Times a Function**

(A) If $f(x) = 3x^2$, then $f'(x) = 3 \cdot 2x^{2-1} = 6x$.

(B) If $y = \frac{t^3}{6} = \frac{1}{6}t^3$, then $\frac{dy}{dt} = \frac{1}{6} \cdot 3t^{3-1} = \frac{1}{2}t^2$.

(C) If $y = \frac{1}{2x^4} = \frac{1}{2}x^{-4}$, then $y' = \frac{1}{2}(-4x^{-4-1}) = -2x^{-5}$, or $\frac{-2}{x^5}$.

(D) $$\frac{d}{dx}\frac{0.4}{\sqrt{x^3}} = \frac{d}{dx}\frac{0.4}{x^{3/2}} = \frac{d}{dx}0.4x^{-3/2} = 0.4\left[-\frac{3}{2}x^{(-3/2)-1}\right]$$
$$= -0.6x^{-5/2}, \text{ or } -\frac{0.6}{\sqrt{x^5}}$$

Matched Problem 4 Find

(A) $f'(x)$ for $f(x) = 4x^5$

(B) $\frac{dy}{dt}$ for $y = \frac{t^4}{12}$

(C) y' for $y = \frac{1}{3x^3}$

(D) $\frac{d}{dx}\frac{0.9}{\sqrt[3]{x}}$

Sum and Difference Properties

Let $f(x) = u(x) + v(x)$, where $u'(x)$ and $v'(x)$ exist. Using the four-step process, we have the following:

Step 1 $f(x + h) = u(x + h) + v(x + h)$

Step 2
$$f(x + h) - f(x) = u(x + h) + v(x + h) - [u(x) + v(x)]$$
$$= u(x + h) - u(x) + v(x + h) - v(x)$$

Step 3
$$\frac{f(x + h) - f(x)}{h} = \frac{u(x + h) - u(x) + v(x + h) - v(x)}{h}$$
$$= \frac{u(x + h) - u(x)}{h} + \frac{v(x + h) - v(x)}{h}$$

Step 4
$$f'(x) = \lim_{h\to 0}\frac{f(x + h) - f(x)}{h}$$
$$= \lim_{h\to 0}\left[\frac{u(x + h) - u(x)}{h} + \frac{v(x + h) - v(x)}{h}\right]$$

$\lim_{x\to c}[g(x) + h(x)] = \lim_{x\to c} g(x) + \lim_{x\to c} h(x)$

$$= \lim_{h\to 0}\frac{u(x + h) - u(x)}{h} + \lim_{h\to 0}\frac{v(x + h) - v(x)}{h}$$
$$= u'(x) + v'(x)$$

Therefore,

The derivative of the sum of two differentiable functions is the sum of the derivatives of the functions.

Similarly, we can show that

The derivative of the difference of two differentiable functions is the difference of the derivatives of the functions.

Together, we have the **sum and difference property** for differentiation:

THEOREM 4 Sum and Difference Property

If $y = f(x) = u(x) \pm v(x)$, then

$$f'(x) = u'(x) \pm v'(x)$$

Also,

$$y' = u' \pm v' \qquad \frac{dy}{dx} = \frac{du}{dx} \pm \frac{dv}{dx}$$

Note: This rule generalizes to the sum and difference of any given number of functions.

With Theorems 1 through 4, we can compute the derivatives of all polynomials and a variety of other functions.

EXAMPLE 5 **Differentiating Sums and Differences**

(A) If $f(x) = 3x^2 + 2x$, then

$$f'(x) \boxed{= (3x^2)' + (2x)' = 3(2x) + 2(1)} = 6x + 2$$

(B) If $y = 4 + 2x^3 - 3x^{-1}$, then

$$y' \boxed{= (4)' + (2x^3)' - (3x^{-1})' = 0 + 2(3x^2) - 3(-1)x^{-2}} = 6x^2 + 3x^{-2}$$

(C) If $y = \sqrt[3]{w} - 3w$, then

$$\frac{dy}{dw} = \frac{d}{dw}w^{1/3} - \frac{d}{dw}3w = \frac{1}{3}w^{-2/3} - 3 = \frac{1}{3w^{2/3}} - 3$$

(D) $$\frac{d}{dx}\left(\frac{5}{3x^2} - \frac{2}{x^4} + \frac{x^3}{9}\right) \boxed{= \frac{d}{dx}\frac{5}{3}x^{-2} - \frac{d}{dx}2x^{-4} + \frac{d}{dx}\frac{1}{9}x^3}$$

$$= \frac{5}{3}(-2)x^{-3} - 2(-4)x^{-5} + \frac{1}{9}\cdot 3x^2$$

$$= -\frac{10}{3x^3} + \frac{8}{x^5} + \frac{1}{3}x^2$$

Matched Problem 5 Find

(A) $f'(x)$ for $f(x) = 3x^4 - 2x^3 + x^2 - 5x + 7$

(B) y' for $y = 3 - 7x^{-2}$

(C) $\frac{dy}{dv}$ for $y = 5v^3 - \sqrt[4]{v}$

(D) $\frac{d}{dx}\left(-\frac{3}{4x} + \frac{4}{x^3} - \frac{x^4}{8}\right)$

Applications

EXAMPLE 6 **Instantaneous Velocity** An object moves along the y axis (marked in feet) so that its position at time x (in seconds) is

$$f(x) = x^3 - 6x^2 + 9x$$

(A) Find the instantaneous velocity function v.

(B) Find the velocity at $x = 2$ and $x = 5$ seconds.

(C) Find the time(s) when the velocity is 0.

SOLUTION

(A) $v = f'(x) = (x^3)' - (6x^2)' + (9x)' = 3x^2 - 12x + 9$

(B) $f'(2) = 3(2)^2 - 12(2) + 9 = -3$ feet per second

$f'(5) = 3(5)^2 - 12(5) + 9 = 24$ feet per second

(C) $v = f'(x) = 3x^2 - 12x + 9 = 0$ — Factor 3 out of each term.

$3(x^2 - 4x + 3) = 0$ — Factor the quadratic term.

$3(x - 1)(x - 3) = 0$ — Use the zero property.

$x = 1, 3$

So, $v = 0$ at $x = 1$ and $x = 3$ seconds.

Matched Problem 6 Repeat Example 6 for $f(x) = x^3 - 15x^2 + 72x$.

EXAMPLE 7 **Tangents** Let $f(x) = x^4 - 6x^2 + 10$.

(A) Find $f'(x)$.

(B) Find the equation of the tangent line at $x = 1$.

(C) Find the values of x where the tangent line is horizontal.

SOLUTION

(A) $f'(x) = (x^4)' - (6x^2)' + (10)'$

$= 4x^3 - 12x$

(B) $y - y_1 = m(x - x_1)$ — $y_1 = f(x_1) = f(1) = (1)^4 - 6(1)^2 + 10 = 5$

$y - 5 = -8(x - 1)$ — $m = f'(x_1) = f'(1) = 4(1)^3 - 12(1) = -8$

$y = -8x + 13$ — Tangent line at $x = 1$

(C) Since a horizontal line has 0 slope, we must solve $f'(x) = 0$ for x:

$f'(x) = 4x^3 - 12x = 0$ — Factor $4x$ out of each term.

$4x(x^2 - 3) = 0$ — Factor the difference of two squares.

$4x(x + \sqrt{3})(x - \sqrt{3}) = 0$ — Use the zero property.

$x = 0, -\sqrt{3}, \sqrt{3}$

Matched Problem 7 Repeat Example 7 for $f(x) = x^4 - 8x^3 + 7$.

Exercises 1-5

A

Find the indicated derivatives in Problems 1–18.

1. $f'(x)$ for $f(x) = 7$
2. $\frac{d}{dx}3$
3. $\frac{dy}{dx}$ for $y = x^9$
4. y' for $y = x^6$
5. $\frac{d}{dx}x^3$
6. $g'(x)$ for $g(x) = x^5$
7. y' for $y = x^{-4}$
8. $\frac{dy}{dx}$ for $y = x^{-8}$
9. $g'(x)$ for $g(x) = x^{8/3}$
10. $f'(x)$ for $f(x) = x^{9/2}$
11. $\frac{dy}{dx}$ for $y = \frac{1}{x^{10}}$
12. y' for $y = \frac{1}{x^{12}}$

13. $f'(x)$ for $f(x) = 5x^2$

14. $\frac{d}{dx}(-2x^3)$

15. y' for $y = 0.4x^7$

16. $f'(x)$ for $f(x) = 0.8x^4$

17. $\frac{d}{dx}\left(\frac{x^3}{18}\right)$

18. $\frac{dy}{dx}$ for $y = \frac{x^5}{25}$

Problems 19–24 refer to functions f and g that satisfy $f'(2) = 3$ and $g'(2) = -1$. In each problem, find $h'(2)$ for the indicated function h.

19. $h(x) = 4f(x)$

20. $h(x) = 5g(x)$

21. $h(x) = f(x) + g(x)$

22. $h(x) = g(x) - f(x)$

23. $h(x) = 2f(x) - 3g(x) + 7$

24. $h(x) = -4f(x) + 5g(x) - 9$

B

Find the indicated derivatives in Problems 25–48.

25. $\frac{d}{dx}(2x - 5)$

26. $\frac{d}{dx}(-4x + 9)$

27. $f'(t)$ if $f(t) = 2t^2 - 3t + 1$

28. $\frac{dy}{dt}$ if $y = 2 + 5t - 8t^3$

29. y' for $y = 5x^{-2} + 9x^{-1}$

30. $g'(x)$ if $g(x) = 5x^{-7} - 2x^{-4}$

31. $\frac{d}{du}(5u^{0.3} - 4u^{2.2})$

32. $\frac{d}{du}(2u^{4.5} - 3.1u + 13.2)$

33. $h'(t)$ if $h(t) = 2.1 + 0.5t - 1.1t^3$

34. $F'(t)$ if $F(t) = 0.2t^3 - 3.1t + 13.2$

35. y' if $y = \frac{2}{5x^4}$

36. w' if $w = \frac{7}{5u^2}$

37. $\frac{d}{dx}\left(\frac{3x^2}{2} - \frac{7}{5x^2}\right)$

38. $\frac{d}{dx}\left(\frac{5x^3}{4} - \frac{2}{5x^3}\right)$

39. $G'(w)$ if $G(w) = \frac{5}{9w^4} + 5\sqrt[3]{w}$

40. $H'(w)$ if $H(w) = \frac{5}{w^6} - 2\sqrt{w}$

41. $\frac{d}{du}(3u^{2/3} - 5u^{1/3})$

42. $\frac{d}{du}(8u^{3/4} + 4u^{-1/4})$

43. $h'(t)$ if $h(t) = \frac{3}{t^{3/5}} - \frac{6}{t^{1/2}}$

44. $F'(t)$ if $F(t) = \frac{5}{t^{1/5}} - \frac{8}{t^{3/2}}$

45. y' if $y = \frac{1}{\sqrt[3]{x}}$

46. w' if $w = \frac{10}{\sqrt[5]{u}}$

47. $\frac{d}{dx}\left(\frac{1.2}{\sqrt{x}} - 3.2x^{-2} + x\right)$

48. $\frac{d}{dx}\left(2.8x^{-3} - \frac{0.6}{\sqrt[3]{x^2}} + 7\right)$

For Problems 49–52, find

(A) $f'(x)$

(B) The slope of the graph of f at $x = 2$ and $x = 4$

(C) The equations of the tangent lines at $x = 2$ and $x = 4$

(D) The value(s) of x where the tangent line is horizontal

49. $f(x) = 6x - x^2$

50. $f(x) = 2x^2 + 8x$

51. $f(x) = 3x^4 - 6x^2 - 7$

52. $f(x) = x^4 - 32x^2 + 10$

If an object moves along the y axis (marked in feet) so that its position at time x (in seconds) is given by the indicated functions in Problems 53–56, find

(A) The instantaneous velocity function $v = f'(x)$

(B) The velocity when $x = 0$ and $x = 3$ seconds

(C) The time(s) when $v = 0$

53. $f(x) = 176x - 16x^2$

54. $f(x) = 80x - 10x^2$

55. $f(x) = x^3 - 9x^2 + 15x$

56. $f(x) = x^3 - 9x^2 + 24x$

Problems 57–64 require the use of a graphing calculator. For each problem, find $f'(x)$ and approximate (to four decimal places) the value(s) of x where the graph of f has a horizontal tangent line.

57. $f(x) = x^2 - 3x - 4\sqrt{x}$

58. $f(x) = x^2 + x - 10\sqrt{x}$

59. $f(x) = 3\sqrt[3]{x^4} - 1.5x^2 - 3x$

60. $f(x) = 3\sqrt[3]{x^4} - 2x^2 + 4x$

61. $f(x) = 0.05x^4 + 0.1x^3 - 1.5x^2 - 1.6x + 3$

62. $f(x) = 0.02x^4 - 0.06x^3 - 0.78x^2 + 0.94x + 2.2$

63. $f(x) = 0.2x^4 - 3.12x^3 + 16.25x^2 - 28.25x + 7.5$

64. $f(x) = 0.25x^4 - 2.6x^3 + 8.1x^2 - 10x + 9$

65. Let $f(x) = ax^2 + bx + c$, $a \neq 0$. Recall that the graph of $y = f(x)$ is a parabola. Use the derivative $f'(x)$ to derive a formula for the x coordinate of the vertex of this parabola.

66. Now that you know how to find derivatives, explain why it is no longer necessary for you to memorize the formula for the x coordinate of the vertex of a parabola.

67. Give an example of a cubic polynomial function that has

(A) No horizontal tangents

(B) One horizontal tangent

(C) Two horizontal tangents

68. Can a cubic polynomial function have more than two horizontal tangents? Explain.

C

Find the indicated derivatives in Problems 69–76.

69. $f'(x)$ if $f(x) = (2x - 1)^2$

70. y' if $y = (2x - 5)^2$

71. $\dfrac{d}{dx}\dfrac{10x + 20}{x}$

72. $\dfrac{dy}{dx}$ if $y = \dfrac{x^2 + 25}{x^2}$

73. $\dfrac{dy}{dx}$ if $y = \dfrac{3x - 4}{12x^2}$

74. $f'(x)$ if $f(x) = \dfrac{2x^5 - 4x^3 + 2x}{x^3}$

In Problems 75-80, discuss the validity of each statement. If the statement is always true, explain why. If not, give a counterexample.

75. The derivative of a sum is the sum of the derivatives.

76. The derivative of a difference is the difference of the derivatives.

77. The derivative of a product is the product of the derivatives.

78. The derivative of a quotient is the quotient of the derivatives.

79. The derivative of a constant is 0.

80. The derivative of a constant times a function is 0.

Applications

81. Sales analysis. A company's total sales (in millions of dollars) t months from now are given by

$$S(t) = 0.03t^3 + 0.5t^2 + 2t + 3$$

(A) Find $S'(t)$.

(B) Find $S(5)$ and $S'(5)$ (to two decimal places). Write a brief verbal interpretation of these results.

(C) Find $S(10)$ and $S'(10)$ (to two decimal places). Write a brief verbal interpretation of these results.

82. Sales analysis. A company's total sales (in millions of dollars) t months from now are given by

$$S(t) = 0.015t^4 + 0.4t^3 + 3.4t^2 + 10t - 3$$

(A) Find $S'(t)$.

(B) Find $S(4)$ and $S'(4)$ (to two decimal places). Write a brief verbal interpretation of these results.

(C) Find $S(8)$ and $S'(8)$ (to two decimal places). Write a brief verbal interpretation of these results.

83. Advertising. A marine manufacturer will sell $N(x)$ power boats after spending $\$x$ thousand on advertising, as given by

$$N(x) = 1{,}000 - \frac{3{,}780}{x} \qquad 5 \le x \le 30$$

(see figure).

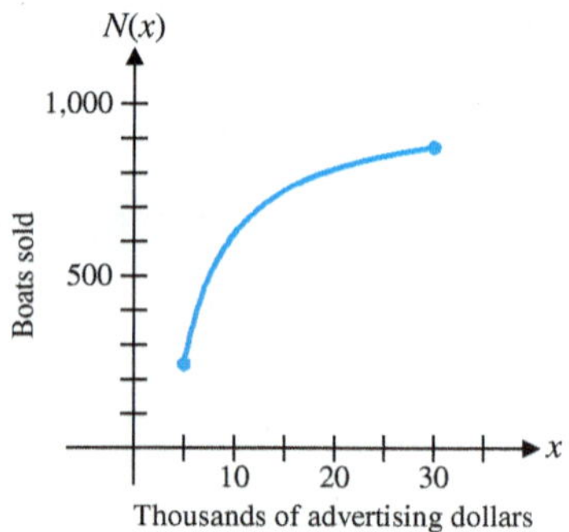

Figure for 83

(A) Find $N'(x)$.

(B) Find $N'(10)$ and $N'(20)$. Write a brief verbal interpretation of these results.

84. Price–demand equation. Suppose that, in a given gourmet food store, people are willing to buy x pounds of chocolate candy per day at $\$p$ per quarter pound, as given by the price–demand equation

$$x = 10 + \frac{180}{p} \qquad 2 \le p \le 10$$

This function is graphed in the figure. Find the demand and the instantaneous rate of change of demand with respect to price when the price is \$5. Write a brief verbal interpretation of these results.

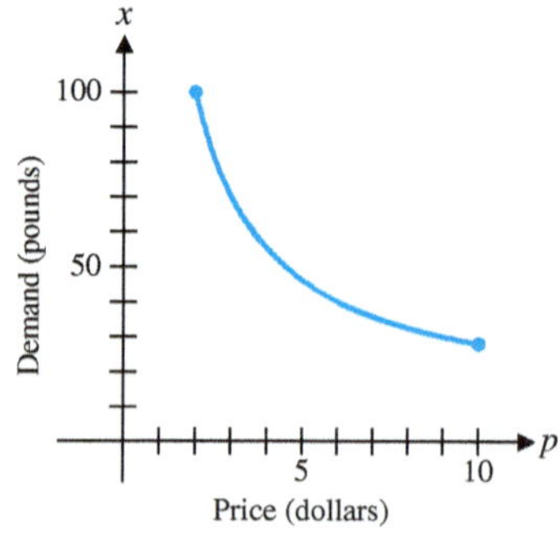

Figure for 84

85. College enrollment. The percentages of male high-school graduates who enrolled in college are given in the second column of Table 1.

Table 1 College enrollment percentages

Year	Male	Female
1970	55.2	48.5
1980	46.7	51.8
1990	58.0	62.2
2000	59.9	66.2
2006	65.8	66.1

(A) Let x represent time (in years) since 1970, and let y represent the corresponding percentage of male high-school graduates who enrolled in college. Enter the data in a graphing calculator and find a cubic regression equation for the data.

 (B) If $y = M(x)$ denotes the regression equation found in part A, find $M(46)$ and $M'(46)$ (to the nearest tenth), and write a brief verbal interpretation of these results.

 **86.** **College enrollment.** The percentages of female high-school graduates who enrolled in college are given in the third column of Table 1.

(A) Let x represent time (in years) since 1970, and let y represent the corresponding percentage of female high-school graduates who enrolled in college. Enter the data in a graphing calculator and find a cubic regression equation for the data.

 (B) If $y = F(x)$ denotes the regression equation found in part A, find $F(46)$ and $F'(46)$ (to the nearest tenth), and write a brief verbal interpretation of these results.

87. **Medicine.** A person x inches tall has a pulse rate of y beats per minute, as given approximately by

$$y = 590x^{-1/2} \qquad 30 \le x \le 75$$

What is the instantaneous rate of change of pulse rate at the

(A) 36-inch level?

(B) 64-inch level?

88. **Ecology.** A coal-burning electrical generating plant emits sulfur dioxide into the surrounding air. The concentration $C(x)$, in parts per million, is given approximately by

$$C(x) = \frac{0.1}{x^2}$$

where x is the distance from the plant in miles. Find the instantaneous rate of change of concentration at

(A) $x = 1$ mile

(B) $x = 2$ miles

89. **Learning.** Suppose that a person learns y items in x hours, as given by

$$y = 50\sqrt{x} \qquad 0 \le x \le 9$$

(see figure). Find the rate of learning at the end of

(A) 1 hour (B) 9 hours

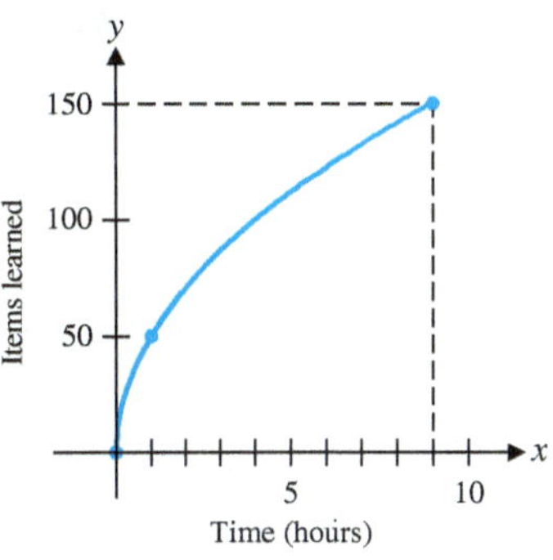

Figure for 89

90. **Learning.** If a person learns y items in x hours, as given by

$$y = 21\sqrt[3]{x^2} \qquad 0 \le x \le 8$$

find the rate of learning at the end of

(A) 1 hour (B) 8 hours

Answers to Matched Problems

1. All are 0.
2. (A) $6x^5$ (B) $30x^{29}$ (C) $-2t^{-3} = -2/t^3$ (D) $\frac{3}{2}x^{1/2}$
3. (A) $-x^{-2}$, or $-1/x^2$ (B) $\frac{2}{3}u^{-1/3}$, or $2/(3\sqrt[3]{u})$ (C) $-\frac{1}{2}x^{-3/2}$, or $-1/(2\sqrt{x^3})$
4. (A) $20x^4$ (B) $t^3/3$ (C) $-x^{-4}$, or $-1/x^4$ (D) $-0.3x^{-4/3}$, or $-0.3/\sqrt[3]{x^4}$
5. (A) $12x^3 - 6x^2 + 2x - 5$ (B) $14x^{-3}$, or $14/x^3$ (C) $15v^2 - \frac{1}{4}v^{-3/4}$, or $15v^2 - 1/(4v^{3/4})$ (D) $3/(4x^2) - (12/x^4) - (x^3/2)$
6. (A) $v = 3x^2 - 30x + 72$ (B) $f'(2) = 24$ ft/s; $f'(5) = -3$ ft/s (C) $x = 4$ and $x = 6$ seconds
7. (A) $f'(x) = 4x^3 - 24x^2$ (B) $y = -20x + 20$ (C) $x = 0$ and $x = 6$

1-6 Differentials

- Increments
- Differentials
- Approximations Using Differentials

In this section, we introduce increments and differentials. Increments are useful and they provide an alternative notation for defining the derivative. Differentials are often easier to compute than increments and can be used to approximate increments.

Increments

In Section 1-4, we defined the derivative of f at x as the limit of the difference quotient

$$f'(x) = \lim_{h \to 0} \frac{f(x + h) - f(x)}{h}$$

We considered various interpretations of this limit, including slope, velocity, and instantaneous rate of change. Increment notation enables us to interpret the numerator and denominator of the difference quotient separately.

Given $y = f(x) = x^3$, if x changes from 2 to 2.1, then y will change from $y = f(2) = 2^3 = 8$ to $y = f(2.1) = 2.1^3 = 9.261$. The change in x is called the *increment in x* and is denoted by Δx (read as "delta x").* Similarly, the change in y is called the *increment in y* and is denoted by Δy. In terms of the given example, we write

$$\Delta x = 2.1 - 2 = 0.1 \quad \text{Change in } x$$

$$\begin{aligned} \Delta y &= f(2.1) - f(2) && f(x) = x^3 \\ &= 2.1^3 - 2^3 && \text{Use a calculator.} \\ &= 9.261 - 8 \\ &= 1.261 && \text{Corresponding change in } y \end{aligned}$$

CONCEPTUAL INSIGHT

The symbol Δx does not represent the product of Δ and x but is the symbol for a single quantity: the *change in x*. Likewise, the symbol Δy represents a single quantity: the *change in y*.

DEFINITION Increments

For $y = f(x)$, $\Delta x = x_2 - x_1$, so $x_2 = x_1 + \Delta x$, and

$$\begin{aligned} \Delta y &= y_2 - y_1 \\ &= f(x_2) - f(x_1) \\ &= f(x_1 + \Delta x) - f(x_1) \end{aligned}$$

Δy represents the change in y corresponding to a change Δx in x.
Δx can be either positive or negative.

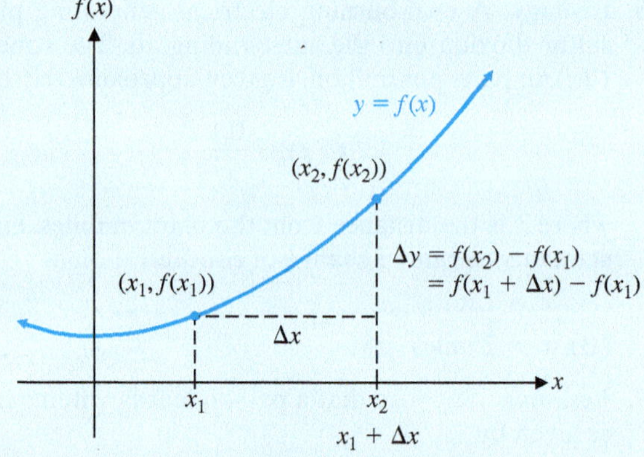

[**Note:** Δy depends on the function f, the input x_1, and the increment Δx.]

EXAMPLE 1 **Increments** Given the function $y = f(x) = \dfrac{x^2}{2}$,

(A) Find Δx, Δy, and $\Delta y/\Delta x$ for $x_1 = 1$ and $x_2 = 2$.

(B) Find $\dfrac{f(x_1 + \Delta x) - f(x_1)}{\Delta x}$ for $x_1 = 1$ and $\Delta x = 2$.

SOLUTION (A) $\Delta x = x_2 - x_1 = 2 - 1 = 1$

$$\begin{aligned} \Delta y &= f(x_2) - f(x_1) \\ &= f(2) - f(1) = \frac{4}{2} - \frac{1}{2} = \frac{3}{2} \end{aligned}$$

$$\frac{\Delta y}{\Delta x} = \frac{f(x_2) - f(x_1)}{x_2 - x_1} = \frac{\frac{3}{2}}{1} = \frac{3}{2}$$

*Δ is the symbol for the Greek letter delta.

(B) $\frac{f(x_1 + \Delta x) - f(x_1)}{\Delta x} = \frac{f(1 + 2) - f(1)}{2}$

$$= \frac{f(3) - f(1)}{2} = \frac{\frac{9}{2} - \frac{1}{2}}{2} = \frac{4}{2} = 2$$

Matched Problem 1 Given the function $y = f(x) = x^2 + 1$,

(A) Find Δx, Δy, and $\Delta y/\Delta x$ for $x_1 = 2$ and $x_2 = 3$.

(B) Find $\frac{f(x_1 + \Delta x) - f(x_1)}{\Delta x}$ for $x_1 = 1$ and $\Delta x = 2$.

In Example 1, we observe another notation for the difference quotient

$$\frac{f(x + h) - f(x)}{h} \tag{1}$$

It is common to refer to h, the change in x, as Δx. Then the difference quotient (1) takes on the form

$$\frac{f(x + \Delta x) - f(x)}{\Delta x} \quad \text{or} \quad \frac{\Delta y}{\Delta x} \qquad \Delta y = f(x + \Delta x) - f(x)$$

and the derivative is defined by

$$f'(x) = \lim_{\Delta x \to 0} \frac{f(x + \Delta x) - f(x)}{\Delta x}$$

or

$$f'(x) = \lim_{\Delta x \to 0} \frac{\Delta y}{\Delta x} \tag{2}$$

if the limit exists.

Explore & Discuss 1 Suppose that $y = f(x)$ defines a function whose domain is the set of all real numbers. If every increment Δy is equal to 0, then what is the range of f?

Differentials

Assume that the limit in equation (2) exists. Then, for small Δx, the difference quotient $\Delta y/\Delta x$ provides a good approximation for $f'(x)$. Also, $f'(x)$ provides a good approximation for $\Delta y/\Delta x$. We write

$$\frac{\Delta y}{\Delta x} \approx f'(x) \qquad \Delta x \text{ is small, but } \neq 0 \tag{3}$$

Multiplying both sides of (3) by Δx gives us

$$\Delta y \approx f'(x)\,\Delta x \qquad \Delta x \text{ is small, but } \neq 0 \tag{4}$$

From equation (4), we see that $f'(x)\Delta x$ provides a good approximation for Δy when Δx is small.

Because of the practical and theoretical importance of $f'(x)\,\Delta x$, we give it the special name **differential** and represent it with the special symbol ***dy*** or ***df***:

$$dy = f'(x)\Delta x \quad \text{or} \quad df = f'(x)\Delta x$$

For example,

$$d(2x^3) = (2x^3)'\,\Delta x = 6x^2\,\Delta x$$

$$d(x) = (x)'\,\Delta x = 1\,\Delta x = \Delta x$$

In the second example, we usually drop the parentheses in $d(x)$ and simply write

$$dx = \Delta x$$

In summary, we have the following:

DEFINITION Differentials

If $y = f(x)$ defines a differentiable function, then the **differential dy, or df**, is defined as the product of $f'(x)$ and dx, where $dx = \Delta x$. Symbolically,

$$dy = f'(x)\,dx, \qquad \text{or} \qquad df = f'(x)\,dx$$

where

$$dx = \Delta x$$

Note: The differential dy (or df) is actually a function involving two independent variables, x and dx. A change in either one or both will affect dy (or df).

EXAMPLE 2 **Differentials** Find dy for $f(x) = x^2 + 3x$. Evaluate dy for

(A) $x = 2$ and $dx = 0.1$

(B) $x = 3$ and $dx = 0.1$

(C) $x = 1$ and $dx = 0.02$

SOLUTION

$$\begin{aligned} dy &= f'(x)\,dx \\ &= (2x + 3)\,dx \end{aligned}$$

(A) When $x = 2$ and $dx = 0.1$,

$$dy = [2(2) + 3]0.1 = 0.7$$

(B) When $x = 3$ and $dx = 0.1$,

$$dy = [2(3) + 3]0.1 = 0.9$$

(C) When $x = 1$ and $dx = 0.02$,

$$dy = [2(1) + 3]0.02 = 0.1$$

Matched Problem 2 Find dy for $f(x) = \sqrt{x} + 3$. Evaluate dy for

(A) $x = 4$ and $dx = 0.1$

(B) $x = 9$ and $dx = 0.12$

(C) $x = 1$ and $dx = 0.01$

We now have two interpretations of the symbol dy/dx. Referring to the function $y = f(x) = x^2 + 3x$ in Example 2 with $x = 2$ and $dx = 0.1$, we have

$$\frac{dy}{dx} = f'(2) = 7 \qquad \text{Derivative}$$

and

$$\frac{dy}{dx} = \frac{0.7}{0.1} = 7 \qquad \text{Ratio of differentials}$$

Approximations Using Differentials

Earlier, we noted that for small Δx,

$$\frac{\Delta y}{\Delta x} \approx f'(x) \qquad \text{and} \qquad \Delta y \approx f'(x)\Delta x$$

Also, since

$$dy = f'(x)\,dx$$

it follows that

$$\Delta y \approx dy$$

and dy can be used to approximate Δy.

To interpret this result geometrically, we need to recall a basic property of the slope. The vertical change in a line is equal to the product of the slope and the horizontal change, as shown in Figure 1.

Now consider the line tangent to the graph of $y = f(x)$, as shown in Figure 2. Since $f'(x)$ is the slope of the tangent line and dx is the horizontal change in the tangent line, it follows that the vertical change in the tangent line is given by $dy = f'(x)\,dx$, as indicated in Figure 2.

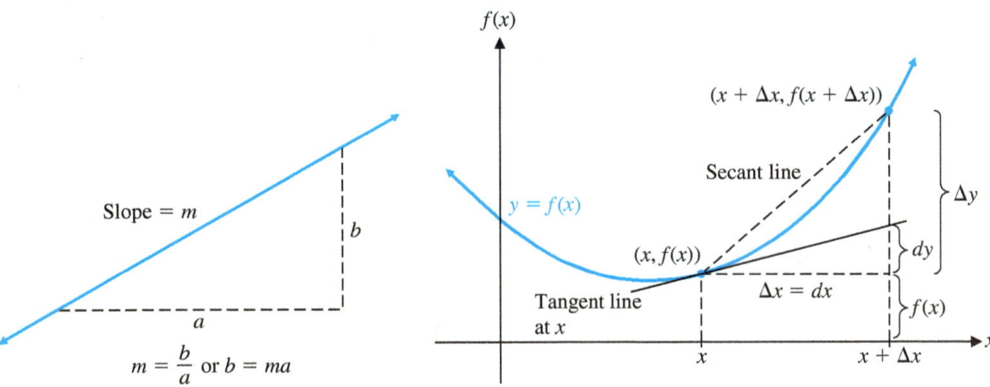

Figure 1

Figure 2

EXAMPLE 3 **Comparing Increments and Differentials** Let $y = f(x) = 6x - x^2$.

(A) Find Δy and dy when $x = 2$.

(B) Graph Δy and dy from part A for $-1 \le \Delta x \le 1$.

(C) Compare Δy and dy from part A for $\Delta x = 0.1, 0.2$, and 0.3.

SOLUTION (A)

$$\begin{aligned}\Delta y &= f(2 + \Delta x) - f(2)\\ &= 6(2 + \Delta x) - (2 + \Delta x)^2 - (6 \cdot 2 - 2^2) && \text{Remove parentheses.}\\ &= 12 + 6\Delta x - 4 - 4\Delta x - \Delta x^2 - 12 + 4 && \text{Collect like terms.}\\ &= 2\Delta x - \Delta x^2\end{aligned}$$

Since $f'(x) = 6 - 2x$, $f'(2) = 2$, and $dx = \Delta x$, $dy = f'(2)\,dx = 2\Delta x$

(B) The graphs of $\Delta y = 2\Delta x - \Delta x^2$ and $dy = 2\Delta x$ are shown in Figure 3. Examining the graphs, we conclude that the differential provides a linear approximation of the increments and that the approximation is better for values of $\Delta x = dx$ close to 0.

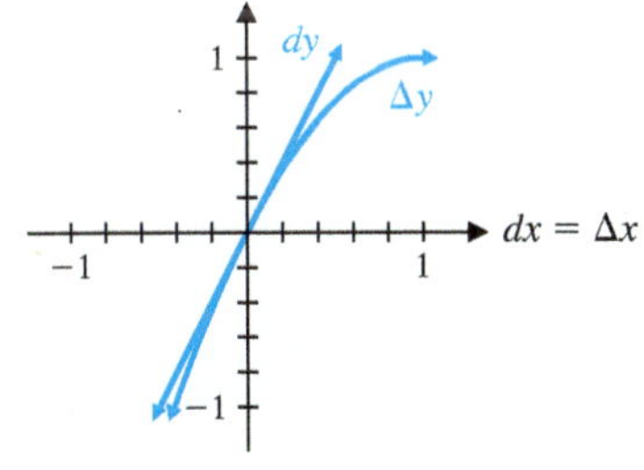

Figure 3 $\Delta y = 2\Delta x - \Delta x^2$, $dy = 2\Delta x$

(C) Table 1 compares the values of Δx and dy for the indicated values of Δx.

Table 1

Δx	Δy	dy
0.1	0.19	0.2
0.2	0.36	0.4
0.3	0.51	0.6

Matched Problem 3 Repeat Example 3 for $x = 4$ and $\Delta x = dx = -0.1, -0.2$, and -0.3.

CONCEPTUAL INSIGHT

The error in the approximation $\Delta y \approx dy$ is usually small when $\Delta x = dx$ is small (see Example 3C), but it can be quite substantial in some cases.

EXAMPLE 4 **Cost–Revenue** A company manufactures and sells x transistor radios per week. If the weekly cost and revenue equations are

$$C(x) = 5{,}000 + 2x \qquad R(x) = 10x - \frac{x^2}{1{,}000} \qquad 0 \le x \le 8{,}000$$

then find the approximate changes in revenue and profit if production is increased from 2,000 to 2,010 units per week.

SOLUTION We will approximate ΔR and ΔP with dR and dP, respectively, using $x = 2{,}000$ and $dx = 2{,}010 - 2{,}000 = 10$.

$$\begin{aligned} R(x) &= 10x - \frac{x^2}{1{,}000} \\ dR &= R'(x)\,dx \\ &= \left(10 - \frac{x}{500}\right)dx \\ &= \left(10 - \frac{2{,}000}{500}\right)10 \\ &= \$60 \text{ per week} \end{aligned}$$

$$\begin{aligned} P(x) = R(x) - C(x) &= 10x - \frac{x^2}{1{,}000} - 5{,}000 - 2x \\ &= 8x - \frac{x^2}{1{,}000} - 5{,}000 \\ dP &= P'(x)\,dx \\ &= \left(8 - \frac{x}{500}\right)dx \\ &= \left(8 - \frac{2{,}000}{500}\right)10 \\ &= \$40 \text{ per week} \end{aligned}$$

Matched Problem 4 Repeat Example 4 with production increasing from 6,000 to 6,010.

Comparing the results in Example 4 and Matched Problem 4, we see that an increase in production results in a revenue and profit increase at the 2,000 production level but a revenue and profit loss at the 6,000 production level.

Exercises 1-6

A

In Problems 1–6, find the indicated quantities for $y = f(x) = 3x^2$.

1. Δx, Δy, and $\Delta y/\Delta x$; given $x_1 = 1$ and $x_2 = 4$
2. Δx, Δy, and $\Delta y/\Delta x$; given $x_1 = 2$ and $x_2 = 5$
3. $\dfrac{f(x_1 + \Delta x) - f(x_1)}{\Delta x}$; given $x_1 = 1$ and $\Delta x = 2$
4. $\dfrac{f(x_1 + \Delta x) - f(x_1)}{\Delta x}$; given $x_1 = 2$ and $\Delta x = 1$
5. $\Delta y/\Delta x$; given $x_1 = 1$ and $x_2 = 3$
6. $\Delta y/\Delta x$; given $x_1 = 2$ and $x_2 = 3$

In Problems 7–12, find dy for each function.

7. $y = 30 + 12x^2 - x^3$
8. $y = 200x - \dfrac{x^2}{30}$
9. $y = x^2\left(1 - \dfrac{x}{9}\right)$
10. $y = x^3(60 - x)$
11. $y = \dfrac{590}{\sqrt{x}}$
12. $y = 52\sqrt{x}$

B

In Problems 13 and 14, find the indicated quantities for $y = f(x) = 3x^2$.

13. (A) $\dfrac{f(2 + \Delta x) - f(2)}{\Delta x}$ (simplify)
    (B) What does the quantity in part (A) approach as Δx approaches 0?
14. (A) $\dfrac{f(3 + \Delta x) - f(3)}{\Delta x}$ (simplify)
    (B) What does the quantity in part (A) approach as Δx approaches 0?

In Problems 15–18, find dy for each function.

15. $y = (2x + 1)^2$
16. $y = (3x + 5)^2$
17. $y = \dfrac{x^2 + 9}{x}$
18. $y = \dfrac{(x - 1)^2}{x^2}$

In Problems 19–22, evaluate dy and Δy for each function for the indicated values.

19. $y = f(x) = x^2 - 3x + 2;\ x = 5,\ dx = \Delta x = 0.2$

20. $y = f(x) = 30 + 12x^2 - x^3;\ x = 2,\ dx = \Delta x = 0.1$

21. $y = f(x) = 75\left(1 - \frac{2}{x}\right);\ x = 5,\ dx = \Delta x = -0.5$

22. $y = f(x) = 100\left(x - \frac{4}{x^2}\right);\ x = 2,\ dx = \Delta x = -0.1$

23. A cube with 10-inch sides is covered with a coat of fiberglass 0.2 inch thick. Use differentials to estimate the volume of the fiberglass shell.

24. A sphere with a radius of 5 centimeters is coated with ice 0.1 centimeter thick. Use differentials to estimate the volume of the ice. [Recall that $V = \frac{4}{3}\pi r^3$.]

C

In Problems 25–28,

(A) Find Δy and dy for the function f at the indicated value of x.

(B) Graph Δy and dy from part A.

(C) Compare the values of Δy and dy from part A at the indicated values of Δx.

25. $f(x) = x^2 + 2x + 3;\ x = -0.5,\ \Delta x = dx = 0.1, 0.2, 0.3$

26. $f(x) = x^2 + 2x + 3;\ x = -2,\ \Delta x = dx = -0.1, -0.2, -0.3$

27. $f(x) = x^3 - 2x^2;\ x = 1,\ \Delta x = dx = 0.05, 0.10, 0.15$

28. $f(x) = x^3 - 2x^2;\ x = 2,\ \Delta x = dx = -0.05, -0.10, -0.15$

In Problems 29–32, discuss the validity of each statement. If the statement is always true, explain why. If not, give a counterexample.

29. If the graph of the function $y = f(x)$ is a line, then the functions Δy and dy (of the independent variable $\Delta x = dx$) for $f(x)$ at $x = 3$ are identical.

30. If the graph of the function $y = f(x)$ is a parabola, then the functions Δy and dy (of the independent variable $\Delta x = dx$) for $f(x)$ at $x = 0$ are identical.

31. Suppose that $y = f(x)$ defines a differentiable function whose domain is the set of all real numbers. If every differential dy at $x = 2$ is equal to 0, then $f(x)$ is a constant function.

32. Suppose that $y = f(x)$ defines a function whose domain is the set of all real numbers. If every increment at $x = 2$ is equal to 0, then $f(x)$ is a constant function.

33. Find dy if $y = (1 - 2x)\sqrt[3]{x^2}$.

34. Find dy if $y = (2x^2 - 4)\sqrt{x}$.

35. Find dy and Δy for $y = 52\sqrt{x}$, $x = 4$, and $\Delta x = dx = 0.3$.

36. Find dy and Δy for $y = 590/\sqrt{x}$, $x = 64$, and $\Delta x = dx = 1$.

Applications

Use differential approximations in the following problems.

37. Advertising. A company will sell N units of a product after spending $\$x$ thousand in advertising, as given by

$$N = 60x - x^2 \qquad 5 \le x \le 30$$

Approximately what increase in sales will result by increasing the advertising budget from \$10,000 to \$11,000? From \$20,000 to \$21,000?

38. Price–demand. Suppose that the daily demand (in pounds) for chocolate candy at $\$x$ per pound is given by

$$D = 1{,}000 - 40x^2 \qquad 1 \le x \le 5$$

If the price is increased from \$3.00 per pound to \$3.20 per pound, what is the approximate change in demand?

39. Average cost. For a company that manufactures tennis rackets, the average cost per racket $\overline{C}$ is

$$\overline{C} = \frac{400}{x} + 5 + \frac{1}{2}x \qquad x \ge 1$$

where x is the number of rackets produced per hour. What will the approximate change in average cost per racket be if production is increased from 20 per hour to 25 per hour? From 40 per hour to 45 per hour?

40. Revenue and profit. A company manufactures and sells x televisions per month. If the cost and revenue equations are

$$C(x) = 72{,}000 + 60x$$

$$R(x) = 200x - \frac{x^2}{30} \qquad 0 \le x \le 6{,}000$$

what will the approximate changes in revenue and profit be if production is increased from 1,500 to 1,510? From 4,500 to 4,510?

41. Pulse rate. The average pulse rate y (in beats per minute) of a healthy person x inches tall is given approximately by

$$y = \frac{590}{\sqrt{x}} \qquad 30 \le x \le 75$$

Approximately how will the pulse rate change for a change in height from 36 to 37 inches? From 64 to 65 inches?

42. Measurement. An egg of a particular bird is nearly spherical. If the radius to the inside of the shell is 5 millimeters and the radius to the outside of the shell is 5.3 millimeters, approximately what is the volume of the shell? [Remember that $V = \frac{4}{3}\pi r^3$.]

43. Medicine. A drug is given to a patient to dilate her arteries. If the radius of an artery is increased from 2 to 2.1 millimeters, approximately how much is the cross-sectional area increased? [Assume that the cross section of the artery is circular; that is, $A = \pi r^2$.]

44. Drug sensitivity. One hour after x milligrams of a particular drug are given to a person, the change in body temperature T (in degrees Fahrenheit) is given by

$$T = x^2\left(1 - \frac{x}{9}\right) \qquad 0 \le x \le 6$$

Approximate the changes in body temperature produced by the following changes in drug dosages:

(A) From 2 to 2.1 milligrams

(B) From 3 to 3.1 milligrams

(C) From 4 to 4.1 milligrams

45. **Learning.** A particular person learning to type has an achievement record given approximately by

$$N = 75\left(1 - \frac{2}{t}\right) \qquad 3 \le t \le 20$$

where N is the number of words per minute typed after t weeks of practice. What is the approximate improvement from 5 to 5.5 weeks of practice?

46. **Learning.** If a person learns y items in x hours, as given approximately by

$$y = 52\sqrt{x} \qquad 0 \le x \le 9$$

what is the approximate increase in the number of items learned when x changes from 1 to 1.1 hours? From 4 to 4.1 hours?

47. **Politics.** In a new city, the voting population (in thousands) is given by

$$N(t) = 30 + 12t^2 - t^3 \qquad 0 \le t \le 8$$

where t is time in years. Find the approximate change in votes for the following changes in time:

(A) From 1 to 1.1 years

(B) From 4 to 4.1 years

(C) From 7 to 7.1 years

Answers to Matched Problems

1. (A) $\Delta x = 1, \Delta y = 5, \Delta y/\Delta x = 5$
(B) 4

2. $dy = \dfrac{1}{2\sqrt{x}}dx$
(A) 0.025
(B) 0.02
(C) 0.005

3. (A) $\Delta y = -2\Delta x - \Delta x^2; dy = -2\Delta x$
(B)

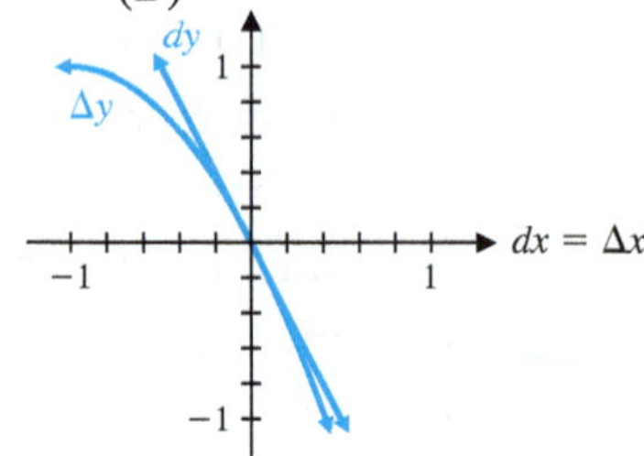

(C)

Δx	Δy	dy
−0.1	0.19	0.2
−0.2	0.36	0.4
−0.3	0.51	0.6

4. $dR = -\$20/\text{wk}; dP = -\$40/\text{wk}$

1-7 Marginal Analysis in Business and Economics

- Marginal Cost, Revenue, and Profit
- Application
- Marginal Average Cost, Revenue, and Profit

Marginal Cost, Revenue, and Profit

One important application of calculus to business and economics involves *marginal analysis.* In economics, the word *marginal* refers to a rate of change—that is, to a derivative. Thus, if $C(x)$ is the total cost of producing x items, then $C'(x)$ is called the *marginal cost* and represents the instantaneous rate of change of total cost with respect to the number of items produced. Similarly, the *marginal revenue* is the derivative of the total revenue function, and the *marginal profit* is the derivative of the total profit function.

DEFINITION Marginal Cost, Revenue, and Profit

If x is the number of units of a product produced in some time interval, then

$$\begin{aligned}
\text{total cost} &= C(x)\\
\textbf{marginal cost} &= C'(x)\\
\text{total revenue} &= R(x)\\
\textbf{marginal revenue} &= R'(x)\\
\text{total profit} &= P(x) = R(x) - C(x)\\
\textbf{marginal profit} &= P'(x) = R'(x) - C'(x)\\
&= (\text{marginal revenue}) - (\text{marginal cost})
\end{aligned}$$

Marginal cost (or revenue or profit) is the instantaneous rate of change of cost (or revenue or profit) relative to production at a given production level.

To begin our discussion, we consider a cost function $C(x)$. It is important to remember that $C(x)$ represents the *total* cost of producing x items, not the cost of producing a *single* item. To find the cost of producing a single item, we use the difference of two successive values of $C(x)$:

$$\begin{aligned} \text{Total cost of producing } x + 1 \text{ items} &= C(x + 1) \\ \text{Total cost of producing } x \text{ items} &= C(x) \\ \text{Exact cost of producing the } (x + 1)\text{st item} &= C(x + 1) - C(x) \end{aligned}$$

EXAMPLE 1 **Cost Analysis** A company manufactures fuel tanks for cars. The total weekly cost (in dollars) of producing x tanks is given by

$$C(x) = 10{,}000 + 90x - 0.05x^2$$

(A) Find the marginal cost function.

(B) Find the marginal cost at a production level of 500 tanks per week.

(C) Interpret the results of part B.

(D) Find the exact cost of producing the 501st item.

SOLUTION (A) $C'(x) = 90 - 0.1x$

(B) $C'(500) = 90 - 0.1(500) = \$40$ Marginal cost

(C) At a production level of 500 tanks per week, the total production costs are increasing at the rate of \$40 per tank.

(D)

$$\begin{aligned} C(501) &= 10{,}000 + 90(501) - 0.05(501)^2 \\ &= \$42{,}539.95 && \text{Total cost of producing 501 tanks per week} \\ C(500) &= 10{,}000 + 90(500) - 0.05(500)^2 \\ &= \$42{,}500.00 && \text{Total cost of producing 500 tanks per week} \\ C(501) - C(500) &= 42{,}539.95 - 42{,}500.00 \\ &= \$39.95 && \text{Exact cost of producing the 501st tank} \end{aligned}$$

Matched Problem 1 A company manufactures automatic transmissions for cars. The total weekly cost (in dollars) of producing x transmissions is given by

$$C(x) = 50{,}000 + 600x - 0.75x^2$$

(A) Find the marginal cost function.

(B) Find the marginal cost at a production level of 200 transmissions per week.

(C) Interpret the results of part B.

(D) Find the exact cost of producing the 201st transmission.

In Example 1, we found that the cost of the 501st tank and the marginal cost at a production level of 500 tanks differ by only a nickel. Increments and differentials will help us understand the relationship between marginal cost and the cost of a single item. If $C(x)$ is any total cost function, then

$$C'(x) \approx \frac{C(x + \Delta x) - C(x)}{\Delta x} \qquad \text{See Section 1-6}$$

$$C'(x) \approx C(x + 1) - C(x) \qquad \Delta x = 1$$

We see that the marginal cost $C'(x)$ approximates $C(x + 1) - C(x)$, the exact cost of producing the $(x + 1)$st item. These observations are summarized next and are illustrated in Figure 1.

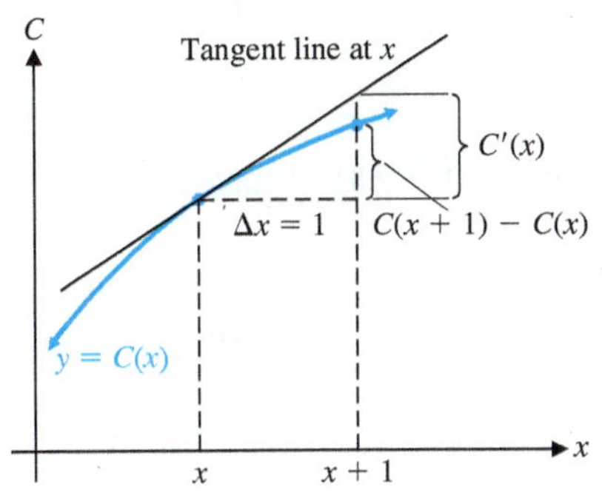

Figure 1 $C'(x) \approx C(x + 1) - C(x)$

THEOREM 1 Marginal Cost and Exact Cost

If $C(x)$ is the total cost of producing x items, then the marginal cost function approximates the exact cost of producing the $(x + 1)$st item:

Marginal cost   Exact cost

$$C'(x) \approx C(x + 1) - C(x)$$

Similar statements can be made for total revenue functions and total profit functions.

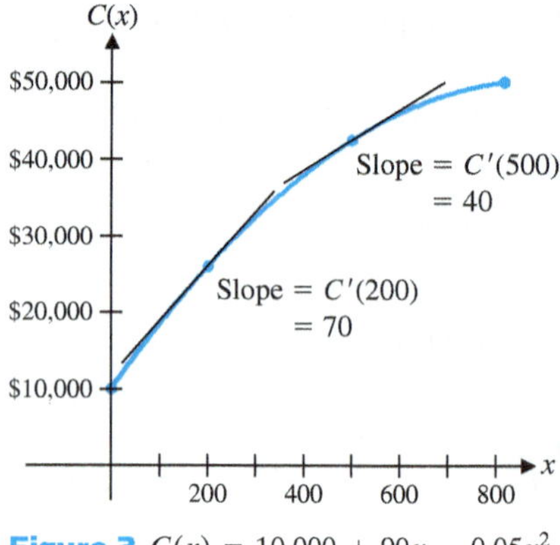

Figure 2 $C(x) = 10{,}000 + 90x - 0.05x^2$

CONCEPTUAL INSIGHT

Theorem 1 states that the marginal cost at a given production level x approximates the cost of producing the $(x + 1)$st, or *next,* item. In practice, the marginal cost is used more frequently than the exact cost. One reason for this is that the marginal cost is easily visualized when one is examining the graph of the total cost function. Figure 2 shows the graph of the cost function discussed in Example 1, with tangent lines added at $x = 200$ and $x = 500$. The graph clearly shows that as production increases, the slope of the tangent line decreases. Thus, the cost of producing the next tank also decreases, a desirable characteristic of a total cost function. We will have much more to say about graphical analysis in Chapter 3.

Application

Now we discuss how price, demand, revenue, cost, and profit are tied together in typical applications. Although either price or demand can be used as the independent variable in a price–demand equation, it is common to use demand as the independent variable when marginal revenue, cost, and profit are also involved.

EXAMPLE 2 **Production Strategy** A company's market research department recommends the manufacture and marketing of a new headphone set for MP3 players. After suitable test marketing, the research department presents the following **price–demand equation**:

$$x = 10{,}000 - 1{,}000p \qquad x \text{ is demand at price } p. \tag{1}$$

Solving (1) for p gives

$$p = 10 - 0.001x \tag{2}$$

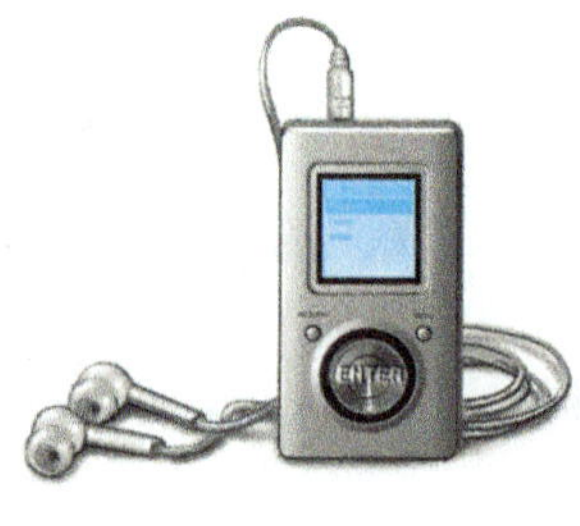

where x is the number of headphones that retailers are likely to buy at \$$p$ per set.

The financial department provides the **cost function**

$$C(x) = 7{,}000 + 2x \tag{3}$$

where \$7,000 is the estimate of fixed costs (tooling and overhead) and \$2 is the estimate of variable costs per headphone set (materials, labor, marketing, transportation, storage, etc.).

(A) Find the domain of the function defined by the price–demand equation (2).

(B) Find and interpret the marginal cost function $C'(x)$.

(C) Find the revenue function as a function of x and find its domain.

(D) Find the marginal revenue at $x = 2{,}000$, $5{,}000$, and $7{,}000$. Interpret these results.

(E) Graph the cost function and the revenue function in the same coordinate system. Find the intersection points of these two graphs and interpret the results.

(F) Find the profit function and its domain and sketch the graph of the function.

(G) Find the marginal profit at $x = 1{,}000$, $4{,}000$, and $6{,}000$. Interpret these results.

SOLUTION (A) Since price p and demand x must be non-negative, we have $x \geq 0$ and

$$\begin{aligned} p = 10 - 0.001x &\geq 0 \\ 10 &\geq 0.001x \\ 10{,}000 &\geq x \end{aligned}$$

Thus, the permissible values of x are $0 \leq x \leq 10{,}000$.

(B) The marginal cost is $C'(x) = 2$. Since this is a constant, it costs an additional \$2 to produce one more headphone set at any production level.

(C) The **revenue** is the amount of money R received by the company for manufacturing and selling x headphone sets at \$$p$ per set and is given by

$$R = (\text{number of headphone sets sold})(\text{price per headphone set}) = xp$$

In general, the revenue R can be expressed as a function of p using equation (1) or as a function of x using equation (2). As we mentioned earlier, when using marginal functions, we will always use the number of items x as the independent variable. Thus, the **revenue function** is

$$\begin{aligned} R(x) &= xp = x(10 - 0.001x) \quad \text{Using equation (2)} \qquad (4) \\ &= 10x - 0.001x^2 \end{aligned}$$

Since equation (2) is defined only for $0 \leq x \leq 10{,}000$, it follows that the domain of the revenue function is $0 \leq x \leq 10{,}000$.

(D) The **marginal revenue** is

$$R'(x) = 10 - 0.002x$$

For production levels of $x = 2{,}000, 5{,}000$, and $7{,}000$, we have

$$R'(2{,}000) = 6 \qquad R'(5{,}000) = 0 \qquad R'(7{,}000) = -4$$

This means that at production levels of 2,000, 5,000, and 7,000, the respective approximate changes in revenue per unit change in production are \$6, \$0, and −\$4. That is, at the 2,000 output level, revenue increases as production increases; at the 5,000 output level, revenue does not change with a "small" change in production; and at the 7,000 output level, revenue decreases with an increase in production.

(E) Graphing $R(x)$ and $C(x)$ in the same coordinate system results in Figure 3. The intersection points are called the **break-even points**, because revenue equals cost at these production levels. The company neither makes nor loses money, but just breaks even. The break-even points are obtained as follows:

$$\begin{aligned} C(x) &= R(x) \\ 7{,}000 + 2x &= 10x - 0.001x^2 \\ 0.001x^2 - 8x + 7{,}000 &= 0 \quad \text{Solve by the quadratic formula} \\ x^2 - 8{,}000x + 7{,}000{,}000 &= 0 \quad \text{(see Appendix A-7).} \\ x &= \frac{8{,}000 \pm \sqrt{8{,}000^2 - 4(7{,}000{,}000)}}{2} \\ &= \frac{8{,}000 \pm \sqrt{36{,}000{,}000}}{2} \\ &= \frac{8{,}000 \pm 6{,}000}{2} \\ &= 1{,}000, \quad 7{,}000 \end{aligned}$$

$$\begin{aligned} R(1{,}000) &= 10(1{,}000) - 0.001(1{,}000)^2 = 9{,}000 \\ C(1{,}000) &= 7{,}000 + 2(1{,}000) = 9{,}000 \\ R(7{,}000) &= 10(7{,}000) - 0.001(7{,}000)^2 = 21{,}000 \\ C(7{,}000) &= 7{,}000 + 2(7{,}000) = 21{,}000 \end{aligned}$$

The break-even points are (1,000, 9,000) and (7,000, 21,000), as shown in Figure 3. Further examination of the figure shows that cost is greater than revenue for production levels between 0 and 1,000 and also between 7,000 and 10,000. Consequently, the company incurs a loss at these levels. By contrast, for production levels between 1,000 and 7,000, revenue is greater than cost, and the company makes a profit.

(F) The **profit function** is

$$\begin{aligned} P(x) &= R(x) - C(x) \\ &= (10x - 0.001x^2) - (7{,}000 + 2x) \\ &= -0.001x^2 + 8x - 7{,}000 \end{aligned}$$

The domain of the cost function is $x \geq 0$, and the domain of the revenue function is $0 \leq x \leq 10{,}000$. The domain of the profit function is the set of x values for which both functions are defined—that is, $0 \leq x \leq 10{,}000$. The graph of the profit function is shown in Figure 4. Notice that the x coordinates of the break-even points in Figure 3 are the x intercepts of the profit function. Furthermore, the intervals on which cost is greater than revenue and on which revenue is greater than cost correspond, respectively, to the intervals on which profit is negative and on which profit is positive.

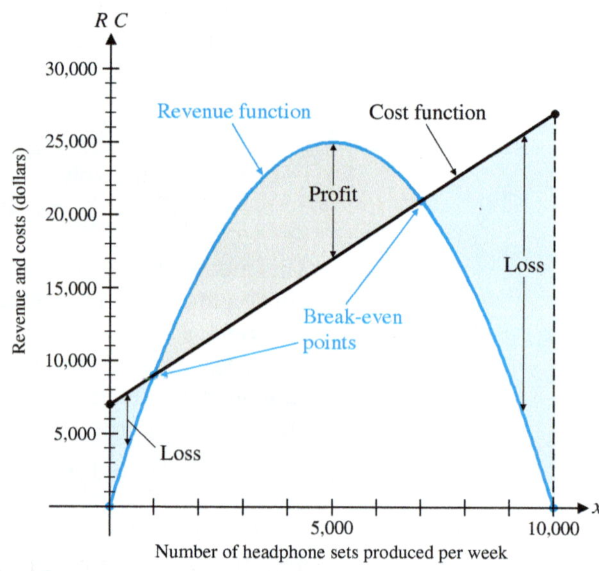

Figure 3

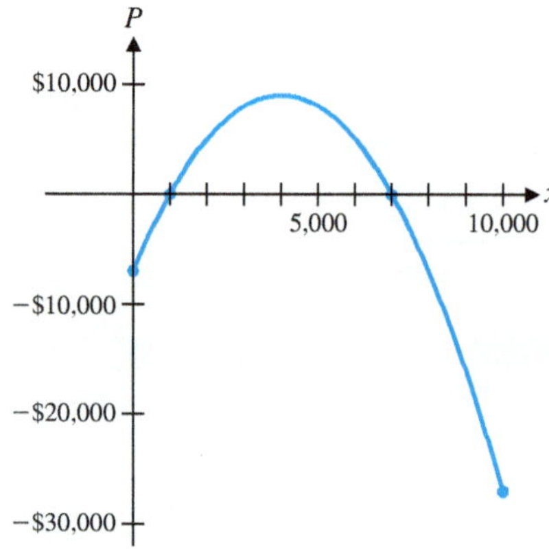

Figure 4

(G) The **marginal profit** is

$$P'(x) = -0.002x + 8$$

For production levels of 1,000, 4,000, and 6,000, we have

$$P'(1{,}000) = 6 \qquad P'(4{,}000) = 0 \qquad P'(6{,}000) = -4$$

This means that at production levels of 1,000, 4,000, and 6,000, the respective approximate changes in profit per unit change in production are \$6, \$0, and −\$4. That is, at the 1,000 output level, profit will be increased if production is increased; at the 4,000 output level, profit does not change for "small" changes in production; and at the 6,000 output level, profits will decrease if production is increased. It seems that the best production level to produce a maximum profit is 4,000.

Example 2 requires careful study since a number of important ideas in economics and calculus are involved. In the next chapter, we will develop a systematic procedure for finding the production level (and, using the demand equation, the selling price) that will maximize profit.

Matched Problem 2 Refer to the revenue and profit functions in Example 2.

(A) Find $R'(3{,}000)$ and $R'(6{,}000)$. Interpret the results.

(B) Find $P'(2{,}000)$ and $P'(7{,}000)$. Interpret the results.

Marginal Average Cost, Revenue, and Profit

Sometimes it is desirable to carry out marginal analysis relative to **average cost (cost per unit), average revenue (revenue per unit),** and **average profit (profit per unit).**

DEFINITION Marginal Average Cost, Revenue, and Profit

If x is the number of units of a product produced in some time interval, then

Cost per unit: **average cost** $= \overline{C}(x) = \dfrac{C(x)}{x}$

marginal average cost $= \overline{C}'(x) = \dfrac{d}{dx}\overline{C}(x)$

Revenue per unit: **average revenue** $= \overline{R}(x) = \dfrac{R(x)}{x}$

marginal average revenue $= \overline{R}'(x) = \dfrac{d}{dx}\overline{R}(x)$

Profit per unit: **average profit** $= \overline{P}(x) = \dfrac{P(x)}{x}$

marginal average profit $= \overline{P}'(x) = \dfrac{d}{dx}\overline{P}(x)$

EXAMPLE 3 **Cost Analysis** A small machine shop manufactures drill bits used in the petroleum industry. The manager estimates that the total daily cost (in dollars) of producing x bits is

$$C(x) = 1{,}000 + 25x - 0.1x^2$$

(A) Find $\overline{C}(x)$ and $\overline{C}'(x)$.

(B) Find $\overline{C}(10)$ and $\overline{C}'(10)$. Interpret these quantities.

(C) Use the results in part (B) to estimate the average cost per bit at a production level of 11 bits per day.

SOLUTION (A) $\overline{C}(x) = \dfrac{C(x)}{x} = \dfrac{1{,}000 + 25x - 0.1x^2}{x}$

$$= \frac{1{,}000}{x} + 25 - 0.1x \qquad \text{Average cost function}$$

$$\overline{C}'(x) = \frac{d}{dx}\overline{C}(x) = -\frac{1{,}000}{x^2} - 0.1 \qquad \text{Marginal average cost function}$$

(B) $\overline{C}(10) = \dfrac{1{,}000}{10} + 25 - 0.1(10) = \$124$

$$\overline{C}'(10) = -\frac{1{,}000}{10^2} - 0.1 = -\$10.10$$

At a production level of 10 bits per day, the average cost of producing a bit is \$124. This cost is decreasing at the rate of \$10.10 per bit.

(C) If production is increased by 1 bit, then the average cost per bit will decrease by approximately \$10.10. So, the average cost per bit at a production level of 11 bits per day is approximately $\$124 - \$10.10 = \$113.90$.

Matched Problem 3 Consider the cost function for the production of headphone sets from Example 2:

$$C(x) = 7{,}000 + 2x$$

(A) Find $\overline{C}(x)$ and $\overline{C}'(x)$.

(B) Find $\overline{C}(100)$ and $\overline{C}'(100)$. Interpret these quantities.

(C) Use the results in part (B) to estimate the average cost per headphone set at a production level of 101 headphone sets.

Explore & Discuss 1 A student produced the following solution to Matched Problem 3:

$$C(x) = 7{,}000 + 2x \quad \text{Cost}$$

$$C'(x) = 2 \quad \text{Marginal cost}$$

$$\frac{C'(x)}{x} = \frac{2}{x} \quad \text{"Average" of the marginal cost}$$

Explain why the last function is not the same as the marginal average cost function.

CAUTION

1. The marginal average cost function is computed by first finding the average cost function and then finding its derivative. As Explore & Discuss 1 illustrates, reversing the order of these two steps produces a different function that does not have any useful economic interpretations.
2. Recall that the marginal cost function has two interpretations: the usual interpretation of any derivative as an instantaneous rate of change and the special interpretation as an approximation to the exact cost of the $(x + 1)$st item. This special interpretation does not apply to the marginal average cost function. Referring to Example 3, we would be incorrect to interpret $\overline{C}'(10) = -\$10.10$ to mean that the average cost of the next bit is approximately $-\$10.10$. In fact, the phrase "average cost of the next bit" does not even make sense. Averaging is a concept applied to a collection of items, not to a single item.

These remarks also apply to revenue and profit functions.

Exercises 1-7

A

In Problems 1–4, find the marginal cost function.

1. $C(x) = 175 + 0.8x$
2. $C(x) = 4{,}500 + 9.5x$
3. $C(x) = 210 + 4.6x - 0.01x^2$
4. $C(x) = 790 + 13x - 0.2x^2$

In Problems 5–8, find the marginal revenue function.

5. $R(x) = 4x - 0.01x^2$
6. $R(x) = 36x - 0.03x^2$
7. $R(x) = x(12 - 0.04x)$
8. $R(x) = x(25 - 0.05x)$

In Problems 9–12, find the marginal profit function if the cost and revenue, respectively, are those in the indicated problems.

9. Problem 1 and Problem 5
10. Problem 2 and Problem 6
11. Problem 3 and Problem 7
12. Problem 4 and Problem 8

B

In Problems 13–20, find the indicated function if cost and revenue are given by $C(x) = 145 + 1.1x$ and $R(x) = 5x - 0.02x^2$, respectively.

13. Average cost function
14. Average revenue function

15. Marginal average cost function

16. Marginal average revenue function

17. Profit function

18. Marginal profit function

19. Average profit function

20. Marginal average profit function

C

In Problems 21–24, discuss the validity of each statement. If the statement is always true, explain why. If not, give a counterexample.

21. If a cost function is linear, then the marginal cost is a constant.

22. If a price–demand equation is linear, then the marginal revenue function is linear.

23. Marginal profit is equal to marginal cost minus marginal revenue.

24. Marginal average cost is equal to average marginal cost.

Applications

25. Cost analysis. The total cost (in dollars) of producing x food processors is

$$C(x) = 2{,}000 + 50x - 0.5x^2$$

(A) Find the exact cost of producing the 21st food processor.

(B) Use marginal cost to approximate the cost of producing the 21st food processor.

26. Cost analysis. The total cost (in dollars) of producing x electric guitars is

$$C(x) = 1{,}000 + 100x - 0.25x^2$$

(A) Find the exact cost of producing the 51st guitar.

(B) Use marginal cost to approximate the cost of producing the 51st guitar.

27. Cost analysis. The total cost (in dollars) of manufacturing x auto body frames is

$$C(x) = 60{,}000 + 300x$$

(A) Find the average cost per unit if 500 frames are produced.

(B) Find the marginal average cost at a production level of 500 units and interpret the results.

(C) Use the results from parts (A) and (B) to estimate the average cost per frame if 501 frames are produced.

28. Cost analysis. The total cost (in dollars) of printing x dictionaries is

$$C(x) = 20{,}000 + 10x$$

(A) Find the average cost per unit if 1,000 dictionaries are produced.

(B) Find the marginal average cost at a production level of 1,000 units and interpret the results.

(C) Use the results from parts (A) and (B) to estimate the average cost per dictionary if 1,001 dictionaries are produced.

29. Profit analysis. The total profit (in dollars) from the sale of x skateboards is

$$P(x) = 30x - 0.3x^2 - 250 \qquad 0 \le x \le 100$$

(A) Find the exact profit from the sale of the 26th skateboard.

(B) Use marginal profit to approximate the profit from the sale of the 26th skateboard.

30. Profit analysis. The total profit (in dollars) from the sale of x calendars is

$$P(x) = 22x - 0.2x^2 - 400 \qquad 0 \le x \le 100$$

(A) Find the exact profit from the sale of the 41st calendar.

(B) Use the marginal profit to approximate the profit from the sale of the 41st calendar.

31. Profit analysis. The total profit (in dollars) from the sale of x DVDs is

$$P(x) = 5x - 0.005x^2 - 450 \qquad 0 \le x \le 1{,}000$$

Evaluate the marginal profit at the given values of x, and interpret the results.

(A) $x = 450$ (B) $x = 750$

32. Profit analysis. The total profit (in dollars) from the sale of x cameras is

$$P(x) = 12x - 0.02x^2 - 1{,}000 \qquad 0 \le x \le 600$$

Evaluate the marginal profit at the given values of x, and interpret the results.

(A) $x = 200$ (B) $x = 350$

33. Profit analysis. The total profit (in dollars) from the sale of x lawn mowers is

$$P(x) = 30x - 0.03x^2 - 750 \qquad 0 \le x \le 1{,}000$$

(A) Find the average profit per mower if 50 mowers are produced.

(B) Find the marginal average profit at a production level of 50 mowers and interpret the results.

(C) Use the results from parts (A) and (B) to estimate the average profit per mower if 51 mowers are produced.

34. Profit analysis. The total profit (in dollars) from the sale of x gas grills is

$$P(x) = 20x - 0.02x^2 - 320 \qquad 0 \le x \le 1{,}000$$

(A) Find the average profit per grill if 40 grills are produced.

(B) Find the marginal average profit at a production level of 40 grills and interpret the results.

(C) Use the results from parts (A) and (B) to estimate the average profit per grill if 41 grills are produced.

35. Revenue analysis. The price p (in dollars) and the demand x for a brand of running shoes are related by the equation

$$x = 4{,}000 - 40p$$

(A) Express the price p in terms of the demand x, and find the domain of this function.

(B) Find the revenue $R(x)$ from the sale of x pairs of running shoes. What is the domain of R?

(C) Find the marginal revenue at a production level of 1,600 pairs and interpret the results.

(D) Find the marginal revenue at a production level of 2,500 pairs, and interpret the results.

36. Revenue analysis. The price p (in dollars) and the demand x for a particular steam iron are related by the equation

$$x = 1{,}000 - 20p$$

(A) Express the price p in terms of the demand x, and find the domain of this function.

(B) Find the revenue $R(x)$ from the sale of x steam irons. What is the domain of R?

(C) Find the marginal revenue at a production level of 400 steam irons and interpret the results.

(D) Find the marginal revenue at a production level of 650 steam irons and interpret the results.

37. Revenue, cost, and profit. The price–demand equation and the cost function for the production of table saws are given, respectively, by

$$x = 6{,}000 - 30p \quad \text{and} \quad C(x) = 72{,}000 + 60x$$

where x is the number of saws that can be sold at a price of $\$p$ per saw and $C(x)$ is the total cost (in dollars) of producing x saws.

(A) Express the price p as a function of the demand x, and find the domain of this function.

(B) Find the marginal cost.

(C) Find the revenue function and state its domain.

(D) Find the marginal revenue.

(E) Find $R'(1{,}500)$ and $R'(4{,}500)$ and interpret these quantities.

(F) Graph the cost function and the revenue function on the same coordinate system for $0 \le x \le 6{,}000$. Find the break-even points, and indicate regions of loss and profit.

(G) Find the profit function in terms of x.

(H) Find the marginal profit.

(I) Find $P'(1{,}500)$ and $P'(3{,}000)$ and interpret these quantities.

38. Revenue, cost, and profit. The price–demand equation and the cost function for the production of HDTVs are given, respectively, by

$$x = 9{,}000 - 30p \quad \text{and} \quad C(x) = 150{,}000 + 30x$$

where x is the number of HDTVs that can be sold at a price of $\$p$ per TV and $C(x)$ is the total cost (in dollars) of producing x TVs.

(A) Express the price p as a function of the demand x, and find the domain of this function.

(B) Find the marginal cost.

(C) Find the revenue function and state its domain.

(D) Find the marginal revenue.

(E) Find $R'(3{,}000)$ and $R'(6{,}000)$ and interpret these quantities.

(F) Graph the cost function and the revenue function on the same coordinate system for $0 \le x \le 9{,}000$. Find the break-even points and indicate regions of loss and profit.

(G) Find the profit function in terms of x.

(H) Find the marginal profit.

(I) Find $P'(1{,}500)$ and $P'(4{,}500)$ and interpret these quantities.

39. Revenue, cost, and profit. A company is planning to manufacture and market a new two-slice electric toaster. After conducting extensive market surveys, the research department provides the following estimates: a weekly demand of 200 toasters at a price of \$16 per toaster and a weekly demand of 300 toasters at a price of \$14 per toaster. The financial department estimates that weekly fixed costs will be \$1,400 and variable costs (cost per unit) will be \$4.

(A) Assume that the relationship between price p and demand x is linear. Use the research department's estimates to express p as a function of x and find the domain of this function.

(B) Find the revenue function in terms of x and state its domain.

(C) Assume that the cost function is linear. Use the financial department's estimates to express the cost function in terms of x.

(D) Graph the cost function and revenue function on the same coordinate system for $0 \le x \le 1{,}000$. Find the break-even points and indicate regions of loss and profit.

(E) Find the profit function in terms of x.

(F) Evaluate the marginal profit at $x = 250$ and $x = 475$ and interpret the results.

40. Revenue, cost, and profit. The company in Problem 39 is also planning to manufacture and market a four-slice toaster. For this toaster, the research department's estimates are a weekly demand of 300 toasters at a price of \$25 per toaster and a weekly demand of 400 toasters at a price of \$20. The financial department's estimates are fixed weekly costs of \$5,000 and variable costs of \$5 per toaster.

(A) Assume that the relationship between price p and demand x is linear. Use the research department's estimates to express p as a function of x, and find the domain of this function.

(B) Find the revenue function in terms of x and state its domain.

(C) Assume that the cost function is linear. Use the financial department's estimates to express the cost function in terms of x.

(D) Graph the cost function and revenue function on the same coordinate system for $0 \le x \le 800$. Find the break-even points and indicate regions of loss and profit.

(E) Find the profit function in terms of x.

(F) Evaluate the marginal profit at $x = 325$ and $x = 425$ and interpret the results.

41. Revenue, cost, and profit. The total cost and the total revenue (in dollars) for the production and sale of x ski jackets are given, respectively, by

$$C(x) = 24x + 21{,}900 \quad \text{and} \quad R(x) = 200x - 0.2x^2 \qquad 0 \le x \le 1{,}000$$

(A) Find the value of x where the graph of $R(x)$ has a horizontal tangent line.

(B) Find the profit function $P(x)$.

(C) Find the value of x where the graph of $P(x)$ has a horizontal tangent line.

(D) Graph $C(x)$, $R(x)$, and $P(x)$ on the same coordinate system for $0 \le x \le 1{,}000$. Find the break-even points. Find the x intercepts of the graph of $P(x)$.

42. Revenue, cost, and profit. The total cost and the total revenue (in dollars) for the production and sale of x hair dryers are given, respectively, by

$$C(x) = 5x + 2{,}340 \quad \text{and} \quad R(x) = 40x - 0.1x^2$$
$$0 \le x \le 400$$

(A) Find the value of x where the graph of $R(x)$ has a horizontal tangent line.

(B) Find the profit function $P(x)$.

(C) Find the value of x where the graph of $P(x)$ has a horizontal tangent line.

(D) Graph $C(x)$, $R(x)$, and $P(x)$ on the same coordinate system for $0 \le x \le 400$. Find the break-even points. Find the x intercepts of the graph of $P(x)$.

43. Break-even analysis. The price–demand equation and the cost function for the production of garden hoses are given, respectively, by

$$p = 20 - \sqrt{x} \quad \text{and} \quad C(x) = 500 + 2x$$

where x is the number of garden hoses that can be sold at a price of $\$p$ per unit and $C(x)$ is the total cost (in dollars) of producing x garden hoses.

(A) Express the revenue function in terms of x.

(B) Graph the cost function and revenue function in the same viewing window for $0 \le x \le 400$. Use approximation techniques to find the break-even points correct to the nearest unit.

44. Break-even analysis. The price–demand equation and the cost function for the production of handwoven silk scarves are given, respectively, by

$$p = 60 - 2\sqrt{x} \quad \text{and} \quad C(x) = 3{,}000 + 5x$$

where x is the number of scarves that can be sold at a price of $\$p$ per unit and $C(x)$ is the total cost (in dollars) of producing x scarves.

(A) Express the revenue function in terms of x.

(B) Graph the cost function and the revenue function in the same viewing window for $0 \le x \le 900$. Use approximation techniques to find the break-even points correct to the nearest unit.

45. Break-even analysis. Table 2 contains price–demand and total cost data for the production of projectors, where p is the wholesale price (in dollars) of a projector for an annual demand of x projectors and C is the total cost (in dollars) of producing x projectors.

Table 2

x	$p(\$)$	$C(\$)$
3,190	581	1,130,000
4,570	405	1,241,000
5,740	181	1,410,000
7,330	85	1,620,000

(A) Find a quadratic regression equation for the price–demand data, using x as the independent variable.

(B) Find a linear regression equation for the cost data, using x as the independent variable. Use this equation to estimate the fixed costs and variable costs per projector. Round answers to the nearest dollar.

(C) Find the break-even points. Round answers to the nearest integer.

(D) Find the price range for which the company will make a profit. Round answers to the nearest dollar.

46. Break-even analysis. Table 3 contains price–demand and total cost data for the production of treadmills, where p is the wholesale price (in dollars) of a treadmill for an annual demand of x treadmills and C is the total cost (in dollars) of producing x treadmills.

Table 3

x	$p(\$)$	$C(\$)$
2,910	1,435	3,650,000
3,415	1,280	3,870,000
4,645	1,125	4,190,000
5,330	910	4,380,000

(A) Find a linear regression equation for the price–demand data, using x as the independent variable.

(B) Find a linear regression equation for the cost data, using x as the independent variable. Use this equation to estimate the fixed costs and variable costs per treadmill. Round answers to the nearest dollar.

(C) Find the break-even points. Round answers to the nearest integer.

(D) Find the price range for which the company will make a profit. Round answers to the nearest dollar.

Answers to Matched Problems

1. (A) $C'(x) = 600 - 1.5x$

(B) $C'(200) = 300$. At a production level of 200 transmissions, total costs are increasing at the rate of \$300 per transmission.

(C) $C(201) - C(200) = \$299.25$

2. (A) $R'(3{,}000) = 4$. At a production level of 3,000, a unit increase in production will increase revenue by approximately \$4.
$R'(6{,}000) = -2$. At a production level of 6,000, a unit increase in production will decrease revenue by approximately \$2.

(B) $P'(2{,}000) = 4$. At a production level of 2,000, a unit increase in production will increase profit by approximately \$4.
$P'(7{,}000) = -6$. At a production level of 7,000, a unit increase in production will decrease profit by approximately \$6.

3. (A) $\overline{C}(x) = \dfrac{7{,}000}{x} + 2$; $\overline{C}'(x) = -\dfrac{7{,}000}{x^2}$

(B) $\overline{C}(100) = \$72$; $\overline{C}'(100) = -\$0.70$. At a production level of 100 headphone sets, the average cost per headphone set is \$72. This average cost is decreasing at a rate of \$0.70 per headphone set.

(C) Approx. \$71.30.

Chapter 1 Review

Important Terms, Symbols, and Concepts

1-1 Introduction to Limits

EXAMPLES

- The graph of the function $y = f(x)$ is the graph of the set of all ordered pairs $(x, f(x))$. Ex. 1, p. 3
- The limit of the function $y = f(x)$ as x approaches c is L, written as $\lim_{x \to c} f(x) = L$, if the functional value $f(x)$ is close to the single real number L whenever x is close, but not equal, to c (on either side of c). Ex. 2, p. 4 Ex. 3, p. 5
- The limit of the function $y = f(x)$ as x approaches c from the left is K, written as $\lim_{x \to c^-} f(x) = K$, if $f(x)$ is close to K whenever x is close to, but to the left of, c on the real-number line. Ex. 4, p. 6 Ex. 5, p. 9
- The limit of the function $y = f(x)$ as x approaches c from the right is L, written as $\lim_{x \to c^+} f(x) = L$, if $f(x)$ is close to L whenever x is close to, but to the right of, c on the real-number line. Ex. 6, p. 10 Ex. 7, p. 10 Ex. 8, p. 11
- The limit of the difference quotient $[f(a + h) - f(a)]/h$ always results in a 0/0 indeterminate form. Algebraic simplification is often required to evaluate this type of limit. Ex. 9, p. 12 Ex. 10, p. 13 Ex. 11, p. 13

1-2 Infinite Limits and Limits at Infinity

- If $f(x)$ increases or decreases without bound as x approaches a from either side of a, then the line $x = a$ is a **vertical asymptote** of the graph of $y = f(x)$. Ex. 1, p. 19 Ex. 2, p. 20
- If $f(x)$ gets close to L as x increases without bound or decreases without bound, then L is called the limit of f at ∞ or $-\infty$. Ex. 3, p. 22
- The end behavior of a polynomial is described in terms of limits at infinity. Ex. 4, p. 23
- If $f(x)$ approaches L as $x \to \infty$ or as $x \to -\infty$, then the line $y = L$ is a **horizontal asymptote** of the graph of $y = f(x)$. Polynomial functions never have horizontal asymptotes. A rational function can have at most one. Ex. 5, p. 25 Ex. 6, p. 25

1-3 Continuity

- Intuitively, the graph of a continuous function can be drawn without lifting a pen off the paper. Algebraically, a function f is **continuous at c** if Ex. 1, p. 31 Ex. 2, p. 32

  1. $\lim_{x \to c} f(x)$ exists, 2. $f(c)$ exists, and 3. $\lim_{x \to c} f(x) = f(c)$
- Continuity properties are useful for determining where a function is continuous and where it is discontinuous. Ex. 3, p. 34
- Continuity properties are also useful for solving inequalities. Ex. 4, p. 35

1-4 The Derivative

- Given a function $y = f(x)$, the **average rate of change** is the ratio of the change in y to the change in x. Ex. 1, p. 41
- The **instantaneous rate of change** is the limit of the average rate of change as the change in x approaches 0. Ex. 2, p. 42
- The slope of the secant line through two points on the graph of a function $y = f(x)$ is the ratio of the change in y to the change in x. The slope of the tangent line at the point $(a, f(a))$ is the limit of the slope of the secant line through the points $(a, f(a))$ and $(a + h, f(a + h))$ as h approaches 0. Ex. 3, p. 45 Ex. 4, p. 47 Ex. 5, p. 48
- The **derivative of $y = f(x)$ at x**, denoted as $f'(x)$, is the limit of the difference quotient $[f(x + h) - f(x)]/h$ as $h \to 0$ (if the limit exists). Ex. 6, p. 49
- The four-step method is used to find derivatives.
- If the limit of the difference quotient does not exist at $x = a$, then f is nondifferentiable at a and $f'(a)$ does not exist.

1-5 Basic Differentiation Properties

- The derivative of a constant function is 0. Ex. 1, p. 55
- For any real number n, the derivative of $f(x) = x^n$ is nx^{n-1}. Ex. 2, p. 56
- If f is a differential function, then the derivative of $kf(x)$ is $kf'(x)$. Ex. 3, p. 57
- The derivative of the sum or difference of two differential functions is the sum or difference of the derivatives of the functions. Ex. 4, p. 58 Ex. 5, p. 59

1-6 Differentials

- Given the function $y = f(x)$, the change in x is also called the **increment of x** and is denoted as Δx. The corresponding change in y is called the **increment of y** and is given by $\Delta y = f(x + \Delta x) - f(x)$.
- If $y = f(x)$ is differentiable at x, then the **differential of x** is $dx = \Delta x$ and the **differential of $y = f(x)$** is $dy = f'(x)dx$, or $df = f'(x)dx$. In this context, x and dx are both independent variables.

Ex. 1, p. 64
Ex. 2, p. 66
Ex. 3, p. 67

1-7 Marginal Analysis in Business and Economics

- If $y = C(x)$ is the total cost of producing x items, then $y = C'(x)$ is the **marginal cost** and $C(x + 1) - C(x)$ is the exact cost of producing item $x + 1$. Furthermore, $C'(x) \approx C(x + 1) - C(x)$. Similar statements can be made regarding total revenue and total profit functions.

Ex. 1, p. 71
Ex. 2, p. 72

- If $y = C(x)$ is the total cost of producing x items, then the **average cost**, or cost per unit, is $\overline{C}(x) = \frac{C(x)}{x}$ and the **marginal average cost** is $\overline{C}'(x) = \frac{d}{dx}\overline{C}(x)$. Similar statements can be made regarding total revenue and total profit functions.

Ex. 3, p. 75

Review Exercises

Work through all the problems in this chapter review, and check your answers in the back of the book. Answers to all review problems are there, along with section numbers in italics to indicate where each type of problem is discussed. Where weaknesses show up, review appropriate sections of the text.

Many of the problems in this exercise set ask you to find a derivative. Most of the answers to these problems contain both an unsimplified form and a simplified form of the derivative. When checking your work, first check that you applied the rules correctly, and then check that you performed the algebraic simplification correctly.

A

1. Find the indicated quantities for $y = f(x) = 2x^2 + 5$:
   (A) The change in y if x changes from 1 to 3
   (B) The average rate of change of y with respect to x if x changes from 1 to 3
   (C) The slope of the secant line through the points $(1, f(1))$ and $(3, f(3))$ on the graph of $y = f(x)$
   (D) The instantaneous rate of change of y with respect to x at $x = 1$
   (E) The slope of the line tangent to the graph of $y = f(x)$ at $x = 1$
   (F) $f'(1)$
2. Use the four-step process to find $f'(x)$ for $f(x) = -3x + 2$.
3. If $\lim_{x\to 1} f(x) = 2$ and $\lim_{x\to 1} g(x) = 4$, find
   (A) $\lim_{x\to 1}(5f(x) + 3g(x))$
   (B) $\lim_{x\to 1}[f(x)g(x)]$
   (C) $\lim_{x\to 1}\frac{g(x)}{f(x)}$
   (D) $\lim_{x\to 1}[5 + 2x - 3g(x)]$

In Problems 4–10, use the graph of f to estimate the indicated limits and function values.

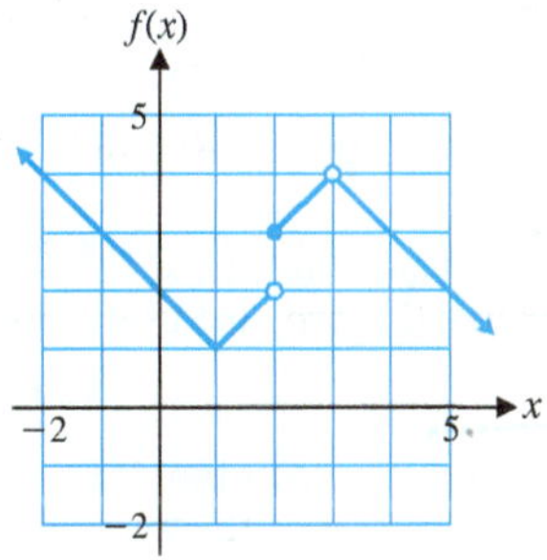

Figure for 4–10

4. $f(1.5)$
5. $f(2.5)$
6. $f(2.75)$
7. $f(3.25)$
8. (A) $\lim_{x\to 1^-} f(x)$ (B) $\lim_{x\to 1^+} f(x)$
   (C) $\lim_{x\to 1} f(x)$ (D) $f(1)$
9. (A) $\lim_{x\to 2^-} f(x)$ (B) $\lim_{x\to 2^+} f(x)$
   (C) $\lim_{x\to 2} f(x)$ (D) $f(2)$
10. (A) $\lim_{x\to 3^-} f(x)$ (B) $\lim_{x\to 3^+} f(x)$
    (C) $\lim_{x\to 3} f(x)$ (D) $f(3)$

In Problems 11–13, use the graph of the function f shown in the figure to answer each question.

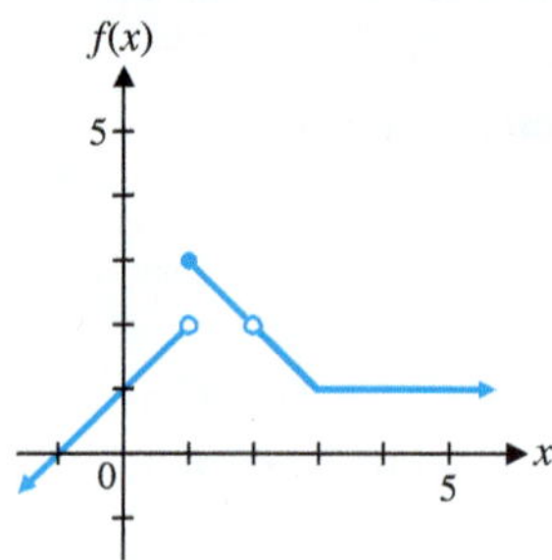

Figure for 11–13

11. (A) $\lim_{x \to 1} f(x) = ?$ (B) $f(1) = ?$
(C) Is f continuous at $x = 1$?

12. (A) $\lim_{x \to 2} f(x) = ?$ (B) $f(2) = ?$
(C) Is f continuous at $x = 2$?

13. (A) $\lim_{x \to 3} f(x) = ?$ (B) $f(3) = ?$
(C) Is f continuous at $x = 3$?

In Problems 14–23, refer to the following graph of $y = f(x)$:

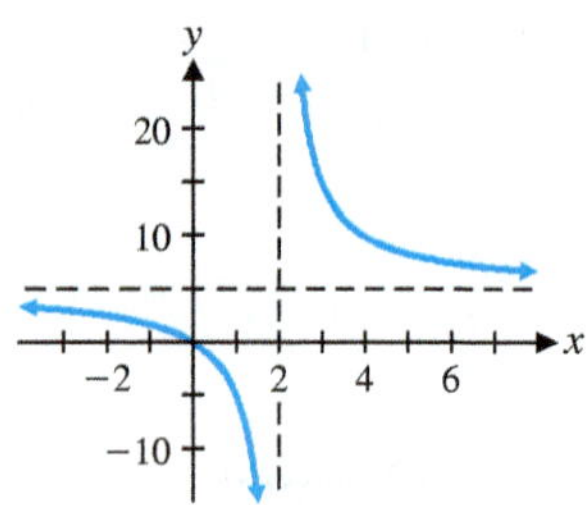

Figure for 14–23

14. $\lim_{x \to \infty} f(x) = ?$

15. $\lim_{x \to -\infty} f(x) = ?$

16. $\lim_{x \to 2^+} f(x) = ?$

17. $\lim_{x \to 2^-} f(x) = ?$

18. $\lim_{x \to 0^-} f(x) = ?$

19. $\lim_{x \to 0^+} f(x) = ?$

20. $\lim_{x \to 0} f(x) = ?$

21. Identify any vertical asymptotes.

22. Identify any horizontal asymptotes.

23. Where is $y = f(x)$ discontinuous?

24. Use the four-step process to find $f'(x)$ for $f(x) = 5x^2$.

25. If $f(5) = 4$, $f'(5) = -1$, $g(5) = 2$, and $g'(5) = -3$, then find $h'(5)$ for each of the following functions:

(A) $h(x) = 3f(x)$

(B) $h(x) = -2g(x)$

(C) $h(x) = 2f(x) + 5$

(D) $h(x) = -g(x) - 1$

(E) $h(x) = 2f(x) + 3g(x)$

In Problems 26–31, find $f'(x)$ and simplify.

26. $f(x) = \frac{1}{3}x^3 - 5x^2 + 1$

27. $f(x) = 2x^{1/2} - 3x$

28. $f(x) = 5$

29. $f(x) = \frac{3}{2x} + \frac{5x^3}{4}$

30. $f(x) = \frac{0.5}{x^4} + 0.25x^4$

31. $f(x) = (3x^3 - 2)(x + 1)$ (*Hint:* Multiply and then differentiate.)

In Problems 32–35, find the indicated quantities for $y = f(x) = x^2 + x$.

32. Δx, Δy, and $\Delta y/\Delta x$ for $x_1 = 1$ and $x_2 = 3$.

33. $[f(x_1 + \Delta x) - f(x_1)]/\Delta x$ for $x_1 = 1$ and $\Delta x = 2$.

34. dy for $x_1 = 1$ and $x_2 = 3$.

35. Δy and dy for $x = 1$, $\Delta x = dx = 0.2$.

B

Problems 36–38 refer to the function.

$$f(x) = \begin{cases} x^2 & \text{if } 0 \le x < 2 \\ 8 - x & \text{if } x \ge 2 \end{cases}$$

which is graphed in the figure.

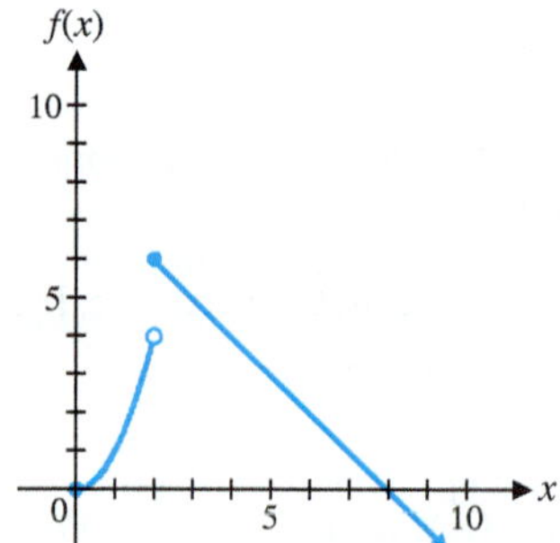

Figure for 36–38

36. (A) $\lim_{x \to 2^-} f(x) = ?$ (B) $\lim_{x \to 2^+} f(x) = ?$
(C) $\lim_{x \to 2} f(x) = ?$ (D) $f(2) = ?$
(E) Is f continuous at $x = 2$?

37. (A) $\lim_{x \to 5^-} f(x) = ?$ (B) $\lim_{x \to 5^+} f(x) = ?$
(C) $\lim_{x \to 5} f(x) = ?$ (D) $f(5) = ?$
(E) Is f continuous at $x = 5$?

38. Solve each inequality. Express answers in interval notation.
(A) $f(x) < 0$ (B) $f(x) \ge 0$

In Problems 39–41, solve each inequality. Express the answer in interval notation. Use a graphing calculator in Problem 41 to approximate partition numbers to four decimal places.

39. $x^2 - x < 12$

40. $\frac{x - 5}{x^2 + 3x} > 0$

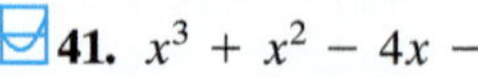
41. $x^3 + x^2 - 4x - 2 > 0$

42. Let $f(x) = 0.5x^2 - 5$.

(A) Find the slope of the secant line through $(2, f(2))$ and $(4, f(4))$.

(B) Find the slope of the secant line through $(2, f(2))$ and $(2 + h, f(2 + h))$, $h \ne 0$.

(C) Find the slope of the tangent line at $x = 2$.

In Problems 43–46, find the indicated derivative and simplify.

43. $\frac{dy}{dx}$ for $y = \frac{1}{3}x^{-3} - 5x^{-2} + 1$

44. y' for $y = \frac{3\sqrt{x}}{2} + \frac{5}{3\sqrt{x}}$

45. $g'(x)$ for $g(x) = 1.8\sqrt[3]{x} + \frac{0.9}{\sqrt[3]{x}}$

46. $\frac{dy}{dx}$ for $y = \frac{2x^3 - 3}{5x^3}$

47. For $y = f(x) = x^2 + 4$, find

(A) The slope of the graph at $x = 1$

(B) The equation of the tangent line at $x = 1$ in the form $y = mx + b$

In Problems 48 and 49, find the value(s) of x where the tangent line is horizontal.

48. $f(x) = 10x - x^2$

49. $f(x) = x^3 + 3x^2 - 45x - 135$

In Problems 50 and 51, approximate (to four decimal places) the value(s) of x where the graph of f has a horizontal tangent line.

50. $f(x) = x^4 - 2x^3 - 5x^2 + 7x$

51. $f(x) = x^5 - 10x^3 - 5x + 10$

52. If an object moves along the y axis (scale in feet) so that it is at $y = f(x) = 8x^2 - 4x + 1$ at time x (in seconds), find

(A) The instantaneous velocity function

(B) The velocity at time $x = 3$ seconds

53. An object moves along the y axis (scale in feet) so that at time x (in seconds) it is at $y = f(x) = -5x^2 + 16x + 3$. Find

(A) The instantaneous velocity function

(B) The time(s) when the velocity is 0

54. Let $f(x) = x^3$, $g(x) = (x - 4)^3$, and $h(x) = (x + 3)^3$.

(A) How are the graphs of f, g, and h related? Illustrate your conclusion by graphing f, g, and h on the same coordinate axes.

(B) How would you expect the graphs of the derivatives of these functions to be related? Illustrate your conclusion by graphing f', g', and h' on the same coordinate axes.

In Problems 55–59, determine where f is continuous. Express the answer in interval notation.

55. $f(x) = x^2 - 4$

56. $f(x) = \frac{x + 1}{x - 2}$

57. $f(x) = \frac{x + 4}{x^2 + 3x - 4}$

58. $f(x) = \sqrt[3]{4 - x^2}$

59. $f(x) = \sqrt{4 - x^2}$

In Problems 60–69, evaluate the indicated limits if they exist.

60. Let $f(x) = \frac{2x}{x^2 - 3x}$. Find

(A) $\lim_{x \to 1} f(x)$ (B) $\lim_{x \to 3} f(x)$ (C) $\lim_{x \to 0} f(x)$

61. Let $f(x) = \frac{x + 1}{(3 - x)^2}$. Find

(A) $\lim_{x \to 1} f(x)$ (B) $\lim_{x \to -1} f(x)$ (C) $\lim_{x \to 3} f(x)$

62. Let $f(x) = \frac{|x - 4|}{x - 4}$. Find

(A) $\lim_{x \to 4^-} f(x)$ (B) $\lim_{x \to 4^+} f(x)$ (C) $\lim_{x \to 4} f(x)$

63. Let $f(x) = \frac{x - 3}{9 - x^2}$. Find

(A) $\lim_{x \to 3} f(x)$ (B) $\lim_{x \to -3} f(x)$ (C) $\lim_{x \to 0} f(x)$

64. Let $f(x) = \frac{x^2 - x - 2}{x^2 - 7x + 10}$. Find

(A) $\lim_{x \to -1} f(x)$ (B) $\lim_{x \to 2} f(x)$ (C) $\lim_{x \to 5} f(x)$

65. Let $f(x) = \frac{2x}{3x - 6}$. Find

(A) $\lim_{x \to \infty} f(x)$ (B) $\lim_{x \to -\infty} f(x)$ (C) $\lim_{x \to 2} f(x)$

66. Let $f(x) = \frac{2x^3}{3(x - 2)^2}$. Find

(A) $\lim_{x \to \infty} f(x)$ (B) $\lim_{x \to -\infty} f(x)$ (C) $\lim_{x \to 2} f(x)$

67. Let $f(x) = \frac{2x}{3(x - 2)^3}$. Find

(A) $\lim_{x \to \infty} f(x)$ (B) $\lim_{x \to -\infty} f(x)$ (C) $\lim_{x \to 2} f(x)$

68. $\lim_{h \to 0} \frac{f(2 + h) - f(2)}{h}$ for $f(x) = x^2 + 4$

69. $\lim_{h \to 0} \frac{f(x + h) - f(x)}{h}$ for $f(x) = \frac{1}{x + 2}$

In Problems 70 and 71, use the definition of the derivative and the four-step process to find $f'(x)$.

70. $f(x) = x^2 - x$

71. $f(x) = \sqrt{x} - 3$

Problems 72–77 refer to the function f in the figure. Determine whether f is differentiable at the indicated value of x.

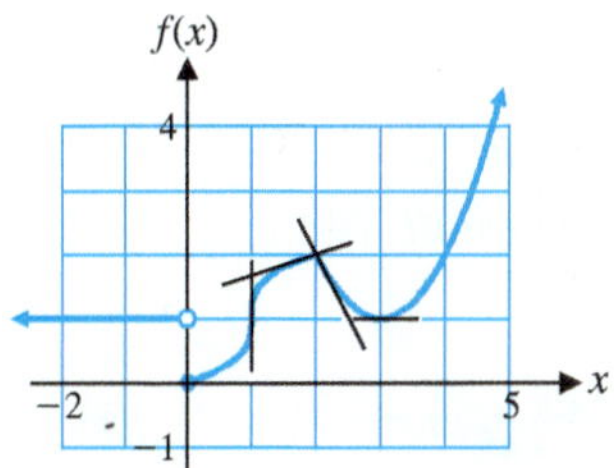

72. $x = -1$

73. $x = 0$

74. $x = 1$

75. $x = 2$

76. $x = 3$

77. $x = 4$

In Problems 78–82, find all horizontal and vertical asymptotes.

78. $f(x) = \frac{5x}{x - 7}$

79. $f(x) = \frac{-2x + 5}{(x - 4)^2}$

80. $f(x) = \frac{x^2 + 9}{x - 3}$

81. $f(x) = \frac{x^2 - 9}{x^2 + x - 2}$

82. $f(x) = \dfrac{x^3 - 1}{x^3 - x^2 - x + 1}$

83. The domain of the power function $f(x) = x^{1/5}$ is the set of all real numbers. Find the domain of the derivative $f'(x)$. Discuss the nature of the graph of $y = f(x)$ for any x values excluded from the domain of $f'(x)$.

84. Let f be defined by

$$f(x) = \begin{cases} x^2 - m & \text{if } x \le 1 \\ -x^2 + m & \text{if } x > 1 \end{cases}$$

where m is a constant.

(A) Graph f for $m = 0$, and find

$$\lim_{x\to 1^-} f(x) \quad \text{and} \quad \lim_{x\to 1^+} f(x)$$

(B) Graph f for $m = 2$, and find

$$\lim_{x\to 1^-} f(x) \quad \text{and} \quad \lim_{x\to 1^+} f(x)$$

(C) Find m so that

$$\lim_{x\to 1^-} f(x) = \lim_{x\to 1^+} f(x)$$

and graph f for this value of m.

(D) Write a brief verbal description of each graph. How does the graph in part (C) differ from the graphs in parts (A) and (B)?

85. Let $f(x) = 1 - |x - 1|, 0 \le x \le 2$ (see the figure).

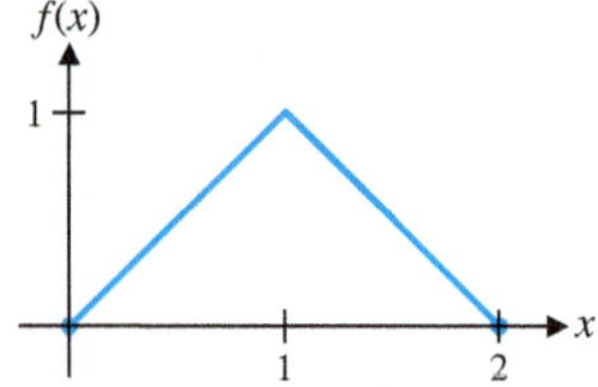

Figure for 85

(A) $\displaystyle\lim_{h\to 0^-} \frac{f(1 + h) - f(1)}{h} = ?$

(B) $\displaystyle\lim_{h\to 0^+} \frac{f(1 + h) - f(1)}{h} = ?$

(C) $\displaystyle\lim_{h\to 0} \frac{f(1 + h) - f(1)}{h} = ?$

(D) Does $f'(1)$ exist?

Applications

86. **Natural-gas rates.** Table 1 shows the winter rates for natural gas charged by the Bay State Gas Company. The customer charge is a fixed monthly charge, independent of the amount of gas used per month.

Table 1 Natural Gas Rates

Monthly customer charge	\$7.47
First 90 therms	\$0.4000 per therm
All usage over 90 therms	\$0.2076 per therm

(A) Write a piecewise definition of the monthly charge $S(x)$ for a customer who uses x therms in a winter month.

(B) Graph $S(x)$.

(C) Is $S(x)$ continuous at $x = 90$? Explain.

87. **Cost analysis.** The total cost (in dollars) of producing x HDTVs is

$$C(x) = 10{,}000 + 200x - 0.1x^2$$

(A) Find the exact cost of producing the 101st TV.

(B) Use the marginal cost to approximate the cost of producing the 101st TV.

88. **Cost analysis.** The total cost (in dollars) of producing x bicycles is

$$C(x) = 5{,}000 + 40x + 0.05x^2$$

(A) Find the total cost and the marginal cost at a production level of 100 bicycles and interpret the results.

(B) Find the average cost and the marginal average cost at a production level of 100 bicycles and interpret the results.

89. **Cost analysis.** The total cost (in dollars) of producing x laser printers per week is shown in the figure. Which is greater, the approximate cost of producing the 201st printer or the approximate cost of producing the 601st printer? Does this graph represent a manufacturing process that is becoming more efficient or less efficient as production levels increase? Explain.

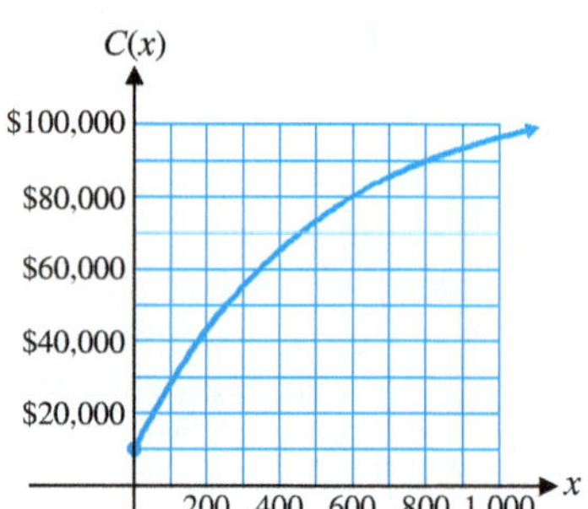

Figure for 89

90. **Cost analysis.** Let

$$p = 25 - 0.01x \quad \text{and} \quad C(x) = 2x + 9{,}000$$
$$0 \le x \le 2{,}500$$

be the price–demand equation and cost function, respectively, for the manufacture of umbrellas.

(A) Find the marginal cost, average cost, and marginal average cost functions.

(B) Express the revenue in terms of x, and find the marginal revenue, average revenue, and marginal average revenue functions.

(C) Find the profit, marginal profit, average profit, and marginal average profit functions.

(D) Find the break-even point(s).

(E) Evaluate the marginal profit at $x = 1{,}000$, 1,150, and 1,400, and interpret the results.

(F) Graph $R = R(x)$ and $C = C(x)$ on the same coordinate system, and locate regions of profit and loss.

91. Employee training. A company producing computer components has established that, on average, a new employee can assemble $N(t)$ components per day after t days of on-the-job training, as given by

$$N(t) = \frac{40t - 80}{t}, t \geq 2$$

(A) Find the average rate of change of $N(t)$ from 2 days to 5 days.

(B) Find the instantaneous rate of change of $N(t)$ at 2 days.

92. Sales analysis. The total number of swimming pools, N (in thousands), sold during a year is given by

$$N(t) = 2t + \frac{1}{3}t^{3/2}$$

where t is the number of months since the beginning of the year. Find $N(9)$ and $N'(9)$, and interpret these quantities.

93. Natural-gas consumption. The data in Table 2 give the U.S. consumption of natural gas in trillions of cubic feet.

Table 2

Year	Natural-Gas Consumption
1960	12.0
1970	21.1
1980	19.9
1990	18.7
2000	21.9

(A) Let x represent time (in years), with $x = 0$ corresponding to 1960, and let y represent the corresponding U.S. consumption of natural gas. Enter the data set in a graphing calculator and find a cubic regression equation for the data.

(B) If $y = N(x)$ denotes the regression equation found in part (A), find $N(50)$ and $N'(50)$, and write a brief verbal interpretation of these results.

94. Break-even analysis. Table 3 contains price–demand and total cost data from a bakery for the production of kringles (a Danish pastry), where p is the price (in dollars) of a kringle for a daily demand of x kringles and C is the total cost (in dollars) of producing x kringles.

Table 3

x	$p(\$)$	$C(\$)$
125	9	740
140	8	785
170	7	850
200	6	900

(A) Find a linear regression equation for the price–demand data, using x as the independent variable.

(B) Find a linear regression equation for the cost data, using x as the independent variable. Use this equation to estimate the fixed costs and variable costs per kringle.

(C) Find the break-even points.

(D) Find the price range for which the bakery will make a profit.

95. Pollution. A sewage treatment plant uses a pipeline that extends 1 mile toward the center of a large lake. The concentration of effluent $C(x)$ in parts per million, x meters from the end of the pipe is given approximately by

$$C(x) = \frac{500}{x^2}, x \geq 1$$

What is the instantaneous rate of change of concentration at 10 meters? At 100 meters?

96. Medicine. The body temperature (in degrees Fahrenheit) of a patient t hours after taking a fever-reducing drug is given by

$$F(t) = 0.16t^2 - 1.6t + 102$$

Find $F(4)$ and $F'(4)$. Write a brief verbal interpretation of these quantities.

97. Learning. If a person learns N items in t hours, as given by

$$N(t) = 20\sqrt{t}$$

find the rate of learning after

(A) 1 hour (B) 4 hours

98. Physics: The coefficient of thermal expansion (CTE) is a measure of the expansion of an object subjected to extreme temperatures. We want to use a Michaelis–Menten function of the form

$$C(T) = \frac{C_{max}T}{M + T}$$

where C = CTE, T is temperature in K (degrees Kelvin), and C_{max} and M are constants. Table 4 lists the coefficients of thermal expansion for titanium at various temperatures.

Table 4 Coefficients of Thermal Expansion

T (K)	Titanium
100	4.5
200	7.4
293	8.6
500	9.9
800	11.1
1100	11.7

(A) Plot the points in columns 1 and 2 of Table 4 on graph paper and estimate C_{max} to the nearest integer. To estimate M, add the horizontal line CTE $= \frac{C_{max}}{2}$ to your graph, connect successive points on the graph with straight-line segments, and estimate the value of T (to the nearest multiple of fifty) that satisfies

$$C(T) = \frac{C_{max}}{2}.$$

(B) Use the constants $\frac{C_{max}}{2}$ and M from part (A) to form a Michaelis–Menten function for the CTE of titanium.

(C) Use the function from part (B) to estimate the CTE of titanium at 600 K and to estimate the temperature when the CTE of titanium is 10.

Additional Derivative Topics

Introduction

In this chapter, we develop techniques for finding derivatives of a wide variety of functions, including exponential and logarithmic functions. There are straightforward procedures—the product rule, quotient rule, and chain rule—for writing down the derivative of any function that is the product, quotient, or composite of functions whose derivatives are known. With the ability to calculate derivatives easily, we consider a wealth of applications involving rates of change. For example, we apply the derivative to study population growth, radioactive decay, elasticity of demand, and environmental crises (see Problem 31 in Section 2-6 or Problem 71 in Section 2-7).

2-1 The Constant e and Continuous Compound Interest

- The Constant e
- Continuous Compound Interest

Both the exponential function with base e and continuous compound interest were introduced informally. Now, with an understanding of limit concepts, we can give precise definitions of e and continuous compound interest.

The Constant e

The irrational number e is a particularly suitable base for both exponential and logarithmic functions. The reasons for choosing this number as a base will become clear as we develop differentiation formulas for the exponential function e^x and the natural logarithmic function $\ln x$.

In precalculus treatments, the number e is defined informally as the irrational number that can be approximated by the expression $[1 + (1/n)]^n$ for n sufficiently large. Now we will use the limit concept to formally define e as either of the following two limits. [*Note:* If $s = 1/n$, then as $n \to \infty$, $s \to 0$.]

DEFINITION The Number e

$$e = \lim_{n\to\infty}\left(1 + \frac{1}{n}\right)^n \quad \text{or, alternatively,} \quad e = \lim_{s\to 0}(1 + s)^{1/s}$$

Both limits are equal to $e = 2.718\,281\,828\,459\ldots$

Proof that the indicated limits exist and represent an irrational number between 2 and 3 is not easy and is omitted.

CONCEPTUAL INSIGHT

The two limits used to define e are unlike any we have encountered so far. Some people reason (incorrectly) that both limits are 1, since $1 + s \to 1$ as $s \to 0$ and 1 to any power is 1. An ordinary scientific calculator with a y^x key can convince you otherwise. Consider the following table of values for s and $f(s) = (1 + s)^{1/s}$ and Figure 1 for s close to 0. Compute the table values with a calculator yourself, and try several values of s even closer to 0. Note that the function is discontinuous at $s = 0$.

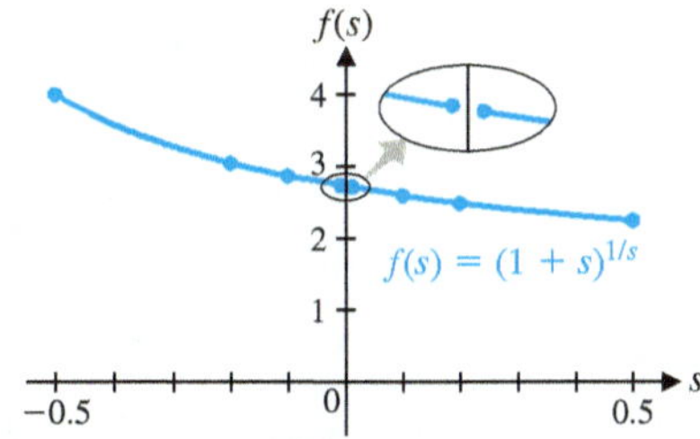

Figure 1

s approaches 0 from the left → 0 ← s approaches 0 from the right

s	−0.5	−0.2	−0.1	−0.01 → 0 ← 0.01	0.1	0.2	0.5
$(1 + s)^{1/s}$	4.0000	3.0518	2.8680	2.7320 → e ← 2.7048	2.5937	2.4883	2.2500

Continuous Compound Interest

Now we can see how e appears quite naturally in the important application of compound interest. Let us start with simple interest, move on to compound interest, and then proceed on to continuous compound interest.

On one hand, if a principal P is borrowed at an annual rate r,* then after t years at simple interest, the borrower will owe the lender an amount A given by

$$A = P + Prt = P(1 + rt) \quad \text{Simple interest} \qquad (1)$$

On the other hand, if interest is compounded n times a year, then the borrower will owe the lender an amount A given by

$$A = P\left(1 + \frac{r}{n}\right)^{nt} \quad \text{Compound interest} \qquad (2)$$

where r/n is the interest rate per compounding period and nt is the number of compounding periods. Suppose that P, r, and t in equation (2) are held fixed and n is increased. Will the amount A increase without bound, or will it tend to approach some limiting value?

Let us perform a calculator experiment before we attack the general limit problem. If $P = \$100$, $r = 0.06$, and $t = 2$ years, then

$$A = 100\left(1 + \frac{0.06}{n}\right)^{2n}$$

We compute A for several values of n in Table 1. The biggest gain appears in the first step, then the gains slow down as n increases. The amount A appears to approach \$112.75 as n gets larger and larger.

Table 1

Compounding Frequency	n	$A = 100\left(1 + \frac{0.06}{n}\right)^{2n}$
Annually	1	\$112.3600
Semiannually	2	112.5509
Quarterly	4	112.6493
Monthly	12	112.7160
Weekly	52	112.7419
Daily	365	112.7486
Hourly	8,760	112.7496

Keeping P, r, and t fixed in equation (2), we compute the following limit and observe an interesting and useful result:

$$\begin{aligned}
\lim_{n\to\infty} P\left(1 + \frac{r}{n}\right)^{nt} &= P \lim_{n\to\infty}\left(1 + \frac{r}{n}\right)^{(n/r)rt} && \text{Insert } r/r \text{ in the exponent and let } s = r/n. \text{ Note that } n \to \infty \text{ implies } s \to 0. \\
&= P \lim_{s\to 0}[(1 + s)^{1/s}]^{rt} && \text{Use a limit property.}^{\dagger} \\
&= P[\lim_{s\to 0}(1 + s)^{1/s}]^{rt} && \lim_{s\to 0}(1 + s)^{1/s} = e \\
&= Pe^{rt}
\end{aligned}$$

The resulting formula is called the **continuous compound interest formula**, a widely used formula in business and economics.

THEOREM 1 Continuous Compound Interest Formula

If a principal P is invested at an annual rate r (expressed as a decimal) compounded continuously, then the amount A in the account at the end of t years is given by

$$A = Pe^{rt}$$

*If r is the interest rate written as a decimal, then $100r\%$ is the rate in percent. For example, if $r = 0.12$, then $100r\% = 100(0.12)\% = 12\%$. The expressions 0.12 and 12% are equivalent. Unless stated otherwise, all formulas in this book use r in decimal form.

†The following new limit property is used: If $\lim_{x\to c} f(x)$ exists, then $\lim_{x\to c}[f(x)]^p = [\lim_{x\to c} f(x)]^p$, provided that the last expression names a real number.

EXAMPLE 1 **Computing Continuously Compounded Interest** If \$100 is invested at 6% compounded continuously,* what amount will be in the account after 2 years? How much interest will be earned?

SOLUTION

$$A = Pe^{rt}$$
$$= 100e^{(0.06)(2)} \quad \text{6\% is equivalent to } r = 0.06.$$
$$\approx \$112.7497$$

Compare this result with the values calculated in Table 1. The interest earned is \$112.7497 − \$100 = \$12.7497.

Matched Problem 1 What amount (to the nearest cent) will be in an account after 5 years if \$100 is invested at an annual nominal rate of 8% compounded annually? Semiannually? Continuously?

EXAMPLE 2 **Graphing the Growth of an Investment** Union Savings Bank offers a 5-year certificate of deposit (CD) that earns 5.75% compounded continuously. If \$1,000 is invested in one of these CDs, graph the amount in the account as a function of time for a period of 5 years.

SOLUTION We want to graph

$$A = 1{,}000e^{0.0575t} \qquad 0 \le t \le 5$$

Using a calculator, we construct a table of values (Table 2). Then we graph the points from the table and join the points with a smooth curve (Fig. 2).

Table 2

t	A(\$)
0	1,000
1	1,059
2	1,122
3	1,188
4	1,259
5	1,333

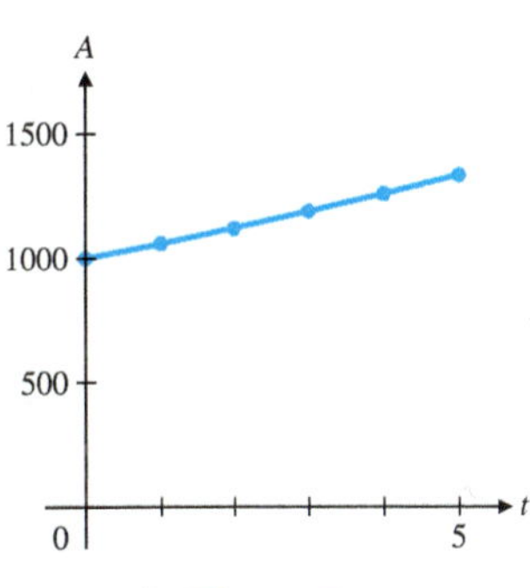

Figure 2

CONCEPTUAL INSIGHT

Depending on the domain, the graph of an exponential function can appear to be linear. Table 2 shows that the graph in Figure 2 is *not* linear. The slope determined by the first two points (for $t = 0$ and $t = 1$) is 59 but the slope determined by the first and third points (for $t = 0$ and $t = 2$) is 61. For a linear graph, the slope determined by any two points is constant.

Matched Problem 2 If \$5,000 is invested in a Union Savings Bank 4-year CD that earns 5.61% compounded continuously, graph the amount in the account as a function of time for a period of 4 years.

*Following common usage, we will often write "at 6% compounded continuously," understanding that this means "at an annual nominal rate of 6% compounded continuously."

EXAMPLE 3 **Computing Growth Time** How long will it take an investment of \$5,000 to grow to \$8,000 if it is invested at 5% compounded continuously?

SOLUTION Starting with the continous compound interest formula $A = Pe^{rt}$, we must solve for t:

$$A = Pe^{rt}$$

$$8{,}000 = 5{,}000e^{0.05t}$$

Divide both sides by 5,000 and reverse the equation.

$$e^{0.05t} = 1.6$$

Take the natural logarithm of both sides—recall that $\log_b b^x = x$.

$$\ln e^{0.05t} = \ln 1.6$$

$$0.05t = \ln 1.6$$

$$t = \frac{\ln 1.6}{0.05}$$

$$t \approx 9.4 \text{ years}$$

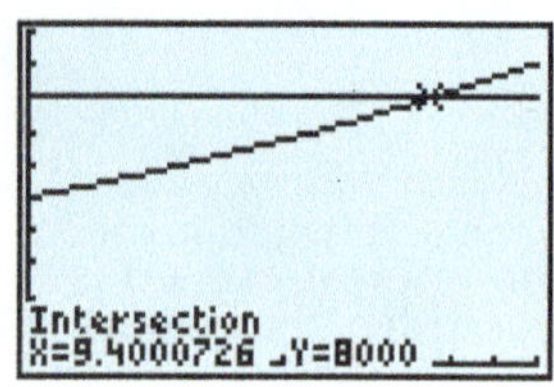

Figure 3
$y_1 = 5{,}000e^{0.05x}$
$y_2 = 8{,}000$

Figure 3 shows an alternative method for solving Example 3 on a graphing calculator.

Matched Problem 3 How long will it take an investment of \$10,000 to grow to \$15,000 if it is invested at 9% compounded continuously?

EXAMPLE 4 **Computing Doubling Time** How long will it take money to double if it is invested at 6.5% compounded continuously?

SOLUTION Starting with the continuous compound interest formula $A = Pe^{rt}$, we solve for t, given $A = 2P$ and $R = 0.065$:

$$2P = Pe^{0.065t}$$

Divide both sides by P and reverse the equation.

$$e^{0.065t} = 2$$

Take the natural logarithm of both sides.

$$\ln e^{0.065t} = \ln 2$$

$$0.065t = \ln 2$$

$$t = \frac{\ln 2}{0.065}$$

$$t \approx 10.66 \text{ years}$$

Matched Problem 4 How long will it take money to triple if it is invested at 5.5% compounded continuously?

EXPLORE & DISCUSS 1

You are considering three options for investing \$10,000: at 7% compounded annually, at 6% compounded monthly, and at 5% compounded continuously.

(A) Which option would be the best for investing \$10,000 for 8 years?

(B) How long would you need to invest your money for the third option to be the best?

Exercises 2-1

A

Use a calculator to evaluate A to the nearest cent in Problems 1 and 2.

1. $A = \$1{,}000e^{0.1t}$ for $t = 2, 5,$ and 8
2. $A = \$5{,}000e^{0.08t}$ for $t = 1, 4,$ and 10
3. If \$6,000 is invested at 10% compounded continuously, graph the amount in the account as a function of time for a period of 8 years.
4. If \$4,000 is invested at 8% compounded continuously, graph the amount in the account as a function of time for a period of 6 years.

B

In Problems 5–10, solve for t or r to two decimal places.

5. $2 = e^{0.06t}$
6. $2 = e^{0.03t}$
7. $3 = e^{0.1r}$
8. $3 = e^{0.25t}$
9. $2 = e^{5r}$
10. $3 = e^{10r}$

C

In Problems 11 and 12, use a calculator to complete each table to five decimal places.

11.

n	$[1 + (1/n)]^n$
10	2.593 74
100	
1,000	
10,000	
100,000	
1,000,000	
10,000,000	
↓	↓
∞	$e = 2.718\ 281\ 828\ 459\ldots$

12.

s	$(1 + s)^{1/s}$
0.01	2.704 81
−0.01	
0.001	
−0.001	
0.000 1	
−0.000 1	
0.000 01	
−0.000 01	
↓	↓
0	$e = 2.718\ 281\ 828\ 459\ldots$

13. Use a calculator and a table of values to investigate

$$\lim_{n\to\infty}(1 + n)^{1/n}$$

Do you think this limit exists? If so, what do you think it is?

14. Use a calculator and a table of values to investigate

$$\lim_{s\to 0^+}\left(1 + \frac{1}{s}\right)^s.$$

Do you think this limit exists? If so, what do you think it is?

15. It can be shown that the number e satisfies the inequality

$$\left(1 + \frac{1}{n}\right)^n < e < \left(1 + \frac{1}{n}\right)^{n+1} \qquad n \ge 1$$

Illustrate this condition by graphing

$$y_1 = (1 + 1/n)^n$$
$$y_2 = 2.718\ 281\ 828 \approx e$$
$$y_3 = (1 + 1/n)^{n+1}$$

in the same viewing window, for $1 \le n \le 20$.

16. It can be shown that

$$e^s = \lim_{n\to\infty}\left(1 + \frac{s}{n}\right)^n$$

for any real number s. Illustrate this equation graphically for $s = 2$ by graphing

$$y_1 = (1 + 2/n)^n$$
$$y_2 = 7.389\ 056\ 099 \approx e^2$$

in the same viewing window, for $1 \le n \le 50$.

Applications

17. **Continuous compound interest.** Provident Bank offers a 10-year CD that earns 4.15% compounded continuously.
    (A) If \$10,000 is invested in this CD, how much will it be worth in 10 years?
    (B) How long will it take for the account to be worth \$18,000?
18. **Continuous compound interest.** Provident Bank also offers a 3-year CD that earns 3.64% compounded continuously.
    (A) If \$10,000 is invested in this CD, how much will it be worth in 3 years?
    (B) How long will it take for the account to be worth \$11,000?
19. **Present value.** A note will pay \$20,000 at maturity 10 years from now. How much should you be willing to pay for the note now if money is worth 5.2% compounded continuously?
20. **Present value.** A note will pay \$50,000 at maturity 5 years from now. How much should you be willing to pay for the note now if money is worth 6.4% compounded continuously?
21. **Continuous compound interest.** An investor bought stock for \$20,000. Five years later, the stock was sold for \$30,000. If interest is compounded continuously, what annual nominal rate of interest did the original \$20,000 investment earn?

22. Continuous compound interest. A family paid \$99,000 cash for a house. Fifteen years later, the house was sold for \$195,000. If interest is compounded continuously, what annual nominal rate of interest did the original \$99,000 investment earn?

23. Present value. Solving $A = Pe^{rt}$ for P, we obtain

$$P = Ae^{-rt}$$

which is the present value of the amount A due in t years if money earns interest at an annual nominal rate r compounded continuously.

(A) Graph $P = 10{,}000e^{-0.08t}$, $0 \le t \le 50$.

(B) $\lim_{t \to \infty} 10{,}000e^{-0.08t} = ?$ [Guess, using part (A).]

[*Conclusion:* The longer the time until the amount A is due, the smaller is its present value, as we would expect.]

24. Present value. Referring to Problem 23, in how many years will the \$10,000 be due in order for its present value to be \$5,000?

25. Doubling time. How long will it take money to double if it is invested at 4% compounded continuously?

26. Doubling time. How long will it take money to double if it is invested at 5% compounded continuously?

27. Doubling rate. At what nominal rate compounded continuously must money be invested to double in 8 years?

28. Doubling rate. At what nominal rate compounded continuously must money be invested to double in 10 years?

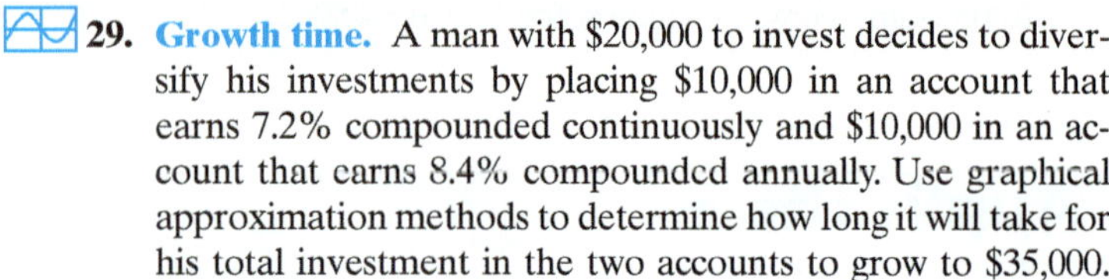

29. Growth time. A man with \$20,000 to invest decides to diversify his investments by placing \$10,000 in an account that earns 7.2% compounded continuously and \$10,000 in an account that earns 8.4% compounded annually. Use graphical approximation methods to determine how long it will take for his total investment in the two accounts to grow to \$35,000.

30. Growth time. A woman invests \$5,000 in an account that earns 8.8% compounded continuously and \$7,000 in an account that earns 9.6% compounded annually. Use graphical approximation methods to determine how long it will take for her total investment in the two accounts to grow to \$20,000.

31. Doubling times

(A) Show that the doubling time t (in years) at an annual rate r compounded continuously is given by

$$t = \frac{\ln 2}{r}$$

(B) Graph the doubling-time equation from part (A) for $0.02 \le r \le 0.30$. Is this restriction on r reasonable? Explain.

(C) Determine the doubling times (in years, to two decimal places) for $r = 5\%, 10\%, 15\%, 20\%, 25\%$, and 30%.

32. Doubling rates

(A) Show that the rate r that doubles an investment at continuously compounded interest in t years is given by

$$r = \frac{\ln 2}{t}$$

(B) Graph the doubling-rate equation from part (A) for $1 \le t \le 20$. Is this restriction on t reasonable? Explain.

(C) Determine the doubling rates for $t = 2, 4, 6, 8, 10$, and 12 years.

33. Radioactive decay. A mathematical model for the decay of radioactive substances is given by

$$Q = Q_0e^{rt}$$

where

Q_0 = amount of the substance at time $t = 0$
r = continuous compound rate of decay
t = time in years
Q = amount of the substance at time t

If the continuous compound rate of decay of radium per year is $r = -0.000\,433\,2$, how long will it take a certain amount of radium to decay to half the original amount? (This period is the *half-life* of the substance.)

34. Radioactive decay. The continuous compound rate of decay of carbon-14 per year is $r = -0.000\,123\,8$. How long will it take a certain amount of carbon-14 to decay to half the original amount? (Use the radioactive decay model in Problem 33.)

35. Radioactive decay. A cesium isotope has a half-life of 30 years. What is the continuous compound rate of decay? (Use the radioactive decay model in Problem 33.)

36. Radioactive decay. A strontium isotope has a half-life of 90 years. What is the continuous compound rate of decay? (Use the radioactive decay model in Problem 33.)

37. World population. A mathematical model for world population growth over short intervals is given by

$$P = P_0e^{rt}$$

where

P_0 = population at time $t = 0$
r = continuous compound rate of growth
t = time in years
P = population at time t

How long will it take world population to double if it continues to grow at its current continuous compound rate of 1.3% per year?

38. U.S. population. How long will it take for the U.S. population to double if it continues to grow at a rate of 0.975% per year?

39. Population growth. Some underdeveloped nations have population doubling times of 50 years. At what continuous compound rate is the population growing? (Use the population growth model in Problem 37.)

40. Population growth. Some developed nations have population doubling times of 200 years. At what continuous compound rate is the population growing? (Use the population growth model in Problem 37.)

Answers to Matched Problems

1. \$146.93; \$148.02; \$149.18
2. $A = 5{,}000e^{0.0561t}$

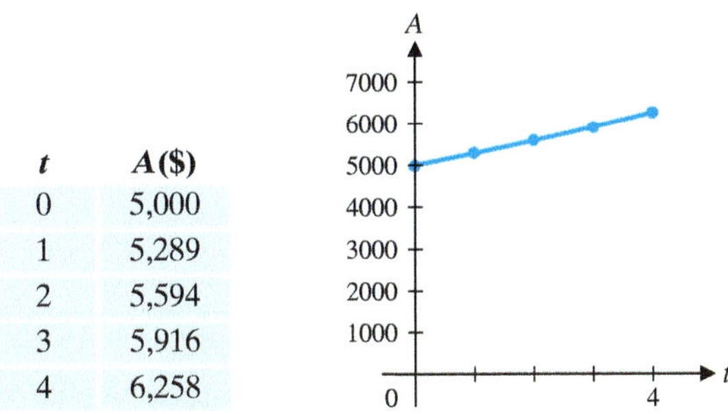

t	A(\$)
0	5,000
1	5,289
2	5,594
3	5,916
4	6,258

3. 4.51 yr
4. 19.97 yr

2-2 Derivatives of Exponential and Logarithmic Functions

- The Derivative of e^x
- The Derivative of $\ln x$
- Other Logarithmic and Exponential Functions
- Exponential and Logarithmic Models

In this section, we find formulas for the derivatives of logarithmic and exponential functions. In particular, recall that $f(x) = e^x$ is the exponential function with base $e \approx 2.718$, and the inverse of the function e^x is the natural logarithm function $\ln x$. More generally, if b is a positive real number, $b \neq 1$,then the exponential function b^x with base b, and the logarithmic function $\log_b x$ with base b, are inverses of each other.

The Derivative of e^x

In the process of finding the derivative of e^x, we use (without proof) the fact that

$$\lim_{h \to 0} \frac{e^h - 1}{h} = 1 \tag{1}$$

EXPLORE & DISCUSS 1

Complete Table 1.

Table 1

h	-0.1	-0.01	-0.001	$\rightarrow 0 \leftarrow$	0.001	0.01	0.1
$\dfrac{e^h - 1}{h}$							

Do your calculations make it reasonable to conclude that

$$\lim_{h \to 0} \frac{e^h - 1}{h} = 1?$$

Discuss.

We now apply the four-step process (Section 1-4) to the exponential function $f(x) = e^x$.

Step 1 Find $f(x + h)$.

$$f(x + h) = e^{x+h} = e^x e^h$$

Step 2 Find $f(x + h) - f(x)$.

$$f(x + h) - f(x) = e^x e^h - e^x \quad \text{Factor out } e^x.$$

$$= e^x(e^h - 1)$$

Step 3 Find $\dfrac{f(x+h)-f(x)}{h}$.

$$\frac{f(x+h)-f(x)}{h} = \frac{e^x(e^h-1)}{h} = e^x\left(\frac{e^h-1}{h}\right)$$

Step 4 Find $f'(x) = \lim_{h\to 0}\dfrac{f(x+h)-f(x)}{h}$.

$$\begin{aligned} f'(x) &= \lim_{h\to 0}\frac{f(x+h)-f(x)}{h} \\ &= \lim_{h\to 0} e^x\left(\frac{e^h-1}{h}\right) \\ &= e^x \lim_{h\to 0}\left(\frac{e^h-1}{h}\right) && \text{Use the limit in (1).} \\ &= e^x \cdot 1 = e^x \end{aligned}$$

Therefore,

$$\frac{d}{dx}e^x = e^x$$ The derivative of the exponential function is the exponential function.

EXAMPLE 1 **Finding Derivatives** Find $f'(x)$ for

(A) $f(x) = 5e^x - 3x^4 + 9x + 16$ (B) $f(x) = -7x^e + 2e^x + e^2$

SOLUTIONS (A) $f'(x) = 5e^x - 12x^3 + 9$ (B) $f'(x) = -7ex^{e-1} + 2e^x$

Remember that e is a real number, so the power rule (Section 1–5) is used to find the derivative of x^e. The derivative of the exponential function e^x, however, is e^x. Note that $e^2 \approx 7.389$ is a constant, so its derivative is 0.

Matched Problem 1 Find $f'(x)$ for

(A) $f(x) = 4e^x + 8x^2 + 7x - 14$ (B) $f(x) = x^7 - x^5 + e^3 - x + e^x$

CAUTION

$$\frac{d}{dx}e^x \neq xe^{x-1} \qquad \frac{d}{dx}e^x = e^x$$

The power rule cannot be used to differentiate the exponential function. The power rule applies to exponential forms x^n, where the exponent is a constant and the base is a variable. In the exponential form e^x, the base is a constant and the exponent is a variable.

The Derivative of ln x

We summarize some important facts about logarithmic functions:

SUMMARY

Recall that the inverse of an exponential function is called a **logarithmic function**. For $b > 0$ and $b \neq 1$,

Logarithmic form		**Exponential form**
$y = \log_b x$	is equivalent to	$x = b^y$
Domain: $(0, \infty)$		Domain: $(-\infty, \infty)$
Range: $(-\infty, \infty)$		Range: $(0, \infty)$

The graphs of $y = \log_b x$ and $y = b^x$ are symmetric with respect to the line $y = x$. (See Figure 1.)

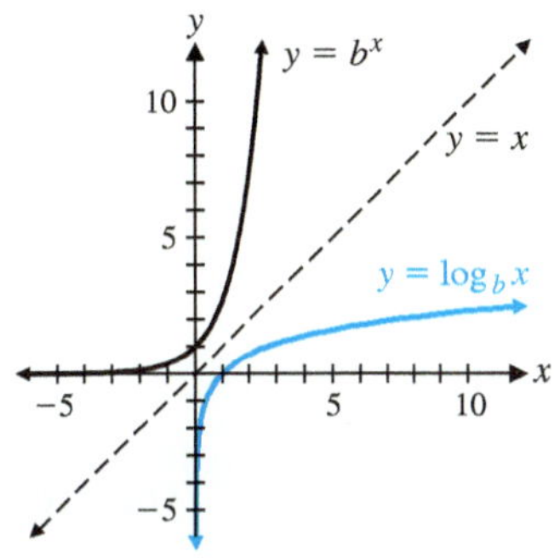

Figure 1

Of all the possible bases for logarithmic functions, the two most widely used are

$\log x = \log_{10} x$ Common logarithm (base 10)

$\ln x = \log_e x$ Natural logarithm (base e)

We are now ready to use the definition of the derivative and the four-step process discussed in Section 1-4 to find a formula for the derivative of $\ln x$. Later we will extend this formula to include $\log_b x$ for any base b.

Let $f(x) = \ln x, x > 0$.

Step 1 Find $f(x + h)$.

$$f(x + h) = \ln(x + h)$$ $\ln(x + h)$ cannot be simplified.

Step 2 Find $f(x + h) - f(x)$.

$$f(x + h) - f(x) = \ln(x + h) - \ln x$$ Use $\ln A - \ln B = \ln \frac{A}{B}$.

$$= \ln \frac{x + h}{x}$$

Step 3 Find $\frac{f(x + h) - f(x)}{h}$.

$$\frac{f(x + h) - f(x)}{h} = \frac{\ln(x + h) - \ln x}{h}$$

$$= \frac{1}{h} \ln \frac{x + h}{x}$$

$$= \frac{x}{x} \cdot \frac{1}{h} \ln \frac{x + h}{x}$$ Multiply by $1 = x/x$ to change form.

$$= \frac{1}{x}\left[\frac{x}{h}\ln\left(1 + \frac{h}{x}\right)\right] \qquad \textit{Use } p \ln A = \ln A^p.$$

$$= \frac{1}{x}\ln\left(1 + \frac{h}{x}\right)^{x/h}$$

Step 4 Find $f'(x) = \lim_{h\to 0}\frac{f(x+h) - f(x)}{h}$.

$$f'(x) = \lim_{h\to 0}\frac{f(x+h) - f(x)}{h}$$

$$= \lim_{h\to 0}\left[\frac{1}{x}\ln\left(1 + \frac{h}{x}\right)^{x/h}\right] \qquad \textit{Let } s = h/x. \textit{ Note that } h \to 0 \textit{ implies } s \to 0.$$

$$= \frac{1}{x}\lim_{s\to 0}\left[\ln(1 + s)^{1/s}\right] \qquad \textit{Use a new limit property.*}$$

$$= \frac{1}{x}\ln\left[\lim_{s\to 0}(1 + s)^{1/s}\right] \qquad \textit{Use the definition of } e.$$

$$= \frac{1}{x}\ln e \qquad \ln e = \log_e e = 1$$

$$= \frac{1}{x}$$

Therefore,

$$\frac{d}{dx}\ln x = \frac{1}{x}$$

CONCEPTUAL INSIGHT

In finding the derivative of ln x, we used the following properties of logarithms:

$$\ln\frac{A}{B} = \ln A - \ln B \qquad \ln A^p = p\ln A$$

We also noted that there is no property that simplifies $\ln(A + B)$.

EXAMPLE 2 **Finding Derivatives** Find y' for

(A) $y = 3e^x + 5\ln x$ (B) $y = x^4 - \ln x^4$

SOLUTIONS (A) $y' = 3e^x + \frac{5}{x}$

(B) Before taking the derivative, we use a property of logarithms to rewrite y.

$$y = x^4 - \ln x^4 \qquad \textit{Use } \ln M^p = p\ln M.$$

$$y = x^4 - 4\ln x \qquad \textit{Now take the derivative of both sides.}$$

$$y' = 4x^3 - \frac{4}{x}$$

Matched Problem 2 Find y' for

(A) $y = 10x^3 - 100\ln x$ (B) $y = \ln x^5 + e^x - \ln e^2$

*The following new limit property is used: If $\lim_{x\to c} f(x)$ exists and is positive, then $\lim_{x\to c}[\ln f(x)] = \ln[\lim_{x\to c} f(x)]$.

Other Logarithmic and Exponential Functions

In most applications involving logarithmic or exponential functions, the number e is the preferred base. However, in some situations it is convenient to use a base other than e. Derivatives of $y = \log_b x$ and $y = b^x$ can be obtained by expressing these functions in terms of the natural logarithmic and exponential functions.

We begin by finding a relationship between $\log_b x$ and $\ln x$ for any base b such that $b > 0$ and $b \neq 1$.

$$y = \log_b x \qquad \text{Change to exponential form.}$$
$$b^y = x \qquad \text{Take the natural logarithm of both sides.}$$
$$\ln b^y = \ln x \qquad \text{Recall that } \ln b^y = y \ln b.$$
$$y \ln b = \ln x \qquad \text{Solve for } y.$$
$$y = \frac{1}{\ln b}\ln x$$

Therefore,

$$\log_b x = \frac{1}{\ln b}\ln x \qquad \text{Change-of-base formula for logarithms*} \qquad (2)$$

Similarly, we can find a relationship between b^x and e^x for any base b such that $b > 0$, $b \neq 1$.

$$y = b^x \qquad \text{Take the natural logarithm of both sides.}$$
$$\ln y = \ln b^x \qquad \text{Recall that } \ln b^x = x \ln b.$$
$$\ln y = x \ln b \qquad \text{Take the exponential function of both sides.}$$
$$y = e^{x \ln b}$$

Therefore,

$$b^x = e^{x \ln b} \qquad \text{Change-of-base formula for exponential functions} \qquad (3)$$

Differentiating both sides of equation (2) gives

$$\frac{d}{dx}\log_b x = \frac{1}{\ln b}\frac{d}{dx}\ln x = \frac{1}{\ln b}\left(\frac{1}{x}\right)$$

It can be shown that the derivative of the function e^{cx}, where c is a constant, is the function ce^{cx} (see Problems 49–50 in Exercise 2-2 or the more general results of Section 2–4). Therefore, differentiating both sides of equation (3), we have

$$\frac{d}{dx}b^x = e^{x \ln b}\ln b = b^x \ln b$$

For convenience, we list the derivative formulas for exponential and logarithmic functions:

Derivatives of Exponential and Logarithmic Functions

For $b > 0$, $b \neq 1$,

$$\frac{d}{dx}e^x = e^x \qquad \frac{d}{dx}b^x = b^x \ln b$$

$$\frac{d}{dx}\ln x = \frac{1}{x} \qquad \frac{d}{dx}\log_b x = \frac{1}{\ln b}\left(\frac{1}{x}\right)$$

*Equation (2) is a special case of the **general change-of-base formula** for logarithms (which can be derived in the same way): $\log_b x = (\log_a x)/(\log_a b)$.

EXAMPLE 3 **Finding Derivatives** Find $g'(x)$ for

(A) $g(x) = 2^x - 3^x$ (B) $g(x) = \log_4 x^5$

SOLUTIONS (A) $g'(x) = 2^x \ln 2 - 3^x \ln 3$

(B) First, use a property of logarithms to rewrite $g(x)$.

$$g(x) = \log_4 x^5 \quad \text{Use } \log_b M^p = p \log_b M.$$

$$g(x) = 5 \log_4 x \quad \text{Take the derivative of both sides.}$$

$$g'(x) = \frac{5}{\ln 4}\left(\frac{1}{x}\right)$$

Matched Problem 3 Find $g'(x)$ for

(A) $g(x) = x^{10} + 10^x$ (B) $g(x) = \log_2 x - 6 \log_5 x$

EXPLORE & DISCUSS 2

(A) The graphs of $f(x) = \log_2 x$ and $g(x) = \log_4 x$ are shown in Figure 2. Which graph belongs to which function?

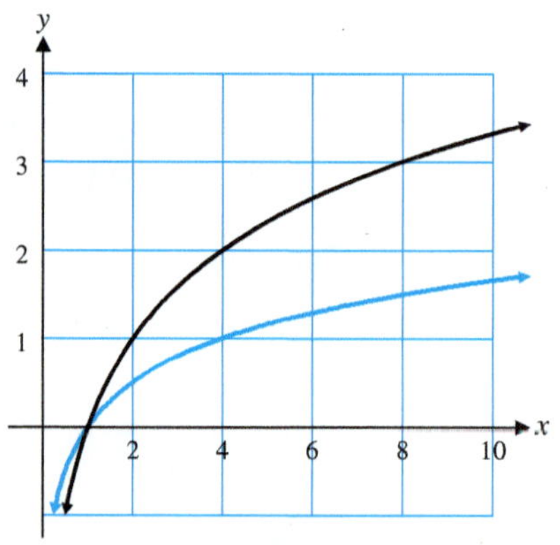

Figure 2

(B) Sketch graphs of $f'(x)$ and $g'(x)$.

(C) The function $f(x)$ is related to $g(x)$ in the same way that $f'(x)$ is related to $g'(x)$. What is that relationship?

Exponential and Logarithmic Models

EXAMPLE 4 **Price–Demand Model** An Internet store sells Australian wool blankets. If the store sells x blankets at a price of \$$p$ per blanket, then the price–demand equation is $p = 350(0.999)^x$. Find the rate of change of price with respect to demand when the demand is 800 blankets and interpret the result.

SOLUTION $$\frac{dp}{dx} = 350(0.999)^x \ln 0.999$$

If $x = 800$, then

$$\frac{dp}{dx} = 350(0.999)^{800} \ln 0.999 \approx -0.157, \text{ or } -\$0.16$$

When the demand is 800 blankets, the price is decreasing by \$0.16 per blanket.

Matched Problem 4 The store in Example 4 also sells a reversible fleece blanket. If the price–demand equation for reversible fleece blankets is $p = 200(0.998)^x$, find the rate of change of price with respect to demand when the demand is 400 blankets and interpret the result.

EXAMPLE 5 **Cable TV Subscribers** A statistician used data from the U.S. Census Bureau to construct the model

$$S(t) = 21 \ln t + 2$$

where $S(t)$ is the number of cable TV subscribers (in millions) in year t ($t = 0$ corresponds to 1980). Use this model to estimate the number of cable TV subscribers in 2015 and the rate of change of the number of subscribers in 2015. Round both to the nearest tenth of a million. Interpret these results.

SOLUTION Since 2015 corresponds to $t = 35$, we must find $S(35)$ and $S'(35)$.

$$S(35) = 21 \ln 35 + 2 = 76.7 \text{ million}$$

$$S'(t) = 21\frac{1}{t} = \frac{21}{t}$$

$$S'(35) = \frac{21}{35} = 0.6 \text{ million}$$

In 2015 there will be approximately 76.7 million subscribers, and this number is growing at the rate of 0.6 million subscribers per year.

Matched Problem 5 A model for newspaper circulation is

$$C(t) = 83 - 9 \ln t$$

where $C(t)$ is newspaper circulation (in millions) in year t ($t = 0$ corresponds to 1980). Use this model to estimate the circulation and the rate of change of circulation in 2015. Round both to the nearest tenth of a million. Interpret these results.

CONCEPTUAL INSIGHT

On most graphing calculators, exponential regression produces a function of the form $y = a \cdot b^x$. Formula (3) on page 97 allows you to change the base b (chosen by the graphing calculator) to the more familiar base e:

$$y = a \cdot b^x = a \cdot e^{x \ln b}$$

On most graphing calculators, logarithmic regression produces a function of the form $y = a + b \ln x$. Formula (2) on page 97 allows you to write the function in terms of logarithms to any base d that you may prefer:

$$y = a + b \ln x = a + b(\ln d) \log_d x$$

Exercises 2-2

A

In Problems 1–14, find $f'(x)$.

1. $f(x) = 5e^x + 3x + 1$
2. $f(x) = -7e^x - 2x + 5$
3. $f(x) = -2 \ln x + x^2 - 4$
4. $f(x) = 6 \ln x - x^3 + 2$
5. $f(x) = x^3 - 6e^x$
6. $f(x) = 9e^x + 2x^2$
7. $f(x) = e^x + x - \ln x$
8. $f(x) = \ln x + 2e^x - 3x^2$
9. $f(x) = \ln x^3$
10. $f(x) = \ln x^8$
11. $f(x) = 5x - \ln x^5$
12. $f(x) = 4 + \ln x^9$
13. $f(x) = \ln x^2 + 4e^x$
14. $f(x) = \ln x^{10} + 2 \ln x$

B

In Problems 15–22, find the equation of the line tangent to the graph of f at the indicated value of x.

15. $f(x) = 3 + \ln x; x = 1$
16. $f(x) - 2 \ln x; x - 1$
17. $f(x) = 3e^x; x = 0$
18. $f(x) = e^x + 1; x = 0$
19. $f(x) = \ln x^3; x = e$
20. $f(x) = 1 + \ln x^4; x = e$
21. $f(x) = 2 + e^x; x = 1$
22. $f(x) = 5e^x; x = 1$
23. A student claims that the line tangent to the graph of $f(x) = e^x$ at $x = 3$ passes through the point $(2, 0)$ (see the figure). Is she correct? Will the line tangent at $x = 4$ pass through $(3, 0)$? Explain.

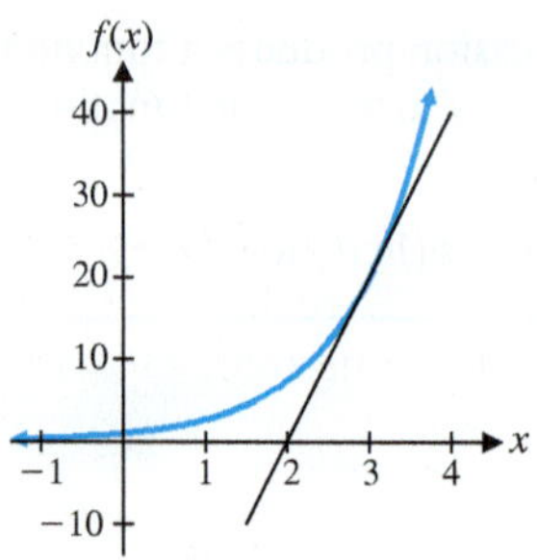

Figure for 23

24. Refer to Problem 23. Does the line tangent to the graph of $f(x) = e^x$ at $x = 1$ pass through the origin? Are there any other lines tangent to the graph of f that pass through the origin? Explain.
25. A student claims that the line tangent to the graph of $g(x) = \ln x$ at $x = 3$ passes through the origin (see the figure). Is he correct? Will the line tangent at $x = 4$ pass through the origin? Explain.

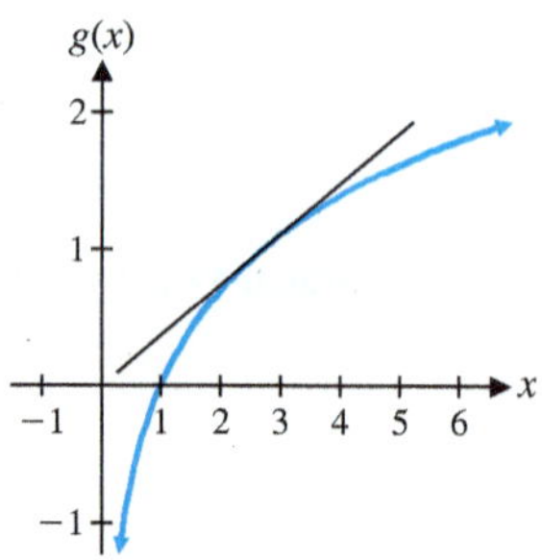

Figure for 25

26. Refer to Problem 25. Does the line tangent to the graph of $f(x) = \ln x$ at $x = e$ pass through the origin? Are there any other lines tangent to the graph of f that pass through the origin? Explain.

In Problems 27–30, first use appropriate properties of logarithms to rewrite f(x), and then find $f'(x)$.

27. $f(x) = 10x + \ln 10x$
28. $f(x) = 2 + 3 \ln \dfrac{1}{x}$
29. $f(x) = \ln \dfrac{4}{x^3}$
30. $f(x) = x + 5 \ln 6x$

C

In Problems 31–42, find $\dfrac{dy}{dx}$ for the indicated function y.

31. $y = \log_2 x$
32. $y = 3 \log_5 x$
33. $y = 3^x$
34. $y = 4^x$
35. $y = 2x - \log x$
36. $y = \log x + 4x^2 + 1$
37. $y = 10 + x + 10^x$
38. $y = x^5 - 5^x$
39. $y = 3 \ln x + 2 \log_3 x$
40. $y = -\log_2 x + 10 \ln x$
41. $y = 2^x + e^2$
42. $y = e^3 - 3^x$

In Problems 43–48, use graphical approximation methods to find the points of intersection of f(x) and g(x) (to two decimal places).

43. $f(x) = e^x; g(x) = x^4$

    [Note that there are three points of intersection and that e^x is greater than x^4 for large values of x.]

44. $f(x) = e^x; g(x) = x^5$

    [Note that there are two points of intersection and that e^x is greater than x^5 for large values of x.]

45. $f(x) = (\ln x)^2$; $g(x) = x$

46. $f(x) = (\ln x)^3$; $g(x) = \sqrt{x}$

47. $f(x) = \ln x$; $g(x) = x^{1/5}$

48. $f(x) = \ln x$; $g(x) = x^{1/4}$

49. Explain why $\lim_{h \to 0} \frac{e^{ch} - 1}{h} = c$.

50. Use the result of Problem 49 and the four-step process to show that if $f(x) = e^{cx}$, then $f'(x) = ce^{cx}$.

Applications

51. **Salvage value.** The estimated salvage value S (in dollars) of a company airplane after t years is given by

$$S(t) = 300{,}000(0.9)^t$$

What is the rate of depreciation (in dollars per year) after 1 year? 5 years? 10 years?

52. **Resale value.** The estimated resale value R (in dollars) of a company car after t years is given by

$$R(t) = 20{,}000(0.86)^t$$

What is the rate of depreciation (in dollars per year) after 1 year? 2 years? 3 years?

53. **Bacterial growth.** A single cholera bacterium divides every 0.5 hour to produce two complete cholera bacteria. If we start with a colony of 5,000 bacteria, then after t hours, there will be

$$A(t) = 5{,}000 \cdot 2^{2t} = 5{,}000 \cdot 4^t$$

bacteria. Find $A'(t)$, $A'(1)$, and $A'(5)$, and interpret the results.

54. **Bacterial growth.** Repeat Problem 53 for a starting colony of 1,000 bacteria such that a single bacterium divides every 0.25 hour.

55. **Blood pressure.** An experiment was set up to find a relationship between weight and systolic blood pressure in children. Using hospital records for 5,000 children, the experimenters found that the systolic blood pressure was given approximately by

$$P(x) = 17.5(1 + \ln x) \qquad 10 \le x \le 100$$

where $P(x)$ is measured in millimeters of mercury and x is measured in pounds. What is the rate of change of blood pressure with respect to weight at the 40-pound weight level? At the 90-pound weight level?

56. **Blood pressure.** Refer to Problem 55. Find the weight (to the nearest pound) at which the rate of change of blood pressure with respect to weight is 0.3 millimeter of mercury per pound.

57. **Psychology: stimulus/response.** In psychology, the Weber–Fechner law for the response to a stimulus is

$$R = k \ln \frac{S}{S_0}$$

where R is the response, S is the stimulus, and S_0 is the lowest level of stimulus that can be detected. Find dR/dS.

58. **Psychology: learning.** A mathematical model for the average of a group of people learning to type is given by

$$N(t) = 10 + 6 \ln t \qquad t \ge 1$$

where $N(t)$ is the number of words per minute typed after t hours of instruction and practice (2 hours per day, 5 days per week). What is the rate of learning after 10 hours of instruction and practice? After 100 hours?

Answers to Matched Problems

1. (A) $4e^x + 16x + 7$
   (B) $7x^6 - 5x^4 - 1 + e^x$
2. (A) $30x^2 - \frac{100}{x}$
   (B) $\frac{5}{x} + e^x$
3. (A) $10x^9 + 10^x \ln 10$
   (B) $\left(\frac{1}{\ln 2} - \frac{6}{\ln 5}\right)\frac{1}{x}$
4. The price is decreasing at the rate of $0.18 per blanket.
5. The circulation in 2015 is approximately 51.0 million and is decreasing at the rate of 0.3 million per year.

2-3 Derivatives of Products and Quotients

- Derivatives of Products
- Derivatives of Quotients

The derivative properties discussed in Section 1-5 add substantially to our ability to compute and apply derivatives to many practical problems. In this and the next two sections, we add a few more properties that will increase this ability even further.

Derivatives of Products

In Section 1-5, we found that the derivative of a sum is the sum of the derivatives. Is the derivative of a product the product of the derivatives?

EXPLORE & DISCUSS 1

Let $F(x) = x^2$, $S(x) = x^3$, and $f(x) = F(x)S(x) = x^5$. Which of the following is $f'(x)$?

(A) $F'(x)S'(x)$ (B) $F(x)S'(x)$

(C) $F'(x)S(x)$ (D) $F(x)S'(x) + F'(x)S(x)$

Comparing the various expressions computed in Explore & Discuss 1, we see that the derivative of a product is not the product of the derivatives.

Using the definition of the derivative and the four-step process, we can show that

The derivative of the product of two functions is the first function times the derivative of the second function, plus the second function times the derivative of the first function.

THEOREM 1 Product Rule

If

$$y = f(x) = F(x)S(x)$$

and if $F'(x)$ and $S'(x)$ exist, then

$$f'(x) = F(x)S'(x) + S(x)F'(x)$$

Using simplified notation,

$$y' = FS' + SF' \qquad \text{or} \qquad \frac{dy}{dx} = F\frac{dS}{dx} + S\frac{dF}{dx}$$

EXAMPLE 1 **Differentiating a Product** Use two different methods to find $f'(x)$ for

$$f(x) = 2x^2(3x^4 - 2).$$

SOLUTION **Method 1.** Use the product rule:

$$\begin{aligned} f'(x) &= 2x^2(3x^4 - 2)' + (3x^4 - 2)(2x^2)' && \text{First times derivative of second, plus second times derivative of first} \\ &= 2x^2(12x^3) + (3x^4 - 2)(4x) \\ &= 24x^5 + 12x^5 - 8x \\ &= 36x^5 - 8x \end{aligned}$$

Method 2. Multiply first; then take derivatives:

$$f(x) = 2x^2(3x^4 - 2) = 6x^6 - 4x^2$$
$$f'(x) = 36x^5 - 8x$$

Matched Problem 1 Use two different methods to find $f'(x)$ for $f(x) = 3x^3(2x^2 - 3x + 1)$.

Some products we encounter can be differentiated by either method illustrated in Example 1. In other situations, the product rule *must* be used. Unless instructed otherwise, you should use the product rule to differentiate all products in this section in order to gain experience with this important differentiation rule.

EXAMPLE 2 **Tangent Lines** Let $f(x) = (2x - 9)(x^2 + 6)$.

(A) Find the equation of the line tangent to the graph of $f(x)$ at $x = 3$.

(B) Find the value(s) of x where the tangent line is horizontal.

SOLUTION (A) First, find $f'(x)$:

$$\begin{aligned} f'(x) &= (2x - 9)(x^2 + 6)' + (x^2 + 6)(2x - 9)' \\ &= (2x - 9)(2x) + (x^2 + 6)(2) \end{aligned}$$

Then, find $f(3)$ and $f'(3)$:

$$f(3) = [2(3) - 9](3^2 + 6) = (-3)(15) = -45$$

$$f'(3) = [2(3) - 9]2(3) + (3^2 + 6)(2) = -18 + 30 = 12$$

Now, find the equation of the tangent line at $x = 3$:

$$\begin{aligned} y - y_1 &= m(x - x_1) && y_1 = f(x_1) = f(3) = -45 \\ y - (-45) &= 12(x - 3) && m = f'(x_1) = f'(3) = 12 \\ y &= 12x - 81 && \text{Tangent line at } x = 3 \end{aligned}$$

(B) The tangent line is horizontal at any value of x such that $f'(x) = 0$, so

$$\begin{aligned} f'(x) = (2x - 9)2x + (x^2 + 6)2 &= 0 \\ 6x^2 - 18x + 12 &= 0 \\ x^2 - 3x + 2 &= 0 \\ (x - 1)(x - 2) &= 0 \\ x &= 1, 2 \end{aligned}$$

The tangent line is horizontal at $x = 1$ and at $x = 2$.

Matched Problem 2 Repeat Example 2 for $f(x) = (2x + 9)(x^2 - 12)$.

CONCEPTUAL INSIGHT

As Example 2 illustrates, the way we write $f'(x)$ depends on what we want to do. If we are interested only in evaluating $f'(x)$ at specified values of x, then the form in part (A) is sufficient. However, if we want to solve $f'(x) = 0$, we must multiply and collect like terms, as we did in part (B).

EXAMPLE 3 **Finding Derivatives** Find $f'(x)$ for

(A) $f(x) = 2x^3e^x$

(B) $f(x) = 6x^4 \ln x$

SOLUTIONS

(A) $$\begin{aligned} f'(x) &= 2x^3(e^x)' + e^x(2x^3)' \\ &= 2x^3e^x + e^x(6x^2) \\ &= 2x^2e^x(x + 3) \end{aligned}$$

(B) $$\begin{aligned} f'(x) &= 6x^4 (\ln x)' + (\ln x)(6x^4)' \\ &= 6x^4 \frac{1}{x} + (\ln x)(24x^3) \\ &= 6x^3 + 24x^3 \ln x \\ &= 6x^3(1 + 4 \ln x) \end{aligned}$$

Matched Problem 3 Find $f'(x)$ for

(A) $f(x) = 5x^8 e^x$

(B) $f(x) = x^7 \ln x$

Derivatives of Quotients

The derivative of a quotient of two functions is not the quotient of the derivatives of the two functions.

EXPLORE & DISCUSS 2

Let $T(x) = x^5$, $B(x) = x^2$, and

$$f(x) = \frac{T(x)}{B(x)} = \frac{x^5}{x^2} = x^3$$

Which of the following is $f'(x)$?

(A) $\dfrac{T'(x)}{B'(x)}$ (B) $\dfrac{T'(x)B(x)}{[B(x)]^2}$ (C) $\dfrac{T(x)B'(x)}{[B(x)]^2}$

(D) $\dfrac{T'(x)B(x)}{[B(x)]^2} - \dfrac{T(x)B'(x)}{[B(x)]^2} = \dfrac{T'(x)B(x) - T(x)B'(x)}{[B(x)]^2}$

The expressions in Explore & Discuss 2 suggest that the derivative of a quotient leads to a more complicated quotient than expected.

If $T(x)$ and $B(x)$ are any two differentiable functions and

$$f(x) = \frac{T(x)}{B(x)}$$

then

$$f'(x) = \frac{B(x)T'(x) - T(x)B'(x)}{[B(x)]^2}$$

Therefore,

The derivative of the quotient of two functions is the denominator function times the derivative of the numerator function, minus the numerator function times the derivative of the denominator function, divided by the denominator function squared.

THEOREM 2 Quotient Rule

If

$$y = f(x) = \frac{T(x)}{B(x)}$$

and if $T'(x)$ and $B'(x)$ exist, then

$$f'(x) = \frac{B(x)T'(x) - T(x)B'(x)}{[B(x)]^2}$$

Using simplified notation,

$$y' = \frac{BT' - TB'}{B^2} \quad \text{or} \quad \frac{dy}{dx} = \frac{B\dfrac{dT}{dx} - T\dfrac{dB}{dx}}{B^2}$$

EXAMPLE 4 **Differentiating Quotients**

(A) If $f(x) = \dfrac{x^2}{2x - 1}$, find $f'(x)$. (B) If $y = \dfrac{t^2 - t}{t^3 + 1}$, find y'.

(C) Find $\dfrac{d}{dx}\dfrac{x^2 - 3}{x^2}$ by using the quotient rule and also by splitting the fraction into two fractions.

SOLUTION (A)
$$f'(x) = \frac{(2x - 1)(x^2)' - x^2(2x - 1)'}{(2x - 1)^2}$$

The denominator times the derivative of the numerator, minus the numerator times the derivative of the denominator, divided by the square of the denominator

$$= \frac{(2x - 1)(2x) - x^2(2)}{(2x - 1)^2}$$
$$= \frac{4x^2 - 2x - 2x^2}{(2x - 1)^2}$$
$$= \frac{2x^2 - 2x}{(2x - 1)^2}$$

(B)
$$y' = \frac{(t^3 + 1)(t^2 - t)' - (t^2 - t)(t^3 + 1)'}{(t^3 + 1)^2}$$
$$= \frac{(t^3 + 1)(2t - 1) - (t^2 - t)(3t^2)}{(t^3 + 1)^2}$$
$$= \frac{2t^4 - t^3 + 2t - 1 - 3t^4 + 3t^3}{(t^3 + 1)^2}$$
$$= \frac{-t^4 + 2t^3 + 2t - 1}{(t^3 + 1)^2}$$

(C) **Method 1.** Use the quotient rule:

$$\frac{d}{dx}\frac{x^2 - 3}{x^2} = \frac{x^2\frac{d}{dx}(x^2 - 3) - (x^2 - 3)\frac{d}{dx}x^2}{(x^2)^2}$$
$$= \frac{x^2(2x) - (x^2 - 3)2x}{x^4}$$
$$= \frac{2x^3 - 2x^3 + 6x}{x^4} = \frac{6x}{x^4} = \frac{6}{x^3}$$

Method 2. Split into two fractions:

$$\frac{x^2 - 3}{x^2} = \frac{x^2}{x^2} - \frac{3}{x^2} = 1 - 3x^{-2}$$
$$\frac{d}{dx}(1 - 3x^{-2}) = 0 - 3(-2)x^{-3} = \frac{6}{x^3}$$

Comparing methods 1 and 2, we see that it often pays to change an expression algebraically before choosing a differentiation formula.

Matched Problem 4 Find

(A) $f'(x)$ for $f(x) = \dfrac{2x}{x^2 + 3}$ (B) y' for $y = \dfrac{t^3 - 3t}{t^2 - 4}$

(C) $\dfrac{d}{dx}\dfrac{2 + x^3}{x^3}$ in two ways

EXAMPLE 5 **Finding Derivatives** Find $f'(x)$ for

(A) $f(x) = \dfrac{3e^x}{1 + e^x}$ (B) $f(x) = \dfrac{\ln x}{2x + 5}$

SOLUTIONS (A)
$$f'(x) = \frac{(1 + e^x)(3e^x)' - 3e^x(1 + e^x)'}{(1 + e^x)^2}$$
$$= \frac{(1 + e^x)3e^x - 3e^xe^x}{(1 + e^x)^2}$$
$$= \frac{3e^x}{(1 + e^x)^2}$$

(B)
$$f'(x) = \frac{(2x + 5)(\ln x)' - (\ln x)(2x + 5)'}{(2x + 5)^2}$$
$$= \frac{(2x + 5) \cdot \frac{1}{x} - (\ln x)(2)}{(2x + 5)^2} \qquad \text{Multiply by } \frac{x}{x}$$
$$= \frac{2x + 5 - 2x \ln x}{x(2x + 5)^2}$$

Matched Problem 5 Find $f'(x)$ for

(A) $f(x) = \dfrac{x^3}{e^x + 2}$ (B) $f(x) = \dfrac{4x}{1 + \ln x}$

EXAMPLE 6 **Sales Analysis** The total sales S (in thousands of games) of a video game t months after the game is introduced are given by

$$S(t) = \frac{125t^2}{t^2 + 100}$$

(A) Find $S'(t)$.

(B) Find $S(10)$ and $S'(10)$. Write a brief interpretation of these results.

(C) Use the results from part (B) to estimate the total sales after 11 months.

SOLUTION (A)
$$S'(t) = \frac{(t^2 + 100)(125t^2)' - 125t^2(t^2 + 100)'}{(t^2 + 100)^2}$$
$$= \frac{(t^2 + 100)(250t) - 125t^2(2t)}{(t^2 + 100)^2}$$
$$= \frac{250t^3 + 25{,}000t - 250t^3}{(t^2 + 100)^2}$$
$$= \frac{25{,}000t}{(t^2 + 100)^2}$$

(B) $S(10) = \dfrac{125(10)^2}{10^2 + 100} = 62.5$ and $S'(10) = \dfrac{25{,}000(10)}{(10^2 + 100)^2} = 6.25.$

Total sales after 10 months are 62,500 games, and sales are increasing at the rate of 6,250 games per month.

(C) Total sales will increase by approximately 6,250 games during the next month, so the estimated total sales after 11 months are 62,500 + 6,250 = 68,750 games.

Matched Problem 6 Refer to Example 6. Suppose that the total sales S (in thousands of games) t months after the game is introduced are given by

$$S(t) = \frac{150t}{t + 3}$$

(A) Find $S'(t)$.

(B) Find $S(12)$ and $S'(12)$. Write a brief interpretation of these results.

(C) Use the results from part (B) to estimate the total sales after 13 months.

Exercises 2-3

Answers to most of the problems in this exercise set contain both an unsimplified form and a simplified form of the derivative. When checking your work, first check that you applied the rules correctly and then check that you performed the algebraic simplification correctly. Unless instructed otherwise, when differentiating a product, use the product rule rather than performing the multiplication first.

A

In Problems 1–26, find $f'(x)$ and simplify.

1. $f(x) = 2x^3(x^2 - 2)$

2. $f(x) = 5x^2(x^3 + 2)$

3. $f(x) = (x - 3)(2x - 1)$

4. $f(x) = (3x + 2)(4x - 5)$

5. $f(x) = \frac{x}{x - 3}$

6. $f(x) = \frac{3x}{2x + 1}$

7. $f(x) = \frac{2x + 3}{x - 2}$

8. $f(x) = \frac{3x - 4}{2x + 3}$

9. $f(x) = 3xe^x$

10. $f(x) = x^2e^x$

11. $f(x) = x^3 \ln x$

12. $f(x) = 5x \ln x$

13. $f(x) = (x^2 + 1)(2x - 3)$

14. $f(x) = (3x + 5)(x^2 - 3)$

15. $f(x) = (0.4x + 2)(0.5x - 5)$

16. $f(x) = (0.5x - 4)(0.2x + 1)$

17. $f(x) = \frac{x^2 + 1}{2x - 3}$

18. $f(x) = \frac{3x + 5}{x^2 - 3}$

19. $f(x) = (x^2 + 2)(x^2 - 3)$

20. $f(x) = (x^2 - 4)(x^2 + 5)$

21. $f(x) = \frac{x^2 + 2}{x^2 - 3}$

22. $f(x) = \frac{x^2 - 4}{x^2 + 5}$

23. $f(x) = \frac{e^x}{x^2 + 1}$

24. $f(x) = \frac{1 - e^x}{1 + e^x}$

25. $f(x) = \frac{\ln x}{1 + x}$

26. $f(x) = \frac{2x}{1 + \ln x}$

In Problems 27–38, find $h'(x)$, where $f(x)$ is an unspecified differentiable function.

27. $h(x) = xf(x)$

28. $h(x) = x^2f(x)$

29. $h(x) = x^3f(x)$

30. $h(x) = \frac{f(x)}{x}$

31. $h(x) = \frac{f(x)}{x^2}$

32. $h(x) = \frac{f(x)}{x^3}$

33. $h(x) = \frac{x}{f(x)}$

34. $h(x) = \frac{x^2}{f(x)}$

35. $h(x) = e^xf(x)$

36. $h(x) = \frac{e^x}{f(x)}$

37. $h(x) = \frac{\ln x}{f(x)}$

38. $h(x) = \frac{f(x)}{\ln x}$

B

In Problems 39–48, find the indicated derivatives and simplify.

39. $f'(x)$ for $f(x) = (2x + 1)(x^2 - 3x)$

40. y' for $y = (x^3 + 2x^2)(3x - 1)$

41. $\frac{dy}{dt}$ for $y = (2.5t - t^2)(4t + 1.4)$

42. $\frac{d}{dt}[(3 - 0.4t^3)(0.5t^2 - 2t)]$

43. y' for $y = \frac{5x - 3}{x^2 + 2x}$

44. $f'(x)$ for $f(x) = \frac{3x^2}{2x - 1}$

45. $\frac{d}{dw}\frac{w^2 - 3w + 1}{w^2 - 1}$

46. $\frac{dy}{dw}$ for $y = \frac{w^4 - w^3}{3w - 1}$

47. y' for $y = (1 + x - x^2)e^x$

48. $\frac{dy}{dt}$ for $y = (1 + e^t)\ln t$

In Problems 49–54, find $f'(x)$ and find the equation of the line tangent to the graph of f at $x = 2$.

49. $f(x) = (1 + 3x)(5 - 2x)$

50. $f(x) = (7 - 3x)(1 + 2x)$

51. $f(x) = \frac{x - 8}{3x - 4}$ **52.** $f(x) = \frac{2x - 5}{2x - 3}$

53. $f(x) = \frac{x}{2^x}$ **54.** $f(x) = (x - 2) \ln x$

In Problems 55–58, find $f'(x)$ and find the value(s) of x where $f'(x) = 0$.

55. $f(x) = (2x - 15)(x^2 + 18)$

56. $f(x) = (2x - 3)(x^2 - 6)$

57. $f(x) = \frac{x}{x^2 + 1}$ **58.** $f(x) = \frac{x}{x^2 + 9}$

In Problems 59–62, find $f'(x)$ in two ways: (1) using the product or quotient rule and (2) simplifying first.

59. $f(x) = x^3(x^4 - 1)$ **60.** $f(x) = x^4(x^3 - 1)$

61. $f(x) = \frac{x^3 + 9}{x^3}$ **62.** $f(x) = \frac{x^4 + 4}{x^4}$

C

In Problems 63–82, find each indicated derivative and simplify.

63. $f(w) = (w + 1)2^w$

64. $g(w) = (w - 5) \log_3 w$ **65.** $\frac{d}{dx}\frac{3x^2 - 2x + 3}{4x^2 + 5x - 1}$

66. y' for $y = \frac{x^3 - 3x + 4}{2x^2 + 3x - 2}$

67. $\frac{dy}{dx}$ for $y = 9x^{1/3}(x^3 + 5)$

68. $\frac{d}{dx}[(4x^{1/2} - 1)(3x^{1/3} + 2)]$

69. y' for $y = \frac{\log_2 x}{1 + x^2}$ **70.** $\frac{dy}{dx}$ for $y = \frac{10^x}{1 + x^4}$

71. $f'(x)$ for $f(x) = \frac{6\sqrt[3]{x}}{x^2 - 3}$

72. y' for $y = \frac{2\sqrt{x}}{x^2 - 3x + 1}$

73. $g'(t)$ if $g(t) = \frac{0.2t}{3t^2 - 1}$

74. $h'(t)$ if $h(t) = \frac{-0.05t^2}{2t + 1}$

75. $\frac{d}{dx}[4x \log x^5]$ **76.** $\frac{d}{dt}[10^t \log t]$

77. $\frac{d}{dx}\frac{x^3 - 2x^2}{\sqrt[3]{x^2}}$

78. $\frac{dy}{dx}$ for $y = \frac{x^2 - 3x + 1}{\sqrt[4]{x}}$

79. $f'(x)$ for $f(x) = \frac{(2x^2 - 1)(x^2 + 3)}{x^2 + 1}$

80. y' for $y = \frac{2x - 1}{(x^3 + 2)(x^2 - 3)}$

81. $\frac{dy}{dt}$ for $y = \frac{t \ln t}{e^t}$ **82.** $\frac{dy}{du}$ for $y = \frac{u^2 e^u}{1 + \ln u}$

Applications

83. Sales analysis. The total sales S (in thousands of DVDs) of a DVD are given by

$$S(t) = \frac{90t^2}{t^2 + 50}$$

where t is the number of months since the release of the DVD.

(A) Find $S'(t)$.

(B) Find $S(10)$ and $S'(10)$. Write a brief interpretation of these results.

(C) Use the results from part (B) to estimate the total sales after 11 months.

84. Sales analysis. A communications company has installed a new cable television system in a city. The total number N (in thousands) of subscribers t months after the installation of the system is given by

$$N(t) = \frac{180t}{t + 4}$$

(A) Find $N'(t)$.

(B) Find $N(16)$ and $N'(16)$. Write a brief interpretation of these results.

(C) Use the results from part (B) to estimate the total number of subscribers after 17 months.

85. Price–demand equation. According to economic theory, the demand x for a quantity in a free market decreases as the price p increases (see the figure). Suppose that the number x of DVD players people are willing to buy per week from a retail chain at a price of \$$p$ is given by

$$x = \frac{4{,}000}{0.1p + 1} \qquad 10 \le p \le 70$$

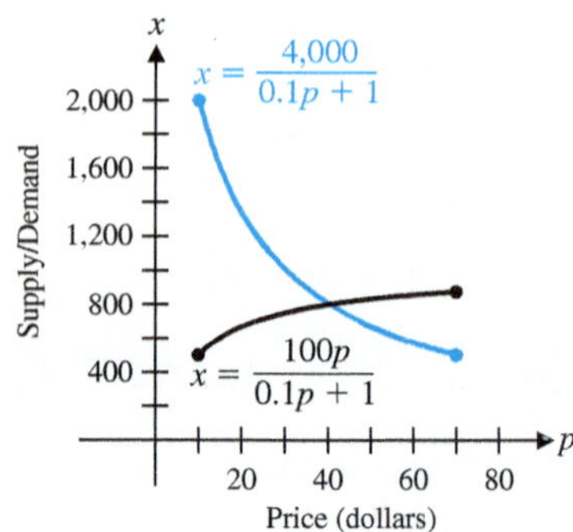

Figure for 85 and 86

(A) Find dx/dp.

(B) Find the demand and the instantaneous rate of change of demand with respect to price when the price is \$40. Write a brief interpretation of these results.

(C) Use the results from part (B) to estimate the demand if the price is increased to \$41.

86. Price–supply equation. According to economic theory, the supply x of a quantity in a free market increases as the price p increases (see the figure). Suppose that the number x of DVD players a retail chain is willing to sell per week at a price of \$$p$ is given by

$$x = \frac{100p}{0.1p + 1} \qquad 10 \le p \le 70$$

(A) Find dx/dp.

(B) Find the supply and the instantaneous rate of change of supply with respect to price when the price is \$40. Write a brief verbal interpretation of these results.

(C) Use the results from part (B) to estimate the supply if the price is increased to \$41.

87. Medicine. A drug is injected into a patient's bloodstream through her right arm. The drug concentration (in milligrams per cubic centimeter) in the bloodstream of the left arm t hours after the injection is given by

$$C(t) = \frac{0.14t}{t^2 + 1}$$

(A) Find $C'(t)$.

(B) Find $C'(0.5)$ and $C'(3)$, and interpret the results.

88. Drug sensitivity. One hour after a dose of x milligrams of a particular drug is administered to a person, the change in body temperature $T(x)$, in degrees Fahrenheit, is given approximately by

$$T(x) = x^2\left(1 - \frac{x}{9}\right) \qquad 0 \le x \le 7$$

The rate $T'(x)$ at which T changes with respect to the size of the dosage x is called the *sensitivity* of the body to the dosage.

(A) Use the product rule to find $T'(x)$.

(B) Find $T'(1)$, $T'(3)$, and $T'(6)$.

89. Learning. In the early twentieth century, L. L. Thurstone found that a given person successfully accomplished $N(x)$ acts after x practice acts, as given by

$$N(x) = \frac{100x + 200}{x + 32}$$

(A) Find the instantaneous rate of change of learning, $N'(x)$, with respect to the number of practice acts, x.

(B) Find $N'(4)$ and $N'(68)$.

Answers to Matched Problems

1. $30x^4 - 36x^3 + 9x^2$
2. (A) $y = 84x - 297$
   (B) $x = -4, x = 1$
3. (A) $5x^8e^x + e^x(40x^7) = 5x^7(x + 8)e^x$
   (B) $x^7 \cdot \frac{1}{x} + \ln x\,(7x^6) = x^6\,(1 + 7\ln x)$
4. (A) $\frac{(x^2 + 3)2 - (2x)(2x)}{(x^2 + 3)^2} = \frac{6 - 2x^2}{(x^2 + 3)^2}$
   (B) $\frac{(t^2 - 4)(3t^2 - 3) - (t^3 - 3t)(2t)}{(t^2 - 4)^2} = \frac{t^4 - 9t^2 + 12}{(t^2 - 4)^2}$
   (C) $-\frac{6}{x^4}$
5. (A) $\frac{(e^x + 2)\,3x^2 - x^3e^x}{(e^x + 2)^2}$
   (B) $\frac{(1 + \ln x)\,4 - 4x\frac{1}{x}}{(1 + \ln x)^2} = \frac{4\ln x}{(1 + \ln x)^2}$
6. (A) $S'(t) = \frac{450}{(t + 3)^2}$
   (B) $S(12) = 120$; $S'(12) = 2$. After 12 months, the total sales are 120,000 games, and sales are increasing at the rate of 2,000 games per month.
   (C) 122,000 games

2-4 The Chain Rule

- Composite Functions
- General Power Rule
- The Chain Rule

The word *chain* in the name "chain rule" comes from the fact that a function formed by composition involves a chain of functions—that is, a function of a function. The *chain rule* enables us to compute the derivative of a composite function in terms of the derivatives of the functions making up the composition. In this section, we review composite functions, introduce the chain rule by means of a special case known as the *general power rule,* and then discuss the chain rule itself.

Composite Functions

The function $m(x) = (x^2 + 4)^3$ is a combination of a quadratic function and a cubic function. To see this more clearly, let

$$y = f(u) = u^3 \qquad \text{and} \qquad u = g(x) = x^2 + 4$$

We can express y as a function of x:

$$y = f(u) = f[g(x)] = [x^2 + 4]^3 = m(x)$$

The function m is the *composite* of the two functions f and g.

DEFINITION Composite Functions

A function m is a **composite** of functions f and g if

$$m(x) = f[g(x)]$$

The domain of m is the set of all numbers x such that x is in the domain of g, and $g(x)$ is in the domain of f.

EXAMPLE 1 **Composite Functions** Let $f(u) = e^u$ and $g(x) = -3x$. Find $f[g(x)]$ and $g[f(u)]$.

SOLUTION

$$f[g(x)] = f(-3x) = e^{-3x}$$

$$g[f(u)] = g(e^u) = -3e^u$$

Matched Problem 1 Let $f(u) = 2u$ and $g(x) = e^x$. Find $f[g(x)]$ and $g[f(u)]$.

EXAMPLE 2 **Composite Functions** Write each function as a composition of two simpler functions.

(A) $y = 100e^{0.04x}$ (B) $y = \sqrt{4 - x^2}$

SOLUTION (A) Let

$$y = f(u) = 100e^u$$

$$u = g(x) = 0.04x$$

Check: $y = f[g(x)] = 100e^{g(x)} = 100e^{0.04x}$

(B) Let

$$y = f(u) = \sqrt{u}$$

$$u = g(x) = 4 - x^2$$

Check: $y = f[g(x)] = \sqrt{g(x)} = \sqrt{4 - x^2}$

Matched Problem 2 Write each function as a composition of two simpler functions.

(A) $y = 50e^{-2x}$ (B) $y = \sqrt[3]{1 + x^3}$

CONCEPTUAL INSIGHT

There can be more than one way to express a function as a composition of simpler functions. Choosing $y = f(u) = 100u$ and $u = g(x) = e^{0.04x}$ in Example 2A produces the same result:

$$y = f[g(x)] = 100g(x) = 100e^{0.04x}$$

Since we will be using composition as a means to an end (finding a derivative), usually it will not matter what functions you choose for the composition.

General Power Rule

We have already made extensive use of the power rule,

$$\frac{d}{dx}x^n = nx^{n-1} \tag{1}$$

Now we can generalize this rule in order to differentiate composite functions of the form $[u(x)]^n$, where $u(x)$ is a differentiable function. Is rule (1) still valid if we replace x with a function $u(x)$?

EXPLORE & DISCUSS 1

Let $u(x) = 2x^2$ and $f(x) = [u(x)]^3 = 8x^6$. Which of the following is $f'(x)$?

(A) $3[u(x)]^2$ (B) $3[u'(x)]^2$ (C) $3[u(x)]^2u'(x)$

The calculations in Explore & Discuss 1 show that we cannot generalize the power rule simply by replacing x with $u(x)$ in equation (1).

How can we find a formula for the derivative of $[u(x)]^n$, where $u(x)$ is an arbitrary differentiable function? Let's begin by considering the derivatives of $[u(x)]^2$ and $[u(x)]^3$ to see if a general pattern emerges. Since $[u(x)]^2 = u(x)u(x)$, we use the product rule to write

$$\begin{aligned}\frac{d}{dx}[u(x)]^2 &= \frac{d}{dx}[u(x)u(x)]\\ &= u(x)u'(x) + u(x)u'(x)\\ &= 2u(x)u'(x)\end{aligned} \tag{2}$$

Because $[u(x)]^3 = [u(x)]^2u(x)$, we use the product rule and the result in equation (2) to write

$$\begin{aligned}\frac{d}{dx}[u(x)]^3 &= \frac{d}{dx}\{[u(x)]^2u(x)\}\\ &= [u(x)]^2\frac{d}{dx}u(x) + u(x)\frac{d}{dx}[u(x)]^2\\ &= [u(x)]^2u'(x) + u(x)[2u(x)u'(x)]\\ &= 3[u(x)]^2u'(x)\end{aligned}$$

Use equation (2) to substitute for $\frac{d}{dx}[u(x)]^2$.

Continuing in this fashion, we can show that

$$\frac{d}{dx}[u(x)]^n = n[u(x)]^{n-1}u'(x) \qquad n \text{ a positive integer} \tag{3}$$

Using more advanced techniques, we can establish formula (3) for all real numbers n, obtaining the **general power rule**.

THEOREM 1 General Power Rule

If $u(x)$ is a differentiable function, n is any real number, and

$$y = f(x) = [u(x)]^n$$

then

$$f'(x) = n[u(x)]^{n-1}u'(x)$$

Using simplified notation,

$$y' = nu^{n-1}u' \qquad \text{or} \qquad \frac{d}{dx}u^n = nu^{n-1}\frac{du}{dx} \qquad \text{where } u = u(x)$$

EXAMPLE 3 **Using the General Power Rule** Find the indicated derivatives:

(A) $f'(x)$ if $f(x) = (3x + 1)^4$ (B) y' if $y = (x^3 + 4)^7$

(C) $\frac{d}{dt}\frac{1}{(t^2 + t + 4)^3}$ (D) $\frac{dh}{dw}$ if $h(w) = \sqrt{3 - w}$

SOLUTION (A) $f(x) = (3x + 1)^4$ Let $u = 3x + 1$, $n = 4$.

$$f'(x) = 4(3x + 1)^3(3x + 1)' \qquad nu^{n-1}\frac{du}{dx}$$

$$= 4(3x + 1)^3\,3 \qquad \frac{du}{dx} = 3$$

$$= 12(3x + 1)^3$$

(B) $y = (x^3 + 4)^7$ Let $u = (x^3 + 4)$, $n = 7$.

$$y' = 7(x^3 + 4)^6(x^3 + 4)' \qquad nu^{n-1}\frac{du}{dx}$$

$$= 7(x^3 + 4)^6\,3x^2 \qquad \frac{du}{dx} = 3x^2$$

$$= 21x^2(x^3 + 4)^6$$

(C) $\frac{d}{dt}\frac{1}{(t^2 + t + 4)^3}$

$$= \frac{d}{dt}(t^2 + t + 4)^{-3} \qquad \text{Let } u = t^2 + t + 4,\ n = -3.$$

$$= -3(t^2 + t + 4)^{-4}(t^2 + t + 4)' \qquad nu^{n-1}\frac{du}{dt}$$

$$= -3(t^2 + t + 4)^{-4}(2t + 1) \qquad \frac{du}{dt} = 2t + 1$$

$$= \frac{-3(2t + 1)}{(t^2 + t + 4)^4}$$

(D) $h(w) = \sqrt{3 - w} = (3 - w)^{1/2}$ Let $u = 3 - w$, $n = \frac{1}{2}$.

$$\frac{dh}{dw} = \frac{1}{2}(3 - w)^{-1/2}(3 - w)' \qquad nu^{n-1}\frac{du}{dw}$$

$$= \frac{1}{2}(3 - w)^{-1/2}(-1) \qquad \frac{du}{dw} = -1$$

$$= -\frac{1}{2(3 - w)^{1/2}} \quad \text{or} \quad -\frac{1}{2\sqrt{3 - w}}$$

Matched Problem 3 Find the indicated derivatives:

(A) $h'(x)$ if $h(x) = (5x + 2)^3$ (B) y' if $y = (x^4 - 5)^5$

(C) $\frac{d}{dt}\frac{1}{(t^2 + 4)^2}$ (D) $\frac{dg}{dw}$ if $g(w) = \sqrt{4 - w}$

Notice that we used two steps to differentiate each function in Example 3. First, we applied the general power rule, and then we found du/dx. As you gain experience with the general power rule, you may want to combine these two steps. If you do this, be certain to multiply by du/dx. For example,

$$\frac{d}{dx}(x^5 + 1)^4 = 4(x^5 + 1)^3 5x^4 \qquad \text{Correct}$$

$$\frac{d}{dx}(x^5 + 1)^4 \neq 4(x^5 + 1)^3 \qquad du/dx = 5x^4 \text{ is missing}$$

CONCEPTUAL INSIGHT

If we let $u(x) = x$, then $du/dx = 1$, and the general power rule reduces to the (ordinary) power rule discussed in Section 1-5. Compare the following:

$$\frac{d}{dx}x^n = nx^{n-1} \qquad \text{Yes—power rule}$$

$$\frac{d}{dx}u^n = nu^{n-1}\frac{du}{dx} \qquad \text{Yes—general power rule}$$

$$\frac{d}{dx}u^n \neq nu^{n-1} \qquad \text{Unless } u(x) = x + k, \text{ so that } du/dx = 1$$

The Chain Rule

We have used the general power rule to find derivatives of composite functions of the form $f(g(x))$, where $f(u) = u^n$ is a power function. But what if f is not a power function? Then a more general rule, the *chain rule,* enables us to compute the derivatives of many composite functions of the form $f(g(x))$.

Suppose that

$$y = m(x) = f[g(x)]$$

is a composite of f and g, where

$$y = f(u) \qquad \text{and} \qquad u = g(x)$$

To express the derivative dy/dx in terms of the derivatives of f and g, we use the definition of a derivative (see Section 1-4).

$$\frac{dy}{dx} = \lim_{h\to 0}\frac{m(x+h) - m(x)}{h} \qquad \text{Substitute } m(x+h) = f[g(x+h)] \text{ and } m(x) = f[g(x)].$$

$$= \lim_{h\to 0}\frac{f[g(x+h)] - f[g(x)]}{h} \qquad \text{Multiply by } 1 = \frac{g(x+h) - g(x)}{g(x+h) - g(x)}.$$

$$= \lim_{h\to 0}\left[\frac{f[g(x+h)] - f[g(x)]}{h}\cdot\frac{g(x+h) - g(x)}{g(x+h) - g(x)}\right]$$

$$= \lim_{h\to 0}\left[\frac{f[g(x+h)] - f[g(x)]}{g(x+h) - g(x)}\cdot\frac{g(x+h) - g(x)}{h}\right] \qquad (4)$$

We recognize the second factor in equation (4) as the difference quotient for $g(x)$. To interpret the first factor as the difference quotient for $f(u)$, we let $k = g(x+h) - g(x)$. Since $u = g(x)$, we write

$$u + k = g(x) + g(x+h) - g(x) = g(x+h)$$

Substituting in equation (1), we have

$$\frac{dy}{dx} = \lim_{h\to 0}\left[\frac{f(u+k) - f(u)}{k}\cdot\frac{g(x+h) - g(x)}{h}\right] \qquad (5)$$

If we assume that $k = [g(x+h) - g(x)] \to 0$ as $h \to 0$, we can find the limit of each difference quotient in equation (5):

$$\frac{dy}{dx} = \left[\lim_{k\to 0}\frac{f(u+k) - f(u)}{k}\right]\left[\lim_{h\to 0}\frac{g(x+h) - g(x)}{h}\right]$$

$$= f'(u)g'(x)$$

$$= \frac{dy}{du}\frac{du}{dx}$$

This result is correct under general conditions and is called the *chain rule*, but our "derivation" is superficial because it ignores some hidden problems. Since a formal proof of the chain rule is beyond the scope of this book, we simply state it as follows:

THEOREM 2 Chain Rule

If $y = f(u)$ and $u = g(x)$ define the composite function

$$y = m(x) = f[g(x)]$$

then

$$\frac{dy}{dx} = \frac{dy}{du}\frac{du}{dx} \quad \text{provided that } \frac{dy}{du} \text{ and } \frac{du}{dx} \text{ exist}$$

or, equivalently,

$$m'(x) = f'[g(x)]g'(x) \quad \text{provided that } f'[g(x)] \text{ and } g'(x) \text{ exist}$$

EXAMPLE 4 **Using the Chain Rule** Find dy/du, du/dx, and dy/dx (express dy/dx as a function of x) for

(A) $y = u^{3/2}$ and $u = 3x^2 + 1$

(B) $y = e^u$ and $u = 2x^3 + 5$

(C) $y = \ln u$ and $u = x^2 - 4x + 2$

SOLUTION

(A) $\frac{dy}{du} = \frac{3}{2}u^{1/2}$ and $\frac{du}{dx} = 6x$ — Basic derivative rules

$\frac{dy}{dx} = \frac{dy}{du}\frac{du}{dx}$ — Chain rule

$= \frac{3}{2}u^{1/2}(6x) = 9x(3x^2 + 1)^{1/2}$ — Since $u = 3x^2 + 1$

(B) $\frac{dy}{du} = e^u$ and $\frac{du}{dx} = 6x^2$ — Basic derivative rules

$\frac{dy}{dx} = \frac{dy}{du}\frac{du}{dx}$ — Chain rule

$= e^u(6x^2) = 6x^2e^{2x^3+5}$ — Since $u = 2x^3 + 5$

(C) $\frac{dy}{du} = \frac{1}{u}$ and $\frac{du}{dx} = 2x - 4$ — Basic derivative rules

$\frac{dy}{dx} = \frac{dy}{du}\frac{du}{dx}$ — Chain rule

$= \frac{1}{u}(2x - 4) = \frac{2x - 4}{x^2 - 4x + 2}$ — Since $u = x^2 - 4x + 2$

Matched Problem 4 Find dy/du, du/dx, and dy/dx (express dy/dx as a function of x) for

(A) $y = u^{-5}$ and $u = 2x^3 + 4$

(B) $y = e^u$ and $u = 3x^4 + 6$

(C) $y = \ln u$ and $u = x^2 + 9x + 4$

EXPLORE & DISCUSS 2

Let $m(x) = f[g(x)]$. Use the chain rule and Figures 1 and 2 to find

(A) $f(4)$ (B) $g(6)$ (C) $m(6)$

(D) $f'(4)$ (E) $g'(6)$ (F) $m'(6)$

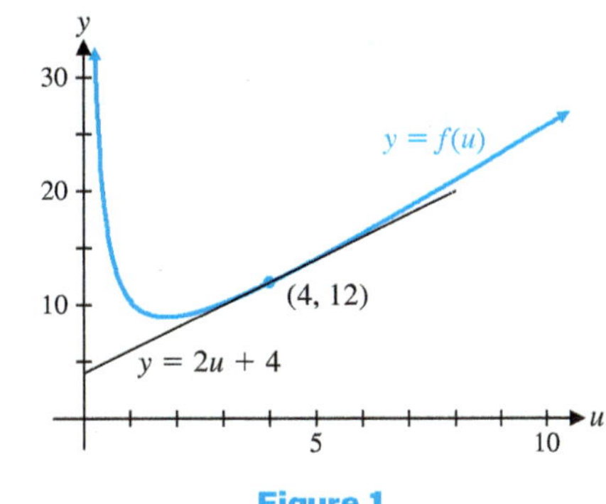

Figure 1

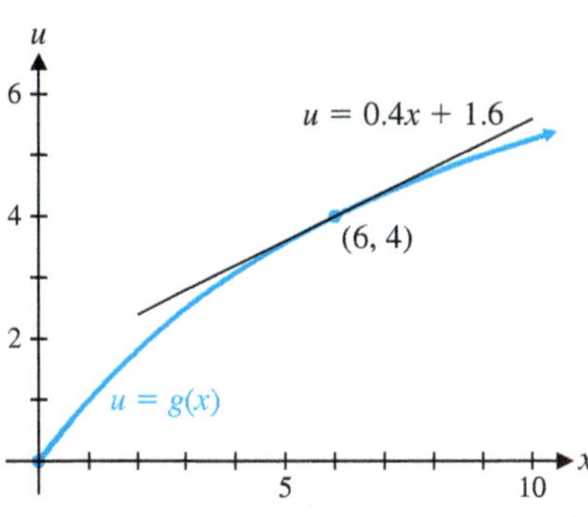

Figure 2

The chain rule can be extended to compositions of three or more functions. For example, if $y = f(w)$, $w = g(u)$, and $u = h(x)$, then

$$\frac{dy}{dx} = \frac{dy}{dw}\frac{dw}{du}\frac{du}{dx}$$

EXAMPLE 5 **Using the Chain Rule** For $y = h(x) = e^{1+(\ln x)^2}$, find dy/dx.

SOLUTION Note that h is of the form $y = e^w$, where $w = 1 + u^2$ and $u = \ln x$.

$$\begin{aligned}\frac{dy}{dx} &= \frac{dy}{dw}\frac{dw}{du}\frac{du}{dx}\\ &= e^w(2u)\left(\frac{1}{x}\right)\\ &= e^{1+u^2}(2u)\left(\frac{1}{x}\right) && \text{Since } w = 1 + u^2\\ &= e^{1+(\ln x)^2}(2\ln x)\left(\frac{1}{x}\right) && \text{Since } u = \ln x\\ &= \frac{2}{x}(\ln x)e^{1+(\ln x)^2}\end{aligned}$$

Matched Problem 5 For $y = h(x) = [\ln(1 + e^x)]^3$, find dy/dx.

The chain rule generalizes basic derivative rules. We list three general derivative rules here for convenient reference [the first, equation (6), is the general power rule of Theorem 1].

General Derivative Rules

$$\frac{d}{dx}[f(x)]^n = n[f(x)]^{n-1}f'(x) \tag{6}$$

$$\frac{d}{dx}\ln[f(x)] = \frac{1}{f(x)}f'(x) \tag{7}$$

$$\frac{d}{dx}e^{f(x)} = e^{f(x)}f'(x) \tag{8}$$

Unless directed otherwise, you now have a choice between the chain rule and the general derivative rules. However, practicing with the chain rule will help prepare you for concepts that appear later in the text. Examples 4 and 5 illustrate the chain rule method, and the next example illustrates the general derivative rules method.

EXAMPLE 6 **Using General Derivative Rules**

(A) $\frac{d}{dx}e^{2x} = e^{2x}\frac{d}{dx}2x$ Using equation (5)

$= e^{2x}(2) = 2e^{2x}$

(B) $\frac{d}{dx}\ln(x^2 + 9) = \frac{1}{x^2 + 9}\frac{d}{dx}(x^2 + 9)$ Using equation (4)

$= \frac{1}{x^2 + 9}2x = \frac{2x}{x^2 + 9}$

(C) $\frac{d}{dx}(1 + e^{x^2})^3 = 3(1 + e^{x^2})^2\frac{d}{dx}(1 + e^{x^2})$ Using equation (3)

$= 3(1 + e^{x^2})^2 e^{x^2}\frac{d}{dx}x^2$ Using equation (5)

$= 3(1 + e^{x^2})^2 e^{x^2}(2x)$

$= 6xe^{x^2}(1 + e^{x^2})^2$

Matched Problem 6 Find

(A) $\frac{d}{dx}\ln(x^3 + 2x)$ (B) $\frac{d}{dx}e^{3x^2+2}$ (C) $\frac{d}{dx}(2 + e^{-x^2})^4$

Exercises 2-4

For many of the problems in this exercise set, the answers in the back of the book include both an unsimplified form and a simplified form. When checking your work, first check that you applied the rules correctly, and then check that you performed the algebraic simplification correctly.

A

In Problems 1–4, find f[g(x)].

1. $f(u) = u^3;\ \ g(x) = 3x^2 + 2$
2. $f(u) = u^4;\ \ g(x) = 1 - 4x^3$
3. $f(u) = e^u;\ \ g(x) = -x^2$
4. $g(u) = e^u;\ \ g(x) = 3x^3$

In Problems 5–8, write each composite function in the form $y = f(u)$ and $u = g(x)$.

5. $y = (3x^2 - x + 5)^4$
6. $y = (2x^3 + x + 3)^5$
7. $y = e^{1+x+x^2}$
8. $y = e^{x^4+2x^2+5}$

In Problems 9–16, replace ? with an expression that will make the indicated equation valid.

9. $\frac{d}{dx}(3x + 4)^4 = 4(3x + 4)^3\underline{\ \ ?\ \ }$
10. $\frac{d}{dx}(5 - 2x)^6 = 6(5 - 2x)^5\underline{\ \ ?\ \ }$
11. $\frac{d}{dx}(4 - 2x^2)^3 = 3(4 - 2x^2)^2\underline{\ \ ?\ \ }$
12. $\frac{d}{dx}(3x^2 + 7)^5 = 5(3x^2 + 7)^4\underline{\ \ ?\ \ }$
13. $\frac{d}{dx}e^{x^2+1} = e^{x^2+1}\underline{\ \ ?\ \ }$
14. $\frac{d}{dx}e^{4x-2} = e^{4x-2}\underline{\ \ ?\ \ }$
15. $\frac{d}{dx}\ln(x^4 + 1) = \frac{1}{x^4 + 1}\underline{\ \ ?\ \ }$
16. $\frac{d}{dx}\ln(x - x^3) = \frac{1}{x - x^3}\underline{\ \ ?\ \ }$

In Problems 17–42, find $f'(x)$ and simplify.

17. $f(x) = (x + 3)^2$
18. $f(x) = (x - 6)^3$
19. $f(x) = (2x + 5)^3$
20. $f(x) = (3x - 7)^5$
21. $f(x) = (5 - 2x)^4$
22. $f(x) = (9 - 5x)^2$
23. $f(x) = (4 + 0.2x)^5$
24. $f(x) = (6 - 0.5x)^4$
25. $f(x) = (3x^2 + 5)^5$
26. $f(x) = (5x^2 - 3)^6$
27. $f(x) = 5e^x$
28. $f(x) = 10 - 4e^x$
29. $f(x) = e^{5x}$
30. $f(x) = 6e^{-2x}$
31. $f(x) = 3e^{-6x}$
32. $f(x) = e^{x^2 + 3x + 1}$
33. $f(x) = (2x - 5)^{1/2}$
34. $f(x) = (4x + 3)^{1/2}$
35. $f(x) = (x^4 + 1)^{-2}$
36. $f(x) = (x^5 + 2)^{-3}$
37. $f(x) = 4 - 2 \ln x$
38. $f(x) = 8 \ln x$
39. $f(x) = 3 \ln(1 + x^2)$
40. $f(x) = 2 \ln(x^2 - 3x + 4)$
41. $f(x) = (1 + \ln x)^3$
42. $f(x) = (x - 2 \ln x)^4$

In Problems 43–48, find $f'(x)$ and the equation of the line tangent to the graph of f at the indicated value of x. Find the value(s) of x where the tangent line is horizontal.

43. $f(x) = (2x - 1)^3;\ x = 1$
44. $f(x) = (3x - 1)^4;\ x = 1$
45. $f(x) = (4x - 3)^{1/2};\ x = 3$
46. $f(x) = (2x + 8)^{1/2};\ x = 4$
47. $f(x) = 5e^{x^2-4x+1};\ x = 0$
48. $f(x) = \ln(1 - x^2 + 2x^4);\ x = 1$

B

In Problems 49–64, find the indicated derivative and simplify.

49. y' if $y = 3(x^2 - 2)^4$
50. y' if $y = 2(x^3 + 6)^5$
51. $\frac{d}{dt}2(t^2 + 3t)^{-3}$
52. $\frac{d}{dt}3(t^3 + t^2)^{-2}$
53. $\frac{dh}{dw}$ if $h(w) = \sqrt{w^2 + 8}$
54. $\frac{dg}{dw}$ if $g(w) = \sqrt[3]{3w - 7}$
55. $g'(x)$ if $g(x) = 4xe^{3x}$
56. $h'(x)$ if $h(x) = \frac{e^{2x}}{x^2 + 9}$
57. $\frac{d}{dx}\frac{\ln(1 + x)}{x^3}$
58. $\frac{d}{dx}[x^4 \ln(1 + x^4)]$
59. $F'(t)$ if $F(t) = (e^{t^2+1})^3$
60. $G'(t)$ if $G(t) = (1 - e^{2t})^2$
61. y' if $y = \ln(x^2 + 3)^{3/2}$
62. y' if $y = [\ln(x^2 + 3)]^{3/2}$
63. $\frac{d}{dw}\frac{1}{(w^3 + 4)^5}$
64. $\frac{d}{dw}\frac{1}{(w^2 - 2)^6}$

In Problems 65–70, find $f'(x)$ and find the equation of the line tangent to the graph of f at the indicated value of x.

65. $f(x) = x(4 - x)^3;\ x = 2$
66. $f(x) = x^2(1 - x)^4;\ x = 2$
67. $f(x) = \frac{x}{(2x - 5)^3};\ x = 3$
68. $f(x) = \frac{x^4}{(3x - 8)^2};\ x = 4$
69. $f(x) = \sqrt{\ln x};\ x = e$
70. $f(x) = e^{\sqrt{x}};\ x = 1$

In Problems 71–76, find $f'(x)$ and find the value(s) of x where the tangent line is horizontal.

71. $f(x) = x^2(x - 5)^3$
72. $f(x) = x^3(x - 7)^4$
73. $f(x) = \frac{x}{(2x + 5)^2}$
74. $f(x) = \frac{x - 1}{(x - 3)^3}$
75. $f(x) = \sqrt{x^2 - 8x + 20}$
76. $f(x) = \sqrt{x^2 + 4x + 5}$

77. A student reasons that the functions $f(x) = \ln[5(x^2 + 3)^4]$ and $g(x) = 4 \ln(x^2 + 3)$ must have the same derivative since he has entered $f(x)$, $g(x)$, $f'(x)$, and $g'(x)$ into a graphing calculator, but only three graphs appear (see the figure). Is his reasoning correct? Are $f'(x)$ and $g'(x)$ the same function? Explain.

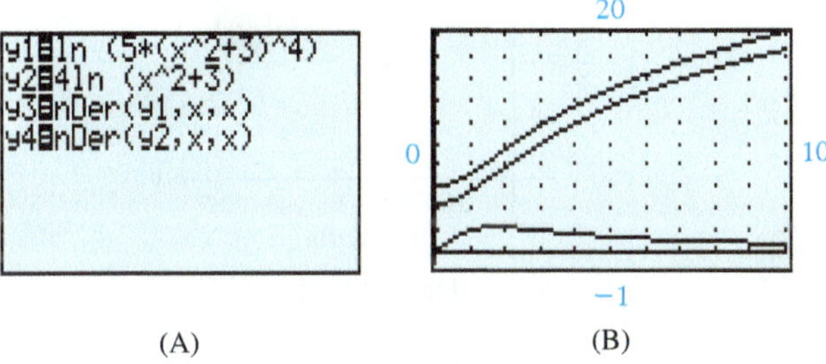

Figure for 77

78. A student reasons that the functions $f(x) = (x + 1) \ln(x + 1) - x$ and $g(x) = (x + 1)^{1/3}$ must have the same derivative since she has entered $f(x)$, $g(x)$, $f'(x)$, and $g'(x)$ into a graphing calculator, but only three graphs appear (see the figure). Is her reasoning correct? Are $f'(x)$ and $g'(x)$ the same function? Explain.

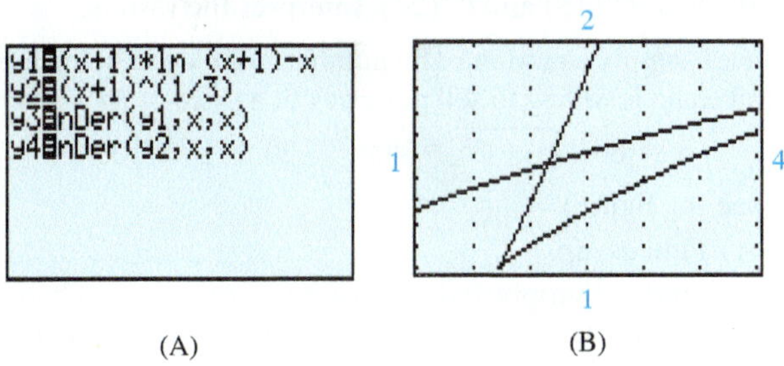

Figure for 78

C

In Problems 79–90, find each derivative and simplify.

79. $\frac{d}{dx}[3x(x^2 + 1)^3]$

80. $\frac{d}{dx}[2x^2(x^3 - 3)^4]$

81. $\frac{d}{dx}\frac{(x^3 - 7)^4}{2x^3}$

82. $\frac{d}{dx}\frac{3x^2}{(x^2 + 5)^3}$

83. $\frac{d}{dx}\log_2(3x^2 - 1)$

84. $\frac{d}{dx}\log(x^3 - 1)$

85. $\frac{d}{dx}10^{x^2+x}$

86. $\frac{d}{dx}8^{1-2x^2}$

87. $\frac{d}{dx}\log_3(4x^3 + 5x + 7)$

88. $\frac{d}{dx}\log_5(5^{x^2-1})$

89. $\frac{d}{dx}2^{x^3-x^2+4x+1}$

90. $\frac{d}{dx}10^{\ln x}$

Applications

91. **Cost function.** The total cost (in hundreds of dollars) of producing x cell phones per day is

$$C(x) = 10 + \sqrt{2x + 16} \qquad 0 \le x \le 50$$

(see the figure).

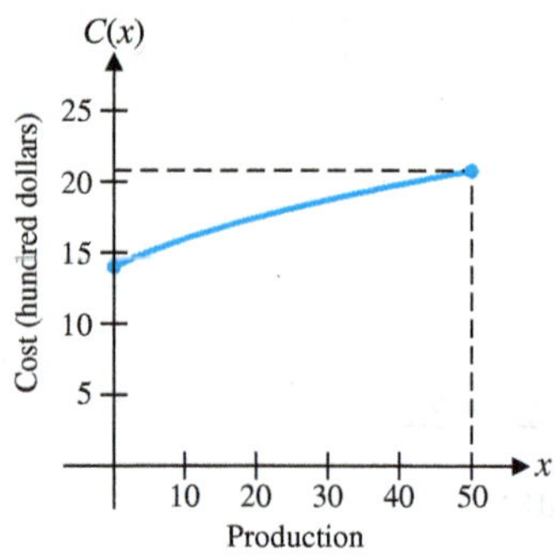

Figure for 91

(A) Find $C'(x)$.

(B) Find $C'(24)$ and $C'(42)$. Interpret the results.

92. **Cost function.** The total cost (in hundreds of dollars) of producing x cameras per week is

$$C(x) = 6 + \sqrt{4x + 4} \qquad 0 \le x \le 30$$

(A) Find $C'(x)$.

(B) Find $C'(15)$ and $C'(24)$. Interpret the results.

93. **Price–supply equation.** The number x of bicycle helmets a retail chain is willing to sell per week at a price of $\$p$ is given by

$$x = 80\sqrt{p + 25} - 400 \qquad 20 \le p \le 100$$

(see the figure).

(A) Find dx/dp.

(B) Find the supply and the instantaneous rate of change of supply with respect to price when the price is \$75. Write a brief interpretation of these results.

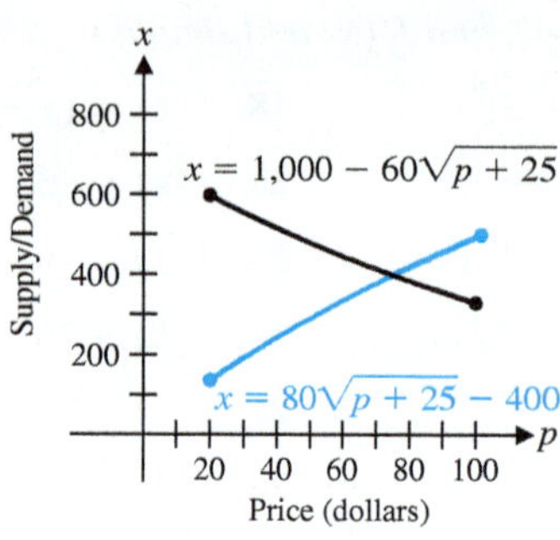

Figure for 93 and 94

94. **Price–demand equation.** The number x of bicycle helmets people are willing to buy per week from a retail chain at a price of $\$p$ is given by

$$x = 1{,}000 - 60\sqrt{p + 25} \qquad 20 \le p \le 100$$

(see the figure).

(A) Find dx/dp.

(B) Find the demand and the instantaneous rate of change of demand with respect to price when the price is \$75. Write a brief interpretation of these results.

95. **Drug concentration.** The drug concentration in the bloodstream t hours after injection is given approximately by

$$C(t) = 4.35e^{-t} \qquad 0 \le t \le 5$$

where $C(t)$ is concentration in milligrams per milliliter.

(A) What is the rate of change of concentration after 1 hour? After 4 hours?

(B) Graph C.

96. **Water pollution.** The use of iodine crystals is a popular way of making small quantities of water safe to drink. Crystals placed in a 1-ounce bottle of water will dissolve until the solution is saturated. After saturation, half of the solution is poured into a quart container of water, and after about an hour, the water is usually safe to drink. The half-empty 1-ounce bottle is then refilled, to be used again in the same way. Suppose that the concentration of iodine in the 1-ounce bottle t minutes after the crystals are introduced can be approximated by

$$C(t) = 250(1 - e^{-t}) \qquad t \ge 0$$

where $C(t)$ is the concentration of iodine in micrograms per milliliter.

(A) What is the rate of change of the concentration after 1 minute? After 4 minutes?

(B) Graph C for $0 \le t \le 5$.

97. **Blood pressure and age.** A research group using hospital records developed the following mathematical model relating systolic blood pressure and age:

$$P(x) = 40 + 25\ln(x + 1) \qquad 0 \le x \le 65$$

$P(x)$ is pressure, measured in millimeters of mercury, and x is age in years. What is the rate of change of pressure at the end of 10 years? At the end of 30 years? At the end of 60 years?

98. **Biology.** A yeast culture at room temperature (68°F) is placed in a refrigerator set at a constant temperature of 38°F. After t hours, the temperature T of the culture is given approximately by

$$T = 30e^{-0.58t} + 38 \qquad t \geq 0$$

What is the rate of change of temperature of the culture at the end of 1 hour? At the end of 4 hours?

99. **Learning.** In 1930, L. L. Thurstone developed the following formula to indicate how learning time T depends on the length of a list n:

$$T = f(n) = \frac{c}{k}n\sqrt{n - a}$$

Here, a, c, and k are empirical constants. Suppose that, for a particular person, the time T (in minutes) required to learn a list of length n is

$$T = f(n) = 2n\sqrt{n - 2}$$

(A) Find dT/dn.

(B) Find $f'(11)$ and $f'(27)$. Interpret the results.

Answers to Matched Problems

1. $f[g(x)] = 2e^x$, $g[f(u)] = e^{2u}$
2. (A) $f(u) = 50e^u$, $u = -2x$
   (B) $f(u) = \sqrt[3]{u}$, $u = 1 + x^3$
   [*Note:* There are other correct answers.]
3. (A) $15(5x + 2)^2$
   (B) $20x^3(x^4 - 5)^4$
   (C) $-4t/(t^2 + 4)^3$
   (D) $-1/(2\sqrt{4 - w})$
4. (A) $\frac{dy}{du} = -5u^{-4}, \frac{du}{dx} = 6x^2, \frac{dy}{dx} = -30x^2(2x^3 + 4)^{-6}$
   (B) $\frac{dy}{du} = e^u, \frac{du}{dx} = 12x^3, \frac{dy}{dx} = 12x^3e^{3x^4+6}$
   (C) $\frac{dy}{du} = \frac{1}{u}, \frac{du}{dx} = 2x + 9, \frac{dy}{dx} = \frac{2x + 9}{x^2 + 9x + 4}$
5. $\frac{3e^x[\ln(1 + e^x)]^2}{1 + e^x}$
6. (A) $\frac{3x^2 + 2}{x^3 + 2x}$ (B) $6xe^{3x^2+2}$ (C) $-8xe^{-x^2}(2 + e^{-x^2})^3$

2-5 Implicit Differentiation

- Special Function Notation
- Implicit Differentiation

Special Function Notation

The equation

$$y = 2 - 3x^2 \tag{1}$$

defines a function f with y as a dependent variable and x as an independent variable. Using function notation, we would write

$$y = f(x) \qquad \text{or} \qquad f(x) = 2 - 3x^2$$

In order to minimize the number of symbols, we will often write equation (1) in the form

$$y = 2 - 3x^2 = y(x)$$

where y is *both* a dependent variable and a function symbol. This is a convenient notation, and no harm is done as long as one is aware of the double role of y. Other examples are

$$x = 2t^2 - 3t + 1 = x(t)$$
$$z = \sqrt{u^2 - 3u} = z(u)$$
$$r = \frac{1}{(s^2 - 3s)^{2/3}} = r(s)$$

Until now, we have considered functions involving only one independent variable. There is no reason to stop there: The concept can be generalized to functions involving two or more independent variables, and this will be done in detail in

Chapter 6. For now, we will "borrow" the notation for a function involving two independent variables. For example,

$$F(x, y) = x^2 - 2xy + 3y^2 - 5$$

specifies a function F involving two independent variables.

Implicit Differentiation

Consider the equation

$$3x^2 + y - 2 = 0 \qquad (2)$$

and the equation obtained by solving equation (2) for y in terms of x,

$$y = 2 - 3x^2 \qquad (3)$$

Both equations define the same function with x as the independent variable and y as the dependent variable. For equation (3), we write

$$y = f(x)$$

where

$$f(x) = 2 - 3x^2 \qquad (4)$$

and we have an **explicit** (directly stated) rule that enables us to determine y for each value of x. On the other hand, the y in equation (2) is the same y as in equation (3), and equation (2) **implicitly** gives (implies, though does not directly express) y as a function of x. We say that equations (3) and (4) define the function f explicitly and equation (2) defines f implicitly.

The direct use of an equation that defines a function implicitly to find the derivative of the dependent variable with respect to the independent variable is called **implicit differentiation**. Let's differentiate equation (2) implicitly and equation (3) directly, and compare results.

Starting with

$$3x^2 + y - 2 = 0$$

we think of y as a function of x and write

$$3x^2 + y(x) - 2 = 0$$

Then we differentiate both sides with respect to x:

$$\frac{d}{dx}[(3x^2 + y(x) - 2)] = \frac{d}{dx}0$$

$$\frac{d}{dx}3x^2 + \frac{d}{dx}y(x) - \frac{d}{dx}2 = 0$$

$$6x + y' - 0 = 0$$

Since y is a function of x, but is not explicitly given, we simply write $\frac{d}{dx}y(x) = y'$ to indicate its derivative.

Now we solve for y':

$$y' = -6x$$

Note that we get the same result if we start with equation (3) and differentiate directly:

$$y = 2 - 3x^2$$
$$y' = -6x$$

Why are we interested in implicit differentiation? Why not solve for y in terms of x and differentiate directly? The answer is that there are many equations of the form

$$F(x, y) = 0 \qquad (5)$$

that are either difficult or impossible to solve for y explicitly in terms of x (try it for $x^2y^5 - 3xy + 5 = 0$ or for $e^y - y = 3x$, for example). But it can be shown that, under fairly general conditions on F, equation (5) will define one or more functions in which y is a dependent variable and x is an independent variable. To find y' under these conditions, we differentiate equation (5) implicitly.

EXPLORE & DISCUSS 1

(A) How many tangent lines are there to the graph in Figure 1 when $x = 0$? When $x = 1$? When $x = 2$? When $x = 4$? When $x = 6$?

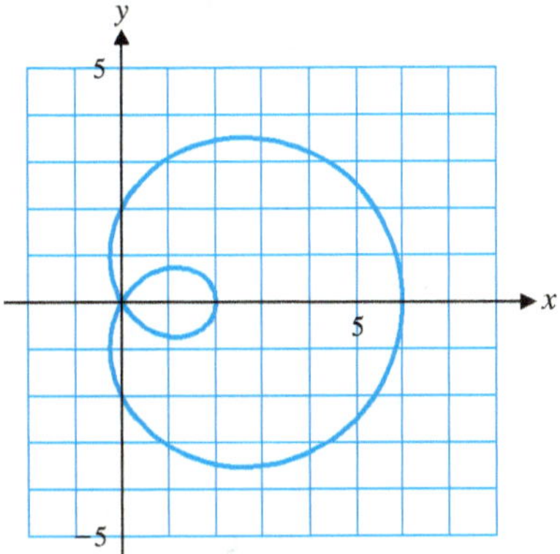

Figure 1

(B) Sketch the tangent lines referred to in part (A), and estimate each of their slopes.

(C) Explain why the graph in Figure 1 is not the graph of a function.

EXAMPLE 1 **Differentiating Implicitly** Given

$$F(x, y) = x^2 + y^2 - 25 = 0 \quad (6)$$

find y' and the slope of the graph at $x = 3$.

SOLUTION We start with the graph of $x^2 + y^2 - 25 = 0$ (a circle, as shown in Fig. 2) so that we can interpret our results geometrically. From the graph, it is clear that equation (6) does not define a function. But with a suitable restriction on the variables, equation (6) can define two or more functions. For example, the upper half and the lower half of the circle each define a function. On each half-circle, a point that corresponds to $x = 3$ is found by substituting $x = 3$ into equation (6) and solving for y:

$$\begin{aligned} x^2 + y^2 - 25 &= 0 \\ (3)^2 + y^2 &= 25 \\ y^2 &= 16 \\ y &= \pm 4 \end{aligned}$$

The point $(3, 4)$ is on the upper half-circle, and the point $(3, -4)$ is on the lower half-circle. We will use these results in a moment. We now differentiate equation (6) implicitly, treating y as a function of x [i.e., $y = y(x)$]:

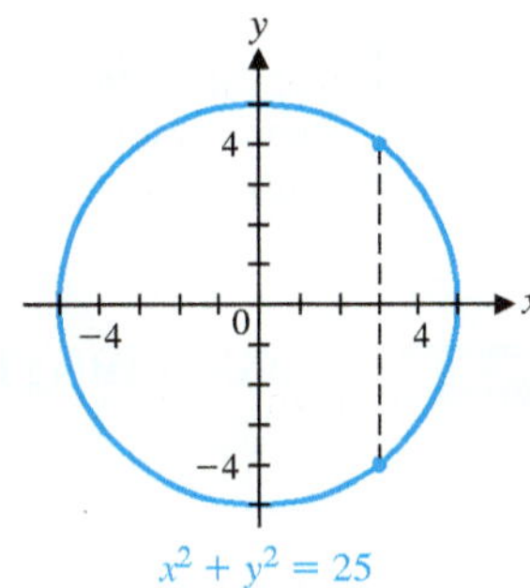

Figure 2

$$x^2 + y^2 - 25 = 0$$

$$x^2 + [y(x)]^2 - 25 = 0$$

$$\frac{d}{dx}\{x^2 + [y(x)]^2 - 25\} = \frac{d}{dx}0$$

$$\frac{d}{dx}x^2 + \frac{d}{dx}[y(x)]^2 - \frac{d}{dx}25 = 0 \quad \text{Use the chain rule.}$$

$$2x + 2[y(x)]^{2-1}y'(x) - 0 = 0$$

$$2x + 2yy' = 0 \quad \text{Solve for } y' \text{ in terms of } x \text{ and } y.$$

$$y' = -\frac{2x}{2y}$$

$$y' = -\frac{x}{y} \quad \text{Leave the answer in terms of } x \text{ and } y.$$

We have found y' without first solving $x^2 + y^2 - 25 = 0$ for y in terms of x. And by leaving y' in terms of x and y, we can use $y' = -x/y$ to find y' for *any* point on the graph of $x^2 + y^2 - 25 = 0$ (except where $y = 0$). In particular, for $x = 3$, we found that (3, 4) and (3, −4) are on the graph. The slope of the graph at (3, 4) is

$$y'|_{(3,4)} = -\tfrac{3}{4} \quad \text{The slope of the graph at (3, 4)}$$

and the slope at (3, −4) is

$$y'|_{(3,-4)} = -\tfrac{3}{-4} = \tfrac{3}{4} \quad \text{The slope of the graph at (3, −4)}$$

The symbol

$$y'|_{(a,b)}$$

is used to indicate that we are evaluating y' at $x = a$ and $y = b$.

The results are interpreted geometrically in Figure 3 on the original graph.

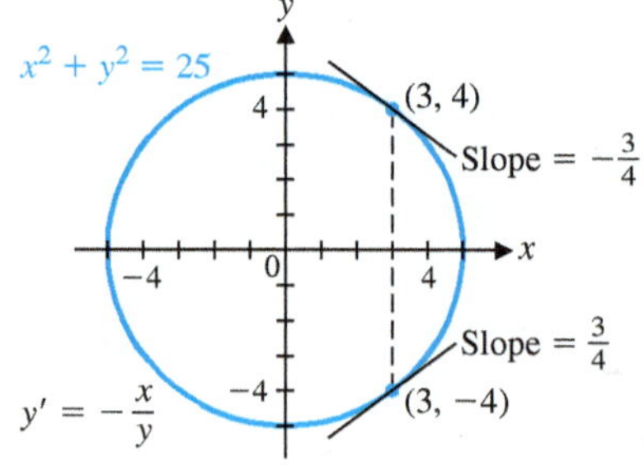

Figure 3

Matched Problem 1 Graph $x^2 + y^2 - 169 = 0$, find y' by implicit differentiation, and find the slope of the graph when $x = 5$.

CONCEPTUAL INSIGHT

When differentiating implicitly, the derivative of y^2 is $2yy'$, not just $2y$. This is because y represents a function of x, so the chain rule applies. Suppose, for example, that y represents the function $y = 5x + 4$. Then

$$(y^2)' = [(5x + 4)^2]' = 2(5x + 4) \cdot 5 = 2yy'$$

So, when differenting implicitly, the derivative of y is y', the derivative of y^2 is $2yy'$, the derivative of y^3 is $3y^2y'$, and so on.

EXAMPLE 2 **Differentiating Implicitly** Find the equation(s) of the tangent line(s) to the graph of

$$y - xy^2 + x^2 + 1 = 0 \qquad (7)$$

at the point(s) where $x = 1$.

SOLUTION We first find y when $x = 1$:

$$\begin{aligned} y - xy^2 + x^2 + 1 &= 0 \\ y - (1)y^2 + (1)^2 + 1 &= 0 \\ y - y^2 + 2 &= 0 \\ y^2 - y - 2 &= 0 \\ (y - 2)(y + 1) &= 0 \\ y &= -1, 2 \end{aligned}$$

So there are two points on the graph of (7) where $x = 1$, namely, $(1, -1)$and $(1, 2)$. We next find the slope of the graph at these two points by differentiating equation (7) implicitly:

$$y - xy^2 + x^2 + 1 = 0$$

$$\frac{d}{dx}y - \frac{d}{dx}xy^2 + \frac{d}{dx}x^2 + \frac{d}{dx}1 = \frac{d}{dx}0$$

Use the product rule and the chain rule for $\frac{d}{dx}xy^2$.

$$y' - (x \cdot 2yy' + y^2) + 2x = 0$$

$$y' - 2xyy' - y^2 + 2x = 0$$

Solve for y' by getting all terms involving y' on one side.

$$\begin{aligned} y' - 2xyy' &= y^2 - 2x \\ (1 - 2xy)y' &= y^2 - 2x \\ y' &= \frac{y^2 - 2x}{1 - 2xy} \end{aligned}$$

Now find the slope at each point:

$$y'|_{(1,-1)} = \frac{(-1)^2 - 2(1)}{1 - 2(1)(-1)} = \frac{1 - 2}{1 + 2} = \frac{-1}{3} = -\frac{1}{3}$$

$$y'|_{(1,2)} = \frac{(2)^2 - 2(1)}{1 - 2(1)(2)} = \frac{4 - 2}{1 - 4} = \frac{2}{-3} = -\frac{2}{3}$$

Equation of tangent line at $(1, -1)$:

$$\begin{aligned} y - y_1 &= m(x - x_1) \\ y + 1 &= -\tfrac{1}{3}(x - 1) \\ y + 1 &= -\tfrac{1}{3}x + \tfrac{1}{3} \\ y &= -\tfrac{1}{3}x - \tfrac{2}{3} \end{aligned}$$

Equation of tangent line at $(1, 2)$:

$$\begin{aligned} y - y_1 &= m(x - x_1) \\ y - 2 &= -\tfrac{2}{3}(x - 1) \\ y - 2 &= -\tfrac{2}{3}x + \tfrac{2}{3} \\ y &= -\tfrac{2}{3}x + \tfrac{8}{3} \end{aligned}$$

Matched Problem 2 Repeat Example 2 for $x^2 + y^2 - xy - 7 = 0$ at $x = 1$.

EXAMPLE 3 **Differentiating Implicitly** Find x' for $x = x(t)$ defined implicitly by

$$t \ln x = xe^t - 1$$

and evaluate x' at $(t, x) = (0, 1)$.

SOLUTION It is important to remember that x is the dependent variable and t is the independent variable. Therefore, we differentiate both sides of the equation with respect to t (using product and chain rules where appropriate) and then solve for x':

$$t \ln x = xe^t - 1 \qquad \text{Differentiate implicitly with respect to } t.$$

$$\frac{d}{dt}(t \ln x) = \frac{d}{dt}(xe^t) - \frac{d}{dt}1$$

$$t\frac{x'}{x} + \ln x = xe^t + x'e^t \qquad \text{Clear fractions.}$$

$$\mathbf{x} \cdot t\frac{x'}{x} + \mathbf{x} \cdot \ln x = \mathbf{x} \cdot xe^t + \mathbf{x} \cdot e^t x' \qquad x \neq 0$$

$$tx' + x \ln x = x^2e^t + xe^tx' \qquad \text{Solve for } x'.$$

$$tx' - xe^tx' = x^2e^t - x \ln x \qquad \text{Factor out } x'.$$

$$(t - xe^t)x' = x^2e^t - x \ln x$$

$$x' = \frac{x^2e^t - x \ln x}{t - xe^t}$$

Now we evaluate x' at $(t, x) = (0, 1)$, as requested:

$$x'|_{(0,1)} = \frac{(1)^2e^0 - 1 \ln 1}{0 - 1e^0} = \frac{1}{-1} = -1$$

Matched Problem 3 Find x' for $x = x(t)$ defined implicitly by

$$1 + x \ln t = te^x$$

and evaluate x' at $(t, x) = (1, 0)$.

Exercises 2-5

A

In Problems 1–4, find y′ in two ways:

(A) Differentiate the given equation implicitly and then solve for y′.

(B) Solve the given equation for y and then differentiate directly.

1. $3x + 5y + 9 = 0$

2. $-2x + 6y - 4 = 0$

3. $3x^2 - 4y - 18 = 0$

4. $2x^3 + 5y - 2 = 0$

In Problems 5–22, use implicit differentiation to find y′ and evaluate y′ at the indicated point.

5. $y - 5x^2 + 3 = 0; (1, 2)$

6. $5x^3 - y - 1 = 0; (1, 4)$

7. $x^2 - y^3 - 3 = 0; (2, 1)$

8. $y^2 + x^3 + 4 = 0; (-2, 2)$

9. $y^2 + 2y + 3x = 0; (-1, 1)$

10. $y^2 - y - 4x = 0; (0, 1)$

B

11. $xy - 6 = 0; (2, 3)$

12. $3xy - 2x - 2 = 0; (2, 1)$

13. $2xy + y + 2 = 0; (-1, 2)$

14. $2y + xy - 1 = 0; (-1, 1)$

15. $x^2y - 3x^2 - 4 = 0; (2, 4)$

16. $2x^3y - x^3 + 5 = 0; (-1, 3)$

17. $e^y = x^2 + y^2; (1, 0)$

18. $x^2 - y = 4e^y; (2, 0)$

19. $x^3 - y = \ln y; (1, 1)$

20. $\ln y = 2y^2 - x; (2, 1)$

21. $x \ln y + 2y = 2x^3; (1, 1)$

22. $xe^y - y = x^2 - 2; (2, 0)$

In Problems 23 and 24, find x′ for x = x(t) defined implicitly by the given equation. Evaluate x′ at the indicated point.

23. $x^2 - t^2x + t^3 + 11 = 0; (-2, 1)$

24. $x^3 - tx^2 - 4 = 0; (-3, -2)$

Problems 25 and 26 refer to the equation and graph shown in the figure.

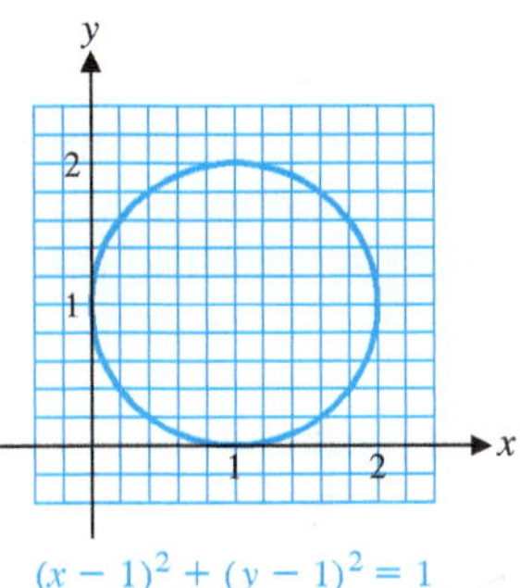

Figure for 25 and 26

25. Use implicit differentiation to find the slopes of the tangent lines at the points on the graph where $x = 1.6$. Check your answers by visually estimating the slopes on the graph in the figure.

26. Find the slopes of the tangent lines at the points on the graph where $x = 0.2$. Check your answers by visually estimating the slopes on the graph in the figure.

In Problems 27–30, find the equation(s) of the tangent line(s) to the graphs of the indicated equations at the point(s) with the given value of x.

27. $xy - x - 4 = 0; x = 2$

28. $3x + xy + 1 = 0; x = -1$

29. $y^2 - xy - 6 = 0; x = 1$

30. $xy^2 - y - 2 = 0; x = 1$

31. If $xe^y = 1$, find y' in two ways, first by differentiating implicitly and then by solving for y explicitly in terms of x. Which method do you prefer? Explain.

32. Explain the difficulty that arises in solving $x^3 + y + xe^y = 1$ for y as an explicit function of x. Find the slope of the tangent line to the graph of the equation at the point (0, 1).

C

In Problems 33–40, find y′ and the slope of the tangent line to the graph of each equation at the indicated point.

33. $(1 + y)^3 + y = x + 7; (2, 1)$

34. $(y - 3)^4 - x = y; (-3, 4)$

35. $(x - 2y)^3 = 2y^2 - 3; (1, 1)$

36. $(2x - y)^4 - y^3 = 8; (-1, -2)$

37. $\sqrt{7 + y^2} - x^3 + 4 = 0; (2, 3)$

38. $6\sqrt{y^3 + 1} - 2x^{3/2} - 2 = 0; (4, 2)$

39. $\ln(xy) = y^2 - 1; (1, 1)$

40. $e^{xy} - 2x = y + 1; (0, 0)$

41. Find the equation(s) of the tangent line(s) at the point(s) on the graph of the equation

$$y^3 - xy - x^3 = 2$$

where $x = 1$. Round all approximate values to two decimal places.

42. Refer to the equation in Problem 41. Find the equation(s) of the tangent line(s) at the point(s) on the graph where $y = -1$. Round all approximate values to two decimal places.

Applications

For the demand equations in Problems 43–46, find the rate of change of p with respect to x by differentiating implicitly (x is the number of items that can be sold at a price of $p).

43. $x = p^2 - 2p + 1{,}000$

44. $x = p^3 - 3p^2 + 200$

45. $x = \sqrt{10{,}000 - p^2}$

46. $x = \sqrt[3]{1{,}500 - p^3}$

47. Biophysics. In biophysics, the equation

$$(L + m)(V + n) = k$$

is called the *fundamental equation of muscle contraction*, where m, n, and k are constants and V is the velocity of the shortening of muscle fibers for a muscle subjected to a load L. Find dL/dV by implicit differentiation.

48. Biophysics. In Problem 47, find dV/dL by implicit differentiation.

49. Speed of sound. The speed of sound in air is given by the formula

$$v = k\sqrt{T}$$

where v is the velocity of sound, T is the temperature of the air, and k is a constant. Use implicit differentiation to find $\frac{dT}{dv}$.

50. Gravity. The equation

$$F = G\frac{m_1m_2}{r^2}$$

is Newton's law of universal gravitation. G is a constant and F is the gravitational force between two objects having masses m_1 and m_2 that are a distance r from each other. Use implicit differentiation to find $\frac{dr}{dF}$. Assume that m_1 and m_2 are constant.

51. Speed of sound. Refer to Problem 49. Find $\frac{dv}{dT}$ and discuss the connection between $\frac{dv}{dT}$ and $\frac{dT}{dv}$.

52. Gravity. Refer to Problem 50. Find $\frac{dF}{dr}$ and discuss the connection between $\frac{dF}{dr}$ and $\frac{dr}{dF}$.

Answers to Matched Problems

1. $y' = -x/y$. When $x = 5$, $y = \pm 12$; thus, $y'|_{(5,12)} = -\frac{5}{12}$ and $y'|_{(5,-12)} = \frac{5}{12}$

2. $y' = \dfrac{y - 2x}{2y - x}$; $y = \frac{4}{5}x - \frac{14}{5}$, $y = \frac{1}{5}x + \frac{14}{5}$

3. $x' = \dfrac{te^x - x}{t\ln t - t^2e^x}$; $x'|_{(1,0)} = -1$

2-6 Related Rates

Union workers are concerned that the rate at which wages are increasing is lagging behind the rate of increase in the company's profits. An automobile dealer wants to predict how badly an anticipated increase in interest rates will decrease his rate of sales. An investor is studying the connection between the rate of increase in the Dow Jones average and the rate of increase in the gross domestic product over the past 50 years.

In each of these situations, there are two quantities—wages and profits, for example—that are changing with respect to time. We would like to discover the precise relationship between the rates of increase (or decrease) of the two quantities. We begin our discussion of such *related rates* by considering familiar situations in which the two quantities are distances and the two rates are velocities.

EXAMPLE 1 **Related Rates and Motion** A 26-foot ladder is placed against a wall (Fig. 1). If the top of the ladder is sliding down the wall at 2 feet per second, at what rate is the bottom of the ladder moving away from the wall when the bottom of the ladder is 10 feet away from the wall?

SOLUTION Many people think that since the ladder is a constant length, the bottom of the ladder will move away from the wall at the rate that the top of the ladder is moving down the wall. This is not the case, however.

At any moment in time, let x be the distance of the bottom of the ladder from the wall and let y be the distance of the top of the ladder from the ground (see Fig. 1). Both x and y are changing with respect to time and can be thought of as functions of time; that is, $x = x(t)$ and $y = y(t)$. Furthermore, x and y are related by the Pythagorean relationship:

Figure 1

$$x^2 + y^2 = 26^2 \tag{1}$$

Differentiating equation (1) implicitly with respect to time t and using the chain rule where appropriate, we obtain

$$2x\frac{dx}{dt} + 2y\frac{dy}{dt} = 0 \tag{2}$$

The rates dx/dt and dy/dt are related by equation (2). This is a **related-rates problem**.

Our problem is to find dx/dt when $x = 10$ feet, given that $dy/dt = -2$ (y is decreasing at a constant rate of 2 feet per second). We have all the quantities we need in equation (2) to solve for dx/dt, except y. When $x = 10$, y can be found from equation (1):

$$10^2 + y^2 = 26^2$$
$$y = \sqrt{26^2 - 10^2} = 24 \text{ feet}$$

Substitute $dy/dt = -2$, $x = 10$, and $y = 24$ into (2). Then solve for dx/dt:

$$2(10)\frac{dx}{dt} + 2(24)(-2) = 0$$
$$\frac{dx}{dt} = \frac{-2(24)(-2)}{2(10)} = 4.8 \text{ feet per second}$$

The bottom of the ladder is moving away from the wall at a rate of 4.8 feet per second.

CONCEPTUAL INSIGHT

In the solution to Example 1, we used equation (1) in two ways: first, to find an equation relating dy/dt and dx/dt, and second, to find the value of y when $x = 10$. These steps must be done in this order. Substituting $x = 10$ and then differentiating does not produce any useful results:

$$x^2 + y^2 = 26^2$$
$$100 + y^2 = 26^2$$

Substituting 10 for x has the effect of stopping the ladder.

$$0 + 2yy' = 0$$
$$y' = 0$$

The rate of change of a stationary object is always 0, but that is not the rate of change of the moving ladder.

Matched Problem 1 Again, a 26-foot ladder is placed against a wall (Fig. 1). If the bottom of the ladder is moving away from the wall at 3 feet per second, at what rate is the top moving down when the top of the ladder is 24 feet above ground?

Explore & Discuss 1

(A) For which values of x and y in Example 1 is dx/dt equal to 2 (i.e., the same rate that the ladder is sliding down the wall)?

(B) When is dx/dt greater than 2? Less than 2?

DEFINITION Suggestions for Solving Related-Rates Problems

Step 1 Sketch a figure if helpful.

Step 2 Identify all relevant variables, including those whose rates are given and those whose rates are to be found.

Step 3 Express all given rates and rates to be found as derivatives.

Step 4 Find an equation connecting the variables identified in step 2.

Step 5 Implicitly differentiate the equation found in step 4, using the chain rule where appropriate, and substitute in all given values.

Step 6 Solve for the derivative that will give the unknown rate.

EXAMPLE 2 **Related Rates and Motion** Suppose that two motorboats leave from the same point at the same time. If one travels north at 15 miles per hour and the other travels east at 20 miles per hour, how fast will the distance between them be changing after 2 hours?

SOLUTION First, draw a picture, as shown in Figure 2.

All variables, x, y, and z, are changing with time. They can be considered as functions of time: $x = x(t)$, $y = y(t)$, and $z = z(t)$, given implicitly. It now makes sense to take derivatives of each variable with respect to time. From the Pythagorean theorem,

$$z^2 = x^2 + y^2 \tag{3}$$

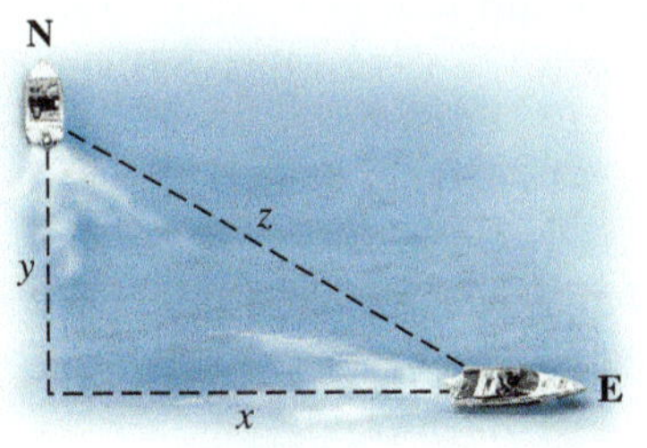

Figure 2

We also know that

$$\frac{dx}{dt} = 20 \text{ miles per hour} \quad \text{and} \quad \frac{dy}{dt} = 15 \text{ miles per hour}$$

We want to find dz/dt at the end of 2 hours—that is, when $x = 40$ miles and $y = 30$ miles. To do this, we differentiate both sides of equation (3) with respect to t and solve for dz/dt:

$$2z\frac{dz}{dt} = 2x\frac{dx}{dt} + 2y\frac{dy}{dt} \tag{4}$$

We have everything we need except z. From equation (3), when $x = 40$ and $y = 30$, we find z to be 50. Substituting the known quantities into equation (4), we obtain

$$2(50)\frac{dz}{dt} = 2(40)(20) + 2(30)(15)$$

$$\frac{dz}{dt} = 25 \text{ miles per hour}$$

The boats will be separating at a rate of 25 miles per hour.

Matched Problem 2 Repeat Example 2 for the same situation at the end of 3 hours.

EXAMPLE 3 **Related Rates and Motion** Suppose that a point is moving along the graph of $x^2 + y^2 = 25$ (Fig. 3). When the point is at $(-3, 4)$, its x coordinate is increasing at the rate of 0.4 unit per second. How fast is the y coordinate changing at that moment?

SOLUTION Since both x and y are changing with respect to time, we can consider each as a function of time, namely,

$$x = x(t) \quad \text{and} \quad y = y(t)$$

but restricted so that

$$x^2 + y^2 = 25 \tag{5}$$

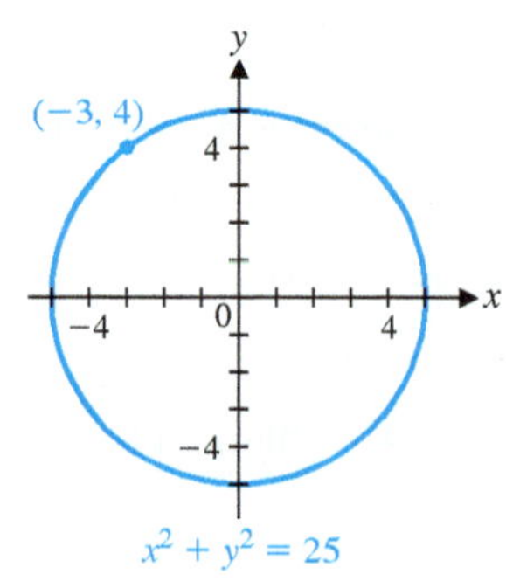

Figure 3

We want to find dy/dt, given $x = -3$, $y = 4$, and $dx/dt = 0.4$. Implicitly differentiating both sides of equation (5) with respect to t, we have

$$x^2 + y^2 = 25$$

$$2x\frac{dx}{dt} + 2y\frac{dy}{dt} = 0 \quad \text{Divide both sides by 2.}$$

$$x\frac{dx}{dt} + y\frac{dy}{dt} = 0 \quad \text{Substitute } x = -3, y = 4, \text{ and } dx/dt = 0.4, \text{ and solve for } dy/dt.$$

$$(-3)(0.4) + 4\frac{dy}{dt} = 0$$

$$\frac{dy}{dt} = 0.3 \text{ unit per second}$$

Matched Problem 3 A point is moving on the graph of $y^3 = x^2$. When the point is at $(-8, 4)$, its y coordinate is decreasing by 2 units per second. How fast is the x coordinate changing at that moment?

EXAMPLE 4 **Related Rates and Business** Suppose that for a company manufacturing flash drives, the cost, revenue, and profit equations are given by

$$C = 5{,}000 + 2x \quad \text{Cost equation}$$

$$R = 10x - 0.001x^2 \quad \text{Revenue equation}$$

$$P = R - C \quad \text{Profit equation}$$

where the production output in 1 week is x flash drives. If production is increasing at the rate of 500 flash drives per week when production is 2,000 flash drives, find the rate of increase in

(A) Cost (B) Revenue (C) Profit

SOLUTION If production x is a function of time (it must be, since it is changing with respect to time), then C, R, and P must also be functions of time. These functions are given implicitly (rather than explicitly). Letting t represent time in weeks, we differentiate both sides of each of the preceding three equations with respect to t and then substitute $x = 2{,}000$ and $dx/dt = 500$ to find the desired rates.

(A) $C = 5{,}000 + 2x$ *Think: $C = C(t)$ and $x = x(t)$.*

$$\frac{dC}{dt} = \frac{d}{dt}(5{,}000) + \frac{d}{dt}(2x) \quad \text{Differentiate both sides with respect to } t.$$

$$\frac{dC}{dt} = 0 + 2\frac{dx}{dt} = 2\frac{dx}{dt}$$

Since $dx/dt = 500$ when $x = 2{,}000$,

$$\frac{dC}{dt} = 2(500) = \$1{,}000 \text{ per week}$$

Cost is increasing at a rate of $1,000 per week.

(B) $R = 10x - 0.001x^2$

$$\frac{dR}{dt} = \frac{d}{dt}(10x) - \frac{d}{dt}0.001x^2$$

$$\frac{dR}{dt} = 10\frac{dx}{dt} - 0.002x\frac{dx}{dt}$$

$$\frac{dR}{dt} = (10 - 0.002x)\frac{dx}{dt}$$

Since $dx/dt = 500$ when $x = 2{,}000$,

$$\frac{dR}{dt} = [10 - 0.002(2{,}000)](500) = \$3{,}000 \text{ per week}$$

Revenue is increasing at a rate of $3,000 per week.

(C) $P = R - C$

$$\frac{dP}{dt} = \frac{dR}{dt} - \frac{dC}{dt} \quad \text{Results from parts (A) and (B)}$$
$$= \$3{,}000 - \$1{,}000$$
$$= \$2{,}000 \text{ per week}$$

Profit is increasing at a rate of $2,000 per week.

Matched Problem 4 Repeat Example 4 for a production level of 6,000 flash drives per week.

Exercises 2-6

A

In Problems 1–6, assume that $x = x(t)$ and $y = y(t)$. Find the indicated rate, given the other information.

1. $y = x^2 + 2$; $dx/dt = 3$ when $x = 5$; find dy/dt
2. $y = x^3 - 3$; $dx/dt = -2$ when $x = 2$; find dy/dt
3. $x^2 + y^2 = 1$; $dy/dt = -4$ when $x = -0.6$ and $y = 0.8$; find dx/dt
4. $x^2 + y^2 = 4$; $dy/dt = 5$ when $x = 1.2$ and $y = -1.6$; find dx/dt
5. $x^2 + 3xy + y^2 = 11$; $dx/dt = 2$ when $x = 1$ and $y = 2$; find dy/dt
6. $x^2 - 2xy - y^2 = 7$; $dy/dt = -1$ when $x = 2$ and $y = -1$; find dx/dt

B

7. A point is moving on the graph of $xy = 36$. When the point is at (4, 9), its x coordinate is increasing by 4 units per second. How fast is the y coordinate changing at that moment?
8. A point is moving on the graph of $4x^2 + 9y^2 = 36$. When the point is at (3, 0), its y coordinate is decreasing by 2 units per second. How fast is its x coordinate changing at that moment?
9. A boat is being pulled toward a dock as shown in the figure. If the rope is being pulled in at 3 feet per second, how fast is the distance between the dock and the boat decreasing when it is 30 feet from the dock?

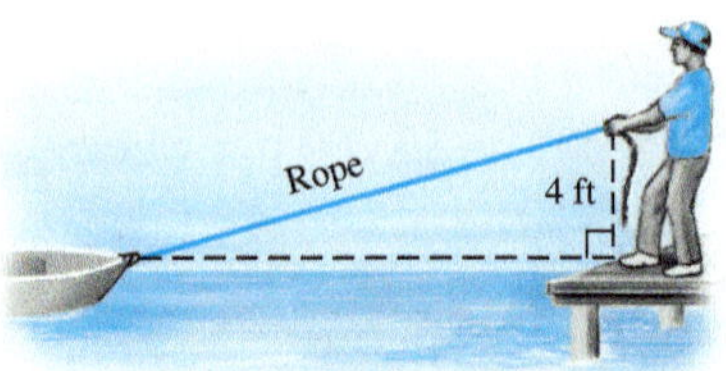

Figure for 9 and 10

10. Refer to Problem 9. Suppose that the distance between the boat and the dock is decreasing by 3.05 feet per second. How fast is the rope being pulled in when the boat is 10 feet from the dock?
11. A rock thrown into a still pond causes a circular ripple. If the radius of the ripple is increasing by 2 feet per second, how fast is the area changing when the radius is 10 feet? [Use $A = \pi R^2$, $\pi \approx 3.14$.]
12. Refer to Problem 11. How fast is the circumference of a circular ripple changing when the radius is 10 feet? [Use $C = 2\pi R$, $\pi \approx 3.14$.]
13. The radius of a spherical balloon is increasing at the rate of 3 centimeters per minute. How fast is the volume changing when the radius is 10 centimeters? [Use $V = \frac{4}{3}\pi R^3$, $\pi \approx 3.14$.]
14. Refer to Problem 13. How fast is the surface area of the sphere increasing when the radius is 10 centimeters? [Use $S = 4\pi R^2$, $\pi \approx 3.14$.]
15. Boyle's law for enclosed gases states that if the volume is kept constant, the pressure P and temperature T are related by the equation

$$\frac{P}{T} = k$$

where k is a constant. If the temperature is increasing at 3 kelvins per hour, what is the rate of change of pressure when the temperature is 250 kelvins and the pressure is 500 pounds per square inch?

16. Boyle's law for enclosed gases states that if the temperature is kept constant, the pressure P and volume V of a gas are related by the equation

$$VP = k$$

where k is a constant. If the volume is decreasing by 5 cubic inches per second, what is the rate of change of pressure when the volume is 1,000 cubic inches and the pressure is 40 pounds per square inch?

17. A 10-foot ladder is placed against a vertical wall. Suppose that the bottom of the ladder slides away from the wall at a constant rate of 3 feet per second. How fast is the top of the ladder sliding down the wall (negative rate) when the bottom is 6 feet from the wall? [*Hint:* Use the Pythagorean theorem, $a^2 + b^2 = c^2$, where c is the length of the hypotenuse of a right triangle and a and b are the lengths of the two shorter sides.]
18. A weather balloon is rising vertically at the rate of 5 meters per second. An observer is standing on the ground 300 meters from where the balloon was released. At what rate is the distance between the observer and the balloon changing when the balloon is 400 meters high?

C

19. A streetlight is on top of a 20-foot pole. A person who is 5 feet tall walks away from the pole at the rate of 5 feet per second. At what rate is the tip of the person's shadow moving away from the pole when he is 20 feet from the pole?
20. Refer to Problem 19. At what rate is the person's shadow growing when he is 20 feet from the pole?
21. Helium is pumped into a spherical balloon at a constant rate of 4 cubic feet per second. How fast is the radius increasing after 1 minute? After 2 minutes? Is there any time at which the radius is increasing at a rate of 100 feet per second? Explain.
22. A point is moving along the x axis at a constant rate of 5 units per second. At which point is its distance from (0, 1) increasing at a rate of 2 units per second? At 4 units per second? At 5 units per second? At 10 units per second? Explain.
23. A point is moving on the graph of $y = e^x + x + 1$ in such a way that its x coordinate is always increasing at a rate of 3 units per second. How fast is the y coordinate changing when the point crosses the x axis?
24. A point is moving on the graph of $x^3 + y^2 = 1$ in such a way that its y coordinate is always increasing at a rate of 2 units per second. At which point(s) is the x coordinate increasing at a rate of 1 unit per second?

Applications

25. Cost, revenue, and profit rates. Suppose that for a company manufacturing calculators, the cost, revenue, and profit equations are given by

$$C = 90{,}000 + 30x \qquad R = 300x - \frac{x^2}{30}$$
$$P = R - C$$

where the production output in 1 week is x calculators. If production is increasing at a rate of 500 calculators per week when production output is 6,000 calculators, find the rate of increase (decrease) in

(A) Cost (B) Revenue (C) Profit

26. Cost, revenue, and profit rates. Repeat Problem 25 for

$$C = 72{,}000 + 60x \qquad R = 200x - \frac{x^2}{30}$$
$$P = R - C$$

where production is increasing at a rate of 500 calculators per week at a production level of 1,500 calculators.

27. Advertising. A retail store estimates that weekly sales s and weekly advertising costs x (both in dollars) are related by

$$s = 60{,}000 - 40{,}000e^{-0.0005x}$$

The current weekly advertising costs are \$2,000, and these costs are increasing at the rate of \$300 per week. Find the current rate of change of sales.

28. Advertising. Repeat Problem 27 for

$$s = 50{,}000 - 20{,}000e^{-0.0004x}$$

29. Price–demand. The price p (in dollars) and demand x for a product are related by

$$2x^2 + 5xp + 50p^2 = 80{,}000$$

(A) If the price is increasing at a rate of \$2 per month when the price is \$30, find the rate of change of the demand.

(B) If the demand is decreasing at a rate of 6 units per month when the demand is 150 units, find the rate of change of the price.

30. Price–demand. Repeat Problem 29 for

$$x^2 + 2xp + 25p^2 = 74{,}500$$

31. Pollution. An oil tanker aground on a reef is forming a circular oil slick about 0.1 foot thick (see the figure). To estimate the rate dV/dt (in cubic feet per minute) at which the oil is leaking from the tanker, it was found that the radius of the slick was increasing at 0.32 foot per minute $(dR/dt = 0.32)$ when the radius R was 500 feet. Find dV/dt, using $\pi \approx 3.14$.

Figure for 31

32. Learning. A person who is new on an assembly line performs an operation in T minutes after x performances of the operation, as given by

$$T = 6\left(1 + \frac{1}{\sqrt{x}}\right)$$

If $dx/dt = 6$ operations per hours, where t is time in hours, find dT/dt after 36 performances of the operation.

Answers to Matched Problems

1. $dy/dt = -1.25$ ft/sec
2. $dz/dt = 25$ mi/hr
3. $dx/dt = 6$ units/sec
4. (A) $dC/dt = \$1{,}000/\text{wk}$
   (B) $dR/dt = -\$1{,}000/\text{wk}$
   (C) $dP/dt = -\$2{,}000/\text{wk}$

2-7 Elasticity of Demand

- Relative Rate of Change
- Elasticity of Demand

When will a price increase lead to an increase in revenue? To answer this question and study relationships among price, demand, and revenue, economists use the notion of *elasticity of demand.* In this section, we define the concepts of *relative rate of change*, *percentage rate of change*, and *elasticity of demand.*

Relative Rate of Change

Explore & Discuss 1

A broker is trying to sell you two stocks: Biotech and Comstat. The broker estimates that Biotech's earnings will increase \$2 per year over the next several years, while Comstat's earnings will increase only \$1 per year. Is this sufficient information for you to choose between the two stocks? What other information might you request from the broker to help you decide?

Interpreting rates of change is a fundamental application of calculus. In Explore & Discuss 1, Biotech's earnings are increasing at twice the rate of Comstat's, but that does not automatically make Biotech the better buy. The obvious information that is missing is the cost of each stock. If Biotech costs \$100 a share and Comstat costs \$25 share, then which stock is the better buy? To answer this question, we introduce two new concepts: *relative rate of change* and *percentage rate of change*.

DEFINITION Relative and Percentage Rates of Change

The **relative rate of change** of a function $f(x)$ is $\dfrac{f'(x)}{f(x)}$.

The **percentage rate of change** is $100 \times \dfrac{f'(x)}{f(x)}$.

Because

$$\frac{d}{dx}\ln f(x) = \frac{f'(x)}{f(x)}$$

the relative rate of change of $f(x)$ is the derivative of the logarithm of $f(x)$. This is also referred to as the **logarithmic derivative** of $f(x)$. Returning to Explore & Discuss 1, we can now write

	Relative rate of change		Percentage rate of change
Biotech	$\dfrac{2}{100} = 0.02$	or	2%
Comstat	$\dfrac{1}{25} = 0.04$	or	4%

EXAMPLE 1 **Percentage Rate of Change** Table 1 lists the GDP (gross domestic product expressed in billions of 2005 dollars) and U.S. population from 2000 to 2008. A model for the GDP is

$$f(t) = 280t + 11{,}147$$

where t is years since 2000. Find and graph the percentage rate of change of $f(t)$ for $0 \le t \le 8$.

Table 1

Year	Real GDP (billions of 2005 dollars)	Population (in millions)
2000	\$11,226	282.2
2002	\$11,553	287.7
2004	\$12,264	292.9
2006	\$12,976	298.4
2008	\$13,312	304.1

SOLUTION If $p(t)$ is the percentage rate of change of $f(t)$, then

$$p(t) = 100 \times \frac{d}{dx}\ln(280t + 11{,}147)$$

$$= \frac{28{,}000}{280t + 11{,}147}$$

The graph of $p(t)$ is shown in Figure 1 (graphing details omitted). Notice that $p(t)$ is decreasing, even though the GDP is increasing.

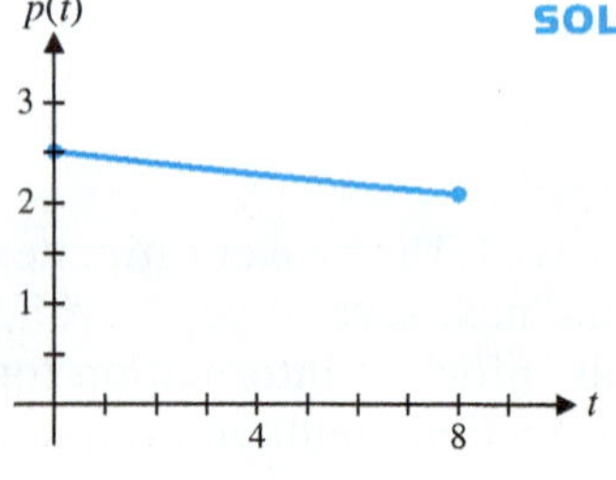

Figure 1

Matched Problem 1 A model for the population data in Table 1 is

$$f(t) = 2.7t + 282$$

where t is years since 2000. Find and graph $p(t)$, the percentage rate of change of $f(t)$ for $0 \le t \le 8$.

Elasticity of Demand

Economists use logarithmic derivatives and relative rates of change to study the relationship among price changes, demand, and revenue. Suppose the price $\$p$ and the demand x for a certain product are related by the price–demand equation

$$x + 500p = 10{,}000 \qquad (1)$$

In problems involving revenue, cost, and profit, it is customary to use the demand equation to express price as a function of demand. Since we want to know the effects that changes in price have on demand, it is more convenient to express demand as a function of price. Solving (1) for x, we have

$$\begin{aligned} x &= 10{,}000 - 500p \\ &= 500(20 - p) \quad \text{Demand as a function of price} \end{aligned}$$

or

$$x = f(p) = 500(20 - p) \qquad 0 \le p \le 20 \qquad (2)$$

Since x and p both represent nonnegative quantities, we must restrict p so that $0 \le p \le 20$. For most products, demand is assumed to be a decreasing function of price. That is, price increases result in lower demand, and price decreases result in higher demand (see Figure 2).

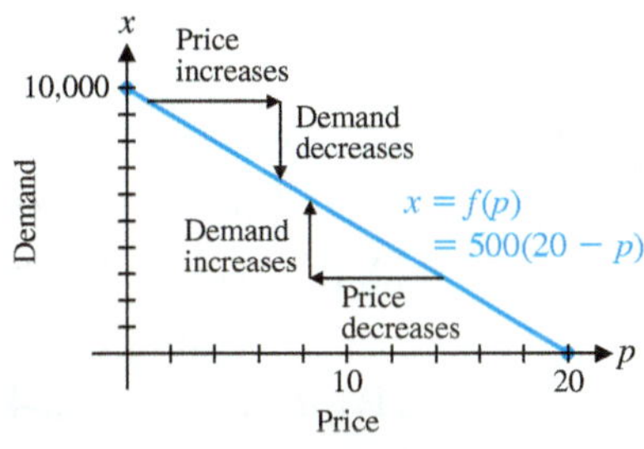

Figure 2

Economists use the *elasticity of demand* to study the relationship between changes in price and changes in demand. The **elasticity of demand** is the negative of the ratio of the relative rate of change of demand to the relative rate of change of price. If price and demand are related by a price–demand equation of the form $x = f(p)$, then the elasticity of demand can be expressed as

$$\begin{aligned} -\frac{\text{relative rate of change of demand}}{\text{relative rate of change of price}} &= -\frac{\dfrac{d}{dp}\ln f(p)}{\dfrac{d}{dp}\ln p} \\ &= -\frac{\dfrac{f'(p)}{f(p)}}{\dfrac{1}{p}} \\ &= -\frac{pf'(p)}{f(p)} \end{aligned}$$

THEOREM 1 Elasticity of Demand

If price and demand are related by $x = f(p)$, then the elasticity of demand is given by

$$E(p) = -\frac{pf'(p)}{f(p)}$$

CONCEPTUAL INSIGHT

Since p and $f(p)$ are nonnegative and $f'(p)$ is negative (remember, demand is usually a decreasing function of price), $E(p)$ is nonnegative. This is why elasticity of demand is defined as the negative of a ratio.

The next example illustrates interpretations of the elasticity of demand.

EXAMPLE 2 **Elasticity of Demand** Find $E(p)$ for the price–demand equation

$$x = f(p) = 500(20 - p)$$

Find and interpret each of the following:

(A) $E(4)$ (B) $E(16)$ (C) $E(10)$

SOLUTION $$E(p) = -\frac{pf'(p)}{f(p)} = -\frac{p(-500)}{500(20 - p)} = \frac{p}{20 - p}$$

In order to interpret values of $E(p)$, we must recall the definition of elasticity:

$$E(p) = -\frac{\text{relative rate of change of demand}}{\text{relative rate of change of price}}$$

or

$$-\begin{pmatrix}\text{relative rate of}\\ \text{change of demand}\end{pmatrix} \approx E(p)\begin{pmatrix}\text{relative rate of}\\ \text{change of price}\end{pmatrix}$$

(A) $E(4) = \frac{4}{16} = 0.25 < 1$. If the \$4 price changes by 10%, then the demand will change by approximately $0.25(10\%) = 2.5\%$.

(B) $E(16) = \frac{16}{4} = 4 > 1$. If the \$16 price changes by 10%, then the demand will change by approximately $4(10\%) = 40\%$.

(C) $E(10) = \frac{10}{10} = 1$. If the \$10 price changes by 10%, then the demand will also change by approximately 10%.

CONCEPTUAL INSIGHT

Do not be concerned with the omission of any negative signs in these interpretations. We already know that if price increases, then demand decreases, and vice versa (Fig. 2 on page 133). $E(p)$ is a measure of *how much* the demand changes for a given change in price.

Matched Problem 2 Find $E(p)$ for the price–demand equation

$$x = f(p) = 1{,}000(40 - p)$$

Find and interpret each of the following:

(A) $E(8)$ (B) $E(30)$ (C) $E(20)$

The three cases illustrated in the solution to Example 2 are referred to as **inelastic demand, elastic demand**, and **unit elasticity**, as indicated in Table 2.

Table 2

$E(p)$	Demand	Interpretation
$0 < E(p) < 1$	Inelastic	Demand is not sensitive to changes in price. A change in price produces a smaller change in demand.
$E(p) > 1$	Elastic	Demand is sensitive to changes in price. A change in price produces a larger change in demand.
$E(p) = 1$	Unit	A change in price produces the same change in demand.

Now we want to see how revenue and elasticity are related. We have used the following model for revenue many times before:

$$\text{Revenue} = (\text{demand}) \times (\text{price})$$

Since we are looking for a connection between $E(p)$ and revenue, we will use a price–demand equation written in the form $x = f(p)$, where x is demand and p is price.

$$\begin{aligned} R(p) &= xp = pf(p) && \text{Revenue as a function of price} \\ R'(p) &= pf'(p) + f(p) \\ &= f(p)\left[\frac{pf'(p)}{f(p)} + 1\right] && E(p) = -\frac{pf'(p)}{f(p)} \\ &= f(p)[1 - E(p)] \end{aligned}$$

Since $x = f(p) > 0$, it follows that $R'(p)$ and $[1 - E(p)]$ always have the same sign (see Table 3).

Table 3 Revenue and Elasticity of Demand

All Are True or All Are False	All Are True or All Are False
$R'(p) > 0$	$R'(p) < 0$
$E(p) < 1$	$E(p) > 1$
Demand is inelastic	Demand is elastic

These facts are interpreted in the following summary and in Figure 3:

SUMMARY Revenue and Elasticity of Demand

When demand is inelastic,

A price increase will increase revenue.
A price decrease will decrease revenue.

When demand is elastic,

A price increase will decrease revenue.
A price decrease will increase revenue.

CONCEPTUAL INSIGHT

We know that a price increase will decrease demand at all price levels (Fig. 2). As Figure 3 illustrates, the effects of a price increase on revenue depends on the price level. If demand is elastic, then a price increase decreases revenue. If demand is inelastic, then a price increase increases revenue.

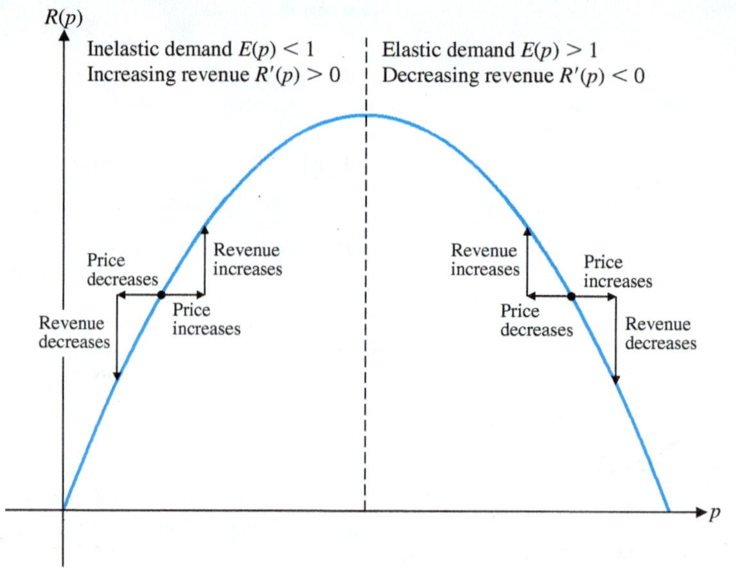

Figure 3 Revenue and elasticity

EXAMPLE 3 **Elasticity and Revenue** A manufacturer of sunglasses currently sells one type for \$15 a pair. The price p and the demand x for these glasses are related by

$$x = f(p) = 9{,}500 - 250p$$

If the current price is increased, will revenue increase or decrease?

SOLUTION

$$\begin{aligned} E(p) &= -\frac{pf'(p)}{f(p)} \\ &= -\frac{p(-250)}{9{,}500 - 250p} \\ &= \frac{p}{38 - p} \\ E(15) &= \frac{15}{23} \approx 0.65 \end{aligned}$$

At the \$15 price level, demand is inelastic and a price increase will increase revenue.

Matched Problem 3 Repeat Example 3 if the current price for sunglasses is \$21 a pair.

Exercises 2-7

A

In Problems 1–6, find the relative rate of change of f(x).

1. $f(x) = 35x - 0.4x^2$
2. $f(x) = 60x - 1.2x^2$
3. $f(x) = 7 + 4e^{-x}$
4. $f(x) = 15 - 3e^{-0.5x}$
5. $f(x) = 12 + 5 \ln x$
6. $f(x) = 25 - 2 \ln x$

In Problems 7–16, find the relative rate of change of f(x) at the indicated value of x. Round to three decimal places.

7. $f(x) = 45;\ x = 100$
8. $f(x) = 580;\ x = 300$
9. $f(x) = 420 - 5x;\ x = 25$
10. $f(x) = 500 - 6x;\ x = 40$

11. $f(x) = 420 - 5x; x = 55$

12. $f(x) = 500 - 6x; x = 75$

13. $f(x) = 4x^2 - \ln x; x = 2$

14. $f(x) = 9x - 5 \ln x; x = 3$

15. $f(x) = 4x^2 - \ln x; x = 5$

16. $f(x) = 9x - 5 \ln x; x = 7$

In Problems 17–24, find the percentage rate of change of f(x) at the indicated value of x. Round to the nearest tenth of a percent.

17. $f(x) = 225 + 65x; x = 5$

18. $f(x) = 75 + 110x; x = 4$

19. $f(x) = 225 + 65x; x = 15$

20. $f(x) = 75 + 110x; x = 16$

21. $f(x) = 5{,}100 - 3x^2; x = 35$

22. $f(x) = 3{,}000 - 8x^2; x = 12$

23. $f(x) = 5{,}100 - 3x^2; x = 41$

24. $f(x) = 3{,}000 - 8x^2; x = 18$

In Problems 25–30, use the price–demand equation to find E(p), the elasticity of demand.

25. $x = f(p) = 25{,}000 - 450p$

26. $x = f(p) = 10{,}000 - 190p$

27. $x = f(p) = 4{,}800 - 4p^2$

28. $x = f(p) = 8{,}400 - 7p^2$

29. $x = f(p) = 98 - 0.6e^p$

30. $x = f(p) = 160 - 35 \ln p$

B

In Problems 31–34, use the price–demand equation to determine whether demand is elastic, is inelastic, or has unit elasticity at the indicated values of p.

31. $x = f(p) = 12{,}000 - 10p^2$
(A) $p = 10$ (B) $p = 20$
(C) $p = 30$

32. $x = f(p) = 1{,}875 - p^2$
(A) $p = 15$ (B) $p = 25$
(C) $p = 40$

33. $x = f(p) = 950 - 2p - 0.1p^2$
(A) $p = 30$ (B) $p = 50$
(C) $p = 70$

34. $x = f(p) = 875 - p - 0.05p^2$
(A) $p = 50$ (B) $p = 70$
(C) $p = 100$

35. Given the price–demand equation

$$p + 0.005x = 30$$

(A) Express the demand x as a function of the price p.
(B) Find the elasticity of demand, $E(p)$.
(C) What is the elasticity of demand when $p = \$10$? If this price is increased by 10%, what is the approximate change in demand?
(D) What is the elasticity of demand when $p = \$25$? If this price is increased by 10%, what is the approximate change in demand?
(E) What is the elasticity of demand when $p = \$15$? If this price is increased by 10%, what is the approximate change in demand?

36. Given the price–demand equation

$$p + 0.01x = 50$$

(A) Express the demand x as a function of the price p.
(B) Find the elasticity of demand, $E(p)$.
(C) What is the elasticity of demand when $p = \$10$? If this price is decreased by 5%, what is the approximate change in demand?
(D) What is the elasticity of demand when $p = \$45$? If this price is decreased by 5%, what is the approximate change in demand?
(E) What is the elasticity of demand when $p = \$25$? If this price is decreased by 5%, what is the approximate change in demand?

37. Given the price–demand equation

$$0.02x + p = 60$$

(A) Express the demand x as a function of the price p.
(B) Express the revenue R as a function of the price p.
(C) Find the elasticity of demand, $E(p)$.
(D) For which values of p is demand elastic? Inelastic?
(E) For which values of p is revenue increasing? Decreasing?
(F) If $p = \$10$ and the price is decreased, will revenue increase or decrease?
(G) If $p = \$40$ and the price is decreased, will revenue increase or decrease?

38. Repeat Problem 37 for the price–demand equation

$$0.025x + p = 50$$

In Problems 39–46, use the price–demand equation to find the values of p for which demand is elastic and the values for which demand is inelastic. Assume that price and demand are both positive.

39. $x = f(p) = 210 - 30p$

40. $x = f(p) = 480 - 8p$

41. $x = f(p) = 3{,}125 - 5p^2$

42. $x = f(p) = 2{,}400 - 6p^2$

43. $x = f(p) = \sqrt{144 - 2p}$

44. $x = f(p) = \sqrt{324 - 2p}$

45. $x = f(p) = \sqrt{2{,}500 - 2p^2}$

46. $x = f(p) = \sqrt{3{,}600 - 2p^2}$

In Problems 47–52, use the demand equation to find the revenue function. Sketch the graph of the revenue function, and indicate the regions of inelastic and elastic demand on the graph.

47. $x = f(p) = 20(10 - p)$

48. $x = f(p) = 10(16 - p)$

49. $x = f(p) = 40(p - 15)^2$

50. $x = f(p) = 10(p - 9)^2$

51. $x = f(p) = 30 - 10\sqrt{p}$

52. $x = f(p) = 30 - 5\sqrt{p}$

C

If a price–demand equation is solved for p, then price is expressed as $p = g(x)$ and x becomes the independent variable. In this case, it can be shown that the elasticity of demand is given by

$$E(x) = -\frac{g(x)}{xg'(x)}$$

In Problems 53–56, use the price–demand equation to find $E(x)$ at the indicated value of x.

53. $p = g(x) = 50 - 0.1x, x = 200$

54. $p = g(x) = 30 - 0.05x, x = 400$

55. $p = g(x) = 50 - 2\sqrt{x}, x = 400$

56. $p = g(x) = 20 - \sqrt{x}, x = 100$

In Problems 57–60, use the price–demand equation to find the values of x for which demand is elastic and for which demand is inelastic.

57. $p = g(x) = 180 - 0.3x$

58. $p = g(x) = 640 - 0.4x$

59. $p = g(x) = 90 - 0.1x^2$

60. $p = g(x) = 540 - 0.2x^2$

61. Find $E(p)$ for $x = f(p) = Ap^{-k}$, where A and k are positive constants.

62. Find $E(p)$ for $x = f(p) = Ae^{-kp}$, where A and k are positive constants.

Applications

63. Rate of change of cost. A fast-food restaurant can produce a hamburger for \$2.50. If the restaurant's daily sales are increasing at the rate of 30 hamburgers per day, how fast is its daily cost for hamburgers increasing?

64. Rate of change of cost. The fast-food restaurant in Problem 63 can produce an order of fries for \$0.80. If the restaurant's daily sales are increasing at the rate of 45 orders of fries per day, how fast is its daily cost for fries increasing?

65. Revenue and elasticity. The price–demand equation for hamburgers at a fast-food restaurant is

$$x + 400p = 3{,}000$$

Currently, the price of a hamburger is \$3.00. If the price is increased by 10%, will revenue increase or decrease?

66. Revenue and elasticity. Refer to Problem 65. If the current price of a hamburger is \$4.00, will a 10% price increase cause revenue to increase or decrease?

67. Revenue and elasticity. The price–demand equation for an order of fries at a fast-food restaurant is

$$x + 1{,}000p = 2{,}500$$

Currently, the price of an order of fries is \$0.99. If the price is decreased by 10%, will revenue increase or decrease?

68. Revenue and elasticity. Refer to Problem 67. If the current price of an order of fries is \$1.29, will a 10% price decrease cause revenue to increase or decrease?

69. Maximum revenue. Refer to Problem 65. What price will maximize the revenue from selling hamburgers?

70. Maximum revenue. Refer to Problem 67. What price will maximize the revenue from selling fries?

71. Population growth. A model for Canada's population growth (Table 4) is

$$f(t) = 0.31t + 18.5$$

where t is years since 1960. Find and graph the percentage rate of change of $f(t)$ for $0 \le t \le 50$.

72. Population growth. A model for Mexico's population growth (Table 4) is

Table 4 Population Growth

Year	Canada (millions)	Mexico (millions)
1960	18	39
1970	22	53
1980	25	68
1990	28	85
2000	31	100
2010	34	112

$$f(t) = 1.49t + 38.8$$

where t is years since 1960. Find and graph the percentage rate of change of $f(t)$ for $0 \le t \le 50$.

73. Crime. A model for the number of robberies in the United States (Table 5) is

$$r(t) = 9.7 - 2.7 \ln t$$

where t is years since 1990. Find the relative rate of change for robberies in 2008.

Table 5 Number of Victimizations per 1,000 Population Age 12 and Over

	Robbery	Aggravated Assault
1995	5.4	9.5
1997	4.3	8.6
1999	3.6	6.7
2001	2.8	5.3
2003	2.5	4.6
2005	2.6	4.3

74. Crime. A model for the number of assaults in the United States (Table 5) is

$$a(t) = 18.2 - 5.2 \ln t$$

where t is years since 1990. Find the relative rate of change for assaults in 2008.

Answers to Matched Problems

1. $p(t) = \frac{270}{2.7t + 282}$

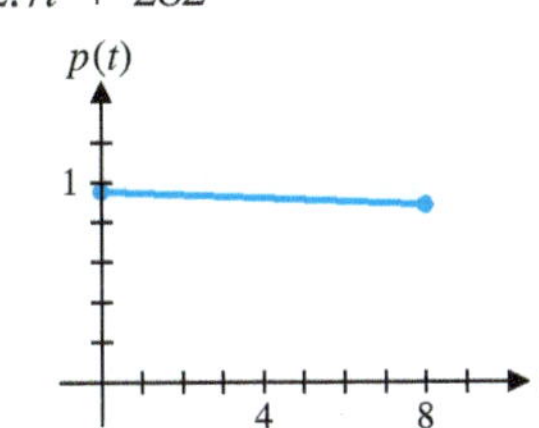

2. $E(p) = \frac{p}{40 - p}$
   (A) $E(8) = 0.25$; demand is inelastic.
   (B) $E(30) = 3$; demand is elastic.
   (C) $E(20) = 1$; demand has unit elasticity.

3. $E(21) = \frac{21}{17} \approx 1.2$; demand is elastic. Increasing price will decrease revenue.

Chapter 2 Review

Important Terms, Symbols, and Concepts

2-1 The Constant *e* and Continuous Compound Interest

EXAMPLES

- The number e is defined as

$$\lim_{x\to\infty}\left(1 + \frac{1}{n}\right)^n = \lim_{x\to 0}(1 + s)^{1/s} = 2.718\,281\,828\,459\ldots$$

- If a principal P is invested at an annual rate r (expressed as a decimal) compounded continuously, then the amount A in the account at the end of t years is given by the **compound interest formula** — Ex. 1, p. 89; Ex. 2, p. 89; Ex. 3, p. 90; Ex. 4, p. 90

$$A = Pe^{rt}$$

2-2 Derivatives of Exponential and Logarithmic Functions

- For $b > 0, b \neq 1$, — Ex. 1, p. 94; Ex. 2, p. 96; Ex. 3, p. 98; Ex. 4, p. 98; Ex. 5, p. 99

$$\frac{d}{dx}e^x = e^x \qquad \frac{d}{dx}b^x = b^x \ln b$$

$$\frac{d}{dx}\ln x = \frac{1}{x} \qquad \frac{d}{dx}\log_b x = \frac{1}{\ln b}\frac{1}{x}$$

- The **change-of-base formulas** allow conversion from base e to any base $b, b > 0, b \neq 1$:

$$b^x = e^{x \ln b} \qquad \log_b x = \frac{\ln x}{\ln b}$$

2-3 Derivatives of Products and Quotients

- Product rule. If $y = f(x) = F(x)\,S(x)$, then $f'(x) = F(x)S'(x) + S(x)F'(x)$, provided that both $F'(x)$ and $S'(x)$ exist. — Ex. 1, p. 102; Ex. 2, p. 103; Ex. 3, p. 103
- Quotient rule. If $y = f(x) = \frac{T(x)}{B(x)}$, then $f'(x) = \frac{B(x)\,T'(x) - T(x)\,B'(x)}{[B(x)]^2}$ provided that both $T'(x)$ and $B'(x)$ exist. — Ex. 4, p. 105; Ex. 5, p. 106; Ex. 6, p. 106

2-4 The Chain Rule

- A function m is a **composite** of functions f and g if $m(x) = f[g(x)]$. — Ex. 1, p. 110
- The **chain rule** gives a formula for the derivative of the composite function $m(x) = f[g(x)]$: — Ex. 2, p. 110; Ex. 4, p. 114; Ex. 5, p. 115

$$m'(x) = f'[g(x)]g'(x)$$

- A special case of the chain rule is called the **general power rule:** — Ex. 3, p. 112

$$\frac{d}{dx}[f(x)]^n = n[f(x)]^{n-1}f'(x)$$

- Other special cases of the chain rule are the following **general derivative rules:** — Ex. 6, p. 116

$$\frac{d}{dx}\ln[f(x)] = \frac{1}{f(x)}f'(x)$$

$$\frac{d}{dx}e^{f(x)} = e^{f(x)}f'(x)$$

2-5 Implicit Differentiation

- If $y = y(x)$ is a function defined by the equation $F(x, y) = 0$, then we use **implicit differentiation** to find an equation in x, y, and y'. Ex. 1, p. 121 Ex. 2, p. 122 Ex. 3, p. 123

2-6 Related Rates

- If x and y represent quantities that are changing with respect to time and are related by the equation $F(x, y) = 0$, then implicit differentiation produces an equation that relates x, y, dy/dt, and dx/dt. Problems of this type are called **related-rates problems**. Ex. 1, p. 126 Ex. 2, p. 127 Ex. 3, p. 128
- Suggestions for solving related-rates problems are given on page 127.

2-7 Elasticity of Demand

- The **relative rate of change**, or the **logarithmic derivative**, of a function $f(x)$ is $f'(x)/f(x)$, and the **percentage rate of change** is $100 \times [f'(x)/f(x)]$. Ex. 1, p. 132
- If price and demand are related by $x = f(p)$, then the **elasticity of demand** is given by Ex. 2, p. 134

$$E(p) = -\frac{pf'(p)}{f(p)} = -\frac{\text{relative rate of change of demand}}{\text{relative rate of change of price}}$$

- **Demand is inelastic** if $0 < E(p) < 1$. (Demand is not sensitive to changes in price; a change in price produces a smaller change in demand.) **Demand is elastic** if $E(p) > 1$. (Demand is sensitive to changes in price; a change in price produces a larger change in demand.) **Demand has unit elasticity** if $E(p) = 1$. (A change in price produces the same change in demand.) Ex. 3, p. 136
- If $R(p) = pf(p)$ is the revenue function, then $R'(p)$ and $[1 - E(p)]$ always have the same sign.

Review Exercises

Work through all the problems in this chapter review, and check your answers in the back of the book. Answers to all review problems are there, along with section numbers in italics to indicate where each type of problem is discussed. Where weaknesses show up, review appropriate sections of the text.

A

1. Use a calculator to evaluate $A = 2{,}000e^{0.09t}$ to the nearest cent for $t = 5$, 10, and 20.

Find the indicated derivatives in Problems 2–4.

2. $\frac{d}{dx}(2 \ln x + 3e^x)$ **3.** $\frac{d}{dx}e^{2x-3}$

4. y' for $y = \ln(2x + 7)$

5. Let $y = \ln u$ and $u = 3 + e^x$.
(A) Express y in terms of x.
(B) Use the chain rule to find dy/dx, and then express dy/dx in terms of x.

6. Find y' for $y = y(x)$ defined implicity by the equation $2y^2 - 3x^3 - 5 = 0$, and evaluate at $(x, y) = (1, 2)$.

7. For $y = 3x^2 - 5$, where $x = x(t)$ and $y = y(t)$, find dy/dt if $dx/dt = 3$ when $x = 12$.

8. Given the demand equation $25p + x = 1{,}000$,
(A) Express the demand x as a function of the price p.
(B) Find the elasticity of demand, $E(p)$.
(C) Find $E(15)$ and interpret.
(D) Express the revenue function as a function of price p.
(E) If $p = \$25$, what is the effect of a price cut on revenue?

B

9. Find the slope of the line tangent to $y = 100e^{-0.1x}$ when $x = 0$.

10. Use a calculator and a table of values to investigate

$$\lim_{n\to\infty}\left(1 + \frac{2}{n}\right)^n$$

Do you think the limit exists? If so, what do you think it is?

Find the indicated derivatives in Problems 11–16.

11. $\frac{d}{dz}[(\ln z)^7 + \ln z^7]$ **12.** $\frac{d}{dx}(x^6 \ln x)$

13. $\frac{d}{dx}\frac{e^x}{x^6}$ **14.** y' for $y = \ln(2x^3 - 3x)$

15. $f'(x)$ for $f(x) = e^{x^3-x^2}$ **16.** dy/dx for $y = e^{-2x} \ln 5x$

17. Find the equation of the line tangent to the graph of $y = f(x) = 1 + e^{-x}$ at $x = 0$. At $x = -1$.

18. Find y' for $y = y(x)$ defined implicitly by the equation $x^2 - 3xy + 4y^2 = 23$, and find the slope of the graph at $(-1, 2)$.

19. Find x' for $x = x(t)$ defined implicitly by $x^3 - 2t^2x + 8 = 0$, and evaluate at $(t, x) = (-2, 2)$.

20. Find y' for $y = y(x)$ defined implicitly by $x - y^2 = e^y$, and evaluate at $(1, 0)$.

21. Find y' for $y = y(x)$ defined implicitly by $\ln y = x^2 - y^2$, and evaluate at $(1, 1)$.

22. A point is moving on the graph of $y^2 - 4x^2 = 12$ so that its x coordinate is decreasing by 2 units per second when $(x, y) = (1, 4)$. Find the rate of change of the y coordinate.

23. A 17-foot ladder is placed against a wall. If the foot of the ladder is pushed toward the wall at 0.5 foot per second, how fast is the top of the ladder rising when the foot is 8 feet from the wall?

24. Water is leaking onto a floor. The resulting circular pool has an area that is increasing at the rate of 24 square inches per minute. How fast is the radius R of the pool increasing when the radius is 12 inches? $[A = \pi R^2]$

C

25. Find the values of p for which demand is elastic and the values for which demand is inelastic if the price–demand equation is

$$x = f(p) = 20(p - 15)^2 \qquad 0 \le p \le 15$$

26. Graph the revenue function and indicate the regions of inelastic and elastic demand of the graph if the price–demand equation is

$$x = f(p) = 5(20 - p) \qquad 0 \le p \le 20$$

27. Let $y = w^3$, $w = \ln u$, and $u = 4 - e^x$.

(A) Express y in terms of x.

(B) Use the chain rule to find dy/dx, and then express dy/dx in terms of x.

Find the indicated derivatives in Problems 28–30.

28. y' for $y = 5^{x^2-1}$

29. $\frac{d}{dx}\log_5(x^2 - x)$

30. $\frac{d}{dx}\sqrt{\ln(x^2 + x)}$

31. Find y' for $y = y(x)$ defined implicitly by the equation $e^{xy} = x^2 + y + 1$, and evaluate at $(0, 0)$.

32. A rock thrown into a still pond causes a circular ripple. The radius is increasing at a constant rate of 3 feet per second. Show that the area does not increase at a constant rate. When is the rate of increase of the area the smallest? The largest? Explain.

33. A point moves along the graph of $y = x^3$ in such a way that its y coordinate is increasing at a constant rate of 5 units per second. Does the x coordinate ever increase at a faster rate than the y coordinate? Explain.

Applications

34. **Doubling time.** How long will it take money to double if it is invested at 5% interest compounded

(A) Annually? (B) Continuously?

35. **Continuous compound interest.** If \$100 is invested at 10% interest compounded continuously, then the amount (in dollars) at the end of t years is given by

$$A = 100e^{0.1t}$$

Find $A'(t)$, $A'(1)$, and $A'(10)$.

36. **Marginal analysis.** The price–demand equation for 14-cubic-foot refrigerators at an appliance store is

$$p(x) = 1{,}000e^{-0.02x}$$

where x is the monthly demand and p is the price in dollars. Find the marginal revenue equation.

37. **Demand equation.** Given the demand equation

$$x = \sqrt{5{,}000 - 2p^3}$$

find the rate of change of p with respect to x by implicit differentiation (x is the number of items that can be sold at a price of \$$p$ per item).

38. **Rate of change of revenue.** A company is manufacturing kayaks and can sell all that it manufactures. The revenue (in dollars) is given by

$$R = 750x - \frac{x^2}{30}$$

where the production output in 1 day is x kayaks. If production is increasing at 3 kayaks per day when production is 40 kayaks per day, find the rate of increase in revenue.

39. **Revenue and elasticity.** The price–demand equation for home-delivered large pizzas is

$$p = 38.2 - 0.002x$$

where x is the number of pizzas delivered weekly. The current price of one pizza is \$21. In order to generate additional revenue from the sale of large pizzas, would you recommend a price increase or a price decrease? Explain.

40. **Average income.** A model for the average income per household before taxes are paid is

$$f(t) = 1{,}700t + 20{,}500$$

where t is years since 1980. Find the relative rate of change of household income in 2015.

41. **Drug concentration.** The drug concentration in the bloodstream t hours after injection is given approximately by

$$C(t) = 5e^{-0.3t}$$

where $C(t)$ is concentration in milligrams per milliliter. What is the rate of change of concentration after 1 hour? After 5 hours?

42. **Wound healing.** A circular wound on an arm is healing at the rate of 45 square millimeters per day (the area of the wound is decreasing at this rate). How fast is the radius R of the wound decreasing when $R = 15$ millimeters? $[A = \pi R^2]$

43. **Psychology: learning.** In a computer assembly plant, a new employee, on the average, is able to assemble

$$N(t) = 10(1 - e^{-0.4t})$$

units after t days of on-the-job training.

(A) What is the rate of learning after 1 day? After 5 days?

(B) Find the number of days (to the nearest day) after which the rate of learning is less than 0.25 unit per day.

44. **Learning.** A new worker on the production line performs an operation in T minutes after x performances of the operation, as given by

$$T = 2\left(1 + \frac{1}{x^{3/2}}\right)$$

If, after performing the operation 9 times, the rate of improvement is $dx/dt = 3$ operations per hour, find the rate of improvement in time dT/dt in performing each operation.

Graphing and Optimization

Introduction

Since the derivative is associated with the slope of the graph of a function at a point, we might expect that it is also related to other properties of a graph. As we will see in this chapter, the derivative can tell us a great deal about the shape of the graph of a function. In particular, we will study methods for finding absolute maximum and minimum values. Manufacturing companies can use these methods to find production levels that will minimize cost or maximize profit, pharmacologists can use them to find levels of drug dosages that will produce maximum sensitivity, and advertisers can use them to determine the number of ads that will maximize the rate of change of sales (see, for example, Problem 93 in Section 3-2).

3-1 First Derivative and Graphs

- Increasing and Decreasing Functions
- Local Extrema
- First-Derivative Test
- Economics Applications

Increasing and Decreasing Functions

Sign charts will be used throughout this chapter. You may find it helpful to review the terminology and techniques for constructing sign charts in Section 1-2.

EXPLORE & DISCUSS 1

Figure 1 shows the graph of $y = f(x)$ and a sign chart for $f'(x)$, where

$$f(x) = x^3 - 3x$$

and

$$f'(x) = 3x^2 - 3 = 3(x + 1)(x - 1)$$

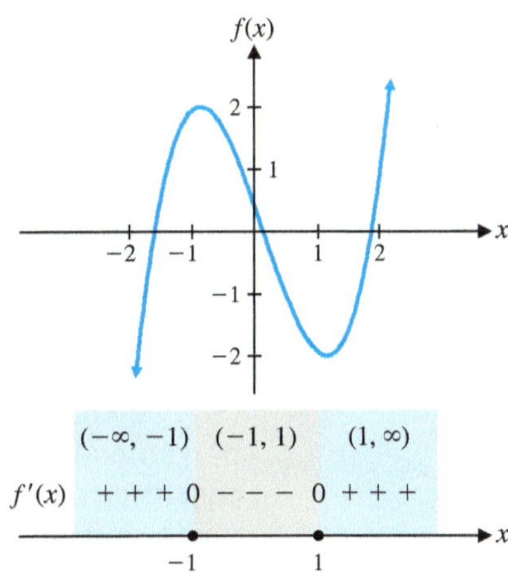

Figure 1

Discuss the relationship between the graph of f and the sign of $f'(x)$ over each interval on which $f'(x)$ has a constant sign. Also, describe the behavior of the graph of f at each partition number for f'.

As they are scanned from left to right, graphs of functions generally have rising and falling sections. If you scan the graph of $f(x) = x^3 - 3x$ in Figure 1 from left to right, you will observe the following:

- On the interval $(-\infty, -1)$, the graph of f is rising, $f(x)$ is increasing,* and the slope of the graph is positive $[f'(x) > 0]$.
- On the interval $(-1, 1)$, the graph of f is falling, $f(x)$ is decreasing, and the slope of the graph is negative $[f'(x) < 0]$.
- On the interval $(1, \infty)$, the graph of f is rising, $f(x)$ is increasing, and the slope of the graph is positive $[f'(x) > 0]$.
- At $x = -1$ and $x = 1$, the slope of the graph is 0 $[f'(x) = 0]$.

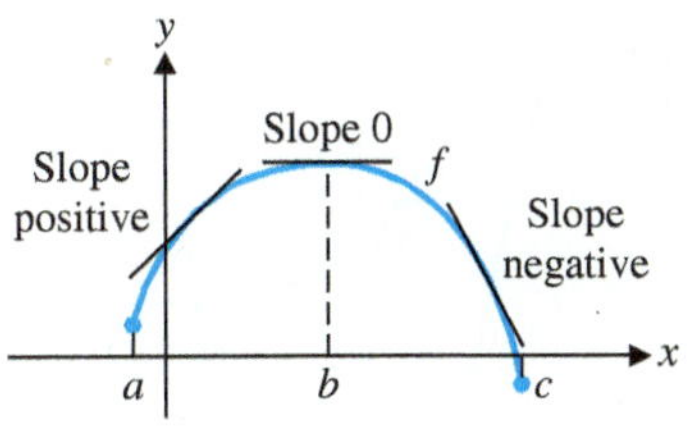

Slope positive | Slope 0 | f | Slope negative | y | x | a | b | c

Figure 2

If $f'(x) > 0$ (is positive) on the interval (a, b) (Fig. 2), then $f(x)$ increases (↗) and the graph of f rises as we move from left to right over the interval. If $f'(x) < 0$ (is negative) on an interval (a, b), then $f(x)$ decreases (↘) and the graph of f falls as we move from left to right over the interval. We summarize these important results in Theorem 1.

*Formally, we say that the function f is **increasing** on an interval (a, b) if $f(x_2) > f(x_1)$ whenever $a < x_1 < x_2 < b$, and f is **decreasing** on (a, b) if $f(x_2) < f(x_1)$ whenever $a < x_1 < x_2 < b$.

THEOREM 1 Increasing and Decreasing Functions

For the interval (a, b),

$f'(x)$	$f(x)$	Graph of f	Examples
+	Increases ↗	Rises ↗	
–	Decreases ↘	Falls ↘	

EXAMPLE 1 **Finding Intervals on Which a Function Is Increasing or Decreasing** Given the function $f(x) = 8x - x^2$,

(A) Which values of x correspond to horizontal tangent lines?

(B) For which values of x is $f(x)$ increasing? Decreasing?

(C) Sketch a graph of f. Add any horizontal tangent lines.

SOLUTION (A) $f'(x) = 8 - 2x = 0$

$$x = 4$$

So, a horizontal tangent line exists at $x = 4$ only.

(B) We will construct a sign chart for $f'(x)$ to determine which values of x make $f'(x) > 0$ and which values make $f'(x) < 0$. Recall from Section 1-2 that the partition numbers for a function are the points where the function is 0 or discontinuous. When constructing a sign chart for $f'(x)$, we must locate all points where $f'(x) = 0$ or $f'(x)$ is discontinuous. From part (A), we know that $f'(x) = 8 - 2x = 0$ at $x = 4$. Since $f'(x) = 8 - 2x$ is a polynomial, it is continuous for all x. So, 4 is the only partition number. We construct a sign chart for the intervals $(-\infty, 4)$ and $(4, \infty)$, using test numbers 3 and 5:

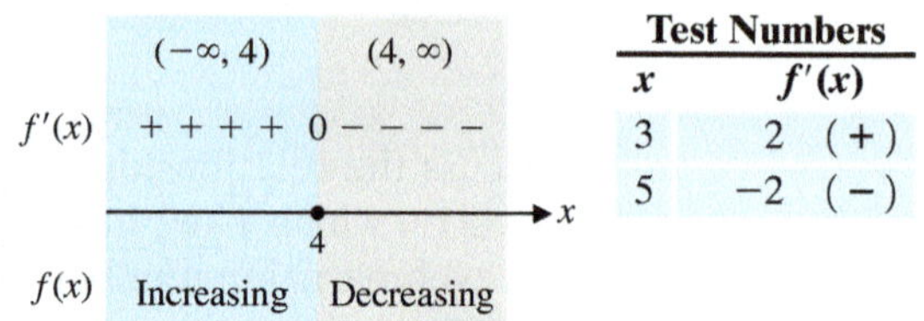

Test Numbers

x	$f'(x)$
3	2 (+)
5	−2 (−)

Therefore, $f(x)$ is increasing on $(-\infty, 4)$ and decreasing on $(4, \infty)$.

x	$f(x)$
0	0
2	12
4	16
6	12
8	0

f(x)
Horizontal tangent line
15
10
5
0
5
10
x
f(x) increasing
f(x) decreasing

Matched Problem 1 Repeat Example 1 for $f(x) = x^2 - 6x + 10$.

As Example 1 illustrates, the construction of a sign chart will play an important role in using the derivative to analyze and sketch the graph of a function *f*. The partition numbers for f' are central to the construction of these sign charts and also to the analysis of the graph of $y = f(x)$. We already know that if $f'(c) = 0$, then the graph of $y = f(x)$ will have a horizontal tangent line at $x = c$. But the partition numbers for f' also include the numbers *c* such that $f'(c)$ does not exist.* There are two possibilities at this type of number: (1) $f(c)$ does not exist; or (2) $f(c)$ exists but the slope of the tangent line at $x = c$ is undefined.

DEFINITION Critical Values

The values of x in the domain of *f* where $f'(x) = 0$ or where $f'(x)$ does not exist are called the **critical values** of *f*.

CONCEPTUAL INSIGHT

The critical values of *f* are always in the domain of *f* and are also partition numbers for f', but f' may have partition numbers that are not critical values.

If *f* is a polynomial, then both the partition numbers for f' and the critical values of *f* are the solutions of $f'(x) = 0$.

EXAMPLE 2 **Partition Numbers and Critical Values** Find the critical values of *f*, the intervals on which *f* is increasing, and those on which *f* is decreasing, for $f(x) = 1 + x^3$.

SOLUTION Begin by finding the partition number for $f'(x)$:

$$f'(x) = 3x^2 = 0, \quad \text{only at } x = 0$$

The partition number 0 is in the domain of *f*, so 0 is the only critical value of *f*.

The sign chart for $f'(x) = 3x^2$ (partition number is 0) is

	$(-\infty, 0)$		$(0, \infty)$
$f'(x)$	+ + + + +	0	+ + + + +
x		0	
$f(x)$	Increasing		Increasing

Test Numbers

x	$f'(x)$
−1	3 (+)
1	3 (+)

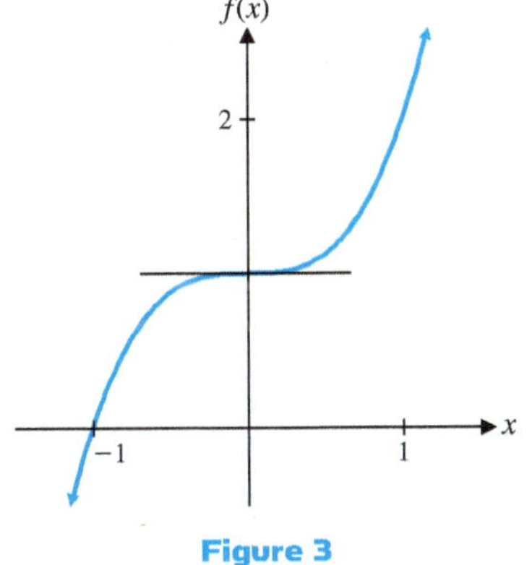

Figure 3

The sign chart indicates that $f(x)$ is increasing on $(-\infty, 0)$ and $(0, \infty)$. Since *f* is continuous at $x = 0$, it follows that $f(x)$ is increasing for all *x*. The graph of *f* is shown in Figure 3.

Matched Problem 2 Find the critical values of *f*, the intervals on which *f* is increasing, and those on which *f* is decreasing, for $f(x) = 1 - x^3$.

EXAMPLE 3 **Partition Numbers and Critical Values** Find the critical values of *f*, the intervals on which *f* is increasing, and those on which *f* is decreasing, for $f(x) = (1 - x)^{1/3}$.

SOLUTION

$$f'(x) = -\frac{1}{3}(1 - x)^{-2/3} = \frac{-1}{3(1 - x)^{2/3}}$$

*We are assuming that $f'(c)$ does not exist at any point of discontinuity of f'. There do exist functions *f* such that f' is discontinuous at $x = c$, yet $f'(c)$ exists. However, we do not consider such functions in this book.

To find partition numbers for f', we note that f' is continuous for all x, except for values of x for which the denominator is 0; that is, $f'(1)$ does not exist and f' is discontinuous at $x = 1$. Since the numerator is the constant -1, $f'(x) \neq 0$ for any value of x. Thus, $x = 1$ is the only partition number for f'. Since 1 is in the domain of f, $x = 1$ is also the only critical value of f. When constructing the sign chart for f' we use the abbreviation ND to note the fact that $f'(x)$ is *not defined* at $x = 1$.

The sign chart for $f'(x) = -1/[3(1 - x)^{2/3}]$ (partition number is 1) is as follows:

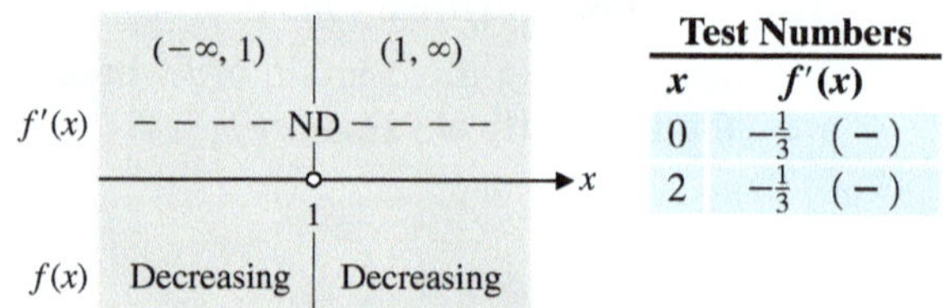

Test Numbers

x	$f'(x)$
0	$-\frac{1}{3}$ (−)
2	$-\frac{1}{3}$ (−)

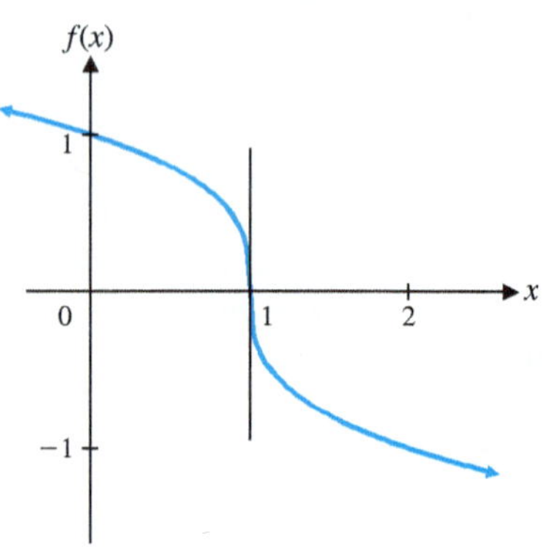

Figure 4

The sign chart indicates that f is decreasing on $(-\infty, 1)$ and $(1, \infty)$. Since f is continuous at $x = 1$, it follows that $f(x)$ is decreasing for all x. **A continuous function can be decreasing (or increasing) on an interval containing values of x where $f'(x)$ does not exist.** The graph of f is shown in Figure 4. Notice that the undefined derivative at $x = 1$ results in a vertical tangent line at $x = 1$. **A vertical tangent will occur at $x = c$ if f is continuous at $x = c$ and if $|f'(x)|$ becomes larger and larger as x approaches c.**

Matched Problem 3 Find the critical values of f, the intervals on which f is increasing, and those on which f is decreasing, for $f(x) = (1 + x)^{1/3}$.

EXAMPLE 4 **Partition Numbers and Critical Values** Find the critical values of f, the intervals on which f is increasing, and those on which f is decreasing, for $f(x) = \dfrac{1}{x - 2}$.

SOLUTION

$$f(x) = \frac{1}{x - 2} = (x - 2)^{-1}$$

$$f'(x) = -(x - 2)^{-2} = \frac{-1}{(x - 2)^2}$$

To find the partition numbers for f', note that $f'(x) \neq 0$ for any x and f' is not defined at $x = 2$. Thus, $x = 2$ is the only partition number for f'. However, $x = 2$ is *not* in the domain of f. Consequently, $x = 2$ is not a critical value of f. This function has no critical values.

The sign chart for $f'(x) = -1/(x - 2)^2$ (partition number is 2) is as follows:

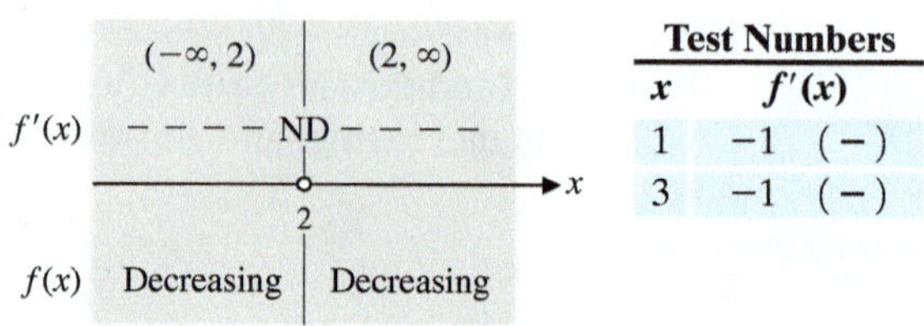

Test Numbers

x	$f'(x)$
1	−1 (−)
3	−1 (−)

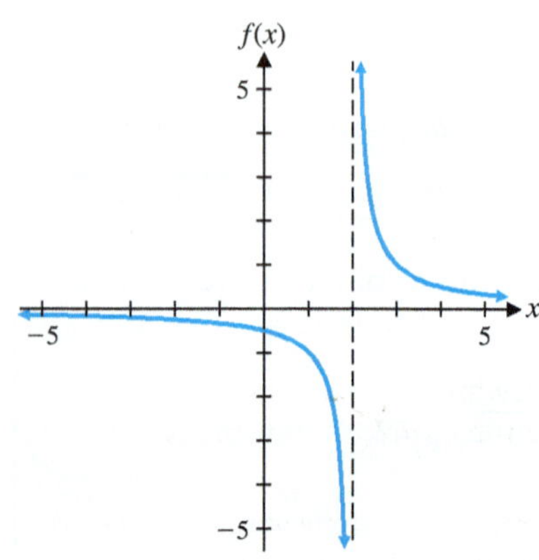

Figure 5

Therefore, f is decreasing on $(-\infty, 2)$ and $(2, \infty)$. The graph of f is shown in Figure 5.

Matched Problem 4 Find the critical values for f, the intervals on which f is increasing, and those on which f is decreasing, for $f(x) = \frac{1}{x}$.

EXAMPLE 5 **Partition Numbers and Critical Values** Find the critical values of f, the intervals on which f is increasing, and those on which f is decreasing, for $f(x) = 8 \ln x - x^2$.

SOLUTION The natural logarithm function $\ln x$ is defined on $(0, \infty)$, or $x > 0$, so $f(x)$ is defined only for $x > 0$.

$$f(x) = 8 \ln x - x^2, \; x > 0$$

$$f'(x) = \frac{8}{x} - 2x \qquad \text{Find a common denominator.}$$

$$= \frac{8}{x} - \frac{2x^2}{x} \qquad \text{Subtract numerators.}$$

$$= \frac{8 - 2x^2}{x} \qquad \text{Factor numerator.}$$

$$= \frac{2(2 - x)(2 + x)}{x}, \quad x > 0$$

Note that $f'(x) = 0$ at -2 and at 2, and $f'(x)$ is discontinuous at 0. These are the partition numbers for $f'(x)$. Since the domain of f is $(0, \infty)$, 0 and -2 are not critical values. The remaining partition number, 2, is the only critical value for $f(x)$.

The sign chart for $f'(x) = \frac{2(2 - x)(2 + x)}{x}$, $x > 0$ (partition number is 2), is as follows:

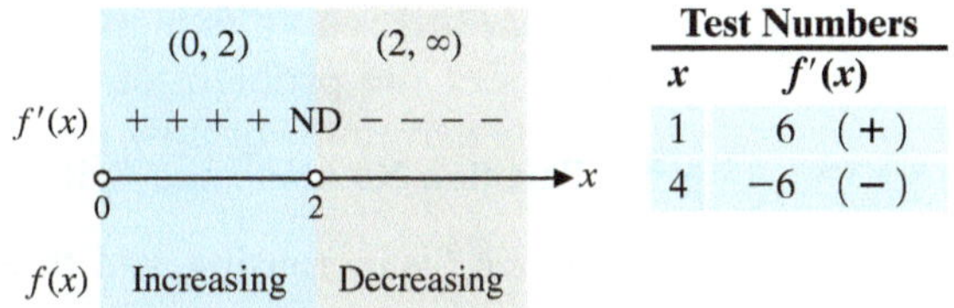

Test Numbers

x	$f'(x)$
1	6 (+)
4	−6 (−)

Therefore, f is increasing on $(0, 2)$ and decreasing on $(2, \infty)$. The graph of f is shown in Figure 6.

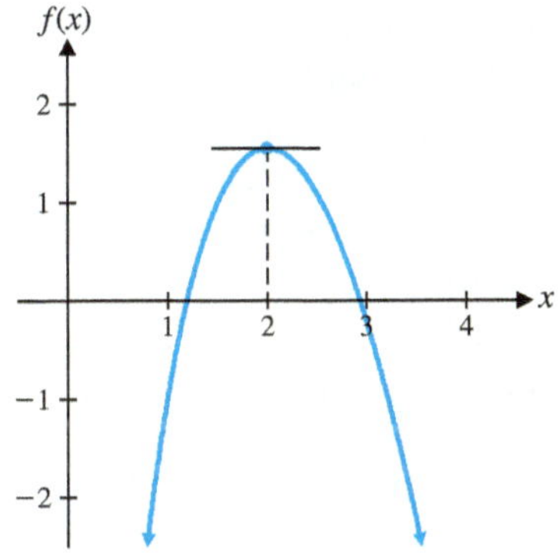

Figure 6

Matched Problem 5 Find the critical values of f, the intervals on which f is increasing, and those on which f is decreasing, for $f(x) = 5 \ln x - x$.

CONCEPTUAL INSIGHT

Examples 4 and 5 illustrate two important ideas:

1. Do not assume that all partition numbers for the derivative f' are critical values of the function f. To be a critical value, a partition number must also be in the domain of f.
2. The values for which a function is increasing or decreasing must always be expressed in terms of open intervals that are subsets of the domain of the function.

Local Extrema

When the graph of a continuous function changes from rising to falling, a high point, or *local maximum,* occurs. When the graph changes from falling to rising, a low point, or *local minimum,* occurs. In Figure 7, high points occur at c_3 and c_6, and low points occur at c_2 and c_4. In general, we call $f(c)$ a **local maximum** if there exists an interval (m, n) containing c such that

$$f(x) \le f(c) \qquad \text{for all } x \text{ in } (m, n)$$

Note that this inequality need hold only for values of x near c, which is why we use the term *local.*

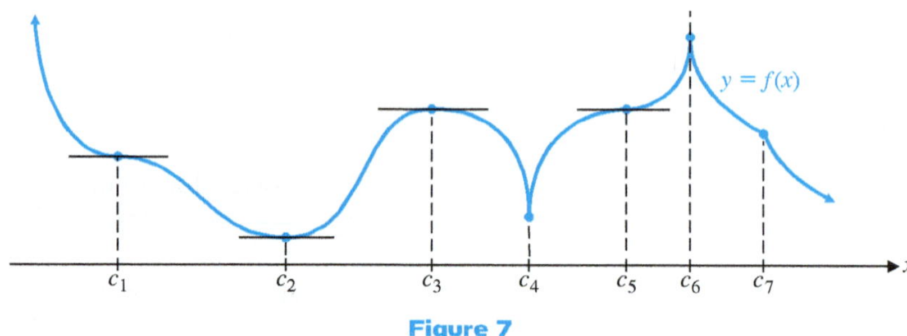

Figure 7

The quantity $f(c)$ is called a **local minimum** if there exists an interval (m, n) containing c such that

$$f(x) \ge f(c) \qquad \text{for all } x \text{ in } (m, n)$$

The quantity $f(c)$ is called a **local extremum** if it is either a local maximum or a local minimum. A point on a graph where a local extremum occurs is also called a **turning point**. In Figure 7 we see that local maxima occur at c_3 and c_6, local minima occur at c_2 and c_4, and all four values produce local extrema. Also, the local maximum $f(c_3)$ is not the highest point on the graph in Figure 7. Later in this chapter, we consider the problem of finding the highest and lowest points on a graph, or absolute extrema. For now, we are concerned only with locating *local* extrema.

EXAMPLE 6 **Analyzing a Graph** Use the graph of f in Figure 8 to find the intervals on which f is increasing, those on which f is decreasing, any local maxima, and any local minima.

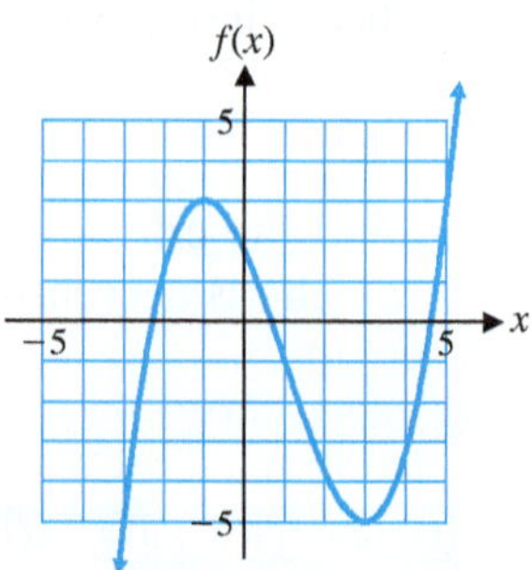

Figure 8

SOLUTION The function f is increasing (the graph is rising) on $(-\infty, -1)$ and on $(3, \infty)$ and is decreasing (the graph is falling) on $(-1, 3)$. Because the graph changes from rising to falling at $x = -1$, $f(-1) = 3$ is a local maximum. Because the graph changes from falling to rising at $x = 3$, $f(3) = -5$ is a local minimum.

Matched Problem 6 Use the graph of g in Figure 9 to find the intervals on which g is increasing, those on which g is decreasing, any local maxima, and any local minima.

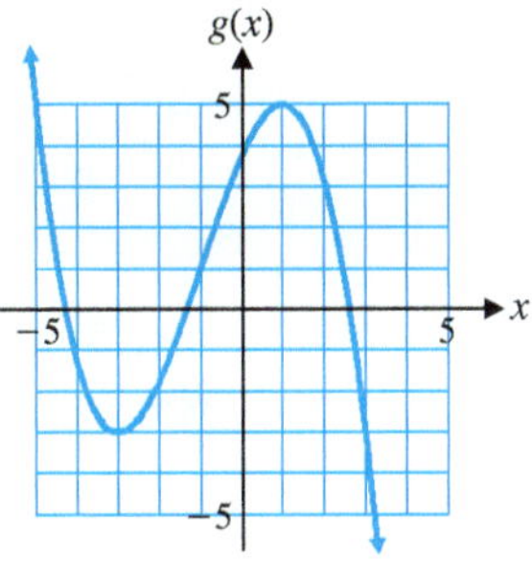

Figure 9

How can we locate local maxima and minima if we are given the equation of a function and not its graph? The key is to examine the critical values of the function. The local extrema of the function f in Figure 7 occur either at points where the derivative is 0 (c_2 and c_3) or at points where the derivative does not exist (c_4 and c_6). In other words, local extrema occur only at critical values of f. Theorem 2 shows that this is true in general.

THEOREM 2 Existence of Local Extrema

If f is continuous on the interval (a, b), c is a number in (a, b), and $f(c)$ is a local extremum, then either $f'(c) = 0$ or $f'(c)$ does not exist (is not defined).

Theorem 2 states that a local extremum can occur only at a critical value, but it does not imply that every critical value produces a local extremum. In Figure 7, c_1 and c_5 are critical values (the slope is 0), but the function does not have a local maximum or local minimum at either of these values.

Our strategy for finding local extrema is now clear: We find all critical values of f and test each one to see if it produces a local maximum, a local minimum, or neither.

First-Derivative Test

If $f'(x)$ exists on both sides of a critical value c, the sign of $f'(x)$ can be used to determine whether the point $(c, f(c))$ is a local maximum, a local minimum, or neither. The various possibilities are summarized in the following box and are illustrated in Figure 10:

PROCEDURE First-Derivative Test for Local Extrema

Let c be a critical value of f [$f(c)$ is defined and either $f'(c) = 0$ or $f'(c)$ is not defined]. Construct a sign chart for $f'(x)$ close to and on either side of c.

Sign Chart	$f(c)$
$f'(x)$ − − − + + + (m c n) x $f(x)$ Decreasing Increasing	$f(c)$ is a local minimum. If $f'(x)$ changes from negative to positive at c, then $f(c)$ is a local minimum.

Sign Chart	$f(c)$
$f'(x)$: + + + | − − − on (m, n), change at c; $f(x)$: Increasing | Decreasing	$f(c)$ is a local maximum. If $f'(x)$ changes from positive to negative at c, then $f(c)$ is a local maximum.
$f'(x)$: + + + | + + + on (m, n), at c; $f(x)$: Increasing | Increasing	$f(c)$ is not a local extremum. If $f'(x)$ does not change sign at c, then $f(c)$ is neither a local maximum nor a local minimum.
$f'(x)$: − − − | − − − on (m, n), at c; $f(x)$: Decreasing | Decreasing	$f(c)$ is not a local extremum. If $f'(x)$ does not change sign at c, then $f(c)$ is neither a local maximum nor a local minimum.

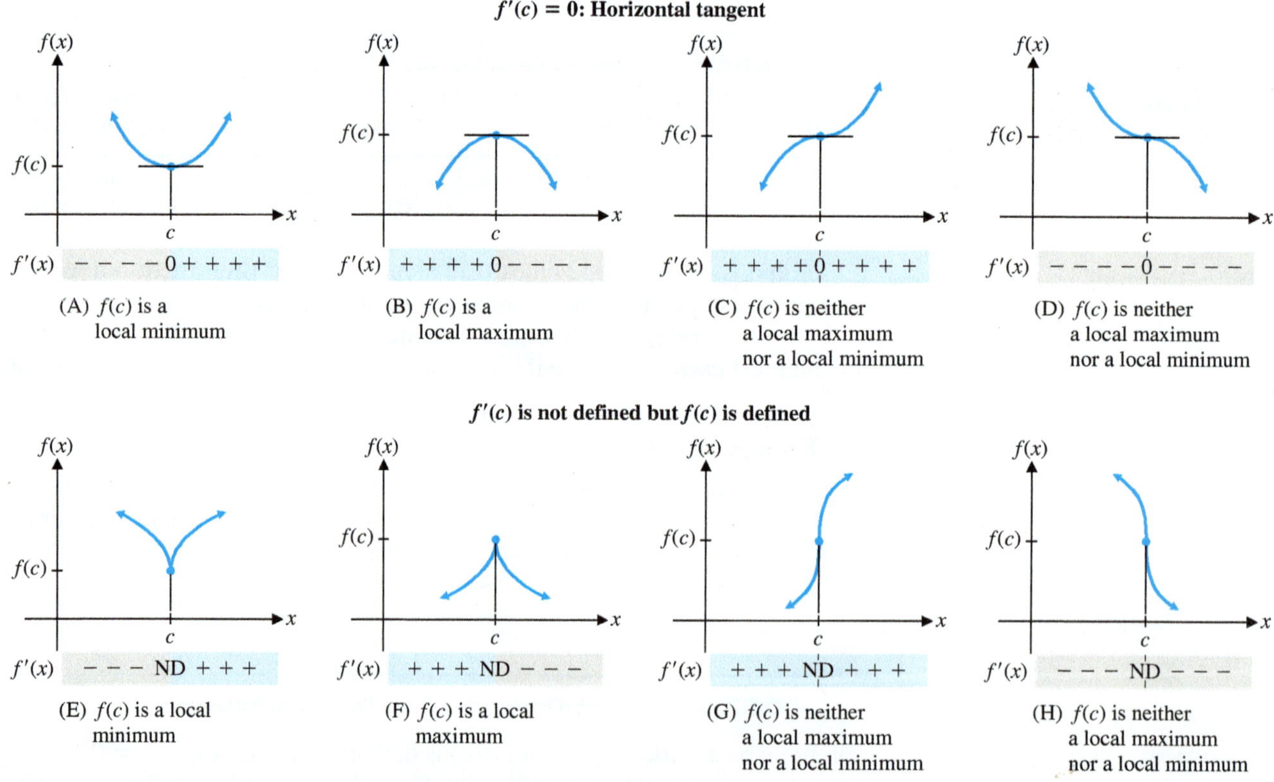

Figure 10 Local extrema

EXAMPLE 7 **Locating Local Extrema** Given $f(x) = x^3 - 6x^2 + 9x + 1$,

(A) Find the critical values of f.

(B) Find the local maxima and minima.

(C) Sketch the graph of f.

SOLUTION (A) Find all numbers x in the domain of f where $f'(x) = 0$ or $f'(x)$ does not exist.

$$f'(x) = 3x^2 - 12x + 9 = 0$$
$$3(x^2 - 4x + 3) = 0$$
$$3(x - 1)(x - 3) = 0$$
$$x = 1 \quad \text{or} \quad x = 3$$

$f'(x)$ exists for all x; the critical values are $x = 1$ and $x = 3$.

(B) The easiest way to apply the first-derivative test for local maxima and minima is to construct a sign chart for $f'(x)$ for all x. Partition numbers for $f'(x)$ are $x = 1$ and $x = 3$ (which also happen to be critical values of f).

Sign chart for $f'(x) = 3(x - 1)(x - 3)$:

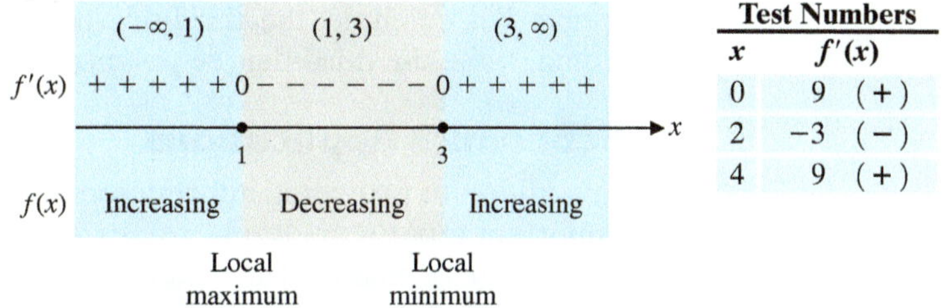

Test Numbers	
x	$f'(x)$
0	9 (+)
2	−3 (−)
4	9 (+)

The sign chart indicates that f increases on $(-\infty, 1)$, has a local maximum at $x = 1$, decreases on $(1, 3)$, has a local minimum at $x = 3$, and increases on $(3, \infty)$. These facts are summarized in the following table:

x	$f'(x)$	$f(x)$	Graph of f
$(-\infty, 1)$	+	Increasing	Rising
$x = 1$	0	Local maximum	Horizontal tangent
$(1, 3)$	−	Decreasing	Falling
$x = 3$	0	Local minimum	Horizontal tangent
$(3, \infty)$	+	Increasing	Rising

(C) We sketch a graph of f, using the information from part (B) and point-by-point plotting.

x	$f(x)$
0	1
1	5
2	3
3	1
4	5

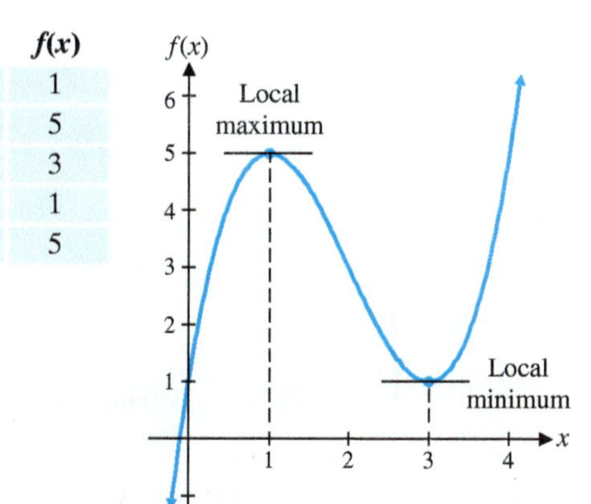

Matched Problem 7 Given $f(x) = x^3 - 9x^2 + 24x - 10$,

(A) Find the critical values of f.

(B) Find the local maxima and minima.

(C) Sketch a graph of f.

How can you tell if you have found all the local extrema of a function? In general, this can be a difficult question to answer. However, in the case of a polynomial function, there is an easily determined upper limit on the number of local extrema. Since the local extrema are the x intercepts of the derivative, this limit is a consequence of the number of x intercepts of a polynomial. The relevant information is summarized in the following theorem, which is stated without proof:

THEOREM 3 Intercepts and Local Extrema of Polynomial Functions
If $f(x) = a_n x^n + a_{n-1}x^{n-1} + \cdots + a_1 x + a_0$, $a_n \neq 0$, is an nth-degree polynomial, then f has at most n x intercepts and at most $n - 1$ local extrema.

Theorem 3 does not guarantee that every nth-degree polynomial has exactly $n - 1$ local extrema; it says only that there can never be more than $n - 1$ local extrema. For example, the third-degree polynomial in Example 7 has two local extrema, while the third-degree polynomial in Example 2 does not have any.

Economics Applications

In addition to providing information for hand-sketching graphs, the derivative is an important tool for analyzing graphs and discussing the interplay between a function and its rate of change. The next two examples illustrate this process in the context of economics applications.

EXAMPLE 8 **Agricultural Exports and Imports** Over the past few decades, the United States has exported more agricultural products than it has imported, maintaining a positive balance of trade in this area. However, the trade balance fluctuated considerably during that period. The graph in Figure 11 approximates the rate of change of the balance of trade over a 15-year period, where $B(t)$ is the balance of trade (in billions of dollars) and t is time (in years).

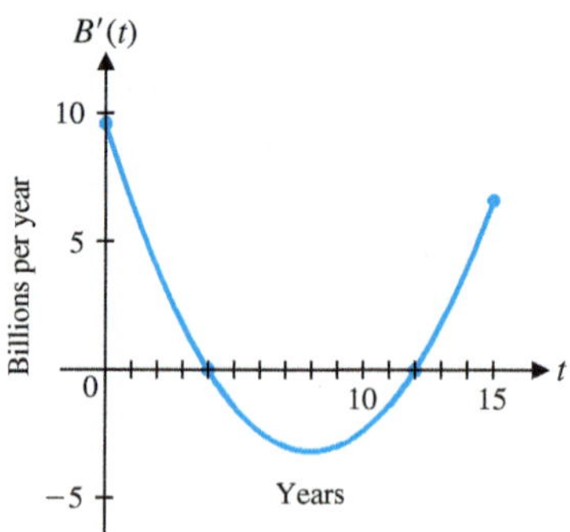

Figure 11 Rate of change of the balance of trade

(A) Write a brief description of the graph of $y = B(t)$, including a discussion of any local extrema.

(B) Sketch a possible graph of $y = B(t)$.

SOLUTION (A) The graph of the derivative $y = B'(t)$ contains the same essential information as a sign chart. That is, we see that $B'(t)$ is positive on $(0, 4)$, 0 at $t = 4$, negative on $(4, 12)$, 0 at $t = 12$, and positive on $(12, 15)$. The trade balance increases for the first 4 years to a local maximum, decreases for the next 8 years to a local minimum, and then increases for the final 3 years.

(B) Without additional information concerning the actual values of $y = B(t)$, we cannot produce an accurate graph. However, we can sketch a possible graph that illustrates the important features, as shown in Figure 12. The absence of a scale on the vertical axis is a consequence of the lack of information about the values of $B(t)$.

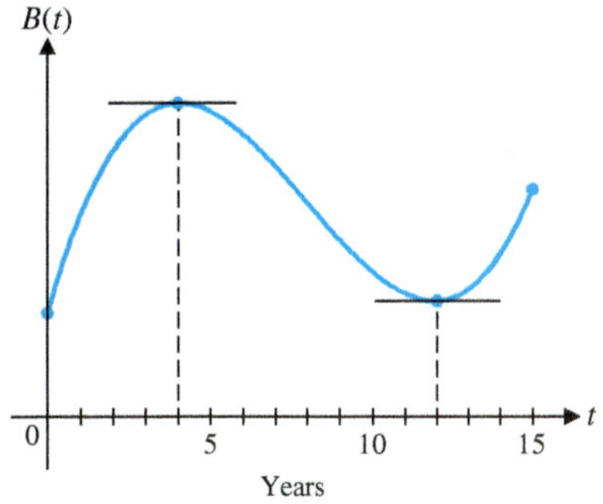

Figure 12 Balance of trade

Matched Problem 8 The graph in Figure 13 approximates the rate of change of the U.S. share of the total world production of motor vehicles over a 20-year period, where $S(t)$ is the U.S. share (as a percentage) and t is time (in years).

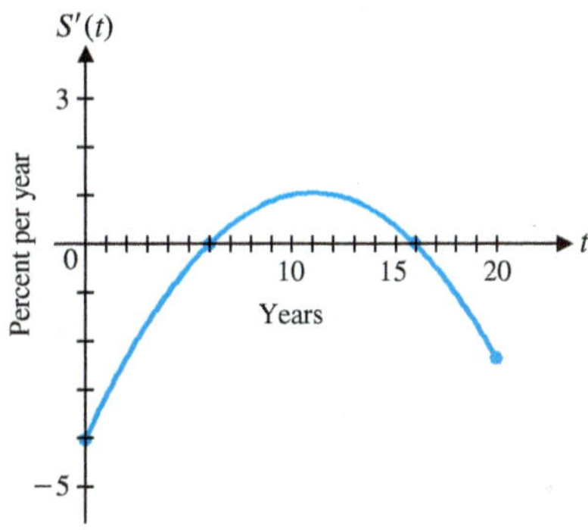

Figure 13

(A) Write a brief description of the graph of $y = S(t)$, including a discussion of any local extrema.

(B) Sketch a possible graph of $y = S(t)$.

EXAMPLE 9 **Revenue Analysis** The graph of the total revenue $R(x)$ (in dollars) from the sale of x bookcases is shown in Figure 14.

(A) Write a brief description of the graph of the marginal revenue function $y = R'(x)$, including a discussion of any x intercepts.

(B) Sketch a possible graph of $y = R'(x)$.

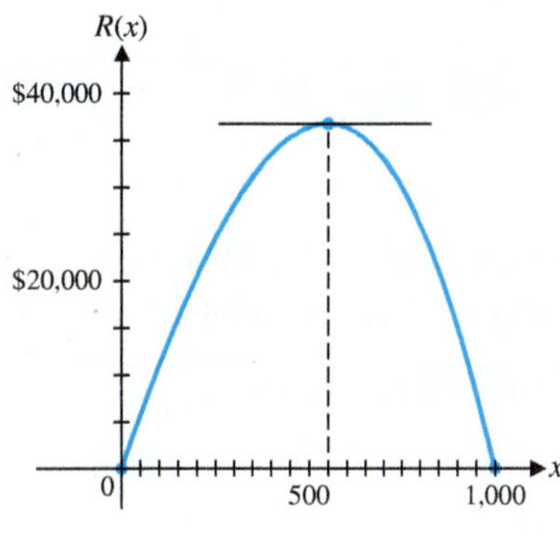

Figure 14 Revenue

SOLUTION (A) The graph of $y = R(x)$ indicates that $R(x)$ increases on $(0, 550)$, has a local maximum at $x = 550$, and decreases on $(550, 1{,}000)$. Consequently, the marginal revenue function $R'(x)$ must be positive on $(0, 550)$, 0 at $x = 550$, and negative on $(550, 1{,}000)$.

(B) A possible graph of $y = R'(x)$ illustrating the information summarized in part (A) is shown in Figure 15.

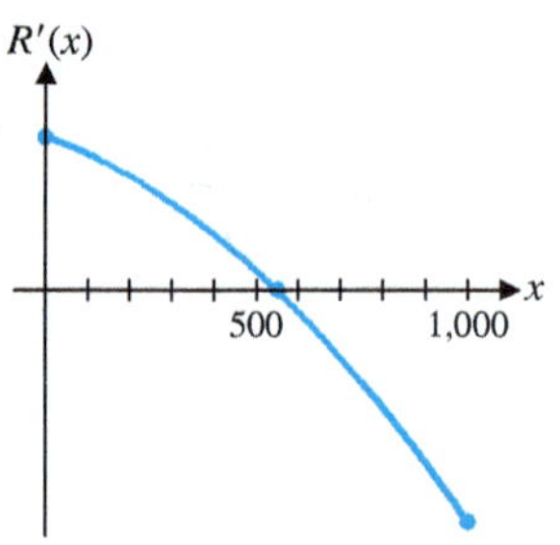

Figure 15 Marginal revenue

Matched Problem 9 The graph of the total revenue $R(x)$ (in dollars) from the sale of x desks is shown in Figure 16.

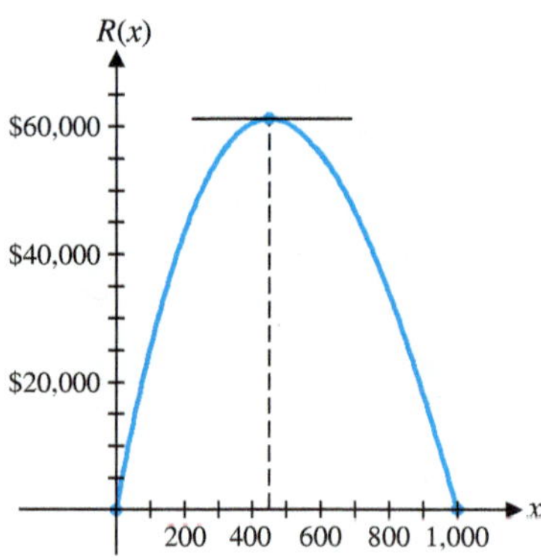

Figure 16

(A) Write a brief description of the graph of the marginal revenue function $y = R'(x)$, including a discussion of any x intercepts.

(B) Sketch a possible graph of $y = R'(x)$.

Comparing Examples 8 and 9, we see that we were able to obtain more information about the function from the graph of its derivative (Example 8) than we were when the process was reversed (Example 9). In the next section, we introduce some ideas that will help us obtain additional information about the derivative from the graph of the function.

Exercises 3-1

A

Problems 1–8 refer to the following graph of $y = f(x)$:

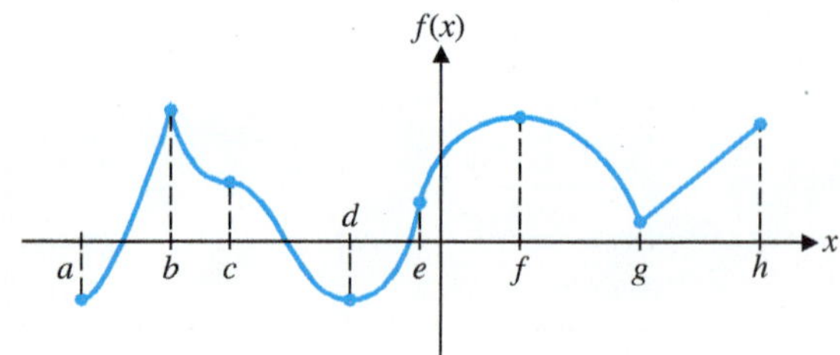

Figure for 1–8

1. Identify the intervals on which $f(x)$ is increasing.
2. Identify the intervals on which $f(x)$ is decreasing.
3. Identify the intervals on which $f'(x) < 0$.
4. Identify the intervals on which $f'(x) > 0$.

5. Identify the x coordinates of the points where $f'(x) = 0$.
6. Identify the x coordinates of the points where $f'(x)$ does not exist.
7. Identify the x coordinates of the points where $f(x)$ has a local maximum.
8. Identify the x coordinates of the points where $f(x)$ has a local minimum.

In Problems 9 and 10, $f(x)$ is continuous on $(-\infty, \infty)$ and has critical values at $x = a, b, c,$ and d. Use the sign chart for $f'(x)$ to determine whether f has a local maximum, a local minimum, or neither at each critical value.

9.

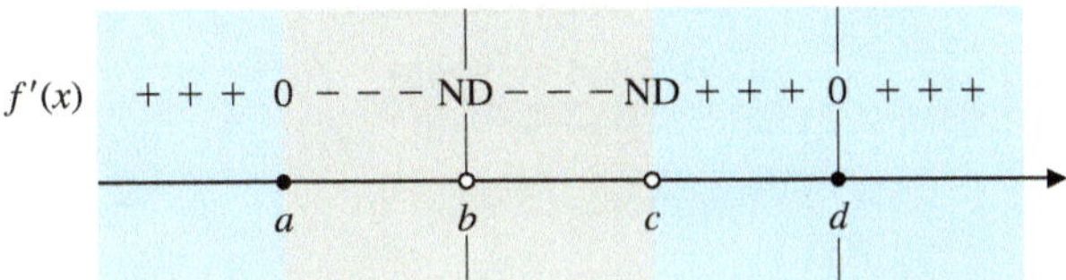

10.

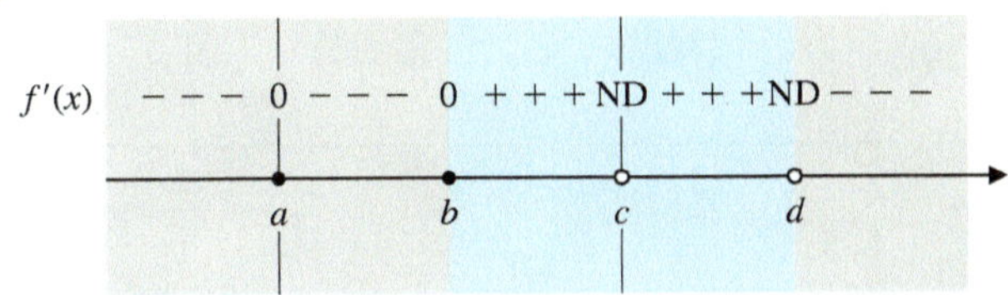

In Problems 11–18, match the graph of f with one of the sign charts a–h in the figure.

11.

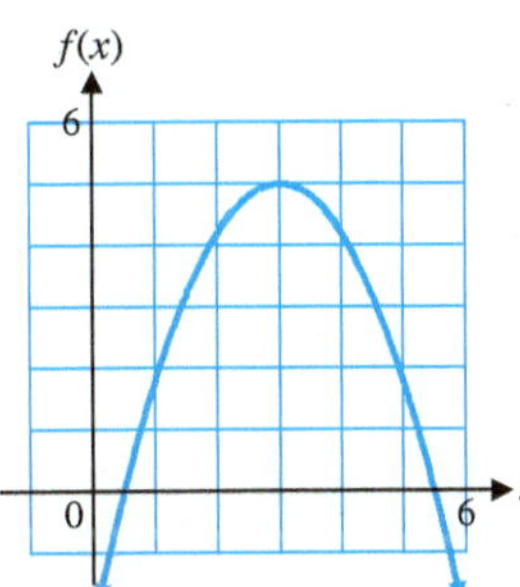

12.

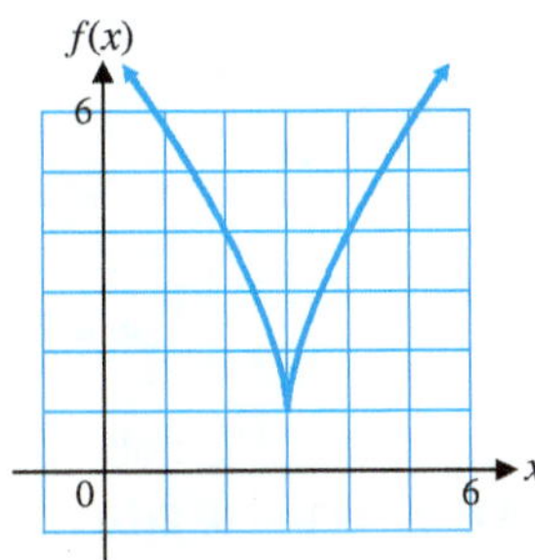

13.

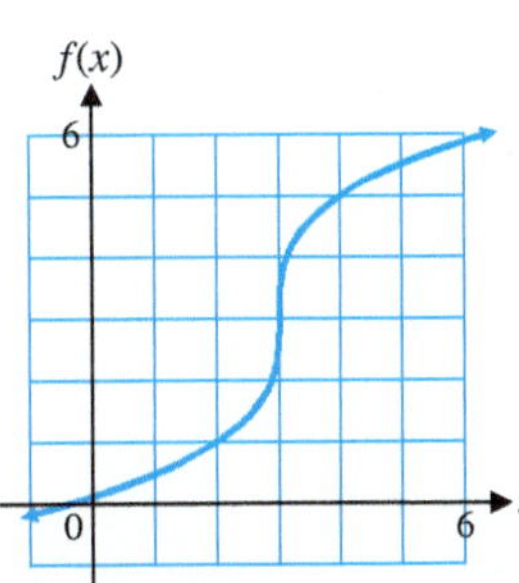

14.

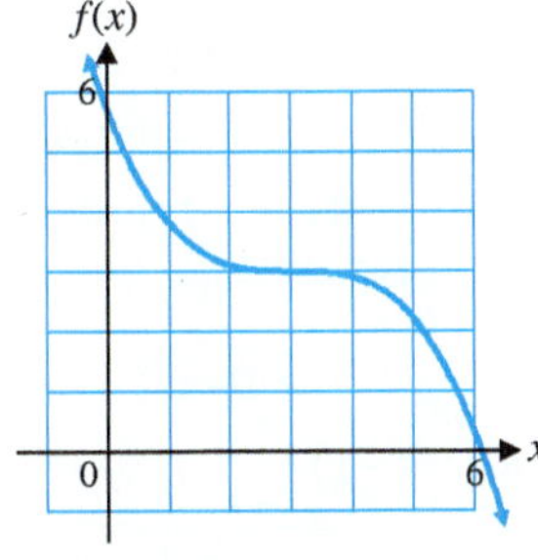

15.

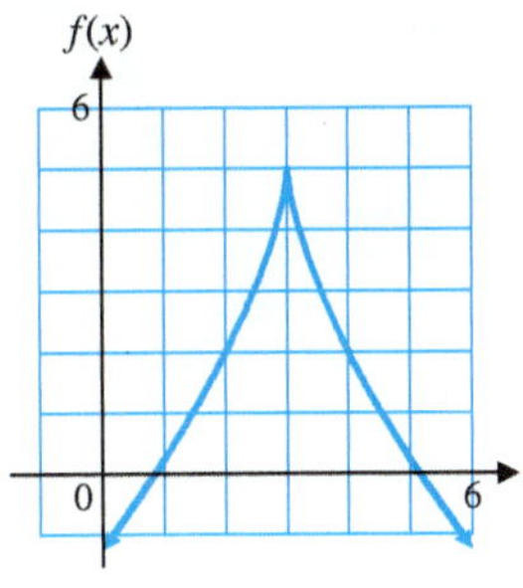

16.

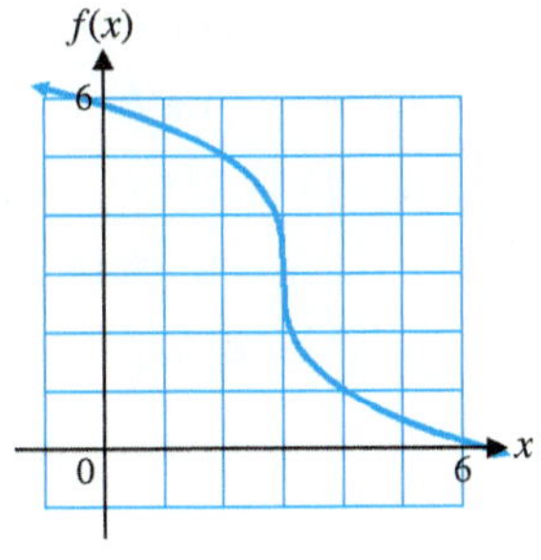

17.

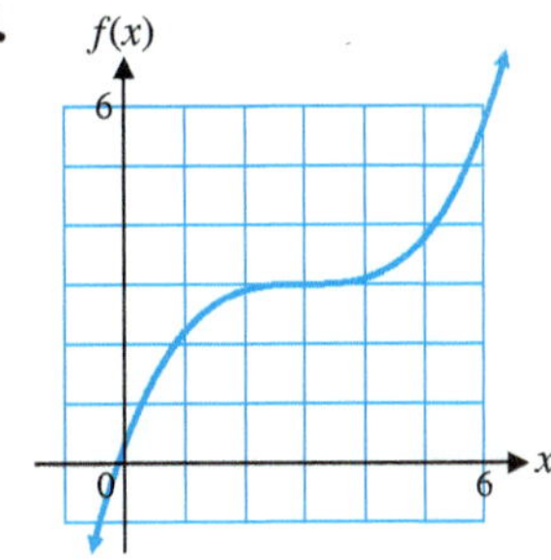

18.

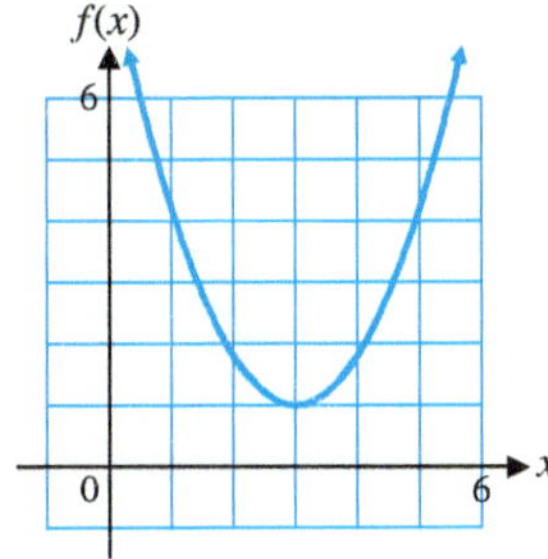

(a) $f'(x)$ – – – – – – – – 0 + + + + + + + +
3

(b)

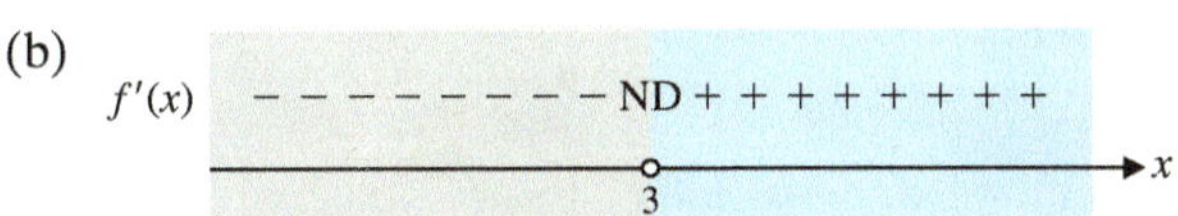

(c) $f'(x)$ + + + + + + + + 0 + + + + + + + +
3

(d)

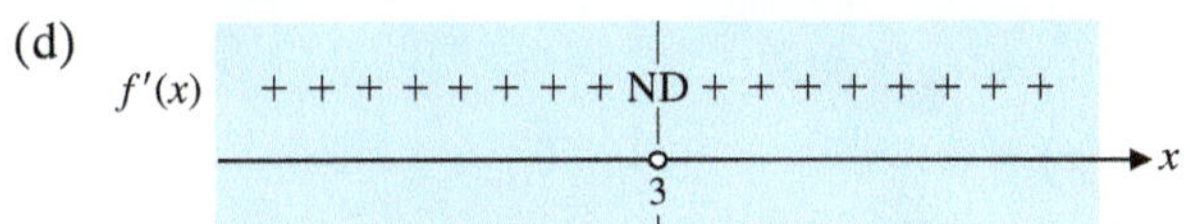

(e) $f'(x)$ + + + + + + + + 0 – – – – – – – –
3

(f)

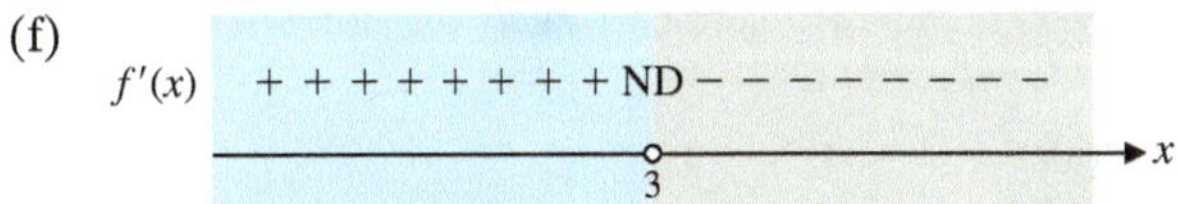

(g) $f'(x)$ – – – – – – – – ND – – – – – – – –
3

(h)

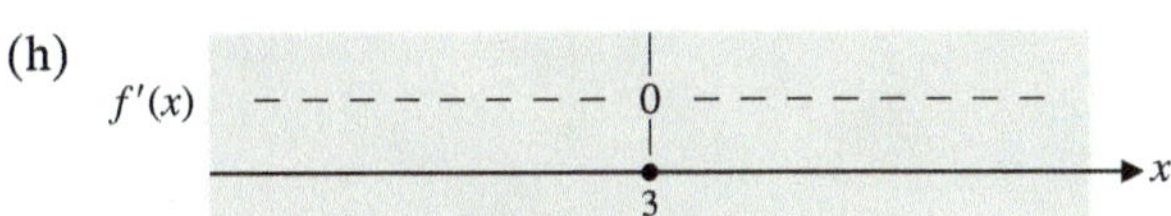

Figure for 11–18

B

In Problems 19–26, find (A) $f'(x)$, (B) the critical values of f, and (C) the partition numbers for f'.

19. $f(x) = x^3 - 12x + 8$
20. $f(x) = x^3 - 27x + 30$
21. $f(x) = (x + 5)^{1/3}$
22. $f(x) = (x - 9)^{2/3}$

23. $f(x) = \dfrac{6}{x+2}$

24. $f(x) = \dfrac{5}{x-4}$

25. $f(x) = |x|$

26. $f(x) = |x+3|$

In Problems 27–40, find the intervals on which f(x) is increasing, the intervals on which f(x) is decreasing, and the local extrema.

27. $f(x) = 2x^2 - 4x$

28. $f(x) = -3x^2 - 12x$

29. $f(x) = -2x^2 - 16x - 25$

30. $f(x) = -3x^2 + 12x - 5$

31. $f(x) = x^3 + 4x - 5$

32. $f(x) = -x^3 - 4x + 8$

33. $f(x) = 2x^3 - 3x^2 - 36x$

34. $f(x) = -2x^3 + 3x^2 + 120x$

35. $f(x) = 3x^4 - 4x^3 + 5$

36. $f(x) = x^4 + 2x^3 + 5$

37. $f(x) = (x-1)e^{-x}$

38. $f(x) = x \ln x - x$

39. $f(x) = 4x^{1/3} - x^{2/3}$

40. $f(x) = (x^2 - 9)^{2/3}$

In Problems 41–46, use a graphing calculator to approximate the critical values of f(x) to two decimal places. Find the intervals on which f(x) is increasing, the intervals on which f(x) is decreasing, and the local extrema.

41. $f(x) = x^4 - 4x^3 + 9x$

42. $f(x) = x^4 + 5x^3 - 15x$

43. $f(x) = x \ln x - (x-2)^3$

44. $f(x) = e^{-x} - 3x^2$

45. $f(x) = e^x - 2x^2$

46. $f(x) = \dfrac{\ln x}{x} - 5x + x^2$

In Problems 47–54, find the intervals on which f(x) is increasing and the intervals on which f(x) is decreasing. Then sketch the graph. Add horizontal tangent lines.

47. $f(x) = 4 + 8x - x^2$

48. $f(x) = 2x^2 - 8x + 9$

49. $f(x) = x^3 - 3x + 1$

50. $f(x) = x^3 - 12x + 2$

51. $f(x) = 10 - 12x + 6x^2 - x^3$

52. $f(x) = x^3 + 3x^2 + 3x$

53. $f(x) = x^4 - 18x^2$

54. $f(x) = -x^4 + 50x^2$

In Problems 55–62, f(x) is continuous on $(-\infty, \infty)$. Use the given information to sketch the graph of f.

55.

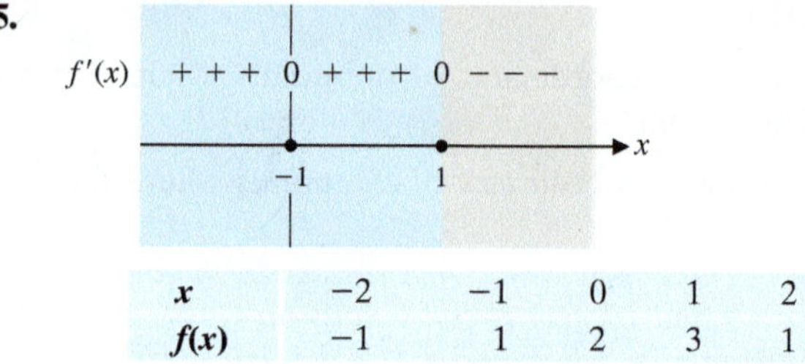

x	−2	−1	0	1	2
$f(x)$	−1	1	2	3	1

56.

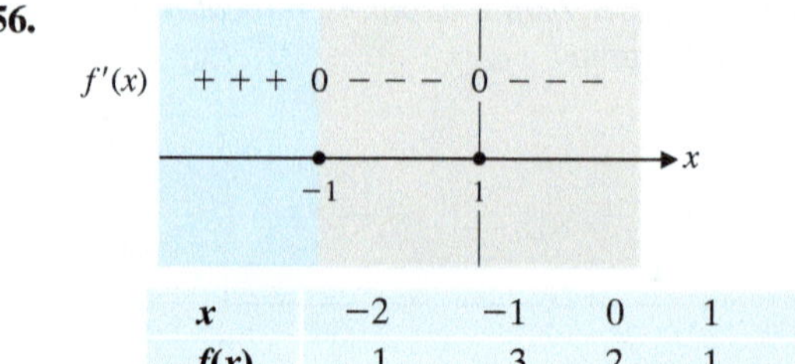

x	−2	−1	0	1	2
$f(x)$	1	3	2	1	−1

57.

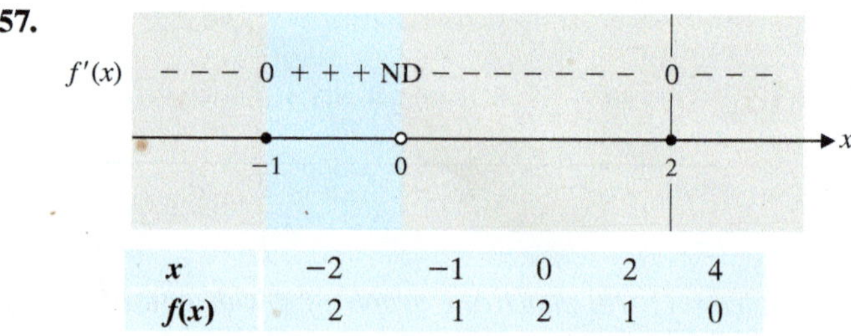

x	−2	−1	0	2	4
$f(x)$	2	1	2	1	0

58.

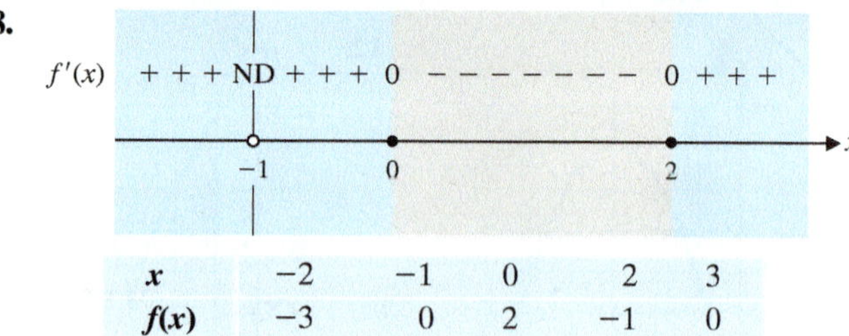

x	−2	−1	0	2	3
$f(x)$	−3	0	2	−1	0

59. $f(-2) = 4, f(0) = 0, f(2) = -4;$
$f'(-2) = 0, f'(0) = 0, f'(2) = 0;$
$f'(x) > 0$ on $(-\infty, -2)$ and $(2, \infty)$;
$f'(x) < 0$ on $(-2, 0)$ and $(0, 2)$

60. $f(-2) = -1, f(0) = 0, f(2) = 1;$
$f'(-2) = 0, f'(2) = 0;$
$f'(x) > 0$ on $(-\infty, -2)$, $(-2, 2)$, and $(2, \infty)$

61. $f(-1) = 2, f(0) = 0, f(1) = -2;$
$f'(-1) = 0, f'(1) = 0$, $f'(0)$ is not defined;
$f'(x) > 0$ on $(-\infty, -1)$ and $(1, \infty)$;
$f'(x) < 0$ on $(-1, 0)$ and $(0, 1)$

62. $f(-1) = 2, f(0) = 0, f(1) = 2;$
$f'(-1) = 0, f'(1) = 0$, $f'(0)$ is not defined;
$f'(x) > 0$ on $(-\infty, -1)$ and $(0, 1)$;
$f'(x) < 0$ on $(-1, 0)$ and $(1, \infty)$

Problems 63–68 involve functions f_1–f_6 and their derivatives, g_1–g_6. Use the graphs shown in figures (A) and (B) to match each function f_i with its derivative g_j.

63. f_1 **64.** f_2 **65.** f_3

66. f_4 **67.** f_5 **68.** f_6

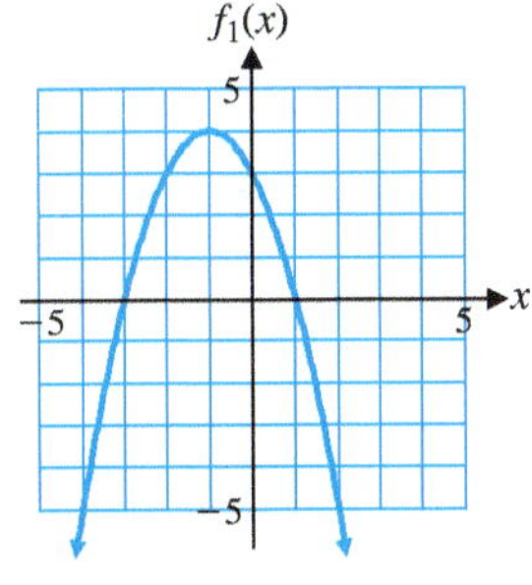

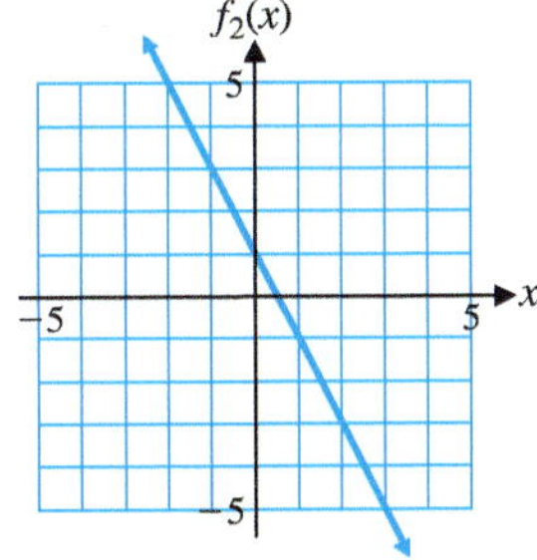

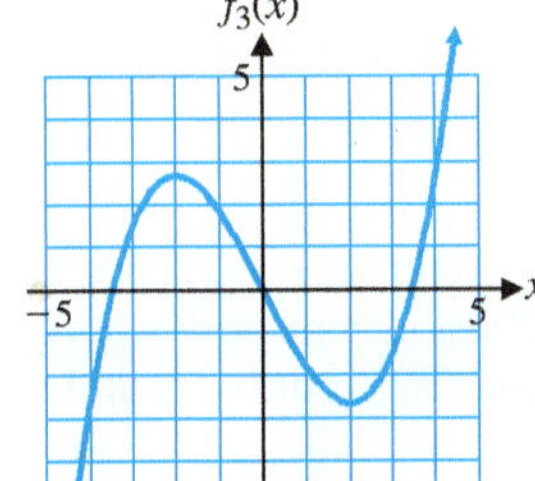

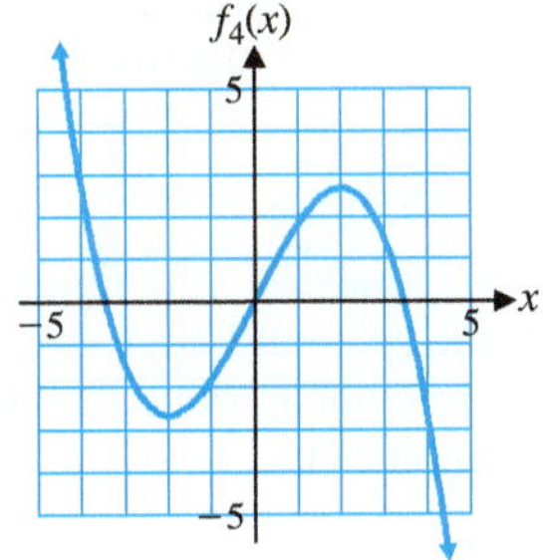

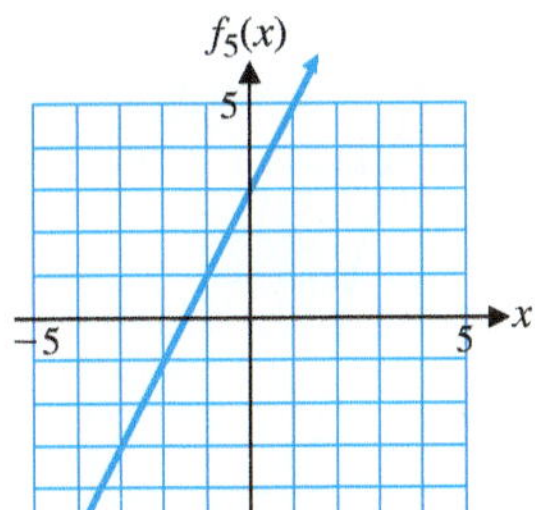

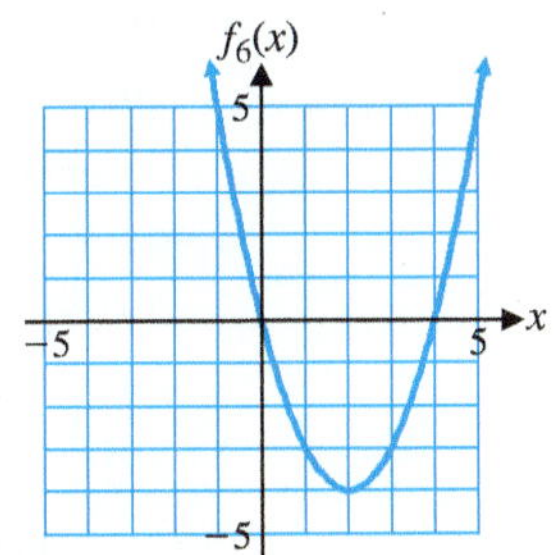

Figure (A) for 63–68

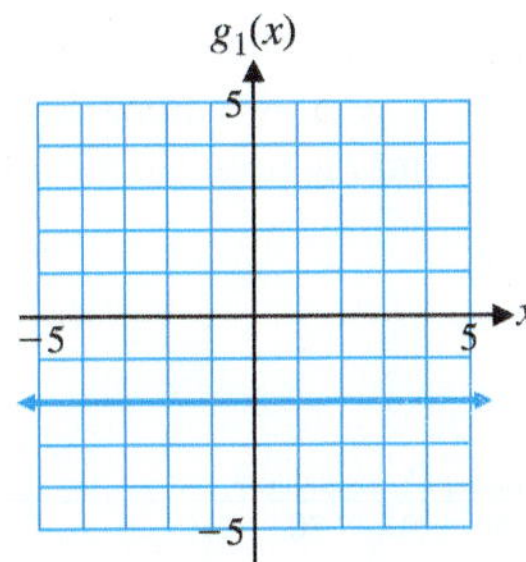

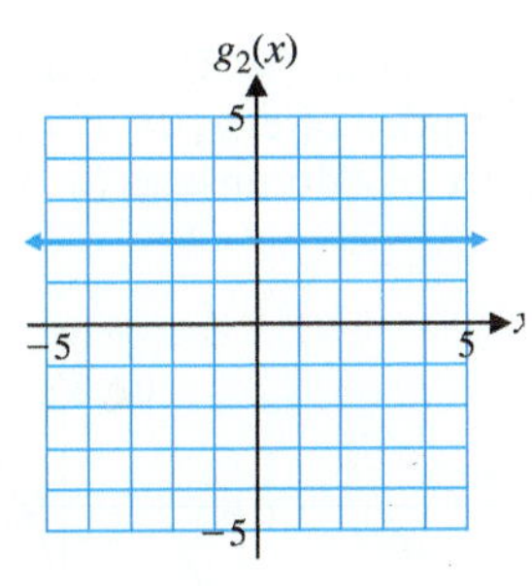

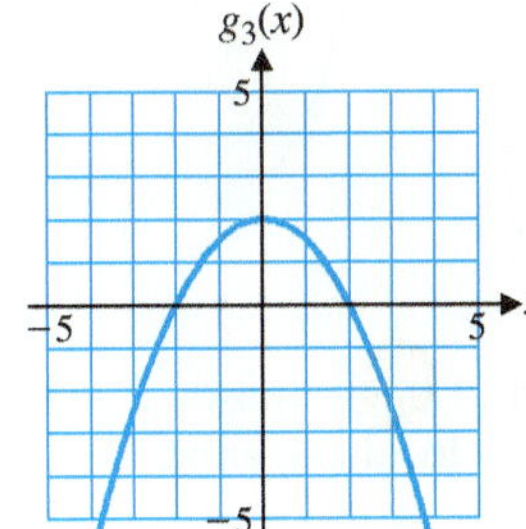

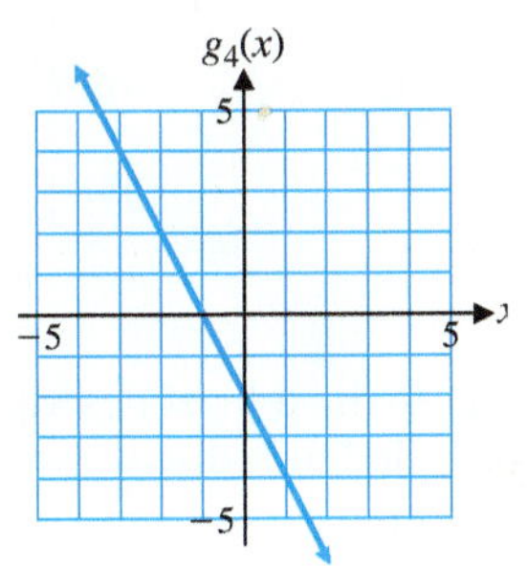

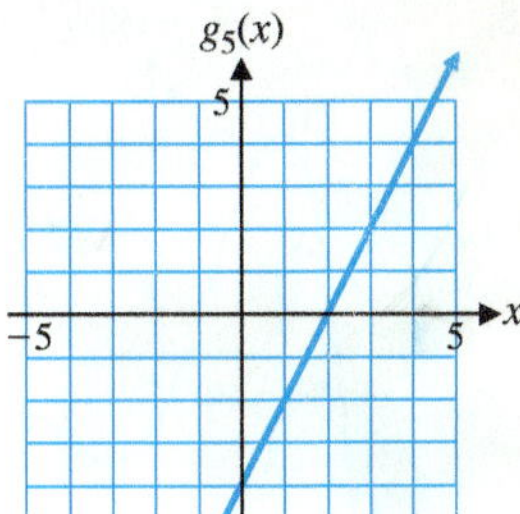

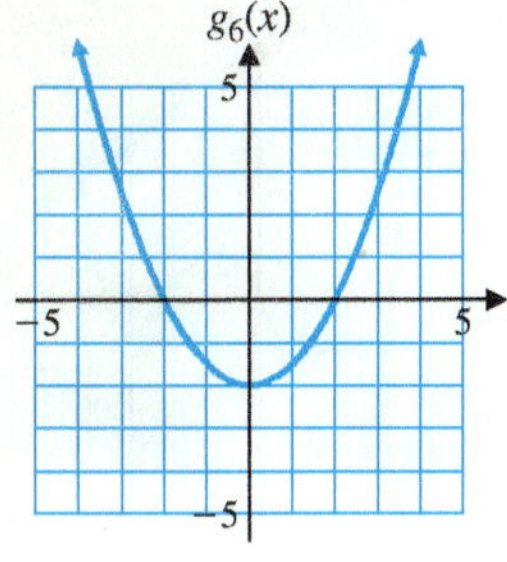

Figure (B) for 63–68

In Problems 69–74, use the given graph of $y = f'(x)$ to find the intervals on which f is increasing, the intervals on which f is decreasing, and the local extrema. Sketch a possible graph of $y = f(x)$.

69.

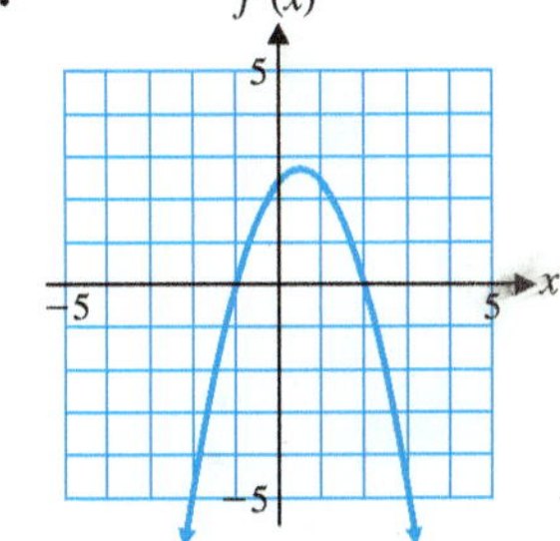

70.

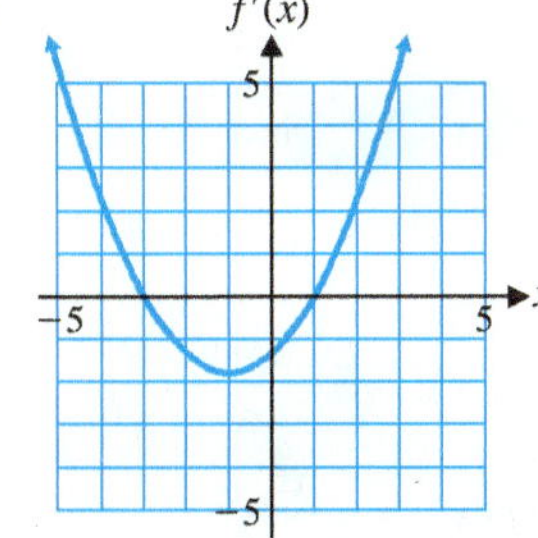

71.

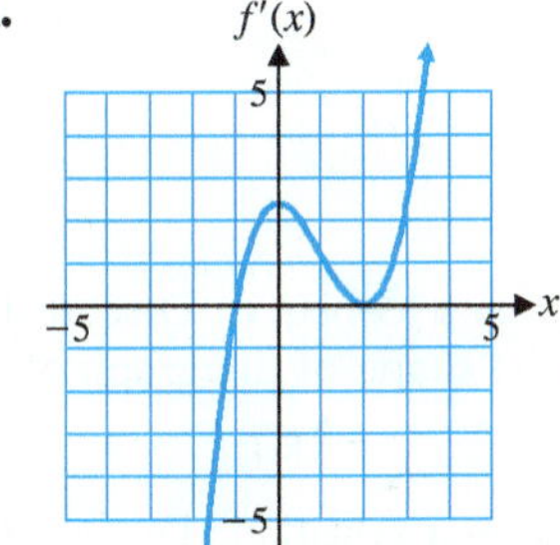

72.

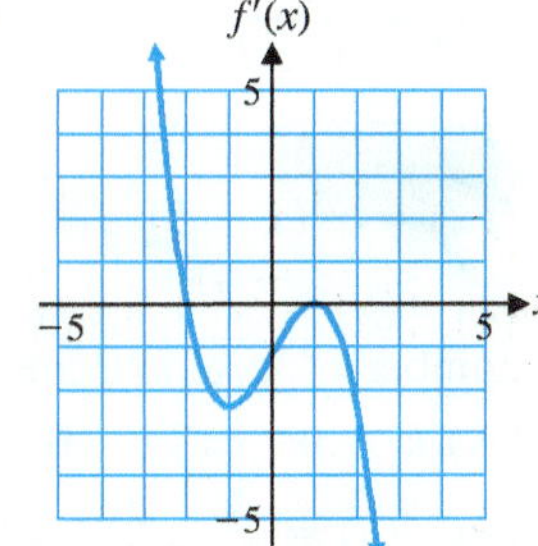

73.

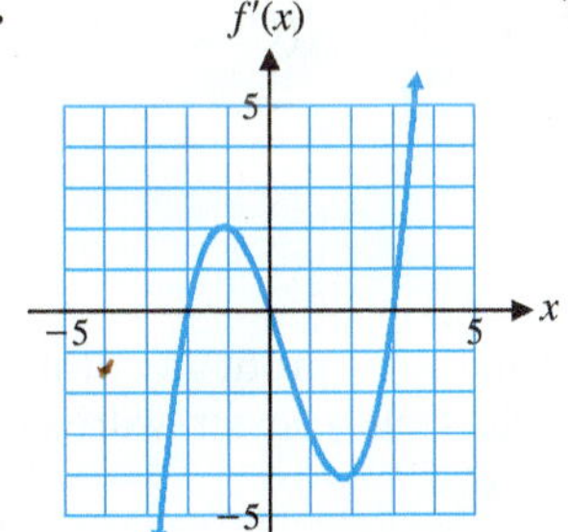

74.

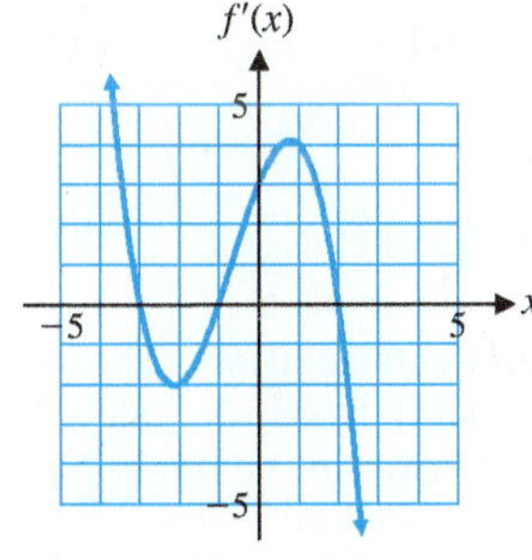

In Problems 75–78, use the given graph of $y = f(x)$ to find the intervals on which $f'(x) > 0$, the intervals on which $f'(x) < 0$, and the values of x for which $f'(x) = 0$. Sketch a possible graph of $y = f'(x)$.

75.

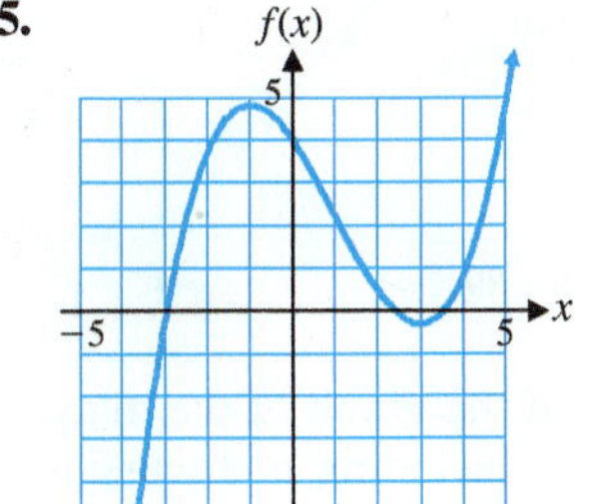

76.

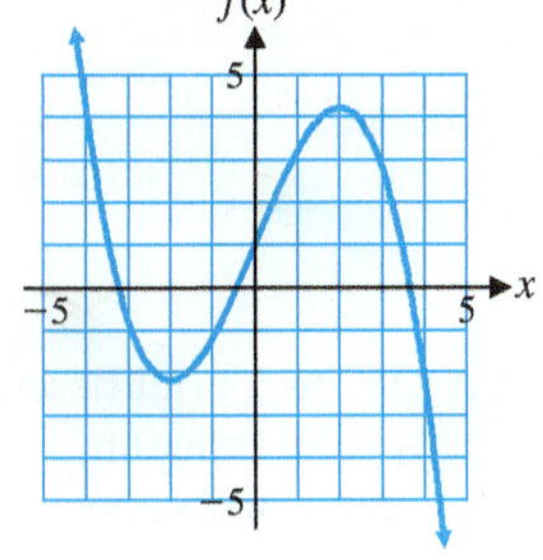

77.

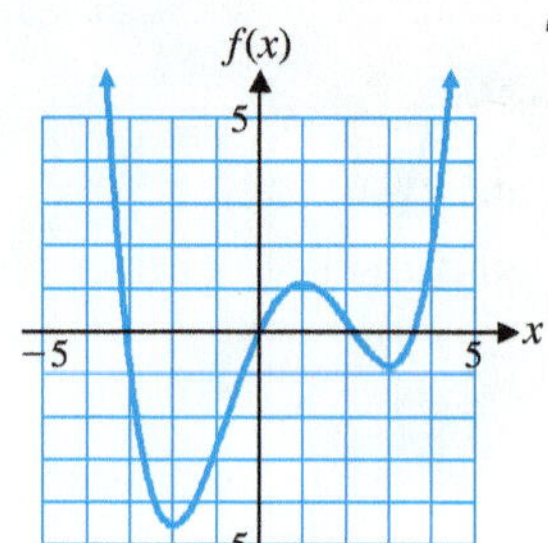

78.

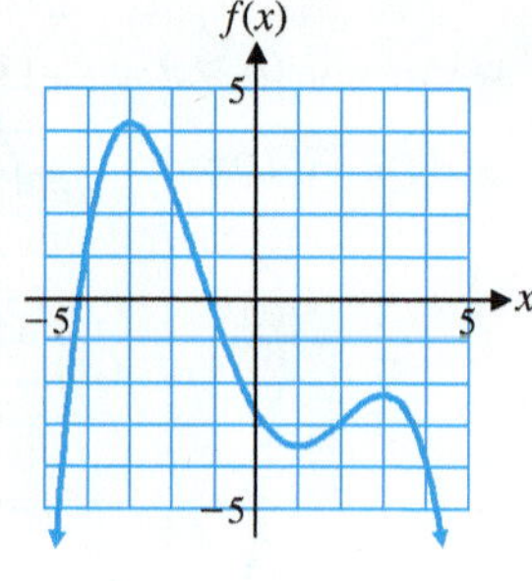

C

In Problems 79–86, find the critical values, the intervals on which f(x) is increasing, the intervals on which f(x) is decreasing, and the local extrema. Do not graph.

79. $f(x) = x + \frac{4}{x}$

80. $f(x) = \frac{9}{x} + x$

81. $f(x) = 1 + \frac{1}{x} + \frac{1}{x^2}$

82. $f(x) = 3 - \frac{4}{x} - \frac{2}{x^2}$

83. $f(x) = \frac{x^2}{x - 2}$

84. $f(x) = \frac{x^2}{x + 1}$

85. $f(x) = x^4(x - 6)^2$

86. $f(x) = x^3(x - 5)^2$

87. Let $f(x) = x^3 + kx$, where k is a constant. Discuss the number of local extrema and the shape of the graph of f if

(A) $k > 0$ (B) $k < 0$ (C) $k = 0$

88. Let $f(x) = x^4 + kx^2$, where k is a constant. Discuss the number of local extrema and the shape of the graph of f if

(A) $k > 0$ (B) $k < 0$ (C) $k = 0$

Applications

89. **Profit analysis.** The graph of the total profit $P(x)$ (in dollars) from the sale of x cordless electric screwdrivers is shown in the figure.

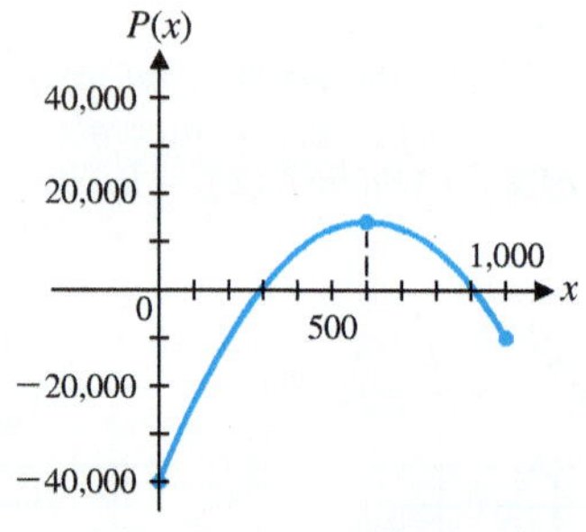

Figure for 89

(A) Write a brief description of the graph of the marginal profit function $y = P'(x)$, including a discussion of any x intercepts.

(B) Sketch a possible graph of $y = P'(x)$.

90. **Revenue analysis.** The graph of the total revenue $R(x)$ (in dollars) from the sale of x cordless electric screwdrivers is shown in the figure.

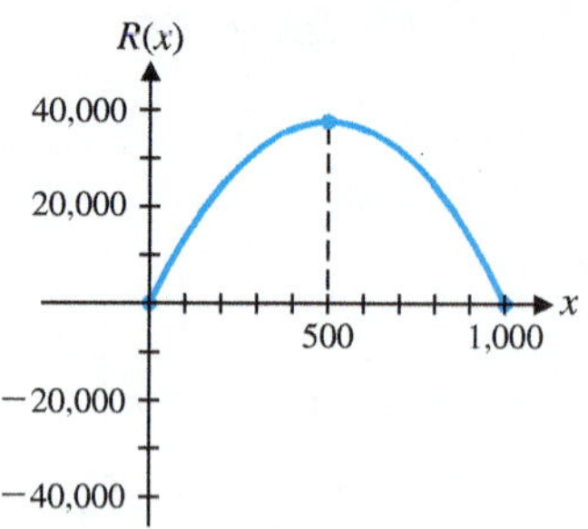

Figure for 90

(A) Write a brief description of the graph of the marginal revenue function $y = R'(x)$, including a discussion of any x intercepts.

(B) Sketch a possible graph of $y = R'(x)$.

91. **Price analysis.** The figure approximates the rate of change of the price of bacon over a 70-month period, where $B(t)$ is the price of a pound of sliced bacon (in dollars) and t is time (in months).

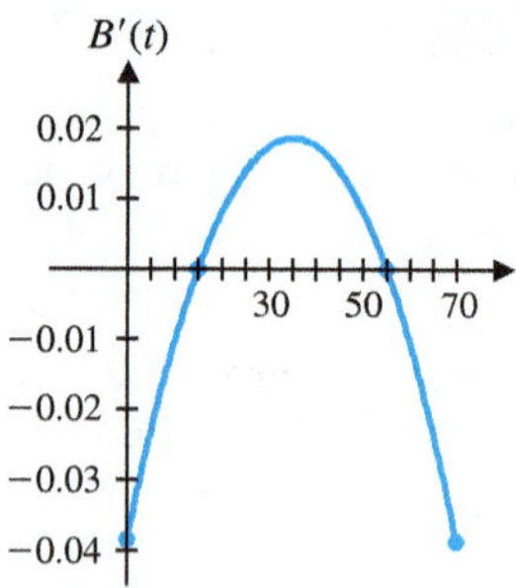

Figure for 91

(A) Write a brief description of the graph of $y = B(t)$, including a discussion of any local extrema.

(B) Sketch a possible graph of $y = B(t)$.

92. **Price analysis.** The figure approximates the rate of change of the price of eggs over a 70-month period, where $E(t)$ is the price of a dozen eggs (in dollars) and t is time (in months).

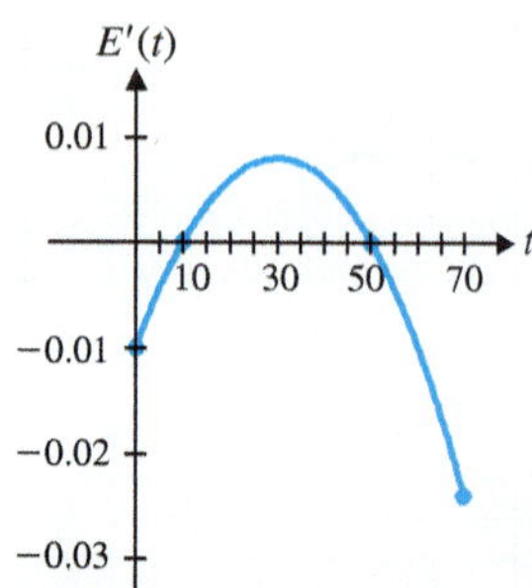

Figure for 92

(A) Write a brief description of the graph of $y = E(t)$, including a discussion of any local extrema.

(B) Sketch a possible graph of $y = E(t)$.

93. **Average cost.** A manufacturer incurs the following costs in producing x water ski vests in one day, for $0 < x < 150$: fixed costs, \$320; unit production cost, \$20 per vest; equipment maintenance and repairs, $0.05x^2$ dollars. So, the cost of manufacturing x vests in one day is given by

$$C(x) = 0.05x^2 + 20x + 320 \qquad 0 < x < 150$$

(A) What is the average cost $\overline{C}(x)$ per vest if x vests are produced in one day?

(B) Find the critical values of $\overline{C}(x)$, the intervals on which the average cost per vest is decreasing, the intervals on which the average cost per vest is increasing, and the local extrema. Do not graph.

94. **Average cost.** A manufacturer incurs the following costs in producing x rain jackets in one day for $0 < x < 200$: fixed costs, \$450; unit production cost, \$30 per jacket; equipment maintenance and repairs, $0.08x^2$ dollars.

(A) What is the average cost $\overline{C}(x)$ per jacket if x jackets are produced in one day?

(B) Find the critical values of $\overline{C}(x)$, the intervals on which the average cost per jacket is decreasing, the intervals on which the average cost per jacket is increasing, and the local extrema. Do not graph.

95. **Marginal analysis.** Show that profit will be increasing over production intervals (a, b) for which marginal revenue is greater than marginal cost. [*Hint:* $P(x) = R(x) - C(x)$]

96. **Marginal analysis.** Show that profit will be decreasing over production intervals (a, b) for which marginal revenue is less than marginal cost.

97. **Medicine.** A drug is injected into the bloodstream of a patient through the right arm. The drug concentration in the bloodstream of the left arm t hours after the injection is approximated by

$$C(t) = \frac{0.28t}{t^2 + 4} \qquad 0 < t < 24$$

Find the critical values of $C(t)$, the intervals on which the drug concentration is increasing, the intervals on which the concentration of the drug is decreasing, and the local extrema. Do not graph.

98. **Medicine.** The concentration $C(t)$, in milligrams per cubic centimeter, of a particular drug in a patient's bloodstream is given by

$$C(t) = \frac{0.3t}{t^2 + 6t + 9} \qquad 0 < t < 12$$

where t is the number of hours after the drug is taken orally. Find the critical values of $C(t)$, the intervals on which the drug concentration is increasing, the intervals on which the drug concentration is decreasing, and the local extrema. Do not graph.

Answers to Matched Problems

1. (A) Horizontal tangent line at $x = 3$.

(B) Decreasing on $(-\infty, 3)$; increasing on $(3, \infty)$

(C)

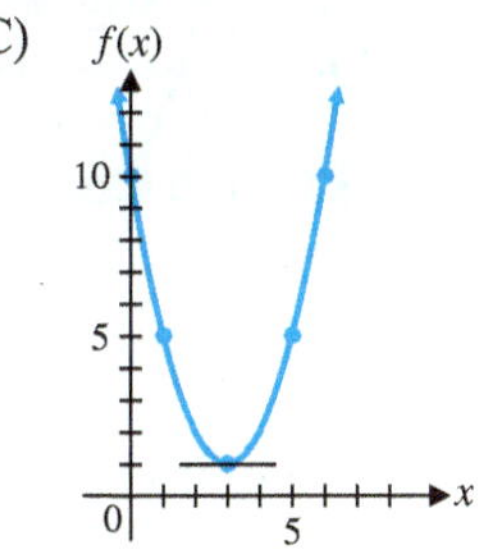

2. Partition number: $x = 0$; critical value: $x = 0$; decreasing for all x

3. Partition number: $x = -1$; critical value: $x = -1$; increasing for all x

4. Partition number: $x = 0$; no critical values; decreasing on $(-\infty, 0)$ and $(0, \infty)$

5. Partition number: $x = 5$; critical value: $x = 5$; increasing on $(0, 5)$; decreasing on $(5, \infty)$

6. Increasing on $(-3, 1)$; decreasing on $(-\infty, -3)$ and $(1, \infty)$; local maximum at $x = 1$; local minimum at $x = -3$

7. (A) Critical values: $x = 2, x = 4$

(B) Local maximum at $x = 2$; local minimum at $x = 4$

(C)

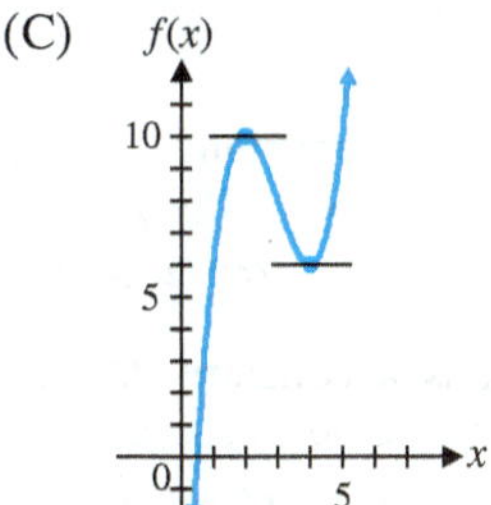

8. (A) The U.S. share of the world market decreases for 6 years to a local minimum, increases for the next 10 years to a local maximum, and then decreases for the final 4 years.

(B)

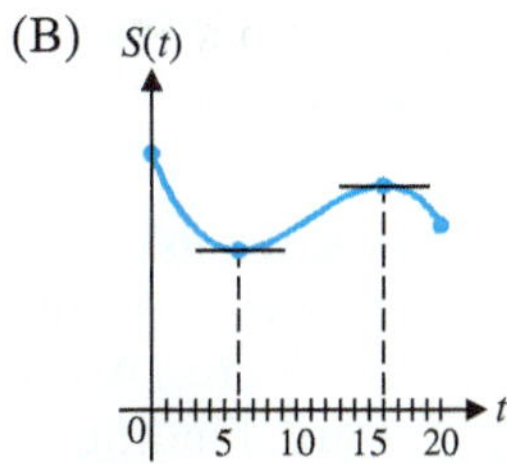

9. (A) The marginal revenue is positive on (0, 450), 0 at $x = 450$, and negative on (450, 1,000).

(B)

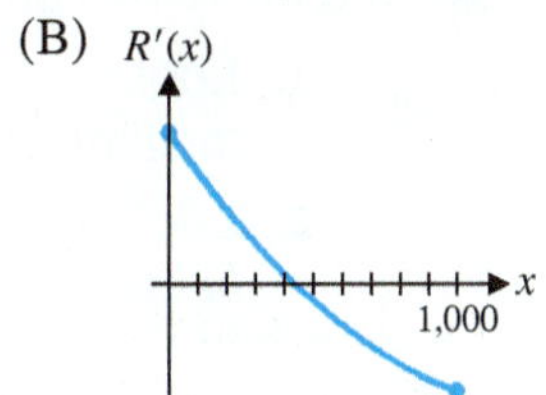

3-2 Second Derivative and Graphs

In Section 3-1, we saw that the derivative can be used when a graph is rising and falling. Now we want to see what the *second derivative* (the derivative of the derivative) can tell us about the shape of a graph.

Using Concavity as a Graphing Tool

Consider the functions

$$f(x) = x^2 \qquad \text{and} \qquad g(x) = \sqrt{x}$$

for x in the interval $(0, \infty)$. Since

$$f'(x) = 2x > 0 \qquad \text{for } 0 < x < \infty$$

and

$$g'(x) = \frac{1}{2\sqrt{x}} > 0 \qquad \text{for } 0 < x < \infty$$

both functions are increasing on $(0, \infty)$.

EXPLORE & DISCUSS 1

(A) Discuss the difference in the shapes of the graphs of f and g shown in Figure 1.

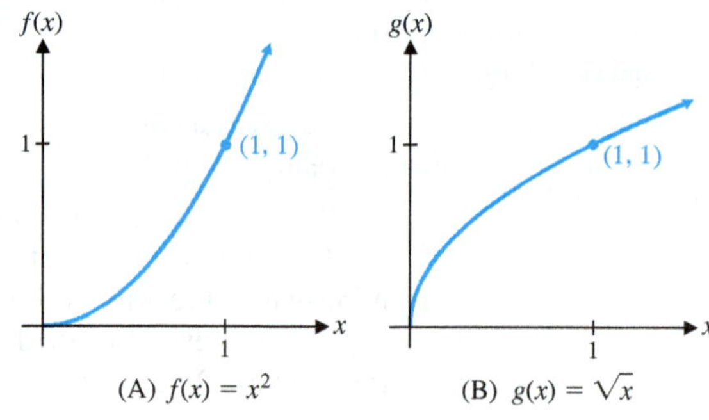

(A) $f(x) = x^2$ (B) $g(x) = \sqrt{x}$

Figure 1

(B) Complete the following table, and discuss the relationship between the values of the derivatives of f and g and the shapes of their graphs:

x	0.25	0.5	0.75	1
$f'(x)$				
$g'(x)$				

We use the term *concave upward* to describe a graph that opens upward and *concave downward* to describe a graph that opens downward. Thus, the graph of f in Figure 1A is concave upward, and the graph of g in Figure 1B is concave downward. Finding a mathematical formulation of concavity will help us sketch and analyze graphs.

We examine the slopes of f and g at various points on their graphs (see Fig. 2) and make two observations about each graph:

1. Looking at the graph of f in Figure 2A, we see that $f'(x)$ (the slope of the tangent line) is *increasing* and that the graph lies *above* each tangent line;
2. Looking at Figure 2B, we see that $g'(x)$ is *decreasing* and that the graph lies *below* each tangent line.

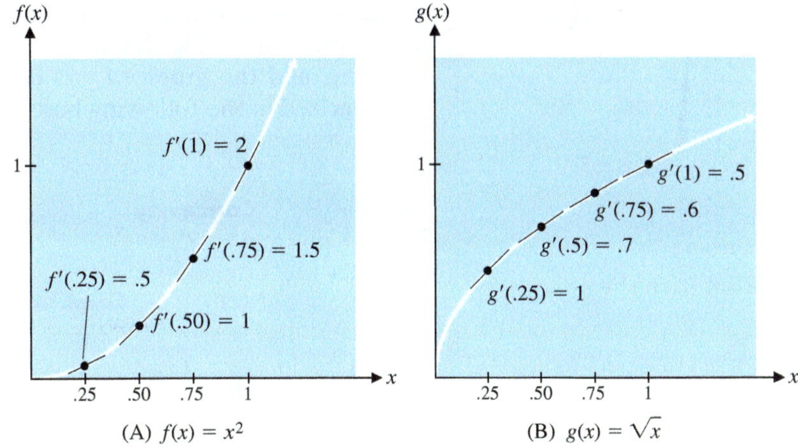

Figure 2

With these ideas in mind, we can state the general definition of concavity.

DEFINITION Concavity

The graph of a function f is **concave upward** on the interval (a, b) if $f'(x)$ is *increasing* on (a, b) and is **concave downward** on the interval (a, b) if $f'(x)$ is *decreasing* on (a, b).

Geometrically, the graph is concave upward on (a, b) if it lies above its tangent lines in (a, b) and is concave downward on (a, b) if it lies below its tangent lines in (a, b).

How can we determine when $f'(x)$ is increasing or decreasing? In Section 3-1, we used the derivative to determine when a function is increasing or decreasing. To determine when the function $f'(x)$ is increasing or decreasing, we use the derivative of $f'(x)$. The derivative of the derivative of a function is called the *second derivative* of the function. Various notations for the second derivative are given in the following box:

NOTATION Second Derivative

For $y = f(x)$, the **second derivative** of f, provided that it exists, is

$$f''(x) = \frac{d}{dx}f'(x)$$

Other notations for $f''(x)$ are

$$\frac{d^2y}{dx^2} \quad \text{and} \quad y''$$

Returning to the functions f and g discussed at the beginning of this section, we have

$$f(x) = x^2 \qquad g(x) = \sqrt{x} = x^{1/2}$$

$$f'(x) = 2x \qquad g'(x) = \frac{1}{2}x^{-1/2} = \frac{1}{2\sqrt{x}}$$

$$f''(x) = \frac{d}{dx}2x = 2 \qquad g''(x) = \frac{d}{dx}\frac{1}{2}x^{-1/2} = -\frac{1}{4}x^{-3/2} = -\frac{1}{4\sqrt{x^3}}$$

For $x > 0$, we see that $f''(x) > 0$; so, $f'(x)$ is increasing, and the graph of f is concave upward (see Fig. 2A). For $x > 0$, we also see that $g''(x) < 0$; so, $g'(x)$ is decreasing, and the graph of g is concave downward (see Fig. 2B). These ideas are summarized in the following box:

SUMMARY Concavity

For the interval (a, b),

$f''(x)$	$f'(x)$	Graph of $y = f(x)$	Examples
+	Increasing	Concave upward	
−	Decreasing	Concave downward	

CONCEPTUAL INSIGHT

Be careful not to confuse concavity with falling and rising. A graph that is concave upward on an interval may be falling, rising, or both falling and rising on that interval. A similar statement holds for a graph that is concave downward. See Figure 3.

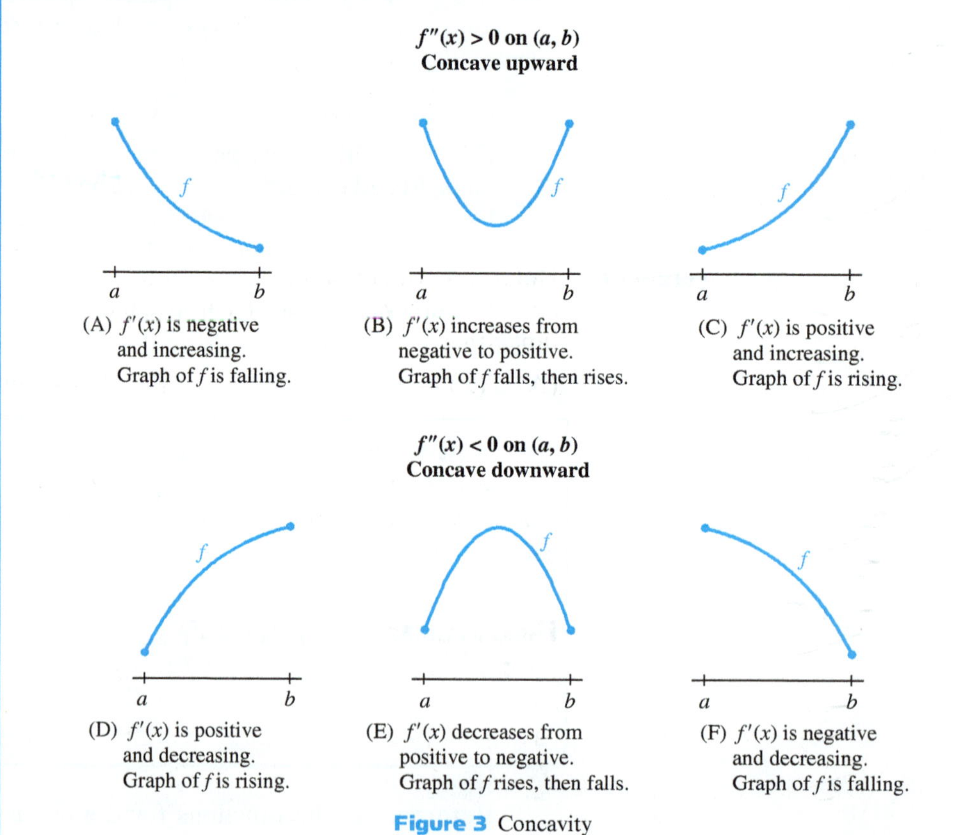

Figure 3 Concavity

EXAMPLE 1 **Concavity of Graphs** Determine the intervals on which the graph of each function is concave upward and the intervals on which it is concave downward. Sketch a graph of each function.

(A) $f(x) = e^x$ (B) $g(x) = \ln x$ (C) $h(x) = x^3$

SOLUTION

(A) $f(x) = e^x$

$f'(x) = e^x$

$f''(x) = e^x$

Since $f''(x) > 0$ on $(-\infty, \infty)$, the graph of $f(x) = e^x$ [Fig. 4(A)] is concave upward on $(-\infty, \infty)$.

(B) $g(x) = \ln x$

$g'(x) = \dfrac{1}{x}$

$g''(x) = -\dfrac{1}{x^2}$

The domain of $g(x) = \ln x$ is $(0, \infty)$ and $g''(x) < 0$ on this interval, so the graph of $g(x) = \ln x$ [Fig. 4(B)] is concave downward on $(0, \infty)$.

(C) $h(x) = x^3$

$h'(x) = 3x^2$

$h''(x) = 6x$

Since $h''(x) < 0$ when $x < 0$ and $h''(x) > 0$ when $x > 0$, the graph of $h(x) = x^3$ [Fig. 4(C)] is concave downward on $(-\infty, 0)$ and concave upward on $(0, \infty)$.

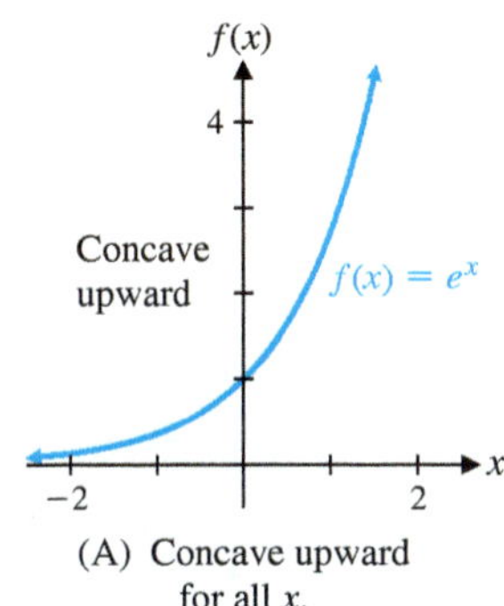

(A) Concave upward for all x.

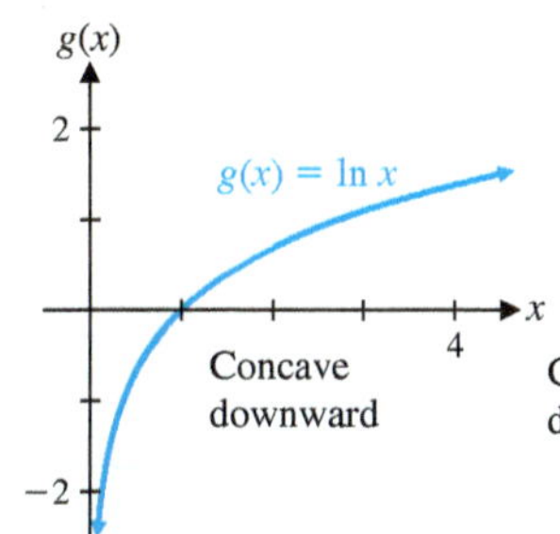

(B) Concave downward for $x > 0$.

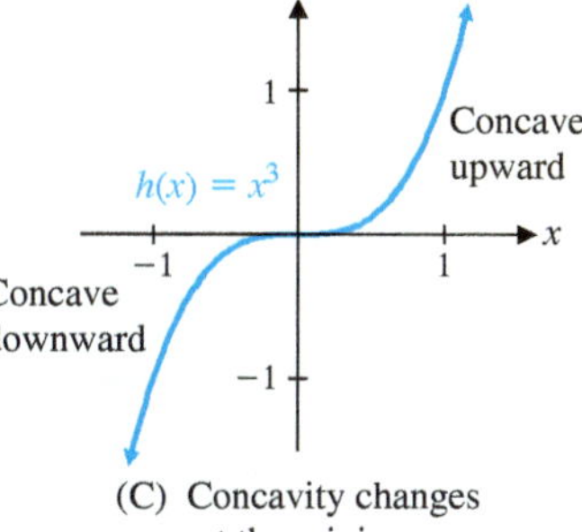

(C) Concavity changes at the origin.

Figure 4

Matched Problem 1 Determine the intervals on which the graph of each function is concave upward and the intervals on which it is cancave downward. Sketch a graph of each function.

(A) $f(x) = -e^{-x}$ (B) $g(x) = \ln \dfrac{1}{x}$ (C) $h(x) = x^{1/3}$

Refer to Example 1. The graphs of $f(x) = e^x$ and $g(x) = \ln x$ never change concavity. But the graph of $h(x) = x^3$ changes concavity at $(0, 0)$. This point is called an *inflection point.*

Finding Inflection Points

An **inflection point** is a point on the graph of the function where the concavity changes (from upward to downward or from downward to upward). For the concavity to change at a point, $f''(x)$must change sign at that point. But in Section 1-2, we saw that the partition numbers* identify the points where a function can change sign.

> **THEOREM 1 Inflection Points**
> If $y = f(x)$ is continuous on (a, b) and has an inflection point at $x = c$, then either $f''(c) = 0$ or $f''(c)$ does not exist.

*As we did with the first derivative, we assume that if f'' is discontinuous at c, then $f''(c)$ does not exist.

Note that inflection points can occur only at partition numbers of f'', but not every partition number of f'' produces an inflection point. Two additional requirements must be satisfied for an inflection point to occur:

A partition number c for f'' produces an inflection point for the graph of f only if

1. $f''(x)$ changes sign at c and

2. c is in the domain of f.

Figure 5 illustrates several typical cases.

If $f'(c)$ exists and $f''(x)$ changes sign at $x = c$, then the tangent line at an inflection point $(c, f(c))$ will always lie below the graph on the side that is concave upward and above the graph on the side that is concave downward (see Fig. 5A, B, and C).

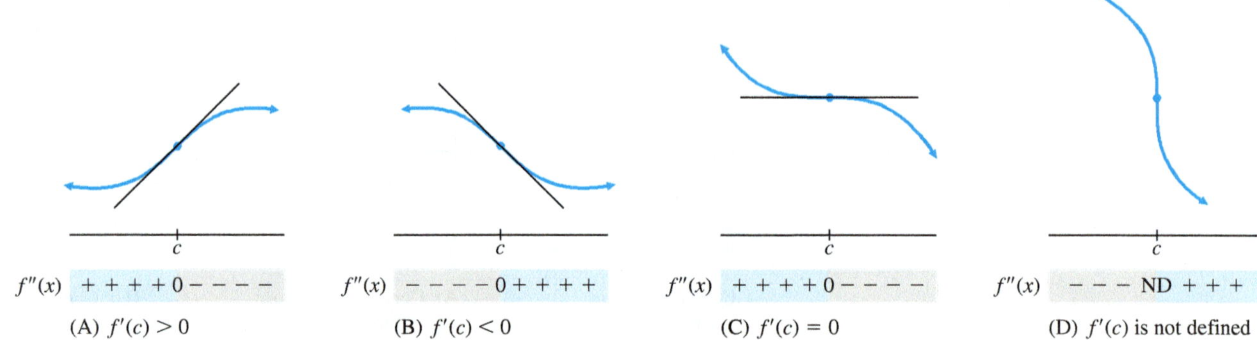

Figure 5 Inflection points

EXAMPLE 2 **Locating Inflection Points** Find the inflection point(s) of

$$f(x) = x^3 - 6x^2 + 9x + 1$$

SOLUTION Since inflection points occur at values of x where $f''(x)$ changes sign, we construct a sign chart for $f''(x)$.

$$f(x) = x^3 - 6x^2 + 9x + 1$$
$$f'(x) = 3x^2 - 12x + 9$$
$$f''(x) = 6x - 12 = 6(x - 2)$$

The sign chart for $f''(x) = 6(x - 2)$ (partition number is 2) is as follows:

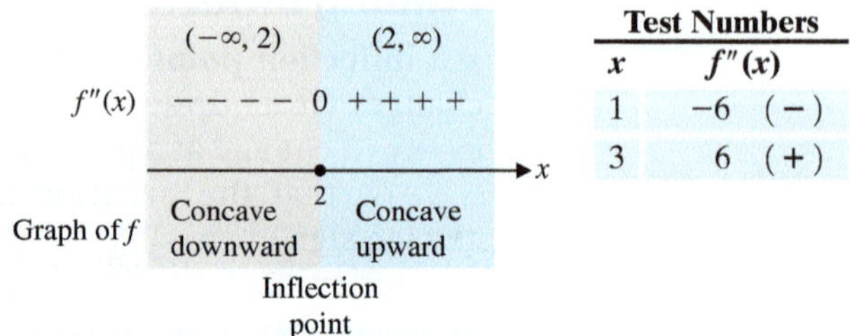

Test Numbers	
x	$f''(x)$
1	−6 (−)
3	6 (+)

From the sign chart, we see that the graph of f has an inflection point at $x = 2$. That is, the point

$$(2, f(2)) = (2, 3) \qquad f(2) = 2^3 - 6 \cdot 2^2 + 9 \cdot 2 + 1 = 3$$

is an inflection point on the graph of f.

Matched Problem 2 Find the inflection point(s) of

$$f(x) = x^3 - 9x^2 + 24x - 10$$

EXAMPLE 3 **Locating Inflection Points** Find the inflection point(s) of

$$f(x) = \ln(x^2 - 4x + 5)$$

SOLUTION First we find the domain of f. Since $\ln x$ is defined only for $x > 0$, f is defined only for

$$x^2 - 4x + 5 > 0 \quad \text{Use completing the square.}$$

$$(x - 2)^2 + 1 > 0 \quad \text{True for all } x.$$

So the domain of f is $(-\infty, \infty)$. Now we find $f''(x)$ and construct a sign chart for it.

$$f(x) = \ln(x^2 - 4x + 5)$$

$$f'(x) = \frac{2x - 4}{x^2 - 4x + 5}$$

$$f''(x) = \frac{(x^2 - 4x + 5)(2x - 4)' - (2x - 4)(x^2 - 4x + 5)'}{(x^2 - 4x + 5)^2}$$

$$= \frac{(x^2 - 4x + 5)2 - (2x - 4)(2x - 4)}{(x^2 - 4x + 5)^2}$$

$$= \frac{2x^2 - 8x + 10 - 4x^2 + 16x - 16}{(x^2 - 4x + 5)^2}$$

$$= \frac{-2x^2 + 8x - 6}{(x^2 - 4x + 5)^2}$$

$$= \frac{-2(x - 1)(x - 3)}{(x^2 - 4x + 5)^2}$$

The partition numbers for $f''(x)$ are $x = 1$ and $x = 3$.
Sign chart for $f''(x)$:

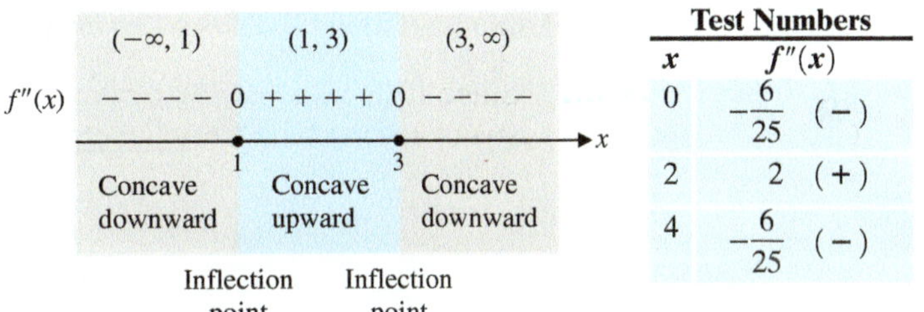

Test Numbers

x	$f''(x)$
0	$-\frac{6}{25}$ (−)
2	2 (+)
4	$-\frac{6}{25}$ (−)

The sign chart shows that the graph of f has inflection points at $x = 1$ and $x = 3$.

Matched Problem 3 Find the inflection point(s) of

$$f(x) = \ln(x^2 - 2x + 5)$$

CONCEPTUAL INSIGHT

It is important to remember that the partition numbers for f'' are only *candidates* for inflection points. The function f must be defined at $x = c$, and the second derivative must change sign at $x = c$ in order for the graph to have an inflection point at $x = c$. For example, consider

$$f(x) = x^4 \qquad g(x) = \frac{1}{x}$$

$$f'(x) = 4x^3 \qquad g'(x) = -\frac{1}{x^2}$$

$$f''(x) = 12x^2 \qquad g''(x) = \frac{2}{x^3}$$

In each case, $x = 0$ is a partition number for the second derivative, but neither the graph of $f(x)$ nor the graph of $g(x)$ has an inflection point at $x = 0$. Function f does not have an inflection point at $x = 0$ because $f''(x)$ does not change sign at $x = 0$ (see Fig. 6A). Function g does not have an inflection point at $x = 0$ because $g(0)$ is not defined (see Fig. 6B).

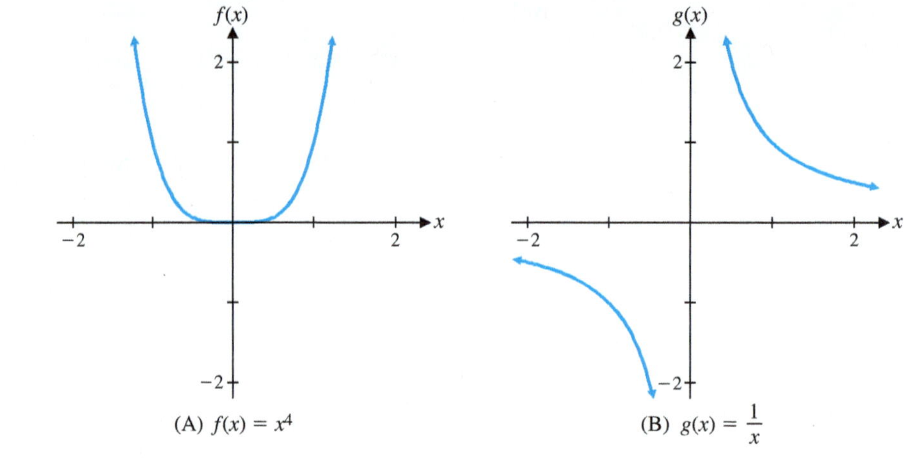

Figure 6

Analyzing Graphs

In the next example, we combine increasing/decreasing properties with concavity properties to analyze the graph of a function.

EXAMPLE 4 **Analyzing a Graph** Figure 7 shows the graph of the derivative of a function f. Use this graph to discuss the graph of f. Include a sketch of a possible graph of f.

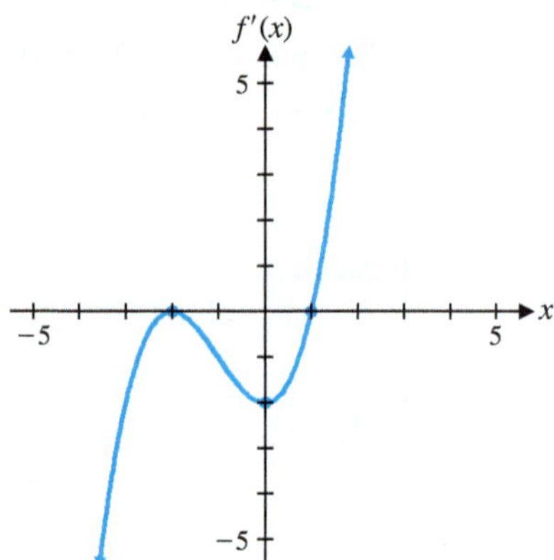

Figure 7

SOLUTION The sign of the derivative determines where the original function is increasing and decreasing, and the increasing/decreasing properties of the derivative determine the concavity of the original function. The relevant information obtained from the graph of f' is summarized in Table 1, and a possible graph of f is shown in Figure 8.

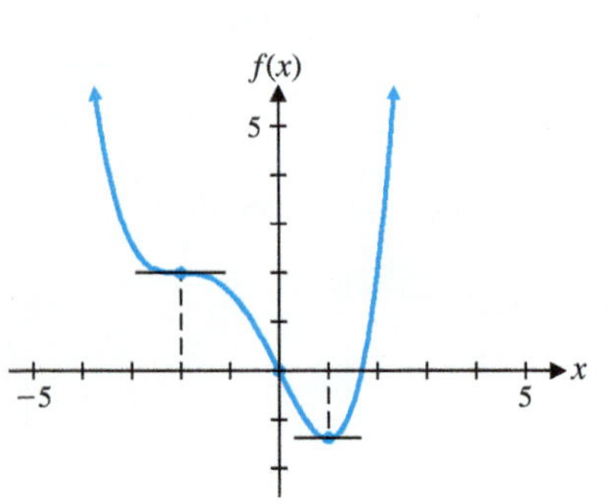

Figure 8

Table 1

x	$f'(x)$ (Fig. 7)	$f(x)$ (Fig. 8)
$-\infty < x < -2$	Negative and increasing	Decreasing and concave upward
$x = -2$	Local maximum	Inflection point
$-2 < x < 0$	Negative and decreasing	Decreasing and concave downward
$x = 0$	Local minimum	Inflection point
$0 < x < 1$	Negative and increasing	Decreasing and concave upward
$x = 1$	x intercept	Local minimum
$1 < x < \infty$	Positive and increasing	Increasing and concave upward

Matched Problem 4 Figure 9 shows the graph of the derivative of a function *f*. Use this graph to discuss the graph of *f*. Include a sketch of a possible graph of *f*.

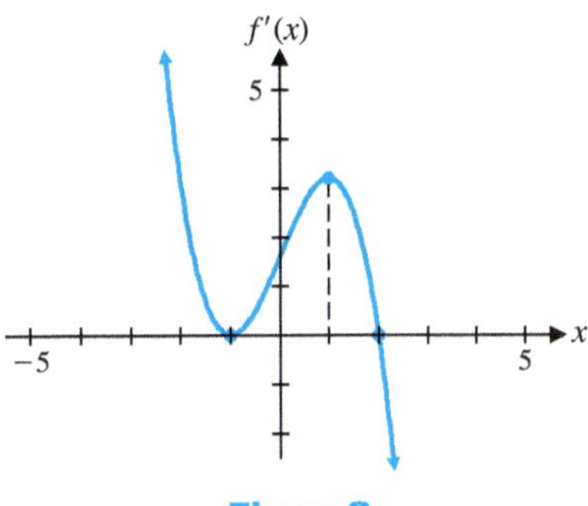

Figure 9

Curve Sketching

Graphing calculators and computers produce the graph of a function by plotting many points. However, important points on a plot many be difficult to identify. Using information gained from the function $f(x)$ and its derivatives, and plotting the important points—intercepts, local extrema, and inflection points—we can sketch by hand a very good representation of the graph of $f(x)$. This graphing process is called **curve sketching**.

PROCEDURE Graphing Strategy (First Version)*

Step 1 *Analyze f(x).* Find the domain and the intercepts. The *x* intercepts are the solutions of $f(x) = 0$, and the *y* intercept is $f(0)$.

Step 2 *Analyze* $f'(x)$. Find the partition numbers for, and critical values of, $f'(x)$. Construct a sign chart for $f'(x)$, determine the intervals on which *f* is increasing and decreasing, and find local maxima and minima.

Step 3 *Analyze* $f''(x)$. Find the partition numbers for $f''(x)$. Construct a sign chart for $f''(x)$, determine the intervals on which the graph of *f* is concave upward and concave downward, and find inflection points.

Step 4 *Sketch the graph of f.* Locate intercepts, local maxima and minima, and inflection points. Sketch in what you know from steps 1–3. Plot additional points as needed and complete the sketch.

EXAMPLE 5 **Using the Graphing Strategy** Follow the graphing strategy and analyze the function

$$f(x) = x^4 - 2x^3$$

State all the pertinent information and sketch the graph of *f*.

SOLUTION **Step 1** *Analyze f(x).* Since *f* is a polynomial, its domain is $(-\infty, \infty)$.

x intercept: $f(x) = 0$

$$x^4 - 2x^3 = 0$$

$$x^3(x - 2) = 0$$

$$x = 0, 2$$

y intercept: $f(0) = 0$

*We will modify this summary in Section 3-4 to include additional information about the graph of *f*.

Step 2 *Analyze* $f'(x)$. $f'(x) = 4x^3 - 6x^2 = 4x^2(x - \frac{3}{2})$
Critical values of $f(x)$: 0 and $\frac{3}{2}$
Partition numbers for $f'(x)$: 0 and $\frac{3}{2}$
Sign chart for $f'(x)$:

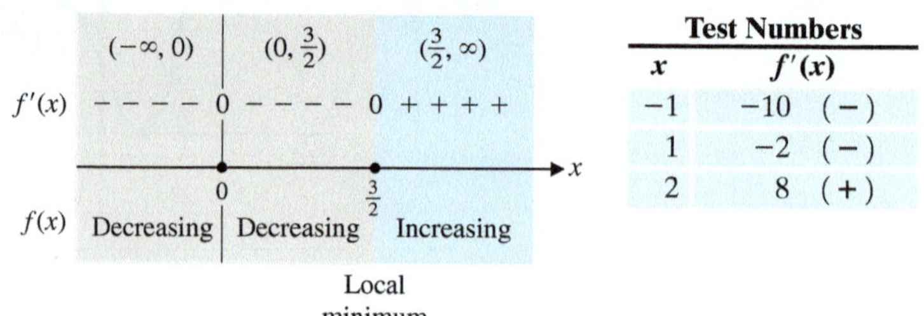

Test Numbers	
x	$f'(x)$
−1	−10 (−)
1	−2 (−)
2	8 (+)

So $f(x)$ is decreasing on $(-\infty, \frac{3}{2})$, is increasing on $(\frac{3}{2}, \infty)$, and has a local minimum at $x = \frac{3}{2}$.

Step 3 *Analyze* $f''(x)$. $f''(x) = 12x^2 - 12x = 12x(x - 1)$

Partition numbers for $f''(x)$: 0 and 1

Sign chart for $f''(x)$:

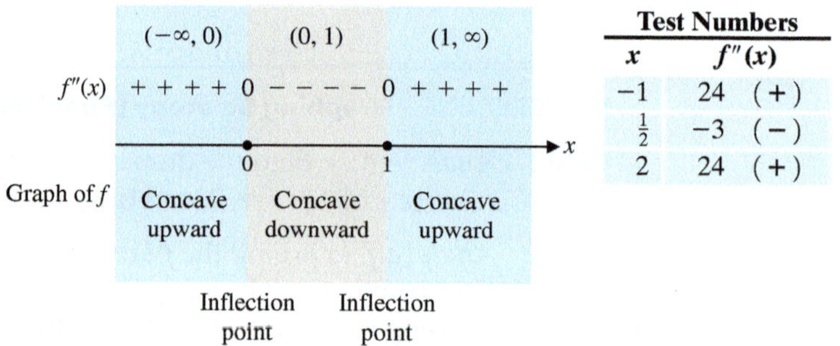

Test Numbers	
x	$f''(x)$
−1	24 (+)
$\frac{1}{2}$	−3 (−)
2	24 (+)

So the graph of f is concave upward on $(-\infty, 0)$ and $(1, \infty)$, is concave downward on $(0, 1)$, and has inflection points at $x = 0$ and $x = 1$.

Step 4 *Sketch the graph of f.*

Test Numbers	
x	$f(x)$
0	0
1	−1
$\frac{3}{2}$	$-\frac{27}{16}$
2	0

Matched Problem 5 Follow the graphing strategy and analyze the function $f(x) = x^4 + 4x^3$. State all the pertinent information and sketch the graph of f.

CONCEPTUAL INSIGHT

Refer to the solution of Example 5. Combining the sign charts for $f'(x)$ and $f''(x)$ (Fig. 10) partitions the real-number line into intervals on which neither $f'(x)$ nor $f''(x)$ changes sign. On each of these intervals, the graph of $f(x)$ must have one of four basic shapes (see also Fig. 3, parts A, C, D, and F on page 286). This reduces sketching the graph of a function to plotting the points identified in the graphing strategy and connecting them with one of the basic shapes.

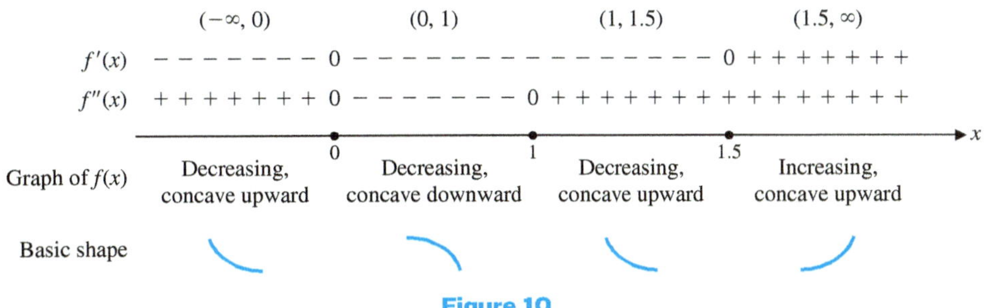

Figure 10

EXAMPLE 6 **Using the Graphing Strategy** Follow the graphing strategy and analyze the function

$$f(x) = 3x^{5/3} - 20x$$

State all the pertinent information and sketch the graph of f. Round any decimal values to two decimal places.

SOLUTION **Step 1** *Analyze f(x).* $f(x) = 3x^{5/3} - 20x$

Since x^p is defined for any x and any positive p, the domain of f is $(-\infty, \infty)$.

$$x \text{ intercepts: Solve } f(x) = 0$$

$$3x^{5/3} - 20x = 0$$

$$3x\left(x^{2/3} - \frac{20}{3}\right) = 0 \qquad (a^2 - b^2) = (a - b)(a + b)$$

$$3x\left(x^{1/3} - \sqrt{\frac{20}{3}}\right)\left(x^{1/3} + \sqrt{\frac{20}{3}}\right) = 0$$

The x intercepts of f are

$$x = 0, \quad x = \left(\sqrt{\frac{20}{3}}\right)^3 \approx 17.21, \quad x = \left(-\sqrt{\frac{20}{3}}\right)^3 \approx -17.21$$

y intercept: $f(0) = 0$.

Step 2 Analyze $f'(x)$.

$$\begin{aligned} f'(x) &= 5x^{2/3} - 20 \\ &= 5(x^{2/3} - 4) \qquad \text{Again, } a^2 - b^2 = (a - b)(a + b) \\ &= 5(x^{1/3} - 2)(x^{1/3} + 2) \end{aligned}$$

Critical values of f: $x = 2^3 = 8$ and $x = (-2)^3 = -8$.
Partition numbers for f: $-8, 8$
Sign chart for $f'(x)$:

	$(-\infty, -8)$		$(-8, 8)$		$(8, \infty)$
$f'(x)$	+ + + +	0	− − − −	0	+ + + +
x		−8		8	
	Increasing		Decreasing		Increasing
		Local maximum		Local minimum	

Test Numbers

x	$f'(x)$
−12	6.21 (+)
0	−20 (−)
12	6.21 (+)

So f is increasing on $(-\infty, -8)$ and $(8, \infty)$ and decreasing on $(-8, 8)$. Therefore, $f(-8)$ is a local maximum, and $f(8)$ is a local minimum.

Step 3 *Analyze $f''(x)$.*

$$f'(x) = 5x^{2/3} - 20$$

$$f''(x) = \frac{10}{3}x^{-1/3} = \frac{10}{3x^{1/3}}$$

Partition number for f'': 0

Sign chart for $f''(x)$:

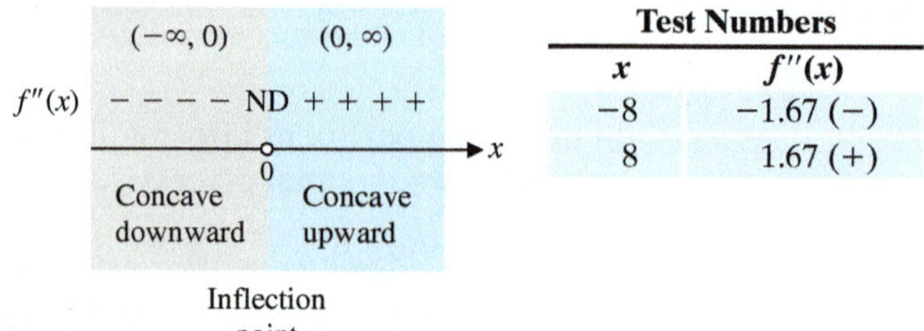

Test Numbers

x	$f''(x)$
−8	−1.67 (−)
8	1.67 (+)

So f is concave downward on $(-\infty, 0)$, is concave upward on $(0, \infty)$, and has an inflection point at $x = 0$.

Step 4 *Sketch the graph of f.*

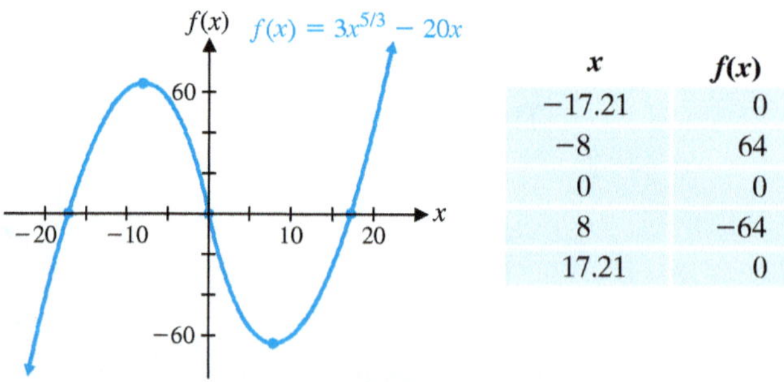

x	$f(x)$
−17.21	0
−8	64
0	0
8	−64
17.21	0

Matched Problem 6 Follow the graphing strategy and analyze the function $f(x) = 3x^{2/3} - x$. State all the pertinent information and sketch the graph of f. Round any decimal values to two decimal places.

Point of Diminishing Returns

If a company decides to increase spending on advertising, it would expect sales to increase. At first, sales will increase at an increasing rate and then increase at a decreasing rate. The value of x where the rate of change of sales goes from increasing to decreasing is called the **point of diminishing returns.** This is also the point where the rate of change has a maximum value. Money spent after this point may increase sales but at a lower rate.

EXAMPLE 7 **Maximum Rate of Change** Currently, a discount appliance store is selling 200 large-screen television sets monthly. If the store invests $\$x$ thousand in an advertising campaign, the ad company estimates that sales will increase to

$$N(x) = 3x^3 - 0.25x^4 + 200 \qquad 0 \le x \le 9$$

When is the rate of change of sales increasing and when is it decreasing? What is the point of diminishing returns and the maximum rate of change of sales? Graph N and N' on the same coordinate system.

SOLUTION The rate of change of sales with respect to advertising expenditures is

$$N'(x) = 9x^2 - x^3 = x^2(9 - x)$$

To determine when $N'(x)$ is increasing and decreasing, we find $N''(x)$, the derivative of $N'(x)$:

$$N''(x) = 18x - 3x^2 = 3x(6 - x)$$

The information obtained by analyzing the signs of $N'(x)$ and $N''(x)$ is summarized in Table 2 (sign charts are omitted).

Table 2

x	$N''(x)$	$N'(x)$	$N'(x)$	$N(x)$
$0 < x < 6$	+	+	Increasing	Increasing, concave upward
$x = 6$	0	+	Local maximum	Inflection point
$6 < x < 9$	−	+	Decreasing	Increasing, concave downward

Examining Table 2, we see that $N'(x)$ is increasing on (0, 6) and decreasing on (6, 9). The point of diminishing returns is $x = 6$ and the maximum rate of change is $N'(6) = 108$. Note that $N'(x)$ has a local maximum and $N(x)$ has an inflection point at $x = 6$.

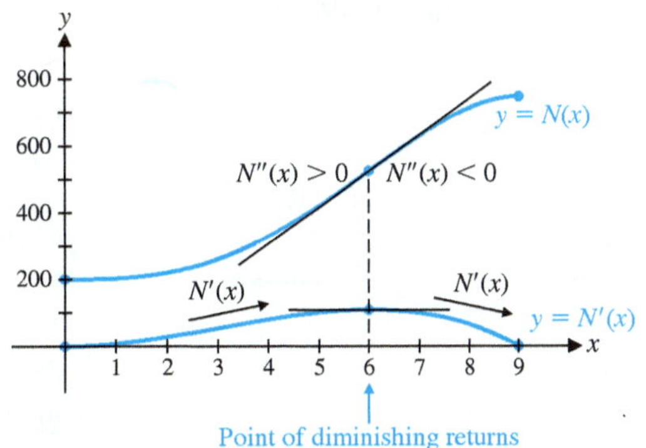

Matched Problem 7 Repeat Example 7 for

$$N(x) = 4x^3 - 0.25x^4 + 500 \qquad 0 \le x \le 12$$

Exercises 3-2

A

1. Use the graph of $y = f(x)$ to identify
   (A) Intervals on which the graph of f is concave upward
   (B) Intervals on which the graph of f is concave downward
   (C) Intervals on which $f''(x) < 0$
   (D) Intervals on which $f''(x) > 0$
   (E) Intervals on which $f'(x)$ is increasing
   (F) Intervals on which $f'(x)$ is decreasing
   (G) The x coordinates of inflection points
   (H) The x coordinates of local extrema for $f'(x)$

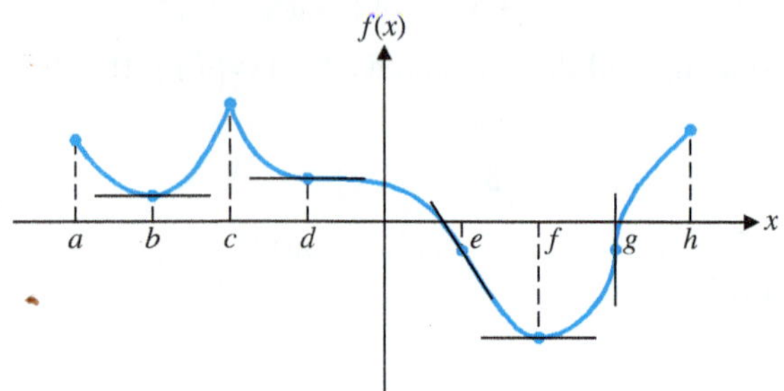

Figure for 1

2. Use the graph of $y = g(x)$ to identify
   (A) Intervals on which the graph of g is concave upward
   (B) Intervals on which the graph of g is concave downward
   (C) Intervals on which $g''(x) < 0$
   (D) Intervals on which $g''(x) > 0$
   (E) Intervals on which $g'(x)$ is increasing
   (F) Intervals on which $g'(x)$ is decreasing
   (G) The x coordinates of inflection points
   (H) The x coordinates of local extrema for $g'(x)$

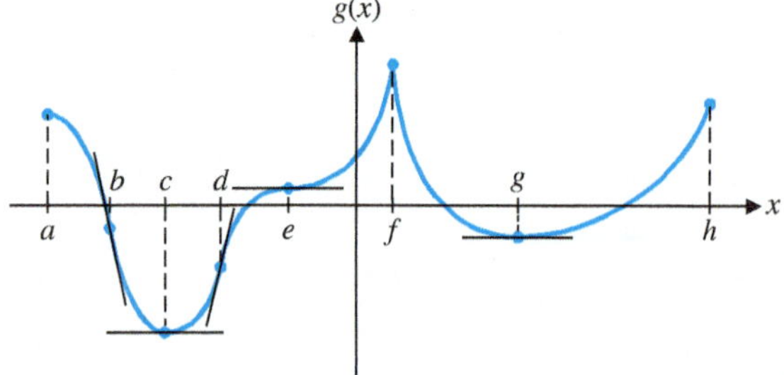

Figure for 2

In Problems 3–6, match the indicated conditions with one of the graphs (A)–(D) shown in the figure.

3. $f'(x) > 0$ and $f''(x) > 0$ on (a, b)
4. $f'(x) > 0$ and $f''(x) < 0$ on (a, b)
5. $f'(x) < 0$ and $f''(x) > 0$ on (a, b)
6. $f'(x) < 0$ and $f''(x) < 0$ on (a, b)

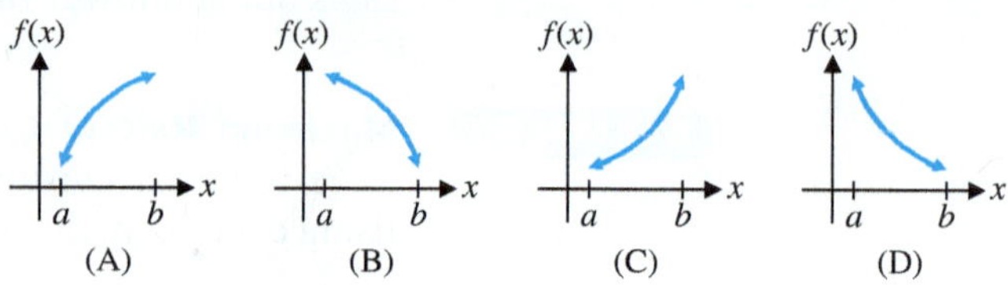

Figure for 3–6

In Problems 7–18, find the indicated derivative for each function.

7. $f''(x)$ for $f(x) = 2x^3 - 4x^2 + 5x - 6$
8. $g''(x)$ for $g(x) = -x^3 + 2x^2 - 3x + 9$
9. $h''(x)$ for $h(x) = 2x^{-1} - 3x^{-2}$
10. $k''(x)$ for $k(x) = -6x^{-2} + 12x^{-3}$
11. d^2y/dx^2 for $y = x^2 - 18x^{1/2}$
12. d^2y/dx^2 for $y = x^3 - 24x^{1/3}$
13. y'' for $y = (x^2 + 9)^4$
14. y'' for $y = (x^2 - 16)^5$
15. $f''(x)$ for $f(x) = e^{-x^2}$
16. $f''(x)$ for $f(x) = xe^{-x}$
17. y'' for $y = \dfrac{\ln x}{x^2}$
18. y'' for $y = x^2 \ln x$

In Problems 19–24, find the x and y coordinates of all inflection points.

19. $f(x) = x^3 + 30x^2$
20. $f(x) = x^3 - 24x^2$
21. $f(x) = x^{5/3} + 2$
22. $f(x) = 5 - x^{4/3}$
23. $f(x) = 1 + x + x^{2/5}$
24. $f(x) = x^{3/5} - 6x + 7$

In Problems 25–34, find the intervals on which the graph of f is concave upward, the intervals on which the graph of f is concave downward, and the inflection points.

25. $f(x) = x^4 + 6x^2$
26. $f(x) = x^4 + 6x$
27. $f(x) = x^3 - 4x^2 + 5x - 2$
28. $f(x) = -x^3 - 5x^2 + 4x - 3$
29. $f(x) = -x^4 + 12x^3 - 12x + 24$
30. $f(x) = x^4 - 2x^3 - 36x + 12$
31. $f(x) = \ln(x^2 - 2x + 10)$
32. $f(x) = \ln(x^2 + 6x + 13)$
33. $f(x) = 8e^x - e^{2x}$
34. $f(x) = e^{3x} - 9e^x$

In Problems 35–42, f(x) is continuous on $(-\infty, \infty)$. Use the given information to sketch the graph of f.

35.

x	−4	−2	−1	0	2	4
$f(x)$	0	3	1.5	0	−1	−3

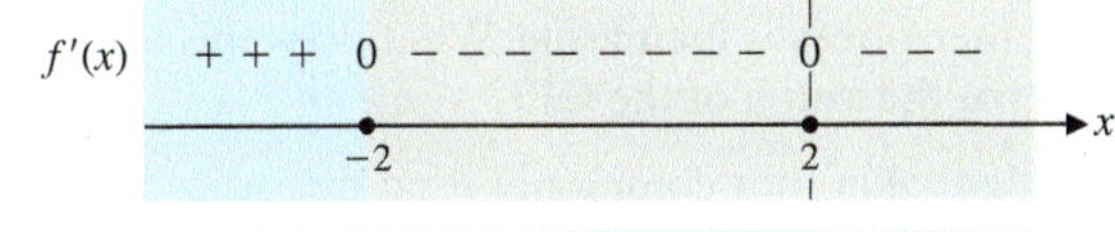

f″(x) − − − − − − − − 0 + + + 0 − − −

−1 2 x

36.

x	−4	−2	−1	0	2	4
$f(x)$	0	−2	−1	0	1	3

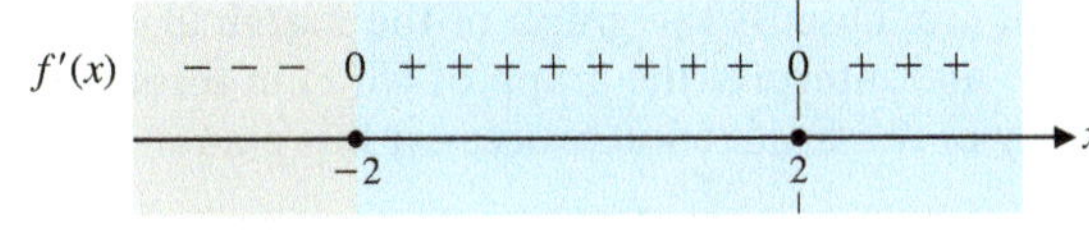

f″(x) + + + + + + + + 0 − − − 0 + + +

−1 2 x

37.

x	−3	0	1	2	4	5
$f(x)$	−4	0	2	1	−1	0

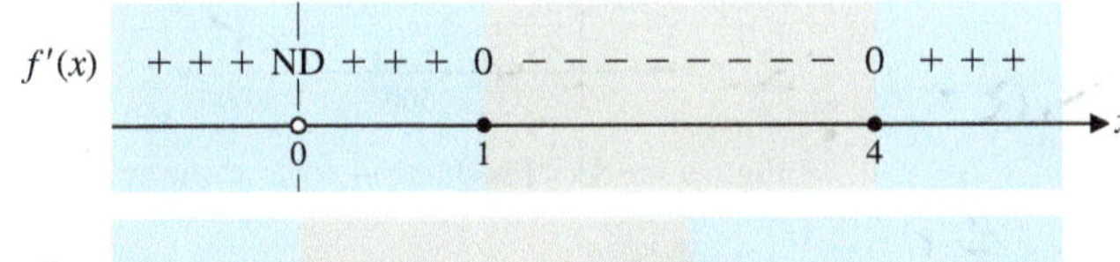

f″(x) + + + ND − − − − − − − − 0 + + + + + + + +

0 2 x

38.

x	−4	−2	0	2	4	6
$f(x)$	0	3	0	−2	0	3

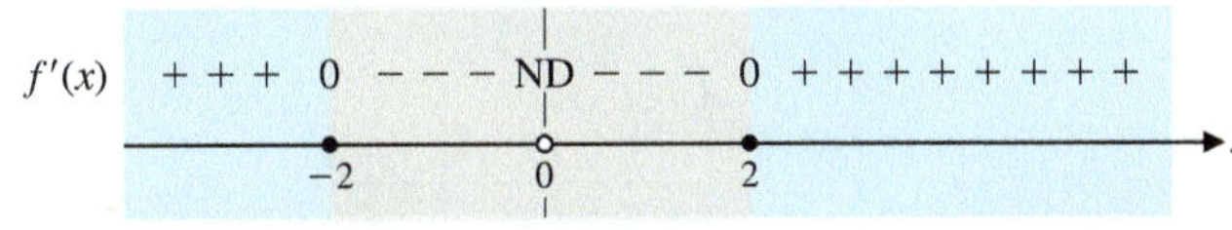

f″(x) − − − − − − − − ND + + + + + + + + 0 − − −

0 4 x

39. $f(0) = 2, f(1) = 0, f(2) = -2;$

$f'(0) = 0, f'(2) = 0;$

$f'(x) > 0$ on $(-\infty, 0)$ and $(2, \infty)$;

$f'(x) < 0$ on $(0, 2)$;

$f''(1) = 0;$

$f''(x) > 0$ on $(1, \infty)$;

$f''(x) < 0$ on $(-\infty, 1)$

40. $f(-2) = -2, f(0) = 1, f(2) = 4;$

$f'(-2) = 0, f'(2) = 0;$

$f'(x) > 0$ on $(-2, 2)$;

$f'(x) < 0$ on $(-\infty, -2)$ and $(2, \infty)$;

$f''(0) = 0;$

$f''(x) > 0$ on $(-\infty, 0)$;

$f''(x) < 0$ on $(0, \infty)$

41. $f(-1) = 0, f(0) = -2, f(1) = 0;$

$f'(0) = 0, f'(-1)$ and $f'(1)$ are not defined;

$f'(x) > 0$ on $(0, 1)$ and $(1, \infty)$;

$f'(x) < 0$ on $(-\infty, -1)$ and $(-1, 0)$;

$f''(-1)$ and $f''(1)$ are not defined;

$f''(x) > 0$ on $(-1, 1)$;

$f''(x) < 0$ on $(-\infty, -1)$ and $(1, \infty)$

42. $f(0) = -2, f(1) = 0, f(2) = 4;$

$f'(0) = 0, f'(2) = 0, f'(1)$ is not defined;

$f'(x) > 0$ on $(0, 1)$ and $(1, 2)$;

$f'(x) < 0$ on $(-\infty, 0)$ and $(2, \infty)$;

$f''(1)$ is not defined;

$f''(x) > 0$ on $(-\infty, 1)$;

$f''(x) < 0$ on $(1, \infty)$

B

In Problems 43–64, summarize the pertinent information obtained by applying the graphing strategy and sketch the graph of $y = f(x)$.

43. $f(x) = (x - 2)(x^2 - 4x - 8)$

44. $f(x) = (x - 3)(x^2 - 6x - 3)$

45. $f(x) = (x + 1)(x^2 - x + 2)$

46. $f(x) = (1 - x)(x^2 + x + 4)$

47. $f(x) = -0.25x^4 + x^3$

48. $f(x) = 0.25x^4 - 2x^3$

49. $f(x) = 16x(x - 1)^3$

50. $f(x) = -4x(x + 2)^3$

51. $f(x) = (x^2 + 3)(9 - x^2)$

52. $f(x) = (x^2 + 3)(x^2 - 1)$

53. $f(x) = (x^2 - 4)^2$

54. $f(x) = (x^2 - 1)(x^2 - 5)$

55. $f(x) = 2x^6 - 3x^5$

56. $f(x) = 3x^5 - 5x^4$

57. $f(x) = 1 - e^{-x}$

58. $f(x) = 2 - 3e^{-2x}$

59. $f(x) = e^{0.5x} + 4e^{-0.5x}$

60. $f(x) = 2e^{0.5x} + e^{-0.5x}$

61. $f(x) = -4 + 2 \ln x$

62. $f(x) = 5 - 3 \ln x$

63. $f(x) = \ln(x + 4) - 2$

64. $f(x) = 1 - \ln(x - 3)$

In Problems 65–68, use the graph of $y = f'(x)$ to discuss the graph of $y = f(x)$. Organize your conclusions in a table (see Example 4), and sketch a possible graph of $y = f(x)$.

65.

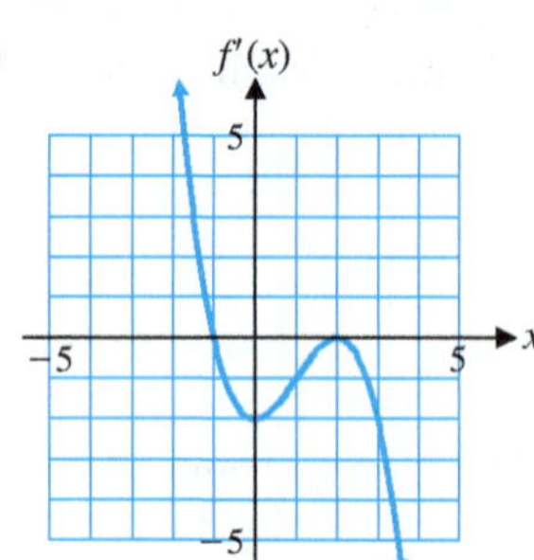

66.

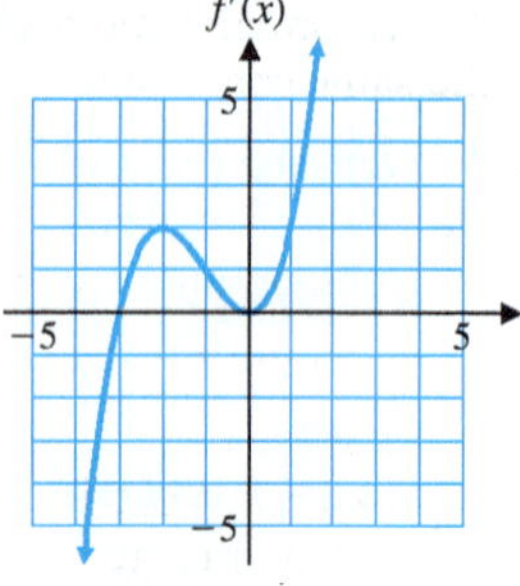

67.

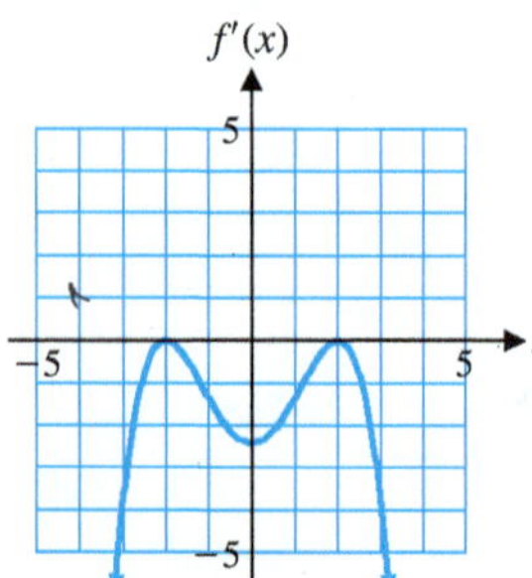

68.

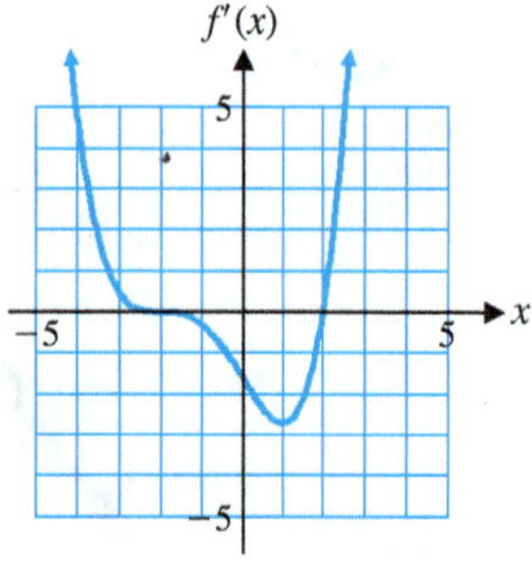

In Problems 69–76, apply steps 1–3 of the graphing strategy to f(x). Use a graphing calculator to approximate (to two decimal places) x intercepts, critical values, and x coordinates of inflection points. Summarize all the pertinent information.

69. $f(x) = x^4 - 5x^3 + 3x^2 + 8x - 5$

70. $f(x) = x^4 + 2x^3 - 5x^2 - 4x + 4$

71. $f(x) = x^4 - 21x^3 + 100x^2 + 20x + 100$

72. $f(x) = x^4 - 12x^3 + 28x^2 + 76x - 50$

73. $f(x) = -x^4 - x^3 + 2x^2 - 2x + 3$

74. $f(x) = -x^4 + x^3 + x^2 + 6$

75. $f(x) = 0.1x^5 + 0.3x^4 - 4x^3 - 5x^2 + 40x + 30$

76. $f(x) = x^5 + 4x^4 - 7x^3 - 20x^2 + 20x - 20$

C

In Problems 77–80, assume that f is a polynomial.

77. Explain how you can locate inflection points for the graph of $y = f(x)$ by examining the graph of $y = f'(x)$.

78. Explain how you can determine where $f'(x)$ is increasing or decreasing by examining the graph of $y = f(x)$.

79. Explain how you can locate local maxima and minima for the graph of $y = f'(x)$ by examining the graph of $y = f(x)$.

80. Explain how you can locate local maxima and minima for the graph of $y = f(x)$ by examining the graph of $y = f'(x)$.

Applications

81. **Inflation.** One commonly used measure of inflation is the annual rate of change of the Consumer Price Index (CPI). A TV news story says that the rate of change of inflation for consumer prices is increasing. What does this say about the shape of the graph of the CPI?

82. **Inflation.** Another commonly used measure of inflation is the annual rate of change of the Producer Price Index (PPI). A government report states that the rate of change of inflation for producer prices is decreasing. What does this say about the shape of the graph of the PPI?

83. **Cost analysis.** A company manufactures a variety of camp stoves at different locations. The total cost $C(x)$ (in dollars) of producing x camp stoves per week at plant A is shown in the figure. Discuss the graph of the marginal cost function $C'(x)$ and interpret the graph of $C'(x)$ in terms of the efficiency of the production process at this plant.

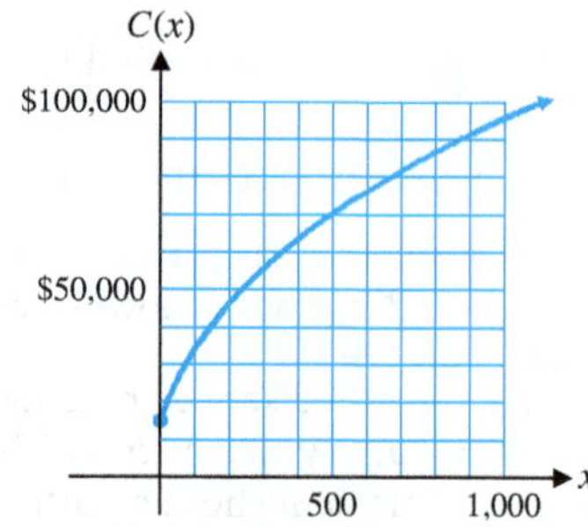

Figure for 83 Production costs at plant A

84. **Cost analysis.** The company in Problem 83 produces the same camp stove at another plant. The total cost $C(x)$ (in dollars) of producing x camp stoves per week at plant B is shown in the figure. Discuss the graph of the marginal cost function $C'(x)$ and interpret the graph of $C'(x)$ in terms of the efficiency of the production process at plant B. Compare the production processes at the two plants.

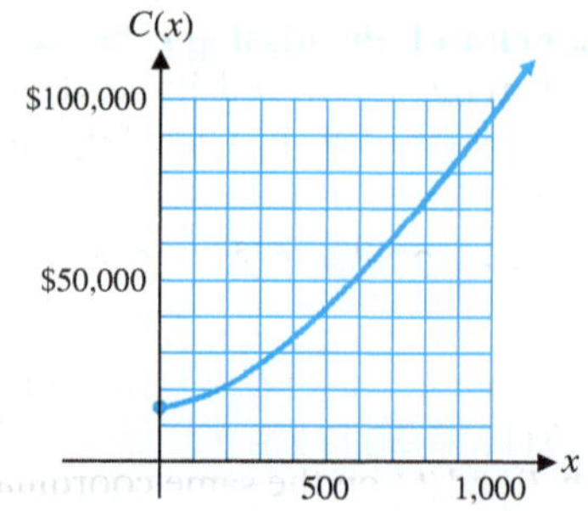

Figure for 84 Production costs at plant B

85. **Revenue.** The marketing research department of a computer company used a large city to test market the firm's new laptop. The department found that the relationship between price p (dollars per unit) and the demand x (units per week) was given approximately by

$$p = 1{,}296 - 0.12x^2 \qquad 0 < x < 80$$

So, weekly revenue can be approximated by

$$R(x) = xp = 1{,}296x - 0.12x^3 \qquad 0 < x < 80$$

(A) Find the local extrema for the revenue function.

(B) On which intervals is the graph of the revenue function concave upward? Concave downward?

86. Profit. Suppose that the cost equation for the company in Problem 85 is

$$C(x) = 830 + 396x$$

(A) Find the local extrema for the profit function.

(B) On which intervals is the graph of the profit function concave upward? Concave downward?

87. Revenue. A dairy is planning to introduce and promote a new line of organic ice cream. After test marketing the new line in a large city, the marketing research department found that the demand in that city is given approximately by

$$p = 10e^{-x} \qquad 0 \le x \le 5$$

where x thousand quarts were sold per week at a price of $\$p$ each.

(A) Find the local extrema for the revenue function.

(B) On which intervals is the graph of the revenue function concave upward? Concave downward?

88. Revenue. A national food service runs food concessions for sporting events throughout the country. The company's marketing research department chose a particular football stadium to test market a new jumbo hot dog. It was found that the demand for the new hot dog is given approximately by

$$p = 8 - 2 \ln x \qquad 5 \le x \le 50$$

where x is the number of hot dogs (in thousands) that can be sold during one game at a price of $\$p$.

(A) Find the local extrema for the revenue function.

(B) On which intervals is the graph of the revenue function concave upward? Concave downward?

89. Production: point of diminishing returns. A T-shirt manufacturer is planning to expand its workforce. It estimates that the number of T-shirts produced by hiring x new workers is given by

$$T(x) = -0.25x^4 + 5x^3 \qquad 0 \le x \le 15$$

When is the rate of change of T-shirt production increasing and when is it decreasing? What is the point of diminishing returns and the maximum rate of change of T-shirt production? Graph T and T' on the same coordinate system.

90. Production: point of diminishing returns. A baseball cap manufacturer is planning to expand its workforce. It estimates that the number of baseball caps produced by hiring x new workers is given by

$$T(x) = -0.25x^4 + 6x^3 \qquad 0 \le x \le 18$$

When is the rate of change of baseball cap production increasing and when is it decreasing? What is the point of diminishing returns and the maximum rate of change of baseball cap production? Graph T and T' on the same coordinate system.

91. Advertising: point of diminishing returns. A company estimates that it will sell $N(x)$ units of a product after spending $\$x$ thousand on advertising, as given by

$$N(x) = -0.25x^4 + 23x^3 - 540x^2 + 80{,}000 \qquad 24 \le x \le 45$$

When is the rate of change of sales increasing and when is it decreasing? What is the point of diminishing returns and the maximum rate of change of sales? Graph N and N' on the same coordinate system.

92. Advertising: point of diminishing returns. A company estimates that it will sell $N(x)$ units of a product after spending $\$x$ thousand on advertising, as given by

$$N(x) = -0.25x^4 + 13x^3 - 180x^2 + 10{,}000 \qquad 15 \le x \le 24$$

When is the rate of change of sales increasing and when is it decreasing? What is the point of diminishing returns and the maximum rate of change of sales? Graph N and N' on the same coordinate system.

93. Advertising. An automobile dealer uses TV advertising to promote car sales. On the basis of past records, the dealer arrived at the following data, where x is the number of ads placed monthly and y is the number of cars sold that month:

Number of Ads x	Number of Cars y
10	325
12	339
20	417
30	546
35	615
40	682
50	795

(A) Enter the data in a graphing calculator and find a cubic regression equation for the number of cars sold monthly as a function of the number of ads.

(B) How many ads should the dealer place each month to maximize the rate of change of sales with respect to the number of ads, and how many cars can the dealer expect to sell with this number of ads? Round answers to the nearest integer.

94. Advertising. A sporting goods chain places TV ads to promote golf club sales. The marketing director used past records to determine the following data, where x is the number of ads placed monthly and y is the number of golf clubs sold that month.

Number of Ads x	Number of Golf Clubs y
10	345
14	488
20	746
30	1,228
40	1,671
50	1,955

(A) Enter the data in a graphing calculator and find a cubic regression equation for the number of golf clubs sold monthly as a function of the number of ads.

(B) How many ads should the store manager place each month to maximize the rate of change of sales with respect to the number of ads, and how many golf clubs can the manager expect to sell with this number of ads? Round answers to the nearest integer.

95. Population growth: bacteria. A drug that stimulates reproduction is introduced into a colony of bacteria. After t minutes, the number of bacteria is given approximately by

$$N(t) = 1{,}000 + 30t^2 - t^3 \qquad 0 \le t \le 20$$

(A) When is the rate of growth, $N'(t)$, increasing? Decreasing?

(B) Find the inflection points for the graph of N.

(C) Sketch the graphs of N and N' on the same coordinate system.

(D) What is the maximum rate of growth?

96. Drug sensitivity. One hour after x milligrams of a particular drug are given to a person, the change in body temperature $T(x)$, in degrees Fahrenheit, is given by

$$T(x) = x^2\left(1 - \frac{x}{9}\right) \qquad 0 \le x \le 6$$

The rate $T'(x)$ at which $T(x)$ changes with respect to the size of the dosage x is called the *sensitivity* of the body to the dosage.

(A) When is $T'(x)$ increasing? Decreasing?

(B) Where does the graph of T have inflection points?

(C) Sketch the graphs of T and T' on the same coordinate system.

(D) What is the maximum value of $T'(x)$?

97. Learning. The time T (in minutes) it takes a person to learn a list of length n is

$$T(n) = 0.08n^3 - 1.2n^2 + 6n \qquad n \ge 0$$

(A) When is the rate of change of T with respect to the length of the list increasing? Decreasing?

(B) Where does the graph of T have inflection points?

(C) Graph T and T' on the same coordinate system.

(D) What is the minimum value of $T'(n)$?

Answers to Matched Problems

1. (A) Concave downward on $(-\infty, \infty)$

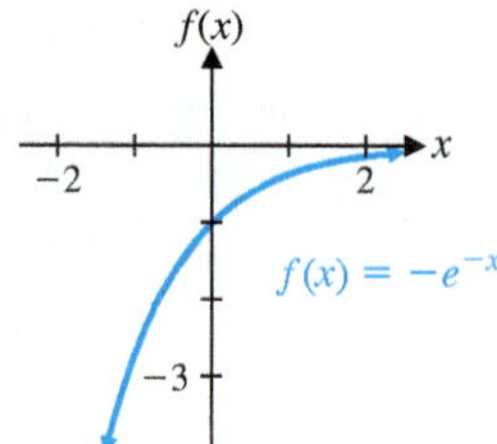

(B) Concave upward on $(0, \infty)$

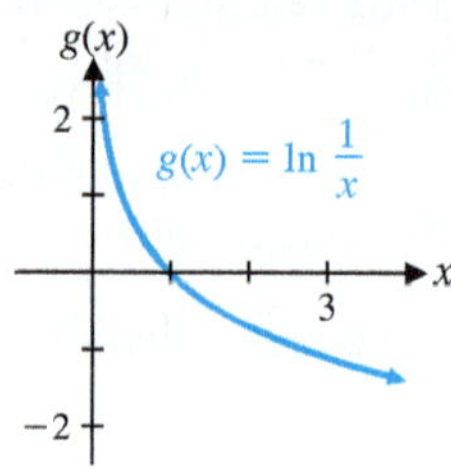

(C) Concave upward on $(-\infty, 0)$ and concave downward on $(0, \infty)$

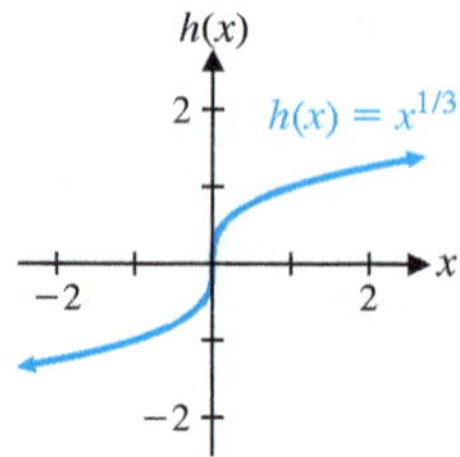

2. The only inflection point is $(3, f(3)) = (3, 8)$.

3. The inflection points are $(-1, f(-1)) = (-1, \ln 8)$ and $(3, f(3)) = (3, \ln 8)$.

4.

x	$f'(x)$	$f(x)$
$-\infty < x < -1$	Positive and decreasing	Increasing and concave downward
$x = -1$	Local minimum	Inflection point
$-1 < x < 1$	Positive and increasing	Increasing and concave upward
$x = 1$	Local maximum	Inflection point
$1 < x < 2$	Positive and decreasing	Increasing and concave downward
$x = 2$	x intercept	Local maximum
$2 < x < \infty$	Negative and decreasing	Decreasing and concave downward

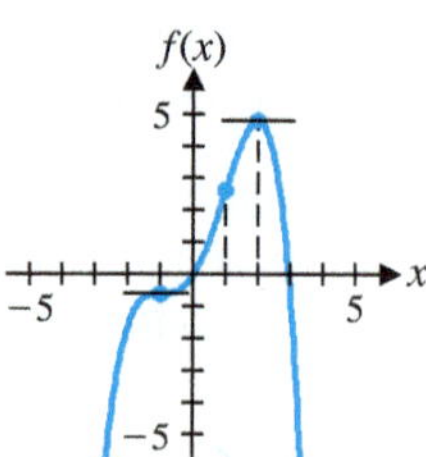

5. x intercepts: $-4, 0$; y intercept: $f(0) = 0$

Decreasing on $(-\infty, -3)$; increasing on $(-3, \infty)$; local minimum at $x = -3$

Concave upward on $(-\infty, -2)$ and $(0, \infty)$; concave downward on $(-2, 0)$

Inflection points at $x = -2$ and $x = 0$

x	$f(x)$
-4	0
-3	-27
-2	-16
0	0

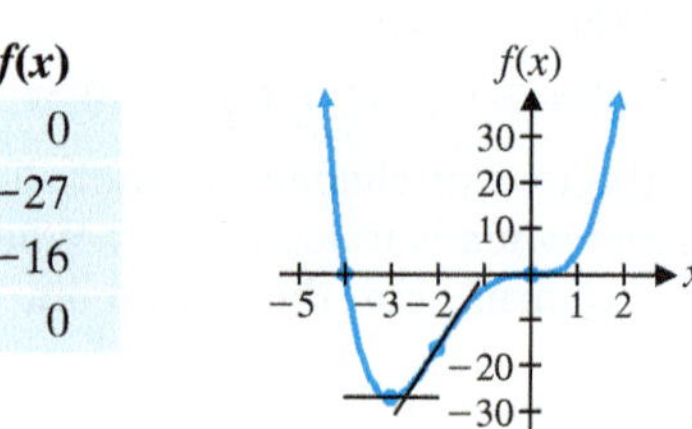

6. x intercepts: 0, 27; y intercept: $f(0) = 0$

Decreasing on $(-\infty, 0)$ and $(8, \infty)$; increasing on $(0, 8)$; local minimum: $f(0) = 0$; local maximum: $f(8) = 4$

Concave downward on $(-\infty, 0)$ and $(0, \infty)$; no inflection points

x	$f(x)$
0	0
8	4
27	0

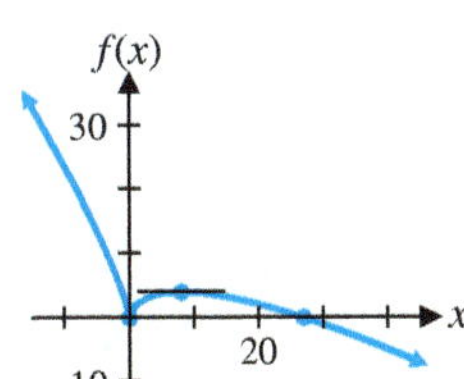

7. $N'(x)$ is increasing on $(0, 8)$ and decreasing on $(8, 12)$. The point of diminishing returns is $x = 8$ and the maximum rate of change is $N'(8) = 256$.

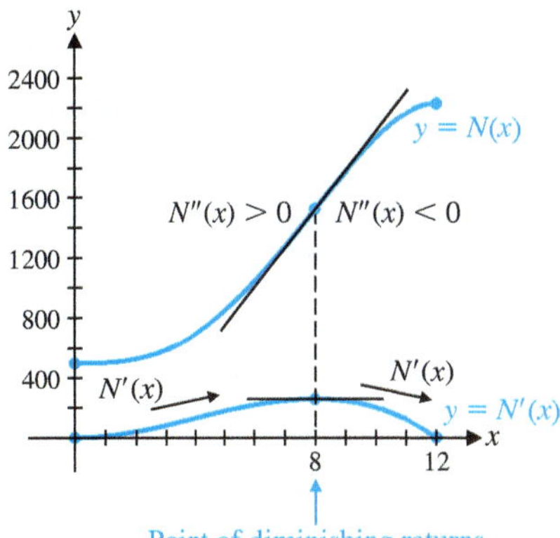

3-3 L'Hôpital's Rule

Introduction

The ability to evaluate a wide variety of different types of limits is one of the skills that are necessary to apply the techniques of calculus successfully. Limits play a fundamental role in the development of the derivative and are an important graphing tool. In order to deal effectively with graphs, we need to develop some more methods for evaluating limits.

In this section, we discuss a powerful technique for evaluating limits of quotients called *L'Hôpital's rule.* The rule is named after the French mathematician Marquis de L'Hôpital (1661–1704). To use L'Hôpital's rule, it is necessary to be familiar with the limit properties of some basic functions. Figure 1 reviews some limits involving powers of x that were discussed earlier.

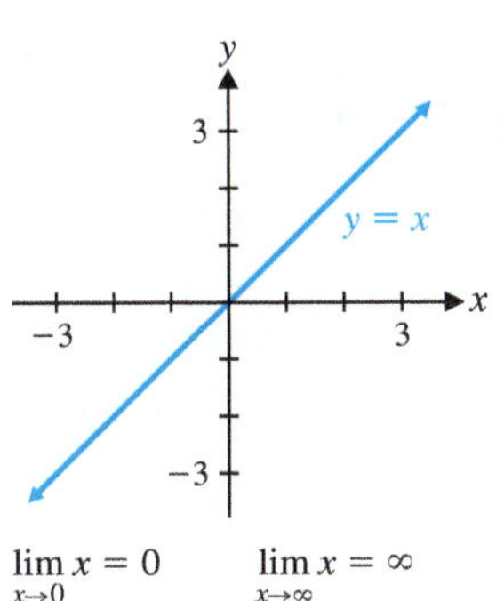

$\lim_{x\to 0} x = 0$ $\quad \lim_{x\to\infty} x = \infty$

$\lim_{x\to-\infty} x = -\infty$

(A) $y = x$

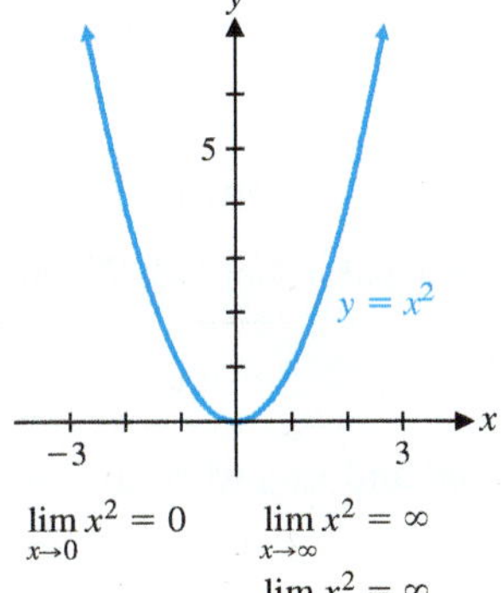

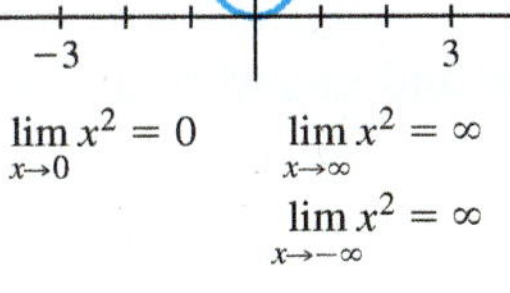

(B) $y = x^2$

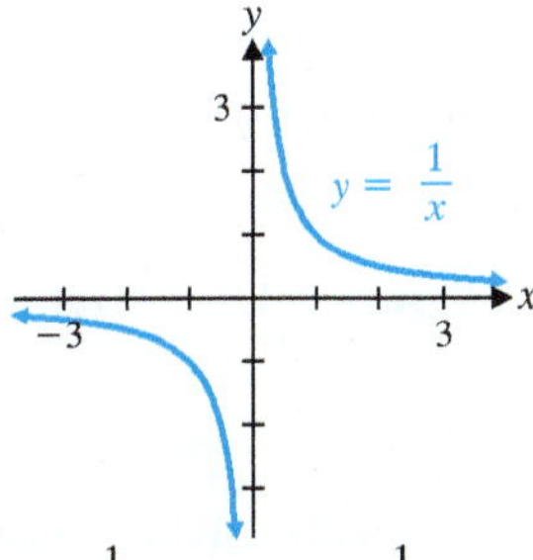

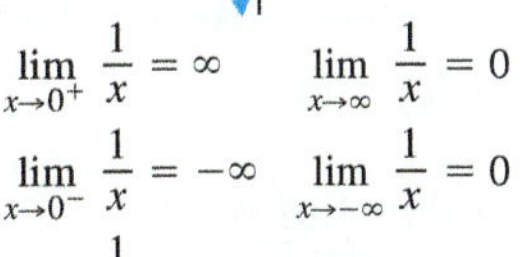

(C) $y = \frac{1}{x}$

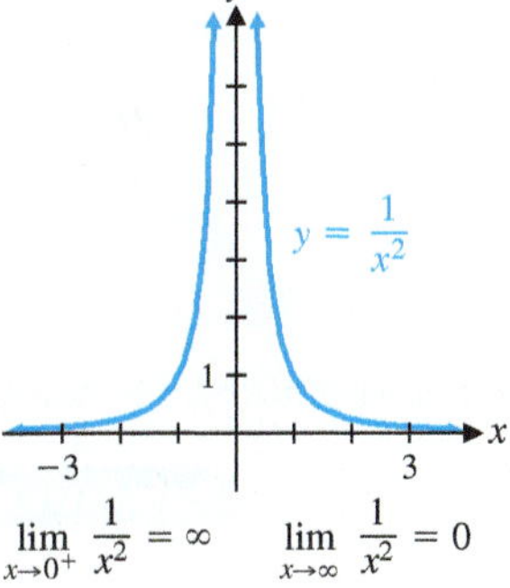

$\lim_{x\to 0^+} \frac{1}{x^2} = \infty$ $\quad \lim_{x\to\infty} \frac{1}{x^2} = 0$

$\lim_{x\to 0^-} \frac{1}{x^2} = \infty$ $\quad \lim_{x\to-\infty} \frac{1}{x^2} = 0$

$\lim_{x\to 0} \frac{1}{x^2} = \infty$

(D) $y = \frac{1}{x^2}$

Figure 1 Limits involving powers of x

The limits in Figure 1 are easily extended to functions of the form $f(x) = (x - c)^n$ and $g(x) = 1/(x - c)^n$. In general, if n is an odd integer, then limits involving $(x - c)^n$ or $1/(x - c)^n$ as x approaches c (or $\pm\infty$) behave, respectively, like the limits of x and $1/x$ as x approaches 0 (or $\pm\infty$). If n is an even integer, then limits involving these expressions behave, respectively, like the limits of x^2 and $1/x^2$ as x approaches 0 (or $\pm\infty$).

EXAMPLE 1 **Limits Involving Powers of $x - c$**

(A) $\lim_{x\to 2} \frac{5}{(x-2)^4} = \infty$ Compare with $\lim_{x\to 0} \frac{1}{x^2}$ in Figure 1.

(B) $\lim_{x\to -1^-} \frac{4}{(x+1)^3} = -\infty$ Compare with $\lim_{x\to 0^-} \frac{1}{x}$ in Figure 1.

(C) $\lim_{x\to \infty} \frac{4}{(x-9)^6} = 0$ Compare with $\lim_{x\to \infty} \frac{1}{x^2}$ in Figure 1.

(D) $\lim_{x\to -\infty} 3x^3 = -\infty$ Compare with $\lim_{x\to -\infty} x$ in Figure 1.

Matched Problem 1 Evaluate each limit.

(A) $\lim_{x\to 3^+} \frac{7}{(x-3)^5}$ (B) $\lim_{x\to -4} \frac{6}{(x+4)^6}$

(C) $\lim_{x\to -\infty} \frac{3}{(x+2)^3}$ (D) $\lim_{x\to \infty} 5x^4$

Figure 2 reviews limits of exponential and logarithmic functions.

The limits in Figure 2 also generalize to other simple exponential and logarithmic forms.

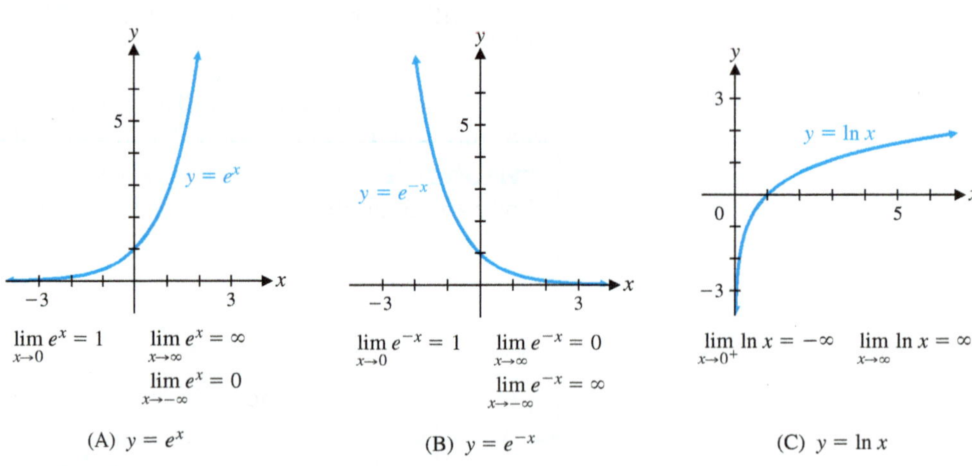

Figure 2 Limits involving exponential and logarithmic functions

EXAMPLE 2 **Limits Involving Exponential and Logarithmic Forms**

(A) $\lim_{x\to \infty} 2e^{3x} = \infty$ Compare with $\lim_{x\to \infty} e^x$ in Figure 2.

(B) $\lim_{x\to \infty} 4e^{-5x} = 0$ Compare with $\lim_{x\to \infty} e^{-x}$ in Figure 2.

(C) $\lim_{x\to \infty} \ln(x+4) = \infty$ Compare with $\lim_{x\to \infty} \ln x$ in Figure 2.

(D) $\lim_{x\to 2^+} \ln(x-2) = -\infty$ Compare with $\lim_{x\to 0^+} \ln x$ in Figure 2.

Matched Problem 2 Evaluate each limit.

(A) $\lim_{x \to -\infty} 2e^{-6x}$ (B) $\lim_{x \to -\infty} 3e^{2x}$

(C) $\lim_{x \to -4^+} \ln(x + 4)$ (D) $\lim_{x \to \infty} \ln(x - 10)$

Now that we have reviewed the limit properties of some basic functions, we are ready to consider the main topic of this section: L'Hôpital's rule.

L'Hôpital's Rule and the Indeterminate Form 0/0

Recall that the limit

$$\lim_{x \to c} \frac{f(x)}{g(x)}$$

is a 0/0 indeterminate form if

$$\lim_{x \to c} f(x) = 0 \quad \text{and} \quad \lim_{x \to c} g(x) = 0$$

The quotient property for limits in Section 1-1 does not apply since $\lim_{x \to c} g(x) = 0$.

If we are dealing with a 0/0 indeterminate form, the limit may or may not exist, and we cannot tell which is true without further investigation.

Each of the following is a 0/0 indeterminate form:

$$\lim_{x \to 2} \frac{x^2 - 4}{x - 2} \quad \text{and} \quad \lim_{x \to 1} \frac{e^x - e}{x - 1}$$

The first limit can be evaluated by performing some algebraic simplifications, such as

$$\lim_{x \to 2} \frac{x^2 - 4}{x - 2} = \lim_{x \to 2} \frac{(x - 2)(x + 2)}{x - 2} = \lim_{x \to 2}(x + 2) = 4$$

The second cannot. Instead, we turn to the powerful **L'Hôpital's rule**, which we now state without proof. This rule can be used whenever a limit is a 0/0 indeterminate form.

THEOREM 1 L'Hôpital's Rule for 0/0 Indeterminate Forms: Version 1

For c a real number,

if $\lim_{x \to c} f(x) = 0$ and $\lim_{x \to c} g(x) = 0$, then

$$\lim_{x \to c} \frac{f(x)}{g(x)} = \lim_{x \to c} \frac{f'(x)}{g'(x)}$$

provided that the second limit exists or is $+\infty$ or $-\infty$.

EXAMPLE 3 **L'Hôpital's Rule** Evaluate $\lim_{x \to 1} \frac{e^x - e}{x - 1}$.

SOLUTION **Step 1** *Check to see if L'Hôpital's rule applies:*

$$\lim_{x \to 1}(e^x - e) = e^1 - e = 0 \quad \text{and} \quad \lim_{x \to 1}(x - 1) = 1 - 1 = 0$$

L'Hôpital's rule does apply.

Step 2 *Apply L'Hôpital's rule:*

0/0 form

$$\lim_{x \to 1} \frac{e^x - e}{x - 1} = \lim_{x \to 1} \frac{\frac{d}{dx}(e^x - e)}{\frac{d}{dx}(x - 1)}$$ Apply L'Hôpital's rule.

$$= \lim_{x \to 1} \frac{e^x}{1}$$ e^x is continuous at $x = 1$.

$$= \frac{e^1}{1} = e$$

Matched Problem 3 Evaluate $\lim_{x \to 4} \frac{e^x - e^4}{x - 4}$.

CONCEPTUAL INSIGHT

In L'Hôpital's rule, the symbol $f'(x)/g'(x)$ represents the derivative of $f(x)$ divided by the derivative of $g(x)$, not the derivative of the quotient $f(x)/g(x)$.

When applying L'Hôpital's rule to a 0/0 indeterminate form, be certain that you differentiate the numerator and denominator separately.

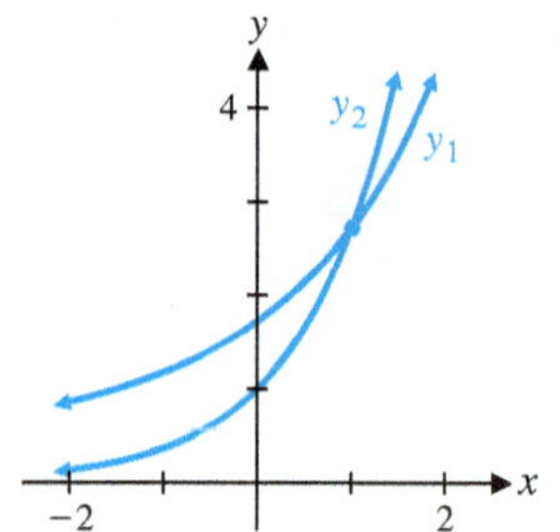

Figure 3

The functions

$$y_1 = \frac{e^x - e}{x - 1} \quad \text{and} \quad y_2 = \frac{e^x}{1}$$

of Example 3 are different functions (see Fig. 3), but both functions have the same limit e as x approaches 1. Although y_1 is undefined at $x = 1$, the graph of y_1 provides a check of the answer to Example 3.

EXAMPLE 4 **L'Hôpital's Rule** Evaluate $\lim_{x \to 0} \frac{\ln(1 + x^2)}{x^4}$.

SOLUTION **Step 1** *Check to see if L'Hôpital's rule applies:*

$$\lim_{x \to 0} \ln(1 + x^2) = \ln 1 = 0 \quad \text{and} \quad \lim_{x \to 0} x^4 = 0$$

L'Hôpital's rule does apply.

Step 2 *Apply L'Hôpital's rule:*

0/0 form

$$\lim_{x \to 0} \frac{\ln(1 + x^2)}{x^4} = \lim_{x \to 0} \frac{\frac{dx}{dx} \ln(1 + x^2)}{\frac{d}{dx} x^4}$$ Apply L'Hôpital's rule.

$$\lim_{x \to 0} \frac{\ln(1 + x^2)}{x^4} = \lim_{x \to 0} \frac{\frac{2x}{1 + x^2}}{4x^3}$$ Multiply numerator and denominator by $1/4x^3$.

$$= \lim_{x \to 0} \frac{\frac{2x}{1 + x^2} \cdot \frac{1}{4x^3}}{4x^3 \cdot \frac{1}{4x^3}}$$ Simplify.

$$= \lim_{x \to 0} \frac{1}{2x^2(1 + x^2)}$$

$$= \infty$$

Apply Theorem 1 in Section 1-3 and compare with Fig. 1(D).

Matched Problem 4 Evaluate $\lim_{x \to 1} \frac{\ln x}{(x-1)^3}$.

EXAMPLE 5 **L'Hôpital's Rule May Not Be Applicable** Evaluate $\lim_{x \to 1} \frac{\ln x}{x}$.

SOLUTION **Step 1** *Check to see if L'Hôpital's rule applies:*

$$\lim_{x \to 1} \ln x = \ln 1 = 0, \qquad \text{but} \qquad \lim_{x \to 1} x = 1 \neq 0$$

L'Hôpital's rule does not apply.

Step 2 *Evaluate by another method.* The quotient property for limits from Section 1-1 does apply, and we have

$$\lim_{x \to 1} \frac{\ln x}{x} = \frac{\lim_{x \to 1} \ln x}{\lim_{x \to 1} x} = \frac{\ln 1}{1} = \frac{0}{1} = 0$$

Note that applying L'Hôpital's rule would give us an incorrect result:

$$\lim_{x \to 1} \frac{\ln x}{x} \neq \lim_{x \to 1} \frac{\frac{d}{dx} \ln x}{\frac{d}{dx} x} = \lim_{x \to 1} \frac{1/x}{1} = 1$$

Matched Problem 5 Evaluate $\lim_{x \to 0} \frac{x}{e^x}$.

CONCEPTUAL INSIGHT

As Example 5 illustrates, all limits involving quotients are not 0/0 indeterminate forms.

You must always check to see if L'Hôpital's rule applies before you use it.

EXAMPLE 6 **Repeated Application of L'Hôpital's Rule** Evaluate

$$\lim_{x \to 0} \frac{x^2}{e^x - 1 - x}$$

SOLUTION **Step 1** *Check to see if L'Hôpital's rule applies:*

$$\lim_{x \to 0} x^2 = 0 \qquad \text{and} \qquad \lim_{x \to 0} (e^x - 1 - x) = 0$$

L'Hôpital's rule does apply.

Step 2 Apply L'Hôpital's rule:

0/0 form

$$\lim_{x \to 0} \frac{x^2}{e^x - 1 - x} = \lim_{x \to 0} \frac{\frac{d}{dx} x^2}{\frac{d}{dx}(e^x - 1 - x)} = \lim_{x \to 0} \frac{2x}{e^x - 1}$$

Since $\lim_{x \to 0} 2x = 0$ and $\lim_{x \to 0}(e^x - 1) = 0$, the new limit obtained is also a 0/0 indeterminate form, and L'Hôpital's rule can be applied again.

Step 3 *Apply L'Hôpital's rule again:*

0/0 form

$$\lim_{x\to 0}\frac{2x}{e^x - 1} = \lim_{x\to 0}\frac{\frac{d}{dx}2x}{\frac{d}{dx}(e^x - 1)} = \lim_{x\to 0}\frac{2}{e^x} = \frac{2}{e^0} = 2$$

Therefore,

$$\lim_{x\to 0}\frac{x^2}{e^x - 1 - x} = \lim_{x\to 0}\frac{2x}{e^x - 1} = \lim_{x\to 0}\frac{2x}{e^x} = 2$$

Matched Problem 6 Evaluate $\lim_{x\to 0}\frac{e^{2x} - 1 - 2x}{x^2}$

One-Sided Limits and Limits at ∞

In addition to examining the limit as x approaches c, we have discussed one-sided limits and limits at ∞ in Chapter 3. L'Hôpital's rule is valid in these cases also.

THEOREM 2 L'Hôpital's Rule for 0/0 Indeterminate Forms: Version 2 (for one-sided limits and limits at infinity)

The first version of L'Hôpital's rule (Theorem 1) remains valid if the symbol $x \to c$ is replaced everywhere it occurs with one of the following symbols:

$$x \to c^+ \qquad x \to c^- \qquad x \to \infty \qquad x \to -\infty$$

For example, if $\lim_{x\to\infty} f(x) = 0$ and $\lim_{x\to\infty} g(x) = 0$, then

$$\lim_{x\to\infty}\frac{f(x)}{g(x)} = \lim_{x\to\infty}\frac{f'(x)}{g'(x)}$$

provided that the second limit exists or is $+\infty$ or $-\infty$. Similar rules can be written for $x \to c^+$, $x \to c^-$, and $x \to -\infty$.

EXAMPLE 7 **L'Hôpital's Rule for One-Sided Limits** Evaluate $\lim_{x\to 1^+}\frac{\ln x}{(x - 1)^2}$.

SOLUTION **Step 1** *Check to see if L'Hôpital's rule applies:*

$$\lim_{x\to 1^+}\ln x = 0 \qquad \text{and} \qquad \lim_{x\to 1^+}(x - 1)^2 = 0$$

L'Hôpital's rule does apply.

Step 2 *Apply L'Hôpital's rule:*

0/0 form

$$\lim_{x\to 1^+}\frac{\ln x}{(x - 1)^2} = \lim_{x\to 1^+}\frac{\frac{d}{dx}(\ln x)}{\frac{d}{dx}(x - 1)^2} \qquad \text{Apply L'Hôpital's rule.}$$

$$= \lim_{x\to 1^+}\frac{1/x}{2(x - 1)} \qquad \text{Simplify.}$$

$$= \lim_{x\to 1^+}\frac{1}{2x(x - 1)}$$

$$= \infty$$

The limit as $x \to 1^+$ is ∞ because $1/2x(x - 1)$ has a vertical asymptote at $x = 1$ (Theorem 1, Section 1-3) and $x(x - 1) > 0$ for $x > 1$.

Matched Problem 7 Evaluate $\lim_{x \to 1^-} \frac{\ln x}{(x - 1)^2}$.

EXAMPLE 8 **L'Hôpital's Rule for Limits at Infinity** Evaluate $\lim_{x \to \infty} \frac{\ln(1 + e^{-x})}{e^{-x}}$.

SOLUTION **Step 1** *Check to see if L'Hôpital's rule applies:*

$$\lim_{x \to \infty} \ln(1 + e^{-x}) = \ln(1 + 0) = \ln 1 = 0 \text{ and } \lim_{x \to \infty} e^{-x} = 0$$

L'Hôpital's rule does apply.

Step 2 *Apply L'Hôpital's rule:*

0/0 form

$$\lim_{x \to \infty} \frac{\ln(1 + e^{-x})}{e^{-x}} = \lim_{x \to \infty} \frac{\frac{d}{dx}[\ln(1 + e^{-x})]}{\frac{d}{dx}e^{-x}}$$ Apply L'Hôpital's rule.

$$= \lim_{x \to \infty} \frac{-e^{-x}/(1 + e^{-x})}{-e^{-x}}$$ Multiply numerator and denominator by $-e^x$.

$$\lim_{x \to \infty} \frac{\ln(1 + e^{-x})}{e^{-x}} = \lim_{x \to \infty} \frac{1}{1 + e^{-x}}$$ $\lim_{x \to \infty} e^{-x} = 0$

$$= \frac{1}{1 + 0} = 1$$

Matched Problem 8 Evaluate $\lim_{x \to -\infty} \frac{\ln(1 + 2e^x)}{e^x}$.

L'Hôpital's Rule and the Indeterminate Form ∞/∞

In Section 1-3, we discussed techniques for evaluating limits of rational functions such as

$$\lim_{x \to \infty} \frac{2x^2}{x^3 + 3} \qquad \lim_{x \to \infty} \frac{4x^3}{2x^2 + 5} \qquad \lim_{x \to \infty} \frac{3x^3}{5x^3 + 6} \qquad (1)$$

Each of these limits is an ∞/∞ *indeterminate form.* In general, if $\lim_{x \to c} f(x) = \pm\infty$ and $\lim_{x \to c} g(x) = \pm\infty$, then

$$\lim_{x \to c} \frac{f(x)}{g(x)}$$

is called an ∞/∞ **indeterminate form**. Furthermore, $x \to c$ can be replaced in all three limits above with $x \to c^+$, $x \to c^-$, $x \to \infty$, or $x \to -\infty$. It can be shown that L'Hôpital's rule also applies to these ∞/∞ indeterminate forms.

THEOREM 3 L'Hôpital's Rule for the Indeterminate Form ∞/∞: Version 3
Versions 1 and 2 of L'Hôpital's rule for the indeterminate form 0/0 are also valid if the limit of f and the limit of g are both infinite; that is, both $+\infty$ and $-\infty$ are permissible for either limit.

For example, if $\lim_{x\to c^+} f(x) = \infty$ and $\lim_{x\to c^+} g(x) = -\infty$, then L'Hôpital's rule can be applied to $\lim_{x\to c^+} [f(x)/g(x)]$.

Explore & Discuss 1

Evaluate each of the limits in (1) in two ways:

1. Use Theorem 4 in Section 3-3.
2. Use L'Hôpital's rule.

Given a choice, which method would you choose? Why?

EXAMPLE 9 **L'Hôpital's Rule for the Indeterminate Form ∞/∞** Evaluate $\lim_{x\to\infty} \frac{\ln x}{x^2}$.

SOLUTION **Step 1** *Check to see if L'Hôpital's rule applies:*

$$\lim_{x\to\infty} \ln x = \infty \quad \text{and} \quad \lim_{x\to\infty} x^2 = \infty$$

L'Hôpital's rule does apply.

Step 2 *Apply L'Hôpital's rule:*

$$\lim_{x\to\infty} \frac{\ln x}{x^2} = \lim_{x\to\infty} \frac{\frac{d}{dx}(\ln x)}{\frac{d}{dx}x^2} \qquad \infty/\infty \text{ form} \qquad \text{Apply L'Hôpital's rule.}$$

$$= \lim_{x\to\infty} \frac{1/x}{2x} \qquad \text{Simplify.}$$

$$\lim_{x\to\infty} \frac{\ln x}{x^2} = \lim_{x\to\infty} \frac{1}{2x^2} \qquad \text{See Figure 1(D).}$$

$$= 0$$

Matched Problem 9 Evaluate $\lim_{x\to\infty} \frac{\ln x}{x}$.

EXAMPLE 10 **L'Hôpital's Rule for the Indeterminate Form ∞/∞** Evaluate $\lim_{x\to\infty} \frac{e^x}{x^2}$.

SOLUTION **Step 1** *Check to see if L'Hôpital's rule applies:*

$$\lim_{x\to\infty} e^x = \infty \quad \text{and} \quad \lim_{x\to\infty} x^2 = \infty$$

L'Hôpital's rule does apply.

Step 2 *Apply L'Hôpital's rule:*

$$\lim_{x\to\infty} \frac{e^x}{x^2} = \lim_{x\to\infty} \frac{\frac{d}{dx}e^x}{\frac{d}{dx}x^2} = \lim_{x\to\infty} \frac{e^x}{2x} \qquad \infty/\infty \text{ form}$$

Since $\lim_{x\to\infty} e^x = \infty$ and $\lim_{x\to\infty} 2x = \infty$, this limit is an ∞/∞ indeterminate form and L'Hôpital's rule can be applied again.

Step 3 *Apply L'Hôpital's rule again:*

$$\lim_{x\to\infty} \frac{e^x}{2x} = \lim_{x\to\infty} \frac{\frac{d}{dx}e^x}{\frac{d}{dx}2x} = \lim_{x\to\infty} \frac{e^x}{2} = \infty \qquad \infty/\infty \text{ form}$$

Therefore,

$$\lim_{x\to\infty}\frac{e^x}{x^2} = \lim_{x\to\infty}\frac{e^x}{2x} = \lim_{x\to\infty}\frac{e^x}{2} = \infty$$

Matched Problem 10 Evaluate $\lim_{x\to\infty}\frac{e^{2x}}{x^2}$.

CONCEPTUAL INSIGHT

The three versions of L'Hôpital's rule cover a multitude of limits—far too many to remember case by case. Instead, we suggest you use the following pattern, common to all versions, as a memory aid:

1. All versions involve three limits: $\lim f(x)/g(x)$, $\lim f(x)$, and $\lim g(x)$.
2. The independent variable x must behave the same way in all three limits. The acceptable behaviors are $x \to c$, $x \to c^+$, $x \to c^-$, $x \to \infty$, or $x \to -\infty$.
3. The form of $\lim f(x)/g(x)$ must be $\frac{0}{0}$ or $\frac{\pm\infty}{\pm\infty}$ and both $\lim f(x)$ and $\lim g(x)$ must approach 0 or both must approach $\pm\infty$.

Exercises 3-3

A

Use L'Hôpital's rule to find each limit in Problems 1–14.

1. $\lim_{x\to 2}\frac{x^4 - 16}{x^3 - 8}$

2. $\lim_{x\to 1}\frac{x^6 - 1}{x^5 - 1}$

3. $\lim_{x\to 2}\frac{x^2 + x - 6}{x^2 + 6x - 16}$

4. $\lim_{x\to 4}\frac{x^2 - 8x + 16}{x^2 - 5x + 4}$

5. $\lim_{x\to 0}\frac{e^x - 1}{x}$

6. $\lim_{x\to 1}\frac{\ln x}{x - 1}$

7. $\lim_{x\to 0}\frac{\ln(1 + 4x)}{x}$

8. $\lim_{x\to 0}\frac{e^{2x} - 1}{x}$

9. $\lim_{x\to\infty}\frac{2x^2 + 7}{5x^3 + 9}$

10. $\lim_{x\to\infty}\frac{3x^4 + 6}{2x^2 + 5}$

11. $\lim_{x\to\infty}\frac{e^{3x}}{x}$

12. $\lim_{x\to\infty}\frac{x}{e^{4x}}$

13. $\lim_{x\to\infty}\frac{x^2}{\ln x}$

14. $\lim_{x\to\infty}\frac{\ln x}{x^4}$

In Problems 15–18, explain why L'Hôpital's rule does not apply. If the limit exists, find it by other means.

15. $\lim_{x\to 1}\frac{x^2 + 5x + 4}{x^3 + 1}$

16. $\lim_{x\to\infty}\frac{e^{-x}}{\ln x}$

17. $\lim_{x\to 2}\frac{x + 2}{(x - 2)^4}$

18. $\lim_{x\to -3}\frac{x^2}{(x + 3)^5}$

In Problems 19–22, the limit can be found in two ways. Use L'Hôpital's rule to find the limit and check your answer using an algebraic simplification.

19. $\lim_{x\to -4}\frac{x + 4}{x^2 - 16}$

20. $\lim_{x\to 5}\frac{x - 5}{x^2 - 25}$

21. $\lim_{x\to 9}\frac{x^2 - x - 72}{x - 9}$

22. $\lim_{x\to 7}\frac{x - 7}{x^2 + 6x - 91}$

B

Find each limit in Problems 23–46. Note that L'Hôpital's rule does not apply to every problem, and some problems will require more than one application of L'Hôpital's rule.

23. $\lim_{x\to 0}\frac{e^{4x} + 1 - 4x}{x^2}$

24. $\lim_{x\to 0}\frac{3x + 1 - e^{3x}}{x^2}$

25. $\lim_{x\to 2}\frac{\ln(x - 1)}{x - 1}$

26. $\lim_{x\to -1}\frac{\ln(x + 2)}{x + 2}$

27. $\lim_{x\to 0^+}\frac{\ln(1 + x^2)}{x^3}$

28. $\lim_{x\to 0^-}\frac{\ln(1 + 2x)}{x^2}$

29. $\lim_{x\to 0^+}\frac{\ln(1 + \sqrt{x})}{x}$

30. $\lim_{x\to 0^+}\frac{\ln(1 + x)}{\sqrt{x}}$

31. $\lim_{x\to -2}\frac{x^2 + 2x + 1}{x^2 + x + 1}$

32. $\lim_{x\to 1}\frac{2x^3 - 3x^2 + 1}{x^3 - 3x + 2}$

33. $\lim_{x\to -1}\frac{x^3 + x^2 - x - 1}{x^3 + 4x^2 + 5x + 2}$

34. $\lim_{x\to 3}\frac{x^3 + 3x^2 - x - 3}{x^2 + 6x + 9}$

35. $\lim_{x\to 2^-}\frac{x^3 - 12x + 16}{x^3 - 6x^2 + 12x - 8}$

36. $\lim_{x\to 1^+}\frac{x^3 + x^2 - x + 1}{x^3 + 3x^2 + 3x - 1}$

37. $\lim_{x\to\infty}\frac{3x^2 + 5x}{4x^3 + 7}$

38. $\lim_{x\to\infty}\frac{4x^2 + 9x}{5x^2 + 8}$

39. $\lim_{x\to\infty} \frac{x^2}{e^{2x}}$

40. $\lim_{x\to\infty} \frac{e^{3x}}{x^3}$

41. $\lim_{x\to\infty} \frac{1 + e^{-x}}{1 + x^2}$

42. $\lim_{x\to-\infty} \frac{1 + e^{-x}}{1 + x^2}$

43. $\lim_{x\to\infty} \frac{e^{-x}}{\ln(1 + 4e^{-x})}$

44. $\lim_{x\to\infty} \frac{\ln(1 + 2e^{-x})}{\ln(1 + e^{-x})}$

45. $\lim_{x\to 0} \frac{e^x - e^{-x} - 2x}{x^3}$

46. $\lim_{x\to 0} \frac{e^{2x} - 1 - 2x - 2x^2}{x^3}$

C

47. Find $\lim_{x\to 0^+}(x \ln x)$.
[*Hint*: Write $x \ln x = (\ln x)/x^{-1}$.]

48. Find $\lim_{x\to 0^+}(\sqrt{x} \ln x)$.
[*Hint*: Write $\sqrt{x} \ln x = (\ln x)/x^{-1/2}$.]

In Problems 49–52, n is a positive integer. Find each limit.

49. $\lim_{x\to\infty} \frac{\ln x}{x^n}$

50. $\lim_{x\to\infty} \frac{x^n}{\ln x}$

51. $\lim_{x\to\infty} \frac{e^x}{x^n}$

52. $\lim_{x\to\infty} \frac{x^n}{e^x}$

In Problems 53–56, show that the repeated application of L'Hôpital's rule does not lead to a solution. Then use algebraic manipulation to evaluate each limit. [Hint: If $x > 0$ and $n > 0$, then $\sqrt[n]{x^n} = x$.]

53. $\lim_{x\to\infty} \frac{\sqrt{1 + x^2}}{x}$

54. $\lim_{x\to-\infty} \frac{x}{\sqrt{4 + x^2}}$

55. $\lim_{x\to-\infty} \frac{\sqrt[3]{x^3 + 1}}{x}$

56. $\lim_{x\to\infty} \frac{x^2}{\sqrt[3]{(x^3 + 1)^2}}$

Answers to Matched Problems

1. (A) ∞
   (B) ∞
   (C) 0
   (D) ∞
2. (A) ∞
   (B) 0
   (C) $-\infty$
   (D) ∞
3. e^4
4. ∞
5. 0
6. 2
7. $-\infty$
8. 2
9. 0
10. ∞

3-4 Curve-Sketching Techniques

- Modifying the Graphing Strategy
- Using the Graphing Strategy
- Modeling Average Cost

When we summarized the graphing strategy in Section 3-2, we omitted one important topic: asymptotes. Polynomial functions do not have any asymptotes. Asymptotes of rational functions were discussed in Section 1-3, but what about all the other functions, such as logarithmic and exponential functions? Since investigating asymptotes always involves limits, we can now use L'Hôpital's rule (Section 3-3) as a tool for finding asymptotes of many different types of functions.

Modifying the Graphing Strategy

The first version of the graphing strategy in Section 3-2 made no mention of asymptotes. Including information about asymptotes produces the following (and final) version of the graphing strategy.

PROCEDURE Graphing Strategy (Final Version)

Step 1 *Analyze $f(x)$.*
(A) Find the domain of f.
(B) Find the intercepts.
(C) Find asymptotes.

Step 2 *Analyze $f'(x)$.* Find the partition numbers for, and critical values of, $f'(x)$. Construct a sign chart for $f'(x)$, determine the intervals on which f is increasing and decreasing, and find local maxima and minima.

Step 3 *Analyze $f''(x)$.* Find the partition numbers of $f''(x)$. Construct a sign chart for $f''(x)$, determine the intervals on which the graph of f is concave upward and concave downward, and find inflection points.

Step 4 *Sketch the graph of f.* Draw asymptotes and locate intercepts, local maxima and minima, and inflection points. Sketch in what you know from steps 1–3. Plot additional points as needed and complete the sketch.

Using the Graphing Strategy

We will illustrate the graphing strategy with several examples. From now on, you should always use the final version of the graphing strategy. If a function does not have any asymptotes, simply state this fact.

EXAMPLE 1 **Using the Graphing Strategy** Use the graphing strategy to analyze the function $f(x) = (x - 1)/(x - 2)$. State all the pertinent information and sketch the graph of f.

SOLUTION **Step 1** *Analyze $f(x)$.* $f(x) = \dfrac{x - 1}{x - 2}$

(A) Domain: All real x, except $x = 2$

(B) y intercept: $f(0) = \dfrac{0 - 1}{0 - 2} = \dfrac{1}{2}$

x intercepts: Since a fraction is 0 when its numerator is 0 and its denominator is not 0, the x intercept is $x = 1$.

(C) Horizontal asymptote: $\dfrac{a_m x^m}{b_n x^n} = \dfrac{x}{x} = 1$

So the line $y = 1$ is a horizontal asymptote.

Vertical asymptote: The denominator is 0 for $x = 2$, and the numerator is not 0 for this value. Therefore, the line $x = 2$ is a vertical asymptote.

Step 2 *Analyze $f'(x)$.* $f'(x) = \dfrac{(x - 2)(1) - (x - 1)(1)}{(x - 2)^2} = \dfrac{-1}{(x - 2)^2}$

Critical values of $f(x)$: None

Partition number for $f'(x)$: $x = 2$

Sign chart for $f'(x)$:

	$(-\infty, 2)$		$(2, \infty)$
$f'(x)$	- - - -	ND	- - - -
x		2	
$f(x)$	Decreasing		Decreasing

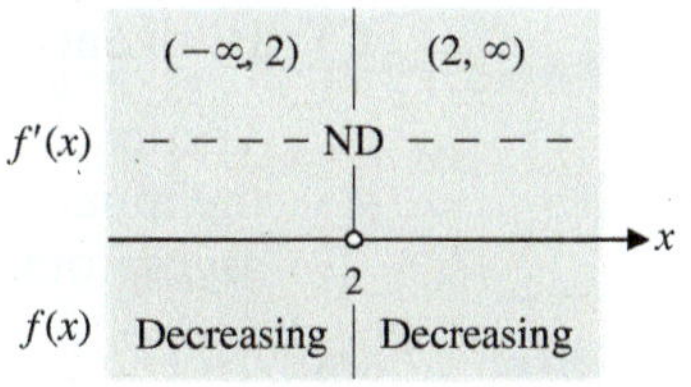

Test Numbers

x	$f'(x)$
1	-1 $(-)$
3	-1 $(-)$

So $f(x)$ is decreasing on $(-\infty, 2)$ and $(2, \infty)$. There are no local extrema.

Step 3 *Analyze $f''(x)$.* $f''(x) = \dfrac{2}{(x - 2)^3}$

Partition number for $f''(x)$: $x = 2$

Sign chart for $f''(x)$:

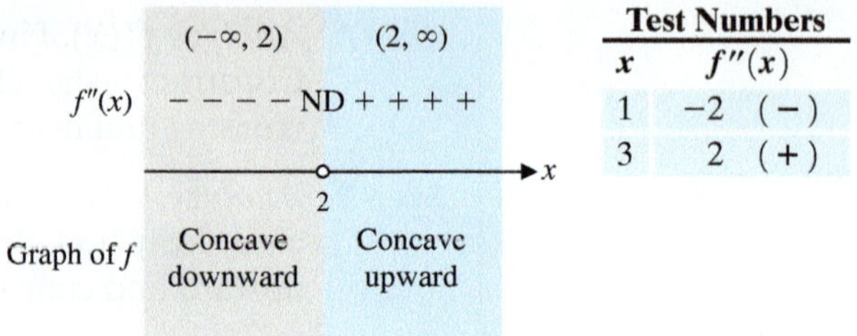

Test Numbers

x	$f''(x)$
1	-2 $(-)$
3	2 $(+)$

The graph of f is concave downward on $(-\infty, 2)$ and concave upward on $(2, \infty)$. Since $f(2)$ is not defined, there is no inflection point at $x = 2$, even though $f''(x)$ changes sign at $x = 2$.

Step 4 *Sketch the graph of f.* Insert intercepts and asymptotes, and plot a few additional points (for functions with asymptotes, plotting additional points is often helpful). Then sketch the graph.

x	$f(x)$
-2	$\frac{3}{4}$
0	$\frac{1}{2}$
1	0
$\frac{3}{2}$	-1
$\frac{5}{2}$	3
3	2
4	$\frac{3}{2}$

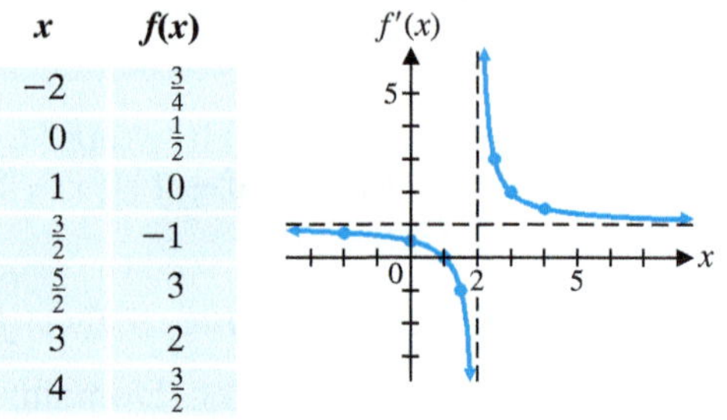

Matched Problem 1 Follow the graphing strategy and analyze the function $f(x) = 2x/(1 - x)$. State all the pertinent information and sketch the graph of f.

EXAMPLE 2 **Using the Graphing Strategy** Use the graphing strategy to analyze the function

$$g(x) = \frac{2x - 1}{x^2}$$

State all pertinent information and sketch the graph of g.

SOLUTION **Step 1** *Analyze* $g(x)$.

(A) Domain: All real x, except $x = 0$

(B) x intercept: $x = \frac{1}{2} = 0.5$

y intercept: Since 0 is not in the domain of g, there is no y intercept.

(C) Horizontal asymptote: $y = 0$ (the x axis)

Vertical asymptote: The denominator of $g(x)$ is 0 at $x = 0$ and the numerator is not. So the line $x = 0$ (the y axis) is a vertical asymptote.

Step 2 *Analyze* $g'(x)$.

$$g(x) = \frac{2x - 1}{x^2} = \frac{2}{x} - \frac{1}{x^2} = 2x^{-1} - x^{-2}$$

$$g'(x) = -2x^{-2} + 2x^{-3} = -\frac{2}{x^2} + \frac{2}{x^3} = \frac{-2x + 2}{x^3}$$

$$= \frac{2(1 - x)}{x^3}$$

Critical values of $g(x)$: $x = 1$

Partition numbers for $g'(x)$: $x = 0, x = 1$

Sign chart for $g'(x)$:

	$(-\infty, 0)$		$(0, 1)$		$(1, \infty)$
$g'(x)$	$- - - -$	ND	$+ + + +$	0	$- - - -$
x		0		1	

Function $f(x)$ is decreasing on $(-\infty, 0)$ and $(1, \infty)$, is increasing on $(0, 1)$, and has a local maximum at $x = 1$.

Step 3 *Analyze* $g''(x)$.

$$g'(x) = -2x^{-2} + 2x^{-3}$$

$$g''(x) = 4x^{-3} - 6x^{-4} = \frac{4}{x^3} - \frac{6}{x^4} = \frac{4x - 6}{x^4} = \frac{2(2x - 3)}{x^4}$$

Partition numbers for $g''(x)$: $x = 0, x = \frac{3}{2} = 1.5$

Sign chart for $g''(x)$:

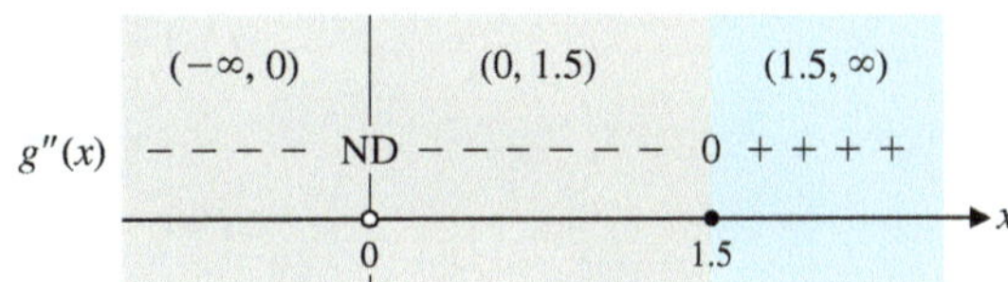

Function $g(x)$ is concave downward on $(-\infty, 0)$ and $(0, 1.5)$, is concave upward on $(1.5, \infty)$, and has an inflection point at $x = 1.5$.

Step 4 *Sketch the graph of* g. Plot key points, note that the coordinate axes are asymptotes, and sketch the graph.

x	$g(x)$
−10	−0.21
−1	−3
0.5	0
1	1
1.5	0.89
10	0.19

g(x)
Horizontal asymptote
Vertical asymptote

Matched Problem 2 Use the graphing strategy to analyze the function

$$h(x) = \frac{4x + 3}{x^2}$$

State all pertinent information and sketch the graph of h.

EXAMPLE 3 **Graphing Strategy** Follow the steps of the graphing strategy and analyze the function $f(x) = xe^x$. State all the pertinent information and sketch the graph of f.

SOLUTION **Step 1** Analyze $f(x)$: $f(x) = xe^x$.

(A) Domain: All real numbers

(B) y intercept: $f(0) = 0$

x intercept: $xe^x = 0$ for $x = 0$ only, since $e^x > 0$ for all x.

(C) Vertical asymptotes: None

Horizontal asymptotes: We use tables to determine the nature of the graph of f as $x \to \infty$ and $x \to -\infty$:

x	1	5	10	$\to \infty$
$f(x)$	2.72	742.07	220,264.66	$\to \infty$

x	−1	−5	−10	$\to -\infty$
$f(x)$	−0.37	−0.03	−0.000 45	$\to 0$

Step 2 *Analyze $f'(x)$:*

$$f'(x) = x\frac{d}{dx}e^x + e^x\frac{d}{dx}x$$
$$= xe^x + e^x = e^x(x+1)$$

Critical value of $f(x)$: -1

Partition number for $f'(x)$: -1

Sign chart for $f'(x)$:

	$(-\infty, -1)$		$(-1, \infty)$
$f'(x)$	− − − −	0	+ + + +
x		−1	
$f(x)$	Decreasing		Increasing

Test Numbers

x	$f'(x)$
−2	$-e^{-2}$ (−)
0	1 (+)

So $f(x)$ decreases on $(-\infty, -1)$, has a local minimum at $x = -1$, and increases on $(-1, \infty)$.

Step 3 *Analyze $f''(x)$:*

$$f''(x) = e^x\frac{d}{dx}(x+1) + (x+1)\frac{d}{dx}e^x$$
$$= e^x + (x+1)e^x = e^x(x+2)$$

Sign chart for $f''(x)$ (partition number is -2):

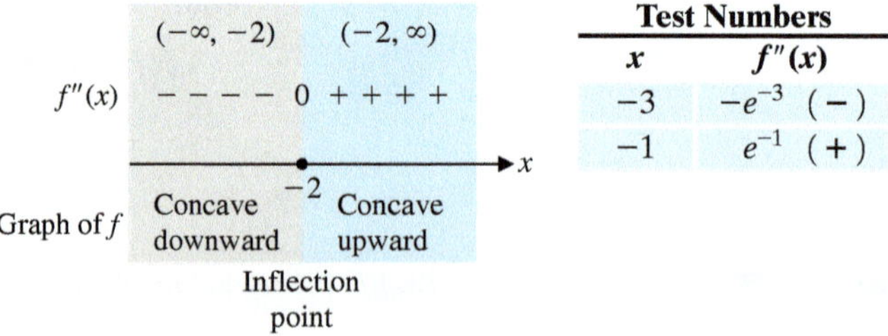

The graph of f is concave downward on $(-\infty, -2)$, has an inflection point at $x = -2$, and is concave upward on $(-2, \infty)$.

Step 4 *Sketch the graph of f, using the information from steps 1 to 3:*

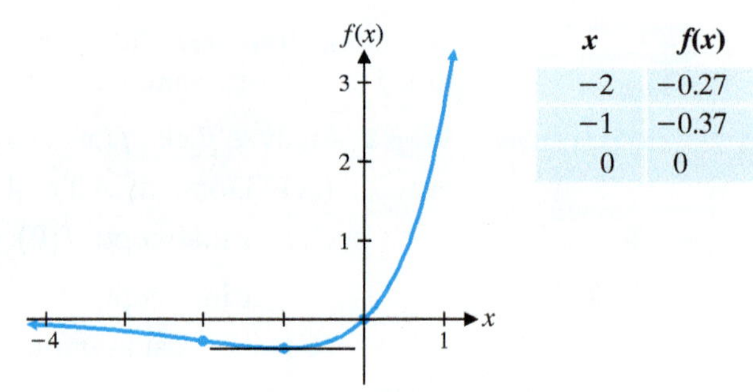

x	$f(x)$
−2	−0.27
−1	−0.37
0	0

Matched Problem 3 Analyze the function $f(x) = xe^{-0.5x}$. State all the pertinent information and sketch the graph of f.

EXPLORE & DISCUSS 1

Refer to the discussion of vertical asymptotes in the solution of Example 3. We used tables of values to estimate limits at infinity and determine horizontal asymptotes. In some cases, the functions involved in these limits can be written in a form that allows us to use L'Hôpital's rule.

$$\begin{aligned}
\lim_{x\to-\infty} f(x) &= \lim_{x\to-\infty} xe^{x} && -\infty \cdot 0 \text{ form} \quad \text{Rewrite as a fraction.} \\
&= \lim_{x\to-\infty} \frac{x}{e^{-x}} && -\infty/\infty \text{ form} \quad \text{Apply L'Hôpital's rule.} \\
&= \lim_{x\to-\infty} \frac{1}{-e^{-x}} && \text{Simplify.} \\
&= \lim_{x\to-\infty} (-e^{x}) && \text{Property of } e^x \\
&= 0
\end{aligned}$$

Use algebraic manipulation and L'Hôpital's rule to verify the value of each of the following limits:

(A) $\lim_{x\to\infty} xe^{-0.5x} = 0$

(B) $\lim_{x\to 0^+} x^2(\ln x - 0.5) = 0$

(C) $\lim_{x\to 0^+} x \ln x = 0$

EXAMPLE 4 **Graphing Strategy** Let $f(x) = x^2 \ln x - 0.5x^2$. Follow the steps in the graphing strategy and analyze this function. State all the pertinent information and sketch the graph of f.

SOLUTION **Step 1** *Analyze* $f(x)$: $f(x) = x^2 \ln x - 0.5x^2 = x^2(\ln x - 0.5)$.

(A) Domain: $(0, \infty)$

(B) y intercept: None [$f(0)$ is not defined.]

x intercept: Solve $x^2(\ln x - 0.5) = 0$

$$\ln x - 0.5 = 0 \quad \text{or} \quad x^2 = 0 \qquad \text{Discard, since 0 is not in the domain of } f.$$

$$\ln x = 0.5 \qquad \ln x = a \text{ if and only if } x = e^a.$$

$$x = e^{0.5} \qquad x \text{ intercept}$$

(C) Asymptotes: None. The following tables suggest the nature of the graph as $x \to 0^+$ and as $x \to \infty$:

x	0.1	0.01	0.001	$\to 0^+$
$f(x)$	−0.0280	−0.00051	−0.00007	$\to 0$

See Explore & Discuss 1(B).

x	10	100	1,000	$\to \infty$
$f(x)$	180	41,000	6,400,000	$\to \infty$

Step 2 *Analyze* $f'(x)$:

$$\begin{aligned} f'(x) &= x^2\frac{d}{dx}\ln x + (\ln x)\frac{d}{dx}x^2 - 0.5\frac{d}{dx}x^2 \\ &= x^2\frac{1}{x} + (\ln x)\,2x - 0.5(2x) \\ &= x + 2x\ln x - x \\ &= 2x\ln x \end{aligned}$$

Critical value of $f(x)$: 1

Partition number for $f'(x)$: 1

Sign chart for $f'(x)$:

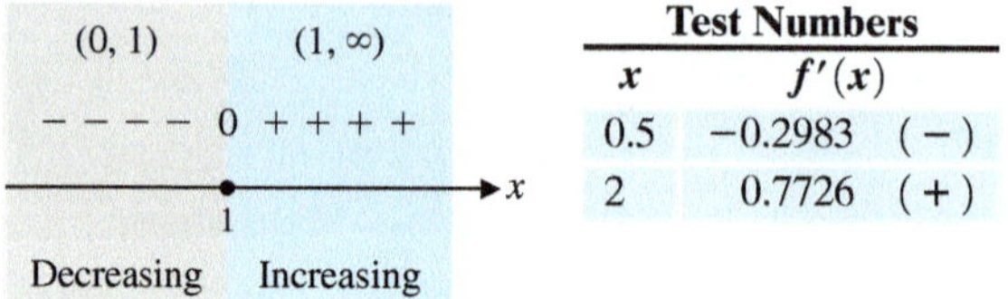

Test Numbers	
x	$f'(x)$
0.5	−0.2983 (−)
2	0.7726 (+)

The function $f(x)$ decreases on $(0, 1)$, has a local minimum at $x = 1$, and increases on $(1, \infty)$.

Step 3 *Analyze* $f''(x)$:

$$\begin{aligned} f''(x) &= 2x\frac{d}{dx}(\ln x) + (\ln x)\frac{d}{dx}(2x) \\ &= 2x\frac{1}{x} + (\ln x)\,2 \\ &= 2 + 2\ln x = 0 \\ 2\ln x &= -2 \\ \ln x &= -1 \\ x &= e^{-1} \approx 0.3679 \end{aligned}$$

Sign chart for $f'(x)$ (partition number is e^{-1}):

$(0, e^{-1})$ (e^{-1}, ∞)

- - - - 0 + + + +

x

e^{-1}

Concave downward Concave upward

Test Numbers	
x	$f'(x)$
0.2	−1.2189 (−)
1	2 (+)

The graph of $f(x)$ is concave downward on $(0, e^{-1})$, has an inflection point at $x = e^{-1}$, and is concave upward on (e^{-1}, ∞).

Step 4 *Sketch the graph of f, using the information from steps 1 to 3:*

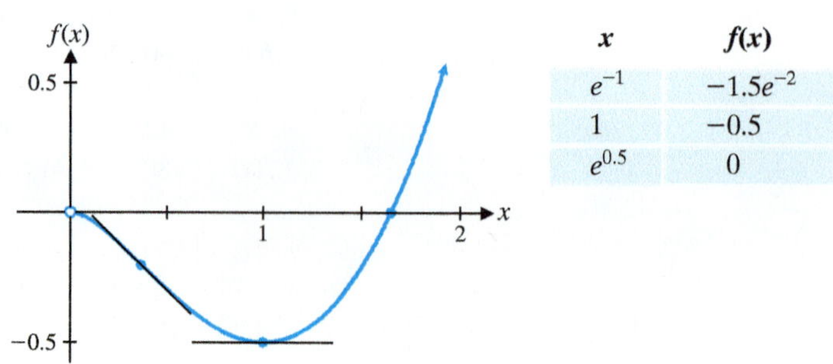

x	$f(x)$
e^{-1}	$-1.5e^{-2}$
1	−0.5
$e^{0.5}$	0

Matched Problem 4 Analyze the function $f(x) = x \ln x$. State all pertinent information and sketch the graph of f.

Modeling Average Cost

EXAMPLE 5 **Average Cost** Given the cost function $C(x) = 5{,}000 + 0.5x^2$, where x is the number of items produced, use the graphing strategy to analyze the graph of the average cost function. State all the pertinent information and sketch the graph of the average cost function. Find the marginal cost function and graph it on the same set of coordinate axes.

SOLUTION The average cost function is

$$\overline{C}(x) = \frac{5{,}000 + 0.5x^2}{x} = \frac{5{,}000}{x} + 0.5x$$

Step 1 *Analyze* $\overline{C}(x)$.

(A) Domain: Since negative values of x do not make sense and $\overline{C}(0)$ is not defined, the domain is the set of positive real numbers.

(B) Intercepts: None

(C) Horizontal asymptote: $\dfrac{a_m x^m}{b_n x^n} = \dfrac{0.5x^2}{x} = 0.5x$

So there is no horizontal asymptote.

Vertical asymptote: The line $x = 0$ is a vertical asymptote since the denominator is 0 and the numerator is not 0 for $x = 0$.

Oblique asymptotes: If a graph approaches a line that is neither horizontal nor vertical as x approaches ∞ or $-\infty$, then that line is called an **oblique asymptote**. If x is a large positive number, then $5{,}000/x$ is very small and

$$\overline{C}(x) = \frac{5{,}000}{x} + 0.5x \approx 0.5x$$

That is,

$$\lim_{x\to\infty} [\overline{C}(x) - 0.5x] = \lim_{x\to\infty} \frac{5{,}000}{x} = 0$$

This implies that the graph of $y = \overline{C}(x)$ approaches the line $y = 0.5x$ as x approaches ∞. That line is an oblique asymptote for the graph of $y = \overline{C}(x)$.*

Step 2 *Analyze* $\overline{C}'(x)$.

$$\begin{aligned}\overline{C}'(x) &= -\frac{5{,}000}{x^2} + 0.5\\ &= \frac{0.5x^2 - 5{,}000}{x^2}\\ &= \frac{0.5(x - 100)(x + 100)}{x^2}\end{aligned}$$

*If $f(x) = n(x)/d(x)$ is a rational function for which the degree of $n(x)$ is 1 more than the degree of $d(x)$, then we can use polynomial long division to write $f(x) = mx + b + r(x)/d(x)$, where the degree of $r(x)$ is less than the degree of $d(x)$. The line $y = mx + b$ is then an oblique asymptote for the graph of $y = f(x)$.

Critical value for $\overline{C}(x)$: 100
Partition numbers for $\overline{C}'(x)$: 0 and 100
Sign chart for $\overline{C}'(x)$:

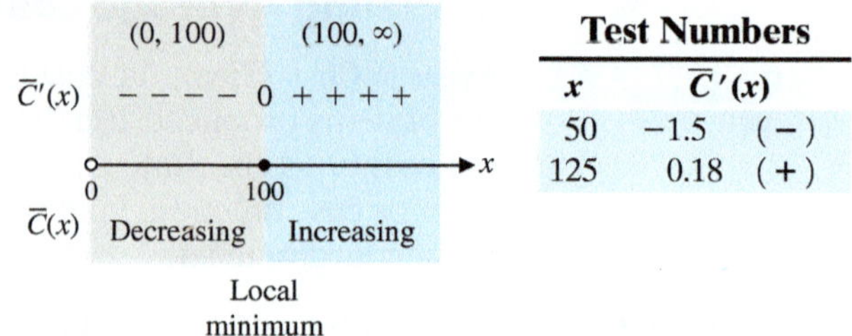

Test Numbers

x	$\overline{C}'(x)$
50	−1.5 (−)
125	0.18 (+)

So $\overline{C}(x)$ is decreasing on (0, 100), is increasing on $(100, \infty)$, and has a local minimum at $x = 100$.

Step 3 *Analyze* $\overline{C}''(x)$: $\overline{C}''(x) = \dfrac{10{,}000}{x^3}$.

$\overline{C}''(x)$ is positive for all positive x, so the graph of $y = \overline{C}(x)$ is concave upward on $(0, \infty)$.

Step 4 *Sketch the graph of* $\overline{C}$. The graph of $\overline{C}$ is shown in Figure 1.

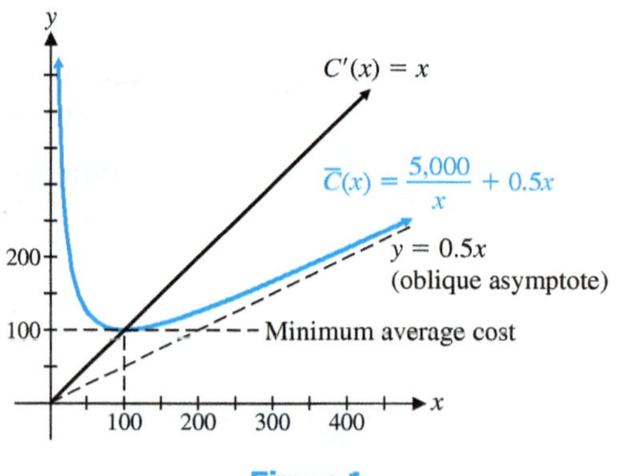

Figure 1

The marginal cost function is $C'(x) = x$. The graph of this linear function is also shown in Figure 1.

Figure 1 illustrates an important principle in economics:

The minimum average cost occurs when the average cost is equal to the marginal cost.

Matched Problem 5 Given the cost function $C(x) = 1{,}600 + 0.25x^2$, where x is the number of items produced,

(A) Use the graphing strategy to analyze the graph of the average cost function. State all the pertinent information and sketch the graph of the average cost function. Find the marginal cost function and graph it on the same set of coordinate axes. Include any oblique asymptotes.

(B) Find the minimum average cost.

Exercises 3-4

A

1. Use the graph of f in the figure to identify the following:

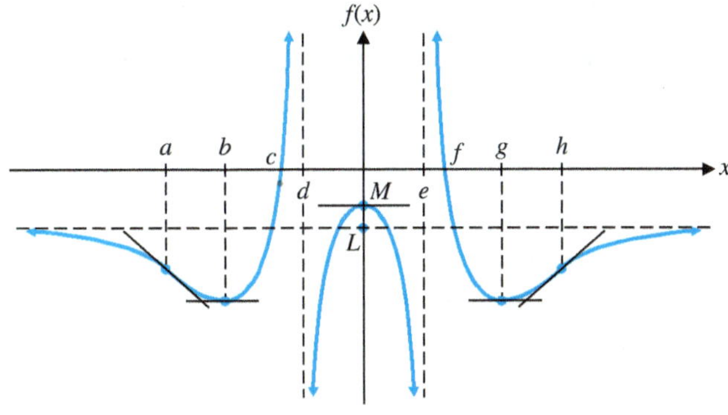

Figure for 1

(A) the intervals on which $f'(x) < 0$

(B) the intervals on which $f'(x) > 0$

(C) the intervals on which $f(x)$ is increasing

(D) the intervals on which $f(x)$ is decreasing

(E) the x coordinate(s) of the point(s) where $f(x)$ has a local maximum

(F) the x coordinate(s) of the point(s) where $f(x)$ has a local minimum

(G) the intervals on which $f''(x) < 0$

(H) the intervals on which $f''(x) > 0$

(I) the intervals on which the graph of f is concave upward

(J) the intervals on which the graph of f is concave downward

(K) the x coordinate(s) of the inflection point(s)

(L) the horizontal asymptote(s)

(M) the vertical asymptote(s)

2. Repeat Problem 1 for the following graph of f:

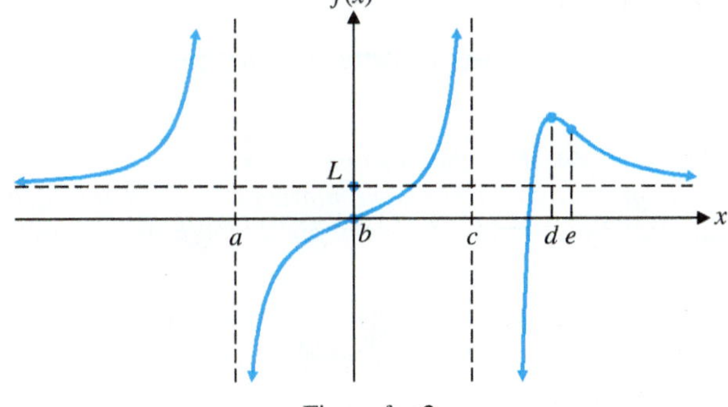

Figure for 2

In Problems 3–10, use the given information to sketch the graph of f. Assume that f is continuous on its domain and that all intercepts are included in the table of values.

3. Domain: All real x; $\lim_{x \to \pm\infty} f(x) = 2$

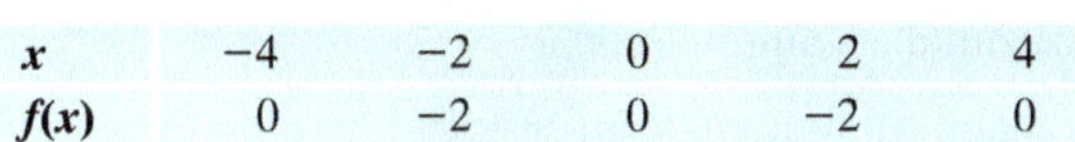

x	-4	-2	0	2	4
$f(x)$	0	-2	0	-2	0

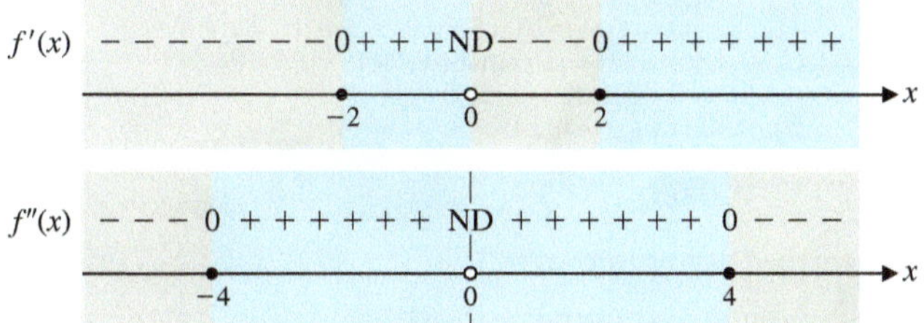

4. Domain: All real x; $\lim_{x \to -\infty} f(x) = -3$; $\lim_{x \to \infty} f(x) = 3$

x	-2	-1	0	1	2
$f(x)$	0	2	0	-2	0

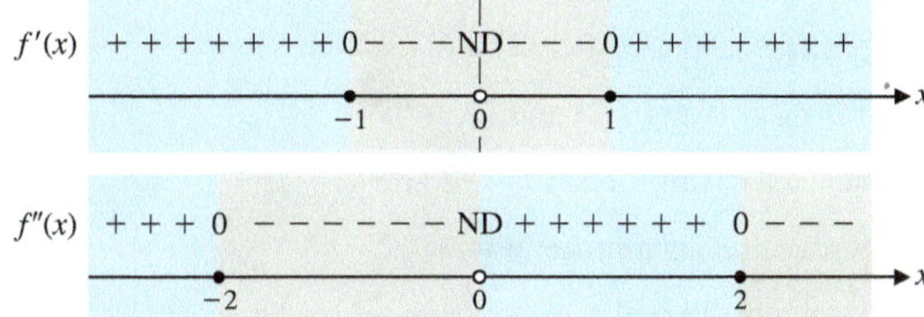

5. Domain: All real x, except $x = -2$; $\lim_{x \to -2^-} f(x) = \infty$; $\lim_{x \to -2^+} f(x) = -\infty$; $\lim_{x \to \infty} f(x) = 1$

x	-4	0	4	6
$f(x)$	0	0	3	2

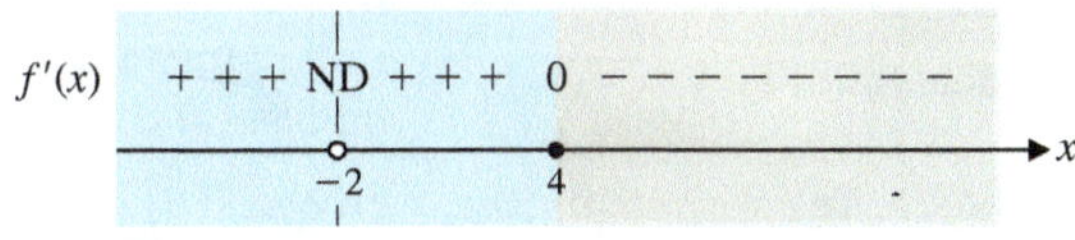

6. Domain: All real x, except $x = 1$; $\lim_{x \to 1^-} f(x) = \infty$; $\lim_{x \to 1^+} f(x) = \infty$; $\lim_{x \to \infty} f(x) = -2$

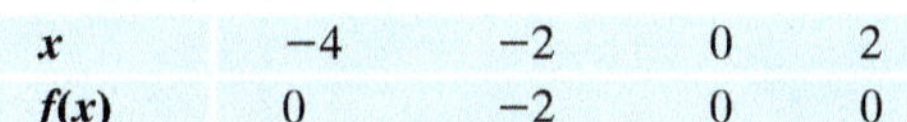

x	-4	-2	0	2
$f(x)$	0	-2	0	0

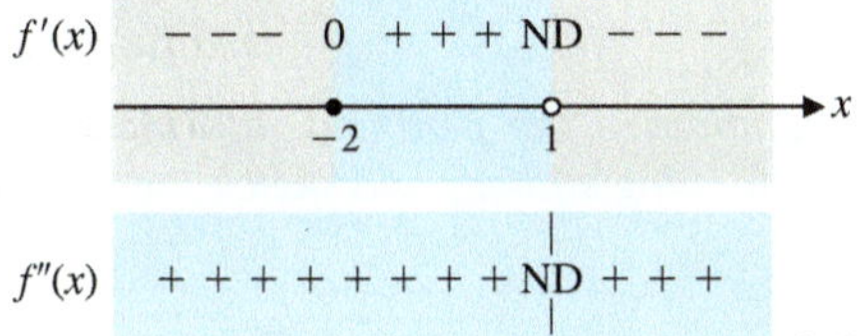

7. Domain: All real x, except $x = -1$;
$f(-3) = 2, f(-2) = 3, f(0) = -1, f(1) = 0$;
$f'(x) > 0$ on $(-\infty, -1)$ and $(-1, \infty)$;
$f''(x) > 0$ on $(-\infty, -1)$; $f''(x) < 0$ on $(-1, \infty)$;
vertical asymptote: $x = -1$;
horizontal asymptote: $y = 1$

8. Domain: All real x, except $x = 1$;
$f(0) = -2, f(2) = 0$;
$f'(x) < 0$ on $(-\infty, 1)$ and $(1, \infty)$;
$f''(x) < 0$ on $(-\infty, 1)$;
$f''(x) > 0$ on $(1, \infty)$;
vertical asymptote: $x = 1$;
horizontal asymptote: $y = -1$

9. Domain: All real x, except $x = -2$ and $x = 2$;
$f(-3) = -1, f(0) = 0, f(3) = 1$;
$f'(x) < 0$ on $(-\infty, -2)$ and $(2, \infty)$;
$f'(x) > 0$ on $(-2, 2)$;
$f''(x) < 0$ on $(-\infty, -2)$ and $(-2, 0)$;
$f''(x) > 0$ on $(0, 2)$ and $(2, \infty)$;
vertical asymptotes: $x = -2$ and $x = 2$;
horizontal asymptote: $y = 0$

10. Domain: All real x, except $x = -1$ and $x = 1$;
$f(-2) = 1, f(0) = 0, f(2) = 1$;
$f'(x) > 0$ on $(-\infty, -1)$ and $(0, 1)$;
$f'(x) < 0$ on $(-1, 0)$ and $(1, \infty)$;
$f''(x) > 0$ on $(-\infty, -1)$, $(-1, 1)$, and $(1, \infty)$;
vertical asymptotes: $x = -1$ and $x = 1$;
horizontal asymptote: $y = 0$

In Problems 11–16, find the domain and intercepts.

11. $f(x) = \sqrt{x + 4}$
12. $f(x) = \sqrt{x + 25}$
13. $f(x) = 75 - 5x$
14. $f(x) = 4x + 52$
15. $f(x) = \dfrac{95}{x - 5}$
16. $f(x) = \dfrac{102}{x + 3}$

B

In Problems 17–56, summarize the pertinent information obtained by applying the graphing strategy and sketch the graph of $y = f(x)$.

17. $f(x) = \dfrac{x + 3}{x - 3}$
18. $f(x) = \dfrac{2x - 4}{x + 2}$
19. $f(x) = \dfrac{x}{x - 2}$
20. $f(x) = \dfrac{2 + x}{3 - x}$
21. $f(x) = 5 + 5e^{-0.1x}$
22. $f(x) = 3 + 7e^{-0.2x}$
23. $f(x) = 5xe^{-0.2x}$
24. $f(x) = 10xe^{-0.1x}$
25. $f(x) = \ln(1 - x)$
26. $f(x) = \ln(2x + 4)$
27. $f(x) = x - \ln x$
28. $f(x) = \ln(x^2 + 4)$
29. $f(x) = \dfrac{x}{x^2 - 4}$
30. $f(x) = \dfrac{1}{x^2 - 4}$
31. $f(x) = \dfrac{1}{1 + x^2}$
32. $f(x) = \dfrac{x^2}{1 + x^2}$
33. $f(x) = \dfrac{2x}{1 - x^2}$
34. $f(x) = \dfrac{2x}{x^2 - 9}$
35. $f(x) = \dfrac{-5x}{(x - 1)^2}$
36. $f(x) = \dfrac{x}{(x - 2)^2}$
37. $f(x) = \dfrac{x^2 + x - 2}{x^2}$
38. $f(x) = \dfrac{x^2 - 5x - 6}{x^2}$
39. $f(x) = \dfrac{x^2}{x - 1}$
40. $f(x) = \dfrac{x^2}{2 + x}$
41. $f(x) = \dfrac{3x^2 + 2}{x^2 - 9}$
42. $f(x) = \dfrac{2x^2 + 5}{4 - x^2}$
43. $f(x) = \dfrac{x^3}{x - 2}$
44. $f(x) = \dfrac{x^3}{4 - x}$
45. $f(x) = (3 - x)e^x$
46. $f(x) = (x - 2)e^x$
47. $f(x) = e^{-0.5x^2}$
48. $f(x) = e^{-2x^2}$
49. $f(x) = x^2 \ln x$
50. $f(x) = \dfrac{\ln x}{x}$
51. $f(x) = (\ln x)^2$
52. $f(x) = \dfrac{x}{\ln x}$
53. $f(x) = \dfrac{1}{x^2 + 2x - 8}$
54. $f(x) = \dfrac{1}{3 - 2x - x^2}$
55. $f(x) = \dfrac{x^3}{3 - x^2}$
56. $f(x) = \dfrac{x^3}{x^2 - 12}$

C

In Problems 57–64, show that the line $y = x$ is an oblique asymptote for the graph of $y = f(x)$, summarize all pertinent information obtained by applying the graphing strategy, and sketch the graph of $y = f(x)$.

57. $f(x) = x + \dfrac{4}{x}$
58. $f(x) = x - \dfrac{9}{x}$
59. $f(x) = x - \dfrac{4}{x^2}$
60. $f(x) = x + \dfrac{32}{x^2}$
61. $f(x) = x - \dfrac{9}{x^3}$
62. $f(x) = x + \dfrac{27}{x^3}$
63. $f(x) = x + \dfrac{1}{x} + \dfrac{4}{x^3}$
64. $f(x) = x - \dfrac{16}{x^3}$

In Problems 65–72, summarize all pertinent information obtained by applying the graphing strategy, and sketch the graph of $y = f(x)$. [Note: These rational functions are not reduced to lowest terms.]

65. $f(x) = \dfrac{x^2 + x - 6}{x^2 - 6x + 8}$

66. $f(x) = \dfrac{x^2 + x - 6}{x^2 - x - 12}$

67. $f(x) = \dfrac{2x^2 + x - 15}{x^2 - 9}$

68. $f(x) = \dfrac{2x^2 + 11x + 14}{x^2 - 4}$

69. $f(x) = \dfrac{x^3 - 5x^2 + 6x}{x^2 - x - 2}$

70. $f(x) = \dfrac{x^3 - 5x^2 - 6x}{x^2 + 3x + 2}$

71. $f(x) = \dfrac{x^2 + x - 2}{x^2 - 2x + 1}$

72. $f(x) = \dfrac{x^2 + x - 2}{x^2 + 4x + 4}$

Applications

73. Revenue. The marketing research department for a computer company used a large city to test market the firm's new laptop. The department found that the relationship between price p (dollars per unit) and demand x (units sold per week) was given approximately by

$$p = 1{,}296 - 0.12x^2 \qquad 0 \le x \le 80$$

So, weekly revenue can be approximated by

$$R(x) = xp = 1{,}296x - 0.12x^3 \qquad 0 \le x \le 80$$

Graph the revenue function R.

74. Profit. Suppose that the cost function $C(x)$ (in dollars) for the company in Problem 73 is

$$C(x) = 830 + 396x$$

(A) Write an equation for the profit $P(x)$.

(B) Graph the profit function P.

75. Pollution. In Silicon Valley, a number of computer firms were found to be contaminating underground water supplies with toxic chemicals stored in leaking underground containers. A water quality control agency ordered the companies to take immediate corrective action and contribute to a monetary pool for the testing and cleanup of the underground contamination. Suppose that the required monetary pool (in millions of dollars) is given by

$$P(x) = \frac{2x}{1 - x} \qquad 0 \le x < 1$$

where x is the percentage (expressed as a decimal fraction) of the total contaminant removed.

(A) Where is $P(x)$ increasing? Decreasing?

(B) Where is the graph of P concave upward? Downward?

(C) Find any horizontal and vertical asymptotes.

(D) Find the x and y intercepts.

(E) Sketch a graph of P.

76. Employee training. A company producing dive watches has established that, on average, a new employee can assemble $N(t)$ dive watches per day after t days of on-the-job training, as given by

$$N(t) = \frac{100t}{t + 9} \qquad t \ge 0$$

(A) Where is $N(t)$ increasing? Decreasing?

(B) Where is the graph of N concave upward? Downward?

(C) Find any horizontal and vertical asymptotes.

(D) Find the intercepts.

(E) Sketch a graph of N.

77. Replacement time. An outboard motor has an initial price of \$3,200. A service contract costs \$300 for the first year and increases \$100 per year thereafter. The total cost of the outboard motor (in dollars) after n years is given by

$$C(n) = 3{,}200 + 250n + 50n^2 \qquad n \ge 1$$

(A) Write an expression for the average cost per year, $\overline{C}(n)$, for n years.

(B) Graph the average cost function found in part (A).

(C) When is the average cost per year minimum? (This time is frequently referred to as the **replacement time** for this piece of equipment.)

78. Construction costs. The management of a manufacturing plant wishes to add a fenced-in rectangular storage yard of 20,000 square feet, using a building as one side of the yard (see the figure). If x is the distance (in feet) from the building to the fence, show that the length of the fence required for the yard is given by

$$L(x) = 2x + \frac{20{,}000}{x} \qquad x > 0$$

Figure for 78

(A) Graph L.

(B) What are the dimensions of the rectangle requiring the least amount of fencing?

79. Average and marginal costs. The total daily cost (in dollars) of producing x mountain bikes is given by

$$C(x) = 1{,}000 + 5x + 0.1x^2$$

(A) Sketch the graphs of the average cost function and the marginal cost function on the same set of coordinate axes. Include any oblique asymptotes.

(B) Find the minimum average cost.

80. Average and marginal costs. The total daily cost (in dollars) of producing x city bikes is given by

$$C(x) = 500 + 2x + 0.2x^2$$

(A) Sketch the graphs of the average cost function and the marginal cost function on the same set of coordinate axes. Include any oblique asymptotes.

(B) Find the minimum average cost.

81. Minimizing average costs. The table gives the total daily costs y (in dollars) of producing x pepperoni pizzas at various production levels.

Number of Pizzas x	Total Costs y
50	395
100	475
150	640
200	910
250	1,140
300	1,450

(A) Enter the data into a graphing calculator and find a quadratic regression equation for the total costs.

(B) Use the regression equation from part (A) to find the minimum average cost (to the nearest cent) and the corresponding production level (to the nearest integer).

82. Minimizing average costs. The table gives the total daily costs y (in dollars) of producing x deluxe pizzas at various production levels.

Number of Pizzas x	Total Costs y
50	595
100	755
150	1,110
200	1,380
250	1,875
300	2,410

(A) Enter the data into a graphing calculator and find a quadratic regression equation for the total costs.

(B) Use the regression equation from part (A) to find the minimum average cost (to the nearest cent) and the corresponding production level (to the nearest integer).

83. Medicine. A drug is injected into the bloodstream of a patient through her right arm. The drug concentration in the bloodstream of the left arm t hours after the injection is given by

$$C(t) = \frac{0.14t}{t^2 + 1}$$

Graph C.

84. Physiology. In a study on the speed of muscle contraction in frogs under various loads, researchers found that the speed of contraction decreases with increasing loads. More precisely, they found that the relationship between speed of contraction, S (in centimeters per second), and load w (in grams) is given approximately by

$$S(w) = \frac{26 + 0.06w}{w} \qquad w \geq 5$$

Graph S.

85. Psychology: retention. Each student in a psychology class is given one day to memorize the same list of 30 special characters. The lists are turned in at the end of the day, and for each succeeding day for 30 days, each student is asked to turn in a list of as many of the symbols as can be recalled. Averages are taken, and it is found that

$$N(t) = \frac{5t + 20}{t} \qquad t \geq 1$$

provides a good approximation of the average number $N(t)$ of symbols retained after t days. Graph N.

Answers to Matched Problems

1. Domain: All real x, except $x = 1$
y intercept: $f(0) = 0$; x intercept: 0
Horizontal asymptote: $y = -2$;
vertical asymptote: $x = 1$
Increasing on $(-\infty, 1)$ and $(1, \infty)$
Concave upward on $(-\infty, 1)$; concave downward on $(1, \infty)$

x	$f(x)$
-1	-1
0	0
$\frac{1}{2}$	2
$\frac{3}{2}$	-6
2	-4
5	$-\frac{5}{2}$

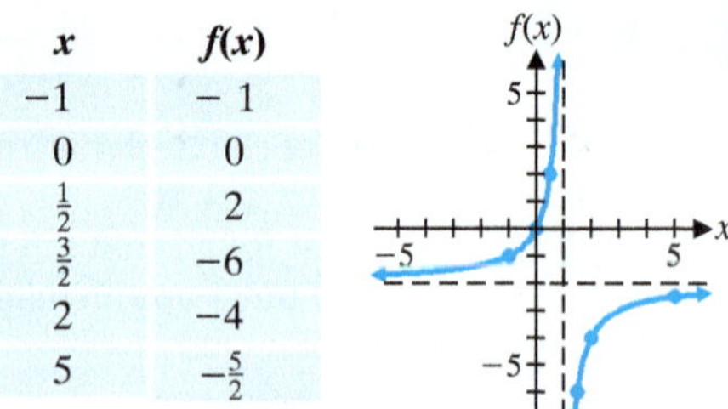

2. Domain: All real x, except $x = 0$
x intercept: $= -\frac{3}{4} = -0.75$
$h(0)$ is not defined
Vertical asymptote: $x = 0$ (the y axis)
Horizontal asymptote: $y = 0$ (the x axis)
Increasing on $(-1.5, 0)$
Decreasing on $(-\infty, -1.5)$ and $(0, \infty)$
Local minimum at $x = 1.5$
Concave upward on $(-2.25, 0)$ and $(0, \infty)$
Concave downward on $(-\infty, -2.25)$
Inflection point at $x = -2.25$

x	$h(x)$
-10	-0.37
-2.25	-1.19
-1.5	-1.33
-0.75	0
2	2.75
10	0.43

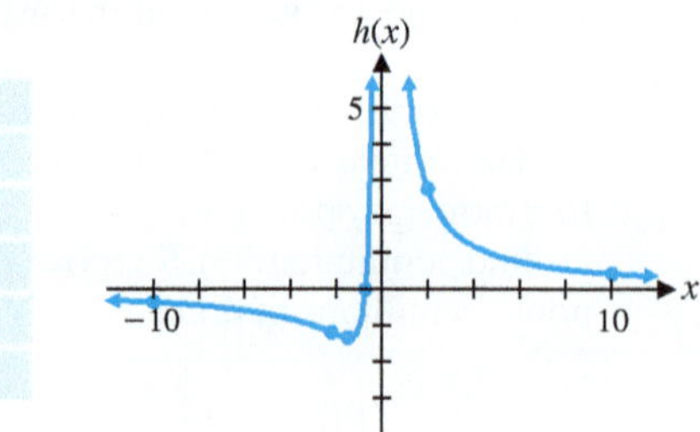

3. Domain: $(-\infty, \infty)$
y intercept: $f(0) = 0$
x intercept: $x = 0$
Horizontal asymptote: $y = 0$ (the x axis)
Increasing on $(-\infty, 2)$
Decreasing on $(2, \infty)$
Local maximum at $x = 2$
Concave downward on $(-\infty, 4)$
Concave upward on $(4, \infty)$
Inflection point at $x = 4$

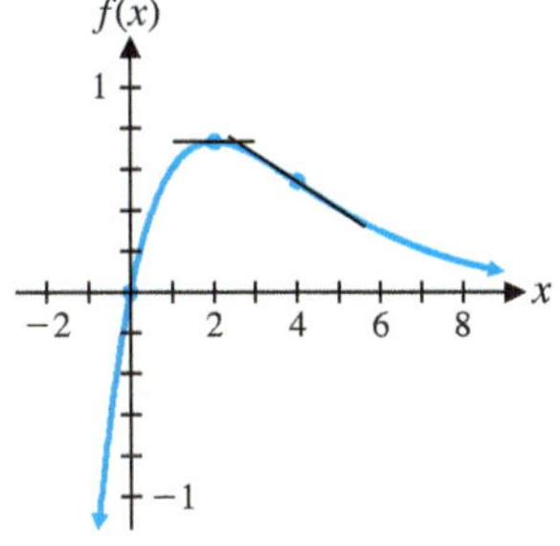

4. Domain: $(0, \infty)$
y intercept: None [$f(0)$ is not defined]
x intercept: $x = 1$
Increasing on (e^{-1}, ∞)
Decreasing on $(0, e^{-1})$
Local minimum at $x = e^{-1} \approx 0.368$
Concave upward on $(0, \infty)$

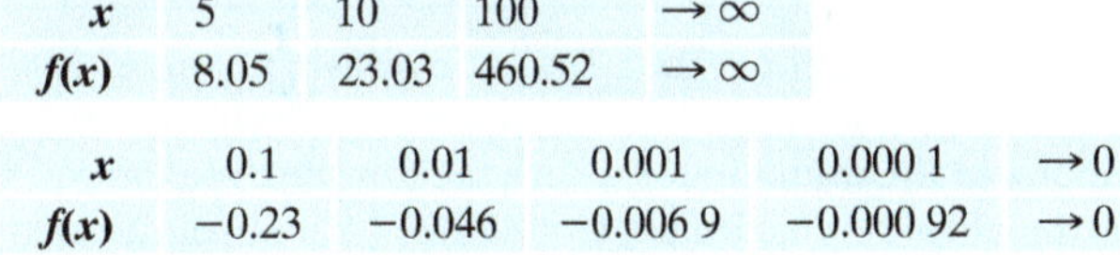

x	5	10	100	$\to \infty$
$f(x)$	8.05	23.03	460.52	$\to \infty$

x	0.1	0.01	0.001	0.000 1	$\to 0$
$f(x)$	−0.23	−0.046	−0.006 9	−0.000 92	$\to 0$

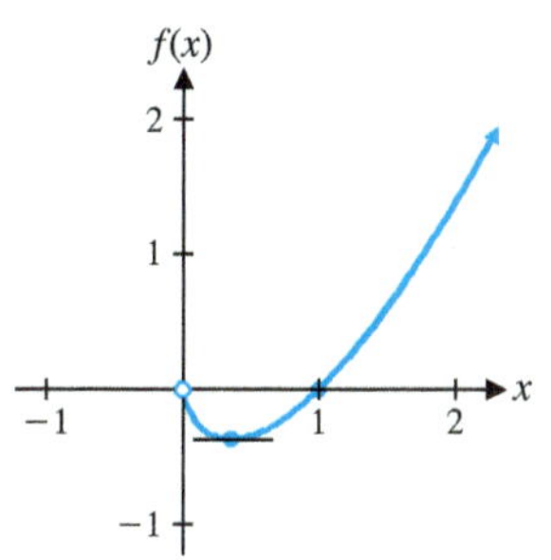

5. (A) Domain: $(0, \infty)$
Intercepts: None
Vertical asymptote: $x = 0$; oblique asymptote: $y = 0.25x$
Decreasing on $(0, 80)$; increasing on $(80, \infty)$; local minimum at $x = 80$
Concave upward on $(0, \infty)$

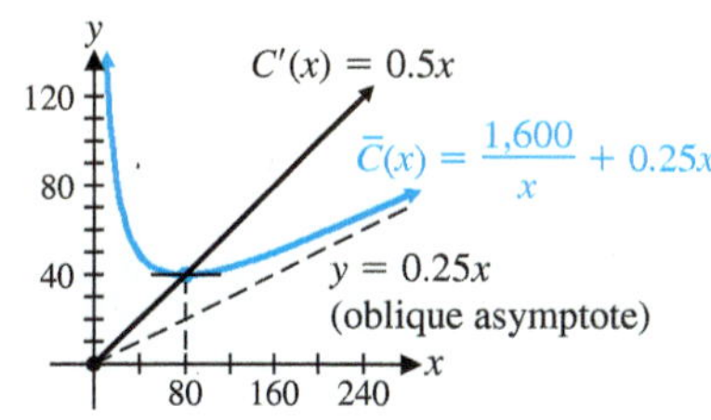

(B) Minimum average cost is 40 at $x = 80$.

3-5 Absolute Maxima and Minima

- Absolute Maxima and Minima
- Second Derivative and Extrema

Now we will consider one of the most important applications of the derivative: finding the absolute maximum or minimum value of a function. An economist may be interested in the price or production level of a commodity that will bring a maximum profit; a doctor may be interested in the time it takes for a drug to reach its maximum concentration in the bloodstream after an injection; and a city planner might be interested in the location of heavy industry in a city in order to produce minimum pollution in residential and business areas. In this section, we develop the procedures needed to find the absolute maximum and absolute minimum values of a function.

Absolute Maxima and Minima

Recall that $f(c)$ is a local maximum value if $f(x) \le f(c)$ for x near c and a local minimum value if $f(x) \ge f(c)$ for x near c. Now we are interested in finding the largest and the smallest values of $f(x)$ throughout its domain.

DEFINITION Absolute Maxima and Minima

If $f(c) \geq f(x)$ for all x in the domain of f, then $f(c)$ is called the **absolute maximum value** of f.

If $f(c) \leq f(x)$ for all x in the domain of f, then $f(c)$ is called the **absolute minimum value** of f.

Figure 1 illustrates some typical examples.

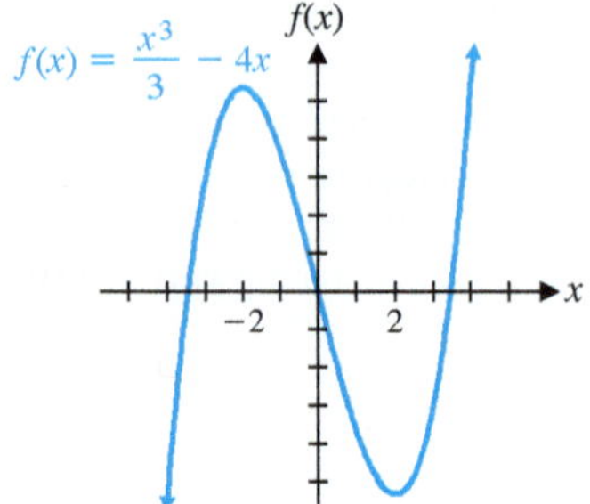

(A) No absolute maximum or minimum
One local maximum at $x = -2$
One local minimum at $x = 2$

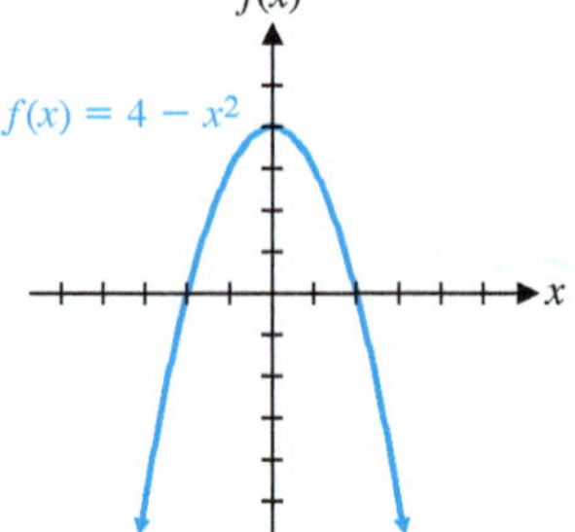

(B) Absolute maximum at $x = 0$
No absolute minimum

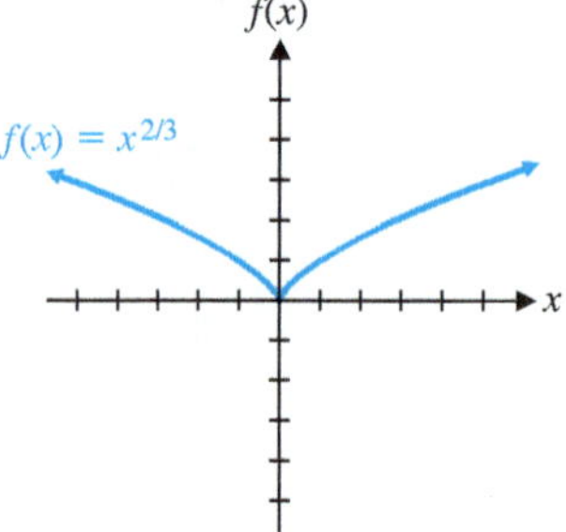

(C) Absolute minimum at $x = 0$
No absolute maximum

Figure 1

CONCEPTUAL INSIGHT

If $f(c)$ is the absolute maximum value of a function f, then $f(c)$ is obviously a "value" of f. It is common practice to omit "value" and to refer to $f(c)$ as the **absolute maximum** of f. In either usage, note that c is a value of x in the domain of f where the absolute maximum occurs. It is incorrect to refer to c as the absolute maximum. Collectively, the absolute maximum and the absolute minimum are referred to as **absolute extrema**.

In many applications, the domain of a function is restricted because of practical or physical considerations. If the domain is restricted to some closed interval, as is often the case, then Theorem 1 applies.

THEOREM 1 Extreme Value Theorem

A function f that is continuous on a closed interval $[a, b]$ has both an absolute maximum value and an absolute minimum value on that interval.

It is important to understand that the absolute maximum and minimum values depend on both the function f and the interval $[a, b]$. Figure 2 illustrates four cases.

In all four cases illustrated in Figure 2, the absolute maximum value and absolute minimum value occur at a critical value or an endpoint. This property is generalized in Theorem 2. Note that both the absolute maximum value and the absolute minimum value are unique, but each can occur at more than one point in the interval (Fig. 2D).

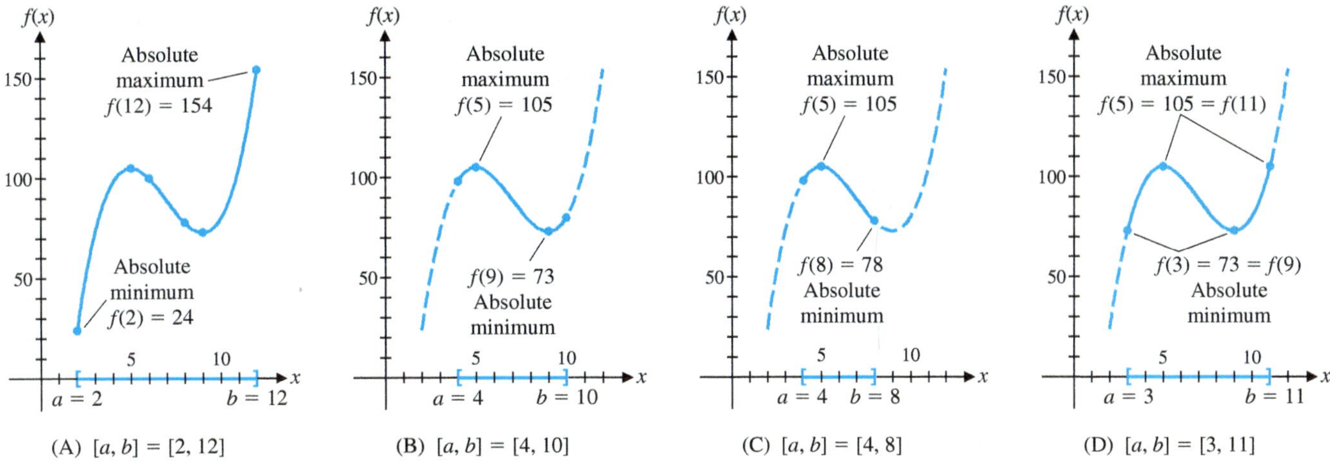

Figure 2 Absolute extrema for $f(x) = x^3 - 21x^2 + 135x - 170$ for various closed intervals

THEOREM 2 Locating Absolute Extrema

Absolute extrema (if they exist) must always occur at critical values or at endpoints.

To find the absolute maximum or minimum value of a continuous function on a closed interval, we simply identify the endpoints and critical values in the interval, evaluate the function at each, and then choose the largest and smallest values out of this group.

PROCEDURE Finding Absolute Extrema on a Closed Interval

Step 1 Check to make certain that f is continuous over $[a, b]$.

Step 2 Find the critical values in the interval (a, b).

Step 3 Evaluate f at the endpoints a and b and at the critical values found in step 2.

Step 4 The absolute maximum $f(x)$ on $[a, b]$ is the largest value found in step 3.

Step 5 The absolute minimum $f(x)$ on $[a, b]$ is the smallest value found in step 3.

EXAMPLE 1 **Finding Absolute Extrema** Find the absolute maximum and absolute minimum values of

$$f(x) = x^3 + 3x^2 - 9x - 7$$

on each of the following intervals:

(A) $[-6, 4]$ (B) $[-4, 2]$ (C) $[-2, 2]$

SOLUTION (A) The function is continuous for all values of x.

$$f'(x) = 3x^2 + 6x - 9 = 3(x - 1)(x + 3)$$

So, $x = -3$ and $x = 1$ are critical values in the interval $(-6, 4)$. Evaluate f at the endpoints and critical values (-6, -3, 1, and 4), and choose the maximum and minimum from these:

$$f(-6) = -61 \quad \text{Absolute minimum}$$
$$f(-3) = 20$$
$$f(1) = -12$$
$$f(4) = 69 \quad \text{Absolute maximum}$$

(B) Interval: $[-4, 2]$

x	$f(x)$	
-4	13	
-3	20	Absolute maximum
1	-12	Absolute minimum
2	-5	

(C) Interval: $[-2, 2]$

x	$f(x)$	
-2	15	Absolute maximum
1	-12	Absolute minimum
2	-5	

The critical value $x = -3$ is not included in this table, because it is not in the interval $[-2, 2]$.

Matched Problem 1 Find the absolute maximum and absolute minimum values of

$$f(x) = x^3 - 12x$$

on each of the following intervals:

(A) $[-5, 5]$ (B) $[-3, 3]$ (C) $[-3, 1]$

Now, suppose that we want to find the absolute maximum or minimum value of a function that is continuous on an interval that is not closed. Since Theorem 1 no longer applies, we cannot be certain that the absolute maximum or minimum value exists. Figure 3 illustrates several ways that functions can fail to have absolute extrema.

In general, the best procedure to follow in searching for absolute extrema on an interval that is not of the form $[a, b]$ is to sketch a graph of the function. However, many applications can be solved with a new tool that does not require any graphing.

Second Derivative and Extrema

The second derivative can be used to classify the local extrema of a function. Suppose that f is a function satisfying $f'(c) = 0$ and $f''(c) > 0$. First, note that if $f''(c) > 0$, then it follows from the properties of limits* that $f''(x) > 0$ in some

*Actually, we are assuming that $f''(x)$ is continuous in an interval containing c. It is unlikely that we will encounter a function for which $f''(c)$ exists but $f''(x)$ is not continuous in an interval containing c.

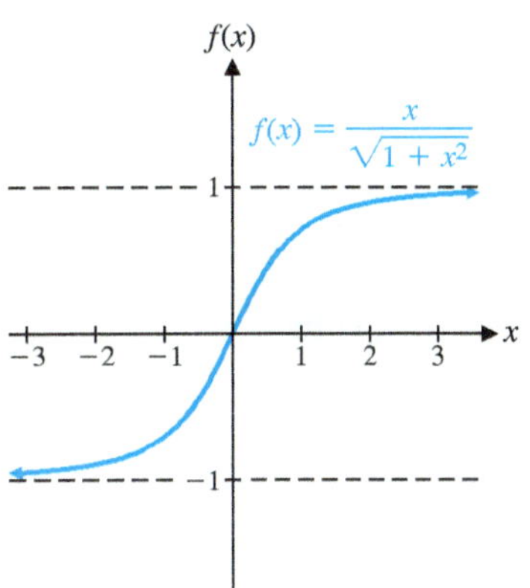

(A) No absolute extrema on $(-\infty, \infty)$: $-1 < f(x) < 1$ for all x [$f(x) \neq 1$ or -1 for any x]

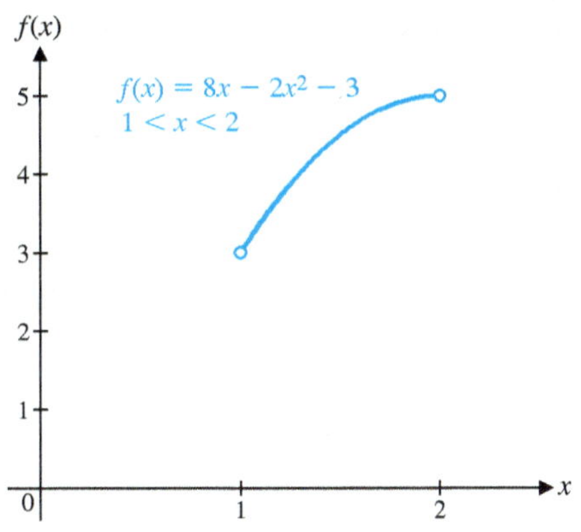

(B) No absolute extrema on $(1, 2)$: $3 < f(x) < 5$ for $x \in (1, 2)$ [$f(x) \neq 3$ or 5 for any $x \in (1, 2)$]

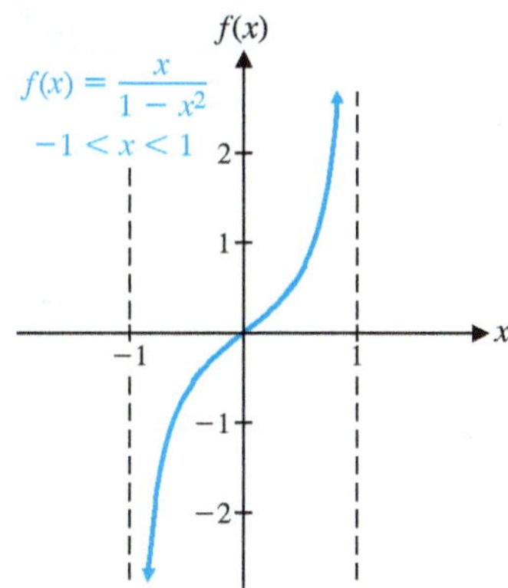

(C) No absolute extrema on $(-1, 1)$: Graph has vertical asymptotes at $x = -1$ and $x = 1$

Figure 3 Functions with no absolute extrema

interval (m, n) containing c. Thus, the graph of f must be concave upward in this interval. But this implies that $f'(x)$ is increasing in the interval. Since $f'(c) = 0$, $f'(x)$ must change from negative to positive at $x = c$, and $f(c)$ is a local minimum (see Fig. 4). Reasoning in the same fashion, we conclude that if $f'(c) = 0$ and $f''(c) < 0$, then $f(c)$ is a local maximum. Of course, it is possible that both $f'(c) = 0$ and $f''(c) = 0$. In this case, the second derivative cannot be used to determine the shape of the graph around $x = c$; $f(c)$ may be a local minimum, a local maximum, or neither.

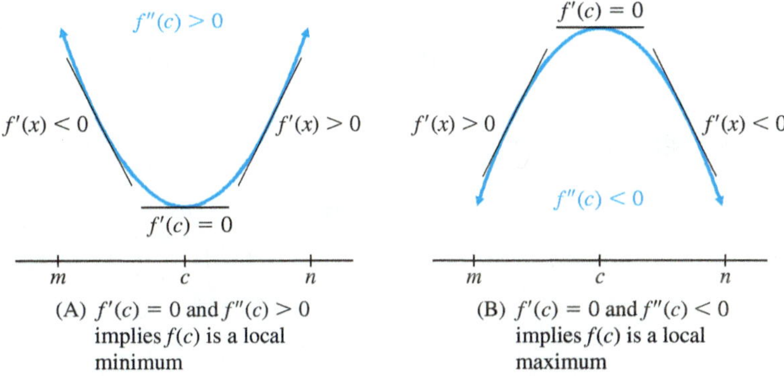

(A) $f'(c) = 0$ and $f''(c) > 0$ implies $f(c)$ is a local minimum

(B) $f'(c) = 0$ and $f''(c) < 0$ implies $f(c)$ is a local maximum

Figure 4 Second derivative and local extrema

The sign of the second derivative provides a simple test for identifying local maxima and minima. This test is most useful when we do not want to draw the graph of the function. If we are interested in drawing the graph and have already constructed the sign chart for $f'(x)$, then the first-derivative test can be used to identify the local extrema.

RESULT Second-Derivative Test

Let c be a critical value of $f(x)$.

$f'(c)$	$f''(c)$	Graph of f is:	$f(c)$	Example
0	+	Concave upward	Local minimum	∪
0	−	Concave downward	Local maximum	∩
0	0	?	Test does not apply	

EXAMPLE 2 **Testing Local Extrema** Find the local maxima and minima for each function. Use the second-derivative test when it applies.

(A) $f(x) = x^3 - 6x^2 + 9x + 1$

(B) $f(x) = xe^{-0.2x}$

(C) $f(x) = \frac{1}{6}x^6 - 4x^5 + 25x^4$

SOLUTION (A) Take first and second derivatives and find critical values:

$$f(x) = x^3 - 6x^2 + 9x + 1$$
$$f'(x) = 3x^2 - 12x + 9 = 3(x - 1)(x - 3)$$
$$f''(x) = 6x - 12 = 6(x - 2)$$

Critical values are $x = 1$ and $x = 3$.

$f''(1) = -6 < 0$ f has a local maximum at $x = 1$.

$f''(3) = 6 > 0$ f has a local minimum at $x = 3$.

(B)
$$f(x) = xe^{-0.2x}$$
$$f'(x) = e^{-0.2x} + xe^{-0.2x}(-0.2)$$
$$= e^{-0.2x}(1 - 0.2x)$$
$$f''(x) = e^{-0.2x}(-0.2)(1 - 0.2x) + e^{-0.2x}(-0.2)$$
$$= e^{-0.2x}(0.04x - 0.4)$$

Critical value: $x = 1/0.2 = 5$

$f''(5) = e^{-1}(-0.2) < 0$

f has a local maximum at $x = 5$.

(C)
$$f(x) = \frac{1}{6}x^6 - 4x^5 + 25x^4$$
$$f'(x) = x^5 - 20x^4 + 100x^3 = x^3(x - 10)^2$$
$$f''(x) = 5x^4 - 80x^3 + 300x^2$$

Critical values are $x = 0$ and $x = 10$.

$f''(0) = 0$ The second-derivative test fails at both critical values, so

$f''(10) = 0$ the first-derivative test must be used.

Sign chart for $f'(x) = x^3(x - 10)^2$ (partition numbers are 0 and 10):

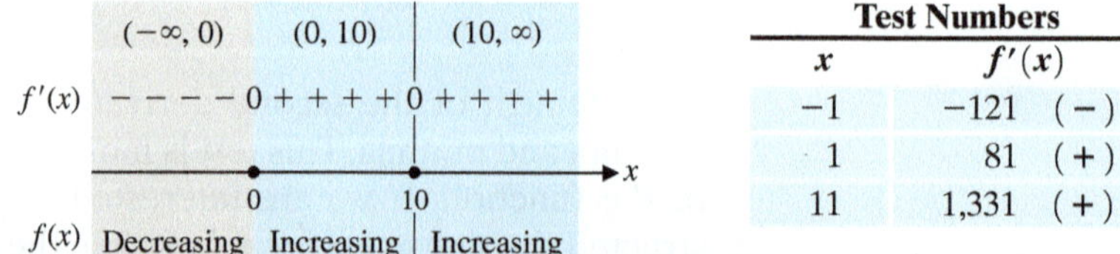

Test Numbers	
x	$f'(x)$
−1	−121 (−)
1	81 (+)
11	1,331 (+)

From the chart, we see that $f(x)$ has a local minimum at $x = 0$ and does not have a local extremum at $x = 10$.

Matched Problem 2 Find the local maxima and minima for each function. Use the second-derivative test when it applies.

(A) $f(x) = x^3 - 9x^2 + 24x - 10$

(B) $f(x) = e^x - 5x$

(C) $f(x) = 10x^6 - 24x^5 + 15x^4$

CONCEPTUAL INSIGHT

The second-derivative test does not apply if $f''(c) = 0$ or if $f''(c)$ is not defined. As Example 2C illustrates, if $f''(c) = 0$, then $f(c)$ may or may not be a local extremum. Some other method, such as the first-derivative test, must be used when $f''(c) = 0$ or $f''(c)$ does not exist.

The solution of many optimization problems involves searching for an absolute extremum. If the function in question has only one critical value, then the second-derivative test not only classifies the local extremum but also guarantees that the local extremum is, in fact, the absolute extremum.

THEOREM 3 Second-Derivative Test for Absolute Extremum

Let f be continuous on an interval I with only one critical value c in I.

If $f'(c) = 0$ and $f''(c) > 0$, then $f(c)$ is the absolute minimum of f on I.

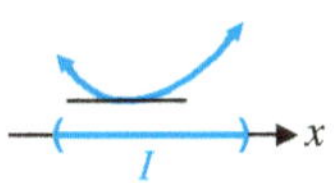

If $f'(c) = 0$ and $f''(c) < 0$, then $f(c)$ is the absolute maximum of f on I.

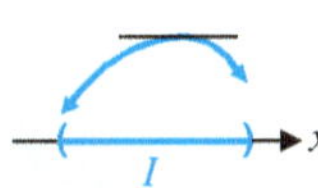

Since the second-derivative test cannot be applied when $f''(c) = 0$ or $f''(c)$ does not exist, Theorem 3 makes no mention of these cases.

EXAMPLE 3 **Finding an Absolute Extremum on an Open Interval** Find the absolute minimum value of each function on $(0, \infty)$.

(A) $f(x) = x + \frac{4}{x}$

(B) $f(x) = (\ln x)^2 - 3 \ln x$

SOLUTION (A) $f(x) = x + \frac{4}{x}$

$$f'(x) = 1 - \frac{4}{x^2} = \frac{x^2 - 4}{x^2} = \frac{(x - 2)(x + 2)}{x^2}$$ Critical values are $x = -2$ and $x = 2$.

$$f''(x) = \frac{8}{x^3}$$

The only critical value in the interval $(0, \infty)$ is $x = 2$. Since $f''(2) = 1 > 0$, $f(2) = 4$ is the absolute minimum value of f on $(0, \infty)$.

(B) $f(x) = (\ln x)^2 - 3 \ln x$

$$f'(x) = (2 \ln x)\frac{1}{x} - \frac{3}{x} = \frac{2 \ln x - 3}{x}$$ Critical value is $x = e^{3/2}$.

$$f''(x) = \frac{x\frac{2}{x} - (2 \ln x - 3)}{x^2} = \frac{5 - 2 \ln x}{x^2}$$

The only critical value in the interval $(0, \infty)$ is $x = e^{3/2}$. Since $f''(e^{3/2}) = 2/e^3 > 0$, $f(e^{3/2}) = -2.25$ is the absolute minimum value of f on $(0, \infty)$.

Matched Problem 3 Find the absolute maximum value of each function on $(0, \infty)$.

(A) $f(x) = 12 - x - \frac{5}{x}$ (B) $f(x) = 5 \ln x - x$

Exercises 3-5

A

Problems 1–10 refer to the graph of $y = f(x)$ shown here. Find the absolute minimum and the absolute maximum over the indicated interval.

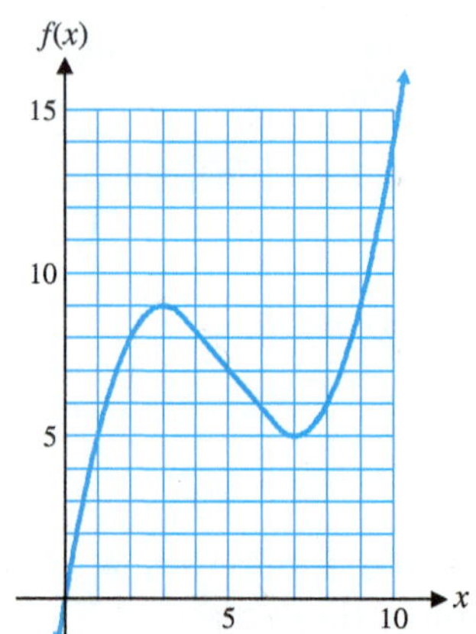

Figure for 1–10

1. $[0, 10]$ **2.** $[2, 8]$ **3.** $[0, 8]$ **4.** $[2, 10]$

5. $[1, 10]$ **6.** $[0, 9]$ **7.** $[1, 9]$ **8.** $[0, 2]$

9. $[2, 5]$ **10.** $[5, 8]$

In Problems 11–16, find the absolute maximum and absolute minimum of each function on the given interval.

11. $f(x) = 4x + 9$ on $[2, 3]$ **12.** $f(x) = -3x + 20$ on $[-2, 6]$

13. $f(x) = e^{-x}$ on $[-1, 1]$ **14.** $f(x) = \ln x$ on $[1, 2]$

15. $f(x) = 9 - x^2$ on $[-4, 4]$

16. $f(x) = x^2 - 6x + 7$ on $[0, 10]$

In Problems 17–32, find the absolute maximum and minimum, if either exists, for each function.

17. $f(x) = x^2 - 2x + 3$ **18.** $f(x) = x^2 + 4x - 3$

19. $f(x) = -x^2 - 6x + 9$ **20.** $f(x) = -x^2 + 2x + 4$

21. $f(x) = x^3 + x$ **22.** $f(x) = -x^3 - 2x$

23. $f(x) = 8x^3 - 2x^4$ **24.** $f(x) = x^4 - 4x^3$

25. $f(x) = x + \frac{16}{x}$ **26.** $f(x) = x + \frac{25}{x}$

27. $f(x) = \frac{x^2}{x^2 + 1}$ **28.** $f(x) = \frac{1}{x^2 + 1}$

29. $f(x) = \frac{2x}{x^2 + 1}$ **30.** $f(x) = \frac{-8x}{x^2 + 4}$

31. $f(x) = \frac{x^2 - 1}{x^2 + 1}$ **32.** $f(x) = \frac{9 - x^2}{x^2 + 4}$

B

In Problems 33–56, find the indicated extremum of each function on the given interval.

33. Absolute minimum value on $[0, \infty)$ for

$$f(x) = 2x^2 - 8x + 6$$

34. Absolute maximum value on $[0, \infty)$ for

$$f(x) = 6x - x^2 + 4$$

35. Absolute maximum value on $[0, \infty)$ for

$$f(x) = 3x^2 - x^3$$

36. Absolute minimum value on $[0, \infty)$ for

$$f(x) = x^3 - 6x^2$$

37. Absolute minimum value on $[0, \infty)$ for

$$f(x) = (x + 4)(x - 2)^2$$

38. Absolute minimum value on $[0, \infty)$ for

$$f(x) = (2 - x)(x + 1)^2$$

39. Absolute maximum value on $(0, \infty)$ for

$$f(x) = 2x^4 - 8x^3$$

40. Absolute maximum value on $(0, \infty)$ for

$$f(x) = 4x^3 - 8x^4$$

41. Absolute maximum value on $(0, \infty)$ for

$$f(x) = 20 - 3x - \frac{12}{x}$$

42. Absolute minimum value on $(0, \infty)$ for

$$f(x) = 4 + x + \frac{9}{x}$$

43. Absolute maximum value on $(0, \infty)$ for

$$f(x) = 10 + 2x + \frac{64}{x^2}$$

44. Absolute maximum value on $(0, \infty)$ for

$$f(x) = 20 - 4x - \frac{250}{x^2}$$

45. Absolute minimum value on $(0, \infty)$ for

$$f(x) = x + \frac{1}{x} + \frac{30}{x^3}$$

46. Absolute minimum value on $(0, \infty)$ for

$$f(x) = 2x + \frac{5}{x} + \frac{4}{x^3}$$

47. Absolute minimum value on $(0, \infty)$ for

$$f(x) = \frac{e^x}{x^2}$$

48. Absolute maximum value on $(0, \infty)$ for

$$f(x) = \frac{x^4}{e^x}$$

49. Absolute maximum value on $(0, \infty)$ for

$$f(x) = \frac{x^3}{e^x}$$

50. Absolute minimum value on $(0, \infty)$ for

$$f(x) = \frac{e^x}{x}$$

51. Absolute maximum value on $(0, \infty)$ for

$$f(x) = 5x - 2x \ln x$$

52. Absolute minimum value on $(0, \infty)$ for

$$f(x) = 4x \ln x - 7x$$

53. Absolute maximum value on $(0, \infty)$ for

$$f(x) = x^2(3 - \ln x)$$

54. Absolute minimum value on $(0, \infty)$ for

$$f(x) = x^3(\ln x - 2)$$

55. Absolute maximum value on $(0, \infty)$ for

$$f(x) = \ln(xe^{-x})$$

56. Absolute maximum value on $(0, \infty)$ for

$$f(x) = \ln(x^2e^{-x})$$

In Problems 57–62, find the absolute maximum and minimum, if either exists, for each function on the indicated intervals.

57. $f(x) = x^3 - 6x^2 + 9x - 6$
(A) $[-1, 5]$ (B) $[-1, 3]$ (C) $[2, 5]$

58. $f(x) = 2x^3 - 3x^2 - 12x + 24$
(A) $[-3, 4]$ (B) $[-2, 3]$ (C) $[-2, 1]$

59. $f(x) = (x - 1)(x - 5)^3 + 1$
(A) $[0, 3]$ (B) $[1, 7]$ (C) $[3, 6]$

60. $f(x) = x^4 - 8x^2 + 16$
(A) $[-1, 3]$ (B) $[0, 2]$ (C) $[-3, 4]$

61. $f(x) = x^4 - 4x^3 + 5$
(A) $[-1, 2]$ (B) $[0, 4]$ (C) $[-1, 1]$

62. $f(x) = x^4 - 18x^2 + 32$
(A) $[-4, 4]$ (B) $[-1, 1]$ (C) $[1, 3]$

In Problems 63–70, describe the graph of f at the given point relative to the existence of a local maximum or minimum with one of the following phrases: "Local maximum," "Local minimum," "Neither," or "Unable to determine from the given information." Assume that f(x) is continuous on $(-\infty, \infty)$.

63. $(2, f(2))$ if $f'(2) = 0$ and $f''(2) > 0$

64. $(4, f(4))$ if $f'(4) = 1$ and $f''(4) < 0$

65. $(-3, f(-3))$ if $f'(-3) = 0$ and $f''(-3) = 0$

66. $(-1, f(-1))$ if $f'(-1) = 0$ and $f''(-1) < 0$

67. $(6, f(6))$ if $f'(6) = 1$ and $f''(6)$ does not exist

68. $(5, f(5))$ if $f'(5) = 0$ and $f''(5)$ does not exist

69. $(-2, f(-2))$ if $f'(-2) = 0$ and $f''(-2) < 0$

70. $(1, f(1))$ if $f'(1) = 0$ and $f''(1) > 0$

Answers to Matched Problems

1. (A) Absolute maximum: $f(5) = 65$; absolute minimum: $f(-5) = -65$
   (B) Absolute maximum: $f(-2) = 16$; absolute minimum: $f(2) = -16$
   (C) Absolute maximum: $f(-2) = 16$; absolute minimum: $f(1) = -11$
2. (A) $f(2)$ is a local maximum; $f(4)$ is a local minimum.
   (B) $f(\ln 5) = 5 - 5 \ln 5$ is a local minimum.
   (C) $f(0)$ is a local minimum; there is no local extremum at $x = 1$.
3. (A) $f(\sqrt{5}) = 12 - 2\sqrt{5}$(B) $f(5) = 5 \ln 5 - 5$

3-6 Optimization

- Area and Perimeter
- Maximizing Revenue and Profit
- Inventory Control

Now we can use calculus to solve **optimization problems**—problems that involve finding the absolute maximum value or the absolute minimum value of a function. As you work through this section, note that the statement of the problem does not usually include the function to be optimized. Often, it is your responsibility to find the function and then to find its absolute extremum.

Area and Perimeter

The techniques used to solve optimization problems are best illustrated through examples.

EXAMPLE 1 **Maximizing Area** A homeowner has \$320 to spend on building a fence around a rectangular garden. Three sides of the fence will be constructed with wire fencing at a cost of \$2 per linear foot. The fourth side will be constructed with wood fencing at a cost of \$6 per linear foot. Find the dimensions and the area of the largest garden that can be enclosed with \$320 worth of fencing.

SOLUTION To begin, we draw a figure (Fig. 1), introduce variables, and look for relationships among the variables.

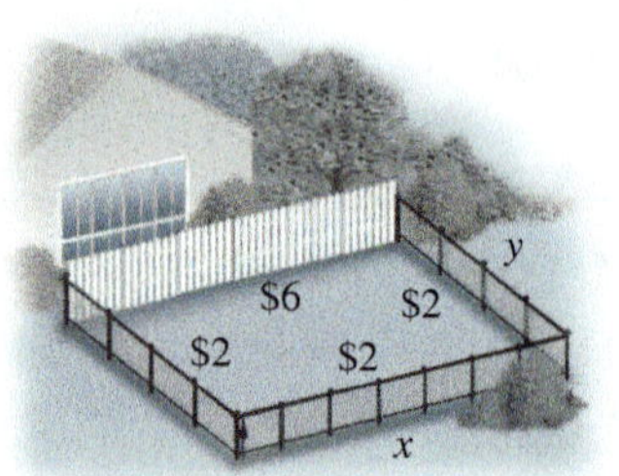

Figure 1

Since we don't know the dimensions of the garden, the lengths of fencing are represented by the variables x and y. The costs of the fencing materials are fixed and are represented by constants.

Now we look for relationships among the variables. The area of the garden is

$$A = xy$$

while the cost of the fencing is

$$\begin{aligned} C &= 2y + 2x + 2y + 6x \\ &= 8x + 4y \end{aligned}$$

The problem states that the homeowner has \$320 to spend on fencing. We assume that enclosing the largest area will use all the money available for fencing. The problem has now been reduced to

$$\text{Maximize} \quad A = xy \quad \text{subject to} \quad 8x + 4y = 320$$

Before we can use calculus to find the maximum area A, we must express A as a function of a single variable. We use the cost equation to eliminate one of the variables in the area expression (we choose to eliminate y—either will work).

$$\begin{aligned} 8x + 4y &= 320 \\ 4y &= 320 - 8x \\ y &= 80 - 2x \end{aligned}$$

$$A = xy = x(80 - 2x) = 80x - 2x^2$$

Now we consider the permissible values of x. Because x is one of the dimensions of a rectangle, x must satisfy

$$x \ge 0 \quad \text{Length is always nonnegative.}$$

And because $y = 80 - 2x$ is also a dimension of a rectangle, y must satisfy

$$\begin{aligned} y = 80 - 2x &\ge 0 \quad \text{Width is always nonnegative.} \\ 80 &\ge 2x \\ 40 &\ge x \quad \text{or} \quad x \le 40 \end{aligned}$$

We summarize the preceding discussion by stating the following model for this optimization problem:

$$\text{Maximize} \quad A(x) = 80x - 2x^2 \quad \text{for } 0 \le x \le 40$$

Next, we find any critical values of A:

$$\begin{aligned} A'(x) = 80 - 4x &= 0 \\ 80 &= 4x \\ x &= \frac{80}{4} = 20 \quad \text{Critical value} \end{aligned}$$

Table 1

x	$A(x)$
0	0
20	800
40	0

Since $A(x)$ is continuous on $[0, 40]$, the absolute maximum value of A, if it exists, must occur at a critical value or an endpoint. Evaluating A at these values (Table 1), we see that the maximum area is 800 when

$$x = 20 \quad \text{and} \quad y = 80 - 2(20) = 40$$

Finally, we must answer the questions posed in the problem. The dimensions of the garden with the maximum area of 800 square feet are 20 feet by 40 feet, with one 20-foot side of wood fencing.

Matched Problem 1 Repeat Example 1 if the wood fencing costs $8 per linear foot and all other information remains the same.

We summarize the steps in the solution of Example 1 in the following box:

PROCEDURE Strategy for Solving Optimization Problems

Step 1 Introduce variables, look for relationships among the variables, and construct a mathematical model of the form

$$\text{Maximize (or minimize) } f(x) \text{ on the interval } I$$

Step 2 Find the critical values of $f(x)$.

Step 3 Use the procedures developed in Section 3-5 to find the absolute maximum (or minimum) value of $f(x)$ on the interval I and the value(s) of x where this occurs.

Step 4 Use the solution to the mathematical model to answer all the questions asked in the problem.

EXAMPLE 2 **Minimizing Perimeter** Refer to Example 1. The homeowner judges that an area of 800 square feet for the garden is too small and decides to increase the area to 1,250 square feet. What is the minimum cost of building a fence that will enclose a garden with an area of 1,250 square feet? What are the dimensions of this garden? Assume that the cost of fencing remains unchanged.

SOLUTION Refer to Figure 1 and the solution of Example 1. This time we want to minimize the cost of the fencing that will enclose 1,250 square feet. The problem can be expressed as

$$\text{Minimize} \quad C = 8x + 4y \quad \text{subject to} \quad xy = 1{,}250$$

Since x and y represent distances, we know that $x \geq 0$ and $y \geq 0$. But neither variable can equal 0 because their product must be 1,250.

$$xy = 1{,}250 \qquad \text{Solve the area equation for } y.$$

$$y = \frac{1{,}250}{x}$$

$$C(x) = 8x + 4\frac{1{,}250}{x} \qquad \text{Substitute for } y \text{ in the cost equation.}$$

$$= 8x + \frac{5{,}000}{x} \qquad x > 0$$

The model for this problem is

$$\text{Minimize } C(x) = 8x + \frac{5{,}000}{x} \quad \text{for } x > 0$$

$$= 8x + 5{,}000x^{-1}$$

$$C'(x) = 8 - 5{,}000x^{-2}$$

$$= 8 - \frac{5{,}000}{x^2} = 0$$

$$8 = \frac{5{,}000}{x^2}$$

$$x^2 = \frac{5{,}000}{8} = 625$$

$$x = \sqrt{625} = 25$$ The negative square root is discarded, since $x > 0$.

We use the second derivative to determine the behavior at $x = 25$.

$$C'(x) = 8 - 5{,}000x^{-2}$$

$$C''(x) = 0 + 10{,}000x^{-3} = \frac{10{,}000}{x^3}$$

$$C''(25) = \frac{10{,}000}{25^3} = 0.64 > 0$$

The second-derivative test shows that $C(x)$ has a local minimum at $x = 25$, and since $x = 25$ is the only critical value of $x > 0$, then $C(25)$ must be the absolute minimum value of $C(x)$ for $x > 0$. When $x = 25$, the cost is

$$C(25) = 8(25) + \frac{5{,}000}{25} = 200 + 200 = \$400$$

and

$$y = \frac{1{,}250}{25} = 50$$

The minimal cost for enclosing a 1,250-square-foot garden is $400, and the dimensions are 25 feet by 50 feet, with one 25-foot side of wood fencing.

Matched Problem 2 Repeat Example 2 if the homeowner wants to enclose an 1,800-square-foot garden and all other data remain unchanged.

CONCEPTUAL INSIGHT

The restrictions on the variables in the solutions of Examples 1 and 2 are typical of problems involving areas or perimeters (or the cost of the perimeter):

$$8x + 4y = 320$$ Cost of fencing (Example 1)

$$xy = 1{,}250$$ Area of garden (Example 2)

The equation in Example 1 restricts the values of x to

$$0 \le x \le 40 \quad \text{or} \quad [0, 40]$$

The endpoints are included in the interval for our convenience (a closed interval is easier to work with than an open one). The area function is defined at each endpoint, so it does no harm to include them.

The equation in Example 2 restricts the values of x to

$$x > 0 \qquad \text{or} \qquad (0, \infty)$$

Neither endpoint can be included in this interval. We cannot include 0 because the area is not defined when $x = 0$, and we can never include ∞ as an endpoint. Remember, ∞ is not a number; it is a symbol that indicates the interval is unbounded.

Maximizing Revenue and Profit

EXAMPLE 3 **Maximizing Revenue** An office supply company sells x permanent markers per year at $\$p$ per marker. The price–demand equation for these markers is $p = 10 - 0.001x$. What price should the company charge for the markers to maximize revenue? What is the maximum revenue?

SOLUTION

$$\text{Revenue} = \text{price} \times \text{demand}$$

$$R(x) = (10 - 0.001x)x$$

$$= 10x - 0.001x^2$$

Both price and demand must be nonnegative, so

$$x \geq 0 \quad \text{and} \quad p = 10 - 0.001x \geq 0$$

$$10 \geq 0.001x$$

$$10{,}000 \geq x$$

The mathematical model for this problem is

$$\text{Maximize} \quad R(x) = 10x - 0.001x^2 \qquad 0 \leq x \leq 10{,}000$$

$$R'(x) = 10 - 0.002x$$

$$10 - 0.002x = 0$$

$$10 = 0.002x$$

$$x = \frac{10}{0.002} = 5{,}000 \qquad \text{Critical value}$$

Use the second-derivative test for absolute extrema:

$$R''(x) = -0.002 < 0 \quad \text{for all } x$$

$$\text{Max} \quad R(x) = R(5{,}000) = \$25{,}000$$

When the demand is $x = 5{,}000$, the price is

$$10 - 0.001(5{,}000) = \$5 \qquad p = 10 - 0.001x$$

The company will realize a maximum revenue of \$25,000 when the price of a marker is \$5.

Matched Problem 3 An office supply company sells x heavy-duty paper shredders per year at $\$p$ per shredder. The price–demand equation for these shredders is

$$p = 300 - \frac{x}{30}$$

What price should the company charge for the shredders to maximize revenue? What is the maximum revenue?

EXAMPLE 4 **Maximizing Profit** The total annual cost of manufacturing x permanent markers for the office supply company in Example 3 is

$$C(x) = 5{,}000 + 2x$$

What is the company's maximum profit? What should the company charge for each marker, and how many markers should be produced?

SOLUTION Using the revenue model in Example 3, we have

$$\text{Profit} = \text{Revenue} - \text{Cost}$$

$$\begin{aligned} P(x) &= R(x) - C(x) \\ &= 10x - 0.001x^2 - 5{,}000 - 2x \\ &= 8x - 0.001x^2 - 5{,}000 \end{aligned}$$

The mathematical model for profit is

$$\text{Maximize} \quad P(x) = 8x - 0.001x^2 - 5{,}000 \qquad 0 \le x \le 10{,}000$$

The restrictions on x come from the revenue model in Example 3.

$$P'(x) = 8 - 0.002x = 0$$

$$8 = 0.002x$$

$$x = \frac{8}{0.002} = 4{,}000 \quad \text{Critical value}$$

$$P''(x) = -0.002 < 0 \quad \text{for all } x$$

Since $x = 4{,}000$ is the only critical value and $P''(x) < 0$,

$$\text{Max } P(x) = P(4{,}000) = \$11{,}000$$

Using the price–demand equation from Example 3 with $x = 4{,}000$, we find that

$$p = 10 - 0.001(4{,}000) = \$6 \quad p = 10 - 0.001x$$

A maximum profit of \$11,000 is realized when 4,000 markers are manufactured annually and sold for \$6 each.

The results in Examples 3 and 4 are illustrated in Figure 2.

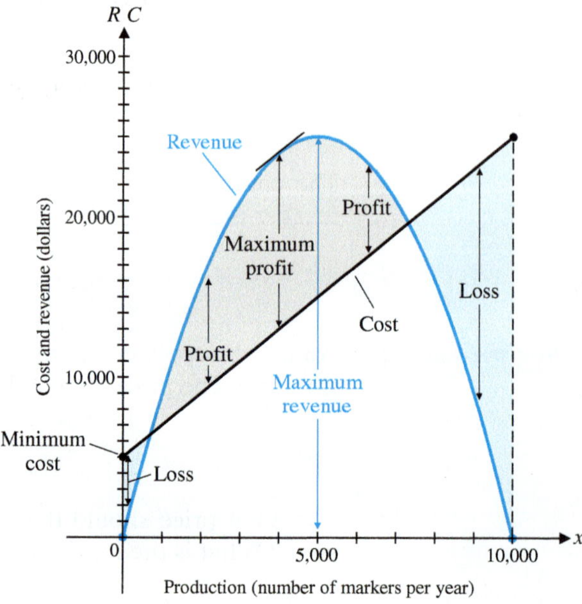

Figure 2

CONCEPTUAL INSIGHT

In Figure 2, notice that the maximum revenue and the maximum profit occur at different production levels. The maximum profit occurs when

$$P'(x) = R'(x) - C'(x) = 0$$

that is, when the marginal revenue is equal to the marginal cost. Notice that the slopes of the revenue function and the cost function are the same at this production level.

Matched Problem 4 The annual cost of manufacturing x paper shredders for the office supply company in Matched Problem 3 is $C(x) = 90{,}000 + 30x$. What is the company's maximum profit? What should it charge for each shredder, and how many shredders should it produce?

EXAMPLE 5 **Maximizing Profit** The government decides to tax the company in Example 4 \$2 for each marker produced. Taking into account this additional cost, how many markers should the company manufacture each week to maximize its weekly profit? What is the maximum weekly profit? How much should the company charge for the markers to realize the maximum weekly profit?

SOLUTION The tax of \$2 per unit changes the company's cost equation:

$$\begin{aligned} C(x) &= \text{original cost} + \text{tax} \\ &= 5{,}000 + 2x + 2x \\ &= 5{,}000 + 4x \end{aligned}$$

The new profit function is

$$\begin{aligned} P(x) &= R(x) - C(x) \\ &= 10x - 0.001x^2 - 5{,}000 - 4x \\ &= 6x - 0.001x^2 - 5{,}000 \end{aligned}$$

So, we must solve the following equation:

$$\begin{aligned} \text{Maximize} \quad P(x) &= 6x - 0.001x^2 - 5{,}000 \qquad 0 \le x \le 10{,}000 \\ P'(x) &= 6 - 0.002x \\ 6 - 0.002x &= 0 \\ x &= 3{,}000 \qquad \text{Critical value} \\ P''(x) &= -0.002 < 0 \quad \text{for all } x \\ \text{Max } P(x) &= P(3{,}000) = \$4{,}000 \end{aligned}$$

Using the price–demand equation (Example 3) with $x = 3{,}000$, we find that

$$p = 10 - 0.001(3{,}000) = \$7 \qquad p = 10 - 0.001x$$

The company's maximum profit is \$4,000 when 3,000 markers are produced and sold weekly at a price of \$7.

Even though the tax caused the company's cost to increase by \$2 per marker, the price that the company should charge to maximize its profit increases by only \$1. The company must absorb the other \$1, with a resulting decrease of \$7,000 in maximum profit.

Matched Problem 5 The government decides to tax the office supply company in Matched Problem 4 \$20 for each shredder produced. Taking into account this additional cost, how many shredders should the company manufacture each week to maximize its weekly profit? What is the maximum weekly profit? How much should the company charge for the shredders to realize the maximum weekly profit?

EXAMPLE 6 **Maximizing Revenue** When a management training company prices its seminar on management techniques at \$400 per person, 1,000 people will attend the seminar. The company estimates that for each \$5 reduction in price, an additional 20 people will attend the seminar. How much should the company charge for the seminar in order to maximize its revenue? What is the maximum revenue?

SOLUTION Let x represent the number of \$5 price reductions.

$$400 - 5x = \text{price per customer}$$

$$1{,}000 + 20x = \text{number of customers}$$

$$\text{Revenue} = (\text{price per customer})(\text{number of customers})$$

$$R(x) = (400 - 5x) \times (1{,}000 + 20x)$$

Since price cannot be negative, we have

$$400 - 5x \geq 0$$

$$400 \geq 5x$$

$$80 \geq x \quad \text{or} \quad x \leq 80$$

A negative value of x would result in a price increase. Since the problem is stated in terms of price reductions, we must restrict x so that $x \geq 0$. Putting all this together, we have the following model:

$$\text{Maximize} \quad R(x) = (400 - 5x)(1{,}000 + 20x) \quad \text{for } 0 \leq x \leq 80$$

$$R(x) = 400{,}000 + 3{,}000x - 100x^2$$

$$R'(x) = 3{,}000 - 200x = 0$$

$$3{,}000 = 200x$$

$$x = 15 \quad \text{Critical value}$$

Table 2

x	$R(x)$
0	400,000
15	422,500
80	0

Since $R(x)$ is continuous on the interval [0, 80], we can determine the behavior of the graph by constructing a table. Table 2 shows that $R(15) = \$422{,}500$ is the absolute maximum revenue. The price of attending the seminar at $x = 15$ is $400 - 5(15) = \$325$. The company should charge \$325 for the seminar in order to receive a maximum revenue of \$422,500.

Matched Problem 6 A walnut grower estimates from past records that if 20 trees are planted per acre, then each tree will average 60 pounds of nuts per year. If, for each additional tree planted per acre, the average yield per tree drops 2 pounds, then how many trees should be planted to maximize the yield per acre? What is the maximum yield?

EXAMPLE 7 **Maximizing Revenue** After additional analysis, the management training company in Example 6 decides that its estimate of attendance was too high. Its new estimate is that only 10 additional people will attend the seminar for each \$5 decrease in price. All other information remains the same. How much should the company charge for the seminar now in order to maximize revenue? What is the new maximum revenue?

SOLUTION Under the new assumption, the model becomes

$$\text{Maximize} \quad R(x) = (400 - 5x)(1{,}000 + 10x) \qquad 0 \le x \le 80$$
$$= 400{,}000 - 1{,}000x - 50x^2$$
$$R'(x) = -1{,}000 - 100x = 0$$
$$-1{,}000 = 100x$$
$$x = -10 \quad \text{Critical value}$$

Table 3

x	$R(x)$
0	400,000
80	0

Note that $x = -10$ is not in the interval [0, 80]. Since $R(x)$ is continuous on [0, 80], we can use a table to find the absolute maximum revenue. Table 3 shows that the maximum revenue is $R(0) = \$400{,}000$. The company should leave the price at \$400. Any \$5 decreases in price will lower the revenue.

Matched Problem 7 After further analysis, the walnut grower in Matched Problem 6 determines that each additional tree planted will reduce the average yield by 4 pounds. All other information remains the same. How many additional trees per acre should the grower plant now in order to maximize the yield? What is the new maximum yield?

CONCEPTUAL INSIGHT

The solution in Example 7 is called an **endpoint solution** because the optimal value occurs at the endpoint of an interval rather than at a critical value in the interior of the interval. It is always important to verify that the optimal value has been found.

Inventory Control

EXAMPLE 8 **Inventory Control** A multimedia company anticipates that there will be a demand for 20,000 copies of a certain DVD during the next year. It costs the company \$0.50 to store a DVD for one year. Each time it must make additional DVDs, it costs \$200 to set up the equipment. How many DVDs should the company make during each production run to minimize its total storage and setup costs?

SOLUTION This type of problem is called an **inventory control problem**. One of the basic assumptions made in such problems is that the demand is uniform. For example, if there are 250 working days in a year, then the daily demand would be $20{,}000 \div 250 = 80$ DVDs. The company could decide to produce all 20,000 DVDs at the beginning of the year. This would certainly minimize the setup costs but would result in very large storage costs. At the other extreme, the company could produce 80 DVDs each day. This would minimize the storage costs but would result in very large setup costs. Somewhere between these two extremes is the optimal solution that will minimize the total storage and setup costs. Let

$$x = \text{number of DVDs manufactured during each production run}$$
$$y = \text{number of production runs}$$

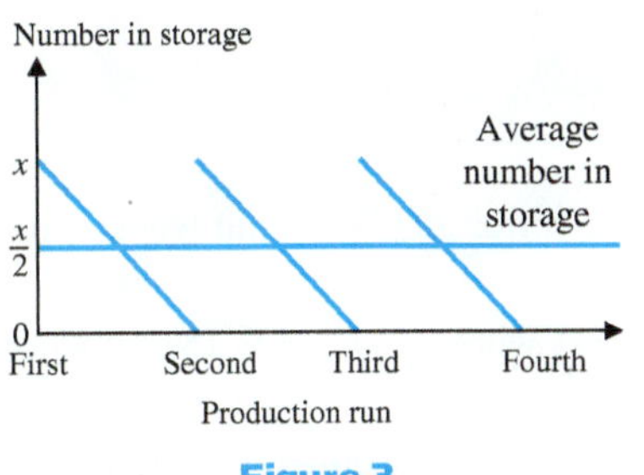

Figure 3

It is easy to see that the total setup cost for the year is $200y$, but what is the total storage cost? If the demand is uniform, then the number of DVDs in storage between production runs will decrease from x to 0, and the average number in storage each day is $x/2$. This result is illustrated in Figure 3.

Since it costs \$0.50 to store a DVD for one year, the total storage cost is $0.5(x/2) = 0.25x$ and the total cost is

$$\text{total cost} = \text{setup cost} + \text{storage cost}$$
$$C = 200y + 0.25x$$

In order to write the total cost C as a function of one variable, we must find a relationship between x and y. If the company produces x DVDs in each of y production runs, then the total number of DVDs produced is xy.

$$xy = 20{,}000$$

$$y = \frac{20{,}000}{x}$$

Certainly, x must be at least 1 and cannot exceed 20,000. We must solve the following equation:

$$\text{Minimize} \quad C(x) = 200\left(\frac{20{,}000}{x}\right) + 0.25x \qquad 1 \le x \le 20{,}000$$

$$C(x) = \frac{4{,}000{,}000}{x} + 0.25x$$

$$C'(x) = -\frac{4{,}000{,}000}{x^2} + 0.25$$

$$-\frac{4{,}000{,}000}{x^2} + 0.25 = 0$$

$$x^2 = \frac{4{,}000{,}000}{0.25}$$

$$x^2 = 16{,}000{,}000$$

$$x = 4{,}000$$

$-4{,}000$ is not a critical value, since $1 \le x \le 20{,}000$.

$$C''(x) = \frac{8{,}000{,}000}{x^3} > 0 \qquad \text{for } x \in (1, 20{,}000)$$

Therefore,

$$\text{Min } C(x) = C(4{,}000) = 2{,}000$$

$$y = \frac{20{,}000}{4{,}000} = 5$$

The company will minimize its total cost by making 4,000 DVDs five times during the year.

Matched Problem 8 Repeat Example 8 if it costs \$250 to set up a production run and \$0.40 to store a DVD for one year.

Exercises 3-6

Preliminary word problems:

1. Find two numbers whose sum is 15 and whose product is a maximum.
2. Find two numbers whose sum is 21 and whose product is a maximum.
3. Find two numbers whose difference is 15 and whose product is a minimum.
4. Find two numbers whose difference is 21 and whose product is a minimum.
5. Find two positive numbers whose product is 15 and whose sum is a minimum.
6. Find two positive numbers whose product is 21 and whose sum is a minimum.
7. Find the dimensions of a rectangle with an area of 200 square feet that has the minimum perimeter.

8. Find the dimensions of a rectangle with an area of 108 square feet that has the minimum perimeter.

9. Find the dimensions of a rectangle with a perimeter of 148 feet that has the maximum area.

10. Find the dimensions of a rectangle with a perimeter of 76 feet that has the maximum area.

11. Maximum revenue and profit. A company manufactures and sells x videophones per week. The weekly price–demand and cost equations are, respectively,

$$p = 500 - 0.5x \quad \text{and} \quad C(x) = 20{,}000 + 135x$$

(A) What price should the company charge for the phones, and how many phones should be produced to maximize the weekly revenue? What is the maximum weekly revenue?

(B) What is the maximum weekly profit? How much should the company charge for the phones, and how many phones should be produced to realize the maximum weekly profit?

12. Maximum revenue and profit. A company manufactures and sells x digital cameras per week. The weekly price–demand and cost equations are, respectively,

$$p = 400 - 0.4x \quad \text{and} \quad C(x) = 2{,}000 + 160x$$

(A) What price should the company charge for the cameras, and how many cameras should be produced to maximize the weekly revenue? What is the maximum revenue?

(B) What is the maximum weekly profit? How much should the company charge for the cameras, and how many cameras should be produced to realize the maximum weekly profit?

13. Maximum revenue and profit. A company manufactures and sells x television sets per month. The monthly cost and price–demand equations are

$$C(x) = 72{,}000 + 60x$$

$$p = 200 - \frac{x}{30} \qquad 0 \le x \le 6{,}000$$

(A) Find the maximum revenue.

(B) Find the maximum profit, the production level that will realize the maximum profit, and the price the company should charge for each television set.

(C) If the government decides to tax the company \$5 for each set it produces, how many sets should the company manufacture each month to maximize its profit? What is the maximum profit? What should the company charge for each set?

14. Maximum revenue and profit. Repeat Problem 13 for

$$C(x) = 60{,}000 + 60x$$

$$p = 200 - \frac{x}{50} \qquad 0 \le x \le 10{,}000$$

15. Maximum profit. The following table contains price–demand and total cost data for the production of extreme-cold sleeping bags, where p is the wholesale price (in dollars) of a sleeping bag for an annual demand of x sleeping bags and C is the total cost (in dollars) of producing x sleeping bags:

x	p	C
950	240	130,000
1,200	210	150,000
1,800	160	180,000
2,050	120	190,000

(A) Find a quadratic regression equation for the price–demand data, using x as the independent variable.

(B) Find a linear regression equation for the cost data, using x as the independent variable.

(C) What is the maximum profit? What is the wholesale price per extreme-cold sleeping bag that should be charged to realize the maximum profit? Round answers to the nearest dollar.

16. Maximum profit. The following table contains price–demand and total cost data for the production of regular sleeping bags, where p is the wholesale price (in dollars) of a sleeping bag for an annual demand of x sleeping bags and C is the total cost (in dollars) of producing x sleeping bags:

x	p	C
2,300	98	145,000
3,300	84	170,000
4,500	67	190,000
5,200	51	210,000

(A) Find a quadratic regression equation for the price–demand data, using x as the independent variable.

(B) Find a linear regression equation for the cost data, using x as the independent variable.

(C) What is the maximum profit? What is the wholesale price per regular sleeping bag that should be charged to realize the maximum profit? Round answers to the nearest dollar.

17. Maximum revenue. A deli sells 640 sandwiches per day at a price of \$8 each.

(A) A market survey shows that for every \$0.10 reduction in price, 40 more sandwiches will be sold. How much should the deli charge for a sandwich in order to maximize revenue?

(B) A different market survey shows that for every \$0.20 reduction in the original \$8 price, 15 more sandwiches will be sold. Now how much should the deli charge for a sandwich in order to maximize revenue?

18. Maximum revenue. A university student center sells 1,600 cups of coffee per day at a price of \$2.40.

(A) A market survey shows that for every \$0.05 reduction in price, 50 more cups of coffee will be sold. How much should the student center charge for a cup of coffee in order to maximize revenue?

(B) A different market survey shows that for every \$0.10 reduction in the original \$2.40 price, 60 more cups of coffee will be sold. Now how much should the student center charge for a cup of coffee in order to maximize revenue?

19. Car rental. A car rental agency rents 200 cars per day at a rate of \$30 per day. For each \$1 increase in rate, 5 fewer cars are rented. At what rate should the cars be rented to produce the maximum income? What is the maximum income?

20. **Rental income.** A 300-room hotel in Las Vegas is filled to capacity every night at \$80 a room. For each \$1 increase in rent, 3 fewer rooms are rented. If each rented room costs \$10 to service per day, how much should the management charge for each room to maximize gross profit? What is the maximum gross profit?

21. **Agriculture.** A commercial cherry grower estimates from past records that if 30 trees are planted per acre, then each tree will yield an average of 50 pounds of cherries per season. If, for each additional tree planted per acre (up to 20), the average yield per tree is reduced by 1 pound, how many trees should be planted per acre to obtain the maximum yield per acre? What is the maximum yield?

22. **Agriculture.** A commercial pear grower must decide on the optimum time to have fruit picked and sold. If the pears are picked now, they will bring 30¢ per pound, with each tree yielding an average of 60 pounds of salable pears. If the average yield per tree increases 6 pounds per tree per week for the next 4 weeks, but the price drops 2¢ per pound per week, when should the pears be picked to realize the maximum return per tree? What is the maximum return?

23. **Manufacturing.** A candy box is to be made out of a piece of cardboard that measures 8 by 12 inches. Squares of equal size will be cut out of each corner, and then the ends and sides will be folded up to form a rectangular box. What size square should be cut from each corner to obtain a maximum volume?

24. **Packaging.** A parcel delivery service will deliver a package only if the length plus girth (distance around) does not exceed 108 inches.

    (A) Find the dimensions of a rectangular box with square ends that satisfies the delivery service's restriction and has maximum volume. What is the maximum volume?

    (B) Find the dimensions (radius and height) of a cylindrical container that meets the delivery service's requirement and has maximum volume. What is the maximum volume?

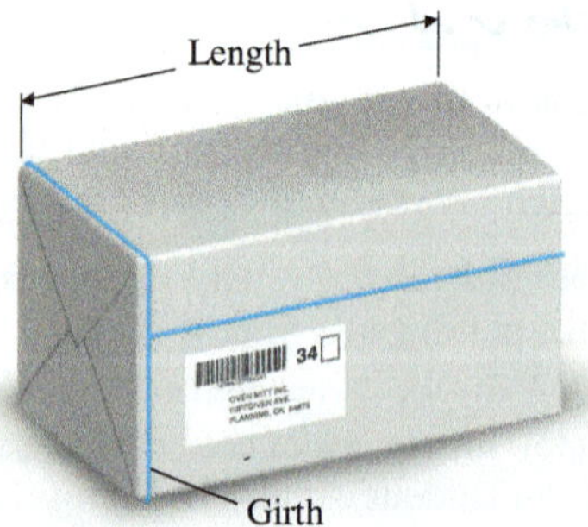

Figure for 24

25. **Construction costs.** A fence is to be built to enclose a rectangular area of 800 square feet. The fence along three sides is to be made of material that costs \$6 per foot. The material for the fourth side costs \$18 per foot. Find the dimensions of the rectangle that will allow for the most economical fence to be built.

26. **Construction costs.** The owner of a retail lumber store wants to construct a fence to enclose an outdoor storage area adjacent to the store, using all of the store as part of one side of the area (see the figure). Find the dimensions that will enclose the largest area if

    (A) 240 feet of fencing material are used.

    (B) 400 feet of fencing material are used.

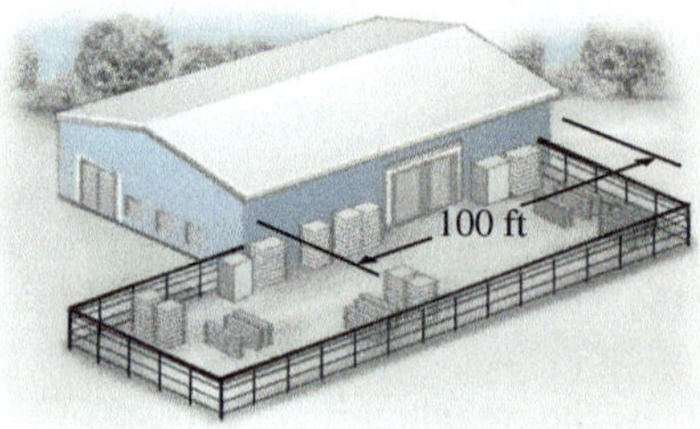

Figure for 26

27. **Inventory control.** A paint manufacturer has a uniform annual demand for 16,000 cans of automobile primer. It costs \$4 to store one can of paint for one year and \$500 to set up the plant for production of the primer. How many times a year should the company produce this primer in order to minimize the total storage and setup costs?

28. **Inventory control.** A pharmacy has a uniform annual demand for 200 bottles of a certain antibiotic. It costs \$10 to store one bottle for one year and \$40 to place an order. How many times during the year should the pharmacy order the antibiotic in order to minimize the total storage and reorder costs?

29. **Inventory control.** A publishing company sells 50,000 copies of a certain book each year. It costs the company \$1 to store a book for one year. Each time that it prints additional copies, it costs the company \$1,000 to set up the presses. How many books should the company produce during each printing in order to minimize its total storage and setup costs?

30. **Operational costs.** The cost per hour for fuel to run a train is $v^2/4$ dollars, where v is the speed of the train in miles per hour. (Note that the cost goes up as the square of the speed.) Other costs, including labor, are \$300 per hour. How fast should the train travel on a 360-mile trip to minimize the total cost for the trip?

31. **Construction costs.** A freshwater pipeline is to be run from a source on the edge of a lake to a small resort community on an island 5 miles offshore, as indicated in the figure.

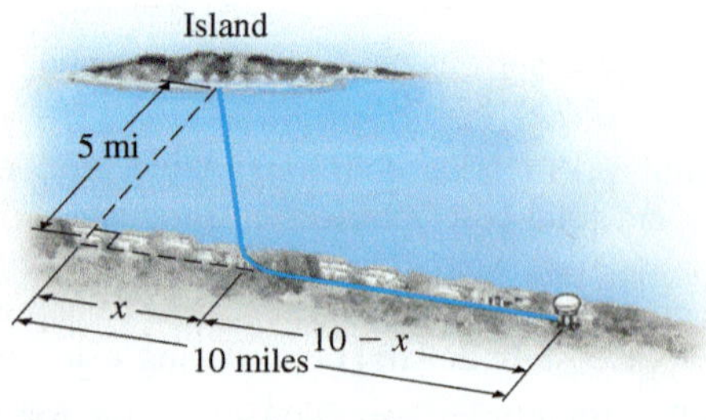

Figure for 31

(A) If it costs 1.4 times as much to lay the pipe in the lake as it does on land, what should x be (in miles) to minimize the total cost of the project?

(B) If it costs only 1.1 times as much to lay the pipe in the lake as it does on land, what should x be to minimize the total cost of the project? [*Note:* Compare with Problem 34.]

32. Drug concentration. The concentration $C(t)$, in milligrams per cubic centimeter, of a particular drug in a patient's bloodstream is given by

$$C(t) = \frac{0.16t}{t^2 + 4t + 4}$$

where t is the number of hours after the drug is taken. How many hours after the drug is taken will the concentration be maximum? What is the maximum concentration?

33. Bacteria control. A lake used for recreational swimming is treated periodically to control harmful bacteria growth. Suppose that t days after a treatment, the concentration of bacteria per cubic centimeter is given by

$$C(t) = 30t^2 - 240t + 500 \qquad 0 \le t \le 8$$

How many days after a treatment will the concentration be minimal? What is the minimum concentration?

34. Bird flights. Some birds tend to avoid flights over large bodies of water during daylight hours. Suppose that an adult bird with this tendency is taken from its nesting area on the edge of a large lake to an island 5 miles offshore and is then released (see the figure).

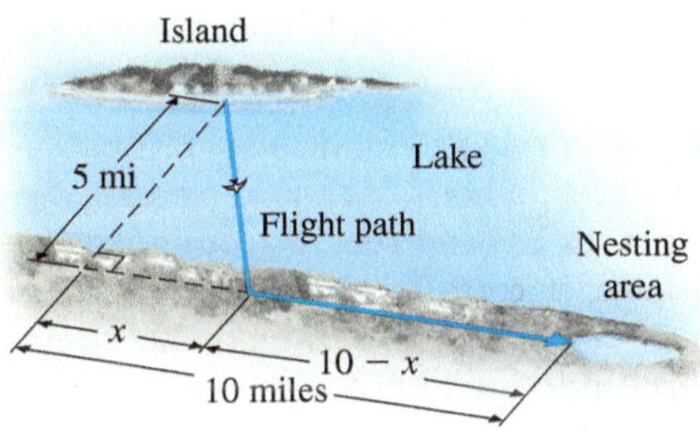

Figure for 34

(A) If it takes 1.4 times as much energy to fly over water as land, how far up the shore (x, in miles) should the bird head to minimize the total energy expended in returning to the nesting area?

(B) If it takes only 1.1 times as much energy to fly over water as land, how far up the shore should the bird head to minimize the total energy expended in returning to the nesting area? [*Note:* Compare with Problem 31.]

35. Botany. If it is known from past experiments that the height (in feet) of a certain plant after t months is given approximately by

$$H(t) = 4t^{1/2} - 2t \qquad 0 \le t \le 2$$

then how long, on average, will it take a plant to reach its maximum height? What is the maximum height?

36. Pollution. Two heavily industrial areas are located 10 miles apart, as shown in the figure. If the concentration of particulate matter (in parts per million) decreases as the reciprocal of the square of the distance from the source, and if area A_1 emits eight times the particulate matter as A_2, then the concentration of particulate matter at any point between the two areas is given by

$$C(x) = \frac{8k}{x^2} + \frac{k}{(10 - x)^2} \qquad 0.5 \le x \le 9.5, \quad k > 0$$

How far from A_1 will the concentration of particulate matter between the two areas be at a minimum?

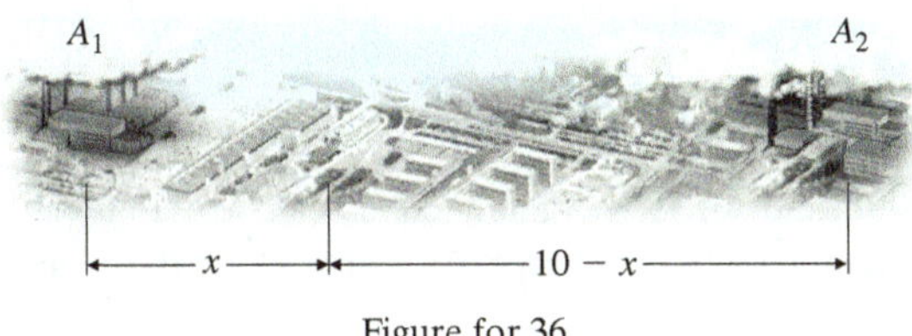

Figure for 36

37. Politics. In a newly incorporated city, it is estimated that the voting population (in thousands) will increase according to

$$N(t) = 30 + 12t^2 - t^3 \qquad 0 \le t \le 8$$

where t is time in years. When will the rate of increase be most rapid?

38. Learning. A large grocery chain found that, on average, a checker can recall $P\%$ of a given price list x hours after starting work, as given approximately by

$$P(x) = 96x - 24x^2 \qquad 0 \le x \le 3$$

At what time x does the checker recall a maximum percentage? What is the maximum?

Answers to Matched Problems

1. The dimensions of the garden with the maximum area of 640 square feet are 16 feet by 40 feet, with one 16-foot side with wood fencing.
2. The minimal cost for enclosing a 1,800-square-foot garden is \$480, and the dimensions are 30 feet by 60 feet, with one 30-foot side with wood fencing.
3. The company will realize a maximum revenue of \$675,000 when the price of a shredder is \$150.
4. A maximum profit of \$456,750 is realized when 4,050 shredders are manufactured annually and sold for \$165 each.
5. A maximum profit of \$378,750 is realized when 3,075 shredders are manufactured annually and sold for \$175 each.
6. The maximum yield is 1,250 pounds per acre when 5 additional trees are planted on each acre.
7. The maximum yield is 1,200 pounds when no additional trees are planted.
8. The company should produce 5,000 DVDs four times a year.

Chapter 3 Review

Important Terms, Symbols, and Concepts

3-1 First Derivative and Graphs

EXAMPLES

- **Increasing and decreasing properties** of a function can be determined by examining a sign chart for the derivative. Ex. 1, p. 144
- A number c is a **partition number** for $f'(x)$ if $f'(c) = 0$ or $f'(x)$ is discontinuous at c. If c is also in the domain of $f(x)$, then c is a **critical value.** Ex. 2, p. 145, Ex. 3, p. 145
- Increasing and decreasing properties and **local extrema** for $f(x)$ can be determined by examining the graph of $f'(x)$. Ex. 4, p. 146, Ex. 5, p. 147, Ex. 6, p. 148
- The **first-derivative test** is used to locate extrema of a function. Ex. 7, p. 150

3-2 Second Derivative and Graphs

- The **second derivative** of a function f can be used to determine the concavity of the graph of f. Ex. 1, p. 162
- **Inflection points** on a graph are points where the concavity changes. Ex. 2, p. 164, Ex. 3, p. 165
- The concavity of the graph of $f(x)$ can also be determined by an examination of the graph of $f'(x)$. Ex. 4, p. 166
- A **graphing strategy** is used to organize the information obtained from the first and second derivatives. Ex. 5, p. 167, Ex. 6, p. 169

3-3 L'Hôpital's Rule

- Limits at infinity and infinite limits involving powers of $x - c$, e^x, and $\ln x$ are reviewed. Ex. 1, p. 178, Ex. 2, p. 178
- The first version of **L'Hôpital's rule** applies to limits involving the indeterminate form $0/0$ as $x \to c$. Ex. 3, p. 179, Ex. 4, p. 180
- You must always check that L'Hôpital's rule applies. Ex. 5, p. 181
- L'Hôpital's rule can be applied more than once. Ex. 6, p. 181
- L'Hôpital's rule applies to one-sided limits and limits at infinity. Ex. 7, p. 182, Ex. 8, p. 183
- L'Hôpital's rule applies to limits involving the indeterminate form ∞/∞. Ex. 9, p. 184, Ex. 10, p. 184

3-4 Curve-Sketching Techniques

- The graphing strategy first used in Section 3-2 is expanded to include horizontal and vertical asymptotes. Ex. 1, p. 187, Ex. 2, p. 188, Ex. 3, p. 189, Ex. 4, p. 191
- If $f(x) = n(x)/d(x)$ is a rational function with the degree of $n(x)$ 1 more than the degree of $d(x)$, then the graph of $f(x)$ has an **oblique asymptote** of the form $y = mx + b$. Ex. 5, p. 193

3-5 Absolute Maxima and Minima

- The steps involved in finding the **absolute maximum and absolute minimum** values of a continuous function on a closed interval are listed in a procedure. Ex. 1, p. 201
- The **second-derivative test for local extrema** can be used to test critical values, but it does not work in all cases. Ex. 2, p. 204
- If a function is continuous on an interval I and has only one critical value in I, then the **second-derivative test for absolute extrema** can be used to find the absolute extrema, but it not does work in all cases. Ex. 3, p. 205

3-6 Optimization

- The methods used to solve **optimization problems** are summarized and illustrated by examples. Ex. 1, p. 208, Ex. 2, p. 209, Ex. 3, p. 211, Ex. 4, p. 212, Ex. 5, p. 213, Ex. 6, p. 214, Ex. 7, p. 214, Ex. 8, p. 215

Review Exercises

Work through all the problems in this chapter review, and check your answers in the back of the book. Answers to all review problems are there, along with section numbers in italics to indicate where each type of problem is discussed. Where weaknesses show up, review appropriate sections in the text.

A

Problems 1–8 refer to the following graph of $y = f(x)$. Identify the points or intervals on the x axis that produce the indicated behavior.

1. $f(x)$ is increasing.

2. $f'(x) < 0$

3. The graph of f is concave downward.

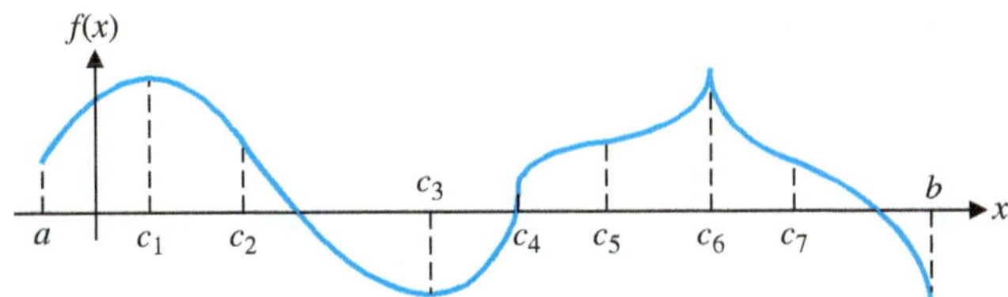

Figure for 1–8

4. Local minima

5. Absolute maxima

6. $f'(x)$ appears to be 0.

7. $f'(x)$ does not exist.

8. Inflection points

In Problems 9 and 10, use the given information to sketch the graph of f. Assume that f is continuous on its domain and that all intercepts are included in the information given.

9. Domain: All real x

x	−3	−2	−1	0	2	3
$f(x)$	0	3	2	0	−3	0

$f'(x)$: + + + 0 − − − − − − − − 0 − − − ND + + + (at $x = -2$, 0, 2)

$f''(x)$: − − − − − − − − 0 + + + 0 − − − ND − − − (at $x = -1$, 0, 2)

10. Domain: All real x

$f(-2) = 1, f(0) = 0, f(2) = 1;$

$f'(0) = 0; f'(x) < 0$ on $(-\infty, 0);$

$f'(x) > 0$ on $(0, \infty);$

$f''(-2) = 0, f''(2) = 0;$

$f''(x) < 0$ on $(-\infty, -2)$ and $(2, \infty);$

$f''(x) > 0$ on $(-2, 2);$

$\lim_{x\to-\infty} f(x) = 2; \lim_{x\to\infty} f(x) = 2$

11. Find $f''(x)$ for $f(x) = x^4 + 5x^3$.

12. Find y'' for $y = 3x + \frac{4}{x}$.

In Problems 13 and 14, find the domain and intercepts.

13. $f(x) = \frac{5 + x}{4 - x}$

14. $f(x) = \ln(x + 2)$

In Problems 15 and 16, find the horizontal and vertical asymptotes.

15. $f(x) = \frac{x + 3}{x^2 - 4}$

16. $f(x) = \frac{2x - 7}{3x + 10}$

In Problems 17 and 18, find the x and y coordinates of all inflection points.

17. $f(x) = x^4 - 12x^2$

18. $f(x) = (2x + 1)^{1/3} - 6$

In Problems 19 and 20, find (A) $f'(x)$, (B) the critical values of f, and (C) the partition numbers for f'.

19. $f(x) = x^{1/5}$

20. $f(x) = x^{-1/5}$

In Problems 21–30, summarize all the pertinent information obtained by applying the final version of the graphing strategy (Section 3-4) to f, and sketch the graph of f.

21. $f(x) = x^3 - 18x^2 + 81x$

22. $f(x) = (x + 4)(x - 2)^2$

23. $f(x) = 8x^3 - 2x^4$

24. $f(x) = (x - 1)^3(x + 3)$

25. $f(x) = \frac{3x}{x + 2}$

26. $f(x) = \frac{x^2}{x^2 + 27}$

27. $f(x) = \frac{x}{(x + 2)^2}$

28. $f(x) = \frac{x^3}{x^2 + 3}$

29. $f(x) = 5 - 5e^{-x}$

30. $f(x) = x^3 \ln x$

Find each limit in Problems 31–40.

31. $\lim_{x\to 0} \frac{e^{3x} - 1}{x}$

32. $\lim_{x\to 2} \frac{x^2 - 5x + 6}{x^2 + x - 6}$

33. $\lim_{x\to 0^-} \frac{\ln(1 + x)}{x^2}$

34. $\lim_{x\to 0} \frac{\ln(1 + x)}{1 + x}$

35. $\lim_{x\to\infty} \frac{e^{4x}}{x^2}$

36. $\lim_{x\to 0} \frac{e^x + e^{-x} - 2}{x^2}$

37. $\lim_{x\to 0^+} \frac{\sqrt{1 + x} - 1}{\sqrt{x}}$

38. $\lim_{x\to\infty} \frac{\ln x}{x^5}$

39. $\lim_{x\to\infty} \frac{\ln(1 + 6x)}{\ln(1 + 3x)}$

40. $\lim_{x\to 0} \frac{\ln(1 + 6x)}{\ln(1 + 3x)}$

41. Use the graph of $y = f'(x)$ shown here to discuss the graph of $y = f(x)$. Organize your conclusions in a table (see Example 4, Section 3-2). Sketch a possible graph of $y = f(x)$.

Figure for 41 and 42

42. Refer to the above graph of $y = f'(x)$. Which of the following could be the graph of $y = f''(x)$?

(A)

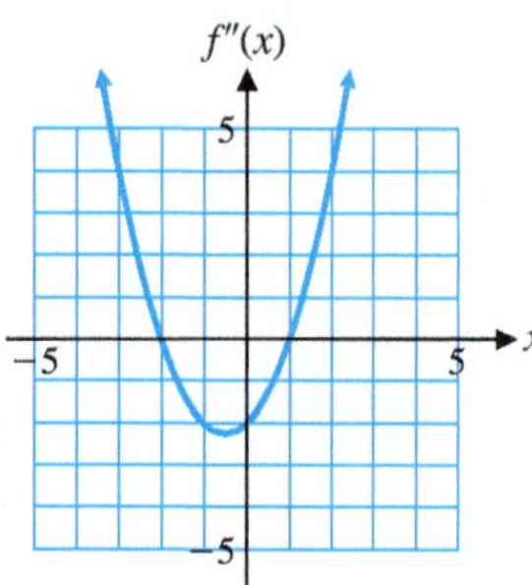

(B)

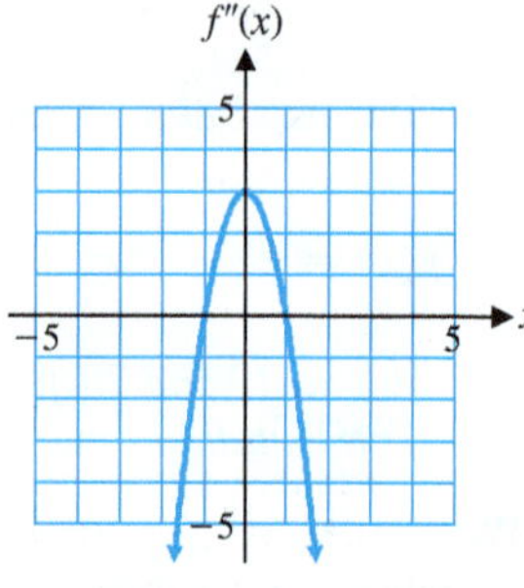

(C)

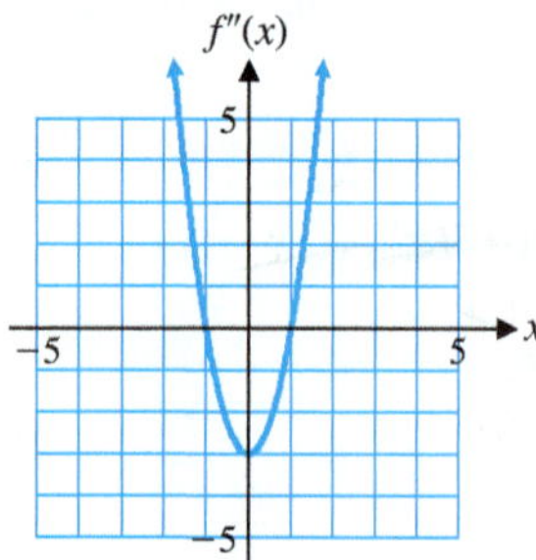

43. Use the second-derivative test to find any local extrema for

$$f(x) = x^3 - 6x^2 - 15x + 12$$

44. Find the absolute maximum and absolute minimum, if either exists, for

$$y = f(x) = x^3 - 12x + 12 \qquad -3 \le x \le 5$$

45. Find the absolute minimum, if it exists, for

$$y = f(x) = x^2 + \frac{16}{x^2} \qquad x > 0$$

46. Find the absolute maximum value, if it exists, for

$$f(x) = 11x - 2x \ln x \qquad x > 0$$

47. Find the absolute maximum value, if it exists, for

$$f(x) = 10xe^{-2x} \qquad x > 0$$

48. Let $y = f(x)$ be a polynomial function with local minima at $x = a$ and $x = b$, $a < b$. Must f have at least one local maximum between a and b? Justify your answer.

49. The derivative of $f(x) = x^{-1}$ is $f'(x) = -x^{-2}$. Since $f'(x) < 0$ for $x \ne 0$, is it correct to say that $f(x)$ is decreasing for all x except $x = 0$? Explain.

50. Discuss the difference between a partition number for $f'(x)$ and a critical value of $f(x)$, and illustrate with examples.

C

51. Find the absolute maximum for $f'(x)$ if

$$f(x) = 6x^2 - x^3 + 8$$

Graph f and f' on the same coordinate system for $0 \le x \le 4$.

52. Find two positive numbers whose product is 400 and whose sum is a minimum. What is the minimum sum?

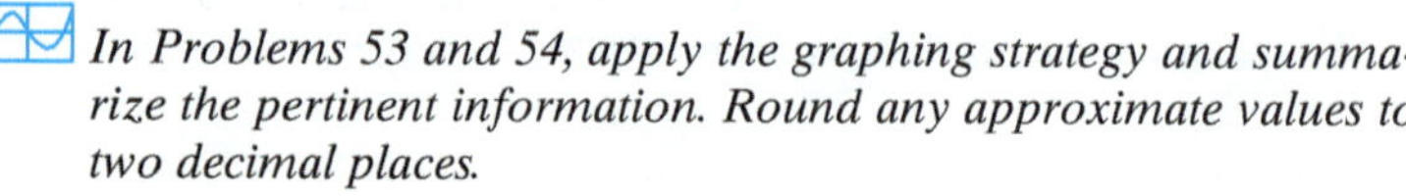

In Problems 53 and 54, apply the graphing strategy and summarize the pertinent information. Round any approximate values to two decimal places.

53. $f(x) = x^4 + x^3 - 4x^2 - 3x + 4$

54. $f(x) = 0.25x^4 - 5x^3 + 31x^2 - 70x$

55. Find the absolute maximum value, if it exists, for

$$f(x) = 3x - x^2 + e^{-x} \quad x > 0$$

56. Find the absolute maximum value, if it exists, for

$$f(x) = \frac{\ln x}{e^x} \quad x > 0$$

Applications

57. Price analysis. The graph in the figure approximates the rate of change of the price of tomatoes over a 60-month period, where $p(t)$ is the price of a pound of tomatoes and t is time (in months).

(A) Write a brief description of the graph of $y = p(t)$, including a discussion of local extrema and inflection points.

(B) Sketch a possible graph of $y = p(t)$.

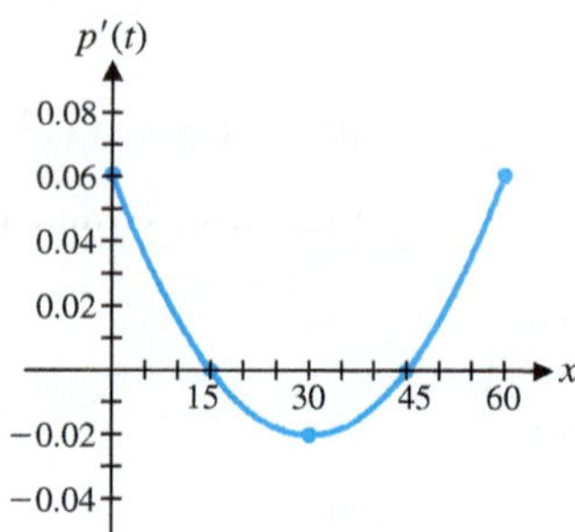

Figure for 57

58. Maximum revenue and profit. A company manufactures and sells x e-book readers per month. The monthly cost and price–demand equations are, respectively,

$$C(x) = 350x + 50{,}000$$

$$p = 500 - 0.025x \qquad 0 \le x \le 20{,}000$$

(A) Find the maximum revenue.

(B) How many readers should the company manufacture each month to maximize its profit? What is the maximum monthly profit? How much should the company charge for each reader?

(C) If the government decides to tax the company \$20 for each reader it produces, how many readers should the company manufacture each month to maximize its profit? What is the maximum monthly profit? How much should the company charge for each reader?

59. Construction. A fence is to be built to enclose a rectangular area. The fence along three sides is to be made of material that costs \$5 per foot. The material for the fourth side costs \$15 per foot.

(A) If the area is 5,000 square feet, find the dimensions of the rectangle that will allow for the most economical fence.

(B) If \$3,000 is available for the fencing, find the dimensions of the rectangle that will enclose the most area.

60. Rental income. A 200-room hotel in Reno is filled to capacity every night at a rate of \$40 per room. For each \$1 increase in the nightly rate, 4 fewer rooms are rented. If each rented room costs \$8 a day to service, how much should the management charge per room in order to maximize gross profit? What is the maximum gross profit?

61. Inventory control. A computer store sells 7,200 boxes of storage disks annually. It costs the store \$0.20 to store a box of disks for one year. Each time it reorders disks, the store must pay a \$5.00 service charge for processing the order. How many times during the year should the store order disks to minimize the total storage and reorder costs?

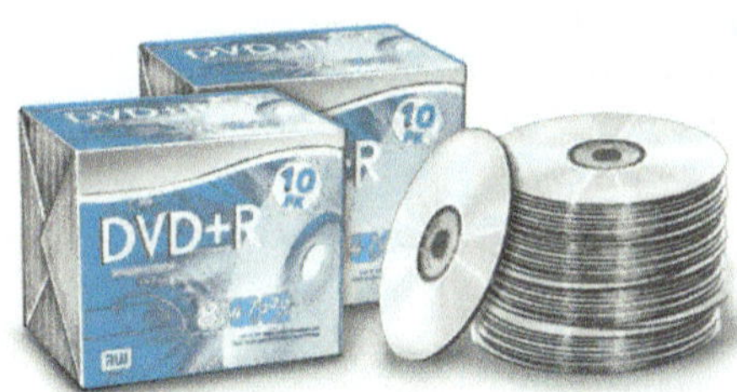

62. Average cost. The total cost of producing x dorm refrigerators per day is given by

$$C(x) = 4{,}000 + 10x + 0.1x^2$$

Find the minimum average cost. Graph the average cost and the marginal cost functions on the same coordinate system. Include any oblique asymptotes.

63. Average cost. The cost of producing x wheeled picnic coolers is given by

$$C(x) = 200 + 50x - 50\ln x \qquad x \ge 1$$

Find the minimum average cost.

64. Marginal analysis. The price–demand equation for a GPS device is

$$p(x) = 1{,}000e^{-0.02x}$$

where x is the monthly demand and p is the price in dollars. Find the production level and price per unit that produce the maximum revenue. What is the maximum revenue?

65. Maximum revenue. Graph the revenue function from Problem 64 for $0 \le x \le 100$.

66. Maximum profit. Refer to Problem 64. If the GPS devices cost the store \$220 each, find the price (to the nearest cent) that maximizes the profit. What is the maximum profit (to the nearest dollar)?

67. Maximum profit. The data in the table show the daily demand x for cream puffs at a state fair at various price levels p. If it costs \$1 to make a cream puff, use logarithmic regression ($p = a + b\ln x$) to find the price (to the nearest cent) that maximizes profit.

Demand x	Price per Cream Puff(\$) p
3,125	1.99
3,879	1.89
5,263	1.79
5,792	1.69
6,748	1.59
8,120	1.49

68. Construction costs. The ceiling supports in a new discount department store are 12 feet apart. Lights are to be hung from these supports by chains in the shape of a "Y." If the lights are 10 feet below the ceiling, what is the shortest length of chain that can be used to support these lights?

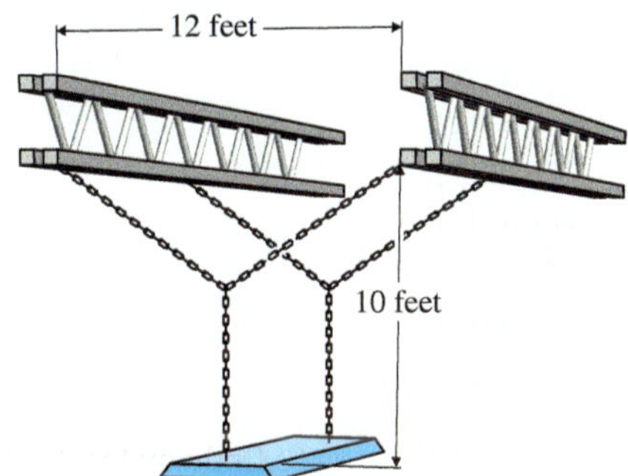

69. Average cost. The table gives the total daily cost y (in dollars) of producing x dozen chocolate chip cookies at various production levels.

Dozens of Cookies x	Total Cost y
50	119
100	187
150	248
200	382
250	505
300	695

(A) Enter the data into a graphing calculator and find a quadratic regression equation for the total cost.

(B) Use the regression equation from part (A) to find the minimum average cost (to the nearest cent) and the corresponding production level (to the nearest integer).

70. Advertising—point of diminishing returns. A company estimates that it will sell $N(x)$ units of a product after spending \$$x$ thousand on advertising, as given by

$$N(x) = -0.25x^4 + 11x^3 - 108x^2 + 3{,}000$$
$$9 \le x \le 24$$

When is the rate of change of sales increasing and when is it decreasing? What is the point of diminishing returns and the maximum rate of change of sales? Graph N and N' on the same coordinate system.

71. Advertising. A chain of appliance stores uses TV ads to promote its HDTV sales. Analyzing past records produced the data in the following table, where x is the number of ads placed monthly and y is the number of HDTVs sold that month:

Number of Ads x	Number of HDTVs y
10	271
20	427
25	526
30	629
45	887
48	917

(A) Enter the data into a graphing calculator, set the calculator to display two decimal places, and find a cubic regression equation for the number of HDTVs sold monthly as a function of the number of ads.

(B) How many ads should be placed each month to maximize the rate of change of sales with respect to the number of ads, and how many HDTVs can be expected to be sold with that number of ads? Round answers to the nearest integer.

72. Bacteria control. If t days after a treatment the bacteria count per cubic centimeter in a body of water is given by

$$C(t) = 20t^2 - 120t + 800 \qquad 0 \le t \le 9$$

then in how many days will the count be a minimum?

73. Politics. In a new suburb, it is estimated that the number of registered voters will grow according to

$$N = 10 + 6t^2 - t^3 \qquad 0 \le t \le 5$$

where t is time in years and N is in thousands. When will the rate of increase be a maximum?

Integration

Introduction

In the preceding three chapters, we studied the *derivative* and its applications. In Chapter 4, we introduce the *integral*, the second key concept of calculus. The integral can be used to calculate areas, volumes, the index of income concentration, and consumers' surplus. At first glance, the integral may appear to be unrelated to the derivative. There is, however, a close connection between these two concepts, which is made precise by the *fundamental theorem of calculus* (Section 4-5). We consider many applications of integrals and differential equations in Chapter 4. See, for example, Problem 67 in Section 4-3, which explores how the age of an archeological site or artifact can be estimated.

4-1 Antiderivatives and Indefinite Integrals

- Antiderivatives
- Indefinite Integrals: Formulas and Properties
- Applications

Many operations in mathematics have reverses—addition and subtraction, multiplication and division, powers and roots. We now know how to find the derivatives of many functions. The reverse operation, *antidifferentiation* (the reconstruction of a function from its derivative), will receive our attention in this and the next two sections.

Antiderivatives

A function F is an **antiderivative** of a function f if $F'(x) = f(x)$.

The function $F(x) = \frac{x^3}{3}$ is an antiderivative of the function $f(x) = x^2$ because

$$\frac{d}{dx}\left(\frac{x^3}{3}\right) = x^2$$

However, $F(x)$ is not the only antiderivative of x^2. Note also that

$$\frac{d}{dx}\left(\frac{x^3}{3} + 2\right) = x^2 \qquad \frac{d}{dx}\left(\frac{x^3}{3} - \pi\right) = x^2 \qquad \frac{d}{dx}\left(\frac{x^3}{3} + \sqrt{5}\right) = x^2$$

Therefore,

$$\frac{x^3}{3} + 2 \qquad \frac{x^3}{3} - \pi \qquad \frac{x^3}{3} + \sqrt{5}$$

are also antiderivatives of x^2 because each has x^2 as a derivative. In fact, it appears that

$$\frac{x^3}{3} + C \qquad \text{for any real number } C$$

is an antiderivative of x^2 because

$$\frac{d}{dx}\left(\frac{x^3}{3} + C\right) = x^2$$

Antidifferentiation of a given function does not give a unique function, but an entire family of functions.

Does the expression

$$\frac{x^3}{3} + C \qquad \text{with } C \text{ any real number}$$

include all antiderivatives of x^2? Theorem 1 (stated without proof) indicates that the answer is yes.

THEOREM 1 Antiderivatives

If the derivatives of two functions are equal on an open interval (a, b), then the functions differ by at most a constant. Symbolically, if F and G are differentiable functions on the interval (a, b) and $F'(x) = G'(x)$ for all x in (a, b), then $F(x) = G(x) + k$ for some constant k.

CONCEPTUAL INSIGHT

Suppose that $F(x)$ is an antiderivative of $f(x)$. If $G(x)$ is any other antiderivative of $f(x)$, then by Theorem 1, the graph of $G(x)$ is a vertical translation of the graph of $F(x)$.

EXAMPLE 1 **A Family of Antiderivatives** Note that

$$\frac{d}{dx}\left(\frac{x^2}{2}\right) = x$$

(A) Find all antiderivatives of $f(x) = x$.

(B) Graph the antiderivative of $f(x) = x$ that passes through the point (0, 0); through the point (0, 1); through the point (0, 2).

(C) How are the graphs of the three antiderivatives in part (B) related?

SOLUTION (A) By Theorem 1, any antiderivative of $f(x)$ has the form

$$F(x) = \frac{x^2}{2} + k$$

where k is a real number.

(B) Because $F(0) = (0^2/2) + k = k$, the functions

$$F_0(x) = \frac{x^2}{2}, \quad F_1(x) = \frac{x^2}{2} + 1, \quad \text{and} \quad F_2(x) = \frac{x^2}{2} + 2$$

pass through the points (0, 0), (0, 1), and (0, 2), respectively (see Fig. 1).

(C) The graphs of the three antiderivatives are vertical translations of each other.

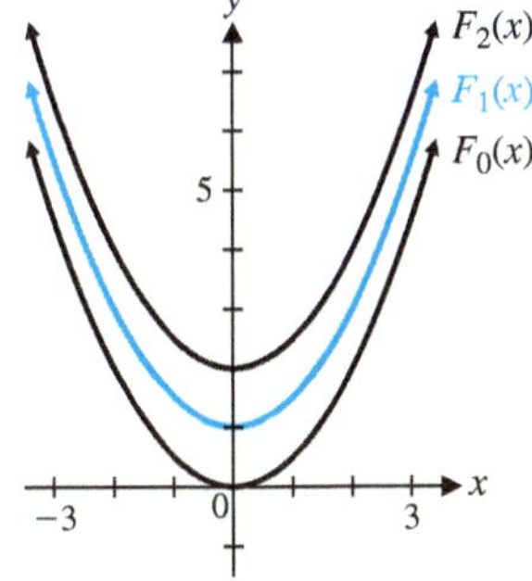

Figure 1

Matched Problem 1 Note that

$$\frac{d}{dx}(x^3) = 3x^2$$

(A) Find all antiderivatives of $f(x) = 3x^2$.

(B) Graph the antiderivative of $f(x) = 3x^2$ that passes through the point (0, 0); through the point (0, 1); through the point (0, 2).

(C) How are the graphs of the three antiderivatives in part (B) related?

Indefinite Integrals: Formulas and Properties

Theorem 1 states that if the derivatives of two functions are equal, then the functions differ by at most a constant. We use the symbol

$$\int f(x)\, dx$$

called the **indefinite integral**, to represent the family of all antiderivatives of $f(x)$, and we write

$$\int f(x)\, dx = F(x) + C \qquad \text{if} \qquad F'(x) = f(x)$$

The symbol $\int$ is called an **integral sign**, and the function $f(x)$ is called the **integrand**. The symbol dx indicates that the antidifferentiation is performed with respect to the variable x. (We will have more to say about the symbols $\int$ and dx later in the chapter.) The arbitrary constant C is called the **constant of integration**. Referring to the preceding discussion, we can write

$$\int x^2\, dx = \frac{x^3}{3} + C \qquad \text{since} \qquad \frac{d}{dx}\left(\frac{x^3}{3} + C\right) = x^2$$

Of course, variables other than x can be used in indefinite integrals. For example,

$$\int t^2\,dt = \frac{t^3}{3} + C \qquad \text{since} \qquad \frac{d}{dt}\left(\frac{t^3}{3} + C\right) = t^2$$

or

$$\int u^2\,du = \frac{u^3}{3} + C \qquad \text{since} \qquad \frac{d}{du}\left(\frac{u^3}{3} + C\right) = u^2$$

The fact that indefinite integration and differentiation are reverse operations, except for the addition of the constant of integration, can be expressed symbolically as

$$\frac{d}{dx}\left[\int f(x)\,dx\right] = f(x)$$ The derivative of the indefinite integral of $f(x)$ is $f(x)$.

and

$$\int F'(x)\,dx = F(x) + C$$ The indefinite integral of the derivative of $F(x)$ is $F(x) + C$.

We can develop formulas for the indefinite integrals of certain basic functions from the formulas for derivatives in Chapters 1 and 2.

FORMULAS Indefinite Integrals of Basic Functions

For C a constant,

1. $\int x^n\,dx = \frac{x^{n+1}}{n+1} + C, \quad n \neq -1$
2. $\int e^x\,dx = e^x + C$
3. $\int \frac{1}{x}\,dx = \ln|x| + C, \quad x \neq 0$

To justify each formula, show that the derivative of the right-hand side is the integrand of the left-hand side (see Problems 77–80 in Exercise 4-1). Note that formula 1 does not give the antiderivative of x^{-1} (because $x^{n+1}/(n+1)$ is undefined when $n = -1$), but formula 3 does.

EXPLORE & DISCUSS 1

Formulas 1, 2, and 3 do *not* provide a formula for the indefinite integral of the function $\ln x$. Show that if $x > 0$, then

$$\int \ln x\,dx = x \ln x - x + C$$

by differentiating the right-hand side.

We can obtain properties of the indefinite integral from derivative properties that were established in Chapter 1.

PROPERTIES Indefinite Integrals

For k a constant,

4. $\int kf(x)\,dx = k\int f(x)\,dx$
5. $\int [f(x) \pm g(x)]\,dx = \int f(x)\,dx \pm \int g(x)\,dx$

Property 4 states that

The indefinite integral of a constant times a function is the constant times the indefinite integral of the function.

Property 5 states that

The indefinite integral of the sum of two functions is the sum of the indefinite integrals, and the indefinite integral of the difference of two functions is the difference of the indefinite integrals.

To establish property 4, let F be a function such that $F'(x) = f(x)$. Then

$$k\int f(x)\,dx = k\int F'(x)\,dx = k[F(x) + C_1] = kF(x) + kC_1$$

and since $[kF(x)]' = kF'(x) = kf(x)$, we have

$$\int kf(x)\,dx = \int kF'(x)\,dx = kF(x) + C_2$$

But $kF(x) + kC_1$ and $kF(x) + C_2$ describe the same set of functions, because C_1 and C_2 are arbitrary real numbers. Property 4 is established. Property 5 can be established in a similar manner (see Problems 81 and 82 in Exercise 4-1).

CAUTION Property 4 states that **a constant factor can be moved across an integral sign. A variable factor cannot be moved across an integral sign:**

Constant Factor	*Variable Factor*
$\int 5x^{1/2}\,dx = 5\int x^{1/2}\,dx$	$\int xx^{1/2}\,dx \neq x\int x^{1/2}\,dx$

Indefinite integral formulas and properties can be used together to find indefinite integrals for many frequently encountered functions. If $n = 0$, then formula 1 gives

$$\int dx = x + C$$

Therefore, by property 4,

$$\int k\,dx = k(x + C) = kx + kC$$

Because kC is a constant, we replace it with a single symbol that denotes an arbitrary constant (usually C), and write

$$\int k\,dx = kx + C$$

In words,

The indefinite integral of a constant function with value k is $kx + C$.

Similarly, using property 5 and then formulas 2 and 3, we obtain

$$\begin{aligned}\int\left(e^x + \frac{1}{x}\right)dx &= \int e^x\,dx + \int \frac{1}{x}\,dx \\ &= e^x + C_1 + \ln|x| + C_2\end{aligned}$$

Because $C_1 + C_2$ is a constant, we replace it with the symbol C and write

$$\int\left(e^x + \frac{1}{x}\right)dx = e^x + \ln|x| + C$$

EXAMPLE 2 **Using Indefinite Integral Properties and Formulas**

(A) $\int 5\,dx = 5x + C$

(B) $\int 9e^x\,dx = 9\int e^x\,dx = 9e^x + C$

(C) $\int 5t^7\,dt = 5\int t^7\,dt = 5\frac{t^8}{8} + C = \frac{5}{8}t^8 + C$

(D) $\int (4x^3 + 2x - 1)\,dx = \int 4x^3\,dx + \int 2x\,dx - \int dx$

$= 4\int x^3\,dx + 2\int x\,dx - \int dx$

$= \frac{4x^4}{4} + \frac{2x^2}{2} - x + C$

$= x^4 + x^2 - x + C$

Property 4 can be extended to the sum and difference of an arbitrary number of functions.

(E) $\int \left(2e^x + \frac{3}{x}\right)dx = 2\int e^x\,dx + 3\int \frac{1}{x}\,dx$

$= 2e^x + 3\ln|x| + C$

To check any of the results in Example 2, we differentiate the final result to obtain the integrand in the original indefinite integral. When you evaluate an indefinite integral, do not forget to include the arbitrary constant C.

Matched Problem 2 Find each indefinite integral:

(A) $\int 2\,dx$

(B) $\int 16e^t\,dt$

(C) $\int 3x^4\,dx$

(D) $\int (2x^5 - 3x^2 + 1)\,dx$

(E) $\int \left(\frac{5}{x} - 4e^x\right)dx$

EXAMPLE 3 **Using Indefinite Integral Properties and Formulas**

(A) $\int \frac{4}{x^3}\,dx = \int 4x^{-3}\,dx = \frac{4x^{-3+1}}{-3+1} + C = -2x^{-2} + C$

(B) $\int 5\sqrt[3]{u^2}\,du = 5\int u^{2/3}\,du = 5\frac{u^{(2/3)+1}}{\frac{2}{3}+1} + C$

$= 5\frac{u^{5/3}}{\frac{5}{3}} + C = 3u^{5/3} + C$

(C) $\int \frac{x^3 - 3}{x^2}\,dx = \int \left(\frac{x^3}{x^2} - \frac{3}{x^2}\right)dx$

$= \int (x - 3x^{-2})\,dx$

$= \int x\,dx - 3\int x^{-2}\,dx$

$= \frac{x^{1+1}}{1+1} - 3\frac{x^{-2+1}}{-2+1} + C$

$= \frac{1}{2}x^2 + 3x^{-1} + C$

(D) $\int \left(\frac{2}{\sqrt[3]{x}} - 6\sqrt{x}\right) dx = \int (2x^{-1/3} - 6x^{1/2})\,dx$

$$= 2\int x^{-1/3}\,dx - 6\int x^{1/2}\,dx$$

$$= 2\frac{x^{(-1/3)+1}}{-\frac{1}{3}+1} - 6\frac{x^{(1/2)+1}}{\frac{1}{2}+1} + C$$

$$= 2\frac{x^{2/3}}{\frac{2}{3}} - 6\frac{x^{3/2}}{\frac{3}{2}} + C$$

$$= 3x^{2/3} - 4x^{3/2} + C$$

(E) $\int x(x^2 + 2)\,dx = \int (x^3 + 2x)\,dx = \frac{x^4}{4} + x^2 + C$

Matched Problem 3 Find each indefinite integral:

(A) $\int \left(2x^{2/3} - \frac{3}{x^4}\right) dx$

(B) $\int 4\sqrt[5]{w^3}\,dw$

(C) $\int \frac{x^4 - 8x^3}{x^2}\,dx$

(D) $\int \left(8\sqrt[3]{x} - \frac{6}{\sqrt{x}}\right) dx$

(E) $\int (x^2 - 2)(x + 3)\,dx$

CAUTION

1. Note from Example 3(E) that

$$\int x(x^2 + 2)\,dx \neq \frac{x^2}{2}\left(\frac{x^3}{3} + 2x\right) + C$$

In general, the **indefinite integral of a product is not the product of the indefinite integrals.** (This is expected because the derivative of a product is not the product of the derivatives.)

2. $$\int e^x\,dx \neq \frac{e^{x+1}}{x+1} + C$$

The power rule applies only to power functions of the form x^n, where the exponent n is a real constant not equal to -1 and the base x is the variable. The function e^x is an exponential function with variable exponent x and constant base e. The correct form is

$$\int e^x\,dx = e^x + C$$

3. Not all elementary functions have elementary antiderivatives. It is impossible, for example, to give a formula for the antiderivative of $f(x) = e^{x^2}$ in terms of elementary functions. Nevertheless, finding such a formula, when it exists, can markedly simplify the solution of certain problems.

Applications

Let's consider some applications of the indefinite integral.

EXAMPLE 4 **Curves** Find the equation of the curve that passes through (2, 5) if the slope of the curve is given by $dy/dx = 2x$ at any point x.

SOLUTION We want to find a function $y = f(x)$ such that

$$\frac{dy}{dx} = 2x \tag{1}$$

and

$$y = 5 \quad \text{when} \quad x = 2 \tag{2}$$

If $dy/dx = 2x$, then

$$y = \int 2x\,dx \tag{3}$$
$$= x^2 + C$$

Since $y = 5$ when $x = 2$, we determine the *particular value of C* so that

$$5 = 2^2 + C$$

So $C = 1$, and

$$y = x^2 + 1$$

is the *particular antiderivative* out of all those possible from equation (3) that satisfies both equations (1) and (2) (see Fig. 2).

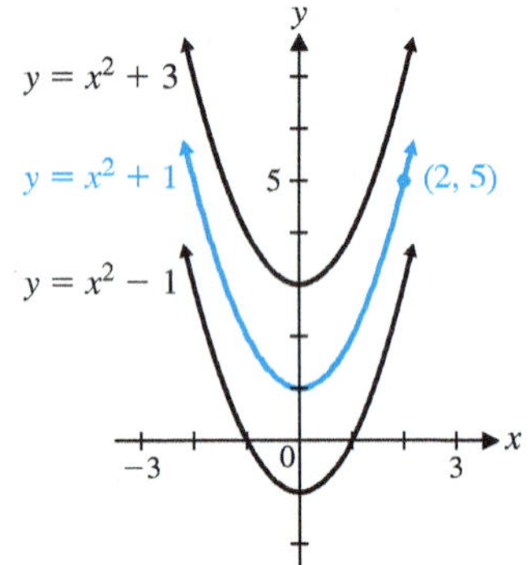

Figure 2 $y = x^2 + C$

Matched Problem 4 Find the equation of the curve that passes through (2, 6) if the slope of the curve is given by $dy/dx = 3x^2$ at any point x.

In certain situations, it is easier to determine the rate at which something happens than to determine how much of it has happened in a given length of time (for example, population growth rates, business growth rates, the rate of healing of a wound, rates of learning or forgetting). If a rate function (derivative) is given and we know the value of the dependent variable for a given value of the independent variable, then we can often find the original function by integration.

EXAMPLE 5 **Cost Function** If the marginal cost of producing x units of a commodity is given by

$$C'(x) = 0.3x^2 + 2x$$

and the fixed cost is \$2,000, find the cost function $C(x)$ and the cost of producing 20 units.

SOLUTION Recall that marginal cost is the derivative of the cost function and that fixed cost is cost at a zero production level. So we want to find $C(x)$, given

$$C'(x) = 0.3x^2 + 2x \qquad C(0) = 2{,}000$$

We find the indefinite integral of $0.3x^2 + 2x$ and determine the arbitrary integration constant using $C(0) = 2{,}000$:

$$C'(x) = 0.3x^2 + 2x$$
$$C(x) = \int (0.3x^2 + 2x)\,dx$$
$$= 0.1x^3 + x^2 + K$$

Since C represents the cost, we use K for the constant of integration.

But

$$C(0) = (0.1)0^3 + 0^2 + K = 2{,}000$$

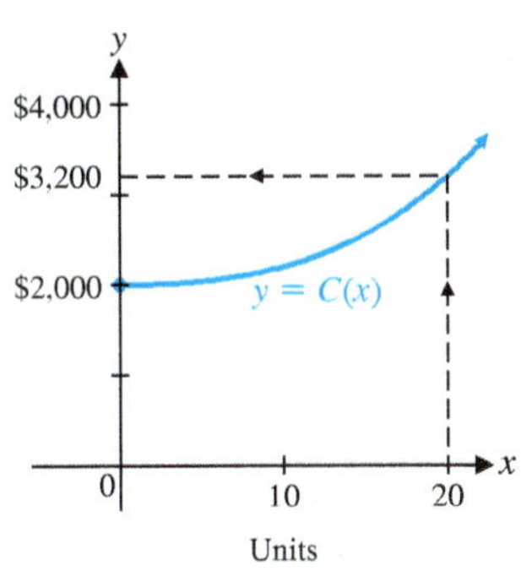

Figure 3

So $K = 2{,}000$, and the cost function is

$$C(x) = 0.1x^3 + x^2 + 2{,}000$$

We now find $C(20)$, the cost of producing 20 units:

$$C(20) = (0.1)20^3 + 20^2 + 2{,}000$$
$$= \$3{,}200$$

See Figure 3 for a geometric representation.

Matched Problem 5 Find the revenue function $R(x)$ when the marginal revenue is

$$R'(x) = 400 - 0.4x$$

and no revenue results at a zero production level. What is the revenue at a production level of 1,000 units?

EXAMPLE 6 **Advertising** A satellite radio station is launching an aggressive advertising campaign in order to increase the number of daily listeners. The station currently has 27,000 daily listeners, and management expects the number of daily listeners, $S(t)$, to grow at the rate of

$$S'(t) = 60t^{1/2}$$

listeners per day, where t is the number of days since the campaign began. How long should the campaign last if the station wants the number of daily listeners to grow to 41,000?

SOLUTION We must solve the equation $S(t) = 41{,}000$ for t, given that

$$S'(t) = 60t^{1/2} \qquad \text{and} \qquad S(0) = 27{,}000$$

First, we use integration to find $S(t)$:

$$\int S(t) = 60t^{1/2}\,dt$$
$$= 60\frac{t^{3/2}}{\frac{3}{2}} + C$$
$$= 40t^{3/2} + C$$

Since

$$S(0) = 40(0)^{3/2} + C = 27{,}000$$

we have $C = 27{,}000$ and

$$S(t) = 40t^{3/2} + 27{,}000$$

Now we solve the equation $S(t) = 41{,}000$ for t:

$$40t^{3/2} + 27{,}000 = 41{,}000$$
$$40t^{3/2} = 14{,}000$$
$$t^{3/2} = 350$$
$$t = 350^{2/3} \qquad \text{Use a calculator.}$$
$$= 49.664\,419\ldots$$

The advertising campaign should last approximately 50 days.

Matched Problem 6 There are 64,000 subscribers to an online fashion magazine. Due to competition from a new magazine, the number $C(t)$ of subscribers is expected to decrease at the rate of

$$C'(t) = -600t^{1/3}$$

subscribers per month, where t is the time in months since the new magazine began publication. How long will it take until the number of subscribers to the online fashion magazine drops to 46,000?

Exercises 4-1

A

In Problems 1–16, find each indefinite integral. Check by differentiating.

1. $\int 7\, dx$
2. $\int 10\, dx$
3. $\int 8x\, dx$
4. $\int 14x\, dx$
5. $\int 9x^2\, dx$
6. $\int 15x^2\, dx$
7. $\int x^5\, dx$
8. $\int x^8\, dx$
9. $\int x^{-3}\, dx$
10. $\int x^{-4}\, dx$
11. $\int 10x^{3/2}\, dx$
12. $\int 8x^{1/3}\, dx$
13. $\int \frac{3}{z}\, dz$
14. $\int \frac{7}{z}\, dz$
15. $\int 16e^u\, du$
16. $\int 5e^u\, du$

In Problems 17–24, find all the antiderivatives for each derivative.

17. $\frac{dy}{dx} = 200x^4$
18. $\frac{dx}{dt} = 42t^5$
19. $\frac{dP}{dx} = 24 - 6x$
20. $\frac{dy}{dx} = 3x^2 - 4x^3$
21. $\frac{dy}{dx} = e^x + 3$
22. $\frac{dy}{dx} = x - e^x$
23. $\frac{dx}{dt} = 5t^{-1} + 1$
24. $\frac{du}{dv} = \frac{4}{v} + \frac{v}{4}$

B

In Problems 25–30, discuss the validity of each statement. If the statement is always true, explain why. If not, give a counterexample.

25. The constant function $f(x) = \pi$ is an antiderivative of the constant function $k(x) = 0$.
26. The constant function $k(x) = 0$ is an antiderivative of the constant function $f(x) = \pi$.
27. If n is an integer, then $x^{n+1}/(n + 1)$ is an antiderivative of x^n.
28. The constant function $k(x) = 0$ is an antiderivative of itself.
29. The function $h(x) = 5e^x$ is an antiderivative of itself.
30. The constant function $g(x) = 5e^{\pi}$ is an antiderivative of itself.

In Problems 31–34, could the three graphs in each figure be antiderivatives of the same function? Explain.

31.

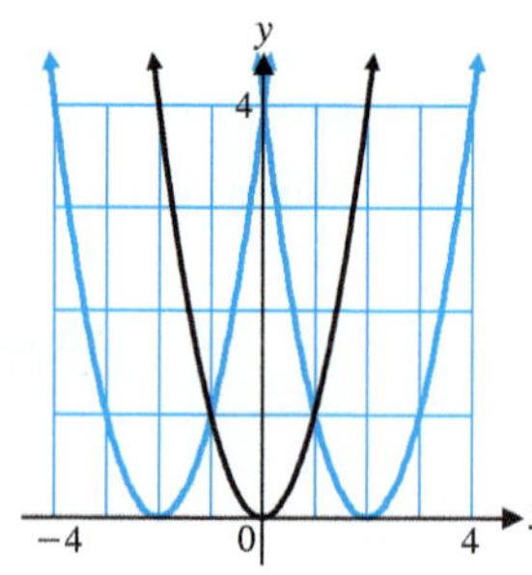

32.

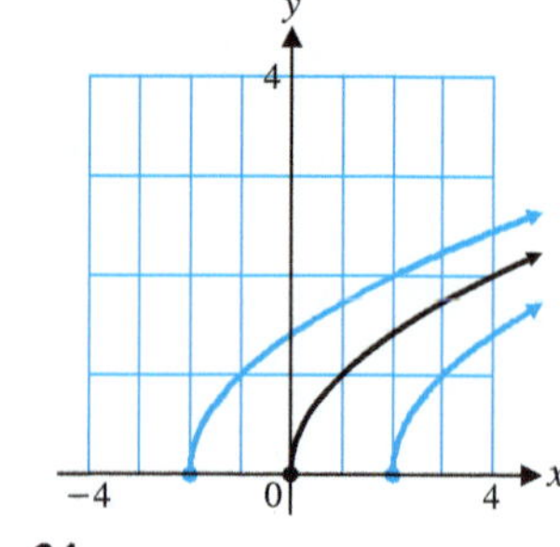

33.

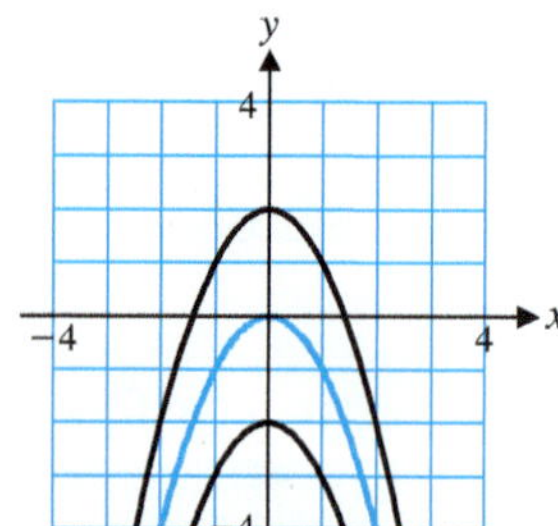

34.

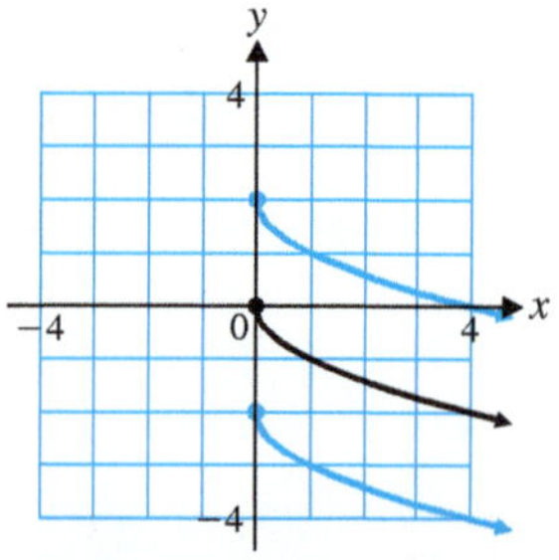

In Problems 35–50, find each indefinite integral. (Check by differentiation.)

35. $\int 5x(1 - x)\, dx$
36. $\int x^2(1 + x^3)\, dx$
37. $\int \frac{du}{\sqrt{u}}$
38. $\int \frac{dt}{\sqrt[3]{t}}$
39. $\int \frac{dx}{4x^3}$
40. $\int \frac{6\, dm}{m^2}$

41. $\int \frac{4 + u}{u} du$

42. $\int \frac{1 - y^2}{3y} dy$

43. $\int (5e^z + 4) dz$

44. $\int \frac{e^t - t}{2} dt$

45. $\int \left(3x^2 - \frac{2}{x^2}\right) dx$

46. $\int \left(4x^3 + \frac{2}{x^3}\right) dx$

47. $\int \left(3\sqrt{x} + \frac{2}{\sqrt{x}}\right) dx$

48. $\int \left(\frac{2}{\sqrt[3]{x}} - \sqrt[3]{x^2}\right) dx$

49. $\int \frac{e^x - 3x}{4} dx$

50. $\int \frac{e^x - 3x^2}{2} dx$

In Problems 51–58, find the particular antiderivative of each derivative that satisfies the given condition.

51. $C'(x) = 6x^2 - 4x; C(0) = 3{,}000$

52. $R'(x) = 600 - 0.6x; R(0) = 0$

53. $\frac{dx}{dt} = \frac{20}{\sqrt{t}}; x(1) = 40$

54. $\frac{dR}{dt} = \frac{100}{t^2}; R(1) = 400$

55. $\frac{dy}{dx} = 2x^{-2} + 3x^{-1} - 1; y(1) = 0$

56. $\frac{dy}{dx} = 3x^{-1} + x^{-2}; y(1) = 1$

57. $\frac{dx}{dt} = 4e^t - 2; x(0) = 1$

58. $\frac{dy}{dt} = 5e^t - 4; y(0) = -1$

59. Find the equation of the curve that passes through (2, 3) if its slope is given by

$$\frac{dy}{dx} = 4x - 3$$

for each x.

60. Find the equation of the curve that passes through (1, 3) if its slope is given by

$$\frac{dy}{dx} = 12x^2 - 12x$$

for each x.

C

In Problems 61–66, find each indefinite integral.

61. $\int \frac{2x^4 - x}{x^3} dx$

62. $\int \frac{x^{-1} - x^4}{x^2} dx$

63. $\int \frac{x^5 - 2x}{x^4} dx$

64. $\int \frac{1 - 3x^4}{x^2} dx$

65. $\int \frac{x^2 e^x - 2x}{x^2} dx$

66. $\int \frac{1 - xe^x}{x} dx$

In Problems 67–72, find the particular antiderivative of each derivative that satisfies the given condition.

67. $\frac{dM}{dt} = \frac{t^2 - 1}{t^2}; M(4) = 5$

68. $\frac{dR}{dx} = \frac{1 - x^4}{x^3}; R(1) = 4$

69. $\frac{dy}{dx} = \frac{5x + 2}{\sqrt[3]{x}}; y(1) = 0$

70. $\frac{dx}{dt} = \frac{\sqrt{t^3} - t}{\sqrt{t^3}}; x(9) = 4$

71. $p'(x) = -\frac{10}{x^2}; p(1) = 20$

72. $p'(x) = \frac{10}{x^3}; p(1) = 15$

In Problems 73–76, find the derivative or indefinite integral as indicated.

73. $\frac{d}{dx}\left(\int x^3 dx\right)$

74. $\frac{d}{dt}\left(\int \frac{\ln t}{t} dt\right)$

75. $\int \frac{d}{dx}(x^4 + 3x^2 + 1) dx$

76. $\int \frac{d}{du}(e^{u^2}) du$

77. Use differentiation to justify the formula

$$\int x^n dx = \frac{x^{n+1}}{n + 1} + C$$

provided that $n \neq -1$.

78. Use differentiation to justify the formula

$$\int e^x dx = e^x + C$$

79. Assuming that $x > 0$, use differentiation to justify the formula

$$\int \frac{1}{x} dx = \ln|x| + C$$

80. Assuming that $x < 0$, use differentiation to justify the formula

$$\int \frac{1}{x} dx = \ln|x| + C$$

81. Show that the indefinite integral of the sum of two functions is the sum of the indefinite integrals.

[*Hint*: Assume that $\int f(x) dx = F(x) + C_1$ and $\int g(x) dx = G(x) + C_2$. Using differentiation, show that $F(x) + C_1 + G(x) + C_2$ is the indefinite integral of the function $s(x) = f(x) + g(x)$.]

82. Show that the indefinite integral of the difference of two functions is the difference of the indefinite integrals.

Applications

83. Cost function. The marginal average cost of producing x sports watches is given by

$$\overline{C}'(x) = -\frac{1{,}000}{x^2} \qquad \overline{C}(100) = 25$$

where $\overline{C}(x)$ is the average cost in dollars. Find the average cost function and the cost function. What are the fixed costs?

84. Renewable energy. In 2007, U.S. consumption of renewable energy was 6.8 quadrillion Btu (or 6.8×10^{15} Btu). Since the 1960s, consumption has been growing at a rate (in quadrillion Btu's per year) given by

$$f'(t) = 0.004t + 0.062$$

where t is years after 1960. Find $f(t)$ and estimate U.S. consumption of renewable energy in 2020.

85. Production costs. The graph of the marginal cost function from the production of x thousand bottles of sunscreen per month [where cost $C(x)$ is in thousands of dollars per month] is given in the figure.

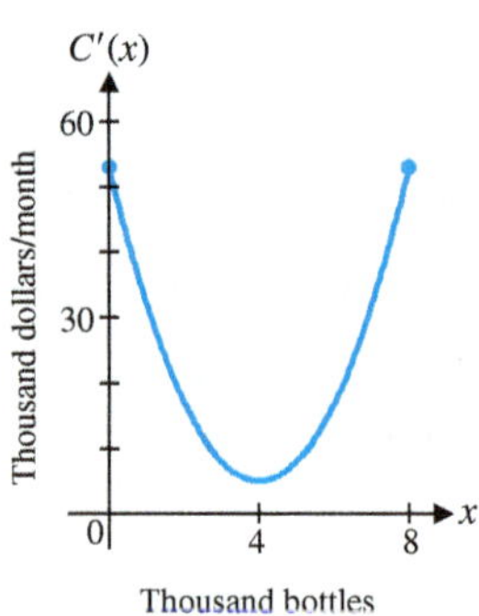

Figure for 85

(A) Using the graph shown, describe the shape of the graph of the cost function $C(x)$ as x increases from 0 to 8,000 bottles per month.

(B) Given the equation of the marginal cost function,

$$C'(x) = 3x^2 - 24x + 53$$

find the cost function if monthly fixed costs at 0 output are \$30,000. What is the cost of manufacturing 4,000 bottles per month? 8,000 bottles per month?

(C) Graph the cost function for $0 \le x \le 8$. [Check the shape of the graph relative to the analysis in part (A).]

(D) Why do you think that the graph of the cost function is steeper at both ends than in the middle?

86. Revenue. The graph of the marginal revenue function from the sale of x sports watches is given in the figure.

(A) Using the graph shown, describe the shape of the graph of the revenue function $R(x)$ as x increases from 0 to 1,000.

(B) Find the equation of the marginal revenue function (the linear function shown in the figure).

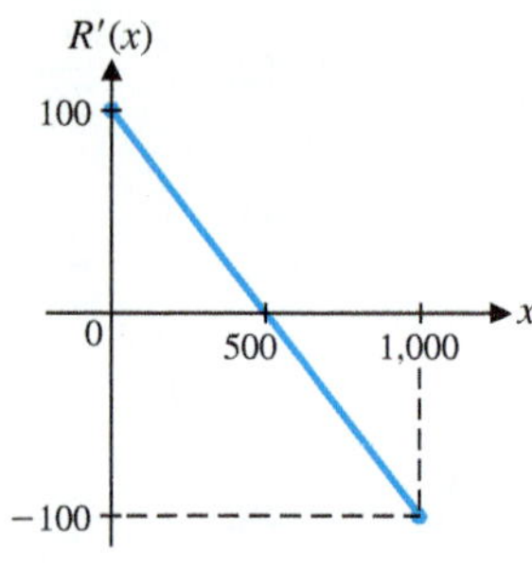

Figure for 86

(C) Find the equation of the revenue function that satisfies $R(0) = 0$. Graph the revenue function over the interval [0, 1,000]. [Check the shape of the graph relative to the analysis in part (A).]

(D) Find the price–demand equation and determine the price when the demand is 700 units.

87. Sales analysis. Monthly sales of an SUV model are expected to decline at the rate of

$$S'(t) = -24t^{1/3}$$

SUVs per month, where t is time in months and $S(t)$ is the number of SUVs sold each month. The company plans to stop manufacturing this model when monthly sales reach 300 SUVs. If monthly sales now ($t = 0$) are 1,200 SUVs, find $S(t)$. How long will the company continue to manufacture this model?

88. Sales analysis. The rate of change of the monthly sales of a newly released football game is given by

$$S'(t) = 500t^{1/4} \qquad S(0) = 0$$

where t is the number of months since the game was released and $S(t)$ is the number of games sold each month. Find $S(t)$. When will monthly sales reach 20,000 games?

 **89. Sales analysis.** Repeat Problem 87 if $S'(t) = -24t^{1/3} - 70$ and all other information remains the same. Use a graphing calculator to approximate the solution of the equation $S(t) = 800$ to two decimal places.

 **90. Sales analysis.** Repeat Problem 88 if $S'(t) = 500t^{1/4} + 300$ and all other information remains the same. Use a graphing calculator to approximate the solution of the equation $S(t) = 20{,}000$ to two decimal places.

91. Labor costs. A defense contractor is starting production on a new missile control system. On the basis of data collected during the assembly of the first 16 control systems, the production manager obtained the following function describing the rate of labor use:

$$g(x) = 2{,}400x^{-1/2}$$

$g(x)$ is the number of labor-hours required to assemble the xth unit of the control system. For example, after assembly of 16 units, the rate of assembly is 600 labor-hours per unit, and after assembly of 25 units, the rate of assembly is 480 labor-hours per unit. The more units assembled, the more efficient the process. If 19,200 labor-hours are required to assemble of the first 16 units, how many labor-hours $L(x)$ will be required to assemble the first x units? The first 25 units?

92. **Labor costs.** If the rate of labor use in Problem 91 is

$$g(x) = 2{,}000x^{-1/3}$$

and if the first 8 control units require 12,000 labor-hours, how many labor-hours, $L(x)$, will be required for the first x control units? The first 27 control units?

93. **Weight–height.** For an average person, the rate of change of weight W (in pounds) with respect to height h (in inches) is given approximately by

$$\frac{dW}{dh} = 0.0015h^2$$

Find $W(h)$ if $W(60) = 108$ pounds. Find the weight of an average person who is 5 feet, 10 inches, tall.

94. **Wound healing.** The area A of a healing wound changes at a rate given approximately by

$$\frac{dA}{dt} = -4t^{-3} \qquad 1 \le t \le 10$$

where t is time in days and $A(1) = 2$ square centimeters. What will the area of the wound be in 10 days?

95. **Urban growth.** The rate of growth of the population $N(t)$ of a new city t years after its incorporation is estimated to be

$$\frac{dN}{dt} = 400 + 600\sqrt{t} \qquad 0 \le t \le 9$$

If the population was 5,000 at the time of incorporation, find the population 9 years later.

96. **Learning.** A college language class was chosen for an experiment in learning. Using a list of 50 words, the experiment involved measuring the rate of vocabulary memorization at different times during a continuous 5-hour study session. It was found that the average rate of learning for the entire class was inversely proportional to the time spent studying and was given approximately by

$$V'(t) = \frac{15}{t} \qquad 1 \le t \le 5$$

If the average number of words memorized after 1 hour of study was 15 words, what was the average number of words memorized after t hours of study for $1 \le t \le 5$? After 4 hours of study? Round answer to the nearest whole number.

Answers to Matched Problems

1. (A) $x^3 + C$

(B)

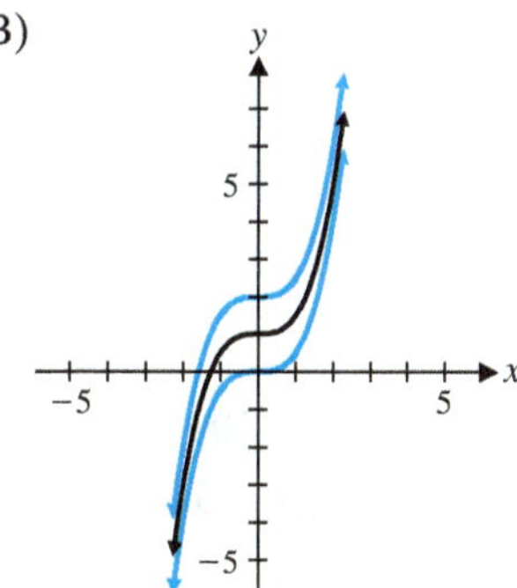

(C) The graphs are vertical translations of each other.

2. (A) $2x + C$

(B) $16e^t + C$

(C) $\frac{3}{5}x^5 + C$

(D) $\frac{1}{3}x^6 - x^3 + x + C$

(E) $5\ln|x| - 4e^x + C$

3. (A) $\frac{6}{5}x^{5/3} + x^{-3} + C$

(B) $\frac{5}{2}w^{8/5} + C$

(C) $\frac{1}{3}x^3 - 4x^2 + C$

(D) $6x^{4/3} - 12x^{1/2} + C$

(E) $\frac{1}{4}x^4 + x^3 - x^2 - 6x + C$

4. $y = x^3 - 2$

5. $R(x) = 400x - 0.2x^2$; $R(1{,}000) = \$200{,}000$

6. $t = (40)^{3/4} \approx 16$ mo

4-2 Integration by Substitution

- Reversing the Chain Rule
- Integration by Substitution
- Additional Substitution Techniques
- Application

Many of the indefinite integral formulas introduced in the preceding section are based on corresponding derivative formulas studied earlier. We now consider indefinite integral formulas and procedures based on the chain rule for differentiation.

Reversing the Chain Rule

Recall the chain rule:

$$\frac{d}{dx}f[g(x)] = f'[g(x)]g'(x)$$

The expression on the right is formed from the expression on the left by taking the derivative of the outside function f and multiplying it by the derivative of the inside function g. If we recognize an integrand as a chain-rule form $f'[g(x)]g'(x)$, we can easily find an antiderivative and its indefinite integral:

$$\int f'[g(x)]g'(x)\,dx = f[g(x)] + C \tag{1}$$

We are interested in finding the indefinite integral

$$\int 3x^2e^{x^3-1}\,dx \tag{2}$$

The integrand appears to be the chain-rule form $e^{g(x)}g'(x)$, which is the derivative of $e^{g(x)}$. Since

$$\frac{d}{dx}e^{x^3-1} = 3x^2e^{x^3-1}$$

it follows that

$$\int 3x^2e^{x^3-1}\,dx = e^{x^3-1} + C \tag{3}$$

How does the following indefinite integral differ from integral (2)?

$$\int x^2e^{x^3-1}\,dx \tag{4}$$

It is missing the constant factor 3. That is, $x^2e^{x^3-1}$ is within a constant factor of being the derivative of e^{x^3-1}. But because a constant factor can be moved across the integral sign, this causes us little trouble in finding the indefinite integral of $x^2e^{x^3-1}$. We introduce the constant factor 3 and at the same time multiply by $\frac{1}{3}$ and move the $\frac{1}{3}$ factor outside the integral sign. This is equivalent to multiplying the integrand in integral (4) by 1:

$$\int x^2e^{x^3-1}\,dx = \int \frac{3}{3}x^2e^{x^3-1}\,dx \tag{5}$$

$$= \frac{1}{3}\int 3x^2e^{x^3-1}\,dx = \frac{1}{3}e^{x^3-1} + C$$

The derivative of the rightmost side of equation (5) is the integrand of the indefinite integral (4). Check this.

How does the following indefinite integral differ from integral (2)?

$$\int 3xe^{x^3-1}\,dx \tag{6}$$

It is missing a variable factor x. This is more serious. As tempting as it might be, we *cannot* adjust integral (6) by introducing the variable factor x and moving $1/x$ outside the integral sign, as we did with the constant 3 in equation (5).

CAUTION A constant factor can be moved across an integral sign, but a variable factor cannot.

There is nothing wrong with educated guessing when you are looking for an antiderivative of a given function. You have only to check the result by differentiation. If you are right, you go on your way; if you are wrong, you simply try another approach.

In Section 2-4, we saw that the chain rule extends the derivative formulas for x^n, e^x, and $\ln x$ to derivative formulas for $[f(x)]^n$, $e^{f(x)}$, and $\ln[f(x)]$. The chain rule can also be used to extend the indefinite integral formulas discussed in Section 4-1. Some general formulas are summarized in the following box:

FORMULAS General Indefinite Integral Formulas

1. $\int [f(x)]^n f'(x)\,dx = \frac{[f(x)]^{n+1}}{n+1} + C, n \neq -1$

2. $\int e^{f(x)} f'(x)\,dx = e^{f(x)} + C$

3. $\int \frac{1}{f(x)} f'(x)\,dx = \ln|f(x)| + C$

We can verify each formula by using the chain rule to show that the derivative of the function on the right is the integrand on the left. For example,

$$\frac{d}{dx}[e^{f(x)} + C] = e^{f(x)} f'(x)$$

verifies formula 2.

EXAMPLE 1 **Reversing the Chain Rule**

(A) $\int (3x+4)^{10}(3)\,dx = \frac{(3x+4)^{11}}{11} + C$ Formula 1 with $f(x) = 3x + 4$ and $f'(x) = 3$

Check:

$$\frac{d}{dx}\frac{(3x+4)^{11}}{11} = 11\frac{(3x+4)^{10}}{11}\frac{d}{dx}(3x+4) = (3x+4)^{10}(3)$$

(B) $\int e^{x^2}(2x)\,dx = e^{x^2} + C$ Formula 2 with $f(x) = x^2$ and $f'(x) = 2x$

Check:

$$\frac{d}{dx}e^{x^2} = e^{x^2}\frac{d}{dx}x^2 = e^{x^2}(2x)$$

(C) $\int \frac{1}{1+x^3}3x^2\,dx = \ln|1+x^3| + C$ Formula 3 with $f(x) = 1 + x^3$ and $f'(x) = 3x^2$

Check:

$$\frac{d}{dx}\ln|1+x^3| = \frac{1}{1+x^3}\frac{d}{dx}(1+x^3) = \frac{1}{1+x^3}3x^2$$

Matched Problem 1 Find each indefinite integral.

(A) $\int (2x^3-3)^{20}(6x^2)\,dx$ (B) $\int e^{5x}(5)\,dx$

(C) $\int \frac{1}{4+x^2}2x\,dx$

Integration by Substitution

The key step in using formulas 1, 2, and 3 is recognizing the form of the integrand. Some people find it difficult to identify $f(x)$ and $f'(x)$ in these formulas and prefer to use a *substitution* to simplify the integrand. The *method of substitution,* which we now discuss, becomes increasingly useful as one progresses in studies of integration.

We start by introducing the idea of the *differential.* We represented the derivative by the symbol dy/dx taken as a whole. We now define dy and dx as two separate quantities with the property that their ratio is still equal to $f'(x)$:

DEFINITION Differentials

If $y = f(x)$ defines a differentiable function, then

1. The **differential** dx of the independent variable x is an arbitrary real number.
2. The **differential** dy of the dependent variable y is defined as the product of $f'(x)$ and dx:

$$dy = f'(x)\,dx$$

Differentials involve mathematical subtleties that are treated carefully in advanced mathematics courses. Here, we are interested in them mainly as a bookkeeping device to aid in the process of finding indefinite integrals. We can always check an indefinite integral by differentiating.

EXAMPLE 2 **Differentials**

(A) If $y = f(x) = x^2$, then

$$dy = f'(x)\,dx = 2x\,dx$$

(B) If $u = g(x) = e^{3x}$, then

$$du = g'(x)\,dx = 3e^{3x}\,dx$$

(C) If $w = h(t) = \ln(4 + 5t)$, then

$$dw = h'(t)\,dt = \frac{5}{4 + 5t}\,dt$$

Matched Problem 2 (A) Find dy for $y = f(x) = x^3$.
(B) Find du for $u = h(x) = \ln(2 + x^2)$.
(C) Find dv for $v = g(t) = e^{-5t}$.

The **method of substitution** is developed through Examples 3–6.

EXAMPLE 3 **Using Substitution** Find $\int (x^2 + 2x + 5)^5(2x + 2)\,dx$.

SOLUTION If

$$u = x^2 + 2x + 5$$

then the differential of u is

$$du = (2x + 2)\,dx$$

Notice that du is one of the factors in the integrand. Substitute u for $x^2 + 2x + 5$ and du for $(2x + 2)\,dx$ to obtain

$$\begin{aligned}\int (x^2 + 2x + 5)^5(2x + 2)\,dx &= \int u^5\,du\\ &= \frac{u^6}{6} + C\\ &= \frac{1}{6}(x^2 + 2x + 5)^6 + C \quad \text{Since } u = x^2 + 2x + 5\end{aligned}$$

Check:

$$\frac{d}{dx}\frac{1}{6}(x^2 + 2x + 5)^6 = \frac{1}{6}(6)(x^2 + 2x + 5)^5 \frac{d}{dx}(x^2 + 2x + 5)$$
$$= (x^2 + 2x + 5)^5(2x + 2)$$

Matched Problem 3 Find $\int (x^2 - 3x + 7)^4(2x - 3)\,dx$ by substitution.

The substitution method is also called the **change-of-variable method** since u replaces the variable x in the process. Substituting $u = f(x)$ and $du = f'(x)\,dx$ in formulas 1, 2, and 3 produces the general indefinite integral formulas 4, 5, and 6:

FORMULAS General Indefinite Integral Formulas

4. $\displaystyle\int u^n\,du = \frac{u^{n+1}}{n + 1} + C, \qquad n \neq -1$

5. $\displaystyle\int e^u\,du = e^u + C$

6. $\displaystyle\int \frac{1}{u}\,du = \ln|u| + C$

These formulas are valid if u is an independent variable, or if u is a function of another variable and du is the differential of u with respect to that variable.

The substitution method for evaluating certain indefinite integrals is outlined as follows:

PROCEDURE Integration by Substitution

Step 1 Select a substitution that appears to simplify the integrand. In particular, try to select u so that du is a factor in the integrand.

Step 2 Express the integrand entirely in terms of u and du, completely eliminating the original variable and its differential.

Step 3 Evaluate the new integral if possible.

Step 4 Express the antiderivative found in step 3 in terms of the original variable.

EXAMPLE 4 **Using Substitution** Use a substitution to find each indefinite integral.

(A) $\displaystyle\int (3x + 4)^6(3)\,dx$

(B) $\displaystyle\int e^{t^2}(2t)\,dt$

SOLUTION (A) If we let $u = 3x + 4$, then $du = 3\,dx$, and

$$\int (3x + 4)^6(3)\,dx = \int u^6\,du \quad \text{Use formula 4.}$$
$$= \frac{u^7}{7} + C$$
$$= \frac{(3x + 4)^7}{7} + C \quad \text{Since } u = 3x + 4$$

Check:

$$\frac{d}{dx}\frac{(3x + 4)^7}{7} = \frac{7(3x + 4)^6}{7}\frac{d}{dx}(3x + 4) = (3x + 4)^6(3)$$

(B) If we let $u = t^2$, then $du = 2t\,dt$, and

$$\int e^{t^2}(2t)\,dt = \int e^u\,du \quad \text{Use formula 5.}$$
$$= e^u + C$$
$$= e^{t^2} + C \quad \text{Since } u = t^2$$

Check:

$$\frac{d}{dt}e^{t^2} = e^{t^2}\frac{d}{dt}t^2 = e^{t^2}(2t)$$

Matched Problem 4 Use a substitution to find each indefinite integral.

(A) $\int (2x^3 - 3)^4(6x^2)\,dx$ (B) $\int e^{5w}(5)\,dw$

Additional Substitution Techniques

In order to use the substitution method, **the integrand must be expressed entirely in terms of u and du.** In some cases, the integrand must be modified before making a substitution and using one of the integration formulas. Example 5 illustrates this process.

EXAMPLE 5 **Substitution Techniques** Integrate.

(A) $\int \frac{1}{4x + 7}\,dx$ (B) $\int te^{-t^2}\,dt$

(C) $\int 4x^2\sqrt{x^3 + 5}\,dx$

SOLUTION (A) If $u = 4x + 7$, then $du = 4\,dx$. We are missing a factor of 4 in the integrand to match formula 6 exactly. Recalling that a constant factor can be moved across an integral sign, we proceed as follows:

$$\int \frac{1}{4x + 7}dx = \int \frac{1}{4x + 7}\frac{4}{4}\,dx$$
$$= \frac{1}{4}\int \frac{1}{4x + 7}4\,dx \quad \text{Substitute } u = 4x + 7 \text{ and } du = 4\,dx.$$
$$= \frac{1}{4}\int \frac{1}{u}\,du \quad \text{Use formula 6.}$$
$$= \tfrac{1}{4}\ln|u| + C$$
$$= \tfrac{1}{4}\ln|4x + 7| + C \quad \text{Since } u = 4x + 7$$

Check:

$$\frac{d}{dx}\frac{1}{4}\ln|4x+7| = \frac{1}{4}\frac{1}{4x+7}\frac{d}{dx}(4x+7) = \frac{1}{4}\frac{1}{4x+7}4 = \frac{1}{4x+7}$$

(B) If $u = -t^2$, then $du = -2t\,dt$. Proceed as in part (A):

$$\int te^{-t^2}\,dt = \int e^{-t^2}\frac{-2}{-2}t\,dt$$

$$= -\frac{1}{2}\int e^{-t^2}(-2t)\,dt \qquad \text{Substitute } u = -t^2 \text{ and } du = -2t\,dt.$$

$$= -\frac{1}{2}\int e^{u}\,du \qquad \text{Use formula 5.}$$

$$= -\tfrac{1}{2}e^{u} + C$$

$$= -\tfrac{1}{2}e^{-t^2} + C \qquad \text{Since } u = -t^2$$

Check:

$$\frac{d}{dt}\left(-\tfrac{1}{2}e^{-t^2}\right) = -\tfrac{1}{2}e^{-t^2}\frac{d}{dt}(-t^2) = -\tfrac{1}{2}e^{-t^2}(-2t) = te^{-t^2}$$

(C) $$\int 4x^2\sqrt{x^3+5}\,dx = 4\int\sqrt{x^3+5}(x^2)\,dx \qquad \text{Move the 4 across the integral sign and proceed as before.}$$

$$= 4\int\sqrt{x^3+5}\frac{3}{3}(x^2)\,dx$$

$$= \frac{4}{3}\int\sqrt{x^3+5}(3x^2)\,dx \qquad \text{Substitute } u = x^3 + 5 \text{ and } du = 3x^2\,dx.$$

$$= \frac{4}{3}\int\sqrt{u}\,du$$

$$= \frac{4}{3}\int u^{1/2}\,du \qquad \text{Use formula 4.}$$

$$= \frac{4}{3}\frac{u^{3/2}}{\frac{3}{2}} + C$$

$$= \tfrac{8}{9}u^{3/2} + C$$

$$= \tfrac{8}{9}(x^3+5)^{3/2} + C \qquad \text{Since } u = x^3 + 5$$

Check:

$$\frac{d}{dx}\left[\tfrac{8}{9}(x^3+5)^{3/2}\right] = \tfrac{4}{3}(x^3+5)^{1/2}\frac{d}{dx}(x^3+5)$$

$$= \tfrac{4}{3}(x^3+5)^{1/2}(3x^2) = 4x^2\sqrt{x^3+5}$$

Matched Problem 5 Integrate.

(A) $\int e^{-3x}\,dx$ (B) $\int \frac{x}{x^2-9}\,dx$

(C) $\int 5t^2(t^3+4)^{-2}\,dt$

Even if it is not possible to find a substitution that makes an integrand match one of the integration formulas exactly, a substitution may simplify the integrand sufficiently so that other techniques can be used.

EXAMPLE 6 **Substitution Techniques** Find $\int \frac{x}{\sqrt{x+2}}\,dx$.

SOLUTION Proceeding as before, if we let $u = x + 2$, then $du = dx$ and

$$\int \frac{x}{\sqrt{x+2}}\,dx = \int \frac{x}{\sqrt{u}}\,du$$

Notice that this substitution is not complete because we have not expressed the integrand entirely in terms of u and du. As we noted earlier, only a constant factor can be moved across an integral sign, so we cannot move x outside the integral sign. Instead, we must return to the original substitution, solve for x in terms of u, and use the resulting equation to complete the substitution:

$$u = x + 2 \qquad \text{Solve for } x \text{ in terms of } u.$$

$$u - 2 = x \qquad \text{Substitute this expression for } x.$$

Thus,

$$\begin{aligned}
\int \frac{x}{\sqrt{x+2}}\,dx &= \int \frac{u-2}{\sqrt{u}}\,du \qquad \text{Simplify the integrand.}\\
&= \int \frac{u-2}{u^{1/2}}\,du\\
&= \int (u^{1/2} - 2u^{-1/2})\,du\\
&= \int u^{1/2}\,du - 2\int u^{-1/2}\,du\\
&= \frac{u^{3/2}}{\frac{3}{2}} - 2\frac{u^{1/2}}{\frac{1}{2}} + C\\
&= \tfrac{2}{3}(x+2)^{3/2} - 4(x+2)^{1/2} + C \qquad \text{Since } u = x + 2
\end{aligned}$$

Check:

$$\begin{aligned}
\frac{d}{dx}\left[\tfrac{2}{3}(x+2)^{3/2} - 4(x+2)^{1/2}\right] &= (x+2)^{1/2} - 2(x+2)^{-1/2}\\
&= \frac{x+2}{(x+2)^{1/2}} - \frac{2}{(x+2)^{1/2}}\\
&= \frac{x}{(x+2)^{1/2}}
\end{aligned}$$

Matched Problem 6 Find $\int x\sqrt{x+1}\,dx$.

We can find the indefinite integral of some functions in more than one way. For example, we can use substitution to find

$$\int x(1 + x^2)^2\,dx$$

by letting $u = 1 + x^2$. As a second approach, we can expand the integrand, obtaining

$$\int (x + 2x^3 + x^5)\,dx$$

for which we can easily calculate an antiderivative. In such a case, choose the approach that you prefer.

There are also some functions for which substitution is not an effective approach to finding the indefinite integral. For example, substitution is not helpful in finding

$$\int e^{x^2}\,dx \qquad \text{or} \qquad \int \ln x\,dx$$

Application

EXAMPLE 7 **Price–Demand** The market research department of a supermarket chain has determined that, for one store, the marginal price $p'(x)$ at x tubes per week for a certain brand of toothpaste is given by

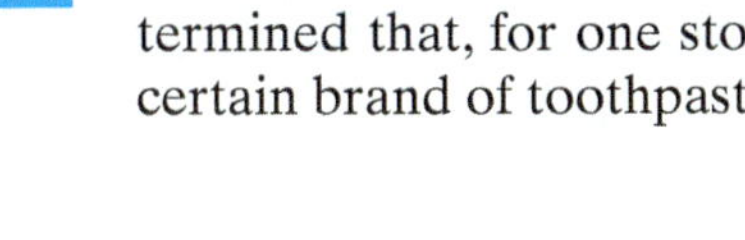

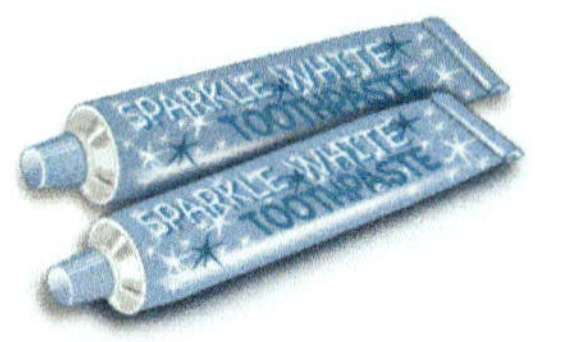

$$p'(x) = -0.015e^{-0.01x}$$

Find the price–demand equation if the weekly demand is 50 tubes when the price of a tube is \$4.35. Find the weekly demand when the price of a tube is \$3.89.

SOLUTION

$$\begin{aligned} p(x) &= \int -0.015e^{-0.01x}\,dx \\ &= -0.015\int e^{-0.01x}\,dx \\ &= -0.015\int e^{-0.01x}\frac{-0.01}{-0.01}\,dx \\ &= \frac{-0.015}{-0.01}\int e^{-0.01x}(-0.01)\,dx && \text{Substitute } u = -0.01x \text{ and } du = -0.01\,dx. \\ &= 1.5\int e^{u}\,du \\ &= 1.5e^{u} + C \\ &= 1.5e^{-0.01x} + C && \text{Since } u = -0.01x \end{aligned}$$

We find C by noting that

$$\begin{aligned} p(50) = 1.5e^{-0.01(50)} + C &= \$4.35 \\ C &= \$4.35 - 1.5e^{-0.5} && \text{Use a calculator.} \\ &= \$4.35 - 0.91 \\ &= \$3.44 \end{aligned}$$

So,

$$p(x) = 1.5e^{-0.01x} + 3.44$$

To find the demand when the price is \$3.89, we solve $p(x) = \$3.89$ for x:

$$\begin{aligned} 1.5e^{-0.01x} + 3.44 &= 3.89 \\ 1.5e^{-0.01x} &= 0.45 \\ e^{-0.01x} &= 0.3 \\ -0.01x &= \ln 0.3 \\ x &= -100 \ln 0.3 \approx 120 \text{ tubes} \end{aligned}$$

Matched Problem 7 The marginal price $p'(x)$ at a supply level of x tubes per week for a certain brand of toothpaste is given by

$$p'(x) = 0.001e^{0.01x}$$

Find the price–supply equation if the supplier is willing to supply 100 tubes per week at a price of \$3.65 each. How many tubes would the supplier be willing to supply at a price of \$3.98 each?

We conclude with two final cautions. The first was stated earlier, but it is worth repeating.

CAUTION

1. A variable cannot be moved across an integral sign.
2. An integral must be expressed entirely in terms of u and du before applying integration formulas 4, 5, and 6.

Exercises 4-2

A

In Problems 1–36, find each indefinite integral and check the result by differentiating.

1. $\int (3x + 5)^2(3)\,dx$

2. $\int (6x - 1)^3(6)\,dx$

3. $\int (x^2 - 1)^5(2x)\,dx$

4. $\int (x^6 + 1)^4(6x^5)\,dx$

5. $\int (5x^3 + 1)^{-3}(15x^2)\,dx$

6. $\int (4x^2 - 3)^{-6}(8x)\,dx$

7. $\int e^{5x}(5)\,dx$

8. $\int e^{x^3}(3x^2)\,dx$

9. $\int \frac{1}{1 + x^2}(2x)\,dx$

10. $\int \frac{1}{5x - 7}(5)\,dx$

11. $\int \sqrt{1 + x^4}(4x^3)\,dx$

12. $\int (x^2 + 9)^{-1/2}(2x)\,dx$

B

13. $\int (x + 3)^{10}\,dx$

14. $\int (x - 3)^{-4}\,dx$

15. $\int (6t - 7)^{-2}\,dt$

16. $\int (5t + 1)^3\,dt$

17. $\int (t^2 + 1)^5\,t\,dt$

18. $\int (t^3 + 4)^{-2}\,t^2\,dt$

19. $\int xe^{x^2}\,dx$

20. $\int e^{-0.01x}\,dx$

21. $\int \frac{1}{5x + 4}\,dx$

22. $\int \frac{x}{1 + x^2}\,dx$

23. $\int e^{1-t}\,dt$

24. $\int \frac{3}{2 - t}\,dt$

25. $\int \frac{t}{(3t^2 + 1)^4}\,dt$

26. $\int \frac{t^2}{(t^3 - 2)^5}\,dt$

27. $\int x\sqrt{x + 4}\,dx$

28. $\int x\sqrt{x - 9}\,dx$

29. $\int \frac{x}{\sqrt{x - 3}}\,dx$

30. $\int \frac{x}{\sqrt{x + 5}}\,dx$

31. $\int x(x - 4)^9\,dx$

32. $\int x(x + 6)^8\,dx$

33. $\int e^{2x}(1 + e^{2x})^3\,dx$

34. $\int e^{-x}(1 - e^{-x})^4\,dx$

35. $\int \frac{1 + x}{4 + 2x + x^2}\,dx$

36. $\int \frac{x^2 - 1}{x^3 - 3x + 7}\,dx$

In Problems 37–42, the indefinite integral can be found in more than one way. First use the substitution method to find the indefinite integral. Then find it without using substitution. Check that your answers are equivalent.

37. $\int 5(5x + 3)\,dx$

38. $\int -7(4 - 7x)\,dx$

39. $\int 2x(x^2 - 1)\,dx$

40. $\int 3x^2(x^3 + 1)\,dx$

41. $\int 5x^4(x^5)^4\,dx$

42. $\int 8x^7(x^8)^3\,dx$

In Problems 43–48, suppose that the indicated "solutions" were given to you by a student whom you are tutoring.

(A) How would you have the student check each solution?

(B) Is the solution right or wrong? If the solution is wrong, explain what is wrong and how it can be corrected.

(C) Show a correct solution for each incorrect solution, and check the result by differentiation.

43. $\int \frac{1}{2x - 3}\,dx = \ln|2x - 3| + C$

44. $\int \frac{x}{x^2 + 5}\,dx = \ln|x^2 + 5| + C$

45. $\int x^3 e^{x^4}\,dx = e^{x^4} + C$

46. $\int e^{4x-5}\,dx = e^{4x-5} + C$

47. $\int 2(x^2-2)^2\,dx = \dfrac{(x^2-2)^3}{3x} + C$

48. $\int (-10x)(x^2-3)^{-4}\,dx = (x^2-3)^{-5} + C$

C

In Problems 49–60, find each indefinite integral and check the result by differentiating.

49. $\int x\sqrt{3x^2+7}\,dx$

50. $\int x^2\sqrt{2x^3+1}\,dx$

51. $\int x(x^3+2)^2\,dx$

52. $\int x(x^2+2)^2\,dx$

53. $\int x^2(x^3+2)^2\,dx$

54. $\int (x^2+2)^2\,dx$

55. $\int \dfrac{x^3}{\sqrt{2x^4+3}}\,dx$

56. $\int \dfrac{x^2}{\sqrt{4x^3-1}}\,dx$

57. $\int \dfrac{(\ln x)^3}{x}\,dx$

58. $\int \dfrac{e^x}{1+e^x}\,dx$

59. $\int \dfrac{1}{x^2}e^{-1/x}\,dx$

60. $\int \dfrac{1}{x\ln x}\,dx$

In Problems 61–66, find the antiderivative of each derivative.

61. $\dfrac{dx}{dt} = 7t^2(t^3+5)^6$

62. $\dfrac{dm}{dn} = 10n(n^2-8)^7$

63. $\dfrac{dy}{dt} = \dfrac{3t}{\sqrt{t^2-4}}$

64. $\dfrac{dy}{dx} = \dfrac{5x^2}{(x^3-7)^4}$

65. $\dfrac{dp}{dx} = \dfrac{e^x+e^{-x}}{(e^x-e^{-x})^2}$

66. $\dfrac{dm}{dt} = \dfrac{\ln(t-5)}{t-5}$

Applications

67. **Price–demand equation.** The marginal price for a weekly demand of x bottles of shampoo in a drugstore is given by

$$p'(x) = \frac{-6{,}000}{(3x+50)^2}$$

Find the price–demand equation if the weekly demand is 150 when the price of a bottle of shampoo is \$8. What is the weekly demand when the price is \$6.50?

68. **Price–supply equation.** The marginal price at a supply level of x bottles of shampoo per week is given by

$$p'(x) = \frac{300}{(3x+25)^2}$$

Find the price–supply equation if the distributor of the shampoo is willing to supply 75 bottles a week at a price of \$5.00 per bottle. How many bottles would the supplier be willing to supply at a price of \$5.15 per bottle?

69. **Cost function.** The weekly marginal cost of producing x pairs of tennis shoes is given by

$$C'(x) = 12 + \frac{500}{x+1}$$

where $C(x)$ is cost in dollars. If the fixed costs are \$2,000 per week, find the cost function. What is the average cost per pair of shoes if 1,000 pairs of shoes are produced each week?

70. **Revenue function.** The weekly marginal revenue from the sale of x pairs of tennis shoes is given by

$$R'(x) = 40 - 0.02x + \frac{200}{x+1} \qquad R(0) = 0$$

where $R(x)$ is revenue in dollars. Find the revenue function. Find the revenue from the sale of 1,000 pairs of shoes.

71. **Marketing.** An automobile company is ready to introduce a new line of hybrid cars through a national sales campaign. After test marketing the line in a carefully selected city, the marketing research department estimates that sales (in millions of dollars) will increase at the monthly rate of

$$S'(t) = 10 - 10e^{-0.1t} \qquad 0 \le t \le 24$$

t months after the campaign has started.

(A) What will be the total sales $S(t)$ t months after the beginning of the national campaign if we assume no sales at the beginning of the campaign?

(B) What are the estimated total sales for the first 12 months of the campaign?

(C) When will the estimated total sales reach \$100 million? Use a graphing calculator to approximate the answer to two decimal places.

72. **Marketing.** Repeat Problem 71 if the monthly rate of increase in sales is found to be approximated by

$$S'(t) = 20 - 20e^{-0.05t} \qquad 0 \le t \le 24$$

73. **Oil production.** Using production and geological data, the management of an oil company estimates that oil will be pumped from a field producing at a rate given by

$$R(t) = \frac{100}{t+1} + 5 \qquad 0 \le t \le 20$$

where $R(t)$ is the rate of production (in thousands of barrels per year) t years after pumping begins. How many barrels of oil $Q(t)$ will the field produce in the first t years if $Q(0) = 0$? How many barrels will be produced in the first 9 years?

74. **Oil production.** Assume that the rate in Problem 73 is found to be

$$R(t) = \frac{120t}{t^2+1} + 3 \qquad 0 \le t \le 20$$

(A) When is the rate of production greatest?

(B) How many barrels of oil $Q(t)$ will the field produce in the first t years if $Q(0) = 0$? How many barrels will be produced in the first 5 years?

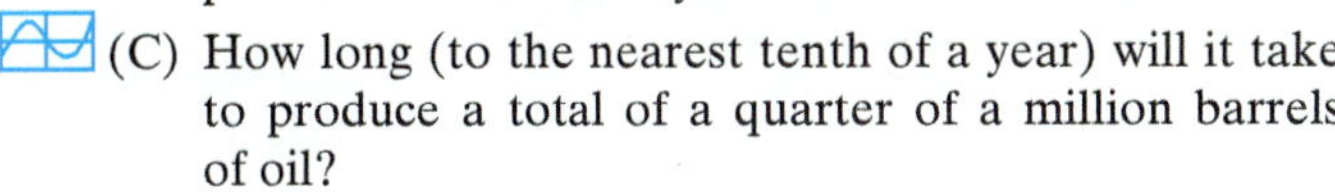

(C) How long (to the nearest tenth of a year) will it take to produce a total of a quarter of a million barrels of oil?

75. Biology. A yeast culture is growing at the rate of $w'(t) = 0.2e^{0.1t}$ grams per hour. If the starting culture weighs 2 grams, what will be the weight of the culture $W(t)$ after t hours? After 8 hours?

76. Medicine. The rate of healing for a skin wound (in square centimeters per day) is approximated by $A'(t) = -0.9e^{-0.1t}$. If the initial wound has an area of 9 square centimeters, what will its area $A(t)$ be after t days? After 5 days?

77. Pollution. A contaminated lake is treated with a bactericide. The rate of increase in harmful bacteria t days after the treatment is given by

$$\frac{dN}{dt} = -\frac{2{,}000t}{1 + t^2} \qquad 0 \le t \le 10$$

where $N(t)$ is the number of bacteria per milliliter of water. Since dN/dt is negative, the count of harmful bacteria is decreasing.

(A) Find the minimum value of dN/dt.

(B) If the initial count was 5,000 bacteria per milliliter, find $N(t)$ and then find the bacteria count after 10 days.

(C) When (to two decimal places) is the bacteria count 1,000 bacteria per milliliter?

78. Pollution. An oil tanker aground on a reef is losing oil and producing an oil slick that is radiating outward at a rate given approximately by

$$\frac{dR}{dt} = \frac{60}{\sqrt{t + 9}} \qquad t \ge 0$$

where R is the radius (in feet) of the circular slick after t minutes. Find the radius of the slick after 16 minutes if the radius is 0 when $t = 0$.

79. Learning. An average student enrolled in an advanced typing class progressed at a rate of $N'(t) = 6e^{-0.1t}$ words per minute per week t weeks after enrolling in a 15-week course. If, at the beginning of the course, a student could type 40 words per minute, how many words per minute $N(t)$ would the student be expected to type t weeks into the course? After completing the course?

80. Learning. An average student enrolled in a stenotyping class progressed at a rate of $N'(t) = 12e^{-0.06t}$ words per minute per week t weeks after enrolling in a 15-week course. If, at the beginning of the course, a student could stenotype at zero words per minute, how many words per minute $N(t)$ would the student be expected to handle t weeks into the course? After completing the course?

81. College enrollment. The projected rate of increase in enrollment at a new college is estimated by

$$\frac{dE}{dt} = 5{,}000(t + 1)^{-3/2} \qquad t \ge 0$$

where $E(t)$ is the projected enrollment in t years. If enrollment is 2,000 now ($t = 0$), find the projected enrollment 15 years from now.

Answers to Matched Problems

1. (A) $\frac{1}{21}(2x^3 - 3)^{21} + C$
(B) $e^{5x} + C$
(C) $\ln|4 + x^2| + C$ or $\ln(4 + x^2) + C$, since $4 + x^2 > 0$

2. (A) $dy = 3x^2\,dx$
(B) $du = \dfrac{2x}{2 + x^2}dx$
(C) $dv = -5e^{-5t}\,dt$

3. $\frac{1}{5}(x^2 - 3x + 7)^5 + C$

4. (A) $\frac{1}{5}(2x^3 - 3)^5 + C$
(B) $e^{5w} + C$

5. (A) $-\frac{1}{3}e^{-3x} + C$
(B) $\frac{1}{2}\ln|x^2 - 9| + C$
(C) $-\frac{5}{3}(t^3 + 4)^{-1} + C$

6. $\frac{2}{5}(x + 1)^{5/2} - \frac{2}{3}(x + 1)^{3/2} + C$

7. $p(x) = 0.1e^{0.01x} + 3.38$; 179 tubes

4-3 Differential Equations; Growth and Decay

- Differential Equations and Slope Fields
- Continuous Compound Interest Revisited
- Exponential Growth Law
- Population Growth, Radioactive Decay, and Learning
- Comparison of Exponential Growth Phenomena

In the preceding section, we considered equations of the form

$$\frac{dy}{dx} = 6x^2 - 4x \qquad p'(x) = -400e^{-0.04x}$$

These are examples of *differential equations.* In general, an equation is a **differential equation** if it involves an unknown function and one or more of its derivatives. Other examples of differential equations are

$$\frac{dy}{dx} = ky \qquad y'' - xy' + x^2 = 5 \qquad \frac{dy}{dx} = 2xy$$

The first and third equations are called **first-order** equations because each involves a first derivative but no higher derivative. The second equation is called a **second-**

order equation because it involves a second derivative and no higher derivatives. Finding solutions of different types of differential equations (functions that satisfy the equation) is the subject matter of entire books and courses on this topic. Here, we consider only a few special first-order equations that have immediate and significant applications.

We start by looking at some first-order equations geometrically, in terms of *slope fields.* We then consider continuous compound interest as modeled by a first-order differential equation. From this treatment, we can generalize our approach to a wide variety of other types of growth phenomena.

Differential Equations and Slope Fields

We introduce the concept of *slope field* through an example. Consider the first-order differential equation

$$\frac{dy}{dx} = 0.2y \qquad (1)$$

A function f is a solution of equation (1) if $y = f(x)$ satisfies equation (1) for all values of x in the domain of f. Geometrically interpreted, equation (1) gives us the slope of a solution curve that passes through the point (x, y). For example, if $y = f(x)$ is a solution of equation (1) that passes through the point $(0, 2)$, then the slope of f at $(0, 2)$ is given by

$$\frac{dy}{dx} = 0.2(2) = 0.4$$

We indicate this relationship by drawing a short segment of the tangent line at the point $(0, 2)$, as shown in Figure 1A. The procedure is repeated for points $(-3, 1)$ and $(2, 3)$. Assuming that the graph of f passes through all three points, we sketch an approximate graph of f in Figure 1B.

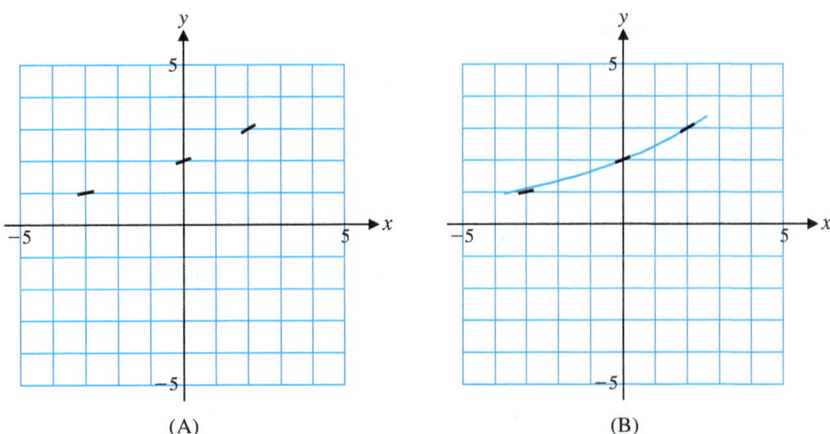

Figure 1

If we continue the process of drawing tangent line segments at each point grid in Figure 1—a task easily handled by computers, but not by hand—we obtain a *slope field.* A slope field for differential equation (1), drawn by a computer, is shown in Figure 2. In general, a **slope field** for a first-order differential equation is obtained by drawing tangent line segments determined by the equation at each point in a grid.

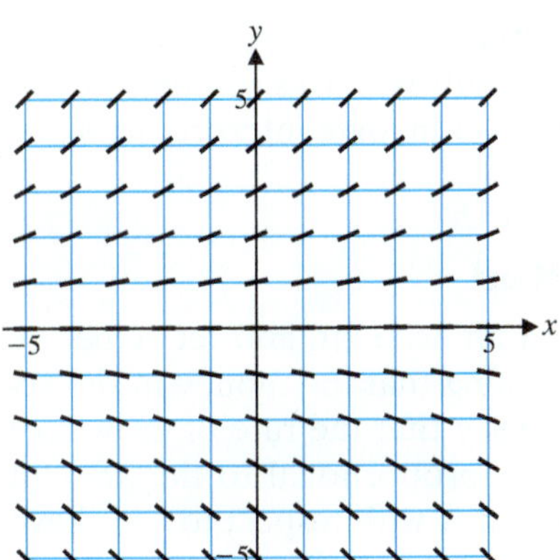

Figure 2

Explore & Discuss 1

(A) In Figure 1A (or a copy), draw tangent line segments for a solution curve of differential equation (1) that passes through $(-3, -1)$, $(0, -2)$, and $(2, -3)$.

(B) In Figure 1B (or a copy), sketch an approximate graph of the solution curve that passes through the three points given in part (A). Repeat the tangent line segments first.

(C) What type of function, of all the elementary functions discussed in the first two chapters, appears to be a solution of differential equation (1)?

In Explore & Discuss 1, if you guessed that all solutions of equation (1) are exponential functions, you are to be congratulated. We now show that

$$y = Ce^{0.2x} \tag{2}$$

is a solution of equation (1) for any real number C. We substitute $y = Ce^{0.2x}$ into equation (1) to see if the left side is equal to the right side for all real x:

$$\frac{dy}{dx} = 0.2y$$

$$\textit{Left side:} \quad \frac{dy}{dx} = \frac{d}{dx}(Ce^{0.2x}) = 0.2Ce^{0.2x}$$

$$\textit{Right side:} \quad 0.2y = 0.2Ce^{0.2x}$$

So equation (2) is a solution of equation (1) for C any real number. Which values of C will produce solution curves that pass through $(0, 2)$ and $(0, -2)$, respectively? Substituting the coordinates of each point into equation (2) and solving for C, we obtain

$$y = 2e^{0.2x} \quad \text{and} \quad y = -2e^{0.2x} \tag{3}$$

The graphs of equations (3) are shown in Figure 3, and they confirm the results shown in Figure 1B.

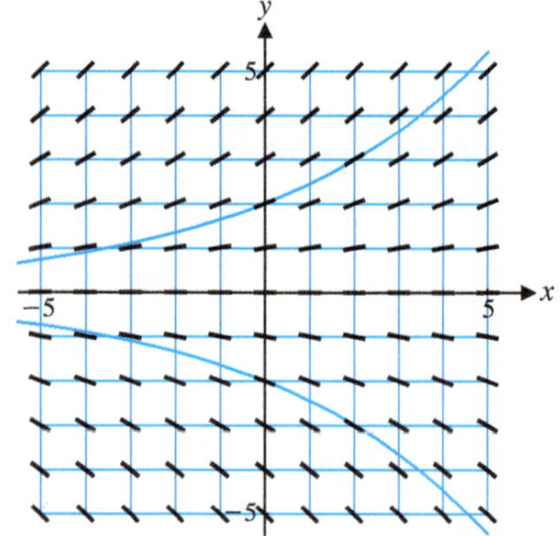

Figure 3

CONCEPTUAL INSIGHT

For a complicated first-order differential equation, say,

$$\frac{dy}{dx} = \frac{3 + \sqrt{xy}}{x^2 - 5y^4}$$

it may be impossible to find a formula analogous to (2) for its solutions. Nevertheless, it is routine to evaluate the right-hand side at each point in a grid. The resulting slope field provides a graphical representation of the solutions of the differential equation.

Drawing slope fields by hand is not a task for human beings: A 20-by-20 grid would require drawing 400 tangent line segments! Repetitive tasks of this type are what computers are for. A few problems in Exercise 4-3 involve interpreting slope fields, not drawing them.

Continuous Compound Interest Revisited

Let P be the initial amount of money deposited in an account, and let A be the amount in the account at any time t. Instead of assuming that the money in the account earns a particular rate of interest, suppose we say that the rate of growth of the amount of money in the account at any time t is proportional to the amount present at that time. Since dA/dt is the rate of growth of A with respect to t, we have

$$\frac{dA}{dt} = rA \qquad A(0) = P \qquad A, P > 0 \tag{4}$$

where r is an appropriate constant. We would like to find a function $A = A(t)$ that satisfies these conditions. Multiplying both sides of equation (4) by $1/A$, we obtain

$$\frac{1}{A}\frac{dA}{dt} = r$$

Now we integrate each side with respect to t:

$$\int \frac{1}{A}\frac{dA}{dt}dt = \int r\,dt \qquad \frac{dA}{dt}dt = A'(t)dt = dA$$

$$\int \frac{1}{A}dA = \int r\,dt$$

$$\ln|A| = rt + C \qquad |A| = A, \text{ since } A > 0$$

$$\ln A = rt + C$$

We convert this last equation into the equivalent exponential form

$$A = e^{rt+C} \qquad \text{Definition of logarithmic function: } y = \ln x \text{ if and only if } x = e^y$$

$$= e^C e^{rt} \qquad \text{Property of exponents: } b^m b^n = b^{m+n}$$

Since $A(0) = P$, we evaluate $A(t) = e^C e^{rt}$ at $t = 0$ and set the result equal to P:

$$A(0) = e^C e^0 = e^C = P$$

Hence, $e^C = P$, and we can rewrite $A = e^C e^{rt}$ in the form

$$A = Pe^{rt}$$

This is the same continuous compound interest formula obtained in Section 3-1, where the principal P is invested at an annual nominal rate r compounded continuously for t years.

Exponential Growth Law

In general, if the rate of change of a quantity Q with respect to time is proportional to the amount of Q present and $Q(0) = Q_0$, then, proceeding in exactly the same way as we just did, we obtain the following theorem:

THEOREM 1 Exponential Growth Law

If $\frac{dQ}{dt} = rQ$ and $Q(0) = Q_0$, then $Q = Q_0 e^{rt}$,

where

Q_0 = amount of Q at $t = 0$
r = relative growth rate (expressed as a decimal)
t = time
Q = quantity at time t

The constant r in the exponential growth law is called the **relative growth rate**. If the relative growth rate is $r = 0.02$, then the quantity Q is growing at a rate $dQ/dt = 0.02Q$ (that is, 2% of the quantity Q per unit of time t). Note the distinction between the relative growth rate r and the rate of growth dQ/dt of the quantity Q. If $r < 0$, then $dQ/dt < 0$ and Q is decreasing. This type of growth is called **exponential decay**.

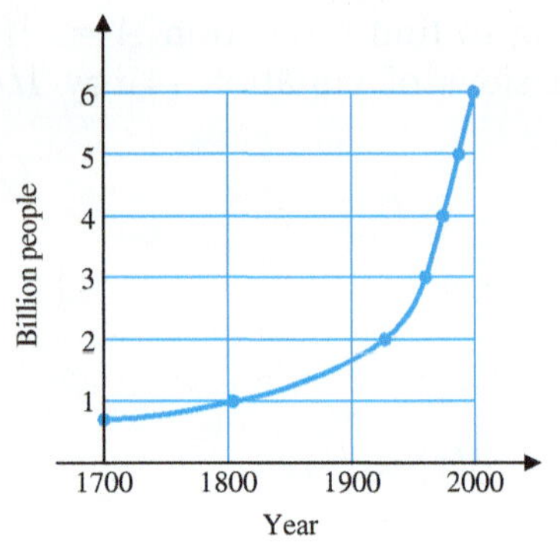

Figure 4 World population growth

Once we know that the rate of growth is proportional to the amount present, we recognize exponential growth and can use Theorem 1 without solving the differential equation each time. The exponential growth law applies not only to money invested at interest compounded continuously, but also to many other types of problems—population growth, radioactive decay, the depletion of a natural resource, and so on.

Population Growth, Radioactive Decay, and Learning

The world population passed 1 billion in 1804, 2 billion in 1927, 3 billion in 1960, 4 billion in 1974, 5 billion in 1987, and 6 billion in 1999, as illustrated in Figure 4. **Population growth** over certain periods often can be approximated by the exponential growth law of Theorem 1.

EXAMPLE 1 **Population Growth** India had a population of about 1.2 billion in 2010 ($t = 0$). Let P represent the population (in billions) t years after 2010, and assume a growth rate of 1.5% compounded continuously.

(A) Find an equation that represents India's population growth after 2010, assuming that the 1.5% growth rate continues.

(B) What is the estimated population (to the nearest tenth of a billion) of India in the year 2030?

(C) Graph the equation found in part (A) from 2000 to 2030.

SOLUTION (A) The exponential growth law applies, and we have

$$\frac{dP}{dt} = 0.015P \qquad P(0) = 1.2$$

Therefore,

$$P = 1.2e^{0.015t} \tag{5}$$

(B) Using equation (5), we can estimate the population in India in 2030 ($t = 20$):

$$P = 1.2e^{0.015(20)} = 1.6 \text{ billion people}$$

(C) The graph is shown in Figure 5.

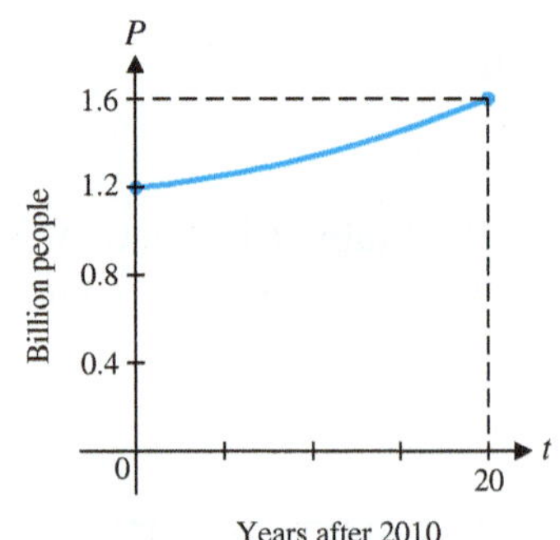

Figure 5 Population of India

Matched Problem 1 Assuming the same continuous compound growth rate as in Example 1, what will India's population be (to the nearest tenth of a billion) in the year 2020?

EXAMPLE 2 **Population Growth** If the exponential growth law applies to Canada's population growth, at what continuous compound growth rate will the population double over the next 100 years?

SOLUTION We must find r, given that $P = 2P_0$ and $t = 100$:

$$P = P_0e^{rt}$$

$$2P_0 = P_0e^{100r}$$

$$2 = e^{100r}$$

Take the natural logarithm of both sides and reverse the equation.

$$100r = \ln 2$$

$$r = \frac{\ln 2}{100}$$

$$\approx 0.0069 \quad \text{or} \quad 0.69\%$$

Matched Problem 2 If the exponential growth law applies to population growth in Nigeria, find the doubling time (to the nearest year) of the population if it grows at 2.1% per year compounded continuously.

We now turn to another type of exponential growth: **radioactive decay**. In 1946, Willard Libby (who later received a Nobel Prize in chemistry) found that as long as a plant or animal is alive, radioactive carbon-14 is maintained at a constant level in its tissues. Once the plant or animal is dead, however, the radioactive carbon-14 diminishes by radioactive decay at a rate proportional to the amount present.

$$\frac{dQ}{dt} = rQ \qquad Q(0) = Q_0$$

This is another example of the exponential growth law. The continuous compound rate of decay for radioactive carbon-14 is 0.000 123 8, so $r = -0.000\,123\,8$, since decay implies a negative continuous compound growth rate.

EXAMPLE 3 **Archaeology** A human bone fragment was found at an archaeological site in Africa. If 10% of the original amount of radioactive carbon-14 was present, estimate the age of the bone (to the nearest 100 years).

SOLUTION By the exponential growth law for

$$\frac{dQ}{dt} = -0.000\,123\,8Q \qquad Q(0) = Q_0$$

we have

$$Q = Q_0e^{-0.0001238t}$$

We must find t so that $Q = 0.1Q_0$ (since the amount of carbon-14 present now is 10% of the amount Q_0 present at the death of the person).

$$0.1Q_0 = Q_0e^{-0.0001238t}$$

$$0.1 = e^{-0.0001238t}$$

$$\ln 0.1 = \ln e^{-0.0001238t}$$

$$t = \frac{\ln 0.1}{-0.000\,123\,8} \approx 18{,}600 \text{ years}$$

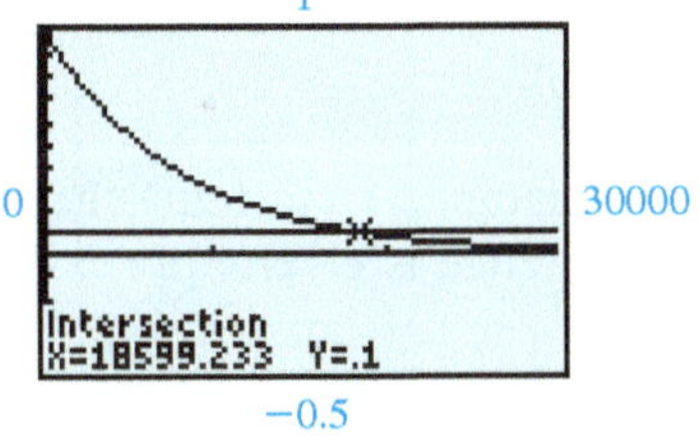

Figure 6 $y_1 = e^{-0.0001238x}$; $y_2 = 0.1$

See Figure 6 for a graphical solution to Example 3.

Matched Problem 3 Estimate the age of the bone in Example 3 (to the nearest 100 years) if 50% of the original amount of carbon-14 is present.

In learning certain skills, such as typing and swimming, one often assumes that there is a maximum skill attainable—say, M—and the rate of improvement is proportional to the difference between what has been achieved y and the maximum attainable M. Mathematically,

$$\frac{dy}{dt} = k(M - y) \qquad y(0) = 0$$

We solve this type of problem with the same technique used to obtain the exponential growth law. First, multiply both sides of the first equation by $1/(M - y)$ to get

$$\frac{1}{M - y}\frac{dy}{dt} = k$$

and then integrate each side with respect to t:

$$\int \frac{1}{M - y}\frac{dy}{dt}\,dt = \int k\,dt$$

$$-\int \frac{1}{M - y}\left(-\frac{dy}{dt}\right)dt = \int k\,dt \qquad \text{Substitute } u = M - y \text{ and}$$

$$-\int \frac{1}{u}\,du = \int k\,dt \qquad du = -dy = -\frac{dy}{dt}dt.$$

$$-\ln|u| = kt + C \qquad \text{Substitute } M - y, \text{ which is } > 0, \text{ for } u.$$

$$-\ln(M - y) = kt + C \qquad \text{Multiply both sides by } -1.$$

$$\ln(M - y) = -kt - C$$

Change this last equation to an equivalent exponential form:

$$M - y = e^{-kt-C}$$

$$M - y = e^{-C}e^{-kt}$$

$$y = M - e^{-C}e^{-kt}$$

Now, $y(0) = 0$; hence,

$$y(0) = M - e^{-C}e^{0} = 0$$

Solving for e^{-C}, we obtain

$$e^{-C} = M$$

and our final solution is

$$y = M - Me^{-kt} = M(1 - e^{-kt})$$

EXAMPLE 4 **Learning** For a particular person learning to swim, the distance y (in feet) that the person is able to swim in 1 minute after t hours of practice is given approximately by

$$y = 50(1 - e^{-0.04t})$$

What is the rate of improvement (to two decimal places) after 10 hours of practice?

SOLUTION

$$y = 50 - 50e^{-0.04t}$$

$$y'(t) = 2e^{-0.04t}$$

$$y'(10) = 2e^{-0.04(10)} \approx 1.34 \text{ feet per hour of practice}$$

Matched Problem 4 In Example 4, what is the rate of improvement (to two decimal places) after 50 hours of practice?

Comparison of Exponential Growth Phenomena

The graphs and equations given in Table 1 compare several widely used growth models. These models are divided into two groups: unlimited growth and limited growth. Following each equation and graph is a short (and necessarily incomplete) list of areas in which the models are used.

Table 1 Exponential Growth

Description	Model	Solution	Graph	Uses
Unlimited growth: Rate of growth is proportional to the amount present	$\frac{dy}{dt} = ky$ $k, t > 0$ $y(0) = c$	$y = ce^{kt}$		• Short-term population growth (people, bacteria, etc.) • Growth of money at continuous compound interest • Price–supply curves
Exponential decay: Rate of growth is proportional to the amount present	$\frac{dy}{dt} = -ky$ $k, t > 0$ $y(0) = c$	$y = ce^{-kt}$		• Depletion of natural resources • Radioactive decay • Absorption of light in water • Price–demand curves • Atmospheric pressure (t is altitude)
Limited growth: Rate of growth is proportional to the difference between the amount present and a fixed limit	$\frac{dy}{dt} = k(M - y)$ $k, t > 0$ $y(0) = 0$	$y = M(1 - e^{-kt})$		• Sales fads (for example, skateboards) • Depreciation of equipment • Company growth • Learning
Logistic growth: Rate of growth is proportional to the amount present and to the difference between the amount present and a fixed limit	$\frac{dy}{dt} = ky(M - y)$ $k, t > 0$ $y(0) = \frac{M}{1 + c}$	$y = \frac{M}{1 + ce^{-kMt}}$		• Long-term population growth • Epidemics • Sales of new products • Spread of a rumor • Company growth

Exercises 4-3

A

In Problems 1–12, find the general or particular solution, as indicated, for each differential equation.

1. $\frac{dy}{dx} = 6x$

2. $\frac{dy}{dx} = 3x^{-2}$

3. $\frac{dy}{dx} = \frac{7}{x}$

4. $\frac{dy}{dx} = e^{0.1x}$

5. $\frac{dy}{dx} = e^{0.02x}$

6. $\frac{dy}{dx} = 8x^{-1}$

7. $\frac{dy}{dx} = x^2 - x;\ y(0) = 0$

8. $\frac{dy}{dx} = \sqrt{x};\ y(0) = 0$

9. $\frac{dy}{dx} = -2xe^{-x^2};\ y(0) = 3$

10. $\frac{dy}{dx} = e^{x-3};\ y(3) = -5$

11. $\frac{dy}{dx} = \frac{2}{1 + x};\ y(0) = 5$

12. $\frac{dy}{dx} = \frac{1}{4(3 - x)};\ y(0) = 1$

B

Problems 13–18 refer to the following slope fields:

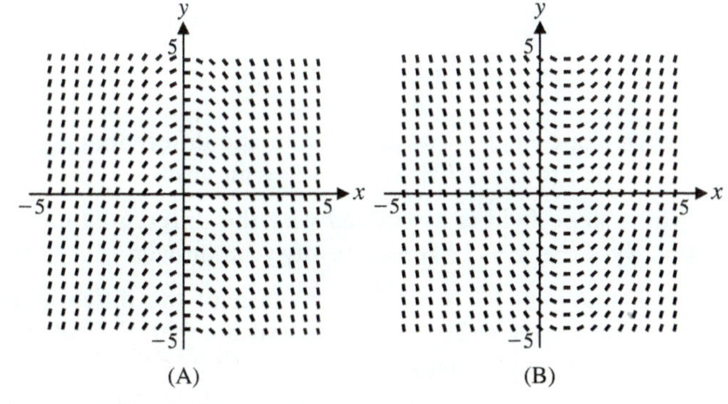

Figure for 13–18

13. Which slope field is associated with the differential equation $dy/dx = x - 1$? Briefly justify your answer.

14. Which slope field is associated with the differential equation $dy/dx = -x$? Briefly justify your answer.

15. Solve the differential equation $dy/dx = x - 1$ and find the particular solution that passes through $(0, -2)$.

16. Solve the differential equation $dy/dx = -x$ and find the particular solution that passes through $(0, 3)$.

17. Graph the particular solution found in Problem 15 in the appropriate figure A or B (or a copy).

18. Graph the particular solution found in Problem 16 in the appropriate figure A or B (or a copy).

In Problems 19–26, find the general or particular solution, as indicated, for each differential equation.

19. $\dfrac{dy}{dt} = 2y$

20. $\dfrac{dy}{dt} = -3y$

21. $\dfrac{dy}{dx} = -0.5y;\ y(0) = 100$

22. $\dfrac{dy}{dx} = 0.1y;\ y(0) = -2.5$

23. $\dfrac{dx}{dt} = -5x$

24. $\dfrac{dx}{dt} = 4t$

25. $\dfrac{dx}{dt} = -5t$

26. $\dfrac{dx}{dt} = 4x$

C

Problems 27–34 refer to the following slope fields:

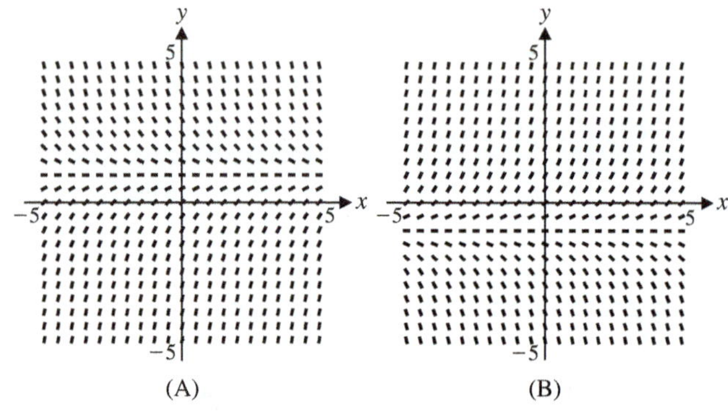

Figure for 27–34

27. Which slope field is associated with the differential equation $dy/dx = 1 - y$? Briefly justify your answer.

28. Which slope field is associated with the differential equation $dy/dx = y + 1$? Briefly justify your answer.

29. Show that $y = 1 - Ce^{-x}$ is a solution of the differential equation $dy/dx = 1 - y$ for any real number C. Find the particular solution that passes through $(0, 0)$.

30. Show that $y = Ce^{x} - 1$ is a solution of the differential equation $dy/dx = y + 1$ for any real number C. Find the particular solution that passes through $(0, 0)$.

31. Graph the particular solution found in Problem 29 in the appropriate figure, Figure A or Figure B (or a copy).

32. Graph the particular solution found in Problem 30 in the appropriate figure A or B (or a copy).

33. Use a graphing calculator to graph $y = 1 - Ce^{-x}$ for $C = -2, -1, 1$, and 2, for $-5 \le x \le 5$, $-5 \le y \le 5$, all in the same viewing window. Observe how the solution curves go with the flow of the tangent line segments in the corresponding slope field shown in figure A or figure B.

34. Use a graphing calculator to graph $y = Ce^{x} - 1$ for $C = -2, -1, 1$, and 2, for $-5 \le x \le 5$, $-5 \le y \le 5$, all in the same viewing window. Observe how the solution curves go with the flow of the tangent line segments in the corresponding slope field shown in figure A or figure B.

35. Show that $y = \sqrt{C - x^2}$ is a solution of the differential equation $dy/dx = -x/y$ for any positive real number C. Find the particular solution that passes through $(3, 4)$.

36. Show that $y = \sqrt{x^2 + C}$ is a solution of the differential equation $dy/dx = x/y$ for any real number C. Find the particular solution that passes through $(-6, 7)$.

37. Show that $y = Cx$ is a solution of the differential equation $dy/dx = y/x$ for any real number C. Find the particular solution that passes through $(-8, 24)$.

38. Show that $y = C/x$ is a solution of the differential equation $dy/dx = -y/x$ for any real number C. Find the particular solution that passes through $(2, 5)$.

39. Show that $y = 1/(1 + ce^{-t})$ is a solution of the differential equation $dy/dt = y(1 - y)$ for any real number c. Find the particular solution that passes through $(0, -1)$.

40. Show that $y = 2/(1 + ce^{-6t})$ is a solution of the differential equation $dy/dt = 3y(2 - y)$ for any real number c. Find the particular solution that passes through $(0, 1)$.

In Problems 41–48, use a graphing calculator to graph the given examples of the various cases in Table 1 on page 255.

41. Unlimited growth:

$y = 1{,}000e^{0.08t}$
$0 \le t \le 15$
$0 \le y \le 3{,}500$

42. Unlimited growth:

$y = 5{,}250e^{0.12t}$
$0 \le t \le 10$
$0 \le y \le 20{,}000$

43. Exponential decay:

$p = 100e^{-0.05x}$
$0 \le x \le 30$
$0 \le p \le 100$

44. Exponential decay:

$p = 1{,}000e^{-0.08x}$
$0 \le x \le 40$
$0 \le p \le 1{,}000$

45. Limited growth:

$N = 100(1 - e^{-0.05t})$
$0 \le t \le 100$
$0 \le N \le 100$

46. Limited growth:

$$N = 1{,}000(1 - e^{-0.07t})$$
$$0 \le t \le 70$$
$$0 \le N \le 1{,}000$$

47. Logistic growth:

$$N = \frac{1{,}000}{1 + 999e^{-0.4t}}$$
$$0 \le t \le 40$$
$$0 \le N \le 1{,}000$$

48. Logistic growth:

$$N = \frac{400}{1 + 99e^{-0.4t}}$$
$$0 \le t \le 30$$
$$0 \le N \le 400$$

49. Show that the rate of logistic growth, $dy/dt = ky(M - y)$, has its maximum value when $y = M/2$.

50. Find the value of t for which the logistic function

$$y = \frac{M}{1 + ce^{-kMt}}$$

is equal to $M/2$.

51. Let $Q(t)$ denote the population of the world at time t. In 1999, the world population was 6.0 billion and increasing at 1.3% per year; in 2009, it was 6.8 billion and increasing at 1.2% per year. In which year, 1999 or 2009, was dQ/dt (the rate of growth of Q with respect to t) greater? Explain.

52. Refer to Problem 51. Explain why the world population function $Q(t)$ does not satisfy an exponential growth law.

Applications

53. Continuous compound interest. Find the amount A in an account after t years if

$$\frac{dA}{dt} = 0.03A \quad \text{and} \quad A(0) = 1{,}000$$

54. Continuous compound interest. Find the amount A in an account after t years if

$$\frac{dA}{dt} = 0.02A \quad \text{and} \quad A(0) = 5{,}250$$

55. Continuous compound interest. Find the amount A in an account after t years if

$$\frac{dA}{dt} = rA \quad A(0) = 8{,}000 \quad A(2) = 8{,}260.14$$

56. Continuous compound interest. Find the amount A in an account after t years if

$$\frac{dA}{dt} = rA \quad A(0) = 5{,}000 \quad A(5) = 5{,}581.39$$

57. Price–demand. The marginal price dp/dx at x units of demand per week is proportional to the price p. There is no weekly demand at a price of \$100 per unit $[p(0) = 100]$, and there is a weekly demand of 5 units at a price of \$77.88 per unit $[p(5) = 77.88]$.

(A) Find the price–demand equation.

(B) At a demand of 10 units per week, what is the price?

(C) Graph the price–demand equation for $0 \le x \le 25$.

58. Price–supply. The marginal price dp/dx at x units of supply per day is proportional to the price p. There is no supply at a price of \$10 per unit $[p(0) = 10]$, and there is a daily supply of 50 units at a price of \$12.84 per unit $[p(50) = 12.84]$.

(A) Find the price–supply equation.

(B) At a supply of 100 units per day, what is the price?

(C) Graph the price–supply equation for $0 \le x \le 250$.

59. Advertising. A company is trying to expose a new product to as many people as possible through TV ads. Suppose that the rate of exposure to new people is proportional to the number of those who have not seen the product out of L possible viewers. No one is aware of the product at the start of the campaign, and after 10 days, 40% of L are aware of the product. Mathematically,

$$\frac{dN}{dt} = k(L - N) \quad N(0) = 0 \quad N(10) = 0.4L$$

(A) Solve the differential equation.

(B) What percent of L will have been exposed after 5 days of the campaign?

(C) How many days will it take to expose 80% of L?

(D) Graph the solution found in part (A) for $0 \le t \le 90$.

60. Advertising. Suppose that the differential equation for Problem 59 is

$$\frac{dN}{dt} = k(L - N) \quad N(0) = 0 \quad N(10) = 0.1L$$

(A) Explain what the equation $N(10) = 0.1L$ means.

(B) Solve the differential equation.

(C) How many days will it take to expose 50% of L?

(D) Graph the solution found in part (B) for $0 \le t \le 300$.

61. Biology. For relatively clear bodies of water, the intensity of light is reduced according to

$$\frac{dI}{dx} = -kI \quad I(0) = I_0$$

where I is the intensity of light at x feet below the surface. For the Sargasso Sea off the West Indies, $k = 0.00942$. Find I in terms of x, and find the depth at which the light is reduced to half of that at the surface.

62. Blood pressure. Under certain assumptions, the blood pressure P in the largest artery in the human body (the aorta) changes between beats with respect to time t according to

$$\frac{dP}{dt} = -aP \quad P(0) = P_0$$

where a is a constant. Find $P = P(t)$ that satisfies both conditions.

63. Drug concentration. A single injection of a drug is administered to a patient. The amount Q in the body then decreases at a rate proportional to the amount present. For a particular drug, the rate is 4% per hour. Thus,

$$\frac{dQ}{dt} = -0.04Q \quad Q(0) = Q_0$$

where t is time in hours.

(A) If the initial injection is 3 milliliters [$Q(0) = 3$], find $Q = Q(t)$ satisfying both conditions.

(B) How many milliliters (to two decimal places) are in the body after 10 hours?

(C) How many hours (to two decimal places) will it take for only 1 milliliter of the drug to be left in the body?

(D) Graph the solution found in part (A).

64. Simple epidemic. A community of 1,000 people is homogeneously mixed. One person who has just returned from another community has influenza. Assume that the home community has not had influenza shots and all are susceptible. One mathematical model assumes that influenza tends to spread at a rate in direct proportion to the number N who have the disease and to the number $1{,}000 - N$ who have not yet contracted the disease. Mathematically,

$$\frac{dN}{dt} = kN(1{,}000 - N) \qquad N(0) = 1$$

where N is the number of people who have contracted influenza after t days. For $k = 0.0004$, $N(t)$ is given by

$$N(t) = \frac{1{,}000}{1 + 999e^{-0.4t}}$$

(A) How many people have contracted influenza after 10 days? After 20 days?

(B) How many days will it take until half the community has contracted influenza?

(C) Find $\lim_{t\to\infty} N(t)$.

(D) Graph $N = N(t)$ for $0 \le t \le 30$.

65. Nuclear accident. One of the dangerous radioactive isotopes detected after the Chernobyl nuclear disaster in 1986 was cesium-137. If 93.3% of the cesium-137 emitted during the disaster was still present 3 years later, find the continuous compound rate of decay of this isotope.

66. Insecticides. Many countries have banned the use of the insecticide DDT because of its long-term adverse effects. Five years after a particular country stopped using DDT, the amount of DDT in the ecosystem had declined to 75% of the amount present at the time of the ban. Find the continuous compound rate of decay of DDT.

67. Archaeology. A skull found in an ancient tomb has 5% of the original amount of radioactive carbon-14 present. Estimate the age of the skull. (See Example 3.)

68. Learning. For a person learning to type, the number N of words per minute that the person could type after t hours of practice was given by

$$N = 100(1 - e^{-0.02t})$$

What is the rate of improvement after 10 hours of practice? After 40 hours of practice?

69. Small-group analysis. In a study on small-group dynamics, sociologists found that when 10 members of a discussion group were ranked according to the number of times each participated, the number $N(k)$ of times that the kth-ranked person participated was given by

$$N(k) = N_1 e^{-0.11(k-1)} \qquad 1 \le k \le 10$$

where N_1 is the number of times that the first-ranked person participated in the discussion. If $N_1 = 180$, in a discussion group of 10 people, estimate how many times the sixth-ranked person participated. How about the 10th-ranked person?

70. Perception. The Weber–Fechner law concerns a person's sensed perception of various strengths of stimulation involving weights, sound, light, shock, taste, and so on. One form of the law states that the rate of change of sensed sensation S with respect to stimulus R is inversely proportional to the strength of the stimulus R. So

$$\frac{dS}{dR} = \frac{k}{R}$$

where k is a constant. If we let R_0 be the threshold level at which the stimulus R can be detected (the least amount of sound, light, weight, and so on, that can be detected), then

$$S(R_0) = 0$$

Find a function S in terms of R that satisfies these conditions.

71. Rumor propagation. Sociologists have found that a rumor tends to spread at a rate in direct proportion to the number x who have heard it and to the number $P - x$ who have not, where P is the total population. If a resident of a 400-student dormitory hears a rumor that there is a case of TB on campus, then $P = 400$ and

$$\frac{dx}{dt} = 0.001x(400 - x) \qquad x(0) = 1$$

where t is time (in minutes). From these conditions, it can be shown that

$$x(t) = \frac{400}{1 + 399e^{-0.4t}}$$

(A) How many people have heard the rumor after 5 minutes? after 20 minutes?

(B) Find $\lim_{t\to\infty} x(t)$.

(C) Graph $x = x(t)$ for $0 \le t \le 30$.

72. Rumor propagation. In Problem 71, how long (to the nearest minute) will it take for half of the group of 400 to have heard the rumor?

Answers to Matched Problems

1. 1.4 billion people
2. 33 yr
3. 5,600 yr
4. 0.27 ft/hr

4-4 The Definite Integral

- Approximating Areas by Left and Right Sums
- The Definite Integral as a Limit of Sums
- Properties of the Definite Integral

The first three sections of this chapter focused on the *indefinite integral.* In this section, we introduce the *definite integral.* The definite integral is used to compute areas, probabilities, average values of functions, future values of continuous income streams, and many other quantities. Initially, the concept of the definite integral may seem unrelated to the notion of the indefinite integral. There is, however, a close connection between the two integrals. The fundamental theorem of calculus, discussed in Section 4-5, makes that connection precise.

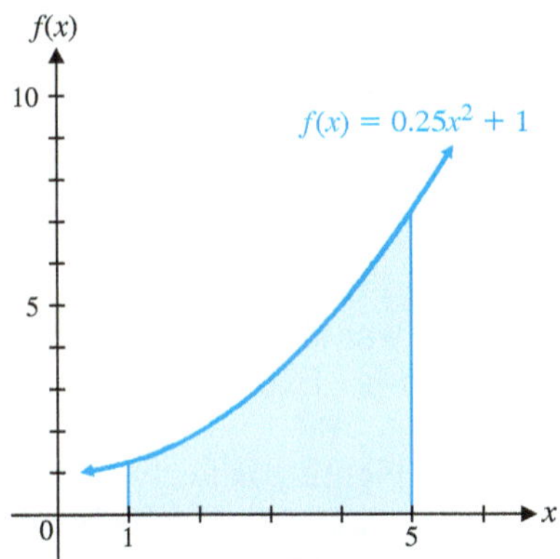

Figure 1 What is the shaded area?

Approximating Areas by Left and Right Sums

How do we find the shaded area in Figure 1? That is, how do we find the area bounded by the graph of $f(x) = 0.25x^2 + 1$, the x axis, and the vertical lines $x = 1$ and $x = 5$? [This cumbersome description is usually shortened to "the area under the graph of $f(x) = 0.25x^2 + 1$ from $x = 1$ to $x = 5$."] Our standard geometric area formulas do not apply directly, but the formula for the area of a rectangle can be used indirectly. To see how, we look at a method of approximating the area under the graph by using rectangles. This method will give us any accuracy desired, which is quite different from finding the area exactly. Our first area approximation is made by dividing the interval [1, 5] on the x axis into four equal parts, each of length

$$\Delta x = \frac{5 - 1}{4} = 1^*$$

We then place a **left rectangle** on each subinterval, that is, a rectangle whose base is the subinterval and whose height is the value of the function at the left endpoint of the subinterval (see Fig. 2).

Summing the areas of the left rectangles in Figure 2 results in a **left sum** of four rectangles, denoted by L_4, as follows:

$$\begin{aligned} L_4 &= f(1)\cdot 1 + f(2)\cdot 1 + f(3)\cdot 1 + f(4)\cdot 1 \\ &= 1.25 + 2.00 + 3.25 + 5 = 11.5 \end{aligned}$$

From Figure 3, since $f(x)$ is increasing, we see that the left sum L_4 underestimates the area, and we can write

$$11.5 = L_4 < \text{Area}$$

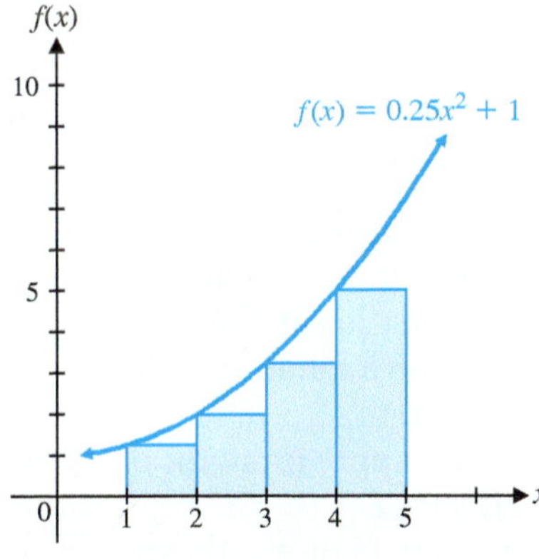

Figure 2 Left rectangles

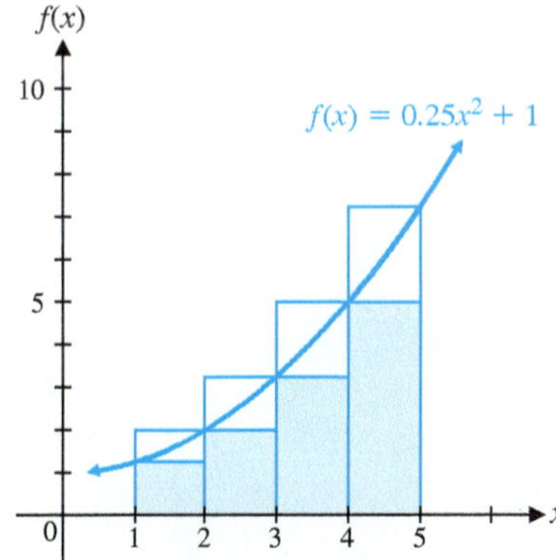

Figure 3 Left and right rectangles

*It is customary to denote the length of the subintervals by Δx, which is read "delta x," since Δ is the Greek capital letter delta.

EXPLORE & DISCUSS 1

If $f(x)$ were decreasing over the interval [1, 5], would the left sum L_4 over- or underestimate the actual area under the curve? Explain.

Similarly, we use the right endpoint of each subinterval to find the height of the **right rectangle** placed on top of it. Superimposing right rectangles on top of Figure 2, we get Figure 3.

Summing the areas of the right rectangles in Figure 3 results in a **right sum** of four rectangles, denoted by R_4, as follows (compare R_4 with L_4 and note that R_4 can be obtained from L_4 by deleting one rectangular area and adding one more):

$$R_4 = f(2)\cdot 1 + f(3)\cdot 1 + f(4)\cdot 1 + f(5)\cdot 1$$
$$= 2.00 + 3.25 + 5.00 + 7.25 = 17.5$$

From Figure 3, since $f(x)$ is increasing, we see that the right sum R_4 overestimates the area, and we conclude that the actual area is between 11.5 and 17.5. That is,

$$11.5 = L_4 < \text{Area} < R_4 = 17.5$$

EXPLORE & DISCUSS 2

If $f(x)$ in Figure 3 were decreasing over the interval [1, 5], would the right sum R_4 overestimate or underestimate the actual area under the curve? Explain.

The first approximation of the area under the curve in Figure 1 is fairly coarse, but the method outlined can be continued with increasingly accurate results by dividing the interval [1, 5] into more and more subintervals of equal horizontal length. Of course, this is not a job for hand calculation, but a job that computers are designed to do.* Figure 4 shows left- and right-rectangle approximations for 16 equal subdivisions.

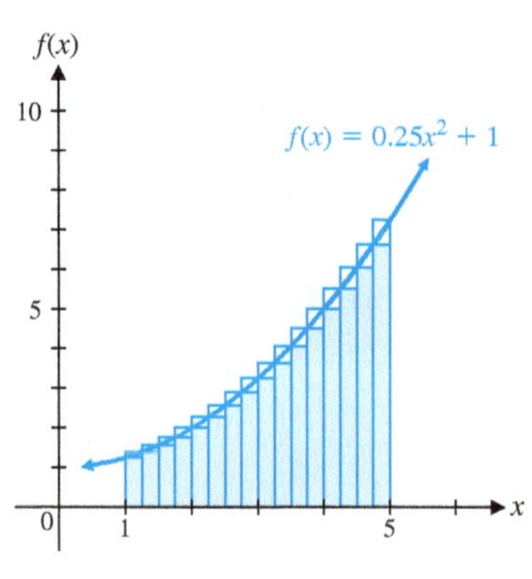

Figure 4

For this case,

$$\Delta x = \frac{5 - 1}{16} = 0.25$$

$$L_{16} = f(1)\cdot\Delta x + f(1.25)\cdot\Delta x + \cdots + f(4.75)\cdot\Delta x$$
$$= 13.59$$

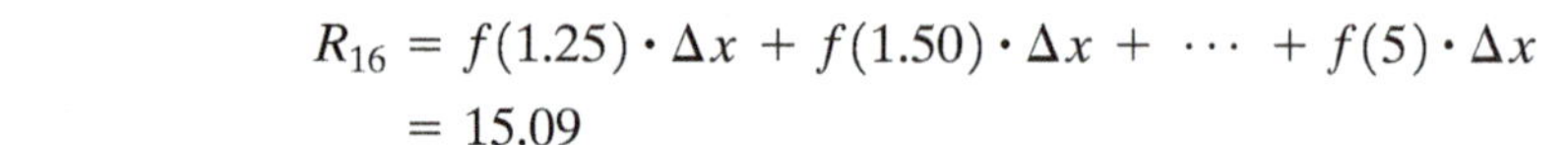

$$R_{16} = f(1.25)\cdot\Delta x + f(1.50)\cdot\Delta x + \cdots + f(5)\cdot\Delta x$$
$$= 15.09$$

Thus, we now know that the area under the curve is between 13.59 and 15.09. That is,

$$13.59 = L_{16} < \text{Area} < R_{16} = 15.09$$

For 100 equal subdivisions, computer calculations give us

$$14.214 = L_{100} < \text{Area} < R_{100} = 14.454$$

The **error in an approximation** is the absolute value of the difference between the approximation and the actual value. In general, neither the actual value nor the error in an approximation is known. However, it is often possible to calculate an **error bound**—a positive number such that the error is guaranteed to be less than or equal to that number.

The error in the approximation of the area under the graph of f from $x = 1$ to $x = 5$ by the left sum L_{16} (or the right sum R_{16}) is less than the sum of the areas of the small rectangles in Figure 4. By stacking those rectangles (see Fig. 5), we see that

$$\text{Error} = |\text{Area} - L_{16}| < |f(5) - f(1)|\cdot\Delta x = 1.5$$

Therefore, 1.5 is an error bound for the approximation of the area under f by L_{16}. We can apply the same stacking argument to any positive function that is increasing on $[a, b]$ or decreasing on $[a, b]$, to obtain the error bound in Theorem 1.

*The computer software that accompanies this book will perform these calculations (see the preface).

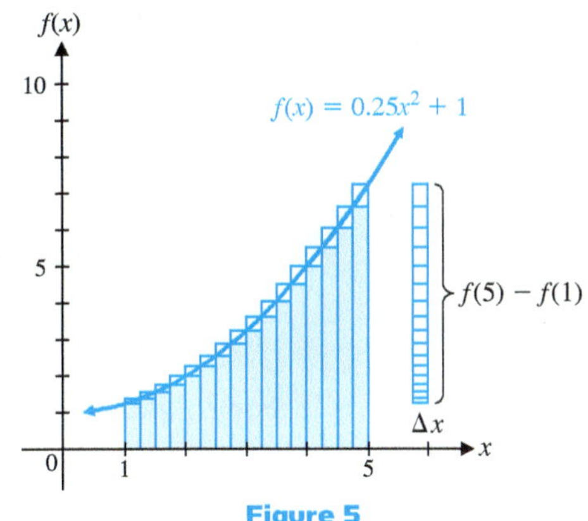

Figure 5

> **THEOREM 1 Error Bounds for Approximations of Area by Left or Right Sums**
> If $f(x) > 0$ and is either increasing on $[a, b]$ or decreasing on $[a, b]$, then
> $$|f(b) - f(a)| \cdot \frac{b - a}{n}$$
> is an error bound for the approximation of the area between the graph of f and the x axis, from $x = a$ to $x = b$, by L_n or R_n.

Because the error bound of Theorem 1 approaches 0 as $n \to \infty$, it can be shown that left and right sums, for certain functions, approach the same limit as $n \to \infty$.

> **THEOREM 2 Limits of Left and Right Sums**
> If $f(x) > 0$ and is either increasing on $[a, b]$ or decreasing on $[a, b]$, then its left and right sums approach the same real number as $n \to \infty$.

The number approached as $n \to \infty$ by the left and right sums in Theorem 2 is the area between the graph of f and the x axis from $x = a$ to $x = b$.

EXAMPLE 1 **Approximating Areas** Given the function $f(x) = 9 - 0.25x^2$, we want to approximate the area under $y = f(x)$ from $x = 2$ to $x = 5$.

(A) Graph the function over the interval [0, 6]. Then draw left and right rectangles for the interval [2, 5] with $n = 6$.

(B) Calculate L_6, R_6, and error bounds for each.

(C) How large should n be in order for the approximation of the area by L_n or R_n to be within 0.05 of the true value?

SOLUTION (A) $\Delta x = 0.5$:

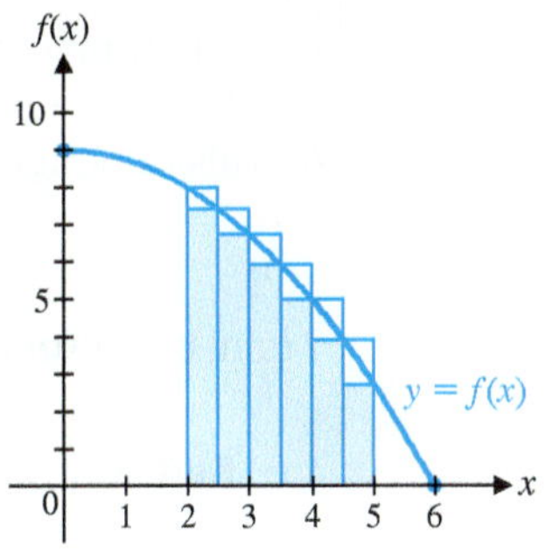

(B) $L_6 = f(2) \cdot \Delta x + f(2.5) \cdot \Delta x + f(3) \cdot \Delta x + f(3.5) \cdot \Delta x + f(4) \cdot \Delta x + f(4.5) \cdot \Delta x = 18.53$

$R_6 = f(2.5) \cdot \Delta x + f(3) \cdot \Delta x + f(3.5) \cdot \Delta x + f(4) \cdot \Delta x + f(4.5) \cdot \Delta x + f(5) \cdot \Delta x = 15.91$

The error bound for L_6 and R_6 is

$$\text{error} \leq |f(5) - f(2)|\frac{5 - 2}{6} = |2.75 - 8|(0.5) = 2.625$$

(C) For L_n and R_n, find n such that error ≤ 0.05:

$$|f(b) - f(a)|\frac{b - a}{n} \leq 0.05$$

$$|2.75 - 8|\frac{3}{n} \leq 0.05$$

$$|-5.25|\frac{3}{n} \leq 0.05$$

$$15.75 \leq 0.05n$$

$$n \geq \frac{15.75}{0.05} = 315$$

Matched Problem 1 Given the function $f(x) = 8 - 0.5x^2$, we want to approximate the area under $y = f(x)$ from $x = 1$ to $x = 3$.

(A) Graph the function over the interval [0, 4]. Then draw left and right rectangles for the interval [1, 3] with $n = 4$.

(B) Calculate L_4, R_4, and error bounds for each.

(C) How large should n be in order for the approximation of the area by L_n or R_n to be within 0.5 of the true value?

CONCEPTUAL INSIGHT

Note from Example 1(C) that a relatively large value of n ($n = 315$) is required to approximate the area by L_n or R_n to within 0.05. In other words, 315 rectangles must be used, and 315 terms must be summed, to guarantee that the error does not exceed 0.05. We can obtain a more efficient approximation of the area (fewer terms are summed to achieve a given accuracy) by replacing rectangles with trapezoids. The resulting **trapezoidal rule**, and other methods for approximating areas, are discussed in Group Activity 1 in this book's Web site.

The Definite Integral as a Limit of Sums

Left and right sums are special cases of more general sums, called *Riemann sums* [named after the German mathematician Georg Riemann (1826–1866)], that are used to approximate areas by means of rectangles.

Let f be a function defined on the interval $[a, b]$. We partition $[a, b]$ into n subintervals of equal length $\Delta x = (b - a)/n$ with endpoints

$$a = x_0 < x_1 < x_2 < \cdots < x_n = b$$

Then, using **summation notation** (see Appendix B-1), we have

$$\textbf{Left sum: } L_n = f(x_0)\Delta x + f(x_1)\Delta x + \cdots + f(x_{n-1})\Delta x = \sum_{k=1}^{n} f(x_{k-1})\Delta x$$

$$\textbf{Right sum: } R_n = f(x_1)\Delta x + f(x_2)\Delta x + \cdots + f(x_n)\Delta x = \sum_{k=1}^{n} f(x_k)\Delta x$$

$$\textbf{Riemann sum: } S_n = f(c_1)\Delta x + f(c_2)\Delta x + \cdots + f(c_n)\Delta x = \sum_{k=1}^{n} f(c_k)\Delta x$$

In a **Riemann sum**,* each c_k is required to belong to the subinterval $[x_{k-1}, x_k]$. Left and right sums are the special cases of Riemann sums in which c_k is the left endpoint or right endpoint, respectively, of the subinterval. If $f(x) > 0$, then each term of a Riemann sum S_n represents the area of a rectangle having height $f(c_k)$ and width Δx (see Fig. 6). If $f(x)$ has both positive and negative values, then some terms of S_n represent areas of rectangles, and others represent the negatives of areas of rectangles, depending on the sign of $f(c_k)$ (see Fig. 7).

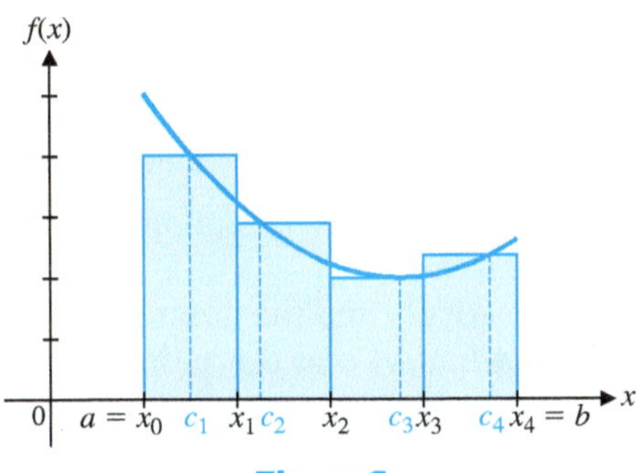

Figure 6

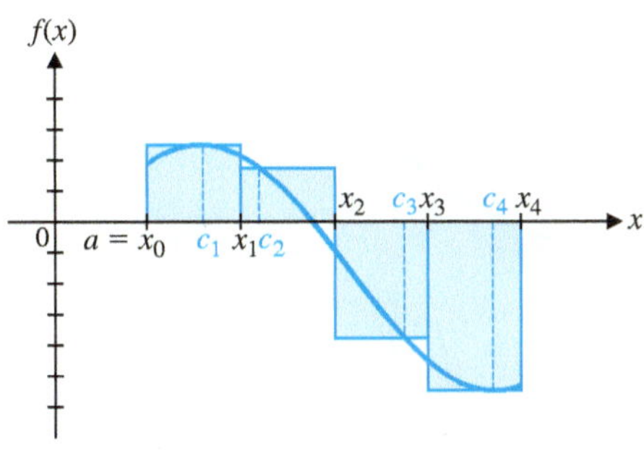

Figure 7

EXAMPLE 2 **Riemann Sums** Consider the function $f(x) = 15 - x^2$ on $[1, 5]$. Partition the interval $[1, 5]$ into four subintervals of equal length. For each subinterval $[x_{k-1}, x_k]$, let c_k be the midpoint. Calculate the corresponding Riemann sum S_4. (Riemann sums for which the c_k are the midpoints of the subintervals are called **midpoint sums**.)

SOLUTION

$$\Delta x = \frac{5-1}{4} = 1$$

$$\begin{aligned} S_4 &= f(c_1)\cdot\Delta x + f(c_2)\cdot\Delta x + f(c_3)\cdot\Delta x + f(c_4)\cdot\Delta x \\ &= f(1.5)\cdot 1 + f(2.5)\cdot 1 + f(3.5)\cdot 1 + f(4.5)\cdot 1 \\ &= 12.75 + 8.75 + 2.75 - 5.25 = 19 \end{aligned}$$

Matched Problem 2 Consider the function $f(x) = x^2 - 2x - 10$ on $[2, 8]$. Partition the interval $[2, 8]$ into three subintervals of equal length. For each subinterval $[x_{k-1}, x_k]$, let c_k be the midpoint. Calculate the corresponding Riemann sum S_3.

By analyzing properties of a continuous function on a closed interval, it can be shown that the conclusion of Theorem 2 is valid if f is continuous. In that case, not just left and right sums, but Riemann sums, have the same limit as $n \to \infty$.

THEOREM 3 Limit of Riemann Sums

If f is a continuous function on $[a, b]$, then the Riemann sums for f on $[a, b]$ approach a real number limit I as $n \to \infty$.†

*The term *Riemann sum* is often applied to more general sums in which the subintervals $[x_{k-1}, x_k]$ are not required to have the same length. Such sums are not considered in this book.

†The precise meaning of this limit statement is as follows: For each $e > 0$, there exists some $d > 0$ such that $|S_n - I| < e$ whenever S_n is a Riemann sum for f on $[a, b]$ for which $\Delta x < d$.

DEFINITION Definite Integral

Let f be a continuous function on $[a, b]$. The limit I of Riemann sums for f on $[a, b]$, guaranteed to exist by Theorem 2, is called the **definite integral** of f from a to b and is denoted as

$$\int_a^b f(x)\,dx$$

The **integrand** is $f(x)$, the **lower limit of integration** is a, and the **upper limit of integration** is b.

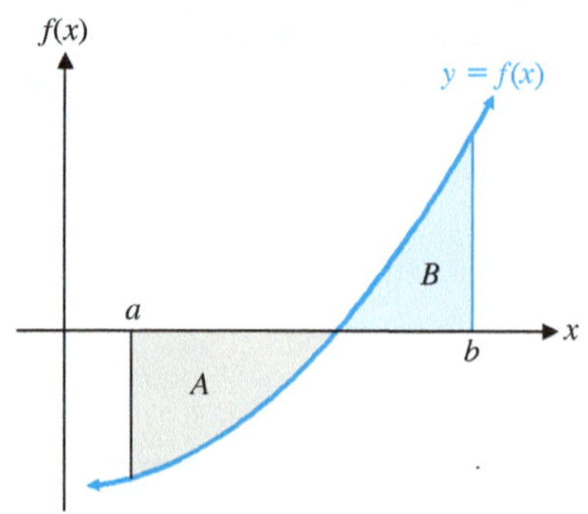

Figure 8 $\int_a^b f(x)\,dx = -A + B$

Because area is a positive quantity, the definite integral has the following geometric interpretation:

$$\int_a^b f(x)\,dx$$

represents the cumulative sum of the signed areas between the graph of f and the x axis from $x = a$ to $x = b$, where the areas above the x axis are counted positively and the areas below the x axis are counted negatively (see Fig. 8, where A and B are the actual areas of the indicated regions).

EXAMPLE 3 **Definite Integrals** Calculate the definite integrals by referring to Figure 9.

(A) $\int_a^b f(x)\,dx$

(B) $\int_a^c f(x)\,dx$

(C) $\int_b^c f(x)\,dx$

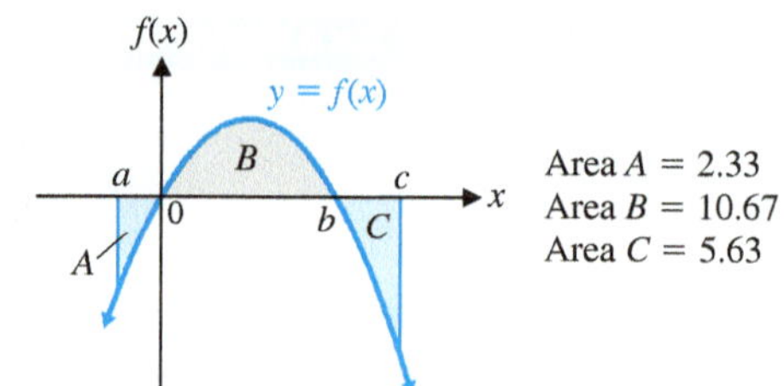

Figure 9

SOLUTION (A) $\int_a^b f(x)\,dx = -2.33 + 10.67 = 8.34$

(B) $\int_a^c f(x)\,dx = -2.33 + 10.67 - 5.63 = 2.71$

(C) $\int_b^c f(x)\,dx = -5.63$

Matched Problem 3 Referring to the figure for Example 3, calculate the definite integrals.

(A) $\int_a^0 f(x)\,dx$ (B) $\int_0^c f(x)\,dx$ (C) $\int_0^b f(x)\,dx$

Properties of the Definite Integral

Because the definite integral is defined as the limit of Riemann sums, many properties of sums are also properties of the definite integral. Note that Properties 3 and 4 are similar to the indefinite integral properties given in Section 4-1.

PROPERTIES Properties of Definite Integrals

1. $\int_a^a f(x)\,dx = 0$
2. $\int_a^b f(x)\,dx = -\int_b^a f(x)\,dx$
3. $\int_a^b kf(x)\,dx = k\int_a^b f(x)\,dx$, k a constant
4. $\int_a^b [f(x) \pm g(x)]\,dx = \int_a^b f(x)\,dx \pm \int_a^b g(x)\,dx$
5. $\int_a^b f(x)\,dx = \int_a^c f(x)\,dx + \int_c^b f(x)\,dx$

EXAMPLE 4 **Using Properties of the Definite Integral** If

$$\int_0^2 x\,dx = 2 \qquad \int_0^2 x^2\,dx = \frac{8}{3} \qquad \int_2^3 x^2\,dx = \frac{19}{3}$$

then

(A) $\int_0^2 12x^2\,dx = 12\int_0^2 x^2\,dx = 12\left(\frac{8}{3}\right) = 32$

(B) $\int_0^2 (2x - 6x^2)\,dx = 2\int_0^2 x\,dx - 6\int_0^2 x^2\,dx = 2(2) - 6\left(\frac{8}{3}\right) = -12$

(C) $\int_3^2 x^2\,dx = -\int_2^3 x^2\,dx = -\frac{19}{3}$

(D) $\int_5^5 3x^2\,dx = 0$

(E) $\int_0^3 3x^2\,dx = 3\int_0^2 x^2\,dx + 3\int_2^3 x^2\,dx = 3\left(\frac{8}{3}\right) + 3\left(\frac{19}{3}\right) = 27$

Matched Problem 4 Using the same integral values given in Example 4, find

(A) $\int_2^3 6x^2\,dx$ (B) $\int_0^2 (9x^2 - 4x)\,dx$ (C) $\int_2^0 3x\,dx$

(D) $\int_{-2}^{-2} 3x\,dx$ (E) $\int_0^3 12x^2\,dx$

Exercises 4-4

A

Problems 1–4 refer to the rectangles A, B, C, D, and E in the following figure.

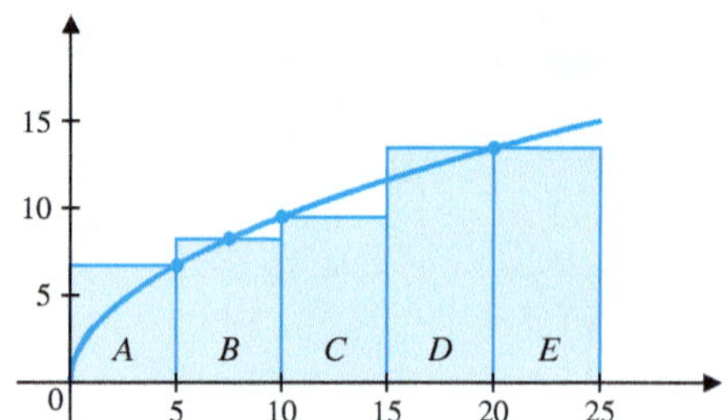

1. Which rectangles are left rectangles?
2. Which rectangles are right rectangles?
3. Which rectangles are neither left nor right rectangles?
4. Which rectangles are both left and right rectangles?

Problems 5–8 refer to the rectangles F, G, H, I, and J in the following figure.

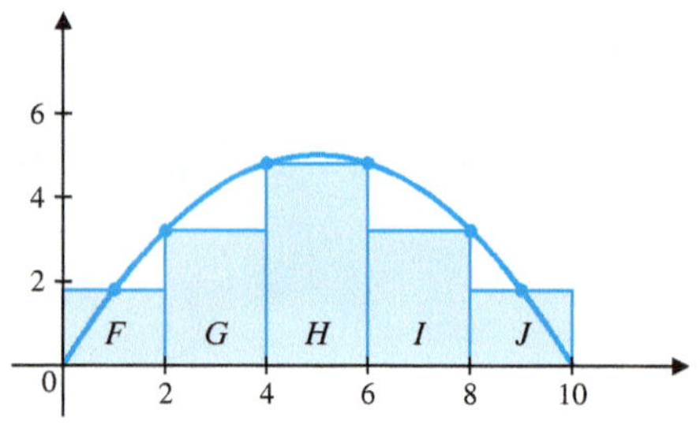

Figure for 5–8

5. Which rectangles are right rectangles?
6. Which rectangles are left rectangles?
7. Which rectangles are both left and right rectangles?
8. Which rectangles are neither left nor right rectangles?

Problems 9–16 involve estimating the area under the curves in Figures A–D from $x = 1$ to $x = 4$. For each figure, divide the interval $[1, 4]$ into three equal subintervals.

9. Draw in left and right rectangles for Figures A and B.
10. Draw in left and right rectangles for Figures C and D.
11. Using the results of Problem 9, compute L_3 and R_3 for Figure A and for Figure B.
12. Using the results of Problem 10, compute L_3 and R_3 for Figure C and for Figure D.
13. Replace the question marks with L_3 and R_3 as appropriate. Explain your choice.

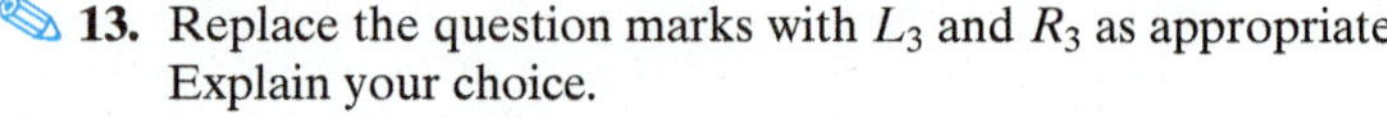

$$? \le \int_1^4 f(x)\,dx \le ? \qquad ? \le \int_1^4 g(x)\,dx \le ?$$

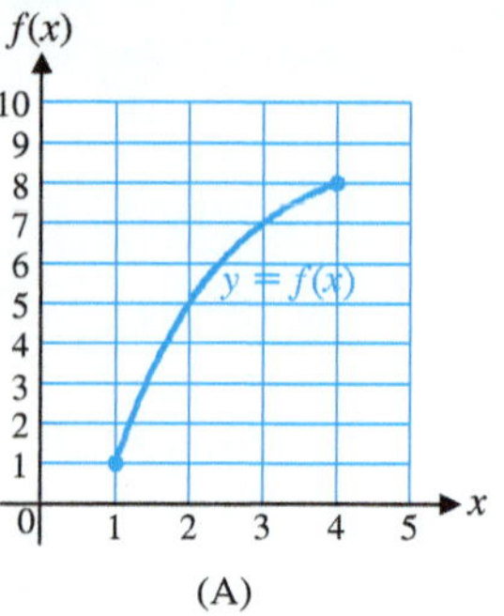

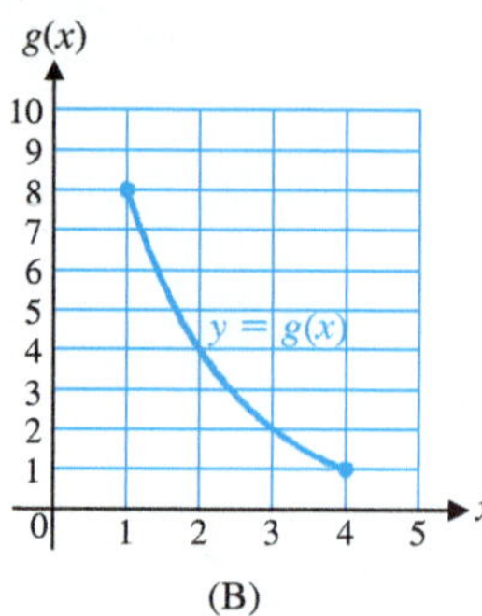

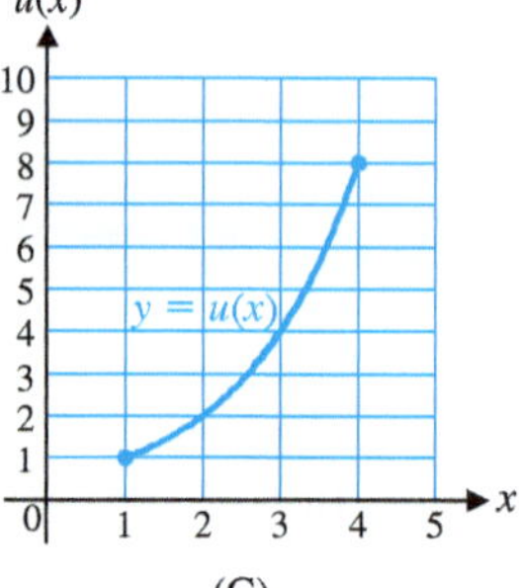

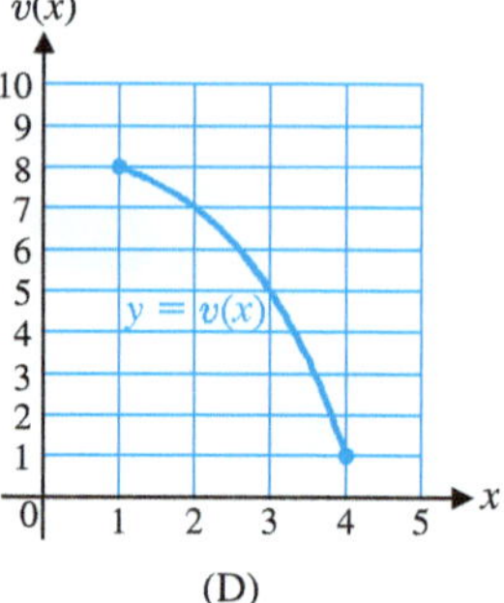

Figure for 9–16

14. Replace the question marks with L_3 and R_3 as appropriate. Explain your choice.

$$? \le \int_1^4 u(x)\,dx \le ? \qquad ? \le \int_1^4 v(x)\,dx \le ?$$

15. Compute error bounds for L_3 and R_3 found in Problem 11 for both figures.
16. Compute error bounds for L_3 and R_3 found in Problem 12 for both figures.

In Problems 17–20, calculate the indicated Riemann sum S_n for the function $f(x) = 25 - 3x^2$.

17. Partition $[-2, 8]$ into five subintervals of equal length, and for each subinterval $[x_{k-1}, x_k]$, let $c_k = (x_{k-1} + x_k)/2$.
18. Partition $[0, 12]$ into four subintervals of equal length, and for each subinterval $[x_{k-1}, x_k]$, let $c_k = (x_{k-1} + 2x_k)/3$.
19. Partition $[0, 12]$ into four subintervals of equal length, and for each subinterval $[x_{k-1}, x_k]$, let $c_k = (2x_{k-1} + x_k)/3$.
20. Partition $[-5, 5]$ into five subintervals of equal length, and for each subinterval $[x_{k-1}, x_k]$, let $c_k = (x_{k-1} + x_k)/2$.

In Problems 21–24, calculate the indicated Riemann sum S_n for the function $f(x) = x^2 - 5x - 6$.

21. Partition $[0, 3]$ into three subintervals of equal length, and let $c_1 = 0.7$, $c_2 = 1.8$, and $c_3 = 2.4$.
22. Partition $[0, 3]$ into three subintervals of equal length, and let $c_1 = 0.2$, $c_2 = 1.5$, and $c_3 = 2.8$.

23. Partition [1, 7] into six subintervals of equal length, and let $c_1 = 1, c_2 = 3, c_3 = 3, c_4 = 5, c_5 = 5$, and $c_6 = 7$.

24. Partition [1, 7] into six subintervals of equal length, and let $c_1 = 2, c_2 = 2, c_3 = 4, c_4 = 4, c_5 = 6$, and $c_6 = 6$.

In Problems 25–36, calculate the definite integral by referring to the figure with the indicated areas.

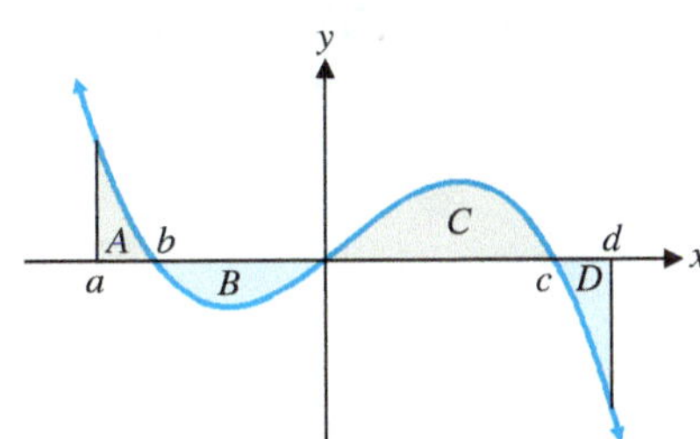

Area $A = 1.408$
Area $B = 2.475$
Area $C = 5.333$
Area $D = 1.792$

Figure for 25–36

25. $\int_b^0 f(x)\,dx$ **26.** $\int_0^c f(x)\,dx$

27. $\int_a^c f(x)\,dx$ **28.** $\int_b^d f(x)\,dx$

29. $\int_a^d f(x)\,dx$ **30.** $\int_0^d f(x)\,dx$

31. $\int_c^0 f(x)\,dx$ **32.** $\int_d^a f(x)\,dx$

33. $\int_0^a f(x)\,dx$ **34.** $\int_c^a f(x)\,dx$

35. $\int_d^b f(x)\,dx$ **36.** $\int_c^b f(x)\,dx$

In Problems 37–48, calculate the definite integral, given that

$$\int_1^4 x\,dx = 7.5 \qquad \int_1^4 x^2\,dx = 21 \qquad \int_4^5 x^2\,dx = \frac{61}{3}$$

37. $\int_1^4 2x\,dx$ **38.** $\int_1^4 3x^2\,dx$

39. $\int_1^4 (5x + x^2)\,dx$ **40.** $\int_1^4 (7x - 2x^2)\,dx$

41. $\int_1^4 (x^2 - 10x)\,dx$ **42.** $\int_1^4 (4x^2 - 9x)\,dx$

43. $\int_1^5 6x^2\,dx$ **44.** $\int_1^5 -4x^2\,dx$

45. $\int_4^4 (7x - 2)^2\,dx$

46. $\int_5^5 (10 - 7x + x^2)\,dx$

47. $\int_5^4 9x^2\,dx$

48. $\int_4^1 x(1 - x)\,dx$

B

In Problems 49–54, discuss the validity of each statement. If the statement is always true, explain why. If it is not always true, give a counterexample.

49. If $\int_a^b f(x)\,dx = 0$, then $f(x) = 0$ for all x in $[a, b]$.

50. If $f(x) = 0$ for all x in $[a, b]$, then $\int_a^b f(x)\,dx = 0$.

51. If $f(x) = 2x$ on [0, 10], then there is a positive integer n for which the left sum L_n equals the exact area under the graph of f from $x = 0$ to $x = 10$.

52. If $f(x) = 2x$ on [0, 10] and n is a positive integer, then there is some Riemann sum S_n that equals the exact area under the graph of f from $x = 0$ to $x = 10$.

53. If the area under the graph of f on $[a, b]$ is equal to both the left sum L_n and the right sum R_n for some positive integer n, then f is constant on $[a, b]$.

54. If f is a decreasing function on $[a, b]$, then the area under the graph of f is greater than the left sum L_n and less than the right sum R_n, for any positive integer n.

Problems 55 and 56 refer to the following figure showing two parcels of land along a river:

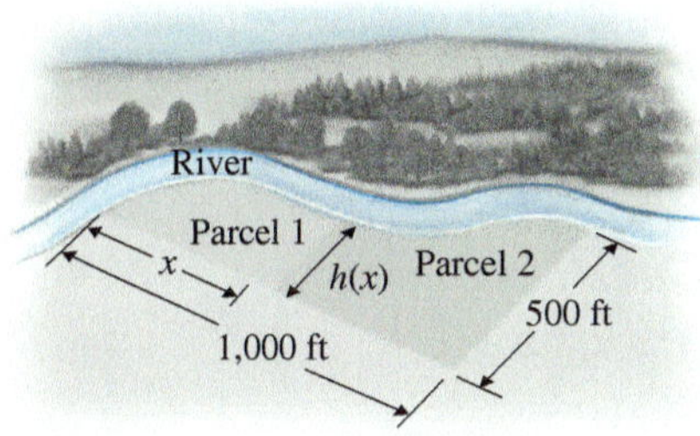

Figure for 55 and 56

55. You want to purchase both parcels of land shown in the figure and make a quick check on their combined area. There is no equation for the river frontage, so you use the average of the left and right sums of rectangles covering the area. The 1,000-foot baseline is divided into 10 equal parts. At the end of each subinterval, a measurement is made from the baseline to the river, and the results are tabulated. Let x be the distance from the left end of the baseline and let $h(x)$ be the distance from the baseline to the river at x. Use L_{10} to estimate the combined area of both parcels, and calculate an error bound for this estimate. How many subdivisions of the baseline would be required so that the error incurred in using L_n would not exceed 2,500 square feet?

x	0	100	200	300	400	500
$h(x)$	0	183	235	245	260	286

x	600	700	800	900	1,000
$h(x)$	322	388	453	489	500

56. Refer to Problem 55. Use R_{10} to estimate the combined area of both parcels, and calculate an error bound for this estimate. How many subdivisions of the baseline would be required so that the error incurred in using R_n would not exceed 1,000 square feet?

C

Problems 57 and 58 refer to the following figure:

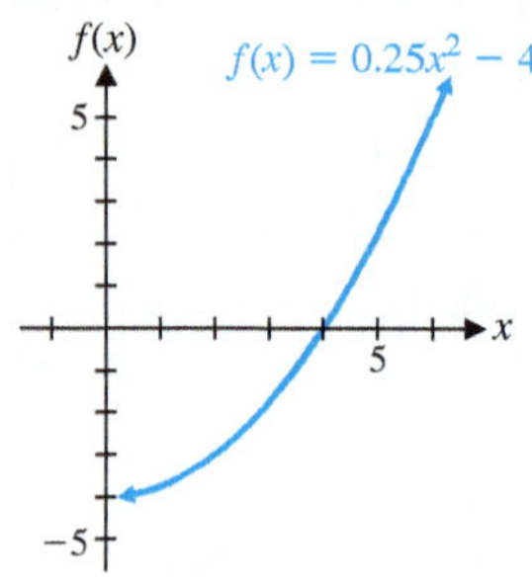

Figure for 57 and 58

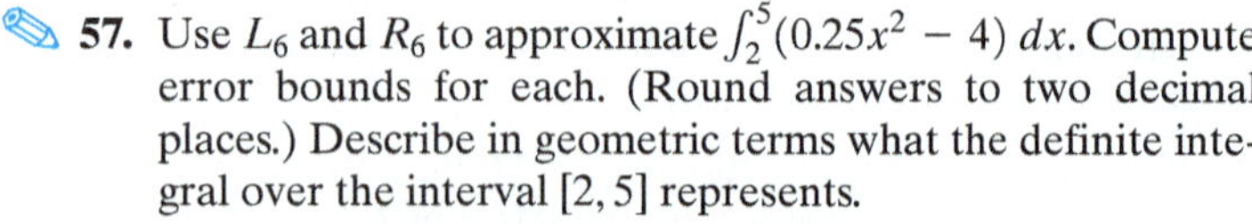

57. Use L_6 and R_6 to approximate $\int_2^5 (0.25x^2 - 4)\, dx$. Compute error bounds for each. (Round answers to two decimal places.) Describe in geometric terms what the definite integral over the interval $[2, 5]$ represents.

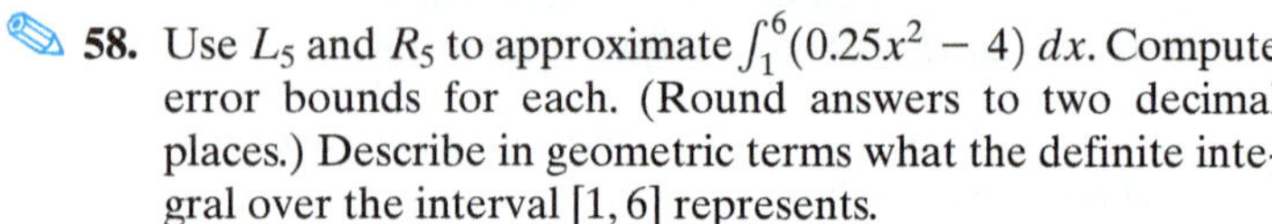

58. Use L_5 and R_5 to approximate $\int_1^6 (0.25x^2 - 4)\, dx$. Compute error bounds for each. (Round answers to two decimal places.) Describe in geometric terms what the definite integral over the interval $[1, 6]$ represents.

For Problems 59–62, use a graphing calculator to determine the intervals on which each function is increasing or decreasing.

59. $f(x) = e^{-x^2}$

60. $f(x) = \dfrac{3}{1 + 2e^{-x}}$

61. $f(x) = x^4 - 2x^2 + 3$

62. $f(x) = e^{x^2}$

In Problems 63–66, the left sum L_n or the right sum R_n is used to approximate the definite integral to the indicated accuracy. How large must n be chosen in each case? (Each function is increasing over the indicated interval.)

63. $\int_1^3 \ln x\, dx = R_n \pm 0.1$

64. $\int_0^{10} \ln(x^2 + 1)\, dx = L_n \pm 0.5$

65. $\int_1^3 x^x\, dx = L_n \pm 0.5$ **66.** $\int_1^4 x^x\, dx = R_n \pm 0.5$

Applications

67. Employee training. A company producing electric motors has established that, on the average, a new employee can assemble $N(t)$ components per day after t days of on-the-job training, as shown in the following table (a new employee's productivity increases continuously with time on the job):

t	0	20	40	60	80	100	120
$N(t)$	10	51	68	76	81	84	86

Use left and right sums to estimate the area under the graph of $N(t)$ from $t = 0$ to $t = 60$. Use three subintervals of equal length for each. Calculate an error bound for each estimate.

68. Employee training. For a new employee in Problem 67, use left and right sums to estimate the area under the graph of $N(t)$ from $t = 20$ to $t = 100$. Use four equal subintervals for each. Replace the question marks with the values of L_4 or R_4 as appropriate:

$$? \le \int_{20}^{100} N(t)\, dt \le ?$$

69. Medicine. The rate of healing, $A'(t)$ (in square centimeters per day), for a certain type of skin wound is given approximately by the following table:

t	0	1	2	3	4	5
$A'(t)$	0.90	0.81	0.74	0.67	0.60	0.55

t	6	7	8	9	10
$A'(t)$	0.49	0.45	0.40	0.36	0.33

(A) Use left and right sums over five equal subintervals to approximate the area under the graph of $A'(t)$ from $t = 0$ to $t = 5$.

(B) Replace the question marks with values of L_5 and R_5 as appropriate:

$$? \le \int_0^5 A'(t)\, dt \le ?$$

70. Medicine. Refer to Problem 69. Use left and right sums over five equal subintervals to approximate the area under the graph of $A'(t)$ from $t = 5$ to $t = 10$. Calculate an error bound for this estimate.

71. Learning. A psychologist found that, on average, the rate of learning a list of special symbols in a code $N'(x)$ after x days of practice was given approximately by the following table values:

x	0	2	4	6	8	10	12
$N'(x)$	29	26	23	21	19	17	15

Use left and right sums over three equal subintervals to approximate the area under the graph of $N'(x)$ from $x = 6$ to $x = 12$. Calculate an error bound for this estimate.

72. Learning. For the data in Problem 71, use left and right sums over three equal subintervals to approximate the area under the graph of $N'(x)$ from $x = 0$ to $x = 6$. Replace the question marks with values of L_3 and R_3 as appropriate:

$$? \le \int_0^6 N'(x)\, dx \le ?$$

Answers to Matched Problems

1. (A) $\Delta x = 0.5$:

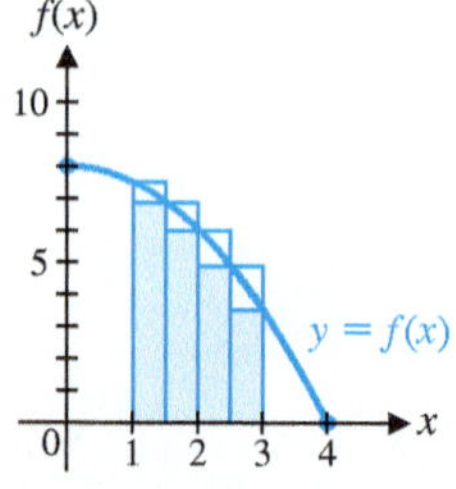

(B) $L_4 = 12.625$, $R_4 = 10.625$; error for L_4 and $R_4 = 2$

(C) $n > 16$ for L_n and R_n

2. $S_3 = 46$

3. (A) -2.33 (B) 5.04 (C) 10.67

4. (A) 38 (B) 16 (C) -6 (D) 0 (E) 108

4-5 The Fundamental Theorem of Calculus

- Introduction to the Fundamental Theorem
- Evaluating Definite Integrals
- Recognizing a Definite Integral: Average Value

The definite integral of a function f on an interval $[a, b]$ is a number, the area (if $f(x) > 0$) between the graph of f and the x axis from $x = a$ to $x = b$. The indefinite integral of a function is a family of antiderivatives. In this section, we explain the connection between these two integrals, a connection made precise by the fundamental theorem of calculus.

Introduction to the Fundamental Theorem

Suppose that the daily cost function for a small manufacturing firm is given (in dollars) by

$$C(x) = 180x + 200 \qquad 0 \le x \le 20$$

Then the marginal cost function is given (in dollars per unit) by

$$C'(x) = 180$$

What is the change in cost as production is increased from $x = 5$ units to $x = 10$ units? That change is equal to

$$\begin{aligned} C(10) - C(5) &= (180 \cdot 10 + 200) - (180 \cdot 5 + 200) \\ &= 180(10 - 5) \\ &= \$900 \end{aligned}$$

Notice that $180(10 - 5)$ is equal to the area between the graph of $C'(x)$ and the x axis from $x = 5$ to $x = 10$. Therefore,

$$C(10) - C(5) = \int_5^{10} 180\,dx$$

In other words, the change in cost from $x = 5$ to $x = 10$ is equal to the area between the marginal cost function and the x axis from $x = 5$ to $x = 10$ (see Fig. 1).

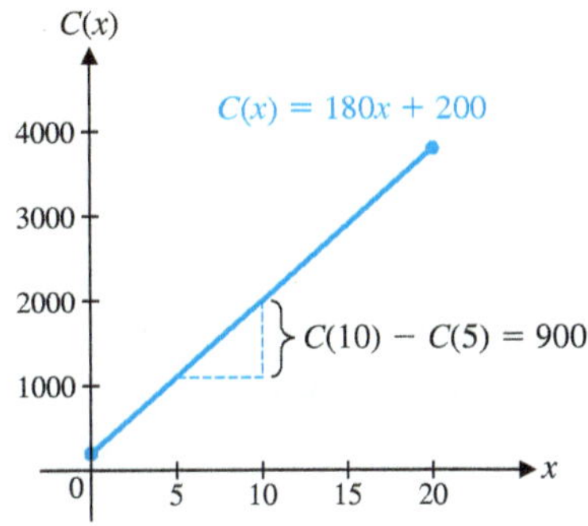

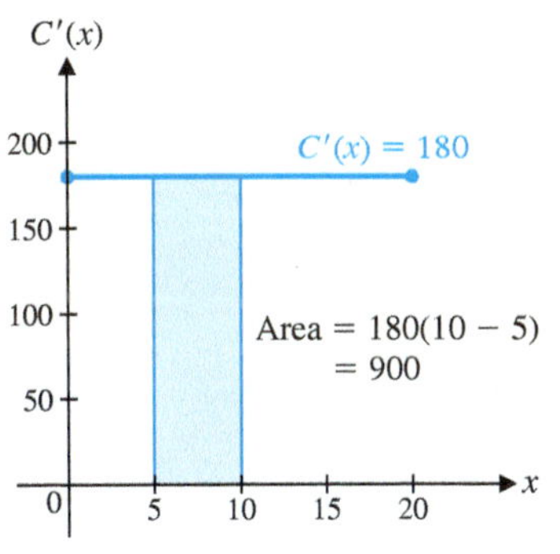

Figure 1

CONCEPTUAL INSIGHT

Consider the formula for the slope of a line:

$$m = \frac{y_2 - y_1}{x_2 - x_1}$$

Multiplying both sides of this equation by $x_2 - x_1$ gives

$$y_2 - y_1 = m(x_2 - x_1)$$

The right-hand side, $m(x_2 - x_1)$, is equal to the area of a rectangle of height m and width $x_2 - x_1$. So the change in y coordinates is equal to the area under the constant function with value m from $x = x_1$ to $x = x_2$.

EXAMPLE 1 **Change in Cost vs Area under Marginal Cost** The daily cost function for a company (in dollars) is given by

$$C(x) = -5x^2 + 210x + 400 \qquad 0 \le x \le 20$$

(A) Graph $C(x)$ for $0 \le x \le 20$, calculate the change in cost from $x = 5$ to $x = 10$, and indicate that change in cost on the graph.

(B) Graph the marginal cost function $C'(x)$ for $0 \le x \le 20$, and use geometric formulas (see Appendix C) to calculate the area between $C'(x)$ and the x axis from $x = 5$ to $x = 10$.

(C) Compare the results of the calculations in parts (A) and (B).

SOLUTION (A) $C(10) - C(5) = 2{,}000 - 1{,}325 = 675$, and this change in cost is indicated in Figure 2A.

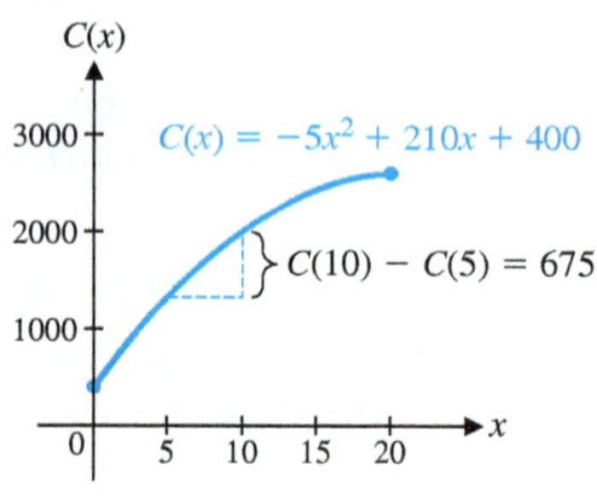

Figure 2(A)

(B) $C'(x) = -10x + 210$, so the area between $C'(x)$ and the x axis from $x = 5$ to $x = 10$ (see Fig. 2B) is the area of a trapezoid (geometric formulas are given in Appendix C):

$$\text{Area} = \frac{C'(5) + C'(10)}{2}(10 - 5) = \frac{160 + 110}{2}(5) = 675$$

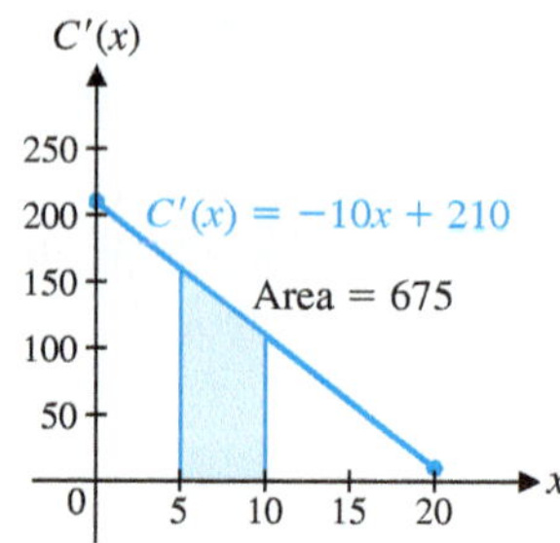

Figure 2(B)

(C) The change in cost from $x = 5$ to $x = 10$ is equal to the area between the marginal cost function and the x axis from $x = 5$ to $x = 10$.

Matched Problem 1 Repeat Example 1 for the daily cost function

$$C(x) = -7.5x^2 + 305x + 625$$

The connection illustrated in Example 1, between the change in a function from $x = a$ to $x = b$ and the area under the derivative of the function, provides the link between antiderivatives (or indefinite integrals) and the definite integral. This link is known as the fundamental theorem of calculus. (See Problems 59 and 60 in Exercise 4-5 for an outline of its proof.)

THEOREM 1 Fundamental Theorem of Calculus

If f is a continuous function on $[a, b]$, and F is any antiderivative of f, then

$$\int_a^b f(x)\,dx = F(b) - F(a)$$

CONCEPTUAL INSIGHT

Because a definite integral is the limit of Riemann sums, we expect that it would be difficult to calculate definite integrals exactly. The fundamental theorem, however, gives us an easy method for evaluating definite integrals, *provided that we can find an antiderivative* $F(x)$ of $f(x)$: Simply calculate the difference $F(b) - F(a)$. But what if we are unable to find an antiderivative of $f(x)$? In that case, we must resort to left sums, right sums, or other approximation methods to approximate the definite integral. However, it is often useful to remember that such an approximation is also an estimate of the change $F(b) - F(a)$.

Evaluating Definite Integrals

By the fundamental theorem, we can evaluate $\int_a^b f(x)\,dx$ easily and exactly whenever we can find an antiderivative $F(x)$ of $f(x)$. We simply calculate the difference $F(b) - F(a)$.

Now you know why we studied techniques of indefinite integration before this section—so that we would have methods of finding antiderivatives of large classes of elementary functions for use with the fundamental theorem. It is important to remember that

Any antiderivative of $f(x)$ can be used in the fundamental theorem. One generally chooses the simplest antiderivative by letting $C = 0$, since any other value of C will drop out in computing the difference $F(b) - F(a)$.

In evaluating definite integrals by the fundamental theorem, it is convenient to use the notation $F(x)|_a^b$, which represents the change in $F(x)$ from $x = a$ to $x = b$, as an intermediate step in the calculation. This technique is illustrated in the following examples.

EXAMPLE 2 **Evaluating Definite Integrals** Evaluate $\int_1^2 \left(2x + 3e^x - \frac{4}{x}\right) dx$.

SOLUTION

$$\begin{aligned}\int_1^2 \left(2x + 3e^x - \frac{4}{x}\right) dx &= 2\int_1^2 x\,dx + 3\int_1^2 e^x\,dx - 4\int_1^2 \frac{1}{x}\,dx \\ &= 2\frac{x^2}{2}\Big|_1^2 + 3e^x\Big|_1^2 - 4\ln|x|\Big|_1^2 \\ &= (2^2 - 1^2) + (3e^2 - 3e^1) - (4\ln 2 - 4\ln 1) \\ &= 3 + 3e^2 - 3e - 4\ln 2 \approx 14.24\end{aligned}$$

Matched Problem 2 Evaluate $\int_1^3 \left(4x - 2e^x + \frac{5}{x}\right) dx$.

The evaluation of a definite integral is a two-step process: First, find an antiderivative. Then find the change in that antiderivative. If *substitution techniques* are required to find the antiderivative, there are two different ways to proceed. The next example illustrates both methods.

EXAMPLE 3 **Definite Integrals and Substitution Techniques** Evaluate

$$\int_0^5 \frac{x}{x^2 + 10}\,dx$$

SOLUTION We solve this problem using substitution in two different ways.

Method 1. Use substitution in an indefinite integral to find an antiderivative as a function of x. Then evaluate the definite integral.

$$\begin{aligned}\int \frac{x}{x^2 + 10}\,dx &= \frac{1}{2}\int \frac{1}{x^2 + 10}2x\,dx && \text{Substitute } u = x^2 + 10 \text{ and } du = 2x\,dx.\\ &= \frac{1}{2}\int \frac{1}{u}\,du\\ &= \tfrac{1}{2}\ln|u| + C\\ &= \tfrac{1}{2}\ln(x^2 + 10) + C && \text{Since } u = x^2 + 10 > 0\end{aligned}$$

We choose $C = 0$ and use the antiderivative $\frac{1}{2}\ln(x^2 + 10)$ to evaluate the definite integral.

$$\begin{aligned}\int_0^5 \frac{x}{x^2 + 10}\,dx &= \frac{1}{2}\ln(x^2 + 10)\Big|_0^5\\ &= \tfrac{1}{2}\ln 35 - \tfrac{1}{2}\ln 10 \approx 0.626\end{aligned}$$

Method 2. Substitute directly into the definite integral, changing both the variable of integration and the limits of integration. In the definite integral

$$\int_0^5 \frac{x}{x^2 + 10}\,dx$$

the upper limit is $x = 5$ and the lower limit is $x = 0$. When we make the substitution $u = x^2 + 10$ in this definite integral, we must change the limits of integration to the corresponding values of u:

$$\begin{aligned}&x = 5 &&\text{implies} && u = 5^2 + 10 = 35 && \text{New upper limit}\\ &x = 0 &&\text{implies} && u = 0^2 + 10 = 10 && \text{New lower limit}\end{aligned}$$

We have

$$\begin{aligned}\int_0^5 \frac{x}{x^2 + 10}\,dx &= \frac{1}{2}\int_0^5 \frac{1}{x^2 + 10}2x\,dx\\ &= \frac{1}{2}\int_{10}^{35} \frac{1}{u}\,du\\ &= \frac{1}{2}\left(\ln|u|\Big|_{10}^{35}\right)\\ &= \tfrac{1}{2}(\ln 35 - \ln 10) \approx 0.626\end{aligned}$$

Matched Problem 3 Use both methods described in Example 3 to evaluate $\int_0^1 \frac{1}{2x + 4}\,dx$.

EXAMPLE 4 **Definite Integrals and Substitution** Use method 2 described in Example 3 to evaluate

$$\int_{-4}^1 \sqrt{5 - t}\,dt$$

SOLUTION If $u = 5 - t$, then $du = -dt$, and

$t = 1$ implies $u = 5 - 1 = 4$ New upper limit

$t = -4$ implies $u = 5 - (-4) = 9$ New lower limit

Notice that the lower limit for u is larger than the upper limit. Be careful not to reverse these two values when substituting into the definite integral:

$$\begin{aligned}\int_{-4}^{1} \sqrt{5 - t}\, dt &= -\int_{-4}^{1} \sqrt{5 - t}\,(-dt) \\ &= -\int_{9}^{4} \sqrt{u}\, du \\ &= -\int_{9}^{4} u^{1/2}\, du \\ &= -\left(\frac{u^{3/2}}{\frac{3}{2}}\bigg|_{9}^{4}\right) \\ &= -\left[\tfrac{2}{3}(4)^{3/2} - \tfrac{2}{3}(9)^{3/2}\right] \\ &= -\left[\tfrac{16}{3} - \tfrac{54}{3}\right] = \tfrac{38}{3} \approx 12.667\end{aligned}$$

Matched Problem 4 Use method 2 described in Example 3 to evaluate $\int_{2}^{5} \frac{1}{\sqrt{6 - t}} dt$.

EXAMPLE 5 **Change in Profit** A company manufactures x HDTVs per month. The monthly marginal profit (in dollars) is given by

$$P'(x) = 165 - 0.1x \qquad 0 \le x \le 4{,}000$$

The company is currently manufacturing 1,500 HDTVs per month, but is planning to increase production. Find the change in the monthly profit if monthly production is increased to 1,600 HDTVs.

SOLUTION

$$\begin{aligned}P(1{,}600) - P(1{,}500) &= \int_{1{,}500}^{1{,}600} (165 - 0.1x)\, dx \\ &= (165x - 0.05x^2)\Big|_{1{,}500}^{1{,}600} \\ &= [165(1{,}600) - 0.05(1{,}600)^2] \\ &\quad - [165(1{,}500) - 0.05(1{,}500)^2] \\ &= 136{,}000 - 135{,}000 \\ &= 1{,}000\end{aligned}$$

Increasing monthly production from 1,500 units to 1,600 units will increase the monthly profit by \$1,000.

Matched Problem 5 Repeat Example 5 if

$$P'(x) = 300 - 0.2x \qquad 0 \le x \le 3{,}000$$

and monthly production is increased from 1,400 to 1,500 HDTVs.

EXAMPLE 6 **Useful Life** An amusement company maintains records for each video game installed in an arcade. Suppose that $C(t)$ and $R(t)$ represent the total accumulated costs and revenues (in thousands of dollars), respectively, t years after a particular game has been installed. Suppose also that

$$C'(t) = 2 \qquad R'(t) = 9e^{-0.5t}$$

The value of t for which $C'(t) = R'(t)$ is called the **useful life** of the game.

(A) Find the useful life of the game, to the nearest year.

(B) Find the total profit accumulated during the useful life of the game.

SOLUTION (A) $R'(t) = C'(t)$

$$9e^{-0.5t} = 2$$

$$e^{-0.5t} = \tfrac{2}{9} \qquad \text{Convert to equivalent logarithmic form.}$$

$$-0.5t = \ln\tfrac{2}{9}$$

$$t = -2\ln\tfrac{2}{9} \approx 3 \text{ years}$$

Thus, the game has a useful life of 3 years. This is illustrated graphically in Figure 3.

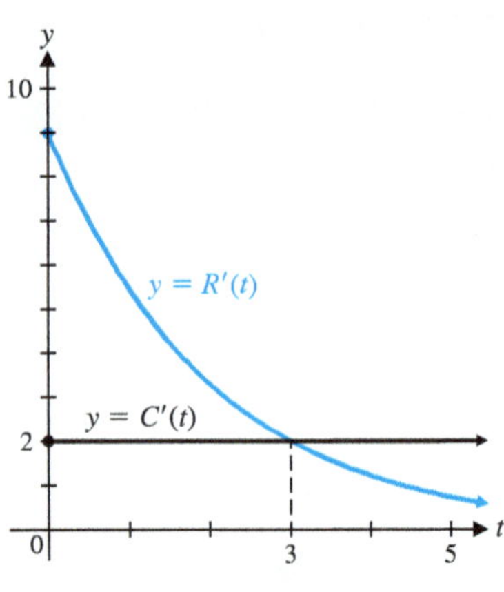

Figure 3 Useful life

(B) The total profit accumulated during the useful life of the game is

$$P(3) - P(0) = \int_0^3 P'(t)\,dt$$

$$= \int_0^3 [R'(t) - C'(t)]\,dt$$

$$= \int_0^3 (9e^{-0.5t} - 2)\,dt$$

$$= \left(\frac{9}{-0.5}e^{-0.5t} - 2t\right)\Big|_0^3 \qquad \text{Recall: } \int e^{ax}\,dx = \frac{1}{a}e^{ax} + C$$

$$= (-18e^{-0.5t} - 2t)\Big|_0^3$$

$$= (-18e^{-1.5} - 6) - (-18e^{0} - 0)$$

$$= 12 - 18e^{-1.5} \approx 7.984 \quad \text{or} \quad \$7{,}984$$

Matched Problem 6 Repeat Example 6 if $C'(t) = 1$ and $R'(t) = 7.5e^{-0.5t}$.

EXAMPLE 7 **Numerical Integration on a Graphing Calculator** Evaluate $\int_{-1}^{2} e^{-x^2}\,dx$ to three decimal places.

SOLUTION The integrand e^{-x^2} does not have an elementary antiderivative, so we are unable to use the fundamental theorem to evaluate the definite integral. Instead, we use a numerical integration routine that has been preprogrammed into a graphing calculator. (Consult your user's manual for specific details.) Such a routine is an approximation algorithm, more powerful than the left-sum and right-sum methods discussed in Section 4-4. From Figure 4,

```
fnInt(e^(-X²),X,
-1,2)
         1.628905524
```

Figure 4

$$\int_{-1}^{2} e^{-x^2}\,dx = 1.629$$

Matched Problem 7 Evaluate $\int_{1.5}^{4.3} \frac{x}{\ln x}\,dx$ to three decimal places.

Recognizing a Definite Integral: Average Value

Recall that the derivative of a function f was defined in Section 1-3 by

$$f'(x) = \lim_{h\to 0} \frac{f(x+h) - f(x)}{h}$$

This form is generally not easy to compute directly but is easy to recognize in certain practical problems (slope, instantaneous velocity, rates of change, and so on). Once we know that we are dealing with a derivative, we proceed to try to compute the derivative with the use of derivative formulas and rules.

Similarly, evaluating a definite integral with the use of the definition

$$\int_a^b f(x)\,dx = \lim_{n\to\infty} [f(c_1)\Delta x_1 + f(c_2)\Delta x_2 + \cdots + f(c_n)\Delta x_n] \tag{1}$$

is generally not easy, but the form on the right occurs naturally in many practical problems. We can use the fundamental theorem to evaluate the definite integral (once it is recognized) if an antiderivative can be found; otherwise, we will approximate it with a rectangle sum. We will now illustrate these points by finding the *average value* of a continuous function.

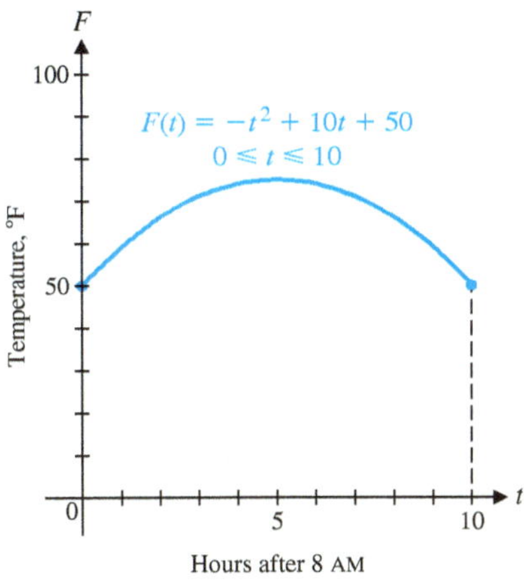

Figure 5

Suppose that the temperature F (in degrees Fahrenheit) in the middle of a small shallow lake from 8 AM ($t = 0$) to 6 PM ($t = 10$) during the month of May is given approximately as shown in Figure 5.

How can we compute the average temperature from 8 AM to 6 PM? We know that the average of a finite number of values $a_1, a_2, \ldots, a_n$ is given by

$$\text{average} = \frac{a_1 + a_2 + \cdots + a_n}{n}$$

But how can we handle a continuous function with infinitely many values? It would seem reasonable to divide the time interval [0, 10] into n equal subintervals, compute the temperature at a point in each subinterval, and then use the average of the temperatures as an approximation of the average value of the continuous function $F = F(t)$ over [0, 10]. We would expect the approximations to improve as n increases. In fact, we would define the limit of the average of n values as $n \to \infty$ as the *average value of F over* [0, 10] if the limit exists. This is exactly what we will do:

$$\begin{pmatrix}\text{average temperature}\\ \text{for } n \text{ values}\end{pmatrix} = \frac{1}{n}[F(t_1) + F(t_2) + \cdots + F(t_n)] \tag{2}$$

Here t_k is a point in the kth subinterval. We will call the limit of equation (2) as $n \to \infty$ the *average temperature over the time interval* [0, 10].

Form (2) resembles form (1), but we are missing the Δt_k. We take care of this by multiplying equation (2) by $(b - a)/(b - a)$, which will change the form of equation (2) without changing its value:

$$\begin{aligned}\frac{b-a}{b-a}\cdot\frac{1}{n}[F(t_1) + F(t_2) + \cdots + F(t_n)] &= \frac{1}{b-a}\cdot\frac{b-a}{n}[F(t_1) + F(t_2) + \cdots + F(t_n)]\\ &= \frac{1}{b-a}\left[F(t_1)\frac{b-a}{n} + F(t_2)\frac{b-a}{n} + \cdots + F(t_n)\frac{b-a}{n}\right]\\ &= \frac{1}{b-a}[F(t_1)\Delta t + F(t_2)\Delta t + \cdots + F(t_n)\Delta t]\end{aligned}$$

Therefore,

$$\begin{aligned}\begin{pmatrix}\text{average temperature}\\ \text{over } [a, b] = [0, 10]\end{pmatrix} &= \lim_{n\to\infty}\left\{\frac{1}{b-a}[F(t_1)\Delta t + F(t_2)\Delta t + \cdots + F(t_n)\Delta t]\right\}\\ &= \frac{1}{b-a}\left\{\lim_{n\to\infty}[F(t_1)\,\Delta t + F(t_2)\,\Delta t + \cdots + F(t_n)\,\Delta t]\right\}\end{aligned}$$

The limit inside the braces is of form (1)—that is, a definite integral. So

$$\begin{pmatrix}\text{average temperature}\\ \text{over } [a, b] = [0, 10]\end{pmatrix} = \frac{1}{b-a}\int_a^b F(t)\,dt$$

$$= \frac{1}{10-0}\int_0^{10}(-t^2 + 10t + 50)\,dt$$

$$= \frac{1}{10}\left(-\frac{t^3}{3} + 5t^2 + 50t\right)\Bigg|_0^{10}$$

$$= \frac{200}{3} \approx 67°\text{F}$$

We now use the fundamental theorem to evaluate the definite integral.

Proceeding as before for an arbitrary continuous function f over an interval $[a, b]$, we obtain the following general formula:

DEFINITION Average Value of a Continuous Function f over $[a, b]$

$$\frac{1}{b-a}\int_a^b f(x)\,dx$$

EXPLORE & DISCUSS 1

In Figure 6, the rectangle shown has the same area as the area under the graph of $y = f(x)$ from $x = a$ to $x = b$. Explain how the average value of $f(x)$ over the interval $[a, b]$ is related to the height of the rectangle.

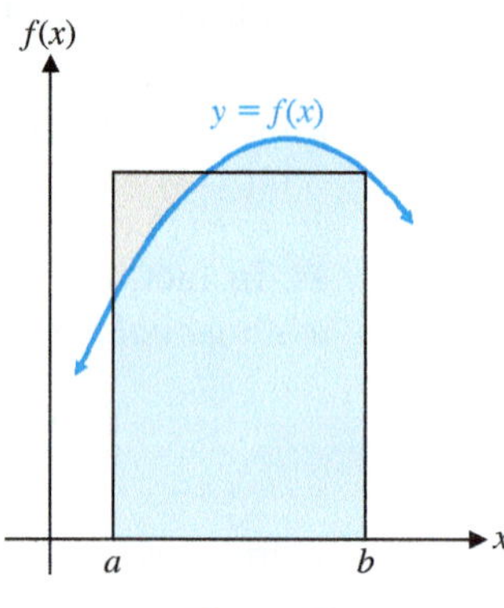

Figure 6

EXAMPLE 8 **Average Value of a Function** Find the average value of $f(x) = x - 3x^2$ over the interval $[-1, 2]$.

SOLUTION

$$\frac{1}{b-a}\int_a^b f(x)\,dx = \frac{1}{2-(-1)}\int_{-1}^{2}(x - 3x^2)\,dx$$

$$= \frac{1}{3}\left(\frac{x^2}{2} - x^3\right)\Bigg|_{-1}^{2} = -\frac{5}{2}$$

Matched Problem 8 Find the average value of $g(t) = 6t^2 - 2t$ over the interval $[-2, 3]$.

EXAMPLE 9 **Average Price** Given the demand function

$$p = D(x) = 100e^{-0.05x}$$

find the average price (in dollars) over the demand interval [40, 60].

SOLUTION

$$\begin{aligned} \text{Average price} &= \frac{1}{b-a}\int_a^b D(x)\,dx \\ &= \frac{1}{60-40}\int_{40}^{60} 100e^{-0.05x}\,dx \\ &= \frac{100}{20}\int_{40}^{60} e^{-0.05x}\,dx \qquad \text{Use } \int e^{ax}\,dx = \frac{1}{a}e^{ax},\ a \neq 0. \\ &= -\frac{5}{0.05}e^{-0.05x}\Big|_{40}^{60} \\ &= 100(e^{-2} - e^{-3}) \approx \$8.55 \end{aligned}$$

Matched Problem 9 Given the supply equation

$$p = S(x) = 10e^{0.05x}$$

find the average price (in dollars) over the supply interval [20, 30].

Exercises 4-5

A

In Problems 1–4,

(A) Calculate the change in $F(x)$ from $x = 10$ to $x = 15$.

(B) Graph $F'(x)$ and use geometric formulas (see Appendix C) to calculate the area between the graph of $F'(x)$ and the x axis from $x = 10$ to $x = 15$.

(C) Verify that your answers to (A) and (B) are equal, as is guaranteed by the fundamental theorem of calculus.

1. $F(x) = 3x^2 + 160$

2. $F(x) = 9x + 120$

3. $F(x) = -x^2 + 42x + 240$

4. $F(x) = x^2 + 30x + 210$

Evaluate the integrals in Problems 5–24.

5. $\int_0^{10} 4\,dx$

6. $\int_0^{8} 9x\,dx$

7. $\int_0^{6} x^2\,dx$

8. $\int_0^{4} x^3\,dx$

9. $\int_1^{4} (5x + 3)\,dx$

10. $\int_2^{5} (2x - 1)\,dx$

11. $\int_0^{1} e^x\,dx$

12. $\int_0^{2} 4e^x\,dx$

13. $\int_1^{2} \frac{1}{x}\,dx$

14. $\int_1^{5} \frac{2}{x}\,dx$

15. $\int_{-2}^{2} (x^3 + 7x)\,dx$

16. $\int_0^{8} (0.25x - 1)\,dx$

17. $\int_2^{5} (2x + 9)\,dx$

18. $\int_1^{4} (6x - 5)\,dx$

19. $\int_5^{2} (2x + 9)\,dx$

20. $\int_4^{1} (6x - 5)\,dx$

21. $\int_2^{3} (6 - x^3)\,dx$

22. $\int_6^{9} (5 - x^2)\,dx$

23. $\int_6^{6} (x^2 - 5x + 1)^{10}\,dx$

24. $\int_{-3}^{-3} (x^2 + 4x + 2)^8\,dx$

B

Evaluate the integrals in Problems 25–40.

25. $\int_1^{2} (2x^{-2} - 3)\,dx$

26. $\int_1^{2} (5 - 16x^{-3})\,dx$

27. $\int_1^{4} 3\sqrt{x}\,dx$

28. $\int_4^{25} \frac{2}{\sqrt{x}}\,dx$

29. $\int_2^{3} 12(x^2 - 4)^5 x\,dx$

30. $\int_0^{1} 32(x^2 + 1)^7 x\,dx$

31. $\int_3^9 \frac{1}{x-1}dx$

32. $\int_2^8 \frac{1}{x+1}dx$

33. $\int_{-5}^{10} e^{-0.05x}\,dx$

34. $\int_{-10}^{25} e^{-0.01x}\,dx$

35. $\int_1^e \frac{\ln t}{t}dt$

36. $\int_e^{e^2} \frac{(\ln t)^2}{t}dt$

37. $\int_0^1 xe^{-x^2}\,dx$

38. $\int_0^1 xe^{x^2}\,dx$

39. $\int_1^1 e^{x^2}\,dx$

40. $\int_{-1}^{-1} e^{-x^2}\,dx$

In Problems 41–48,

(A) Find the average value of each function over the indicated interval.

*(B) Use a graphing calculator to graph the function and its average value over the indicated interval in the same viewing window.*

41. $f(x) = 500 - 50x; [0, 10]$

42. $g(x) = 2x + 7; [0, 5]$

43. $f(t) = 3t^2 - 2t; [-1, 2]$

44. $g(t) = 4t - 3t^2; [-2, 2]$

45. $f(x) = \sqrt{x}; [1, 8]$

46. $g(x) = \sqrt{x+1}; [3, 8]$

47. $f(x) = 4e^{-0.2x}; [0, 10]$

48. $f(x) = 64e^{0.08x}; [0, 10]$

C

Evaluate the integrals in Problems 49–54.

49. $\int_2^3 x\sqrt{2x^2-3}\,dx$

50. $\int_0^1 x\sqrt{3x^2+2}\,dx$

51. $\int_0^1 \frac{x-1}{x^2-2x+3}dx$

52. $\int_1^2 \frac{x+1}{2x^2+4x+4}dx$

53. $\int_{-1}^1 \frac{e^{-x}-e^x}{(e^{-x}+e^x)^2}dx$

54. $\int_6^7 \frac{\ln(t-5)}{t-5}dt$

*Use a numerical integration routine to evaluate each definite integral in Problems 55–58 (to three decimal places).*

55. $\int_{1.7}^{3.5} x\ln x\,dx$

56. $\int_{-1}^1 e^{x^2}\,dx$

57. $\int_{-2}^2 \frac{1}{1+x^2}dx$

58. $\int_0^3 \sqrt{9-x^2}\,dx$

59. The **mean value theorem** states that if $F(x)$ is a differentiable function on the interval $[a, b]$, then there exists some number c between a and b such that

$$F'(c) = \frac{F(b) - F(a)}{b - a}$$

Explain why the mean value theorem implies that if a car averages 60 miles per hour in some 10-minute interval, then the car's instantaneous velocity is 60 miles per hour at least once in that interval.

60. The fundamental theorem of calculus can be proved by showing that, for every positive integer n, there is a Riemann sum for f on $[a, b]$ that is equal to $F(b) - F(a)$. By the mean value theorem (see Problem 59), within each subinterval $[x_{k-1}, x_k]$ that belongs to a partition of $[a, b]$, there is some c_k such that

$$f(c_k) = F'(c_k) = \frac{F(x_k) - F(x_{k-1})}{x_k - x_{k-1}}$$

Multiplying by the denominator $x_k - x_{k-1}$, we get

$$f(c_k)(x_k - x_{k-1}) = F(x_k) - F(x_{k-1})$$

Show that the Riemann sum

$$S_n = \sum_{k=1}^{n} f(c_k)(x_k - x_{k-1})$$

is equal to $F(b) - F(a)$.

Applications

61. **Cost.** A company manufactures mountain bikes. The research department produced the marginal cost function

$$C'(x) = 500 - \frac{x}{3} \qquad 0 \le x \le 900$$

where $C'(x)$ is in dollars and x is the number of bikes produced per month. Compute the increase in cost going from a production level of 300 bikes per month to 900 bikes per month. Set up a definite integral and evaluate it.

62. **Cost.** Referring to Problem 61, compute the increase in cost going from a production level of 0 bikes per month to 600 bikes per month. Set up a definite integral and evaluate it.

63. **Salvage value.** A new piece of industrial equipment will depreciate in value, rapidly at first and then less rapidly as time goes on. Suppose that the rate (in dollars per year) at which the book value of a new milling machine changes is given approximately by

$$V'(t) = f(t) = 500(t - 12) \qquad 0 \le t \le 10$$

where $V(t)$ is the value of the machine after t years. What is the total loss in value of the machine in the first 5 years? In the second 5 years? Set up appropriate integrals and solve.

64. **Maintenance costs.** Maintenance costs for an apartment house generally increase as the building gets older. From past records, the rate of increase in maintenance costs (in

dollars per year) for a particular apartment complex is given approximately by

$$M'(x) = f(x) = 90x^2 + 5{,}000$$

where x is the age of the apartment complex in years and $M(x)$ is the total (accumulated) cost of maintenance for x years. Write a definite integral that will give the total maintenance costs from the end of the second year to the end of the seventh year, and evaluate the integral.

65. Employee training. A company producing computer components has established that, on the average, a new employee can assemble $N(t)$ components per day after t days of on-the-job training, as indicated in the following table (a new employee's productivity usually increases with time on the job, up to a leveling-off point):

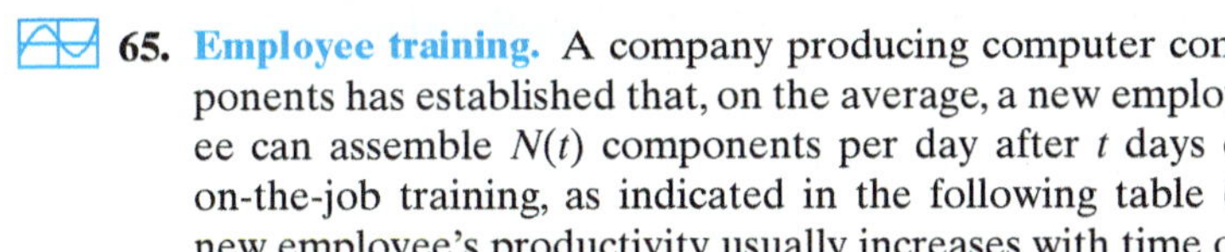

t	0	20	40	60	80	100	120
$N(t)$	10	51	68	76	81	84	85

(A) Find a quadratic regression equation for the data, and graph it and the data set in the same viewing window.

(B) Use the regression equation and a numerical integration routine on a graphing calculator to approximate the number of units assembled by a new employee during the first 100 days on the job.

66. Employee training. Refer to Problem 65.

(A) Find a cubic regression equation for the data, and graph it and the data set in the same viewing window.

(B) Use the regression equation and a numerical integration routine on a graphing calculator to approximate the number of units assembled by a new employee during the second 60 days on the job.

67. Useful life. The total accumulated costs $C(t)$ and revenues $R(t)$ (in thousands of dollars), respectively, for a photocopying machine satisfy

$$C'(t) = \tfrac{1}{11}t \qquad \text{and} \qquad R'(t) = 5te^{-t^2}$$

where t is time in years. Find the useful life of the machine, to the nearest year. What is the total profit accumulated during the useful life of the machine?

68. Useful life. The total accumulated costs $C(t)$ and revenues $R(t)$ (in thousands of dollars), respectively, for a coal mine satisfy

$$C'(t) = 3 \qquad \text{and} \qquad R'(t) = 15e^{-0.1t}$$

where t is the number of years that the mine has been in operation. Find the useful life of the mine, to the nearest year. What is the total profit accumulated during the useful life of the mine?

69. Average cost. The total cost (in dollars) of manufacturing x auto body frames is $C(x) = 60{,}000 + 300x$.

(A) Find the average cost per unit if 500 frames are produced. [*Hint*: Recall that $\overline{C}(x)$ is the average cost per unit.]

(B) Find the average value of the cost function over the interval $[0, 500]$.

(C) Discuss the difference between parts (A) and (B).

70. Average cost. The total cost (in dollars) of printing x dictionaries is $C(x) = 20{,}000 + 10x$.

(A) Find the average cost per unit if 1,000 dictionaries are produced.

(B) Find the average value of the cost function over the interval $[0, 1{,}000]$.

(C) Discuss the difference between parts (A) and (B).

71. Cost. The marginal cost at various levels of output per month for a company that manufactures sunglasses is shown in the following table, with the output x given in thousands of units per month and the total cost $C(x)$ given in thousands of dollars per month:

x	0	1	2	3	4	5	6	7	8
$C'(x)$	58	30	18	9	5	7	17	33	51

(A) Find a quadratic regression equation for the data, and graph it and the data set in the same viewing window.

(B) Use the regression equation and a numerical integration routine on a graphing calculator to approximate (to the nearest dollar) the increased cost in going from a production level of 2 thousand sunglasses per month to 8 thousand sunglasses per month.

72. Cost. Refer to Problem 71.

(A) Find a cubic regression equation for the data, and graph it and the data set in the same viewing window.

(B) Use the regression equation and a numerical integration routine on a graphing calculator to approximate (to the nearest dollar) the increased cost in going from a production level of 1 thousand sunglasses per month to 7 thousand sunglasses per month.

73. Supply function. Given the supply function

$$p = S(x) = 10(e^{0.02x} - 1)$$

find the average price (in dollars) over the supply interval $[20, 30]$.

74. Demand function. Given the demand function

$$p = D(x) = \frac{1{,}000}{x}$$

find the average price (in dollars) over the demand interval $[400, 600]$.

75. Labor costs and learning. A defense contractor is starting production on a new missile control system. On the basis of data collected during assembly of the first 16 control systems, the production manager obtained the following function for the rate of labor use:

$$g(x) = 2{,}400x^{-1/2}$$

$g(x)$ is the number of labor-hours required to assemble the xth unit of a control system. Approximately how many labor-hours will be required to assemble the 17th through the 25th control units? [*Hint*: Let $a = 16$ and $b = 25$.]

76. Labor costs and learning. If the rate of labor use in Problem 75 is

$$g(x) = 2{,}000x^{-1/3}$$

then approximately how many labor-hours will be required to assemble the 9th through the 27th control units? [*Hint*: Let $a = 8$ and $b = 27$.]

77. Inventory. A store orders 600 units of a product every 3 months. If the product is steadily depleted to 0 by the end of each 3 months, the inventory on hand I at any time t during the year is shown in the following figure:

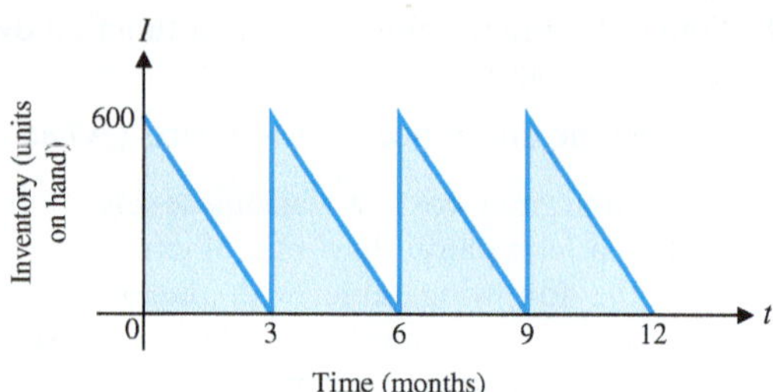

Figure for 77

(A) Write an inventory function (assume that it is continuous) for the first 3 months. [The graph is a straight line joining $(0, 600)$ and $(3, 0)$.]

(B) What is the average number of units on hand for a 3-month period?

78. Repeat Problem 77 with an order of 1,200 units every 4 months.

79. Oil production. Using production and geological data, the management of an oil company estimates that oil will be pumped from a producing field at a rate given by

$$R(t) = \frac{100}{t + 1} + 5 \qquad 0 \le t \le 20$$

where $R(t)$ is the rate of production (in thousands of barrels per year) t years after pumping begins. Approximately how many barrels of oil will the field produce during the first 10 years of production? From the end of the 10th year to the end of the 20th year of production?

80. Oil production. In Problem 79, if the rate is found to be

$$R(t) = \frac{120t}{t^2 + 1} + 3 \qquad 0 \le t \le 20$$

then approximately how many barrels of oil will the field produce during the first 5 years of production? The second 5 years of production?

81. Biology. A yeast culture weighing 2 grams is expected to grow at the rate of $W'(t) = 0.2e^{0.1t}$ grams per hour at a higher controlled temperature. How much will the weight of the culture increase during the first 8 hours of growth? How much will the weight of the culture increase from the end of the 8th hour to the end of the 16th hour of growth?

82. Medicine. The rate of healing of a skin wound (in square centimeters per day) is given approximately by $A'(t) = -0.9e^{-0.1t}$. The initial wound has an area of 9 square centimeters. How much will the area change during the first 5 days? The second 5 days?

83. Temperature. If the temperature $C(t)$ in an aquarium changes according to

$$C(t) = t^3 - 2t + 10 \qquad 0 \le t \le 2$$

(in degrees Celsius) over a 2-hour period, what is the average temperature over this period?

84. Medicine. A drug is injected into the bloodstream of a patient through her right arm. The drug concentration in the bloodstream of the left arm t hours after the injection is given by

$$C(t) = \frac{0.14t}{t^2 + 1}$$

What is the average drug concentration in the bloodstream of the left arm during the first hour after the injection? During the first 2 hours after the injection?

85. Politics. Public awareness of a congressional candidate before and after a successful campaign was approximated by

$$P(t) = \frac{8.4t}{t^2 + 49} + 0.1 \qquad 0 \le t \le 24$$

where t is time in months after the campaign started and $P(t)$ is the fraction of the number of people in the congressional district who could recall the candidate's name. What is the average fraction of the number of people who could recall the candidate's name during the first 7 months of the campaign? During the first 2 years of the campaign?

86. Population composition. The number of children in a large city was found to increase and then decrease rather drastically. If the number of children over a 6-year period was given by

$$N(t) = -\tfrac{1}{4}t^2 + t + 4 \qquad 0 \le t \le 6$$

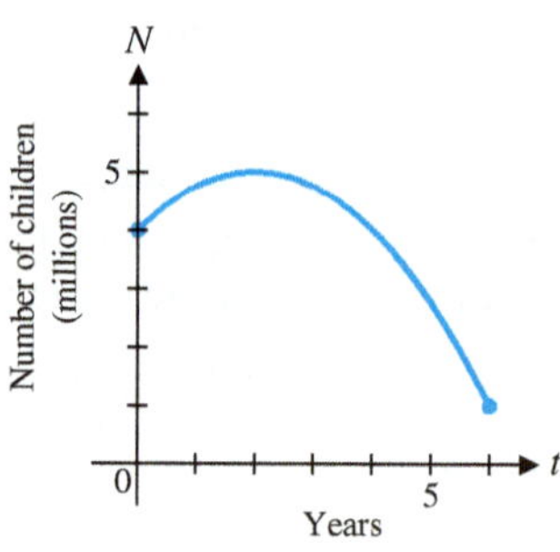

Figure for 86

what was the average number of children in the city over the 6-year period? [Assume that $N = N(t)$ is continuous.]

Answers to Matched Problems

1. (A)

$C(x) = -7.5x^2 + 305x + 625$

$C(10) - C(5) = 962.5$

(B)

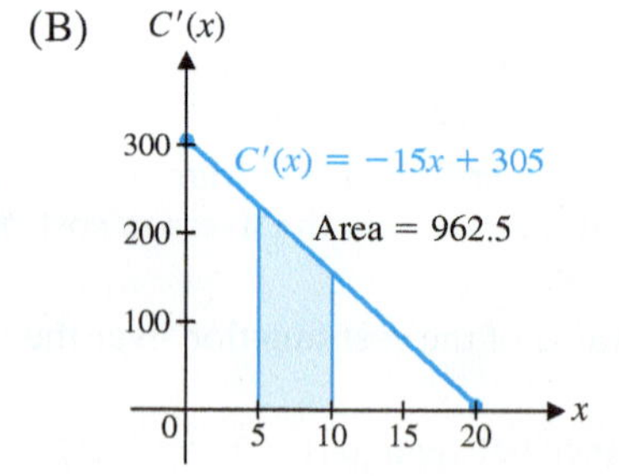

(C) The change in cost from $x = 5$ to $x = 10$ is equal to the area between the marginal cost function and the x axis from $x = 5$ to $x = 10$.

2. $16 + 2e - 2e^3 + 5 \ln 3 \approx -13.241$

3. $\frac{1}{2}(\ln 6 - \ln 4) \approx 0.203$

4. 2

5. $1,000

6. (A) $-2 \ln \frac{2}{15} \approx 4$ yr

(B) $11 - 15e^{-2} \approx 8.970$ or $8,970

7. 8.017

8. 13

9. $35.27

Chapter 4 Review

Examples

4-1 Antiderivatives and Indefinite Integrals

- A function F is an **antiderivative** of a function f if $F'(x) = f(x)$. Ex. 1, p. 227
- If F and G are both antiderivatives of f, then F and G differ by a constant; that is, $F(x) = G(x) + k$ for some constant k.
- We use the symbol $\int f(x)\,dx$, called an **indefinite integral,** to represent the family of all antiderivatives of f, and we write

$$\int f(x)\,dx = F(x) + C$$

The symbol $\int$ is called an **integral sign,** $f(x)$ is the **integrand,** and C is the **constant of integration.**

- Indefinite integrals of basic functions are given by the formulas on page 228.
- Properties of indefinite integrals are given on page 228; in particular, a constant factor can be moved across an integral sign. However, a variable factor *cannot* be moved across an integral sign.

Ex. 2, p. 230
Ex. 3, p. 230
Ex. 4, p. 232
Ex. 5, p. 232
Ex. 6, p. 233

4-2 Integration by Substitution

- The **method of substitution** (also called the **change-of-variable method**) is a technique for finding indefinite integrals. It is based on the following formula, which is obtained by reversing the chain rule: Ex. 1, p. 239

$$\int f'[g(x)]g'(x)\,dx = f[g(x)] + C$$

- This formula implies the general indefinite integral formulas on page 241.
- When using the method of substitution, it is helpful to use differentials as a bookkeeping device:
  1. The **differential dx** of the independent variable x is an arbitrary real number.
  2. The **differential dy** of the dependent variable y is defined by $dy = f'(x)\,dx$.
- Guidelines for using the substitution method are given by the procedure on page 241.

Ex. 2, p. 240
Ex. 3, p. 240
Ex. 4, p. 241
Ex. 5, p. 242
Ex. 6, p. 244
Ex. 7, p. 245

4-3 Differential Equations; Growth and Decay

- An equation is a **differential equation** if it involves an unknown function and one or more of its derivatives.
- The equation

$$\frac{dy}{dx} = 3x(1 + xy^2)$$

is a **first-order** differential equation because it involves the first derivative of the unknown function y but no second or higher order derivative.

- A **slope field** can be constructed for the preceding differential equation by drawing a tangent line segment with slope $3x(1 + xy^2)$ at each point (x, y) of a grid. The slope field gives a graphical representation of the functions that are solutions of the differential equation.
- The differential equation

$$\frac{dQ}{dt} = rQ$$

(in words, the rate at which the unknown function Q increases is proportional to Q) is called the **exponential growth law.** The constant r is called the **relative growth rate.** The solutions of the exponential growth law are the functions

$$Q(t) = Q_0 e^{rt}$$

where Q_0 denotes $Q(0)$, the amount present at time $t = 0$. These functions can be used to solve problems in population growth, continuous compound interest, radioactive decay, blood pressure, and light absorption.

Ex. 1, p. 252
Ex. 2, p. 252
Ex. 3, p. 253

- Table 1 on page 255 gives the solutions of other first-order differential equations that can be used to model the limited or logistic growth of epidemics, sales, and corporations. Ex. 4, p. 254

4-4 The Definite Integral

- If the function f is positive on $[a, b]$, then the area between the graph of f and the x axis from $x = a$ to $x = b$ can be approximated by partitioning $[a, b]$ into n subintervals $[x_{k-1}, x_k]$ of equal length $\Delta x = (b - a)/n$ and summing the areas of n rectangles. This can be done using **left sums, right sums,** or, more generally, **Riemann sums:**

Ex. 1, p. 261
Ex. 2, p. 263

Left sum: $L_n = \sum_{k=1}^{n} f(x_{k-1})\Delta x$

Right sum: $R_n = \sum_{k=1}^{n} f(x_k)\Delta x$

Riemann sum: $S_n = \sum_{k=1}^{n} f(c_k)\Delta x$

In a Riemann sum, each c_k is required to belong to the subinterval $[x_{k-1}, x_k]$. Left sums and right sums are the special cases of Riemann sums in which c_k is the left endpoint and right endpoint, respectively, of the subinterval.

- The **error in an approximation** is the absolute value of the difference between the approximation and the actual value. An **error bound** is a positive number such that the error is guaranteed to be less than or equal to that number.
- Theorem 1 on page 261 gives error bounds for the approximation of the area between the graph of a positive function f and the x axis from $x = a$ to $x = b$, by left sums or right sums, if f is either increasing or decreasing.
- If $f(x) > 0$ and is either increasing on $[a, b]$ or decreasing on $[a, b]$, then the left and right sums of $f(x)$ approach the same real number as $n \to \infty$ (Theorem 2, page 261).
- If f is a continuous function on $[a, b]$, then the Riemann sums for f on $[a, b]$ approach a real-number limit I as $n \to \infty$ (Theorem 3, page 263).
- Let f be a continuous function on $[a, b]$. Then the limit I of Riemann sums for f on $[a, b]$, guaranteed to exist by Theorem 3, is called the **definite integral** of f from a to b and is denoted

$$\int_a^b f(x)\,dx$$

The **integrand** is $f(x)$, the **lower limit of integration** is a, and the **upper limit of integration** is b.

- Geometrically, the definite integral

Ex. 3, p. 264

$$\int_a^b f(x)\,dx$$

represents the cumulative sum of the signed areas between the graph of f and the x axis from $x = a$ to $x = b$.

- Properties of the definite integral are given on page 265.

Ex. 4, p. 265

4-5 The Fundamental Theorem of Calculus

- If f is a continuous function on $[a, b]$ and F is any antiderivative of f, then

Ex. 1, p. 270
Ex. 2, p. 271

$$\int_a^b f(x)\,dx = F(b) - F(a)$$

This is the fundamental theorem of calculus (see page 271).

- The fundamental theorem gives an easy and exact method for evaluating definite integrals, provided that we can find an antiderivative $F(x)$ of $f(x)$. In practice, we first find an antiderivative $F(x)$ (when possible), using techniques for computing indefinite integrals. Then we calculate the difference $F(b) - F(a)$. If it is impossible to find an antiderivative, we must resort to left or right sums, or other approximation methods, to evaluate the definite integral. Graphing calculators have a built-in numerical approximation routine, more powerful than left- or right-sum methods, for this purpose.

Ex. 3, p. 272
Ex. 4, p. 272
Ex. 5, p. 273
Ex. 6, p. 273
Ex. 7, p. 274

- If f is a continuous function on $[a, b]$, then the **average value** of f over $[a, b]$ is defined to be

Ex. 8, p. 276

$$\frac{1}{b-a}\int_a^b f(x)\,dx$$

Review Exercises

Work through all the problems in this chapter review and check your answers in the back of the book. Answers to all review problems are there, along with section numbers in italics to indicate where each type of problem is discussed. Where weaknesses show up, review appropriate sections of the text.

A

Find each integral in Problems 1–6.

1. $\int (6x + 3)\,dx$

2. $\int_{10}^{20} 5\,dx$

3. $\int_0^9 (4 - t^2)\,dt$

4. $\int (1 - t^2)^3 t\,dt$

5. $\int \frac{1 + u^4}{u}\,du$

6. $\int_0^1 xe^{-2x^2}\,dx$

In Problems 7 and 8, find the derivative or indefinite integral as indicated.

7. $\frac{d}{dx}\left(\int e^{-x^2}\,dx\right)$

8. $\int \frac{d}{dx}\left(\sqrt{4 + 5x}\right)dx$

9. Find a function $y = f(x)$ that satisfies both conditions:

$$\frac{dy}{dx} = 3x^2 - 2 \qquad f(0) = 4$$

10. Find all antiderivatives of

(A) $\frac{dy}{dx} = 8x^3 - 4x - 1$ (B) $\frac{dx}{dt} = e^t - 4t^{-1}$

11. Approximate $\int_1^5 (x^2 + 1)\,dx$, using a right sum with $n = 2$. Calculate an error bound for this approximation.

12. Evaluate the integral in Problem 11, using the fundamental theorem of calculus, and calculate the actual error $|I - R_2|$ produced by using R_2.

13. Use the following table of values and a left sum with $n = 4$ to approximate $\int_1^{17} f(x)\,dx$:

x	1	5	9	13	17
$f(x)$	1.2	3.4	2.6	0.5	0.1

14. Find the average value of $f(x) = 6x^2 + 2x$ over the interval $[-1, 2]$.

15. Describe a rectangle that has the same area as the area under the graph of $f(x) = 6x^2 + 2x$ from $x = -1$ to $x = 2$ (see Problem 14).

In Problems 16 and 17, calculate the indicated Riemann sum S_n for the function $f(x) = 100 - x^2$.

16. Partition [3, 11] into four subintervals of equal length, and for each subinterval $[x_{i-1}, x_i]$, let $c_i = (x_{i-1} + x_i)/2$.

17. Partition $[-5, 5]$ into five subintervals of equal length and let $c_1 = -4$, $c_2 = -1$, $c_3 = 1$, $c_4 = 2$, and $c_5 = 5$.

B

Use the graph and actual areas of the indicated regions in the figure to evaluate the integrals in Problems 18–25:

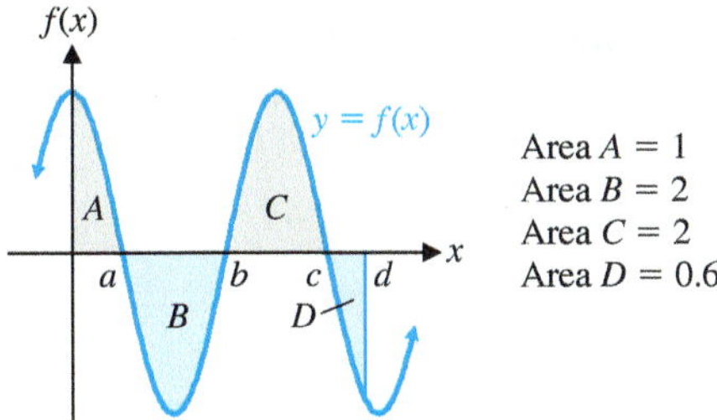

Figure for 18–25

18. $\int_a^b 5f(x)\,dx$

19. $\int_b^c \frac{f(x)}{5}\,dx$

20. $\int_b^d f(x)\,dx$

21. $\int_a^c f(x)\,dx$

22. $\int_0^d f(x)\,dx$

23. $\int_b^a f(x)\,dx$

24. $\int_c^b f(x)\,dx$

25. $\int_d^0 f(x)\,dx$

Problems 26–31 refer to the slope field shown in the figure:

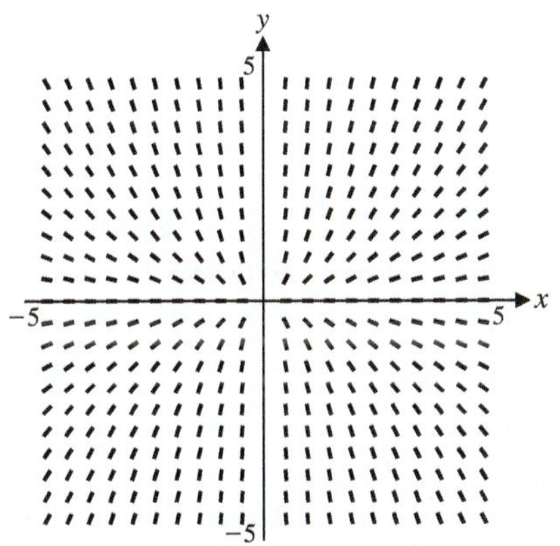

Figure for 26–31

26. (A) For $dy/dx = (2y)/x$, what is the slope of a solution curve at (2, 1)? At (−2, −1)?

(B) For $dy/dx = (2x)/y$, what is the slope of a solution curve at (2, 1)? At (−2, −1)?

27. Is the slope field shown in the figure for $dy/dx = (2x)/y$ or for $dy/dx = (2y)/x$? Explain.

28. Show that $y = Cx^2$ is a solution of $dy/dx = (2y)/x$ for any real number C.

29. Referring to Problem 28, find the particular solution of $dy/dx = (2y)/x$ that passes through (2, 1). Through (−2, −1).

30. Graph the two particular solutions found in Problem 29 in the slope field shown (or a copy).

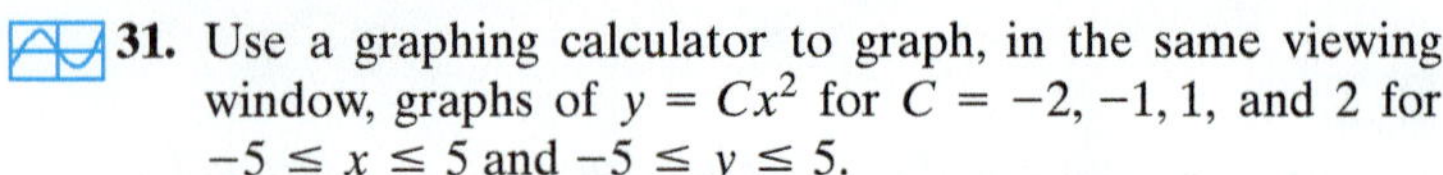

31. Use a graphing calculator to graph, in the same viewing window, graphs of $y = Cx^2$ for $C = -2, -1, 1,$ and 2 for $-5 \le x \le 5$ and $-5 \le y \le 5$.

Find each integral in Problems 32–42.

32. $\int_{-1}^{1} \sqrt{1 + x}\, dx$

33. $\int_{-1}^{0} x^2(x^3 + 2)^{-2}\, dx$

34. $\int 5e^{-t}\, dt$

35. $\int_{1}^{e} \frac{1 + t^2}{t}\, dt$

36. $\int xe^{3x^2}\, dx$

37. $\int_{-3}^{1} \frac{1}{\sqrt{2 - x}}\, dx$

38. $\int_{0}^{3} \frac{x}{1 + x^2}\, dx$

39. $\int_{0}^{3} \frac{x}{(1 + x^2)^2}\, dx$

40. $\int x^3(2x^4 + 5)^5\, dx$

41. $\int \frac{e^{-x}}{e^{-x} + 3}\, dx$

42. $\int \frac{e^x}{(e^x + 2)^2}\, dx$

43. Find a function $y = f(x)$ that satisfies both conditions:

$$\frac{dy}{dx} = 3x^{-1} - x^{-2} \qquad f(1) = 5$$

44. Find the equation of the curve that passes through $(2, 10)$ if its slope is given by

$$\frac{dy}{dx} = 6x + 1$$

for each x.

45. (A) Find the average value of $f(x) = 3\sqrt{x}$ over the interval $[1, 9]$.

(B) Graph $f(x) = 3\sqrt{x}$ and its average over the interval $[1, 9]$ in the same coordinate system.

C

Find each integral in Problems 46–50.

46. $\int \frac{(\ln x)^2}{x}\, dx$

47. $\int x(x^3 - 1)^2\, dx$

48. $\int \frac{x}{\sqrt{6 - x}}\, dx$

49. $\int_{0}^{7} x\sqrt{16 - x}\, dx$

50. $\int_{1}^{1} (x + 1)^9 dx$

51. Find a function $y = f(x)$ that satisfies both conditions:

$$\frac{dy}{dx} = 9x^2e^{x^3} \qquad f(0) = 2$$

52. Solve the differential equation

$$\frac{dN}{dt} = 0.06N \qquad N(0) = 800 \qquad N > 0$$

Graph Problems 53–56 on a graphing calculator, and identify each curve as unlimited growth, exponential decay, limited growth, or logistic growth:

53. $N = 50(1 - e^{-0.07t}); 0 \le t \le 80, 0 \le N \le 60$

54. $p = 500e^{-0.03x}; 0 \le x \le 100, 0 \le p \le 500$

55. $A = 200e^{0.08t}; 0 \le t \le 20, 0 \le A \le 1{,}000$

56. $N = \frac{100}{1 + 9e^{-0.3t}}; 0 \le t \le 25, 0 \le N \le 100$

Use a numerical integration routine to evaluate each definite integral in Problems 57–59 (to three decimal places).

57. $\int_{-0.5}^{0.6} \frac{1}{\sqrt{1 - x^2}}\, dx$

58. $\int_{-2}^{3} x^2e^x\, dx$

59. $\int_{0.5}^{2.5} \frac{\ln x}{x^2}\, dx$

Applications

60. Cost. A company manufactures downhill skis. The research department produced the marginal cost graph shown in the accompanying figure, where $C'(x)$ is in dollars and x is the number of pairs of skis produced per week. Estimate the increase in cost going from a production level of 200 to 600 pairs of skis per week. Use left and right sums over two equal subintervals. Replace the question marks with the values of L_2 and R_2 as appropriate:

$$? \le \int_{200}^{600} C'(x)\, dx \le ?$$

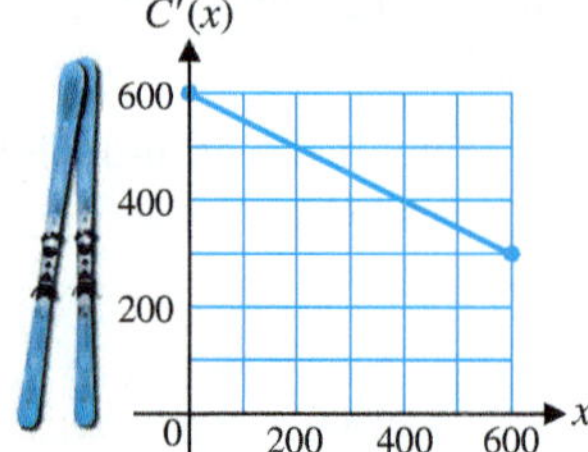

Figure for 60

61. Cost. Assuming that the marginal cost function in Problem 60 is linear, find its equation and write a definite integral that represents the increase in costs going from a production level of 200 to 600 pairs of skis per week. Evaluate the definite integral.

62. Profit and production. The weekly marginal profit for an output of x units is given approximately by

$$P'(x) = 150 - \frac{x}{10} \qquad 0 \le x \le 40$$

What is the total change in profit for a change in production from 10 units per week to 40 units? Set up a definite integral and evaluate it.

63. Profit function. If the marginal profit for producing x units per day is given by

$$P'(x) = 100 - 0.02x \qquad P(0) = 0$$

where $P(x)$ is the profit in dollars, find the profit function P and the profit on 10 units of production per day.

64. Resource depletion. An oil well starts out producing oil at the rate of 60,000 barrels of oil per year, but the production rate is expected to decrease by 4,000 barrels per year. Thus, if $P(t)$ is the total production (in thousands of barrels) in t years, then

$$P'(t) = f(t) = 60 - 4t \qquad 0 \le t \le 15$$

Write a definite integral that will give the total production after 15 years of operation, and evaluate the integral.

65. Inventory. Suppose that the inventory of a certain item t months after the first of the year is given approximately by

$$I(t) = 10 + 36t - 3t^2 \qquad 0 \le t \le 12$$

What is the average inventory for the second quarter of the year?

66. Price–supply. Given the price–supply function

$$p = S(x) = 8(e^{0.05x} - 1)$$

find the average price (in dollars) over the supply interval [40, 50].

67. Useful life. The total accumulated costs C(t) and revenues $R(t)$ (in thousands of dollars), respectively, for a coal mine satisfy

$$C'(t) = 3 \qquad \text{and} \qquad R'(t) = 20e^{-0.1t}$$

where t is the number of years that the mine has been in operation. Find the useful life of the mine, to the nearest year. What is the total profit accumulated during the useful life of the mine?

68. Marketing. The market research department for an automobile company estimates that sales (in millions of dollars) of a new electric car will increase at the monthly rate of

$$S'(t) = 4e^{-0.08t} \qquad 0 \le t \le 24$$

t months after the introduction of the car. What will be the total sales $S(t)$ t months after the car is introduced if we assume that there were 0 sales at the time the car entered the marketplace? What are the estimated total sales during the first 12 months after the introduction of the car? How long will it take for the total sales to reach $40 million?

69. Wound healing. The area of a healing skin wound changes at a rate given approximately by

$$\frac{dA}{dt} = -5t^{-2} \qquad 1 \le t \le 5$$

where t is time in days and $A(1) = 5$ square centimeters. What will be the area of the wound in 5 days?

70. Pollution. An environmental protection agency estimates that the rate of seepage of toxic chemicals from a waste dump (in gallons per year) is given by

$$R(t) = \frac{1{,}000}{(1 + t)^2}$$

where t is the time in years since the discovery of the seepage. Find the total amount of toxic chemicals that seep from the dump during the first 4 years of its discovery.

71. Population. The population of Mexico was 111 million in 2009 and was growing at a rate of 1.13% per year, compounded continuously.

(A) Assuming that the population continues to grow at this rate, estimate the population of Mexico in the year 2025.

(B) At the growth rate indicated, how long will it take the population of Mexico to double?

72. Archaeology. The continuous compound rate of decay for carbon-14 is $r = -0.000\,123\,8$. A piece of animal bone found at an archaeological site contains 4% of the original amount of carbon-14. Estimate the age of the bone.

73. Learning. An average student enrolled in a typing class progressed at a rate of $N'(t) = 7e^{-0.1t}$ words per minute t weeks after enrolling in a 15-week course. If a student could type 25 words per minute at the beginning of the course, how many words per minute $N(t)$ would the student be expected to type t weeks into the course? After completing the course?

Additional Integration Topics

Introduction

In Chapter 5 we explore additional applications and techniques of integration. We use the integral to find probabilities and to calculate several quantities that are important in business and economics: the total income and future value produced by a continuous income stream, consumers' and producers' surplus, and the Gini index of income concentration. The Gini index is a single number that measures the equality of a country's income distribution (see Problems 87 and 88, for example, in Section 5-1).

5-1 Area Between Curves

- Area Between Two Curves
- Application: Income Distribution

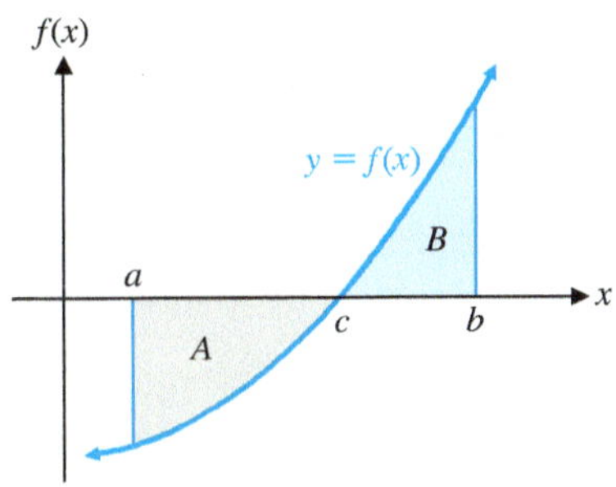

Figure 1 $\int_a^b f(x)\,dx = -A + B$

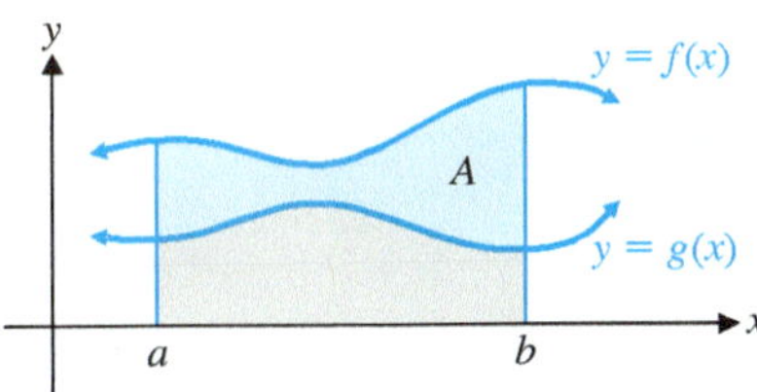

Figure 2

In Chapter 4, we found that the definite integral $\int_a^b f(x)\,dx$ represents the sum of the signed areas between the graph of $y = f(x)$ and the x axis from $x = a$ to $x = b$, where the areas above the x axis are counted positively and the areas below the x axis are counted negatively (see Fig. 1). In this section, we are interested in using the definite integral to find the actual area between a curve and the x axis or the actual area between two curves. These areas are always nonnegative quantities—**area measure is never negative**.

Area Between Two Curves

Consider the area bounded by $y = f(x)$ and $y = g(x)$, where $f(x) \geq g(x) \geq 0$, for $a \leq x \leq b$, as shown in Figure 2.

$$\begin{pmatrix}\text{Area } A \text{ between}\\ f(x) \text{ and } g(x)\end{pmatrix} = \begin{pmatrix}\text{area}\\ \text{under } f(x)\end{pmatrix} - \begin{pmatrix}\text{area}\\ \text{under } g(x)\end{pmatrix}$$

Areas are from $x = a$ to $x = b$ above the x axis.

$$= \int_a^b f(x)\,dx - \int_a^b g(x)\,dx$$

Use definite integral property 4 (Section 4-4).

$$= \int_a^b [f(x) - g(x)]\,dx$$

It can be shown that the preceding result does not require $f(x)$ or $g(x)$ to remain positive over the interval $[a, b]$. A more general result is stated in the following box:

THEOREM 1 Area Between Two Curves

If f and g are continuous and $f(x) \geq g(x)$ over the interval $[a, b]$, then the area bounded by $y = f(x)$ and $y = g(x)$ for $a \leq x \leq b$ is given exactly by

$$A = \int_a^b [f(x) - g(x)]\,dx$$

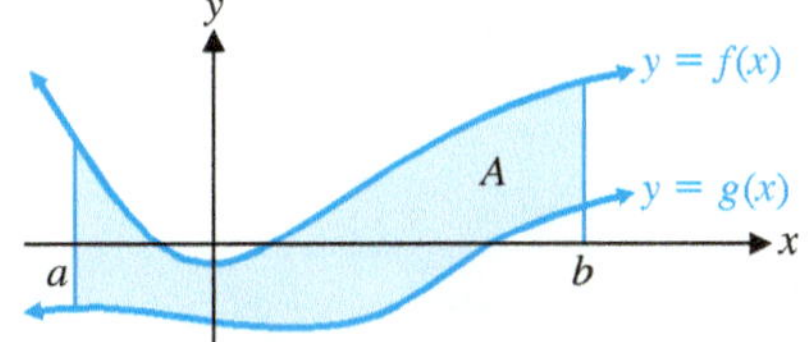

CONCEPTUAL INSIGHT

Theorem 1 requires the graph of f to be *above* (or equal to) the graph of g throughout $[a, b]$, but f and g can be either positive, negative, or 0. In Section 4-4, we considered the special cases of Theorem 1 in which (1) f is positive and g is the zero function on $[a, b]$; and (2) f is the zero function and g is negative on $[a, b]$:

Special case 1. If f is continuous and positive over $[a, b]$, then the area bounded by the graph of f and the x axis for $a \leq x \leq b$ is given exactly by

$$\int_a^b f(x)\,dx$$

Special case 2. If g is continuous and negative over $[a, b]$, then the area bounded by the graph of g and the x axis for $a \leq x \leq b$ is given exactly by

$$\int_a^b -g(x)\,dx$$

EXAMPLE 1 **Area Between a Curve and the x Axis** Find the area bounded by $f(x) = 6x - x^2$ and $y = 0$ for $1 \le x \le 4$.

SOLUTION We sketch a graph of the region first (Fig. 3). The solution of every area problem should begin with a sketch. Since $f(x) \ge 0$ on $[1, 4]$,

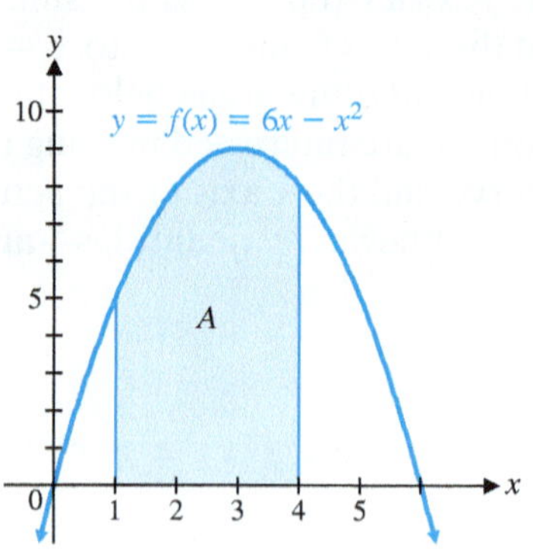

Figure 3

$$A = \int_1^4 (6x - x^2)\,dx = \left(3x^2 - \frac{x^3}{3}\right)\Big|_1^4$$

$$= \left[3(4)^2 - \frac{(4)^3}{3}\right] - \left[3(1)^2 - \frac{(1)^3}{3}\right]$$

$$= 48 - \tfrac{64}{3} - 3 + \tfrac{1}{3}$$

$$= 48 - 21 - 3$$

$$= 24$$

Matched Problem 1 Find the area bounded by $f(x) = x^2 + 1$ and $y = 0$ for $-1 \le x \le 3$.

EXAMPLE 2 **Area Between a Curve and the x Axis** Find the area between the graph of $f(x) = x^2 - 2x$ and the x axis over the indicated intervals:

(A) $[1, 2]$ (B) $[-1, 1]$

SOLUTION We begin by sketching the graph of f, as shown in Figure 4.

(A) From the graph, we see that $f(x) \le 0$ for $1 \le x \le 2$, so we integrate $-f(x)$:

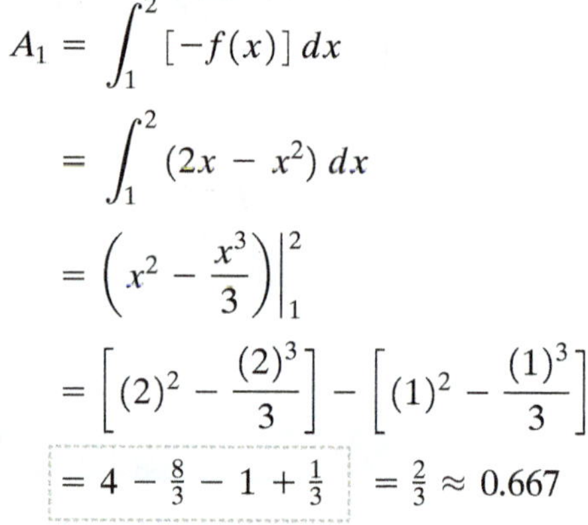

$$A_1 = \int_1^2 [-f(x)]\,dx$$

$$= \int_1^2 (2x - x^2)\,dx$$

$$= \left(x^2 - \frac{x^3}{3}\right)\Big|_1^2$$

$$= \left[(2)^2 - \frac{(2)^3}{3}\right] - \left[(1)^2 - \frac{(1)^3}{3}\right]$$

$$= 4 - \tfrac{8}{3} - 1 + \tfrac{1}{3} \quad = \tfrac{2}{3} \approx 0.667$$

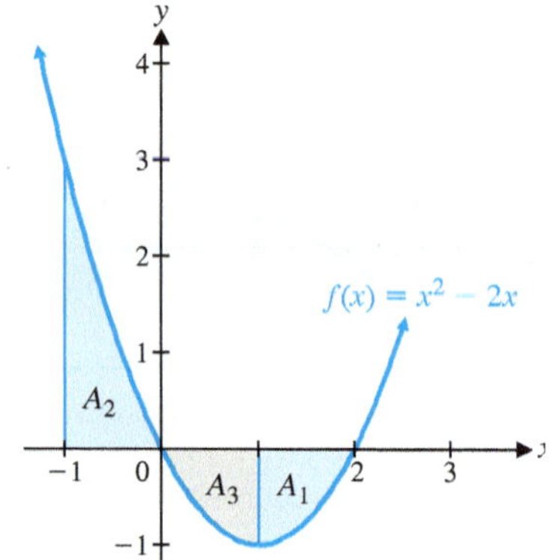

Figure 4

(B) Since the graph shows that $f(x) \ge 0$ on $[-1, 0]$ and $f(x) \le 0$ on $[0, 1]$, the computation of this area will require two integrals:

$$A = A_2 + A_3$$

$$= \int_{-1}^0 f(x)\,dx + \int_0^1 [-f(x)]\,dx$$

$$= \int_{-1}^0 (x^2 - 2x)\,dx + \int_0^1 (2x - x^2)\,dx$$

$$= \left(\frac{x^3}{3} - x^2\right)\Big|_{-1}^0 + \left(x^2 - \frac{x^3}{3}\right)\Big|_0^1$$

$$= \tfrac{4}{3} + \tfrac{2}{3} \quad = 2$$

Matched Problem 2 Find the area between the graph of $f(x) = x^2 - 9$ and the x axis over the indicated intervals:

(A) $[0, 2]$ (B) $[2, 4]$

EXAMPLE 3 **Area Between Two Curves** Find the area bounded by the graphs of $f(x) = \frac{1}{2}x + 3$, $g(x) = -x^2 + 1$, $x = -2$, and $x = 1$.

SOLUTION We first sketch the area (Fig. 5) and then set up and evaluate an appropriate definite integral. We observe from the graph that $f(x) \ge g(x)$ for $-2 \le x \le 1$, so

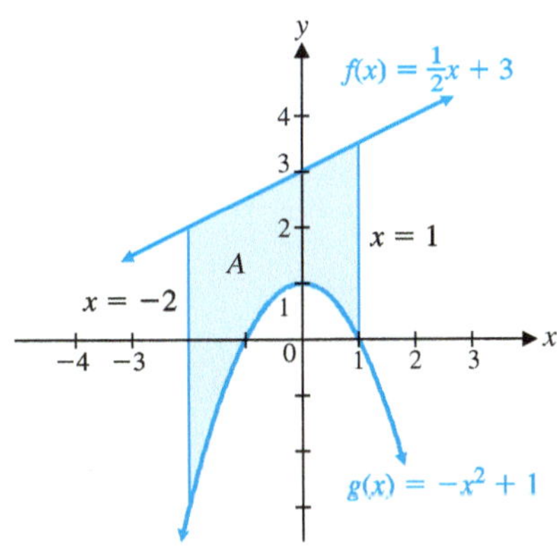

Figure 5

$$A = \int_{-2}^{1} [f(x) - g(x)]\,dx = \int_{-2}^{1} \left[\left(\frac{x}{2} + 3\right) - (-x^2 + 1)\right] dx$$

$$= \int_{-2}^{1} \left(x^2 + \frac{x}{2} + 2\right) dx$$

$$= \left(\frac{x^3}{3} + \frac{x^2}{4} + 2x\right)\Bigg|_{-2}^{1}$$

$$= \left(\frac{1}{3} + \frac{1}{4} + 2\right) - \left(\frac{-8}{3} + \frac{4}{4} - 4\right) = \frac{33}{4} = 8.25$$

Matched Problem 3 Find the area bounded by $f(x) = x^2 - 1$, $g(x) = -\frac{1}{2}x - 3$, $x = -1$, and $x = 2$.

EXAMPLE 4 **Area Between Two Curves** Find the area bounded by $f(x) = 5 - x^2$ and $g(x) = 2 - 2x$.

SOLUTION First, graph f and g on the same coordinate system, as shown in Figure 6. Since the statement of the problem does not include any limits on the values of x, we must determine the appropriate values from the graph. The graph of f is a parabola and the graph of g is a line. The area bounded by these two graphs extends from the intersection point on the left to the intersection point on the right. To find these intersection points, we solve the equation $f(x) = g(x)$ for x:

$$f(x) = g(x)$$
$$5 - x^2 = 2 - 2x$$
$$x^2 - 2x - 3 = 0$$
$$x = -1, 3$$

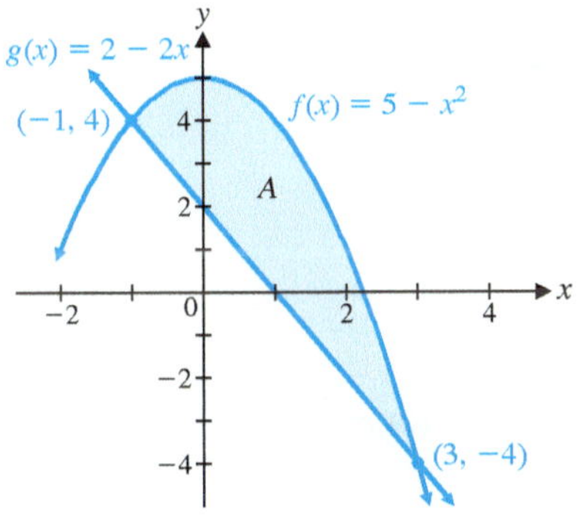

Figure 6

You should check these values in the original equations. (Note that the area between the graphs for $x < -1$ is unbounded on the left, and the area between the graphs for $x > 3$ is unbounded on the right.) Figure 6 shows that $f(x) \ge g(x)$ over the interval $[-1, 3]$, so we have

$$A = \int_{-1}^{3} [f(x) - g(x)]\,dx = \int_{-1}^{3} [5 - x^2 - (2 - 2x)]\,dx$$

$$= \int_{-1}^{3} (3 + 2x - x^2)\,dx$$

$$= \left(3x + x^2 - \frac{x^3}{3}\right)\Bigg|_{-1}^{3}$$

$$= \left[3(3) + (3)^2 - \frac{(3)^3}{3}\right] - \left[3(-1) + (-1)^2 - \frac{(-1)^3}{3}\right] = \frac{32}{3} \approx 10.667$$

Matched Problem 4 Find the area bounded by $f(x) = 6 - x^2$ and $g(x) = x$.

EXAMPLE 5 **Area Between Two Curves** Find the area bounded by $f(x) = x^2 - x$ and $g(x) = 2x$ for $-2 \le x \le 3$.

SOLUTION The graphs of f and g are shown in Figure 7. Examining the graph, we see that $f(x) \ge g(x)$ on the interval $[-2, 0]$, but $g(x) \ge f(x)$ on the interval $[0, 3]$. Thus, two integrals are required to compute this area:

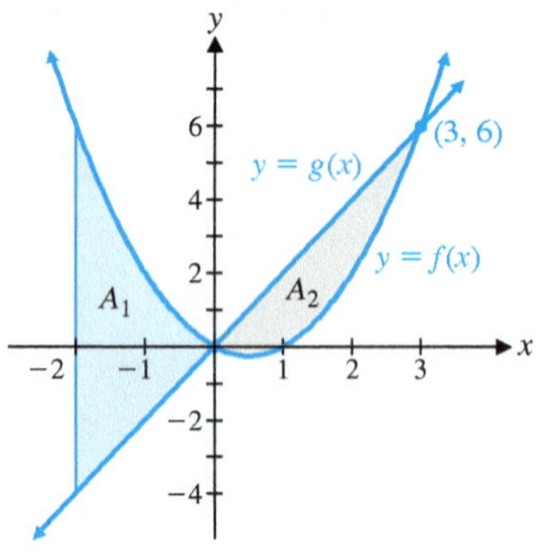

Figure 7

$$A_1 = \int_{-2}^{0} [f(x) - g(x)]\,dx \qquad f(x) \ge g(x) \text{ on } [-2, 0]$$

$$= \int_{-2}^{0} [x^2 - x - 2x]\,dx$$

$$= \int_{-2}^{0} (x^2 - 3x)\,dx$$

$$= \left(\frac{x^3}{3} - \frac{3}{2}x^2\right)\Big|_{-2}^{0}$$

$$= (0) - \left[\frac{(-2)^3}{3} - \frac{3}{2}(-2)^2\right] = \frac{26}{3} \approx 8.667$$

$$A_2 = \int_{0}^{3} [g(x) - f(x)]\,dx \qquad g(x) \ge f(x) \text{ on } [0, 3]$$

$$= \int_{0}^{3} [2x - (x^2 - x)]\,dx$$

$$= \int_{0}^{3} (3x - x^2)\,dx$$

$$= \left(\frac{3}{2}x^2 - \frac{x^3}{3}\right)\Big|_{0}^{3}$$

$$= \left[\frac{3}{2}(3)^2 - \frac{(3)^3}{3}\right] - (0) = \frac{9}{2} = 4.5$$

The total area between the two graphs is

$$A = A_1 + A_2 = \tfrac{26}{3} + \tfrac{9}{2} = \tfrac{79}{6} \approx 13.167$$

Matched Problem 5 Find the area bounded by $f(x) = 2x^2$ and $g(x) = 4 - 2x$ for $-2 \le x \le 2$.

EXAMPLE 6 **Computing Areas with a Numerical Integration Routine** Find the area (to three decimal places) bounded by $f(x) = e^{-x^2}$ and $g(x) = x^2 - 1$.

SOLUTION First, we use a graphing calculator to graph the functions f and g and find their intersection points (see Fig. 8A). We see that the graph of f is bell shaped and the graph of g is a parabola. We note that $f(x) \ge g(x)$ on the interval $[-1.131, 1.131]$ and compute the area A by a numerical integration routine (see Fig. 8B):

$$A = \int_{-1.131}^{1.131} \left[e^{-x^2} - (x^2 - 1)\right] dx = 2.876$$

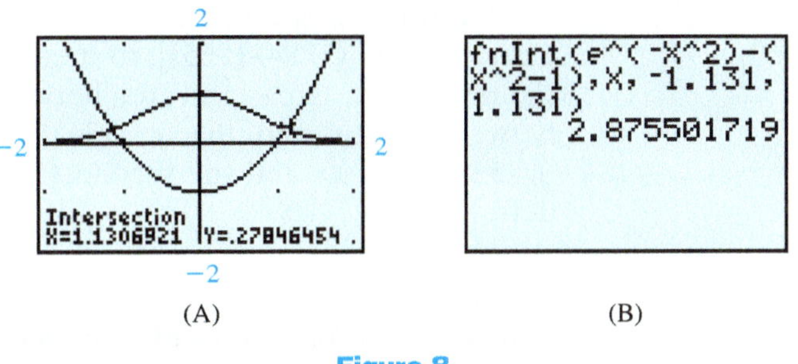

Figure 8

Matched Problem 6 Find the area (to three decimal places) bounded by the graphs of $f(x) = x^2 \ln x$ and $g(x) = 3x - 3$.

Application: Income Distribution

The U.S. Census Bureau compiles and analyzes a great deal of data having to do with the distribution of income among families in the United States. For 2006, the Bureau reported that the lowest 20% of families received 3% of all family income and the top 20% received 50%. Table 1 and Figure 9 give a detailed picture of the distribution of family income in 2006.

The graph of $y = f(x)$ in Figure 9 is called a **Lorenz curve** and is generally found by using *regression analysis,* a technique for fitting a function to a data set over a given interval. The variable ***x* represents the cumulative percentage of families at or below a given income level,** and ***y* represents the cumulative percentage of total family income received.** For example, data point (0.40, 0.12) in Table 1 indicates that the bottom 40% of families (those with incomes under $38,000) received 12% of the total income for all families in 2006, data point (0.60, 0.27) indicates that the bottom 60% of families received 27% of the total income for all families that year, and so on.

Table 1 Family Income Distribution in the United States, 2006

Income Level	x	y
Under $20,000	0.20	0.03
Under $38,000	0.40	0.12
Under $60,000	0.60	0.27
Under $97,000	0.80	0.49

Source: U.S. Census Bureau

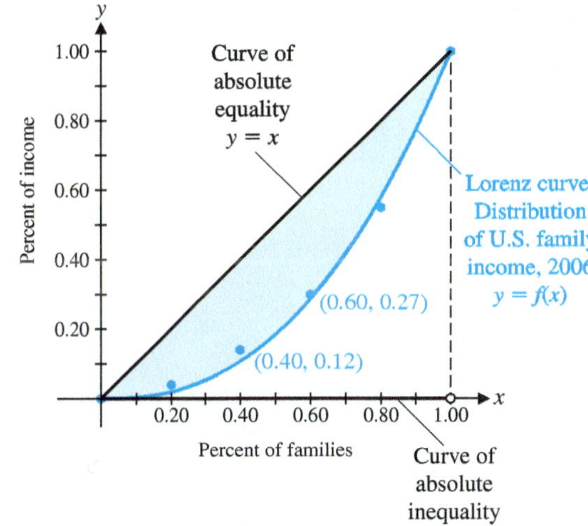

Figure 9 Lorenz curve

Absolute equality of income would occur if the area between the Lorenz curve and $y = x$ were 0. In this case, the Lorenz curve would be $y = x$ and all families would receive equal shares of the total income. That is, 5% of the families would receive 5% of the income, 20% of the families would receive 20% of the income, 65% of the families would receive 65% of the income, and so on. The maximum possible area between a Lorenz curve and $y = x$ is $\frac{1}{2}$, the area of the triangle below $y = x$. In this case, we would have **absolute inequality**: All the income would be in the hands of one family and the rest would have none. In actuality, Lorenz curves lie between these two extremes. But as the shaded area increases, the greater is the inequality of income distribution.

We use a single number, the **Gini index** [named after the Italian sociologist Corrado Gini (1884–1965)], to measure income concentration. The Gini index is the ratio of two areas: the area between $y = x$ and the Lorenz curve, and the area between $y = x$ and the x axis, from $x = 0$ to $x = 1$. The first area equals $\int_0^1 [x - f(x)]\,dx$ and the second (triangular) area equals $\frac{1}{2}$, giving the following definition:

DEFINITION Gini Index of Income Concentration

If $y = f(x)$ is the equation of a Lorenz curve, then

$$\textbf{Gini index} = 2\int_0^1 [x - f(x)]\,dx$$

The Gini index is always a number between 0 and 1:

A Gini index of 0 indicates absolute equality—all people share equally in the income. A Gini index of 1 indicates absolute inequality—one person has all the income and the rest have none.

The closer the index is to 0, the closer the income is to being equally distributed. The closer the index is to 1, the closer the income is to being concentrated in a few hands. The Gini index of income concentration is used to compare income distributions at various points in time, between different groups of people, before and after taxes are paid, between different countries, and so on.

EXAMPLE 7 **Distribution of Income** The Lorenz curve for the distribution of income in a certain country in 2010 is given by $f(x) = x^{2.6}$. Economists predict that the Lorenz curve for the country in the year 2025 will be given by $g(x) = x^{1.8}$. Find the Gini index of income concentration for each curve, and interpret the results.

SOLUTION The Lorenz curves are shown in Figure 10.

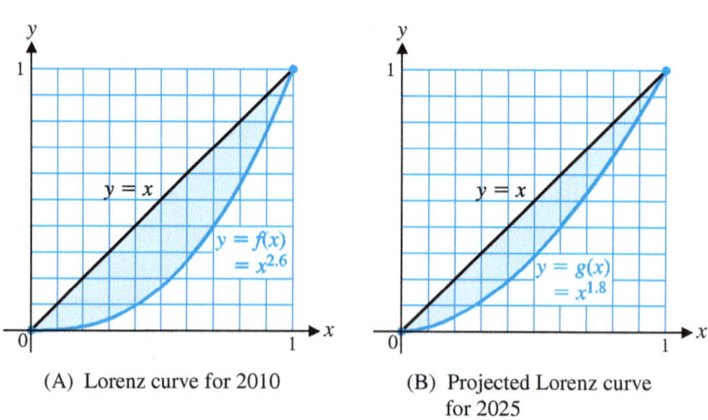

(A) Lorenz curve for 2010

(B) Projected Lorenz curve for 2025

Figure 10

The Gini index in 2010 is (see Fig. 10A)

$$2\int_0^1 [x - f(x)]\,dx = 2\int_0^1 [x - x^{2.6}]\,dx = 2\left(\frac{1}{2}x^2 - \frac{1}{3.6}x^{3.6}\right)\Big|_0^1$$

$$= 2\left(\frac{1}{2} - \frac{1}{3.6}\right) \approx 0.444$$

The projected Gini index in 2025 is (see Fig. 10B)

$$2\int_0^1 [x - g(x)]\,dx = 2\int_0^1 [x - x^{1.8}]\,dx = 2\left(\frac{1}{2}x^2 - \frac{1}{2.8}x^{2.8}\right)\Big|_0^1$$

$$= 2\left(\frac{1}{2} - \frac{1}{2.8}\right) \approx 0.286$$

If this projection is correct, the Gini index will decrease, and income will be more equally distributed in the year 2025 than in 2010.

Matched Problem 7 Repeat Example 7 if the projected Lorenz curve in the year 2025 is given by $g(x) = x^{3.8}$.

EXPLORE & DISCUSS 1 Do you agree or disagree with each of the following statements (explain your answers by referring to the data in Table 2):

(A) In countries with a low Gini index, there is little incentive for individuals to strive for success, and therefore productivity is low.

(B) In countries with a high Gini index, it is almost impossible to rise out of poverty, and therefore productivity is low.

Table 2

Country	Gini Index	Per Capita Gross Domestic Product
Brazil	0.57	$8,402
China	0.47	6,757
France	0.33	30,386
Germany	0.28	29,461
Japan	0.25	31,267
Jordan	0.39	5,530
Mexico	0.46	10,751
Russia	0.40	10,845
Sweden	0.25	32,525
United States	0.41	41,890

Source: The World Bank

Exercises 5-1

A

Problems 1–6 refer to Figures A–D on page 294. Set up definite integrals in Problems 1–4 that represent the indicated shaded area.

1. Shaded area in Figure B
2. Shaded area in Figure A
3. Shaded area in Figure C
4. Shaded area in Figure D

5. Explain why $\int_a^b h(x)\,dx$ does not represent the area between the graph of $y = h(x)$ and the x axis from $x = a$ to $x = b$ in Figure C.
6. Explain why $\int_a^b [-h(x)]\,dx$ represents the area between the graph of $y = h(x)$ and the x axis from $x = a$ to $x = b$ in Figure C.

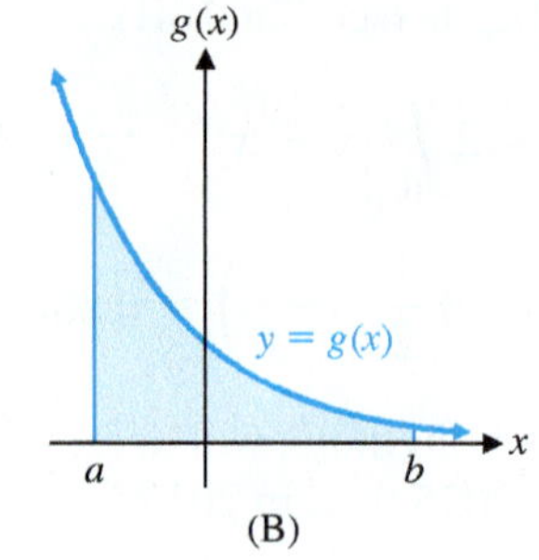

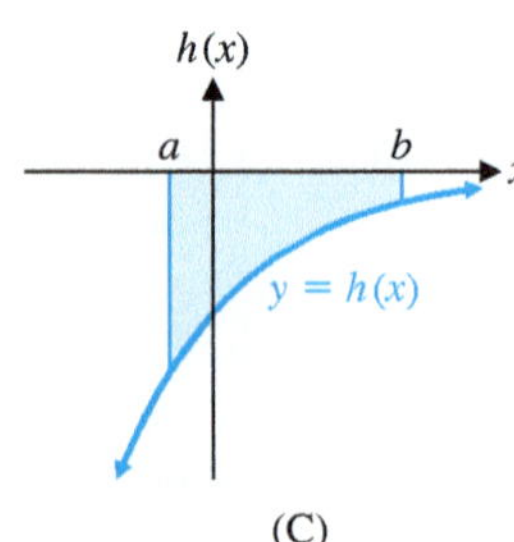

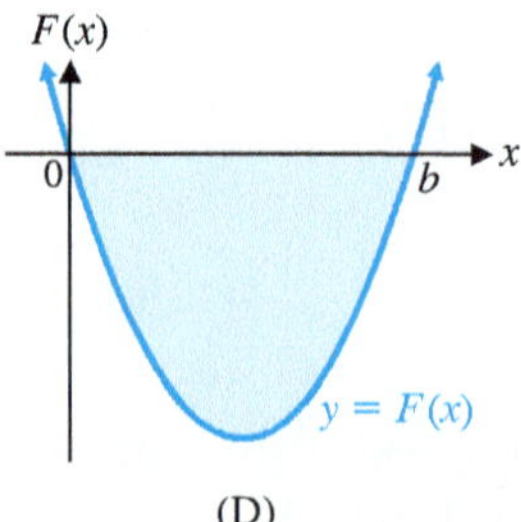

Figures for 1–6

In Problems 7–20, find the area bounded by the graphs of the indicated equations over the given interval. Compute answers to three decimal places.

7. $y = x + 4; y = 0; 0 \le x \le 4$

8. $y = -x + 10; y = 0; -2 \le x \le 2$

9. $y = x^2 - 20; y = 0; -3 \le x \le 0$

10. $y = x^2 + 2; y = 0; 0 \le x \le 3$

11. $y = -x^2 + 10; y = 0; -3 \le x \le 3$

12. $y = -2x^2; y = 0; -6 \le x \le 0$

13. $y = x^3 + 1; y = 0; 0 \le x \le 2$

14. $y = -x^3 + 3; y = 0; -2 \le x \le 1$

15. $y = x(1 - x); y = 0; -1 \le x \le 0$

16. $y = -x(3 - x); y = 0; 1 \le x \le 2$

17. $y = -e^x; y = 0; -1 \le x \le 1$

18. $y = e^x; y = 0; 0 \le x \le 1$

19. $y = \frac{1}{x}; y = 0; 1 \le x \le e$

20. $y = -\frac{1}{x}; y = 0; -1 \le x \le -\frac{1}{e}$

B

In Problems 21–26, use a definite integral to find the area bounded by the graphs of the indicated equations over the given interval. Then check your answer by finding the area without using a definite integral. [Hint: Partition the region into triangles and/or rectangles].

21. $y = x; y = -2; 0 \le x \le 3$

22. $y = -x; y = 8; 0 \le x \le 5$

23. $y = 2x + 3; y = x - 1; 4 \le x \le 6$

24. $y = 3x + 6; y = \frac{1}{2}x + 1; 8 \le x \le 10$

25. $y = -4x + 1; y = \frac{3}{2}x + 1; 0 \le x \le 4$

26. $y = -2x - 3; y = x - 3; 0 \le x \le 20$

Problems 27–36 refer to Figures A and B. Set up definite integrals in Problems 27–34 that represent the indicated shaded areas over the given intervals.

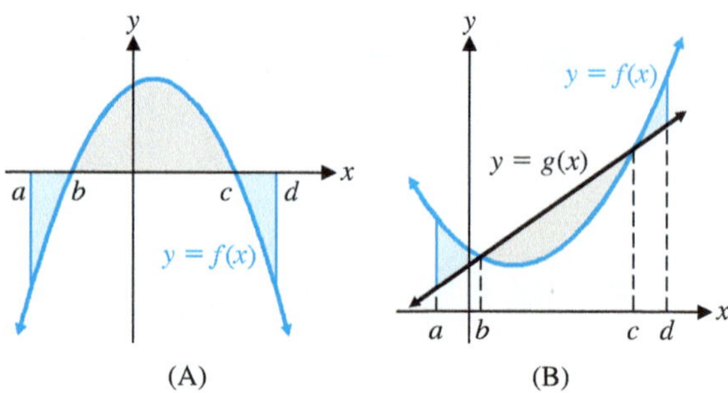

Figures for 27–34

27. Over interval $[a, b]$ in Figure A

28. Over interval $[c, d]$ in Figure A

29. Over interval $[b, d]$ in Figure A

30. Over interval $[a, c]$ in Figure A

31. Over interval $[c, d]$ in Figure B

32. Over interval $[a, b]$ in Figure B

33. Over interval $[a, c]$ in Figure B

34. Over interval $[b, d]$ in Figure B

35. Referring to Figure B, explain how you would use definite integrals and the functions f and g to find the area bounded by the two functions from $x = a$ to $x = d$.

36. Referring to Figure A, explain how you would use definite integrals to find the area between the graph of $y = f(x)$ and the x axis from $x = a$ to $x = d$.

In Problems 37–52, find the area bounded by the graphs of the indicated equations over the given intervals (when stated). Compute answers to three decimal places.

37. $y = -x; y = 0; -2 \le x \le 1$

38. $y = -x + 1; y = 0; -1 \le x \le 2$

39. $y = x^2 - 4; y = 0; 0 \le x \le 3$

40. $y = 4 - x^2; y = 0; 0 \le x \le 4$

41. $y = x^2 - 3x; y = 0; -2 \le x \le 2$

42. $y = -x^2 - 2x; y = 0; -2 \le x \le 1$

43. $y = -2x + 8; y = 12; -1 \le x \le 2$

44. $y = 2x + 6; y = 3; -1 \le x \le 2$

45. $y = 3x^2; y = 12$

46. $y = x^2; y = 9$

47. $y = 4 - x^2; y = -5$

48. $y = x^2 - 1; y = 3$

49. $y = x^2 + 1; y = 2x - 2; -1 \le x \le 2$

50. $y = x^2 - 1; y = x - 2; -2 \le x \le 1$

51. $y = e^{0.5x}; y = -\frac{1}{x}; 1 \le x \le 2$

52. $y = \frac{1}{x}; y = -e^x; 0.5 \le x \le 1$

In Problems 53–58, set up a definite integral that represents the area bounded by the graphs of the indicated equations over the given interval. Find the areas to three decimal places. [Hint: A circle of radius r, with center at the origin, has equation $x^2 + y^2 = r^2$ and area πr^2].

53. $y = \sqrt{9 - x^2}; y = 0; -3 \le x \le 3$

54. $y = \sqrt{25 - x^2}; y = 0; -5 \le x \le 5$

55. $y - -\sqrt{16 - x^2}; y - 0; 0 \le x \le 4$

56. $y = -\sqrt{36 - x^2}; y = 0; -6 \le x \le 0$

57. $y = -\sqrt{4 - x^2}; y = \sqrt{4 - x^2}; -2 \le x \le 2$

58. $y = -\sqrt{100 - x^2}; y = \sqrt{100 - x^2}; -10 \le x \le 10$

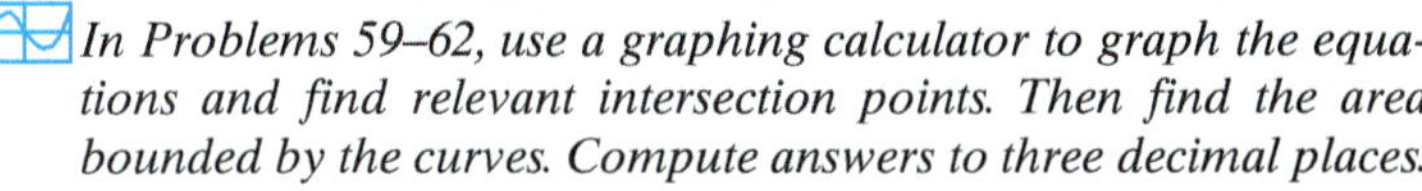
In Problems 59–62, use a graphing calculator to graph the equations and find relevant intersection points. Then find the area bounded by the curves. Compute answers to three decimal places.

59. $y = 3 - 5x - 2x^2; y = 2x^2 + 3x - 2$

60. $y = 3 - 2x^2; y = 2x^2 - 4x$

61. $y = -0.5x + 2.25; y = \frac{1}{x}$

62. $y = x - 4.25; y = -\frac{1}{x}$

C

In Problems 63–68, find the area bounded by the graphs of the indicated equations over the given intervals (when stated). Compute answers to three decimal places.

63. $y = e^x; y = e^{-x}; 0 \le x \le 4$

64. $y = e^x; y = -e^{-x}; 1 \le x \le 2$

65. $y = x^3; y = 4x$

66. $y = x^3 + 1; y = x + 1$

67. $y = x^3 - 3x^2 - 9x + 12; y = x + 12$

68. $y = x^3 - 6x^2 + 9x; y = x$

In Problems 69–74, use a graphing calculator to graph the equations and find relevant intersection points. Then find the area bounded by the curves. Compute answers to three decimal places.

69. $y = x^3 - x^2 + 2; y = -x^3 + 8x - 2$

70. $y = 2x^3 + 2x^2 - x; y = -2x^3 - 2x^2 + 2x$

71. $y = e^{-x}; y = 3 - 2x$

72. $y = 2 - (x + 1)^2; y = e^{x+1}$

73. $y = e^x; y = 5x - x^3$

74. $y = 2 - e^x; y = x^3 + 3x^2$

In Problems 75–78, use a numerical integration routine on a graphing calculator to find the area bounded by the graphs of the indicated equations over the given interval (when stated). Compute answers to three decimal places.

75. $y = e^{-x}; y = \sqrt{\ln x}; 2 \le x \le 5$

76. $y = x^2 + 3x + 1; y = e^{e^x}; -3 \le x \le 0$

77. $y = e^{x^2}; y = x + 2$

78. $y = \ln(\ln x); y = 0.01x$

Applications

In the applications that follow, it is helpful to sketch graphs to get a clearer understanding of each problem and to interpret results. A graphing calculator will prove useful if you have one, but it is not necessary.

79. Oil production. Using production and geological data, the management of an oil company estimates that oil will be pumped from a producing field at a rate given by

$$R(t) = \frac{100}{t + 10} + 10 \qquad 0 \le t \le 15$$

where $R(t)$ is the rate of production (in thousands of barrels per year) t years after pumping begins. Find the area between the graph of R and the t axis over the interval $[5, 10]$ and interpret the results.

80. Oil production. In Problem 79, if the rate is found to be

$$R(t) = \frac{100t}{t^2 + 25} + 4 \qquad 0 \le t \le 25$$

then find the area between the graph of R and the t axis over the interval $[5, 15]$ and interpret the results.

81. Useful life. An amusement company maintains records for each video game it installs in an arcade. Suppose that $C(t)$ and $R(t)$ represent the total accumulated costs and revenues (in thousands of dollars), respectively, t years after a particular game has been installed. If

$$C'(t) = 2 \qquad \text{and} \qquad R'(t) = 9e^{-0.3t}$$

then find the area between the graphs of C' and R' over the interval on the t axis from 0 to the useful life of the game and interpret the results.

82. Useful life. Repeat Problem 81 if

$$C'(t) = 2t \qquad \text{and} \qquad R'(t) = 5te^{-0.1t^2}$$

83. Income distribution. In a study on the effects of World War II on the U.S. economy, an economist used data from the U.S. Census Bureau to produce the following Lorenz curves for the distribution of U.S. income in 1935 and in 1947:

$$f(x) = x^{2.4} \qquad \text{Lorenz curve for 1935}$$

$$g(x) = x^{1.6} \qquad \text{Lorenz curve for 1947}$$

Find the Gini index of income concentration for each Lorenz curve and interpret the results.

84. Income distribution. Using data from the U.S. Census Bureau, an economist produced the following Lorenz curves for the distribution of U.S. income in 1962 and in 1972:

$$f(x) = \tfrac{3}{10}x + \tfrac{7}{10}x^2 \quad \text{Lorenz curve for 1962}$$

$$g(x) = \tfrac{1}{2}x + \tfrac{1}{2}x^2 \quad \text{Lorenz curve for 1972}$$

Find the Gini index of income concentration for each Lorenz curve and interpret the results.

85. Distribution of wealth. Lorenz curves also can provide a relative measure of the distribution of a country's total assets. Using data in a report by the U.S. Congressional Joint Economic Committee, an economist produced the following Lorenz curves for the distribution of total U.S. assets in 1963 and in 1983:

$$f(x) = x^{10} \quad \text{Lorenz curve for 1963}$$

$$g(x) = x^{12} \quad \text{Lorenz curve for 1983}$$

Find the Gini index of income concentration for each Lorenz curve and interpret the results.

86. Income distribution. The government of a small country is planning sweeping changes in the tax structure in order to provide a more equitable distribution of income. The Lorenz curves for the current income distribution and for the projected income distribution after enactment of the tax changes are as follows:

$$f(x) = x^{2.3} \quad \text{Current Lorenz curve}$$

$$g(x) = 0.4x + 0.6x^2 \quad \text{Projected Lorenz curve after changes in tax laws}$$

Find the Gini index of income concentration for each Lorenz curve. Will the proposed changes provide a more equitable income distribution? Explain.

87. Distribution of wealth. The data in the following table describe the distribution of wealth in a country:

x	0	0.20	0.40	0.60	0.80	1
y	0	0.12	0.31	0.54	0.78	1

(A) Use quadratic regression to find the equation of a Lorenz curve for the data.

(B) Use the regression equation and a numerical integration routine to approximate the Gini index of income concentration.

88. Distribution of wealth. Refer to Problem 87.

(A) Use cubic regression to find the equation of a Lorenz curve for the data.

(B) Use the cubic regression equation you found in Part (A) and a numerical integration routine to approximate the Gini index of income concentration.

89. Biology. A yeast culture is growing at a rate of $W'(t) = 0.3e^{0.1t}$ grams per hour. Find the area between the graph of W' and the t axis over the interval $[0, 10]$ and interpret the results.

90. Natural resource depletion. The instantaneous rate of change in demand for U.S. lumber since 1970 ($t = 0$), in billions of cubic feet per year, is given by

$$Q'(t) = 12 + 0.006t^2 \qquad 0 \le t \le 50$$

Find the area between the graph of Q' and the t axis over the interval $[15, 20]$, and interpret the results.

91. Learning. A college language class was chosen for a learning experiment. Using a list of 50 words, the experiment measured the rate of vocabulary memorization at different times during a continuous 5-hour study session. The average rate of learning for the entire class was inversely proportional to the time spent studying and was given approximately by

$$V'(t) = \frac{15}{t} \qquad 1 \le t \le 5$$

Find the area between the graph of V' and the t axis over the interval $[2, 4]$, and interpret the results.

92. Learning. Repeat Problem 91 if $V'(t) = 13/t^{1/2}$ and the interval is changed to $[1, 4]$.

Answers to Matched Problems

1. $A = \int_{-1}^{3} (x^2 + 1)\,dx = \frac{40}{3} \approx 13.333$
2. (A) $A = \int_{0}^{2} (9 - x^2)\,dx = \frac{46}{3} \approx 15.333$

   (B) $A = \int_{2}^{3} (9 - x^2)\,dx + \int_{3}^{4} (x^2 - 9)\,dx = 6$
3. $A = \displaystyle\int_{-1}^{2} \left[(x^2 - 1) - \left(-\frac{x}{2} - 3\right)\right] dx = \frac{39}{4} = 9.75$
4. $A = \int_{-3}^{2} [(6 - x^2) - x]\,dx = \frac{125}{6} \approx 20.833$
5. $A = \int_{-2}^{1} [(4 - 2x) - 2x^2]\,dx + \int_{1}^{2} [2x^2 - (4 - 2x)]\,dx = \frac{38}{3} \approx 12.667$
6. 0.443
7. Gini index of income concentration ≈ 0.583; income will be less equally distributed in 2025.

5-2 Applications in Business and Economics

- Probability Density Functions
- Continuous Income Stream
- Future Value of a Continuous Income Stream
- Consumers' and Producers' Surplus

This section contains important applications of the definite integral to business and economics. Included are three independent topics: probability density functions, continuous income streams, and consumers' and producers' surplus. Any of the three may be covered in any order as time and interests dictate.

Probability Density Functions

We now take a brief, informal look at the use of the definite integral to determine probabilities. A more formal treatment of the subject requires the use of the special "improper" integral form $\int_{-\infty}^{\infty} f(x)\,dx$, which we will not discuss.

Suppose that an experiment is designed in such a way that any real number x on the interval $[c, d]$ is a possible outcome. For example, x may represent an IQ score, the height of a person in inches, or the life of a lightbulb in hours. Technically, we refer to x as a *continuous random variable.*

In certain situations, we can find a function f with x as an independent variable such that the function f can be used to determine the probability that the outcome x of an experiment will be in the interval $[c, d]$. Such a function, called a **probability density function**, must satisfy the following three conditions (see Fig. 1):

1. $f(x) \geq 0$ for all real x.
2. The area under the graph of $f(x)$ over the interval $(-\infty, \infty)$ is exactly 1.
3. If $[c, d]$ is a subinterval of $(-\infty, \infty)$, then

$$\text{Probability } (c \leq x \leq d) = \int_c^d f(x)\,dx$$

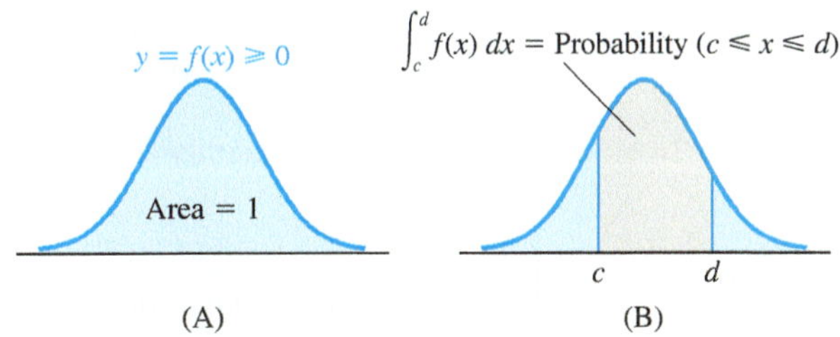

Figure 1 Probability density function

EXAMPLE 1 **Duration of Telephone Calls** Suppose that the length of telephone calls (in minutes) is a continuous random variable with the probability density function shown in Figure 2:

$$f(t) = \begin{cases} \frac{1}{4}e^{-t/4} & \text{if } t \geq 0 \\ 0 & \text{otherwise} \end{cases}$$

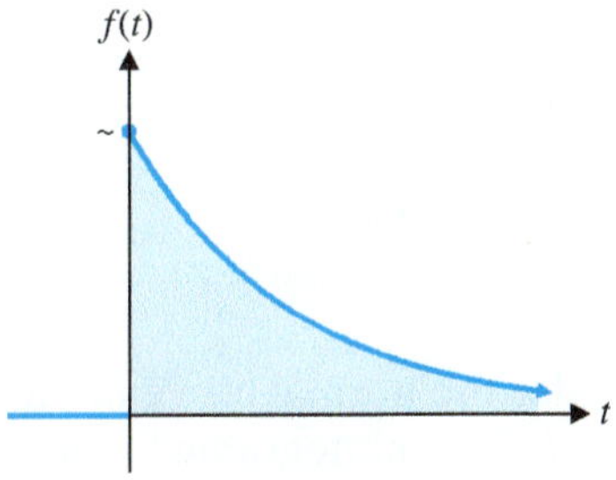

(A) Determine the probability that a call selected at random will last between 2 and 3 minutes.

(B) Find b (to two decimal places) so that the probability of a call selected at random lasting between 2 and b minutes is .5.

Figure 2

SOLUTION (A)
$$\text{Probability } (2 \leq t \leq 3) = \int_2^3 \tfrac{1}{4}e^{-t/4}\,dt$$
$$= (-e^{-t/4})\Big|_2^3$$
$$= -e^{-3/4} + e^{-1/2} \approx .13$$

(B) We want to find b such that Probability $(2 \le t \le b) = .5$.

$$\int_2^b \tfrac{1}{4}e^{-t/4}\,dt = .5$$

$$-e^{-b/4} + e^{-1/2} = .5 \qquad \text{Solve for } b.$$

$$e^{-b/4} = e^{-.5} - .5$$

$$-\frac{b}{4} = \ln(e^{-.5} - .5)$$

$$b = 8.96 \text{ minutes}$$

So the probability of a call selected at random lasting from 2 to 8.96 minutes is .5.

Matched Problem 1 (A) In Example 1, find the probability that a call selected at random will last 4 minutes or less.

(B) Find b (to two decimal places) so that the probability of a call selected at random lasting b minutes or less is .9

CONCEPTUAL INSIGHT

The probability that a phone call in Example 1 lasts exactly 2 minutes (not 1.999 minutes, not 1.999 999 minutes) is given by

$$\text{Probability } (2 \le t \le 2) = \int_2^2 \tfrac{1}{4}e^{-t/4}\,dt \qquad \text{Use Property 1, Section 4-4}$$

$$= 0$$

In fact, for any *continuous* random variable x with probability density function $f(x)$, the probability that x is exactly equal to a constant c is equal to 0:

$$\text{Probability } (c \le x \le c) = \int_c^c f(x)\,dx \qquad \text{Use Property 1, Section 4-4}$$

$$= 0$$

In this respect, a *continuous* random variable differs from a *discrete* random variable. If x, for example, is the discrete random variable that represents the number of dots that appear on the top face when a fair die is rolled, then

$$\text{Probability } (2 \le x \le 2) = \tfrac{1}{6}$$

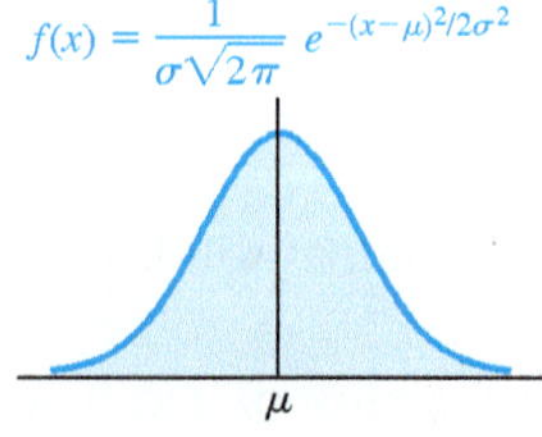

Figure 3 Normal curve

One of the most important probability density functions, the **normal probability density function**, is defined as follows and graphed in Figure 3:

$$f(x) = \frac{1}{\sigma\sqrt{2\pi}} e^{-(x-\mu)^2/2\sigma^2} \qquad \begin{array}{l}\mu \text{ is the mean.}\\ \sigma \text{ is the standard deviation.}\end{array}$$

It can be shown (but not easily) that the area under the normal curve in Figure 3 over the interval $(-\infty, \infty)$ is exactly 1. Since $\int e^{-x^2}\,dx$ is nonintegrable in terms of elementary functions (that is, the antiderivative cannot be expressed as a finite combination of simple functions), probabilities such as

$$\text{Probability } (c \le x \le d) = \frac{1}{\sigma\sqrt{2\pi}} \int_c^d e^{-(x-\mu)^2/2\sigma^2}\,dx$$

can be determined by making an appropriate substitution in the integrand and then using a table of areas under the standard normal curve (that is, the normal curve with $\mu = 0$ and $\sigma = 1$). As an alternative to a table, calculators and computers can be used to compute areas under normal curves.

Continuous Income Stream

We start with a simple example having an obvious solution and generalize the concept to examples having less obvious solutions.

Suppose that an aunt has established a trust that pays you \$2,000 a year for 10 years. What is the total amount you will receive from the trust by the end of the 10th year? Since there are 10 payments of \$2,000 each, you will receive

$$10 \times \$2{,}000 = \$20{,}000$$

We now look at the same problem from a different point of view. Let's assume that the income stream is continuous at a rate of \$2,000 per year. In Figure 4, the area under the graph of $f(t) = 2{,}000$ from 0 to t represents the income accumulated t years after the start. For example, for $t = \frac{1}{4}$ year, the income would be $\frac{1}{4}(2{,}000) = \$500$; for $t = \frac{1}{2}$ year, the income would be $\frac{1}{2}(2{,}000) = \$1{,}000$; for $t = 1$ year, the income would be $1(2{,}000) = \$2{,}000$; for $t = 5.3$ years, the income would be $5.3(2{,}000) = \$10{,}600$; and for $t = 10$ years, the income would be $10(2{,}000) = \$20{,}000$. The total income over a 10-year period—that is, the area under the graph of $f(t) = 2{,}000$ from 0 to 10—is also given by the definite integral

$$\int_0^{10} 2{,}000\, dt = 2{,}000t\Big|_0^{10} = 2{,}000(10) - 2{,}000(0) = \$20{,}000$$

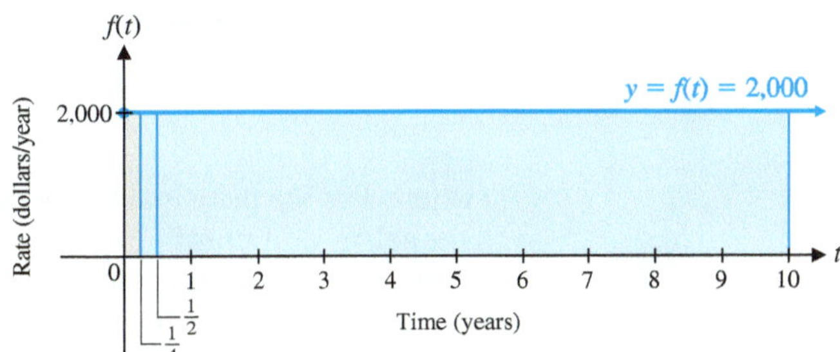

Figure 4 Continuous income stream

EXAMPLE 2 **Continuous Income Stream** The rate of change of the income produced by a vending machine is given by

$$f(t) = 5{,}000e^{0.04t}$$

where t is time in years since the installation of the machine. Find the total income produced by the machine during the first 5 years of operation.

SOLUTION The area under the graph of the rate-of-change function from 0 to 5 represents the total change in income over the first 5 years (Fig. 5), and is given by a definite integral:

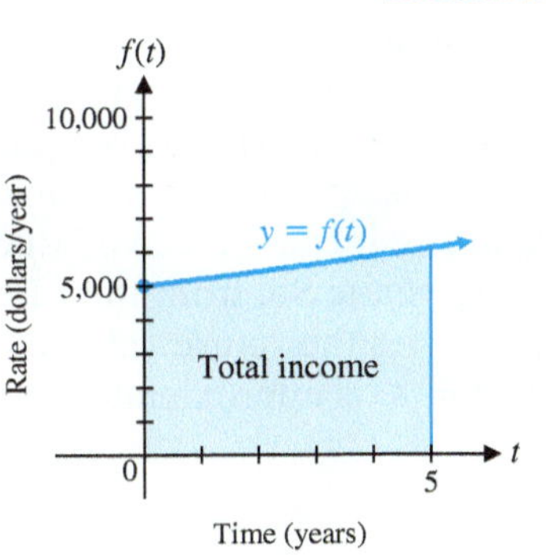

Figure 5 Continuous income stream

$$\begin{aligned}\text{Total income} &= \int_0^5 5{,}000e^{0.04t}\, dt \\ &= 125{,}000e^{0.04t}\Big|_0^5 \\ &= 125{,}000e^{0.04(5)} - 125{,}000e^{0.04(0)} \\ &= 152{,}675 - 125{,}000 \\ &= \$27{,}675 \qquad \text{Rounded to the nearest dollar}\end{aligned}$$

The vending machine produces a total income of \$27,675 during the first 5 years of operation.

Matched Problem 2 Referring to Example 2, find the total income produced (to the nearest dollar) during the second 5 years of operation.

In reality, income from a vending machine is not usually received as a single payment at the end of each year, even though the rate is given as a yearly rate. Income is usually collected on a daily or weekly basis. In problems of this type, it is convenient to assume that income is actually received in a **continuous stream**; that is, we assume that the rate at which income is received is a continuous function of time. The rate of change is called the **rate of flow** of the continuous income stream. In general, we have the following definition:

DEFINITION Total Income for a Continuous Income Stream

If $f(t)$ is the rate of flow of a continuous income stream, then the **total income** produced during the period from $t = a$ to $t = b$ is

$$\text{Total income} = \int_a^b f(t)\,dt$$

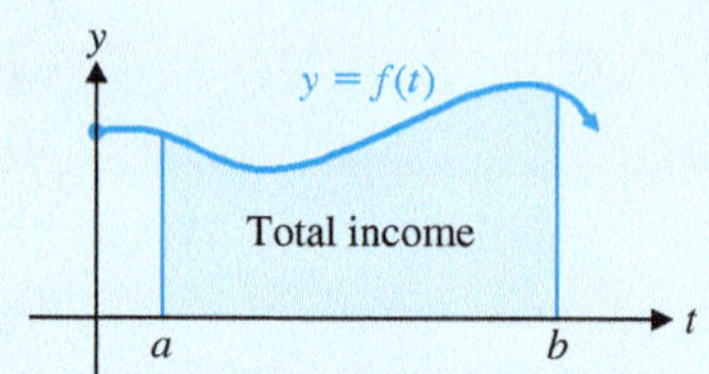

Future Value of a Continuous Income Stream

In Section 2-1, we discussed the continuous compound interest formula

$$A = Pe^{rt}$$

where P is the principal (or present value), A is the amount (or future value), r is the annual rate of continuous compounding (expressed as a decimal), and t is time in years. For example, if money is worth 12% compounded continuously, then the future value of a \$10,000 investment in 5 years is (to the nearest dollar)

$$A = 10{,}000e^{0.12(5)} = \$18{,}221$$

We want to apply the future value concept to the income produced by a continuous income stream. Suppose that $f(t)$ is the rate of flow of a continuous income stream, and the income produced by this continuous income stream is invested as soon as it is received at a rate r, compounded continuously. We already know how to find the total income produced after T years, but how can we find the total of the income produced and the interest earned by this income? Since the income is received in a continuous flow, we cannot just use the formula $A = Pe^{rt}$. This formula is valid only for a single deposit P, not for a continuous flow of income. Instead, we use a Riemann sum approach that will allow us to apply the formula $A = Pe^{rt}$ repeatedly. To begin, we divide the time interval $[0, T]$ into n equal subintervals of length Δt and choose an arbitrary point c_k in each subinterval, as shown in Figure 6.

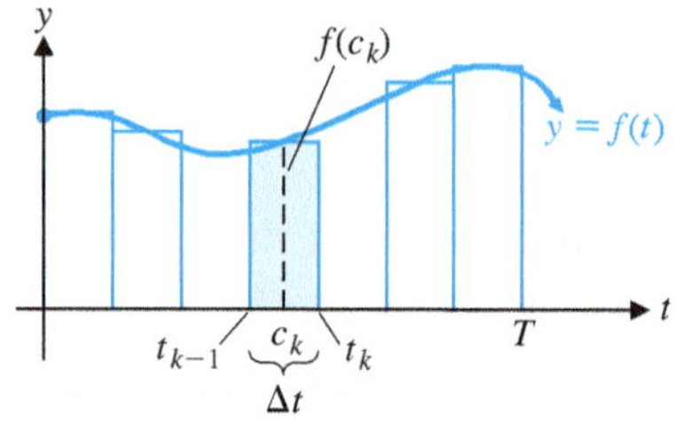

Figure 6

The total income produced during the period from $t = t_{k-1}$ to $t = t_k$ is equal to the area under the graph of $f(t)$ over this subinterval and is approximately equal to $f(c_k)\,\Delta t$, the area of the shaded rectangle in Figure 6. The income received during this period will earn interest for approximately $T - c_k$ years. So, from the future-value formula $A = Pe^{rt}$ with $P = f(c_k)\,\Delta t$ and $t = T - c_k$, the future value of the income produced during the period from $t = t_{k-1}$ to $t = t_k$ is approximately equal to

$$f(c_k)\,\Delta t\,e^{(T-c_k)r}$$

The total of these approximate future values over n subintervals is then

$$f(c_1)\,\Delta t\,e^{(T-c_1)r} + f(c_2)\,\Delta t\,e^{(T-c_2)r} + \cdots + f(c_n)\,\Delta t\,e^{(T-c_n)r} = \sum_{k=1}^{n} f(c_k)e^{r(T-c_k)}\,\Delta t$$

This equation has the form of a Riemann sum, the limit of which is a definite integral. (See the definition of the definite integral in Section 4-4.) Therefore, the *future value FV* of the income produced by the continuous income stream is given by

$$FV = \int_0^T f(t)e^{r(T-t)}\,dt$$

Since r and T are constants, we also can write

$$FV = \int_0^T f(t)e^{rT}e^{-rt}\,dt = e^{rT}\int_0^T f(t)e^{-rt}\,dt \qquad (1)$$

This last form is preferable, since the integral is usually easier to evaluate than the first form.

DEFINITION Future Value of a Continuous Income Stream

If $f(t)$ is the rate of flow of a continuous income stream, $0 \le t \le T$, and if the income is continuously invested at a rate r, compounded continuously, then the **future value FV** at the end of T years is given by

$$FV = \int_0^T f(t)e^{r(T-t)}\,dt = e^{rT}\int_0^T f(t)e^{-rt}\,dt$$

The future value of a continuous income stream is the total value of all money produced by the continuous income stream (income and interest) at the end of T years.

We return to the trust that your aunt set up for you. Suppose that the \$2,000 per year you receive from the trust is invested as soon as it is received at 8%, compounded continuously. We consider the trust income to be a continuous income stream with a flow rate of \$2,000 per year. What is its future value (to the nearest dollar) by the end of the 10th year? Using the definite integral for future value from the preceding box, we have

$$FV = e^{rT}\int_0^T f(t)e^{-rt}\,dt$$

$$FV = e^{0.08(10)}\int_0^{10} 2{,}000e^{-0.08t}\,dt \qquad r = 0.08,\ T = 10,\ f(t) = 2{,}000$$

$$= 2{,}000e^{0.8}\int_0^{10} e^{-0.08t}\,dt$$

$$= 2{,}000e^{0.8}\left[\frac{e^{-0.08t}}{-0.08}\right]\Bigg|_0^{10}$$

$$= 2{,}000e^{0.8}[-12.5e^{-0.8} + 12.5] = \$30{,}639$$

At the end of 10 years, you will have received \$30,639, including interest. How much is interest? Since you received \$20,000 in income from the trust, the interest is the difference between the future value and income. So,

$$\$30{,}639 - \$20{,}000 = \$10{,}639$$

is the interest earned by the income received from the trust over the 10-year period.

EXAMPLE 3 **Future Value of a Continuous Income Stream** Using the continuous income rate of flow for the vending machine in Example 2, namely,

$$f(t) = 5{,}000e^{0.04t}$$

find the future value of this income stream at 12%, compounded continuously for 5 years, and find the total interest earned. Compute answers to the nearest dollar.

SOLUTION Using the formula

$$FV = e^{rT}\int_0^T f(t)e^{-rt}\,dt$$

with $r = 0.12$, $T = 5$, and $f(t) = 5{,}000e^{0.04t}$, we have

$$\begin{aligned} FV &= e^{0.12(5)}\int_0^5 5{,}000e^{0.04t}e^{-0.12t}\,dt \\ &= 5{,}000e^{0.6}\int_0^5 e^{-0.08t}\,dt \\ &= 5{,}000e^{0.6}\left(\frac{e^{-0.08t}}{-0.08}\right)\Bigg|_0^5 \\ &= 5{,}000e^{0.6}(-12.5e^{-0.4} + 12.5) \\ &= \$37{,}545 \quad \text{Rounded to the nearest dollar} \end{aligned}$$

The future value of the income stream at 12% compounded continuously at the end of 5 years is \$37,545.

In Example 2, we saw that the total income produced by this vending machine over a 5-year period was \$27,675. The difference between future value and income is interest. So,

$$\$37{,}545 - \$27{,}675 = \$9{,}870$$

is the interest earned by the income produced by the vending machine during the 5-year period.

Matched Problem 3 Repeat Example 3 if the interest rate is 9%, compounded continuously.

Consumers' and Producers' Surplus

Figure 7

Let $p = D(x)$ be the price–demand equation for a product, where x is the number of units of the product that consumers will purchase at a price of \$$p$ per unit. Suppose that $\overline{p}$, is the current price and $\overline{x}$ is the number of units that can be sold at that price. Then the price–demand curve in Figure 7 shows that if the price is higher than $\overline{p}$, the demand x is less than $\overline{x}$, but some consumers are still willing to pay the higher price. Consumers who are willing to pay more than $\overline{p}$, but who are still able to buy the product at $\overline{p}$, have saved money. We want to determine the total amount saved by all the consumers who are willing to pay a price higher than $\overline{p}$ for the product.

To do this, consider the interval $[c_k, c_k + \Delta x]$, where $c_k + \Delta x < \overline{x}$. If the price remained constant over that interval, the savings on each unit would be the difference between $D(c_k)$, the price consumers are willing to pay, and $\overline{p}$, the price they actually pay. Since Δx represents the number of units purchased by consumers over the interval, the total savings to consumers over this interval is approximately equal to

$$[D(c_k) - \overline{p}]\,\Delta x \quad \text{(savings per unit)} \times \text{(number of units)}$$

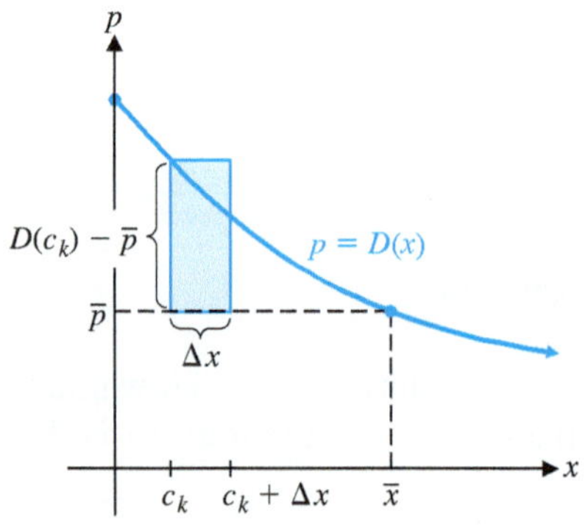

which is the area of the shaded rectangle shown in Figure 7. If we divide the interval $[0, \overline{x}]$ into n equal subintervals, then the total savings to consumers is approximately equal to

$$[D(c_1) - \overline{p}]\,\Delta x + [D(c_2) - \overline{p}]\,\Delta x + \cdots + [D(c_n) - \overline{p}]\,\Delta x = \sum_{k=1}^{n}[D(c_k) - \overline{p}]\,\Delta x$$

which we recognize as a Riemann sum for the integral

$$\int_0^{\overline{x}} [D(x) - \overline{p}]\,dx$$

We define the *consumers' surplus* to be this integral.

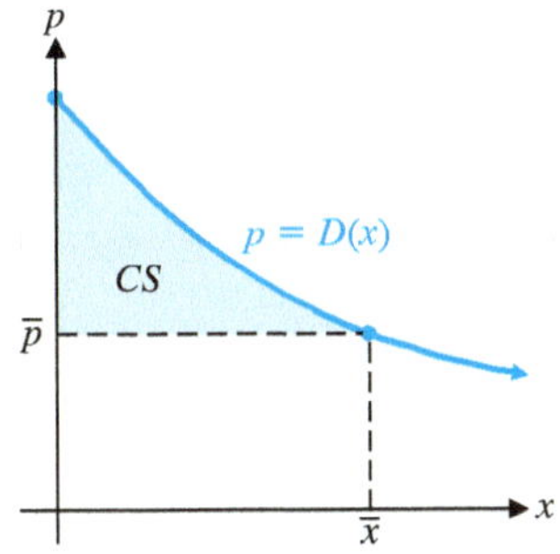

Figure 8

DEFINITION Consumers' Surplus

If $(\overline{x}, \overline{p})$ is a point on the graph of the price–demand equation $p = D(x)$ for a particular product, then the **consumers' surplus *CS*** at a price level of $\overline{p}$ is

$$CS = \int_0^{\overline{x}} [D(x) - \overline{p}]\,dx$$

which is the area between $p = \overline{p}$ and $p = D(x)$ from $x = 0$ to $x = \overline{x}$, as shown in Figure 8.

The consumers' surplus represents the total savings to consumers who are willing to pay more than $\overline{p}$ for the product but are still able to buy the product for $\overline{p}$.

EXAMPLE 4 **Consumers' Surplus** Find the consumers' surplus at a price level of \$8 for the price–demand equation

$$p = D(x) = 20 - 0.05x$$

SOLUTION **Step 1** Find $\overline{x}$, the demand when the price is $\overline{p} = 8$:

$$\begin{aligned}\overline{p} &= 20 - 0.05\overline{x}\\ 8 &= 20 - 0.05\overline{x}\\ 0.05\overline{x} &= 12\\ \overline{x} &= 240\end{aligned}$$

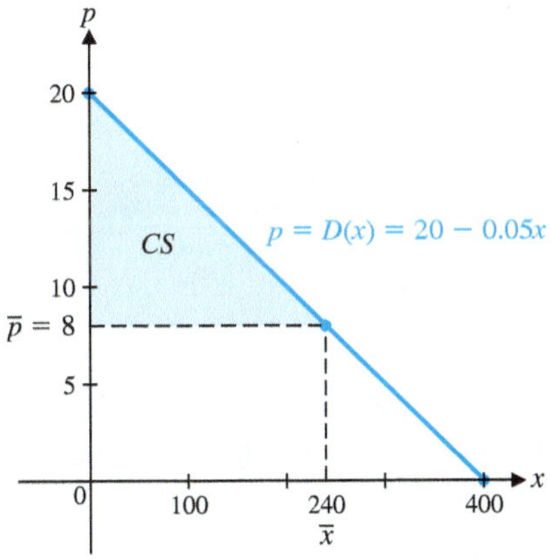

Step 2 Sketch a graph, as shown in Figure 9.

Figure 9

Step 3 Find the consumers' surplus (the shaded area in the graph):

$$\begin{aligned}CS &= \int_0^{\overline{x}} [D(x) - \overline{p}]\,dx\\ &= \int_0^{240} (20 - 0.05x - 8)\,dx\\ &= \int_0^{240} (12 - 0.05x)\,dx\\ &= (12x - 0.025x^2)\Big|_0^{240}\\ &= 2{,}880 - 1{,}440 = \$1{,}440\end{aligned}$$

The total savings to consumers who are willing to pay a higher price for the product is \$1,440.

Matched Problem 4 Repeat Example 4 for a price level of \$4.

If $p = S(x)$ is the price–supply equation for a product, $\overline{p}$ is the current price, and $\overline{x}$ is the current supply, then some suppliers are still willing to supply some units at a lower price than $\overline{p}$. The additional money that these suppliers gain from the higher price is called the *producers' surplus* and can be expressed in terms of a definite integral (proceeding as we did for the consumers' surplus).

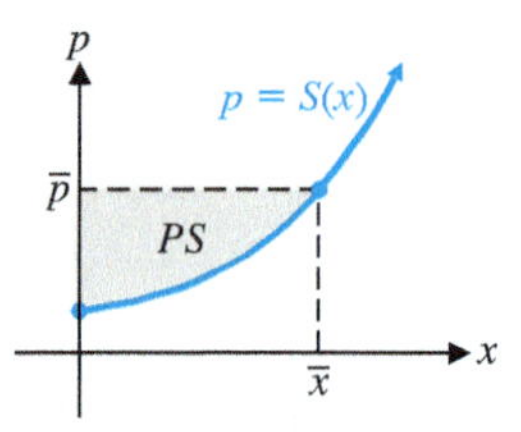

Figure 10

DEFINITION Producers' Surplus

If $(\overline{x}, \overline{p})$ is a point on the graph of the price–supply equation $p = S(x)$, then the **producers' surplus *PS*** at a price level of $\overline{p}$ is

$$PS = \int_0^{\overline{x}} [\overline{p} - S(x)]\, dx$$

which is the area between $p = \overline{p}$ and $p = S(x)$ from $x = 0$ to $x = \overline{x}$, as shown in Figure 10.

The producers' surplus represents the total gain to producers who are willing to supply units at a lower price than $\overline{p}$ but are still able to supply units at $\overline{p}$.

EXAMPLE 5 **Producers' Surplus** Find the producers' surplus at a price level of \$20 for the price–supply equation

$$p = S(x) = 2 + 0.0002x^2$$

SOLUTION **Step 1** Find $\overline{x}$, the supply when the price is $\overline{p} = 20$:

$$\overline{p} = 2 + 0.0002\overline{x}^2$$
$$20 = 2 + 0.0002\overline{x}^2$$
$$0.0002\overline{x}^2 = 18$$
$$\overline{x}^2 = 90{,}000$$
$$\overline{x} = 300 \qquad \text{There is only one solution, since } \overline{x} \geq 0.$$

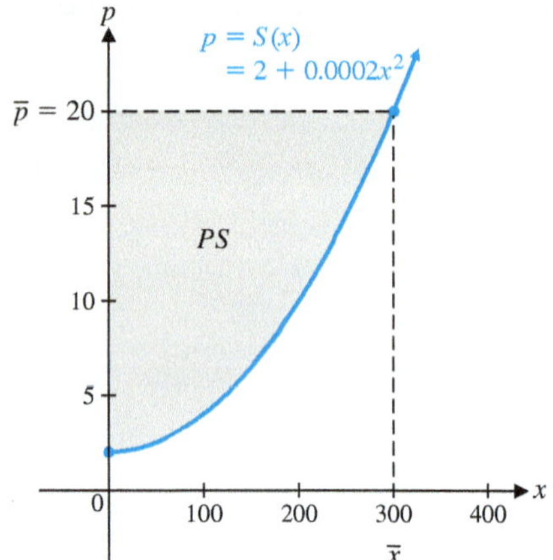

Figure 11

Step 2 Sketch a graph, as shown in Figure 11.

Step 3 Find the producers' surplus (the shaded area in the graph):

$$PS = \int_0^{\overline{x}} [\overline{p} - S(x)]\, dx = \int_0^{300} [20 - (2 + 0.0002x^2)]\, dx$$

$$= \int_0^{300} (18 - 0.0002x^2)\, dx = \left(18x - 0.0002\frac{x^3}{3}\right)\Big|_0^{300}$$

$$= 5{,}400 - 1{,}800 = \$3{,}600$$

The total gain to producers who are willing to supply units at a lower price is \$3,600.

Matched Problem 5 Repeat Example 5 for a price level of \$4.

In a free competitive market, the price of a product is determined by the relationship between supply and demand. If $p = D(x)$ and $p = S(x)$ are the price–demand and price–supply equations, respectively, for a product and if $(\overline{x}, \overline{p})$ is the point of intersection of these equations, then $\overline{p}$ is called the **equilibrium price** and $\overline{x}$ is called the **equilibrium quantity**. If the price stabilizes at the equilibrium price $\overline{p}$, then this is the price level that will determine both the consumers' surplus and the producers' surplus.

EXAMPLE 6 **Equilibrium Price and Consumers' and Producers' Surplus** Find the equilibrium price and then find the consumers' surplus and producers' surplus at the equilibrium price level, if

$$p = D(x) = 20 - 0.05x \qquad \text{and} \qquad p = S(x) = 2 + 0.0002x^2$$

SOLUTION **Step 1** Find the equilibrium point. Set $D(x)$ equal to $S(x)$ and solve:

$$\begin{aligned} D(x) &= S(x) \\ 20 - 0.05x &= 2 + 0.0002x^2 \\ 0.0002x^2 + 0.05x - 18 &= 0 \\ x^2 + 250x - 90{,}000 &= 0 \\ x &= 200, -450 \end{aligned}$$

Figure 12

Since x cannot be negative, the only solution is $x = 200$. The equilibrium price can be determined by using $D(x)$ or $S(x)$. We will use both to check our work:

$$\begin{aligned} \overline{p} &= D(200) \\ &= 20 - 0.05(200) = 10 \end{aligned} \qquad \begin{aligned} \overline{p} &= S(200) \\ &= 2 + 0.0002(200)^2 = 10 \end{aligned}$$

The equilibrium price is $\overline{p} = 10$, and the equilibrium quantity is $\overline{x} = 200$.

Step 2 Sketch a graph, as shown in Figure 12.

Step 3 Find the consumers' surplus:

$$\begin{aligned} CS &= \int_0^{\overline{x}} [D(x) - \overline{p}]\, dx = \int_0^{200} (20 - 0.05x - 10)\, dx \\ &= \int_0^{200} (10 - 0.05x)\, dx \\ &= (10x - 0.025x^2)\Big|_0^{200} \\ &= 2{,}000 - 1{,}000 = \$1{,}000 \end{aligned}$$

Step 4 Find the producers' surplus:

$$\begin{aligned} PS &= \int_0^{\overline{x}} [\overline{p} - S(x)]\, dx \\ &= \int_0^{200} [10 - (2 + 0.0002x^2)]\, dx \\ &= \int_0^{200} (8 - 0.0002x^2)\, dx \\ &= \left(8x - 0.0002\frac{x^3}{3}\right)\Bigg|_0^{200} \\ &= 1{,}600 - \tfrac{1{,}600}{3} \approx \$1{,}067 \end{aligned}$$

Rounded to the nearest dollar

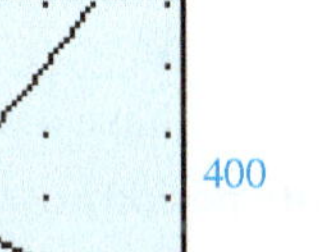

(A)

(B)

Figure 13

A graphing calculator offers an alternative approach to finding the equilibrium point for Example 6 (Fig. 13A). A numerical integration command can then be used to find the consumers' and producers' surplus (Fig. 13B).

Matched Problem 6 Repeat Example 6 for

$$p = D(x) = 25 - 0.001x^2 \qquad \text{and} \qquad p = S(x) = 5 + 0.1x$$

Exercises 5-2

A

In Problems 1–10, evaluate each definite integral to two decimal places.

1. $\int_0^1 e^{-2t}\,dt$

2. $\int_1^3 e^{-t}\,dt$

3. $\int_0^2 e^{4(2-t)}\,dt$

4. $\int_0^1 e^{3(1-t)}\,dt$

5. $\int_0^8 e^{0.06(8-t)}\,dt$

6. $\int_1^{10} e^{0.07(10-t)}\,dt$

7. $\int_0^{20} e^{0.08t}e^{0.12(20-t)}\,dt$

8. $\int_0^{15} e^{0.05t}e^{0.06(15-t)}\,dt$

9. $\int_0^{30} 500e^{0.02t}e^{0.09(30-t)}\,dt$

10. $\int_0^{25} 900e^{0.03t}e^{0.04(25-t)}\,dt$

B

In Problems 11 and 12, explain which of (A), (B), and (C) are equal before evaluating the expressions. Then evaluate each expression to two decimal places.

11. (A) $\int_0^8 e^{0.07(8-t)}\,dt$

(B) $\int_0^8 (e^{0.56} - e^{0.07t})\,dt$

(C) $e^{0.56}\int_0^8 e^{-0.07t}\,dt$

12. (A) $\int_0^{10} 2{,}000e^{0.05t}e^{0.12(10-t)}\,dt$

(B) $2{,}000e^{1.2}\int_0^{10} e^{-0.07t}\,dt$

(C) $2{,}000e^{0.05}\int_0^{10} e^{0.12(10-t)}\,dt$

Applications

Unless stated to the contrary, compute all monetary answers to the nearest dollar.

13. The life expectancy (in years) of a microwave oven is a continuous random variable with probability density function

$$f(x) = \begin{cases} 2/(x+2)^2 & \text{if } x \ge 0 \\ 0 & \text{otherwise} \end{cases}$$

(A) Find the probability that a randomly selected microwave oven lasts at most 6 years.

(B) Find the probability that a randomly selected microwave oven lasts from 6 to 12 years.

(C) Graph $y = f(x)$ for $[0, 12]$ and show the shaded region for part (A).

14. The shelf life (in years) of a laser pointer battery is a continuous random variable with probability density function

$$f(x) = \begin{cases} 1/(x+1)^2 & \text{if } x \ge 0 \\ 0 & \text{otherwise} \end{cases}$$

(A) Find the probability that a randomly selected laser pointer battery has a shelf life of 3 years or less.

(B) Find the probability that a randomly selected laser pointer battery has a shelf life of from 3 to 9 years.

(C) Graph $y = f(x)$ for $[0, 10]$ and show the shaded region for part (A).

15. In Problem 13, find d so that the probability of a randomly selected microwave oven lasting d years or less is .8.

16. In Problem 14, find d so that the probability of a randomly selected laser pointer battery lasting d years or less is .5.

17. A manufacturer guarantees a product for 1 year. The time to failure of the product after it is sold is given by the probability density function

$$f(t) = \begin{cases} .01e^{-.01t} & \text{if } t \ge 0 \\ 0 & \text{otherwise} \end{cases}$$

where t is time in months. What is the probability that a buyer chosen at random will have a product failure

(A) During the warranty period?

(B) During the second year after purchase?

18. In a certain city, the daily use of water (in hundreds of gallons) per household is a continuous random variable with probability density function

$$f(x) = \begin{cases} .15e^{-.15x} & \text{if } x \ge 0 \\ 0 & \text{otherwise} \end{cases}$$

Find the probability that a household chosen at random will use

(A) At most 400 gallons of water per day

(B) Between 300 and 600 gallons of water per day

19. In Problem 17, what is the probability that the product will last at least 1 year? [*Hint:* Recall that the total area under the probability density function curve is 1.]

20. In Problem 18, what is the probability that a household will use more than 400 gallons of water per day? [See the hint in Problem 19.]

21. Find the total income produced by a continuous income stream in the first 5 years if the rate of flow is $f(t) = 2{,}500$.

22. Find the total income produced by a continuous income stream in the first 10 years if the rate of flow is $f(t) = 3{,}000$.

23. Interpret the results of Problem 21 with both a graph and a description of the graph.

24. Interpret the results of Problem 22 with both a graph and a description of the graph.

25. Find the total income produced by a continuous income stream in the first 3 years if the rate of flow is $f(t) = 400e^{0.05t}$.

26. Find the total income produced by a continuous income stream in the first 2 years if the rate of flow is $f(t) = 600e^{0.06t}$.

27. Interpret the results of Problem 25 with both a graph and a description of the graph.

28. Interpret the results of Problem 26 with both a graph and a description of the graph.

29. Starting at age 25, you deposit \$2,000 a year into an IRA account. Treat the yearly deposits into the account as a continuous income stream. If money in the account earns 5%, compounded continuously, how much will be in the account 40 years later, when you retire at age 65? How much of the final amount is interest?

30. Suppose in Problem 29 that you start the IRA deposits at age 30, but the account earns 6%, compounded continuously. Treat the yearly deposits into the account as a continuous income stream. How much will be in the account 35 years later when you retire at age 65? How much of the final amount is interest?

31. Find the future value at 3.25% interest, compounded continuously for 4 years, of the continuous income stream with rate of flow $f(t) = 1{,}650e^{-0.02t}$.

32. Find the future value, at 2.95% interest, compounded continuously for 6 years, of the continuous income stream with rate of flow $f(t) = 2{,}000e^{0.06t}$.

33. Compute the interest earned in Problem 31.

34. Compute the interest earned in Problem 32.

35. An investor is presented with a choice of two investments: an established clothing store and a new computer store. Each choice requires the same initial investment and each produces a continuous income stream of 4%, compounded continuously. The rate of flow of income from the clothing store is $f(t) = 12{,}000$, and the rate of flow of income from the computer store is expected to be $g(t) = 10{,}000e^{0.05t}$. Compare the future values of these investments to determine which is the better choice over the next 5 years.

36. Refer to Problem 35. Which investment is the better choice over the next 10 years?

37. An investor has \$10,000 to invest in either a bond that matures in 5 years or a business that will produce a continuous stream of income over the next 5 years with rate of flow $f(t) = 2{,}150$. If both the bond and the continuous income stream earn 3.75%, compounded continuously, which is the better investment?

38. Refer to Problem 37. Which is the better investment if the rate of the income from the business is $f(t) = 2{,}250$?

39. A business is planning to purchase a piece of equipment that will produce a continuous stream of income for 8 years with rate of flow $f(t) = 9{,}000$. If the continuous income stream earns 6.95%, compounded continuously, what single deposit into an account earning the same interest rate will produce the same future value as the continuous income stream? (This deposit is called the **present value** of the continuous income stream.)

40. Refer to Problem 39. Find the present value of a continuous income stream at 7.65%, compounded continuously for 12 years, if the rate of flow is $f(t) = 1{,}000e^{0.03t}$.

41. Find the future value at a rate r, compounded continuously for T years, of a continuous income stream with rate of flow $f(t) = k$, where k is a constant.

42. Find the future value at a rate r, compounded continuously for T years, of a continuous income stream with rate of flow $f(t) = ke^{ct}$, where c and k are constants, $c \neq r$.

43. Find the consumers' surplus at a price level of $\overline{p} = \$150$ for the price–demand equation

$$p = D(x) = 400 - 0.05x$$

44. Find the consumers' surplus at a price level of $\overline{p} = \$120$ for the price–demand equation

$$p = D(x) = 200 - 0.02x$$

45. Interpret the results of Problem 43 with both a graph and a description of the graph.

46. Interpret the results of Problem 44 with both a graph and a description of the graph.

47. Find the producers' surplus at a price level of $\overline{p} = \$67$ for the price–supply equation

$$p = S(x) = 10 + 0.1x + 0.0003x^2$$

48. Find the producers' surplus at a price level of $\overline{p} = \$55$ for the price–supply equation

$$p = S(x) = 15 + 0.1x + 0.003x^2$$

49. Interpret the results of Problem 47 with both a graph and a description of the graph.

50. Interpret the results of Problem 48 with both a graph and a description of the graph.

In Problems 51–58, find the consumers' surplus and the producers' surplus at the equilibrium price level for the given price–demand and price–supply equations. Include a graph that identifies the consumers' surplus and the producers' surplus. Round all values to the nearest integer.

51. $p = D(x) = 50 - 0.1x;\ p = S(x) = 11 + 0.05x$

52. $p = D(x) = 25 - 0.004x^2;\ p = S(x) = 5 + 0.004x^2$

53. $p = D(x) = 80e^{-0.001x};\ p = S(x) = 30e^{0.001x}$

54. $p = D(x) = 185e^{-0.005x};\ p = S(x) = 25e^{0.005x}$

55. $p = D(x) = 80 - 0.04x;\ p = S(x) = 30e^{0.001x}$

56. $p = D(x) = 190 - 0.2x;\ p = S(x) = 25e^{0.005x}$

57. $p = D(x) = 80e^{-0.001x};\ p = S(x) = 15 + 0.0001x^2$

58. $p = D(x) = 185e^{-0.005x};\ p = S(x) = 20 + 0.002x^2$

59. The following tables give price–demand and price–supply data for the sale of soybeans at a grain market, where x is the number of bushels of soybeans (in thousands of bushels) and p is the price per bushel (in dollars):

Tables for 59–60

Price–Demand		Price–Supply	
x	$p = D(x)$	x	$p = S(x)$
0	6.70	0	6.43
10	6.59	10	6.45
20	6.52	20	6.48
30	6.47	30	6.53
40	6.45	40	6.62

Use quadratic regression to model the price–demand data and linear regression to model the price–supply data.

(A) Find the equilibrium quantity (to three decimal places) and equilibrium price (to the nearest cent).

(B) Use a numerical integration routine to find the consumers' surplus and producers' surplus at the equilibrium price level.

60. Repeat Problem 59, using quadratic regression to model both sets of data.

Answers to Matched Problems

1. (A) .63 (B) 9.21 min

2. $33,803

3. $FV = \$34,691$; interest $= \$7,016$

4. $2,560

5. $133

6. $\overline{p} = 15$; $CS = \$667$; $PS = \$500$

5-3 Integration by Parts

In Section 4-1, we promised to return later to the indefinite integral

$$\int \ln x \, dx$$

since none of the integration techniques considered up to that time could be used to find an antiderivative for $\ln x$. We now develop a very useful technique, called *integration by parts,* that will enable us to find not only the preceding integral, but also many others, including integrals such as

$$\int x \ln x \, dx \qquad \text{and} \qquad \int xe^x \, dx$$

The method of integration by parts is based on the product formula for derivatives. If f and g are differentiable functions, then

$$\frac{d}{dx}[f(x)g(x)] = f(x)g'(x) + g(x)f'(x)$$

which can be written in the equivalent form

$$f(x)g'(x) = \frac{d}{dx}[f(x)g(x)] - g(x)f'(x)$$

Integrating both sides, we obtain

$$\int f(x)g'(x)\, dx = \int \frac{d}{dx}[f(x)g(x)]\, dx - \int g(x)f'(x)\, dx$$

The first integral to the right of the equal sign is $f(x)g(x) + C$. Why? We will leave out the constant of integration for now, since we can add it after integrating the second integral to the right of the equal sign. So,

$$\int f(x)g'(x)\, dx = f(x)g(x) - \int g(x)f'(x)\, dx$$

This equation can be transformed into a more convenient form by letting $u = f(x)$ and $v = g(x)$; then $du = f'(x)\, dx$ and $dv = g'(x)\, dx$. Making these substitutions, we obtain the **integration-by-parts formula**:

Integration-by-Parts Formula

$$\int u\,dv = uv - \int v\,du$$

This formula can be very useful when the integral on the left is difficult or impossible to integrate with standard formulas. If u and dv are chosen with care—this is the crucial part of the process—then the integral on the right side may be easier to integrate than the one on the left. The formula provides us with another tool that is helpful in many, but not all, cases. We are able to easily check the results by differentiating to get the original integrand, a good habit to develop.

EXAMPLE 1 **Integration by Parts** Find $\int xe^x\,dx$, using integration by parts, and check the result.

SOLUTION First, write the integration-by-parts formula:

$$\int u\,dv = uv - \int v\,du \tag{1}$$

Now try to identify u and dv in $\int xe^x\,dx$ so that $\int v\,du$ on the right side of (1) is easier to integrate than $\int u\,dv = \int xe^x\,dx$ on the left side. There are essentially two reasonable choices in selecting u and dv in $\int xe^x\,dx$:

Choice 1	Choice 2
$\int \overbrace{x}^{u}\,\overbrace{e^x\,dx}^{dv}$	$\int \overbrace{e^x}^{u}\,\overbrace{x\,dx}^{dv}$

We pursue choice 1 and leave choice 2 for you to explore (see Explore & Discuss 1 following this example).

From choice 1, $u = x$ and $dv = e^x\,dx$. Looking at formula (1), we need du and v to complete the right side. Let

$$u = x \qquad dv = e^x\,dx$$

Then,

$$du = dx \qquad \int dv = \int e^x\,dx$$

$$v = e^x$$

Any constant may be added to v, but we will always choose 0 for simplicity. The general arbitrary constant of integration will be added at the end of the process.

Substituting these results into formula (1), we obtain

$$\int u\,dv = uv - \int v\,du$$

$$\int xe^x\,dx = xe^x - \int e^x\,dx \quad \text{The right integral is easy to integrate.}$$

$$= xe^x - e^x + C \quad \text{Now add the arbitrary constant } C.$$

Check:

$$\frac{d}{dx}(xe^x - e^x + C) = xe^x + e^x - e^x = xe^x$$

Explore & Discuss 1 Pursue choice 2 in Example 1, using the integration-by-parts formula, and explain why this choice does not work out.

Matched Problem 1 Find $\int xe^{2x}\,dx$.

EXAMPLE 2 **Integration by Parts** Find $\int x \ln x\,dx$.

SOLUTION As before, we have essentially two choices in choosing u and dv:

$$\text{Choice 1: } \int \overbrace{x}^{u}\ \overbrace{\ln x\,dx}^{dv} \qquad \text{Choice 2: } \int \overbrace{\ln x}^{u}\ \overbrace{x\,dx}^{dv}$$

Choice 1 is rejected since we do not yet know how to find an antiderivative of ln x. So we move to choice 2 and choose $u = \ln x$ and $dv = x\,dx$. Then we proceed as in Example 1. Let

$$u = \ln x \qquad dv = x\,dx$$

Then,

$$du = \frac{1}{x}dx \qquad \int dv = \int x\,dx$$

$$v = \frac{x^2}{2}$$

Substitute these results into the integration-by-parts formula:

$$\int u\,dv = uv - \int v du$$

$$\int x \ln x\,dx = (\ln x)\left(\frac{x^2}{2}\right) - \int \left(\frac{x^2}{2}\right)\left(\frac{1}{x}\right)dx$$

$$= \frac{x^2}{2}\ln x - \int \frac{x}{2}dx \qquad \text{An easy integral to evaluate}$$

$$= \frac{x^2}{2}\ln x - \frac{x^2}{4} + C$$

Check:

$$\frac{d}{dx}\left(\frac{x^2}{2}\ln x - \frac{x^2}{4} + C\right) = x \ln x + \left(\frac{x^2}{2}\cdot\frac{1}{x}\right) - \frac{x}{2} = x \ln x$$

Matched Problem 2 Find $\int x \ln 2x\,dx$.

CONCEPTUAL INSIGHT

As you may have discovered in Explore & Discuss 1, some choices for u and dv will lead to integrals that are more complicated than the original integral. This does not mean that there is an error in either the calculations or the integration-by-parts formula. It simply means that the particular choice of u and dv does not change the problem into one we can solve. When this happens, we must look for a different choice of u and dv. In some problems, it is possible that no choice will work.

Guidelines for selecting u and dv for integration by parts are summarized in the following box:

SUMMARY Integration by Parts: Selection of u and dv

For $\int u\,dv = uv - \int v\,du$,

1. The product $u\,dv$ must equal the original integrand.
2. It must be possible to integrate dv (preferably by using standard formulas or simple substitutions).
3. The new integral $\int v\,du$ should not be more complicated than the original integral $\int u\,dv$.
4. For integrals involving $x^p e^{ax}$, try

$$u = x^p \qquad \text{and} \qquad dv = e^{ax}\,dx$$

5. For integrals involving $x^p(\ln x)^q$, try

$$u = (\ln x)^q \qquad \text{and} \qquad dv = x^p\,dx$$

In some cases, repeated use of the integration-by-parts formula will lead to the evaluation of the original integral. The next example provides an illustration of such a case.

EXAMPLE 3 **Repeated Use of Integration by Parts** Find $\int x^2 e^{-x}\,dx$.

SOLUTION Following suggestion 4 in the box, we choose

$$u = x^2 \qquad dv = e^{-x}\,dx$$

Then,

$$du = 2x\,dx \qquad v = -e^{-x}$$

and

$$x^2e^{-x}\,dx = x^2(-e^{-x}) - (-e^{-x})2x\,dx$$
$$= -x^2e^{-x} + 2\,xe^{-x}\,dx \qquad (2)$$

The new integral is not one we can evaluate by standard formulas, but it is simpler than the original integral. Applying the integration-by-parts formula to it will produce an even simpler integral. For the integral $\int xe^{-x}\,dx$, we choose

$$u = x \qquad dv = e^{-x}\,dx$$

Then,

$$du = dx \qquad v = -e^{-x}$$

and

$$\int xe^{-x}\,dx = x(-e^{-x}) - \int(-e^{-x})\,dx$$
$$= -xe^{-x} + \int e^{-x}\,dx$$
$$= -xe^{-x} - e^{-x} \qquad \text{Choose 0 for the constant.} \qquad (3)$$

Substituting equation (3) into equation (2), we have

$$\int x^2e^{-x}\,dx = -x^2e^{-x} + 2(-xe^{-x} - e^{-x}) + C \qquad \text{Add an arbitrary constant here.}$$
$$= -x^2e^{-x} - 2xe^{-x} - 2e^{-x} + C$$

Check:

$$\frac{d}{dx}(-x^2e^{-x} - 2xe^{-x} - 2e^{-x} + C) = x^2e^{-x} - 2xe^{-x} + 2xe^{-x} - 2e^{-x} + 2e^{-x}$$
$$= x^2e^{-x}$$

Matched Problem 3 Find $\int x^2e^{2x}\,dx$.

EXAMPLE 4 **Using Integration by Parts** Find $\int_1^e \ln x\,dx$ and interpret the result geometrically.

SOLUTION First, we find $\int \ln x\,dx$. Then we return to the definite integral. Following suggestion 5 in the box (with $p = 0$), we choose

$$u = \ln x \qquad dv = dx$$

Then,

$$du = \frac{1}{x}dx \qquad v = x$$

$$\int \ln x\,dx = (\ln x)(x) - \int (x)\frac{1}{x}dx$$
$$= x\ln x - x + C$$

This is the important result we mentioned at the beginning of this section. Now we have

$$\int_1^e \ln x\,dx = (x\ln x - x)\Big|_1^e$$
$$= (e\ln e - e) - (1\ln 1 - 1)$$
$$= (e - e) - (0 - 1)$$
$$= 1$$

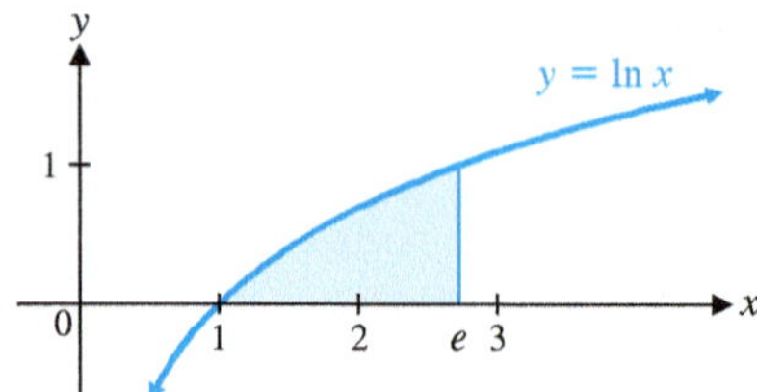

Figure 1

The integral represents the area under the curve $y = \ln x$ from $x = 1$ to $x = e$, as shown in Figure 1.

Matched Problem 4 Find $\int_1^2 \ln 3x\,dx$.

EXPLORE & DISCUSS 2 Try using the integration-by-parts formula on $\int e^{x^2}\,dx$, and explain why it does not work.

Exercises 5-3

A

In Problems 1–4, integrate by parts. Assume that $x > 0$ whenever the natural logarithm function is involved.

1. $\int xe^{3x}\,dx$

2. $\int xe^{4x}\,dx$

3. $\int x^2\ln x\,dx$

4. $\int x^3\ln x\,dx$

B

5. If you want to use integration by parts to find $\int (x+1)^5(x+2)\,dx$, which is the better choice for u: $u = (x+1)^5$ or $u = x + 2$? Explain your choice and then integrate.

6. If you want to use integration by parts to find $\int (5x-7)(x-1)^4\,dx$, which is the better choice for u: $u = 5x - 7$ or $u = (x-1)^4$? Explain your choice and then integrate.

Problems 7–20 are mixed—some require integration by parts, and others can be solved with techniques considered earlier. Integrate as indicated, assuming $x > 0$ whenever the natural logarithm function is involved.

7. $\int xe^{-x}\,dx$

8. $\int (x-1)e^{-x}\,dx$

9. $\int xe^{x^2}\,dx$

10. $\int xe^{-x^2}\,dx$

11. $\int_0^1 (x-3)e^x\,dx$

12. $\int_0^1 (x+1)e^x\,dx$

13. $\int_1^3 \ln 2x\,dx$

14. $\int_1^2 \ln\left(\frac{x}{2}\right)dx$

15. $\int \frac{2x}{x^2+1}\,dx$

16. $\int \frac{x^2}{x^3+5}\,dx$

17. $\int \frac{\ln x}{x}\,dx$

18. $\int \frac{e^x}{e^x+1}\,dx$

19. $\int \sqrt{x}\ln x\,dx$

20. $\int \frac{\ln x}{\sqrt{x}}\,dx$

In Problems 21–24, the integral can be found in more than one way. First use integration by parts, then use a method that does not involve integration by parts. Which method do you prefer?

21. $\int (x-3)(x+1)^2\,dx$

22. $\int (x+2)(x-1)^2\,dx$

23. $\int (2x+1)(x-2)^2\,dx$

24. $\int (5x-1)(x+2)^2\,dx$

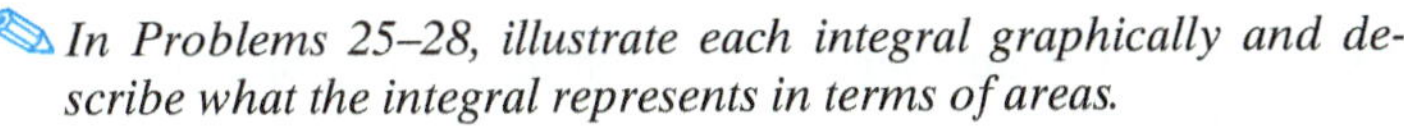
In Problems 25–28, illustrate each integral graphically and describe what the integral represents in terms of areas.

25. Problem 11

26. Problem 12

27. Problem 13

28. Problem 14

C

Problems 29–50 are mixed—some may require use of the integration-by-parts formula along with techniques we have considered earlier; others may require repeated use of the integration-by-parts formula. Assume that $g(x) > 0$ whenever $\ln g(x)$ is involved.

29. $\int x^2e^x\,dx$

30. $\int x^3e^x\,dx$

31. $\int xe^{ax}\,dx,\ a \neq 0$

32. $\int \ln(ax)\,dx,\ a > 0$

33. $\int_1^e \frac{\ln x}{x^2}\,dx$

34. $\int_1^2 x^3e^{x^2}\,dx$

35. $\int_0^2 \ln(x+4)\,dx$

36. $\int_0^2 \ln(4-x)\,dx$

37. $\int xe^{x-2}\,dx$

38. $\int xe^{x+1}\,dx$

39. $\int x\ln(1+x^2)\,dx$

40. $\int x\ln(1+x)\,dx$

41. $\int e^x\ln(1+e^x)\,dx$

42. $\int \frac{\ln(1+\sqrt{x})}{\sqrt{x}}\,dx$

43. $\int (\ln x)^2\,dx$

44. $\int x(\ln x)^2\,dx$

45. $\int (\ln x)^3\,dx$

46. $\int x(\ln x)^3\,dx$

47. $\int_1^e \ln(x^2)\,dx$

48. $\int_1^e \ln(x^4)\,dx$

49. $\int_0^1 \ln(e^{x^2})\,dx$

50. $\int_1^2 \ln(xe^x)\,dx$

In Problems 51–54, use a graphing calculator to graph each equation over the indicated interval and find the area between the curve and the x axis over that interval. Find answers to two decimal places.

51. $y = x - 2 - \ln x;\ 1 \le x \le 4$

52. $y = 6 - x^2 - \ln x;\ 1 \le x \le 4$

53. $y = 5 - xe^x;\ 0 \le x \le 3$

54. $y = xe^x + x - 6;\ 0 \le x \le 3$

Applications

55. Profit. If the marginal profit (in millions of dollars per year) is given by

$$P'(t) = 2t - te^{-t}$$

use an appropriate definite integral to find the total profit (to the nearest million dollars) earned over the first 5 years of operation.

56. Production. An oil field is estimated to produce oil at a rate of $R(t)$ thousand barrels per month t months from now, as given by

$$R(t) = 10te^{-0.1t}$$

Use an appropriate definite integral to find the total production (to the nearest thousand barrels) in the first year of operation.

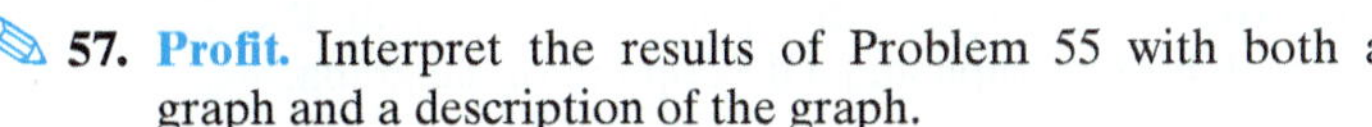
57. Profit. Interpret the results of Problem 55 with both a graph and a description of the graph.

58. Production. Interpret the results of Problem 56 with both a graph and a description of the graph.

59. Continuous income stream. Find the future value at 3.95%, compounded continuously, for 5 years of a continuous income stream with a rate of flow of

$$f(t) = 1{,}000 - 200t$$

60. Continuous income stream. Find the interest earned at 4.15%, compounded continuously, for 4 years for a continuous income stream with a rate of flow of

$$f(t) = 1{,}000 - 250t$$

61. Income distribution. Find the Gini index of income concentration for the Lorenz curve with equation

$$y = xe^{x-1}$$

62. **Income distribution.** Find the Gini index of income concentration for the Lorenz curve with equation

$$y = x^2 e^{x-1}$$

63. **Income distribution.** Interpret the results of Problem 61 with both a graph and a description of the graph.

64. **Income distribution.** Interpret the results of Problem 62 with both a graph and a description of the graph.

65. **Sales analysis.** Monthly sales of a particular personal computer are expected to decline at the rate of

$$S'(t) = -4te^{0.1t}$$

computers per month, where t is time in months and $S(t)$ is the number of computers sold each month. The company plans to stop manufacturing this computer when monthly sales reach 800 computers. If monthly sales now ($t = 0$) are 2,000 computers, find $S(t)$. How long, to the nearest month, will the company continue to manufacture the computer?

66. **Sales analysis.** The rate of change of the monthly sales of a new basketball game is given by

$$S'(t) = 350 \ln(t + 1) \qquad S(0) = 0$$

where t is the number of months since the game was released and $S(t)$ is the number of games sold each month. Find $S(t)$. When, to the nearest month, will monthly sales reach 15,000 games?

67. **Consumers' surplus.** Find the consumers' surplus (to the nearest dollar) at a price level of $\overline{p} = \$2.089$ for the price–demand equation

$$p = D(x) = 9 - \ln(x + 4)$$

Use $\overline{x}$ computed to the nearest higher unit.

68. **Producers' surplus.** Find the producers' surplus (to the nearest dollar) at a price level of $\overline{p} = \$26$ for the price–supply equation

$$p = S(x) = 5 \ln(x + 1)$$

Use $\overline{x}$ computed to the nearest higher unit.

69. **Consumers' surplus.** Interpret the results of Problem 67 with both a graph and a description of the graph.

70. **Producers' surplus.** Interpret the results of Problem 68 with both a graph and a description of the graph.

71. **Pollution.** The concentration of particulate matter (in parts per million) t hours after a factory ceases operation for the day is given by

$$C(t) = \frac{20 \ln(t + 1)}{(t + 1)^2}$$

Find the average concentration for the period from $t = 0$ to $t = 5$.

72. **Medicine.** After a person takes a pill, the drug contained in the pill is assimilated into the bloodstream. The rate of assimilation t minutes after taking the pill is

$$R(t) = te^{-0.2t}$$

Find the total amount of the drug that is assimilated into the bloodstream during the first 10 minutes after the pill is taken.

73. **Learning.** A student enrolled in an advanced typing class progressed at a rate of

$$N'(t) = (t + 6)e^{-0.25t}$$

words per minute per week t weeks after enrolling in a 15-week course. If a student could type 40 words per minute at the beginning of the course, then how many words per minute $N(t)$ would the student be expected to type t weeks into the course? How long, to the nearest week, should it take the student to achieve the 70-word-per-minute level? How many words per minute should the student be able to type by the end of the course?

74. **Learning.** A student enrolled in a stenotyping class progressed at a rate of

$$N'(t) = (t + 10)e^{-0.1t}$$

words per minute per week t weeks after enrolling in a 15-week course. If a student had no knowledge of stenotyping (that is, if the student could stenotype at 0 words per minute) at the beginning of the course, then how many words per minute $N(t)$ would the student be expected to handle t weeks into the course? How long, to the nearest week, should it take the student to achieve 90 words per minute? How many words per minute should the student be able to handle by the end of the course?

75. **Politics.** The number of voters (in thousands) in a certain city is given by

$$N(t) = 20 + 4t - 5te^{-0.1t}$$

where t is time in years. Find the average number of voters during the period from $t = 0$ to $t = 5$.

Answers to Matched Problems

1. $\frac{x}{2}e^{2x} - \frac{1}{4}e^{2x} + C$

2. $\frac{x^2}{2}\ln 2x - \frac{x^2}{4} + C$

3. $\frac{x^2}{2}e^{2x} - \frac{x}{2}e^{2x} + \frac{1}{4}e^{2x} + C$

4. $2 \ln 6 - \ln 3 - 1 \approx 1.4849$

5-4 Integration Using Tables

A **table of integrals** is a list of integration formulas used to evaluate integrals. Table II of Appendix C on pages 722–724 contains a list of integral formulas. Some of these formulas can be derived with the integration techniques discussed earlier, while others require techniques we have not considered. However, it is possible to verify each formula by differentiating the right side.

Using a Table of Integrals

The formulas in Table II on pages 722–724 are organized by categories, such as "Integrals Involving $a + bu$," "Integrals Involving $\sqrt{u^2 - a^2}$," and so on. The variable u is the variable of integration. All other symbols represent constants. To use a table to evaluate an integral, you must first find the category that most closely agrees with the form of the integrand and then find a formula in that category that you can make to match the integrand exactly by assigning values to the constants in the formula.

EXAMPLE 1 **Integration Using Tables** Use Table II to find

$$\int \frac{x}{(5 + 2x)(4 - 3x)} dx$$

SOLUTION Since the integrand

$$f(x) = \frac{x}{(5 + 2x)(4 - 3x)}$$

is a rational function involving terms of the form $a + bu$ and $c + du$, we examine formulas 15 to 20 in Table II on page 722 to see if any of the integrands in these formulas can be made to match $f(x)$ exactly. Comparing the integrand in formula 16 with $f(x)$, we see that this integrand will match $f(x)$ if we let $a = 5$, $b = 2$, $c = 4$, and $d = -3$. Letting $u = x$ and substituting for a, b, c, and d in formula 16, we have

$$\int \frac{u}{(a + bu)(c + du)} du = \frac{1}{ad - bc}\left(\frac{a}{b}\ln|a + bu| - \frac{c}{d}\ln|c + du|\right) \quad \text{Formula 16}$$

$$\int \frac{x}{(\underset{a}{5} + \underset{b}{2}x)(\underset{c}{4} - \underset{d}{3}x)} dx = \frac{1}{5\cdot(-3) - 2\cdot 4}\left(\frac{5}{2}\ln|5 + 2x| - \frac{4}{-3}\ln|4 - 3x|\right) + C$$

$a \cdot d - b \cdot c = 5 \cdot (-3) - 2 \cdot 4 = -23$

$$= -\tfrac{5}{46}\ln|5 + 2x| - \tfrac{4}{69}\ln|4 - 3x| + C$$

Notice that the constant of integration, C, is not included in any of the formulas in Table II. However, you must still include C in all antiderivatives.

Matched Problem 1 Use Table II to find $\displaystyle\int \frac{1}{(5 + 3x)^2(1 + x)} dx$.

EXAMPLE 2 **Integration Using Tables** Evaluate $\displaystyle\int_3^4 \frac{1}{x\sqrt{25 - x^2}} dx$.

SOLUTION First, we use Table II to find

$$\int \frac{1}{x\sqrt{25 - x^2}} dx$$

Since the integrand involves the expression $\sqrt{25 - x^2}$, we examine formulas 29 to 31 in Table II and select formula 29 with $a^2 = 25$ and $a = 5$:

$$\int \frac{1}{u\sqrt{a^2 - u^2}}\,du = -\frac{1}{a}\ln\left|\frac{a + \sqrt{a^2 - u^2}}{u}\right| \qquad \text{Formula 29}$$

$$\int \frac{1}{x\sqrt{25 - x^2}}\,dx = -\frac{1}{5}\ln\left|\frac{5 + \sqrt{25 - x^2}}{x}\right| + C$$

So

$$\int_3^4 \frac{1}{x\sqrt{25 - x^2}}\,dx = -\frac{1}{5}\ln\left|\frac{5 + \sqrt{25 - x^2}}{x}\right|\Bigg|_3^4$$

$$= -\frac{1}{5}\ln\left|\frac{5 + 3}{4}\right| + \frac{1}{5}\ln\left|\frac{5 + 4}{3}\right|$$

$$= -\tfrac{1}{5}\ln 2 + \tfrac{1}{5}\ln 3 = \tfrac{1}{5}\ln 1.5 \approx 0.0811$$

Matched Problem 2 Evaluate $\displaystyle\int_6^8 \frac{1}{x^2\sqrt{100 - x^2}}\,dx$.

Substitution and Integral Tables

As Examples 1 and 2 illustrate, if the integral we want to evaluate can be made to match one in the table exactly, then evaluating the indefinite integral consists simply of substituting the correct values of the constants into the formula. But what happens if we cannot match an integral with one of the formulas in the table? In many cases, a substitution will change the given integral into one that corresponds to a table entry.

EXAMPLE 3 **Integration Using Substitution and Tables** Find $\displaystyle\int \frac{x^2}{\sqrt{16x^2 - 25}}\,dx$.

SOLUTION In order to relate this integral to one of the formulas involving $\sqrt{u^2 - a^2}$ (formulas 40 to 45 in Table II), we observe that if $u = 4x$, then

$$u^2 = 16x^2 \qquad \text{and} \qquad \sqrt{16x^2 - 25} = \sqrt{u^2 - 25}$$

So, we will use the substitution $u = 4x$ to change this integral into one that appears in the table:

$$\int \frac{x^2}{\sqrt{16x^2 - 25}}\,dx = \frac{1}{4}\int \frac{\frac{1}{16}u^2}{\sqrt{u^2 - 25}}\,du \qquad \text{Substitution: } u = 4x,\ du = 4\,dx,\ x = \tfrac{1}{4}u$$

$$= \frac{1}{64}\int \frac{u^2}{\sqrt{u^2 - 25}}\,du$$

This last integral can be evaluated with the aid of formula 44 in Table II with $a = 5$:

$$\int \frac{u^2}{\sqrt{u^2 - a^2}}\,du = \frac{1}{2}\left(u\sqrt{u^2 - a^2} + a^2\ln\left|u + \sqrt{u^2 - a^2}\right|\right) \qquad \text{Formula 44}$$

$$\int \frac{x^2}{\sqrt{16x^2 - 25}}\,dx = \frac{1}{64}\int \frac{u^2}{\sqrt{u^2 - 25}}\,du \qquad \text{Use formula 44 with } a = 5.$$

$$= \tfrac{1}{128}\left(u\sqrt{u^2 - 25} + 25\ln\left|u + \sqrt{u^2 - 25}\right|\right) + C \qquad \text{Substitute } u = 4x.$$

$$= \tfrac{1}{128}\left(4x\sqrt{16x^2 - 25} + 25\ln\left|4x + \sqrt{16x^2 - 25}\right|\right) + C$$

Matched Problem 3 Find $\int \sqrt{9x^2 - 16}\, dx$.

EXAMPLE 4 **Integration Using Substitution and Tables** Find $\displaystyle\int \frac{x}{\sqrt{x^4 + 1}} dx$.

SOLUTION None of the formulas in Table II involve fourth powers; however, if we let $u = x^2$, then

$$\sqrt{x^4 + 1} = \sqrt{u^2 + 1}$$

and this form does appear in formulas 32 to 39. Thus, we substitute $u = x^2$:

$$\int \frac{1}{\sqrt{x^4 + 1}} x\, dx = \frac{1}{2}\int \frac{1}{\sqrt{u^2 + 1}} du \qquad \text{Substitution: } u = x^2,\ du = 2x\, dx$$

We recognize the last integral as formula 36 with $a = 1$:

$$\int \frac{1}{\sqrt{u^2 + a^2}} du = \ln|u + \sqrt{u^2 + a^2}| \qquad \text{Formula 36}$$

$$\int \frac{x}{\sqrt{x^4 + 1}} dx = \frac{1}{2}\int \frac{1}{\sqrt{u^2 + 1}} du \qquad \text{Use formula 36 with } a = 1.$$

$$= \tfrac{1}{2}\ln|u + \sqrt{u^2 + 1}| + C \qquad \text{Substitute } u = x^2.$$

$$= \tfrac{1}{2}\ln|x^2 + \sqrt{x^4 + 1}| + C$$

Matched Problem 4 Find $\int x\sqrt{x^4 + 1}\, dx$.

Reduction Formulas

EXAMPLE 5 **Using Reduction Formulas** Use Table II to find $\int x^2 e^{3x}\, dx$.

SOLUTION Since the integrand involves the function e^{3x}, we examine formulas 46–48 and conclude that formula 47 can be used for this problem. Letting $u = x$, $n = 2$, and $a = 3$ in formula 47, we have

$$\int u^n e^{au}\, du = \frac{u^n e^{au}}{a} - \frac{n}{a}\int u^{n-1}e^{au}\, du \qquad \text{Formula 47}$$

$$\int x^2 e^{3x}\, dx = \frac{x^2 e^{3x}}{3} - \frac{2}{3}\int xe^{3x}\, dx$$

Notice that the expression on the right still contains an integral, but the exponent of x has been reduced by 1. Formulas of this type are called **reduction formulas** and are designed to be applied repeatedly until an integral that can be evaluated is obtained. Applying formula 47 to $\int xe^{3x}\, dx$ with $n = 1$, we have

$$\int x^2 e^{3x}\, dx = \frac{x^2 e^{3x}}{3} - \frac{2}{3}\left(\frac{xe^{3x}}{3} - \frac{1}{3}\int e^{3x}\, dx\right)$$

$$= \frac{x^2 e^{3x}}{3} - \frac{2xe^{3x}}{9} + \frac{2}{9}\int e^{3x}\, dx$$

This last expression contains an integral that is easy to evaluate:

$$\int e^{3x}\, dx = \tfrac{1}{3}e^{3x}$$

After making a final substitution and adding a constant of integration, we have

$$\int x^2 e^{3x}\, dx = \frac{x^2 e^{3x}}{3} - \frac{2xe^{3x}}{9} + \frac{2}{27}e^{3x} + C$$

Matched Problem 5 Use Table II to find $\int (\ln x)^2\, dx$.

Application

EXAMPLE 6 **Producers' Surplus** Find the producers' surplus at a price level of \$20 for the price–supply equation

$$p = S(x) = \frac{5x}{500 - x}$$

SOLUTION **Step 1** Find $\bar{x}$, the supply when the price is $\bar{p} = 20$:

$$\begin{aligned} \bar{p} &= \frac{5\bar{x}}{500 - \bar{x}} \\ 20 &= \frac{5\bar{x}}{500 - \bar{x}} \\ 10{,}000 - 20\bar{x} &= 5\bar{x} \\ 10{,}000 &= 25\bar{x} \\ \bar{x} &= 400 \end{aligned}$$

Step 2 Sketch a graph, as shown in Figure 1.

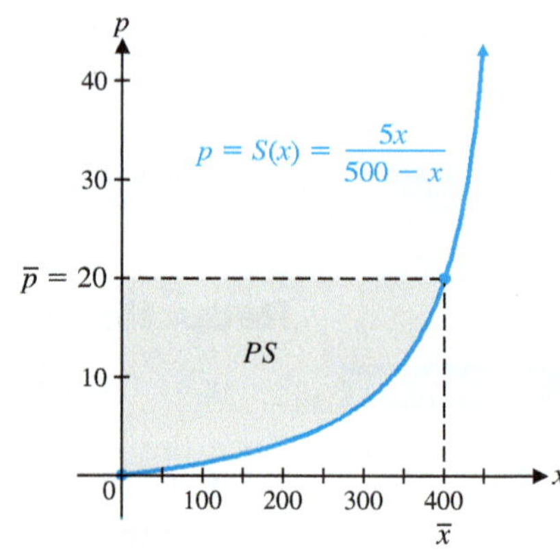

Figure 1

Step 3 Find the producers' surplus (the shaded area of the graph):

$$\begin{aligned} PS &= \int_0^{\bar{x}} [\bar{p} - S(x)]\, dx \\ &= \int_0^{400} \left(20 - \frac{5x}{500 - x}\right) dx \\ &= \int_0^{400} \frac{10{,}000 - 25x}{500 - x}\, dx \end{aligned}$$

Use formula 20 with $a = 10{,}000$, $b = -25$, $c = 500$, and $d = -1$:

$$\int \frac{a + bu}{c + du}\, du = \frac{bu}{d} + \frac{ad - bc}{d^2} \ln|c + du| \qquad \text{Formula 20}$$

$$\begin{aligned} PS &= (25x + 2{,}500 \ln|500 - x|)\Big|_0^{400} \\ &= 10{,}000 + 2{,}500 \ln|100| - 2{,}500 \ln|500| \\ &\approx \$5{,}976 \end{aligned}$$

Matched Problem 6 Find the consumers' surplus at a price level of \$10 for the price–demand equation

$$p = D(x) = \frac{20x - 8{,}000}{x - 500}$$

Exercises 5-4

A

Use Table II on pages 722–724 to find each indefinite integral in Problems 1–14.

1. $\int \frac{1}{x(1+x)}dx$

2. $\int \frac{1}{x^2(1+x)}dx$

3. $\int \frac{1}{(3+x)^2(5+2x)}dx$

4. $\int \frac{x}{(5+2x)^2(2+x)}dx$

5. $\int \frac{x}{\sqrt{16+x}}dx$

6. $\int \frac{1}{x\sqrt{16+x}}dx$

7. $\int \frac{1}{x\sqrt{1-x^2}}dx$

8. $\int \frac{\sqrt{9-x^2}}{x}dx$

9. $\int \frac{1}{x\sqrt{x^2+4}}dx$

10. $\int \frac{1}{x^2\sqrt{x^2-16}}dx$

11. $\int x^2 \ln x\, dx$

12. $\int x^3 \ln x\, dx$

13. $\int \frac{1}{1+e^x}dx$

14. $\int \frac{1}{5+2e^{3x}}dx$

Evaluate each definite integral in Problems 15–20. Use Table II on pages 722–724 to find the antiderivative.

15. $\int_1^3 \frac{x^2}{3+x}dx$

16. $\int_2^6 \frac{x}{(6+x)^2}dx$

17. $\int_0^7 \frac{1}{(3+x)(1+x)}dx$

18. $\int_0^7 \frac{x}{(3+x)(1+x)}dx$

19. $\int_0^4 \frac{1}{\sqrt{x^2+9}}dx$

20. $\int_4^5 \sqrt{x^2-16}\, dx$

B

In Problems 21–32, use substitution techniques and Table II to find each indefinite integral.

21. $\int \frac{\sqrt{4x^2+1}}{x^2}dx$

22. $\int x^2\sqrt{9x^2-1}\, dx$

23. $\int \frac{x}{\sqrt{x^4-16}}dx$

24. $\int x\sqrt{x^4-16}\, dx$

25. $\int x^2\sqrt{x^6+4}\, dx$

26. $\int \frac{x^2}{\sqrt{x^6+4}}dx$

27. $\int \frac{1}{x^3\sqrt{4-x^4}}dx$

28. $\int \frac{\sqrt{x^4+4}}{x}dx$

29. $\int \frac{e^x}{(2+e^x)(3+4e^x)}dx$

30. $\int \frac{e^x}{(4+e^x)^2(2+e^x)}dx$

31. $\int \frac{\ln x}{x\sqrt{4+\ln x}}dx$

32. $\int \frac{1}{x \ln x\sqrt{4+\ln x}}dx$

C

In Problems 33–38, use Table II to find each indefinite integral.

33. $\int x^2e^{5x}\, dx$

34. $\int x^2e^{-4x}\, dx$

35. $\int x^3e^{-x}\, dx$

36. $\int x^3e^{2x}\, dx$

37. $\int (\ln x)^3\, dx$

38. $\int (\ln x)^4\, dx$

Problems 39–46 are mixed—some require the use of Table II, and others can be solved with techniques considered earlier.

39. $\int_3^5 x\sqrt{x^2-9}\, dx$

40. $\int_3^5 x^2\sqrt{x^2-9}\, dx$

41. $\int_2^4 \frac{1}{x^2-1}dx$

42. $\int_2^4 \frac{x}{(x^2-1)^2}dx$

43. $\int \frac{\ln x}{x^2} dx$

44. $\int \frac{(\ln x)^2}{x} dx$

45. $\int \frac{x}{\sqrt{x^2 - 1}} dx$

46. $\int \frac{x^2}{\sqrt{x^2 - 1}} dx$

In Problems 47–50, find the area bounded by the graphs of $y = f(x)$ and $y = g(x)$ to two decimal places. Use a graphing calculator to approximate intersection points to two decimal places.

47. $f(x) = \frac{10}{\sqrt{x^2 + 1}}$; $g(x) = x^2 + 3x$

48. $f(x) = \sqrt{1 + x^2}$; $g(x) = 5x - x^2$

49. $f(x) = x\sqrt{4 + x}$; $g(x) = 1 + x$

50. $f(x) = \frac{x}{\sqrt{x + 4}}$; $g(x) = x - 2$

Applications

Use Table II to evaluate all integrals involved in any solutions of Problems 51–74.

51. **Consumers' surplus.** Find the consumers' surplus at a price level of $\overline{p} = \$15$ for the price–demand equation

$$p = D(x) = \frac{7{,}500 - 30x}{300 - x}$$

52. **Producers' surplus.** Find the producers' surplus at a price level of $\overline{p} = \$20$ for the price–supply equation

$$p = S(x) = \frac{10x}{300 - x}$$

53. **Consumers' surplus.** Graph the price–demand equation and the price-level equation $\overline{p} = 15$ of Problem 51 in the same coordinate system. What region represents the consumers' surplus?

54. **Producers' surplus.** Graph the price-supply equation and the price-level equation $\overline{p} = 20$ of Problem 52 in the same coordinate system. What region represents the producers' surplus?

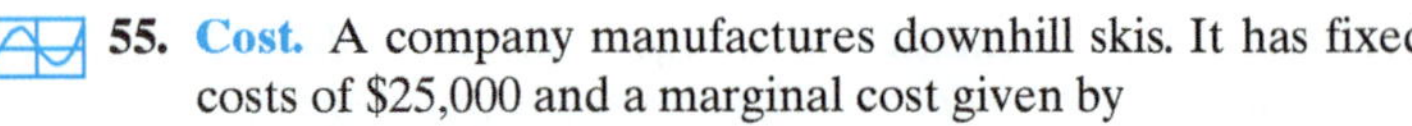

55. **Cost.** A company manufactures downhill skis. It has fixed costs of \$25,000 and a marginal cost given by

$$C'(x) = \frac{250 + 10x}{1 + 0.05x}$$

where $C(x)$ is the total cost at an output of x pairs of skis. Find the cost function $C(x)$ and determine the production level (to the nearest unit) that produces a cost of \$150,000. What is the cost (to the nearest dollar) for a production level of 850 pairs of skis?

56. **Cost.** A company manufactures a portable DVD player. It has fixed costs of \$11,000 per week and a marginal cost given by

$$C'(x) = \frac{65 + 20x}{1 + 0.4x}$$

where $C(x)$ is the total cost per week at an output of x players per week. Find the cost function $C(x)$ and determine the production level (to the nearest unit) that produces a cost of \$52,000 per week. What is the cost (to the nearest dollar) for a production level of 700 players per week?

57. **Continuous income stream.** Find the future value at 4.4%, compounded continuously, for 10 years for the continuous income stream with rate of flow $f(t) = 50t^2$.

58. **Continuous income stream.** Find the interest earned at 3.7%, compounded continuously, for 5 years for the continuous income stream with rate of flow $f(t) = 200t$.

59. **Income distribution.** Find the Gini index of income concentration for the Lorenz curve with equation

$$y = \tfrac{1}{2}x\sqrt{1 + 3x}$$

60. **Income distribution.** Find the Gini index of income concentration for the Lorenz curve with equation

$$y = \tfrac{1}{2}x^2\sqrt{1 + 3x}$$

61. **Income distribution.** Graph $y = x$ and the Lorenz curve of Problem 59 over the interval $[0, 1]$. Discuss the effect of the area bounded by $y = x$ and the Lorenz curve getting smaller relative to the equitable distribution of income.

62. **Income distribution.** Graph $y = x$ and the Lorenz curve of Problem 60 over the interval $[0, 1]$. Discuss the effect of the area bounded by $y = x$ and the Lorenz curve getting larger relative to the equitable distribution of income.

63. **Marketing.** After test marketing a new high-fiber cereal, the market research department of a major food producer estimates that monthly sales (in millions of dollars) will grow at the monthly rate of

$$S'(t) = \frac{t^2}{(1 + t)^2}$$

t months after the cereal is introduced. If we assume 0 sales at the time the cereal is introduced, find $S(t)$, the total sales, t months after the cereal is introduced. Find the total sales during the first 2 years that the cereal is on the market.

64. **Average price.** At a discount department store, the price–demand equation for premium motor oil is given by

$$p = D(x) = \frac{50}{\sqrt{100 + 6x}}$$

where x is the number of cans of oil that can be sold at a price of \$$p$. Find the average price over the demand interval $[50, 250]$.

65. **Marketing.** For the cereal of Problem 63, show the sales over the first 2 years geometrically, and describe the geometric representation.

66. Price–demand. For the motor oil of Problem 64, graph the price–demand equation and the line representing the average price in the same coordinate system over the interval $[50, 250]$. Describe how the areas under the two curves over the interval $[50, 250]$ are related.

67. Profit. The marginal profit for a small car agency that sells x cars per week is given by

$$P'(x) = x\sqrt{2 + 3x}$$

where $P(x)$ is the profit in dollars. The agency's profit on the sale of only 1 car per week is $-\$2{,}000$. Find the profit function and the number of cars that must be sold (to the nearest unit) to produce a profit of \$13,000 per week. How much weekly profit (to the nearest dollar) will the agency have if 80 cars are sold per week?

68. Revenue. The marginal revenue for a company that manufactures and sells x graphing calculators per week is given by

$$R'(x) = \frac{x}{\sqrt{1 + 2x}} \qquad R(0) = 0$$

where $R(x)$ is the revenue in dollars. Find the revenue function and the number of calculators that must be sold (to the nearest unit) to produce \$10,000 in revenue per week. How much weekly revenue (to the nearest dollar) will the company have if 1,000 calculators are sold per week?

69. Pollution. An oil tanker is producing an oil slick that is radiating outward at a rate given approximately by

$$\frac{dR}{dt} = \frac{100}{\sqrt{t^2 + 9}} \qquad t \ge 0$$

where R is the radius (in feet) of the circular slick after t minutes. Find the radius of the slick after 4 minutes if the radius is 0 when $t = 0$.

70. Pollution. The concentration of particulate matter (in parts per million) during a 24-hour period is given approximately by

$$C(t) = t\sqrt{24 - t} \qquad 0 \le t \le 24$$

where t is time in hours. Find the average concentration during the period from $t = 0$ to $t = 24$.

71. Learning. A person learns N items at a rate given approximately by

$$N'(t) = \frac{60}{\sqrt{t^2 + 25}} \qquad t \ge 0$$

where t is the number of hours of continuous study. Determine the total number of items learned in the first 12 hours of continuous study.

72. Politics. The number of voters (in thousands) in a metropolitan area is given approximately by

$$f(t) = \frac{500}{2 + 3e^{-t}} \qquad t \ge 0$$

where t is time in years. Find the average number of voters during the period from $t = 0$ to $t = 10$.

73. Learning. Interpret Problem 71 geometrically. Describe the geometric interpretation.

74. Politics. For the voters of Problem 72, graph $y = f(t)$ and the line representing the average number of voters over the interval $[0, 10]$ in the same coordinate system. Describe how the areas under the two curves over the interval $[0, 10]$ are related.

Answers to Matched Problems

1. $\frac{1}{2}\left(\frac{1}{5 + 3x}\right) + \frac{1}{4}\ln\left|\frac{1 + x}{5 + 3x}\right| + C$
2. $\frac{7}{1{,}200} \approx 0.0058$
3. $\frac{1}{6}(3x\sqrt{9x^2 - 16} - 16\ln|3x + \sqrt{9x^2 - 16}|) + C$
4. $\frac{1}{4}(x^2\sqrt{x^4 + 1} + \ln|x^2 + \sqrt{x^4 + 1}|) + C$
5. $x(\ln x)^2 - 2x\ln x + 2x + C$
6. $3{,}000 + 2{,}000\ln 200 - 2{,}000\ln 500 \approx \$1{,}167$

Chapter 5 Review

Important Terms, Symbols, and Concepts

5-1 Area Between Curves

EXAMPLES

- If f and g are continuous and $f(x) \ge g(x)$ over the interval $[a, b]$, then the area bounded by $y = f(x)$ and $y = g(x)$ for $a \le x \le b$ is given exactly by Ex. 1, p. 288; Ex. 2, p. 288; Ex. 3, p. 289; Ex. 4, p. 289; Ex. 5, p. 290; Ex. 6, p. 290

$$A = \int_a^b [f(x) - g(x)]\,dx$$

- A graphical representation of the distribution of income among a population can be found by plotting data points (x, y), where **x represents the cumulative percentage of families at or below a given income level** and **y represents the cumulative percentage of total family income received.** Regression analysis can be used to find a particular function $y = f(x)$, called a **Lorenz curve,** that best fits the data.
- A single number, the **Gini index,** measures income concentration: Ex. 7, p. 292

$$\text{Gini index} = 2\int_0^1 [x - f(x)]\,dx$$

A Gini index of 0 indicates **absolute equality:** All families share equally in the income. A Gini index of 1 indicates **absolute inequality:** One family has all of the income and the rest have none.

5-2 Applications in Business and Economics

- ***Probability Density Functions.*** If any real number x in an interval is a possible outcome of an experiment, then x is said to be a **continuous random variable.** The probability distribution of a continuous random variable is described by a **probability density function** f that satisfies the following conditions: Ex. 1, p. 297

  1. $f(x) \geq 0$ for all real x.

  2. The area under the graph of $f(x)$ over the interval $(-\infty, \infty)$ is exactly 1.

  3. If $[c, d]$ is a subinterval of $(-\infty, \infty)$, then

$$\text{Probability} \quad (c \leq x \leq d) = \int_c^d f(x)\,dx$$

- ***Continuous Income Stream*** If the rate at which income is received—its **rate of flow**—is a continuous function $f(t)$ of time, then the income is said to be a **continuous income stream.** The **total income** produced by a continuous income stream from $t = a$ to $t = b$ is Ex. 2, p. 299

$$\text{Total income} = \int_a^b f(t)\,dt$$

  The **future value** of a continuous income stream that is invested at rate r, compounded continuously, for $0 \leq t \leq T$, is Ex. 3, p. 302

$$FV = \int_0^T f(t)e^{r(T-t)}\,dt$$

- ***Consumers' and Producers' Surplus*** If $(\overline{x}, \overline{p})$ is a point on the graph of a price–demand equation $p = D(x)$, then the **consumers' surplus** at a price level of $\overline{p}$ is Ex. 4, p. 303

$$CS = \int_0^{\overline{x}} [D(x) - \overline{p}]\,dx$$

  The consumers' surplus represents the total savings to consumers who are willing to pay more than $\overline{p}$ but are still able to buy the product for $\overline{p}$.

  Similarly, for a point $(\overline{x}, \overline{p})$ on the graph of a price–supply equation $p = S(x)$, the **producers' surplus** at a price level of $\overline{p}$ is Ex. 5, p. 304

$$PS = \int_0^{\overline{x}} [\overline{p} - S(x)]\,dx$$

  The producers' surplus represents the total gain to producers who are willing to supply units at a lower price $\overline{p}$, but are still able to supply units at $\overline{p}$.

  If $(\overline{x}, \overline{p})$ is the intersection point of a price–demand equation $p = D(x)$ and a price–supply equation $p = S(x)$, then $\overline{p}$ is called the **equilibrium price** and $\overline{x}$ is called the **equilibrium quantity**. Ex. 6, p. 305

5-3 Integration by Parts

- Some indefinite integrals, but not all, can be found by means of the **integration-by-parts formula** Ex. 1, p. 309 Ex. 2, p. 310 Ex. 3, p. 311 Ex. 4, p. 312

$$\int u\,dv = uv - \int v\,du$$

- Select u and dv with the help of the guidelines in the summary on page 311.

5-4 Integration Using Tables

- A **table of integrals** is a list of integration formulas that can be used to find indefinite or definite integrals of frequently encountered functions. Such a list appears in Table II of Appendix C on pages 722–724. Ex. 1, p. 315 Ex. 2, p. 315 Ex. 3, p. 316 Ex. 4, p. 317 Ex. 5, p. 317 Ex. 6, p. 318

…review and check
…ers to all review prob-
…ers in italics to indicate
…ed. Where weaknesses show
…ne text.
…nswers to three decimal places

…et up definite integrals that represent the shad-
…gure over the indicated intervals.

Figure for 1–3

1. Interval $[a, b]$

2. Interval $[b, c]$

3. Interval $[a, c]$

4. Sketch a graph of the area between the graphs of $y = \ln x$ and $y = 0$ over the interval $[0.5, e]$ and find the area.

In Problems 5–10, evaluate each integral.

5. $\int xe^{4x}\,dx$

6. $\int x \ln x\,dx$

7. $\int \frac{\ln x}{x}\,dx$

8. $\int \frac{x}{1 + x^2}\,dx$

9. $\int \frac{1}{x(1 + x)^2}\,dx$

10. $\int \frac{1}{x^2\sqrt{1 + x}}\,dx$

In Problems 11–16, find the area bounded by the graphs of the indicated equations over the given interval.

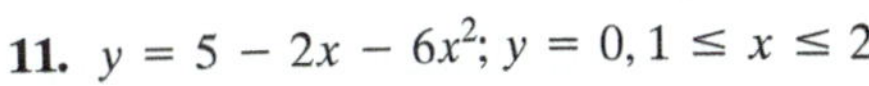

11. $y = 5 - 2x - 6x^2$; $y = 0$, $1 \le x \le 2$

12. $y = 5x + 7$; $y = 12$, $-3 \le x \le 1$

13. $y = -x + 2$; $y = x^2 + 3$, $-1 \le x \le 4$

14. $y = \frac{1}{x}$; $y = -e^{-x}$, $1 \le x \le 2$

15. $y = x$; $y = -x^3$, $-2 \le x \le 2$

16. $y = x^2$; $y = -x^4$; $-2 \le x \le 2$

B

In Problems 17–20, set up definite integrals that represent the shaded areas in the figure over the indicated intervals.

17. Interval $[a, b]$

18. Interval $[b, c]$

Figure for 17–20

19. Interval $[b, d]$

20. Interval $[a, d]$

21. Sketch a graph of the area bounded by the graphs of $y = x^2 - 6x + 9$ and $y = 9 - x$ and find the area.

In Problems 22–27, evaluate each integral.

22. $\int_0^1 xe^x\,dx$

23. $\int_0^3 \frac{x^2}{\sqrt{x^2 + 16}}\,dx$

24. $\int \sqrt{9x^2 - 49}\,dx$

25. $\int te^{-0.5t}\,dt$

26. $\int x^2 \ln x\,dx$

27. $\int \frac{1}{1 + 2e^x}\,dx$

28. Sketch a graph of the area bounded by the indicated graphs, and find the area. In part (B), approximate intersection points and area to two decimal places.
(A) $y = x^3 - 6x^2 + 9x$; $y = x$
(B) $y = x^3 - 6x^2 + 9x$; $y = x + 1$

C

In Problems 29–36, evaluate each integral.

29. $\int \frac{(\ln x)^2}{x}\,dx$

30. $\int x(\ln x)^2\,dx$

31. $\int \frac{x}{\sqrt{x^2 - 36}}\,dx$

32. $\int \frac{x}{\sqrt{x^4 - 36}}\,dx$

33. $\int_0^4 x \ln(10 - x)\,dx$

34. $\int (\ln x)^2\,dx$

35. $\int xe^{-2x^2}\,dx$

36. $\int x^2 e^{-2x}\,dx$

37. Use a numerical integration routine on a graphing calculator to find the area in the first quadrant that is below the graph of

$$y = \frac{6}{2 + 5e^{-x}}$$

and above the graph of $y = 0.2x + 1.6$.

Applications

38. **Product warranty.** A manufacturer warrants a product for parts and labor for 1 year and for parts only for a second year. The time to a failure of the product after it is sold is given by the probability density function

$$f(t) = \begin{cases} 0.21e^{-0.21t} & \text{if } t \ge 0 \\ 0 & \text{otherwise} \end{cases}$$

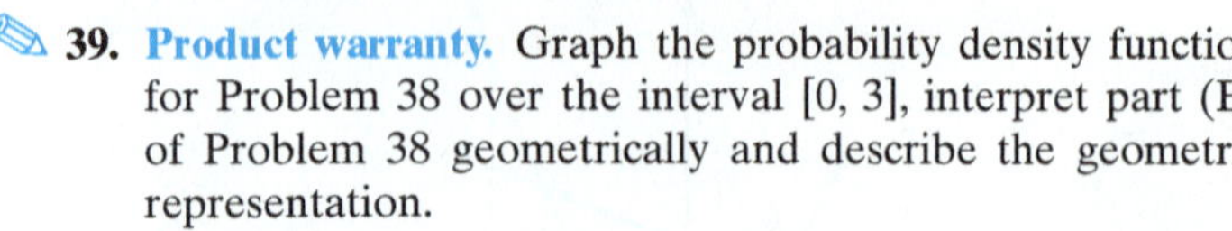

What is the probability that a buyer chosen at random will have a product failure

(A) During the first year of warranty?

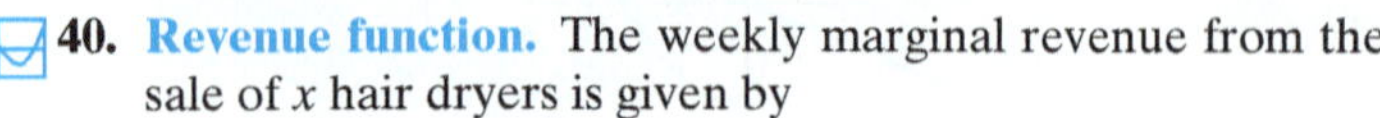

(B) During the second year of warranty?

39. Product warranty. Graph the probability density function for Problem 38 over the interval [0, 3], interpret part (B) of Problem 38 geometrically and describe the geometric representation.

40. Revenue function. The weekly marginal revenue from the sale of x hair dryers is given by

$$R'(x) = 65 - 6\ln(x + 1) \qquad R(0) = 0$$

where $R(x)$ is the revenue in dollars. Find the revenue function and the production level (to the nearest unit) for a revenue of \$20,000 per week. What is the weekly revenue (to the nearest dollar) at a production level of 1,000 hair dryers per week?

41. Continuous income stream. The rate of flow (in dollars per year) of a continuous income stream for a 5-year period is given by

$$f(t) = 2{,}500e^{0.05t} \qquad 0 \le t \le 5$$

(A) Graph $y = f(t)$ over [0, 5] and shade the area that represents the total income received from the end of the first year to the end of the fourth year.

(B) Find the total income received, to the nearest dollar, from the end of the first year to the end of the fourth year.

42. Future value of a continuous income stream. The continuous income stream in Problem 41 is invested at 4%, compounded continuously.

(A) Find the future value (to the nearest dollar) at the end of the 5-year period.

(B) Find the interest earned (to the nearest dollar) during the 5-year period.

43. Income distribution. An economist produced the following Lorenz curves for the current income distribution and the projected income distribution 10 years from now in a certain country:

$$f(x) = 0.1x + 0.9x^2 \qquad \text{Current Lorenz curve}$$

$$g(x) = x^{1.5} \qquad \text{Projected Lorenz curve}$$

(A) Graph $y = x$ and the current Lorenz curve on one set of coordinate axes for [0, 1] and graph $y = x$ and the projected Lorenz curve on another set of coordinate axes over the same interval.

(B) Looking at the areas bounded by the Lorenz curves and $y = x$, can you say that the income will be more or less equitably distributed 10 years from now?

(C) Compute the Gini index of income concentration (to one decimal place) for the current and projected curves. What can you say about the distribution of income 10 years from now? Is it more equitable or less?

44. Consumers' and producers' surplus. Find the consumers' surplus and the producers' surplus at the equilibrium price level for each pair of price–demand and price–supply equations. Include a graph that identifies the consumers' surplus and the producers' surplus. Round all values to the nearest integer.

(A) $p = D(x) = 70 - 0.2x;$
$p = S(x) = 13 + 0.0012x^2$

(B) $p = D(x) = 70 - 0.2x; \; p = S(x) = 13e^{0.006x}$

45. Producers' surplus. The accompanying table gives price–supply data for the sale of hogs at a livestock market, where x is the number of pounds (in thousands) and p is the price per pound (in cents):

x
0
10
20
30
40

(A) Using quadratic regression to model the data, f demand at a price of 52.50 cents per pound.

(B) Use a numerical integration routine to find the pro ers' surplus (to the nearest dollar) at a price level 52.50 cents per pound.

46. Drug assimilation. The rate at which the body eliminates a certain drug (in milliliters per hour) is given by

$$R(t) = \frac{60t}{(t + 1)^2(t + 2)}$$

where t is the number of hours since the drug was administered. How much of the drug is eliminated during the first hour after it was administered? During the fourth hour?

47. With the aid of a graphing calculator, illustrate Problem 46 geometrically.

48. Medicine. For a particular doctor, the length of time (in hours) spent with a patient per office visit has the probability density function

$$f(t) = \begin{cases} \dfrac{\frac{4}{3}}{(t + 1)^2} & \text{if } 0 \le t \le 3 \\ 0 & \text{otherwise} \end{cases}$$

(A) What is the probability that this doctor will spend less than 1 hour with a randomly selected patient?

(B) What is the probability that this doctor will spend more than 1 hour with a randomly selected patient?

49. Medicine. Illustrate part (B) in Problem 48 geometrically. Describe the geometric interpretation.

50. Politics. The rate of change of the voting population of a city with respect to time t (in years) is estimated to be

$$N'(t) = \frac{100t}{(1 + t^2)^2}$$

where $N(t)$ is in thousands. If $N(0)$ is the current voting population, how much will this population increase during the next 3 years?

51. Psychology. Rats were trained to go through a maze by rewarding them with a food pellet upon successful completion of the run. After the seventh successful run, the probability density function for length of time (in minutes) until success on the eighth trial was given by

$$f(t) = \begin{cases} .5e^{-.5t} & \text{if } t \ge 0 \\ 0 & \text{otherwise} \end{cases}$$

What is the probability that a rat selected at random after seven successful runs will take 2 or more minutes to complete the eighth run successfully? [Recall that the area under a probability density function curve from $-\infty$ to ∞ is 1.]

Multivariable Calculus

Introduction

In previous chapters, we have applied the key concepts of calculus, the derivative and the integral, to functions with one independent variable. The graph of such a function is a curve in the plane. In Chapter 6, we extend the key concepts of calculus to functions with two independent variables. Graphs of such functions are surfaces in a three-dimensional coordinate system. We use functions with two independent variables to study how production depends on both labor and capital; how braking distance depends on both the weight and speed of a car; how resistance in a blood vessel depends on both its length and radius. In Section 6-5, we justify the method of least squares and use the method to construct linear models (see, for example, Problem 31 in Section 6-5 on global warming).

6-1 Functions of Several Variables

Functions of Two or More Independent Variables

We introduced the concept of a function with one independent variable. Now we broaden the concept to include functions with more than one independent variable.

A small manufacturing company produces a standard type of surfboard. If fixed costs are \$500 per week and variable costs are \$70 per board produced, the weekly cost function is given by

$$C(x) = 500 + 70x \tag{1}$$

where x is the number of boards produced per week. The cost function is a function of a single independent variable x. For each value of x from the domain of C, there exists exactly one value of $C(x)$ in the range of C.

Now, suppose that the company decides to add a high-performance competition board to its line. If the fixed costs for the competition board are \$200 per week and the variable costs are \$100 per board, then the cost function (1) must be modified to

$$C(x, y) = 700 + 70x + 100y \tag{2}$$

where $C(x, y)$ is the cost for a weekly output of x standard boards and y competition boards. Equation (2) is an example of a function with two independent variables x and y. Of course, as the company expands its product line even further, its weekly cost function must be modified to include more and more independent variables, one for each new product produced.

In general, an equation of the form

$$z = f(x, y)$$

describes a **function of two independent variables** if, for each permissible ordered pair (x, y), there is one and only one value of z determined by $f(x, y)$. The variables x and y are **independent variables**, and the variable z is a **dependent variable**. The set of all ordered pairs of permissible values of x and y is the **domain** of the function, and the set of all corresponding values $f(x, y)$ is the **range** of the function. Unless otherwise stated, we will assume that the domain of a function specified by an equation of the form $z = f(x, y)$ is the set of all ordered pairs of real numbers (x, y) such that $f(x, y)$ is also a real number. It should be noted, however, that certain conditions in practical problems often lead to further restrictions on the domain of a function.

We can similarly define functions of three independent variables, $w = f(x, y, z)$; of four independent variables, $u = f(w, x, y, z)$; and so on. In this chapter, we concern ourselves primarily with functions of two independent variables.

EXAMPLE 1 **Evaluating a Function of Two Independent Variables** For the cost function $C(x, y) = 700 + 70x + 100y$ described earlier, find $C(10, 5)$.

SOLUTION

$$\begin{aligned} C(10, 5) &= 700 + 70(10) + 100(5) \\ &= \$1{,}900 \end{aligned}$$

Matched Problem 1 Find $C(20, 10)$ for the cost function in Example 1.

EXAMPLE 2 **Evaluating a Function of Three Independent Variables** For the function $f(x, y, z) = 2x^2 - 3xy + 3z + 1$, find $f(3, 0, -1)$.

SOLUTION

$$f(3, 0, -1) = 2(3)^2 - 3(3)(0) + 3(-1) + 1$$
$$= 18 - 0 - 3 + 1 = 16$$

Matched Problem 2 Find $f(-2, 2, 3)$ for f in Example 2.

EXAMPLE 3 **Revenue, Cost, and Profit Functions** Suppose the surfboard company discussed earlier has determined that the demand equations for its two types of boards are given by

$$p = 210 - 4x + y$$
$$q = 300 + x - 12y$$

where p is the price of the standard board, q is the price of the competition board, x is the weekly demand for standard boards, and y is the weekly demand for competition boards.

(A) Find the weekly revenue function $R(x, y)$, and evaluate $R(20, 10)$.

(B) If the weekly cost function is

$$C(x, y) = 700 + 70x + 100y$$

find the weekly profit function $P(x, y)$ and evaluate $P(20, 10)$.

SOLUTION (A)

$$\text{Revenue} = \begin{pmatrix}\text{demand for}\\ \text{standard}\\ \text{boards}\end{pmatrix} \times \begin{pmatrix}\text{price of a}\\ \text{standard}\\ \text{board}\end{pmatrix} + \begin{pmatrix}\text{demand for}\\ \text{competition}\\ \text{boards}\end{pmatrix} \times \begin{pmatrix}\text{price of a}\\ \text{competition}\\ \text{board}\end{pmatrix}$$

$$R(x, y) = xp + yq$$
$$= x(210 - 4x + y) + y(300 + x - 12y)$$
$$= 210x + 300y - 4x^2 + 2xy - 12y^2$$
$$R(20, 10) = 210(20) + 300(10) - 4(20)^2 + 2(20)(10) - 12(10)^2$$
$$= \$4{,}800$$

(B) Profit = revenue − cost

$$P(x, y) = R(x, y) - C(x, y)$$
$$= 210x + 300y - 4x^2 + 2xy - 12y^2 - 700 - 70x - 100y$$
$$= 140x + 200y - 4x^2 + 2xy - 12y^2 - 700$$
$$P(20, 10) = 140(20) + 200(10) - 4(20)^2 + 2(20)(10) - 12(10)^2 - 700$$
$$= \$1{,}700$$

Matched Problem 3 Repeat Example 3 if the demand and cost equations are given by

$$p = 220 - 6x + y$$
$$q = 300 + 3x - 10y$$
$$C(x, y) = 40x + 80y + 1{,}000$$

Examples of Functions of Several Variables

A number of concepts can be considered as functions of two or more variables.

Area of a rectangle	$A(x, y) = xy$	A = area; x; y
Volume of a box	$V(x, y, z) = xyz$	V = volume; x; y; z
Volume of a right circular cylinder	$V(r, h) = \pi r^2 h$	r; h
Simple interest	$A(P, r, t) = P(1 + rt)$	A = amount P = principal r = annual rate t = time in years
Compound interest	$A(P, r, t, n) = P\left(1 + \frac{r}{n}\right)^{nt}$	A = amount P = principal r = annual rate t = time in years n = number of compound periods per year
IQ	$Q(M, C) = \frac{M}{C}(100)$	Q = IQ = intelligence quotient M = MA = mental age C = CA = chronological age
Resistance for blood flow in a vessel (Poiseuille's law)	$R(L, r) = k\frac{L}{r^4}$	R = resistance L = length of vessel r = radius of vessel k = constant

EXAMPLE 4 **Package Design** A company uses a box with a square base and an open top for a bath assortment (see figure). If x is the length (in inches) of each side of the base and y is the height (in inches), find the total amount of material $M(x, y)$ required to construct one of these boxes, and evaluate $M(5, 10)$.

SOLUTION

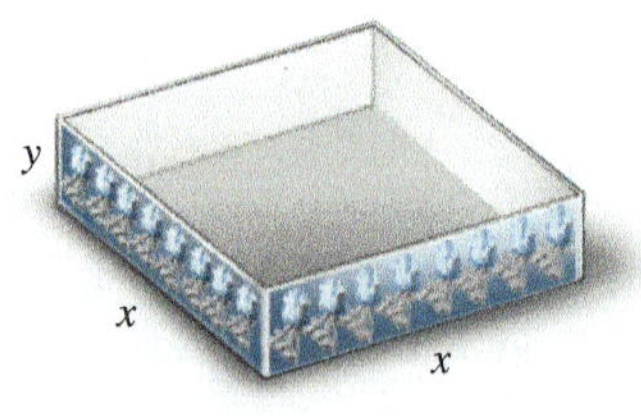

$$\begin{aligned}\text{Area of base} &= x^2\\ \text{Area of one side} &= xy\\ \text{Total material} &= (\text{area of base}) + 4(\text{area of one side})\\ M(x, y) &= x^2 + 4xy\\ M(5, 10) &= (5)^2 + 4(5)(10)\\ &= 225 \text{ square inches}\end{aligned}$$

Matched Problem 4 For the box in Example 4, find the volume $V(x, y)$ and evaluate $V(5, 10)$.

The next example concerns the **Cobb–Douglas production function**

$$f(x, y) = kx^m y^n$$

where k, m, and n are positive constants with $m + n = 1$. Economists use this function to describe the number of units $f(x, y)$ produced from the utilization of x units of labor and y units of capital (for equipment such as tools, machinery, buildings, and so on). Cobb–Douglas production functions are also used to describe the productivity of a single industry, of a group of industries producing the same product, or even of an entire country.

EXAMPLE 5 **Productivity** The productivity of a steel-manufacturing company is given approximately by the function

$$f(x, y) = 10x^{0.2}y^{0.8}$$

with the utilization of x units of labor and y units of capital. If the company uses 3,000 units of labor and 1,000 units of capital, how many units of steel will be produced?

SOLUTION The number of units of steel produced is given by

$$f(3{,}000, 1{,}000) = 10(3{,}000)^{0.2}(1{,}000)^{0.8} \quad \text{Use a calculator.}$$
$$\approx 12{,}457 \text{ units}$$

Matched Problem 5 Refer to Example 5. Find the steel production if the company uses 1,000 units of labor and 2,000 units of capital.

Three-Dimensional Coordinate Systems

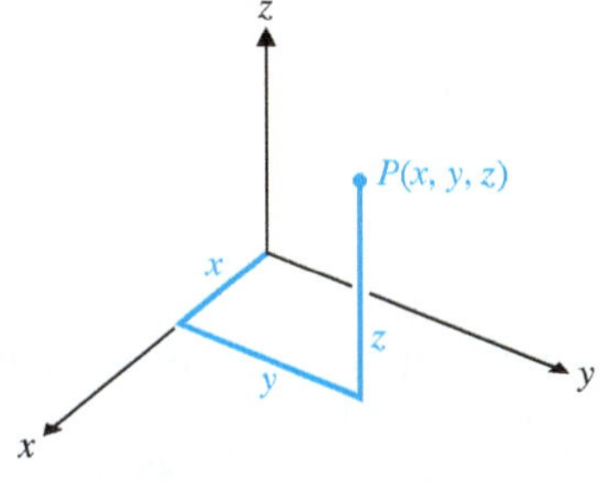

Figure 1 Rectangular coordinate system

We now take a brief look at graphs of functions of two independent variables. Since functions of the form $z = f(x, y)$ involve two independent variables x and y, and one dependent variable z, we need a *three-dimensional coordinate system* for their graphs. A **three-dimensional coordinate system** is formed by three mutually perpendicular number lines intersecting at their origins (see Fig. 1). In such a system, every ordered **triplet of numbers (x, y, z)** can be associated with a unique point, and conversely.

EXAMPLE 6 **Three-Dimensional Coordinates** Locate $(-3, 5, 2)$ in a rectangular coordinate system.

SOLUTION

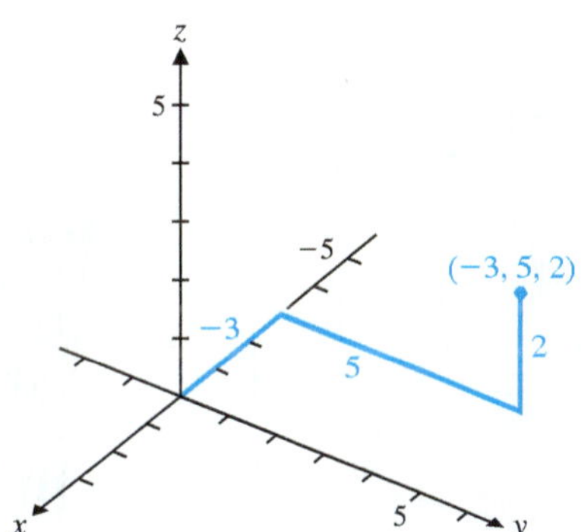

Matched Problem 6 Find the coordinates of the corners A, C, G, and D of the rectangular box shown in the following figure.

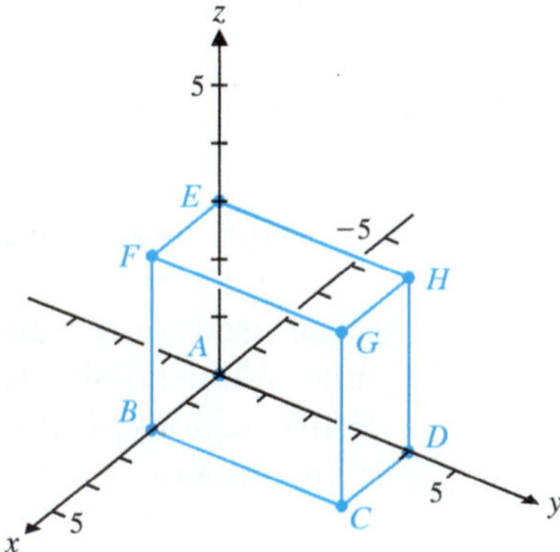

EXPLORE & DISCUSS 1

Imagine that you are facing the front of a classroom whose rectangular walls meet at right angles. Suppose that the point of intersection of the floor, front wall, and left-side wall is the origin of a three-dimensional coordinate system in which every point in the room has nonnegative coordinates. Then the plane $z = 0$ (or, equivalently, the xy plane) can be described as "the floor," and the plane $z = 2$ can be described as "the plane parallel to, but 2 units above, the floor." Give similar descriptions of the following planes:

(A) $x = 0$ (B) $x = 3$ (C) $y = 0$ (D) $y = 4$ (E) $x = -1$

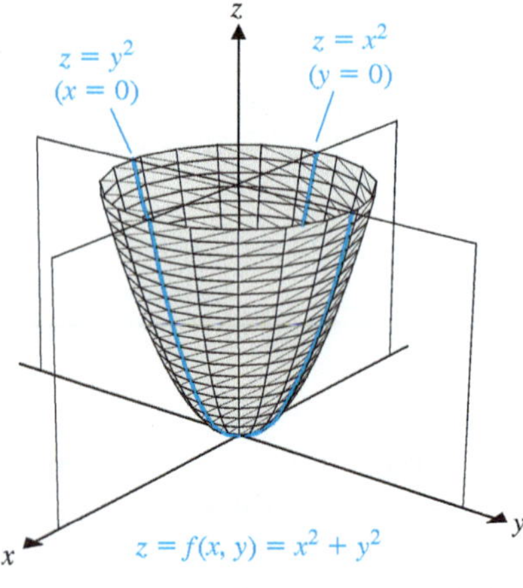

Figure 2 Paraboloid

What does the graph of $z = x^2 + y^2$ look like? If we let $x = 0$ and graph $z = 0^2 + y^2 = y^2$ in the yz plane, we obtain a parabola; if we let $y = 0$ and graph $z = x^2 + 0^2 = x^2$ in the xz plane, we obtain another parabola. The graph of $z = x^2 + y^2$ is either one of these parabolas rotated around the z axis (see Fig. 2). This cup-shaped figure is a *surface* and is called a **paraboloid**.

In general, the graph of any function of the form $z = f(x, y)$ is called a **surface**. The graph of such a function is the graph of all ordered triplets of numbers (x, y, z) that satisfy the equation. Graphing functions of two independent variables is a difficult task, and the general process will not be dealt with in this book. We present only a few simple graphs to suggest extensions of earlier geometric interpretations of the derivative and local maxima and minima to functions of two variables. Note that $z = f(x, y) = x^2 + y^2$ appears (see Fig. 2) to have a local minimum at $(x, y) = (0, 0)$. Figure 3 shows a local maximum at $(x, y) = (0, 0)$.

Figure 4 shows a point at $(x, y) = (0, 0)$, called a **saddle point**, that is neither a local minimum nor a local maximum. Note that if $x = 0$, the saddle point is a local minimum, and if $y = 0$, the saddle point is a local maximum. More will be said about local maxima and minima in Section 6-3.

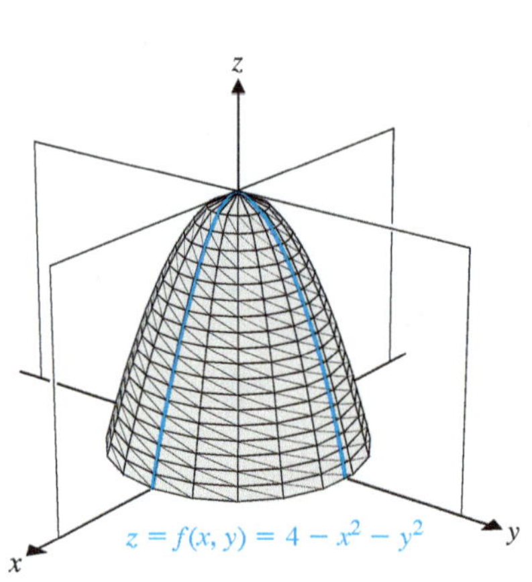

Figure 3 Local maximum: $f(0, 0) = 4$

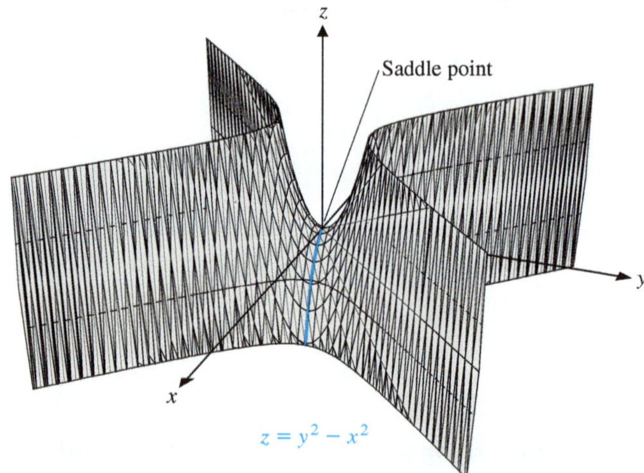

Figure 4 Saddle point at $(0, 0, 0)$

Some graphing calculators are designed to draw graphs (like those of Figs. 2, 3, and 4) of functions of two independent variables. Others, such as the graphing calculator used for the displays in this book, are designed to draw graphs of functions of one independent variable. When using the latter type of calculator, we can graph cross sections produced by cutting surfaces with planes parallel to the xz plane or yz plane to gain insight into the graph of a function of two independent variables.

EXAMPLE 7 **Graphing Cross Sections**

(A) Describe the cross sections of $f(x, y) = 2x^2 + y^2$ in the planes $y = 0$, $y = 1$, $y = 2$, $y = 3$, and $y = 4$.

(B) Describe the cross sections of $f(x, y) = 2x^2 + y^2$ in the planes $x = 0$, $x = 1$, $x = 2$, $x = 3$, and $x = 4$.

SOLUTION (A) The cross section of $f(x, y) = 2x^2 + y^2$ produced by cutting it with the plane $y = 0$ is the graph of the function $f(x, 0) = 2x^2$ in this plane. We can examine the shape of this cross section by graphing $y_1 = 2x^2$ on a graphing calculator (Fig. 5). Similarly, the graphs of $y_2 = f(x, 1) = 2x^2 + 1$, $y_3 = f(x, 2) = 2x^2 + 4$, $y_4 = f(x, 3) = 2x^2 + 9$, and $y_5 = f(x, 4) = 2x^2 + 16$ show the shapes of the other four cross sections (see Fig. 5). Each of these is a parabola that opens upward. Note the correspondence between the graphs in Figure 5 and the actual cross sections of $f(x, y) = 2x^2 + y^2$ shown in Figure 6.

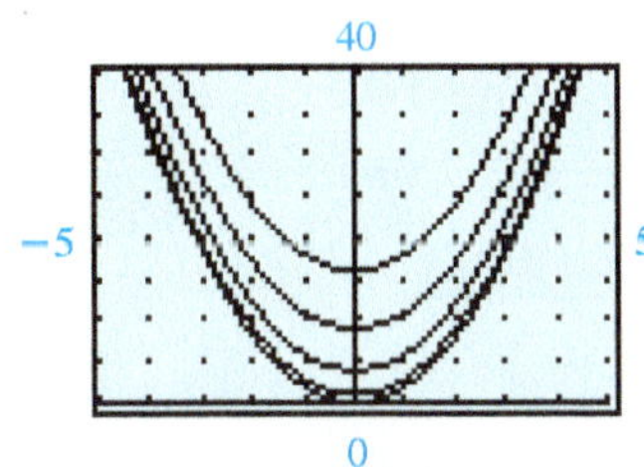

Figure 5

$y_1 = 2x^2$ $y_4 = 2x^2 + 9$
$y_2 = 2x^2 + 1$ $y_5 = 2x^2 + 16$
$y_3 = 2x^2 + 4$

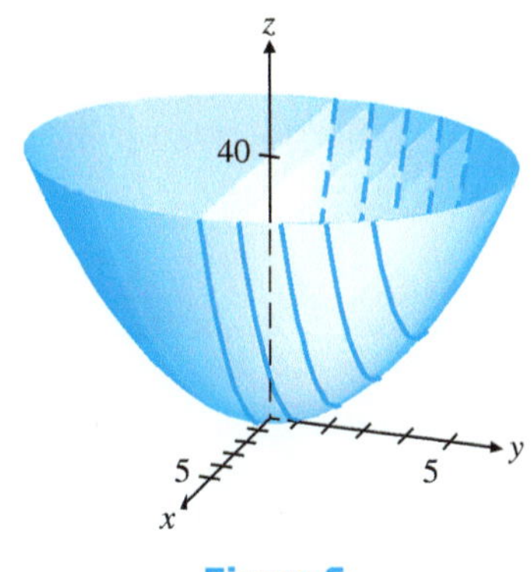

Figure 6

40
−5 5
0

Figure 7

$y_1 = x^2$ $y_4 = 18 + x^2$
$y_2 = 2 + x^2$ $y_5 = 32 + x^2$
$y_3 = 8 + x^2$

(B) The five cross sections are represented by the graphs of the functions $f(0, y) = y^2$, $f(1, y) = 2 + y^2$, $f(2, y) = 8 + y^2$, $f(3, y) = 18 + y^2$, and $f(4, y) = 32 + y^2$. These five functions are graphed in Figure 7. (Note that changing the name of the independent variable from y to x for graphing purposes does not affect the graph displayed.) Each of the five cross sections is a parabola that opens upward.

Matched Problem 7 (A) Describe the cross sections of $g(x, y) = y^2 - x^2$ in the planes $y = 0$, $y = 1$, $y = 2$, $y = 3$, and $y = 4$.

(B) Describe the cross sections of $g(x, y) = y^2 - x^2$ in the planes $x = 0$, $x = 1$, $x = 2$, $x = 3$, and $x = 4$.

CONCEPTUAL INSIGHT

The graph of the *equation*

$$x^2 + y^2 + z^2 = 4 \qquad (3)$$

is the graph of all ordered triplets of numbers (x, y, z) that satisfy the equation. The Pythagorean theorem can be used to show that the distance from the point (x, y, z) to the origin $(0, 0, 0)$ is equal to

$$\sqrt{x^2 + y^2 + z^2}$$

Therefore, the graph of (3) consists of all points that are at a distance 2 from the origin—that is, all points on the sphere of radius 2 and with center at the origin. Recall that a circle in the plane is *not* the graph of a function $y = f(x)$, because it fails the vertical-line test. Similarly, a sphere is *not* the graph of a *function* $z = f(x, y)$ of two variables.

Exercises 6-1

A

In Problems 1–10, find the indicated values of the functions

$$f(x, y) = 2x + 7y - 5 \quad \text{and} \quad g(x, y) = \frac{88}{x^2 + 3y}$$

1. $f(4, -1)$

2. $f(0, 10)$

3. $f(8, 0)$

4. $f(5, 6)$

5. $g(1, 7)$

6. $g(-2, 0)$

7. $g(3, -3)$

8. $g(0, 0)$

9. $3f(-2, 2) + 5g(-2, 2)$

10. $2f(10, -4) - 7g(10, -4)$

In Problems 11–14, find the indicated values of

$$f(x, y, z) = 2x - 3y^2 + 5z^3 - 1$$

11. $f(0, 0, 0)$

12. $f(0, 0, 2)$

13. $f(6, -5, 0)$

14. $f(-10, 4, -3)$

B

In Problems 15–24, find the indicated value of the given function.

15. $P(13, 5)$ for $P(n, r) = \dfrac{n!}{(n - r)!}$

16. $C(13, 5)$ for $C(n, r) = \dfrac{n!}{r!(n - r)!}$

17. $V(4, 12)$ for $V(R, h) = \pi R^2 h$

18. $T(4, 12)$ for $T(R, h) = 2\pi R(R + h)$

19. $S(3, 10)$ for $S(R, h) = \pi R\sqrt{R^2 + h^2}$

20. $W(3, 10)$ for $W(R, h) = \dfrac{1}{3}\pi R^2 h$

21. $A(100, 0.06, 3)$ for $A(P, r, t) = P + Prt$

22. $A(10, 0.04, 3, 2)$ for $A(P, r, t, n) = P\left(1 + \dfrac{r}{n}\right)^{tn}$

23. $P(0.05, 12)$ for $P(r, T) = \displaystyle\int_0^T 4{,}000e^{-rt}\,dt$

24. $F(0.07, 10)$ for $F(r, T) = \displaystyle\int_0^T 4{,}000e^{r(T-t)}\,dt$

In Problems 25–30, find the indicated function f of a single variable.

25. $f(x) = G(x, 0)$ for $G(x, y) = x^2 + 3xy + y^2 - 7$

26. $f(y) = H(0, y)$ for $H(x, y) = x^2 - 5xy - y^2 + 2$

27. $f(y) = K(4, y)$ for $K(x, y) = 10xy + 3x - 2y + 8$

28. $f(x) = L(x, -2)$ for $L(x, y) = 25 - x + 5y - 6xy$

29. $f(y) = M(y, y)$ for $M(x, y) = x^2y - 3xy^2 + 5$

30. $f(x) = N(x, 2x)$ for $N(x, y) = 3xy + x^2 - y^2 + 1$

31. Let $F(x, y) = 2x + 3y - 6$. Find all values of y such that $F(0, y) = 0$.

32. Let $F(x, y) = 5x - 4y + 12$. Find all values of x such that $F(x, 0) = 0$.

33. Let $F(x, y) = 2xy + 3x - 4y - 1$. Find all values of x such that $F(x, x) = 0$.

34. Let $F(x, y) = xy + 2x^2 + y^2 - 25$. Find all values of y such that $F(y, y) = 0$.

35. Let $F(x, y) = x^2 + e^x y - y^2$. Find all values of x such that $F(x, 2) = 0$.

36. Let $G(a, b, c) = a^3 + b^3 + c^3 - (ab + ac + bc) - 6$. Find all values of b such that $G(2, b, 1) = 0$.

C

37. For the function $f(x, y) = x^2 + 2y^2$, find

$$\frac{f(x + h, y) - f(x, y)}{h}$$

38. For the function $f(x, y) = x^2 + 2y^2$, find

$$\frac{f(x, y + k) - f(x, y)}{k}$$

39. For the function $f(x, y) = 2xy^2$, find

$$\frac{f(x + h, y) - f(x, y)}{h}$$

40. For the function $f(x, y) = 2xy^2$, find

$$\frac{f(x, y + k) - f(x, y)}{k}$$

41. Find the coordinates of E and F in the figure for Matched Problem 6 on page 329.

42. Find the coordinates of B and H in the figure for Matched Problem 6 on page 329.

In Problems 43–48, use a graphing calculator as necessary to explore the graphs of the indicated cross sections.

43. Let $f(x, y) = x^2$.

(A) Explain why the cross sections of the surface $z = f(x, y)$ produced by cutting it with planes parallel to $y = 0$ are parabolas.

(B) Describe the cross sections of the surface in the planes $x = 0$, $x = 1$, and $x = 2$.

(C) Describe the surface $z = f(x, y)$.

44. Let $f(x, y) = \sqrt{4 - y^2}$.

(A) Explain why the cross sections of the surface $z = f(x, y)$ produced by cutting it with planes parallel to $x = 0$ are semicircles of radius 2.

(B) Describe the cross sections of the surface in the planes $y = 0$, $y = 2$, and $y = 3$.

(C) Describe the surface $z = f(x, y)$.

45. Let $f(x, y) = \sqrt{36 - x^2 - y^2}$.

(A) Describe the cross sections of the surface $z = f(x, y)$ produced by cutting it with the planes $y = 1$, $y = 2$, $y = 3$, $y = 4$, and $y = 5$.

(B) Describe the cross sections of the surface in the planes $x = 0$, $x = 1$, $x = 2$, $x = 3$, $x = 4$, and $x = 5$.

(C) Describe the surface $z = f(x, y)$.

46. Let $f(x, y) = 100 + 10x + 25y - x^2 - 5y^2$.

(A) Describe the cross sections of the surface $z = f(x, y)$ produced by cutting it with the planes $y = 0$, $y = 1$, $y = 2$, and $y = 3$.

(B) Describe the cross sections of the surface in the planes $x = 0$, $x = 1$, $x = 2$, and $x = 3$.

(C) Describe the surface $z = f(x, y)$.

47. Let $f(x, y) = e^{-(x^2+y^2)}$.

(A) Explain why $f(a, b) = f(c, d)$ whenever (a, b) and (c, d) are points on the same circle centered at the origin in the xy plane.

(B) Describe the cross sections of the surface $z = f(x, y)$ produced by cutting it with the planes $x = 0$, $y = 0$, and $x = y$.

(C) Describe the surface $z = f(x, y)$.

48. Let $f(x, y) = 4 - \sqrt{x^2 + y^2}$.

(A) Explain why $f(a, b) = f(c, d)$ whenever (a, b) and (c, d) are points on the same circle with center at the origin in the xy plane.

(B) Describe the cross sections of the surface $z = f(x, y)$ produced by cutting it with the planes $x = 0$, $y = 0$, and $x = y$.

(C) Describe the surface $z = f(x, y)$.

Applications

49. Cost function. A small manufacturing company produces two models of a surfboard: a standard model and a competition model. If the standard model is produced at a variable cost of \$210 each and the competition model at a variable cost of \$300 each, and if the total fixed costs per month are \$6,000, then the monthly cost function is given by

$$C(x, y) = 6{,}000 + 210x + 300y$$

where x and y are the numbers of standard and competition models produced per month, respectively. Find $C(20, 10)$, $C(50, 5)$, and $C(30, 30)$.

50. Advertising and sales. A company spends \$$x$ thousand per week on online advertising and \$$y$ thousand per week on TV advertising. Its weekly sales are found to be given by

$$S(x, y) = 5x^2y^3$$

Find $S(3, 2)$ and $S(2, 3)$.

51. Revenue function. A supermarket sells two brands of coffee: brand A at \$$p$ per pound and brand B at \$$q$ per pound. The daily demand equations for brands A and B are, respectively,

$$x = 200 - 5p + 4q$$
$$y = 300 + 2p - 4q$$

(both in pounds). Find the daily revenue function $R(p, q)$. Evaluate $R(2, 3)$ and $R(3, 2)$.

52. Revenue, cost, and profit functions. A company manufactures 10- and 3-speed bicycles. The weekly demand and cost equations are

$$p = 230 - 9x + y$$
$$q = 130 + x - 4y$$
$$C(x, y) = 200 + 80x + 30y$$

where \$$p$ is the price of a 10-speed bicycle, \$$q$ is the price of a 3-speed bicycle, x is the weekly demand for 10-speed bicycles, y is the weekly demand for 3-speed bicycles, and $C(x, y)$ is the cost function. Find the weekly revenue function $R(x, y)$ and the weekly profit function $P(x, y)$. Evaluate $R(10, 15)$ and $P(10, 15)$.

53. Productivity. The Cobb–Douglas production function for a petroleum company is given by

$$f(x, y) = 20x^{0.4}y^{0.6}$$

where x is the utilization of labor and y is the utilization of capital. If the company uses 1,250 units of labor and 1,700 units of capital, how many units of petroleum will be produced?

54. Productivity. The petroleum company in Problem 53 is taken over by another company that decides to double both the units of labor and the units of capital utilized in the production of petroleum. Use the Cobb–Douglas production function given in Problem 53 to find the amount of petroleum that will be produced by this increased utilization of labor and capital. What is the effect on productivity of doubling both the units of labor and the units of capital?

55. Future value. At the end of each year, \$5,000 is invested into an IRA earning 3% compounded annually.

(A) How much will be in the account at the end of 30 years? Use the annuity formula

$$F(P, i, n) = P\frac{(1 + i)^n - 1}{i}$$

where

P = periodic payment
i = rate per period
n = number of payments (periods)
F = FV = future value

(B) Use graphical approximation methods to determine the rate of interest that would produce \$300,000 in the account at the end of 30 years.

56. Package design. The packaging department in a company has been asked to design a rectangular box with no top and a partition down the middle (see the figure on the next page). Let x, y, and z be the dimensions of the box (in inches).

Figure for 56

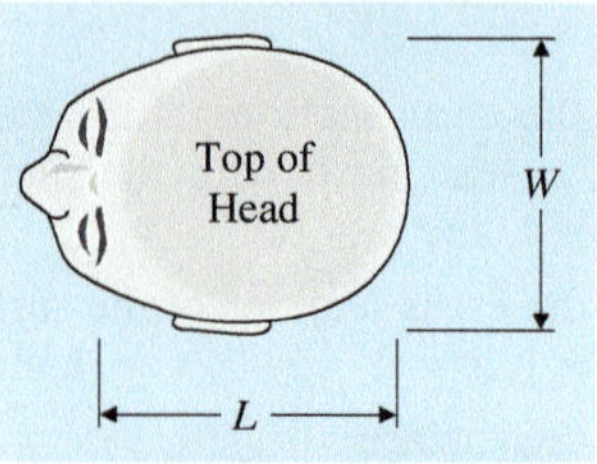

Figure for 59

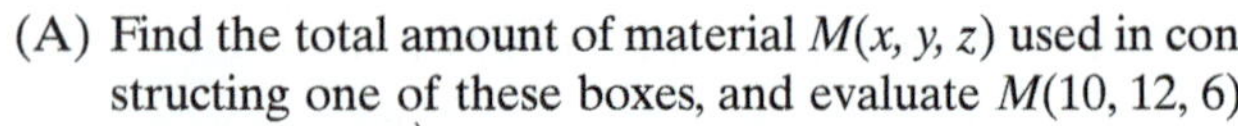

(A) Find the total amount of material $M(x, y, z)$ used in constructing one of these boxes, and evaluate $M(10, 12, 6)$.

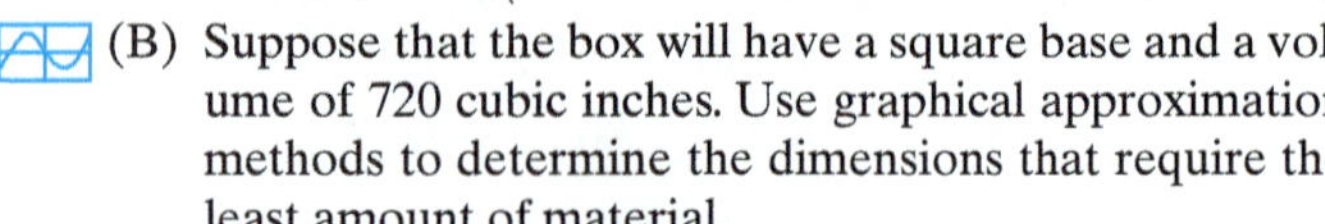

(B) Suppose that the box will have a square base and a volume of 720 cubic inches. Use graphical approximation methods to determine the dimensions that require the least amount of material.

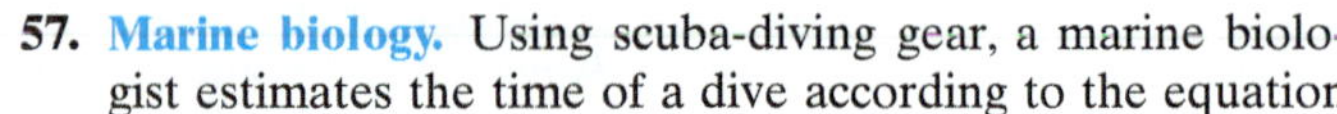

57. **Marine biology.** Using scuba-diving gear, a marine biologist estimates the time of a dive according to the equation

$$T(V, x) = \frac{33V}{x + 33}$$

where

T = time of dive in minutes
V = volume of air, at sea level pressure, compressed into tanks
x = depth of dive in feet

Find $T(70, 47)$ and $T(60, 27)$.

58. **Blood flow.** Poiseuille's law states that the resistance R for blood flowing in a blood vessel varies directly as the length L of the vessel and inversely as the fourth power of its radius r. Stated as an equation,

$$R(L, r) = k\frac{L}{r^4} \qquad k \text{ a constant}$$

Find $R(8, 1)$ and $R(4, 0.2)$.

59. **Physical anthropology.** Anthropologists use an index called the *cephalic index*. The cephalic index C varies directly as the width W of the head and inversely as the length L of the head (both viewed from the top). In terms of an equation,

$$C(W, L) = 100\frac{W}{L}$$

where

W = width in inches
L = length in inches

Find $C(6, 8)$ and $C(8.1, 9)$.

60. **Safety research.** Under ideal conditions, if a person driving a car slams on the brakes and skids to a stop, the length of the skid marks (in feet) is given by the formula

$$L(w, v) = kwv^2$$

where

k = constant
w = weight of car in pounds
v = speed of car in miles per hour

For $k = 0.000\,013\,3$, find $L(2{,}000, 40)$ and $L(3{,}000, 60)$.

61. **Psychology.** The intelligence quotient (IQ) is defined to be the ratio of mental age (MA), as determined by certain tests, to chronological age (CA), multiplied by 100. Stated as an equation,

$$Q(M, C) = \frac{M}{C} \cdot 100$$

where

Q = IQ $\qquad M$ = MA $\qquad C$ = CA

Find $Q(12, 10)$ and $Q(10, 12)$.

Answers to Matched Problems

1. \$3,100

2. 30

3. (A) $R(x, y) = 220x + 300y - 6x^2 + 4xy - 10y^2$; $R(20, 10) = \$4{,}800$

(B) $P(x, y) = 180x + 220y - 6x^2 + 4xy - 10y^2 - 1{,}000$; $P(20, 10) = \$2{,}200$

4. $V(x, y) = x^2y$; $V(5, 10) = 250$ in.^{3}

5. 17,411 units

6. $A\,(0, 0, 0)$; $C(2, 4, 0)$; $G(2, 4, 3)$; $D(0, 4, 0)$

7. (A) Each cross section is a parabola that opens downward.

(B) Each cross section is a parabola that opens upward.

6-2 Partial Derivatives

- Partial Derivatives
- Second-Order Partial Derivatives

Partial Derivatives

We know how to differentiate many kinds of functions of one independent variable and how to interpret the derivatives that result. What about functions with two or more independent variables? Let's return to the surfboard example considered on page 326.

For the company producing only the standard board, the cost function was

$$C(x) = 500 + 70x$$

Differentiating with respect to x, we obtain the marginal cost function

$$C'(x) = 70$$

Since the marginal cost is constant, \$70 is the change in cost for a 1-unit increase in production at any output level.

For the company producing two types of boards—a standard model and a competition model—the cost function was

$$C(x, y) = 700 + 70x + 100y$$

Now suppose that we differentiate with respect to x, holding y fixed, and denote the resulting function by $C_x(x, y)$; or suppose we differentiate with respect to y, holding x fixed, and denote the resulting function by $C_y(x, y)$. Differentiating in this way, we obtain

$$C_x(x, y) = 70 \qquad C_y(x, y) = 100$$

Each of these functions is called a **partial derivative**, and, in this example, each represents marginal cost. The first is the change in cost due to a 1-unit increase in production of the standard board with the production of the competition model held fixed. The second is the change in cost due to a 1-unit increase in production of the competition board with the production of the standard board held fixed.

In general, if $z = f(x, y)$, then the **partial derivative of f with respect to x**, denoted $\partial z/\partial x$, f_x, or $f_x(x, y)$, is defined by

$$\frac{\partial z}{\partial x} = \lim_{h \to 0} \frac{f(x + h, y) - f(x, y)}{h}$$

provided that the limit exists. We recognize this formula as the ordinary derivative of f with respect to x, holding y constant. We can continue to use all the derivative rules and properties discussed in Chapters 1 to 3 and apply them to partial derivatives.

Similarly, the **partial derivative of f with respect to y**, denoted $\partial z/\partial y$, f_y, or $f_y(x, y)$, is defined by

$$\frac{\partial z}{\partial y} = \lim_{k \to 0} \frac{f(x, y + h) - f(x, y)}{k}$$

which is the ordinary derivative with respect to y, holding x constant.

Parallel definitions and interpretations hold for functions with three or more independent variables.

EXAMPLE 1 **Partial Derivatives** For $z = f(x, y) = 2x^2 - 3x^2y + 5y + 1$, find

(A) $\partial z/\partial x$ (B) $f_x(2, 3)$

SOLUTION (A) $z = 2x^2 - 3x^2y + 5y + 1$

Differentiating with respect to x, holding y constant (that is, treating y as a constant), we obtain

$$\frac{\partial z}{\partial x} = 4x - 6xy$$

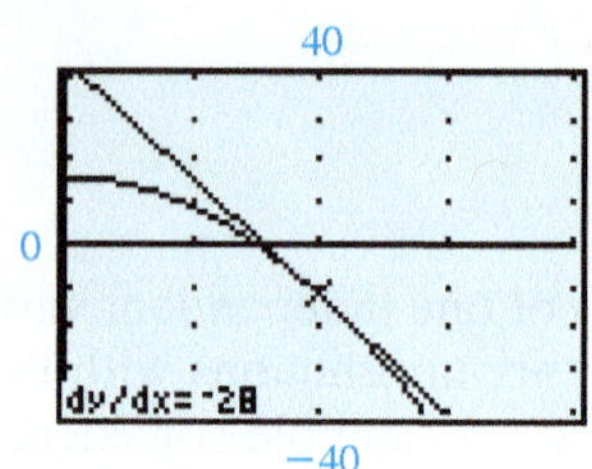

Figure 1 $y_1 = -7x^2 + 16$

(B) $f(x, y) = 2x^2 - 3x^2y + 5y + 1$

First, differentiate with respect to x. From part (A), we have

$$f_x(x, y) = 4x - 6xy$$

Then evaluate this equation at (2, 3):

$$f_x(2, 3) = 4(2) - 6(2)(3) = -28$$

In part 1(B), an alternative approach would be to substitute $y = 3$ into $f(x, y)$ and graph the function $f(x, 3) = -7x^2 + 16$, which represents the cross section of the surface $z = f(x, y)$ produced by cutting it with the plane $y = 3$. Then determine the slope of the tangent line when $x = 2$. Again, we conclude that $f_x(2, 3) = -28$ (see Fig. 1).

Matched Problem 1 For f in Example 1, find

(A) $\partial z/\partial y$ (B) $f_y(2, 3)$

EXAMPLE 2 **Partial Derivatives Using the Chain Rule** For $z = f(x, y) = e^{x^2+y^2}$, find

(A) $\partial z/\partial x$ (B) $f_y(2, 1)$

SOLUTION (A) Using the chain rule [thinking of $z = e^u$, $u = u(x)$; y is held constant], we obtain

$$\frac{\partial z}{\partial x} = e^{x^2+y^2}\frac{\partial(x^2 + y^2)}{\partial x}$$

$$= 2xe^{x^2+y^2}$$

(B)

$$f_y(x, y) = e^{x^2+y^2}\frac{\partial(x^2 + y^2)}{\partial y} = 2ye^{x^2+y^2}$$

$$f_y(2, 1) = 2(1)e^{(2)^2+(1)^2}$$

$$= 2e^5$$

Matched Problem 2 For $z = f(x, y) = (x^2 + 2xy)^5$, find

(A) $\partial z/\partial y$ (B) $f_x(1, 0)$

EXAMPLE 3 **Profit** The profit function for the surfboard company in Example 3 of Section 6-1 was

$$P(x, y) = 140x + 200y - 4x^2 + 2xy - 12y^2 - 700$$

Find $P_x(15, 10)$ and $P_x(30, 10)$, and interpret the results.

SOLUTION

$$P_x(x, y) = 140 - 8x + 2y$$

$$P_x(15, 10) = 140 - 8(15) + 2(10) = 40$$

$$P_x(30, 10) = 140 - 8(30) + 2(10) = -80$$

At a production level of 15 standard and 10 competition boards per week, increasing the production of standard boards by 1 unit and holding the production of competition boards fixed at 10 will increase profit by approximately $40. At a production level of 30 standard and 10 competition boards per week, increasing the production of standard boards by 1 unit and holding the production of competition boards fixed at 10 will decrease profit by approximately $80.

Matched Problem 3 For the profit function in Example 3, find $P_y(25, 10)$ and $P_y(25, 15)$, and interpret the results.

EXAMPLE 4 **Productivity** The productivity of a major computer manufacturer is given approximately by the Cobb–Douglas production function

$$f(x, y) = 15x^{0.4}y^{0.6}$$

with the utilization of x units of labor and y units of capital. The partial derivative $f_x(x, y)$ represents the rate of change of productivity with respect to labor and is called the **marginal productivity of labor**. The partial derivative $f_y(x, y)$ represents the rate of change of productivity with respect to capital and is called the **marginal productivity of capital**. If the company is currently utilizing 4,000 units of labor and 2,500 units of capital, find the marginal productivity of labor and the marginal productivity of capital. For the greatest increase in productivity, should the management of the company encourage increased use of labor or increased use of capital?

SOLUTION

$$f_x(x, y) = 6x^{-0.6}y^{0.6}$$

$$f_x(4{,}000, 2{,}500) = 6(4{,}000)^{-0.6}(2{,}500)^{0.6}$$

$$\approx 4.53 \quad \text{Marginal productivity of labor}$$

$$f_y(x, y) = 9x^{0.4}y^{-0.4}$$

$$f_y(4{,}000, 2{,}500) = 9(4{,}000)^{0.4}(2{,}500)^{-0.4}$$

$$\approx 10.86 \quad \text{Marginal productivity of capital}$$

At the current level of utilization of 4,000 units of labor and 2,500 units of capital, each 1-unit increase in labor utilization (keeping capital utilization fixed at 2,500 units) will increase production by approximately 4.53 units, and each 1-unit increase in capital utilization (keeping labor utilization fixed at 4,000 units) will increase production by approximately 10.86 units. The management of the company should encourage increased use of capital.

Matched Problem 4 The productivity of an airplane-manufacturing company is given approximately by the Cobb–Douglas production function

$$f(x, y) = 40x^{0.3}y^{0.7}$$

(A) Find $f_x(x, y)$ and $f_y(x, y)$.

(B) If the company is currently using 1,500 units of labor and 4,500 units of capital, find the marginal productivity of labor and the marginal productivity of capital.

(C) For the greatest increase in productivity, should the management of the company encourage increased use of labor or increased use of capital?

Partial derivatives have simple geometric interpretations, as shown in Figure 2. If we hold x fixed at $x = a$, then $f_y(a, y)$ is the slope of the curve obtained by intersecting the surface $z = f(x, y)$ with the plane $x = a$. A similar interpretation is given to $f_x(x, b)$.

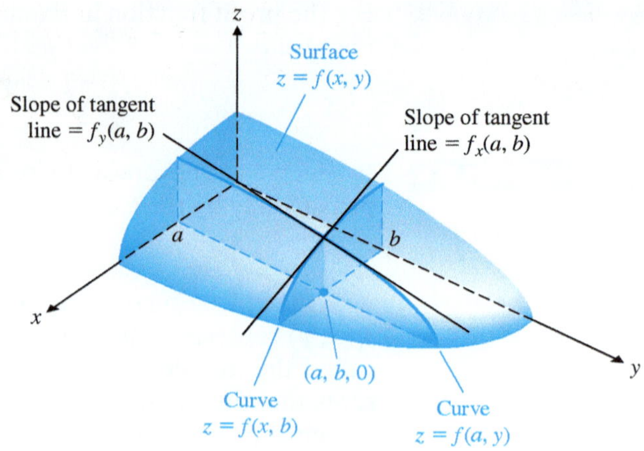

Figure 2

Second-Order Partial Derivatives

The function

$$z = f(x, y) = x^4y^7$$

has two **first-order partial derivatives**:

$$\frac{\partial z}{\partial x} = f_x = f_x(x, y) = 4x^3y^7 \qquad \text{and} \qquad \frac{\partial z}{\partial y} = f_y = f_y(x, y) = 7x^4y^6$$

Each of these partial derivatives, in turn, has two partial derivatives called **second-order partial derivatives** of $z = f(x, y)$. Generalizing the various notations we have for first-order partial derivatives, we write the four second-order partial derivatives of $z = f(x, y) = x^4y^7$ as

Equivalent notations

$$f_{xx} = f_{xx}(x, y) = \frac{\partial^2 z}{\partial x^2} = \frac{\partial}{\partial x}\left(\frac{\partial z}{\partial x}\right) = \frac{\partial}{\partial x}(4x^3y^7) = 12x^2y^7$$

$$f_{xy} = f_{xy}(x, y) = \frac{\partial^2 z}{\partial y\,\partial x} = \frac{\partial}{\partial y}\left(\frac{\partial z}{\partial x}\right) = \frac{\partial}{\partial y}(4x^3y^7) = 28x^3y^6$$

$$f_{yx} = f_{yx}(x, y) = \frac{\partial^2 z}{\partial x\,\partial y} = \frac{\partial}{\partial x}\left(\frac{\partial z}{\partial y}\right) = \frac{\partial}{\partial x}(7x^4y^6) = 28x^3y^6$$

$$f_{yy} = f_{yy}(x, y) = \frac{\partial^2 z}{\partial y^2} = \frac{\partial}{\partial y}\left(\frac{\partial z}{\partial y}\right) = \frac{\partial}{\partial y}(7x^4y^6) = 42x^4y^5$$

In the mixed partial derivative $\partial^2 z/\partial y\,\partial x = f_{xy}$, we started with $z = f(x, y)$ and first differentiated with respect to x (holding y constant). Then we differentiated with respect to y (holding x constant). In the other mixed partial derivative, $\partial^2 z/\partial x\,\partial y = f_{yx}$, the order of differentiation was reversed; however, the final result was the same—that is, $f_{xy} = f_{yx}$. Although it is possible to find functions for which $f_{xy} \neq f_{yx}$, such functions rarely occur in applications involving partial derivatives. For all the functions in this book, we will assume that $f_{xy} = f_{yx}$.

In general, we have the following definitions:

DEFINITION Second-Order Partial Derivatives

If $z = f(x, y)$, then

$$f_{xx} = f_{xx}(x, y) = \frac{\partial^2 z}{\partial x^2} = \frac{\partial}{\partial x}\left(\frac{\partial z}{\partial x}\right)$$

$$f_{xy} = f_{xy}(x, y) = \frac{\partial^2 z}{\partial y\, \partial x} = \frac{\partial}{\partial y}\left(\frac{\partial z}{\partial x}\right)$$

$$f_{yx} = f_{yx}(x, y) = \frac{\partial^2 z}{\partial x\, \partial y} = \frac{\partial}{\partial x}\left(\frac{\partial z}{\partial y}\right)$$

$$f_{yy} = f_{yy}(x, y) = \frac{\partial^2 z}{\partial y^2} = \frac{\partial}{\partial y}\left(\frac{\partial z}{\partial y}\right)$$

EXAMPLE 5 **Second-Order Partial Derivatives** For $z = f(x, y) = 3x^2 - 2xy^3 + 1$, find

(A) $\dfrac{\partial^2 z}{\partial x\, \partial y}, \dfrac{\partial^2 z}{\partial y\, \partial x}$ (B) $\dfrac{\partial^2 z}{\partial x^2}$ (C) $f_{yx}(2, 1)$

SOLUTION (A) First differentiate with respect to y and then with respect to x:

$$\frac{\partial z}{\partial y} = -6xy^2 \qquad \frac{\partial^2 z}{\partial x\, \partial y} = \frac{\partial}{\partial x}\left(\frac{\partial z}{\partial y}\right) = \frac{\partial}{\partial x}(-6xy^2) = -6y^2$$

Now differentiate with respect to x and then with respect to y:

$$\frac{\partial z}{\partial x} = 6x - 2y^3 \qquad \frac{\partial^2 z}{\partial y\, \partial x} = \frac{\partial}{\partial y}\left(\frac{\partial z}{\partial x}\right) = \frac{\partial}{\partial y}(6x - 2y^3) = -6y^2$$

(B) Differentiate with respect to x twice:

$$\frac{\partial z}{\partial x} = 6x - 2y^3 \qquad \frac{\partial^2 z}{\partial x^2} = \frac{\partial}{\partial x}\left(\frac{\partial z}{\partial x}\right) = 6$$

(C) First find $f_{yx}(x, y)$; then evaluate the resulting equation at (2, 1). Again, remember that f_{yx} signifies differentiation first with respect to y and then with respect to x.

$$f_y(x, y) = -6xy^2 \qquad f_{yx}(x, y) = -6y^2$$

and

$$f_{yx}(2, 1) = -6(1)^2 = -6$$

Matched Problem 5 For $z = f(x, y) = x^3y - 2y^4 + 3$, find

(A) $\dfrac{\partial^2 z}{\partial y\, \partial x}$ (B) $\dfrac{\partial^2 z}{\partial y^2}$

(C) $f_{xy}(2, 3)$ (D) $f_{yx}(2, 3)$

CONCEPTUAL INSIGHT

Although the mixed second-order partial derivatives f_{xy} and f_{yx} are equal for all functions considered in this book, it is a good idea to compute both of them, as in Example 5(A), as a check on your work. By contrast, the other two second-order partial derivatives, f_{xx} and f_{yy}, are generally not equal to each other. For example, for the function

$$f(x, y) = 3x^2 - 2xy^3 + 1$$

of Example 5,

$$f_{xx} = 6 \quad \text{and} \quad f_{yy} = -12xy$$

Exercises 6-2

A

In Problems 1–18, find the indicated first-order partial derivative for each function $z = f(x, y)$.

1. $f_x(x, y)$ if $f(x, y) = 4x - 3y + 6$

2. $f_x(x, y)$ if $f(x, y) = 7x + 8y - 2$

3. $f_y(x, y)$ if $f(x, y) = x^2 - 3xy + 2y^2$

4. $f_y(x, y)$ if $f(x, y) = 3x^2 + 2xy - 7y^2$

5. $\dfrac{\partial z}{\partial x}$ if $z = x^3 + 4x^2y + 2y^3$

6. $\dfrac{\partial z}{\partial y}$ if $z = 4x^2y - 5xy^2$

7. $\dfrac{\partial z}{\partial y}$ if $z = (5x + 2y)^{10}$

8. $\dfrac{\partial z}{\partial x}$ if $z = (2x - 3y)^8$

In Problems 9–16, find the indicated value.

9. $f_x(1, 3)$ if $f(x, y) = 5x^3y - 4xy^2$

10. $f_x(4, 1)$ if $f(x, y) = x^2y^2 - 5xy^3$

11. $f_y(1, 0)$ if $f(x, y) = 3xe^y$

12. $f_y(2, 4)$ if $f(x, y) = x^4 \ln y$

13. $f_y(2, 1)$ if $f(x, y) = e^{x^2} - 4y$

14. $f_y(3, 3)$ if $f(x, y) = e^{3x} - y^2$

15. $f_x(1, -1)$ if $f(x, y) = \dfrac{2xy}{1 + x^2y^2}$

16. $f_x(-1, 2)$ if $f(x, y) = \dfrac{x^2 - y^2}{1 + x^2}$

In Problems 17–22, $M(x,y) = 68 + 0.3x - 0.8y$ gives the mileage (in mpg) of a new car as a function of tire pressure x (in psi) and speed (in mph). Find the indicated quantity (include the appropriate units) and explain what it means.

17. $M(32, 40)$

18. $M(22, 40)$

19. $M(32, 50)$

20. $M(22, 50)$

21. $M_x(32, 50)$

22. $M_y(32, 50)$

B

In Problems 23–34, find the indicated second-order partial derivative for each function $f(x, y)$.

23. $f_{xx}(x, y)$ if $f(x, y) = 6x - 5y + 3$

24. $f_{yx}(x, y)$ if $f(x, y) = -2x + y + 8$

25. $f_{xy}(x, y)$ if $f(x, y) = 4x^2 + 6y^2 - 10$

26. $f_{yy}(x, y)$ if $f(x, y) = x^2 + 9y^2 - 4$

27. $f_{xy}(x, y)$ if $f(x, y) = e^{xy^2}$

28. $f_{yx}(x, y)$ if $f(x, y) = e^{3x+2y}$

29. $f_{yy}(x, y)$ if $f(x, y) = \dfrac{\ln x}{y}$

30. $f_{xx}(x, y)$ if $f(x, y) = \dfrac{3 \ln x}{y^2}$

31. $f_{xx}(x, y)$ if $f(x, y) = (2x + y)^5$

32. $f_{yx}(x, y)$ if $f(x, y) = (3x - 8y)^6$

33. $f_{xy}(x, y)$ if $f(x, y) = (x^2 + y^4)^{10}$

34. $f_{yy}(x, y)$ if $f(x, y) = (1 + 2xy^2)^8$

In Problems 35–44, find the indicated function or value if $C(x, y) = 3x^2 + 10xy - 8y^2 + 4x - 15y - 120$.

35. $C_x(x, y)$

36. $C_y(x, y)$

37. $C_x(3, -2)$

38. $C_y(3, -2)$

39. $C_{xx}(x, y)$

40. $C_{yy}(x, y)$

41. $C_{xy}(x, y)$

42. $C_{yx}(x, y)$

43. $C_{xx}(3, -2)$

44. $C_{yy}(3, -2)$

In Problems 45–56, find the indicated function or value if $S(x, y) = x^3 \ln y + 4y^2e^x$.

45. $S_y(x, y)$

46. $S_x(x, y)$

47. $S_y(-1, 1)$

48. $S_x(-1, 1)$

49. $S_{yx}(x, y)$

50. $S_{xy}(x, y)$

51. $S_{yy}(x, y)$

52. $S_{xx}(x, y)$

53. $S_{yx}(-1, 1)$

54. $S_{xy}(-1, 1)$

55. $S_{yy}(-1, 1)$

56. $S_{xx}(-1, 1)$

In Problems 57–62, $S(T, r) = 50(T - 40)(5 - r)$ gives an ice cream shop's daily sales as a function of temperature T (in °F) and rain r (in inches). Find the indicated quantity (include the appropriate units) and explain what it means.

57. $S(60, 2)$ **58.** $S(80, 0)$

59. $S_r(90, 1)$ **60.** $S_T(90, 1)$

61. $S_{Tr}(90, 1)$ **62.** $S_{rT}(90, 1)$

63. (A) Let $f(x, y) = y^3 + 4y^2 - 5y + 3$. Show that $\partial f/\partial x = 0$.

(B) Explain why there are an infinite number of functions $g(x, y)$ such that $\partial g/\partial x = 0$.

64. (A) Find an example of a function $f(x, y)$ such that $\partial f/\partial x = 3$ and $\partial f/\partial y = 2$.

(B) How many such functions are there? Explain.

In Problems 65–70, find $f_{xx}(x, y)$, $f_{xy}(x, y)$, $f_{yx}(x, y)$, and $f_{yy}(x, y)$ for each function f.

65. $f(x, y) = x^2y^2 + x^3 + y$

66. $f(x, y) = x^3y^3 + x + y^2$

67. $f(x, y) = \dfrac{x}{y} - \dfrac{y}{x}$

68. $f(x, y) = \dfrac{x^2}{y} - \dfrac{y^2}{x}$

69. $f(x, y) = xe^{xy}$

70. $f(x, y) = x \ln(xy)$

C

71. For

$$P(x, y) = -x^2 + 2xy - 2y^2 - 4x + 12y - 5$$

find all values of x and y such that

$$P_x(x, y) = 0 \quad \text{and} \quad P_y(x, y) = 0$$

simultaneously.

72. For

$$C(x, y) = 2x^2 + 2xy + 3y^2 - 16x - 18y + 54$$

find all values of x and y such that

$$C_x(x, y) = 0 \quad \text{and} \quad C_y(x, y) = 0$$

simultaneously.

73. For

$$F(x, y) = x^3 - 2x^2y^2 - 2x - 4y + 10$$

find all values of x and y such that

$$F_x(x, y) = 0 \quad \text{and} \quad F_y(x, y) = 0$$

simultaneously.

74. For

$$G(x, y) = x^2 \ln y - 3x - 2y + 1$$

find all values of x and y such that

$$G_x(x, y) = 0 \quad \text{and} \quad G_y(x, y) = 0$$

simultaneously

75. Let $f(x, y) = 3x^2 + y^2 - 4x - 6y + 2$.

(A) Find the minimum value of $f(x, y)$ when $y = 1$.

(B) Explain why the answer to part (A) is not the minimum value of the function $f(x, y)$.

76. Let $f(x, y) = 5 - 2x + 4y - 3x^2 - y^2$.

(A) Find the maximum value of $f(x, y)$ when $x = 2$.

(B) Explain why the answer to part (A) is not the maximum value of the function $f(x, y)$.

77. Let $f(x, y) = 4 - x^4y + 3xy^2 + y^5$.

(A) Use graphical approximation methods to find c (to three decimal places) such that $f(c, 2)$ is the maximum value of $f(x, y)$ when $y = 2$.

(B) Find $f_x(c, 2)$ and $f_y(c, 2)$.

78. Let $f(x, y) = e^x + 2e^y + 3xy^2 + 1$.

(A) Use graphical approximation methods to find d (to three decimal places) such that $f(1, d)$ is the minimum value of $f(x, y)$ when $x = 1$.

(B) Find $f_x(1, d)$ and $f_y(1, d)$.

In Problems 79 and 80, show that the function f satisfies $f_{xx}(x, y) + f_{yy}(x, y) = 0$.

79. $f(x, y) = \ln(x^2 + y^2)$

80. $f(x, y) = x^3 - 3xy^2$

81. For $f(x, y) = x^2 + 2y^2$, find

(A) $\displaystyle\lim_{h \to 0} \frac{f(x + h, y) - f(x, y)}{h}$

(B) $\displaystyle\lim_{k \to 0} \frac{f(x, y + k) - f(x, y)}{k}$

82. For $f(x, y) = 2xy^2$, find

(A) $\displaystyle\lim_{h \to 0} \frac{f(x + h, y) - f(x, y)}{h}$

(B) $\displaystyle\lim_{k \to 0} \frac{f(x, y + k) - f(x, y)}{k}$

Applications

83. Profit function. A firm produces two types of calculators each week, x of type A and y of type B. The weekly revenue and cost functions (in dollars) are

$$R(x, y) = 80x + 90y + 0.04xy - 0.05x^2 - 0.05y^2$$

$$C(x, y) = 8x + 6y + 20{,}000$$

Find $P_x(1{,}200, 1{,}800)$ and $P_y(1{,}200, 1{,}800)$, and interpret the results.

84. Advertising and sales. A company spends \$x per week on online advertising and \$y per week on TV advertising. Its weekly sales were found to be given by

$$S(x, y) = 10x^{0.4}y^{0.8}$$

Find $S_x(3{,}000, 2{,}000)$ and $S_y(3{,}000, 2{,}000)$, and interpret the results.

85. Demand equations. A supermarket sells two brands of coffee: brand A at $\$p$ per pound and brand B at $\$q$ per pound. The daily demands x and y (in pounds) for brands A and B, respectively, are given by

$$x = 200 - 5p + 4q$$

$$y = 300 + 2p - 4q$$

Find $\partial x/\partial p$ and $\partial y/\partial p$, and interpret the results.

86. Revenue and profit functions. A company manufactures 10- and 3-speed bicycles. The weekly demand and cost functions are

$$p = 230 - 9x + y$$

$$q = 130 + x - 4y$$

$$C(x, y) = 200 + 80x + 30y$$

where $\$p$ is the price of a 10-speed bicycle, $\$q$ is the price of a 3-speed bicycle, x is the weekly demand for 10-speed bicycles, y is the weekly demand for 3-speed bicycles, and $C(x, y)$ is the cost function. Find $R_x(10, 5)$ and $P_x(10, 5)$, and interpret the results.

87. Productivity. The productivity of a certain third-world country is given approximately by the function

$$f(x, y) = 10x^{0.75}y^{0.25}$$

with the utilization of x units of labor and y units of capital.

(A) Find $f_x(x, y)$ and $f_y(x, y)$.

(B) If the country is now using 600 units of labor and 100 units of capital, find the marginal productivity of labor and the marginal productivity of capital.

(C) For the greatest increase in the country's productivity, should the government encourage increased use of labor or increased use of capital?

88. Productivity. The productivity of an automobile-manufacturing company is given approximately by the function

$$f(x, y) = 50\sqrt{xy} = 50x^{0.5}y^{0.5}$$

with the utilization of x units of labor and y units of capital.

(A) Find $f_x(x, y)$ and $f_y(x, y)$.

(B) If the company is now using 250 units of labor and 125 units of capital, find the marginal productivity of labor and the marginal productivity of capital.

(C) For the greatest increase in the company's productivity, should the management encourage increased use of labor or increased use of capital?

Problems 89–92 refer to the following: If a decrease in demand for one product results in an increase in demand for another product, the two products are said to be **competitive**, *or* **substitute, products**. *(Real whipping cream and imitation whipping cream are examples of competitive, or substitute, products.) If a decrease in demand for one product results in a decrease in demand for another product, the two products are said to be* **complementary products**. *(Fishing boats and outboard motors are examples of complementary products.) Partial derivatives can be used to test whether two products are competitive, complementary, or neither. We start with demand functions for two products such that the demand for either depends on the prices for both:*

$$x = f(p, q) \quad \text{Demand function for product A}$$

$$y = g(p, q) \quad \text{Demand function for product B}$$

The variables x and y represent the number of units demanded of products A and B, respectively, at a price p for 1 unit of product A and a price q for 1 unit of product B. Normally, if the price of A increases while the price of B is held constant, then the demand for A will decrease; that is, $f_p(p, q) < 0$. *Then, if A and B are competitive products, the demand for B will increase; that is,* $g_p(p, q) > 0$. *Similarly, if the price of B increases while the price of A is held constant, the demand for B will decrease; that is,* $g_q(p, q) < 0$. *Then, if A and B are competitive products, the demand for A will increase; that is,* $f_q(p, q) > 0$. *Reasoning similarly for complementary products, we arrive at the following test:*

Test for Competitive and Complementary Products

Partial Derivatives	Products A and B
$f_q(p, q) > 0$ and $g_p(p, q) > 0$	Competitive (substitute)
$f_q(p, q) < 0$ and $g_p(p, q) < 0$	Complementary
$f_q(p, q) \geq 0$ and $g_p(p, q) \leq 0$	Neither
$f_q(p, q) \leq 0$ and $g_p(p, q) \geq 0$	Neither

Use this test in Problems 89–92 to determine whether the indicated products are competitive, complementary, or neither.

89. Product demand. The weekly demand equations for the sale of butter and margarine in a supermarket are

$$x = f(p, q) = 8{,}000 - 0.09p^2 + 0.08q^2 \quad \text{Butter}$$

$$y = g(p, q) = 15{,}000 + 0.04p^2 - 0.3q^2 \quad \text{Margarine}$$

90. Product demand. The daily demand equations for the sale of brand A coffee and brand B coffee in a supermarket are

$$x = f(p, q) = 200 - 5p + 4q \quad \text{Brand A coffee}$$

$$y = g(p, q) = 300 + 2p - 4q \quad \text{Brand B coffee}$$

91. Product demand. The monthly demand equations for the sale of skis and ski boots in a sporting goods store are

$$x = f(p, q) = 800 - 0.004p^2 - 0.003q^2 \quad \text{Skis}$$

$$y = g(p, q) = 600 - 0.003p^2 - 0.002q^2 \quad \text{Ski boots}$$

92. Product demand. The monthly demand equations for the sale of tennis rackets and tennis balls in a sporting goods store are

$$x = f(p, q) = 500 - 0.5p - q^2 \quad \text{Tennis rackets}$$

$$y = g(p, q) = 10{,}000 - 8p - 100q^2 \quad \text{Tennis balls (cans)}$$

93. Medicine. The following empirical formula relates the surface area A (in square inches) of an average human body to its weight w (in pounds) and its height h (in inches):

$$A = f(w, h) = 15.64w^{0.425}h^{0.725}$$

(A) Find $f_w(w, h)$ and $f_h(w, h)$.

(B) For a 65-pound child who is 57 inches tall, find $f_w(65, 57)$ and $f_h(65, 57)$, and interpret the results.

94. **Blood flow.** Poiseuille's law states that the resistance R for blood flowing in a blood vessel varies directly as the length L of the vessel and inversely as the fourth power of its radius r. Stated as an equation,

$$R(L, r) = k\frac{L}{r^4} \qquad k \text{ a constant}$$

Find $R_L(4, 0.2)$ and $R_r(4, 0.2)$, and interpret the results.

95. **Physical anthropology.** Anthropologists use the cephalic index C, which varies directly as the width W of the head and inversely as the length L of the head (both viewed from the top). In terms of an equation,

$$C(W, L) = 100\frac{W}{L}$$

where

W = width in inches $\qquad L$ = length in inches

Find $C_W(6, 8)$ and $C_L(6, 8)$, and interpret the results.

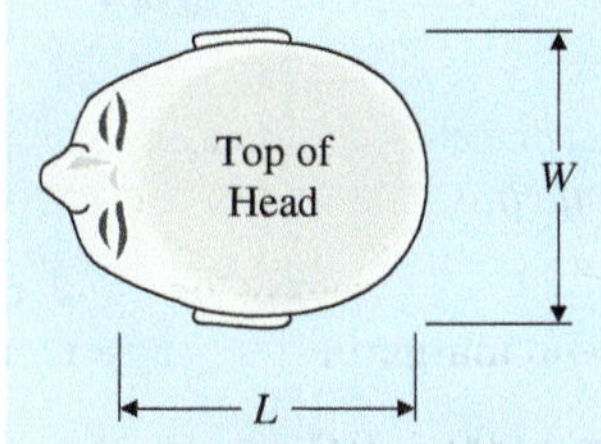

Figure for 95

96. **Safety research.** Under ideal conditions, if a person driving a car slams on the brakes and skids to a stop, the length of the skid marks (in feet) is given by the formula

$$L(w, v) = kwv^2$$

where

k = constant
w = weight of car in pounds
v = speed of car in miles per hour

For $k = 0.000\,013\,3$, find $L_w(2{,}500, 60)$ and $L_v(2{,}500, 60)$, and interpret the results.

Answers to Matched Problems

1. (A) $\partial z/\partial y = -3x^2 + 5$
(B) $f_y(2, 3) = -7$

2. (A) $10x(x^2 + 2xy)^4$
(B) 10

3. $P_y(25, 10) = 10$: At a production level of $x = 25$ and $y = 10$, increasing y by 1 unit and holding x fixed at 25 will increase profit by approximately \$10; $P_y(25, 15) = -110$: At a production level of $x = 25$ and $y = 15$, increasing y by 1 unit and holding x fixed at 25 will decrease profit by approximately \$110

4. (A) $f_x(x, y) = 12x^{-0.7}y^{0.7}$; $f_y(x, y) = 28x^{0.3}y^{-0.3}$
(B) Marginal productivity of labor ≈ 25.89; marginal productivity of capital ≈ 20.14
(C) Labor

5. (A) $3x^2$ (B) $-24y^2$
(C) 12 (D) 12

6-3 Maxima and Minima

We are now ready to undertake a brief, but useful, analysis of local maxima and minima for functions of the type $z = f(x, y)$. We will extend the second-derivative test developed for functions of a single independent variable. We assume that all second-order partial derivatives exist for the function f in some circular region in the xy plane. This guarantees that the surface $z = f(x, y)$ has no sharp points, breaks, or ruptures. In other words, we are dealing only with smooth surfaces with no edges (like the edge of a box), breaks (like an earthquake fault), or sharp points (like the bottom point of a golf tee). (See Figure 1.)

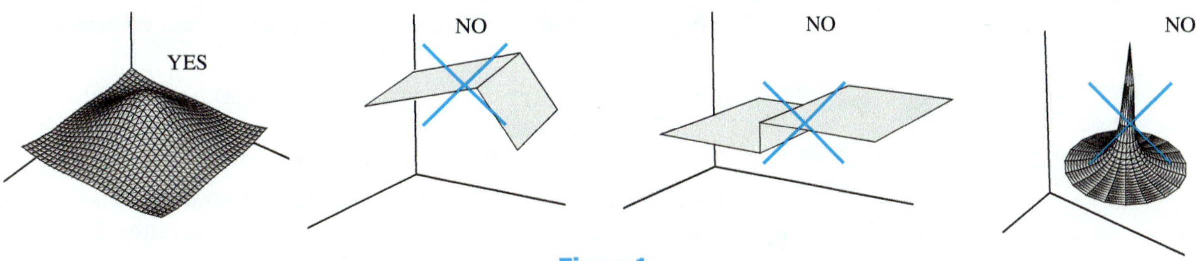

Figure 1

In addition, we will not concern ourselves with boundary points or absolute maxima–minima theory. Despite these restrictions, the procedure we will describe will help us solve a large number of useful problems.

What does it mean for $f(a, b)$ to be a local maximum or a local minimum? We say that **$f(a, b)$ is a local maximum** if there exists a circular region in the domain of f with (a, b) as the center, such that

$$f(a, b) \geq f(x, y)$$

for all (x, y) in the region. Similarly, we say that **$f(a, b)$ is a local minimum** if there exists a circular region in the domain of f with (a, b) as the center, such that

$$f(a, b) \leq f(x, y)$$

for all (x, y) in the region. Figure 2A illustrates a local maximum, Figure 2B a local minimum, and Figure 2C a **saddle point**, which is neither a local maximum nor a local minimum.

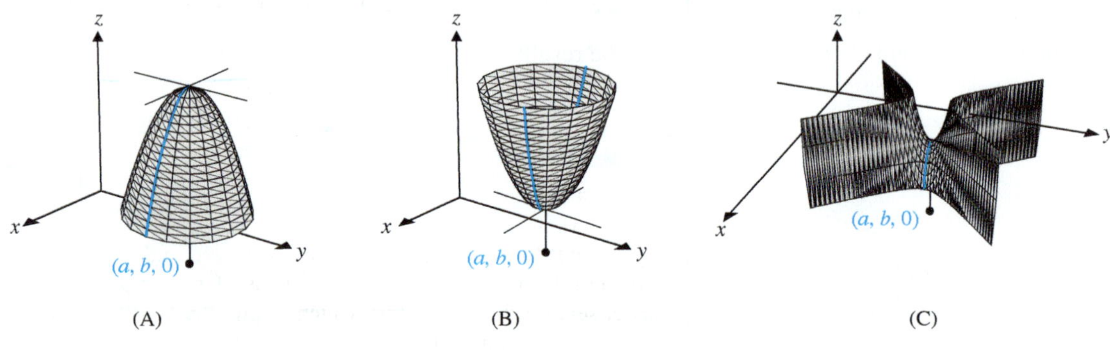

Figure 2

What happens to $f_x(a, b)$ and $f_y(a, b)$ if $f(a, b)$ is a local minimum or a local maximum and the partial derivatives of f exist in a circular region containing (a, b)? Figure 2 suggests that $f_x(a, b) = 0$ and $f_y(a, b) = 0$, since the tangent lines to the given curves are horizontal. Theorem 1 indicates that our intuitive reasoning is correct.

THEOREM 1 Local Extrema and Partial Derivatives
Let $f(a, b)$ be a local extremum (a local maximum or a local minimum) for the function f. If both f_x and f_y exist at (a, b), then

$$f_x(a, b) = 0 \quad \text{and} \quad f_y(a, b) = 0 \tag{1}$$

The converse of this theorem is false. If $f_x(a, b) = 0$ and $f_y(a, b) = 0$, then $f(a, b)$ may or may not be a local extremum; for example, the point $(a, b, f(a, b))$ may be a saddle point (see Fig. 2C).

Theorem 1 gives us *necessary* (but not *sufficient*) conditions for $f(a, b)$ to be a local extremum. We find all points (a, b) such that $f_x(a, b) = 0$ and $f_y(a, b) = 0$ and test these further to determine whether $f(a, b)$ is a local extremum or a saddle point. Points (a, b) such that conditions (1) hold are called **critical points**.

EXPLORE & DISCUSS 1

(A) Let $f(x, y) = y^2 + 1$. Explain why $f(x, y)$ has a local minimum at every point on the x axis. Verify that every point on the x axis is a critical point. Explain why the graph of $z = f(x, y)$ could be described as a trough.

(B) Let $g(x, y) = x^3$. Show that every point on the y axis is a critical point. Explain why no point on the y axis is a local extremum. Explain why the graph of $z = g(x, y)$ could be described as a slide.

The next theorem, using second-derivative tests, gives us *sufficient* conditions for a critical point to produce a local extremum or a saddle point.

THEOREM 2 Second-Derivative Test for Local Extrema

If

1. $z = f(x, y)$
2. $f_x(a, b) = 0$ and $f_y(a, b) = 0$ [(a, b) is a critical point]
3. All second-order partial derivatives of f exist in some circular region containing (a, b) as center.
4. $A = f_{xx}(a, b), \quad B = f_{xy}(a, b), \quad C = f_{yy}(a, b)$

Then

Case 1. If $AC - B^2 > 0$ and $A < 0$, then $f(a, b)$ is a local maximum.

Case 2. If $AC - B^2 > 0$ and $A > 0$, then $f(a, b)$ is a local minimum.

Case 3. If $AC - B^2 < 0$, then f has a saddle point at (a, b).

Case 4. If $AC - B^2 = 0$, the test fails.

CONCEPTUAL INSIGHT

The condition $A = f_{xx}(a, b) < 0$ in case 1 of Theorem 2 is analogous to the condition $f''(c) < 0$ in the second-derivative test for a function of one variable (Section 3-5), which implies that the function is concave downward and therefore has a local maximum. Similarly, the condition $A = f_{xx}(a, b) > 0$ in case 2 is analogous to the condition $f''(c) > 0$ in the earlier second-derivative test, which implies that the function is concave upward and therefore has a local minimum.

To illustrate the use of Theorem 2, we find the local extremum for a very simple function whose solution is almost obvious: $z = f(x, y) = x^2 + y^2 + 2$. From the function f itself and its graph (Fig. 3), it is clear that a local minimum is found at $(0, 0)$. Let us see how Theorem 2 confirms this observation.

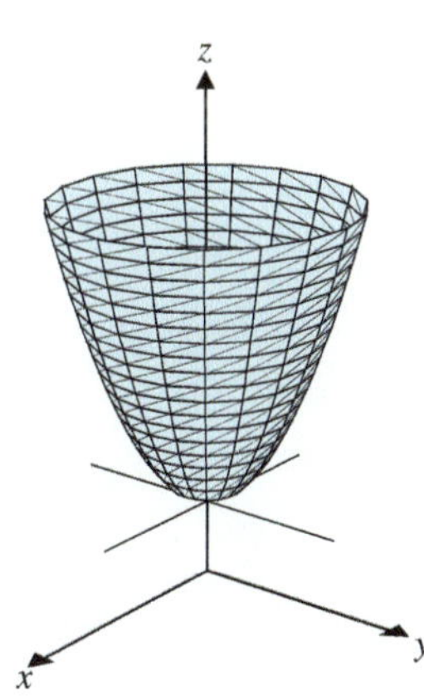

Figure 3

Step 1 Find critical points: Find (x, y) such that $f_x(x, y) = 0$ and $f_y(x, y) = 0$ simultaneously:

$$f_x(x, y) = 2x = 0 \qquad f_y(x, y) = 2y = 0$$
$$x = 0 \qquad\qquad y = 0$$

The only critical point is $(a, b) = (0, 0)$.

Step 2 Compute $A = f_{xx}(0, 0)$, $B = f_{xy}(0, 0)$, and $C = f_{yy}(0, 0)$:

$$f_{xx}(x, y) = 2; \quad \text{so,} \quad A = f_{xx}(0, 0) = 2$$
$$f_{xy}(x, y) = 0; \quad \text{so,} \quad B = f_{xy}(0, 0) = 0$$
$$f_{yy}(x, y) = 2; \quad \text{so,} \quad C = f_{yy}(0, 0) = 2$$

Step 3 Evaluate $AC - B^2$ and try to classify the critical point $(0, 0)$ by using Theorem 2:

$$AC - B^2 = (2)(2) - (0)^2 = 4 > 0 \quad \text{and} \quad A = 2 > 0$$

Therefore, case 2 in Theorem 2 holds. That is, $f(0, 0) = 2$ is a local minimum.

We will now use Theorem 2 to analyze extrema without the aid of graphs.

EXAMPLE 1 **Finding Local Extrema** Use Theorem 2 to find local extrema of

$$f(x, y) = -x^2 - y^2 + 6x + 8y - 21$$

SOLUTION **Step 1** Find critical points: Find (x, y) such that $f_x(x, y) = 0$ and $f_y(x, y) = 0$ simultaneously:

$$f_x(x, y) = -2x + 6 = 0 \qquad f_y(x, y) = -2y + 8 = 0$$
$$x = 3 \qquad\qquad y = 4$$

The only critical point is $(a, b) = (3, 4)$.

Step 2 Compute $A = f_{xx}(3, 4)$, $B = f_{xy}(3, 4)$, and $C = f_{yy}(3, 4)$:

$$f_{xx}(x, y) = -2; \quad \text{so,} \quad A = f_{xx}(3, 4) = -2$$
$$f_{xy}(x, y) = 0; \quad \text{so,} \quad B = f_{xy}(3, 4) = 0$$
$$f_{yy}(x, y) = -2; \quad \text{so,} \quad C = f_{yy}(3, 4) = -2$$

Step 3 Evaluate $AC - B^2$ and try to classify the critical point $(3, 4)$ by using Theorem 2:

$$AC - B^2 = (-2)(-2) - (0)^2 = 4 > 0 \quad \text{and} \quad A = -2 < 0$$

Therefore, case 1 in Theorem 2 holds, and $f(3, 4) = 4$ is a local maximum.

Matched Problem 1 Use Theorem 2 to find local extrema of

$$f(x, y) = x^2 + y^2 - 10x - 2y + 36$$

EXAMPLE 2 **Finding Local Extrema: Multiple Critical Points** Use Theorem 2 to find local extrema of

$$f(x, y) = x^3 + y^3 - 6xy$$

SOLUTION **Step 1** Find critical points of $f(x, y) = x^3 + y^3 - 6xy$:

$$f_x(x, y) = 3x^2 - 6y = 0 \qquad \text{Solve for } y.$$
$$6y = 3x^2$$
$$y = \tfrac{1}{2}x^2 \qquad (2)$$
$$f_y(x, y) = 3y^2 - 6x = 0$$
$$3y^2 = 6x \qquad \text{Use equation (2) to eliminate } y.$$
$$3(\tfrac{1}{2}x^2)^2 = 6x$$
$$\tfrac{3}{4}x^4 = 6x \qquad \text{Solve for } x.$$
$$3x^4 - 24x = 0$$
$$3x(x^3 - 8) = 0$$
$$x = 0 \quad \text{or} \quad x = 2$$
$$y = 0 \qquad y = \tfrac{1}{2}(2)^2 = 2$$

The critical points are $(0, 0)$ and $(2, 2)$.

Since there are two critical points, steps 2 and 3 must be performed twice.

Test (0, 0) **Step 2** Compute $A = f_{xx}(0, 0)$, $B = f_{xy}(0, 0)$, and $C = f_{yy}(0, 0)$:

$$f_{xx}(x, y) = 6x; \quad \text{so,} \quad A = f_{xx}(0, 0) = 0$$

$$f_{xy}(x, y) = -6; \quad \text{so,} \quad B = f_{xy}(0, 0) = -6$$
$$f_{yy}(x, y) = 6y; \quad \text{so,} \quad C = f_{yy}(0, 0) = 0$$

Step 3 Evaluate $AC - B^2$ and try to classify the critical point $(0, 0)$ by using Theorem 2:

$$AC - B^2 = (0)(0) - (-6)^2 = -36 < 0$$

Therefore, case 3 in Theorem 2 applies. That is, f has a saddle point at $(0, 0)$.

Now we will consider the second critical point, $(2, 2)$:

Test (2, 2) **Step 2** Compute $A = f_{xx}(2, 2)$, $B = f_{xy}(2, 2)$, and $C = f_{yy}(2, 2)$:

$$f_{xx}(x, y) = 6x; \quad \text{so,} \quad A = f_{xx}(2, 2) = 12$$
$$f_{xy}(x, y) = -6; \quad \text{so,} \quad B = f_{xy}(2, 2) = -6$$
$$f_{yy}(x, y) = 6y; \quad \text{so,} \quad C = f_{yy}(2, 2) = 12$$

Step 3 Evaluate $AC - B^2$ and try to classify the critical point $(2, 2)$ by using Theorem 2:

$$AC - B^2 = (12)(12) - (-6)^2 = 108 > 0 \quad \text{and} \quad A = 12 > 0$$

So, case 2 in Theorem 2 applies, and $f(2, 2) = -8$ is a local minimum.

Our conclusions in Example 2 may be confirmed geometrically by graphing cross sections of the function f. The cross sections of f in the planes $y = 0$, $x = 0$, $y = x$, and $y = -x$ [each of these planes contains $(0, 0)$] are represented by the graphs of the functions $f(x, 0) = x^3$, $f(0, y) = y^3$, $f(x, x) = 2x^3 - 6x^2$, and $f(x, -x) = 6x^2$, respectively, as shown in Figure 4A (note that the first two functions have the same graph). The cross sections of f in the planes $y = 2$, $x = 2$, $y = x$, and $y = 4 - x$ [each of these planes contains $(2, 2)$] are represented by the graphs of $f(x, 2) = x^3 - 12x + 8$, $f(2, y) = y^3 - 12y + 8$, $f(x, x) = 2x^3 - 6x^2$, and $f(x, 4 - x) = x^3 + (4 - x)^3 + 6x^2 - 24x$, respectively, as shown in Figure 4B (the first two functions have the same graph). Figure 4B illustrates the fact that since f has a local minimum at $(2, 2)$, each of the cross sections of f through $(2, 2)$ has a local minimum of -8 at $(2, 2)$. Figure 4A, by contrast, indicates that some cross sections of f through $(0, 0)$ have a local minimum, some a local maximum, and some neither one, at $(0, 0)$.

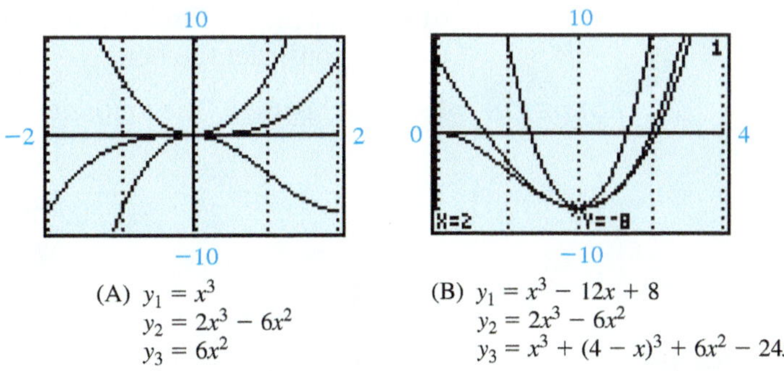

(A) $y_1 = x^3$
$y_2 = 2x^3 - 6x^2$
$y_3 = 6x^2$

(B) $y_1 = x^3 - 12x + 8$
$y_2 = 2x^3 - 6x^2$
$y_3 = x^3 + (4 - x)^3 + 6x^2 - 24x$

Figure 4

Matched Problem 2 Use Theorem 2 to find local extrema for $f(x, y) = x^3 + y^2 - 6xy$.

EXAMPLE 3 **Profit** Suppose that the surfboard company discussed earlier has developed the yearly profit equation

$$P(x, y) = -22x^2 + 22xy - 11y^2 + 110x - 44y - 23$$

where x is the number (in thousands) of standard surfboards produced per year, y is the number (in thousands) of competition surfboards produced per year, and P is profit (in thousands of dollars). How many of each type of board should be produced per year to realize a maximum profit? What is the maximum profit?

SOLUTION **Step 1** Find critical points:

$$P_x(x, y) = -44x + 22y + 110 = 0$$

$$P_y(x, y) = 22x - 22y - 44 = 0$$

Solving this system, we obtain (3, 1) as the only critical point.

Step 2 Compute $A = P_{xx}(3, 1)$, $B = P_{xy}(3, 1)$, and $C = P_{yy}(3, 1)$:

$$P_{xx}(x, y) = -44; \quad \text{so,} \quad A = P_{xx}(3, 1) = -44$$

$$P_{xy}(x, y) = 22; \quad \text{so,} \quad B = P_{xy}(3, 1) = 22$$

$$P_{yy}(x, y) = -22; \quad \text{so,} \quad C = P_{yy}(3, 1) = -22$$

Step 3 Evaluate $AC - B^2$ and try to classify the critical point (3, 1) by using Theorem 2:

$$AC - B^2 = (-44)(-22) - 22^2 = 484 > 0 \quad \text{and} \quad A = -44 < 0$$

Therefore, case 1 in Theorem 2 applies. That is, $P(3, 1) = 120$ is a local maximum. A maximum profit of \$120,000 is obtained by producing and selling 3,000 standard boards and 1,000 competition boards per year.

Matched Problem 3 Repeat Example 3 with

$$P(x, y) = -66x^2 + 132xy - 99y^2 + 132x - 66y - 19$$

EXAMPLE 4 **Package Design** The packaging department in a company is to design a rectangular box with no top and a partition down the middle. The box must have a volume of 48 cubic inches. Find the dimensions that will minimize the amount of material used to construct the box.

SOLUTION Refer to Figure 5. The amount of material used in constructing this box is

Base: xy; Front, back: $2xz$; Sides, partition: $3yz$

$$M = xy + 2xz + 3yz \tag{3}$$

The volume of the box is

$$V = xyz = 48 \tag{4}$$

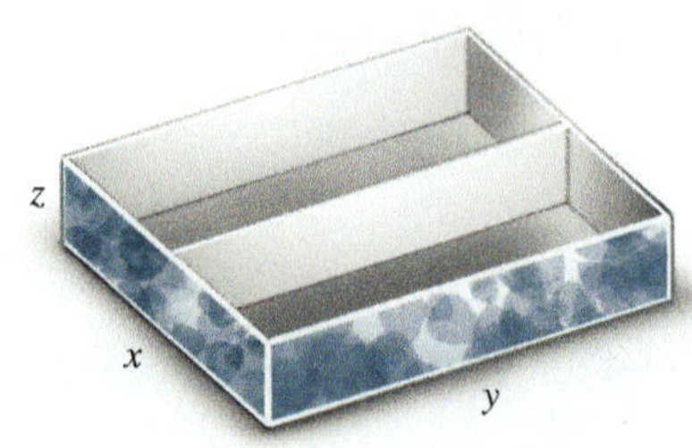

Figure 5

Since Theorem 2 applies only to functions with two independent variables, we must use equation (4) to eliminate one of the variables in equation (3):

$$M = xy + 2xz + 3yz \qquad \text{Substitute } z = 48/xy.$$

$$= xy + 2x\left(\frac{48}{xy}\right) + 3y\left(\frac{48}{xy}\right)$$

$$= xy + \frac{96}{y} + \frac{144}{x}$$

So, we must find the minimum value of

$$M(x, y) = xy + \frac{96}{y} + \frac{144}{x} \qquad x > 0 \qquad \text{and} \qquad y > 0$$

Step 1 Find critical points:

$$M_x(x, y) = y - \frac{144}{x^2} = 0$$

$$y = \frac{144}{x^2} \tag{5}$$

$$M_y(x, y) = x - \frac{96}{y^2} = 0$$

$$x = \frac{96}{y^2}$$ Solve for y^2.

$$y^2 = \frac{96}{x}$$ Use equation (5) to eliminate y and solve for x.

$$\left(\frac{144}{x^2}\right)^2 = \frac{96}{x}$$

$$\frac{20{,}736}{x^4} = \frac{96}{x}$$ Multiply both sides by $x^4/96$ (recall that $x > 0$).

$$x^3 = \frac{20{,}736}{96} = 216$$ Use equation (5) to find y.

$$x = 6$$

$$y = \frac{144}{36} = 4$$

Therefore, (6, 4) is the only critical point.

Step 2 Compute $A = M_{xx}(6, 4)$, $B = M_{xy}(6, 4)$, and $C = M_{yy}(6, 4)$:

$$M_{xx}(x, y) = \frac{288}{x^3}; \quad \text{so,} \quad A = M_{xx}(6, 4) = \tfrac{288}{216} = \tfrac{4}{3}$$

$$M_{xy}(x, y) = 1; \quad \text{so,} \quad B = M_{xy}(6, 4) = 1$$

$$M_{yy}(x, y) = \frac{192}{y^3}; \quad \text{so,} \quad C = M_{yy}(6, 4) = \tfrac{192}{64} = 3$$

Step 3 Evaluate $AC - B^2$ and try to classify the critical point (6, 4) by using Theorem 2:

$$AC - B^2 = \left(\tfrac{4}{3}\right)(3) - (1)^2 = 3 > 0 \qquad \text{and} \qquad A = \tfrac{4}{3} > 0$$

Case 2 in Theorem 2 applies, and $M(x, y)$ has a local minimum at (6, 4). If $x = 6$ and $y = 4$, then

$$z = \frac{48}{xy} = \frac{48}{(6)(4)} = 2$$

The dimensions that will require the minimum amount of material are 6 inches by 4 inches by 2 inches (see Fig. 6).

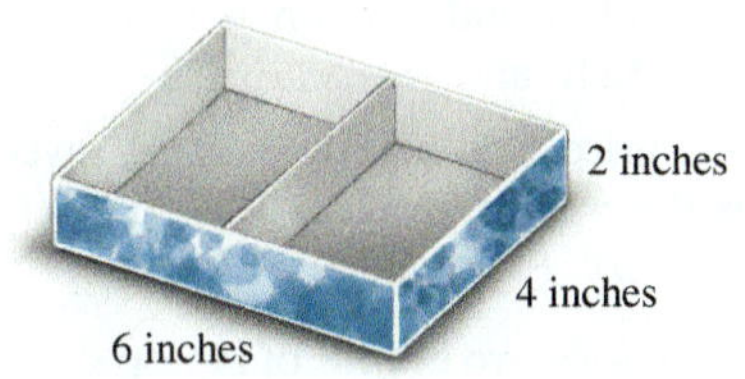

Figure 6

Matched Problem 4 If the box in Example 4 must have a volume of 384 cubic inches, find the dimensions that will require the least amount of material.

Exercises 6-3

A

In Problems 1–4, find $f_x(x, y)$ and $f_y(x, y)$, and explain, using Theorem 1, why $f(x, y)$ has no local extrema.

1. $f(x, y) = 4x + 5y - 6$

2. $f(x, y) = 10 - 2x - 3y + x^2$

3. $f(x, y) = 3.7 - 1.2x + 6.8y + 0.2y^3 + x^4$

4. $f(x, y) = x^3 - y^2 + 7x + 3y + 1$

Use Theorem 2 to find local extrema in Problems 5–24.

5. $f(x, y) = 6 - x^2 - 4x - y^2$

6. $f(x, y) = 3 - x^2 - y^2 + 6y$

7. $f(x, y) = x^2 + y^2 + 2x - 6y + 14$

8. $f(x, y) = x^2 + y^2 - 4x + 6y + 23$

B

9. $f(x, y) = xy + 2x - 3y - 2$

10. $f(x, y) = x^2 - y^2 + 2x + 6y - 4$

11. $f(x, y) = -3x^2 + 2xy - 2y^2 + 14x + 2y + 10$

12. $f(x, y) = -x^2 + xy - 2y^2 + x + 10y - 5$

13. $f(x, y) = 2x^2 - 2xy + 3y^2 - 4x - 8y + 20$

14. $f(x, y) = 2x^2 - xy + y^2 - x - 5y + 8$

C

15. $f(x, y) = e^{xy}$

16. $f(x, y) = x^2y - xy^2$

17. $f(x, y) = x^3 + y^3 - 3xy$

18. $f(x, y) = 2y^3 - 6xy - x^2$

19. $f(x, y) = 2x^4 + y^2 - 12xy$

20. $f(x, y) = 16xy - x^4 - 2y^2$

21. $f(x, y) = x^3 - 3xy^2 + 6y^2$

22. $f(x, y) = 2x^2 - 2x^2y + 6y^3$

23. $f(x, y) = y^3 + 2x^2y^2 - 3x - 2y + 8$

24. $f(x, y) = x \ln y + x^2 - 4x - 5y + 3$

25. Explain why $f(x, y) = x^2$ has an infinite number of local extrema.

26. (A) Find the local extrema of the functions $f(x, y) = x + y$, $g(x, y) = x^2 + y^2$, and $h(x, y) = x^3 + y^3$.

(B) Discuss the local extrema of the function $k(x, y) = x^n + y^n$, where n is a positive integer.

27. (A) Show that (0, 0) is a critical point of the function $f(x, y) = x^4e^y + x^2y^4 + 1$, but that the second-derivative test for local extrema fails.

(B) Use cross sections, as in Example 2, to decide whether f has a local maximum, a local minimum, or a saddle point at (0, 0).

28. (A) Show that (0, 0) is a critical point of the function $g(x, y) = e^{xy^2} + x^2y^3 + 2$, but that the second-derivative test for local extrema fails.

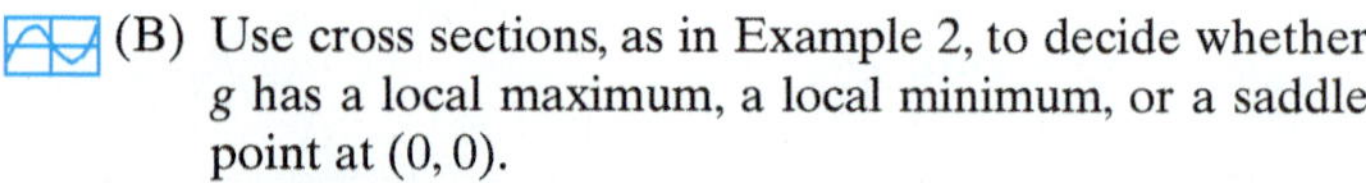

(B) Use cross sections, as in Example 2, to decide whether g has a local maximum, a local minimum, or a saddle point at (0, 0).

Applications

29. Product mix for maximum profit. A firm produces two types of earphones per year: x thousand of type A and y thousand of type B. If the revenue and cost equations for the year are (in millions of dollars)

$$R(x, y) = 2x + 3y$$

$$C(x, y) = x^2 - 2xy + 2y^2 + 6x - 9y + 5$$

determine how many of each type of earphone should be produced per year to maximize profit. What is the maximum profit?

30. Automation–labor mix for minimum cost. The annual labor and automated equipment cost (in millions of dollars) for a company's production of HDTV's is given by

$$C(x, y) = 2x^2 + 2xy + 3y^2 - 16x - 18y + 54$$

where x is the amount spent per year on labor and y is the amount spent per year on automated equipment (both in millions of dollars). Determine how much should be spent on each per year to minimize this cost. What is the minimum cost?

31. Maximizing profit. A store sells two brands of camping chairs. The store pays \$60 for each brand A chair and \$80 for each brand B chair. The research department has estimated the weekly demand equations for these two competitive products to be

$x = 260 - 3p + q$ Demand equation for brand A

$y = 180 + p - 2q$ Demand equation for brand B

where p is the selling price for brand A and q is the selling price for brand B.

(A) Determine the demands x and y when $p = \$100$ and $q = \$120$; when $p = \$110$ and $q = \$110$.

(B) How should the store price each chair to maximize weekly profits? What is the maximum weekly profit? [*Hint:* $C = 60x + 80y$, $R = px + qy$, and $P = R - C$.]

32. Maximizing profit. A store sells two brands of laptop sleeves. The store pays \$25 for each brand A sleeve and \$30 for each brand B sleeve. A consulting firm has estimated the daily demand equations for these two competitive products to be

$x = 130 - 4p + q$ Demand equation for brand A

$y = 115 + 2p - 3q$ Demand equation for brand B

where p is the selling price for brand A and q is the selling price for brand B.

(A) Determine the demands x and y when $p = \$40$ and $q = \$50$; when $p = \$45$ and $q = \$55$.

(B) How should the store price each brand of sleeve to maximize daily profits? What is the maximum daily profit? [*Hint:* $C = 25x + 30y$, $R = px + qy$, and $P = R - C$.]

33. Minimizing cost. A satellite TV station is to be located at $P(x, y)$ so that the sum of the squares of the distances from P to the three towns A, B, and C is a minimum (see the figure). Find the coordinates of P, the location that will minimize the cost of providing satellite TV for all three towns.

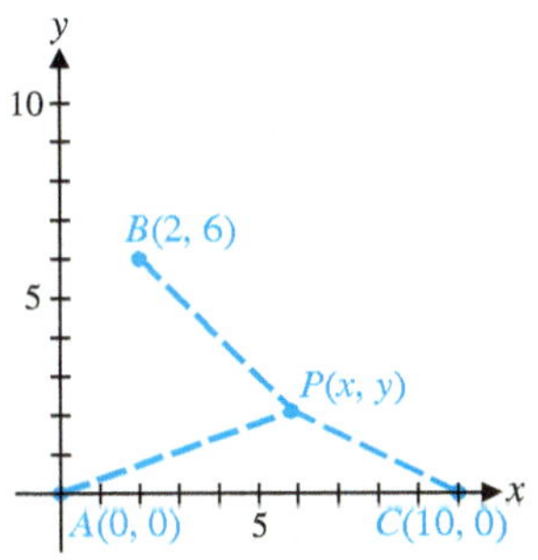

Figure for 33

34. Minimizing cost. Repeat Problem 33, replacing the coordinates of B with $B(6, 9)$ and the coordinates of C with $C(9, 0)$.

35. Minimum material. A rectangular box with no top and two parallel partitions (see the figure) must hold a volume of 64 cubic inches. Find the dimensions that will require the least amount of material.

Figure for 35

36. Minimum material. A rectangular box with no top and two intersecting partitions (see the figure) must hold a volume of 72 cubic inches. Find the dimensions that will require the least amount of material.

Figure for 36

37. Maximum volume. A mailing service states that a rectangular package cannot have the sum of its length and girth exceed 120 inches (see the figure). What are the dimensions of the largest (in volume) mailing carton that can be constructed to meet these restrictions?

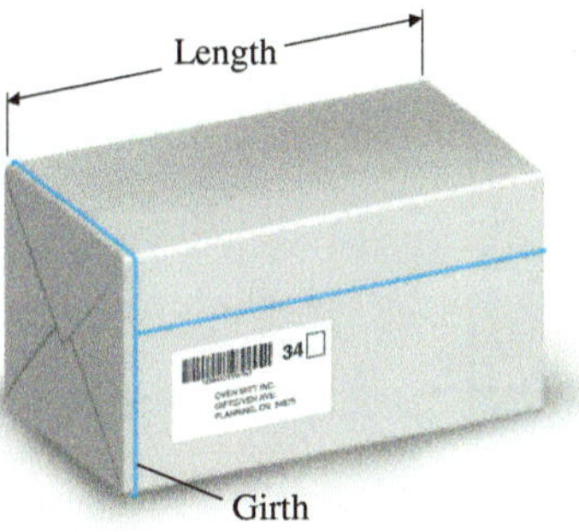

Figure for 37

38. Maximum shipping volume. A shipping box is to be reinforced with steel bands in all three directions, as shown in the figure. A total of 150 inches of steel tape is to be used, with 6 inches of waste because of a 2-inch overlap in each direction. Find the dimensions of the box with maximum volume that can be taped as described.

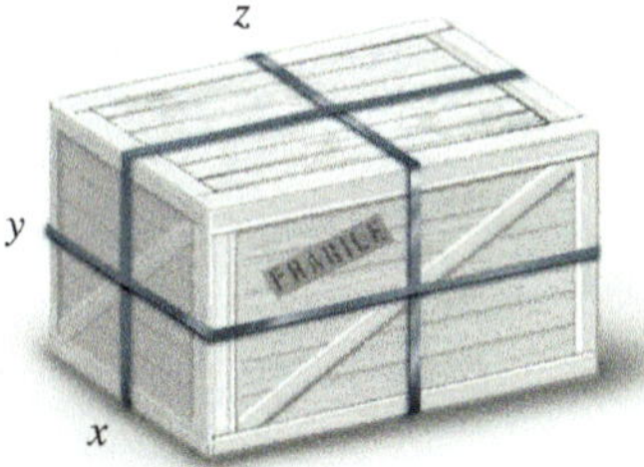

Figure for 38

Answers to Matched Problems

1. $f(5, 1) = 10$ is a local minimum
2. f has a saddle point at $(0, 0)$; $f(6, 18) = -108$ is a local minimum
3. Local maximum for $x = 2$ and $y = 1$; $P(2, 1) = 80$; a maximum profit of \$80,000 is obtained by producing and selling 2,000 standard boards and 1,000 competition boards
4. 12 in. by 8 in. by 4 in.

6-4 Maxima and Minima Using Lagrange Multipliers

- Functions of Two Independent Variables
- Functions of Three Independent Variables

Functions of Two Independent Variables

We now consider a powerful method of solving a certain class of maxima–minima problems. Joseph Louis Lagrange (1736–1813), an eighteenth-century French mathematician, discovered this method, called the **method of Lagrange multipliers**. We introduce the method through an example.

A rancher wants to construct two feeding pens of the same size along an existing fence (see Fig. 1). If the rancher has 720 feet of fencing materials available, how long should x and y be in order to obtain the maximum total area? What is the maximum area?

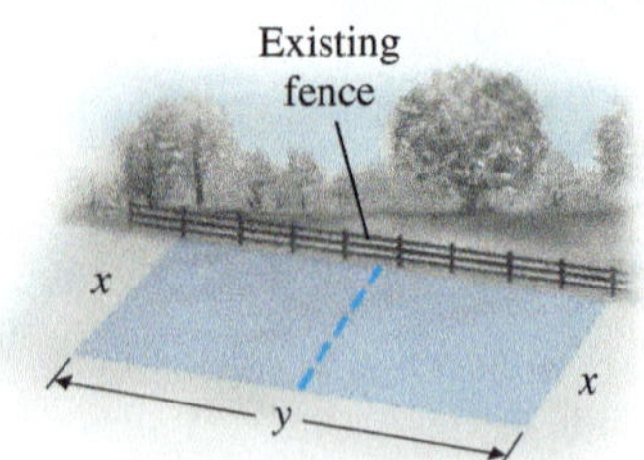

Figure 1

The total area is given by

$$f(x, y) = xy$$

which can be made as large as we like, provided that there are no restrictions on x and y. But there are restrictions on x and y, since we have only 720 feet of fencing. The variables x and y must be chosen so that

$$3x + y = 720$$

This restriction on x and y, called a **constraint**, leads to the following maxima–minima problem:

$$\text{Maximize} \quad f(x, y) = xy \tag{1}$$

$$\text{subject to} \quad 3x + y = 720, \quad \text{or} \quad 3x + y - 720 = 0 \tag{2}$$

This problem is one of a general class of problems of the form

$$\text{Maximize (or minimize)} \quad z = f(x, y) \tag{3}$$

$$\text{subject to} \quad g(x, y) = 0 \tag{4}$$

Of course, we could try to solve equation (4) for y in terms of x, or for x in terms of y, then substitute the result into equation (3), and use methods developed in Section 3-5 for functions of a single variable. But what if equation (4) is more complicated than equation (2), and solving for one variable in terms of the other is either very difficult or impossible? In the method of Lagrange multipliers, we will work with $g(x, y)$ directly and avoid solving equation (4) for one variable in terms of the other. In addition, the method generalizes to functions of arbitrarily many variables subject to one or more constraints.

Now to the method: We form a new function F, using functions f and g in equations (3) and (4), as follows:

$$F(x, y, \lambda) = f(x, y) + \lambda g(x, y) \tag{5}$$

Here, λ (the Greek lowercase letter lambda) is called a **Lagrange multiplier**. Theorem 1 gives the basis for the method.

THEOREM 1 Method of Lagrange Multipliers for Functions of Two Variables
Any local maxima or minima of the function $z = f(x, y)$ subject to the constraint $g(x, y) = 0$ will be among those points (x_0, y_0) for which (x_0, y_0, λ_0) is a solution of the system

$$F_x(x, y, \lambda) = 0$$
$$F_y(x, y, \lambda) = 0$$
$$F_\lambda(x, y, \lambda) = 0$$

where $F(x, y, \lambda) = f(x, y) + \lambda g(x, y)$, provided that all the partial derivatives exist.

We now use the method of Lagrange multipliers to solve the fence problem.

Step 1 Formulate the problem in the form of equations (3) and (4):

$$\text{Maximize} \quad f(x, y) = xy$$
$$\text{subject to} \quad g(x, y) = 3x + y - 720 = 0$$

Step 2 Form the function F, introducing the Lagrange multiplier λ:

$$\begin{aligned} F(x, y, \lambda) &= f(x, y) + \lambda g(x, y) \\ &= xy + \lambda(3x + y - 720) \end{aligned}$$

Step 3 Solve the system $F_x = 0, F_y = 0, F_\lambda = 0$ (the solutions are **critical points** of F):

$$\begin{aligned} F_x &= y + 3\lambda = 0 \\ F_y &= x + \lambda = 0 \\ F_\lambda &= 3x + y - 720 = 0 \end{aligned}$$

From the first two equations, we see that

$$\begin{aligned} y &= -3\lambda \\ x &= -\lambda \end{aligned}$$

Substitute these values for x and y into the third equation and solve for λ:

$$\begin{aligned} -3\lambda - 3\lambda &= 720 \\ -6\lambda &= 720 \\ \lambda &= -120 \end{aligned}$$

So,

$$\begin{aligned} y &= -3(-120) = 360 \text{ feet} \\ x &= -(-120) = 120 \text{ feet} \end{aligned}$$

and $(x_0, y_0, \lambda_0) = (120, 360, -120)$ is the only critical point of F.

Step 4 According to Theorem 1, if the function $f(x, y)$, subject to the constraint $g(x, y) = 0$, has a local maximum or minimum, that maximum or minimum must occur at $x = 120$, $y = 360$. Although it is possible to develop a test similar to Theorem 2 in Section 6-3 to determine the nature of this local extremum, we will not do so. [Note that Theorem 2 cannot be applied to $f(x, y)$ at (120, 360), since this point is not a critical point of the unconstrained function $f(x, y)$.] We simply assume that the maximum value of $f(x, y)$ must occur for $x = 120$, $y = 360$.

$$\begin{aligned} \text{Max } f(x, y) &= f(120, 360) \\ &= (120)(360) = 43{,}200 \text{ square feet} \end{aligned}$$

The key steps in applying the method of Lagrange multipliers are as follows:

PROCEDURE Method of Lagrange Multipliers: Key Steps

Step 1 Write the problem in the form

$$\text{Maximize (or minimize)} \quad z = f(x, y)$$
$$\text{subject to} \quad g(x, y) = 0$$

Step 2 Form the function F:

$$F(x, y, \lambda) = f(x, y) + \lambda g(x, y)$$

Step 3 Find the critical points of F; that is, solve the system

$$F_x(x, y, \lambda) = 0$$
$$F_y(x, y, \lambda) = 0$$
$$F_\lambda(x, y, \lambda) = 0$$

Step 4 If (x_0, y_0, λ_0) is the only critical point of F, we assume that (x_0, y_0) will always produce the solution to the problems we consider. If F has more than one critical point, we evaluate $z = f(x, y)$ at (x_0, y_0) for each critical point (x_0, y_0, λ_0) of F. For the problems we consider, we assume that the largest of these values is the maximum value of $f(x, y)$, subject to the constraint $g(x, y) = 0$, and the smallest is the minimum value of $f(x, y)$, subject to the constraint $g(x, y) = 0$.

EXAMPLE 1 **Minimization Subject to a Constraint** Minimize $f(x, y) = x^2 + y^2$ subject to $x + y = 10$.

SOLUTION **Step 1**

$$\text{Minimize} \quad f(x, y) = x^2 + y^2$$
$$\text{subject to} \quad g(x, y) = x + y - 10 = 0$$

Step 2

$$F(x, y, \lambda) = x^2 + y^2 + \lambda(x + y - 10)$$

Step 3

$$F_x = 2x + \lambda = 0$$
$$F_y = 2y + \lambda = 0$$
$$F_\lambda = x + y - 10 = 0$$

From the first two equations, $x = -\lambda/2$ and $y = -\lambda/2$. Substituting these values into the third equation, we obtain

$$-\frac{\lambda}{2} - \frac{\lambda}{2} = 10$$
$$-\lambda = 10$$
$$\lambda = -10$$

The only critical point is $(x_0, y_0, \lambda_0) = (5, 5, -10)$.

Step 4 Since $(5, 5, -10)$ is the only critical point of F, we conclude that (see step 4 in the box)

$$\text{Min } f(x, y) = f(5, 5) = (5)^2 + (5)^2 = 50$$

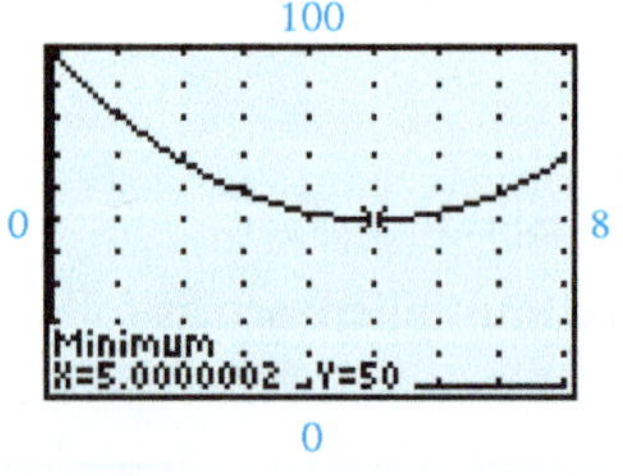

Figure 2 $h(x) = x^2 + (10 - x)^2$

Since $g(x, y)$ in Example 1 has a relatively simple form, an alternative to the method of Lagrange multipliers is to solve $g(x, y) = 0$ for y and then substitute into $f(x, y)$ to obtain the function $h(x) = f(x, 10 - x) = x^2 + (10 - x)^2$ in the single variable x. Then we minimize h (see Fig. 2). From Figure 2, we conclude that Min $f(x, y) = f(5, 5) = 50$. This technique depends on being able to solve the constraint for one of the two variables and so is not always available as an alternative to the method of Lagrange multipliers.

Matched Problem 1 Maximize $f(x, y) = 25 - x^2 - y^2$ subject to $x + y = 4$.

Figures 3 and 4 illustrate the results obtained in Example 1 and Matched Problem 1, respectively.

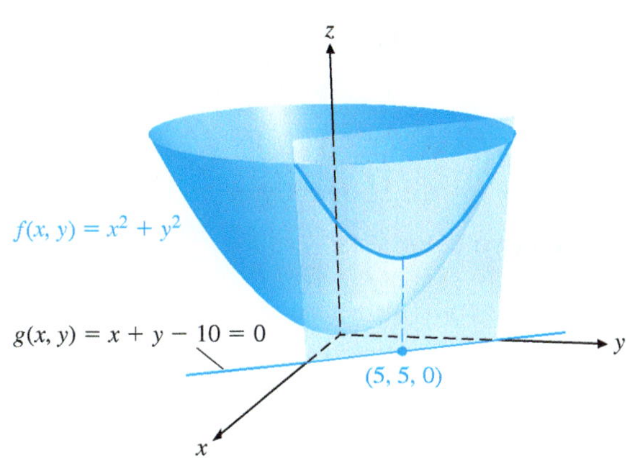

Figure 3

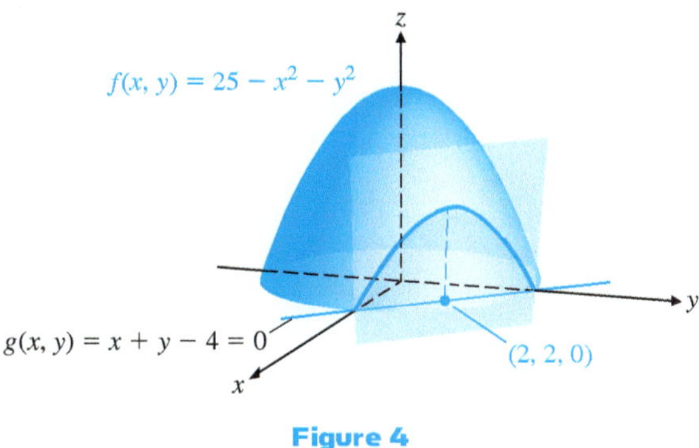

Figure 4

EXPLORE & DISCUSS 1

Consider the problem of minimizing $f(x, y) = 3x^2 + 5y^2$ subject to the constraint $g(x, y) = 2x + 3y - 6 = 0$.

(A) Compute the value of $f(x, y)$ when x and y are integers, $0 \le x \le 3$, $0 \le y \le 2$. Record your answers in the empty boxes next to the points (x, y) in Figure 5.

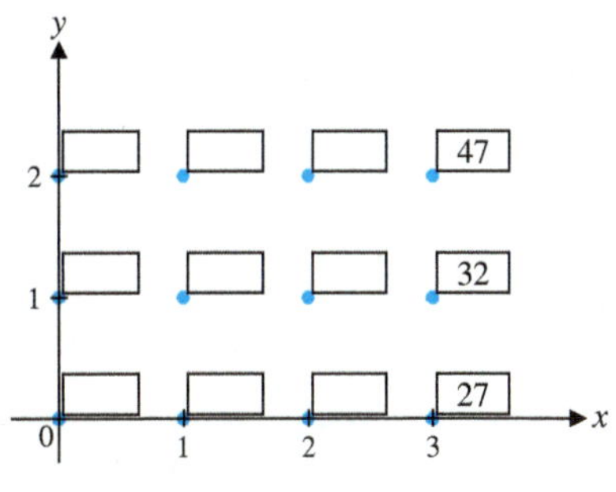

Figure 5

(B) Graph the constraint $g(x, y) = 0$.

(C) Estimate the minimum value of f on the basis of your graph and the computations from part (A).

(D) Use the method of Lagrange multipliers to solve the minimization problem.

EXAMPLE 2 **Productivity** The Cobb–Douglas production function for a new product is given by

$$N(x, y) = 16x^{0.25}y^{0.75}$$

where x is the number of units of labor and y is the number of units of capital required to produce $N(x, y)$ units of the product. Each unit of labor costs \$50 and each unit of capital costs \$100. If \$500,000 has been budgeted for the production of this product, how should that amount be allocated between labor and capital in order to maximize production? What is the maximum number of units that can be produced?

SOLUTION The total cost of using x units of labor and y units of capital is $50x + 100y$. Thus, the constraint imposed by the \$500,000 budget is

$$50x + 100y = 500{,}000$$

Step 1

$$\text{Maximize} \quad N(x, y) = 16x^{0.25}y^{0.75}$$
$$\text{subject to} \quad g(x, y) = 50x + 100y - 500{,}000 = 0$$

Step 2 $$F(x, y, \lambda) = 16x^{0.25}y^{0.75} + \lambda(50x + 100y - 500{,}000)$$

Step 3
$$F_x = 4x^{-0.75}y^{0.75} + 50\lambda = 0$$
$$F_y = 12x^{0.25}y^{-0.25} + 100\lambda = 0$$
$$F_\lambda = 50x + 100y - 500{,}000 = 0$$

From the first two equations,

$$\lambda = -\tfrac{2}{25}x^{-0.75}y^{0.75} \quad \text{and} \quad \lambda = -\tfrac{3}{25}x^{0.25}y^{-0.25}$$

Therefore,

$$-\tfrac{2}{25}x^{-0.75}y^{0.75} = -\tfrac{3}{25}x^{0.25}y^{-0.25}$$ Multiply both sides by $x^{0.75}y^{0.25}$.

$$-\tfrac{2}{25}y = -\tfrac{3}{25}x$$ (We can assume that $x \neq 0$ and $y \neq 0$.)

$$y = \tfrac{3}{2}x$$

Now substitute for y in the third equation and solve for x:

$$50x + 100\left(\tfrac{3}{2}x\right) - 500{,}000 = 0$$
$$200x = 500{,}000$$
$$x = 2{,}500$$

So,

$$y = \tfrac{3}{2}(2{,}500) = 3{,}750$$

and

$$\lambda = -\tfrac{2}{25}(2{,}500)^{-0.75}(3{,}750)^{0.75} \approx -0.1084$$

The only critical point of F is $(2{,}500, 3{,}750, -0.1084)$.

Step 4 Since F has only one critical point, we conclude that maximum productivity occurs when 2,500 units of labor and 3,750 units of capital are used (see step 4 in the method of Lagrange multipliers).

$$\text{Max } N(x, y) = N(2{,}500, 3{,}750)$$
$$= 16(2{,}500)^{0.25}(3{,}750)^{0.75}$$
$$\approx 54{,}216 \text{ units}$$

The negative of the value of the Lagrange multiplier found in step 3 is called the **marginal productivity of money** and gives the approximate increase in production for each additional dollar spent on production. In Example 2, increasing the production budget from \$500,000 to \$600,000 would result in an approximate increase in production of

$$0.1084(100{,}000) = 10{,}840 \text{ units}$$

Note that simplifying the constraint equation

$$50x + 100y - 500{,}000 = 0$$

to

$$x + 2y - 10{,}000 = 0$$

before forming the function $F(x, y, \lambda)$ would make it difficult to interpret $-\lambda$ correctly. **In marginal productivity problems, the constraint equation should not be simplified.**

Matched Problem 2 The Cobb–Douglas production function for a new product is given by

$$N(x, y) = 20x^{0.5}y^{0.5}$$

where x is the number of units of labor and y is the number of units of capital required to produce $N(x, y)$ units of the product. Each unit of labor costs \$40 and each unit of capital costs \$120.

(A) If \$300,000 has been budgeted for the production of this product, how should that amount be allocated in order to maximize production? What is the maximum production?

(B) Find the marginal productivity of money in this case, and estimate the increase in production if an additional \$40,000 is budgeted for production.

Explore & Discuss 2

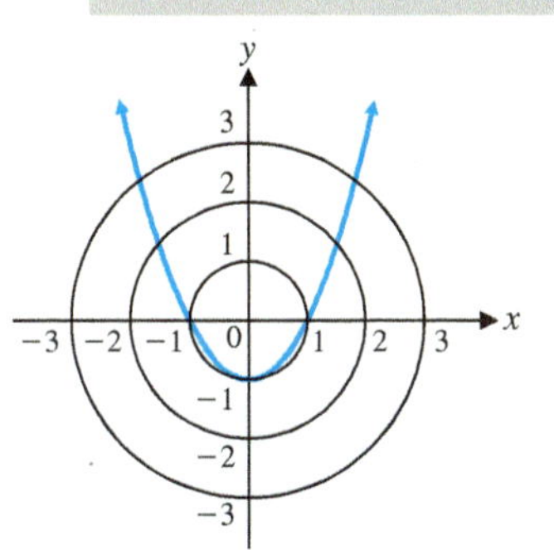

Figure 6

Consider the problem of maximizing $f(x, y) = 4 - x^2 - y^2$ subject to the constraint $g(x, y) = y - x^2 + 1 = 0$.

(A) Explain why $f(x, y) = 3$ whenever (x, y) is a point on the circle of radius 1 centered at the origin. What is the value of $f(x, y)$ when (x, y) is a point on the circle of radius 2 centered at the origin? On the circle of radius 3 centered at the origin? (See Fig. 6.)

(B) Explain why some points on the parabola $y - x^2 + 1 = 0$ lie inside the circle $x^2 + y^2 = 1$.

(C) In light of part (B), would you guess that the maximum value of $f(x, y)$ subject to the constraint is greater than 3? Explain.

(D) Use Lagrange multipliers to solve the maximization problem.

Functions of Three Independent Variables

The method of Lagrange multipliers can be extended to functions with arbitrarily many independent variables with one or more constraints. We now state a theorem for functions with three independent variables and one constraint, and we consider an example that demonstrates the advantage of the method of Lagrange multipliers over the method used in Section 6-3.

THEOREM 2 Method of Lagrange Multipliers for Functions of Three Variables

Any local maxima or minima of the function $w = f(x, y, z)$, subject to the constraint $g(x, y, z) = 0$, will be among the set of points (x_0, y_0, z_0) for which $(x_0, y_0, z_0, \lambda_0)$ is a solution of the system

$$F_x(x, y, z, \lambda) = 0$$
$$F_y(x, y, z, \lambda) = 0$$
$$F_z(x, y, z, \lambda) = 0$$
$$F_\lambda(x, y, z, \lambda) = 0$$

where $F(x, y, z, \lambda) = f(x, y, z) + \lambda g(x, y, z)$, provided that all the partial derivatives exist.

EXAMPLE 3 **Package Design** A rectangular box with an open top and one partition is to be constructed from 162 square inches of cardboard (Fig. 7). Find the dimensions that result in a box with the largest possible volume.

SOLUTION We must maximize

$$V(x, y, z) = xyz$$

subject to the constraint that the amount of material used is 162 square inches. So x, y, and z must satisfy

$$xy + 2xz + 3yz = 162$$

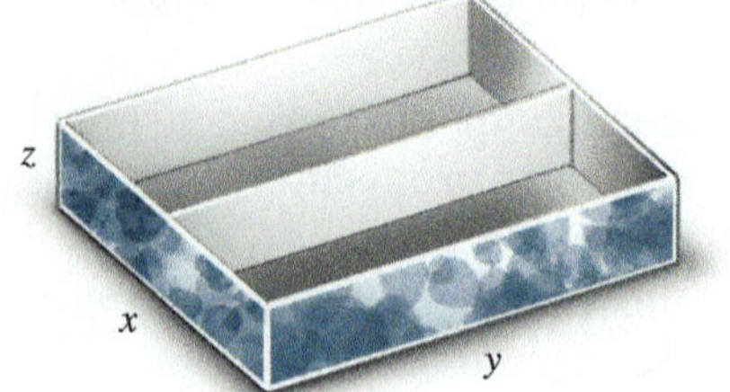

Figure 7

Step 1 Maximize $V(x, y, z) = xyz$
subject to $g(x, y, z) = xy + 2xz + 3yz - 162 = 0$

Step 2 $F(x, y, z, \lambda) = xyz + \lambda(xy + 2xz + 3yz - 162)$

Step 3

$$F_x = yz + \lambda(y + 2z) = 0$$
$$F_y = xz + \lambda(x + 3z) = 0$$
$$F_z = xy + \lambda(2x + 3y) = 0$$
$$F_\lambda = xy + 2xz + 3yz - 162 = 0$$

From the first two equations, we can write

$$\lambda = \frac{-yz}{y + 2z} \qquad \lambda = \frac{-xz}{x + 3z}$$

Eliminating λ, we have

$$\frac{-yz}{y + 2z} = \frac{-xz}{x + 3z}$$
$$-xyz - 3yz^2 = -xyz - 2xz^2$$
$$3yz^2 = 2xz^2 \qquad \text{We can assume that } z \neq 0.$$
$$3y = 2x$$
$$x = \tfrac{3}{2}y$$

From the second and third equations,

$$\lambda = \frac{-xz}{x + 3z} \qquad \lambda = \frac{-xy}{2x + 3y}$$

Eliminating λ, we have

$$\frac{-xz}{x + 3z} = \frac{-xy}{2x + 3y}$$
$$-2x^2z - 3xyz = -x^2y - 3xyz$$
$$2x^2z = x^2y \qquad \text{We can assume that } x \neq 0.$$
$$2z = y$$
$$z = \tfrac{1}{2}y$$

Substituting $x = \frac{3}{2}y$ and $z = \frac{1}{2}y$ into the fourth equation, we have

$$\left(\tfrac{3}{2}y\right)y + 2\left(\tfrac{3}{2}y\right)\left(\tfrac{1}{2}y\right) + 3y\left(\tfrac{1}{2}y\right) - 162 = 0$$
$$\tfrac{3}{2}y^2 + \tfrac{3}{2}y^2 + \tfrac{3}{2}y^2 = 162$$
$$y^2 = 36 \qquad \text{We can assume that } y > 0.$$
$$y = 6$$
$$x = \tfrac{3}{2}(6) = 9 \qquad \text{Using } x = \tfrac{3}{2}y$$
$$z = \tfrac{1}{2}(6) = 3 \qquad \text{Using } z = \tfrac{1}{2}y$$

and finally,

$$\lambda = \frac{-(6)(3)}{6 + 2(3)} = -\frac{3}{2} \quad \text{Using } \lambda = \frac{-yz}{y + 2z}$$

The only critical point of F with x, y, and z all positive is $(9, 6, 3, -\frac{3}{2})$.

Step 4 The box with the maximum volume has dimensions 9 inches by 6 inches by 3 inches (see Fig. 8).

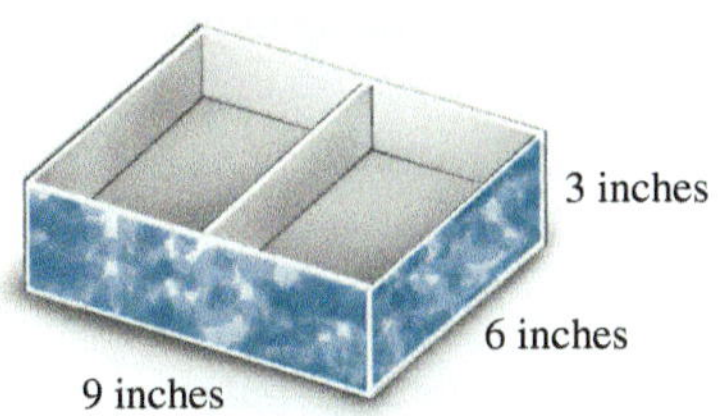

Figure 8

Matched Problem 3 A box of the same type as in Example 3 is to be constructed from 288 square inches of cardboard. Find the dimensions that result in a box with the largest possible volume.

CONCEPTUAL INSIGHT

An alternative to the method of Lagrange multipliers would be to solve Example 3 by means of Theorem 2 (the second-derivative test for local extrema) in Section 6-3. That approach involves solving the material constraint for one of the variables, say, z:

$$z = \frac{162 - xy}{2x + 3y}$$

Then we would eliminate z in the volume function to obtain a function of two variables:

$$V(x, y) = xy\frac{162 - xy}{2x + 3y}$$

The method of Lagrange multipliers allows us to avoid the formidable tasks of calculating the partial derivatives of V and finding the critical points of V in order to apply Theorem 2.

Exercises 6-4

A

Use the method of Lagrange multipliers in Problems 1–4.

1. Maximize $f(x, y) = 2xy$
subject to $x + y = 6$

2. Minimize $f(x, y) = 6xy$
subject to $y - x = 6$

3. Minimize $f(x, y) = x^2 + y^2$
subject to $3x + 4y = 25$

4. Maximize $f(x, y) = 25 - x^2 - y^2$
subject to $2x + y = 10$

B

In Problems 5 and 6, use Theorem 1 to explain why no maxima or minima exist.

5. Minimize $f(x, y) = 4y - 3x$
subject to $2x + 5y = 3$

6. Maximize $f(x, y) = 6x + 5y + 24$
subject to $3x + 2y = 4$

Use the method of Lagrange multipliers in Problems 7–16.

7. Find the maximum and minimum of $f(x, y) = 2xy$ subject to $x^2 + y^2 = 18$.

8. Find the maximum and minimum of $f(x, y) = x^2 - y^2$ subject to $x^2 + y^2 = 25$.

9. Maximize the product of two numbers if their sum must be 10.

10. Minimize the product of two numbers if their difference must be 10.

C

11. Minimize $f(x, y, z) = x^2 + y^2 + z^2$
subject to $2x - y + 3z = -28$

12. Maximize $f(x, y, z) = xyz$
subject to $2x + y + 2z = 120$

13. Maximize and minimize $f(x, y, z) = x + y + z$
subject to $x^2 + y^2 + z^2 = 12$

14. Maximize and minimize $f(x, y, z) = 2x + 4y + 4z$
subject to $x^2 + y^2 + z^2 = 9$

15. Maximize $f(x, y) = y + xy^2$
subject to $x + y^2 = 1$

16. Maximize and minimize $f(x, y) = x + e^y$
subject to $x^2 + y^2 = 1$

In Problems 17 and 18, use Theorem 1 to explain why no maxima or minima exist.

17. Maximize $f(x, y) = e^x + 3e^y$
subject to $x - 2y = 6$

18. Minimize $f(x, y) = x^3 + 2y^3$
subject to $6x - 2y = 1$

19. Consider the problem of maximizing $f(x, y)$ subject to $g(x, y) = 0$, where $g(x, y) = y - 5$. Explain how the maximization problem can be solved without using the method of Lagrange multipliers.

20. Consider the problem of minimizing $f(x, y)$ subject to $g(x, y) = 0$, where $g(x, y) = 4x - y + 3$. Explain how the minimization problem can be solved without using the method of Lagrange multipliers.

21. Consider the problem of maximizing $f(x, y) = e^{-(x^2+y^2)}$ subject to the constraint $g(x, y) = x^2 + y - 1 = 0$.
(A) Solve the constraint equation for y, and then substitute into $f(x, y)$ to obtain a function $h(x)$ of the single variable x. Solve the original maximization problem by maximizing h (round answers to three decimal places).
(B) Confirm your answer by the method of Lagrange multipliers.

22. Consider the problem of minimizing

$$f(x, y) = x^2 + 2y^2$$

subject to the constraint $g(x, y) = ye^{x^2} - 1 = 0$.
(A) Solve the constraint equation for y, and then substitute into $f(x, y)$ to obtain a function $h(x)$ of the single variable x. Solve the original minimization problem by minimizing h (round answers to three decimal places).
(B) Confirm your answer by the method of Lagrange multipliers.

Applications

23. **Budgeting for least cost.** A manufacturing company produces two models of an HDTV per week, x units of model A and y units of model B at a cost (in dollars) of

$$C(x, y) = 6x^2 + 12y^2$$

If it is necessary (because of shipping considerations) that

$$x + y = 90$$

how many of each type of set should be manufactured per week to minimize cost? What is the minimum cost?

24. **Budgeting for maximum production.** A manufacturing firm has budgeted \$60,000 per month for labor and materials. If \$$x$ thousand is spent on labor and \$$y$ thousand is spent on materials, and if the monthly output (in units) is given by

$$N(x, y) = 4xy - 8x$$

then how should the \$60,000 be allocated to labor and materials in order to maximize N? What is the maximum N?

25. **Productivity.** A consulting firm for a manufacturing company arrived at the following Cobb–Douglas production function for a particular product:

$$N(x, y) = 50x^{0.8}y^{0.2}$$

In this equation, x is the number of units of labor and y is the number of units of capital required to produce $N(x, y)$ units of the product. Each unit of labor costs \$40 and each unit of capital costs \$80.
(A) If \$400,000 is budgeted for production of the product, determine how that amount should be allocated to maximize production, and find the maximum production.
(B) Find the marginal productivity of money in this case, and estimate the increase in production if an additional \$50,000 is budgeted for the production of the product.

26. **Productivity.** The research department of a manufacturing company arrived at the following Cobb–Douglas production function for a particular product:

$$N(x, y) = 10x^{0.6}y^{0.4}$$

In this equation, x is the number of units of labor and y is the number of units of capital required to produce $N(x, y)$ units of the product. Each unit of labor costs \$30 and each unit of capital costs \$60.
(A) If \$300,000 is budgeted for production of the product, determine how that amount should be allocated to maximize production, and find the maximum production.
(B) Find the marginal productivity of money in this case, and estimate the increase in production if an additional \$80,000 is budgeted for the production of the product.

27. **Maximum volume.** A rectangular box with no top and two intersecting partitions is to be constructed from 192 square inches of cardboard (see the figure). Find the dimensions that will maximize the volume.

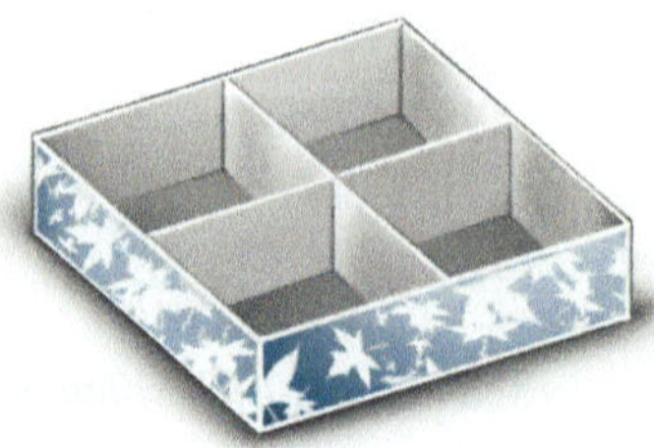

Figure for 27

28. **Maximum volume.** A mailing service states that a rectangular package shall have the sum of its length and girth not to exceed 120 inches (see the figure). What are the dimensions of the largest (in volume) mailing carton that can be constructed to meet these restrictions?

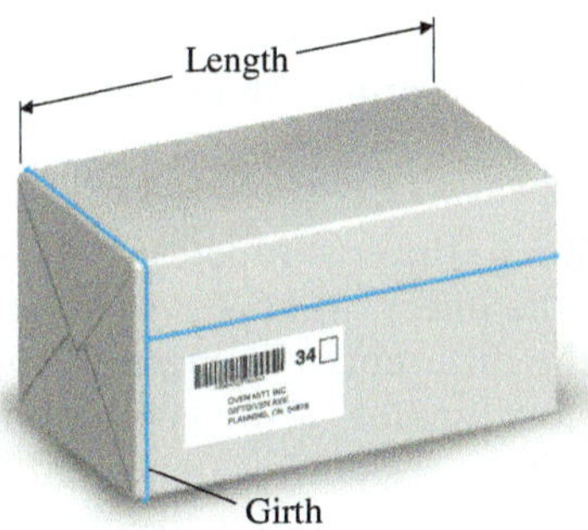

Figure for 28

29. **Agriculture.** Three pens of the same size are to be built along an existing fence (see the figure). If 400 feet of fencing is available, what length should x and y be to produce the maximum total area? What is the maximum area?

Figure for 29

30. **Diet and minimum cost.** A group of guinea pigs is to receive 25,600 calories per week. Two available foods produce $200xy$ calories for a mixture of x kilograms of type M food and y kilograms of type N food. If type M costs \$1 per kilogram and type N costs \$2 per kilogram, how much of each type of food should be used to minimize weekly food costs? What is the minimum cost?

Note: $x \ge 0$, $y \ge 0$

Answers to Matched Problems

1. Max $f(x, y) = f(2, 2) = 17$ (see Fig. 4)
2. (A) 3,750 units of labor and 1,250 units of capital; Max $N(x, y) = N(3{,}750, 1{,}250) \approx 43{,}301$ units

   (B) Marginal productivity of money ≈ 0.1443; increase in production $\approx 5{,}774$ units
3. 12 in. by 8 in. by 4 in.

6-5 Method of Least Squares

- Least Squares Approximation
- Applications

Least Squares Approximation

Regression analysis is the process of fitting an elementary function to a set of data points by the **method of least squares.** Now, using the optimization techniques of Section 6-3, we can develop and explain the mathematical foundation of the method of least squares. We begin with **linear regression**, the process of finding the equation of the line that is the "best" approximation to a set of data points.

Suppose that a manufacturer wants to approximate the cost function for a product. The value of the cost function has been determined for certain levels of production, as listed in Table 1. Although these points do not all lie on a line (see Fig. 1),

Table 1

Number of Units x (hundreds)	Cost y (thousand \$)
2	4
5	6
6	7
9	8

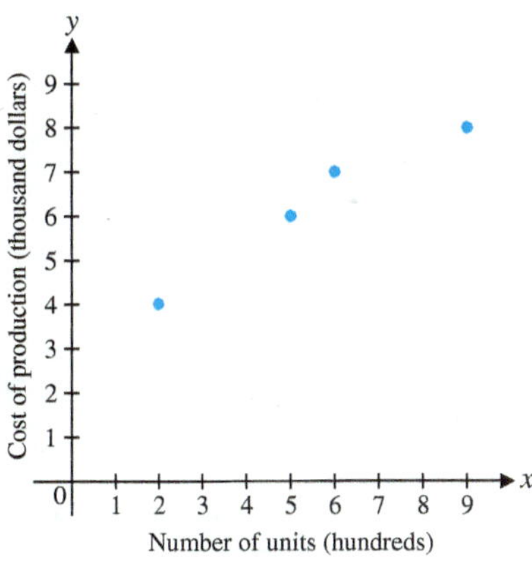

Figure 1

they are very close to being linear. The manufacturer would like to approximate the cost function by a linear function—that is, determine values a and b so that the line

$$y = ax + b$$

is, in some sense, the "best" approximation to the cost function.

What do we mean by "best"? Since the line $y = ax + b$ will not go through all four points, it is reasonable to examine the differences between the y coordinates of the points listed in the table and the y coordinates of the corresponding points on the line. Each of these differences is called the **residual** at that point (see Fig. 2). For example, at $x = 2$, the point from Table 1 is (2, 4) and the point on the line is $(2, 2a + b)$, so the residual is

$$4 - (2a + b) = 4 - 2a - b$$

All the residuals are listed in Table 2.

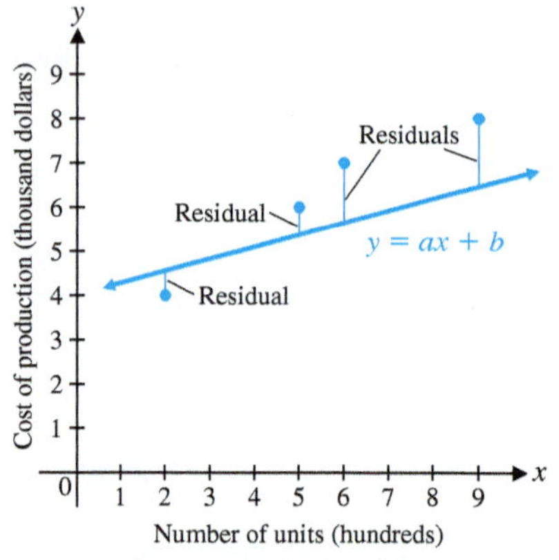

Figure 2

Table 2

x	y	$ax + b$	Residual
2	4	$2a + b$	$4 - 2a - b$
5	6	$5a + b$	$6 - 5a - b$
6	7	$6a + b$	$7 - 6a - b$
9	8	$9a + b$	$8 - 9a - b$

Our criterion for the "best" approximation is the following: Determine the values of a and b that *minimize the sum of the squares* of the residuals. The resulting line is called the **least squares line**, or the **regression line**. To this end, we minimize

$$F(a, b) = (4 - 2a - b)^2 + (6 - 5a - b)^2 + (7 - 6a - b)^2 + (8 - 9a - b)^2$$

Step 1 Find critical points:

$$\begin{aligned} F_a(a, b) &= 2(4 - 2a - b)(-2) + 2(6 - 5a - b)(-5) \\ &\quad + 2(7 - 6a - b)(-6) + 2(8 - 9a - b)(-9) \\ &= -304 + 292a + 44b = 0 \\ F_b(a, b) &= 2(4 - 2a - b)(-1) + 2(6 - 5a - b)(-1) \\ &\quad + 2(7 - 6a - b)(-1) + 2(8 - 9a - b)(-1) \\ &= -50 + 44a + 8b = 0 \end{aligned}$$

After dividing each equation by 2, we solve the system

$$\begin{aligned} 146a + 22b &= 152 \\ 22a + 4b &= 25 \end{aligned}$$

obtaining $(a, b) = (0.58, 3.06)$ as the only critical point.

Step 2 Compute $A = F_{aa}(a, b)$, $B = F_{ab}(a, b)$, and $C = F_{bb}(a, b)$:

$$\begin{aligned} F_{aa}(a, b) &= 292; \quad \text{so,} \quad A = F_{aa}(0.58, 3.06) = 292 \\ F_{ab}(a, b) &= 44; \quad \text{so,} \quad B = F_{ab}(0.58, 3.06) = 44 \\ F_{bb}(a, b) &= 8; \quad \text{so,} \quad C = F_{bb}(0.58, 3.06) = 8 \end{aligned}$$

Step 3 Evaluate $AC - B^2$ and try to classify the critical point (a, b) by using Theorem 2 in Section 6-3:

$$AC - B^2 = (292)(8) - (44)^2 = 400 > 0 \qquad \text{and} \qquad A = 292 > 0$$

Therefore, case 2 in Theorem 2 applies, and $F(a, b)$ has a local minimum at the critical point (0.58, 3.06).

So, the least squares line for the given data is

$$y = 0.58x + 3.06 \qquad \text{Least squares line}$$

The sum of the squares of the residuals is minimized for this choice of a and b (see Fig. 3).

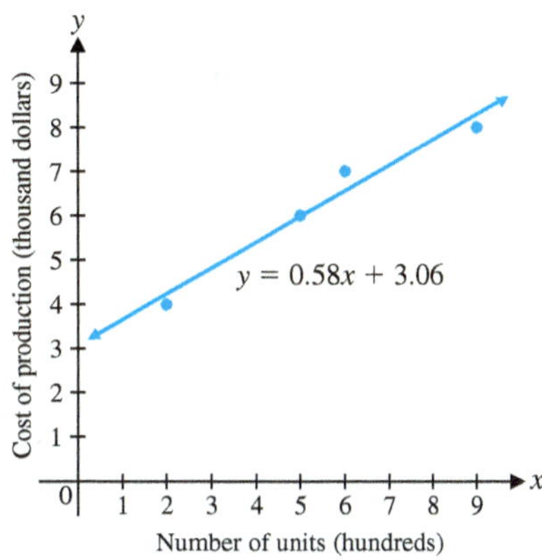

Figure 3

This linear function can now be used by the manufacturer to estimate any of the quantities normally associated with the cost function—such as costs, marginal costs, average costs, and so on. For example, the cost of producing 2,000 units is approximately

$$y = (0.58)(20) + 3.06 = 14.66, \qquad \text{or} \qquad \$14{,}660$$

The marginal cost function is

$$\frac{dy}{dx} = 0.58$$

The average cost function is

$$\bar{y} = \frac{0.58x + 3.06}{x}$$

In general, if we are given a set of n points $(x_1, y_1), (x_2, y_2), \ldots, (x_n, y_n)$, we want to determine the line $y = ax + b$ for which the sum of the squares of the residuals is minimized. Using summation notation, we find that the sum of the squares of the residuals is given by

$$F(a, b) = \sum_{k=1}^{n} (y_k - ax_k - b)^2$$

Note that in this expression the variables are a and b, and the x_k and y_k are all known values. To minimize $F(a, b)$, we thus compute the partial derivatives with respect to a and b and set them equal to 0:

$$F_a(a, b) = \sum_{k=1}^{n} 2(y_k - ax_k - b)(-x_k) = 0$$

$$F_b(a, b) = \sum_{k=1}^{n} 2(y_k - ax_k - b)(-1) = 0$$

Dividing each equation by 2 and simplifying, we see that the coefficients a and b of the least squares line $y = ax + b$ must satisfy the following system of *normal equations:*

$$\left(\sum_{k=1}^{n} x_k^2\right)a + \left(\sum_{k=1}^{n} x_k\right)b = \sum_{k=1}^{n} x_k y_k$$
$$\left(\sum_{k=1}^{n} x_k\right)a + nb = \sum_{k=1}^{n} y_k$$

Solving this system for a and b produces the formulas given in Theorem 1.

THEOREM 1 Least Squares Approximation

For a set of n points $(x_1, y_1), (x_2, y_2), \ldots, (x_n, y_n)$, the coefficients of the least squares line $y = ax + b$ are the solutions of the system of **normal equations**

$$\left(\sum_{k=1}^{n} x_k^2\right)a + \left(\sum_{k=1}^{n} x_k\right)b = \sum_{k=1}^{n} x_k y_k \tag{1}$$
$$\left(\sum_{k=1}^{n} x_k\right)a + nb = \sum_{k=1}^{n} y_k$$

and are given by the formulas

$$a = \frac{n\left(\sum_{k=1}^{n} x_k y_k\right) - \left(\sum_{k=1}^{n} x_k\right)\left(\sum_{k=1}^{n} y_k\right)}{n\left(\sum_{k=1}^{n} x_k^2\right) - \left(\sum_{k=1}^{n} x_k\right)^2} \tag{2}$$

$$b = \frac{\sum_{k=1}^{n} y_k - a\left(\sum_{k=1}^{n} x_k\right)}{n} \tag{3}$$

Now we return to the data in Table 1 and tabulate the sums required for the normal equations and their solution in Table 3.

Table 3

	x_k	y_k	$x_k y_k$	x_k^2
	2	4	8	4
	5	6	30	25
	6	7	42	36
	9	8	72	81
Totals	22	25	152	146

The normal equations (1) are then

$$146a + 22b = 152$$
$$22a + 4b = 25$$

The solution of the normal equations given by equations (2) and (3) is

$$a = \frac{4(152) - (22)(25)}{4(146) - (22)^2} = 0.58$$

$$b = \frac{25 - 0.58(22)}{4} = 3.06$$

Compare these results with step 1 on page 362. Note that Table 3 provides a convenient format for the computation of step 1.

Many graphing calculators have a linear regression feature that solves the system of normal equations obtained by setting the partial derivatives of the sum of squares of the residuals equal to 0. Therefore, in practice, we simply enter the given data points and use the linear regression feature to determine the line $y = ax + b$ that best fits the data (see Fig. 4). There is no need to compute partial derivatives or even to tabulate sums (as in Table 3).

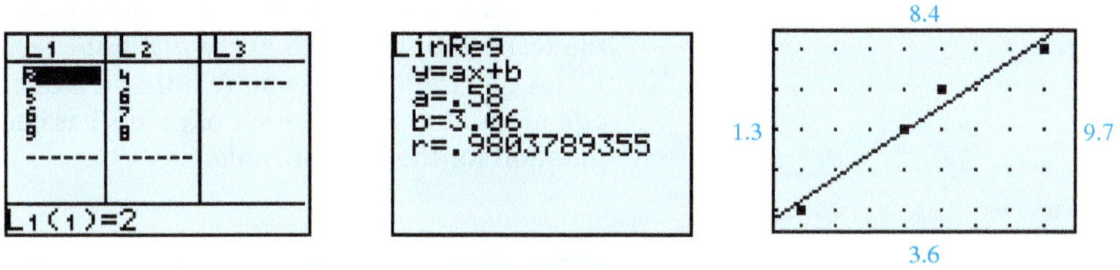

L1	L2	L3
2	4	------
5	6	
6	7	
9	8	
------	------	

L1(1)=2

(A)

LinReg
y=ax+b
a=.58
b=3.06
r=.9803789355

(B)

8.4
1.3
9.7
3.6

(C) $y_1 = 0.58x + 3.06$

Figure 4

Explore & Discuss 1

(A) Plot the four points (0, 0), (0, 1), (10, 0), and (10, 1). Which line would you guess "best" fits these four points? Use formulas (2) and (3) to test your conjecture.

(B) Plot the four points (0, 0), (0, 10), (1, 0) and (1, 10). Which line would you guess "best" fits these four points? Use formulas (2) and (3) to test your conjecture.

(C) If either of your conjectures was wrong, explain how your reasoning was mistaken.

CONCEPTUAL INSIGHT

Formula (2) for a is undefined if the denominator equals 0. When can this happen? Suppose $n = 3$. Then

$$
\begin{aligned}
n\left(\sum_{k=1}^{n} x_k^2\right) - \left(\sum_{k=1}^{n} x_k\right)^2
&= 3\left(x_1^2 + x_2^2 + x_3^2\right) - (x_1 + x_2 + x_3)^2 \\
&= 3\left(x_1^2 + x_2^2 + x_3^2\right) - \left(x_1^2 + x_2^2 + x_3^2 + 2x_1x_2 + 2x_1x_3 + 2x_2x_3\right) \\
&= 2\left(x_1^2 + x_2^2 + x_3^2\right) - (2x_1x_2 + 2x_1x_3 + 2x_2x_3) \\
&= \left(x_1^2 + x_2^2\right) + \left(x_1^2 + x_3^2\right) + \left(x_2^2 + x_3^2\right) - (2x_1x_2 + 2x_1x_3 + 2x_2x_3) \\
&= \left(x_1^2 - 2x_1x_2 + x_2^2\right) + \left(x_1^2 - 2x_1x_3 + x_3^2\right) + \left(x_2^2 - 2x_2x_3 + x_3^2\right) \\
&= (x_1 - x_2)^2 + (x_1 - x_3)^2 + (x_2 - x_3)^2
\end{aligned}
$$

and the last expression is equal to 0 if and only if $x_1 = x_2 = x_3$ (i.e., if and only if the three points all lie on the same vertical line). A similar algebraic manipulation works for any integer $n > 1$, showing that, in formula (2) for a, the denominator equals 0 if and only if all n points lie on the same vertical line.

The method of least squares can also be applied to find the quadratic equation $y = ax^2 + bx + c$ that best fits a set of data points. In this case, the sum of the squares of the residuals is a function of three variables:

$$F(a, b, c) = \sum_{k=1}^{n}\left(y_k - ax_k^2 - bx_k - c\right)^2$$

There are now three partial derivatives to compute and set equal to 0:

$$F_a(a, b, c) = \sum_{k=1}^{n} 2\left(y_k - ax_k^2 - bx_k - c\right)\left(-x_k^2\right) = 0$$

$$F_b(a, b, c) = \sum_{k=1}^{n} 2\left(y_k - ax_k^2 - bx_k - c\right)(-x_k) = 0$$

$$F_c(a, b, c) = \sum_{k=1}^{n} 2\left(y_k - ax_k^2 - bx_k - c\right)(-1) = 0$$

The resulting set of three linear equations in the three variables a, b, and c is called the *set of normal equations for quadratic regression.*

A quadratic regression feature on a calculator is designed to solve such normal equations after the given set of points has been entered. Figure 5 illustrates the computation for the data of Table 1.

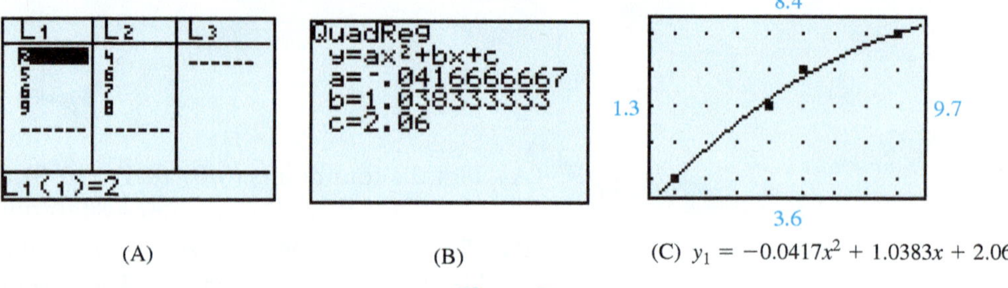

(A) (B) (C) $y_1 = -0.0417x^2 + 1.0383x + 2.06$

Figure 5

EXPLORE & DISCUSS 2

(A) Use the graphs in Figures 4 and 5 to predict which technique, linear regression or quadratic regression, yields the smaller sum of squares of the residuals for the data of Table 1. Explain.

(B) Confirm your prediction by computing the sum of squares of the residuals in each case.

The method of least squares can also be applied to other regression equations—for example, cubic, quartic, logarithmic, exponential, and power regression models. Details are explored in some of the exercises at the end of this section.

Applications

EXAMPLE 1 **Exam Scores** Table 4 lists the midterm and final examination scores of 10 students in a calculus course.

Table 4

Midterm	Final	Midterm	Final
49	61	78	77
53	47	83	81
67	72	85	79
71	76	91	93
74	68	99	99

(A) Use formulas (1), (2), and (3) to find the normal equations and the least squares line for the data given in Table 4.

(B) Use the linear regression feature on a graphing calculator to find and graph the least squares line.

(C) Use the least squares line to predict the final examination score of a student who scored 95 on the midterm examination.

SOLUTION (A) Table 5 shows a convenient way to compute all the sums in the formulas for a and b.

Table 5

	x_k	y_k	$x_k y_k$	x_k^2
	49	61	2,989	2,401
	53	47	2,491	2,809
	67	72	4,824	4,489
	71	76	5,396	5,041
	74	68	5,032	5,476
	78	77	6,006	6,084
	83	81	6,723	6,889
	85	79	6,715	7,225
	91	93	8,463	8,281
	99	99	9,801	9,801
Totals	750	753	58,440	58,496

From the last line in Table 5, we have

$$\sum_{k=1}^{10} x_k = 750 \qquad \sum_{k=1}^{10} y_k = 753 \qquad \sum_{k=1}^{10} x_k y_k = 58{,}440 \qquad \sum_{k=1}^{10} x_k^2 = 58{,}496$$

and the normal equations are

$$\begin{aligned} 58{,}496a + 750b &= 58{,}440 \\ 750a + 10b &= 753 \end{aligned}$$

Using formulas (2) and (3), we obtain

$$a = \frac{10(58{,}440) - (750)(753)}{10(58{,}496) - (750)^2} = \frac{19{,}650}{22{,}460} \approx 0.875$$

$$b = \frac{753 - 0.875(750)}{10} = 9.675$$

The least squares line is given (approximately) by

$$y = 0.875x + 9.675$$

(B) We enter the data and use the linear regression feature, as shown in Figure 6. [The discrepancy between values of a and b in the preceding calculations and those in Figure 6B is due to rounding in part (A).]

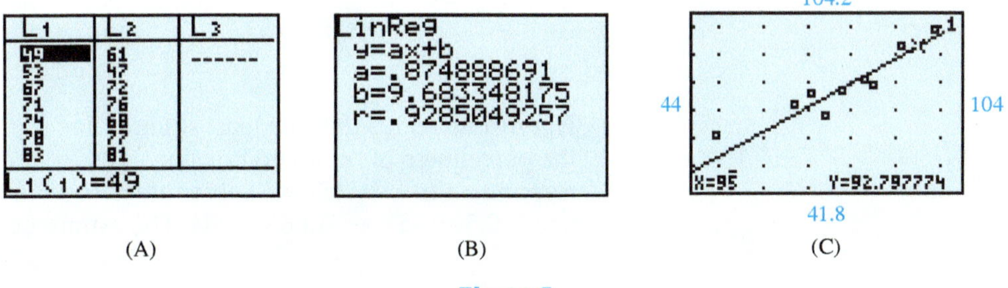

(A) (B) (C)

Figure 6

(C) If $x = 95$, then $y = 0.875(95) + 9.675 \approx 92.8$ is the predicted score on the final exam. This is also indicated in Figure 6C. If we assume that the exam score must be an integer, then we would predict a score of 93.

Matched Problem 1 Repeat Example 1 for the scores listed in Table 6.

Table 6

Midterm	Final	Midterm	Final
54	50	84	80
60	66	88	95
75	80	89	85
76	68	97	94
78	71	99	86

EXAMPLE 2

Energy Consumption The use of fuel oil for home heating in the United States has declined steadily for several decades. Table 7 lists the percentage of occupied housing units in the United States that were heated by fuel oil for various years between 1960 and 2000. Use the data in the table and linear regression to estimate the percentage of occupied housing units in the United States that were heated by fuel oil in the year 2005.

Table 7 Occupied Housing Units Heated by Fuel Oil

Year	Percent	Year	Percent
1960	32.4	1990	13.2
1970	26.0	2000	9.0
1980	18.1		

SOLUTION We enter the data, with $x = 0$ representing 1960, $x = 10$ representing 1970, and so on, and use linear regression as shown in Figure 7.

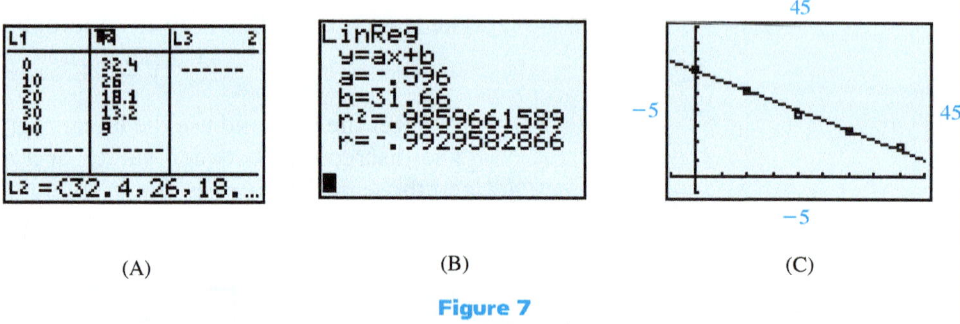

(A) (B) (C)

Figure 7

Figure 7 indicates that the least squares line is $y = -0.596x + 31.66$. To estimate the percentage of occupied housing units heated by fuel oil in the year 2005 (corresponding to $x = 45$), we substitute $x = 45$ in the equation of the least squares line: $-0.596(45) + 31.66 = 4.84$. The estimated percentage for 2005 is 4.84%.

Matched Problem 2

In 1950, coal was still a major source of fuel for home energy consumption, and the percentage of occupied housing units heated by fuel oil was only 22.1%. Add the data for 1950 to the data for Example 2, and compute the new least squares line and the new estimate for the percentage of occupied housing units heated by fuel oil in the year 2005. Discuss the discrepancy between the two estimates. (As in Example 2, let $x = 0$ represent 1960.)

Exercises 6-5

A

In Problems 1–6, find the least squares line. Graph the data and the least squares line.

1.

x	y
1	1
2	3
3	4
4	3

2.

x	y
1	−2
2	−1
3	3
4	5

3.

x	y
1	8
2	5
3	4
4	0

4.

x	y
1	20
2	14
3	11
4	3

5.

x	y
1	3
2	4
3	5
4	6

6.

x	y
1	2
2	3
3	3
4	2

B

In Problems 7–14, find the least squares line and use it to estimate y for the indicated value of x. Round answers to two decimal places.

7.

x	y
1	3
2	1
2	2
3	0

Estimate y when $x = 2.5$.

8.

x	y
1	0
3	1
3	6
3	4

Estimate y when $x = 3$.

9.

x	y
0	10
5	22
10	31
15	46
20	51

Estimate y when $x = 25$.

10.

x	y
−5	60
0	50
5	30
10	20
15	15

Estimate y when $x = 20$.

11.

x	y
−1	14
1	12
3	8
5	6
7	5

Estimate y when $x = 2$.

12.

x	y
2	−4
6	0
10	8
14	12
18	14

Estimate y when $x = 15$.

13.

x	y	x	y
0.5	25	9.5	12
2	22	11	11
3.5	21	12.5	8
5	21	14	5
6.5	18	15.5	1

Estimate y when $x = 8$.

14.

x	y	x	y
0	−15	12	11
2	−9	14	13
4	−7	16	19
6	−7	18	25
8	−1	20	33

Estimate y when $x = 10$.

C

15. To find the coefficients of the parabola

$$y = ax^2 + bx + c$$

that is the "best" fit to the points (1, 2), (2, 1), (3, 1), and (4, 3), minimize the sum of the squares of the residuals

$$F(a, b, c) = (a + b + c - 2)^2 + (4a + 2b + c - 1)^2 + (9a + 3b + c - 1)^2 + (16a + 4b + c - 3)^2$$

by solving the system of normal equations

$$F_a(a, b, c) = 0 \qquad F_b(a, b, c) = 0 \qquad F_c(a, b, c) = 0$$

for a, b, and c. Graph the points and the parabola.

16. Repeat Problem 15 for the points (−1, −2), (0, 1), (1, 2), and (2, 0).

Problems 17 and 18 refer to the system of normal equations and the formulas for a and b given on page 364.

17. Verify formulas (2) and (3) by solving the system of normal equations (1) for a and b.

18. If

$$\bar{x} = \frac{1}{n}\sum_{k=1}^{n} x_k \quad \text{and} \quad \bar{y} = \frac{1}{n}\sum_{k=1}^{n} y_k$$

are the averages of the x and y coordinates, respectively, show that the point $(\bar{x}, \bar{y})$ satisfies the equation of the least squares line, $y = ax + b$.

19. (A) Suppose that $n = 5$ and the x coordinates of the data points $(x_1, y_1), (x_2, y_2), \ldots, (x_n, y_n)$ are −2, −1, 0, 1, 2. Show that system (1) in the text implies that

$$a = \frac{\sum x_k y_k}{\sum x_k^2}$$

and that b is equal to the average of the values of y_k.

(B) Show that the conclusion of part (A) holds whenever the average of the x coordinates of the data points is 0.

20. (A) Give an example of a set of six data points such that half of the points lie above the least squares line and half lie below.

(B) Give an example of a set of six data points such that just one of the points lies above the least squares line and five lie below.

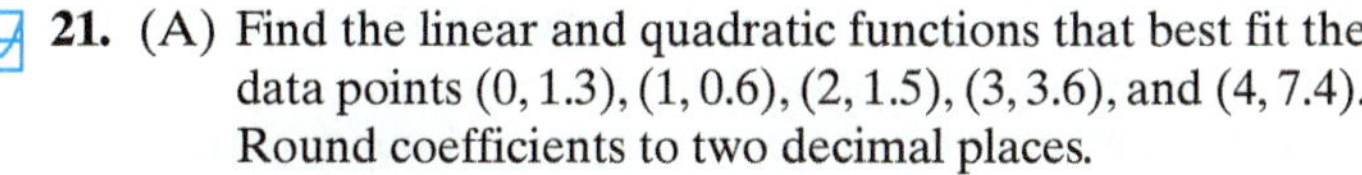

21. (A) Find the linear and quadratic functions that best fit the data points (0, 1.3), (1, 0.6), (2, 1.5), (3, 3.6), and (4, 7.4). Round coefficients to two decimal places.

(B) Which of the two functions best fits the data? Explain.

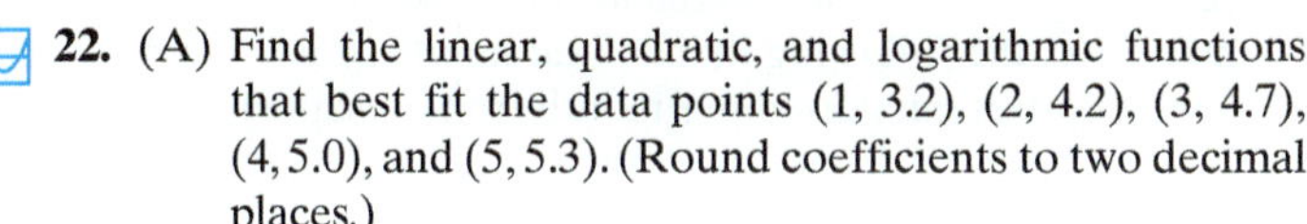

22. (A) Find the linear, quadratic, and logarithmic functions that best fit the data points (1, 3.2), (2, 4.2), (3, 4.7), (4, 5.0), and (5, 5.3). (Round coefficients to two decimal places.)

(B) Which of the three functions best fits the data? Explain.

23. Describe the normal equations for cubic regression. How many equations are there? What are the variables? What techniques could be used to solve the equations?

24. Describe the normal equations for quartic regression. How many equations are there? What are the variables? What techniques could be used to solve the equations?

Applications

25. **Crime rate.** Data on U.S. property crimes (in number of crimes per 100,000 population) are given in the table for the years 2001 through 2006.

U.S. Property Crime Rates

Year	Rate
2001	3,658
2002	3,631
2003	3,591
2004	3,514
2005	3,432
2006	3,335

(A) Find the least squares line for the data, using $x = 0$ for 2000.

(B) Use the least squares line to predict the property crime rate in 2016.

26. **Cable TV revenue.** Data for cable TV revenue are given in the table for the years 2003 through 2007.

Cable TV Revenue

Year	Thousands of dollars
2003	53,991
2004	59,428
2005	65,041
2006	71,843
2007	78,886

(A) Find the least squares line for the data, using $x = 0$ for 2000.

(B) Use the least squares line to predict cable TV revenue in 2017.

27. **Maximizing profit.** The market research department for a drugstore chain chose two summer resort areas to test market a new sunscreen lotion packaged in 4-ounce plastic bottles. After a summer of varying the selling price and recording the monthly demand, the research department arrived at the following demand table, where y is the number of bottles purchased per month (in thousands) at x dollars per bottle:

x	y
5.0	2.0
5.5	1.8
6.0	1.4
6.5	1.2
7.0	1.1

(A) Use the method of least squares to find a demand equation.

(B) If each bottle of sunscreen costs the drugstore chain $4, how should the sunscreen be priced to achieve a maximum monthly profit? [*Hint:* Use the result of part (A), with $C = 4y$, $R = xy$, and $P = R - C$.]

28. **Maximizing profit.** A market research consultant for a supermarket chain chose a large city to test market a new brand of mixed nuts packaged in 8-ounce cans. After a year of varying the selling price and recording the monthly demand, the consultant arrived at the following demand table, where y is the number of cans purchased per month (in thousands) at x dollars per can:

x	y
4.0	4.2
4.5	3.5
5.0	2.7
5.5	1.5
6.0	0.7

(A) Use the method of least squares to find a demand equation.

(B) If each can of nuts costs the supermarket chain $3, how should the nuts be priced to achieve a maximum monthly profit?

29. **Olympic Games.** The table gives the winning heights in the pole vault in the Olympic Games from 1980 to 2008.

Olympic Pole Vault Winning Height

Year	Height (ft)
1980	18.96
1984	18.85
1988	19.35
1992	19.02
1996	19.42
2000	19.35
2004	19.52
2008	19.56

(A) Use a graphing calculator to find the least squares line for the data, letting $x = 0$ for 1980.

(B) Estimate the winning height in the pole vault in the Olympic Games of 2020.

30. **Biology.** In biology, there is an approximate rule, called the *bioclimatic rule for temperate climates*. This rule states that in spring and early summer, periodic phenomena such as the blossoming of flowers, the appearance of insects, and the

ripening of fruit usually come about 4 days later for each 500 feet of altitude. Stated as a formula, the rule becomes

$$d = 8h \qquad 0 \le h \le 4$$

where d is the change in days and h is the altitude (in thousands of feet). To test this rule, an experiment was set up to record the difference in blossoming times of the same type of apple tree at different altitudes. A summary of the results is given in the following table:

h	d
0	0
1	7
2	18
3	28
4	33

(A) Use the method of least squares to find a linear equation relating h and d. Does the bioclimatic rule $d = 8h$ appear to be approximately correct?

(B) How much longer will it take this type of apple tree to blossom at 3.5 thousand feet than at sea level? [Use the linear equation found in part (A).]

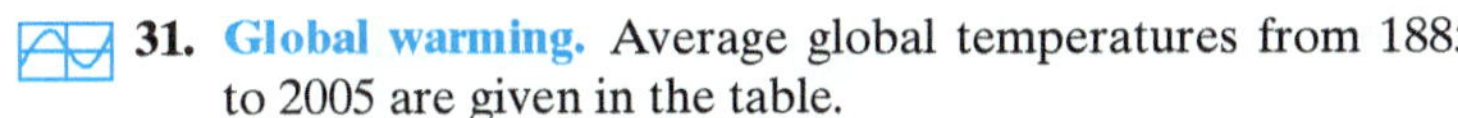

31. **Global warming.** Average global temperatures from 1885 to 2005 are given in the table.

Average Global Temperatures

Year	°F	Year	°F
1885	56.65	1955	57.06
1895	56.64	1965	57.05
1905	56.52	1975	57.04
1915	56.57	1985	57.36
1925	56.74	1995	57.64
1935	57.00	2005	58.59
1945	57.13		

(A) Find the least squares line for the data, using $x = 0$ for 1885.

(B) Use the least squares line to estimate the average global temperature in 2085.

32. **Air pollution.** Data for emissions of sulfur dioxide in the United States from selected years from 1972 to 2007 are given in the table.

Emissions of Sulfur Dioxide in the United States

Year	Million Short Tons	Year	Million Short Tons
1972	30.4	1992	22.1
1977	28.6	1997	18.8
1982	23.2	2002	15.4
1987	22.2	2007	12.4

(A) Find the least squares line for the data, using $x = 0$ for 1972.

(B) Use the least squares line to estimate the emissions of sulfur dioxide in the United States in 2017.

Answers to Matched Problems

1. (A) $y = 0.85x + 9.47$

(B)

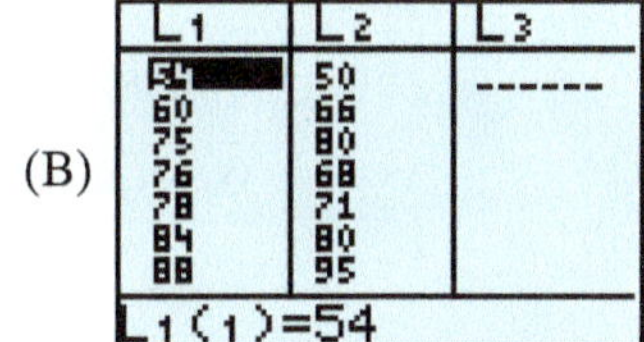

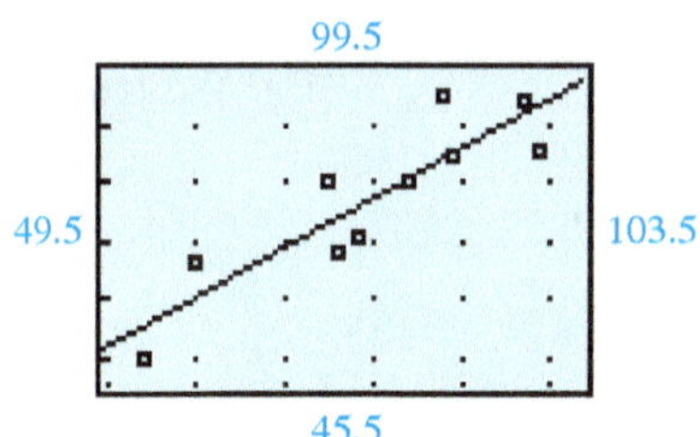

(C) 90.3

2. $y = -0.37x + 25.75$; 9.1%

6-6 Double Integrals over Rectangular Regions

- Introduction
- Definition of the Double Integral
- Average Value over Rectangular Regions
- Volume and Double Integrals

Introduction

We have generalized the concept of differentiation to functions with two or more independent variables. How can we do the same with integration, and how can we interpret the results? Let's look first at the operation of antidifferentiation. We can antidifferentiate a function of two or more variables with respect to one of the variables by treating all the other variables as though they were constants. Thus, this operation is the reverse operation of partial differentiation, just as ordinary antidifferentiation is the reverse operation of ordinary differentiation. We write $\int f(x, y)\,dx$ to indicate that we are to antidifferentiate $f(x, y)$ with respect to x, holding y fixed; we write $\int f(x, y)\,dy$ to indicate that we are to antidifferentiate $f(x, y)$ with respect to y, holding x fixed.

EXAMPLE 1 **Partial Antidifferentiation** Evaluate:

(A) $\int (6xy^2 + 3x^2)\, dy$ (B) $\int (6xy^2 + 3x^2)\, dx$

SOLUTION (A) Treating x as a constant and using the properties of antidifferentiation from Section 4-1, we have

$$\begin{aligned}\int (6xy^2 + 3x^2)\, dy &= \int 6xy^2\, dy + \int 3x^2\, dy \\ &= 6x \int y^2\, dy + 3x^2 \int dy \\ &= 6x\left(\frac{y^3}{3}\right) + 3x^2(y) + C(x) \\ &= 2xy^3 + 3x^2y + C(x)\end{aligned}$$

The dy tells us that we are looking for the antiderivative of $6xy^2 + 3x^2$ with respect to y only, holding x constant.

Note that the constant of integration can be *any function of x alone* since for any such function,

$$\frac{\partial}{\partial y} C(x) = 0$$

Check:

We can verify that our answer is correct by using partial differentiation:

$$\begin{aligned}\frac{\partial}{\partial y}[2xy^3 + 3x^2y + C(x)] &= 6xy^2 + 3x^2 + 0 \\ &= 6xy^2 + 3x^2\end{aligned}$$

(B) We treat y as a constant:

$$\begin{aligned}\int (6xy^2 + 3x^2)\, dx &= \int 6xy^2\, dx + \int 3x^2\, dx \\ &= 6y^2 \int x\, dx + 3 \int x^2\, dx \\ &= 6y^2\left(\frac{x^2}{2}\right) + 3\left(\frac{x^3}{3}\right) + E(y) \\ &= 3x^2y^2 + x^3 + E(y)\end{aligned}$$

The antiderivative contains an arbitrary function $E(y)$ of y alone.

Check:

$$\begin{aligned}\frac{\partial}{\partial x}[3x^2y^2 + x^3 + E(y)] &= 6xy^2 + 3x^2 + 0 \\ &= 6xy^2 + 3x^2\end{aligned}$$

Matched Problem 1 Evaluate (A) $\int (4xy + 12x^2y^3)\, dy$ (B) $\int (4xy + 12x^2y^3)\, dx$

Now that we have extended the concept of antidifferentiation to functions with two variables, we also can evaluate definite integrals of the form

$$\int_a^b f(x, y)\, dx \quad \text{or} \quad \int_c^d f(x, y)\, dy$$

EXAMPLE 2 **Evaluating a Partial Antiderivative** Evaluate, substituting the limits of integration in y if dy is used and in x if dx is used:

(A) $\int_0^2 (6xy^2 + 3x^2)\,dy$ (B) $\int_0^1 (6xy^2 + 3x^2)\,dx$

SOLUTION (A) From Example 1A, we know that

$$\int (6xy^2 + 3x^2)\,dy = 2xy^3 + 3x^2y + C(x)$$

According to properties of the definite integral for a function of one variable, we can use any antiderivative to evaluate the definite integral. Thus, choosing $C(x) = 0$, we have

$$\int_0^2 (6xy^2 + 3x^2)\,dy = (2xy^3 + 3x^2y)\Big|_{y=0}^{y=2}$$
$$= [2x(2)^3 + 3x^2(2)] - [2x(0)^3 + 3x^2(0)]$$
$$= 16x + 6x^2$$

(B) From Example 1B, we know that

$$\int (6xy^2 + 3x^2)\,dx = 3x^2y^2 + x^3 + E(y)$$

Choosing $E(y) = 0$, we have

$$\int_0^1 (6xy^2 + 3x^2)\,dx = (3x^2y^2 + x^3)\Big|_{x=0}^{x=1}$$
$$= [3y^2(1)^2 + (1)^3] - [3y^2(0)^2 + (0)^3]$$
$$= 3y^2 + 1$$

Matched Problem 2 Evaluate:

(A) $\int_0^1 (4xy + 12x^2y^3)\,dy$ (B) $\int_0^3 (4xy + 12x^2y^3)\,dx$

Integrating and evaluating a definite integral with integrand $f(x, y)$ with respect to y produces a function of x alone (or a constant). Likewise, integrating and evaluating a definite integral with integrand $f(x, y)$ with respect to x produces a function of y alone (or a constant). Each of these results, involving at most one variable, can now be used as an integrand in a second definite integral.

EXAMPLE 3 **Evaluating Integrals** Evaluate:

(A) $\int_0^1 \left[\int_0^2 (6xy^2 + 3x^2)\,dy\right] dx$ (B) $\int_0^2 \left[\int_0^1 (6xy^2 + 3x^2)\,dx\right] dy$

SOLUTION (A) Example 2A showed that

$$\int_0^2 (6xy^2 + 3x^2)\,dy = 16x + 6x^2$$

Therefore,

$$\int_0^1 \left[\int_0^2 (6xy^2 + 3x^2)\,dy\right] dx = \int_0^1 (16x + 6x^2)\,dx$$
$$= (8x^2 + 2x^3)\Big|_{x=0}^{x=1}$$
$$= [8(1)^2 + 2(1)^3] - [8(0)^2 + 2(0)^3] = 10$$

(B) Example 2B showed that

$$\int_0^1 (6xy^2 + 3x^2)\,dx = 3y^2 + 1$$

Therefore,

$$\int_0^2\left[\int_0^1 (6xy^2 + 3x^2)\,dx\right]dy = \int_0^2 (3y^2 + 1)\,dy$$
$$= (y^3 + y)\Big|_{y=0}^{y=2}$$
$$= [(2)^3 + 2] - [(0)^3 + 0] = 10$$

```
fnInt(16*X+6*X^2
,X,0,1)
              10
fnInt(3*Y^2+1,Y,
0,2)
              10
```

Figure 1

A numerical integration command can be used as an alternative to the fundamental theorem of calculus to evaluate the last integrals in Examples 3A and 3B, $\int_0^1 (16x + 6x^2)\,dx$ and $\int_0^2 (3y^2 + 1)\,dy$, since the integrand in each case is a function of a single variable (see Fig. 1).

Matched Problem 3 Evaluate

(A) $\int_0^3\left[\int_0^1 (4xy + 12x^2y^3)\,dy\right]dx$ (B) $\int_0^1\left[\int_0^3 (4xy + 12x^2y^3)\,dx\right]dy$

Definition of the Double Integral

Notice that the answers in Examples 3A and 3B are identical. This is not an accident. In fact, it is this property that enables us to define the *double integral,* as follows:

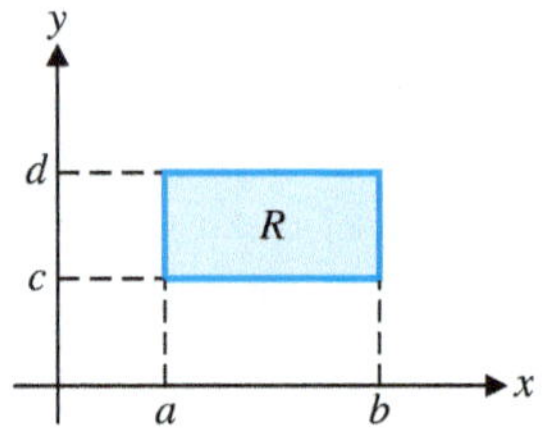

DEFINITION Double Integral

The **double integral** of a function $f(x, y)$ over a rectangle

$$R = \{(x, y) \mid a \le x \le b,\ c \le y \le d\}$$

is

$$\iint_R f(x, y)\,dA = \int_a^b\left[\int_c^d f(x, y)\,dy\right]dx$$
$$= \int_c^d\left[\int_a^b f(x, y)\,dx\right]dy$$

In the double integral $\iint_R f(x, y)\,dA$, $f(x, y)$ is called the **integrand**, and R is called the **region of integration**. The expression dA indicates that this is an integral over a two-dimensional region. The integrals

$$\int_a^b\left[\int_c^d f(x, y)\,dy\right]dx \quad\text{and}\quad \int_c^d\left[\int_a^b f(x, y)\,dx\right]dy$$

are referred to as **iterated integrals** (the brackets are often omitted), and the order in which dx and dy are written indicates the order of integration. This is not the most general definition of the double integral over a rectangular region; however, it is equivalent to the general definition for all the functions we will consider.

EXAMPLE 4 **Evaluating a Double Integral** Evaluate

$$\iint_R (x + y)\, dA \qquad \text{over} \qquad R = \{(x, y) | 1 \le x \le 3, \quad -1 \le y \le 2\}$$

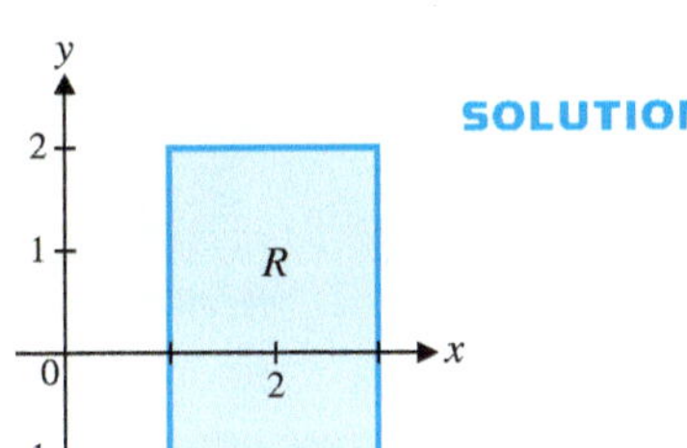

Figure 2

SOLUTION Region R is illustrated in Figure 2. We can choose either order of iteration. As a check, we will evaluate the integral both ways:

$$\begin{aligned}\iint_R (x + y)\, dA &= \int_1^3 \int_{-1}^2 (x + y)\, dy\, dx \\ &= \int_1^3 \left[\left(xy + \frac{y^2}{2}\right)\Big|_{y=-1}^{y=2}\right] dx \\ &= \int_1^3 \left[(2x + 2) - \left(-x + \tfrac{1}{2}\right)\right] dx \\ &= \int_1^3 \left(3x + \tfrac{3}{2}\right) dx \\ &= \left(\tfrac{3}{2}x^2 + \tfrac{3}{2}x\right)\Big|_{x=1}^{x=3} \\ &= \left(\tfrac{27}{2} + \tfrac{9}{2}\right) - \left(\tfrac{3}{2} + \tfrac{3}{2}\right) = 18 - 3 = 15\end{aligned}$$

$$\begin{aligned}\iint_R (x + y)\, dA &= \int_{-1}^2 \int_1^3 (x + y)\, dx\, dy \\ &= \int_{-1}^2 \left[\left(\frac{x^2}{2} + xy\right)\Big|_{x=1}^{x=3}\right] dy \\ &= \int_{-1}^2 \left[\left(\tfrac{9}{2} + 3y\right) - \left(\tfrac{1}{2} + y\right)\right] dy \\ &= \int_{-1}^2 (4 + 2y)\, dy \\ &= (4y + y^2)\Big|_{y=-1}^{y=2} \\ &= (8 + 4) - (-4 + 1) = 12 - (-3) = 15\end{aligned}$$

Matched Problem 4 Evaluate

$$\iint_R (2x - y)\, dA \qquad \text{over} \qquad R = \{(x, y) | -1 \le x \le 5, \quad 2 \le y \le 4\}$$

both ways.

EXAMPLE 5 **Double Integral of an Exponential Function** Evaluate

$$\iint_R 2xe^{x^2+y}\, dA \qquad \text{over} \qquad R = \{(x, y) | 0 \le x \le 1, \quad -1 \le y \le 1\}$$

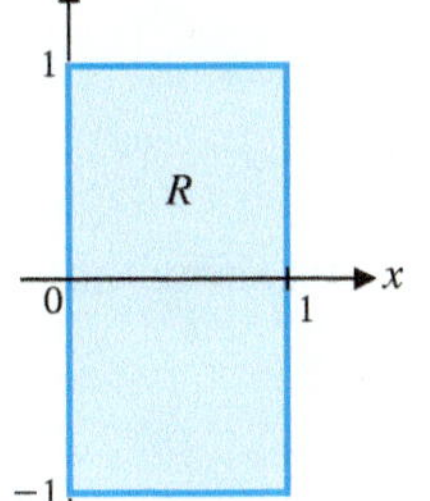

Figure 3

SOLUTION Region R is illustrated in Figure 3.

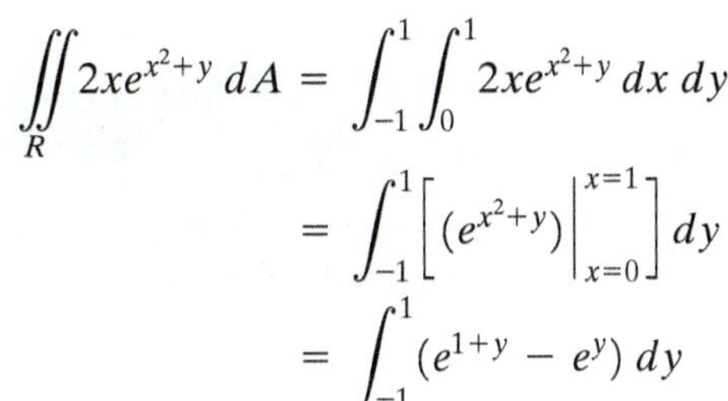

$$\begin{aligned}\iint_R 2xe^{x^2+y}\, dA &= \int_{-1}^1 \int_0^1 2xe^{x^2+y}\, dx\, dy \\ &= \int_{-1}^1 \left[\left(e^{x^2+y}\right)\Big|_{x=0}^{x=1}\right] dy \\ &= \int_{-1}^1 (e^{1+y} - e^y)\, dy\end{aligned}$$

$$= (e^{1+y} - e^y)\Big|_{y=-1}^{y=1}$$
$$= (e^2 - e) - (e^0 - e^{-1})$$
$$= e^2 - e - 1 + e^{-1}$$

Matched Problem 5 Evaluate $\displaystyle\iint_R \frac{x}{y^2} e^{x/y}\, dA$ over $R = \{(x, y) | 0 \le x \le 1,\ \ 1 \le y \le 2\}$.

Average Value over Rectangular Regions

In Section 4-5, the average value of a function $f(x)$ over an interval $[a, b]$ was defined as

$$\frac{1}{b - a} \int_a^b f(x)\, dx$$

This definition is easily extended to functions of two variables over rectangular regions as follows (notice that the denominator $(b - a)(d - c)$ is simply the area of the rectangle R):

DEFINITION Average Value over Rectangular Regions

The **average value** of the function $f(x, y)$ over the rectangle

$$R = \{(x, y) | a \le x \le b,\ \ c \le y \le d\}$$

is

$$\frac{1}{(b - a)(d - c)} \iint_R f(x, y)\, dA$$

EXAMPLE 6 **Average Value** Find the average value of $f(x, y) = 4 - \frac{1}{2}x - \frac{1}{2}y$ over the rectangle $R = \{(x, y) | 0 \le x \le 2,\ \ 0 \le y \le 2\}$.

SOLUTION Region R is illustrated in Figure 4. We have

$$\frac{1}{(b - a)(d - c)} \iint_R f(x, y)\, dA = \frac{1}{(2 - 0)(2 - 0)} \iint_R \left(4 - \frac{1}{2}x - \frac{1}{2}y\right) dA$$
$$= \tfrac{1}{4} \int_0^2 \int_0^2 \left(4 - \tfrac{1}{2}x - \tfrac{1}{2}y\right) dy\, dx$$
$$= \tfrac{1}{4} \int_0^2 \left[\left(4y - \tfrac{1}{2}xy - \tfrac{1}{4}y^2\right)\Big|_{y=0}^{y=2}\right] dx$$
$$= \tfrac{1}{4} \int_0^2 (7 - x)\, dx$$
$$= \tfrac{1}{4}\left(7x - \tfrac{1}{2}x^2\right)\Big|_{x=0}^{x=2}$$
$$= \tfrac{1}{4}(12) = 3$$

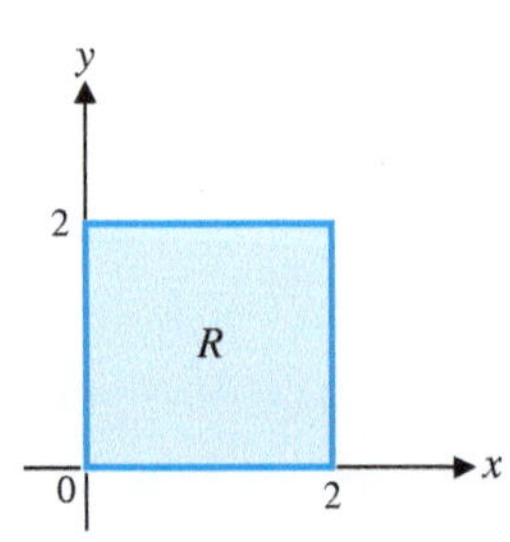

Figure 4

Figure 5 illustrates the surface $z = f(x, y)$, and our calculations show that 3 is the average of the z values over the region R.

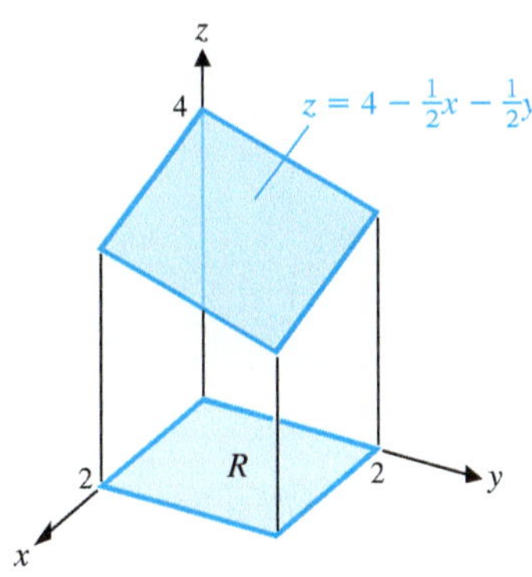

Figure 5

Matched Problem 6 Find the average value of $f(x, y) = x + 2y$ over the rectangle

$$R = \{(x, y) | 0 \le x \le 2, \quad 0 \le y \le 1\}$$

EXPLORE & DISCUSS 1

(A) Which of the functions $f(x, y) = 4 - x^2 - y^2$ and $g(x, y) = 4 - x - y$ would you guess has the greater average value over the rectangle $R = \{(x, y) | 0 \le x \le 1, \quad 0 \le y \le 1\}$? Explain.

(B) Use double integrals to check the correctness of your guess in part (A).

Volume and Double Integrals

One application of the definite integral of a function with one variable is the calculation of areas, so it is not surprising that the definite integral of a function of two variables can be used to calculate volumes of solids.

THEOREM 1 Volume under a Surface

If $f(x, y) \ge 0$ over a rectangle $R = \{(x, y) | a \le x \le b, \quad c \le y \le d\}$, then the volume of the solid formed by graphing f over the rectangle R is given by

$$V = \iint_R f(x, y)\, dA$$

EXAMPLE 7 **Volume** Find the volume of the solid under the graph of $f(x, y) = 1 + x^2 + y^2$ over the rectangle $R = \{(x, y) | 0 \le x \le 1, \quad 0 \le y \le 1\}$.

SOLUTION Figure 6 shows the region R, and Figure 7 illustrates the volume under consideration.

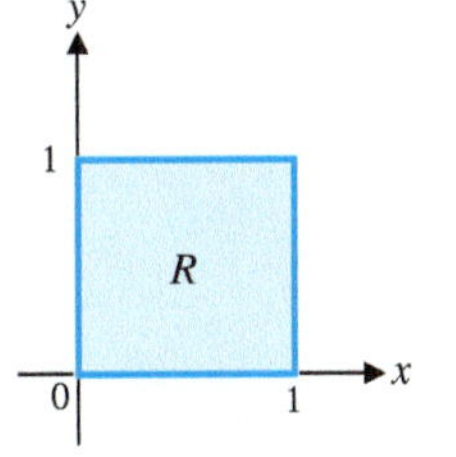

Figure 6

$$V = \iint_R (1 + x^2 + y^2)\, dA$$

$$= \int_0^1 \int_0^1 (1 + x^2 + y^2)\, dx\, dy$$

$$= \int_0^1 \left[\left(x + \tfrac{1}{3}x^3 + xy^2 \right) \Big|_{x=0}^{x=1} \right] dy$$

$$= \int_0^1 \left(\tfrac{4}{3} + y^2 \right) dy$$

$$= \left(\tfrac{4}{3}y + \tfrac{1}{3}y^3 \right) \Big|_{y=0}^{y=1} = \tfrac{5}{3} \text{ cubic units}$$

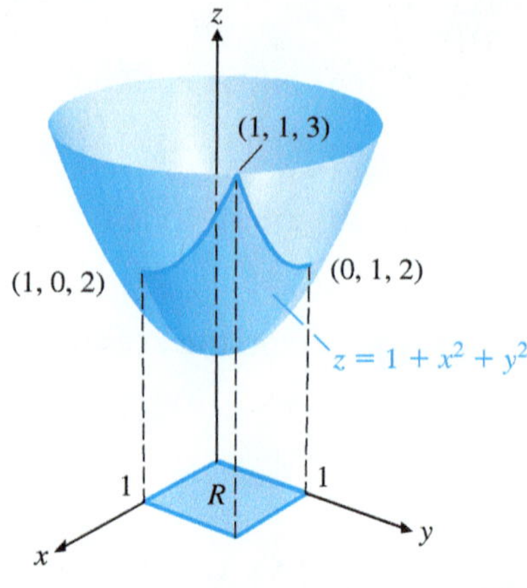

Figure 7

Matched Problem 7 Find the volume of the solid under the graph of $f(x, y) = 1 + x + y$ over the rectangle $R = \{(x, y) | 0 \le x \le 1, \quad 0 \le y \le 2\}$.

CONCEPTUAL INSIGHT

Double integrals can be defined over regions that are more general than rectangles. For example, let $R > 0$. Then the function $f(x, y) = \sqrt{R^2 - (x^2 + y^2)}$ can be integrated over the circular region $C = \{(x, y) | x^2 + y^2 \le R^2\}$. In fact, it can be shown that

$$\iint_C \sqrt{R^2 - (x^2 + y^2)}\, dx\, dy = \frac{2\pi R^3}{3}$$

Because $x^2 + y^2 + z^2 = R^2$ is the equation of a sphere of radius R centered at the origin, the double integral over C represents the volume of the upper hemisphere. Therefore, the volume of a sphere of radius R is given by

$$V = \frac{4\pi R^3}{3} \quad \text{Volume of sphere of radius } R$$

Double integrals can also be used to obtain volume formulas for other geometric figures (see Table 1, Appendix C).

Exercises 6-6

A

In Problems 1–8, find each antiderivative. Then use the antiderivative to evaluate the definite integral.

1. (A) $\int 12x^2y^3\, dy$ (B) $\int_0^1 12x^2y^3\, dy$

2. (A) $\int 12x^2y^3\, dx$ (B) $\int_{-1}^2 12x^2y^3\, dx$

3. (A) $\int (4x + 6y + 5)\, dx$
(B) $\int_{-2}^3 (4x + 6y + 5)\, dx$

4. (A) $\int (4x + 6y + 5)\, dy$
(B) $\int_1^4 (4x + 6y + 5)\, dy$

5. (A) $\int \frac{x}{\sqrt{y + x^2}} dx$ (B) $\int_0^2 \frac{x}{\sqrt{y + x^2}} dx$

6. (A) $\int \frac{x}{\sqrt{y + x^2}} dy$ (B) $\int_1^5 \frac{x}{\sqrt{y + x^2}} dy$

7. (A) $\int \frac{\ln x}{xy} dy$ (B) $\int_1^{e^2} \frac{\ln x}{xy} dy$

8. (A) $\int \frac{\ln x}{xy} dx$ (B) $\int_1^{e} \frac{\ln x}{xy} dx$

B

In Problems 9–16, evaluate each iterated integral. (See the indicated problem for the evaluation of the inner integral.)

9. $\int_{-1}^{2} \int_0^1 12x^2y^3 \, dy \, dx$

(See Problem 1.)

10. $\int_0^1 \int_{-1}^{2} 12x^2y^3 \, dx \, dy$

(See Problem 2.)

11. $\int_1^4 \int_{-2}^{3} (4x + 6y + 5) \, dx \, dy$

(See Problem 3.)

12. $\int_{-2}^{3} \int_1^4 (4x + 6y + 5) \, dy \, dx$

(See Problem 4.)

13. $\int_1^5 \int_0^2 \frac{x}{\sqrt{y + x^2}} dx \, dy$

(See Problem 5.)

14. $\int_0^2 \int_1^5 \frac{x}{\sqrt{y + x^2}} dy \, dx$

(See Problem 6.)

15. $\int_1^e \int_1^{e^2} \frac{\ln x}{xy} dy \, dx$

(See Problem 7.)

16. $\int_1^{e^2} \int_1^e \frac{\ln x}{xy} dx \, dy$

(See Problem 8.)

Use both orders of iteration to evaluate each double integral in Problems 17–20.

17. $\iint_R xy \, dA; R = \{(x, y)|0 \le x \le 2, \quad 0 \le y \le 4\}$

18. $\iint_R \sqrt{xy} \, dA; R = \{(x, y)|1 \le x \le 4, \quad 1 \le y \le 9\}$

19. $\iint_R (x + y)^5 \, dA; R = \{(x, y)|-1 \le x \le 1, \quad 1 \le y \le 2\}$

20. $\iint_R xe^y \, dA; R = \{(x, y)|-2 \le x \le 3, \quad 0 \le y \le 2\}$

In Problems 21–24, find the average value of each function over the given rectangle.

21. $f(x, y) = (x + y)^2$;

$R = \{(x, y)|1 \le x \le 5, \quad -1 \le y \le 1\}$

22. $f(x, y) = x^2 + y^2$;

$R = \{(x, y)|-1 \le x \le 2, \quad 1 \le y \le 4\}$

23. $f(x, y) = x/y$; $R = \{(x, y)|1 \le x \le 4, \quad 2 \le y \le 7\}$

24. $f(x, y) = x^2y^3$;

$R = \{(x, y)|-1 \le x \le 1, \quad 0 \le y \le 2\}$

In Problems 25–28, find the volume of the solid under the graph of each function over the given rectangle.

25. $f(x, y) = 2 - x^2 - y^2$;

$R = \{(x, y)|0 \le x \le 1, \quad 0 \le y \le 1\}$

26. $f(x, y) = 5 - x$;

$R = \{(x, y)|0 \le x \le 5, \quad 0 \le y \le 5\}$

27. $f(x, y) = 4 - y^2$;

$R = \{(x, y)|0 \le x \le 2, \quad 0 \le y \le 2\}$

28. $f(x, y) = e^{-x-y}$;

$R = \{(x, y)|0 \le x \le 1, \quad 0 \le y \le 1\}$

C

Evaluate each double integral in Problems 29–32. Select the order of integration carefully; each problem is easy to do one way and difficult the other.

29. $\iint_R xe^{xy} \, dA; R = \{(x, y)|0 \le x \le 1, \quad 1 \le y \le 2\}$

30. $\iint_R xye^{x^2y} \, dA; R = \{(x, y)|0 \le x \le 1, \quad 1 \le y \le 2\}$

31. $\iint_R \frac{2y + 3xy^2}{1 + x^2} dA$;

$R = \{(x, y)|0 \le x \le 1, \quad -1 \le y \le 1\}$

32. $\iint_R \frac{2x + 2y}{1 + 4y + y^2} dA$;

$R = \{(x, y)|1 \le x \le 3, \quad 0 \le y \le 1\}$

33. Show that $\int_0^2 \int_0^2 (1 - y) \, dx \, dy = 0$. Does the double integral represent the volume of a solid? Explain.

34. (A) Find the average values of the functions $f(x, y) = x + y$, $g(x, y) = x^2 + y^2$, and $h(x, y) = x^3 + y^3$ over the rectangle

$R = \{(x, y)|0 \le x \le 1, \quad 0 \le y \le 1\}$

(B) Does the average value of $k(x, y) = x^n + y^n$ over the rectangle

$R_1 = \{(x, y)|0 \le x \le 1, \quad 0 \le y \le 1\}$

increase or decrease as n increases? Explain.

(C) Does the average value of $k(x, y) = x^n + y^n$ over the rectangle

$R_2 = \{(x, y)|0 \le x \le 2, \quad 0 \le y \le 2\}$

increase or decrease as n increases? Explain.

35. Let $f(x, y) = x^3 + y^2 - e^{-x} - 1$.

(A) Find the average value of $f(x, y)$ over the rectangle $R = \{(x, y) | -2 \le x \le 2, \quad -2 \le y \le 2\}$.

(B) Graph the set of all points (x, y) in R for which $f(x, y) = 0$.

(C) For which points (x, y) in R is $f(x, y)$ greater than 0? Less than 0? Explain.

36. Find the dimensions of the square S centered at the origin for which the average value of $f(x, y) = x^2e^y$ over S is equal to 100.

Applications

37. Multiplier principle. Suppose that Congress enacts a one-time-only 10% tax rebate that is expected to infuse \$$y$ billion, $5 \le y \le 7$, into the economy. If every person and every corporation is expected to spend a proportion x, $0.6 \le x \le 0.8$, of each dollar received, then, by the **multiplier principle** in economics, the total amount of spending S (in billions of dollars) generated by this tax rebate is given by

$$S(x, y) = \frac{y}{1 - x}$$

What is the average total amount of spending for the indicated ranges of the values of x and y? Set up a double integral and evaluate it.

38. Multiplier principle. Repeat Problem 37 if $6 \le y \le 10$ and $0.7 \le x \le 0.9$.

39. Cobb–Douglas production function. If an industry invests x thousand labor-hours, $10 \le x \le 20$, and \$$y$ million, $1 \le y \le 2$, in the production of N thousand units of a certain item, then N is given by

$$N(x, y) = x^{0.75}y^{0.25}$$

What is the average number of units produced for the indicated ranges of x and y? Set up a double integral and evaluate it.

40. Cobb–Douglas production function. Repeat Problem 39 for

$$N(x, y) = x^{0.5}y^{0.5}$$

where $10 \le x \le 30$ and $1 \le y \le 3$.

41. Population distribution. In order to study the population distribution of a certain species of insect, a biologist has constructed an artificial habitat in the shape of a rectangle 16 feet long and 12 feet wide. The only food available to the insects in this habitat is located at its center. The biologist has determined that the concentration C of insects per square foot at a point d units from the food supply (see the figure) is given approximately by

$$C = 10 - \tfrac{1}{10}d^2$$

What is the average concentration of insects throughout the habitat? Express C as a function of x and y, set up a double integral, and evaluate it.

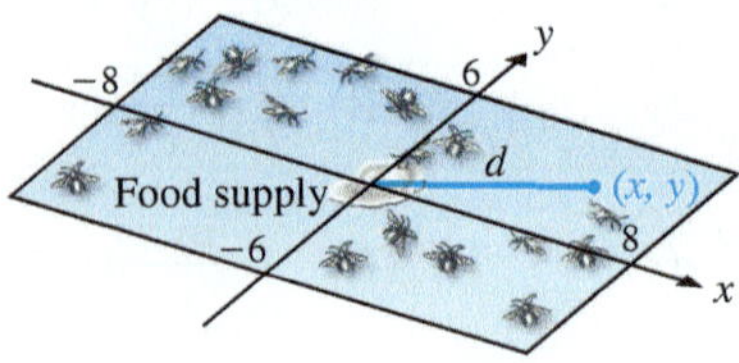

Figure for 41

42. Population distribution. Repeat Problem 41 for a square habitat that measures 12 feet on each side, where the insect concentration is given by

$$C = 8 - \tfrac{1}{10}d^2$$

43. Pollution. A heavy industrial plant located in the center of a small town emits particulate matter into the atmosphere. Suppose that the concentration of particulate matter (in parts per million) at a point d miles from the plant (see the figure) is given by

$$C = 100 - 15d^2$$

If the boundaries of the town form a rectangle 4 miles long and 2 miles wide, what is the average concentration of particulate matter throughout the town? Express C as a function of x and y, set up a double integral, and evaluate it.

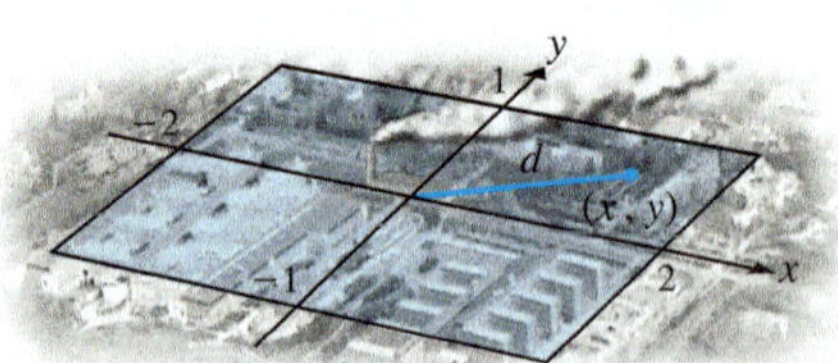

Figure for 43

44. Pollution. Repeat Problem 43 if the boundaries of the town form a rectangle 8 miles long and 4 miles wide and the concentration of particulate matter is given by

$$C = 100 - 3d^2$$

45. Safety research. Under ideal conditions, if a person driving a car slams on the brakes and skids to a stop, the length of the skid marks (in feet) is given by the formula

$$L = 0.000\,013\,3xy^2$$

where x is the weight of the car (in pounds) and y is the speed of the car (in miles per hour). What is the average length of the skid marks for cars weighing between 2,000 and 3,000 pounds and traveling at speeds between 50 and 60 miles per hour? Set up a double integral and evaluate it.

46. Safety research. Repeat Problem 45 for cars weighing between 2,000 and 2,500 pounds and traveling at speeds between 40 and 50 miles per hour.

47. **Psychology.** The intelligence quotient Q for a person with mental age x and chronological age y is given by

$$Q(x, y) = 100\frac{x}{y}$$

In a group of sixth-graders, the mental age varies between 8 and 16 years and the chronological age varies between 10 and 12 years. What is the average intelligence quotient for this group? Set up a double integral and evaluate it.

48. **Psychology.** Repeat Problem 47 for a group with mental ages between 6 and 14 years and chronological ages between 8 and 10 years.

Answers to Matched Problems

1. (A) $2xy^2 + 3x^2y^4 + C(x)$ (B) $2x^2y + 4x^3y^3 + E(y)$
2. (A) $2x + 3x^2$ (B) $18y + 108y^3$
3. (A) 36 (B) 36
4. 12
5. $e - 2e^{1/2} + 1$
6. 2
7. 5 cubic units

6-7 Double Integrals over More General Regions

- Regular Regions
- Double Integrals over Regular Regions
- Reversing the Order of Integration
- Volume and Double Integrals

In this section, we extend the concept of double integration discussed in Section 6-6 to nonrectangular regions. We begin with an example and some new terminology.

Regular Regions

Let R be the region graphed in Figure 1. We can describe R with the following inequalities:

$$R = \{(x, y)|x \le y \le 6x - x^2,\ 0 \le x \le 5\}$$

The region R can be viewed as a union of vertical line segments. For each x in the interval $[0, 5]$, the line segment from the point $(x, g(x))$ to the point $(x, f(x))$ lies in the region R. Any region that can be covered by vertical line segments in this manner is called a *regular x region.*

Now consider the region S in Figure 2. It can be described with the following inequalities:

$$S = \{(x, y)|y^2 \le x \le y + 2,\ -1 \le y \le 2\}$$

The region S can be viewed as a union of horizontal line segments going from the graph of $h(y) = y^2$ to the graph of $k(y) = y + 2$ on the interval $[-1, 2]$. Regions that can be described in this manner are called *regular y regions.*

In general, *regular regions* are defined as follows:

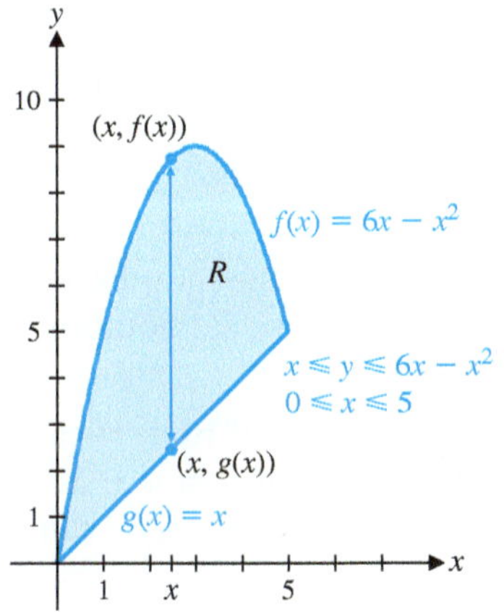

Figure 1

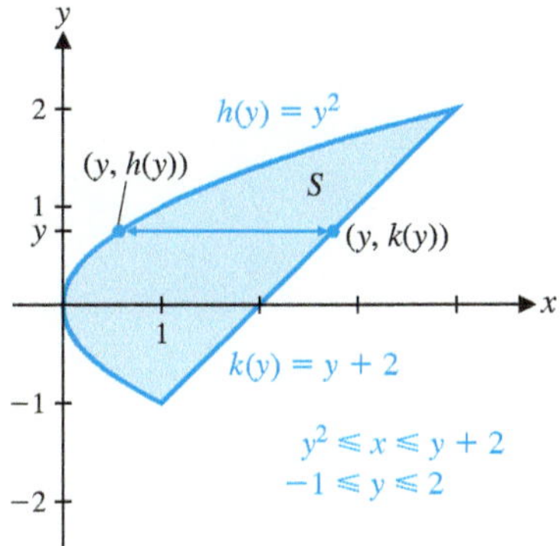

Figure 2

DEFINITION Regular Regions

A region R in the xy plane is a **regular x region** if there exist functions $f(x)$ and $g(x)$ and numbers a and b such that

$$R = \{(x, y)|\, g(x) \le y \le f(x),\ \ a \le x \le b\}$$

A region R is a **regular y region** if there exist functions $h(y)$ and $k(y)$ and numbers c and d such that

$$R = \{(x, y)|\, h(y) \le x \le k(y),\ \ c \le y \le d\}$$

See Figure 3 for a geometric interpretation.

CONCEPTUAL INSIGHT

If, for some region R, there is a horizontal line that has a nonempty intersection I with R, and if I is neither a closed interval nor a point, then R is *not* a regular y region. Similarly, if, for some region R, there is a vertical line that has a nonempty intersection I with R, and if I is neither a closed interval nor a point, then R is *not* a regular x region (see Fig. 3).

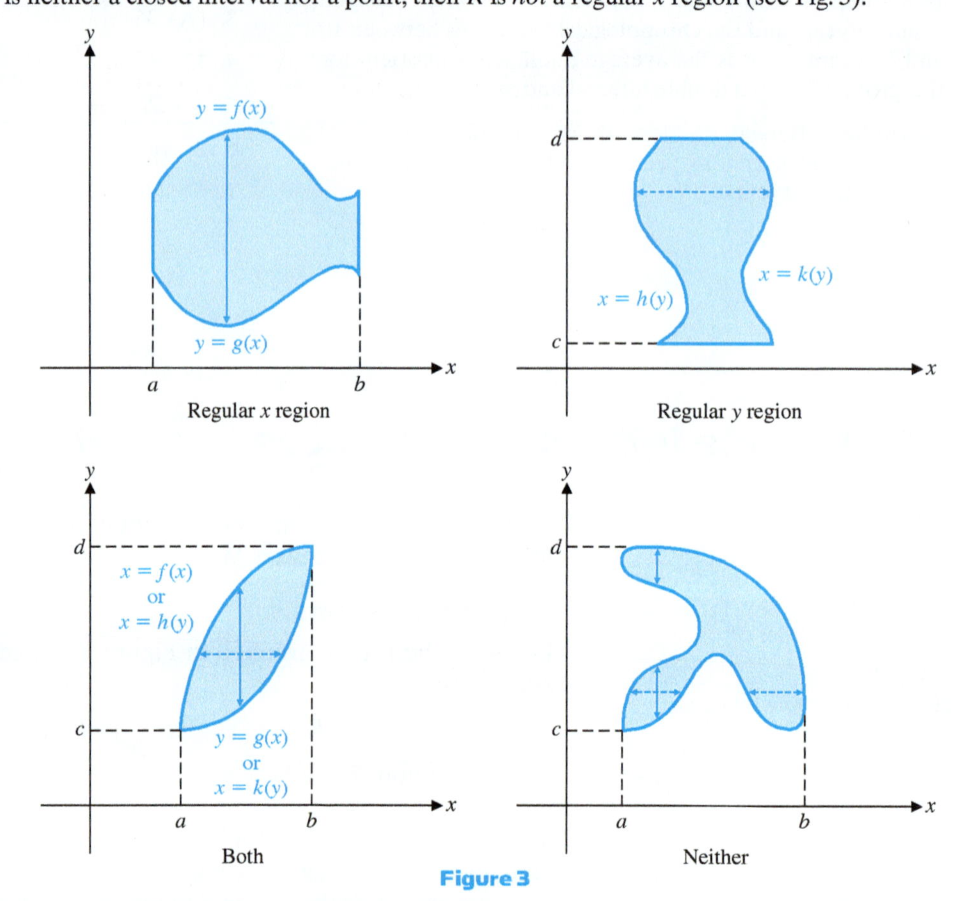

Figure 3

EXAMPLE 1 **Describing a Regular x Region** The region R is bounded by the graphs of $y = 4 - x^2$ and $y = x - 2$, $x \geq 0$, and the y axis. Graph R and use set notation and double inequalities to describe R as a regular x region.

SOLUTION As the solid line in the following figure indicates, R can be covered by vertical line segments that go from the graph of $y = x - 2$ to the graph of $y = 4 - x^2$. So, R is a regular x region. In terms of set notation and double inequalities, we can write

$$R = \{(x, y) \mid x - 2 \leq y \leq 4 - x^2,\ 0 \leq x \leq 2\}$$

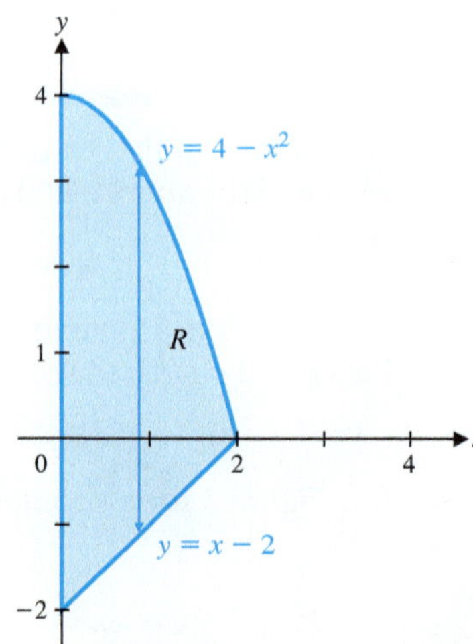

CONCEPTUAL INSIGHT

The region R of Example 1 is also a regular y region, since $R = \{(x, y)|0 \le x \le k(y), -2 \le y \le 4\}$, where

$$k(y) = \begin{cases} 2 + y & \text{if } -2 \le y \le 0 \\ \sqrt{4 - y} & \text{if } 0 \le y \le 4 \end{cases}$$

But because $k(y)$ is piecewise defined, this description is more complicated than the description of R in Example 1 as a regular x region.

Matched Problem 1 Describe the region R bounded by the graphs of $x = 6 - y$ and $x = y^2$, $y \ge 0$, and the x axis as a regular y region.

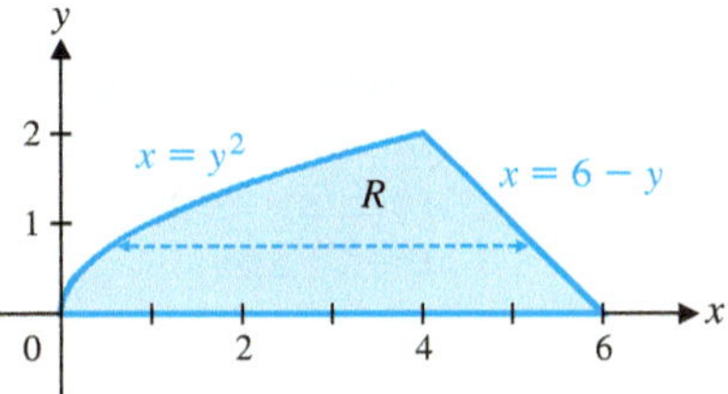

EXAMPLE 2 **Describing Regular Regions** The region R is bounded by the graphs of $x + y^2 = 9$ and $x + 3y = 9$. Graph R and describe R as a regular x region, a regular y region, both, or neither. Represent R in set notation and with double inequalities.

SOLUTION

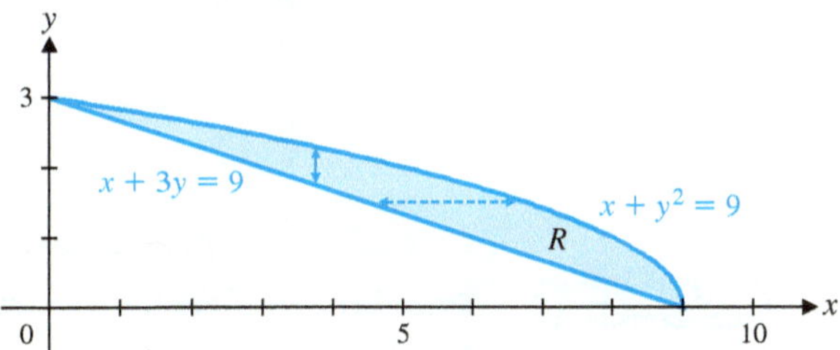

Region R can be covered by vertical line segments that go from the graph of $x + 3y = 9$ to the graph of $x + y^2 = 9$. Thus, R is a regular x region. In order to describe R with inequalities, we must solve each equation for y in terms of x:

$$\begin{aligned} x + 3y &= 9 \\ 3y &= 9 - x \\ y &= 3 - \tfrac{1}{3}x \end{aligned} \qquad \begin{aligned} x + y^2 &= 9 \\ y^2 &= 9 - x \\ y &= \sqrt{9 - x} \end{aligned}$$

We use the positive square root, since the graph is in the first quadrant.

So,

$$R = \{(x, y)|3 - \tfrac{1}{3}x \le y \le \sqrt{9 - x},\ 0 \le x \le 9\}$$

Since region R also can be covered by horizontal line segments (see the dashed line in the preceding figure) that go from the graph of $x + 3y = 9$ to the graph of $x + y^2 = 9$, it is a regular y region. Now we must solve each equation for x in terms of y:

$$\begin{aligned} x + 3y &= 9 \\ x &= 9 - 3y \end{aligned} \qquad \begin{aligned} x + y^2 &= 9 \\ x &= 9 - y^2 \end{aligned}$$

Therefore,

$$R = \{(x, y)|9 - 3y \le x \le 9 - y^2,\ 0 \le y \le 3\}$$

Matched Problem 2 Repeat Example 2 for the region bounded by the graphs of $2y - x = 4$ and $y^2 - x = 4$, as shown in the following figure:

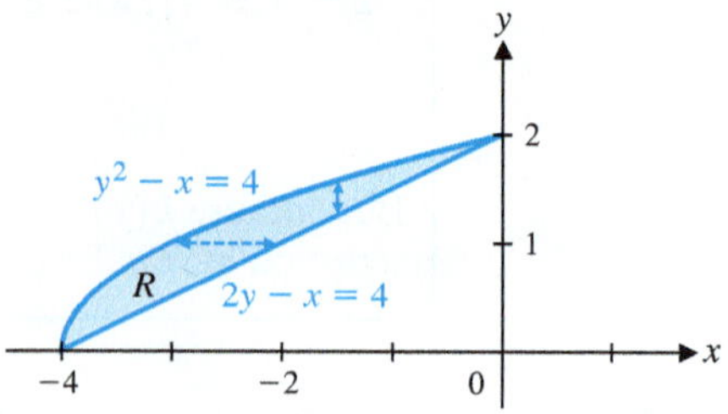

Explore & Discuss 1

A E I O U

Consider the vowels A, E, I, O, U, written in block letters as shown in the margin, to be regions of the plane. One of the vowels is a regular x region, but not a regular y region; one is a regular y region, but not a regular x region; one is both; two are neither. Explain.

Double Integrals over Regular Regions

Now we want to extend the definition of double integration to include regular x regions and regular y regions. The order of integration now depends on the nature of the region R. If R is a regular x region, we integrate with respect to y first, while if R is a regular y region, we integrate with respect to x first.

Note that the variable limits of integration (when present) are always on the inner integral, and the constant limits of integration are always on the outer integral.

DEFINITION Double Integration over Regular Regions

Regular x Region

If $R = \{(x, y) | g(x) \le y \le f(x), \quad a \le x \le b\}$, then

$$\iint_R F(x, y)\,dA = \int_a^b \left[\int_{g(x)}^{f(x)} F(x, y)\,dy\right] dx$$

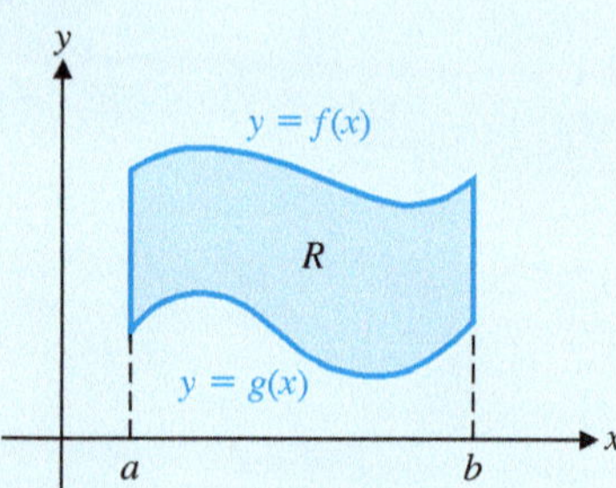

Regular y Region

If $R = \{(x, y) \mid h(y) \le x \le k(y), \quad c \le y \le d\}$, then

$$\iint_R F(x, y)\,dA = \int_c^d \left[\int_{h(y)}^{k(y)} F(x, y)\,dx\right] dy$$

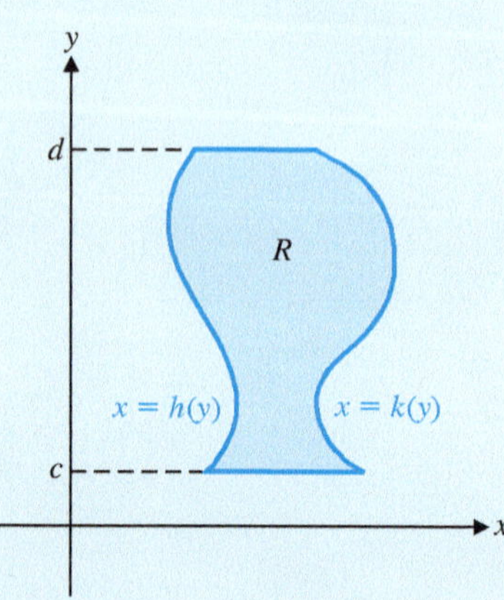

EXAMPLE 3 **Evaluating a Double Integral** Evaluate $\iint_R 2xy\,dA$, where R is the region bounded by the graphs of $y = -x$ and $y = x^2$, $x \ge 0$, and the graph of $x = 1$.

SOLUTION From the graph, we can see that R is a regular x region described by

$$R = \{(x, y) \mid -x \le y \le x^2, \quad 0 \le x \le 1\}$$

$$\begin{aligned}
\iint_R 2xy\,dA &= \int_0^1 \left[\int_{-x}^{x^2} 2xy\,dy\right] dx \\
&= \int_0^1 \left[xy^2 \Big|_{y=-x}^{y=x^2}\right] dx \\
&= \int_0^1 [x(x^2)^2 - x(-x)^2]\,dx \\
&= \int_0^1 (x^5 - x^3)\,dx \\
&= \left(\frac{x^6}{6} - \frac{x^4}{4}\right)\Bigg|_{x=0}^{x=1} \\
&= \left(\tfrac{1}{6} - \tfrac{1}{4}\right) - (0 - 0) = -\tfrac{1}{12}
\end{aligned}$$

Matched Problem 3 Evaluate $\iint_R 3xy^2\,dA$, where R is the region in Example 3.

EXAMPLE 4 **Evaluating a Double Integral** Evaluate $\iint_R (2x + y)\,dA$, where R is the region bounded by the graphs of $y = \sqrt{x}$, $x + y = 2$, and $y = 0$.

SOLUTION From the graph, we can see that R is a regular y region. After solving each equation for x, we can write

$$R = \{(x, y) \mid y^2 \le x \le 2 - y, \quad 0 \le y \le 1\}$$

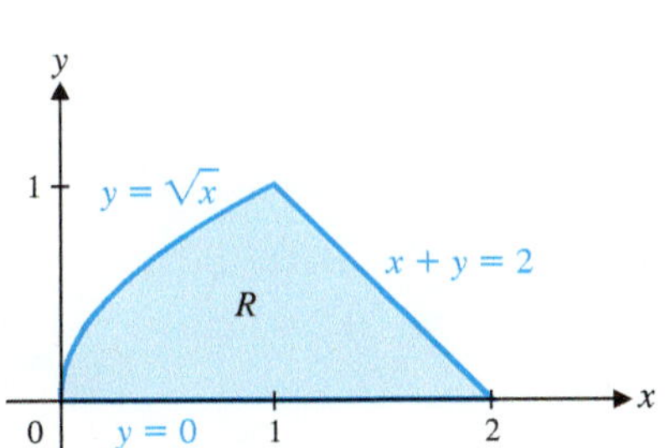

$$\begin{aligned}
\iint_R (2x + y)dA &= \int_0^1 \left[\int_{y^2}^{2-y} (2x + y)dx\right] dy \\
&= \int_0^1 \left[(x^2 + yx)\Big|_{x=y^2}^{x=2-y}\right] dy \\
&= \int_0^1 \{[(2 - y)^2 + y(2 - y)] - [(y^2)^2 + y(y^2)]\}\,dy \\
&= \int_0^1 (4 - 2y - y^3 - y^4)\,dy \\
&= \left(4y - y^2 - \tfrac{1}{4}y^4 - \tfrac{1}{5}y^5\right)\Big|_{y=0}^{y=1} \\
&= \left(4 - 1 - \tfrac{1}{4} - \tfrac{1}{5}\right) - 0 = \tfrac{51}{20}
\end{aligned}$$

Matched Problem 4 Evaluate $\iint_R (y - 4x)\,dA$, where R is the region in Example 4.

EXAMPLE 5 **Evaluating a Double Integral** The region R is bounded by the graphs of $y = \sqrt{x}$ and $y = \frac{1}{2}x$. Evaluate $\iint_R 4xy^3\,dA$ two different ways.

SOLUTION Region R is both a regular x region and a regular y region:

$$R = \{(x, y) \mid \tfrac{1}{2}x \le y \le \sqrt{x},\quad 0 \le x \le 4\} \qquad \text{Regular } x \text{ region}$$

$$R = \{(x, y) \mid y^2 \le x \le 2y,\quad 0 \le y \le 2\} \qquad \text{Regular } y \text{ region}$$

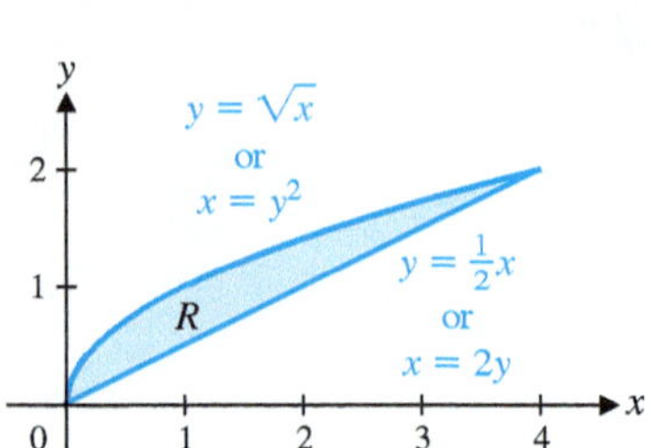

Using the first representation (a regular x region), we obtain

$$\begin{aligned}
\iint_R 4xy^3\,dA &= \int_0^4 \left[\int_{x/2}^{\sqrt{x}} 4xy^3\,dy\right] dx \\
&= \int_0^4 \left[xy^4 \Big|_{y=x/2}^{y=\sqrt{x}}\right] dx \\
&= \int_0^4 [x(\sqrt{x})^4 - x(\tfrac{1}{2}x)^4]\,dx \\
&= \int_0^4 \left(x^3 - \tfrac{1}{16}x^5\right) dx \\
&= \left(\tfrac{1}{4}x^4 - \tfrac{1}{96}x^6\right)\Big|_{x=0}^{x=4} \\
&= \left(64 - \tfrac{128}{3}\right) - 0 = \tfrac{64}{3}
\end{aligned}$$

Using the second representation (a regular y region), we obtain

$$\begin{aligned}
\iint_R 4xy^3\,dA &= \int_0^2 \left[\int_{y^2}^{2y} 4xy^3\,dx\right] dy \\
&= \int_0^2 \left[2x^2y^3 \Big|_{x=y^2}^{x=2y}\right] dy \\
&= \int_0^2 [2(2y)^2y^3 - 2(y^2)^2y^3]\,dy \\
&= \int_0^2 (8y^5 - 2y^7)\,dy \\
&= \left(\tfrac{4}{3}y^6 - \tfrac{1}{4}y^8\right)\Big|_{y=0}^{y=2} \\
&= \left(\tfrac{256}{3} - 64\right) - 0 = \tfrac{64}{3}
\end{aligned}$$

Matched Problem 5 The region R is bounded by the graphs of $y = x$ and $y = \frac{1}{2}x^2$. Evaluate $\iint_R 4xy^3\,dA$ two different ways.

Reversing the Order of Integration

Example 5 shows that

$$\iint_R 4xy^3\,dA = \int_0^4 \left[\int_{x/2}^{\sqrt{x}} 4xy^3\,dy \right] dx = \int_0^2 \left[\int_{y^2}^{2y} 4xy^3\,dx \right] dy$$

In general, if R is both a regular x region and a regular y region, then the two iterated integrals are equal. In rectangular regions, reversing the order of integration in an iterated integral was a simple matter. As Example 5 illustrates, the process is more complicated in nonrectangular regions. The next example illustrates how to start with an iterated integral and reverse the order of integration. Since we are interested in the reversal process and not in the value of either integral, the integrand will not be specified.

EXAMPLE 6 **Reversing the Order of Integration** Reverse the order of integration in

$$\int_1^3 \left[\int_0^{x-1} f(x, y)\,dy \right] dx$$

SOLUTION The order of integration indicates that the region of integration is a regular x region:

$$R = \{(x, y) | 0 \le y \le x - 1,\ \ 1 \le x \le 3\}$$

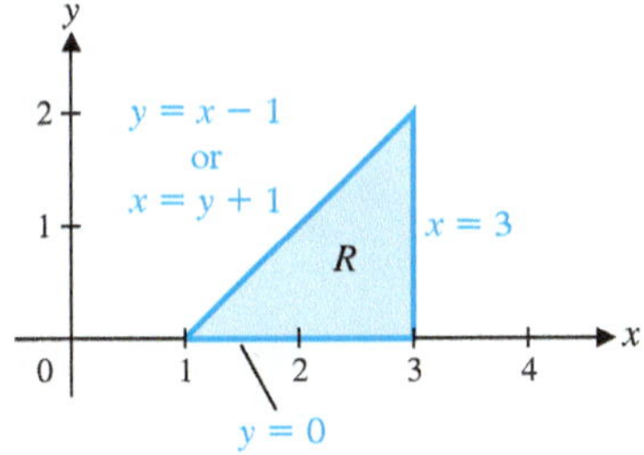

Graph region R to determine whether it is also a regular y region. The graph shows that R is also a regular y region, and we can write

$$R = \{(x, y) | y + 1 \le x \le 3,\ \ 0 \le y \le 2\}$$

$$\int_1^3 \left[\int_0^{x-1} f(x, y)\,dy \right] dx = \int_0^2 \left[\int_{y+1}^3 f(x, y)\,dx \right] dy$$

Matched Problem 6 Reverse the order of integration in $\int_2^4 [\int_0^{4-x} f(x, y)\,dy]\,dx$.

EXPLORE & DISCUSS 2 Explain the difficulty in evaluating $\int_0^2 \int_{x^2}^4 xe^{y^2}dy\,dx$ and how it can be overcome by reversing the order of integration.

Volume and Double Integrals

In Section 6-6, we used the double integral to calculate the volume of a solid with a rectangular base. In general, if a solid can be described by the graph of a positive function $f(x, y)$ over a regular region R (not necessarily a rectangle), then the double integral of the function f over the region R still represents the volume of the corresponding solid.

EXAMPLE 7 **Volume** The region R is bounded by the graphs of $x + y = 1$, $y = 0$, and $x = 0$. Find the volume of the solid under the graph of $z = 1 - x - y$ over the region R.

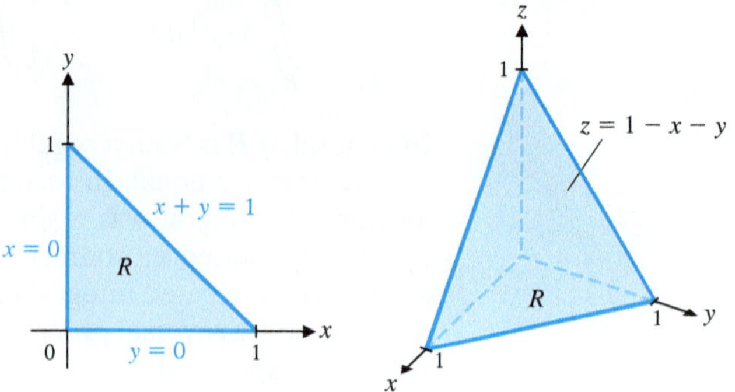

SOLUTION The graph of R shows that R is both a regular x region and a regular y region. We choose to use the regular x region:

$$R = \{(x, y) | 0 \le y \le 1 - x,\quad 0 \le x \le 1\}$$

The volume of the solid is

$$\begin{aligned} V = \iint_R (1 - x - y)\, dA &= \int_0^1 \left[\int_0^{1-x} (1 - x - y)\, dy\right] dx \\ &= \int_0^1 \left[\left(y - xy - \tfrac{1}{2}y^2\right)\Big|_{y=0}^{y=1-x}\right] dx \\ &= \int_0^1 \left[(1 - x) - x(1 - x) - \tfrac{1}{2}(1 - x)^2\right] dx \\ &= \int_0^1 \left(\tfrac{1}{2} - x + \tfrac{1}{2}x^2\right) dx \\ &= \left(\tfrac{1}{2}x - \tfrac{1}{2}x^2 + \tfrac{1}{6}x^3\right)\Big|_{x=0}^{x=1} \\ &= \left(\tfrac{1}{2} - \tfrac{1}{2} + \tfrac{1}{6}\right) - 0 = \tfrac{1}{6} \end{aligned}$$

Matched Problem 7 The region R is bounded by the graphs of $y + 2x = 2$, $y = 0$, and $x = 0$. Find the volume of the solid under the graph of $z = 2 - 2x - y$ over the region R. [*Hint*: Sketch the region first; the solid does not have to be sketched.]

Exercises 6-7

A

In Problems 1–6, graph the region R bounded by the graphs of the equations. Use set notation and double inequalities to describe R as a regular x region and a regular y region in Problems 1 and 2, and as a regular x region or a regular y region, whichever is simpler, in Problems 3–6.

1. $y = 4 - x^2, y = 0, 0 \le x \le 2$

2. $y = x^2, y = 9, 0 \le x \le 3$

3. $y = x^3, y = 12 - 2x, x = 0$

4. $y = 5 - x, y = 1 + x, y = 0$

5. $y^2 = 2x, y = x - 4$

6. $y = 4 + 3x - x^2, x + y = 4$

Evaluate each integral in Problems 7–10.

7. $\int_0^1 \int_0^x (x + y)\, dy\, dx$

8. $\int_0^2 \int_0^y xy\, dx\, dy$

9. $\int_0^1 \int_{y^3}^{\sqrt{y}} (2x + y)\, dx\, dy$

10. $\int_1^4 \int_x^{x^2} (x^2 + 2y)\, dy\, dx$

B

In Problems 11–14, give a verbal description of the region R and determine whether R is a regular x region, a regular y region, both, or neither.

11. $R = \{(x, y) \mid |x| \le 2, \;\; |y| \le 3\}$

12. $R = \{(x, y) \mid 1 \le x^2 + y^2 \le 4\}$

13. $R = \{(x, y) \mid x^2 + y^2 \ge 1, \;\; |x| \le 2, \;\; 0 \le y \le 2\}$

14. $R = \{(x, y) \mid |x| + |y| \le 1\}$

In Problems 15–20, use the description of the region R to evaluate the indicated integral.

15. $\iint_R (x^2 + y^2)\, dA;$
$R = \{(x, y) \mid 0 \le y \le 2x, \;\; 0 \le x \le 2\}$

16. $\iint_R 2x^2 y\, dA;$
$R = \{(x, y) \mid 0 \le y \le 9 - x^2, \;\; -3 \le x \le 3\}$

17. $\iint_R (x + y - 2)^3\, dA;$
$R = \{(x, y) \mid 0 \le x \le y + 2, \;\; 0 \le y \le 1\}$

18. $\iint_R (2x + 3y)\, dA;$
$R = \{(x, y) \mid y^2 - 4 \le x \le 4 - 2y, \;\; 0 \le y \le 2\}$

19. $\iint_R e^{x+y}\, dA;$
$R = \{(x, y) \mid -x \le y \le x, \;\; 0 \le x \le 2\}$

20. $\iint_R \dfrac{x}{\sqrt{x^2 + y^2}}\, dA;$
$R = \{(x, y) \mid 0 \le x \le \sqrt{4y - y^2}, \;\; 0 \le y \le 2\}$

In Problems 21–26, graph the region R bounded by the graphs of the indicated equations. Describe R in set notation with double inequalities, and evaluate the indicated integral.

21. $y = x + 1, y = 0, x = 0, x = 1;$
$\iint_R \sqrt{1 + x + y}\, dA$

22. $y = x^2, \;\; y = \sqrt{x}; \;\; \iint_R 12xy\, dA$

23. $y = 4x - x^2, \;\; y = 0; \;\; \iint_R \sqrt{y + x^2}\, dA$

24. $x = 1 + 3y, \;\; x = 1 - y, \;\; y = 1;$
$\iint_R (x + y + 1)^3\, dA$

25. $y = 1 - \sqrt{x}, \;\; y = 1 + \sqrt{x}, \;\; x = 4;$
$\iint_R x(y - 1)^2\, dA$

26. $y = \frac{1}{2}x, y = 6 - x, \;\; y = 1; \;\; \iint_R \dfrac{1}{x + y}\, dA$

In Problems 27–32, evaluate each integral. Graph the region of integration, reverse the order of integration, and then evaluate the integral with the order reversed.

27. $\int_0^3 \int_0^{3-x} (x + 2y)\, dy\, dx$

28. $\int_0^2 \int_0^y (y - x)^4\, dx\, dy$

29. $\int_0^1 \int_0^{1-x^2} x\sqrt{y}\, dy\, dx$

30. $\int_0^2 \int_{x^3}^{4x} (1 + 2y)\, dy\, dx$

31. $\int_0^4 \int_{x/4}^{\sqrt{x}/2} x\, dy\, dx$

32. $\int_0^4 \int_{y^2/4}^{2\sqrt{y}} (1 + 2xy)\, dx\, dy$

In Problems 33–36, find the volume of the solid under the graph of f (x, y) over the region R bounded by the graphs of the indicated equations. Sketch the region R; the solid does not have to be sketched.

33. $f(x, y) = 4 - x - y$; R is the region bounded by the graphs of $x + y = 4$, $y = 0$, $x = 0$

34. $f(x, y) = (x - y)^2$; R is the region bounded by the graphs of $y = x$, $y = 2$, $x = 0$

35. $f(x, y) = 4$; R is the region bounded by the graphs of $y = 1 - x^2$ and $y = 0$ for $0 \le x \le 1$

36. $f(x, y) = 4xy$; R is the region bounded by the graphs of $y = \sqrt{1 - x^2}$ and $y = 0$ for $0 \le x \le 1$

C

In Problems 37–40, reverse the order of integration for each integral. Evaluate the integral with the order reversed. Do not attempt to evaluate the integral in the original form.

37. $\displaystyle\int_0^2 \int_{x^2}^4 \frac{4x}{1 + y^2}\, dy\, dx$

38. $\displaystyle\int_0^1 \int_y^1 \sqrt{1 - x^2}\, dx\, dy$

39. $\displaystyle\int_0^1 \int_{y^2}^1 4ye^{x^2}\, dx\, dy$

40. $\displaystyle\int_0^4 \int_{\sqrt{x}}^2 \sqrt{3x + y^2}\, dy\, dx$

In Problems 41–46, use a graphing calculator to graph the region R bounded by the graphs of the indicated equations. Use approximation techniques to find intersection points correct to two decimal places. Describe R in set notation with double inequalities, and evaluate the indicated integral correct to two decimal places.

41. $y = 1 + \sqrt{x}, \quad y = x^2, \;\; x = 0; \quad \iint_R x\, dA$

42. $y = 1 + \sqrt[3]{x}, y = x, x = 0; \;\; \iint_R x\, dA$

43. $y = \sqrt[3]{x}, y = 1 - x, y = 0; \;\; \iint_R 24xy\, dA$

44. $y = x^3, y = 1 - x, y = 0; \;\; \iint_R 48xy\, dA$

45. $y = e^{-x}, \quad y = 3 - x; \quad \iint_R 4y\, dA$

46. $y = e^{x}, \quad y = 2 + x; \quad \iint_R 8y\, dA$

Answers to Matched Problems

1. $R = \{(x, y) | y^2 \le x \le 6 - y, \quad 0 \le y \le 2\}$

2. R is both a regular x region and a regular y region:

$R = \{(x, y) \mid \frac{1}{2}x + 2 \le y \le \sqrt{x + 4}, \quad -4 \le x \le 0\}$

$R = \{(x, y) | y^2 - 4 \le x \le 2y - 4, \quad 0 \le y \le 2\}$

3. $\frac{13}{40}$

4. $-\frac{77}{20}$

5. $\frac{16}{3}$

6. $\int_0^2 \int_2^{4-y} f(x, y)\, dx\, dy$

7. $\frac{2}{3}$

Chapter 6 Review

Important Terms, Symbols, and Concepts

6-1 Functions of Several Variables

EXAMPLES

- An equation of the form $z = f(x, y)$ describes a **function of two independent variables** if, for each permissible ordered pair (x, y), there is one and only one value of z determined by $f(x, y)$. The variables x and y are **independent variables**, and z is a **dependent variable**. The set of all ordered pairs of permissible values of x and y is the **domain** of the function, and the set of all corresponding values $f(x, y)$ is the **range**. Functions of more than two independent variables are defined similarly.
- The graph of $z = f(x, y)$ consists of all triples (x, y, z) in a **three-dimensional coordinate system** that satisfy the equation. The graphs of the functions $z = f(x, y) = x^2 + y^2$ and $z = g(x, y) = x^2 - y^2$, for example, are **surfaces**; the first has a local minimum, and the second has a **saddle point**, at $(0, 0)$.

Ex. 1, p. 326
Ex. 2, p. 327
Ex. 3, p. 327
Ex. 4, p. 328
Ex. 5, p. 329
Ex. 6, p. 329
Ex. 7, p. 331

6-2 Partial Derivatives

- If $z = f(x, y)$, then the **partial derivative of f with respect to x**, denoted as $\partial z/\partial x$, f_x, or $f_x(x, y)$, is

$$\frac{\partial z}{\partial x} = \lim_{h \to 0} \frac{f(x + h, y) - f(x, y)}{h}$$

Similarly, the **partial derivative of f with respect to y**, denoted as $\partial z/\partial y$, f_y, or $f_y(x, y)$, is

$$\frac{\partial z}{\partial y} = \lim_{k \to 0} \frac{f(x, y + k) - f(x, y)}{k}$$

The partial derivatives $\partial z/\partial x$ and $\partial z/\partial y$ are said to be **first-order partial derivatives.**

Ex. 1, p. 335
Ex. 2, p. 336
Ex. 3, p. 336
Ex. 4, p. 337

- There are four **second-order partial derivatives** of $z = f(x, y)$:

Ex. 5, p. 339

$$f_{xx} = f_{xx}(x, y) = \frac{\partial^2 z}{\partial x^2} = \frac{\partial}{\partial x}\left(\frac{\partial z}{\partial x}\right)$$

$$f_{xy} = f_{xy}(x, y) = \frac{\partial^2 z}{\partial y\, \partial x} = \frac{\partial}{\partial y}\left(\frac{\partial z}{\partial x}\right)$$

$$f_{yx} = f_{yx}(x, y) = \frac{\partial^2 z}{\partial x\, \partial y} = \frac{\partial}{\partial x}\left(\frac{\partial z}{\partial y}\right)$$

$$f_{yy} = f_{yy}(x, y) = \frac{\partial^2 z}{\partial y^2} = \frac{\partial}{\partial y}\left(\frac{\partial z}{\partial y}\right)$$

6-3 Maxima and Minima

- If $f(a, b) \ge f(x, y)$ for all (x, y) in a circular region in the domain of f with (a, b) as center, then $f(a, b)$ is a **local maximum**. If $f(a, b) \le f(x, y)$ for all (x, y) in such a region, then $f(a, b)$ is a **local minimum**.
- If a function $f(x, y)$ has a local maximum or minimum at the point (a, b), and f_x and f_y exist at (a, b), then both first-order partial derivatives equal 0 at (a, b) [Theorem 1, p. 344].
- The second-derivative test for local extrema (Theorem 2, p. 345) gives conditions on the first- and second-order partial derivatives of $f(x, y)$, which guarantee that $f(a, b)$ is a local maximum, local minimum, or saddle point.

Ex. 1, p. 346
Ex. 2, p. 346
Ex. 3, p. 348
Ex. 4, p. 348

6-4 Maxima and Minima Using Lagrange Multipliers

- The **method of Lagrange multipliers** can be used to find local extrema of a function $z = f(x, y)$ subject to the constraint $g(x, y) = 0$. A procedure that lists the key steps in the method is given on page 353.
- The method of Lagrange multipliers can be extended to functions with arbitrarily many independent variables with one or more constraints (see Theorem 1, p. 352, and Theorem 2, p. 357, for the method when there are two and three independent variables, respectively).

Ex. 1, p. 354
Ex. 2, p. 355
Ex. 3, p. 357

6-5 Method of Least Squares

- **Linear regression** is the process of fitting a line $y = ax + b$ to a set of data points $(x_1, y_1), (x_2, y_2), \ldots, (x_n, y_n)$ by using the **method of least squares**.

Ex. 1, p. 366

- We minimize $F(a, b) = \sum_{k=1}^{n} (y_k - ax_k - b)^2$, the **sum of the squares of the residuals**, by computing the first-order partial derivatives of F and setting them equal to 0. Solving for a and b gives the formulas

$$a = \frac{n\left(\sum_{k=1}^{n} x_k y_k\right) - \left(\sum_{k=1}^{n} x_k\right)\left(\sum_{k=1}^{n} y_k\right)}{n\left(\sum_{k=1}^{n} x_k^2\right) - \left(\sum_{k=1}^{n} x_k\right)^2}$$

$$b = \frac{\sum_{k=1}^{n} y_k - a\left(\sum_{k=1}^{n} x_k\right)}{n}$$

- Graphing calculators have built-in routines to calculate linear—as well as quadratic, cubic, quartic, logarithmic, exponential, power, and trigonometric—regression equations.

Ex. 2, p. 368

6-6 Double Integrals over Rectangular Regions

- The **double integral** of a function $f(x, y)$ over a rectangle

$$R = \{(x, y) | a \le x \le b, \quad c \le y \le d\}$$

is

$$\iint_R f(x, y)\, dA = \int_a^b \left[\int_c^d f(x, y)\, dy\right] dx$$

$$= \int_c^d \left[\int_a^b f(x, y)\, dx\right] dy$$

Ex. 1, p. 372
Ex. 2, p. 373
Ex. 3, p. 373
Ex. 4, p. 375
Ex. 5, p. 375

- In the double integral $\iint_R f(x, y)\, dA$, $f(x, y)$ is called the **integrand** and R is called the **region of integration.** The expression dA indicates that this is an integral over a two-dimensional region. The integrals

$$\int_a^b \left[\int_c^d f(x, y)\, dy\right] dx \quad \text{and} \quad \int_c^d \left[\int_a^b f(x, y)\, dx\right] dy$$

are referred to as **iterated integrals** (the brackets are often omitted), and the order in which dx and dy are written indicates the order of integration.

- The **average value** of the function $f(x, y)$ over the rectangle

Ex. 6, p. 376

$$R = \{(x, y) | a \le x \le b, \quad c \le y \le d\}$$

is

$$\frac{1}{(b - a)(d - c)} \iint_R f(x, y)\, dA$$

- If $f(x, y) \ge 0$ over a rectangle $R = \{(x, y) | a \le x \le b, c \le y \le d\}$, then the volume of the solid formed by graphing f over the rectangle R is given by

$$V = \iint_R f(x, y)\, dA$$

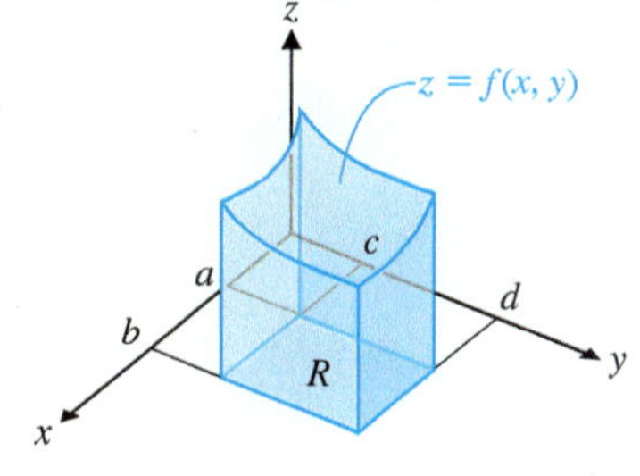

Ex. 7, p. 377

6-7 Double Integrals over More General Regions

- A region R in the xy plane is a **regular x region** if there exist functions $f(x)$ and $g(x)$ and numbers a and b such that Ex. 1, p. 382

$$R = \{(x, y) | g(x) \le y \le f(x), \quad a \le x \le b\}$$

- A region R in the xy plane is a **regular y region** if there exist functions $h(y)$ and $k(y)$ and numbers c and d such that Ex. 2, p. 383

$$R = \{(x, y) | h(y) \le x \le k(y), \quad c \le y \le d\}$$

- The double integral of a function $F(x, y)$ over a regular x region $R = \{(x, y) | g(x) \le y \le f(x), \ a \le x \le b\}$ is Ex. 3, p. 385
Ex. 4, p. 385
Ex. 5, p. 386
Ex. 6, p. 387
Ex. 7, p. 388

$$\iint_R F(x, y)\, dA = \int_a^b \left[\int_{g(x)}^{f(x)} F(x, y)\, dy \right] dx$$

- The double integral of a function $F(x, y)$ over a regular y region $R = \{(x, y) | h(y) \le x \le k(y), \ c \le y \le d\}$ is

$$\iint_R F(x, y)\, dA = \int_c^d \left[\int_{h(y)}^{k(y)} F(x, y)\, dx \right] dy$$

Review Exercises

Work through all the problems in this chapter review and check your answers in the back of the book. Answers to all review problems are there, along with section numbers in italics to indicate where each type of problem is discussed. Where weaknesses show up, review appropriate sections of the text.

A

1. For $f(x, y) = 2{,}000 + 40x + 70y$, find $f(5, 10)$, $f_x(x, y)$, and $f_y(x, y)$.
2. For $z = x^3y^2$, find $\partial^2 z/\partial x^2$ and $\partial^2 z/\partial x\, \partial y$.
3. Evaluate $\int (6xy^2 + 4y)\, dy$.
4. Evaluate $\int (6xy^2 + 4y)\, dx$.
5. Evaluate $\int_0^1 \int_0^1 4xy\, dy\, dx$.
6. For $f(x, y) = 6 + 5x - 2y + 3x^2 + x^3$, find $f_x(x, y)$, and $f_y(x, y)$, and explain why $f(x, y)$ has no local extrema.

B

7. For $f(x, y) = 3x^2 - 2xy + y^2 - 2x + 3y - 7$, find $f(2, 3)$, $f_y(x, y)$, and $f_y(2, 3)$.
8. For $f(x, y) = -4x^2 + 4xy - 3y^2 + 4x + 10y + 81$, find $[f_{xx}(2, 3)][f_{yy}(2, 3)] - [f_{xy}(2, 3)]^2$.
9. If $f(x, y) = x + 3y$ and $g(x, y) = x^2 + y^2 - 10$, find the critical points of $F(x, y, \lambda) = f(x, y) + \lambda g(x, y)$.
10. Use the least squares line for the data in the following table to estimate y when $x = 10$.

x	y
2	12
4	10
6	7
8	3

11. For $R = \{(x, y) | -1 \le x \le 1, \quad 1 \le y \le 2\}$, evaluate the following in two ways:

$$\iint_R (4x + 6y)\, dA$$

12. For $R = \{(x, y) | \sqrt{y} \le x \le 1, \quad 0 \le y \le 1\}$, evaluate

$$\iint_R (6x + y)\, dA$$

C

13. For $f(x, y) = e^{x^2 + 2y}$, find f_x, f_y, and f_{xy}.
14. For $f(x, y) = (x^2 + y^2)^5$, find f_x and f_{xy}.
15. Find all critical points and test for extrema for

$$f(x, y) = x^3 - 12x + y^2 - 6y$$

16. Use Lagrange multipliers to maximize $f(x, y) = xy$ subject to $2x + 3y = 24$.
17. Use Lagrange multipliers to minimize $f(x, y, z) = x^2 + y^2 + z^2$ subject to $2x + y + 2z = 9$.
18. Find the least squares line for the data in the following table.

x	y	x	y
10	50	60	80
20	45	70	85
30	50	80	90
40	55	90	90
50	65	100	110

19. Find the average value of $f(x, y) = x^{2/3}y^{1/3}$ over the rectangle

$$R = \{(x, y) | -8 \le x \le 8, \quad 0 \le y \le 27\}$$

20. Find the volume of the solid under the graph of $z = 3x^2 + 3y^2$ over the rectangle

$$R = \{(x, y) | 0 \le x \le 1, \quad -1 \le y \le 1\}$$

21. Without doing any computation, predict the average value of $f(x, y) = x + y$ over the rectangle $R = \{(x, y) | -10 \le x \le 10, \quad -10 \le y \le 10\}$. Then check the correctness of your prediction by evaluating a double integral.

 **22.** (A) Find the dimensions of the square S centered at the origin such that the average value of

$$f(x, y) = \frac{e^x}{y + 10}$$

over S is equal to 5.

 (B) Is there a square centered at the origin over which

$$f(x, y) = \frac{e^x}{y + 10}$$

has average value 0.05? Explain.

23. Explain why the function $f(x, y) = 4x^3 - 5y^3$, subject to the constraint $3x + 2y = 7$, has no maxima or minima.

24. Find the volume of the solid under the graph of $F(x, y) = 60x^2y$ over the region R bounded by the graph of $x + y = 1$ and the coordinate axes.

Applications

25. **Maximizing profit.** A company produces x units of product A and y units of product B (both in hundreds per month). The monthly profit equation (in thousands of dollars) is given by

$$P(x, y) = -4x^2 + 4xy - 3y^2 + 4x + 10y + 81$$

 (A) Find $P_x(1, 3)$ and interpret the results.

(B) How many of each product should be produced each month to maximize profit? What is the maximum profit?

26. **Minimizing material.** A rectangular box with no top and six compartments (see the figure) is to have a volume of 96 cubic inches. Find the dimensions that will require the least amount of material.

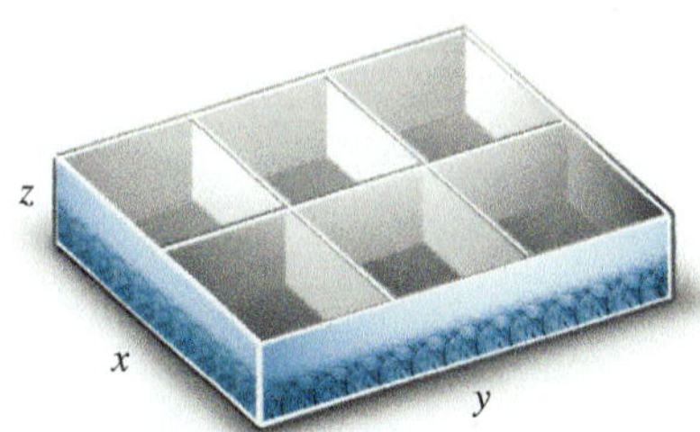

Figure for 26

27. **Profit.** A company's annual profits (in millions of dollars) over a 5-year period are given in the following table. Use the least squares line to estimate the profit for the sixth year.

Year	Profit
1	2
2	2.5
3	3.1
4	4.2
5	4.3

28. **Productivity.** The Cobb–Douglas production function for a product is

$$N(x, y) = 10x^{0.8}y^{0.2}$$

where x is the number of units of labor and y is the number of units of capital required to produce N units of the product.

(A) Find the marginal productivity of labor and the marginal productivity of capital at $x = 40$ and $y = 50$. For the greatest increase in productivity, should management encourage increased use of labor or increased use of capital?

(B) If each unit of labor costs \$100, each unit of capital costs \$50, and \$10,000 is budgeted for production of this product, use the method of Lagrange multipliers to determine the allocations of labor and capital that will maximize the number of units produced and find the maximum production. Find the marginal productivity of money and approximate the increase in production that would result from an increase of \$2,000 in the amount budgeted for production.

(C) If $50 \le x \le 100$ and $20 \le y \le 40$, find the average number of units produced. Set up a double integral, and evaluate it.

29. **Marine biology.** The function used for timing dives with scuba gear is

$$T(V, x) = \frac{33V}{x + 33}$$

where T is the time of the dive in minutes, V is the volume of air (in cubic feet, at sea-level pressure) compressed into tanks, and x is the depth of the dive in feet. Find $T_x(70, 17)$ and interpret the results.

30. **Pollution.** A heavy industrial plant located in the center of a small town emits particulate matter into the atmosphere. Suppose that the concentration of particulate matter (in parts per million) at a point d miles from the plant is given by

$$C = 100 - 24d^2$$

If the boundaries of the town form a square 4 miles long and 4 miles wide, what is the average concentration of particulate matter throughout the town? Express C as a function of x and y, and set up a double integral and evaluate it.

31. **Sociology.** A sociologist found that the number n of long-distance telephone calls between two cities during a given period varied (approximately) jointly as the populations P_1 and P_2 of the two cities and varied inversely as the distance d between the cities. An equation for a period of 1 week is

$$n(P_1, P_2, d) = 0.001\frac{P_1P_2}{d}$$

Find $n(100{,}000, 50{,}000, 100)$.

32. **Education.** At the beginning of the semester, students in a foreign language course take a proficiency exam. The same exam is given at the end of the semester. The results for 5 students are shown in the following table. Use the least squares line to estimate the second exam score of a student who scored 40 on the first exam.

First Exam	Second Exam
30	60
50	75
60	80
70	85
90	90

33. Population density. The following table gives the U.S. population per square mile for the years 1900–2000:

U.S. Population Density

Year	Population (per Square Mile)	Year	Population (per Square Mile)
1900	25.6	1960	50.6
1910	31.0	1970	57.4
1920	35.6	1980	64.0
1930	41.2	1990	70.4
1940	44.2	2000	77.8
1950	50.7		

(A) Find the least squares line for the data, using $x = 0$ for 1900.

(B) Use the least squares line to estimate the population density in the United States in the year 2020.

(C) Now use quadratic regression and exponential regression to obtain the estimate of part (B).

34. Life expectancy. The following table gives life expectancies for males and females in a sample of Central and South American countries:

Life Expectancies for Central and South American Countries

Males	Females	Males	Females
62.30	67.50	70.15	74.10
68.05	75.05	62.93	66.58
72.40	77.04	68.43	74.88
63.39	67.59	66.68	72.80
55.11	59.43		

(A) Find the least squares line for the data.

(B) Use the least squares line to estimate the life expectancy of a female in a Central or South American country in which the life expectancy for males is 60 years.

(C) Now use quadratic regression and logarithmic regression to obtain the estimate of part (B).

Differential Equations

We have considered differential equations of the form

$$\frac{dy}{dt} = ky, \qquad \frac{dy}{dt} = k(M - y) \qquad \text{and} \qquad \frac{dy}{dt} = ky(M - y)$$

Each of these equations establishes a relationship between a quantity y and its rate of growth dy/dt. Their solutions provide models for several types of exponential growth laws. Many different relationships in business, economics and the sciences can be stated in terms of differential equations. Unfortunately, there is no single method that will solve all the differential equations that may be encountered—even in every simple applications. In this chapter, after discussing some basic concepts in the first section, we will consider methods for solving several types of differential equations that have significant applications.

7-1 Basic Concepts

- Solutions of Differential Equations
- Implicit Solutions
- Application

In this section we introduce the concept of a differential equation and discuss what is meant by a solution, including explicit and implicit representations. Actual techniques for finding solutions of differential equations will be discussed in the following sections.

Solutions of Differential Equations

A **differential equation** is an equation involving an unknown function, usually denoted by y, and one or more of its derivatives. For example,

$$y' = 2xy \tag{1}$$

is a differential equation. Since only the first derivative of the unknown function y appears in this equation, it is called a **first-order** differential equation. In general, the order of a differential equation is the highest derivative of the unknown function present in the equation.

Now consider the function

$$y = 4e^{x^2}$$

whose derivative is

$$y\leq = 4e^{x^2}(2x) = 8xe^{x^2}$$

Taken from *Applied Calculus for Business, Economics, Life Sciences, and Social Sciences,* Fifth Edition, by Raymond A. Barnett and Michael R. Ziegler.

Substituting for y and y' in equation (1) gives

$$y' = 2xy$$
$$8xe^{x^2} = 2x(4e^{x^2})$$
$$8xe^{x^2} = 8xe^{x^2}$$

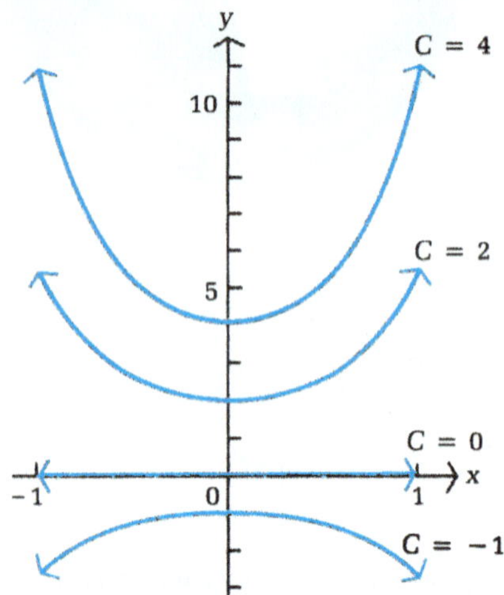

Figure 1 Particular solutions of $y' = 2xy$

which is certainly true for all values of x. This shows that the function $y = 4e^{x^2}$ is a **solution** of equation (1). But this function is not the only solution. In fact, if C is any constant, substituting $y = Ce^{x^2}$ and $y' = 2xCe^{x^2}$ in equation (1) yields the identity

$$y' = 2xy$$
$$2xCe^{x^2} = 2xCe^{x^2}$$

It turns out that all solutions of $y' = 2xy$ can be obtained from $y = Ce^{x^2}$ by assigning C appropriate values: hence, $y = Ce^{x^2}$ is called the **general solution** of equation (1). The collection of all functions of the form $y = Ce^{x^2}$ is called the **family of solutions** of equation (1). The function $y = 4e^{x^2}$, obtained by letting $C = 4$ in the general solution, is called a **particular solution** of equation (1). Other particular solutions are

$$y = -e^{x^2} \qquad C = -1$$
$$y = 0 \qquad C = 0$$
$$y = 2e^{x^2} \qquad C = 2$$

These particular solutions are graphed in Figure 1.

Remark—Recognizing General Solutions

Most, but not all, first-order differential equations have general solutions that involve one arbitrary constant. Whenever we find a function, such as $y = Ce^{x^2}$, that satisfies a given differential equation for any value of the constant C, we will assume this function is the general solution.

EXAMPLE 1 Show that

$$y = Cx^2 + 1$$

is the general solution of the differential equation

$$xy' = 2y - 2$$

On the same set of axes, graph the particular solutions obtained by letting $C = -2, -1, 0, 1,$ and 2.

SOLUTION Substituting $y = Cx^2 + 1$ and $y' = 2Cx$ in the differential equation, we have

$$xy' = 2y - 2$$
$$x(2Cx) = 2(Cx^2 + 1) - 2$$
$$2Cx^2 = 2Cx^2 + 2 - 2$$
$$2Cx^2 = 2Cx^2$$

which shows that $y = Cx^2 + 1$ is the general solution. The particular solutions corresponding to $C = -2, -1, 0, 1,$ and 2 are graphed in Figure 2.

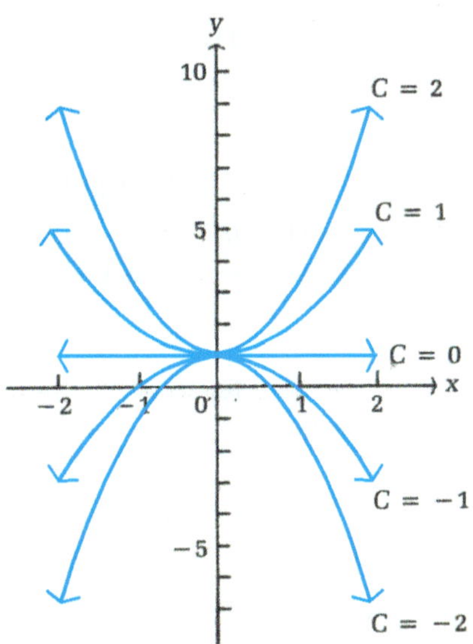

Figure 2 Particular solutions of $xy' = 2y - 2$

Matched Problem 1 Show that

$$y = Cx + 1$$

is the general solution of the differential equation

$$xy' = y - 1$$

On the same set of axes, graph the particular solutions obtained by letting $C = -2, -1, 0, 1$, and 2.

In many applications, we will be interested in finding a particular solution $y(x)$ that satisfies an **initial condition** of the form $y(x_0) = y_0$. The value of the constant C in the general solution must then be selected so that this initial condition is satisfied; that is, so that the solution curve (from the family of solution curves) passes through (x_0, y_0).

EXAMPLE 2 Using the general solution in Example 1, find a particular solution of differential equation

$$xy' = 2y - 2$$

that satisfies the indicated initial condition, if such a solutions exists.

(A) $y(1) = 3$ (B) $y(0) = 3$ (C) $y(0) = 1$

SOLUTION (A) *Initial Condition* $y(1) = 3$. From Example 1, the general solution of the differential equation is

$$y = Cx^2 + 1$$

Substituting $x = 1$ and $y = 3$ in this general solution yields

$$3 = C(1)^2 + 1$$
$$= C + 1$$
$$C = 2$$

Thus, the particular solution satisfying the initial conditions $y(1) = 3$ is

$$y = 2x^2 + 1$$

See the graph labeled C = 2 in Figure 2.

(B) *Initial Condition y(0)* = 3. Substituting x = 0 and y = 3 in the general solutions yields

$$3 = C(0)^2 + 1$$

No matter what value of C we select, $C(0)^2 = 0$ and this equation reduces to

$$3 = 1$$

which is *never* valid. Thus, we must conclude that there is no particular solution of this differential equation that will satisfy the initial condition $y(0) = 3$.

(C) *Initial Conditions y(0)* = *1*. Substituting $x = 0$ and $y = 1$ in the general solution, we have

$$1 = C(0)^2 + 1$$
$$1 = 1$$

This equation is valid for *all* values of C. Thus, all solutions of this differential equation satisfy the initial condition $y(0) = 1$ (see Figure 2).

Remark—Nature of Particular Solutions

The situation illustrated in Example 2A is the most common. Most first-order differential equations you will encounter will have exactly one particular solution that satisfies a given initial condition. However, we must always be aware of the possibility that a differential equation may not have any particular solutions that satisfy a given initial condition, or the possibility that it may have many particular solutions satisfying a given initial condition.

Matched Problem 2 Using the general solution in Problem 1, find a particular solution of the differential equation

$$xy' = y - 1$$

that satisfies the indicated initial condition, if such a solution exists.
(A) $y(1) = 2$ (B) $y(0) = 2$ (C) $y(0) = 1$

Implicit Solutions

In many cases, the solution of a differential equation is defined by an implicit equation rather than by an explicit formula. As with all implicitly defined functions, it may not be possible to derive an explicit formula for an implicitly defined solution.

EXAMPLE 3 If y is defined implicitly by the equation

$$y^3 + e^y - x^4 = C \tag{2}$$

show that y satisfies the differential equation

$$(3y^2 + e^y)y' = 4x^3$$

SOLUTION We use implicit differentiation to show that y satisfies the given differential equation:

$$y^3 + e^y - x^4 = C$$

$$D_x(y^3 + e^y - x^4) = D_xC$$

$$D_xy^3 + D_xe^y - D_xx^4 = 0$$

Remember, $D_xx^n = nx^{n-1}$ and $D_xe^x = e^x$, but $D_xy^n = ny^{n-1}y'$ and $D_xe^y = e^yy'$ since y is a function of x.

$$3y^2y' + e^yy' - 4x^3 = 0$$

$$(3y^2 + e^y)y' = 4x^3$$

Since the last equation is the given differential equation, our calculations show that any function y defined implicitly by equation (2) is a solution of this differential equation. We cannot find an explicit formula for y, since none exists in terms of finite combinations of elementary functions.

Matched Problem 3 If y defined implicitly by the equation

$$y + e^{y^2} - x^2 = C$$

show that y satisfies the differential equation

$$(1 + 2ye^{y^2})y' = 2x$$

EXAMPLE 4 If y is defined implicitly by the equation

$$y^2 - x^2 = C$$

show that y satisfies the differential equation

$$yy' = x$$

Find an explicit expression for the particular solution that satisfies the initial conditions $y(0) = 2$.

SOLUTION Using implicit differentiation, we have

$$y^2 - x^2 = C$$

$$D_x(y^2 - x^2) = D_xC$$

$$2yy' - 2x = 0$$

$$2yy' = 2x$$

$$yy' = x$$

which shows that y satisfies the given differential equation. Substituting $x = 0$ and $y = 2$ in $y^2 - x^2 = C$, we have

$$y^2 - x^2 = C$$

$$(2)^2 - (0)^2 = C$$

$$4 = C$$

Thus, the particular solution satisfying $y(0) = 2$ is a solution of the equation

$$y^2 - x^2 = 4$$

or

$$y^2 = 4 + x^2$$

Solving the last equation for y yields two explicit solutions,

$$y_1(x) = \sqrt{4 + x^2} \qquad \text{and} \qquad y_2(x) = -\sqrt{4 + x^2}$$

The first of these two solutions satisfies

$$y_1(0) = \sqrt{4 + (0)^2} = \sqrt{4} = 2$$

while the second satisfies

$$y_2(0) = -\sqrt{4 + (0)^2} = -\sqrt{4} = -2$$

Thus, the particular solution of the differential equation $yy' = x$ that satisfies the initial condition $y(0) = 2$ is

$$y(x) = \sqrt{4 + x^2}$$

CHECK

$$y'(x) = \frac{1}{2}(4 + x^2)^{-1/2}2x \qquad\qquad yy' = x$$

$$= \frac{x}{\sqrt{4 + x^2}} \qquad\qquad \sqrt{4 + x^2}\left(\frac{x}{\sqrt{4 + x^2}}\right) \stackrel{?}{=} x$$

$$x \stackrel{\checkmark}{=} x$$

Matched Problem 4 If y is defined implicitly by the equation

$$y^2 - x = C$$

show that y satisfies that differential equation

$$2yy' = 1$$

Find an explicit expression for the particular solution that satisfies the initial conditions $y(0) = (3)$.

Application

In economics the price of a product is often studied over a period of time, and so it is natural to view price as a function of time. Let $p(t)$ be the price of a particular product at time t. If $p(t)$ approaches a limiting value $\overline{p}$ as t approaches infinity, then the price for this product is said to be **dynamically stable** and $\overline{p}$ is referred to as the **equilibrium price**. In order to study the behavior of price as a function of time, economists often assume that the price satisfies a differential equation. This approach is illustrated in the next example.

EXAMPLE 5 **Dynamic Price Stability** The price $p(t)$ of a product is assumed to satisfy the differential equation

$$\frac{dp}{dt} = 10 - 0.5p$$

Show that

$$p(t) = 20 - Ce^{-0.5t}$$

is the general solution of this differential equation and evaluate

$$\overline{p} = \lim_{t \to \infty} p(t)$$

On the same set of axes, graph the three particular solutions that satisfy the initial conditions $p(0) = 40, p(0) = 10$, and $p(0) = 20$.

SOLUTION

$$p(t) = 20 - Ce^{-0.5t}$$
$$p'(t) = 0.5Ce^{-0.5t}$$

Substituting in the given differential equation, we have

$$\frac{dp}{dt} = 10 - 0.5p$$

$$0.5Ce^{0.5t} = 10 = 0.5(20 - Ce^{-0.5t}$$

$$= 10 - 10 + 0.5Ce^{-0.5t}$$

$$= 0.5Ce^{-0.5t}$$

which shows that $p(t) = 20 - Ce^{-0.5t}$ is the general solution of this differential equation. To find the equilibrium price, we must evaluate $\lim_{t\to\infty} p(t)$.

$$\begin{aligned} \overline{p} &= \lim_{t\to\infty} p(t) \\ &= \lim_{t\to\infty} (20 - Ce^{0.5t}) \\ &= 20 - C \lim_{t\to\infty} 2e^{-0.5t} \qquad \lim_{t\to\infty} e^{-0.5t} = 0 \\ &= 20 - C \cdot 0 \\ &= 20 \end{aligned}$$

Before we can graph the three particular solutions, we must evaluate the constant C for each of the indicated initial conditions. In each case, we will make use of the equation

$$p(0) = 20 - Ce^0 = 20 - C$$

$p(0) = 40$	$p(0) = 10$	$p(0) = 20$
$20 - C = 40$	$20 - C = 10$	$20 - C = 20$
$C = -20$	$C = 10$	$C = 0$
$p(t) = 20 + 20e^{-0.5t}$	$p(t) = 20 - 10e^{-0.5t}$	$p(t) = 20$

Notice that the equilibrium price $\overline{p}$ does not depend on the constant C and consequently does not depend on the initial value of the price function $p(0)$ (see Figure 3). If $p(0) > \overline{p}$, then the price decreases and approaches $\overline{p}$ as a limit. If $p(0) < \overline{p}$, then the price increases and approaches $\overline{p}$ as a limit. If $p(0) < \overline{p}$, then the price remains constant for all t.

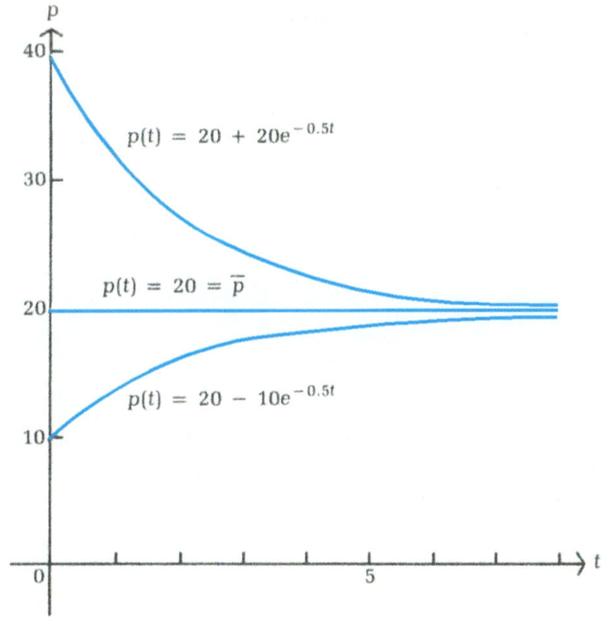

Figure 3 Particular solutions of $\frac{dp}{dt} = 10 - 0.5p$

Matched Problem 5 Show that the price function

$$p(t) = 25 - Ce^{-0.2t}$$

is the general solution of the differential equation

$$\frac{dp}{dt} = 5 - 0.2p$$

Find the equilibrium price $\overline{p}$. On the same set of axes, graph the three particular solutions that satisfy the initial conditions $p(0) = 40$, $p(0) = 5$, and $p(0) = 25$.

Answers to Matched Problems

1.

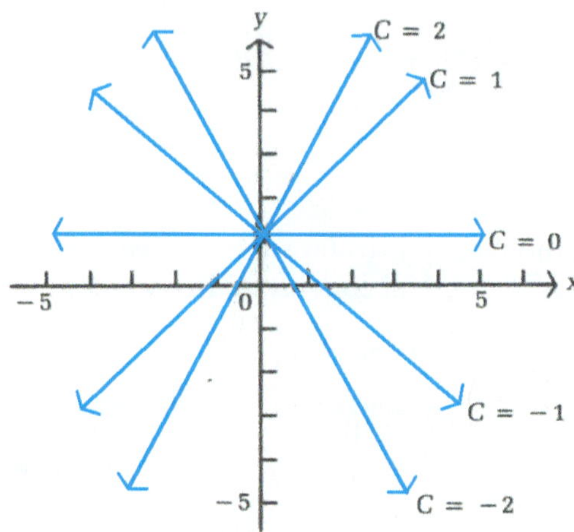

2. (A) $y = x + 1$
(B) No particular solution exists
(C) $y = Cx + 1$ for any C

3. $D_x y + D_x e^{y^2} - D_x x^2 = D_x C$
$y' + e^{y^2} 2yy' - 2x = 0$
$(1 + 2ye^{y^2})y' = 2x$

4. $y(x) = \sqrt{9 + x}$

5. $\overline{p} = 25$

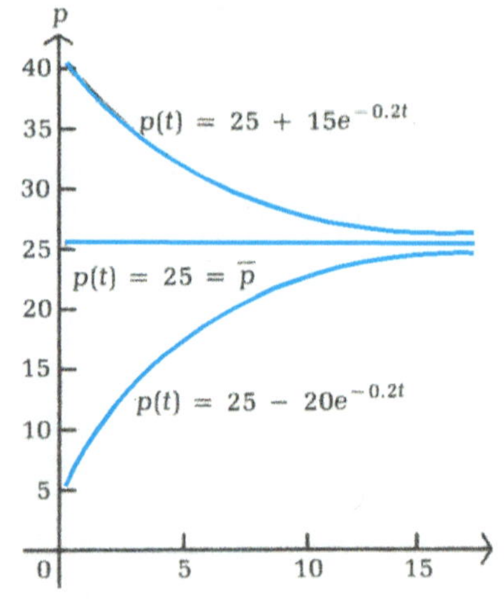

Exercises 7-1

A

Show that the given function y is the general solution of the indicated differential equation. On the same set of axes, graph the particular solutions obtained by letting C = −2, −1, 0, 1, and 2.

1. $y = Cx^2$; $xy' = 2y$
2. $y = C(x - 1)^2$; $(x - 1)y' = 2y$
3. $y = \frac{C}{x}$; $xy' = -y$
4. $y = C(1 + x^2)$; $(1 + x^2)y' = 2xy$

B

Show that the given function y is the general solution of the indicated differential equation. Find a particular solution satisfying the given initial condition.

5. $y = Ce^x - 5x - 5$; $y' = y + 5x$; $y(0) = 2$
6. $y = Ce^{-x} + 2x - 2$; $y' = 2x - y$; $y(0) = 3$
7. $y = e^x + Ce^{2x}$; $y' = 2y - e^x$; $y(0) = -1$
8. $y = e^x + Ce^{-x}$; $y' = 2e^x - y$; $y(0) = 1$

9. $y = x + \frac{C}{x}$; $xy' = 2x - y$; $y(2) = 3$

10. $y = Cx + \frac{2}{x}$; $x^2y' = xy - 4$; $y(2) = -3$

If y is defined implicitly by the given equation, use implicit differentiation to show that y satisfies the indicated differential equation.

11. $y^3 + xy - x^3 = C$; $(3y^2 + x)y' = 3x^2 - y$

12. $y + x^3y^3 - 2x = C$; $(1 + 3x^3y^2)y' = 2 - 3x^2y^3$

13. $xy + e^{y^2} - x^2 = C$; $(x + 2ye^{y^2})y' = 2x - y$

14. $y + e^{xy} - x = C$; $(1 + xe^{xy})y' = 1 - ye^{xy}$

If y is defined implicitly by the given equation, use implicit differentiation to show that y satisfies the indicated differential equation. Find an explicit expression for the particular solution that satisfies the given initial conditions.

15. $y^2 + x^2 = C$; $yy' = -x$; $y(0) = 3$

16. $y^2 + 2x^2 = C$; $yy' = -2x$; $y(0) = -4$

17. $\ln(2 - y) = x + C$; $y' = y - 2$; $y(0) = 1$

18. $\ln(5 - y) = 2x + C$; $y' = 2(y - 5)$; $y(0) = 2$

Use the general solution y of each differential equation to find a particular solution that satisfies the indicated initial condition. Graph the particular solution for $x \leqslant 0$.

19. $y = 2 + Ce^{-x}$; $y' = 2 - y$

(A) $y(0) = 1$ (B) $y(0) = 2$ (C) $y(0) = 3$

20. $y = 4 + Ce^{-x}$; $y' = 4 - y$

(A) $y(0) = 2$ (B) $y(0) = 4$ (C) $y(0) = 5$

21. $y = 2 + Ce^{x}$; $y' = y - 2$

(A) $y(0) = 1$ (B) $y(0) = 2$ (C) $y(0) = 3$

22. $y = 4 + Ce^{x}$; $y' = y - 4$

(A) $y(0) = 2$ (B) $y(0) = 4$ (C) $y(0) = 5$

C

Use the general solution y of each differential equation to find a particular solution that satisfies the indicated initial condition. Graph the particular solutions for $x \leqslant 0$.

23. $y = \frac{10}{1 + Ce^{-x}}$; $y' = 0.1y(10 - y)$

(A) $y(0) = 1$ (B) $y(0) = 10$ (C) $y(0) = 20$

[**Hint:** The particular solution in part A has an inflection point at $x = \ln 9$.]

24. $y = \frac{5}{1 + Ce^{-x}}$; $y' = 0.2y(5 - y)$

(A) $y(0) = 1$ (B) $y(0) = 5$ (C) $y(0) = 10$

[**Hint:** The particular solution in part A has an inflection point at $x = \ln 4$.]

Use the general solution y of each differential equation to find particular solution that satisfies the indicated initial condition.

25. $y = Cx^3 + 2$; $xy' = 3y - 6$

(A) $y(0) = 2$ (B) $y(0) = 0$ (C) $y(1) = 1$

26. $y\ Cx^4 + 1$; $xy' = 4y - 4$

(A) $y(0) = 1$ (B) $y(0) = 0$ (C) $y(1) = 2$

Problems 27 and 28 require the use of a graphic calculator or a computer. In each problem, use [−5, 5] for both the x range and the y range.

27. Given that $y = x + Ce^{-x}$ is the general solution of the differential equation $y' + y = 1 + x$:

(A) In the same viewing rectangle, graph the particular solutions obtained by letting $C = 0, 1, 2$, and 3.

(B) What do the graphs of the solutions for $C > 0$ have in common?

(C) In the same viewing rectangle, graph the particular solutions obtained by letting $C = 0, -1, -2$, and -3.

(D) What do the graphs of the solutions for $C < 0$ have in common?

28. Repeat Problem 27, given that $y = x + (C/x)$ is the general solution of the differential equation $xy' - 2x\quad y$.

Applications

Business Economics

29. Price stability. The price $p(t)$ of a product is assumed to satisfy the differential equation

$$\frac{dp}{dt} = 0.5 - 0.1p$$

Show that

$p(t) = 5 - Ce^{-0.1t}$

is the general solution of this differential equation and evaluate

$$\bar{p} = \lim_{t\to\infty} p(t)$$

Graph the particular solutions that satisfy the initial conditions $p(0) = 1$ and $p(0) = 10$.

30. Price stability. The price $p(t)$ of a product is assumed to satisfy the differential equation

$$\frac{dp}{dt} = 0.8 - 0.2p$$

Show that

$$p(t) = 4 - Ce^{-0.2t}$$

is the general solution of this differential equation and evaluate

$$\bar{p} = \lim_{t\to\infty} p(t)$$

Graph the particular solutions that satisfy the initial conditions $p(0) = 2$ and $p(0) = 8$.

31. Continuous compound interest. If money is deposited at the continuous rate of \$200 per year into an account earning 8% compounded continuously, then the amount A in the account after t years satisfies the differential equation

$$\frac{dA}{dt} = 0.08A + 200$$

Show that

$$A = Ce^{0.08t} - 2{,}500$$

is the general solution of this differential equation. Graph the particular solutions satisfying $A(0) = 0$ and $A(0) = 1{,}000$.

32. Continuous compound interest. If money is withdrawn at the continuous rate of \$500 per year from an account earning 10% compounded continuously, then the amount A in the account after t years satisfies the differential equation

$$\frac{dA}{dt} = 0.1A - 500$$

Show that

$$A = 5{,}000 + Ce^{0.1t}$$

is the general solution of this differential equation. Graph the particular solutions satisfying $A(0) = 4{,}000$ and $A(0) = 6{,}000$.

Life Sciences

33. Population growth—Verhulst growth law. The number $N(t)$ of bacteria in a culture at time t is assumed to satisfy the Verhulst growth law

$$\frac{dN}{dt} = 100 - 0.5N$$

Show that

$$N(t) = 200 - Ce^{-0.5t}$$

is the general solution of this differential equation and evaluate

$$\overline{N} = \lim_{t\to\infty} N(t)$$

where $\overline{N}$ is the equilibrium size of the population. Graph the particular solutions that satisfy $N(0) = 50$ and $N(0) = 300$. (Note the growth can be "negative," that is, populations can decrease in size.)

34. Population growth—logistic growth. The population $N(t)$ of a certain species of animal in a controlled habitat at time t is assumed to satisfy the logistic growth law

$$\frac{dN}{dt} = \frac{1}{500}N(1{,}000 - N)$$

Show that

$$N(t) = \frac{1{,}000}{1 + Ce^{-2t}}$$

Amwest

is the general solution of this differential equation and evaluate

$$\overline{N} = \lim_{t\to\infty} N(t)$$

Graph the particular solution that satisfies the initial condition $N(0) = 200$. [**Hint:** The particular solution has an inflection point at $t = \ln 2$.]

Social Sciences

35. Rumor spread—Gompertz growth law. The rate of propagation of a rumor is assumed to satisfy the Gompertz growth law

$$\frac{dN}{dt} = Ne^{-0.5t}$$

where $N(t)$ is the number of individuals who have heard the rumor at time t. Show that

$$N(t) = Ce^{-2e^{-.05t}}$$

is the general solution of this differential equation. Find the particular solution that satisfies the initial condition $N(0) = 200$, evaluate

$$\overline{N} = \lim_{t\to\infty} N(t)$$

and sketch the graph of this particular solution. [**Hint:** $N(t)$ has an inflection point at $t = \ln 4$.]

7-2 Separation of Variables

- Separation of Variables
- Exponential Growth
- Limited Growth
- Logistic Growth

Separation of Variables

In this section we will develop a technique called *separation of variables,* which can be used to solve differential equations that can be expressed in the form

$$f(y)y' = g(x) \qquad (1)$$

The solution of differential equations by the method of separating the variables is based on the substitution formula for indefinite integrals. If $y = y(x)$ is a differentiable function of x, then

$$\int f(y)\,dy = \int f[y(y)]y'(x)\,dx \qquad \text{Substitution formula for indefinite integrals} \qquad (2)$$

If $y(x)$ is also the solution of (1), and $y(x)$ and $y'(x)$ must satisfy (1). That is,

$$f[y(x)]y'(x) = g(x)$$

Substituting $g(x)$ for $f[y(x)]y'(x)$ in (2), we have

$$\int f(y)\,dy = \int g(x)\,dx$$

Thus, the solution $y(x)$ of (1) is given implicitly by this equation. Of course, for this method to be useful, we must be able to evaluate both of these indefinite integrals.

This discussion is summarized in Theorem 1:

THEOREM 1 Separation of Variables
The solution of the differential equation

$$f(y)y' = g(x)$$

is given implicitly by the equation

$$\int f(y)\,dy = \int g(x)\,dx$$

EXAMPLE 6 Solve: $y' = 2xy^2$

SOLUTION

$$y' = 2xy^2 \qquad \text{Multiply by } 1/y^2 \text{ to separate the variables.}$$

$$\int \frac{1}{y^2}\,dy = \int 2x\,dx \qquad \text{Convert to an equation involving indefinite integrals.}$$

Evaluate each indefinite integral.

$$-\frac{1}{y} + C_1 = x^2 + C_2 \qquad \text{Combine the two constants of integration into a single arbitrary constant.}$$

$$-\frac{1}{y} = x^2 + C \qquad \text{Solve for } y.$$

General solution of $y' = 2xy^2$

$$y = -\frac{1}{x^2 + C}$$

CHECK

$$y' = D_x\left(-\frac{1}{x^2 + C}\right) = \frac{2x}{(x^2 + C)^2}$$

$$y' = 2xy^2$$

Substitute $y' = 2/(x^2 + C)^2$ and $y = -1/(x^2 + C)$.

$$\frac{2x}{(x^2 + C)^2} \stackrel{?}{=} 2x\left(-\frac{1}{x^2 + C}\right)^2$$

$$\frac{2x}{(x^2 + C)^2} \stackrel{\checkmark}{=} \frac{2x}{(x^2 + C)^2}$$

This verifies that our general solution is correct. (You should develop the habit of checking the solution of each differential equation you solve, as we have done here. From now on, we will leave it up to you to check most of the examples worked in the text.)

Matched Problem 6 Solve: $y' = 4x^3y^2$

In some cases the general solution obtained by the technique of separation of variables does not include all the solutions to a differential equation. For example, the function $y = 0$ also satisfies the differential equation in Example 6 (verify this). Yet the solution $y = 0$ cannot be obtained from the expression

$$y = -\frac{1}{x^2 + C}$$

for any choice of the constant C. Solutions of this type are referred to as **singular solutions** and are usually discussed in more advanced courses. We will not attempt to find the singular solutions of any of the differential equations we consider.

EXAMPLE 7 Find the general solution of

$$(1 + x^2)y' = 2x(y - 1)$$

Then find the particular solution that satisfies the initial condition $y(0) = 3$.

SOLUTION First, we find the general solution:

$$(1 + x^2)y' = 2x(y - 1) \quad \text{Multiply by } 1/(1 + x^2) \text{ and } 1/(y - 1) \text{ to separate the variables.}$$

$$\frac{y'}{y - 1} = \frac{2x}{1 + x^2} \quad \text{Convert to an equation involving indefinite integrals.}$$

$$\int \frac{dy}{y - 1} = \int \frac{2x\,dx}{1 + x^2} \quad \text{Evaluate each indefinite integral.}$$

$$\ln|y - 1| = \ln(1 + x^2) + C$$

where C is it an arbitrary constant. Notice that the two constants of integration always can be combined to form a single arbitrary constant. Also note that we can use $1 + x^2$ instead of $|1 + x^2|$, since $1 + x^2$ is always positive. In order to solve for y, we convert this last equation to exponential form:

$$|y - 1| = e^{\ln(1+x^2)+C}$$

$$= e^C e^{\ln(1+x^2)} \quad \text{Use the property } e^{\ln r} = r.$$

$$= e^C(1 + x^2)$$

It can be shown that if we replace e^C with an arbitrary constant K, then we can omit the absolute value signs on the left side of the last equation.* The resulting equation can then be used to find the general solution to the original differential equation. Thus,

$$y - 1 = K(1 + x^2)$$

$$y = 1 + K(1 + x^2) \quad \text{General solution}$$

To find the particular solution that satisfies $y(0) = 3$, we substitute $x = 0$ and $y = 3$ in the general solution and solve for K:

$$3 = 1 + K(1 + 0)$$

$$K = 2$$

$$y = 1 + 2(1 + x^2) = 3 + 2x^2 \quad \text{Particular solution}$$

Matched Problem 7 Find the general solution of

$$(2 + x^4)y' = 4x^3(y - 3)$$

Then find the particular solution that satisfies the initial condition $y(0) = 5$.

---

*In any problem involving separation of variables, you may assume that the equations

$$e^{\ln|f(y)|} = e^C g(x) \quad \text{and} \quad f(y) = Kg(x)$$

are equivalent (both C and K are arbitrary constants). Justification of this assumption involves properties of the absolute value and exponential functions.

Exponential Growth

We now return to the study of exponential growth laws. This time we will place more emphasis on determining the relevant growth law for a particular application and on using separation of variables to solve the corresponding differential equation in each problem. We begin with the familiar exponential growth law.

> **Exponential Growth Law**
>
> If the rate of change with respect to time t of a quantity y is proportional to the amount present, then y satisfies the differential equation
>
> $$\frac{dy}{dt} = ky.$$

Exponential growth includes both the case where y is increasing and the case where y is decreasing, or decaying.

EXAMPLE 8 **Product Analysis** A certain brand of mothballs evaporate at a rate proportional to their volume, losing half their volume every 4 weeks. If the volume of each mothball is initially 15 cubic centimeters and a mothball becomes ineffective when its volume reaches 1 cubic centimeter, how long will these mothballs be effective?

SOLUTION The volume of each mothball is decaying at a rate proportional to its volume. If V is the volume of a mothball after t weeks, then

$$\frac{dV}{dt} = kV$$

Since the initial volume is 15 cubic centimeters, we know that $V(0) = 15$. After 4 weeks, the volume will be half the original volume, so $V(4) = 7.5$. Summarizing these requirements, we have the following exponential decay model:

$$\frac{dV}{dt} = kV$$

$$V(0) = 15 \qquad V(4) = 7.5$$

We want to determine the value of t that satisfies the equation $V(t) = 1$. First, we use separation of variables to find the general solution of the differential equation:

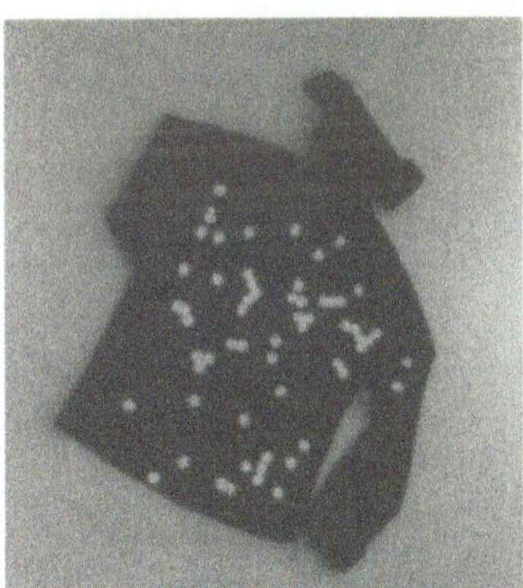

$$\frac{dV}{dt} = kV$$

$$\frac{1}{V}\frac{dV}{dt} = k$$

$$\int \frac{dV}{V} = \int k\,dt$$

$$\ln V = kt + C \qquad \text{We can write } \ln V \text{ in place of } \ln|V|, \text{ since } V > 0.$$

$$V = e^{kt+c} = c^{C}e^{kt} = Ae^{kt} \qquad \text{General solution}$$

where $A = e^{C}$ is a positive constant. Now we use the initial condition to determine the value of the constant A:

$$V(0) = Ae^{0} = A = 15$$

$$V(t) = 15e^{kt}$$

Next, we apply the condition $V(4) = 7.5$ to determine the constant k:

$$V\,(4) = 15e^{4k} = 7.5$$

$$e^{4k} = \frac{7.5}{15} = 0.5$$

$$4k = \ln 0.5$$

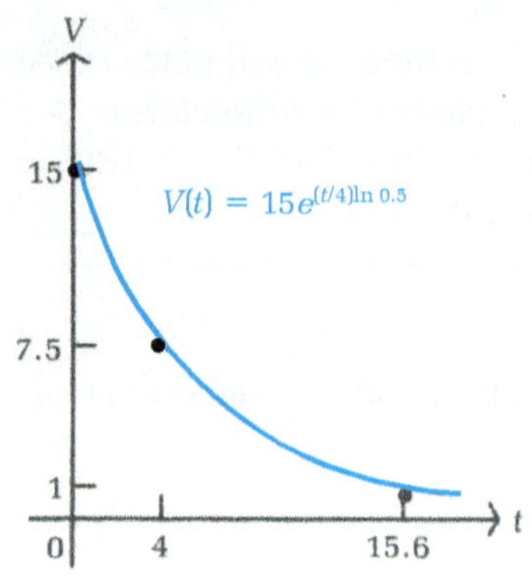

Exponential decay

$$k = \frac{\ln 0.5}{4} \approx -0.1733$$

$$V(t) = 15e^{(t/4)\ln 0.5} \qquad \text{Particular solution}$$

The graph of $V(t)$ is shown in the figure in the margin. To determine how long the mothballs will be effective, we find t when $V = 1$:

$$\begin{aligned} V(t) &= 1 \\ 15e^{(t/4)\ln 0.5} &= 1 \\ e^{(t/4)\ln 0.5} &= \frac{1}{15} \\ \frac{t}{4}\ln 0.5 &= \ln\frac{1}{15} \\ t &= \frac{4\ln\frac{1}{15}}{\ln 0.5} \; / \; 15.6 \text{ weeks} \end{aligned}$$

Matched Problem 8 Repeat Example 8 if the mothballs lose half their volume every 5 weeks.

Limited Growth

In certain situations there is an upper limit (or a lower limit), say M, on the values a variable can assume. This limiting value leads to the limited growth law.

Limited Growth Law

If the rate of change with respect to time t of a quantity y is proportional to the difference between y and a limiting value M, then y satisfies the differential equation

$$\frac{dy}{dt} = k(M - y)$$

When we speak of limited growth, we will include both the case where y increases and approaches M from below and the case where y decreases and approaches M from above.

EXAMPLE 9 **Sales Growth** The annual sales of a new company are expected to grow at a rate proportional to the difference between the sales and an upper limit of $20 million. The sales are 0 initially and $4 million for the second year of operation.

(A) What should the company expect the sales to be during the tenth year?

(B) In what year should the sales be expected to reach $15 million?

SOLUTION If S is the annual sales in millions of dollars during year t, then the model for this problem is

$$\frac{dS}{dt} = k(20 - S)$$

$$S(0) = 0 \qquad S(2) = 4 \tag{3}$$

This is a limited growth model. For part A we want to find $S(10)$, and for part B we want to solve $S(t) = 15$ for t. First, separating the variables in (3) and integrating both sides, we obtain

$$\int \frac{dS}{20 - S} = \int k\,dt$$

$$-\ln(20 - S) = kt + C$$ We can write $-\ln(20 - S)$ in place of $\ln|20 - S|$, since $0 < S < 20$.

$$\ln(20 - S) = -kt - C$$

$$20 - S = e^{-kt-C} = e^{-C}e^{-kt} = Ae^{-kt}$$ $A = e^{-C}$ General solution

$$S = 20 - Ae^{-kt}$$

Now we use the conditions $S(0) = 0$ and $S(2) = 4$ to determine the constants A and k:

$$S(0) = 20 - Ae^0$$
$$= 20 - A = 0$$
$$A = 20$$
$$S(t) = 20 - 20e^{-kt}$$
$$S(2) = 20 - 20e^{-2k} = 4$$
$$20e^{-2k} = 16$$
$$e^{-2k} = \frac{16}{20} = 0.8$$
$$-2k = \ln 0.8$$
$$k = -\frac{\ln 0.8}{2} \approx 0.1116$$
$$S(t) = 20 - 20e^{(t/2)\ln 0.8}$$ Particular solution

The graph of $S(t)$ is shown in the figure in the margin. Notice that the upper limit of \$20 million is a horizontal asymptote.

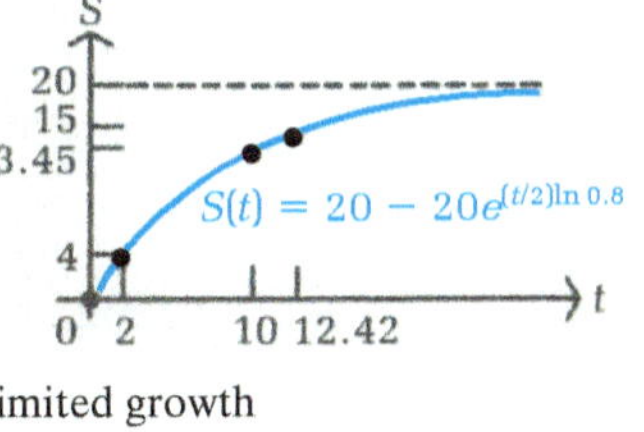

Limited growth

(A) $S(10) = 20 - 20e^{5 \ln 0.8} \approx \$13.45$ million

(B) $S(t) = 20 - 20^{e(t/2) \ln 0.8} = 15$

$$20e^{(t/2) \ln 0.8} = 5$$
$$e^{(t/2) \ln 0.8} = \frac{5}{20} = 0.25$$
$$\frac{t}{2} \ln 0.8 = \ln 0.25$$
$$t = \frac{2 \ln 0.25}{\ln 0.8} \approx 12.43 \text{ years}$$

The annual sales will exceed \$15 million in the thirteenth year.

Matched Problem 9 Repeat Example 9 if the sales during the second year are \$3 million.

Logistic Growth

If a quantity first begins to grow exponentially, but then starts to approach a limiting value, it is said to exhibit logistic growth. More formally, we have:

Logistic Growth Law

If the rate of change with respect to time t of a quantity y is proportional to both the amount present and the difference between y and a limiting value M, then y satisfies the differential equation.

$$\frac{dy}{dt} = ky(M - y)$$

Theoretically, functions satisfying logistic growth laws can be increasing or decreasing, just as was the case for the exponential and limited growth laws. However, decreasing functions are seldom encountered in actual practice.

EXAMPLE 10 **Population Growth** In a study of ciliate protozoans, it has been shown that the rate of growth of the number of *Paramecium caudatum* in a medium with fixed volume is proportional to the product of the number present and the difference between an upper limit of 375 and the number present. The medium initially contains 25 paramecia. After 1 hour there are 125 paramecia. How many paramecia are present after 2 hours?

SOLUTION If p is the number of paramecia in the medium at time t, then the model for this problem is the following logistic growth model:

$$\frac{dP}{dt} = kP(375 - P)$$

$$P(0) = 25 \qquad P(1) = 125$$

We want to find $P(2)$. First, we separate the variables and convert to an equation involving indefinite integrals:

$$\frac{1}{P(375 - P)}\frac{dP}{dt} = k$$

$$\int \frac{1}{P(375 - P)}\,dP = \int k\,dt$$

The integral on the left side of (4) can be evaluated by using an algebraic identity* to derive the next formula. We will use this formula with $u = P$,

$$\int \frac{1}{u(a + bu)}\,du = \frac{1}{a}\ln\left|\frac{u}{a + bu}\right|$$

$$\int \frac{1}{P(375 - P)}\,dP = \frac{1}{375}\ln\left|\frac{P}{375 - P}\right| \qquad \text{Since } 0 < P < 375 \text{, the absolute value signs can be omitted.}$$

$$= \frac{1}{375}\ln\left(\frac{P}{375 - P}\right)$$

Returning to equation (4), we have

$$\frac{1}{375}\ln\left(\frac{P}{375 - P}\right) = \int k\,dt = kt + D \qquad D \text{ is a constant.}$$

$$\ln\left(\frac{P}{375 - P}\right) = 375kt + 375D$$

$$\frac{P}{375 - P} = e^{375kt + 375D} = e^{Bt}e^{C} \qquad B = 375k,\ C = 375D$$

$$P = 375e^{Bt}e^{C} - Pe^{Bt}e^{C}$$

$$P(e^{Bt}e^{C} + 1) = 375e^{Bt}e^{C}$$

$$P = \frac{375e^{Bt}e^{C}}{e^{Bt}e^{C} + 1} \qquad \text{Multiply numerator and denominator by } e^{-Bt}e^{-C} \text{ and let } A = e^{-C}.$$

$$= \frac{375}{1 + Ae^{-Bt}} \qquad \text{General solution}$$

*The algebraic identity

$\frac{1}{x(a - x)} = \frac{1}{a}\left[\frac{(a - x) + x}{x(a - x)}\right] = \frac{1}{a}\left(\frac{1}{x} + \frac{1}{a - x}\right)$ was used to evaluate this integral.

Now we use the conditions $P(0) = 25$ and $P(1) = 125$ to evaluate the constant A and B:

$$P(0) = \frac{375}{1 + A} = 25$$

$$375 = 25 + 25A$$

$$A = 14$$

$$P(t) = \frac{375}{1 + 14e^{-Bt}}$$

$$P(1) = \frac{375}{1 + 14e^{-B}} = 125$$

$$375 = 125 + 1{,}750e^{-B}$$

$$e^{-B} = \frac{250}{1{,}750} = \frac{1}{7}$$

$$-B = \ln\frac{1}{7} = -\ln 7$$

$$B = \ln 7$$

$$P(t) = \frac{375}{1 + 14e^{-2\ln 7}}$$

To determine the population after 2 hours, we evaluate $P(2)$:

$$P(2) = \frac{375}{1 + 14e^{-2\ln 7}} \approx 292 \text{ paramecia}$$

The graph of $P(t)$ is shown in the figure in the margin. This S-shaped curve is typical of logistic growth functions. Notice that the upper limit of 375 is a horizontal asymptote.

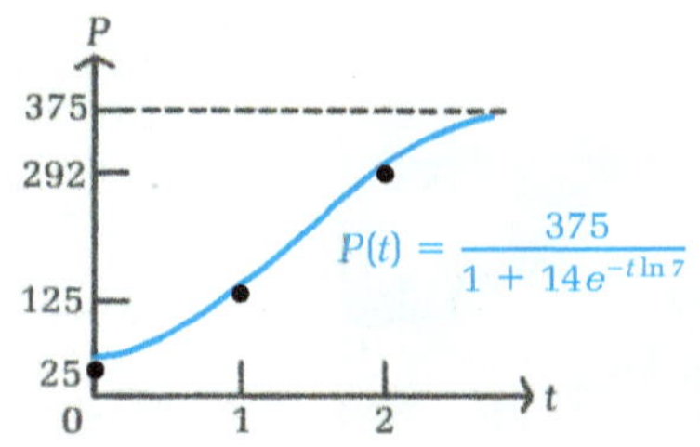

Logistic growth

Matched Problem 10 Repeat Example 10 for 15 paramecia initially and 150 paramecia after 1 hour.

Answers to Matched Problems

6. $y = -1/(x^4 + C)$
7. General solution: $y = 3 + K(2 + x^4)$; particular solution: $y = 5 + x^4$
8. 19.5 weeks
9. (A) \$11.13 million (B) Seventeenth year ($t \approx 17.06$ yr)
10. 343 paramecia

Exercises 7-2

A

Find the general solution for each differential equation. Then find the particular solution satisfying the initial condition.

1. $y' = 1;\quad y(0) = 2$
2. $y' = 2x;\quad y(0) = 4$
3. $y' = \frac{1}{x^{1/2}};\quad y(1) = -2$
4. $y' = 3x^{1/2};\quad y(0) = 7$
5. $y' = y;\quad y(0) = 10$
6. $y' = y - 10;\quad y(0) = 15$

7. $y' = 25 - y \quad y(0) = 5$

8. $y' = 3x^2y; \quad y(0) = \frac{1}{2}$

9. $y' = \frac{y}{x}; \quad y(1) = 5; \quad x > 0$

10. $y' = \frac{y}{x^2}; \quad y(-1) = 2e$

B

Find the general solution for each differential equation. Then find the particular solution satisfying the initial condition.

11. $y' = \frac{1}{y^2}; \quad y(1) = 3$

12. $y' = \frac{x^2}{y^2}; \quad y(0) = 2$

13. $y' = ye^x; \quad y(0) = 3e$

14. $y' = -y^2e^x; \quad y(0) = \frac{1}{2}$

15. $y' = \frac{e^x}{e^y}; \quad y(0) = \ln 2$

16. $y' = y^2(2x + 1); \quad y(0) = -\frac{1}{5}$

17. $y' = xy + x; \quad y(0) = 2$

18. $y' = (2x + 4)(y - 3); \quad y(0) = 2$

19. $y' = (2 - y)^2e^x; \quad y(0) = 1$

20. $y' = \frac{x^{1/2}}{(y - 5)^2}; \quad y(1) = 7$

Find the general solution for each differential equation. Do not attempt to find an explicit expression for the solution.

21. $y' = \frac{1 + x^2}{1 + y^2}$

22. $y' = \frac{6x - 9x^2}{2y + 4}$

23. $xyy' = (1 + x^2)(1 + y^2)$

24. $(xy^2 - x)y' = x^2y - x^2$

25. $x^2e^yy' = x^3 + x^3e^y$

26. $y' = \frac{xe^x}{\ln y}$

C

Find the explicit expression for the particular solution for each differential equation.

27. $xyy' = \ln x; \quad y(1) = 1$

28. $y' = \frac{xe^{x^2}}{y}; \quad y(0) = 2$

29. $xy' = x\sqrt{y} + 2\sqrt{y}; \quad y(1) = 4$

30. $y' = x(x - 1)^{1/2}(y - 1)^{1/2}; \quad y(1) = 1$

31. $yy' = xe^{-y^2}; \quad y(0) = 1$

32. $yy' = x(1 + y^2); \quad y(0) = 1$

Applications

Business & Economics

33. Continuous compound interest. If $5,000 is invested at 12% compounded continuously, how much will be in the account at the end of 10 years?

34. Continuous compound interest. If a sum of money is invested at 9% compounded continuously, how long will it take for the sum to double?

35. Advertising. A company is using radio advertising to introduce a new product to a community of 100,000 people. Suppose the rate at which people learn about the new product is proportional to the number who have not yet heard of it. If no one is aware of the product at the start of the advertising campaign and after 7 days 20,000 people are aware of the product, how long will it take for 50,000 people to become aware of the product?

36. Advertising. Prior to the beginning of an advertising campaign, 10% of the potential users of a certain brand are aware of the brand name. After the first week of the campaign, 20% of the consumers are aware of the brand name. If the percentage of informed consumers is growing at a rate proportional to the product of the percentage of informed consumers and the percentage of uninformed consumers, what percentage of consumers will be aware of the brand name after 5 weeks of advertising?

37. Product analysis. A company wishes to analyze a new room deodorizer. The active ingredient evaporates at a rate proportional to the amount present. Half of the ingredient evaporates in the first 30 days after the deodorizer is installed. If the deodorizer becomes ineffective after 90% of the active ingredient evaporated, how long will one of these deodorizers remain effective?

38. Natural resources. According to the U.S. Department of Agriculture Forest Service, the total consumption of wood in the United States was 11.6 billion cubic feet in 1960 and 13.4 billion cubic feet in 1970. If the consumption is increasing at a rate proportional to the total consumption, how much wood will be used in the year 2000?

39. Sales growth. The annual sales of a new company are expected to grow at a rate proportional to the difference between the sales and an upper limit of $5 million. If the sales are 0 initially and $1 million during the fourth year of operation, when will the sales reach $4 million?

40. Sales analysis. The annual sales of a company have declined from $8 million 2 years ago to $6 million today. If the annual sales continue to decline at a rate proportional to the difference between the annual sales and a lower limit of $3 million, find the annual sales 3 years from now.

Newton's law of cooling states that the rate of change of the temperature of an object is proportional to the difference between the temperature of the object and the temperature of the surrounding medium. Use this law to formulate models for Problems 41–44, and then solve using the techniques discussed in this section.

41. Manufacturing. *As part of a manufacturing process, a metal bar is to be heated in an oven until its temperature reaches 500°F.* The oven is maintained at a constant temperature of 800°F. Before it is placed in the oven, the temperature of the bar is 80°F. After 2 minutes in the oven, the temperature of the bar is 200°F. How long should the bar be left in the oven?

42. Manufacturing. The next step in the manufacturing process described in Problem 41 calls for the heated bar to be cooled in a vat of water until its temperature reaches 100°F.

The water in the vat is maintained at a constant temperature of 50°F. If the temperature of the bar is 500°F when it is first placed in the water and the bar has cooled to 400°F after 5 minutes in the water, how long should the bar be left in the water?

Life Sciences

43. **Food preparation.** A pie is removed from an oven where the temperature is 325°F and placed in a freezer with a constant temperature of 25°F. After 1 hour in the freezer, the temperature of the pie is 225°F. What is the temperature of the pie after 4 hours in the freezer?

44. **Food preparation.** A roast is taken from a freezer where the temperature is 20°F and placed in an oven with a constant temperature of 350°F. After 1 hour in the oven, the temperature of the roast is 185°F. What is the temperature of the roast after 3 hours in the oven?

45. **Population growth.** A culture of bacteria is growing at a rate proportional to the number present. The culture initially contains 100 bacteria. After 1 hour there are 140 bacteria in the culture.

(A) How many bacteria will be present after 5 hours?

(B) When will the culture contain 1,000 bacteria?

46. **Population growth.** A culture of bacteria is growing in a medium that can support a maximum of 1,100 bacteria. The rate of change of the number of bacteria is proportional to the product of the number present and the difference between 1,100 and the number present. The culture initially contains 100 bacteria. After 1 hour there are 140 bacteria.

(A) How many bacteria are present after 5 hours?

(B) When will the culture contain 1,000 bacteria?

47. **Simply epidemic.** An influenza epidemic has spread throughout a community of 50,000 people at a rate proportional to the product of the number of people who have been infected and the number who have not been infected. If 100 individuals were infected initially and 500 were infected 10 days later:

(A) How many people will be infected after 20 days?

(B) When will half the community be infected?

48. **Ecology.** A fish population in a large lake is declining at a rate proportional to the difference between the population and a lower limit of 5,000 fish. If the population has declined from 15,000 fish 3 years ago to 10,000 today, find the population 6 years from now.

Social Sciences

49. **Sensory perception.** A person is subjected to a physical stimulus that has a measurable magnitude, but the intensity of the resulting sensation is difficult to measure. If s is the magnitude of the stimulus and $I(s)$ is the intensity of sensation, experimental evidence suggests that

$$\frac{dI}{ds} = k\frac{I}{s}$$

for some constant k. Express I as a function of s.

50. **Learning.** The number of words per minute, N, a person can type increases with practice. Suppose the rate of change of N is proportional to the difference between N and an upper limit of 140. It is reasonable to assume that a beginner cannot type at all. Thus, $N = 0$ when $t = 0$. If a person can type 35 words per minute after 10 hours of practice:

(A) How many words per minute can that individual type after 20 hours of practice?

(B) How many hours must that individual practice to be able to type 105 words per minute?

51. **Rumor spread.** A rumor spreads through a population of 1,000 people at a rate proportional to the product of the number who have heard it and the number who have not heard it. If 5 people initiate a rumor and 10 people had heard it after 1 day:

(A) How many people will have heard the rumor after 7 days?

(B) How long will it take for 850 people to hear the rumor?

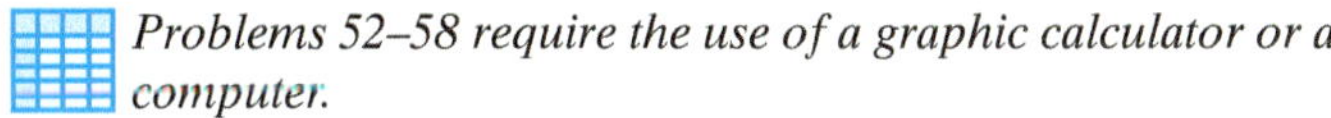
Problems 52–58 require the use of a graphic calculator or a computer.

52. **Rumor spread.** Refer to Problem 51. Graph the particular solution and use approximation techniques to determine how long it will take for 600 people to hear the rumor. Compute the answer to the nearest day.

Business & Economics

53. **Advertising.** Refer to Problem 35. Graph the particular solution and use approximation techniques to determine how long it will take for 70,000 people to become aware of the product. Compute the answer to the nearest day.

54. **Advertising.** Refer to Problem 36. Graph the particular solution and use approximation techniques to determine how long it will take for 60% of the consumers to become aware of the product. Compute the answer correct to one decimal place.

55. **Sales analysis.** A new company has 0 sales initially, sales of \$2 million during the first year, and sales of \$5 million during the third year. If the annual sales S are assumed to be growing at a rate proportional to the difference between the sales and an unknown upper limit M, then by the limited growth law,

$$S(t) = M(1 - e^{kt})$$

where t is time (in years) and $S(t)$ represents sales (in millions of dollars.) Use approximation techniques to find k to one decimal place and M to the nearest million [*Note:* An extraneous solution must be discarded.]

56. **Sales analysis.** Refer to Problem 55. Approximate k to one decimal place and M to the nearest million if the sales during the third year are \$4 million and all other information is unchanged.

Life Sciences

57. **Simple epidemic.** Refer to Problem 47. Graph the particular solution and use approximation techniques to determine how long it will take for 40,000 people to be infected. Compute the answer to the nearest day.

58. **Ecology.** Refer to Problem 48. Graph the particular solution and use approximation techniques to determine how long it will take for the fish population to decline to 5,500. Compute the answer to the nearest year.

7-3 First-Order Linear Differential Equations

Solutions of First-Order Linear Differential Equations

A differential equation that can be expressed in the form

$$y' + f(x)y = g(x)$$

is called a **first-order linear differential equation.**

For example,

$$y' + \frac{2}{x}y = x \tag{1}$$

is a first-order linear differential equation with $f(x) = 2/x$ and $g(x) = x$. This equation cannot be solved by the method of separation of variables. (Try to separate the variables to convince yourself that this is true.) Instead, we will change the form of the equation by multiplying both sides by x2:

$$x^2y' + 2xy = x^3$$

How was x^2 chosen? We will discuss that below. Let us first see how this choice leads to a solution of the problem.

Recall that the product rule for differentiation can be written as

$$FS' + F'S = (FS)'$$

Notice the similarity between the left-hand sides of the last two equations. In fact, if we equate F with x^2 and S with y, then the two expressions are identical:

$$x^2y' + 2xy = FS' + F'S = (FS)' = (x^2y)'$$

Thus, making use of the product rule, we can write the differential equation as

$$(x^2y)' = x^3$$

Now we can integrate both sides:

$$\int (x^2y)'\,dx = \int x^3\,dx$$

$$x^2y = \frac{x^4}{4} + C$$

solving for y, we obtain the general solution

$$y = \frac{x^2}{4} + \frac{C}{x^2}$$

The function x^2, which we used to transform the original equation into one we could solve as illustrated, is called an *integrating factor.* It turns out that there is a specific formula for determining the integrating factor for any first-order linear differential equation. Furthermore, this integrating factor can then be used to find the solution of the differential equation, just as we used x^2 to find the solution to (1). The formula for the integrating factor and a step-by-step summary of the solution process are given in the box.

Solving First-Order Linear Differential Equations

Step 1 Write the equation in the **standard form:**

$$y' + f(x)y = g(x)$$

Step 2 Compute the **integrating factor:**

$$I(x) = e^{\int f(x)\,dx}$$

(When evaluating $\int f(x)\,dx$ choose 0 for the constant of integration.)

Step 3 Multiply both sides of the standard form by the integrating factor $I(x)$. The left side should now be in the form $[I(x)y]'$:

$$[I(x)y]' = I(x)g(x)$$

Step 4 Integrate both sides:

$$I(x)y = \int I(x)g(x)\,dx$$

(When evaluating $\int I(x)g(x)\,dx$, include an arbitrary constant of integration.)

Step 5 Solve for y to obtain the **general solution:**

$$y = \frac{1}{I(x)}\int I(x)g(x)\,dx$$

EXAMPLE 11 Solve: $2xy' + y = 10x^2$

SOLUTION **Step 1** Multiply both sides by $1/(2x)$ to obtain the standard form:

$$y' + \frac{1}{2x}y = 5x \qquad f(x) = \frac{1}{2x} \text{ and } g(x) = 5x$$

Step 2 Find the integrating factor:

$$\begin{aligned} I(x) &= e^{\int f(x)\,dx} \\ &= e^{\int [1/(2x)]\,dx} && \text{Assume } x > 0 \text{ and choose 0 for the constant of integration.} \\ &= e^{(1/2)\ln x} && r\ln t = \ln t^r \\ &= e^{\ln x^{1/2}} && e^{\ln r} = r, r > 0 \\ &= x^{1/2} && \text{Integrating factor} \end{aligned}$$

Step 3 Multiply both sides of the standard form by the integrating factor:

$$x^{1/2}\left(y' + \frac{1}{2x}y\right) = x^{1/2}(5x)$$

$$x^{1/2}y' + \tfrac{1}{2}x^{-1/2}y = 5x^{3/2} \qquad \text{The left side should have the form } [I(x)y]'.$$

$$(x^{1/2}y)' = 5x^{3/2}$$

Step 4 Integrate both sides:

$$\int (x^{1/2}y)'\,dx = \int 5x^{3/2}\,dx \qquad \text{Include an arbitrary constant of integration on the right side.}$$

$$x^{1/2}y = 2x^{5/2} + C$$

Step 5 Solve for y:

$$\begin{aligned} y &= \frac{1}{x^{1/2}}(2x^{5/2} + C) \\ &= 2x^2 + \frac{C}{x^{1/2}} && \text{General solution} \end{aligned}$$

Matched Problem 11 Solve: $xy' + 3y = 4x$

In Example 11, notice that we assumed $x > 0$ to avoid introducing absolute value signs in the integrating factor. Many of the problems in this section will require the evaluation of expressions of the form $e^{\int [h'(x)/h(x)]\,dx}$. In order to avoid the complications caused by the introduction of absolute value signs, we will assume that the domain of $h(x)$ has been restricted so that $h(x) > 0$. This will simplify the solution process. We state the following familiar formula for convenient reference:

> If the domain of $h(x)$ is restricted so that $h(x) > 0$, then
>
> $$\int \frac{h'(x)}{h(x)}\,dx = \ln h(x) \qquad \text{and} \qquad e^{\ln h(x)} = h(x)$$

If a first-order linear differential equation is written in standard form, then multiplying both sides of the equation by its integrating factor will always convert the left side of the equation into the derivative of $I(x)y$. Thus, it is possible to omit steps 3–4 and proceed directly to step 5. This approach is illustrated in the next example. You decide which is easier to use—the step-by-step procedure or the formula in step 5.

EXAMPLE 12 Find the particular solution of the equation

$$y' + 2xy = 4x$$

satisfying the initial condition $y(0) = 5$.

SOLUTION Since the equation is already in standard form, we begin by finding the integrating factor:

$$I(x) = e^{\int 2x\,dx} = e^{x^2} \qquad f(x) = 2x$$

Proceeding directly to step 5, we have

$$\begin{aligned} y &= \frac{1}{I(x)} \int I(x)g(x)\,dx && g(x) = 4x \\ &= \frac{1}{e^{x^2}} \int e^{x^2}(4x)\,dx \\ &= e^{-x^2} \int 4xe^{x^2}\,dx && \text{Let } u = x^2,\ du = 2x\,dx. \\ &= e^{-x^2}(2e^{x^2} + C) \\ &= 2 + Ce^{-x^2} && \text{General solution} \end{aligned}$$

Substituting $x = 0$ and $y = 5$ in the general solution, we have

$$\begin{aligned} 5 &= 2 + C \\ C &= 3 \\ y &= 2 + 3e^{-x^2} && \text{Particular solution} \end{aligned}$$

Matched Problem 12 Find the particular solution of $y' + 3x^2y = 9x^2$ satisfying the initial condition $y(0) = 7$

Common Errors

1. When integrating both sides of an equation such as

$$(x^{1/2}y)' = 5x^{3/2}$$

remember that

$$x^{1/2}y = \int 5x^{3/2}\,dx \neq 2x^{5/2} \qquad \text{"}+C\text{" is missing in the antiderivative.}$$

Remember: A constant of integration must be included when evaluating $\int I(x)g(x)\,dx$. If you omit this constant, you will not be able to find the general solution of the differential equation. See step 4 of Example 11 for the correct procedure.

2. $\frac{1}{e^{x^2}}\int 4xe^{x^2}\,dx \neq \frac{1}{\cancel{e^{x^2}}}\int 4x\cancel{e^{x^2}}\,dx = \int 4x\,dx$

Just as a variable factor cannot be moved across the integral sign, **a variable factor outside the integral sign cannot be used to cancel a factor inside the integral sign.** See Example 12 for the correct procedure.

Applications

If P is the initial amount deposited into an account earning $100r\%$ compounded continuously, and A is the amount in the account after t years, then A satisfies the exponential growth equation

$$\frac{dA}{dt} = rA \qquad A(0) = P$$

Now suppose that money is continuously withdrawn from this account at a rate of $\$m$ per year. Then the amount A in the account at time t must satisfy

$$\begin{pmatrix}\text{Rate of change}\\ \text{of amount } A\end{pmatrix} = \begin{pmatrix}\text{Rate of growth}\\ \text{from continuous}\\ \text{compounding}\end{pmatrix} - \begin{pmatrix}\text{Rate of}\\ \text{withdrawal}\end{pmatrix}$$

$$\frac{dA}{dt} \quad = \quad rA \quad - \quad m$$

or

$$\frac{dA}{dt} - rA = -m$$

EXAMPLE 13 **Continuous Compound Interest** An initial deposit of \$10,000 is made into an account earning 8% compounded continuously. Money is then continuously withdrawn at a constant rate of \$1,000 a year until the account is depleted. Find the amount in the account at any time t. When will the amount be 0? What is the total amount withdrawn from this account?

SOLUTION The amount A in the account at any time t must satisfy

$$\frac{dA}{dt} - 0.08A = -1{,}000 \qquad A(0) = 10{,}000$$

The integrating factor for this equation is

$$I(t) = e^{\int -0.08\,dt} = e^{-0.08t} \qquad f(t) = -0.08$$

Multiplying both sides of the differential equation by $I(t)$ and following the step-by-step procedure, we have

$$e^{-0.008t}\frac{dA}{dt} - 0.08e^{-0.008t}A = -1{,}000e^{-0.008t}$$

$$(e^{-0.08t}A)' = -1{,}000e^{0.08t}$$

$$e^{-0.08t}A = \int -1{,}000e^{-0.08t}\,dt$$

$$= 12{,}500e^{-0.08t} + C$$

$$A = 12{,}500 + Ce^{0.08t} \qquad \text{General solution}$$

Applying the initial condition $A(0) = 10{,}000$ yields

$$A(0) = 12{,}500 + C = 10{,}000$$

$$C = -2{,}500$$

$$A(t) = 12{,}500 - 2{,}500e^{0.08t} \qquad \text{Amount in the account at any time } t$$

To determine when the amount in the account is 0, we solve $A(t) = 0$ for t:

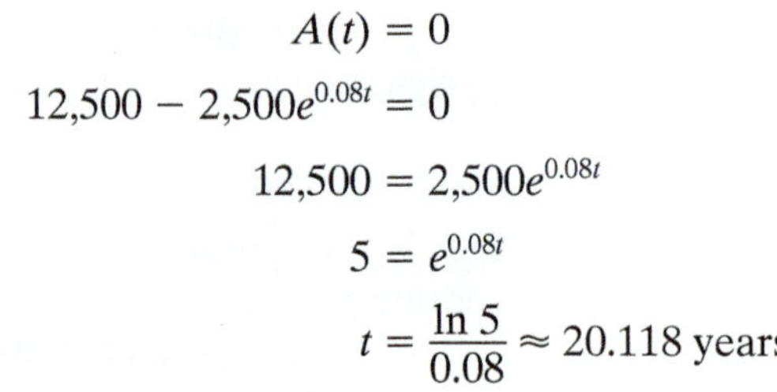

$$A(t) = 0$$

$$12{,}500 - 2{,}500e^{0.08t} = 0$$

$$12{,}500 = 2{,}500e^{0.08t}$$

$$5 = e^{0.08t}$$

$$t = \frac{\ln 5}{0.08} \approx 20.118 \text{ years}$$

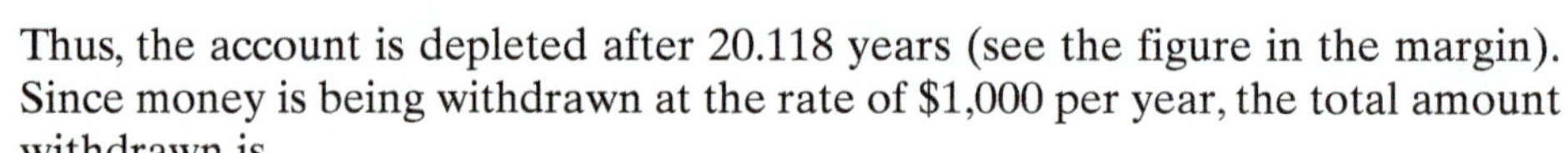

Thus, the account is depleted after 20.118 years (see the figure in the margin). Since money is being withdrawn at the rate of $1,000 per year, the total amount withdrawn is

$$1{,}000(20.118) = \$20{,}118$$

Matched Problem 13 Repeat Example 13 if the account earns 5% compounded continuously.

EXAMPLE 14 **Equilibrium Price** In economics, the supply S and the demand D for a commodity often can be considered as functions of both the price, $p(t)$, and the rate of change of the price, $p'(t)$. (Thus, S and D are ultimately functions of time t.) The **equilibrium price at time t** is the solution of the equation $S = D$. If $p(t)$ is the solution of this equation, then the **long-range equilibrium price** is

$$\overline{p} = \lim_{t\to\infty} p(t)$$

For example, if

$$D = 50 - 2p(t) + 2p'(t)$$

$$S = 20 + 4p(t) + 5p'(t)$$

And $p(0) = 15$, then the equilibrium price at time t is the solution of the equation

$$50 - 2p(t) + 2p'(t) = 20 + 4p(t) + 5p'(t)$$

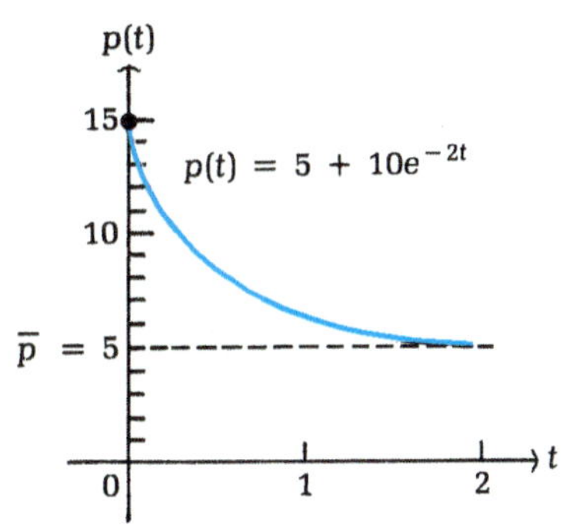

This simplifies to

$$p'(t) + 2p(t) = 10 \qquad f(t) = 2 \quad \text{and} \quad g(t) = 10$$

which is a first-order linear equation with integrating factor

$$I(t) = e^{\int 2\,dt} = e^{2t}$$

Proceeding directly to step 5, we have

$$p(t) = \frac{1}{e^{2t}} \int 10e^{2t}\,dt$$

$$= e^{-2t}(5e^{2t} + C)$$

$$p(t) = 5 + Ce^{-2t}$$ General solution

$$p(0) = 5 + C = 15$$

$$C = 10$$

$$p(t) = 5 + 10e^{-2t}$$ Equilibrium price at time t (see the figure in the margin)

$$\overline{p} = \quad (5 + 10e^{-2t}) = 5$$ Long-range equilibrium price

Matched Problem 14 If $D = 70 + 2p(t) + 2p'(t)$, $S = 30 + 6p(t) + 3p'(t)$, and $p(0) = 25$, find the equilibrium price at time t and find the long-range equilibrium price.

EXAMPLE 15 **Pollution Control** A company has a 1,000 gallon holding tank which is used to control the release of pollutants into a sewage system. Initially, the tank contains 500 gallons of water. Each gallon of water contains 2 pounds of pollutants. Additional polluted water containing 5 pounds of pollutants per gallon is pumped into the tank at the rate of 100 gallons per hour and is thoroughly mixed with the water already present in the tank. At the same time, the uniformly mixed water in the tank is released into the sewage system at a rate of 50 gallons per hour. This process continues for 5 hours. At the end of this 5 hour period, determine:

(A) The total amount of pollutants in the tank.

(B) The rate (in pounds per gallon) at which pollutants are being released into the sewage system.

SOLUTION Let $p(t)$ be the total amount (in pounds) of pollutants in the tank t hours after this process begins. Since the tank initially contains 500 gallons of water and each gallon of water contains 2 pounds of pollutants,

$$p(0) = 2 \cdot 500 = 1{,}000 \qquad \text{Initial amount of pollutants in the tank}$$

Since polluted water is entering and leaving the tank at different rates and with different concentrations of pollutants, the rate of change of the amount of pollutants in the tank will depend on the rate at which pollutants enter the tank and the rate at which they leave the tank:

$$\begin{pmatrix}\text{Rate of change}\\ \text{of pollutants}\end{pmatrix} = \begin{pmatrix}\text{Rate pollutants}\\ \text{enter the tank}\end{pmatrix} - \begin{pmatrix}\text{Rate pollutants}\\ \text{leave the tank}\end{pmatrix}$$

Finding expressions for the two rates on the right side of this equation will produce a differential equation involving $p'(t)$

Since water containing 5 pounds of pollutants per gallon is entering the tank at a constant rate of 100 gallons per hour, pollutants are entering the tank at a constant rate of

$$\begin{pmatrix}\text{5 pounds}\\ \text{per gallon}\end{pmatrix} \times \begin{pmatrix}\text{100 gallons}\\ \text{per hour}\end{pmatrix} = \text{500 pounds per hour}$$

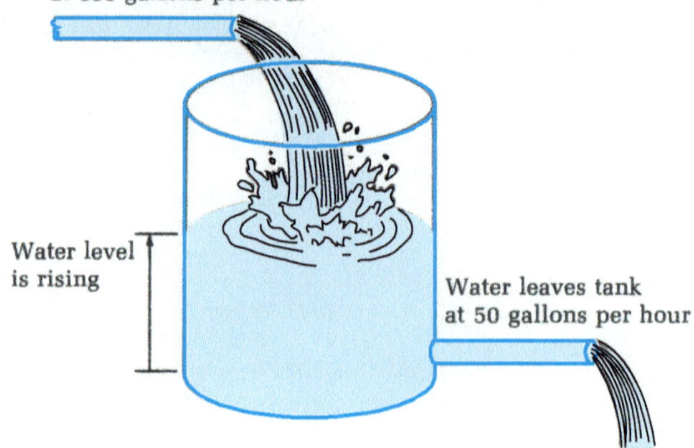

How fast are the pollutants leaving the tank? Since the total amount of pollutants in the tank is increasing, the rate at which the pollutants leave the tank will depend on the amount of pollutants in the tank at time t and the amount of water in the tank at time t. Since 100 gallons of water enter the tank each hour and only 50 gallons leave each hour, the amount of water in the tank increases at the rate of 50 gallons per hour. Thus, the amount of water in the tank at time t is

$$\begin{pmatrix}\text{Initial amount}\\ \text{of water}\end{pmatrix} + \begin{pmatrix}\text{Gallons}\\ \text{per hour}\end{pmatrix} \times \begin{pmatrix}\text{Number}\\ \text{of hours}\end{pmatrix}$$

$$500 \qquad + \qquad 50t$$

The amount of pollutants in each gallon of water at any time t is

$$\frac{p(t)}{500 + 50t} \qquad \frac{\begin{pmatrix}\text{Total amount}\\ \text{of pollutants}\end{pmatrix}}{\begin{pmatrix}\text{Total amount}\\ \text{of water}\end{pmatrix}} = \text{Pollutants per gallon}$$

Since water is leaving the tank at the rate of 50 gallons per hour, the rate at which the pollutants are leaving the tank is

$$\frac{50p(t)}{500 + 50t} = \frac{p(t)}{10 + t}$$

Thus, the rate of change of $p(t)$ must satisfy

$$\begin{pmatrix}\text{Rate of change}\\ \text{of pollutants}\end{pmatrix} = \begin{pmatrix}\text{Rate pollutants}\\ \text{enter the tank}\end{pmatrix} - \begin{pmatrix}\text{Rate pollutants}\\ \text{leave the tank}\end{pmatrix}$$

$$p'(t) \qquad = \qquad 500 \qquad - \qquad \frac{p(t)}{10 + t}$$

This gives the following model for this problem:

$$p'(t) = 500 - \frac{p(t)}{10 + t} \qquad p(0) = 1{,}000$$

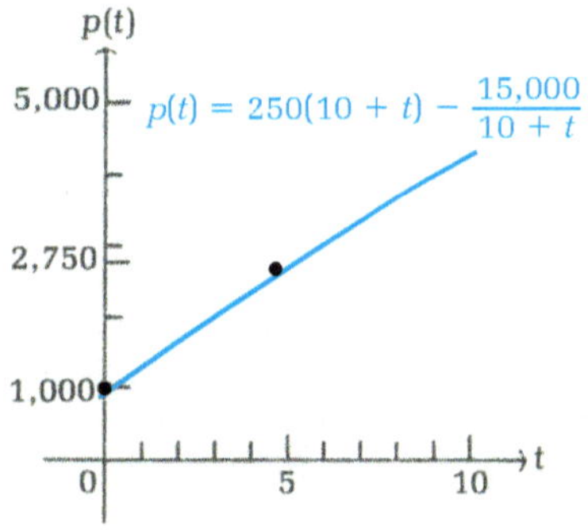

or

$$p'(t) + \frac{1}{10 + t}p(t) = 500$$ First-order linear equation with $f(t) = 1/(10 + t)$ and $g(t) = 500$

$$I(t) = e^{\int dt/(10+t)} = e^{\ln(10+t)} = 10 + t$$ Integrating factor

$$p(t) = \frac{1}{10 + t}\int 500(10 + t)\,dt$$ Proceeding directly to step 5

$$= \frac{1}{10 + t}[250(10 + t)^2 + C]$$

$$p(t) = 250(10 + t) + \frac{C}{10 + t}$$ General solution

$$1{,}000 = 250(10) + \frac{C}{10}$$ Initial condition: $p(0) = 1{,}000$

$$\frac{C}{10} = -1{,}500$$

$$C = -15{,}000$$

$$p(t) = 250(10 + t) - \frac{15{,}000}{10 + t}$$ Particular solution (see the figure in the margin)

(A) To find the total amount of pollutants after 5 hours, we evaluate $p(5)$:

$$p(5) = 250(15) - \frac{15{,}000}{15} = 2{,}750 \text{ pounds}$$

(B) After 5 hours, the tank contains 750 gallons of water. The rate at which pollutants are being released into the sewage system is

$$\frac{2{,}750}{750} \approx 3.67 \text{ pounds per gallon}$$

Matched Problem 15 Repeat Example 15 if water is released from the tank at the rate of 75 gallons per hour.

Answers to Matched Problems

11. $I(x) = x^3;\quad y = x + (C/x^3)$

12. $I(x) = e^{x^3}; y = 3 + 4e^{x^3}$

13. $A(t) = 20{,}000 - 10{,}000e^{0.05t}$; $A = 0$ when $t = (\ln 2)/0.05 \approx 13.83$ yr; total withdrawals $= \$13{,}863$

14. $p(t) = 10 + 15e^{-4t}; \bar{p} = 10$

15. $p(t) = 125(20 + t) - \frac{12{,}000{,}000}{(20 + t)^3}$ (A) 2,357 lb (B) Approx. 3.77 lb/gal

Exercise 7-3

A

In all problems, assume $h(x) > 0$ whenever $\ln h(x)$ is involved.

In Problems 1–12, find the integrating factor, the general solution, and the particular solution satisfying the given initial condition.

1. $y' + 2y = 4$; $y(0) = 1$
2. $y' - 3y = 3$; $y(0) = -1$
3. $y' + y = e^{-2x}$; $y(0) = 3$
4. $y' - 2y = e^{3x}$; $y(0) = 2$
5. $y' - y = 2e^{x}$; $y(0) = -4$
6. $y' + 4y = 3e^{-4x}$; $y(0) = 5$
7. $y' + y = 9x^2e^{-x}$; $y(0) = 2$
8. $y' - 3y = 6\sqrt{x}\, e^{3x}$; $y(0) = -2$
9. $y' + \frac{1}{x}y = 2$; $y(1) = 1$
10. $y' - \frac{3}{x}y = 4$; $y(1) = 1$
11. $y' + \frac{2}{x}y = 10x^2$; $y(2) = 8$
12. $y' - \frac{1}{x}y = \frac{9}{x^3}$; $y(2) = 3$

B

Find the integrating factor $I(x)$ for each equation, and then find the general solution.

13. $y' + xy = 5x$
14. $y' - 2xy = 6x$
15. $y' - 2y = 4x$
16. $y' + y = x^2$
17. $y' + \frac{1}{x}y = e^{x}$
18. $y' + \frac{2}{x}y = e^{3x}$
19. $y' + \frac{1}{x}y = \ln x$
20. $y' - \frac{1}{x}y = \ln x$

C

In Problems 21–26, find the general solution two ways. First, use an integrating factor and then use separation of variables.

21. $y' = \frac{1 - y}{x}$
22. $y' = \frac{y + 2}{x + 1}$
23. $y' = \frac{2x + 2xy}{1 + x^2}$
24. $y' = \frac{4x + 2xy}{4 + x^2}$
25. $y' = 2x(y + 1)$
26. $y' = 3x^2(y + 2)$
27. Use an integrating factor to find the general solution of the unlimited growth model,

$$\frac{dy}{dt} = ky$$

[**Hint:** Remember, the antiderivative of the constant function 0 is an arbitrary constant C.]

28. Use an integrating factor to find the general solution of the limited growth model,

$$\frac{dy}{dt} = k(L - y)$$

Applications

Business & Economics

29. **Continuous compound interest.** An initial deposit of \$20,000 is made into an account that earns 4% compounded continuously. Money is then withdrawn at a constant rate of \$4,000 a year until the amount in the account is 0. Find the amount in the account at any time t. When is the amount 0? What is the total amount withdrawn from the account?

30. **Continuous compound interest.** An initial deposit of \$50,000 is made into an account that earns 10% compounded continuously. Money is then withdrawn at a constant rate of \$6,000 a year until the amount in the account is 0. Find the amount in the account at any time t. When is the amount 0? What is the total amount withdrawn from the account?

31. **Continuous compound interest.** An initial deposit of \$$P$ is made into an account that earns 5% compounded continuously. Money is then withdrawn at a constant rate of \$1,500 a year. After 10 years of continuous withdrawals, the amount in the account is 0. Find the initial deposit P.

32. **Continuous compound interest.** An initial deposit of \$$P$ is made into an account that earns 8% compounded continuously. Money is then withdrawn at a constant rate of \$3,000 a year. After 5 years of continuous withdrawals, the amount in the account is 0. Find the initial deposit P.

33. **Continuous compound interest.** An initial deposit of \$7,000 is made into an account earning 8% compounded continuously. Thereafter, money is deposited into the account at a constant rate of \$2,000 per year. Find the amount in this account at any time t. How much is in this account after 5 years?

34. **Continuous compound interest.** An initial deposit of \$10,000 is made into an account earning 6% compounded continuously. Thereafter, money is deposited into the account at a constant rate of \$6,000 per year. Find the amount in this account at any time t. How much is in this account after 10 years?

35. **Supply–demand.** The supply S and demand D for a certain commodity satisfy the equations

$$S = 35 - 2p(t) + 3p'(t) \text{ and } D = 95 - 5p(t) + 2p'(t)$$

If $p(0) = 30$, find the equilibrium price at time t and the long-range equilibrium price.

36. **Supply–demand.** The supply S and demand D for a certain commodity satisfy the equations

$$S = 70 - 3p(t) + 2p'(t) \text{ and } D = 100 - 5p(t) + p'(t)$$

If $p(0) = 5$, find the equilibrium price at time t and the long-range equilibrium price.

Life Sciences

37. **Pollution.** A 1,000 gallon holding tank contains 200 gallons of water. Initially, each gallon of water in the tank contains 2 pounds of pollutants. Water containing 3 pounds of pollutants per gallon enters the tank at a rate of 75 gallons per hour, and

the uniformly mixed water is released from the tank at a rate of 50 gallons per hour. How many pounds of pollutants are in the tank after 2 hours? At what rate (in pounds per gallon) are the pollutants being released after 2 hours?

38. Pollution. Rework Problem 37 if water is entering the tank at the rate of 100 gallons per hour.

39. Pollution. Rework Problem 37 if water is entering the tank at the rate of 50 gallons per hour.

40. Pollution. Rework Problem 37 if water is entering the tank at the rate of 10 gallons per hour.

In a recent article in the College Mathematics Journal *(January 1987, 18:1), Arthur Segal proposes the following model for weight loss or gain:*

$$\frac{dw}{dt} + 0.005w = \frac{1}{3{,}500}C$$

where $w(t)$ is a person's weight (in pounds) after t days of consuming exactly C calories per day. Use this model to solve Problems 41–44.

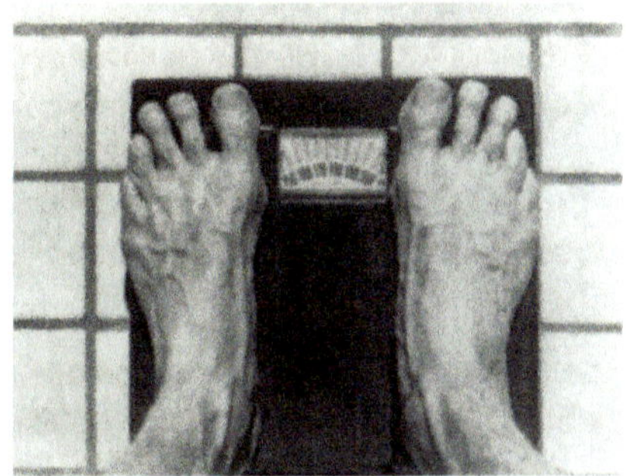

41. Weight loss. A person weighing 160 pounds goes on a diet of 2,100 calories per day. How much will this person weigh after 30 days on this diet? How long will it take this person to lose 10 pounds? Find $\lim_{t\to\infty} w(t)$ and interpret the results.

42. Weight loss. A person weighing 200 pounds goes on a diet of 2,800 calories per day. How much will this person weigh after 90 days on this diet? How long will it take this person to lose 25 pounds? Find $\lim_{t\to\infty} w(t)$ and interpret the result.

43. Weight loss. A person weighing 130 pounds would like to lose 5 pounds during a 30 day period. How many calories per day should this person consume to reach this goal?

44. Weight loss. A person weighing 175 pounds would like to lose 10 pounds during a 45 day period. How many calories per day should this person consume to reach this goal?

Social Sciences

In 1960, William K. Estes proposed the following model for measuring a student's performance in the classroom:

$$\frac{dk}{dt} + Ik = \lambda I$$

where $k(t)$ is the student's knowledge after t weeks (expressed as a percentage and measured by performance on examinations), I is a constant called the coefficient of learning and representing the student's ability to learn (expressed as a percentage and determined by IQ or some similar general intelligence predictor), and λ is a constant representing the fraction of available time the student spends performing helpful acts that should increase knowledge of the subject (studying, going to class, and so on). Use this model to solve Problem 45.

45. Learning theory. Students enrolled in a beginning Spanish class are given a pretest the first day of class in order to determine their initial knowledge of the subject. The results of the pretest, the coefficient of learning, and the fraction of time spent performing helpful acts for two students in the class are given in the table. Use the Estes model to predict the knowledge of each student after 6 weeks in the class.

	score on pretest	coefficient of learning	fraction of helpful acts
STUDENT A	0.1(10%)	0.8	0.9
STUDENT B	0.4(40%)	0.8	0.7

Problems 46–50 require the use of a graphic calculator or a computer. Compute all approximations correct to one decimal place.

46. Learning theory. Refer to Problem 45. Graph both particular solutions in the same viewing rectangle and use approximation techniques to approximate the time when both students have the same level of knowledge.

Business & Economics

47. Continuous compound interest. Refer to Problem 33. Graph the particular solution and use approximation techniques to determine when the account will contain $50,000.

48. Continuous compound interest. Refer to Problem 34. Graph the particular solution and use approximation techniques to determine when the account will contain $200,000.

Life Sciences

49. Pollution. Refer to Problem 37. Graph the particular solution and use approximation techniques to determine when the tank will contain 1,000 pounds of pollutants.

50. Pollution. Refer to Problem 40. Graph the particular solution and use approximation techniques to determine when the tank will contain 1,500 pounds of pollutants.

7-4 Second-Order Differential Equations

- Homogeneous Second-Order Equations
- Nonhomogeneous Second-Order Equations

An equation that can be written in the form

$$a(x)y'' + b(x)y' + c(x)y = d(x)$$

is called a **second-order linear differential equation.** Unfortunately, there is no simple general procedure that can be used to solve every equation of this form. However, there are some special cases of (1) that are easy to solve. In this section we will consider two of these special cases: **homogeneous second-order equations** of the form

$$ay'' + by' + c = 0$$

and **nonhomogeneous second-order equations** of the form

$$ay'' + by' + c = d$$

where a, b, c, and d are constants.

Homogeneous Second-Order Equations

The homogeneous first-order equation

$$ay' + by = 0$$

has a general solution of the form

$$y = Ce^{mx} \qquad m = -\frac{b}{a}$$

where C is an arbitrary constant (verify this). What should we expect the general solution of a homogeneous second-order equation to look like? It is not unreasonable to expect that the general solution of a second-order equation will involve two functions and two arbitrary constants. Theorem 2, which will form the basis for most of our work with second-order equations, states that this is the case.

THEOREM 2 General Solution of Homogeneous Second-Order Equations

If f and g are two functions that satisfy the differential equation

$$ay'' + by' + cy = 0$$

and if g is not a constant multiple of f, then the general solution of this differential equation is

$$y = C_1f(x) + C_2g(x)$$

where C_1 and C_2 are arbitrary constants.

The condition that g not be a constant multiple of f is necessary to ensure that y is in fact the general solution of this equation.

Theorem 2 provides a strategy for solving homogeneous second-order equations. First, we will find two different functions that satisfy the equation, and then we will combine these two functions to form the general solution.

EXAMPLE 16 Solve: $y'' - 2y' - 3y = 0$

SOLUTION Since the general solution of a homogeneous first-order equation involves a function of the form $y = e^{mx}$, we will try to determine whether this second-order equation has any solutions of this form. We begin by substituting $y = e^{mx}$, $y' = me^{mx}$, and $y'' = m^2e^{mx}$ in the given differential equation:

$$y'' - 2y' - 3y = 0$$

$$m^2e^{mx} - 2me^{mx} - 3e^{mx} = 0$$
$$e^{mx}(m^2 - 2m - 3) = 0$$

Since e^{mx} is never 0, in order for this equation to be satisfied, we must have

$$m^2 - 2m - 3 = 0$$
$$(m - 3)(m + 1) = 0$$
$$m_1 = 3 \qquad m_2 = -1$$

Thus, the functions

$$f(x) = e^{3x} \qquad \text{and} \qquad g(x) = e^{-x}$$

both satisfy the given differential equation, and e^{-x} is not a constant multiple of e^{3x}. According to Theorem 2, the general solution is

$$y = C_1e^{3x} + C_2e^{-x}$$

CHECK

$$y = C_1e^{3x} + C_2e^{-x}$$
$$y' = 3C_1e^{3x} - C_2e^{-x}$$
$$y'' = 9C_1e^{3x} + C_2e^{-x}$$
$$\begin{aligned}(y'' - 2y' - 3y) &= (9C_1e^{3x} + C_2e^{-x}) - 2(3C_1e^{3x} - C_2e^{-x}) - 3(C_1e^{3x} + C_2e^{-x})\\ &= (9 - 6 - 3)C_1e^{3x} + (1 + 2 - 3)C_2e^{-x}\\ &\stackrel{\checkmark}{=} 0\end{aligned}$$

Matched Problem 16 Solve: $y'' + y' - 6y = 0$

The quadratic equation

$$m^2 - 2m - 3 = 0$$

which we obtained by substituting $y = e^{mx}$ in the differential equation

$$y'' - 2y' - 3y = 0$$

is called the *characteristic equation* for this differential equation.

The Characteristic Equation

The **characteristic equation for**

$$ay'' + by' + cy = 0$$

is

$$am^2 + bm + c = 0$$

If m is a real root of the characteristic equation, then

$$f(x) = e^{mx}$$

is a solution of the differential equation.

EXAMPLE 17 Solve: $y'' - 16y = 0$

SOLUTION

$$m^2 - 16 = 0 \quad \text{Characteristic equation}$$
$$(m - 4)(m + 4) = 0$$
$$m_1 = 4 \qquad m_2 = -4 \quad \text{Root of characteristic equation}$$
$$f(x) = e^{4x} \qquad g(x) = e^{-4x} \quad \text{Two different solutions of the differential equation}$$
$$y = C_1e^{4x} + C_2e^{-4x} \quad \text{General solution}$$

Matched Problem 17 Solve: $y'' - y' - 2y = 0$

Since the general solution of a second-order equation involves two arbitrary constants, two conditions are required to determine a particular solution. If the value of y and the value of y' are both given for the same value of x, then both conditions are called initial conditions.

EXAMPLE 18 Find the particular solution of $2y'' + 3y' - 2y = 0$ that satisfies the initial conditions $y(0) = 2$ and $y'(0) = -1$.

SOLUTION

$$2m^2 + 3m - 2 = 0 \quad \text{Characteristic equation}$$
$$(2m - 1)(m + 2) = 0$$
$$m_1 = \tfrac{1}{2} \qquad m_2 = -2 \quad \text{Roots of characteristic equation}$$
$$y = C_1e^{x/2} + C_2e^{-2x} \quad \text{General solution}$$
$$y(0) = C_1 + C_2 = 2 \quad \text{First initial condition}$$
$$y' = \tfrac{1}{2}C_1e^{x/2} - 2C_2e^{-2x}$$
$$y'(0) = \tfrac{1}{2}C_1 - 2C_2 - 1 \quad \text{Second initial condition}$$

In order to determine the values of C_1 and C_2, we must solve the system of equations:

$$C_1 + C_2 = 2$$
$$\tfrac{1}{2}C_1 - 2C_2 = -1$$
$$C_1 = 2 - C_2 \quad \text{Solve the first equation for } C_1.$$
$$\tfrac{1}{2}(2 - C_2) - 2C_2 = -1 \quad \text{Substitute for } C_1 \text{ in the second equation and solve for } C_2.$$
$$C_2 = \tfrac{4}{5}$$
$$C_1 = 2 - \tfrac{4}{5} \quad \text{Substitute the value for } C_2 \text{ and solve for } C_1.$$
$$= \tfrac{6}{5}$$

Thus, the particular solution satisfying the given initial conditions is

$$y = \tfrac{6}{5}e^{x/2} + \tfrac{4}{5}e^{-2x}$$

Matched Problem 18 Find the particular solution of $3y'' - 7y' - 6y = 0$ that satisfies the initial condition $y(0) = 1$ and $y'(0) = 2$.

EXAMPLE 19 Solve $y'' - 2y' + y = 0$

SOLUTION

$$\begin{aligned} m^2 - 2m + 1 &= 0 \quad \text{Characteristic equation} \\ (m-1)(m-1) &= 0 \\ m &= 1 \quad \text{Single repeated root of characteristic equation} \end{aligned}$$

This characteristic equation has a single repeated root, $m = 1$, which provides us with one solution of the differential equation $f(x) = e^x$. How can we find a second solution? The answer to this question is given in Theorem 3:

THEOREM 3 Characteristic Equations with a Repeated Root

If the characteristic equation

$$am^2 + bm + c = 0$$

has a single repeated root m, then

$$g(x) = xe^{mx}$$

is a solution of the differential equation

$$ay'' + by' + cy = 0$$

Returning to the differential equation

$$y'' - 2y' + y = 0$$

whose characteristic equation has the single repeated root $m = 1$, we can now conclude that $f(x) = e^x$ and $g(x) = xe^x$ are two different solutions to this equation. Thus, the general solution is

$$y = C_1e^x + C_2xe^x$$

You should check this solution.

Matched Problem 19 Solve: $y'' + 6y' + 9y = 0$

We now know how to find the solution of a homogeneous second-order equation if the characteristic equation has two distinct real roots or one repeated real root. But there is a third possibility. If the discriminant $b^2 - 4ac$ of the characteristic equation is negative, the characteristic equation has two imaginary roots. In this case, the general solution involves both exponential functions and trigonometric functions. For completeness, the imaginary roots case is included in the box (also, see Problems 27–30 in Exercise 7-4).

Homogeneous Second-Order Differential Equations

Differential Equation	Characteristic Equation
$ay'' + by' + cy = 0$	$am^2 + bm + c = 0$
Solution	Nature of Roots
$y = C_1e^{m_1x} + C_2e^{m_2x}$	Two distinct real roots, m_1 and m_2
$y = C_1e^{mx} + C_2xe^{mx}$	A single repeated root, m
$y = e^{px}(C_1 \cos qx + C_2 \sin qx)$	Two distinct imaginary roots, $m_1 = p + qi$ and $m_2 = p - qi$

Nonhomogeneous Second-Order Equations

Now we want to consider nonhomogeneous second-order equations of the form

$$ay'' + by' + cy = d$$

where d is a nonzero constant. The following example illustrates how to solve such an equation.

EXAMPLE 20 Solve: $y'' - 2y' - 8y = 16$

SOLUTION We use a three-step process to solve the equation.

Step 1. Find a constant function satisfying the nonhomogeneous equation. If $y = k$ is a constant function, then $y' = 0$ and $y'' = 0$. Substituting in the equation, we have

$$y'' - 2y' - 8y = 16$$

$$0 - 2(0) - 8k = 16$$

$$k = -2$$

Thus, $y = -2$ is a constant function satisfying the nonhomogeneous equation.

Step 2. Solve the associated homogeneous equation (the equation formed by replacing d with 0):

$$y'' - 2y' - 8y = 0 \quad \text{Associated homogeneous equation}$$

$$m^2 - 2m - 8 = 0 \quad \text{Characteristic equation}$$

$$(m - 4)(m + 2) = 0$$

$$m_1 = 4 \qquad m_2 = -2 \quad \text{Roots of characteristic equation}$$

$$y = C_1e^{4x} + C_2e^{-2x} \quad \text{Solution of homogeneous equation}$$

Step 3. Add together the solutions from steps 1 and 2 to form the general solution of the nonhomogeneous equation:

$$y = -2 + C_1e^{4x} + C_2e^{-2x} \quad \text{General solution of the nonhomogeneous equation}$$

You should check this solution.

Matched Problem 20 Solve: $y'' + y' - 12y = 6$

The three-step process outlined in the solution of Example 20 can be used to solve any nonhomogeneous equation of the form

$$ay'' + by' + cy = d$$

provided $c \neq 0$. If $c = 0$, then there are no solutions of the form $y = k$. In this case, the first step consists of finding any solutions of the nonhomogeneous equation of the form $y = kx$, where k is a constant. The remaining steps are unchanged (see Problems 19–22 in Exercise 7-4).

This three-step process is a simple application of an important theorem studied in more advanced courses. This theorem states that if y_p is any particular solution of a nonhomogeneous differential equation and y_h is the general solution of the associated homogeneous equation, then $y = y_p + y_h$ is the general solution of the nonhomogeneous equation.

Answers to Matched Problems

16. $y = C_1e^{-3x} + C_2e^{2x}$

17. $y = C_1e^{2x} + C_2e^{-x}$

18. $y = \frac{8}{11}e^{3x} + \frac{3}{11}e^{-2x/3}$

19. $y = C_1e^{-3x} + C_2xe^{-3x}$

20. $y = C_1e^{-4x} + C_2e^{3x} - \frac{1}{2}$

Exercise 7-4

A

Find the general solution for each equation.

1. $y'' + 3y' + 2y = 0$

2. $y'' - 6y' + 8y = 0$

3. $y'' + 2y' - 15y = 0$

4. $y'' - 25y = 0$

5. $y'' + 6y' = 0$

6. $y'' - 3y' = 0$

7. $y'' - 4y' + 4y = 0$

8. $y'' + 10y' + 25y = 0$

B

Find the particular solution for each equation that satisfies the initial conditions.

9. $y'' - y = 0;\quad y(0) = 3;\quad y'(0) = 1$

10. $y'' - y' - 2y = 0;\quad y(0) = 1;\quad y'(0) = 2$

11. $3y'' - 10y' + 3y = 0;\quad y(0) = 1;\quad y'(0) = -1$

12. $y'' - 4y' = 0;\quad y(0) = 0;\quad y'(0) = 3$

13. $y'' + 2y' + y = 0;\quad y(0) = 2;\quad y'(0) = 4$

14. $y'' - 2y = 0;\quad y(0) = 0;\quad y'(0) = 1$

Find the general solution for each equation.

15. $y'' - 3y' - 4y = 12$

16. $y'' + 2y' + y = 5$

Find the particular solution that satisfies the initial conditions.

17. $y'' + y' - 2y = 6;\quad y(0) = 0;\quad y'(0) = 0$

18. $y'' - 3y' - 10y = 100;\quad y(0) = -10;\quad y'(0) = 0$

C

Use the following strategy to find the general solution in Problems 19–22:

Step 1 *Find a solution of the nonhomogeneous equation of the form $y = kx$, k a constant.*
Step 2 *Find the general solution of the associated homogeneous equation.*
Step 3 *Add together the solutions from steps 1 and 2.*

19. $2y'' + y' = 1$

20. $y'' + y' = 2$

21. $y'' - 2y' = 3$

22. $2y'' - 4y' = 5$

If the solution of a second-order equation is required to satisfy two conditions of the form $y(a) = y_1$ and $y(b) = y_2$, then these conditions are referred to as boundary conditions. Find the particular solution for each equation that satisfies the given boundary conditions.

23. $y'' - 8y' + 16y = 0;\quad y(0) = 1;\quad y(1) = e^4$

24. $4y'' + 4y' + y = 0;\quad y(0) = 0;\quad y(1) = 2e^{-0.5}$

25. $2y'' - 5y' + 2y = 0;\quad y(0) = 0;\quad y(2) = e^4 - e$

26. $y'' - 3y' = 0;\quad y(0) = 3;\quad y(1) = 2 + e^3$

Problems 27–30 are optional (trigonometric functions are involved.) Find the general solution for each equation.

27. $y'' + y = 0$

28. $y'' + 4y = 0$

29. $y'' - 4y' + 13y = 0$

30. $y'' + 2y' + 3y = 0$

Applications

Business & Economics

31. Supply–demand. In earlier exercises, supply and demand were considered as functions of the price, $p(t)$, and the rate of change of the price, $p'(t)$. In studying certain markets, economists include the second derivative $p''(t)$ in the differential equation to reflect whether the rate of change of $p(t)$ is increasing or decreasing. Suppose that S and D satisfy the equations

$$S = 3 + 0.2p' - 0.05p - p''$$

and

$$D = 2 + 0.8p' - 0.01p + p''$$

and that $p(0) = 75$ and $p'(0) = -15$. The equilibrium price at time t is the solution of the equation $S = D$. Find the equilibrium price at time t. Find the long-range equilibrium price.

32. Public debt. According to the Domar burden-of-debt model, the total public debt $D(t)$ can be modeled by the equation

$$D''(t) - \beta D(t) = 0$$

where β is the constant relative growth rate of income $(0 < \beta < 1)$.

(A) Find the general solution of this equation for any constant β.

(B) Find the particular solution satisfying $D(0) = 1$ and $D'(0) = -\sqrt{\beta}$.

(C) Find the limit of this particular solution as $t \to \infty$.

Social Sciences

33. Learning theory. The differential equation

$$y'' + 5y' + 4y = 8$$

is typical of the equations that occur in the study of learning curves of rats in certain types of psychological experiments. Find the particular solution satisfying $y(0) = 1$ and $y'(0) = 1$. Find the limit of this particular solution as $t \to \infty$.

7-5 Systems of Differential Equations

- Solution of Systems of Differential Equations
- Applications

Solutions of Systems of Differential Equations

In many applications we are interested in the relationship between two quantities, both of which are changing with respect to time. This often leads to a system of differential equations of the type illustrated in Example 21.

EXAMPLE 21 Solve:

$$\frac{dx}{dt} = x + y \qquad (1)$$

$$\frac{dy}{dt} = 4x - 2y \qquad (2)$$

where $x = x(t)$ and $y = y(t)$ are functions of t.

SOLUTION Equations (1) and (2) form a **first-order linear system of differential equations.** Systems of this form can be solved by eliminating one of the variables, in the same way systems of linear algebraic equations are solved.

$$\frac{dx}{dt} = x + y \qquad (1)$$

Solve equation (1) for y.

$$y = \frac{dx}{dt} - x \qquad (3)$$

Differentiate with respect to t.

$$\frac{dy}{dt} = \frac{d^2x}{dt^2} - \frac{dx}{dt} \qquad (4)$$

$$\frac{dy}{dt} = 4x - 2y \qquad (2)$$

Use equations (3) and (4) to substitute for dy/dt and y in equations (2).

$$\frac{d^2x}{dt^2} - \frac{dx}{dt} = 4x - 2\left(\frac{dx}{dt} - x\right)$$

Simplify and collect like terms to obtain a second-order homogeneous equation.

$$\frac{d^2x}{dt^2} + \frac{dx}{dt} - 6x = 0$$

Thus, we have completely eliminated y and dy/dt, so we now have a second-order differential equation involving x alone. The characteristic equation is $m^2 + m - 6 = 0$ which has roots $m_1 = -3$ and $m_2 = 2$. The general solution is

$$x = C_1e^{-3t} + C_2e^{2t}$$

To determine y, we must first compute dx/dt:

$$\frac{dx}{dt} = -3C_1e^{-3t} + 2C_2e^{2t}$$

Substituting for dx/dt and x in equation (3) gives

$$\begin{aligned} y &= -3C_1e^{-3t} + 2C_2e^{2t} - C_1e^{-3t} - C_2e^{2t} \\ &= -4C_1e^{-3t} + C_2e^{2t} \end{aligned}$$

Thus, the solution to this system is

$$x = C_1e^{-3t} + C_2e^{2t} \quad \text{and} \quad y = -4C_1e^{-3t} + C_2e^{2t}$$

You should check this solution.

Matched Problem 21 Solve:

$$\frac{dx}{dt} = x + y$$

$$\frac{dy}{dt} = 3x - y$$

Now, suppose we want to find a particular solution of the system in Example 21. Since there are two arbitrary constants in the general solution, this will require two initial conditions, one for x and one for y.

EXAMPLE 22 Use the general solution in Example 21 to find the particular solution of the system

$$\frac{dx}{dt} = x + y$$

$$\frac{dy}{dt} = 4x - y$$

that satisfies the initial conditions $x(0) = 3$ and $y(0) = -2$.

SOLUTION Substituting $x = 3$, $y = -2$, and $t = 0$ in the general solution from Example 21

$$x = C_1e^{-3t} + C_2e^{2t}$$

$$y = -4C_1e^{-3t} + C_2e^{2t}$$

we obtain the following system of equations:

$$C_1 + C_2 = 3$$

$$-4C_1 + C_2 = -2$$

The solution to this system is $C_1 = 1$ and $C_2 = 2$. Thus, the particular solution we are seeking is

$$x = e^{-3t} + 2e^{2t} \quad \text{and} \quad y = -4e^{-3t} + 2e^{2t}$$

Matched Problem 22 Use the general solution in Problem 21 to find the particular solution of the system

$$\frac{dx}{dt} = x + y$$

$$\frac{dy}{dt} = 3x - y$$

that satisfies the initial conditions $x(0) = 5$ and $y(0) = -3$.

Applications

In studying the relationship between the prices of two commodities in an interrelated market, economists often assume that the rate of change of each price is proportional to a linear combination of both prices. This leads to a system of first-order linear differential equations.

EXAMPLE 23 **Interrelated Markets** The prices p and q (in dollars) of two commodities in an interrelated market satisfy the following system of differential equations, where p and q are functions of time t (in years):

$$p' = p + 2q - 300 \qquad p(0) = 150 \qquad (5)$$

$$q' = -4p - 5q + 900 \qquad q(0) = 25 \qquad (6)$$

Find the solution to this system and analyze the long-term behavior of p and q, that is, the behavior of p and q as t gets larger and larger.

SOLUTION

$$p' = p + 2q - 300 \qquad (5)$$

Solve equation (5) for q.

$$q = \tfrac{1}{2}p' - \tfrac{1}{2}p + 150 \qquad (7)$$

Differentiate with respect to t.

$$q' = \tfrac{1}{2}p'' - \tfrac{1}{2}p'$$

Use equations (7) and (8) to substitute for q' and q in equation (6).

$$q' = -4p - 5q + 900 \qquad (6)$$

$$\tfrac{1}{2}p'' - \tfrac{1}{2}p' = -4p - 5(\tfrac{1}{2}p' - \tfrac{1}{2}p + 150) + 900$$

Simplify.

$$p'' + 4p' + 3p = 300$$

We have eliminated q and q', but the resulting equation is nonhomogeneous. Using the method illustrated in Example 20, we find (details omitted) the general solution of this nonhomogeneous equation to be

$$p = C_1e^{-t} + C_2e^{-3t} + 100$$

Substituting for p and p' in (7) determines q:

$$q = -C_1e^{-t} - 2C_2e^{-3t} + 100$$

Applying the initial conditions $p = 150$ and $q = 25$ when $t = 0$ produces the following system of equations:

$$C_1 + C_2 = 50$$
$$C_1 + 2C_2 = 75$$

The solution to this system is $C_1 = 25$ and $C_2 = 25$. Thus,

$$p = 25e^{-t} + 25e^{-3t} + 100$$
$$q = -25e^{-t} - 50e^{-3t} + 100$$

To determine the behavior of p and q for large values of t, first note that

$$p' = -25e^{-t} - 75e^{-3t} < 0 \qquad \text{for all } t$$

and

$$q' = 25e^{-t} + 150e^{-3t} > 0 \qquad \text{for all } t$$

Thus, p is always decreasing and q is always increasing. Furthermore,

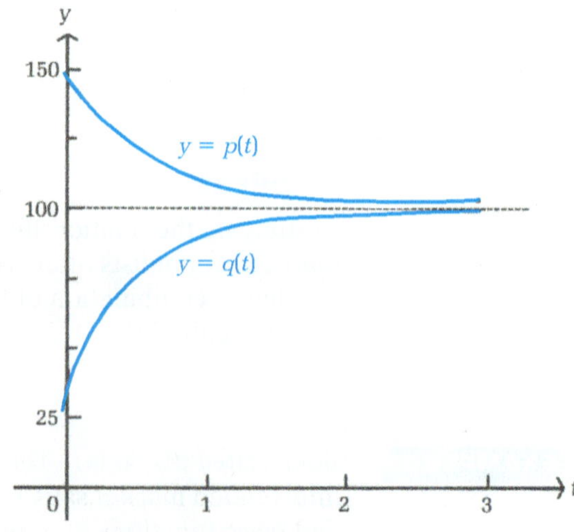

$$\lim_{t\to\infty} p = \lim_{t\to\infty} (25e^{-t} - 25e^{-3t} + 100)$$
$$= 0 + 0 + 100 = 100$$

and

$$\lim_{t\to\infty} q = \lim_{t\to\infty} (-25e^{-t} - 50e^{-3t} + 100)$$
$$= 0 + 0 + 100 = 100$$

The graph of $y = p(t)$ is always falling and approaches the line $y = 100$ from above, while the graph of $y = q(t)$ is always rising and approaches the line $y = 100$ below (see the figure at the top of the next page).

Matched Problem 23 Repeat Example 23 for the system

$$p' = -3p + q + 225 \qquad p(0) = 150$$
$$q' = 2p - 4q + 100 \qquad q(0) = 35$$

In studying organs in the body, scientists often must deal with the relationship between different organs or different parts of the same organ. Each organ or part of an organ is considered to be a compartment. A given substance, such as a drug, may be able to enter or leave a compartment at certain rates and may be able to pass back, and forth between adjacent compartments.

EXAMPLE 24 **Compartment Analysis** Suppose an organ has two compartments separated by a membrane, and assume that a drug injected into compartment 1 can move back and forth between the compartments and can also leave the organ from compartment 2 (see the figure).

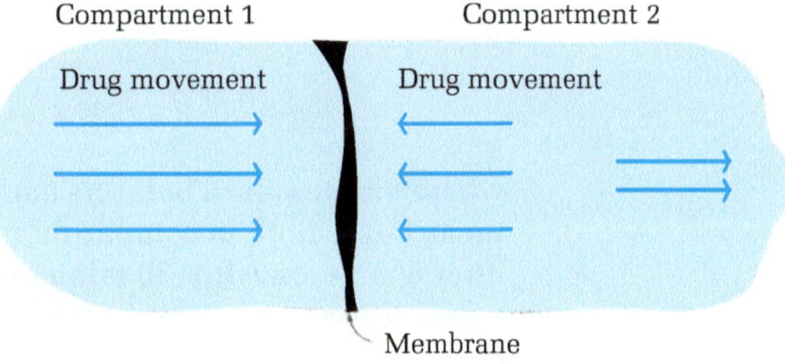

If x is the drug concentration in compartment 1 and y is the drug concentration in compartment 2 at any time t, experimental evidence suggests that x and y satisfy the following system of differential equations:

$$\frac{dx}{dt} = y - 6x$$

$$\frac{dx}{dt} = 6 - 7y$$

where x and y are functions of time t in hours. If there was no drug present in either compartment prior to the injection and if 40 units are injected into compartment 1, then we have initial conditions $x(0) = 40$ and $y(0) = 0$. Find the solution of the system, and find the amount of the drug present in each compartment after 6 minutes.

SOLUTION Proceeding as before (details omitted), we arrive at the second-order equation

$$\frac{d^2x}{dt^2} + 13\frac{dx}{dt} + 36x = 0$$

which has the solution

$$x = -C_1e^{-9t} + C_2e^{-4t}$$

This gives

$$y = -3C_1e^{-9t} + 2C_2e^{-4t}$$

Applying the initial conditions produces the following system of equations:

$$C_1 + C_2 = 40$$

$$-3C_1 + 2C_2 = 0$$

which has the solution $C_1 = 16$ and $C_2 = 24$. Thus, the amount of the drug present in each compartment at time t is

$$x = 16e^{-9t} + 24e^{-4t} \quad \text{Compartment 1}$$

$$y = -48e^{-9t} + 48e^{-4t} \quad \text{Compartment 2}$$

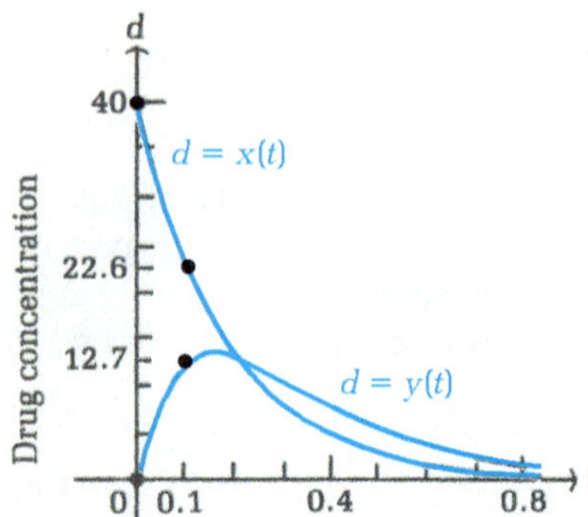

If t is measured in hours, then after 6 minutes, or 0.1 hour, the amount of the drug present in each compartment is (see the figure in the margin)

$$x(0, 1) = 16e^{-0.9} + 24e^{-0.4} \approx 22.6 \text{ units}$$

$$y(0.1) = -48e^{-0.9} + 48e^{-0.4} \approx 12.7 \text{ units}$$

Of the original 40 units injected into compartment 1, 22.6 units are still present, 12.7 units are now in compartment 2, and so 4.7 units have left the organ completely.

Matched Problem 24 If the drug concentrations x and y are governed by the system of equations

$$\frac{dx}{dt} = y - 2x$$

$$\frac{dy}{dt} = 2x - 3y$$

where t is measured in hours and the initial concentration is 30 units in compartment 1 and more in compartment 2, find the concentration in each compartment after 6 minutes. After 30 minutes. After 1 hour.

The population growth of many different species over short periods of time is often governed by the exponential growth law $dx/dt = kx$. If we consider the situation where two species compete for the same food supply, it is reasonable to assume that the rate of growth of each species will be affected by the size of the populations of both species. If x and y are the populations of the two species, then the following system of differential equations can be used as a model for this situation:

$$\frac{dx}{dt} = ax - by$$

$$\frac{dy}{dt} = -cx + dy$$

where a, b, c, and d are positive constants. Notice that the first equation indicates that the rate of growth of x increases as x increases but decreases as y increases. The term $-by$ introduces the competition between the two species. The second equation may be interpreted in a similar manner.

EXAMPLE 25 **Growth of Interacting Species** The populations x and y (in thousands) of two species satisfy the system of differential equations

$$\frac{dx}{dt} = 0.125x - 0.05y$$

$$\frac{dy}{dt} = -0.05x + 0.05y$$

where t is measured in years and the initial conditions are $x(0) = 90$ and $y(0) = 130$. Find the solution to this system and analyze the long-term growth of each species.

SOLUTION Eliminating y and dy/dt (details omitted) leads to the second-order differential equation

$$\frac{d^2x}{dt^2} - 0.175\frac{dx}{dt} + 0.00375x = 0$$

which has the solution

$$x = C_1e^{0.15t} + C_2e^{0.025t}$$

Substituting for dx/dt and x in the first equation and solving for y gives

$$y = -0.5C_1e^{0.15t} + 2C_2e^{0.025t}$$

Applying the initial conditions gives the system

$$C_1 + C_2 = 90$$
$$-\tfrac{1}{2}C_1 + 2C_2 = 130$$

which has the solution $C_1 = 20$ and $C_2 = 70$. Thus,

$$x = 20e^{0.15t} + 70e^{0.025t} \quad \text{and} \quad y = -10e^{0.15t} + 140e^{0.025t}$$

For example, the population of each species after 10 years is

$$x(10) = 20e^{1.5} + 70e^{0.25} \approx 180$$
$$y(10) = -10e^{1.5} + 140e^{0.25} \approx 135$$

We see that x has increased from 90 to 180 thousand and y has increased from 130 to 135 thousand. Since x is the sum of two increasing functions, the first species will continue to increase. On the other hand, y is the difference between two increasing functions. Is it possible that the second species will die out? That is, is y ever 0? To find out, we set the solution for y equal to 0 and try to solve for t:

$$y = -10e^{0.15t} + 140e^{0.025t} = 0$$
$$10e^{0.15t} = 140e^{0.025t}$$
$$\frac{Pe^{0.15t}}{e^{0.025t}} = \frac{140}{10}$$
$$e^{0.125t} = 14$$
$$0.125t = \ln 14$$
$$t = \frac{\ln 14}{0.125} \approx 21$$

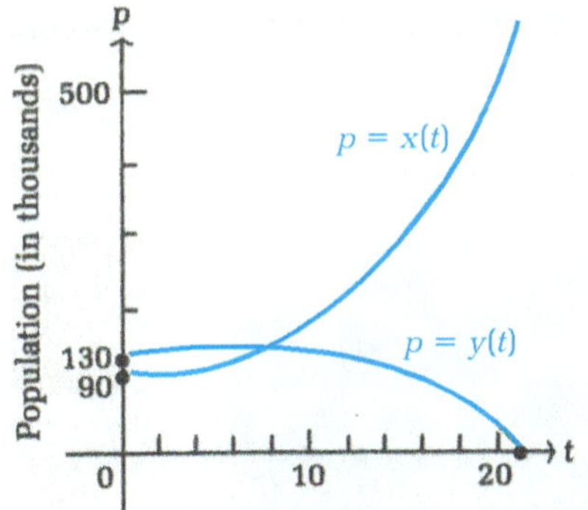

This indicates that the second species will die out after 21 years (see the figure in the margin), and since negative populations do not make any sense, the solution to this system should not be used for larger values of t.

Matched Problem 25 Repeat Example 25 for the system

$$\frac{dx}{dt} = 0.04x - 0.02y \qquad x(0) = 50$$
$$\frac{dy}{dt} = -0.01x + 0.03y \qquad y(0) = 35$$

Answers to Matched Problems

21. $x = C_1e^{2t} + C_2e^{-2t}$; $y = C_1e^{2t} - 3C_2e^{-2t}$
22. $x = 3e^{2t} + 2e^{-2t}$; $y = 3e^{2t} - 6e^{-2t}$
23. $p = 20e^{-2t} + 30e^{-5t} + 100, q = 20e^{-2t} - 60e^{-5t} + 75; \lim_{t\to\infty} p = 100, \lim_{t\to\infty} q = 75$
24. Solution to the system: $x = 10e^{-4t} + 20e^{-t}, y = -20e^{-4t} + 20e^{-t}$

    Concentrations are (approx.) 24.8 units in compartment 1 and 4.7 units in compartment 2 after 6 min, 13.5 units in compartment 1 and 9.4 units in compartment 2 after 30 min, 7.5 units in compartment 1 and 7.0 units in compartment 2 after 1 hr
25. $x(t) = 40e^{0.02t} + 10e^{0.05t}, y(t) = 40e^{0.02t} - 5e^{0.05t}$; the first species grows without bound; the second species will die out after approx. 69 yr.

Exercise 7-5

A

Find the general solution for each system of differential equations, and find the particular solution when initial conditions are given.

1. $x' = -x + y$; $y' = 2x$
2. $x' = 3x + y$; $y' = -2x$
3. $x' = 2x - y$; $y' = 3x - 2y$; $x(0) = 1$; $y(0) = -1$
4. $x' = 5x + 2y$; $y' = 2x + 2y$; $x(0) = 3$; $y(0) = -1$
5. $x' = 2x + y$; $y' = 2x + y$; $x(0) = 2$; $y(0) = -1$
6. $x' = x - y$; $y' = x - y$; $x(0) = 1$; $y(0) = 2$

B

Find the general solution for each system of differential equations, and find the particular solution when initial conditions are given.

7. $x' = -2x + y$; $y' = -3x + 2y$
8. $x' = 3x - y$; $y' = 4x - y$
9. $x' = 5x - 3y + 2$; $y' = 6x - 4y + 4$
10. $x' = y - 3$; $y' = 4x - 16$; $x(0) = 3$; $y(0) = 1$

Applications

Business & Economics

11. **Interrelated markets.** The prices p and q (in dollars) of two commodities in an interrelated market satisfy the following system of equations:

$$p' = -4p + q + 260 \qquad p(0) = 100$$
$$q' = -2p - q + 250 \qquad q(0) = 100$$

Find the solution to this system and analyze the long-term behavior of p and q.

12. **Interrelated markets.** Repeat Problem 11 for the following system:

$$p' = -26p - 40q + 1{,}320 \qquad p(0) = 33$$
$$q' = 15p + 23q - 760 \qquad q(0) = 12$$

Life Sciences

13. **Compartment analysis.** An organ has two compartments separated by a membrane. The drug concentration in the first compartment is represented by $x(t)$ and in the second compartment by $y(t)$ at any time t (measured in hours). Initially, there are no traces of a drug in either compartment. Fifty units of a drug are injected into compartment 1. The flow of the drug is governed by the equations

$$x' = y - 5x$$
$$y' = x - 5y$$

Find the concentration in each compartment after 6 minutes. After 30 minutes. After 1 hour.

14. **Population growth.** The populations x and y (in thousands) of two competing species satisfy the system of differential equations

$$\frac{dx}{dt} = 0.09x - 0.02y \qquad x(0) = 200$$
$$\frac{dy}{dt} = -0.02x + 0.06y \qquad y(0) = 150$$

where t is measured in years. Find the solution of this system and analyze the long-term growth of each species.

In Problems 15–18, use a graphic calculator or a computer to graph the particular solution of the indicated system in the given viewing rectangle.

Business & Economics

15. **Interrelated markets.** System from Problem 11
t range: [0, 2]
p and q range: [75, 100]

16. **Interrelated markets.** System from Problem 12
t range: [0, 3]
p and q range: [0, 40]

Life Sciences

17. **Compartment analysis.** System from Problem 13
t range: [0, 1]
x and y range: [0, 50]

18. **Population growth.** System from Problem 14
t range: [0, 30]
x and y range: [0, 2,000]

Chapter 7 Review

Important Terms and Symbols

7-1 Basic Concepts

Differential equation; order; solution; general solution; family of solutions; particular solution; initial condition; implicit solution; dynamically stable price; equilibrium price

7-2 Separation of Variables

Separation of variables; singular solution; growth laws

$$\int f(y)\,dy = \int g(x)\,dx$$

7-3 First-Order Linear Differential Equations

First-order linear differential equation; standard form; integrating factor; general solution; equilibrium price at time t; long-range equilibrium price

$$y' + f(x)y = g(x); \quad I(x) = e^{\int f(x)\,dx};$$
$$y = \frac{1}{I(x)} \int I(x)g(x)\,dx$$

7-4 Second-Order Differential Equations

Second-order linear differential equation; homogeneous second-order equation; nonhomogeneous second-order equation; characteristic equation; distinct roots; repeated root

$$ay'' + by' + c = 0; \quad ay'' + by' + c = d$$

7-5 Systems of Differential Equations

First-order linear system of differential equations; interrelated markets; compartment analysis; interacting species

Review Exercise

Work through all the problems in this chapter review and check your answers in the back of the book. (Answers to all review problems are there.) Where weaknesses show up, review appropriate sections in the text.

Show that the function y is the general solution of the differential equation. On the same set of axes, graph the particular solutions obtained by letting C = −2, −1, 0, 11, and 2.

1. $y = C\sqrt{x} \quad 2xy' = y$

2. $y = 1 + Ce-x; \quad y' + y = 1$

Find the general solution.

3. $y' = -\dfrac{4y}{x}$

4. $y' = -\dfrac{4y}{x+4}$

5. $y' = 3x^2y^2$

6. $y' = 2y - e^x$

7. $y' = \dfrac{5}{x}y + x^6$

8. $y' = \dfrac{3+y}{2+x} \quad x > -2$

9. $y' = 10 - y; \quad y(0) = 0$

10. $y'' + 12y' + 36y = 0$

B

Find the particular solution that satisfies the given condition(s).

11. $y' = 10 - y;\quad y(0) = 0$

12. $y' + y = x;\quad y(0) = 0$

13. $y' = 2ye^{-x};\quad y(0) = 1$

14. $y' = \dfrac{2x - y}{x + 4};\quad y(0) = 1$

15. $y' = \dfrac{x}{y + 4};\quad y(0) = 0$

16. $y' + \dfrac{2}{x}y = \ln x;\quad y(1) = 2$

17. $yy' = \dfrac{x(1 + y^2)}{1 + x^2};\quad y(0) = 1$

18. $y' + 2xy = 2e^{-x^2};\quad y(0) = 1$

19. $y'' + 4y' = 0;\quad y(0) = 1;\quad y'(0) = -2$

20. $y'' - 16y = 16;\quad y(0) = 1;\quad y'(0) = 0$

C

Find the particular solution that satisfies the initial conditions.

21. $x' = -2x + y$
$y' = -4x + 2y$
$x(0) = 1;\quad y(0) = 1$

22. $x' = -2x + 2y + 6$
$y' = 4x - 8$
$x(0) = 2;\quad y(0) = 2$

Applications

Business & Economics

23. Depreciation. A refrigerator costs $500 when it is new. The value of the refrigerator depreciates to $25 over a 20 year period. If the rate of change of the value is proportional to the value, find the value of the refrigerator 5 years after it was purchased.

24. Sales growth. A new company has sales of $50,000 during the first year of operation. The rate of growth of the annual sales s is proportional to the difference between s and an upper limit of $200,000. Assuming $s = 0$ at $t = 0$, how long will it take for the annual sales to reach $150,000?

25. Supply–demand. The supply S and demand D for a certain commodity satisfy the equations

$$S = 100 + p + p' \qquad \text{and} \qquad D = 200 - p' - p$$

If $p = 75$ when $t = 0$, find the equilibrium price at time t, and find the long-range equilibrium price.

26. Continuous compound interest. An initial deposit of $60,000 is made into an account that earn 5% compounded continuously. Money is then withdrawn at a constant rate of $5,000 a year until the amount in the account is 0. Find the amount in the account at any time t. When is the amount 0? What is the total amount withdrawn from the account?

27. Interrelated markets. The price p and q (in dollars) of two commodities of an interrelated market satisfy the following system of equations:

$$\begin{aligned} p' &= -2p + q + 125 \qquad & p(0) &= 50 \\ q' &= -2p - 5p + 575 \qquad & q(0) &= 150 \end{aligned}$$

Find the solution of this system and analyze the long-term behavior of p and q.

Life Sciences

28. Crop yield. The yield per acre, $y(t)$, of a corn crop satisfies the equation

$$\frac{dy}{dt} = 100 + e^{-t} - y$$

If $y(0) = 0$, find y at any time t. (The yield is in bushels per acre.)

29. Pollution. A 1,000 gallon holding tank contains 100 gallons of unpolluted water. Water containing 2 pounds of pollutant per gallon is pumped into the tank at the rate of 100 gallons per hour. The uniformly mixed water is released from the tank at 50 gallons per hour. Find the total amount of pollutants in the tank after 2 hours.

30. Population growth. The populations x and y (in thousands) of two competing species satisfy the following system of differential equations:

$$\begin{aligned} \frac{dx}{dt} &= 0.03x - 0.01y \qquad & x(0) &= 75 \\ \frac{dy}{dt} &= -0.02x + 0.02y \qquad & y(0) &= 75 \end{aligned}$$

where t is measured in years. Find the solution of this system. Analyze the long-term behavior of each species.

Social Sciences

31. Rumor spread. A single individual starts a rumor in a community of 200 people. The rumor spreads at a rate proportional to the number of people who have not yet heard the rumor. After 2 days, 10 people have heard the rumor

(A) How many people will have heard the rumor after 5 days?

(B) How long will it take for the rumor to spread to 100 people?

Supplementary Problems

1. Find the general solution of the differential equation

$$y'' + 4y' + 4y = 8.$$

2. Find the general solution of the equation

$$y' - 6xy - 24x = 0.$$

3. Find the solution $y = y(x)$ of the differential equation

$$y\frac{dy}{dx} = 2x(y + 1)$$

for which $y(0) = 0$.

4. Find the particular solution of the equation

$$y' = 2y - 8x + 16$$

that satisfies $y(0) = -4$.

5. The supply S and the demand D for soccer balls satisfy the respective equations

$$S = 35 - 2p(t) + 3p'(t), \quad D = 95 - 5p(t) + 2p'(t)$$

where $p = p(t)$ is the price at time t. Find the equilibrium price at time t if $p(0) = 30$. The value $\overline{p}$ of $\lim_{t\to\infty} p(t)$ is called the *long range equilibrium* price. Determine $\overline{p}$ for this example.

6. Repeat problem 5 when

$$S = 70 - 3p(t) + 2p'(t), \quad D = 100 - 5p(t) + p'(t).$$

7. A flea collar for dogs contains an active ingredient that evaporates at a rate proportional to the amount present. Half of the ingredient evaporates in the first 25 days after the collar is removed from its protective packaging. If the collar becomes ineffective after 80% of the active ingredient has evaporated, how long will the collar remain effective?

8. A company wishes to set aside funds for future expansion and so arranges to make continuous deposits into a savings account at the rate of \$10,000 per year. The savings account earns 5% interest compounded continuously.
   (A) Set up the differential equation that is satisfied by the amount $f(t)$ of money in the account at time t.
   (B) Solve the differential equation in part (A), assuming that the balance of the account is zero at time $t = 0$. Determine how much money will be in the account at the end of 5 years.

9. A savings account earns 6% interest per year, compounded continuously, and continuous withdrawals are made from the account at the rate of \$900 per year. If $f(t)$ is the money in the account at time t, write down the differential equation that is satisfied by the account. Solve this differential equation and sketch typical solutions of it.

10. It is a comforting thought to a parachutist that no matter how far he/she falls, his/her velocity never exceeds 176 feet per second, but it does approach that value; this is called the parachutist's *terminal* velocity. If the velocity in feet per second is given by

$$v'(t) = 32 - kv(t)$$

where k is a constant, determine the value of k.

Taylor Polynomials and Infinite Series

Objectives

1. Calculate the nth-degree Taylor polynomial for a given function f at a specified real number a.
2. Calculate the Taylor series for a function f at a specified real number a.
3. Find the interval of convergence of a given Taylor series.
4. Perform various operations—addition, multiplication, differentiation, integration—on Taylor series.
5. Estimate the error in approximating a function by its nth-degree Taylor polynomial.
6. Approximate certain definite integrals using Taylor series.

Chapter Problem

The sum of the infinite series

$$3x - \frac{3^2}{2}x^2 + \frac{3^3}{3}x^3 - \cdots + \frac{(-1)^{n-1}3^n}{n}x^n + \cdots$$

is equal to $\ln(1 + 3x)$ for some, but not all, values of x. For example, if $x = 0$, the sum of the infinite series is 0 and the value of $\ln(1 + 3 \cdot 0)$ is also 0. However, if $x = 1$, the terms of the infinite series get larger and larger in absolute value, and the sum of the series is undefined. For which values of x is the sum of the series equal to $\ln(1 + 3x)$?

Introduction

The circuits inside a calculator are capable only of performing the basic operations of addition, subtraction, multiplication, and division. Yet, many calculators have keys that allow you to evaluate functions such as e^x, ln x, and sin x. How is this

Taken from *Additional Calculus Topics*, by Raymond A. Barnett, Michael R. Ziegler, and Karl E. Byleen.

done? In most cases, the values of these functions are *approximated* by using a carefully selected polynomial, and, of course, polynomials can be evaluated by using the basic arithmetic operations. Thus, the approximating polynomials give the calculator the capability of evaluating nonpolynomial functions. In this chapter we will study one type of approximating polynomial, called a Taylor polynomial after the English mathematician Brook Taylor (1685–1731). For a given function *f*, we will determine the coefficients of the Taylor polynomial, the values of *x* where the Taylor polynomial approximates $f(x)$, and the accuracy of this approximation. We will also consider various applications involving Taylor polynomials, including the approximation of definite integrals.

8-1 Taylor Polynomials

Higher-Order Derivatives

Up to this point, we have considered only the first and second derivatives of a function. In the work that follows, we will need to find higher-order derivatives. For example, if we start with the function *f* defined by

$$f(x) = x^{-1}$$

then the first derivative of *f* is

$$f'(x) = -x^{-2}$$

and the second derivative of *f* is

$$f''(x) = \frac{d}{dx}f'(x) = \frac{d}{dx}(-x^{-2}) = 2x^{-3}$$

Additional higher-order derivatives are found by successive differentiation. Thus, the third derivative is

$$f^{(3)}(x) = \frac{d}{dx}f''(x) = \frac{d}{dx}(2x^{-3}) = -6x^{-4}$$

the fourth derivative is

$$f^{(4)}(x) = \frac{d}{dx}f^{(3)}(x) = \frac{d}{dx}(-6x^{-4}) = 24x^{-5}$$

and so on.

In general, the symbol $f^{(n)}$ is used to represent the ***n*th derivative of the function *f*.** When stating formulas involving a function and its higher-order derivatives, it is convenient to let $f^{(0)}$ represent the function *f* (that is, the zeroth derivative of a function is just the function itself).

The order of the derivative must be enclosed in parentheses, because in most contexts, $f^n(x)$ is interpreted to mean the *n*th power of *f*, not the *n*th derivative. Thus, $f^2(x) = [f(x)]^2 = f(x)f(x)$, while $f^{(2)}(x) = f''(x)$.

Finding a particular higher-order derivative is a routine calculation. However, finding a formula for the nth derivative for arbitrary n requires careful observation of the patterns that develop as each successive derivative is found. Study the next example carefully. Many problems in this chapter will involve similar concepts.

EXAMPLE 1 **Finding nth Derivatives** Find the nth derivative of

$$f(x) = \frac{1}{1+x} = (1+x)^{-1}$$

SOLUTION We begin by finding the first four derivatives of f:

$$f(x) = (1+x)^{-1}$$
$$f'(x) = (-1)(1+x)^{-2}$$
$$f''(x) = (-1)(-2)(1+x)^{-3}$$
$$f^{(3)}(x) = (-1)(-2)(-3)(1+x)^{-4}$$
$$f^{(4)}(x) = (-1)(-2)(-3)(-4)(1+x)^{-5}$$

Notice that we did not multiply out the coefficient of $(1+x)^{-k}$ in each derivative. Our objective is to look for a pattern in the form of each derivative. Multiplying out these coefficients would tend to obscure any pattern that is developing. Instead, we observe that each coefficient is a product of successive negative integers that can be written as follows:

$$(-1) = (-1)^1(1)$$
$$(-1)(-2) = (-1)^2(1)(2)$$
$$(-1)(-2)(-3) = (-1)^3(1)(2)(3)$$
$$(-1)(-2)(-3)(-4) = (-1)^4(1)(2)(3)(4)$$

Next we note that the product of natural numbers in each expression on the right can be written in terms of a factorial:*

$$(-1) = (-1)^1 1! \qquad 1! = 1$$
$$(-1)(-2) = (-1)^2 2! \qquad 2! = 1 \cdot 2$$
$$(-1)(-2)(-3) = (-1)^3 3! \qquad 3! = 1 \cdot 2 \cdot 3$$
$$(-1)(-2)(-3)(-4) = (-1)^4 4! \qquad 4! = 1 \cdot 2 \cdot 3 \cdot 4$$

Substituting these last expressions in the derivatives of f, we have

$$f'(x) = (-1)^1 1!(1+x)^{-2} \qquad n = 1$$
$$f''(x) = (-1)^2 2!(1+x)^{-3} \qquad n = 2$$
$$f^{(3)}(x) = (-1)^3 3!(1+x)^{-4} \qquad n = 3$$
$$f^{(4)}(x) = (-1)^4 4!(1+x)^{-5} \qquad n = 4$$

This suggests that

$$f^{(n)}(x) = (-1)^n n!(1+x)^{-(n+1)} \qquad \text{Arbitrary } n$$

*For n a natural number, $n! = n(n-1)! = n(n-1)(n-2) \cdot \cdots \cdot 2 \cdot 1$ and $0! = 1$.

Matched Problem 1 Find the nth derivative of $f(x) = \ln x$.

EXPLORE & DISCUSS 1

(A) Compute the first six derivatives of the polynomial function $p(x) = 7x^3 - 9x^2 + 4x - 15$.

(B) Let $q(x)$ be a polynomial of degree n. For which orders are the higher-order derivatives of $q(x)$ equal to the constant function $f(x) = 0$?

Approximating e^x with Polynomials

We have already seen that the irrational number e and the exponential function e^x play important roles in many applications, including continuous compound interest, population growth, and exponential decay, to name a few. Now, given the function

$$f(x) = e^x$$

we would like to construct a polynomial function p whose values are close to the values of f, at least for some values of x. If we are successful, then we can use the values of p (which are easily computed) to approximate the values of f. We begin by trying to approximate f for values of x near 0 with a first-degree polynomial of the form

$$p_1(x) = a_0 + a_1 x \qquad (1)$$

We want to place conditions on p_1 that will enable us to determine the unknown coefficients a_0 and a_1. Since we want to approximate f for values near 0, it is reasonable to require that f and p_1 agree at 0. Thus,

$$a_0 = p_1(0) = f(0) = e^0 = 1$$

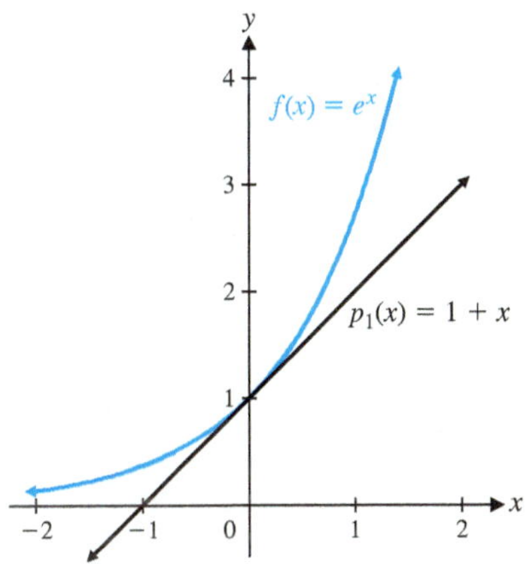

Figure 1

This determines the value of a_0. To determine the value of a_1, we require that both functions have the same slope at 0. Since $p_1'(x) = a_1$ and $f'(x) = e^x$, this implies that

$$a_1 = p_1'(0) = f'(0) = e^0 = 1$$

Thus, after substituting $a_0 = 1$ and $a_1 = 1$ into (1), we obtain

$$p_1(x) = 1 + x$$

which is a first-degree polynomial satisfying

$$p_1(0) = f(0) \qquad \text{and} \qquad p_1'(0) = f'(0)$$

How well does $1 + x$ approximate e^x? Examining the graph in Figure 1, it appears that $1 + x$ is a good approximation to e^x for x very close to 0. However, as x moves away from 0, in either direction, the distance between the values of $1 + x$ and e^x increases and the accuracy of the approximation decreases.

Now we will try to improve this approximation by using a second-degree polynomial of the form

$$p_2(x) = a_0 + a_1 x + a_2 x^2 \qquad (2)$$

We need three conditions to determine the coefficients a_0, a_1, and a_2. We still require that $p_2(0) = f(0)$ and $p_2'(0) = f'(0)$, and add the condition that $p_2''(0) = f''(0)$. This ensures that the graphs of p_2 and f have the same concavity at $x = 0$. Proceeding as before, we compute the first and second derivatives of p_2 and f and apply these conditions:

$$\begin{aligned} p_2(x) &= a_0 + a_1 x + a_2 x^2 \qquad & f(x) &= e^x \\ p_2'(x) &= \phantom{a_0 +{}} a_1 + 2a_2 x & f'(x) &= e^x \\ p_2''(x) &= \phantom{a_0 + a_1 +{}} 2a_2 & f''(x) &= e^x \end{aligned}$$

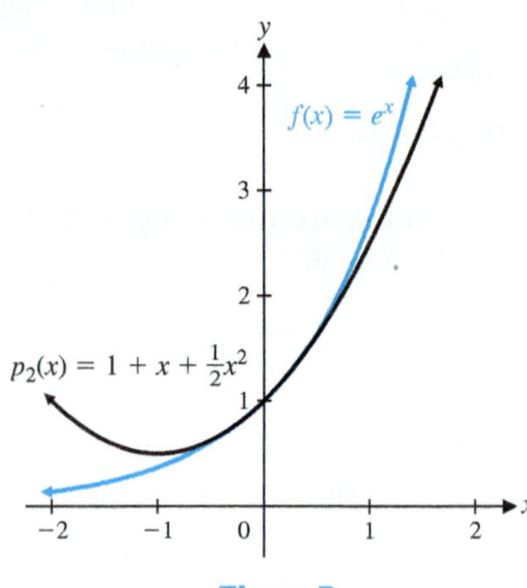

Figure 2

Thus,

$$\begin{aligned} a_0 &= p_2(0) = f(0) = e^0 = 1 &&\text{implies} && a_0 = 1 \\ a_1 &= p_2'(0) = f'(0) = e^0 = 1 &&\text{implies} && a_1 = 1 \\ 2a_2 &= p_2''(0) = f''(0) = e^0 = 1 &&\text{implies} && a_2 = \tfrac{1}{2} \end{aligned}$$

and, substituting in (2), we obtain

$$p_2(x) = 1 + x + \tfrac{1}{2}x^2$$

The graph is shown in Figure 2.

Comparing Figures 1 and 2, we see that for values of x near 0, the graph of $p_2(x) = 1 + x + \frac{1}{2}x^2$ is closer to the graph of $f(x) = e^x$ than was the graph of $p_1(x) = 1 + x$. Thus, the polynomial $1 + x + \frac{1}{2}x^2$ can be used to approximate e^x over a larger range of x values than the polynomial $1 + x$.

It seems reasonable to assume that a third-degree polynomial would yield a still better approximation. Using the notation for higher-order derivatives discussed earlier in this section, we can state the required condition as

Find

$$p_3(x) = a_0 + a_1x + a_2x^2 + a_3x^3 \tag{3}$$

Satisfying

$$p_3^{(k)}(0) = f^{(k)}(0) \qquad k = 0, 1, 2, 3$$

As before, we obtain the value of the additional coefficient a_3 by adding the requirement that

$$p_3^{(3)}(0) = f^{(3)}(0)$$

Thus,

$$\begin{aligned} p_3(x) &= a_0 + a_1x + a_2x^2 + a_3x^3 && f(x) = e^x \\ p_3'(x) &= a_1 + 2a_2x + 3a_3x^2 && f'(x) = e^x \\ p_3''(x) &= 2a_2 + 6a_3x && f''(x) = e^x \\ p_3^{(3)}(x) &= 6a_3 && f^{(3)}(x) = e^x \end{aligned}$$

Applying the conditions $p_3^{(k)}(0) = f^{(k)}(0)$, we have

$$\begin{aligned} a_0 &= p_3(0) = f(0) = e^0 = 1 &&\text{implies} && a_0 = 1 \\ a_1 &= p_3'(0) = f'(0) = e^0 = 1 &&\text{implies} && a_1 = 1 \\ 2a_2 &= p_3''(0) = f''(0) = e^0 = 1 &&\text{implies} && a_2 = \tfrac{1}{2} \\ 6a_3 &= p_3^{(3)}(0) = f^{(3)}(0) = e^0 = 1 &&\text{implies} && a_3 = \tfrac{1}{6} \end{aligned}$$

Substituting in (3), we obtain

$$p_3(x) = 1 + x + \tfrac{1}{2}x^2 + \tfrac{1}{6}x^3$$

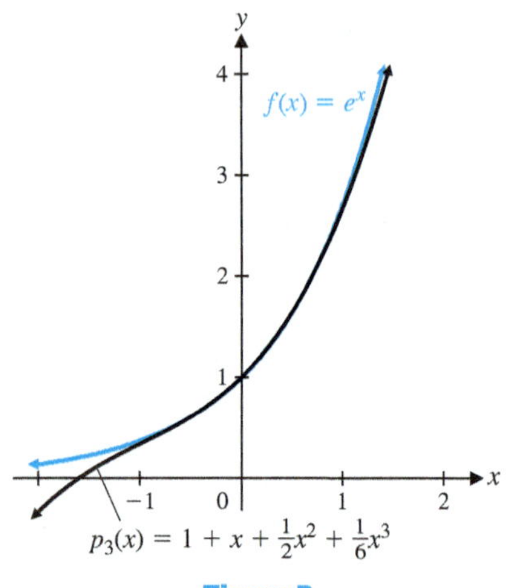

Figure 3

Figure 3 indicates that the approximations provided by p_3 are an improvement over those provided by p_2 and p_1.

In order to compare all three approximations, Table 1 lists the values of p_1, p_2, p_3, and e^x at selected values of x, and Table 2 lists the absolute value of the difference between these polynomials and e^x for the same x values. The values of e^x were obtained by using a calculator and are rounded to six decimal places. Comparing the columns in Table 2, we see that increasing the degree of the approximating polynomial decreases the difference between the polynomial and e^x. Thus, p_2 provides a better approximation than p_1, and p_3 provides a better approximation than p_2. We will have more to say about the accuracy of these approximations later in this chapter.

Table 1 $p_1(x) = 1 + x$, $p_2(x) = 1 + x + \frac{1}{2}x^2$, $p_3(x) = 1 + x + \frac{1}{2}x^2 + \frac{1}{6}x^3$

x	$p_1(x)$	$p_2(x)$	$p_3(x)$	e^x
−0.2	0.8	0.820	0.818 667	0.818 731
−0.1	0.9	0.905	0.904 833	0.904 837
0	1	1	1	1
0.1	1.1	1.105	1.105 167	1.105 171
0.2	1.2	1.220	1.221 333	1.221 403

Table 2

x	$\lvert p_1(x) - e^x\rvert$	$\lvert p_2(x) - e^x\rvert$	$\lvert p_3(x) - e^x\rvert$
−0.2	0.018 731	0.001 269	0.000 064
−0.1	0.004 837	0.000 163	0.000 004
0	0	0	0
0.1	0.005 171	0.000 171	0.000 004
0.2	0.021 403	0.001 403	0.000 069

EXPLORE & DISCUSS 2

(A) Let $p(x)$ be a polynomial of degree $n \geq 1$. Explain why each of the following is true:

(1) The graph of $p(x)$ intersects any vertical line in exactly one point.

(2) The graph of $p(x)$ intersects any horizontal line in at most n points.

(3) The graph of $p(x)$ has neither horizontal nor vertical asymptotes.

(B) Discuss the graphs of

$$f(x) = e^x \qquad g(x) = \frac{5x^2 - 3x + 1}{2x^2 + 3} \qquad h(x) = \frac{1}{x - 2} \qquad k(x) = e^{-x^2}$$

Use part (A) to explain why none of the graphs of f, g, h, or k is the graph of a polynomial function.

Taylor Polynomials at 0

The process we have used to determine p_1, p_2, and p_3 can be continued. Given any positive integer n, we define

$$p_n(x) = a_0 + a_1x + a_2x^2 + \cdots + a_nx^n$$

and require that

$$p_n^{(k)}(0) = f^{(k)}(0) \qquad k = 0, 1, 2, \ldots, n$$

The polynomial p_n is called a *Taylor polynomial.* Before determining p_n for $f(x) = e^x$, it will be convenient to make some general statements concerning the relationship between a_k, $p_n^{(k)}(0)$, and $f^{(k)}(0)$ for an arbitrary function f. First, $p_n(x)$ is differentiated n times to obtain the following relationships:

$$\begin{aligned}
p_n(x) &= a_0 + a_1x + a_2x^2 + a_3x^3 + \cdots + a_nx^n \\
p_n'(x) &= a_1 + 2a_2x + 3a_3x^2 + \cdots + na_nx^{n-1} \\
p_n''(x) &= 2a_2 + (3)(2)a_3x + \cdots + n(n-1)a_nx^{n-2} \\
p_n^{(3)}(x) &= (3)(2)a_3 + \cdots + n(n-1)(n-2)a_nx^{n-3} \\
&\vdots \\
p_n^{(n)}(x) &= n(n-1)(n-2)\cdot\cdots\cdot(1)a_n = n!a_n
\end{aligned}$$

Evaluating each derivative at 0 and applying the requirement that $p_n^{(k)}(0) = f^{(k)}(0)$leads to the following equations:

$$
\begin{aligned}
a_0 &= p_n(0) &&= f(0) \\
a_1 &= p_n'(0) &&= f'(0) \\
2a_2 &= p_n''(0) &&= f''(0) \\
(3)(2)a_3 &= p_n^{(3)}(0) &&= f^{(3)}(0) \\
&\vdots && \vdots \\
n!a_n &= p_n^{(n)}(0) &&= f^{(n)}(0)
\end{aligned}
$$

Solving each equation for a_k, we have

$$a_k = \frac{p_n^{(k)}(0)}{k!} = \frac{f^{(k)}(0)}{k!}$$

This relationship enables us to state the general definition of a Taylor polynomial.

DEFINITION Taylor Polynomial at 0*

If f has n derivatives at 0, then the ***n*th-degree Taylor polynomial for *f* at 0** is

$$p_n(x) = a_0 + a_1x + a_2x^2 + \cdots + a_nx^n = \sum_{k=0}^{n} a_kx^k$$

where

$$p_n^{(k)}(0) = f^{(k)}(0) \qquad \text{and} \qquad a_k = \frac{f^{(k)}(0)}{k!} \qquad k = 0, 1, 2, \ldots, n$$

* Taylor polynomials at 0 are also often referred to as *Maclaurin polynomials,* but we will not use this terminology.

This result can be stated in a form that is more readily remembered as follows:

DEFINITION Taylor Polynomial at 0: Concise Form

The nth-degree Taylor polynomial for f at 0 is

$$p_n(x) = f(0) + f'(0)x + \frac{f''(0)}{2!}x^2 + \cdots + \frac{f^{(n)}(0)}{n!}x^n = \sum_{k=0}^{n} \frac{f^{(k)}(0)}{k!}x^k$$

provided f has n derivatives at 0.

In each of the preceding definitions, notice that we have used both the **expanded notation**

$$
\begin{aligned}
p_n(x) &= a_0 + a_1x + a_2x^2 + \cdots + a_nx^n \\
&= f(0) + f'(0)x + \frac{f''(0)}{2!}x^2 + \cdots + \frac{f^{(n)}(0)}{n!}x^n
\end{aligned}
$$

and the more compact **summation notation**

$$p_n(x) = \sum_{k=0}^{n} a_kx^k = \sum_{k=0}^{n} \frac{f^{(k)}(0)}{k!}x^k$$

to represent a finite sum. In this chapter we will place more emphasis on the expanded notation, since it is usually easier to visualize a polynomial written in this notation, but we will continue to include the summation notation where appropriate.

CONCEPTUAL INSIGHT

The nth-degree Taylor polynomial for f at 0 is always a polynomial, but it does not necessarily have degree n, because it is possible that $f^{(n)}(0) = 0$. For example, if $g(x) = x^3 + 2x + 1$, then $g'(x) = 3x^2 + 2$ and $g''(x) = 6x$, so $g(0) = 1, g'(0) = 2$, and $g''(0) = 0$. Therefore, for this function g at 0, the second-degree Taylor polynomial is $p_2(x) = 1 + 2x$, which actually has degree 1, not 2.

Returning to our original function $f(x) = e^x$, it is now an easy matter to find the nth-degree Taylor polynomial for this function. Since $(d/dx)e^x = e^x$, it follows that

$$f^{(k)}(x) = e^x \qquad f^{(k)}(0) = e^0 = 1 \qquad a_k = \frac{f^{(k)}(0)}{k!} = \frac{1}{k!}$$

for all values of k. Thus, for any n, the nth-degree Taylor polynomial for e^x is

$$p_n(x) = 1 + x + \frac{1}{2!}x^2 + \frac{1}{3!}x^3 + \cdots + \frac{1}{n!}x^n = \sum_{k=0}^{n} \frac{1}{k!}x^k$$

EXAMPLE 2 **Approximation Using a Taylor Polynomial** Find the third-degree Taylor polynomial at 0 for $f(x) = \sqrt{x + 4}$. Use p_3 to approximate $\sqrt{5}$.

SOLUTION **Step 1** **Find the derivatives:**

$$\begin{aligned} f(x) &= (x + 4)^{1/2} \\ f'(x) &= \tfrac{1}{2}(x + 4)^{-1/2} \\ f''(x) &= -\tfrac{1}{4}(x + 4)^{-3/2} \\ f^{(3)}(x) &= \tfrac{3}{8}(x + 4)^{-5/2} \end{aligned}$$

Step 2 **Evaluate the derivatives at 0:**

$$\begin{aligned} f(0) &= 4^{1/2} = 2 \\ f'(0) &= \tfrac{1}{2}(4^{-1/2}) = \tfrac{1}{4} \\ f''(0) &= -\tfrac{1}{4}(4^{-3/2}) = -\tfrac{1}{32} \\ f^{(3)}(0) &= \tfrac{3}{8}(4^{-5/2}) = \tfrac{3}{256} \end{aligned}$$

Step 3 **Find the coefficients of the Taylor polynomial:**

$$\begin{aligned} a_0 &= \frac{f(0)}{0!} = f(0) = 2 \\ a_1 &= \frac{f'(0)}{1!} = f'(0) = \frac{1}{4} \\ a_2 &= \frac{f''(0)}{2!} = \frac{-\frac{1}{32}}{2} = -\frac{1}{64} \\ a_3 &= \frac{f^{(3)}(0)}{3!} = \frac{\frac{3}{256}}{6} = \frac{1}{512} \end{aligned}$$

Step 4 **Write down the Taylor polynomial:**

$$p_3(x) = 2 + \tfrac{1}{4}x - \tfrac{1}{64}x^2 + \tfrac{1}{512}x^3 \quad \text{Taylor polynomial}$$

To use p_3 to approximate $\sqrt{5}$, we must first determine the appropriate value of x:

$$\begin{aligned} f(x) = \sqrt{x+4} &= \sqrt{5} && \text{Square both sides.} \\ x + 4 &= 5 && \text{Solve for } x. \\ x &= 1 \end{aligned}$$

[**Note:** The value of $\sqrt{5}$ obtained by using a calculator is, to six decimal places, 2.236 068.]

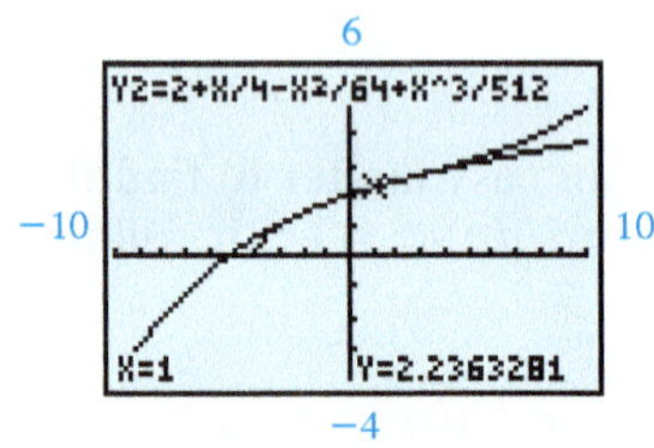

Figure 4 Graphs of $f(x) = \sqrt{x+4}$ and $p_3(x)$

Thus,

$$\sqrt{5} = f(1) \approx p_3(1) = 2 + \tfrac{1}{4} - \tfrac{1}{64} + \tfrac{1}{512} \approx 2.236\,328\,1$$

Compare the evaluation of $p_3(1)$ shown in Figure 4 to the value obtained in Example 2. The figure illustrates graphically that $p_3(x)$ is a good approximation to $f(x)$ for values of x near 0.

Matched Problem 2 Find the second-degree Taylor polynomial at 0 for $f(x) = \sqrt{x+9}$. Use $p_2(x)$ to approximate $\sqrt{10}$.

Many of the problems in this section involve approximating the values of functions that also can be evaluated with the use of a calculator. In such problems you should compare the Taylor polynomial approximation with the calculator value (which is also an approximation). This will give you some indication of the accuracy of the Taylor polynomial approximation. Later in this chapter we will discuss methods for determining the accuracy of any Taylor polynomial approximation.

EXAMPLE 3 **Taylor Polynomials at 0** Find the nth-degree Taylor polynomial at 0 for $f(x) = e^{2x}$.

SOLUTION

Step 1	Step 2	Step 3
$f(x) = e^{2x}$	$f(0) = 1$	$a_0 = f(0) = 1$
$f'(x) = 2e^{2x}$	$f'(0) = 2$	$a_1 = f'(0) = 2$
$f''(x) = 2^2e^{2x}$	$f''(0) = 2^2$	$a_2 = \frac{f''(0)}{2!} = \frac{2^2}{2!}$
$f^{(3)}(x) = 2^3e^{2x}$	$f^{(3)}(0) = 2^3$	$a_3 = \frac{f^{(3)}(0)}{3!} = \frac{2^3}{3!}$
$\vdots$	$\vdots$	$\vdots$
$f^{(n)}(x) = 2^ne^{2x}$	$f^{(n)}(0) = 2^n$	$a_n = \frac{f^{(n)}(0)}{n!} = \frac{2^n}{n!}$

Step 4 *Write down the Taylor polynomial:*

$$p_n(x) = 1 + 2x + \frac{2^2}{2!}x^2 + \frac{2^3}{3!}x^3 + \cdots + \frac{2^n}{n!}x^n = \sum_{k=0}^{n} \frac{2^k}{k!}x^k$$

Matched Problem 3 Find the nth-degree Taylor polynomial at 0 for $f(x) = e^{x/3}$.

Taylor Polynomials at *a*

Suppose we want to approximate the function $f(x) = \sqrt[4]{x}$ with a polynomial. Since $f^{(k)}(0)$is undefined for $k = 1, 2, \ldots$, this function does not have a Taylor polynomial at 0. However, all the derivatives of f exist at any $x > 0$. How can we generalize the definition of the Taylor polynomial to approximate functions such as this one? Before answering this question, we need to review a basic property of polynomials.

We are used to expressing polynomials in powers of x. However, it is also possible to express any polynomial in powers of $x - a$ for an arbitrary number a. For example, the following three expressions all represent the same polynomial:

$$\begin{aligned} p(x) &= 3 + 2x + x^2 && \text{Powers of } x \\ &= 6 + 4(x-1) + (x-1)^2 && \text{Powers of } x - 1 \\ &= 3 - 2(x+2) + (x+2)^2 && \text{Powers of } x + 2 \end{aligned}$$

To verify this statement, we expand the second and third expressions:

$$\begin{aligned} 6 + 4(x-1) + (x-1)^2 &= 6 + 4x - 4 + x^2 - 2x + 1 \\ &= 3 + 2x + x^2 \\ 3 - 2(x+2) + (x+2)^2 &= 3 - 2x - 4 + x^2 + 4x + 4 \\ &= 3 + 2x + x^2 \end{aligned}$$

Now we return to the problem of generalizing the definition of the Taylor polynomial. Proceeding as we did before, given a function f with n derivatives at a number a, we want to find an nth-degree polynomial p_n with the property that

$$p^{(k)}(a) = f^{(k)}(a) \qquad k = 0, 1, \ldots, n$$

That is, we require that p_n and its first n derivatives agree with f and its first n derivatives at the number a. It turns out that it is much easier to find p_n when it is expressed in powers of $x - a$. The general expression for an nth-degree polynomial in powers of $x - a$ and its first n derivatives are as follows:

$$\begin{aligned} p_n(x) &= a_0 + a_1(x-a) + a_2(x-a)^2 + a_3(x-a)^3 + \cdots + a_n(x-a)^n \\ p'_n(x) &= a_1 + 2a_2(x-a) + 3a_3(x-a)^2 + \cdots + na_n(x-a)^{n-1} \\ p''_n(x) &= 2a_2 + (3)(2)a_3(x-a) + \cdots + n(n-1)a_n(x-a)^{n-2} \\ p_n^{(3)}(x) &= (3)(2)(1)a_3 + \cdots + n(n-1)(n-2)a_n(x-a)^{n-3} \\ &\vdots \\ p_n^{(n)}(x) &= n(n-1) \cdot \cdots \cdot (1)a_n = n!a_n \end{aligned}$$

Now we evaluate each function at a and apply the appropriate condition:

$$\begin{aligned} a_0 &= p_n(a) &&= f(a) && k = 0 \\ a_1 &= p'_n(a) &&= f'(a) && k = 1 \\ 2a_2 &= p''_n(a) &&= f''(a) && k = 2 \\ (3)(2)(1)a_3 &= p_n^{(3)}(a) &&= f^{(3)}(a) && k = 3 \\ &\vdots && \vdots && \vdots \\ n!a_n &= p_n^{(n)}(a) &&= f^{(n)}(a) && k = n \end{aligned}$$

DEFINITION Taylor Polynomial at a

The ***n*th-degree Taylor polynomial at *a*** for a function f is

$$p_n(x) = f(a) + f'(a)(x - a) + \frac{f''(a)}{2!}(x - a)^2 + \cdots + \frac{f^{(n)}(a)}{n!}(x - a)^n$$
$$= \sum_{k=0}^{n} \frac{f^{(k)}(a)}{k!}(x - a)^k$$

provided f has n derivatives at a.

Thus, each coefficient of p_n satisfies

$$a_k = \frac{f^{(k)}(a)}{k!} \qquad k = 0, 1, 2, \ldots, n$$

Theoretically, a function f has an nth-degree Taylor polynomial at any value a where it has n derivatives. In practice, we usually choose a so that f and its derivatives are easy to evaluate at $x = a$. For example, if $f(x) = \sqrt{x}$,then good choices for a are 1, 4, 9, 16, and so on.

EXAMPLE 4 **Taylor Polynomials at *a*** Find the third-degree Taylor polynomial at $a = 1$ for $f(x) = \sqrt[4]{x}$. Use $p_3(x)$ to approximate $\sqrt[4]{2}$.

SOLUTION **Step 1 Find the derivatives:**

$$f(x) = x^{1/4}$$
$$f'(x) = \tfrac{1}{4}x^{-3/4}$$
$$f''(x) = -\tfrac{3}{16}x^{-7/4}$$
$$f^{(3)}(x) = \tfrac{21}{64}x^{-11/4}$$

Step 2 Evaluate the derivatives at a = 1:

$$f(1) = 1$$
$$f'(1) = \tfrac{1}{4}$$
$$f''(1) = -\tfrac{3}{16}$$
$$f^{(3)}(1) = \tfrac{21}{64}$$

Step 3 Find the coefficients of the Taylor polynomial:

$$a_0 = f(1) = 1$$
$$a_1 = f'(1) = \tfrac{1}{4}$$
$$a_2 = \frac{f''(1)}{2!} = \frac{-\frac{3}{16}}{2} = -\frac{3}{32}$$
$$a_3 = \frac{f^{(3)}(1)}{3!} = \frac{\frac{21}{64}}{6} = \frac{7}{128}$$

Step 4 Write down the Taylor polynomial:

$$p_3(x) = 1 + \tfrac{1}{4}(x - 1) - \tfrac{3}{32}(x - 1)^2 + \tfrac{7}{128}(x - 1)^3$$

Now we use the Taylor polynomial to approximate $\sqrt[4]{2}$:

$$\sqrt[4]{2} = f(2) \approx p_3(2) = 1 + \tfrac{1}{4} - \tfrac{3}{32} + \tfrac{7}{128} = 1.2109375$$

[**Note:** The value obtained by using a calculator is $\sqrt[4]{2} \approx 1.1892071$.]

Matched Problem 4 Find the second-degree Taylor polynomial at $a = 8$ for $f(x) = \sqrt[3]{x}$. Use $p_2(x)$ to approximate $\sqrt[3]{9}$.

Application

EXAMPLE 5 **Average Price** Given the demand function

$$p = D(x) = \sqrt{2{,}500 - x^2}$$

use the second-degree Taylor polynomial at 0 to approximate the average price (in dollars) over the demand interval $[10, 40]$.

SOLUTION The average price over the demand interval $[10, 40]$ is given by

$$\text{Average price} = \frac{1}{30}\int_{10}^{40} \sqrt{2{,}500 - x^2}\,dx$$

This integral cannot be evaluated by any of the techniques we have discussed. However, we can use a Taylor polynomial to approximate the value of the integral. Omitting the details (which you should supply), the second-degree Taylor polynomial at 0 for $D(x) = \sqrt{2{,}500 - x^2}$ is

$$p_2(x) = 50 - \tfrac{1}{100}x^2$$

Assuming that $D(x) \approx p_2(x)$ for $10 \le x \le 40$ (see Fig. 5), we have

$$\begin{aligned}\text{Average price} &\approx \frac{1}{30}\int_{10}^{40}\left(50 - \frac{1}{100}x^2\right)dx\\ &= \frac{1}{30}\left(50x - \frac{1}{300}x^3\right)\Big|_{10}^{40}\\ &= \$43\end{aligned}$$

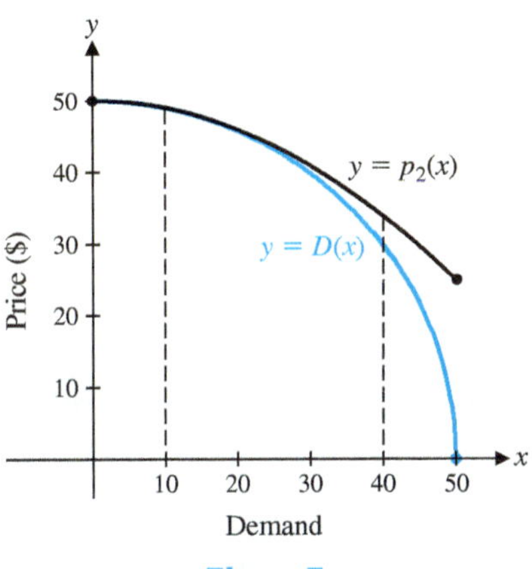

Figure 5

Matched Problem 5 Given the demand function

$$p = D(x) = \sqrt{400 - x^2}$$

use the second-degree Taylor polynomial at 0 to approximate the average price (in dollars) over the demand interval $[5, 15]$.

Example 5 illustrates the convenience of using Taylor polynomials to approximate more complicated functions, but it also raises several important questions. First, how can we determine whether it is reasonable to use the Taylor polynomial to approximate a function on a given interval? Second, is it possible to find the Taylor polynomial without resorting to successive differentiation of the function? Finally, can we determine the accuracy of an approximation such as the approximate average price found in Example 5? These questions will be discussed in the next three sections.

Answers to Matched Problems

1. $f^{(n)}(x) = (-1)^{n-1}(n-1)!x^{-n}$
2. $p_2(x) = 3 + \frac{1}{6}x - \frac{1}{216}x^2$; $\sqrt{10} \approx 3.162\,037$
3. $p_n(x) = 1 + \frac{1}{3}x + \frac{1}{3^2}\frac{1}{2!}x^2 + \frac{1}{3^3}\frac{1}{3!}x^3 + \cdots + \frac{1}{3^n}\frac{1}{n!}x^n = \sum_{k=0}^{n}\frac{1}{3^k}\frac{1}{k!}x^k$
4. $p_2(x) = 2 + \frac{1}{12}(x-8) - \frac{1}{288}(x-8)^2$; $\sqrt[3]{9} \approx 2.079\,861\,1$
5. $\frac{1}{10}\int_5^{15}\left(20 - \frac{x^2}{40}\right)dx = \frac{415}{24} \approx \$17.29$

Exercises 8-1

A

In Problems 1–4, find $f^{(3)}(x)$.

1. $f(x) = \dfrac{1}{x}$

2. $f(x) = \ln(1 + x)$

3. $f(x) = e^{-x}$

4. $f(x) = \sqrt{x}$

In Problems 5–8, find $f^{(4)}(x)$.

5. $f(x) = \ln(1 + 3x)$

6. $f(x) = e^{5x}$

7. $f(x) = \sqrt{1 + x}$

8. $f(x) = \dfrac{1}{2 + x}$

In Problems 9–16, find the indicated Taylor polynomial at 0.

9. $f(x) = e^{-x}; p_4(x)$

10. $f(x) = e^{4x}; p_3(x)$

11. $f(x) = (x + 1)^3; p_4(x)$

12. $f(x) = (1 - x)^4; p_3(x)$

13. $f(x) = \ln(1 + 2x); p_3(x)$

14. $f(x) = \ln(1 + \frac{1}{2}x); p_4(x)$

15. $f(x) = \sqrt[3]{x + 1}; p_3(x)$

16. $f(x) = \sqrt[4]{x + 16}; p_2(x)$

17. (A) Find the third-degree Taylor polynomial $p_3(x)$ for $f(x) = x^4 - 1$ at 0. For which values of x is $|p_3(x) - f(x)| < 0.1$?

(B) Find the fourth-degree Taylor polynomial $p_4(x)$ for $f(x) = x^4 - 1$ at 0. For which values of x is $|p_4(x) - f(x)| < 0.1$?

18. (A) Find the fourth-degree Taylor polynomial $p_4(x)$ for $f(x) = x^5$ at 0. For which values of x is $|p_4(x) - f(x)| < 0.01$?

(B) Find the fifth-degree Taylor polynomial $p_5(x)$ for $f(x) = x^5$ at 0. For which values of x is $|p_5(x) - f(x)| < 0.01$?

In Problems 19–24, find the indicated Taylor polynomial at the given value of a.

19. $f(x) = e^{x-1}; p_4(x)$ at 1

20. $f(x) = e^{2x}; p_3(x)$ at $\frac{1}{2}$

21. $f(x) = x^3; p_3(x)$ at 1

22. $f(x) = x^2 - 6x + 10; p_2(x)$ at $x = 3$

23. $f(x) = \ln(3x); p_3(x)$ at $\frac{1}{3}$

24. $f(x) = \ln(2 - x); p_4(x)$ at 1

25. Use the third-degree Taylor polynomial at 0 for $f(x) = e^{-2x}$ and $x = 0.25$ to approximate $e^{-0.5}$.

26. Use the fourth-degree Taylor polynomial at 0 for $f(x) = \ln(1 - x)$ and $x = 0.1$ to approximate $\ln 0.9$.

27. Use the second-degree Taylor polynomial at 0 for $f(x) = \sqrt{x + 16}$ and $x = 1$ to approximate $\sqrt{17}$.

28. Use the third-degree Taylor polynomial at 0 for $f(x) = \sqrt{(x + 4)^3}$ and $x = 1$ to approximate $\sqrt{125}$.

29. Use the fourth-degree Taylor polynomial at 1 for $f(x) = \sqrt{x}$ and $x = 1.2$ to approximate $\sqrt{1.2}$.

30. Use the third-degree Taylor polynomial at 4 for $f(x) = \sqrt{x}$ and $x = 3.95$ to approximate $\sqrt{3.95}$.

B

In Problems 31–36, find $f^{(n)}(x)$.

31. $f(x) = \dfrac{1}{4 - x}$

32. $f(x) = \dfrac{4}{1 + x}$

33. $f(x) = e^{3x}$

34. $f(x) = \ln(2x + 1)$

35. $f(x) = \ln(6 - x)$

36. $f(x) = e^{x/2}$

In Problems 37–42, find the nth-degree Taylor polynomial at 0. Write the answer in expanded notation.

37. $f(x) = \dfrac{1}{4 - x}$

38. $f(x) = \dfrac{4}{1 + x}$

39. $f(x) = e^{3x}$

40. $f(x) = \ln(2x + 1)$

41. $f(x) = \ln(6 - x)$

42. $f(x) = e^{x/2}$

In Problems 43–50, find the nth-degree Taylor polynomial at the indicated value of a. Write the answer in expanded notation.

43. $f(x) = \dfrac{1}{x}$; at -1

44. $f(x) = \dfrac{2}{x}$; at 1

45. $f(x) = \ln x$; at 1

46. $f(x) = e^x$; at -2

47. $f(x) = e^{-x}$; at 2

48. $f(x) = \ln(2 - x)$; at -1

49. $f(x) = x \ln x$; at 3

50. $f(x) = xe^x$; at 1

51. Let $f(x) = x^4 + 5x^2 + 1$.

(A) Find the third-degree Taylor polynomial $p_3(x)$ for f at 0.

(B) What is the degree of the polynomial $p_3(x)$?

52. Let $f(x) = x^5 + 2x^3 + 8x^2 + 1$.

(A) Find the fourth-degree Taylor polynomial $p_4(x)$ for f at 0.

(B) What is the degree of the polynomial $p_4(x)$?

53. Let $f(x) = x^6 + 2x^3 + 1$. For which values of n does $p_n(x)$, the nth-degree Taylor polynomial for f at 0, have degree n?

54. Let $f(x) = x^4 - 1$. For which values of n does $p_n(x)$, the nth-degree Taylor polynomial for f at 0, have degree n?

C

55. Find the first three Taylor polynomials at 0 for $f(x) = \ln(1 + x)$, and use these polynomials to complete the following two tables. Round all table entries to six decimal places.

x	$p_1(x)$	$p_2(x)$	$p_3(x)$	$f(x)$
-0.2				
-0.1				
0				
0.1				
0.2				

x	$\lvert p_1(x) - f(x)\rvert$	$\lvert p_2(x) - f(x)\rvert$	$\lvert p_3(x) - f(x)\rvert$
-0.2			
-0.1			
0			
0.1			
0.2			

56. Repeat Problem 55 for $f(x) = \sqrt{1 + x}$.

57. Refer to Problem 55. Graph $f(x) = \ln(1 + x)$ and its first three Taylor polynomials in the same viewing window. Use $[-3, 3]$ for both the x range and the y range.

58. Refer to Problem 56. Graph $f(x) = \sqrt{1 + x}$ and its first three Taylor polynomials in the same viewing window. Use $[-3, 3]$ for both the x range and the y range.

59. Consider $f(x) = e^x$ and its third-degree Taylor polynomial $p_3(x)$ at 0. Use graphical approximation techniques to find all values of x such that $|p_3(x) - e^x| < 0.1$.

60. Consider $f(x) = \ln(1 + x)$ and its third-degree Taylor polynomial $p_3(x)$ at 0. Use graphical approximation techniques to find all values of x such that

$$|p_3(x) - \ln(1 + x)| < 0.1$$

In Problems 61–64, write the answer in both expanded notation and summation notation.

61. Find the nth-degree Taylor polynomial for $f(x) = e^x$ at any point a.

62. Find the nth-degree Taylor polynomial for $f(x) = \ln x$ at any $a > 0$.

63. Find the nth-degree Taylor polynomial for $f(x) = 1/x$ at any $a \neq 0$.

64. Find the nth-degree Taylor polynomial at 0 for $f(x) = (x + c)^n$, where c is a constant.

65. Let $f(x)$ be a polynomial. For which values of n is the nth-degree Taylor polynomial for f at 0 equal to $f(x)$? Explain.

66. Explain how any polynomial expressed in powers of x may be rewritten in powers of $x - 1$ using Taylor polynomials. Illustrate your method for $f(x) = 1 + x^2 + x^4$.

67. Let $p_{10}(x)$ be the tenth-degree Taylor polynomial for $f(x) = e^x$ at 0. Do values of x exist for which $|p_{10}(x) - f(x)| \geq 100$? Explain.

68. Let $p_{12}(x)$ be the twelfth-degree Taylor polynomial for $f(x) = 1/x$ at 1. Do values of $x \neq 0$ exist for which $|p_{12}(x) - f(x)| \geq 100$? Explain.

Applications

Business & Economics

69. Average price. Given the demand equation

$$p = D(x) = \tfrac{1}{10}\sqrt{10{,}000 - x^2}$$

use the second-degree Taylor polynomial at 0 to approximate the average price (in dollars) over the demand interval $[0, 30]$.

70. Average price. Given the demand equation

$$p = D(x) = \tfrac{1}{5}\sqrt{1{,}600 - x^2}$$

use the second-degree Taylor polynomial at 0 to approximate the average price (in dollars) over the demand interval $[0, 15]$.

71. Average price. Refer to Problem 69. Use the second-degree Taylor polynomial at $a = 60$ to approximate the average price over the demand interval $[60, 80]$.

72. Average price. Refer to Problem 70. Use the second-degree Taylor polynomial at $a = 24$ to approximate the average price over the demand interval $[24, 32]$.

73. Production. The rate of production of a mine (in millions of dollars per year) is given by

$$R(t) = 2 + 8e^{-0.1t^2}$$

Use the second-degree Taylor polynomial at 0 to approximate the total production during the first 2 years of operation of the mine.

74. Production. The rate of production of an oil well (in millions of dollars per year) is given by

$$R(t) = 5 + 10e^{-0.05t^2}$$

Use the second-degree Taylor polynomial at 0 to approximate the total production during the first 2 years of operation of the well.

Life Sciences

75. Medicine. The rate of healing for a skin wound (in square centimeters per day) is given by

$$A'(t) = \frac{-75}{t^2 + 25}$$

The initial wound has an area of 12 square centimeters. Use the second-degree Taylor polynomial at 0 for $A'(t)$ to approximate the area of the wound after 2 days.

76. Medicine. Rework Problem 75 for $A'(t) = \dfrac{-60}{t^2 + 20}$.

77. Pollution. On an average summer day in a particular large city the air pollution level (in parts per million) is given by

$$P(x) = 20\sqrt{3x^2 + 25} - 80$$

where x is the number of hours elapsed since 8:00 A.M. Use the second-degree Taylor polynomial for $P(x)$ at $a = 5$ to approximate the average pollution level during the 10-hour period from 8:00 A.M. to 6:00 P.M.

78. Pollution. On an average summer day in another large city the air pollution level (in parts per million) is given by

$$P(x) = 10\sqrt{4x^2 + 36} - 50$$

where x is the number of hours elapsed since 8:00 A.M. Use the second-degree Taylor polynomial for $P(x)$ at $a = 4$ to approximate the average pollution level during the 8-hour period from 8:00 A.M. to 4:00 P.M.

Social Sciences

79. Learning. In a particular business college, it was found that an average student enrolled in an advanced typing class progresses at a rate of $N'(t) = 6e^{-0.01t^2}$ words per minute per week, t weeks after enrolling in a 15-week course. At the beginning of the course an average student could type 40 words per minute. Use the second-degree Taylor polynomial at 0 for $N'(t)$ to approximate the improvement in typing after 5 weeks in the course.

80. Learning. In the same business college, it was also found that an average student enrolled in a beginning shorthand class progressed at a rate of $N'(t) = 12e^{-0.005t^2}$ words per minute per week, t weeks after enrolling in a 15-week course. At the beginning of the course none of the students could take any dictation by shorthand. Use the second-degree Taylor polynomial at 0 for $N'(t)$ to approximate the improvement after 5 weeks in the course.

In Problems 81–88, use a graphing utility to graph the given function and its second-degree Taylor polynomial at a in the indicated viewing window.

Business & Economics

81. Average price. From Problem 69,

$$p = D(x) = \tfrac{1}{10}\sqrt{10{,}000 - x^2}$$
$$a = 0$$
x range: $[0, 100]$
p range: $[0, 10]$

82. Average price. From Problem 70,

$$p = D(x) = \tfrac{1}{5}\sqrt{1{,}600 - x^2}$$
$$a = 0$$
x range: $[0, 40]$
p range: $[0, 8]$

83. Production. From Problem 73,

$$y = R(t) = 2 + 8e^{-0.1t^2}$$
$$a = 0$$
t range: $[0, 5]$
y range: $[0, 10]$

84. Production. From Problem 74,

$$y = R(t) = 5 + 10e^{-0.05t^2}$$
$$a = 0$$
t range: $[0, 5]$
y range: $[0, 15]$

Life Sciences

85. Pollution. From Problem 77,

$$y = P(x) = 20\sqrt{3x^2 + 25} - 80$$
$$a = 5$$
x range: $[0, 10]$
y range: $[0, 300]$

86. Pollution. From Problem 78,

$$y = P(x) = 10\sqrt{4x^2 + 36} - 50$$
$$a = 4$$
x range: $[0, 10]$
y range: $[0, 200]$

Social Sciences

87. Learning. From Problem 79,

$$y = N'(t) = 6e^{-0.01t^2}$$
$$a = 0$$
t range: $[0, 10]$
y range: $[0, 6]$

88. Learning. From Problem 80,

$$y = N'(t) = 12e^{-0.005t^2}$$
$$a = 0$$
t range: $[0, 15]$
y range: $[0, 12]$

8-2 Taylor Series

- Introduction
- Taylor Series
- Representation of Functions by Taylor Series

Introduction

If f is a function with derivatives of all order at a point a, then we can construct the Taylor polynomial p_n at a for any integer n. Now we are interested in the relationship between the original function f and the corresponding Taylor polynomial p_n as n

assumes larger and larger values. If the Taylor polynomial is to be a useful tool for approximating functions, then the accuracy of the approximation should improve as we increase the size of n. Indeed, Figures 1, 2, and 3 and Tables 1 and 2 in the preceding section indicate that this is the case for the function e^x.

For a given value of x, we would like to know whether we can make $p_n(x)$ arbitrarily close to $f(x)$ by making n sufficiently large. In other words, we want to know whether

$$\lim_{n\to\infty} p_n(x) = f(x) \tag{1}$$

It turns out that for most functions with derivatives of all order at a point, there is a set of values of x for which equation (1) is valid. We will begin by considering a specific example to illustrate this.

Let

$$f(x) = \frac{1}{1-x} \quad \text{and} \quad a = 0$$

First, we find p_n:

Step 1	**Step 2**	**Step 3**
$f(x) = (1-x)^{-1}$	$f(0) = 1$	$a_0 = f(0) = 1$
$f'(x) = (-1)(1-x)^{-2}(-1) = (1-x)^{-2}$	$f'(0) = 1$	$a_1 = f'(0) = 1$
$f''(x) = (-2)(1-x)^{-3}(-1) = 2(1-x)^{-3}$	$f''(0) = 2$	$a_2 = \frac{f''(0)}{2!} = \frac{2}{2!} = 1$
$f^{(3)}(x) = (-3)(2)(1-x)^{-4}(-1) = 3!(1-x)^{-4}$	$f^{(3)}(0) = 3!$	$a_3 = \frac{f^{(3)}(0)}{3!} = \frac{3!}{3!} = 1$
$\vdots$	$\vdots$	$\vdots$
$f^{(n)}(x) = n!(1-x)^{-n-1}$	$f^{(n)}(0) = n!$	$a_n = \frac{f^{(n)}(0)}{n!} = \frac{n!}{n!} = 1$

Step 4 **Write down the Taylor polynomial:**

$$\begin{aligned} p_n(x) &= 1 + x + x^2 + x^3 + \cdots + x^n \\ &= \sum_{k=0}^{n} x^k \end{aligned} \tag{2}$$

Now we want to evaluate

$$\lim_{n\to\infty} p_n(x) = \lim_{n\to\infty} \sum_{k=0}^{n} x^k$$

It is not possible to evaluate limits written in this form. The difficulty lies in the fact that the number of terms is increasing as n increases. First, we must find a closed form for the summation that does not involve a sum of n terms. [You may recognize that $p_n(x)$ is a finite geometric series with common ratio x.]

If we multiply both sides of (2) by x and subtract this new equation from the original, we obtain an equation that we can solve for $p_n(x)$. Notice how many terms drop out when the subtraction is performed.

$$\begin{aligned} p_n(x) &= 1 + x + x^2 + \cdots + x^{n-1} + x^n \\ xp_n(x) &= \phantom{1 +{}} x + x^2 + \cdots + x^{n-1} + x^n + x^{n+1} \\ \hline p_n(x) - xp_n(x) &= 1 \qquad\qquad\qquad\qquad\qquad\qquad - x^{n+1} \\ p_n(x)(1-x) &= 1 - x^{n+1} \end{aligned}$$

$$p_n(x) = \frac{1 - x^{n+1}}{1 - x} \qquad x \neq 1$$

Since we had to exclude the value $x = 1$ to avoid division by 0, we must consider $x = 1$ as a special case:

$$p_n(1) = \overbrace{1 + 1 + 1^2 + \cdots + 1^n}^{n+1 \text{ terms}} = n + 1$$

Thus,

$$p_n(x) = \begin{cases} \dfrac{1 - x^{n+1}}{1 - x} & \text{if } x \neq 1 \\ n + 1 & \text{if } x = 1 \end{cases}$$

If $x = 1$, it is clear that

$$\lim_{n \to \infty} p_n(1) = \lim_{n \to \infty} (n + 1) = \infty$$

That is, $\lim_{n \to \infty} p_n(1)$ does not exist. What happens to this limit for other values of x? It can be shown that

$$\lim_{n \to \infty} x^{n+1} = \begin{cases} 0 & \text{if } -1 < x < 1 \\ 1 & \text{if } x = 1 \\ \text{Does not exist} & \text{if } x \leq -1 \text{ or } x > 1 \end{cases}$$

Experimenting with a calculator for various values of x and large values of n should convince you that this statement is valid.

Thus,

$$\lim_{n \to \infty} p_n(x) = \begin{cases} \dfrac{1}{1 - x} & \text{if } -1 < x < 1 & \lim_{n \to \infty} x^{n+1} = 0 \\ \text{Does not exist} & \text{if } x = 1 & \lim_{n \to \infty} (n + 1) \text{ does not exist} \\ \text{Does not exist} & \text{if } x \leq -1 \text{ or } x > 1 & \lim_{n \to \infty} x^{n+1} \text{ does not exist} \end{cases}$$

Explore & Discuss 1

(A) Construct a table with eight rows labeled with values of n from 1 to 8, and seven columns labeled with the following values of x: $-1.5, -1, -0.5, 0, 0.5, 1$, and 1.5. Complete the table by computing each entry $p_n(x)$.

(B) Explain how the table from part (A) provides supporting evidence for the statement that $\lim_{n \to \infty} p_n(x)$ equals $1/(1 - x)$ if $-1 < x < 1$, but does not exist if $x \leq -1$ or $x \geq 1$.

We can now conclude that

$$\begin{aligned} f(x) &= \frac{1}{1 - x} \\ &= \lim_{n \to \infty} p_n(x) \qquad -1 < x < 1 \\ &= \lim_{n \to \infty} \sum_{k=0}^{n} x^k \end{aligned}$$

and that the limit does not exist for any other values of x. This is often abbreviated by writing

$$\begin{aligned} \frac{1}{1 - x} &= 1 + x + x^2 + \cdots + x^n + \cdots \qquad -1 < x < 1 \qquad (3) \\ &= \sum_{k=0}^{\infty} x^k \end{aligned}$$

CONCEPTUAL INSIGHT

Equation (3) is often described in words as follows: The sum of an infinite geometric series, having first term 1 and common ratio x, is equal to $1/(1 - x)$, provided that the absolute value of the common ratio is less than 1.

The expression $\Sigma_{k=0}^{\infty} x^k$ is called the *Taylor series for f at* 0, and the expanded notation

$$1 + x + x^2 + \cdots + x^n + \cdots$$

is just another way to write this series. The Taylor series is said to *converge at x* if $\lim_{n\to\infty} p_n(x)$ exists and to *diverge at x* if this limit fails to exist. Thus, $\Sigma_{k=0}^{\infty} x^k$ converges for $-1 < x < 1$ and diverges for $x \le -1$ or $x \ge 1$. The set of values $\{x| -1 < x < 1\} = (-1, 1)$ where this Taylor series converges is called the *interval of convergence.* Since $f(x)$ is equal to the Taylor series for any x in the interval of convergence, we say that f is represented by its Taylor series throughout this interval. That is,

$$\begin{aligned} f(x) &= \frac{1}{1-x} \\ &= 1 + x + x^2 + \cdots + x^n + \cdots \qquad -1 < x < 1 \end{aligned}$$

There can be no relationship between f and its Taylor series outside the interval of convergence since the series is not defined there (see Fig. 1).

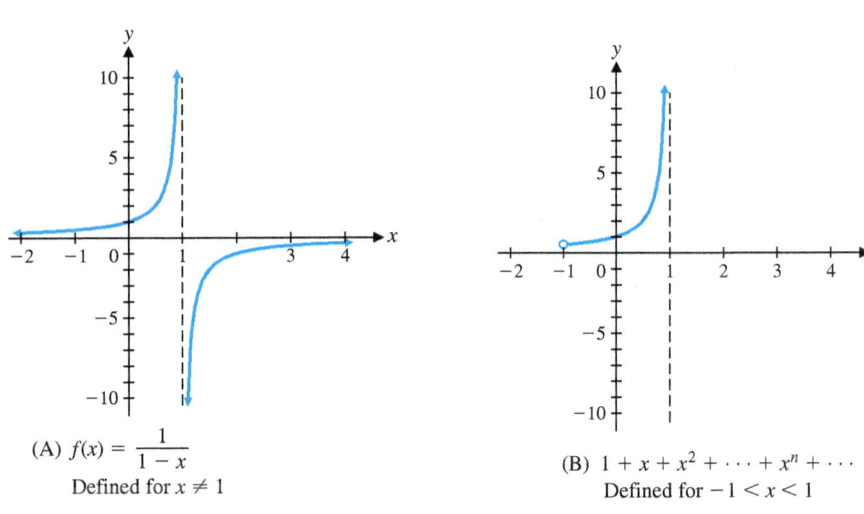

Figure 1

EXPLORE & DISCUSS 2

(A) The six functions $p_n(x) = 1 + x + \cdots + x^n, n = 1, 2, \ldots, 6$, are graphed in Figure 2. Which graph belongs to which function?

(B) Based solely on the graphs, would you conjecture that the interval of convergence of the Taylor series representation

$$\frac{1}{1-x} = 1 + x + x^2 + \cdots + x^n + \cdots$$

is $-1 < x < 1$? Explain.

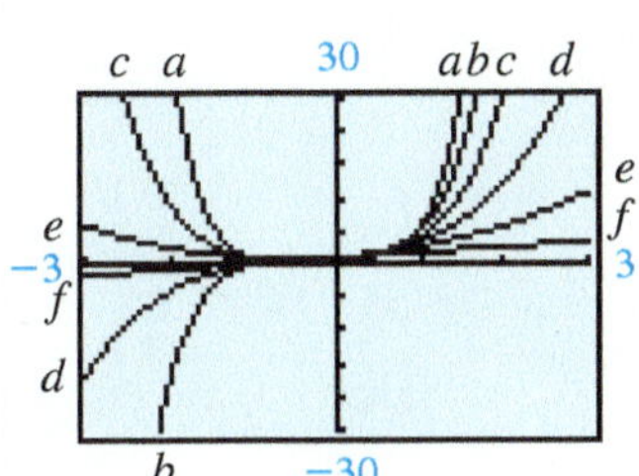

Figure 2

Taylor Series

We now generalize the concepts introduced in the preceding discussion to arbitrary functions.

DEFINITION Taylor Series

If f is a function with derivatives of all order at a point a, $p_n(x)$ is the nth-degree Taylor polynomial for f at a, and $a_n = f^{(n)}(a)/n!$, $n = 0, 1, 2, \ldots$, then

$$\sum_{k=0}^{\infty} a_k(x-a)^k = a_0 + a_1(x-a) + a_2(x-a)^2 + \cdots + a_n(x-a)^n + \cdots$$

is the **Taylor series for f at a.** The Taylor series **converges at x** if

$$\lim_{n\to\infty} p_n(x) = \lim_{n\to\infty} \sum_{k=0}^{n} a_k(x-a)^k$$

exists and **diverges at x** if this limit does not exist. The set of values of x for which this limit exists is called the **interval of convergence.**

We must emphasize that we are not really adding up an infinite number of terms in a Taylor series, nor is a Taylor series an "infinite" polynomial. Rather, for x in the interval of convergence,

$$\sum_{k=0}^{\infty} a_k(x-a)^k = \lim_{n\to\infty} \sum_{k=0}^{n} a_k(x-a)^k$$

Thus, when we write

$$\frac{1}{1-x} = 1 + x + x^2 + \cdots + x^n + \cdots \qquad -1 < x < 1$$

we mean

$$\frac{1}{1-x} = \lim_{n\to\infty}(1 + x + x^2 + \cdots + x^n) \qquad -1 < x < 1$$

In general, it is very difficult to find the interval of convergence by directly evaluating

$$\lim_{n\to\infty} \sum_{k=0}^{n} a_k(x-a)^k$$

For many Taylor series, the following theorem can be used to find the interval of convergence.

THEOREM 1 Interval of Convergence

Let f be a function with derivatives of all order at a point a, let$a_n = f^{(n)}(a)/n!$, $n = 0, 1, 2, \ldots$, and let

$$\sum_{k=0}^{\infty} a_k(x-a)^k = a_0 + a_1(x-a) + a_2(x-a)^2 + \cdots + a_n(x-a)^n + \cdots$$

be the Taylor series for f at a. If $a_n \neq 0$ for $n \geqslant n_0$, then:

Case 1. If

$$\lim_{n\to\infty}\left|\frac{a_{n+1}}{a_n}\right| = L > 0 \qquad \text{and} \qquad R = \frac{1}{L}$$

then the series converges for $|x-a| < R$ and diverges for $|x-a| > R$.

Case 2. If

$$\lim_{n\to\infty}\left|\frac{a_{n+1}}{a_n}\right| = 0$$

then the series converges for all values of x.

Case 3. If

$$\lim_{n\to\infty}\left|\frac{a_{n+1}}{a_n}\right| = \infty$$

then the series converges only at $x = a$.

The various possibilities for the interval of convergence are illustrated in Figure 3. In case 1, the series may or may not converge at the end points $x = a - R$ and $x = a + R$. (Determination of the behavior of a series at the end points of the interval of convergence requires techniques that we will not discuss.)

Case 1. $\lim_{n\to\infty}\left|\frac{a_{n+1}}{a_n}\right| = L > 0$

$R = 1/L$

Case 2. $\lim_{n\to\infty}\left|\frac{a_{n+1}}{a_n}\right| = 0$

Case 3. $\lim_{n\to\infty}\left|\frac{a_{n+1}}{a_n}\right| = \infty$

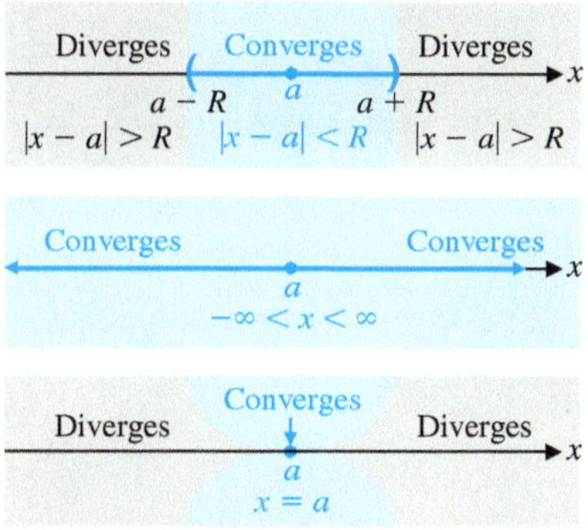

Figure 3 Intervals of convergence

Notice that the set of values where a series converges can be expressed in interval notation as $(a - R, a + R)$ in case 1 and as $(-\infty, \infty)$ in case 2. We were anticipating this result when we used the term *interval of convergence* to describe the set of points where a Taylor series converges.

Finally, the requirement that all the coefficients be nonzero from some point on ensures that the ratio a_{n+1}/a_n is well defined.

Remark—It is possible that

$$\lim_{n\to\infty}\left|\frac{a_{n+1}}{a_n}\right|$$

fails to exist so that none of the cases in Theorem 1 hold. In that event, the interval of convergence will still have one of the forms illustrated by Figure 3, but other techniques are required to determine which one.

CONCEPTUAL INSIGHT

Assume that $a_n \neq 0$ for $n \geq n_0$ and consider case 1 of Theorem 1. The ratio of consecutive terms of the Taylor series is equal to

$$\frac{a_{n+1}(x - a)^{n+1}}{a_n(x - a)^n}$$

The absolute value of this ratio for large n, by the limit in case 1, is approximately $L|x - a|$. Therefore, the Taylor series is approximately an infinite geometric series with common ratio having absolute value $L|x - a|$. Such a series converges if $L|x - a| < 1$, or equivalently, if $|x - a| < 1/L = R$.

The next two examples illustrate the application of Theorem 1. The details of finding the Taylor series are omitted. In the next section we will discuss methods that greatly simplify the process of finding Taylor series. For now, we want to concentrate on understanding the significance of the interval of convergence.

EXAMPLE 1 **Finding the Interval of Convergence** The Taylor series at 0 for $f(x) = e^{2x}$ is

$$1 + 2x + \frac{2^2}{2!}x^2 + \cdots + \frac{2^n}{n!}x^n + \cdots$$

Find the interval of convergence.

SOLUTION

$$a_n = \frac{2^n}{n!} \qquad \text{Notice that } a_n > 0 \text{ for } n \geq 0.$$

$$a_{n+1} = \frac{2^{n+1}}{(n+1)!}$$

$$\frac{a_{n+1}}{a_n} = \frac{\frac{2^{n+1}}{(n+1)!}}{\frac{2^n}{n!}} \qquad \text{Form the ratio } a_{n-1}/a_n \text{ and simplify.}$$

$$= \frac{2^{n+1}}{(n+1)!} \cdot \frac{n!}{2^n} \qquad \frac{n!}{(n+1)!} = \frac{n!}{(n+1)n!} = \frac{1}{n+1}$$

$$= \frac{2}{n+1}$$

$$\lim_{n\to\infty}\left|\frac{a_{n+1}}{a_n}\right| = \lim_{n\to\infty}\left|\frac{2}{n+1}\right| \qquad \text{Absolute value signs can be dropped since } a_{n+1}/a_n > 0.$$

$$= \lim_{n\to\infty}\frac{2}{n+1} = 0^*$$

Case 2 in Theorem 1 applies, and the series converges for all values of x.

Matched Problem 1 The Taylor series at 0 for $f(x) = e^{x/3}$ is

$$1 + \frac{1}{3}x + \frac{1}{3^2 2!}x^2 + \cdots + \frac{1}{3^n n!}x^n + \cdots$$

Find the interval of convergence.

EXAMPLE 2 **Finding the Interval of Convergence** The Taylor series at 0 for $f(x) = 1/(1 + 5x)^2$ is

$$1 - 2 \cdot 5x + 3 \cdot 5^2x^2 - \cdots + (-1)^n(n+1)5^nx^n + \cdots$$

Find the interval of convergence.

SOLUTION

$$a_n = (-1)^n(n+1)5^n \qquad \text{Notice that } a_n \neq 0 \text{ for } n \geq 0.$$

$$a_{n+1} = (-1)^{n+1}(n+2)5^{n+1}$$

$$\frac{a_{n+1}}{a_n} = \frac{(-1)^{n+1}(n+2)5^{n+1}}{(-1)^n(n+1)5^n}$$

$$= \frac{-5(n+2)}{n+1}$$

---

*The limit expressions $\lim_{x\to\infty} g(x)$ and $\lim_{n\to\infty} g(n)$ are very similar. If $\lim_{x\to\infty} g(x) = L$ [that is, $g(x)$ can be made arbitrarily close to L by making the *real number* x sufficiently large], then $\lim_{n\to\infty} g(n) = L$ [that is, $g(n)$ can be made arbitrarily close to L by making the *integer* n sufficiently large]. In particular,

$$\lim_{n\to\infty}\frac{2}{n+1} = \lim_{x\to\infty}\frac{2}{x+1} = 0$$

Thus, the results of Section 2-3 on limits at infinity for rational functions can be helpful in evaluating limits of the form $\lim_{n\to\infty} g(n)$.

$$\lim_{n\to\infty}\left|\frac{a_{n+1}}{a_n}\right| = \lim_{n\to\infty}\left|\frac{-5(n+2)}{n+1}\right|$$

$$= \lim_{n\to\infty}\frac{5(n+2)}{n+1}$$

$$= \lim_{n\to\infty}\frac{5n+10}{n+1} \qquad \text{Divide numerator and denominator by } n.$$

$$= \lim_{n\to\infty}\frac{5+\frac{10}{n}}{1+\frac{1}{n}} = 5$$

Applying case 1 in Theorem 1, we have $L = 5$, $R = \frac{1}{5}$, and the series converges for $-\frac{1}{5} < x < \frac{1}{5}$.

Matched Problem 2 The Taylor series at 0 for $f(x) = 4/(2+x)^2$ is

$$1 - \frac{2}{2}x + \frac{3}{2^2}x^2 - \cdots + \frac{(-1)^n(n+1)}{2^n}x^n + \cdots$$

Find the interval of convergence.

Representation of Functions by Taylor Series

Referring to the function $f(x) = 1/(1-x)$ discussed earlier in this section, our calculations showed that the Taylor series at 0 for f is

$$\lim_{n\to\infty} p_n(x) = 1 + x + x^2 + \cdots + x^n + \cdots$$

with interval of convergence $(-1, 1)$. Furthermore, we showed that f is represented by its Taylor series throughout the interval of convergence. That is,

$$f(x) = \frac{1}{1-x} = \lim_{n\to\infty} p_n(x)$$
$$= 1 + x + x^2 + \cdots + x^n + \cdots \qquad -1 < x < 1$$

A similar statement can be made for most functions that have derivatives of all order at a point a, but not for all. For example, the absolute value function $g(x) = |x|$ has Taylor series (verify this)

$$1 + (x - 1) = x$$

at the point $a = 1$. Clearly, this very simple Taylor series converges for all x, but $g(x) = x$ only for $x \ge 0$. So g is not represented by its Taylor series throughout the interval of convergence of the series. The problem with g is that it is not differentiable at 0. Even if we require that a function f has derivatives of all order at every point in the interval of convergence of its Taylor series, it is still possible that f is not represented by its Taylor series. Since we will not consider such functions, we will make the following assumption.

AGREEMENT **Assumption on the Representation of Functions by Taylor Series**

If f is a function with derivatives of all order throughout the interval of convergence of its Taylor series at a, then **f is represented by its Taylor series throughout the interval of convergence.** Thus, if $p_n(x)$ is the nth-degree Taylor polynomial at a for f and $a_n = f^{(n)}(a)/n!$, $n = 0, 1, 2, \ldots$, then

$$f(x) = \lim_{n\to\infty} p_n(x) = \sum_{k=0}^{\infty} a_k(x-a)^k$$

for any x in the interval of convergence of the Taylor series at a for f.

Consequently, the interval of convergence determines the values of x for which the Taylor polynomials $p_n(x)$ can be used to approximate the values of the function f.

Referring to Examples 1 and 2, we can now write

$$e^{2x} = 1 + 2x + \frac{2^2}{2!}x^2 + \cdots + \frac{2^n}{n!}x^n + \cdots \qquad -\infty < x < \infty$$

and

$$\frac{1}{(1+5x)^2} = 1 - 2 \cdot 5x + 3 \cdot 5^2x^2 - \cdots + (-1)^n(n+1)5^nx^n + \cdots \qquad -\tfrac{1}{5} < x < \tfrac{1}{5}$$

EXAMPLE 3 **Representation of a Function by Its Taylor Series** Let $f(x) = \ln x$.

(A) Find the nth-degree Taylor polynomial at $a = 1$ for f.
(B) Find the Taylor series at $a = 1$ for f.
(C) Determine the values of x for which $f(x) = \lim_{n\to\infty} p_n(x)$.

SOLUTION (A) We use the four-step process from the preceding section to find the nth-degree Taylor polynomial:

Step 1	**Step 2**	**Step 3**
$f(x) = \ln x$	$f(1) = \ln 1 = 0$	$a_0 = f(1) = 0$
$f'(x) = \frac{1}{x} = x^{-1}$	$f'(1) = 1$	$a_1 = f'(1) = 1$
$f''(x) = (-1)x^{-2}$	$f''(1) = -1$	$a_2 = \frac{f''(1)}{2!} = -\frac{1}{2}$
$f^{(3)}(x) = (-1)(-2)x^{-3} = (-1)^2 2! x^{-3}$	$f^{(3)}(1) = (-1)^2 2!$	$a_3 = \frac{f^{(3)}(1)}{3!} = \frac{(-1)^2 2!}{-3!^2} = \frac{1}{3}$
$f^{(4)}(x) = (-1)(-2)(-3)x^{-4} = (-1)^3 3! x^{-4}$	$f^{(4)}(1) = (-1)^3 3!$	$a_4 = \frac{f^{(4)}(1)}{4!} = \frac{(-1)^3 3!}{4!} = \frac{-1}{4}$
$\vdots$	$\vdots$	$\vdots$
$f^{(n)}(x) = (-1)^{n-1}(n-1)!x^{-n}$	$f^{(n)}(1) = (-1)^{n-1}(n-1)!$	$a_n = \frac{f^{(n)}(1)}{n!} = \frac{(-1)^{n-1}(n-1)!}{n!} = \frac{(-1)^{n-1}}{n}$

Step 4 The nth-degree Taylor polynomial at $a = 1$ for $f(x) = \ln x$ is

$$p_n(x) = (x-1) - \frac{1}{2}(x-1)^2 + \frac{1}{3}(x-1)^3 - \cdots + \frac{(-1)^{n-1}}{n}(x-1)^n$$

(B) Once the nth-degree Taylor polynomial for a function has been determined, finding the Taylor series is simply a matter of using the notation correctly. Using the Taylor polynomial from part (A), the Taylor series at $a = 1$ for $f(x) = \ln x$ can be written as

$$\lim_{n\to\infty} p_n(x) = (x-1) - \frac{1}{2}(x-1)^2 + \frac{1}{3}(x-1)^3 - \cdots + \frac{(-1)^{n-1}}{n}(x-1)^n + \cdots$$

or, using summation notation,* as

$$\lim_{n\to\infty} p_n(x) = \sum_{k=0}^{\infty} \frac{(-1)^{k-1}}{k}(x-1)^k$$

(C) To determine the values for which $f(x) = \lim_{n\to\infty} p_n(x)$, we use Theorem 1 to find the interval of convergence of the Taylor series in part (B):

$$a_n = \frac{(-1)^{n-1}}{n}$$

$$a_{n+1} = \frac{(-1)^n}{n+1}$$

$$\frac{a_{n+1}}{a_n} = \frac{\dfrac{(-1)^n}{n+1}}{\dfrac{(-1)^{n-1}}{n}} = \frac{(-1)^n}{n+1} \cdot \frac{n}{(-1)^{n-1}} = \frac{-n}{n+1}$$

$$\lim_{n\to\infty}\left|\frac{a_{n+1}}{a_n}\right| = \lim_{n\to\infty}\left|\frac{-n}{n+1}\right|$$

$$= \lim_{n\to\infty}\frac{n}{n+1} \quad \text{Divide numerator and denominator by } n.$$

$$= \lim_{n\to\infty}\frac{1}{1+\dfrac{1}{n}} = 1$$

From case 1 of Theorem 1, $L = 1$, $R = 1/L = 1$, and the series converges for $|x - 1| < 1$. Converting this to double inequalities, we have

$$|x-1| < 1 \quad |y| < c \text{ is equivalent to } -c < y < c.$$

$$-1 < x - 1 < 1 \quad \text{Add 1 to each side.}$$

$$0 < x < 2 \quad \text{Interval of convergence}$$

Thus, $f(x) = \lim_{n\to\infty} p_n(x)$ for $0 < x < 2$ or, equivalently,

$$\ln x = (x-1) - \frac{1}{2}(x-1)^2 + \frac{1}{3}(x-1)^3 - \cdots + \frac{(-1)^{n-1}}{n}(x-1)^n + \cdots \qquad 0 < x < 2$$

Matched Problem 3 Repeat Example 3 for $f(x) = 1/x$.

Table 1 displays the values of the Taylor polynomials for $f(x) = \ln x$ at $x = 1.5$ and at $x = 3$ for $n = 1, 2, \ldots, 10$ (see Example 3). Notice that $x = 1.5$ is in the interval of convergence of the Taylor series at 1 for $f(x)$ and that the values of $p_n(1.5)$ are approaching ln 1.5 as n increases. On the other hand, $x = 3$ is *not* in the interval of convergence and the values of $p_n(3)$ do not approach ln 3. These calculations illustrate the importance of the interval of convergence.

The values of x must be in the interval of convergence in order to use Taylor polynomials to approximate the values of a function.

*From this point on, we will use the expanded notation exclusively when writing Taylor series.

Table 1

$f(x) = \ln x$

$$p_n(x) = (x-1) - \frac{1}{2}(x-1)^2 + \frac{1}{3}(x-1)^3 - \cdots + \frac{(-1)^{n-1}}{n}(x-1)^n$$

n	$p_n(1.5)$	$p_n(3)$
1	0.5	2
2	0.375	0
3	0.416 667	2.666 667
4	0.401 042	−1.333 333
5	0.407 292	5.066 667
6	0.404 688	−5.6
7	0.405 804	12.685 714
8	0.405 315	−19.314 286
9	0.405 532	37.574 603
10	0.405 435	−64.825 397
	ln 1.5 = 0.405 465	ln 3 = 1.098 612

Answers to Matched Problems

1. $-\infty < x < \infty$
2. $-2 < x < 2$
3. (A) $1 - (x-1) + (x-1)^2 - \cdots + (-1)^n(x-1)^n$
   (B) $1 - (x-1) + (x-1)^2 - \cdots + (-1)^n(x-1)^n + \cdots$
   (C) $0 < x < 2$

Exercises 8-2

A

In Problems 1–12, find the interval of convergence of the given Taylor series representation.

1. $\dfrac{4}{1-x} = 4 + 4x + 4x^2 + \cdots + 4x^n + \cdots$

2. $\dfrac{3}{1-x} = 3 + 3x + 3x^2 + \cdots + 3x^n + \cdots$

3. $\dfrac{1}{1+7x} = 1 - 7x + 7^2x^2 - \cdots + (-1)^n 7^n x^n + \cdots$

4. $\dfrac{1}{1+2x} = 1 - 2x + 2^2x^2 - \cdots + (-1)^n 2^n x^n + \cdots$

5. $\dfrac{1}{6-x} = \dfrac{1}{6} + \dfrac{1}{6^2}x + \dfrac{1}{6^3}x^2 + \cdots + \dfrac{1}{6^{n+1}}x^n + \cdots$

6. $\dfrac{2}{3+x} = \dfrac{2}{3} - \dfrac{2}{3^2}x + \dfrac{2}{3^3}x^2 - \cdots + \dfrac{2(-1)^n}{3^{n+1}}x^n + \cdots$

7. $\ln(1+x) = x - \dfrac{1}{2}x^2 + \dfrac{1}{3}x^3 - \cdots + \dfrac{(-1)^{n-1}}{n}x^n + \cdots$

8. $\ln(1-x) = -x - \dfrac{1}{2}x^2 - \dfrac{1}{3}x^3 - \cdots - \dfrac{1}{n}x^n - \cdots$

9. $\dfrac{1}{(1+x)^2} = 1 - 2x + 3x^2 - \cdots + (-1)^n(n+1)x^n + \cdots$

10. $\dfrac{x}{(1-x)^2} = x + 2x^2 + 3x^3 + \cdots + nx^n + \cdots$

11. $e^x = 1 + x + \dfrac{1}{2!}x^2 + \cdots + \dfrac{1}{n!}x^n + \cdots$

12. $e^{-x} = 1 - x + \dfrac{1}{2!}x^2 - \cdots + \dfrac{(-1)^n}{n!}x^n + \cdots$

13. The Taylor series for e^x at 0 is

$$e^x = 1 + x + \frac{1}{2!}x^2 + \frac{1}{3!}x^3 + \cdots + \frac{1}{n!}x^n + \cdots$$

(A) Find the smallest value of n such that

$$|p_n(2) - e^2| < 0.1$$

(B) Does a value of n exist such that

$|p_n(100) - e^{100}| < 0.1$? Explain.

14. Suppose the function f has the following Taylor series representation at 0:

$$f(x) = 1 + x + 2!x^2 + 3!x^3 + \cdots + n!x^n + \cdots$$

(A) Find the interval of convergence.

(B) Compute $p_n(0.2)$ for $n = 1, 2, 3, \ldots, 10$, where $p_n(x)$ is the nth-degree Taylor polynomial for f at 0. Explain why your computations are consistent with your answer to part (A).

B

In Problems 15–22, find the interval of convergence of the given Taylor series representation.

15. $\ln(4 + x) = (x + 3) - \frac{1}{2}(x + 3)^2 + \frac{1}{3}(x + 3)^3 - \cdots + \frac{(-1)^{n+1}}{n}(x + 3)^n + \cdots$

16. $\ln(9 - 2x) = -2(x - 4) - \frac{2^2}{2}(x - 4)^2 - \frac{2^3}{3}(x - 4)^3 - \cdots - \frac{2^n}{n}(x - 4)^n - \cdots$

17. $e^{2x} = 1 + 2x + \frac{2^2}{2!}x^2 + \cdots + \frac{2^n}{n!}x^n + \cdots$

18. $e^{-x/3} = 1 - \frac{1}{3}x + \frac{1}{2!3^2}x^2 - \cdots + \frac{(-1)^n}{n!3^n}x^n + \cdots$

19. $\frac{1}{6 - x} = \frac{1}{4} + \frac{1}{4^2}(x - 2) + \frac{1}{4^3}(x - 2)^2 + \cdots + \frac{1}{4^{n+1}}(x - 2)^n + \cdots$

20. $\frac{3}{2 + 3x} = -3 - 3^2(x + 1) - 3^3(x + 1)^2 - \cdots - 3^{n+1}(x + 1)^n - \cdots$

21. $\frac{1}{6 - x} = -\frac{1}{2} + \frac{1}{2^2}(x - 8) - \frac{1}{2^3}(x - 8)^2 + \cdots + \frac{(-1)^n}{2^{n+1}}(x - 8)^n + \cdots$

22. $\frac{3}{2 + 3x} = \frac{3}{5} - \frac{3^2}{5^2}(x - 1) + \frac{3^3}{5^3}(x - 1)^2 - \cdots + \frac{(-1)^n 3^{n+1}}{5^{n+1}}(x - 1)^n + \cdots$

23. (A) Graph the nth-degree Taylor polynomials at 0 for $f(x) = e^x$, $n = 1, 2, 3, 4, 5$. Use the viewing window Xmin = −3, Xmax = 3, Ymin = −5, and Ymax = 15. Based solely on the graphs, what would you conjecture the interval of convergence to be? Explain.

(B) What is the actual interval of convergence?

24. (A) Graph the nth-degree Taylor polynomials at 0 for $f(x) = \ln(1 + x)$, $n = 1, 2, 3, 4, 5$. Use the viewing window Xmin = −3, Xmax = 3, Ymin = −15, and Ymax = 15. Based solely on the graphs, what would you conjecture the interval of convergenceto be? Explain.

(B) What is the actual interval of convergence?

C

In Problems 25–30, find the nth-degree Taylor polynomial at 0 for f, find the Taylor series at 0 for f, and determine the values of x for which $f(x) = \lim_{n\to\infty} p_n(x)$.

25. $f(x) = e^{4x}$

26. $f(x) = e^{-x/2}$

27. $f(x) = \ln(1 + 2x)$

28. $f(x) = \ln(1 - 4x)$

29. $f(x) = \frac{2}{2 - x}$

30. $f(x) = \frac{3}{3 + x}$

In Problems 31–34, find the nth-degree Taylor polynomial at the indicated value of a for f, find the Taylor series at a for f, and determine the values of x for which $f(x) = \lim_{n\to\infty} p_n(x)$.

31. $f(x) = \ln(2 - x)$; $a = 1$

32. $f(x) = \ln(2 + x)$; $a = -1$

33. $f(x) = \frac{1}{1 - x}$; $a = 2$

34. $f(x) = \frac{1}{4 - x}$; $a = 3$

35. (A) Find the interval of convergence of the Taylor series representation

$$\frac{1}{4 - x} = \frac{1}{4} + \frac{1}{4^2}x + \frac{1}{4^3}x^2 + \cdots + \frac{1}{4^{n+1}}x^n + \cdots$$

(B) Explain why Theorem 1 is not directly applicable to the Taylor series representation

$$\frac{1}{4 - x^2} = \frac{1}{4} + \frac{1}{4^2}x^2 + \frac{1}{4^3}x^4 + \cdots + \frac{1}{4^{n+1}}x^{2n} + \cdots$$

(C) Use part (A) to find the interval of convergence of the Taylor series of part (B).

36. (A) Explain why Theorem 1 is not directly applicable to the Taylor series representation

$$e^{x^2} = 1 + x^2 + \frac{1}{2!}x^4 + \frac{1}{3!}x^6 + \cdots + \frac{1}{n!}x^{2n} + \cdots$$

(B) Use the Taylor series for e^x at 0 to find the interval of convergence of the Taylor series of part (A).

37. (A) Explain why the interval of convergence of the Taylor series

$$\frac{1}{1 - x} + \frac{1}{1 - x^2} = 2 + x + 2x^2 + x^3 + 2x^4 + x^5 + \cdots$$

cannot be determined by applying Theorem 1.

(B) If two Taylor series have the same interval of convergence, then so does their term-by-term sum. Use this fact to find the interval of convergence of the Taylor series of part (A).

38. (A) Explain why the interval of convergence of the Taylor series

$$e^x - e^{x^2} = x + \left(\frac{1}{2!} - 1\right)x^2 + \frac{1}{3!}x^3 + \left(\frac{1}{4!} - \frac{1}{2!}\right)x^4 + \frac{1}{5!}x^5 + \cdots$$

cannot be determined by applying Theorem 1.

(B) If two Taylor series have the same interval of convergence, then so does their term-by-term difference. Use this fact to find the interval of convergence of the Taylor series of part (A).

8-3 Operations on Taylor Series

- Basic Taylor Series
- Addition and Multiplication
- Differentiation and Integration
- Substitution

Basic Taylor Series

In the preceding sections, we used repeated differentiation and the formula $a_n = f^{(n)}(a)/n!$ to find Taylor polynomials and Taylor series for a variety of simple functions. In this section we will see how to use the series for these simple functions and some basic properties of Taylor series to find the Taylor series for more complicated functions. Table 1 lists the series we will use in this process for convenient reference.

Most of the discussion in this section involves Taylor series at 0. At the end of the section we will see that Taylor series at points other than 0 can be obtained from Taylor series at 0 by using a simple substitution process.

Table 1 Basic Taylor Series

Function $f(x)$	Taylor Series at 0	Interval of Convergence
$\frac{1}{1-x}$	$= 1 + x + x^2 + \cdots + x^n + \cdots$	$-1 < x < 1$
$\frac{1}{1+x}$	$= 1 - x + x^2 - \cdots + (-1)^n x^n + \cdots$	$-1 < x < 1$
$\ln(1-x)$	$= -x - \frac{1}{2}x^2 - \frac{1}{3}x^3 - \cdots - \frac{1}{n}x^n - \cdots$	$-1 < x < 1$
$\ln(1+x)$	$= x - \frac{1}{2}x^2 + \frac{1}{3}x^3 - \cdots + \frac{(-1)^{n-1}}{n}x^n + \cdots$	$-1 < x < 1$
e^x	$= 1 + x + \frac{1}{2!}x^2 + \cdots + \frac{1}{n!}x^n + \cdots$	$-\infty < x < \infty$
e^{-x}	$= 1 - x + \frac{1}{2!}x^2 - \cdots + \frac{(-1)^n}{n!}x^n + \cdots$	$-\infty < x < \infty$

Addition and Multiplication

PROPERTY 1 Addition

Two Taylor series at 0 can be added term by term:
If

$$f(x) = a_0 + a_1x + a_2x^2 + \cdots + a_nx^n + \cdots$$

and

$$g(x) = b_0 + b_1x + b_2x^2 + \cdots + b_nx^n + \cdots$$

then

$$f(x) + g(x) = (a_0 + b_0) + (a_1 + b_1)x + (a_2 + b_2)x^2 + \cdots + (a_n + b_n)x^n + \cdots$$

This operation is valid in the intersection of the intervals of convergence of the series for f and g.

EXAMPLE 1 **Addition of Taylor Series** Find the Taylor series at 0 for $f(x) = e^x + \ln(1 + x)$, and find the interval of convergence.

SOLUTION From Table 1,

$$e^x = 1 + x + \frac{1}{2!}x^2 + \frac{1}{3!}x^3 + \cdots + \frac{1}{n!}x^n + \cdots \qquad -\infty < x < \infty$$

and

$$\ln(1 + x) = x - \frac{1}{2}x^2 + \frac{1}{3}x^3 - \cdots + \frac{(-1)^{n-1}}{n}x^n + \cdots \qquad -1 < x < 1$$

Adding the coefficients of corresponding powers of x, we have

$$\begin{aligned} f(x) &= e^x + \ln(1 + x) \\ &= 1 + 2x + \left(\frac{1}{2!} - \frac{1}{2}\right)x^2 + \left(\frac{1}{3!} + \frac{1}{3}\right)x^3 + \cdots \\ &\qquad + \left[\frac{1}{n!} + \frac{(-1)^{n-1}}{n}\right]x^n + \cdots \end{aligned}$$

The series for e^x converges for all x; however, the series for $\ln(1 + x)$ converges only for $-1 < x < 1$. Thus, the combined series converges for $-1 < x < 1$, the intersection of these two intervals of convergence.

Matched Problem 1 Find the Taylor series at 0 for $f(x) = e^x + 1/(1 - x)$, and find the interval of convergence.

PROPERTY 2 Multiplication

A Taylor series at 0 can be multiplied term by term by an expression of the form cx^r, where c is a nonzero constant and r is a non-negative integer:

If

$$f(x) = a_0 + a_1x + a_2x^2 + \cdots + a_nx^n + \cdots$$

then

$$cx^rf(x) = ca_0x^r + ca_1x^{r+1} + ca_2x^{r+2} + \cdots + ca_nx^{r+n} + \cdots$$

The Taylor series for $cx^rf(x)$ has the same interval of convergence as the Taylor series for f.

EXAMPLE 2 **Multiplication of Taylor Series** Find the Taylor series at 0 for $f(x) = 2x^2e^{-x}$, and find the interval of convergence.

SOLUTION From Table 1,

$$e^{-x} = 1 - x + \frac{1}{2!}x^2 - \cdots + \frac{(-1)^n}{n!}x^n + \cdots \qquad -\infty < x < \infty$$

Multiplying each term of this series by $2x^2$, the series for $f(x)$ is

$$f(x) = 2x^2e^{-x}$$

$$= 2x^2\left[1 - x + \frac{1}{2!}x^2 - \cdots + \frac{(-1)^n}{n!}x^n + \cdots\right]$$

$$= 2x^2 - 2x^3 + \frac{2}{2!}x^4 - \cdots + \frac{2(-1)^n}{n!}x^{n+2} + \cdots$$

Since the series for e^{-x} converges for all x, the series for $f(x)$ also converges for $-\infty < x < \infty$.

Matched Problem 2 Find the Taylor series at 0 for $f(x) = 3x^3 \ln(1 - x)$, and find the interval of convergence.

Differentiation and Integration

PROPERTY 3 Differentiation

A Taylor series at 0 can be differentiated term by term:

If

$$f(x) = a_0 + a_1x + a_2x^2 + a_3x^3 + \cdots + a_nx^n + \cdots$$

then

$$f'(x) = a_1 + 2a_2x + 3a_3x^2 + \cdots + na_nx^{n-1} + \cdots$$

The Taylor series for f' has the same interval of convergence as the Taylor series for f.

CONCEPTUAL INSIGHT

Assume that $a_n \neq 0$ for $n \geq n_0$. Then the ratio of consecutive coefficients in the Taylor series for f' is

$$\frac{(n+1)a_{n+1}}{na_n}$$

and

$$\lim_{n\to\infty}\left|\frac{(n+1)a_{n+1}}{na_n}\right| = \lim_{n\to\infty}\left|\frac{a_{n+1}}{a_n}\right| \qquad \lim_{n\to\infty}\frac{n+1}{n} = 1$$

Therefore, by Theorem 1 of the preceding section, the Taylor series for f and f' have the same interval of convergence.

EXAMPLE 3 **Differentiation of Taylor Series** Find the Taylor series at 0 for $f(x) = 1/(1 - x)^2$, and find the interval of convergence.

SOLUTION We want to relate f to the derivative of one of the functions in Table 1, so that we can apply property 3. Since

$$\begin{aligned}\frac{d}{dx}\left(\frac{1}{1-x}\right) &= \frac{d}{dx}(1-x)^{-1}\\ &= -(1-x)^{-2}(-1)\\ &= \frac{1}{(1-x)^2}\\ &= f(x)\end{aligned}$$

we can apply property 3 to the series for $1/(1 - x)$. From Table 1,

$$\frac{1}{1-x} = 1 + x + x^2 + x^3 + \cdots + x^n + \cdots \qquad -1 < x < 1$$

Differentiating both sides of this equation,

$$\frac{d}{dx}\left(\frac{1}{1-x}\right) = \frac{d}{dx}(1 + x + x^2 + x^3 + \cdots + x^n + \cdots)$$

$$\frac{1}{(1-x)^2} = \frac{d}{dx}1 + \frac{d}{dx}x + \frac{d}{dx}x^2 + \frac{d}{dx}x^3 + \cdots + \frac{d}{dx}x^n + \cdots$$

$$f(x) = 1 + 2x + 3x^2 + \cdots + nx^{n-1} + \cdots$$

Since the series for $1/(1 - x)$ converges for $-1 < x < 1$, the series for $f(x)$ also converges for $-1 < x < 1$. [Verify this statement by applying Theorem 1 in the preceding section to the series for $f(x)$.]

Matched Problem 3 Find the Taylor series at 0 for $f(x) = 1/(1 + x)^2$, and find the interval of convergence.

Explore & Discuss 1 The Taylor series representation for a function f at 0 is

$$f(x) = x - \frac{1}{3!}x^3 + \frac{1}{5!}x^5 - \frac{1}{7!}x^7 + \cdots + \frac{(-1)^n}{(2n+1)!}x^{2n+1} + \cdots \qquad -\infty < x < \infty$$

(A) Find the Taylor series for $f'(x)$.

(B) Find the Taylor series for $f''(x)$. How are f and f'' related?

PROPERTY 4 Integration

A Taylor series at 0 can be integrated term by term:

If

$$f(x) = a_0 + a_1x + a_2x^2 + \cdots + a_nx^n + \cdots$$

then

$$\int f(x)\,dx = C + a_0x + \frac{1}{2}a_1x^2 + \frac{1}{3}a_2x^3 + \cdots + \frac{1}{n+1}a_nx^{n+1} + \cdots$$

where C is the constant of integration. The Taylor series for $\int f(x)\,dx$ has the same interval of convergence as the Taylor series for f.

EXAMPLE 4 **Integration of Taylor Series** If f is a function that satisfies $f'(x) = x^2e^x$ and $f(0) = 2$, find the Taylor series at 0 for f. Find the interval of convergence.

SOLUTION Using the series for e^x from Table 1 and property 2, the series for f' is

$$f'(x) = x^2e^x$$

$$= x^2\left(1 + x + \frac{1}{2!}x^2 + \cdots + \frac{1}{n!}x^n + \cdots\right) \qquad -\infty < x < \infty$$

$$= x^2 + x^3 + \frac{1}{2!}x^4 + \cdots + \frac{1}{n!}x^{n+2} + \cdots$$

Integrating term by term produces a series for f:

$$\begin{aligned} f(x) &= \int f'(x)\,dx \\ &= \int \left(x^2 + x^3 + \frac{1}{2!}x^4 + \cdots + \frac{1}{n!}x^{n+2} + \cdots\right) dx \\ &= \int x^2\,dx + \int x^3\,dx + \frac{1}{2!}\int x^4\,dx + \cdots + \frac{1}{n!}\int x^{n+2}\,dx + \cdots \\ &= C + \frac{1}{3}x^3 + \frac{1}{4}x^4 + \frac{1}{(5)2!}x^5 + \cdots + \frac{1}{(n+3)n!}x^{n+3} + \cdots \end{aligned}$$

Now we use the condition $f(0) = 2$ to evaluate the constant of integration C:

$$\begin{aligned} 2 &= f(0) \\ &= C + 0 + 0 + \cdots + 0 + \cdots \\ &= C \end{aligned}$$

Thus,

$$f(x) = 2 + \frac{1}{3}x^3 + \frac{1}{4}x^4 + \cdots + \frac{1}{(n+3)n!}x^{n+3} + \cdots$$

Since the series for f' converges for $-\infty < x < \infty$, the series for f also converges for $-\infty < x < \infty$.

Matched Problem 4 If f is a function that satisfies $f'(x) = x\ln(1 + x)$ and $f(0) = 4$, find the Taylor series at 0 for f. Find the interval of convergence.

Substitution

EXAMPLE 5 **Using Substitution to Find Taylor Series** Find the Taylor series at 0 for $f(x) = e^{-x^2}$, and find the interval of convergence.

SOLUTION Suppose we try to solve this problem by finding the general form of the nth derivative of f:

$$\begin{aligned} f(x) &= e^{-x^2} \\ f'(x) &= -2xe^{-x^2} \\ f''(x) &= -2e^{-x^2} + 4x^2e^{-x^2} \\ f^{(3)}(x) &= 4xe^{-x^2} + 8xe^{-x^2} - 8x^3e^{-x^2} \\ &= 12xe^{-x^2} - 8x^3e^{-x^2} \\ f^{(4)}(x) &= 12e^{-x^2} - 24x^2e^{-x^2} - 24x^2e^{-x^2} + 16x^4e^{-x^2} \\ &= 12e^{-x^2} - 48x^2e^{-x^2} + 16x^4e^{-x^2} \end{aligned}$$

Since the higher-order derivatives are becoming very complicated and no general pattern is emerging, we will try another approach. How can we relate f to one of the functions in Table 1? If we let $g(x) = e^{-x}$, then f and g are related by

$$f(x) = e^{-x^2} = g(x^2)$$

From Table 1,

$$\begin{aligned} g(x) &= e^{-x} \\ &= 1 - x + \frac{1}{2!}x^2 - \frac{1}{3!}x^3 + \cdots + \frac{(-1)^n}{n!}x^n + \cdots \qquad -\infty < x < \infty \end{aligned}$$

Substituting x^2 for x in the series for g, we have

$$f(x) = g(x^2)$$

$$= 1 - x^2 + \frac{1}{2!}(x^2)^2 - \frac{1}{3!}(x^2)^3 + \cdots + \frac{(-1)^n}{n!}(x^2)^n + \cdots$$

$$= 1 - x^2 + \frac{1}{2!}x^4 - \frac{1}{3!}x^6 + \cdots + \frac{(-1)^n}{n!}x^{2n} + \cdots$$

Since the series for g converges for all values of x, the series for f must also converge for all values of x.

EXPLORE & DISCUSS 1

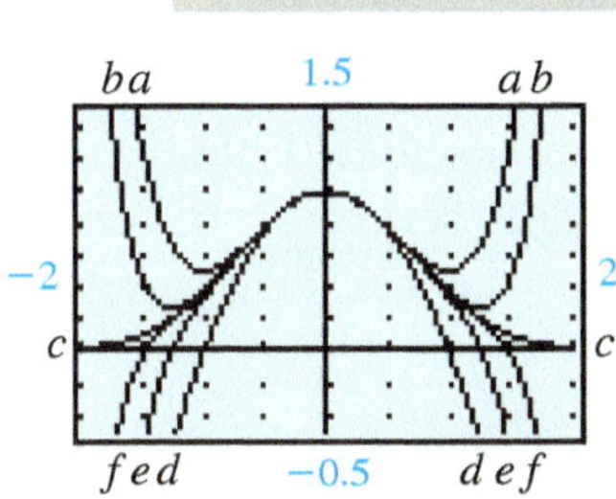

Figure 1

(A) The function $f(x) = e^{-x^2}$ and the Taylor polynomials

$$1 - x^2 + \frac{1}{2!}x^4 - \cdots + \frac{(-1)^n}{n!}x^{2n}$$

are graphed for $n = 1, 2, 3, 4, 5$ in Figure 1. Which curve belongs to which function?

(B) Does Figure 1 provide supporting evidence for the statement that the Taylor series for f at 0 converges for all values of x? Explain.

Matched Problem 5 Find the Taylor series at 0 for $f(x) = e^{x^3}$, and find the interval of convergence.

Making a substitution in a known Taylor series to obtain a new series is a very useful technique. Since there are many different substitutions that can be used, it is difficult to make a general statement concerning the effect of a substitution on the interval of convergence. The following examples illustrate some of the possibilities that may occur.

EXAMPLE 6 **Using Substitution to Find Taylor Series** Find the Taylor series at 0 for $f(x) = 1/(4 - x)$, and find the interval of convergence.

SOLUTION If we factor a 4 out of the denominator of f, we can establish a relationship between $f(x)$ and $g(x) = 1/(1 - x)/(1 - x)$. Thus,

$$f(x) = \frac{1}{4 - x}$$

$$= \left(\frac{1}{4}\right)\frac{1}{1 - (x/4)} \qquad g(x) = \frac{1}{1 - x}$$

$$= \frac{1}{4}g\left(\frac{x}{4}\right)$$

From Table 1,

$$g(x) = \frac{1}{1 - x}$$

$$= 1 + x + x^2 + \cdots + x^n + \cdots \qquad -1 < x < 1$$

Substituting $x/4$ for x in this series and multiplying by $\frac{1}{4}$, we have

$$f(x) = \frac{1}{4}g\left(\frac{x}{4}\right)$$

$$= \frac{1}{4}\left[1 + \left(\frac{x}{4}\right) + \left(\frac{x}{4}\right)^2 + \cdots + \left(\frac{x}{4}\right)^n + \cdots\right]$$

$$= \frac{1}{4} + \frac{1}{4^2}x + \frac{1}{4^3}x^2 + \cdots + \frac{1}{4^{n+1}}x^n + \cdots$$

Since the original series for g converges for $-1 < x < 1$ and we substituted $x/4$ for x in that series, the series for f converges for

$$-1 < \frac{x}{4} < 1 \quad \text{Multiply each member by 4.}$$

$$-4 < x < 4$$

Matched Problem 6 Find the Taylor series at 0 for $f(x) = 1/(3 + x)$, and find the interval of convergence.

EXAMPLE 7 **Using Substitution to Find Taylor Series** Find the Taylor series at 0 for $f(x) = \ln(1 + 4x^2)$, and find the interval of convergence.

SOLUTION If we let $g(x) = \ln(1 + x)$, then

$$\begin{aligned} f(x) &= \ln(1 + 4x^2) \\ &= g(4x^2) \end{aligned}$$

From Table 1, the series for g is

$$g(x) = x - \frac{1}{2}x^2 + \frac{1}{3}x^3 - \cdots + \frac{(-1)^{n-1}}{n}x^n + \cdots \qquad -1 < x < 1$$

Substituting $4x^2$ for x in this series, we have

$$\begin{aligned} f(x) &= g(4x^2) \\ &= 4x^2 - \frac{1}{2}(4x^2)^2 + \frac{1}{3}(4x^2)^3 - \cdots + \frac{(-1)^{n-1}}{n}(4x^2)^n + \cdots \\ &= 4x^2 - \frac{4^2}{2}x^4 + \frac{4^3}{3}x^6 - \cdots + \frac{(-1)^{n-1}4^n}{n}x^{2n} + \cdots \end{aligned}$$

The series for g converges for $-1 < x < 1$. Since we substituted $4x^2$ for x, the series for f converges for

$$-1 < 4x^2 < 1$$

Inequalities of this type are easier to solve if we use absolute value notation:

$$\begin{aligned} -1 < 4x^2 &< 1 && \text{Change to absolute value notation.} \\ |4x^2| &< 1 && \text{Multiply by } \tfrac{1}{4}. \\ |x^2| &< \tfrac{1}{4} \\ |x|^2 &< \tfrac{1}{4} \\ |x| &< \tfrac{1}{2} && \text{Convert to double inequalities.} \\ -\tfrac{1}{2} < x &< \tfrac{1}{2} && \text{See Figure 2 for a graphical solution.} \end{aligned}$$

Thus, the series for f converges for $-\frac{1}{2} < x < \frac{1}{2}$.

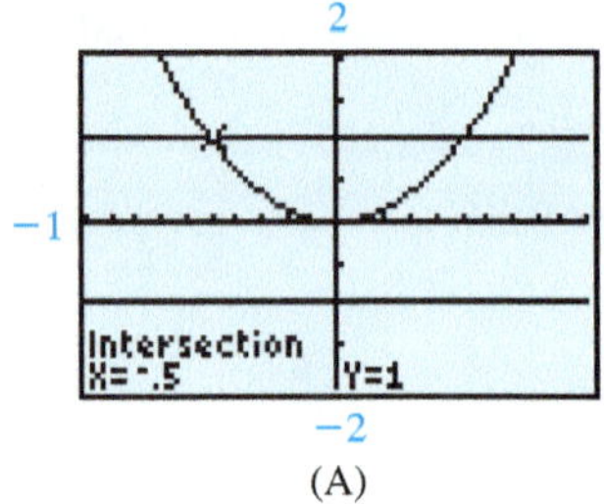

(A)

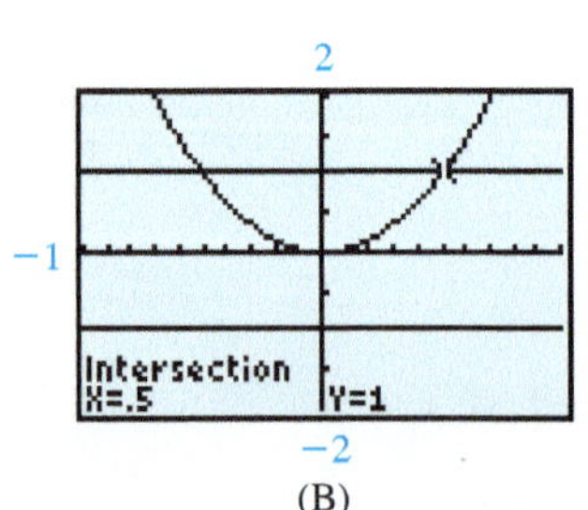

(B)

Figure 2 Graphical solution of $-1 < 4x^2 < 1$

Matched Problem 7 Find the Taylor series at 0 for $f(x) = 1/(1 + 9x^2)$, and find the interval of convergence.

Up to this point in this section we have restricted our attention to Taylor series at 0. How can we use the techniques discussed to find a Taylor series at a point $a \neq 0$? Properties 1–4 could be stated in terms of Taylor series at an arbitrary point a; however, there is an easier way to proceed. The method of substitution allows us to use Taylor series at 0 to find Taylor series at other points. The following example illustrates this technique.

EXAMPLE 8 **Using Substitution to Find Taylor Series** Find the Taylor series at 1 for $f(x) = 1/(2 - x)$, and find the interval of convergence.

SOLUTION In order to find a Taylor series for f in powers of $x - 1$, we will use the substitution $t = x - 1$ to express f as a function of t. If we find the Taylor series at 0 for this new function and then replace t with $x - 1$, we will have obtained the Taylor series at 1 for f.

$$t = x - 1 \qquad \text{Solve for } x.$$

$$x = t + 1 \qquad \text{Substitute for } x \text{ in } f(x) = 1/(2 - x).$$

$$\frac{1}{2 - x} = \frac{1}{2 - (t + 1)}$$

$$= \frac{1}{1 - t} \qquad \text{Find the Taylor series at 0 for this function of } t.$$

$$= 1 + t + t^2 + \cdots + t^n + \cdots \qquad \text{Substitute } x - 1 \text{ for } t.$$

$$-1 < t < 1$$

$$= 1 + (x - 1) + (x - 1)^2 + \cdots + (x - 1)^n + \cdots$$

Since $t = x - 1$ and the series in powers of t converges for $-1 < t < 1$, the series in powers of $x - 1$ converges for

$$-1 < x - 1 < 1 \qquad \text{Add 1 to each member.}$$

$$0 < x < 2$$

Matched Problem 8 Find the Taylor series at 1 for $f(x) = 1/x$, and find the interval of convergence.

Answers to Matched Problems

1. $2 + 2x + \left(\frac{1}{2!} + 1\right)x^2 + \cdots + \left(\frac{1}{n!} + 1\right)x^n + \cdots, -1 < x < 1$
2. $-3x^4 - \frac{3}{2}x^5 - x^6 - \cdots - \frac{3}{n}x^{n+3} - \cdots, -1 < x < 1$
3. $1 - 2x + 3x^2 - \cdots + (-1)^{n+1}nx^{n-1} + \cdots, -1 < x < 1$
4. $4 + \frac{1}{3}x^3 - \frac{1}{8}x^4 + \frac{1}{15}x^5 - \cdots + \frac{(-1)^{n-1}}{n(n+2)}x^{n+2} + \cdots, -1 < x < 1$
5. $1 + x^3 + \frac{1}{2!}x^6 + \cdots + \frac{1}{n!}x^{3n} + \cdots, -\infty < x < \infty$
6. $\frac{1}{3} - \frac{1}{3^2}x + \frac{1}{3^3}x^2 - \cdots + \frac{(-1)^n}{3^{n+1}}x^n + \cdots, -3 < x < 3$
7. $1 - 9x^2 + 9^2x^4 - \cdots + (-1)^n 9^n x^{2n} + \cdots, -\frac{1}{3} < x < \frac{1}{3}$
8. $1 - (x - 1) + (x - 1)^2 - \cdots + (-1)^n(x - 1)^n + \cdots, 0 < x < 2$

Exercises 8-3

Solve all the problems in this exercise by performing operations on the Taylor series in Table 1. State the interval of convergence for each series you find.

A

In Problems 1–12, find the Taylor series at 0.

1. $f(x) = \frac{1}{1-x} + \frac{1}{1+x}$

2. $f(x) = e^x + e^{-x}$

3. $f(x) = \frac{1}{1-x} + e^{-x}$

4. $f(x) = \frac{1}{1+x} + \ln(1+x)$

5. $f(x) = \frac{x^3}{1+x}$

6. $f(x) = xe^x$

7. $f(x) = x^2 \ln(1-x)$

8. $f(x) = \frac{5x^4}{1-x}$

9. $f(x) = e^{x^2}$

10. $f(x) = \ln(1-x^3)$

11. $f(x) = \ln(1+3x)$

12. $f(x) = \frac{x}{1+x^2}$

B

In Problems 13–20, find the Taylor series at 0.

13. $f(x) = \frac{1}{2-x}$

14. $f(x) = \frac{1}{5+x}$

15. $f(x) = \frac{1}{1-8x^3}$

16. $f(x) = \ln(1+16x^2)$

17. $f(x) = 10^x$

18. $f(x) = \log(1+x)$

19. $f(x) = \log_2(1-x)$

20. $f(x) = 3^x$

21. $f(x) = \frac{1}{4+x^2}$

22. $f(x) = \frac{1}{9-x^2}$

23. Substituting $\sqrt{x}$ for x in the Taylor series at 0 for e^x gives the formula

$$e^{\sqrt{x}} = 1 + x^{1/2} + \frac{1}{2!}x + \frac{1}{3!}x^{3/2} + \cdots + \frac{1}{n!}x^{n/2} + \cdots$$

(A) For which values of x is the formula valid?

(B) Is the formula a Taylor series at 0 for $e^{\sqrt{x}}$? Explain.

24. Substituting $1/x$ for x in the Taylor series at 0 for $1/(1-x)$ gives the formula

$$\frac{1}{1-\frac{1}{x}} = 1 + x^{-1} + x^{-2} + \cdots + x^{-n} + \cdots$$

(A) For which values of x is the formula valid?

(B) Explain why the formula is not a Taylor series at 0 for

$$f(x) = \frac{1}{1-\frac{1}{x}}$$

(C) Find a Taylor series at 0 that equals

$$\frac{1}{1-\frac{1}{x}}$$

for all $x \neq 0$ in the interval $-1 < x < 1$.

25. Find the Taylor series at 0 for

(A) $f(x) = \frac{1}{1-x^2}$

(B) $g(x) = \frac{2x}{(1-x^2)^2}$

[*Hint:* Compare $f'(x)$ and $g(x)$.]

26. Find the Taylor series at 0 for

(A) $f(x) = \frac{x}{1-x^2}$

(B) $g(x) = \frac{1+x^2}{(1-x^2)^2}$

[*Hint:* Compare $f'(x)$ and $g(x)$.]

27. If $f(x)$ satisfies $f'(x) = 1/(1+x^2)$ and $f(0) = 0$, find the Taylor series at 0 for $f(x)$.

28. If $f(x)$ satisfies $f'(x) = \ln(1+x^2)$ and $f(0) = 1$, find the Taylor series at 0 for $f(x)$.

29. If $f(x)$ satisfies $f'(x) = x^2 \ln(1-x)$ and $f(0) = 5$, find the Taylor series at 0 for $f(x)$.

30. If $f(x)$ satisfies $f'(x) = xe^x$ and $f(0) = -3$, find the Taylor series at 0 for $f(x)$.

31. Most graphing calculators include the hyperbolic sine function, $f(x) = \sinh x$, as a built-in function. The Taylor series for the hyperbolic sine at 0 is

$$\sinh x = x + \frac{1}{3!}x^3 + \frac{1}{5!}x^5 + \cdots + \frac{1}{(2n+1)!}x^{2n+1} + \cdots \qquad -\infty < x < \infty$$

(A) Graph f and its Taylor polynomials $p_1(x)$, $p_3(x)$, and $p_5(x)$ in the viewing window $-5 \le x \le 5$, $-30 \le y \le 30$.

(B) Explain why $\frac{d^2}{dx^2}(\sinh x) = \sinh x$.

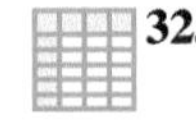

32. Most graphing calculators include the hyperbolic cosine function, $g(x) = \cosh x$, as a built-in function. The Taylor series for the hyperbolic cosine at 0 is

$$\cosh x = 1 + \frac{1}{2!}x^2 + \frac{1}{4!}x^4 + \frac{1}{6!}x^6 + \cdots + \frac{1}{(2n)!}x^{2n} + \cdots \quad -\infty < x < \infty$$

(A) Graph g and its Taylor polynomials $p_2(x)$, $p_4(x)$, and $p_6(x)$ in the viewing window $-5 \le x \le 5, 0 \le y \le 60$.

(B) Explain why $\frac{d^2}{dx^2}(\cosh x) = \cosh x$.

In Problems 33–38, use the substitution $t = x - a$ to find the Taylor series at the indicated value of a.

33. $f(x) = \frac{1}{4 - x}$; at 3

34. $f(x) = \frac{1}{3 + x}$; at -2

35. $f(x) = \ln x$; at 1

36. $f(x) = \ln(3 - x)$; at 2

37. $f(x) = \frac{1}{4 - 3x}$; at 1

38. $f(x) = \frac{1}{5 - 2x}$; at 2

C

39. Use the Taylor series at 0 for $1/(1 - x)$ and repeated applications of property 3 to find the Taylor series at 0 for

$$f(x) = \frac{1}{(1 - x)^3}$$

40. Use the Taylor series at 0 for $1/(1 + x)$ and repeated applications of property 3 to find the Taylor series at 0 for

$$f(x) = \frac{1}{(1 + x)^4}$$

41. Suppose that the Taylor series for f at 0 is

$$f(x) = x + \frac{1}{3}x^3 + \frac{1}{5}x^5 + \cdots + \frac{1}{2n + 1}x^{2n+1} + \cdots \quad -1 < x < 1$$

(A) Find the Taylor series for f'.

(B) Explain why $f'(x) = \frac{1}{1 - x^2}$ for $-1 < x < 1$.

42. Suppose that the Taylor series for g at 0 is

$$g(x) = x - \frac{1}{3}x^3 + \frac{1}{5}x^5 - \cdots + \frac{(-1)^n}{2n + 1}x^{2n+1} + \cdots \quad -1 < x < 1$$

(A) Find the Taylor series for g'.

(B) Explain why $g'(x) = \frac{1}{1 + x^2}$ for $-1 < x < 1$.

43. Find the Taylor series at 0 for $f(x) = \frac{1 + x}{1 - x}$.

$\left[\textbf{Note: } \frac{1 + x}{1 - x} = \frac{1}{1 - x} + \frac{x}{1 - x}\right]$

44. Find the Taylor series at 0 for $f(x) = \frac{1 - 2x}{1 + x}$.

$\left[\textbf{Note: } \frac{1 - 2x}{1 + x} = \frac{1}{1 + x} - \frac{2x}{1 + x}\right]$

45. Find the Taylor series at 0 for $f(x) = \frac{1}{2}\ln\left(\frac{1 + x}{1 - x}\right)$.

$\left[\textbf{Note: } \ln\left(\frac{1 + x}{1 - x}\right) = \ln(1 + x) - \ln(1 - x)\right]$

46. Find the Taylor series at 0 for $f(x) = \ln(1 - 2x + x^2)$.

47. Find the Taylor series at 0 for $f(x) = \frac{e^x + e^{-x}}{2}$.

48. Find the Taylor series at 0 for $f(x) = \frac{e^x - e^{-x}}{2}$.

49. If $f(x)$ satisfies $f''(x) = \ln(1 + x)$, $f'(0) = 3$, and $f(0) = -2$, find the Taylor series at 0 for $f(x)$.

50. If $f(x)$ satisfies $f''(x) = \frac{1}{1 - x^2}$, $f'(0) = 4$, and $f(0) = 5$, find the Taylor series at 0 for $f(x)$.

51. Find the Taylor series at any $a > 0$ for $f(x) = \ln x$.

52. Find the Taylor series at any a for $f(x) = e^x$.

53. If a and b are constants ($b \ne 0, a \ne b^{-1}$), find the Taylor series at a for

$$f(x) = \frac{1}{1 - bx}$$

54. If a and b are constants ($a \ne b$), find the Taylor series at a for

$$f(x) = \frac{1}{b - x}$$

8-4 Approximations Using Taylor Series

- The Remainder
- Taylor's Formula for the Remainder
- Taylor Series with Alternating Terms
- Approximating Definite Integrals

The Remainder

Now that we can find Taylor series for a variety of functions, we return to our original goal: approximating the values of a function.

If x is in the interval of convergence of the Taylor series for a function f and $p_n(x)$ is the nth-degree Taylor polynomial, then

$$f(x) = \lim_{n\to\infty} p_n(x)$$

and $p_n(x)$ can be used to approximate $f(x)$. We want to consider two questions:

1. If we select a particular value of n, how accurate is the approximation $f(x) \approx p_n(x)$?
2. If we want the approximation $f(x) \approx p_n(x)$ to have a specified accuracy, how do we select the proper value of n?

It turns out that both of these questions can be answered by examining the difference between $f(x)$ and $p_n(x)$. This difference is called the *remainder* and is defined in the following box.

DEFINITION The Remainder of a Taylor Series

If p_n is the nth-degree Taylor polynomial for f, then the **remainder** is

$$R_n(x) = f(x) - p_n(x)$$

The **error** in the approximation $f(x) \approx p_n(x)$ is

$$|f(x) - p_n(x)| = |R_n(x)|$$

If the Taylor series at 0 for f is

$$f(x) = \overbrace{a_0 + a_1x + a_2x^2 + \cdots + a_nx^n}^{p_n(x)} + \overbrace{a_{n+1}x^{n-1} + \cdots}^{R_n(x)}$$

then

$$p_n(x) = a_0 + a_1x + a_2x^2 + \cdots + a_nx^n$$

and

$$R_n(x) = a_{n+1}x^{n+1} + a_{n+2}x^{n+2} + \cdots$$

Similar statements can be made for Taylor series at a.

It follows from our basic assumption for the functions we consider that

$$\lim_{n\to\infty} R_n(x) = \lim_{n\to\infty}[f(x) - p_n(x)] = f(x) - f(x) = 0$$

if and only if x is in the interval of convergence of f.*

In general, it is difficult to find the exact value of $R_n(x)$. In fact, since $f(x) = p_n(x) + R_n(x)$, this is equivalent to finding the exact value of $f(x)$. Instead, we will discuss two methods for *estimating* the value of $R_n(x)$. The first method works in all cases, but can be difficult to apply, while the second method is easy to apply, but does not work in all cases.

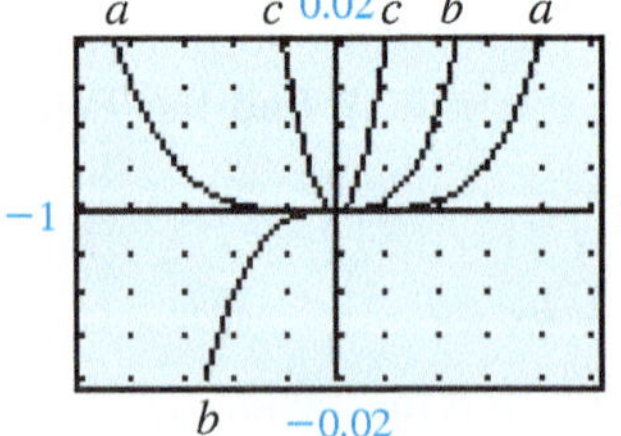

Figure 1

EXPLORE & DISCUSS 1

Let $p_1(x), p_2(x)$, and $p_3(x)$ be Taylor polynomials for $f(x) = e^x$ at 0. The corresponding remainder functions $R_1(x), R_2(x)$, and $R_3(x)$ are graphed in Figure 1.

(A) Which curve belongs to which function?

(B) Use graphical approximation techniques to estimate those values of x for which the error in each of the approximations $f(x) \approx p_n(x)$ is less than 0.01.

---

* In more advanced texts, the statement $\lim_{n\to\infty} R_n(x) = 0$ for x in the interval of convergence is actually proved for functions such as e^x, making our basic assumption unnecessary.

Taylor's Formula for the Remainder

The first method for estimating the remainder of a Taylor series is based on *Taylor's formula for the remainder.*

FORMULA Taylor's Formula for the Remainder

If f has derivatives of all order at 0, then

$$R_n(x) = \frac{f^{(n+1)}(t)x^{n+1}}{(n+1)!}$$

for some number t between 0 and x.

A similar formula can be stated for the remainder of a Taylor series at an arbitrary point a. (We will not use this formula in this text.)

In most applications of the remainder formula, it is not possible to find the value of t. However, if we can find a number M satisfying

$$|f^{(n+1)}(t)| < M \qquad \text{for all } t \text{ between 0 and } x$$

then we can estimate $R_n(x)$ as follows:

$$|R_n(x)| = \left|\frac{f^{(n+1)}(t)x^{n+1}}{(n+1)!}\right| < \frac{M|x|^{n+1}}{(n+1)!}$$

This technique is illustrated in the next example.

EXAMPLE 1 **Taylor's Formula for the Remainder** Use the second-degree Taylor polynomial at 0 for $f(x) = e^x$ to approximate $e^{0.1}$. Use Taylor's formula for the remainder to estimate the error in this approximation.

SOLUTION Since $f^{(n)}(x) = e^x$ for any n, we can write

$$f(x) = e^x = p_2(x) + R_2(x)$$

$$= \overbrace{1 + x + \frac{1}{2!}x^2}^{p_2(x)} + \overbrace{\frac{e^t x^3}{3!}}^{R_2(x)} \qquad f^{(3)}(t) = e^t$$

for some number t between 0 and x. Thus,

$$\begin{aligned} f(0.1) &= e^{0.1} \\ &= p_2(0.1) + R_2(0.1) \\ &\approx p_2(0.1) \\ &= 1 + 0.1 + \tfrac{1}{2}(0.1)^2 \\ &= 1.105 \end{aligned}$$

To estimate the error in this approximation, we must estimate

$$|R_2(0.1)| = \left|\frac{e^t(0.1)^3}{3!}\right| = \frac{e^t}{6{,}000}$$

where $0 \le t \le 0.1$. In order to estimate $|R_2(0.1)|$, we must estimate $g(t) = e^t$ for $0 \le t \le 0.1$. Since e^t is always increasing, $e^t \le e^{0.1}$ for $0 \le t \le 0.1$. However, $e^{0.1}$ is the number we are trying to approximate. We do not want to use this number in our estimate of the error. (This situation occurs frequently in approximation problems involving the exponential function.) Instead, we will use the following rough estimate for e^t (see Fig. 2):

$$\text{If} \quad t \le 1, \quad \text{then} \quad e^t \le 3. \tag{1}$$

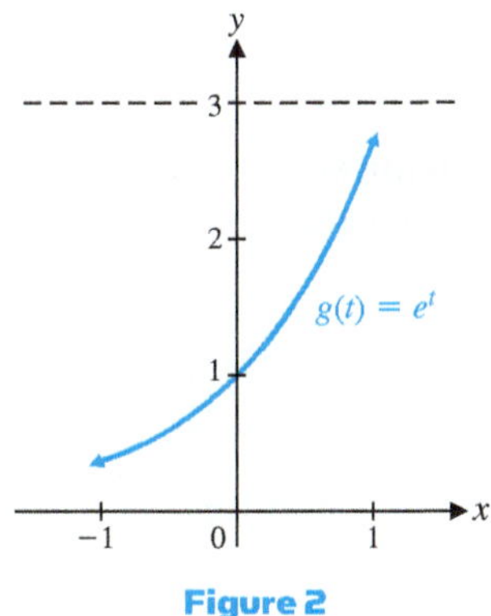

Figure 2

Since estimate (1) holds for $t \le 1$, it certainly holds for $0 \le t \le 0.1$. Thus,

$$|R_2(0.1)| = \frac{e^t}{6{,}000} \le \frac{3}{6{,}000} = 0.0005$$

and we can conclude that the approximate value 1.105 is within ± 0.0005 of the exact value of $e^{0.1}$.

Matched Problem 1 Use the second-degree Taylor polynomial at 0 for $f(x) = e^x$ to approximate $e^{0.2}$. Use Taylor's formula for the remainder to estimate the error in the approximation.

CONCEPTUAL INSIGHT

If f is a polynomial of degree n and $k \ge n$, then $f^{(k+1)}(t) = 0$ for all t. So by Taylor's formula for the remainder, $R_k(x) = 0$. Therefore, in the special case in which f is a polynomial of degree n, each Taylor polynomial p_k, for $k \ge n$, is equal to f.

Taylor Series with Alternating Terms

Taylor's formula for the remainder can be difficult to apply for large values of n. For most functions, the formula for the nth derivative becomes very complicated as n increases. For a certain type of problem, there is another method that does not require estimation of the nth derivative. If the series of numbers that is formed by evaluating a Taylor series at a given number x_0 *alternates in sign and decreases in absolute value*, then the remainder can be estimated by simply examining the numbers in the series. Series of numbers whose terms alternate in sign are called **alternating series.** The estimate for the remainder in an alternating series is given in Theorem 1.

THEOREM 1 Error Estimation for Alternating Series

If x_0 is a number in the interval of convergence for

$$f(x) = a_0 + a_1x + a_2x^2 + \cdots + a_kx^k + \cdots$$

and the terms in the series

$$f(x_0) = a_0 + a_1x_0 + a_2x_0^2 + \cdots + a_kx_0^k + \cdots$$

are alternating in sign and decreasing in absolute value, then the error in the approximation

$$f(x_0) \approx a_0 + a_1x_0 + a_2x_0^2 + \cdots + a_nx_0^n$$

is strictly less than the absolute value of the next term. That is,

$$|R_n(x_0)| < |a_{n+1}x_0^{n+1}|$$

EXAMPLE 2 **Estimating the Remainder for Alternating Series** Use the Taylor series at 0 for $f(x) = e^{-x}$ to approximate $e^{-0.3}$ with an error of no more than 0.0005.

SOLUTION The Taylor series at 0 for $f(x) = e^{-x}$ is

$$e^{-x} = 1 - x + \frac{1}{2!}x^2 - \frac{1}{3!}x^3 + \cdots + \frac{(-1)^k}{k!}x^k + \cdots \qquad -\infty < x < \infty$$

If we substitute $x = 0.3$ in this series, we obtain

$$\begin{aligned} f(0.3) &= e^{-0.3} \\ &= 1 - 0.3 + \tfrac{1}{2}(0.3)^2 - \tfrac{1}{6}(0.3)^3 + \tfrac{1}{24}(0.3)^4 - \cdots \\ &= 1 - 0.3 + 0.045 - 0.0045 + 0.0003375 - \cdots \end{aligned}$$

Since the terms in this series are alternating in sign and decreasing in absolute value, Theorem 1 applies. If we use the first four terms in this series to approximate $e^{-0.3}$, then the error in this approximation is less than the absolute value of the fifth term. That is,

$$|R_3(0.3)| < 0.000\,337\,5 < 0.0005$$

Thus,

$$e^{-0.3} \approx 1 - 0.3 + 0.045 - 0.0045 = 0.7405$$

and the error in this approximation is less than the specified accuracy of 0.0005.

Matched Problem 2 Use the Taylor series at 0 for $f(x) = e^{-x}$ to approximate $e^{-0.1}$ with an error of no more than 0.0005.

As Example 2 illustrates, Theorem 1 is much easier to use than Taylor's formula for the remainder. Notice that we did not have to find an estimate for $f^{(n+1)}(t)$. However, it is very important to understand that Theorem 1 can be applied only if the terms in the series *alternate in sign after they have been evaluated at* x_0. For example, if we try to use the series for e^{-x} to approximate $e^{0.3}$ by substituting $x_0 = -0.3$, we have

$$\begin{aligned} e^{0.3} &= 1 - (-0.3) + \tfrac{1}{2}(-0.3)^2 - \tfrac{1}{6}(-0.3)^3 + \cdots \\ &= 1 + 0.3 + \tfrac{1}{2}(0.3)^2 + \tfrac{1}{6}(0.3)^3 + \cdots \end{aligned}$$

Since these numbers do not alternate in sign, Theorem 1 does not apply. Taylor's formula for the remainder would have to be used to estimate the error in approximations obtained from this series.

Approximating Definite Integrals

In order to find the exact value of a definite intergral $\int_a^b f(x)\,dx$,we must first find an antiderivative of the function *f.* But suppose we cannot find an anti-derivative of *f* (it may not even exist in a convenient form). We saw that Riemann sums can be used to approximate definite integrals. Taylor series techniques provide an alternative method for approximating definite integrals that is often more efficient, and, in the case of alternating series, automatically determines the accuracy of the approximation.

EXAMPLE 3 **Using Taylor Series to Approximate Definite Integrals** Approximate $\int_0^1 e^{-x^2}\,dx$ with a maximum error of 0.005.

SOLUTION If $F(x)$is an antiderivative of e^{-x^2},then

$$\int_0^1 e^{-x^2}\,dx = F(1) - F(0)$$

It is not possible to express $F(x)$as a finite combination of simple functions; however, it is possible to find a Taylor series for *F.* This series can be used to approximate the values of *F* and, consequently, of the definite integral.

Step 1 **Find a Taylor series for the integrand:**

$$e^x = 1 + x + \frac{1}{2!}x^2 + \cdots + \frac{1}{n!}x^n + \cdots \qquad -\infty < x < \infty$$

Thus,

$$\begin{aligned} e^{-x^2} &= 1 + (-x^2) + \frac{1}{2!}(-x^2)^2 + \cdots + \frac{1}{n!}(-x^2)^n + \cdots \\ &= 1 - x^2 + \frac{1}{2!}x^4 - \cdots + \frac{(-1)^n}{n!}x^{2n} + \cdots \qquad -\infty < x < \infty \end{aligned}$$

Step 2 **Find the Taylor series for the antiderivative:** Integrating term by term, we have

$$F(x) = \int e^{-x^2}\,dx$$

$$= \int \left[1 - x^2 + \frac{1}{2!}x^4 - \cdots + \frac{(-1)^n}{n!}x^{2n} + \cdots\right]dx$$

$$= C + x - \frac{1}{3}x^3 + \frac{1}{10}x^5 - \cdots + \frac{(-1)^n}{n!(2n+1)}x^{2n+1} + \cdots \qquad -\infty < x < \infty$$

Step 3 **Approximate the definite integral:** If we choose $C = 0$, then $F(0) = 0$ and

$$\int_0^1 e^{-x^2}\,dx = F(1) - F(0)$$

$$= \left[x - \frac{1}{3}x^3 + \frac{1}{10}x^5 - \cdots + \frac{(-1)^n}{n!(2n+1)}x^{2n+1} + \cdots\right]\Bigg|_0^1$$

$$= \left[1 - \frac{1}{3} + \frac{1}{10} - \cdots + \frac{(-1)^n}{n!(2n+1)} + \cdots\right] - 0$$

$$= 1 - \frac{1}{3} + \frac{1}{10} - \frac{1}{42} + \frac{1}{216} - \cdots$$

$$= 1 - 0.333\,333 + 0.1 - 0.023\,809 + 0.004\,630 - \cdots$$

Since the Taylor series for $F(x)$ converges for all values of x, we can use this last series of numbers to approximate $F(1)$. Notice that the terms in this series are alternating in sign and decreasing in absolute value. Applying Theorem 1, we conclude that the error introduced by approximating $F(1)$ with the first four terms of this series will be no more than 0.004 630, the absolute value of the fifth term. Since this is less than the specified error of 0.005, we have

$$\int_0^1 e^{-x^2}\,dx = F(1) - F(0)$$

$$\approx 1 - 0.333\,333 + 0.1 - 0.023\,809$$

$$\approx 0.743 \quad \text{Rounded to three decimal places}$$

Matched Problem 3 Approximate $\int_0^{0.5} e^{-x^2}$ with a maximum error of 0.005.

Explore & Discuss 2

Suppose you wish to use a Taylor series for

$$f(x) = \frac{1}{1 - x^2} \quad \text{to approximate} \quad \int_2^3 \frac{dx}{1 - x^2}$$

(A) Explain why you would not use the Taylor series for f at 0.

(B) If you use the Taylor series for f at a, which value of a would you expect to yield the best approximation to the integral? Explain.

EXAMPLE 4 **Income Distribution** Approximate the index of income concentration for the Lorenz curve given by

$$f(x) = \frac{11x^4}{10 + x^2}$$

with an error of no more than 0.005.

SOLUTION The index of income concentration for a Lorenz curve is twice the area between the graph of the Lorenz curve and the graph of the line $y = x$ (see Fig. 3). Thus, we must evaluate the integral

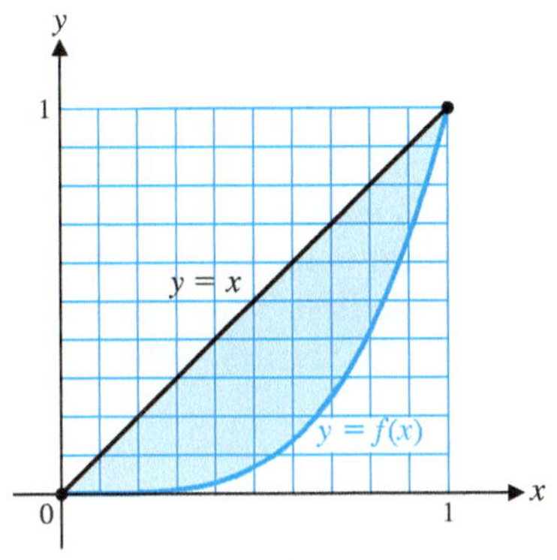

Figure 3

$$2\int_0^1 [x - f(x)]\,dx = 2\int_0^1 x\,dx - 2\int_0^1 f(x)\,dx$$

$$= \int_0^1 2x\,dx - \int_0^1 \frac{22x^4}{10 + x^2}\,dx \qquad (2)$$

The first integral in (2) is easy to evaluate:

$$\int_0^1 2x\,dx = x^2\Big|_0^1 = 1 - 0 = 1$$

Since the second integral in (2) cannot be evaluated by any of the techniques we have discussed, we will use a Taylor series to approximate this integral.

Step 1 **Find a Taylor series for the integrand:**

$$\frac{22x^4}{10 + x^2} = \frac{22x^4}{10}\left[\frac{1}{1 + (x^2/10)}\right] \qquad \text{Substitute } x^2/10 \text{ for } x \text{ in the series for } 1/(1 + x).$$

$$= 2.2x^4\left[1 - \frac{x^2}{10} + \left(\frac{x^2}{10}\right)^2 - \cdots + (-1)^n\left(\frac{x^2}{10}\right)^n + \cdots\right] \qquad -1 < \frac{x^2}{10} < 1$$

$$= 2.2x^4 - \frac{2.2}{10}x^6 + \frac{2.2}{10^2}x^8 - \cdots + \frac{2.2(-1)^n}{10^n}x^{2n+4} + \cdots$$

To find the interval of convergence, we solve $-1 < x^2/10 < 1$ for x:

$$-1 < \frac{x^2}{10} < 1 \qquad \text{Change to absolute value notation.}$$

$$\left|\frac{x^2}{10}\right| < 1 \qquad \text{Multiply by 10.}$$

$$|x^2| < 10 \qquad \text{Take the square root of both sides.}$$

$$|x| < \sqrt{10} \qquad \text{Convert to double inequalities.}$$

$$-\sqrt{10} < x < \sqrt{10} \qquad \text{Interval of convergence}$$

Step 2 **Find the Taylor series for the antiderivative:** Using property 4,

$$\int \frac{22x^4}{10 + x^2}\,dx = \int\left[2.2x^4 - \frac{2.2}{10}x^6 + \frac{2.2}{10^2}x^8 - \cdots + \frac{2.2(-1)^n}{10^n}x^{2n+4} + \cdots\right]dx$$

$$= C + \frac{2.2}{5}x^5 - \frac{2.2}{7\cdot 10}x^7 + \frac{2.2}{9\cdot 10^2}x^9 - \cdots$$

$$+ \frac{2.2(-1)^n}{(2n + 5)10^n}x^{2n+5} + \cdots$$

This series also converges for $-\sqrt{10} < x < \sqrt{10}$.

Step 3 **Approximate the definite integral:** Choosing $C = 0$ in the antiderivative, we have

$$\int_0^1 \frac{22x^4}{10 + x^2}\,dx = \left[\frac{2.2}{5}x^5 - \frac{2.2}{7 \cdot 10}x^7 + \frac{2.2}{9 \cdot 10^2}x^9 - \cdots + \frac{2.2(-1)^n}{(2n+5)10^n}x^{2n+5} + \cdots\right]\Bigg|_0^1$$

$$= \left[\frac{2.2}{5} - \frac{2.2}{7 \cdot 10} + \frac{2.2}{9 \cdot 10^2} - \cdots + \frac{2.2(-1)^n}{(2n+5)10^n} + \cdots\right] - 0$$

$$= 0.44 - 0.031\,429 + 0.002\,444 - \cdots$$

Since the limits of integration, 0 and 1, are within the interval of convergence of the Taylor series for the antiderivative, we can use this series to approximate the value of the definite integral. Notice that the numbers in this series are alternating in sign and decreasing in absolute value. Since the absolute value of the third term is less than 0.005, Theorem 1 implies that we can use the first two terms of this series to approximate the definite integral to the specified accuracy. Thus,

$$\int_0^1 \frac{22x^4}{10 + x^2}\,dx \approx 0.44 - 0.031\,429$$

$$\approx 0.409 \qquad \text{Rounded to three decimal places}$$

Returning to (2), we have

$$2\int_0^1 [x - f(x)]\,dx = 2\int_0^1 x\,dx - 2\int_0^1 f(x)\,dx$$

$$\approx 1 - 0.409$$

$$= 0.591 \qquad \text{Index of income concentration}$$

Matched Problem 4 Repeat Example 4 for $f(x) = \dfrac{11x^5}{10 + x^2}$.

Answers to Matched Problems

1. $e^{0.2} \approx 1 + 0.2 + \frac{1}{2}(0.2)^2 = 1.22$; $|R_2(0.2)| < 0.004$

2. $e^{-0.1} \approx 1 - 0.1 + 0.005 = 0.905$ within $\pm\, 0.000\,167$

3. $\int_0^{0.5} e^{-x}\,dx \approx 0.5 - \frac{1}{3}(0.5)^3 \approx 0.458$ within $\pm 0.003\,125$

4. $\int_0^1 \frac{22x^5}{10 + x^2}\,dx \approx \frac{2.2}{6} - \frac{2.2}{80} \approx 0.339$ within $\pm\, 0.0022$; index of income concentration ≈ 0.661

Exercises 8-4

In Problems 1–30, use Theorem 1 to perform the indicated error estimations.

A

Evaluating the Taylor series at 0 for $f(x) = e^{-x}$ at $x = 0.6$ produces the following series:

$$e^{-0.6} = 1 - 0.6 + 0.18 - 0.036 + 0.0054 - 0.000\,648 + \cdots$$

In Problems 1–4, use the indicated number of terms in this series to approximate $e^{-0.6}$, and then estimate the error in this approximation.

1. Two terms
2. Three terms
3. Four terms
4. Five terms

Evaluating the Taylor series at 0 for $f(x) = \ln(1 + x)$ at $x = 0.3$ produces the following series:

$$\ln 1.3 = 0.3 - 0.045 + 0.009 - 0.002\,025 + 0.000\,486 - 0.000\,121\,5 + \cdots$$

In Problems 5–8, use the indicated number of terms in this series to approximate ln 1.3, and then estimate the error in this approximation.

5. Two terms
6. Three terms
7. Four terms
8. Five terms

In Problems 9–12, use the third-degree Taylor polynomial at 0 for $f(x) = e^{-x}$ to approximate each expression, and then estimate the error in the approximation.

9. $e^{-0.2}$
10. $e^{-0.5}$
11. $e^{-0.03}$
12. $e^{-0.06}$

In Problems 13–16, use the third-degree Taylor polynomial at 0 for $f(x) = \ln(1 + x)$ to approximate each expression, and then estimate the error in the approximation.

13. ln 1.6
14. ln 1.8
15. ln 1.06
16. ln 1.08

B

In Problems 17–24, use a Taylor polynomial at 0 to approximate each expression with an error of no more than 0.000 005. Select the polynomial of lowest degree that can be used to obtain this accuracy and state the degree of this polynomial.

17. $e^{-0.1}$
18. $e^{-0.2}$
19. $e^{-0.01}$
20. $e^{-0.02}$
21. ln 1.2
22. ln 1.1
23. ln 1.02
24. ln 1.01

In Problems 25–30, use a Taylor series at 0 to approximate each integral with an error of no more than 0.0005.

25. $\displaystyle\int_0^{0.2} \frac{1}{1 + x^2}\,dx$

26. $\displaystyle\int_0^{0.5} \frac{x}{1 + x^4}\,dx$

27. $\displaystyle\int_0^{0.6} \ln(1 + x^2)\,dx$

28. $\displaystyle\int_0^{0.7} x\ln(1 + x^4)\,dx$

29. $\displaystyle\int_0^{0.4} x^2 e^{-x^2}\,dx$

30. $\displaystyle\int_0^{0.8} x^4 e^{-x^2}\,dx$

In Problems 31–34, assume that $f(x)$ is a function such that $|f^{(n)}(x)| \le 1$ for all n and all x [the trigonometric functions $f(x) = \sin x$ and $f(x) = \cos x$, for example, satisfy that property] and let $p_n(x)$ be the nth-degree Taylor polynomial for f at 0.

31. Use Taylor's formula for the remainder to find the smallest value of n such that the error in the approximation of $f(2)$ by $p_n(2)$ is guaranteed to be less than 0.001.

32. Use Taylor's formula for the remainder to find the smallest value of n such that the error in the approximation of $f(10)$ by $p_n(10)$ is guaranteed to be less than 0.001.

33. Use Taylor's formula for the remainder to determine the values of x such that the error in the approximation of $f(x)$ by $p_5(x)$ is guaranteed to be less than 0.05.

34. Use Taylor's formula for the remainder to determine the values of x such that the error in the approximation of $f(x)$ by $p_{10}(x)$ is guaranteed to be less than 0.05.

35. Let $f(x) = xe^x$ and consider the third-degree Taylor polynomial $p_3(x)$ for f at 0. Use graphical approximation techniques to determine those values of x for which the error in the approximation $f(x) \approx p_3(x)$ is less than 0.005.

36. Let $f(x) = 1/(4 - x^2)$ and consider the fourth-degree Taylor polynomial $p_4(x)$ for f at 0. Use graphical approximation techniques to determine those values of x for which the error in the approximation $f(x) \approx p_4(x)$ is less than 0.0001.

C

In Problems 37–40, use the second-degree Taylor polynomial at 0 for $f(x) = e^x$ to approximate the given number. Use Taylor's formula for the remainder to estimate the error in each approximation.

37. $e^{-0.3}$
38. $e^{-0.4}$
39. $e^{0.05}$
40. $e^{0.01}$

41. The Taylor series at 0 for $f(x) = \sqrt{16 + x}$ converges for $-16 < x < 16$. Use the second-degree Taylor polynomial at 0 to approximate $\sqrt{17}$. Use Taylor's formula for the remainder to estimate the error in this approximation.

42. The Taylor series at 0 for $f(x) = \sqrt{(4 + x)^3}$ converges for $-4 < x < 4$. Use the third-degree Taylor polynomial at 0 to approximate $\sqrt{125}$. Use Taylor's formula for the remainder to estimate the error in this approximation.

43. To estimate

$$\int_{-2}^{0} \frac{1}{1 - x}\,dx$$

a student takes the first five terms of the Taylor series for $f(x) = 1/(1 - x)$ at 0 and integrates term by term. He obtains the estimate 5.067. A second student estimates the integral by noting that $F(x) = -\ln(1 - x)$ is an antiderivative for $f(x)$. She evaluates $F(x)$ between -2 and 0 to obtain the estimate 1.099. Is either computation correct? How do you account for the large discrepancy between the two estimates?

44. To estimate

$$\int_0^{1.5} \frac{1}{1+x^2}\,dx$$

a student takes the first five nonzero terms of the Taylor series for $f(x) = 1/(1+x^2)$ at 0 and integrates term by term. He obtains the estimate 3.724. A second student doubts the estimate. She claims that since $1/(1+x^2) \le 1$ for $0 \le x \le 1.5$, the value of the integral must be less than 1.5. Is either student correct? How do you account for the large discrepancy between their estimates?

There are different ways to approximate a function f by polynomials. If, for example, $f(a), f'(a)$, and $f''(a)$ are known, then we can construct the second-degree Taylor polynomial $p_2(x)$ at a for $f(x)$; $p_2(x)$ and $f(x)$ will have the same value at a and the same first and second derivatives at a. If, on the other hand, $f(x_1), f(x_2)$, and $f(x_3)$ are known, then we can compute the quadratic regression polynomial $q_2(x)$ for the points $(x_1, f(x_1))$, $(x_2, f(x_2))$, $(x_3, f(x_3))$; $q_2(x)$ and $f(x)$ will have the same values at x_1, x_2, x_3. Problems 45 and 46 concern these contrasting methods of approximation by polynomials.

45. (A) Find the second-degree Taylor polynomial $p(x)$ at 0 for $f(x) = e^x$, and use a graphing utility to compute the quadratic regression polynomial $q_2(x)$ for the points $(-0.1, e^{-0.1})$, $(0, e^0)$, and $(0.1, e^{0.1})$.
(B) Use graphical approximation techniques to find the maximum error for $-0.1 \le x \le 0.1$ in approximating $f(x) = e^x$ by $p_2(x)$ and by $q_2(x)$.
(C) Which polynomial, $p_2(x)$ or $q_2(x)$, gives the better approximation to

$$\int_{-0.1}^{0.1} e^x\,dx?$$

46. (A) Find the fourth-degree Taylor polynomial $p_4(x)$ at 0 for $f(x) = \ln(1+x)$, and use a graphing utility to compute the quartic regression polynomial $q_4(x)$ for the points $(0, \ln 1)$,

$$(-\tfrac{1}{2}, \ln(1-\tfrac{1}{2})), (-\tfrac{3}{4}, \ln(1-\tfrac{3}{4})),$$
$$(-\tfrac{7}{8}, \ln(1-\tfrac{7}{8})), (-\tfrac{15}{16}, \ln(1-\tfrac{15}{16})).$$

(B) Use graphical approximation techniques to find the maximum error for $-\frac{15}{16} \le x \le 0$ in approximating $f(x) = \ln(1+x)$ by $p_4(x)$ and by $q_4(x)$.
(C) Which polynomial, $p_4(x)$ or $q_4(x)$, gives the better approximation to

$$\int_{-15/16}^{0} \ln(1+x)\,dx?$$

Applications

In Problems 47–58, use Theorem 1 to perform the indicated error estimations.

Business & Economics

47. **Income distribution.** The income distribution for a certain country is represented by the Lorenz curve with the equation

$$f(x) = \frac{5x^6}{4+x^2}$$

Approximate the index of income concentration to within ± 0.005.

48. **Income distribution.** Repeat Problem 47 for

$$f(x) = \frac{10x^4}{9+x^2}$$

49. **Marketing.** A soft drink manufacturer is ready to introduce a new diet soda by a national sales campaign. After test marketing the soda in a carefully selected city, the market research department estimates that sales (in millions of dollars) will increase at the monthly rate of

$$S'(t) = 10 - 10e^{-0.01t^2} \qquad 0 \le t \le 12$$

t months after the national campaign is started. Use the fourth-degree Taylor polynomial at 0 for $S'(t)$ to approximate the total sales during the first 4 months of the campaign, and estimate the error in this approximation.

50. **Marketing.** Repeat Problem 49 if the monthly rate of increase in sales is given by

$$S'(t) = 10 - 10e^{-0.005t^2} \qquad 0 \le t \le 12$$

51. **Useful life.** A computer store rents time on desktop publishing systems. The total accumulated costs $C(t)$ and revenues $R(t)$ (in thousands of dollars) from a particular system satisfy

$$C'(t) = 4 \qquad \text{and} \qquad R'(t) = \frac{80}{16+t^2}$$

where t is the time in years that the system has been available for rental. Find the useful life of the system, and approximate the total profit during the useful life to within ± 0.005.

52. **Average price.** Given the demand equation

$$p = D(x) = 10 - 20\ln\left(1 + \frac{x^2}{2{,}500}\right) \qquad 0 \le x \le 40$$

approximate the average price (in dollars) over the demand interval $[0, 20]$ to within ± 0.005.

Life Sciences

53. **Temperature.** The temperature (in degrees Celsius) in an artificial habitat is made to change according to the equation

$$C(t) = 20 + 800\ln\left(1 + \frac{t^2}{100}\right) \qquad 0 \le t \le 2$$

Use a Taylor series at 0 to approximate the average temperature over the time interval $[0, 2]$ to within ± 0.005.

54. **Temperature.** Repeat Problem 53 for

$$C(t) = 10 + 200 \ln\left(1 + \frac{t^2}{50}\right) \quad 0 \le t \le 2$$

55. **Medicine.** The rate of healing for a skin wound (in square centimeters per day) is given by

$$A'(t) = \frac{-75}{t^2 + 25}$$

The initial wound has an area of 12 square centimeters. Use the second-degree Taylor polynomial at 0 for $A'(t)$ to approximate the area of the wound after 2 days, and estimate the error in this approximation.

56. **Medicine.** Repeat Problem 55 for

$$A'(t) = \frac{-60}{t^2 + 20}$$

Social Sciences

57. **Learning.** In a particular business college, it was found that an average student enrolled in an advanced typing class progresses at a rate of $N'(t) = 6e^{-0.01t^2}$ words per minute per week, t weeks after enrolling in a 15-week course. At the beginning of the course an average student could type 40 words per minute. Use the second-degree Taylor polynomial at 0 for $N'(t)$ to approximate the improvement in typing after 5 weeks in the course, and estimate the error in this approximation.

58. **Learning.** In the same business college, it was found that an average student in a beginning shorthand class progressed at a rate of $N'(t) = 12e^{-0.005t^2}$ words per minute per week, t weeks after enrolling in a 15-week course. At the beginning of the course none of the students could take any dictation by shorthand. Use the second-degree Taylor polynomial at 0 for $N'(t)$ to approximate the improvement after 5 weeks in the course, and estimate the error in this approximation.

Chapter 8 Review

Important Terms, Symbols, and Concepts

8-1 Taylor Polynomials

If f is a function that has n derivatives at 0, then the **nth-degree Taylor polynomial for *f* at 0** is

$$p_n(x) = f(0) + f'(0)x + \frac{f''(0)}{2!}x^2 + \cdots + \frac{f^{(n)}(0)}{n!}x^n$$

If *f* has n derivatives at a, then the **nth-degree Taylor polynomial for *f* at *a*** is

$$p_n(x) = f(a) + f'(a)(x - a) + \frac{f''(a)}{2!}(x - a)^2 + \cdots + \frac{f^{(n)}(a)}{n!}(x - a)^n$$

8-2 Taylor Series

If *f* is a function that has derivatives of all order at a point a and $p_n(x)$ is the nth-degree Taylor polynomial for *f* at a, then the **Taylor series for *f* at *a*** is

$$f(a) + f'(a)(x - a) + \frac{f''(a)}{2!}(x - a)^2 + \cdots + \frac{f^{(n)}(a)}{n!}(x - a)^n + \cdots$$

The Taylor series **converges at *x*** if $\lim_{n\to\infty} p_n(x)$ exists, and **diverges at *x*** if the limit does not exist. The set of values of x for which this limit exists is called the **interval of convergence.**

- ***Theorem 1 Interval of Convergence***

Let *f* be a function with derivatives of all order at a point a, let $a_n = f^{(n)}(a)/n!$ for $n = 0, 1, 2, \ldots$, and let $a_0 + a_1(x - a) + a_2(x - a)^2 + \cdots + a_n(x - a)^n + \cdots$ be the Taylor series for *f* at a. If $a_n \ne 0$ for $n \ge n_0$, then:

Case 1. If $\lim_{n\to\infty}\left|\frac{a_{n+1}}{a_n}\right| = L > 0$ and $R = \frac{1}{L}$ then the series converges for $|x - a| < R$ and diverges for $|x - a| > R$.

Case 2. If $\lim_{n\to\infty}\left|\frac{a_{n+1}}{a_n}\right| = 0$, then the series converges for all values of x.

Case 3. If $\lim_{n\to\infty}\left|\frac{a_{n+1}}{a_n}\right| = \infty$, then the series converges only at $x = a$.

We assume that functions are represented by their Taylor series throughout the interval of convergence (there exist functions that do not satisfy this condition, but they are not considered in this supplement).

8-3 Operations on Taylor Series

Property 1. Two Taylor series can be added term by term. This operation is valid in the intersection of the intervals of convergence of the series for f and g.

Property 2. A Taylor series for f at 0 can be multiplied term by term by an expression of the form cx^r, where c is a nonzero constant and r is a non-negative integer. The resulting series has the same interval of convergence as the Taylor series for f.

Property 3. A Taylor series for f at 0 can be differentiated term by term to obtain a Taylor series for f'. Both series have the same interval of convergence.

Property 4. A Taylor series at 0 can be integrated term by term to obtain a Taylor series for $\int f(x)\,dx$. Both series have the same interval of convergence.

Making a substitution in a known Taylor series to obtain a new series is a useful technique. A series for e^{x^2}, for example, can be obtained by substituting x^2 for x in the Taylor series for e^x at 0.

8-4 Approximations Using Taylor Series

If $p_n(x)$ is the nth-degree Taylor polynomial for f, then the **remainder** is $R_n(x) = f(x) - p_n(x)$. The **error** in the approximation $f(x) \approx p_n(x)$ is $|f(x) - p_n(x)| = |R_n(x)|$

- ***Taylor's Formula for the Remainder***

If f has derivatives of all order at 0, then

$$R_n(x) = \frac{f^{(n+1)}(t)\,x^{n+1}}{(n+1)!} \text{ for some number } t \text{ between 0 and } x.$$

THEOREM 1 Error Estimation for Alternating Series

If x_0 is a number in the interval of convergence for

$$f(x) = a_0 + a_1x + a_2x^2 + \cdots + a_kx^k + \cdots$$

and the terms in the series

$$f(x_0) = a_0 + a_1x_0 + a_2x_0^2 + \cdots + a_kx_0^k + \cdots$$

are alternating in sign and decreasing in absolute value, then

$$|R_n(x_0)| < |a_{n+1}x_0^{n+1}|$$

Taylor series techniques provide a method for approximating definite integrals which, in the case of alternating series, automatically determines the accuracy of the approximation.

Review Exercises

Work through all the problems in this chapter review and check your answers in the back of this supplement. Answers to all review problems are there along with section numbers in italics to indicate where each type of problem is discussed. Where weaknesses show up, review appropriate sections in the text.

Unless directed otherwise, use Theorem 1 of Section 8-4 in all problems involving error estimation.

A

1. Find $f^{(4)}(x)$ for $f(x) = \ln(x + 5)$.
2. Use the third-degree Taylor polynomial at 0 for $f(x) = \sqrt[3]{1 + x}$ and $x = 0.01$ to approximate $\sqrt[3]{1.01}$.
3. Use the third-degree Taylor polynomial at $a = 3$ for $f(x) = \sqrt{1 + x}$ and $x = 2.9$ to approximate $\sqrt{3.9}$.
4. Use the second-degree Taylor polynomial at 0 for $f(x) = \sqrt{9 + x^2}$ and $x = 0.1$ to approximate $\sqrt{9.01}$.

Use Theorem 1 of Section 2-2 to find the interval of convergence of each Taylor series representation given in Problems 5–8.

5. $\frac{1}{1 - 4x} = 1 + 4x + 4^2x^2 + \cdots + 4^nx^n + \cdots$

6. $\dfrac{5}{x-1} = 1 - \frac{1}{5}(x-6) + \frac{1}{5^2}(x-6)^2 - \cdots + \dfrac{(-1)^n}{5^n}(x-6)^n + \cdots$

7. $\dfrac{2x}{(1-x)^3} = 1 \cdot 2x + 2 \cdot 3x^2 + 3 \cdot 4\, x^3 + \cdots + n(n+1)x^n + \cdots$

8. $e^{10x} = 1 + 10x + \dfrac{10^2}{2!}x + \cdots + \dfrac{10^n}{n!}x^n + \cdots.$

9. Find the nth derivative of $f(x) = e^{-9x}$.

B

In Problems 10 and 11, use the formula $a_n = f^{(n)}(a)/n!$ to find the Taylor series at the indicated value of a. Use Theorem 1 of Section 2-2 to find the interval of convergence.

10. $f(x) = \dfrac{1}{7-x}$; at 0 11. $f(x) = \ln x$; at 2

In Problems 12–16, use Table 1 of Section 2-3 and the properties of Taylor series to find the Taylor series of each function at the indicated value of a. Find the interval of convergence.

12. $f(x) = \dfrac{1}{10+x}$; at 0 13. $f(x) = \dfrac{x^2}{4-x^2}$; at 0

14. $f(x) = x^2e^{3x}$; at 0 15. $f(x) = x\ln(e+x)$; at 0

16. $f(x) = \dfrac{1}{4-x}$; at 2

17. (A) Explain why Theorem 1 of Section 2-2 is not directly applicable to the Taylor series representation

$$\ln(1-5x^2) = -5\,x^2 - \frac{5^2}{2}\,x^4 - \frac{5^3}{3}\,x^6 - \cdots - \frac{5^n}{n}\,x^{2n} - \cdots$$

(B) Use another method to find the interval of convergence of the Taylor series in part (A).

18. Substituting $\sqrt{x}$ for x in the Taylor series at 0 for $1/(1+x)$ gives the formula

$$\frac{1}{1+\sqrt{x}} = 1 - x^{1/2} + x - x^{3/2} + \cdots + (-1)^n x^{n/2} + \cdots$$

(A) For which values of x is the formula valid?

(B) Is the formula a Taylor series at 0 for $1/(1+\sqrt{x})$? Explain.

In Problems 19 and 20, find the Taylor series at 0 for $f(x)$ and use the relationship $g(x) = f'(x)$ to find the Taylor series at 0 for $g(x)$. Find the interval of convergence for both series.

19. $f(x) = \dfrac{1}{2-x}$; $g(x) = \dfrac{1}{(2-x)^2}$

20. $f(x) = \dfrac{x^2}{1+x^2}$; $g(x) = \dfrac{2x}{(1+x^2)^2}$

In Problems 21 and 22, find the Taylor series at 0, and find the interval of convergence.

21. $f(x) = \displaystyle\int_0^x \frac{t^2}{9+t^2}\,dt$ 22. $f(x) = \displaystyle\int_0^x \frac{t^4}{16-t^2}\,dt$

23. (A) Compute the Taylor series for $f(x) = x^3 - 3x^2 + 4$ at $a = 1$ and at $a = -1$.

(B) Explain how the two series are related.

24. (A) Explain why $f(x) = |x|$ does not have any Taylor polynomials at 0.

(B) If $a \neq 0$, find the Taylor polynomials for $f(x) = |x|$ at a.

In Problems 25 and 26, use the second-degree Taylor polynomial at 0 for $f(x) = e^x$ to approximate the indicated quantity. Use Taylor's formula for the remainder to estimate the error in the approximation.

25. $e^{0.6}$ 26. $e^{0.06}$

In Problems 27 and 28, use a Taylor polynomial at 0 for $f(x) = \ln(1+x)$ to estimate the indicated quantity to within $\pm$ 0.0005. Give the degree of the Taylor polynomial of lowest degree that will provide this accuracy.

27. ln 1.3 28. ln 1.03

29. If $f(x)$ satisfies $f'(x) = x\ln(1-x)$ and $f(0) = 5$, find the Taylor series at 0 for $f(x)$.

30. If $f(x)$ satisfies $f''(x) = xe^{-x}$, $f'(0) = -4$, and $f(0) = 3$, find the Taylor series at 0 for $f(x)$.

C

In Problems 31 and 32, approximate the integral to within $\pm$ 0.0005.

31. $\displaystyle\int_0^1 \frac{1}{16+x^2}\,dx$ 32. $\displaystyle\int_0^1 x^2e^{-0.1x^2}\,dx$

In Problems 33 and 34, assume that $f(x)$ is a function such that $|f^{(n)}(x)| \le 10$ for all n and all x, and let $p_n(x)$ be the nth-degree Taylor polynomial for f at 0.

33. Use Taylor's formula for the remainder to find the smallest value of n such that the error in the approximation of $f(0.5)$ by $p_n(0.5)$ is guaranteed to be less than 10^{-6}.

34. Use Taylor's formula for the remainder to determine the values of x such that the error in the approximation of $f(x)$ by $p_8(x)$ is guaranteed to be less than 10^{-6}.

35. Let $f(x) = e^{x^2}$ and consider the fourth-degree Taylor polynomial $p_4(x)$ for f at 0. Use graphical approximation techniques to determine those values of x for which the error in the approximation $f(x) \approx p_4(x)$ is less than 0.01.

36. Let $f(x) = \ln(1-x^2)$ and consider the sixth-degree Taylor polynomial $p_6(x)$ for f at 0. Use graphical approximation techniques to determine those values of x for which the error in the approximation $f(x) \approx p_6(x)$ is less than 0.001.

Applications

Business & Economics

37. **Average price.** Given the demand equation

$$p = D(x) = \tfrac{1}{10}\sqrt{2{,}500 - x^2}$$

use the second-degree Taylor polynomial at 0 to approximate the average price (in dollars) over the demand interval [0, 15].

38. **Production.** The rate of production of an oil well (in thousands of dollars per year) is given by

$$R(t) = 6 + 3e^{-0.01t^2}$$

Use the second-degree Taylor polynomial at 0 to approximate the total production during the first 10 years of operation of the well.

39. Income distribution. The income distribution for a certain country is represented by the Lorenz curve with the equation

$$f(x) = \frac{9x^3}{8 + x^2}$$

Approximate the index of income concentration to within ± 0.005.

40. Marketing. A cereal manufacturer is ready to introduce a new high-fiber cereal by a national sales campaign. After test-marketing the cereal in a carefully selected city, the market research department estimates that sales (in millions of dollars) will increase at the monthly rate of

$$S'(t) = 20 - 20e^{-0.001t^2} \qquad 0 \le t \le 12$$

t months after the national campaign is started. Use the fourth-degree Taylor polynomial at 0 for $S'(t)$ to approximate the total sales during the first 8 months of the campaign, and estimate the error in this approximation.

Life Sciences

41. Medicine. The rate of healing for a skin wound (in square centimeters per day) is given by

$$A'(t) = \frac{-100}{t^2 + 40}$$

The initial wound has an area of 15 square centimeters. Use the second-degree Taylor polynomial at 0 for $A'(t)$ to approximate the area of the wound after 2 days, and estimate the error in this approximation.

42. Medicine. A large injection of insulin is administered to a patient. The level of insulin in the blood-stream t minutes after the injection is given approximately by

$$L(t) = \frac{5{,}000t^2}{10{,}000 + t^4}$$

Express the average insulin level over the time interval $[0, 5]$ as a definite integral, and use a Taylor series at 0 to approximate this integral to within ± 0.005.

Social Sciences

43. Politics. In a newly incorporated city, the number of voters (in thousands) t years after incorporation is given by

$$N(t) = 10 + 2t - 5e^{-0.01t^2} \qquad 0 \le t \le 5$$

Express the average number of voters over the time interval $[0, 5]$ as a definite integral, and use a Taylor series at 0 to approximate this integral to within ± 0.05.

Group Activity 1 L'Hôpital's Rule

The expressions

$$\lim_{x\to 2} \frac{x^2 - 4}{x - 2} \qquad \text{and} \qquad \lim_{x\to 1} \frac{e^x - e}{x^2 + 3x - 4}$$

are examples of $0/0$ indeterminate forms. The numerator and denominator both have limit 0, and the quotient property for limits does not apply. But the first limit expression can be evaluated by performing some algebraic simplifications:

$$\lim_{x\to 2} \frac{x^2 - 4}{x - 2} = \lim_{x\to 2} \frac{(x - 2)(x + 2)}{x - 2} = \lim_{x\to 2}(x + 2) = 4$$

To evaluate the second limit expression, we can write the numerator $f(x) = e^x - e$ and the denominator $g(x) = x^2 + 3x - 4$ as Taylor series at $a = 1$:

$$\begin{aligned}
\lim_{x\to 1} \frac{f(x)}{g(x)} &= \lim_{x\to 1} \frac{e^x - e}{x^2 + 3x - 4} \\
&= \lim_{x\to 1} \frac{e(x-1) + \frac{e}{2!}(x-1)^2 + \frac{e}{3!}(x-1)^3 + \cdots}{5(x-1) + (x-1)^2} \\
&= \lim_{x\to 1} \frac{(x-1)\left[e + \frac{e}{2!}(x-1) + \frac{e}{3!}(x-1)^2 + \cdots\right]}{(x-1)[5 + (x-1)]} \\
&= \lim_{x\to 1} \frac{\left[e + \frac{e}{2!}(x-1) + \frac{e}{3!}(x-1)^2 + \cdots\right]}{[5 + (x-1)]} = \frac{e}{5}
\end{aligned}$$

The value of the limit, $e/5$, is thus the ratio of the coefficients of $x - 1$ in the Taylor series for f and g. But the coefficient of $x - 1$ in the Taylor series for f is $f'(1)$, and the coefficient of $x - 1$ in the Taylor series for g is $g'(1)$, so the limit is $f'(1)/g'(1)$.

Our reasoning suggests a simple method for computing each of the above limits. We compute the limit of the derivative of the numerator divided by the derivative of the denominator (without calculating Taylor series) as follows:

$$\lim_{x \to 2} \frac{x^2 - 4}{x - 2} = \lim_{x \to 2} \frac{\frac{d}{dx}(x^2 - 4)}{\frac{d}{dx}(x - 2)} = \lim_{x \to 2} \frac{2x}{1} = 4$$

$$\lim_{x \to 1} \frac{e^x - e}{x^2 + 3x - 4} = \lim_{x \to 1} \frac{\frac{d}{dx}(e^x - e)}{\frac{d}{dx}(x^2 + 3x - 4)} = \lim_{x \to 1} \frac{e^x}{2x + 3} = \frac{e}{5}$$

Note that the derivative of the numerator and the derivative of the denominator are computed separately (we are not using the quotient rule!).

The simple computations above are illustrations of a powerful technique for evaluating limits of quotients called *L'Hôpital's rule.* The rule is named after the French mathematician Marquis de L'Hôpital (1661–1704), who is generally credited with writing the first calculus textbook.

L'Hôpital's Rule for 0/0 Indeterminate Forms

For c a real number, if $\lim_{x \to c} f(x) = 0$ and $\lim_{x \to c} g(x) = 0$, then

$$\lim_{x \to c} \frac{f(x)}{g(x)} = \lim_{x \to c} \frac{f'(x)}{g'(x)}$$

provided the second limit exists or is $+\infty$ or $-\infty$. The rule remains valid if the symbol $x \to c$ is replaced everywhere it occurs with one of the following symbols:

$$x \to c^+ \qquad x \to c^- \qquad x \to \infty \qquad x \to -\infty$$

1. Find each limit in two ways: by using algebraic simplification and by using L'Hôpital's rule.

(A) $\lim_{x \to 2} \frac{x^4 - 16}{x^3 - 8}$ (B) $\lim_{x \to 1} \frac{x^6 - 1}{x^5 - 1}$

(C) $\lim_{x \to 2} \frac{x^2 + x - 6}{x^2 + 6x - 16}$ (D) $\lim_{x \to 4} \frac{x^2 - 8x + 16}{x^2 - 5x + 4}$

2. Find each limit in two ways: by using Taylor series expansions and by using L'Hôpital's rule.

(A) $\lim_{x \to 0} \frac{e^x - 1}{x}$ (B) $\lim_{x \to 1} \frac{\ln x}{x - 1}$

(C) $\lim_{x \to 0} \frac{\ln(1 + 4x)}{x}$ (D) $\lim_{x \to 0} \frac{e^{2x} - 1}{x}$

Group Activity 2 Taylor Series Solutions to Differential Equations

Since Taylor series at 0 can be differentiated term by term, they can be used to solve differential equations. The idea is simple: Substitute

$$f(x) = a_0 + a_1 x + a_2 x^2 + \cdots + a_n x^n + \cdots \tag{1}$$

in the differential equation and solve for the coefficients $a_0, a_1, a_2, \ldots$. The following examples will illustrate the technique.

EXAMPLE 1 **Taylor Series Solution** Find the particular solution of $y' = y$ that satisfies $y(0) = 1$.

SOLUTION Substituting $f(x)$ from (1) for y in the differential equation and $f'(x)$ for y', gives

$$a_1 + 2a_2x + 3a_3x^2 + 4a_4x^3 + \cdots = a_0 + a_1x + a_2x^2 + a_3x^3 + \cdots$$

By equating the constant terms, then the coefficients of x, then the coefficients of x^2, and so on, we obtain the following sequence of equations:

$$\begin{aligned} a_1 &= a_0 \\ 2a_2 &= a_1 \\ 3a_3 &= a_2 \\ 4a_4 &= a_3 \\ &\vdots \end{aligned}$$

Since $f(0) = a_0$, the initial condition $y(0) = 1$ implies that $a_0 = 1$. Now the sequence of equations can be used to solve successively for a_1, a_2, a_3, a_4:

$$a_1 = 1, \qquad a_2 = \tfrac{1}{2}, \qquad a_3 = \tfrac{1}{6}, \qquad a_4 = \tfrac{1}{24}$$

Therefore,

$$f(x) = 1 + x + \tfrac{1}{2}x^2 + \tfrac{1}{6}x^3 + \tfrac{1}{24}x^4 + \cdots$$

(A) Identify a well-known function whose Taylor series at 0 agrees with these first five terms of $f(x)$, and show that your function satisfies the differential equation and initial condition of Example 1.

In general, we will not recognize the Taylor series solution to a differential equation as a known function, but we will be able to compute as many terms of the Taylor series solution as desired.

EXAMPLE 2 **Taylor Series Solution** Find the first six nonzero terms of the Taylor series solution to $y'' + xy' = 3x^2 - 2y$, $y(0) = 5$, $y'(0) = -3$.

SOLUTION Substituting $f(x)$ from (1) for y, $f'(x)$ for y', and $f''(x)$ for y'', gives

$$\begin{aligned} &(2a_2 + 6a_3x + 12a_4x^2 + 20a_5x^3 + \cdots) + x(a_1 + 2a_2x + 3a_3x^2 + 4a_4x^3 + \cdots) \\ &= 3x^2 - 2(a_0 + a_1x + a_2x^2 + a_3x^3 + \cdots) \end{aligned}$$

By equating the constant terms, then the coefficients of x, then the coefficients of x^2, and so on, we obtain the following sequence of equations:

$$\begin{aligned} 2a_2 &= -2a_0 \\ 6a_3 + a_1 &= -2a_1 \\ 12a_4 + 2a_2 &= 3 - 2a_2 \\ 20a_5 + 3a_3 &= -2a_3 \end{aligned}$$

Since $f(0) = a_0$, the initial condition $y(0) = 5$ implies that $a_0 = 5$. Since $f'(0) = a_1$, the initial condition $y'(0) = -3$ implies that $a_1 = -3$. Now the sequence of equations can be used to solve successively for a_2, a_3, a_4, and a_5:

$$a_2 = -5, \qquad a_3 = \tfrac{3}{2}, \qquad a_4 = \tfrac{23}{12}, \qquad a_5 = \tfrac{-3}{8}$$

Therefore,

$$f(x) = 5 - 3x - 5x^2 + \tfrac{3}{2}x^3 + \tfrac{23}{12}x^4 - \tfrac{3}{8}x^5 + \cdots$$

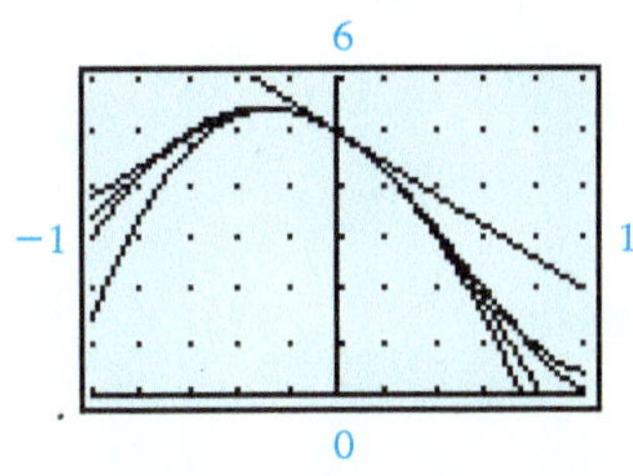

Figure 1

The first-, second-, third-, fourth-, and fifth-degree Taylor polynomials for $f(x)$ in Example 2 are graphed in Figure 1. Note from the graphs that each polynomial satisfies the initial conditions of the differential equation.

(B) Find the coefficients of x^6 and x^7 in the Taylor series solution in Example 2.

(C) Find the first four nonzero terms of the Taylor series solution to $y' + 4xy = 6x$, $y(0) = 1$.

(D) Find the first four nonzero terms of the Taylor series solution to $y'' = -y$, $y(0) = 1$, $y'(0) = 0$.

(E) It is known that the function

$$g(x) = \tfrac{3}{2} - \tfrac{1}{2}e^{-2x^2}$$

is a solution to the first-order linear differential equation and initial condition of part (C). Use a graphing utility to graph $g(x)$ and the sum of the first four nonzero terms of the Taylor series of part (C). Discuss how the graphs differ.

(F) It is known that the function $h(x) = \cos x$ is a solution to the differential equation and initial condition of part (D). Use a graphing utility to graph $h(x)$ and the sum of the first four nonzero terms of the Taylor series of part (D). Discuss how the graphs differ.

Numerical Techniques

Calculus is a powerful tool for solving a wide variety of problems, as evidenced by all the applications we have discussed in this text. However, many real-world applications of calculus cannot be solved by the techniques we have discussed so far. For instance, not all equations can be solved exactly, not all definite integrals can be evaluated by finding an antiderivative, and not all differential equations have solutions that can be expressed in terms of familiar functions. In such cases, numerical methods often are used to approximate solutions that cannot be found exactly. In this chapter we will develop methods for approximating the roots of an equation, the value of a definite integral, and a particular solution of a differential equation. In actual practice, computers are usually used to perform the calculations involved in these approximations. Certain models of graphic calculators also have some built-in approximation routines and can be programmed to perform other routines. However, it is instructive for the beginning student to perform some of these calculations directly using nothing more than an ordinary scientific calculator. We have used a scientific calculator to perform our calculations, and you will need a calculator with similar capabilities in order to work the problems and exercises in this chapter.

Calculator Variation

Do not be alarmed if your calculator results do not agree exactly with those printed in this chapter or in the answer section. The large variety of scientific calculators on the market do not all perform these operations in exactly the same way, so you can expect slight variations in the last two or three decimal places when the calculators are used to their full capacities. Slight variations in results may even occur in the same calculator when a given calculation is performed in two different ways.

Taken from *Applied Calculus for Business, Economics, Life Sciences, and Social Sciences*, Fifth Edition by Raymond A. Barnett and Michael R. Ziegler.

9-1 Newton's Method for Approximating Roots

Zeros, Roots, and Intercepts

A number r is a **zero** of a function f, or a **root** of the equations $f(x) = 0$, if $f(r) = 0$. If $f(x) = ax^2 + bx + c$ is any second-degree polynomial, then its zeros are given by the *quadratic formula*,

$$r = \frac{-b \pm \sqrt{b^2 - 4ac}}{2a}$$

EXAMPLE 1 Find the zeros of $f(x) = x^2 - 8x + 13$.

SOLUTION Since f is a second-degree polynomial and $a = 1, b = -8, c = 13$, the zeros are

$$r_1 = \frac{8 - \sqrt{64 - 52}}{2} = \frac{8 - 2\sqrt{3}}{2} = 4 - \sqrt{3} \approx 2.267\,949\,2$$

$$r_2 = \frac{8 + \sqrt{64 - 52}}{2} = 4 + \sqrt{3} \approx 5.732\,050\,8$$

Matched Problem 1 Use the quadratic formula to find the zeros of $f(x) = x^2 - 6x + 4$.

Notice the use of the symbols = and ≈ in Example 1. The *exact zero* r_1 of f is the irrational number $4-\sqrt{3}$. However, when we use a calculator to evaluate the irrational number $4-\sqrt{3}$, we obtain the finite decimal 2.267 949 2, which is a rational number that *approximates* r_1.

If we graph $y = f(x)$, then the zeros are often referred to as the ***x* intercepts**, since these are the points where the graph crosses the x axis. Figure 1 illustrates this for the function in Example 1.

Using algebraic methods to find the exact zeros of more complicated functions can be very difficult, if not impossible. Many methods have been developed to approximate the zeros of a function. For example, computers and calculators with graphing capabilities can be used to approximate the intercepts of a function by repeatedly expanding the graph near the intercept and displaying the coordinates of the point on the screen that is closest to the actual intercept. Figure 2 illustrates this process for the function $f(x)$ in Example 1.

The coordinates of the point at the location of the cursor are displayed at the bottom of each graph. The graph if Figure 1(C) indicates that $f(x)$ has an intercept at approximately $x = 2.265\,921\,1$. Comparing this result with r_1 in Example 1, we see that this approximation is accurate to two decimal places. Repeated expansions of the graph will improve the accuracy of this approximation.

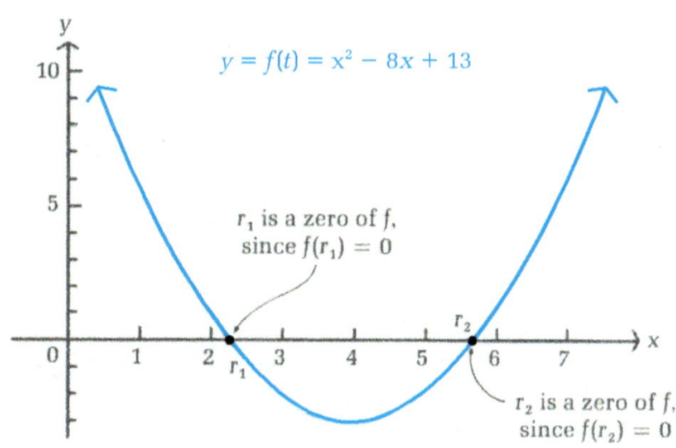

Figure 1

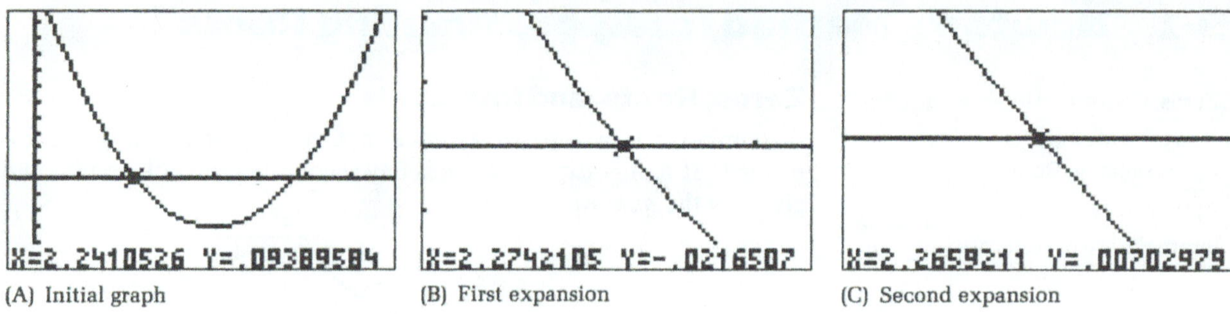

(A) Initial graph (B) First expansion (C) Second expansion

Figure 2 Graphical approximation of the smaller intercept of $f(x) = x^2 - 8x + 13$

Graphical approximation techniques are easy to understand and to apply, but they are not practical for all purposes. They require user interaction at each step, and the number of steps can be large. For example, eight more expansions of the graph in Figure 2 are required to approximate the intercept to seven decimal places. Numerical methods that can be performed automatically by a computer or programmable calculator are generally used in actual practice. One of these—usually called *Newton's method*—produces very accurate approximations in relatively few steps.

Newton's Method for Approximating Roots

Given a differentiable function f, suppose we are able to determine that f has a zero at some number r, but we are unable to find the exact value of r by algebraic methods. In order to approximate the value of r, we begin by selecting an initial value x_1, which we suspect is close to r. (We will have more to say about the selection of this initial value later.) If $f'(x_1) \neq 0$, then the line tangent to the graph of $y = f(x)$ at the point $(x_1, f(x))$ must intersect the x axis (see Figure 3).

Let x_2 be the x intercept of this tangent line. Examining Figure 3, it appears that x_2 is closer to r than x_1 is. How can we find the value of x_2? The equation for the tangent line at $(x_1, f(x_1))$ is

$$y - f(x_1) = f'(x_1)(x - x_1) \tag{1}$$

If x_2 is the x intercept of this line, then the point $(x_2, 0)$ must satisfy the equation of the line. Substituting x_2 for x and 0 for y in (1) and solving for x_2, we have

$$0 - f(x_1) = f'(x_1)(x_2 - x_1)$$

$$x_2 - x_1 = -\frac{f(x_1)}{f'(x_1)} \qquad f'(x_1) \neq 0$$

$$x_2 = x_1 - \frac{f(x_1)}{f'(x_1)} \tag{2}$$

Figure 3

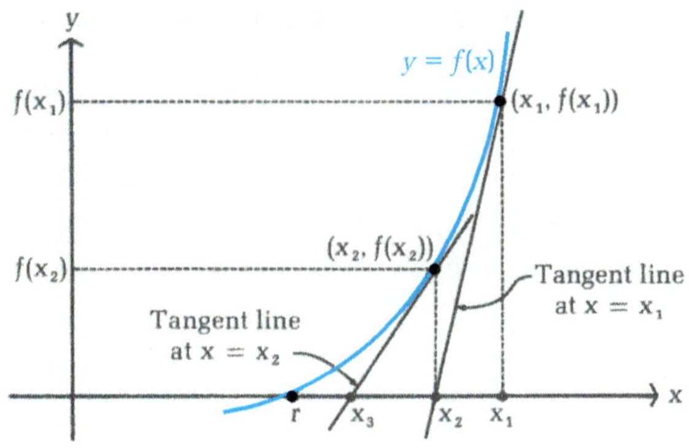

Figure 4

Notice that if $f'(x_1) = 0$, then the tangent line at $(x_1, f(x_1))$ is horizontal and there is no x intercept. To avoid this case, we will assume that $f'(x) \neq 0$ for all values of x near the zero r. Thus, given an initial approximation x_1, we can use (2) to computer x_2, a second approximation to r. For most functions, x_2 will be a better approximation than x_1. However, x_2 still might not be a sufficiently accurate approximation. If we repeat this process, beginning now with the tangent line at the point $(x_2, f(x_2))$, we should obtain an even better approximation to r (see Figure 4).

Let x_3 denote the x intercept of the tangent line at $(x_2, f(x_2))$. The equation of this line is

$$y - f(x_2) = f'(x_2)(x - x_2)$$

Substituting x_3 for x and 0 for y and solving for x_3, we have

$$0 - f(x_2) = f'(x_2)(x_3 - x_2)$$

$$x_3 = x_2 - \frac{f(x_2)}{f'(x_2)} \tag{3}$$

Notice the similarity between the expressions for x_2 in equation (2) and the one for x_3 in equation (3). Continued repetition of this process will produce a sequence of number $x_1, x_2, \ldots, x_n, \ldots,$ where each number in the sequence (after x_1) is obtained by using the preceding number in the **recursion formula**

$$x_n = x_{n-1} - \frac{f(x_{n-1})}{f'(x_{n-1})} \qquad n > 1$$

If the numbers $x_1, x_2, \ldots, x_n, \ldots,$ approach a limit r, then it can be shown that $f(r) = 0$. Hence, r is a zero of f. The process described above is referred to as *Newton's method for approximating roots.*

Newton's Method

Given the function f and the initial approximation x_1, define

$$x_n = x_{n-1} - \frac{f(x_{n-1})}{f'(x_{n-1})} \qquad n > 1, \qquad f'(x_{n-1}) \neq 0$$

If $\lim_{n\to\infty}$ exists, then

$$r = \lim_{n\to\infty} x_n$$

is a zero of f.

EXAMPLE 2 Use Newton's method to approximate the smaller zero of $f(x) = x^2 - 8x + 13$. Compare the result with that obtained by use of the quadratic formula in Example 1.

SOLUTION **Step 1** Find the recursion formula for x_n:

$$f(x) = x^2 - 8x + 13$$

$$f'(x) = 2x - 8$$

$$\begin{aligned} x_n &= x_{n-1} - \frac{f(x_{n-1})}{f'(x_{n-1})} \\ &= x_{n-1} - \frac{x_{n-1}^2 - 8x_{n-1} + 13}{2x_{n-1} - 8} \\ &= \frac{x_{n-1}(2x_{n-1} - 8) - (x_{n-1}^2 - 8x_{n-1} + 13)}{2x_{n-1} - 8} \\ x_n &= \frac{x_{n-1}^2 - 13}{2x_{n-1} - 8} \end{aligned} \tag{4}$$

Step 2 Approximate the root using equation (4). Examining the graph of $y = f(x)$ in Figure 1 (and ignoring the fact that we know the exact root), we see that $x_1 = 2$ is a reasonable first approximation to the root r_1. Using equation (4) and $x_1 = 2$ as a first approximation, we obtain the following successive approximations for r_1:

$$\begin{aligned} x_1 &= 2 \\ x_2 &= \frac{(\mathbf{2})^2 - 13}{2(\mathbf{2}) - 8} = 2.25 \\ x_3 &= \frac{(\mathbf{2.25})^2 - 13}{2(\mathbf{2.25}) - 8} \approx 2.267\,857\,1 \\ x_4 &= \frac{(\mathbf{2.267\,857\,1})^2 - 13}{2(\mathbf{2.267\,857\,1}) - 8} \approx 2.267\,949\,2 \\ x_5 &= \frac{(\mathbf{2.267\,949\,2})^2 - 13}{2(\mathbf{2.267\,949\,2}) - 8} \approx 2.267\,949\,2 \end{aligned}$$

Since $x_4 = x_5$ in the display in our calculator, we have reached the limits of accuracy for our calculator and assume $r_1 \approx 2.267\,949\,2$. Notice that this is the *same approximation to the exact zero* that we obtained in Example 1 using the quadratic formula and our calculator directly.

Matched Problem 2 Use Newton's method to approximate the larger zero of $f(x) = x^2 - 8x + 13$. First examine Figure 1 to obtain a first approximation to the larger zero of f. Then use Newton's method to approximate the larger root to the limit of accuracy of your calculator—that is, until two successive approximation are the same in the display of your calculator.

In Example 2, it took us only three calculations to approximate the zero correct to seven decimal places. We were able to obtain this much accuracy with so few computations because the initial approximation was reasonably close to the actual zero. It is always important to try to select a good first approximation. Figures 5 and 6 illustrate two situations that may occur if x_1 is too far away from r. In Figure 5, each x_n is larger than the preceding value and further away from r. This type of behavior is very common if the initial approximation x_1 is not selected carefully. In Figure 6, the approximating values oscillate between x_1 and x_2, never getting any closer to r. If either of these situations occur when you are using Newton's method, you must select a better initial value for x_1.

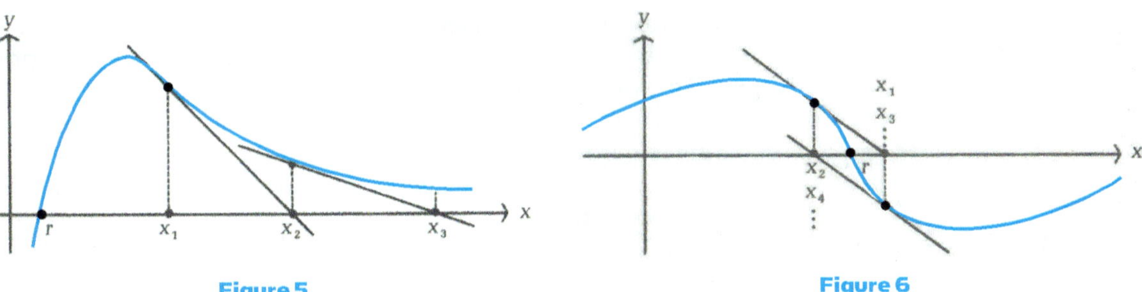

Figure 5

Figure 6

EXAMPLE 3 Sketch the graph of $f(x) = x^3 - 9x^2 + 15x + 10$ and approximate the x intercepts.

SOLUTION **Step 1** Sketch the graph of f:

$$f'(x) = 3x^2 - 18x + 15 = 3(x - 5)(x - 1)$$

Critical values: $x = 1, 5$

The graph of f rises for x in the intervals $(-\infty, 1)$ and $(5, \infty)$ and falls for x in the interval $(1, 5)$. The second derivative,

$$f''(x) = 6x - 18 = 6(x - 3)$$

tells us the graph of f is concave downward in the interval $(-\infty, 3)$ and concave upward in the interval $(3, \infty)$. Using this information and point-by-point plotting to sketch the graph of f, as shown in the figure at the top the next page, we conclude that f has three zeros: r_1 in $(-1, 0)$, r_2 in $(3, 4)$, and r_3 in $(6, 7)$. Notice that the intervals containing intercepts also can be identified by examining the table in the margin for changes in the sign of $f(x)$. Drawing the graph assures us that there is only one zero in each of these intervals and that there are no zeros anywhere else.

Step 2 Write the recursion formula for x_n:

$$x_n = x_{n-1} - \frac{f(x_{n-1})}{f'(x_{n-1})} = x_{n-1} - \frac{x_{n-1}^3 - 9x_{n-1}^2 + 15x_{n-1} + 10}{3x_{n-1}^2 - 18x_{n-1} + 15} \tag{5}$$

This time we will not simplify the recursion formula.

x	$f(x)$	
-1	-15	} Sign change
0	10	
1	17	
2	12	
3	1	} Sign change
4	-10	
5	-15	
6	-8	} Sign change
7	17	

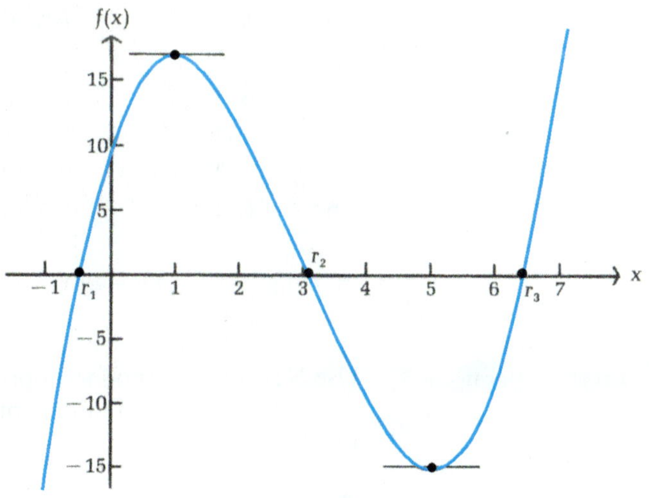

Step 3 Approximate r_1: We will use $x_1 = -1$ for an initial approximation and use a calculator to repeatedly evaluate the expression in (5). At first you may want to evaluate $f(x_{n-1})$ and $f'(x_{n-1})$ separately, recording each value, and then compute x_n. To assist in this process, we have listed both partial and final results of these calculations in the table below. After some practice, you should be able

to store x_{n-1} in the memory of your calculator and evaluate (5) directly, recalling the stored value of x_{n-1} each time you need it. It is good practice to record each value of x_n as you compute it so that you can reenter a previous value in case of an error in a subsequent calculation. If you are using a graphic calculator, you can enter the recursion formula (5) (omitting the subscripts) in the function definition screen of the calculator and repeatedly evaluate this expression for successive value of x.

n	x_{n-1}	$f(x_{n-1})$	$f'(x_{n-1})$	$x_n = x_{n-1} - \frac{f(x_{n-1})}{f'(x_{n-1})}$
2	−1	−15	36	$-1 - \frac{-15}{36} \approx -0.583\ 333\ 33$
3	−0.583 333 33	−2.010 955 3	26.520 833	$-0.583\ 333\ 33 - \frac{-2.010\ 995\ 3}{26.520\ 833} \approx -0.507\ 506\ 33$
4	−0.507 506 33	−0.061 373 72	24.907 803	$-0.507\ 506\ 33 - \frac{-0.061\ 373\ 72}{24.907\ 803} \approx -0.505\ 042\ 29$
5	−0.505 042 29	−0.000 063 77	24.855 964	$-0.505\ 042\ 29 - \frac{-0.000\ 063\ 77}{24.855\ 964} \approx -0.505\ 039\ 72$
6	−0.505 039 72	0.000 000 11	24.855 910	$-0.505\ 039\ 72 - \frac{0.000\ 000\ 11}{24.855\ 910} \approx -0.505\ 039\ 73$

Since x_5 and x_6 are very nearly the same, we conclude that $r_1 \approx -0.505\ 039\ 7$ is a good approximation.

Step 4 Approximate r_2: Initial approximation: $x_1 = 3$.

n	x_{n-1}	$f(x_{n-1})$	$f'(x_{n-1})$	$x_n = x_{n-1} - \frac{f(x_{n-1})}{f'(x_{n-1})}$
2	3	1	212	3.083 333 3
3	3.083 333 3	0.000 578 77	211.979 167	3.083 381 6
4	3.083 381 6	0.000 000 54	211.979 143	3.083 381 6

Thus, $f_2 \approx 3.083\ 381\ 6$.

Step 5 Approximate r_3: Initial approximation: $x_1 = 6$.

n	x_{n-1}	$f(x_{n-1})$	$f'(x_{n-1})$	$x_n = x_{n-1} - \frac{f(x_{n-1})}{f'(x_{n-1})}$
2	6	−8	15	6.533 333 3
3	6.533 333 3	2.711 703 6	25.453 333	6.426 797
4	6.426 797	0.119 100 63	23.228 814	6.421 669 8
5	6.421 669 8	0.000 270 04	23.123 472	6.421 648 1
6	6.421 658 1	0.000 000 47	23.123 232	6.421 658 1

Thus, $r_3 \approx 6.421\ 658\ 1$.

Matched Problem 3 Use Newton's method to approximate the zeros of $f(x) = x^4 - 4x^3 + 10$. (Sketch a graph of f to obtain initial approximations and to be certain all zeros have been located.)

Applications

Many applications in earlier sections involved solving equations in order to find quantities such as the value of x that produces a given functional value, the points of intersection of two graphs, or the critical values of a function. All the problems we considered were carefully selected so that the solutions could be found by algebraic methods. Now that we can use Newton's method to find the roots of equations that can be written in the form $f(x) = 0$, we can consider problems involving equations that do not lend themselves to algebraic solutions. Since Newton's method generally produces an accurate approximation in a relatively small number of steps, most problems can be solved with a scientific calculator. However, performing all these calculations does become tedious, and it is convenient to use a graphic calculator or a computer if one is available. The tables in Examples 4, 5, and 6 were produced using a graphic calculator. All the values in these tables were rounded to seven decimal places.

EXAMPLE 4 **Pollution** The pollution level (in parts per million) in a river t days after an industrial accident near the river is given by

$$P(t) = 50te^{-0.1t}$$

The water is considered to be contaminated as long as the pollution level is above 125 parts per million. Graph $P(t)$ and determine the time period during which the river will be contaminated. Round values of t to one decimal place.

SOLUTION The graph of $P(t)$ is shown in the figure (graphing details omitted):

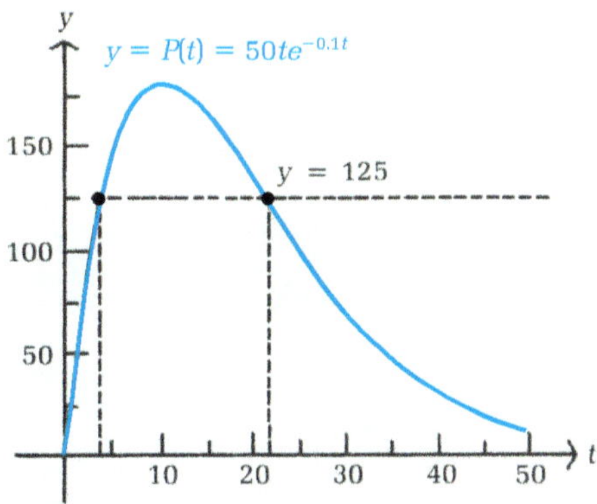

To determine the time interval where $P(t) > 125$ we must first solve the equation

$$P(t) = 125$$

for t. The graph shows that this equation has two solutions, one near 5 and another near 20. Since Newton's method can be used only to solve equations of the form $F(t) = 0$, we define

$$\begin{aligned} F(t) &= P(t) - 125 \\ &= 50te^{-0.1t} - 125 \end{aligned}$$

and apply Newton's method to $F(t)$, first with an initial approximation of 5 and then with an initial approximation of 20. The table gives the results:

Approximation of the Zeros of F(t)

(A) NEAR $t = 5$		(B) NEAR $t = 20$	
n	t_n	n	t_n
1	5	1	20
2	3.243 606 4	2	21.527 359 8
3	3.560 726 3	3	21.532 923 4
4	3.574 007	4	21.532 923 6
5	3.574 029 6		

According to the table, $F(t) = 0$ at $t \approx 3.6$ and $t \approx 21.5$ (rounded to one decimal place). Examining the figure, we see that the graph of $P(t)$ is above the (dashed) horizontal line $y = 125$ for $3.6 < t < 21.5$. Thus,

$$P(t) > 125 \qquad \text{for} \qquad 3.6 < t < 21.5$$

and the river is contaminated for t in the interval $(3.6, 21.5)$.

Matched Problem 4 Repeat Example 4 if the river is considered to be contaminated when the pollution level is above 160 parts per million.

EXAMPLE 5 **Equilibrium Price** Find the equilibrium price correct to two decimal places, given the demand and supply functions

$$p = D(x) = 100 - 10\sqrt{x} \qquad \text{and} \qquad p = S(x) = 10 + \tfrac{1}{50}x^2$$

where p is the price in dollars and x is the number of units in thousands.

SOLUTION Recall that the equilibrium price $\bar{p}$ is the common value of $S(x)$ and $D(x)$ at the points where their graphs intersect. Thus, we must first solve the equation $S(x) = D(x)$ for x. Proceeding as before, we define

$$F(x) = D(x) - S(x) = 90 - 10\sqrt{x} - \tfrac{1}{50}x^2$$

The zeros of F will be the x coordinates of the points of intersection of the graphs of D and S. The figure at the top of the next page shows that the graphs of D and S have one point of intersection and that the x coordinate of this point is close to 40. We apply Newton's method to F with $x_1 = 40$.

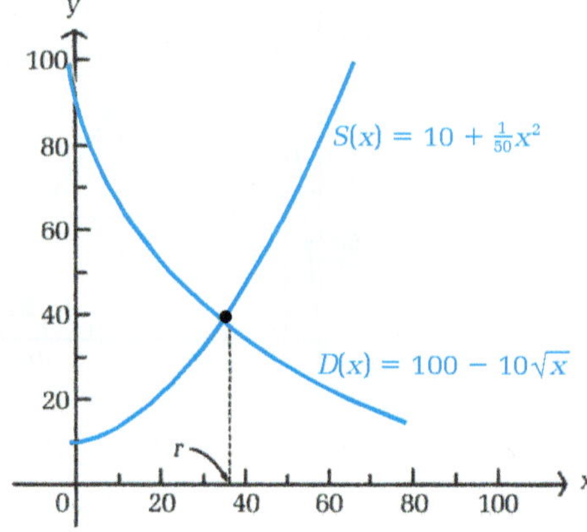

n	x_n
1	40
2	37.805 730 6
3	37.774 841 5
4	37.774 835 5

According to the table, the zero of F is approximately 37.775. We find the equilibrium price (to two decimal places) by using a calculator to evaluate either the demand function or the supply function at $x = 37.775$:

$$D(37.775) \approx 38.54 \approx S(37.775)$$

Thus, the equilibrium price is $\bar{p} \approx \$38.54$.

Matched Problem 5 Repeat Example 5 if

$$p = D(x) = 36 - \tfrac{1}{16}x^2 \qquad \text{and} \qquad p = S(x) = \sqrt{100 + x^2}$$

Internal Rate of Return

If a \$10,000 investment returns a single payment of \$13,000 at the end of 3 years, then the annual rate of return r on this investment can be found by using the compound interest formula:

$$A = P\left(1 + \frac{r}{m}\right)^{mt}$$

Substituting 10,000 for the present value P, 13,000 for the future value A, 1 for the number of compounding periods m in a year, and 3 for the number of years t, we obtain an equation we can solve for r:

$$13{,}000 = 10{,}000(1 + r)^3$$

$$1.3 = (1 + r)^3$$
$$\sqrt[3]{1.3} = 1 + r$$
$$r = \sqrt[3]{1.3} - 1 \approx 0.0914$$

Thus, the rate of return is 9.14% compounded annually.

In the next example we consider an investment that returns a sequence of payments, resulting in a more complicated equation for the rate of return r, and we use Newton's method to solve the equation.

EXAMPLE 6 **Internal Rate of Return** A $10,000 investment returns payments of $5,000 at the end of the first year, $4,000 at the end of the second year, and a final payment of $3,000 at the end of the third year. The present values of these three payments are

$$P_1 = \frac{5{,}000}{1 + r} \qquad P_2 = \frac{4{,}000}{(1 + r)^2} \qquad P_3 = \frac{3{,}000}{(1 + r)^3}$$

The **internal rate of return** of this investment is the annual rate r that produces three present values whose sum equals the original investment of $10,000. Find the internal rate of return. Express the answer as a percentage correct to two decimal places.

SOLUTION The internal rate of return r is the solution of the equation

$$\begin{pmatrix}\text{Initial}\\ \text{investment}\end{pmatrix} = \begin{pmatrix}\text{Sum of present values}\\ \text{of all payments}\end{pmatrix}$$
$$10{,}000 = \frac{5{,}000}{1 + r} + \frac{4{,}000}{(1 + r)^2} + \frac{3{,}000}{(1 + r)^3}$$

or, equivalently, the root of the equation

$$10{,}000 - \frac{5{,}000}{1 + r} - \frac{4{,}000}{(1 + r)^2} - \frac{3{,}000}{(1 + r)^3} = 0$$

Multiplying both sides of this equation by 0.001, we have

$$10 - \frac{5}{1 + r} - \frac{4}{(1 + r)^2} - \frac{3}{(1 + r)^3} = 0$$

If we let $x = 1 + r$ and use negative exponents, then this last equation can be written more simply as

$$f(x) = 10 - 5x^{-1} - 4x^{-2} - 3x^{-3} = 0$$

Since r represents an interest rate, we expect its value to be between 0 and 1. Hence, we expect to find to zero of $f(x)$ between $x = 1$ and $x = 2$. Evaluating f at these values, we have

$$f(1) = -2 \qquad \text{and} \qquad f(2) = 61.25$$

Thus, f does have a zero between $x = 1$ and $x = 2$. Furthermore, since

$$f'(x) = 5x^{-2} + 8x^{-3} + 9x^{-4} > 0$$

for $x > 0$, f is increasing on $(0, \infty)$ and can have only one positive zero. We apply Newton's method to $f(x)$ with $x_1 = 1$ to find this zero.

n	x_n
1	1
2	1.090 909 1
3	1.106 174 5
4	1.106 516 6
5	1.106 516 8

The table indicates that the zero of f (to four decimal places) is $x = 1.1065$. Thus, $r = x - 1 = 0.1065$ and the internal rate of return is 10.65% compounded annually.

Matched Problem 6 An investment of $15,000 returns payments of $5,000 at the end of the first year, $6,000 at the end of the second year, and a final payment of $7,000 at the end of the third year. Find the internal rate of return. Express the answer as a percentage correct to two decimal places.

Remarks

1. The internal rate of return is used by businesses to help make decisions involving long-term capital investments such as equipment replacement, facility expansion, mergers, or new product development.
2. As long as all the payments are nonnegative and their sum exceeds the initial investment, the internal rate of return will be a nonnegative number that can be determined using Newton's method as illustrated in Example 6. Situations where some of the payments are negative (indicating an additional amount was invested) or where the sum of the payments is less than the initial investment (indicating a loss to the investor) require a more detailed analysis. We will not consider these situations.
3. This procedure also can be applied to investments involving payment periods other than annual payments.

Answers to Matched Problems

1. $r_1 = 3 + \sqrt{5} \approx 5.236\,068$; $r_2 = 3 - \sqrt{5} \approx 0.763\,932\,02$
2. $r_2 \approx 5.732\,050\,8$ **3.** $r_1 \approx 1.611\,793\,4$; $r_2 \approx 3.820\,704\,4$
4. $5.6 < t < 16.2$ **5.** \$19.20 **6.** 9.15%

Exercise 9-1

A

In Problems 1–10, find and simplify the recursion formula for x_n. Use this formula and the given value of x_1 to compute x_n for the indicated value of n.

1. $f(x) = x^2 - 4;\quad x_1 = 1;\quad n = 5$

2. $f(x) = x^2 - 2;\quad x_1 = 1;\quad n = 5$

3. $f(x) = x^3 - 8;\quad x_1 = 3;\quad n = 5$

4. $f(x) = x^3 - 2:\quad x_1 = 4;\quad n = 7$

5. $f(x) = e^x + x;\quad x_1 = 0;\quad n = 5$

6. $f(x) = e^x + 2x;\quad x_1 = 1;\quad n = 5$

7. $f(x) = \ln x + x;\quad x_1 = 1;\quad n = 5$

8. $f(x) = \ln x + 2x - 8;\quad x_1 = 10;\quad n = 4$

9. $f(x) = \ln x + x^2;\quad x_1 = 2;\quad n = 5$

10. $f(x) = \ln x - e^{-x};\quad x_1 = 3;\quad n = 7$

B

In Problems 11–18, use Newton's method to find all the zeros of each function. Sketch the graph of the function to obtain initial approximations and to be certain that all the zeros have been located.

11. $f(x) = x^2 - 7x + 2$

12. $f(x) = x^2 + 5x + 3$

13. $f(x) = x^3 + 4x + 10$

14. $f(x) = x^3 - 6x^2 + 12x - 4$

15. $f(x) = x^3 - 12x^2 + 22$

16. $f(x) = x^3 - 18x^2 + 60x - 11$

17. $f(x) = x^4 - 4x^3 - 8x^2 + 4$

18. $f(x) = x^4 - 4x^3 + 4x^2 + 1$

In Problems 19–24, graph f and g on the same set of axes and use Newton's method to find the x coordinate of all points of intersection of the two graphs.

19. $f(x) = x^3; g(x) = x + 4$

20. $f(x) 5 \sqrt{x + 1}; g(x) = x^2 - 4$

21. $f(x) = x^{1/3}; g(x) = 12 - x$

22. $f(x) = x^3; g(x) 5 \dfrac{1}{1 + x^2}$

23. $f(x) = e^x; g(x) = x + 2$

24. $f(x) = \ln x; g(x) = x^2 - 4$

C

25. Newton's algorithm for approximating the square root of a positive number A is often stated as

$$x_n = \frac{x_{n-1}}{2} + \frac{A}{2x_{n-1}}$$

Show that this recursion formula can be derived by applying Newton's method to the function $f(x) = x^2 - A$.

26. Apply Newton's method to the function $f(x) = x^3 - A$ and derive a recursion formula for approximating cube roots.

27. Apply Newton's method to the function $f(x) = x^p - A$, where p is a positive integer, and derive a recursion formula for approximating the pth root of A.

28. Apply Newton's method to the function

$$f(x) = \frac{1}{x} - A$$

and derive a recursion formula for approximating reciprocals without the use of division.

29. Apply Newton's method to $f(x) = x/\sqrt{1 + x^2}$ with $x_1 = 1$. Why does Newton's method fail in this case?

30. Apply Newton's method to $f(x) = 17 + 8x^2 - x^4$ with $x_1 = 1$. Why does Newton's method fail in this case?

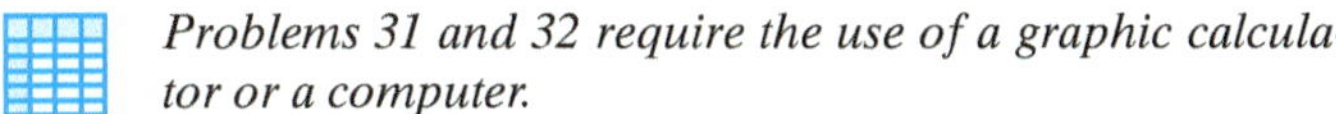

Problems 31 and 32 require the use of a graphic calculator or a computer.

31. Refer to Problem 29. Find the equations of the lines tangent to the graph of $f(x) + x/\sqrt{1 + x^2}$ at $x = 1$ and $x = -1$. Graph f and both tangent lines in the same viewing rectangle. Use $[-2, 2]$ for both the x range and the y range.

32. Refer to Problem 30. Find the equations of the lines tangent to the graph of $f(x) = 17 + 8x^2 - x^4$ at $x = 1$ and $x = -1$. Graph f and both tangent lines in the same viewing rectangle. Use $[-5, 5]$ for the x range and $[-35, 35]$ for the y range.

Applications

Business & Economics

33. Revenue. The revenue (in thousands of dollars) from the sale of x thousand units of a food processor is given by

$$R(x) = 10\,xe^{-0.05x}$$

How many food processors must be sold to generate a revenue of $600,000?

34. Revenue. The revenue (in thousands of dollars) from the sale of x thousand units of a compact disk is given by

$$R(x) = 20\,xe^{-0.01x}$$

How many disks must be sold to generate a revenue of $500,000?

35. Supply–demand. The supply and demand equations for premium motor oil in a particular market are given by

$$p = S(x) = e^{0.02x} \quad \text{and} \quad p = D(x) = 20 - e^{0.08x}$$

where p is the price of a can of oil (in dollars) and x is the number of cans of motor oil (in thousands). Find the equilibrium price (correct to two decimal places).

36. Supply–demand. The supply and demand equations for 20 pound bags of dog food in a certain city are given by

$$p = S(x) = e^{0.1x} \quad \text{and} \quad p = D(x) = 10 - e^{0.05x}$$

where p is the price of a bag of dog food (in dollars) and x is the number of bags of dog food (in thousands). Find the equilibrium price (correct to two decimal places).

37. Internal rate of return. A motion picture distributor pays $6 million for exclusive rights to a film for 2 years, after which

Marty Haigney

all rights revert back to the film's producer. The distributor realizes a profit of $5 million for the first year of distributing the film and $3 million for the second year. Find the internal rate of return of the distributor's investment. Express the answer as a percentage correct to two decimal places.

38. Internal rate of return. A library buys a coin-operated photocopying machine for $5,000. The copier generates annual profits of 2,000 for the library over a 3 year period, after which the copier is sold for $1,000. Find the internal rate of return of this investment. Express the answer as a percentage correct to two decimal places.

39. Internal rate of return. An individual pays $160,000 for a two-family housing unit which generates annual profits of $10,000. After 4 years, the property is sold for $200,000. Find the internal rate of return of this investment. Express the answer as a percentage correct to two decimal places.

40. Internal rate of return. A company spends $200,000 on a piece of equipment with a 5 year life span. The annual profits contributed by the equipment (including any salvage value at the end of the fifth year) are shown in the table in the margin. Find the internal rate of return of the investment in this equipment. Express the answer as a percentage correct to two decimal places.

YEAR	PROFIT
1	$50,000
2	$70,000
3	$80,000
4	$60,000
5	$30,000

Life Sciences

41. Pollution. A swimming pool is treated with a bactericide each morning. The number of bacteria per milliliter of water t hours after the treatment if given by

$$N(t) = 1{,}750 - 800te^{-0.2t}$$

The water is considered safe for swimming as long as the bacteria count is less than 600 bacteria per milliliter.

Marty Haigney

Determine the time period during which this water is safe for swimming. Round values of t to one decimal place.

42. Pollution. In a certain city the amount of a pollutant (in parts per million) in the air t houre after 6 AM is given by

$$P(t) = 100te^{-0.2t}$$

The air quality is classified as poor whenever the amount of pollutant is above 150 parts per million. Determine the time interval (in hours after 6 AM) during which the air quality is poor. Round values of t to one decimal place.

43. Drug concentration. The concentration of a drug in the bloodstream t hours after injection is given approximately by

$$C(t) = 10e^{-0.1t} + 15e^{-0.2t}$$

where $C(t)$ is concentration in milligrams per milliliter. The initial concentration is $C(0) = 25$ milligrams per milliliter. A second injection is to be given when the concentration reaches 15 milligrams per milliliter. Find the time (correct to one decimal place) when the second injection should be administered.

44. Drug sensitivity. One hour after x milligrams of a particular drug is given to a person, the change in body temperature T(x) in degrees Fahrenheit is given by

$$T(x) = x^2\left(1 - \frac{x}{9}\right) \qquad 0 \le x \le 6$$

Find the dosage (correct to two decimal places) that would produce a temperature change of 4°F.

Social Sciences

45. Learning. The relationship between the number of units N(t) a worker can assemble in 1 day and the number of hours t of training the worker has received is given by

$$N(t) = t^3 - 6t^2 + 25t$$

How many hours (correct to one decimal place) of training are required in order for a worker to be able to assemble 500 units a day?

46. Urban growth. The population of a newly incorporated city (in thousands) t years after incorporation is given by

$$N(t) = 12 + t^2 - \tfrac{1}{15}t^3 \qquad 0 \le t \le 10$$

The city will qualify for certain types of federal aid when its population reaches 40,000. When will this city qualify for federal aid? Round the value of t to one decimal place.

9-2 Numerical Integration

Introduction

If f is continuous on the interval $[a, b]$ and F is an antiderivative of f, then the definite integral of f from $x = a$ to $x = b$ is given by

$$\int_a^b f(x)\,dx = F(b) - F(a)$$

However, the antiderivative of f may be difficult to find, or it may not even exist in a convenient closed form. In this case, numerical integration can be used to approximate $\int_a^b f(x)\,dx$.

The rectangle rule for numerical integration, which uses Riemann sums to approximate definite integrals. That is,

$$\int_a^b f(x)\,dx \approx \sum_{k=1}^{n} f(c_k)\Delta x_k \qquad n \text{ sufficiently large}$$

Recall that if $f(x) \ge 0$ for $a \le x \le b$, then each term in the Riemann sum can be interpreted as the area of a rectangle with width Δx and altitude $f(C_k)$; hence the name "rectangle rule" (see Figure 7).

In this section we will develop two other important methods for approximating definite integrals, the *trapezoid rule* and *Simpson's rule*. Each of these methods is based on using a sum of areas of nonrectangular regions to approximate the area under a curve. Both methods are used extensively to approximate definite integrals on computers. Some calculators also have the capability to approximate integrals using one or the other of these methods.

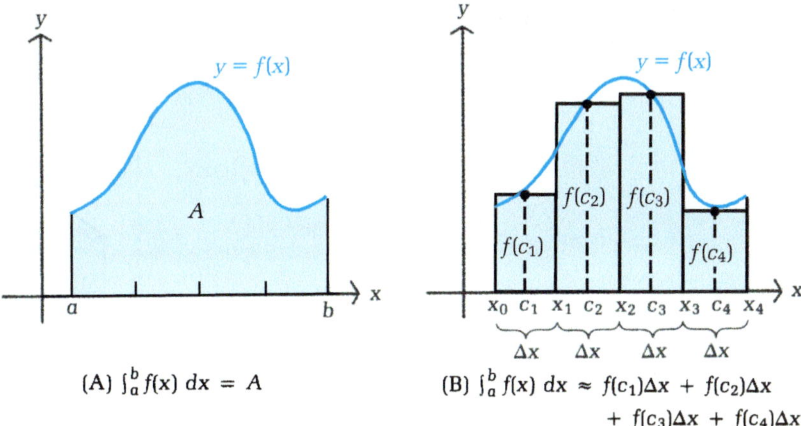

(A) $\int_a^b f(x)\,dx = A$ (B) $\int_a^b f(x)\,dx \approx f(c_1)\Delta x + f(c_2)\Delta x + f(c_3)\Delta x + f(c_4)\Delta x$

Figure 7

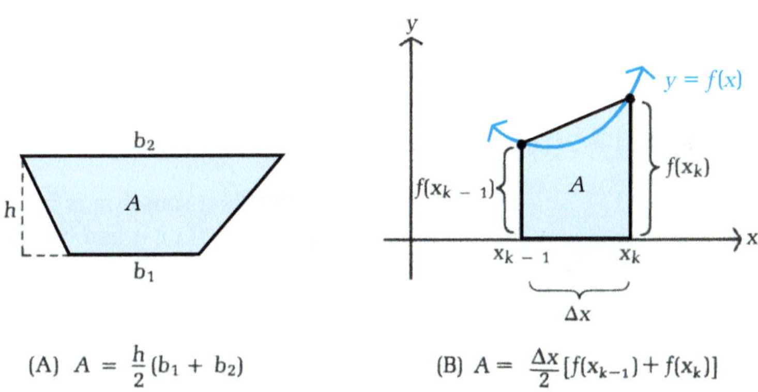

(A) $A = \frac{h}{2}(b_1 + b_2)$ (B) $A = \frac{\Delta x}{2}[f(x_{k-1}) + f(x_k)]$

Figure 8

The Trapezoid Rule

In the rectangle rule, the area under the graph of a positive function $f(x)$ over the interval $[x_{k-1}, x_k]$ was approximated by the area of the rectangle with width $\Delta x = x_k - x_{k-1}$ and altitude $f(c_k)$. Now we will approximate the area under the graph of $f(x)$ from $x = x_{k-1}$ to $x = x_k$ by using the area of a trapezoid. Recall that the area of a trapezoid with altitude h and bases b_1 and b_2 (see Figure 8A) is given by

$$A = \frac{h}{2}(b_1 + b_2) \tag{1}$$

If the points $(x_{k-1}, f(x_{k-1}))$ and $(x_k, f(x_k))$ on the graph of $y = f(x)$ are connected with a straight line segment, a trapezoid is formed (see Figure 8B). The area of this trapezoid is given by

$$A = \frac{\Delta x}{2}[f(x_{k-1}) + f(x_k)]$$

Let $h = \Delta x$, $b_1 = f(x_{k-1})$, and $b_2 = f(x_k)$ in (1).

Connecting the points $(x_0, f(x_0)), (x_1, f(x_1)), \ldots, (x_n, f(x_n))$ with straight line segments forms n trapezoids (as illustrated in Figure 9 for $n = 4$). If T_n denotes the sum of the area of these trapezoids, then T_n is an approximation to the area under the graph of f.

$$\int_a^b f(x)\,dx \approx T_4$$

$$= \frac{\Delta x}{2}[f(x_0) + f(x_1)] + \frac{\Delta x}{2}[f(x_1) + f(x_2)] + \frac{\Delta x}{2}[f(x_2 + f(x_3)] + \frac{\Delta x}{2}[f(x_3) + f(x_4)]$$

Using formula (2) for the area of each trapezoid, we have

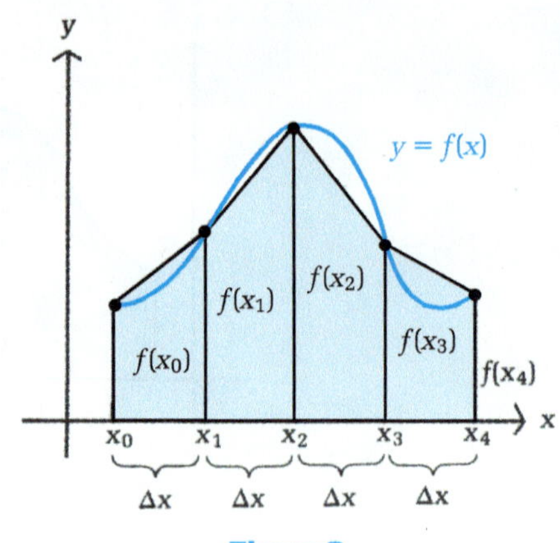

Figure 9

$$T_n = \frac{\Delta x}{2}[f(x_0) + f(x_1)] + \frac{\Delta x}{2}[f(x_1) + f(x_2)] + \frac{\Delta x}{2}[f(x_2 + f(x_3)] + \cdots + \frac{\Delta x}{2}[f(x_{n-1}) + f(x_n)]$$

Factor out $\Delta x/2$.

$$= \frac{\Delta x}{2}[f(x_0 + f(x_1) + f(x_1) + f(x_2) + f(x_2) + f(x_3) + \cdots + f(x_{n-1}) + f(x_n)]$$

$$= \frac{\Delta x}{2}[f(x_0) + 2f(x_1) + 2f(x_2) + \cdots + 2f(x_{n-1}) + f(x_n)]$$

Note: Excluding $f(x_0)$ and $f(x_n)$, each term $f(x_k)$ occurs twice in this sum.

Although this discussion is based on approximating the area under the graph of a positive function $f(x)$, it can be shown that

$$\lim_{n\to\infty} T_n = \int_a^b f(x)\,dx$$

for any function $f(x)$ that is continuous on $[a, b]$. This leads to the *trapezoid rule for approximating definite integrals.*

Trapezoid Rule

Divide the interval from $x = a$ to $x = b$ into n equal subintervals of length $\Delta x = (b - a)/n$. Let $a = x_0 < x_1 < \cdots < x_n = b$ be the end points of these subintervals. Then

$$\int_a^b f(x)\,dx \approx T_n = \frac{\Delta x}{2}[f(x_0) + 2f(x_1) + \cdots + 2f(x_{n-1}) + f(x_n)]$$

EXAMPLE 7 Use the trapezoid rule with the indicated values of n to approximate

$$\int_1^3 \ln x\,dx$$

Round each approximation to three decimal places.

(A) $n = 4$ (B) $n = 10$

SOLUTION (A) **Step 1** Compute Δx:

$$\Delta x = \frac{b - 1}{n} = \frac{3 - 1}{4} = 0.5$$

Step 2 Find the end points of the subintervals:

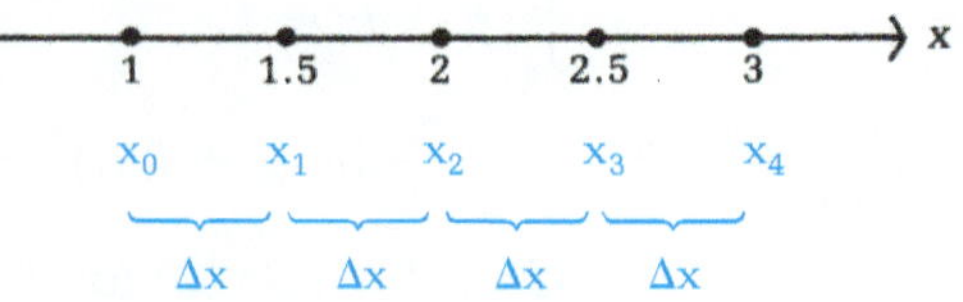

Step 3 Compute T_4:

$$T_4 = \frac{\Delta x}{2}[f(x_0) + 2f(x_1) + 2f(x_2) + 2f(x_3) + f(x_4)]$$

$$= \frac{0.5}{2}[f(1) + 2f(1.5) + 2f(2) + 2f(2.5) + f(3)]$$

$$= 0.25[(\ln 1) + 2(\ln 1.5) + 2(\ln 2) + 2(\ln 2.5) + (\ln 3)]$$

$$\approx 0.25[5.128\,418\,3] \approx 1.282\,104\,6$$

We used a calculator to evaluate T_4. Rounding this value to three decimal places, we can conclude that

$$\int_1^3 \ln x\, dx \approx T_4 \approx 1.282$$

(B) **Step 1** Compute Δx:

$$\Delta x = \frac{3 - 1}{10} = 0.2$$

Step 2 Find the points of the subintervals:

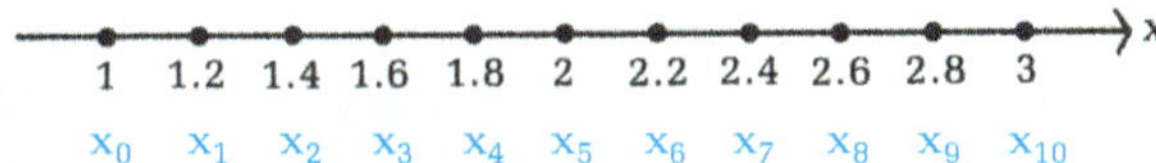

Step 3 Compute T_{10}:

$$T_{10} = \frac{\Delta x}{2}[f(x_0) + 2f(x_1) + \cdots + 2f(x_9) + f(x_{10})]$$

$$= \frac{0.2}{2}[f(1) + 2f(1.2) + \cdots + 2f(\ln 2.8) + f(3)]$$

$$= 0.1[(\ln 1) + 2(\ln 1.2) + \cdots + 2(\ln 2.8) + (\ln 3)]$$

$$\approx 0.1[12.936\,189] \approx 1.293\,618\,9$$

Thus,

$$\int_1^3 \ln x\, dx \approx T_4 \approx 1.294$$

Matched Problem 7 Use the trapezoid rule with the indicated values of n to approximate

$$\int_2^4 \frac{1}{x}\, dx$$

Round each approximation to three decimal places.

(A) $n = 4$ (B) $n = 10$

Referring to Example 7, which approximation is more accurate, T_4 or T_{10}? In order to be able to answer this question, we purposely selected an integral that can be evaluated directly. (In general, we apply approximation methods to integrals that cannot be evaluated directly.) Using integration by parts or a table of integrals to find an antiderivative for $\ln x$, we have

$$\int_1^3 \ln x\, dx = (x \ln x - x)\Big|_1^3$$

$$= 3 \ln 3 - 2$$

$$\approx 1.296 \qquad \text{Rounded to three decimal places}$$

Comparing this value with the approximation obtained in Example 7 we have

$$\int_1^3 \ln x \, dx - T_4 \approx 1.296 - 1.282 = 0.014$$

and

$$\int_1^3 \ln x \, dx - t_{10} \approx 1.296 - 1.294 = 0.002$$

Thus, T_{10} provides the more accurate approximation. In most cases, increasing the value of n increases the accuracy of the approximation provided by the trapezoid rule. We will have more to say about the error in this approximation later in this section.

Simpson's Rule

In the development of the trapezoid rule, we connected each pair of points $(x_{k-1}, f(x_{k-1}))$ and $(x_k, f(x_k))$ with a straight line segment and used the area under the graph of this line segment to approximate the area under the graph of $f(x)$ from $x = x_{k-1}$ to $x = x_k$.

In order to obtain a more accurate approximation, we will now use the area under the graph of a parabola that passes through three successive points on the graph of f. First, we need a formula for the area under the graph of a parabola passing through three noncollinear points. The formula is surprisingly simple.

Let (c, y_0), $(c + h, y_1)$ and $(c + 2h, y_2)$ be three points on the graph of a parabola P (see Figure 10). Notice that the x coordinates of these points are equally spaced. Using integration, it can be shown that the area under the graph of P from $x = c$ to $x = c + 2h$ is

$$A = \frac{h}{3}(y_0 + 4y_1 + y_2) \tag{3}$$

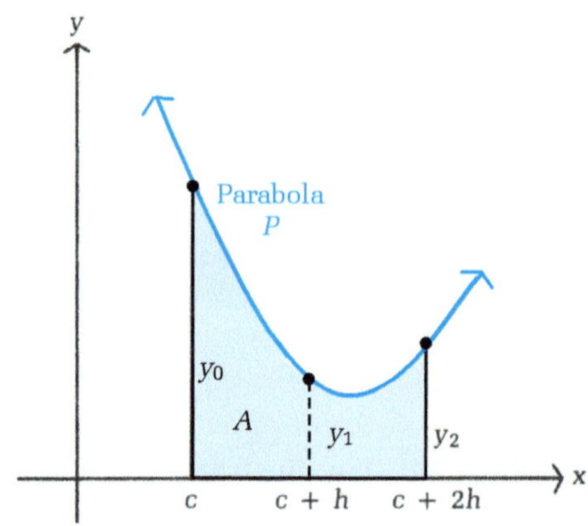

Figure 10

$$A = \frac{h}{3}(y_0 + 4y_1 + y_2)$$

If the graph of the parabola P passes through the points $(x_{k-1}, f(x_{k-1}))$, $(x_k, f(x_k))$ and $(x_{k+1}, f(x_{k+1}))$, then the area under the graph of P from $x = x_{k-1}$ to $x = x_{k+1}$ is

$$A = \frac{\Delta x}{3}[f(x_{k-1}) + 4f(x_k) + f(x_{k+1})] \tag{4}$$

Let $h = \Delta x$, $y_0 = f(x_{k-1})$, $y_1 = f(x_k)$, and $y_2 = f(x_{k+1})$ in (3).

Since A is the area under a parabola over two subintervals of $[a, b]$, we must assume n is even to use this process. If S_n denotes the sum of the areas under the parabolas over the intervals $[x_0, x_2], [x_2, x_4], \ldots, [x_{n-2}, x_n]$, then S_n is an approximation to the area under the graph of f_2. This is illustrated in Figure 11 at the top of the next page using $n = 4$ and two parabolas, P_1 and P_2.

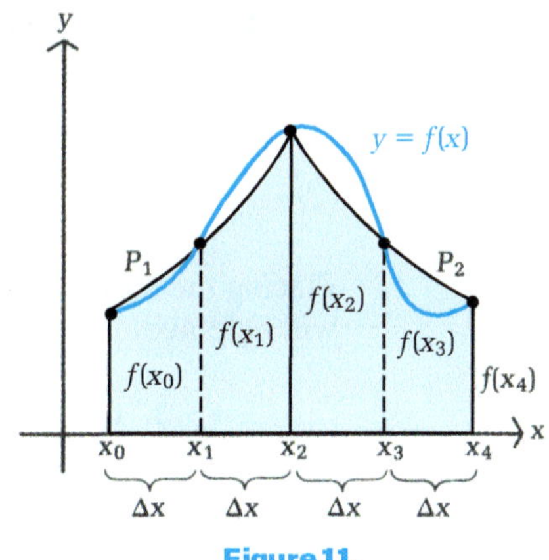

Figure 11

$$\int_a^b f(x)\,dx \approx S_4 = \underbrace{\frac{\Delta x}{3}[f(x_0) + 4f(x_1) + f(x_2)]}_{\text{Area under } P_1} + \underbrace{\frac{\Delta x}{3}[f(x_2) + 4f(x_3) + f(x_4)]}_{\text{Area under } P_2}$$

For arbitrary n, using formula (4) for the area under each parabola, we have

$$S_n = \frac{\Delta x}{3}[f(x_0) + 4f(x_1) + f(x_2)] + \frac{\Delta x}{3}[f(x_2) + 4f(x_3) + f(x_4)]$$

$$+ \frac{\Delta x}{3}[f(x_4) + 4f(x_5) + f(x_6)] + \cdots + \frac{\Delta x}{3}[f(x_{n-2}) + 4f(x_{n-1}) + f(x_n)]$$

Excluding $f(x_0)$ and $f(x_n)$, each term $f(x_k)$ with an *even* subscript occurs twice in this sum. Factoring out $\Delta x/3$ and combining like terms, we can write

$$S_n = \frac{\Delta x}{3}[f(x_0) + 4f(x_1) + 2f(x_2) + 4f(x_3) + 2f(x_4) + \cdots + 2f(x_{n-2}) + 4f(x_{n-1}) + f(x_n)]$$

If f is continuous on $[a, b]$, it can be shown that

$$\int_a^b f(x)\,dx = \lim_{n\to\infty} S_n$$

and S_n can be used to approximate the definite integral. The formula for S_n is referred to as *Simpson's rule*.

Simpson's Rule

Given an even integer n, divide the interval from $x = a$ to $x = b$ into n subintervals of length $\Delta x = (b - a)/n$. Let $a = x_0 < x_1 < \cdots < x_n = b$ be at the end points of these subintervals. Then

$$\int_a^b f(x)\,dx \approx S_n = \frac{\Delta x}{3}[f(x_0) + 4f(x_1) + 2f(x_2) + 4f(x_3) + \cdots + 2f(x_{n-2}) + 4f(x_{n-1}) + f(x_n)]$$

EXAMPLE 8 Use Simpson's rule with the indicated values of n to approximate

$$\int_1^3 \ln x\,dx$$

Round each approximation to three decimal places.

(A) $n = 4$ (B) $n = 10$

SOLUTION (A) **Step 1** Compute Δx:

$$\Delta x = \frac{3 - 1}{4} = 0.5$$

Step 2 Find the end points of the subintervals:

Multiplication factors	1	4	2	4	1
x	1	1.5	2	2.5	3
	x_0	x_1	x_2	x_3	x_4

Placing the correct multiplication factor (1, 2, or 4) for $f(x_k)$ above each end point will help avoid errors in computing S_n).

Step 3 Compute S_4:

$$S_4 = \frac{\Delta x}{3}[f(x_0) + 4f(x_1) + 2f(x_2) + 4f(x_3) + f(x_4)]$$

$$= \frac{0.5}{3}[(\ln 1) + 4(\ln 1.5) + 2(\ln 2) + 4(\ln 2.5) + \ln 3)]$$

$$\approx \frac{0.5}{3}[7.771\,93] \approx 1.295\,321\,7$$

Thus,

$$\int_1^3 \ln x \, dx \approx S_4 \approx 1.295\,32 \qquad \text{Rounded to five decimal places}$$

(B) **Step 1** Compute Δx:

$$\Delta x = \frac{3-1}{10} = 0.2$$

Step 2 Find the end points of the subintervals:

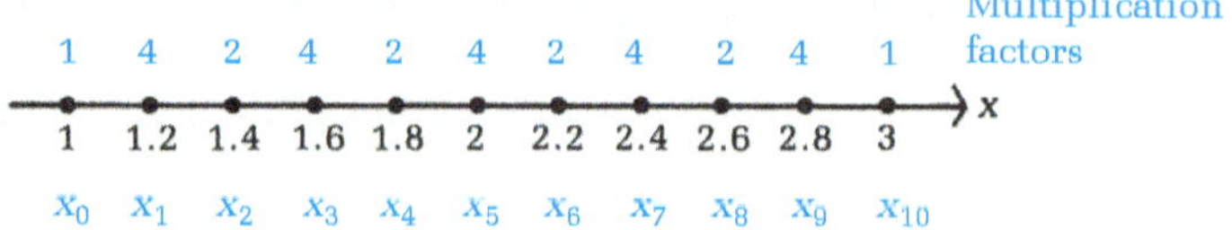

Step 3 Compute S_{10}:

$$S_{10} = \frac{\Delta x}{3}[f(x_0) + 4f(x_1) + 2f(x_2) + 4f(x_3) + 2f(x_4) + 4f(x_5) + 2f(x_6)$$
$$+ 4f(x_7) + 2f(x_8) + 4f(x_9) + f(x_{10})]$$

$$= \frac{0.2}{3}[(\ln 1) + 4(\ln 1.2) + 2(\ln 1.4) + 4(\ln 1.6) + 2(\ln 1.8) + 4(\ln 2)$$
$$+ 2(\ln 2.2) + 4(\ln 2.4) + 2(\ln 2.6) + 4(\ln 2.8) + (\ln 3)]$$

$$\approx \frac{0.2}{3}[19.473\,31] \approx 1.295\,820\,7$$

Thus,

$$\int_1^3 \ln x \, dx \approx S_{10} \approx 1.295\,82 \qquad \text{Rounded to five decimal places}$$

Matched Problem 8 Use Simpson's rule with the indicated values of n to approximate

$$\int_2^4 \frac{1}{x} \, dx$$

Round each approximation to five decimal places.

(A) $n = 4$ (B) $n = 10$

Referring to the discussion following Example 7 the value of $\int_1^3 \ln x\,dx$ rounded to five decimal places is

$$\int_1^3 \ln x\,dx = 3\ln 3 - 2 \approx 1.295\,84$$

Comparing this value with the approximations obtained in Example 12, we have

$$\int_1^3 \ln x\,dx - S_4 \approx 1.295\,84 - 1.295\,32 = 0.000\,52$$

and

$$\int_1^3 \ln x\,dx - s_{10} \approx 1.295\,84 - 1.295\,82 = 0.000\,02$$

Once again, using a larger value of n provides a more accurate approximation. Later in this section we will compare in more detail the accuracy of the approximations obtained by using Simpson's rule with those obtained from the trapezoid rule.

Applications

In the preceding two examples, we purposely selected a definite integral that could be evaluated directly so that we could compare approximate and exact values. The next two examples involve integrals that cannot be evaluated directly and must be approximate.

EXAMPLE 9 **Consumer's and Producers' Surplus** Find the equilibrium price and then use Simpson's rule with $n = 4$ to approximate the consumers' surplus at the equilibrium price level for

$$p = D(x) = \sqrt{100 - 8x^3} \qquad \text{and} \qquad p = S(x) = \sqrt{4 + 4x^3}$$

SOLUTION To find the equilibrium price, set $\sqrt{100 - 8x^3}$ equal to $\sqrt{4 + 4x^3}$ and solve for x:

$$\sqrt{100 - 8x^3} = \sqrt{4 + 4x^3} \quad \text{Square both sides.}$$

$$100 - 8x^3 = 4 + 4x^3$$

$$-12x^3 = -96$$

$$x^3 = 8$$

$$x = 2$$

CHECK $D(2) = \sqrt{100 - 8(2^3)} = \sqrt{36} = 6$

$S(2) = \sqrt{4 + 4(2^3)} = \sqrt{36} = 6$

$D(2) \stackrel{\checkmark}{=} S(2)$

Thus, the equilibrium price is $\bar{p} = 6$, and the corresponding supply and demand is $\bar{x} = 2$. Sketch a graph:

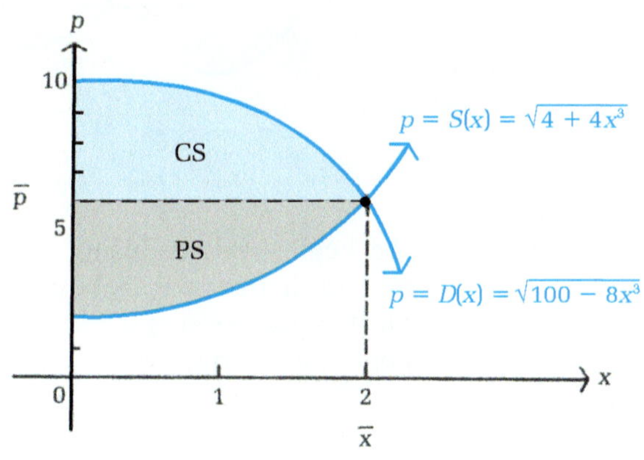

Now,

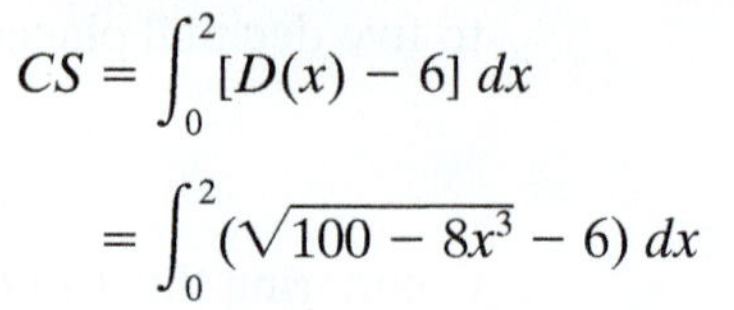

$$CS = \int_0^2 [D(x) - 6]\,dx$$

$$= \int_0^2 (\sqrt{100 - 8x^3} - 6)\,dx$$

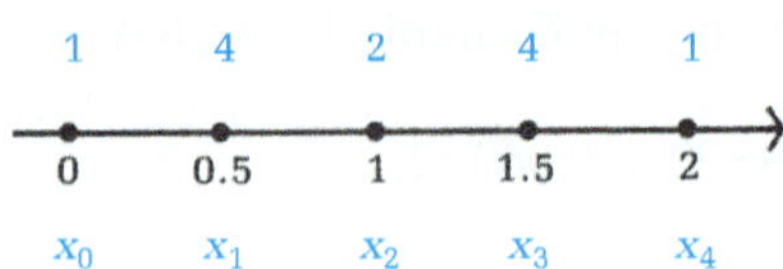

Use Simpson's rule with $n = 4$ to approximate this integral:

$$\Delta x = \frac{2 - 0}{4} = 0.5$$

$$S_4 = \frac{\Delta x}{3}[f(0) + 4f(0.5) + 2f(1) + 4f(1.5) + f(2)$$

$$= \frac{0.5}{3}[(\sqrt{100} - 6) + 4(\sqrt{99} - 6) + 2(\sqrt{92} - 6) + 4(\sqrt{73} - 6) + (\sqrt{36} - 6)]$$

$$\approx \frac{0.5}{3}[37.158\,84] \approx 6.139\,14$$

Thus,

$$CS = \int_0^2 (\sqrt{100 - 8x^3} - 6)\,dx$$

$$\approx 6.193\,14$$

Matched Problem 9 Use Simpson's rule with $n = 4$ to approximate the producers' surplus at the equilibrium price level for Example 9.

Remark — The integral in Example 9 cannot be evaluated directly. The function $D(x) = \sqrt{100 - 8x^3}$ does not have an antiderivative that can be expressed in terms of any of the functions we have studied. You are not expected to recognize functions that do not have antiderivatives. Instead, if the instructions for a problem call for the use of an approximation method, you should assume that the integral may be one that cannot be evaluated directly and you should not spend time trying to find an antiderivative.

EXAMPLE 10 **Biology** A biologist studying the aquatic life in small pond needs to determine the area of the surface of the pond, so that distance across the pond is measured at 10 foot intervals. Use the data in the figure at the top of the next page and the trapezoid rule to approximate the area of the pond.

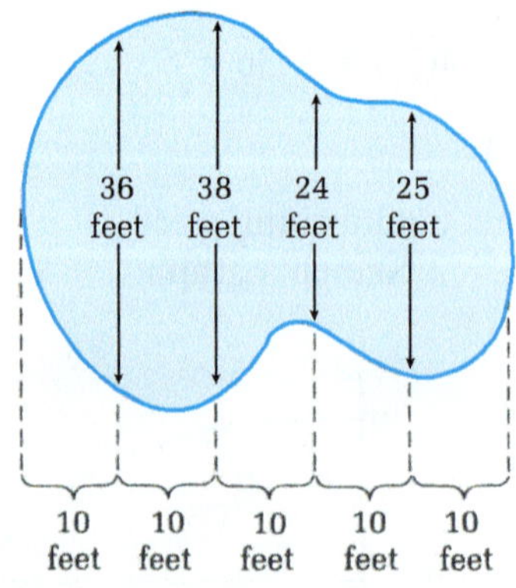

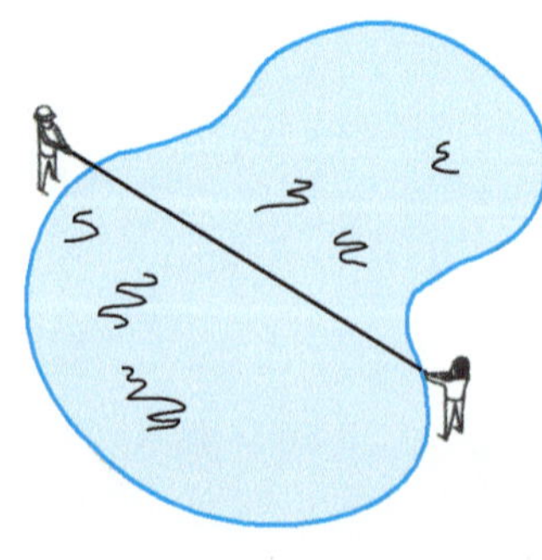

SOLUTION We begin by introducing a coordinate system, as shown in the figure in the margin. If the pond is bounded by the graphs of $y = f(x)$ and $y = g(x)$, then $D(x) = f(x) - g(x)$ is the distance across the pond. The surface area A of the pond is given by

$$A = \int_0^{50} [f(x) - g(x)]\,dx = \int_0^{50} D(x)\,dx$$

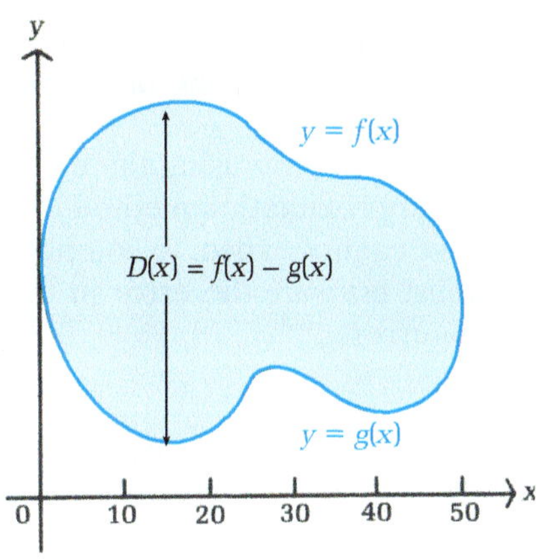

Since there are no equations defining f, g, or D, we cannot evaluate the integral directly. However, we can use the measurements taken by the biologist to approximate the integral. We consider D to be defined by the following table:

x	0	10	20	30	40	50
$D(x)$	0	36	38	24	25	0

Applying the trapezoid rule with $n = 5$ and $\Delta x = 10$, we have

$$T_5 = \tfrac{10}{2}[D(0) + 2D(10) + 2D(20) + 2D(30) + 2D(40) + D(50)]$$

$$= 5(0 + 72 + 76 + 48 + 50 + 0]$$

$$= 1{,}230$$

Thus, the surface area of the pond is approximately 1,230 square feet.

Matched Problem 10 Repeat Example 10 for the following measurements:

x	0	10	20	30	40	50
$D(x)$	0	26	42	33	17	0

Error Estimates

In many actual applications involving numerical integration, it is necessary to approximate the definite integral to a specified accuracy. The difference between the exact value of a definite integral and the approximate value obtained by a numerical method is called the **error in the approximation.**

Error in Approximations

If T_n and S_n are approximations to $\int_a^b f(x)\,dx$ obtained by using the trapezoid rule and Simpson's rule, respectively, then

TRAPEZOID RULE ERROR $\quad \text{Err}(T_n) = \int_a^b f(x)\,dx - T_n$

SIMPSON'S RULE ERROR $\quad \text{Err}(S_n) = \int_a^b f(x)\,dx - S_n$

Earlier in this section, we computed four different approximations to $\int_1^3 \ln x\,dx$ and compared the approximations to the exact value $3 \ln 3 - 2$. These results are summarized in Table 1.

Table 1 Approximations to $\int_1^3 \ln x\,dx$

n	T_n	ERR(T_n)	S_n	ERR(S_n)
4	1.282	0.014	1.295 32	0.000 52
10	1.294	0.002	1.295 82	0.000 02

Examining the column for $\text{Err}(T_n)$ in Table 1, we see that increasing n decreases the error in the approximation. This is also true for $\text{Err}(S_n)$. Although exceptions may occur, in most applications of either the trapezoid rule or Simpson's rule, **increasing n decreases the error in the approximation**. Next, comparing $\text{Err}(T_n)$ and $\text{Err}(S_n)$ in each row to Table 1, we see the $\text{Err}(S_n)$ is much smaller than $\text{Err}(T_n)$.

In fact, $\text{Err}(S_4)$ is even smaller than $\text{Err}(T_{10})$. Again, exceptions may occur, but in most cases, **Simpson's rule provides a more accurate approximation than the trapezoid rule does (and with very little extra work).**

The values of $\text{Err}(T_n)$ and $\text{Err}(S_n)$ in Table 1 were obtained by using the exact value of the integral being approximated. If we apply an approximation method to a definite integral whose exact value is not known, then we cannot expect to compute that actual error in the approximation. Instead, we must *estimate* the error in the approximation. Theorem 1 provides estimates for these errors.

THEOREM 1 Error Estimates for the Trapezoid Rule and Simpson's Rule

Given the definite integral $\int_a^b f(x)\,dx$:

ERROR ESTIMATE FOR THE TRAPEZOID RULE

If $|f''(x)| \leq M_1$ for $a \leq x \leq b$, then $|\text{Err}(T_n)| \leq \dfrac{(b-a)^3 M_1}{12n^2}$

ERROR ESTIMATE FOR SIMPSON'S RULE

If $|f^{(4)}(x)| \leq a \leq x \leq b$, then $|\text{Err}(S_n)| \leq \dfrac{(b-a)^5 M_2}{180n^4}$

EXAMPLE 11 Use Theorem 1 to estimate $\text{Err}(T_n)$ and $\text{Err}(S_n)$ for the definite integral

$$\int_1^3 \ln x\,dx$$

if $n = 10, 20, 30, 40$, or 50

SOLUTION To use Theorem 1, we must maximize the absolute value for the second and fourth derivatives of $f(x) = \ln x$ for $1 \leq x \leq 3$. Thus, we find

$$f(x) = \ln x \qquad f'(x) = \frac{1}{x} \qquad f''(x) = -\frac{1}{x^2}$$
$$f^{(3)}(x) = \frac{2}{x^3} \qquad f^{(4)}(x) = -\frac{6}{x^4}$$

Examining the graphs in figures A and B, we see that

$$M_1 = \text{Max}\,|f''(x)| = |f''(1)| = 1 \qquad 1 \leq x \leq 3$$

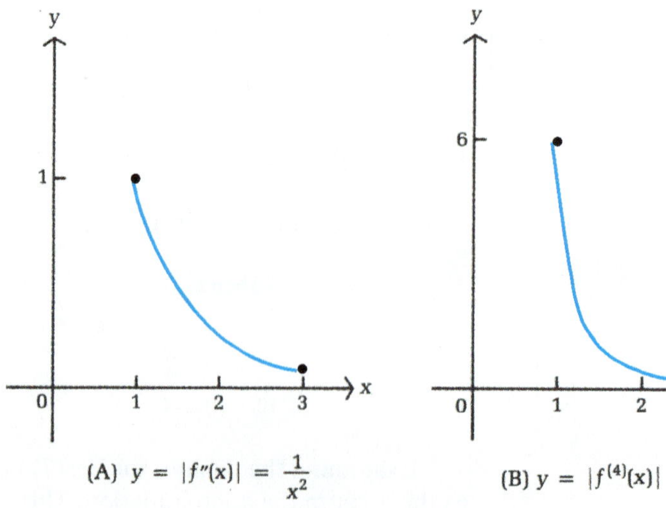

(A) $y = |f''(x)| = \frac{1}{x^2}$ (B) $y = |f^{(4)}(x)| = \frac{6}{x^4}$

and

$$M_2 = \text{Max}|f^{(4)}(x)| = |f^{(4)}(1)| = 6 \qquad 1 \leq x \leq 3$$

Error Estimates for $\int_1^3 \ln x\, dx$

n	ESTIMATE FOR ERR(T_n) $2/(3n^2)$	ESTIMATE FOR ERR(S_n) $16/(15n^4)$
10	0.007	0.000 1
20	0.002	0.000 007
30	0.0007	0.000 001
40	0.0004	0.000 000 4
50	0.0003	0.000 000 2

Thus, according to Theorem 1,

$$|\text{Err}(T_n)| \le \frac{(b-a)^3 M_1}{12n^2} = \frac{(3-1)^3 \cdot 1}{12n^2} = \frac{2}{3n^2}$$

and

$$|\text{Err}(S_n)| \le \frac{(b-a)^5 M_2}{180n^4} = \frac{(3-1)^5 \cdot 6}{180n^4} = \frac{16}{15n^4}$$

The estimates for specific values of n are listed in the table in the margin. All values in this table have been rounded to the first nonzero decimal place.

Matched Problem 11 Repeat Example 11 for: $\int_2^4 \frac{81}{x}\, dx$

In theory, the estimates in Theorem 1 can be used to determine the value of n required for either rule to provide an approximation of a specified accuracy. For example, the table in Example 11 indicates that using Simpson's rule with $n = 30$ will provide an approximation correct to within $\pm 0.000\,001$. We were able to obtain this estimate because the function $f(x) = \ln x$ has a very simple fourth derivative. In most applications, maximizing the absolute value of the second or fourth derivative of the integrand can be a formidable task. For example, to estimate the error in using Simpson's rule to approximate

$$\int_0^2 f(x)\, dx = \int_0^2 (\sqrt{100 - 8x^3} - 6)\, dx \tag{5}$$

(see Example 9), we would have to maximize the absolute value of

$$f^{(4)}(x) = \frac{2.304x^2(x^7 - 700x^3 - 12{,}500)}{(100 - 8x^3)^{7/2}}$$

In actual practice, computers are used to compute approximations to definite integrals. Theorem 1 can be used to devise a variety of numerical techniques for estimating the error in computer-generated approximations to definite integrals. One of the simplest techniques is to continue to double the number of subintervals until the difference between S_n and S_{2n} is less than the desired accuracy. Computer-generated values of S_n for the integral in (5) are listed in Table 2 for $n = 4, 8, \dots, 128$. Each value is rounded to six decimal places. Since S_{64} and S_{128} are the same, it is reasonable to conclude that the approximate value of the integral in (5) is 6.204 489, correct to six decimal places.

Thus, in practice, it is possible to compute approximations by Simpson's rule to a specified accuracy without maximizing the absolute value of the fourth derivative of the integrand.

TABLE 2
Computer Approximation of $\int_0^2 (\sqrt{100 - 8x^3} - 6)\, dx$

–SIMPSONS RULE–
A = 0   B = 2
F(x) = SQR(100 – 8* × ^3) – 6

N	S(N)
4	6.19314
8	6.203338
16	6.204398
32	6.204483
64	6.204489
128	6.204489

Answers to Matched Problems

7. (A) 0.697 (B) 0.694 **8.** (A) 0.693 25 (B) 0.693 15
9. 5.520 78 **10.** Approx. 1,180 ft^2
11.

n	$1/(6n^2)$	$2/(5n^4)$
10	0.002	0.000 01
20	0.000 4	0.000 000 8
30	0.000 2	0.000 000 2
40	0.000 1	0.000 000 05
50	0.000 07	0.000 000 02

Exercise 9-2

A

Write out all the terms in the trapezoid rule for the indicated value of n.

1. $n = 5$

2. $n = 6$

Write out all the terms in Simpson's rule for the indicated value of n.

3. $n = 6$

4. $n = 8$

Given a, b, and n, divide the interval [a, b] into n equal subintervals and graph the end points of these subintervals on a number line. Use the values of these end points to write a formula for T_n.

5. $a = 0, b = 2, n = 5$

6. $a = 0, b = 3, n = 4$

Given a, b, and n, divide the interval [a, b] into n equal subintervals and graph the end points of these subintervals. Use the values of these end points to write a formula for S_n.

7. $a = 0, b = 3, n = 4$

8. $a = 0, b = 9, n = 6$

B

Use the trapezoid rule with n 5 5 to approximate each integral. Round each result to four decimal places.

9. $\int_0^1 \sqrt{1 + x^3}\, dx$

10.

11. $\int_2^4 \ln(1 + x^2)dx$

12. $\int_2^4 \ln(1 + x^3)dx$

Use Simpson's rule with $n = 4$ to approximate each definite integral. Round each result to four decimal places.

13. $\int_0^1 e^{x^2} dx$

14. $\int_0^1 e^{-x^2} dx$

15. $\int_{-1}^1 \frac{1}{\sqrt{4 + x^3}} dx$

16. $\int_{-1}^1 \frac{1}{\sqrt{4 + 4x^4}} dx$

Use the values in each table and the trapezoid rule to approximate the area bounded by the graph of the function and the x axis over the interval of x values in the table.

17.

x	0	0.5	1	1.5	2
$f(x)$	5	7	6	4	2

18.

x	0	0.5	1	1.5	2
$f(x)$	4	3	4	6	9

19.

x	−3	−1	−1	0	1	2	3
$f(x)$	10	12	14	15	13	11	9

20.

x	−6	−4	−2	0	2	4	6
$f(x)$	25	20	15	5	10	25	35

Use Simpson's rule with the indicated value of n to approximate the area of the region bounded by the graphs of the given equations. Round each result to five decimal places.

21. $y = \frac{10}{4 + x^2}; y = 0, 0 \le x \le 4, n = 8$

22. $y = \frac{10}{1 + x^2}; y = 0, -2 \le x \le 2, n = 8$

23. $y = \sqrt{x - x^2}; y = 0, 0 \le x \le 1, n = 10$

24. $y = \sqrt{2x - x^2}; y = 0, 0 \le x \le 2, n = 10$

C

Find the exact value of each integral; then construct a table containing T_n, $Err(T_n)$, S_n and $Err(S_n)$ for $n = 4$ and $n = 10$. Round all values in the tables to six decimal places.

25. $\int_0^1 e^x\, dx$

26. $\int_1^2 \frac{1}{\sqrt{x}} dx$

In Problems 27 and 28, estimate $ERR(T_n)$ and $Err(S_n)$ for each integral. Construct a table containing the values of these estimates rounded to the first nonzero decimal place for n = 10, 20, 30, 40, and 50.

27. $\int_1^3 (x \ln x)\, dx$

28. $\int_0^4 \sqrt{x + 4}\, dx$

29. Show that the trapezoid rule gives the exact value of

$$\int_a^b (Ax + B)\, dx$$

for any constants A and B. [*Hint:* Use Theorem 1 to estimate $\text{Err}(T_n)$ for $f(x) = Ax + B$.]

30. Show that Simpson's rule give the exact value of

$$\int_a^b (Ax^3 + Bx^2 + Cx + D)\, dx$$

for any constants A, B, C, and D. [*Hint:* Use Theorem 1 to estimate $\text{Err}(S_n)$ for $f(x) = Ax^3 + Bx^2 + Cx + D$.]

Problems 31–38 are related to Problems 9–16. In each problem, use a numerical integration routine on a calculator or a computer to approximate the integral to four decimal places.

31. $\int_0^1 \sqrt{1 + x^3}\, dx$

32. $\int_0^1 \sqrt{1 + x^4}\, dx$

33. $\int_2^4 \ln(1 + x^2)\, dx$

34. $\int_2^4 \ln(1 + x^3)\, dx$

35. $\int_0^1 e^{x^2}\, dx$

36. $\int_0^1 e^{-x^2}\, dx$

37. $\int_{-1}^1 \frac{1}{\sqrt{4 + x^3}} dx$

38. $\int_{-1}^1 \frac{1}{\sqrt{4 + x^4}} dx$

Applications

In Problems 39–52, use Simpson's rule with n = 4 to approximate the required definite integrals. round each result to two decimal places.

Business & Economics

39. Consumer's and producers' surplus. Find the equilibrium price (in dollars), and then find the consumers' and producers' surplus for

$$p = D(x) = \sqrt{400 - 4x^3}$$

and

$$p = S(x) = \sqrt{16 + 2x^3}$$

40. Oil production. The management of an oil company estimates that oil will be pumped from a producing field at a rate given by

$$R(t) = \frac{100}{\sqrt{t^2 + 1}} \qquad 0 \le t \le 8$$

where $R(t)$ is the rate of production in thousands of barrels per year t years after pumping begins. Approximately how much oil will the field produce during the first 8 years of production?

41. Marginal revenue. The total accumulated revenues and costs (in thousands of dollars) t months after the opening of a new restaurant are denoted by $R(t)$ and $C(t)$, respectively. The graphs of $R'(t)$ and $C'(t)$ are shown in the figure. Typically, newly opened restaurants lose money until they develop clientele. How much did this restaurant lose during the first 8 months of operation? Set up an appropriate definite integral and use the graph to estimate the necessary values of $R'(t)$ and $C'(t)$.

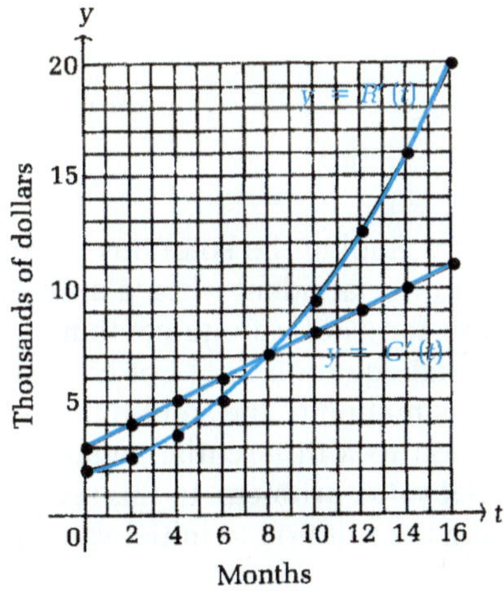

42. Marginal revenue. Refer to Problem 41. How much profit did the restaurant make during the second 8 months of operation? What was the total profit for the first 16 months of operation?

The results of surveying two irregularly shaped pieces of river-front property are given in the figure. Use the measurements in Problems 43 and 44.

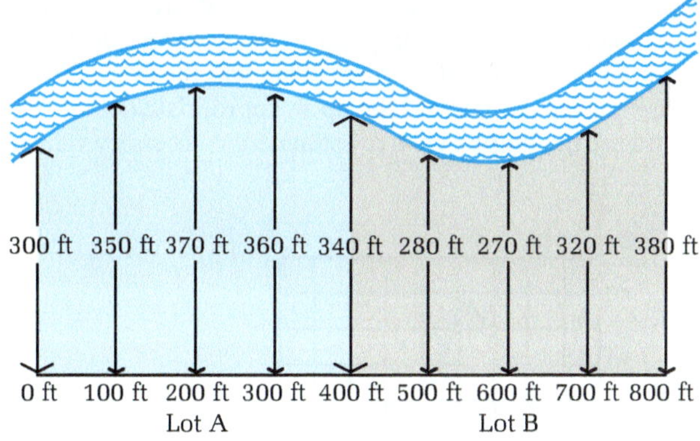

43. Real estate. Find the area of lot A.

44. Real estate. Find the area of lot B.

45. Construction. A city park commission is planning to construct ponds in several parks. All the ponds will have the same basic shape, but the surface area will depend on the location selected for the pond. The city architect designed the basic shape of the pond using an unspecified dimension d (see the figure in the margin). Determine the value of d for a pond with a surface area of 1,650 square meters.

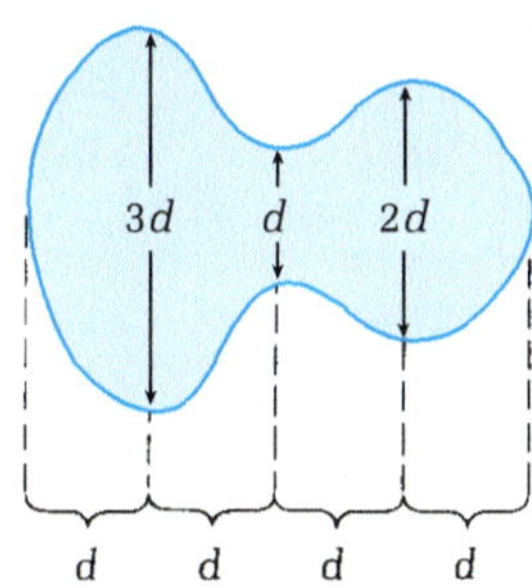

46. Construction. Repeat Problem 45 for a pond with a surface area of 2,400 square meters whose basic shape is given in the figure below

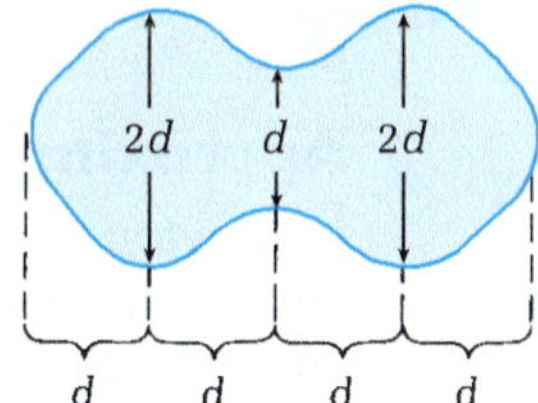

Life Sciences

47. Medicine. The body assimilates a drug at a rate given by

$$R'(t) = 7 - 3\ln(t^2 - 6t + 10) \qquad 0 \le t \le 3$$

where t is the time (in hours) since the drug was administered. Find the total amount of the drug assimilated in the first 3 hours.

48. Medicine. The level of concentration of a certain drug in the bloodstream t minutes after it is administered is given by

$$L(t) = 10 - 6\ln(t^2 - 4t + 5) \qquad 0 \le t \le 4$$

Find the average level of concentration for t in the interval $[0, 4]$.

49. Temperature. The temperature C (in degrees Celsius) in an artificial habitat was graphed by a recording device over a 2 hour period (see the figure). What was the average temperature during this period? Set up an appropriate definite integral and use the graph below to estimate the necessary values of C.

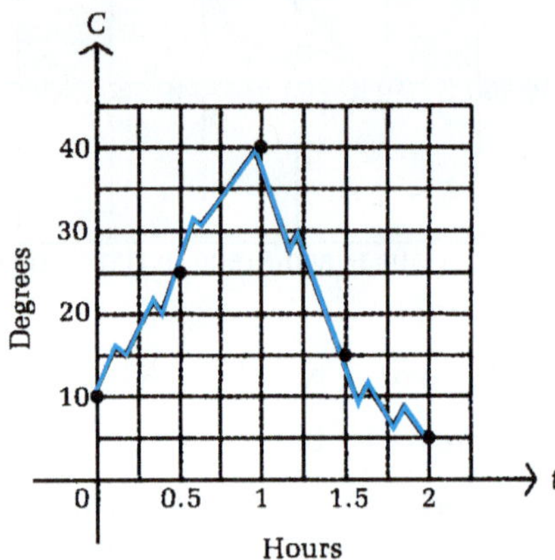

50. Ecology. A biologist is studying the ecosystem on a small island. In order to determine the area of the island, a grid was imposed over an aerial photograph of the island (see the figure). Find the area of the island. Set up an appropriate definite integral and use the graph to estimate the necessary distances.

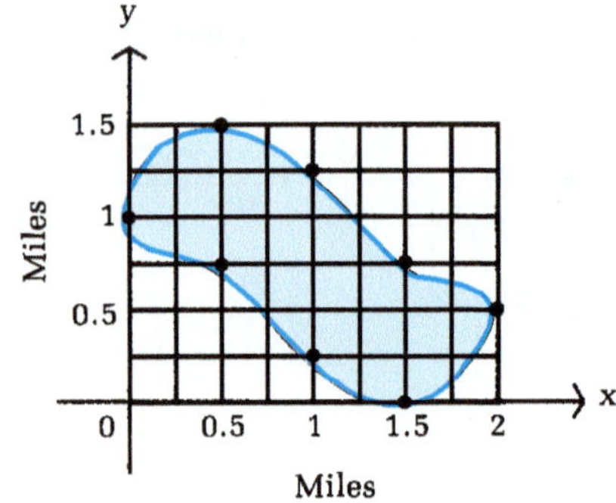

Social Sciences

51. Learning. A person learns N items at a rate given by

$$N'(t) = \sqrt{24 + \frac{1}{t^2}} \qquad 1 \le t \le 6$$

where t is the number of hours of continuous study. Find the total number of items N learned from $t = 1$ to $t = 6$ hours of study.

52. Politics. In a newly incorporated city, the rate of change of the voting population, $N'(t)$, with respect to time t (in years) is shown in the figure, where $N(t)$ is in thousands. What is the total increase in the voting population during the first 4 years? Set up an appropriate definite integral and estimate the necessary values from the graph.

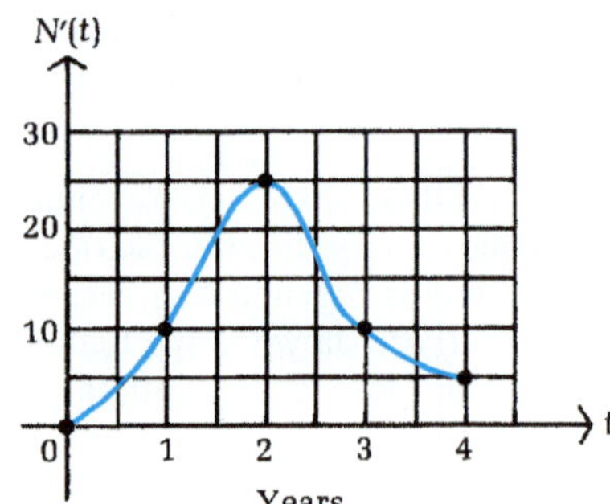

9-3 Euler's Method

- Approximate Solutions
- Euler's method
- Comparisons of Exact and Approximate solutions
- Application
- Euler's Method for Systems of Differential Equations

Approximate Solutions

In previous chapters, we developed techniques for finding the solutions to certain types of differential equations. But it is just a fortunate accident when a differential equation has a simple, neat solution expressed in terms of a finite combination of familiar functions. Most do not. When a differential equation does not have such a nice solution, it is necessary to settle for an approximation to the solution.

The problem of finding the particular solution to a differential equation that satisfies a given initial condition is called an **initial value problem**. In this section we develop a numerical technique for approximating the solution to initial value problems of the form

$$y' = f(x, y) \qquad y(a) = C \qquad a \le x \le b \tag{1}$$

To avoid confusion between the actual solution of an initial value problem and an approximation to that solution, we will often refer to the actual solution as the *exact solution*. That is, a function $y = y(x)$ is the **exact solution** of (1) if

$$y'(x) = f(x, y(x)) \qquad a \le x \le b \qquad \text{and} \qquad y(a) = C$$

Note that an exact solution is always a particular solution. General solutions involving arbitrary constants cannot be approximated numerically.

Given any initial value problem of the form (1), we want to find a function $p(x)$ whose values can be used to approximate the values of the exact solution $y(x)$ for $a \leq x \leq b$. Furthermore, we want to be able to find this approximate solution $p(x)$ without first finding the exact solution $y(x)$. The procedure we will use to find an approximate solution is outlined in the box.

Approximating the Solution of:

$$y' = f(x, y) \qquad y(a) = C \qquad a \leq x \leq b$$

Step 1 Divide the interval $a \leq x \leq b$ into n equal subintervals of length $\Delta x = (b - a)/n$ with end points

$$a = x_0 < x_1 < \ldots < x_n = b$$

Step 2 Let $y_0 = y(a) = C$ and find a sequence of values $y_2, y_2, \ldots, y_n$ that are approximations to the values of the exact solution $y(x)$ at $x = x_1, x_2, \ldots, x_n$. That is

$$y(x_k) \approx y_k \qquad k = 1, 2, \ldots, n$$

Step 3 Connect the points $(x_0, y_0), (x_1, y_1), \ldots, (x_n, y_n)$ with straight line segments to form a continuous piecewise linear function $p(x)$ that approximates $y(x)$ for $a \leq x \leq b$.

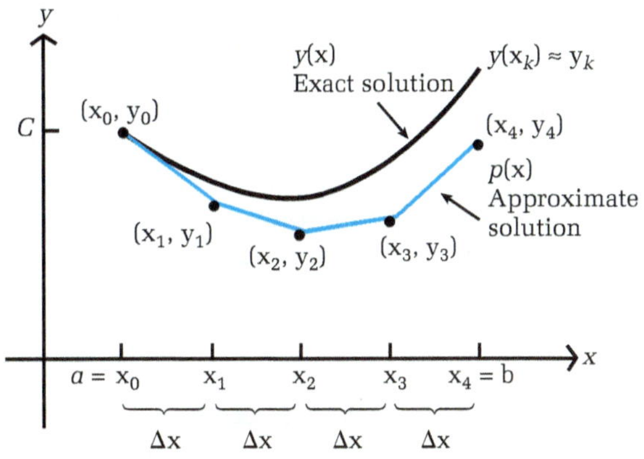

Many different methods can be used to find the approximate values $y_1, y_2, \ldots, y_n$ mentioned in step 2 in the box. One of the oldest and easiest to use is *Euler's method.*

Euler's Method

Euler's method of approximating exact solutions applies only to initial value problems of the form

$$y' = f(x, y) \qquad y(a) = C \qquad a \leq x \leq b \tag{1}$$

However, this type of problem occurs frequently in applications, and in most cases the problem does not have a simple exact solution. Note that $y' = f(x, y)$ gives us the slope of a tangent line to a solution curve at point (x, y) on the curve. Euler's method for computing the approximate values $y_1, y_2, \ldots, y_n$ is based on the use of actual and approximate tangent lines to the exact solution curve. The slopes of these tangent lies are determined by $y' = f(x, y)$.

The method starts by finding the line tangent to the graph of the exact solution $y(x)$ at $x = x_0$. The equation of this tangent line is

$$y = y(x_0) + y'(x_0)(x - x_0) \qquad \text{Tangent line at } x = x_0 \tag{2}$$

Since $x_0 = a$, the initial condition in (1) implies that $y(x_0) = y(a) = C$. In the box outlining the approximation procedure, we have denoted this initial value by y_0 instead of C for consistency in notation.

Now, how can we determine the value of $y'(x_0)$? Remember that $y(x)$ is the solution of (1) and must satisfy

$$y'(x) = f(x, y(x))$$

Substituting x_0 for x in this equation, we have

$$\begin{aligned} y'(x_0) &= f(x_0, y(x_0)) \qquad y(x_0) = y_0 \\ &= f(x_0, y_0) \end{aligned}$$

Substituting y_0 for $y(x_0)$ and $f(x_0, y_0)$ for $y'(x_0)$ in (2) enables us to write

$$y = y_0 + f(x_0, y_0)(x - x_0) \quad \text{Tangent line at } x = x_0$$

This line can be used to approximate $y(x)$ for the values of x near x_0. In particular, we can use this line to approximate $y(x_1)$.

Let

$$\begin{aligned} y_1 &= y_0 + f(x_0, y_0)(x_1 - x_0) \quad \text{Tangent line at } x_0 \text{ evaluated at } x = x_1 \\ &= y_0 + f(x_0, y_0)\Delta x \end{aligned}$$

As Figure 12 illustrates, y_1 can be used to approximate $y(x_1)$. That is,

$$y(x_1) \approx y_1 = y_0 + f(x_0, y_0)\Delta x$$

Notice that y_1 is computed by using the given function $f(x, y)$, the initial values $x_0 = a$ and $y_0 = C$, and $\Delta x = x_1 - x_0$. In particular, we do not need to know that actual solution $y(x)$ in order to compute y_1.

Suppose we try to use the tangent at $x = x_1$ to find a approximate value for $y(x_2)$. The equation of this tangent line is

$$y = y(x_1) + y'(x_1)(x - x_1) \quad \text{Tangent line at } x = x_0$$

Since we do not know the exact values of $y(x_1)$ or $y'(x_1)$, we cannot use this line to approximate $y(x_2)$. However, we have already concluded that $y(x_1) \approx y_1$. Furthermore, since $y(x)$ satisfies the differential equation in (1) for $a \leqslant x \leqslant b$, we know that $y'(x_1) = f(x_1, y(x_1))$. It is reasonable to assume that $y(x_1) \approx y_1$ implies

$$y'(x_1) = f(x_1, y(x_1)) \approx f(x_1, y_1)$$

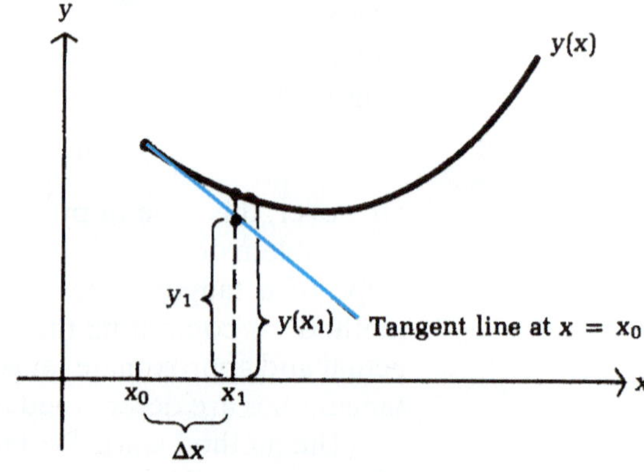

Figure 12

$y(x_1) \approx y_1 = y_0 + f(x_0, y_0)\Delta x$

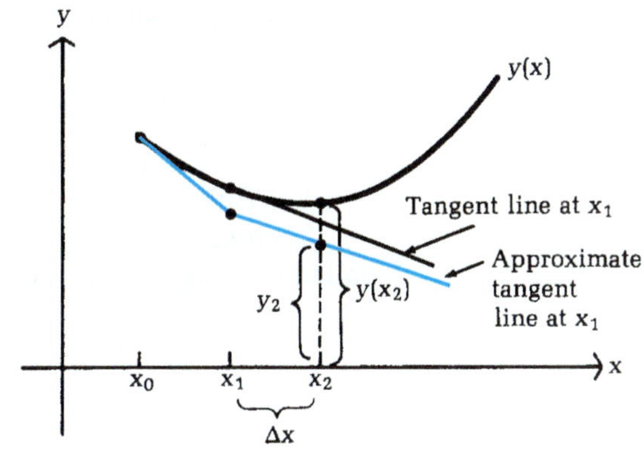

Figure 13
$y(x_2) \approx y_2 = y_1 + f(x_1, y_1)\Delta x$

Thus, we can use the approximation $y(x_1) \approx y_1$ and $y'(x_1) \approx f(x_1, y_1)$ to find the equation of a line that approximates the tangent line at x_1. That is,

$$y = \underbrace{y(x_1)}_{\approx} + \underbrace{y'(x_1)}_{\approx}(x - x_1) \quad \text{Tangent line at } x_1$$

$$y = \overbrace{y_1} + \overbrace{f(x_1, y_1)}(x - x_1) \quad \text{Approximate tangent line at } x_1$$

Now we can use this approximate tangent line to approximate $y(x;i2)$.

Let

$$y_2 = y_1 + f(x_1, y_1)(x_2 - x_1) \quad \text{Approximate tangent line evaluated at } x = x_2$$

$$= y_1 + f(x_1, y_1)\Delta x$$

Referring to Figure 13 we see that

$$y(x_2) \approx y_2 = y_1 + f(x_1, y_1)\Delta x$$

Notice that the calculation of y_2 depends on the given function $f(x, y)$, the known values of x_1 and Δx_1, and the previously computed approximate value y_1.

If we find the approximate tangent line at x_2 and use this line to approximate $y(x_3)$ (see Figure 14), we obtain

$$y(x_3) \approx y_3 = y_2 + f(x_2, y_2)\Delta x$$

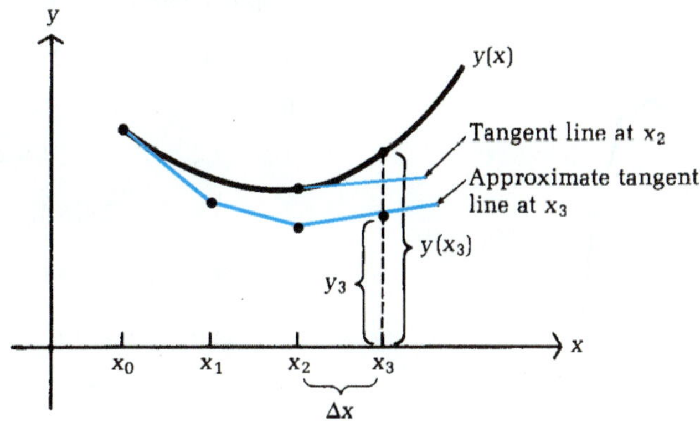

Figure 14
$y(x_3) \approx y_3 = y_2 + f(x_2, y_2)\Delta x$

Continuing this process, we are led to the **recursion formula** (each approximate value is calculated in terms of the previous one)

$$y(x_k) \approx y_k = y_{k-1} + f(x_{k-1}, y_{k-1})\Delta x \qquad k = 1, 2, \ldots, n$$

which is ideal for calculator or computer application. This recursion formula is the heart of Euler's method.

Each calculation in using this formula is called a **step** and $\Delta x = x_k - x_{k-1}$ is called the **step size** (It is convenient, but not necessary, to use equal step sizes, as we have done here.) We summarize *Euler's method* for computing the approximate values of the solution of an initial value problem in the box.

Euler's Method

If $y(x)$ is the (exact) solution to the initial value problem

$$y' = f(x, y) \qquad y(x_0) = y_0$$

$x_0, x_1, \ldots, x_n$ is the sequence of values defined by

$$x_k = x_{k-1} + \Delta x \qquad k = 1, 2, \ldots, n$$

and $y_1, y_2, \ldots, y_n$ is the sequence of values defined by

$$y_k = y_{k-1} + f(x_{k-1}, y_{k-1})\Delta x \qquad k = 1, 2, \ldots, n$$

then

$$y(x_k) \approx y_k \qquad k = 1, 2, \ldots, n$$

EXAMPLE 12 Use Euler's method with a step size of $\Delta x = 0.2$ to approximate the solution of

$$y' = 10\sqrt{y} - 12x \qquad y(1) = 2$$

for $1 \leqslant x \leqslant 2$. Round each approximate value to two decimal places and graph the approximate solution.

SOLUTION Let $x_0 = 1, y_0 = 2, \Delta x = 0.2$, and

$$f(x, y) = 10\sqrt{y} - 12x$$

Then Euler's formula for y_k is

$$\begin{aligned} y_k &= y_{k-1} + f(x_{k-1}, y_{k-1})\Delta x \\ &= y_{k-1} + (10\sqrt{y_{k-1}} - 12x_{k-1})(0.2) \end{aligned}$$

We can use this recursion formula to obtain the sequence of approximate values listed in the table.

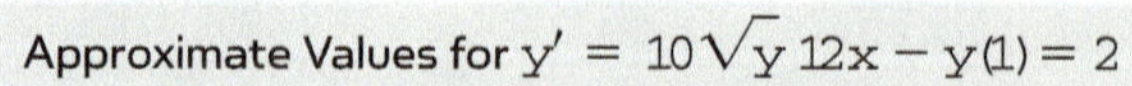
Approximate Values for $y' = 10\sqrt{y}$ $12x - y(1) = 2$

k	x_k	$y_k = y_{k-1} + (10\sqrt{y_{k-1}} - 12x_{k-1})(0.2)$
0	1	2
1	1.2	$2 + [10\sqrt{2} - 12(1)](0.2) \approx 2.428\,472\,1$
2	1.4	$2.428\,427\,1 + [10\sqrt{2.428\,427\,1} - 12(1.2)](0.2) \approx 2.665\,109\,4$
3	1.6	$2.665\,109\,4 + [10\sqrt{2.665\,109\,4} - 12(1.4)](0.2) \approx 2.570\,142$
4	1.8	$2.570\,142 + [10\sqrt{2.570\,142} - 12(1.6)](0.2) \approx 1.936\,474\,4$
5	2	$1.936\,474\,4 + [10\sqrt{1.936\,474\,4} - 12(1.8)](0.2) \approx 0.399\,619\,75$

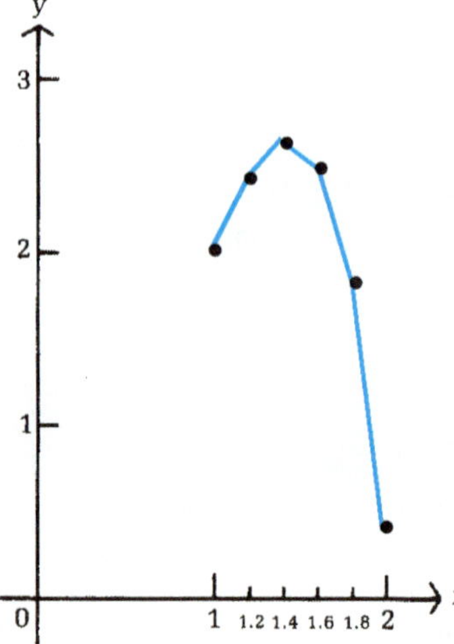

Rounding the values of *yk* from this table to 2.43, 2.67, 2.57, 1.94, and 0.40 we graph the approximate solution, as shown in the figure.

Matched Problem 12 Use Euler's method with a step size of $\Delta x = 0.2$ to approximate the solution of

$$y' = 10x - 9\sqrt{y} \qquad y(1) = 3$$

for $1 \leq x \leq 2$. Round each approximate value to two decimal places and graph the approximate solution.

A calculator with a memory will be very helpful in computing the approximate values $y_1, \ldots, y_n$. In most problems, each value of y_k can be stored in the memory and used to compute the next value without reentering the previously computed value. We used this procedure in Example 12 to calculate the values of y_n in the table (to the limit of accuracy provided by our calculator) and then rounded these values to two decimal places for graphing purposes. Unless otherwise noted, we will follow this procedure in all the examples and exercises in this section.

Comparison of Exact and Approximate Solutions

It is very difficult to make any estimates concerning the error in approximations obtained by using Euler's method and we will not do so in this text. However, it is instructive to apply Euler's method to a problem whose solution is known, so that we can compare the graphs of the exact and approximate solutions.

EXAMPLE 13 Approximate the solution of

$$y' = y + 7x - 3 \qquad y(0) = 1$$

for $0 \leq x \leq 1$ using Euler's method with a step size of:

(A) $\Delta x = 0.2$ (B) $\Delta x = 0.1$

(C) The exact solution of this equation is

$$y = 5e^x - 7x - 4$$

Graph the exact solution and both approximate solutions from parts A and B on the same set of axes.

SOLUTIONS (A) $x_0 = 0, y_0 = 1, \Delta x = 0.2, f(x, y) = y + 7x - 3$, and $y_k = y_{k-1} + (y_{k-1} + 7x_{k-1} - 3)(0.2)$

k	x_k	$y_k = y_{k-1} + (y_{k-1} + 7x_{k-1} - 3)(0.2)$
0	0	1
1	0.2	0.6
2	0.4	0.4
3	0.6	0.44
4	0.8	$0.768 \approx 0.77$
5	1	$1.4416 \approx 1.44$

(B) $x_0 = 0, y_0 = 1, \Delta x = 0.1, f(x, y) = y + 7x - 3$, and $y_k = y_{k-1} + (y_{k-1} + 7x_{k-1} - 3)(0.1)$

k	x_k	$y_k = y_{k-1} + (y_{k-1} + 7x_{k-1} - 3)(0.1)$
0	0	1
1	0.1	0.8
2	0.2	0.65
3	0.3	$0.555 \approx 0.56$
4	0.4	$0.520\,5 \approx 0.52$
5	0.5	$0.552\,55 \approx 0.55$
6	0.6	$0.667\,805 \approx 0.66$
7	0.7	$0.843\,585\,5 \approx 0.84$
8	0.8	$1.117\,944\,1 \approx 1.12$
9	0.9	$1.489\,738\,5 \approx 1.49$
10	1	$1.968\,712\,3 \approx 1.97$

(C) In order to make an accurate comparison, we used point-by-point plotting to graph $y(x) = 5e^x - 7x - 4$ in the following figure:

x	$y(x)$
0	1
0.1	0.83
0.2	0.71
0.3	0.65
0.4	0.66
0.5	0.74
0.6	0.91
0.7	1.17
0.8	1.53
0.9	2.00
1	2.59

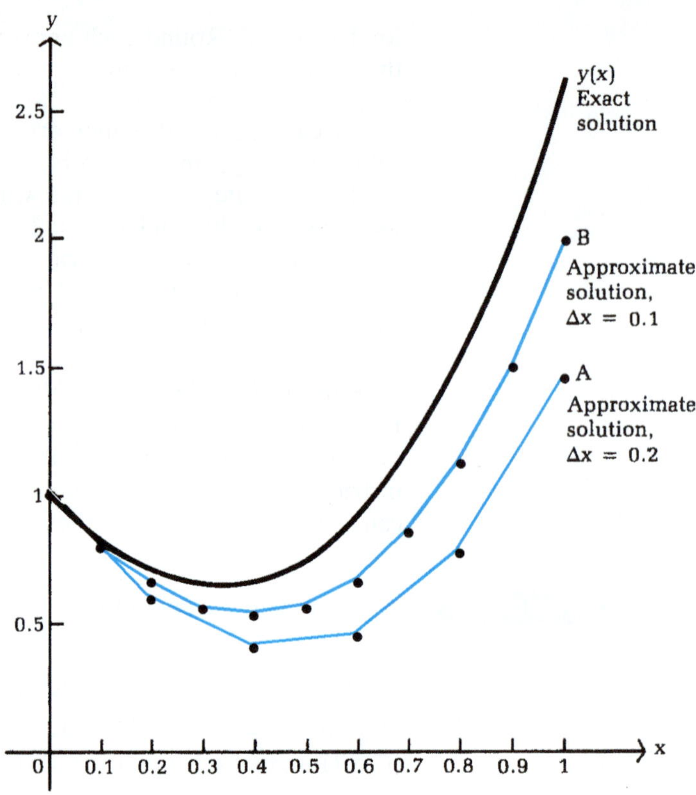

Matched Problem 13 Approximate the solution of

$$y' = y - 8x + 1 \qquad y(0) = 1$$

for $0 \leq x \leq 1$ using Euler's method with a step size of:

(A) $\Delta x = 0.2$ (B) $\Delta x = 0.1$

(C) The exact solution of this equation is

$$y = -6e^x + 8x + 7$$

Graph the exact solution and both approximate solutions from parts A and B on the same set of axes.

The figure in Example 13 illustrates fairly typical behavior for approximate solutions obtained by Euler's method. Notice that the distance between the graph of the actual solution and the graph of each approximate solution increases with each step. For most initial value problems, **the error in the approximation increases with each step.**

Comparing the graphs of the two approximate solutions, we see that the graph of the approximate solution for $\Delta x = 0.1$ is closer to the graph of the exact solution than is the graph of the approximate solution for $\Delta x = 0.3$. Thus, **decreasing the step size usually increases the accuracy in the approximation.**

Application

Up to this point, we have been concerned with approximating the solution of an initial value problem over an interval $a \leq x \leq b$. In many applications, we are interested in approximating the value of the solution only at $x = b$. Euler's method is a step-by-step process, with each approximate value depending on the previous one. Thus, to approximate the solution at $x = b$, we must compute all the intermediate approximate values.

EXAMPLE 14 **Sales Growth** The annual sales S (in millions of dollars) of a new company are expected to satisfy the growth equation

$$\frac{dS}{dt} = S(2 - \sqrt{S})$$

where t is the number of years the company has been in operation. If the annual sales for the first year ($t = 1$) are \$1.5 million, approximate the annual sales for the third year by using Euler's method with a step size of:

(A) $\Delta t = 1$ (B) $\Delta t = 0.5$ (C) $\Delta t = 0.25$

SOLUTIONS If $S(t)$ is the solution of

$$\frac{dS}{dt} = S(2 - \sqrt{S}) \qquad S(1) = 1.5$$

we must approximate $S(3)$. Using successively smaller step sizes will give us some information about the accuracy of our approximations.

(A) $\Delta t = 1, t_0 = 1, S_0 = 1.5, f(t, S) = S(2 - \sqrt{S},$
and $S_k = S_{k-1} + [S_{k-1}(2 - \sqrt{S_{K-1}})](1)$

k	t_k	$S_k = S_{k-1} + [S_{k-1}(2 - \sqrt{S_{k-1}})](1)$
0	1	1.5
1	2	2.662 882 7
2	3	3.643 265 2

(B) $\Delta t = 0.5$

k	t_k	$S_k = S_{k-1} + [S_{k-1}(2 - \sqrt{S_{k-1}})](0.5)$
0	1	1.5
1	1.5	2.081 441 3
2	2	2.661 414
3	2.5	3.151 933 8
4	3	3.505 945 4

(C) $\Delta t = 0.25$

k	t_k	$S_k = S_{k-1} + [S_{k-1}(2 - \sqrt{S_{k-1}})](0.25)$
0	1	1.5
1	1.25	1.790 720 7
2	1.5	2.087 005 2
3	1.75	2.376 761 3
4	2	2.649 093 7
5	2.25	2.895 721 9
6	2.5	3.111 683 3
7	2.75	3.295 275 4
8	3	3.447 443 5

The last line of each of these tables contains an approximation to $S(3)$. These values, rounded to one decimal place, are displayed in the table below. Since decreasing the step size did not produce a large change in the approximate value of $S(3)$, we can conclude that $S(3) \approx 3.4$ is a reasonable approximation. Thus, we would expect the annual sales for the third year to be approximately \$3.4 million.

Approximations to $S(3)$

Δt	1	0.5	0.25
Approximation	3.6	3.5	3.4

Matched Problem 14 Repeat Example 14 if the annual sales during the first year are \$2.5 million.

Refer to the table in Example 14, where the approximations to $S(3)$ are summarized. How can we be sure that 3.4 is close to the exact value of $S(3)$? Perhaps the approximations keep decreasing as the step size decreases and the exact value of $S(3)$ is nowhere near 3.4. One way to answer this question is to continue to compute approximate values of $S(3)$, using successively smaller step sizes. A common procedure is to halve the step size each time, as we did in Example 19. Since halving the step size doubles the number of steps and the number of intermediate values that must be calculated, it is convenient to use a computer to carry out these calculations.

Table 3 contains the approximate values found in Example 19, along with the approximate values obtained by using a computer program with step sizes of 0.125 and 0.0625, all rounded to one decimal place. After examining the values in Table 8, we can feel more confident in concluding that $S(3) \approx 3.4$.

TABLE 3 Approximations to $S(3)$

Δt	1	0.5	0.25	0.125	0.0625
Approximation	3.6	3.5	3.4	3.4	3.4

Euler's Method For Systems of Differential Equations

Euler's method is easily modified to approximate the solution to a system of differential equations. We simply compute approximate values for each variable in the system, as described in the box at the top of the next page.

Euler's Method for a System of Differential Equations

If $x = x(t)$ and $y = y(t)$ is the solution to the system of differential equations

$$\frac{dx}{dt} = f(x, y) \qquad x(t_0) = x_0$$

$$\frac{dy}{dt} = g(x, y) \qquad y(t_0) = y_0$$

$t_0, t_1, \ldots, t_n$ is the sequence of values defined by

$$t_k = t_{k-1} + \Delta t \qquad k = 1, 2, \ldots, n$$

and $x_1, x_2, \ldots, x_n$ and $y_1, y_2, \ldots, y_n$ are defined by

$$x_k = x_{k-1} + f(x_{k-1}, y_{k-1})\Delta t$$

$$y_k = y_{k-1} + g(x_{k-1}, y_{k-1})\Delta t \qquad k = 1, 2, \ldots, n$$

then

$$x(t_k) \approx x_k \quad \text{and} \quad y(t_k) \approx y_k \qquad k = 1, 2, \ldots, n$$

EXAMPLE 15 **Interrelated Prices** The prices (in dollar) of two interrelated commodities are p and q. The rate of change of each price is related to both prices by

$$\frac{dp}{dt} = -0.15p + 0.1q + 10$$

$$\frac{dq}{dt} = 0.1p - 0.2q + 15$$

where t is time in months. Initially, $p = 150$ and $q = 200$. Use Euler's method with a step size of $\Delta t = 1$ and five steps to approximate the solution to this system. Round the approximate values to one decimal place after each step and use these rounded values in the next step.

SOLUTION Let $\Delta t = 1$, $t_0 = 0$, $p_0 = 150$, $q_0 = 200$,

$$f(p, q) = -0.15p + 0.1q + 10$$

and

$$g(p, q) = 0.1p - 0.2q + 15$$

Then

$$p_k = p_{k-1} + f(p_{k-1}, q_{k-1})(1)$$

and

$$q_k = q_{k-1} + g(p_{k-1}, q_{k-1})(1)$$

Notice that we cannot compute these approximations independently. Instead, we must compute p_1 and q_1, then p_2 and q_2, and so on. Also note that we can omit the factor Δt, since $\Delta t = 1$ for this approximation.

$p_k = p_{k-1} + f(p_{k-1}, q_{k-1})$

$p_0 = 150$

$p_1 = 150 + f(150, 200)$

$\quad = 150 + 7.5 = 157.5$

$p_2 = 157.5 + f(157.5, 190)$

$\quad = 157.5 + 5.375 \approx 162.9$

$p_3 = 162.9 + f(162.9, 182.8)$

$\quad = 162.9 + 3.485 \approx 166.7$

$p_4 = 166.7 + f(166.7, 177.5)$

$\quad = 166.7 + 2.745 \approx 169.7.4$

$p_5 = 169.4 + f(169.4, 173.7)$

$\quad = 169.4 + 1.96 = 171.4$

$q_k = q_{k-1} + g(p_{k-1}, q_{k-1})$

$q_0 = 200$

$q_1 = 200 + g(150, 200)$

$\quad = 200 - 10 = 190$

$q_2 = 190 + g(157.5, 190)$

$\quad = 190 - 7.25 \approx 182.8$

$q_3 = 182.8 + g(162.9, 182.8)$

$\quad = 182.8 - 5.27 = 177.5$

$q_4 = 177.5 + g(166.7, 177.5)$

$\quad = 177.5 - 3.83 \approx 173.7$

$q_5 = 173.7 + g(169.4, 173.7)$

$\quad = 173.7 - 2.8 \approx 170.9$

In summary:

p_k	150	157.5	162.9	166.7	169.4	171.4
q_k	200	190	182.8	177.5	173.7	170.9

Thus, over a 5 month period, the price p increases from \$150 to approximately \$171.40 and the price q decreases from \$200 to approximately \$170.90.

Matched Problem 15 Repeat Example 15 if the initial values are $p = 200$ and $q = 150$.

Answers to Matched Problems

12.

k	x_k	y_k
0	1	3
1	1.2	1.88
2	1.4	1.81
3	1.6	2.19
4	1.8	2.73
5	2	3.35

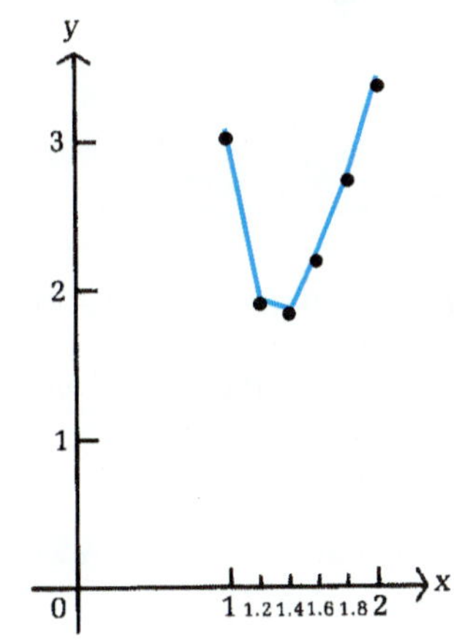

13. (A)

k	x_k	y_k
0	0	1
1	0.2	1.4
2	0.4	1.56
3	0.6	1.43
4	0.8	0.96
5	1	0.07

(B)

k	x_k	y_k
0	0	1
1	0.1	1.2
2	0.2	1.34
3	0.3	1.41
4	0.4	1.42
5	0.5	1.34
6	0.6	1.17
7	0.7	0.91
8	0.8	0.54
9	0.9	0.05
10	1	−0.56

(C)

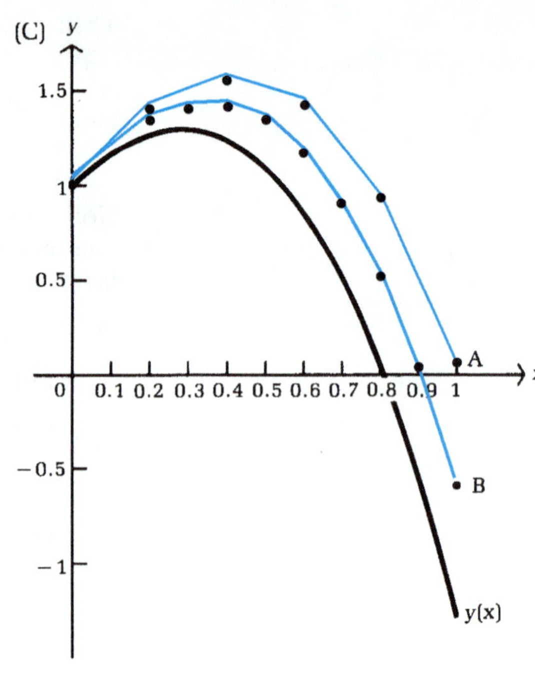

14.

Δt	1	0.5	0.25
Approximation	4.0	3.8	3.8

Approximately $3.8 million

15.

p_k	200	195	191.3	188.5	186.3	184.4
q_k	150	155	158.5	160.9	162.6	163.7

Exercise 9-3

A

In Problems 1–6, use Euler's method with a step size of $\Delta x = 0.2$ to approximate the solution for $0 \leq x \leq 1$. Round each approximate value to two decimal places and graph the approximate solution.

1. $y' = x + y; y(0) = 2$

2. $y' = x - y; y(0) = 2$

3. $y' = y; y(0) = 1$

4. $y' = \frac{1}{2y}, y(0) = 1$

5. $y' = 2\sqrt{y} - 8x; y(0) = 1$

6. $y' = x - 2y^2; y(0) = 1$

B

In Problems 7–10, use Euler's method with a step size of $\Delta x = 0.2$ to approximate the solution for $0 \leq x \leq 1$. Round each approximate value to two decimal places. Graph the approximate solution and the exact solution of the same set of axes:

7. $y' = -y; y(0) = 1$; exact solution: $y(x) = e^{-x}$

8. $y' = -y^2; y(0) = 1$; exact solution: $y(x) = 1/(1 + x)$

9. $y' = 1 + y; y(0) = 0$; exact solution: $y(x = e^x - 1$

10. $y' = 1 - y; y(0) = 0$; exact solution: $y(x) = 1 - e^{-x}$

In Problems 11 and 12, use Euler's method with a step size of $\Delta x = 0.2$ and $\Delta x = 0.1$ to find two approximate solutions for $0 \leq x \leq 1$. Round each approximate value to two decimal places. Graph both approximate solutions and the exact solution on the same set of axes.

11. $y' = y - 6x; y(0) = 1$; exact solution: $y(x) = -5e^x + 6x + 6$

12. $y' = y - 7x + 1; y(0) = 1$; exact solution: $y(x) = -5e^x + 7x + 6$

In Problems 13–20, approximate the indicated value of y by using Euler's method with the given step size. Round the approximate value to two decimal places.

13. $y' = 1 + y^2; y(0) = 0$, use $\Delta x = 0.2$ to approximate $y(1)$

14. $y' = \sqrt{1 + y^2}; y(0) = 0$; use $\Delta x = 0.2$ to approximate $y(1)$

15. $y' = y^2 - x^2; y(1) = -1$; use $\Delta x = 0.4$ to approximate $y(3)$

16. $y' = \sqrt{x^2 + y^2}$; $y(0) = 0$;
use $\Delta x = 0.1$ to approximate $y(0.5)$

17. $y' = x + e^{-y}$; $y(0) = 0$;
use $\Delta x = 0.1$ to approximate $y(1)$

18. $y' = x + \ln y$; $y(0) = 1$;
use $\Delta x = 0.3$ to approximate $y(1.2)$

19. $y' = \ln(x + y)$; $y(1) = 0$;
use $\Delta x = 0.2$ to approximate $y(3)$

20. $y' = e^{-xy}$; $y(0) = 0$;
use $\Delta x = 0.5$ to approximate $y(3)$

In Problems 21–24, use Euler's method with $\Delta t = 0.2$ and five steps to approximate the solution. Round the approximate values to two decimal places after each step and use these rounded values in the next step.

21. $\frac{dx}{dt} = 2x - y, \frac{dy}{dt} = 3x - 2y; x(0) = 1, y(0) = -1$

22. $\frac{dx}{dt} = 5x + 2y, \frac{dy}{dt} = 2x + 2y; x(0) = 3, y(0) = -1$

23. $\frac{dx}{dt} = 2x + y, \frac{dy}{dt} = 2x + y; x(0) = 2, y(0) = -1$

24. $\frac{dx}{dt} = x - y, \frac{dy}{dt} = x - y; x(0) = 1, y(0) = 2$

C

25. Use Euler's method with a step size of $\delta x = 0.2$ to approximate the solution of

$$y' = 5x - 2\sqrt{y}$$

on the interval $0 \leq x \leq 1$ if:

(A) $y(0) = 1$ (B) $y(0) = 2$ (C) $y(0) = 3$

Round all approximate values to two decimal places and graph all three approximate solutions on the same set of axes.

26. Repeat Problem 25 for the equation $y' = 4x - \sqrt{y}$.

27. Use step sizes of $\Delta x = 1$, $\Delta x = 0.5$, and $\Delta x = 0.25$ to approximate $y(3)$ if $y(x)$ is the solution of

$$y' = \sqrt{y}(2 - \sqrt{y}) \qquad y(1) = 2$$

Round each approximate value to one decimal place.

28. Repeat Problem 27 for

$$y' = \sqrt{y}(4 - \sqrt{y}) \qquad y(1) = 2$$

Applications

Business & Economics

29. Advertising. A company is using an extensive newspaper advertising campaign to introduce a new product. The number of people N (in thousands) who have heard of the product satisfies

$$\frac{dN}{dt} = 0.1N(10 - \sqrt{N})$$

If $N(0) = 1$, use Euler's method with $\Delta t = 1$ to approximate $N(5)$.

30. Sales growth. The annual sales s (in millions of dollars) of a company satisfy

$$\frac{ds}{dt} = 0.01s(50 - s^{1.5})$$

If the annual sales now ($t = 0$) are \$2 million, use Euler's method with a step size of 1 to approximate the annual sales 5 years from now.

31. Interrelated markets. The prices p and q (in dollars) of two commodities in an interrelated market satisfy the following system of differential equations:

$$\frac{dp}{dt} = 2p - q$$

$$\frac{dq}{dt} = -p + q$$

where t is time in years. If $p(0) = 100$ and $q(0) = 160$, use Euler's method with a step size of 0.5 to approximate $p(2)$ and $q(2)$. Round the approximate values to one decimal place after each step and use the rounded values in the next step.

32. Interrelated markets. Repeat Problem 31 for the system

$$\frac{dp}{dt} = p - q \; p(0) = 150$$

$$\frac{dp}{dt} = p - 2q \qquad q(0) = 100$$

Life Sciences

33. Epidemic. The rate at which a disease spreads through a community is given by

$$\frac{dN}{dt} = 0.1\sqrt{N}(10 - \sqrt{N})$$

where N is the total number (in thousands) of infected individuals at time t. If $N(0) = 1$, use Euler's method with a step size of 1 to approximate $N(4)$.

34. Population growth. The population y (in tens of thousands) of a certain species of bird satisfies the growth equation

$$\frac{dy}{dt} = -y + y^2 - 0.1y^3$$

where t is measured in years. If the initial population is 5,000, use Euler's method with a step size of 0.5 to approximate the population after 2 years.

35. Population growth. If an ecological system contains two species (say, rabbits and foxes) that interact with each other in a predator-prey relationship, then the rate of growth of each species is related to the current population of both species. Suppose that the number of rabbits, y (in thousands), and the number of foxes, x (in thousands), satisfy the system of equations

$$\frac{dy}{dt} = y(0.8 - 0.04x)$$

$$\frac{dx}{dt} = x(0.006y - 0.3)$$

where t is time in years. If $y(0) = 55$ and $x(0) = 15$, use Euler's method with a step size of 1 to approximate $y(4)$

and $x(4)$. Round the approximate values to one decimal place after each step and use the rounded values in the next step.

36. Population growth. Repeat Problem 35 for the system

$$\frac{dy}{dt} = y(0.5 - 0.02x) \qquad y(0) = 60$$

$$\frac{dx}{dt} = x(0.01y - 0.5) \qquad x(0) = 20$$

Social Sciences

37. Rumor spread. A rumor spreads through a community at a rate given by

$$\frac{dN}{dt} = 0.1N^{2/3}(100 - n^{1/3})$$

where N is the total number of individuals who have heard the rumor at time t. If $N(0) = 1{,}000$, use Euler's method with a step size of 1 to approximation $N(4)$.

Chapter 9 Review

Important Terms and Symbols

9-1 Newton's Method for Approximating Roots

Zero; root; $x$intercept; Newton's method; recursion formula for x_n; internal rate of return

$$x_n = x_{n-1} - \frac{f(x_{n-1})}{f'(x_{n-1})}$$

9-2 Numerical Integration

Trapezoid rule; Simpson's rule; error estimates; Err(T_n; Err(S_n)

$$T_n = \frac{\Delta x}{2}[f(x_0) + 2f(x_1) + \cdots + 2f(x_{n-1}) + f(x_n)]$$

$$S_n = \frac{\Delta x}{3}[f(x_0) + 4f(x_1) + 2f(x_2) + 4f(x_3) + \cdots + 2f(x_{n-2}) + 4f(x_{n-1}) + f(x^n)]$$

9-3 Euler's Method

Initial value problem; exact solution; approximate solution; recursion formula step; step size; Euler's method; Euler's method for a system of differential equations

$$y_k = y_{k-1} + f(x_{k-1}, y_{k-1})\Delta x$$

Review Exercises

Work through all the problems in this chapter review and check your answers in the back of the book. (Answers to all review problems are there.) Where weaknesses show up, review appropriate sections in the text.

In Problems 1 and 2, find and simplify Newton's recursion formula for x_n. Use this expression and the given value of x_1 to compute x_n for the indicated value of n.

1. $f(x) = x^2 - 10; \quad x_1 = 3, \quad n = 4$

2. $f(x) = \ln x + x + 1; \quad x_1 = 0.5, \quad n = 5$

3. Use the trapezoid rule with $n = 5$ to approximate

$$\int_0^2 e^{x^2}\,dx$$

Round the result to three decimal places.

4. Use Simpson's rule with $n = 4$ to approximate

$$\int_0^3 \sqrt{4 + x^3}\, dx$$

Round the result to three decimal places.

In Problems 5 and 6, use Euler's method with a step size of $\Delta x = 0.2$ to approximate the solution for $0 \leqslant x \leqslant 1$. Round each approximate value to two decimal places and graph the approximate solution.

5. $y' = -2y;\ \ y(0) = 3$

6. $y' = y - 5x;\ \ y(0) = 1$

In Problems 7 and 8, sketch the graph of the function and use Newton's method to find the x intercept(s).

7. $f(x) = x^3 - 12x + 3$

8. $f(x) = x^3 - 9x^2 + 15x + 30$

In Problems 9 and 10, graph both functions on the same set of axes and find their point(s) of intersection.

9. $f(x) = e^{-x};\ \ g(x) = x^3$

10. $f(x) = \frac{1}{x^3};\ \ g(x)\ 5\ x + 1$

11. Use the trapezoid rule with $n = 10$ to approximate

$$\int_0^2 \frac{1}{e^x + e^{-x}} dx$$

Round the result to five decimal places.

12. Use Simpson's rule with $n = 8$ to approximate

$$\int_1^3 [\ln (1 + x^2)]^2\, dx$$

Round the result to five decimal places.

13. Use the value in the table and the trapezoid rule to approximate the area bounded by the graph of the function f and the x axis over the interval of x values in the table.

x	−3	−2	−1	0	1	2	3
$f(x)$	5	7	8	12	11	6	4

14. Use Simpson's rule with $n = 10$ to approximate the area of the region bounded by the graphs of

$$y = \frac{1}{\ln x} \quad \text{and} \quad y = 0 \quad 2 \leqslant x \leqslant 3$$

Round the result to five decimal places.

15. Use Euler's method with $\Delta x = 0.2$ to approximate the solution to

$$y' = -\frac{x}{y} \qquad y(0) = 1$$

for $0 \leqslant x \leqslant 1$. Round each approximate value to two decimal places. The exact solution to this equation is

$$y = \sqrt{1 - x^2}$$

Graph the approximate solution and the exact solution on the same set of axes.

In Problems 16 and 17, use Euler's method and the given step size to approximate the indicated value of the solution.

16. $y' = x + \sqrt{y}; y(0) = 0$; use $\Delta x = 0.2$ to approximate $y(1)$

17. $y' = \frac{1}{y} + \frac{1}{x}; y(1) = 1$;
use $\Delta x = 0.4$ to approximate $y(3)$

18. Use Euler's method with $\Delta t = 0.2$ and five steps to approximate the solution to

$$\frac{dx}{dt} = x + 2y \qquad x(0) = 1$$

$$\frac{dy}{dt} = y + rx \qquad y(0) = 0$$

Round the approximate values to one decimal place after each step and use the rounded values in the next step.

C

19. Apply Newton's method to $f(x) = ax^2 + bx + c$ and derive a recursion formula that approximates the roots of a quadratic polynomial.

20. Find the exact value of

$$\int_0^2 x^4\, dx$$

Then construct a table containing the values of T_n, $\text{Err}(T_n)$, S_n, and $\text{Err}(S_n)$ for $n = 4$ and $n = 8$. Round all values in the tables to three decimal places.

21. Use Euler's method with step sizes of $\Delta x = 0.2$ and $\Delta x = 0.1$ to find two approximate solutions to

$$y' = 2y - 10x + 1 \qquad y(0) = 1$$

for $0 \leqslant y \leqslant 1$. Round each approximate value to two decimal places. The exact solution is

$$y = -e^{2x}. + 5x + 2$$

Graph both approximate solutions and the exact solution on the same set of axes.

Applications

Business & Economics

22. Break-even point. The cost and revenue function for a product are

$$C(x) = 10 + 0.5x$$

and

$$R(x) = 2x + \frac{10x}{1 + x^2} \qquad x \geqslant 1$$

Use Newton's method to approximate the break-even point.

23. Internal rate of return. A small business pays \$20,000 to lease a delivery van for 3 years. The annual profits generated by this van during the leasing period are \$8,000, \$10,000, and \$9,000, respectively. Find the internal rate of return of this investment. Express the answer as a percentage correct to two decimal places.

24. Consumer's and producers' surplus. Find the equilibrium price (in dollars) and then find the consumers' and producers' surplus for

$$p = D(x) = \sqrt{144 - 4x^3}$$

and

$$p = S(x) = \sqrt{9 + x^3}$$

Use Simpson's rule with $n = 4$, and round each result to two decimal places.

25. **Marginal analysis.** The marginal cost and revenue functions for a machine are

$$C'(t) = \frac{2t^2}{4 + t^2} \quad \text{and} \quad R'(t) = \frac{8}{r + t^2}$$

where $C(t)$ and $R(t)$ are the total cost and revenue functions (in thousands of dollars) t years after the machine has been put into use. Find the useful life of the machine, and use Simpson's rule with $n = 4$ to approximate the total profit earned by the machine during its useful life. Round the result to the nearest hundred dollars.

26. **Sales growth.** The annual sales s (in thousands of dollars) of a company satisfy the equation

$$\frac{ds}{dt} = 0.1\sqrt{s}(100 - \sqrt{s})$$

If $s(0) = 50$ use Euler's method with $\Delta t = 1$ to approximate $s(5)$. Round the result to the nearest thousand dollars.

Life Sciences

27. **Medicine.** A patient is given a drug containing a particular hormone. The level of the hormone in the bloodstream (in milligrams per milliliter) t hours after the drug was administered is given by

$$L(t) = 100te^{-0.4t}$$

Approximate the time interval during which the hormone level is greater than 50 milligrams per milliliter. Round the values of t to one decimal place.

28. **Medicine.** The blood pressure B of an individual was graphed by a recording device over a 1 second interval (see the figure). What was the average blood pressure during this time period? Set up an appropriate definite integral, use the trapezoid rule with n = 5 to approximate this integral, and use the graph to estimate the necessary values of B, where B is expressed in a unit of measurement for pressure called a torr.

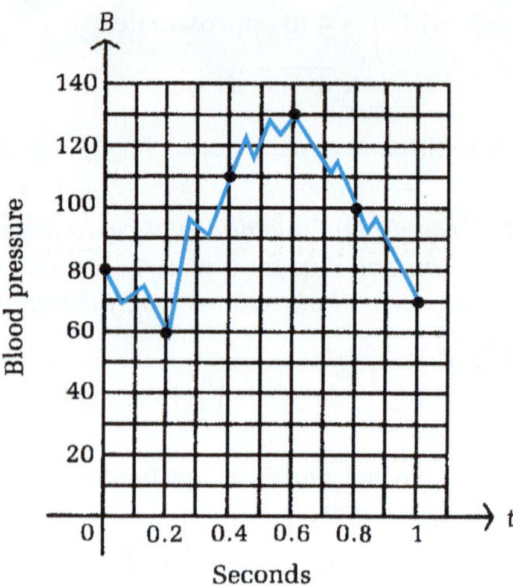

29. **Population growth.** Suppose the number of foxes, x (in thousands), and the number of rabbits, y (in thousands), in an ecological system satisfy

$$\frac{dy}{dt} = y(0.2 - 0.05x) \qquad y(0) = 50$$

$$\frac{dx}{dt} = x(0.005y - 0.2) \qquad x(0) = 15$$

where t is measured in years. Use Euler's method with $\Delta t = 1$ to approximate the population of each species after 5 years. Round the approximate values to one decimal place after each step and use the rounded values in the next step.

Social Sciences

30. **Psychology.** The rate at which workers can assemble electronic components t hours after they begin work is given by

$$N'(t) = \frac{4t^2 + 4t + 15}{4t^2 + 4t + 5}$$

Use Simpson's rule with $n = 4$ to approximate the total number of components a worker will assemble during the first 4 hours at work. Round the result to the nearest integer.

Probability and Calculus

Introduction

In this chapter we discuss some basic probability concepts, with an emphasis on probability distributions. We first consider some important classes of probability distributions associated with discrete sample spaces and then discuss several different probability distributions associated with continuous sample spaces. As we will see, calculus techniques play an important role in extending probability concepts from discrete to continuous sample spaces.

10-0 Improper Integrals

- Improper Integrals
- Application: Capital Value

Improper Integrals

We are now going to consider an integral form that has wide application in probability studies as well as other areas. Earlier, when we introduced the idea of a definite integral,

$$\int_a^b f(x)\,dx \tag{1}$$

we required f to be continuous over a closed interval $[a, b]$. Now we are going to extend the meaning of (1) so that the interval $[a, b]$ may become infinite in length.

EXPLORE & DISCUSS 1

For each of the following functions, find the area under the graph from $x = 1$ to $x = b, b > 1$ (see Fig. 1). Discuss the behavior of this area as b assumes larger and larger values.

(A) $f(x) = \dfrac{1}{x^{5/4}} = x^{-5/4}$ (B) $g(x) = \dfrac{1}{x^{4/5}} = x^{-4/5}$

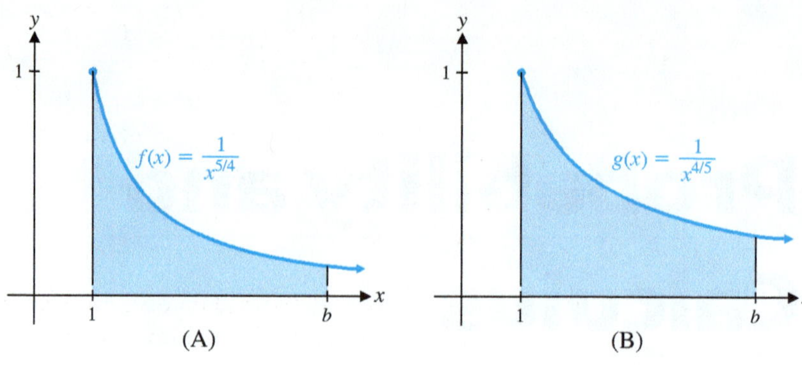

Figure 1

Let us investigate a particular example that will motivate several general definitions. What would be a reasonable interpretation for the following expression?

$$\int_1^{\infty} \frac{dx}{x^2}$$

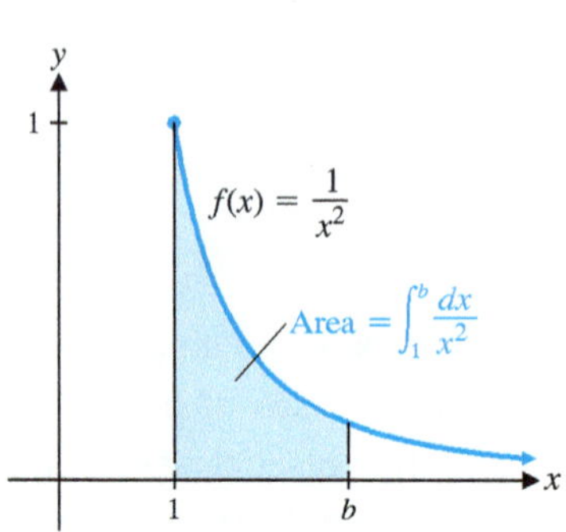

Figure 2

Sketching a graph of $f(x) = 1/x^2, x \ge 1$ (see Fig. 2), we note that for any fixed $b > 1$, $\int_1^b f(x)\,dx$ is the area between the graph of $y = 1/x^2$, the x axis, $x = 1$, and $x = b$.

Let us see what happens when we let $b \to \infty$; that is, when we compute the following limit:

$$\lim_{b\to\infty} \int_1^b \frac{dx}{x^2} = \lim_{b\to\infty} \left[(-x^{-1}) \Big|_1^b \right]$$

$$= \lim_{b\to\infty} \left(-\frac{1}{b} + 1 \right) = 1$$

Did you expect this result? No matter how large b is taken, the area under the graph from $x = 1$ to $x = b$ never exceeds 1, and in the limit it *is* 1 (see Fig. 3). This suggests that we write

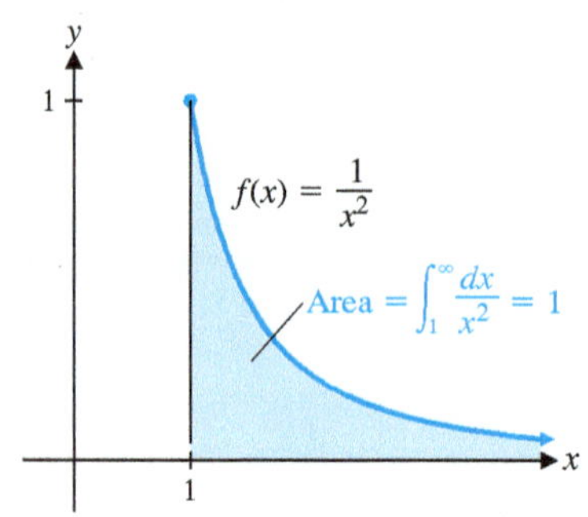

Figure 3

$$\int_1^{\infty} \frac{dx}{x^2} = \lim_{b\to\infty} \int_1^b \frac{dx}{x^2} = 1$$

This integral is an example of an *improper integral*. In general, the forms

$$\int_{-\infty}^{b} f(x)\,dx \qquad \int_a^{\infty} f(x)\,dx \qquad \int_{-\infty}^{\infty} f(x)\,dx$$

where f is continuous over the indicated interval, are called **improper integrals.** (These integrals are "improper" because the interval of integration is unbounded, as indicated by the use of ∞ for one or both limits of integration. There are other types of improper integrals involving certain types of points of discontinuity within the interval of integration, but these will not be considered here.) Each type of improper integral above is formally defined in the following box:

DEFINITION Improper Integrals

If f is continuous over the indicated interval and the limit exists, then

1. $\int_a^{\infty} f(x)\,dx = \lim_{b\to\infty} \int_a^b f(x)\,dx$

2. $\int_{-\infty}^{b} f(x)\,dx = \lim_{a\to-\infty} \int_a^b f(x)\,dx$

3. $\displaystyle\int_{-\infty}^{\infty} f(x)\,dx = \int_{-\infty}^{c} f(x)\,dx + \int_{c}^{\infty} f(x)\,dx$

where c is any point on $(-\infty, \infty)$, provided *both* improper integrals on the right exist.

If the indicated limit exists, then the improper integral is said to exist or to **converge;** if the limit does not exist, then the improper integral is said not to exist or to **diverge** (and no value is assigned to it).

EXAMPLE 1 **Evaluating an Improper Integral** Evaluate the following, if it converges:

$$\int_{2}^{\infty} \frac{dx}{x}$$

SOLUTION

$$\begin{aligned}\int_{2}^{\infty} \frac{dx}{x} &= \lim_{b\to\infty} \int_{2}^{b} \frac{dx}{x}\\ &= \lim_{b\to\infty}\left[(\ln x)\Big|_{2}^{b}\right]\\ &= \lim_{b\to\infty}(\ln b - \ln 2)\end{aligned}$$

Since $\ln b \to \infty$ as $b \to \infty$, the limit does not exist. Hence, the improper integral diverges.

Matched Problem 1 Evaluate the following, if it converges:

$$\int_{3}^{\infty} \frac{dx}{(x-1)^2}.$$

EXAMPLE 2 **Evaluating an Improper Integral** Evaluate the following, if it converges:

$$\int_{-\infty}^{2} e^x\,dx$$

SOLUTION

$$\begin{aligned}\int_{-\infty}^{2} e^x\,dx &= \lim_{a\to-\infty} \int_{a}^{2} e^x\,dx\\ &= \lim_{a\to-\infty}\left(e^x\Big|_{a}^{2}\right)\\ &= \lim_{a\to-\infty}(e^2 - e^a) = e^2 - 0 = e^2 \quad \text{The integral converges.}\end{aligned}$$

Matched Problem 2 Evaluate the following, if it converges:

$$\int_{-\infty}^{-1} x^{-2}\,dx.$$

EXAMPLE 3 **Evaluating an Improper Integral** Evaluate the following, if it converges:

$$\int_{-\infty}^{\infty} \frac{2x}{(1+x^2)^2}\,dx$$

SOLUTION

$$\int_{-\infty}^{\infty} \frac{2x}{(1+x^2)^2}\,dx = \int_{-\infty}^{0} (1+x^2)^{-2}2x\,dx + \int_{0}^{\infty} (1+x^2)^{-2}2x\,dx$$

$$= \lim_{a \to -\infty} \int_a^0 (1 + x^2)^{-2} 2x\, dx + \lim_{b \to \infty} \int_0^b (1 + x^2)^{-2} 2x\, dx$$

$$= \lim_{a \to -\infty} \left[\frac{(1 + x^2)^{-1}}{-1} \Big|_a^0 \right] + \lim_{b \to \infty} \left[\frac{(1 + x^2)^{-1}}{-1} \Big|_0^b \right]$$

$$= \lim_{a \to -\infty} \left[-1 + \frac{1}{1 + a^2} \right] + \lim_{b \to \infty} \left[-\frac{1}{1 + b^2} + 1 \right]$$

$$= -1 + 1 = 0 \quad \text{The integral converges.}$$

Matched Problem 3 Evaluate the following, if it converges:

$$\int_{-\infty}^{\infty} \frac{dx}{e^x}.$$

CONCEPTUAL INSIGHT

It is impossible, in general, solely by inspecting the graph of the function, to determine whether an improper integral converges or diverges. For example, the functions

$$f(x) = \frac{1}{x} \quad \text{and} \quad g(x) = \frac{1}{x^{1.001}}$$

have graphs that are nearly indistinguishable on $[1, \infty)$. But by evaluating the appropriate limits, it can be shown that $\int_0^\infty f(x)\, dx$ diverges and $\int_0^\infty g(x)\, dx$ converges.

Explore & Discuss 2 Let

$$f(x) = \begin{cases} 2 - x & \text{if } 0 \le x \le 1 \\ 0 & \text{otherwise} \end{cases}$$

(A) Graph $y = f(x)$ and explain why f is not continuous at $x = 0$ and at $x = 1$.

(B) Find $\int_0^1 f(x)\, dx$.

(C) Discuss possible interpretations of the improper integral $\int_{-\infty}^{\infty} f(x)\, dx$.

In probability studies, we frequently encounter functions that are 0 over one or more intervals on the real axis. An improper integral involving such a function can be simplified by deleting the intervals where the function is 0 and considering the integral over the remaining intervals. The next example illustrates this process.

EXAMPLE 4 **Evaluating an Improper Integral** Evaluate $\int_{-\infty}^{\infty} f(x)\, dx$ for

$$f(x) = \begin{cases} \dfrac{4}{(x + 2)^2} & \text{if } x \ge 0 \\ 0 & \text{otherwise} \end{cases}$$

SOLUTION The function f is discontinuous at $x = 0$ (see Fig. 4). However, since $f(x) = 0$ for $x < 0$, we can proceed as follows:

$$\int_{-\infty}^{\infty} f(x)\, dx = \int_0^{\infty} f(x)\, dx \qquad f(x) = 0 \text{ for } x < 0.$$

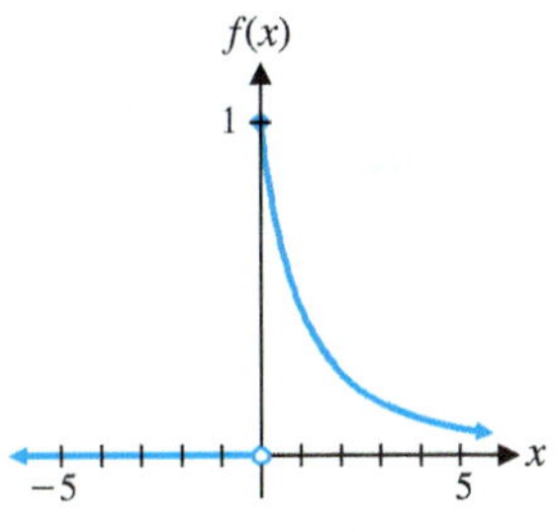

Figure 4

$$= \int_0^{\infty} \frac{4}{(x+2)^2}\,dx$$

$$= \lim_{b\to\infty} \int_0^{b} \frac{4}{(x+2)^2}\,dx$$

$$= \lim_{b\to\infty} \left[-4(x+2)^{-1} \Big|_0^b \right]$$

$$= \lim_{b\to\infty} \left(\frac{-4}{b+2} + 2 \right)$$

$$= 2$$ The integral converges.

Matched Problem 4 Evaluate $\int_{-\infty}^{\infty} f(x)\,dx$ for $f(x) = \begin{cases} 4x - x^2 & \text{if } 0 \le x \le 4 \\ 0 & \text{otherwise.} \end{cases}$

EXAMPLE 5 **Oil Production** It is estimated that an oil well will produce oil at a rate of $R(t)$ million barrels per year t years from now, as given by

$$R(t) = 8e^{-0.05t} - 8e^{-0.1t}$$

Estimate the total amount of oil that will be produced by this well.

SOLUTION The total amount of oil produced in T years of operation is $\int_0^T R(t)\,dt$ (see Fig. 5). At some point in the future, the annual production rate will become so low that it will no longer be economically feasible to operate the well. However, since we

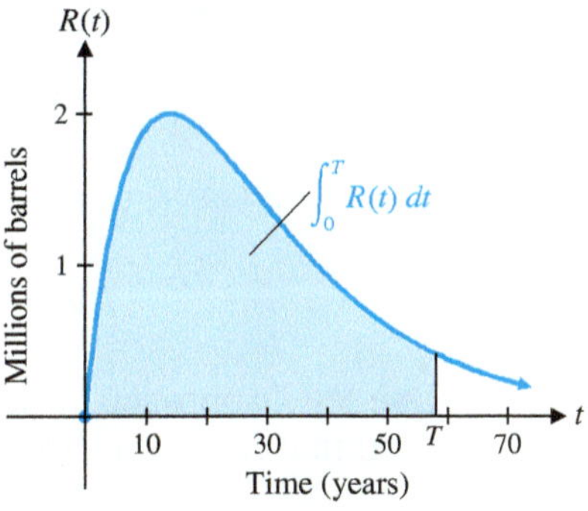

Figure 5

do not know when this will occur, it is convenient to assume that the well is operated indefinitely so we can use an improper integral. Thus, the total amount of oil produced is approximately

$$\int_0^{\infty} R(t)\,dt = \lim_{T\to\infty} \int_0^{T} R(t)\,dt$$

$$= \lim_{T\to\infty} \int_0^{T} (8e^{-0.05t} - 8e^{-0.1t})\,dt$$

$$= \lim_{T\to\infty} \left[(-160e^{-0.05t} + 80e^{-0.1t}) \Big|_0^T \right]$$

$$= \lim_{T\to\infty} (-160e^{-0.05T} + 80e^{-0.1T} + 80)$$

$$= 80 \quad \text{or} \quad 80{,}000{,}000 \text{ barrels}$$

X	Y1
50	67.405
60	72.232
70	75.241
80	77.096
90	78.232
100	78.926
110	79.347

Y1=-160e^(-.05X...

Figure 6 Total production over $[0, x]$

The total production over any finite time period $[0, T]$ is less than 80,000,000 barrels, and the larger T is, the closer the total production will be to 80,000,000 barrels. Figure 6 displays the total production over $[0, T]$ for values of $T(= X)$ from 50 to 110 years.

Matched Problem 5 The annual production rate (in millions of barrels) for an oil well is given by

$$R(t) = 4e^{-0.1t} - e^{-0.2t}$$

Assuming that the well is operated indefinitely, find the total production.

Application: Capital Value

Recall that if money is invested at a rate r compounded continuously for T years, then the present value, PV, and the future value, FV, are related by the continuous compound interest formula (Section 11-1 of *College Mathematics* 11/e)

$$FV = PVe^{rT}$$

which also can be written as

$$PV = e^{-rT}FV$$

This relationship is valid for both single deposits and amounts generated by continuous income streams. Thus, if $f(t)$ is a continuous income stream, the future value is given by (see Section 7-2 of Calculus 11/e)

$$FV = e^{rT}\int_0^T f(t)e^{-rt}\,dt \quad \text{Future value}$$

and, using the compound interest formula, the present value is given by

$$PV = e^{-rT}FV = e^{-rT}e^{rT}\int_0^T f(t)e^{-rt}\,dt$$

Since $e^{-rT}e^{rT} = 1$, we have

$$PV = \int_0^T f(t)e^{-rt}\,dt \quad \text{Present value}$$

If we let T approach ∞ in this formula for present value, we obtain an improper integral that represents the *capital value* of the income stream.

DEFINITION Capital Value of a Perpetual Income Stream

A continuous income stream is called **perpetual** if it never stops producing income. The **capital value, CV,** of a perpetual income stream $f(t)$ at a rate r compounded continuously is the present value over the time interval $[0, \infty)$. That is,

$$CV = \int_0^\infty f(t)e^{-rt}\,dt$$

Capital value provides a method for expressing the worth (in terms of today's dollars) of an investment that will produce income for an indefinite period of time.

EXAMPLE 6 **Capital Value** A family has leased the oil rights of a property to a petroleum company in return for a perpetual annual payment of $1,200. Find the capital value of this lease at 5% compounded continuously.

SOLUTION The annual payments from the oil company produce a continuous income stream with rate of flow $f(t) = 1{,}200$ that continues indefinitely. (It is common practice to treat a sequence of equal periodic payments as a continuous income stream with a constant rate of flow, even if the income is received only at the end of each period.) Thus, the capital value is

$$CV = \int_0^\infty f(t)e^{-rt}\,dt$$

$$\begin{aligned} CV &= \int_0^\infty 1{,}200e^{-0.05t}\,dt \\ &= \lim_{T\to\infty}\int_0^T 1{,}200e^{-0.05t}\,dt \\ &= \lim_{T\to\infty}\left[-24{,}000e^{-0.05t}\Big|_0^T\right] \\ &= \lim_{T\to\infty}(-24{,}000e^{-0.05T} + 24{,}000) = \$24{,}000 \end{aligned}$$

Matched Problem 6 Repeat Example 6 if the interest rate is 6% compounded continuously.

Answers to Matched Problems

1. $\frac{1}{2}$ **2.** 1 **3.** Diverges **4.** $\frac{32}{3}$ **5.** 35,000,000 barrels **6.** \$20,000

Exercises 10-O

A

In Problems 1–16, find the value of each improper integral that converges.

1. $\int_1^\infty \frac{dx}{x^3}$

2. $\int_1^\infty x\,dx$

3. $\int_1^\infty x^2\,dx$

4. $\int_1^\infty \frac{dx}{x^4}$

5. $\int_{-\infty}^0 e^{x/2}\,dx$

6. $\int_{-\infty}^{-1} \frac{dx}{x}$

7. $\int_4^\infty \frac{dx}{100\sqrt{x}}$

8. $\int_1^\infty e^{-2x}\,dx$

9. $\int_9^\infty \frac{dx}{x\sqrt{x}}$

10. $\int_8^\infty \frac{dx}{\sqrt[3]{x}}$

11. $\int_0^\infty \frac{dx}{(x+1)^{2/3}}$

12. $\int_0^\infty \frac{dx}{\sqrt{x+1}}$

13. $\int_1^\infty \frac{dx}{x^{0.99}}$

14. $\int_1^\infty \frac{dx}{x^{1.01}}$

15. $0.3\int_0^\infty e^{-0.3x}\,dx$

16. $0.01\int_0^\infty e^{-0.1x}\,dx$

In Problems 17–22, graph $y = f(x)$ and find the value of $\int_{-\infty}^{\infty} f(x)\,dx$ if it converges.

17. $f(x) = \begin{cases} 1 + x^2 & \text{if } 0 \le x \le 2 \\ 0 & \text{otherwise} \end{cases}$

18. $f(x) = \begin{cases} 2 + x & \text{if } 0 \le x \le 3 \\ 0 & \text{otherwise} \end{cases}$

19. $f(x) = \begin{cases} e^{-0.1x} & \text{if } x \ge 0 \\ 0 & \text{otherwise} \end{cases}$

20. $f(x) = \begin{cases} 2 - e^{-0.2x} & \text{if } x \ge 0 \\ 0 & \text{otherwise} \end{cases}$

21. $f(x) = \begin{cases} 4/(x+2) & \text{if } x \ge 2 \\ 0 & \text{otherwise} \end{cases}$

22. $f(x) = \begin{cases} 4/(x+1)^2 & \text{if } x \ge 1 \\ 0 & \text{otherwise} \end{cases}$

In Problems 23–26, discuss the validity of each statement. If the statement is always true, explain why. If not, give a counterexample.

23. If f is a continuous increasing function on $[0, \infty)$ such that $f(x) > 0$ for all x, then $\int_0^\infty f(x)\,dx$ diverges.

24. If f is a continuous decreasing function on $[0, \infty)$ such that $f(x) > 0$ for all x, then $\int_0^\infty f(x)\,dx$ converges.

25. If f is a continuous function on $[1, \infty)$ such that $f(x) > 0$ for all x and $f(x) \ge 10$ for $1 \le x \le 1{,}000$, then $\int_1^\infty f(x)\,dx$ diverges.

26. If f is a continuous decreasing function on $[1, \infty)$ such that $0 < f(x) \le 0.001$ for all x, then $\int_0^\infty f(x)\,dx$ converges.

In Problems 27–36, find $F(b)$, use a graphing utility to graph $F(b)$, and use the graph to estimate $\lim_{b\to\infty} F(b)$ or, in Problems 31 and 32, $\lim_{b\to-\infty} F(b)$.

27. From Problem 1, $F(b) = \int_1^b \frac{dx}{x^3}$

28. From Problem 2, $F(b) = \int_1^b x\,dx$

29. From Problem 3, $F(b) = \int_1^b x^2\,dx$

30. From Problem 4, $F(b) = \int_1^b \frac{dx}{x^4}$

31. From Problem 5, $F(b) = \int_b^0 e^{x/2}\,dx$

32. From Problem 6, $F(b) = \int_b^{-1} \frac{dx}{x}$

33. From Problem 7, $F(b) = \int_4^b \frac{dx}{100\sqrt{x}}$

34. From Problem 8, $F(b) = \int_1^b e^{-2x}\,dx$

35. From Problem 9, $F(b) = \int_9^b \frac{dx}{x\sqrt{x}}$

36. From Problem 10, $F(b) = \int_8^b \frac{dx}{\sqrt[3]{x}}$

37. If f is continuous on $[0, \infty)$ and $\int_0^\infty f(x)\,dx$ converges, does $\int_1^\infty f(x)\,dx$ converge? Explain.

38. If f is continuous on $[0, \infty)$ and $\int_0^\infty f(x)\,dx$ diverges, does $\int_1^\infty f(x)\,dx$ diverge? Explain.

B

In Problems 39–48, find the value of each improper integral that converges.

39. $\int_0^\infty \frac{1}{k}e^{-x/k}\,dx,\ k > 0$

40. $\int_1^\infty \frac{1}{x^p}\,dx,\ p > 1$

41. $\int_{-\infty}^\infty \frac{x}{1+x^2}\,dx$

42. $\int_{-\infty}^\infty xe^{-x^2}\,dx$

43. $\int_0^\infty (e^{-x} - e^{-2x})\,dx$

44. $\int_0^\infty (e^{-x} + 2e^{x})\,dx$

45. $\int_{-\infty}^0 \frac{1}{\sqrt{1-x}}\,dx$

46. $\int_{-\infty}^0 \frac{1}{\sqrt[3]{(1-x)^4}}\,dx$

47. $\int_1^\infty \frac{\ln x}{x}\,dx$

48. $\int_0^\infty \frac{e^{-x}}{1+e^{-x}}\,dx$

Applications

49. Capital value. The perpetual annual rent for a property is \$6,000. Find the capital value at 5% compounded continuously.

50. Capital value. The perpetual annual rent for a property is \$10,000. Find the capital value at 4% compounded continuously.

51. Capital value. A trust fund produces a perpetual stream of income with rate of flow

$$f(t) = 1{,}500e^{0.04t}$$

Find the capital value at 7% compounded continuously.

52. Capital value. A trust fund produces a perpetual stream of income with rate of flow

$$f(t) = 1{,}000e^{0.02t}$$

Find the capital value at 6% compounded continuously.

53. Capital value. Refer to Problem 49. Discuss the effect on the capital value if the interest rate is increased to 6%, and then if decreased to 4%. Interpret these results.

54. Capital value. Refer to Problem 50. Discuss the effect on the capital value if the interest rate is increased to 5%, and then if decreased to 2%. Interpret these results.

55. Production. The rate of production of a natural gas well (in billions of cubic feet per year) is given by (see figure below)

$$R(t) = 3e^{-0.2t} - 3e^{-0.4t}$$

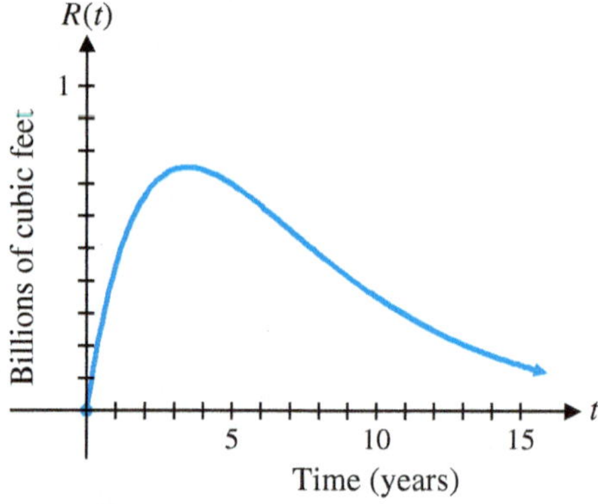

(A) Assuming that the well is operated indefinitely, find the total production.

(B) When will the output from the well reach 50% of the total production? Round answer to two decimal places.

56. Production. The rate of production of an oil well (in millions of barrels per year) is given by

$$R(t) = \frac{1{,}000t}{(50 + t^2)^2}$$

(A) Assuming that the well is operated indefinitely, find the total production.

(B) When will the output from the well reach 50% of the total production? Round answer to two decimal places.

57. **Pollution.** It has been estimated that the rate of seepage of toxic chemicals from a waste dump is $R(t)$ gallons per year t years from now, where

$$R(t) = \frac{500}{(1 + t)^2}$$

Assuming that this seepage continues indefinitely, find the total amount of toxic chemicals that seep from the dump.

58. **Drug assimilation.** When a person takes a drug, the body does not assimilate all of the drug. One way to determine the amount of the drug assimilated is to measure the rate at which the drug is eliminated from the body. If the rate of elimination of the drug (in milliliters per minute) is given by

$$R(t) = 3e^{-0.1t} - 3e^{-0.3t}$$

where t is the time in minutes since the drug was administered, how much of the drug is eliminated from the body?

59. **Immigration.** In order to control rising population, a country is planning to institute a new policy for immigration. Analysts estimate that the rate of immigration into the country t years after the policy is instituted (in millions of immigrants per year) will be given by

$$R(t) = \frac{400}{(5 + t)^3}$$

Find the total number of immigrants that will enter the country under this policy, assuming that the policy is applied indefinitely.

10-1 Finite Probability Models

- Sample Space
- Random Variable; Probability Distribution
- Expected Value of a Random Variable
- Mean and Standard Deviation

This section provides a relatively brief and informal introduction to probability. More detailed and formal treatments can be found in books and courses devoted entirely to the subject.

Sample Space

Probability studies are based on experiments that do not yield the same results each time they are performed, no matter how carefully they are repeated under the same conditions. These experiments are called **random experiments**. Familiar examples of random experiments are flipping coins, rolling dice, observing the frequency of defective items from an assembly line, or observing the frequency of deaths in a certain age group. Probability theory is a branch of mathematics that has been developed to deal with outcomes of random experiments, both real and conceptual. In the work that follows, we will simply use the word **experiment** to mean a random experiment.

If we identify a set S of outcomes of an experiment in such a way that in each trial of the experiment one and only one of the outcomes in the set will occur, then we call the set S a **sample space** for the experiment. Each element in S is called a **simple outcome**. For example, if 3 distinct coins are tossed, then a sample space for this experiment is

$$S = \{\text{HHH, HHT, HTH, HTT, THH THT, TTH, TTT}\}$$

where H is heads and T is tails.

An **event** E is defined to be any subset of the sample space (including the empty set, $\varnothing$, and the entire sample space S). An event is called a **simple event** if it contains only one element and a **compound event** if it contains more than one element. We say that **an event E occurs** if any of the simple outcomes in E occurs. Referring to the sample space S given above, the event that all 3 coins turn up heads and the event that all 3 coins turn up the same can be described, respectively, as

$$E_1 = \{\text{HHH}\} \qquad \text{and} \qquad E_2 = \{\text{HHH, TTT}\}$$

Event E_1 is a simple event,* while event E_2 is a compound event.

*Technically, there is a logical distinction between the simple outcome HHH (a single element in the sample space) and the simple event {HHH} (a subset of the sample space consisting of a single element). But we will just keep this in mind and use the terms simple *outcome* and *simple event* interchangeably.

If each of these 3 coins is fair (a head is as likely to occur as a tail), then each simple event in the sample space is **equally likely** to occur. Since there are 8 different simple events in S, the probability of any one of these simple events occurring is

$$P(e) = \tfrac{1}{8} \qquad e \in S$$

More generally, we have the following result.

THEOREM 1 Probability in Equally Likely Sample Spaces

If we assume each simple event in sample space S is as likely as any other to occur, then the probability of an arbitrary event E in S is given by

$$P(E) = \frac{\text{Number of elements in } E}{\text{Number of elements in } S} = \frac{n(E)}{n(S)}$$

Let's apply this theorem to the events E_1 and E_2 in the coin-tossing example discussed above:

$$P(E_1) = \frac{n(E_1)}{n(S)} = \frac{1}{8} \qquad E_1 = \{\text{HHH}\}$$

$$P(E_2) = \frac{n(E_2)}{n(S)} = \frac{2}{8} = \frac{1}{4} \quad E_2 = \{\text{HHH, TTT}\}$$

Random Variable Probability Distribution

When performing a random experiment, a sample space S is selected in such a way that all probability problems of interest relative to the experiment can be solved. In many situations we may not be interested in each simple event in the sample space S but in some numerical value associated with the event. For example, if 3 coins are tossed, we may be interested in the number of heads that turn up rather than in the particular pattern that turns up. Or, in selecting a random sample of students, we may be interested in the proportion that are women rather than which particular students are women. In the same way, a "craps" player is usually interested in the sum of the dots on the showing faces rather than the pattern of dots on each face.

In each of these examples, we have a rule that assigns to each simple event in S a single real number. Mathematically speaking, we are dealing with a function. Historically, this particular type of function has been called a "random variable."

Random Variable

A **random variable** is a function that assigns a numerical value to each simple event in a sample space S.

The term *random variable* is an unfortunate choice, since it is neither random nor a variable—it is a function with a numerical value and it is defined on a sample space. But the terminology has stuck and is now standard, so we shall have to live with it. Capital letters, such as X, are used to represent random variables.

Let us return to the experiment of tossing 3 coins. A sample space S of equally likely simple events is indicated in Table 1. Suppose we are interested in the number of heads (0, 1, 2, or 3) appearing on each toss of the 3 coins and the probability of each of these events. We introduce a random variable X (a function) that indicates the number of heads for each simple event in S (see the second column in Table 1). For example, $X(E_1) = 0$, $X(e_2) = 1$, and so on. The random variable X assigns a numerical value to each simple event in the sample space S.

Table 1 Number of Heads in the Toss of 3 Coins

Sample Space S	Number of Heads $X(e_i)$
e_1: TTT	0
e_2: TTH	1
e_3: THT	1
e_4: HTT	1
e_5: THH	2
e_6: HTH	2
e_7: HHT	2
e_8: HHH	3

We are interested in the probability of the occurrence of each image value of X; that is, in the probability of the occurrence of 0 heads, 1 head, 2 heads, or 3 heads in the single toss of 3 coins. We indicate this probability by

$$P(x) \qquad \text{where} \qquad x \in \{0, 1, 2, 3\}$$

Table 2 Probability Distribution

NUMBER OF HEADS x	0	1	2	3
PROBABILITY $P(x)$	$\frac{1}{8}$	$\frac{3}{8}$	$\frac{3}{8}$	$\frac{1}{8}$

The function P is called the **probability function* of the random variable X.**

What is $P(2)$, the probability of getting exactly 2 heads on the single toss of 3 coins? "Exactly 2 heads occur" is the event

$$E = \{THH, HTH, HHT\}$$

Thus,

$$P(2) = \frac{n(E)}{n(S)} = \frac{3}{8}$$

Proceeding similarly for $P(0)$, $P(1)$, and $P(3)$, we obtain the results in Table 2. This table is called a **probability distribution for the random variable X.** Probability distributions are also represented graphically, as shown in Figure 1. The graph of a probability distribution is often called a **histogram.**

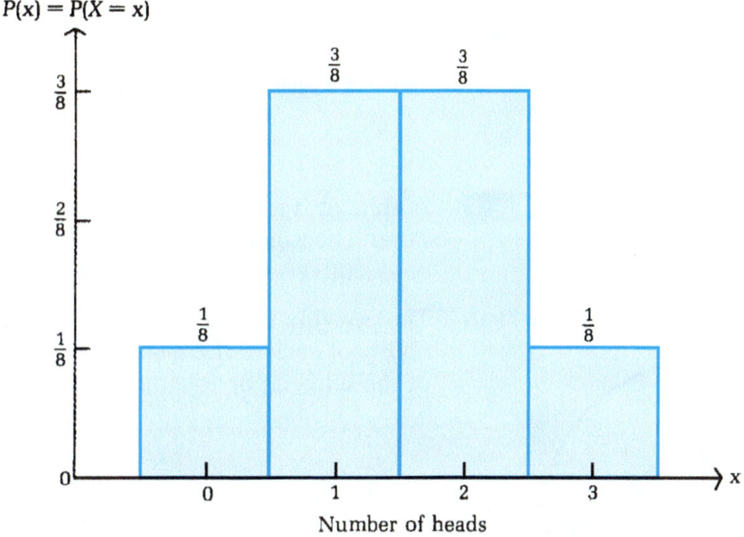

Figure 1 Histogram for a probability distribution

Note from Table 2 or Figure 1 that

1. $0 \leq P(x) \leq 1, x \in \{0, 1, 2, 3\}$
2. $P(0) + P(1) + P(2) + P(3) = \frac{1}{8} + \frac{3}{8} + \frac{3}{8} + \frac{1}{8} = 1$

These are general properties that any probability distribution of a random variable X associated with a finite sample space must have.

Probability Distribution of a Random Variable X

A probability function $P(X = x) = P(x)$ is a **probability distribution of the random variable X if**

1. $0 \leq P(x) \leq 1, x \in \{x_1, x_2, \ldots, x_n\}$
2. $P(x_1) + P(x_2) + \cdots + P(x_n) = 1$

where $\{x_1, x_2, \ldots, x_n\}$ are the (range) values of X (see Figure 2).

Figure 2 illustrates the process of forming a probability distribution of a random variable.

*Formally, the probability function P of the random variable X is defined by $P(x) = P(\{e_i \in S | X(e_i) = x\})$ which, because of its cumbersome nature, is usually simplified to $P(X = x)$ or, simply, $P(x)$. We will use the simplified notation.

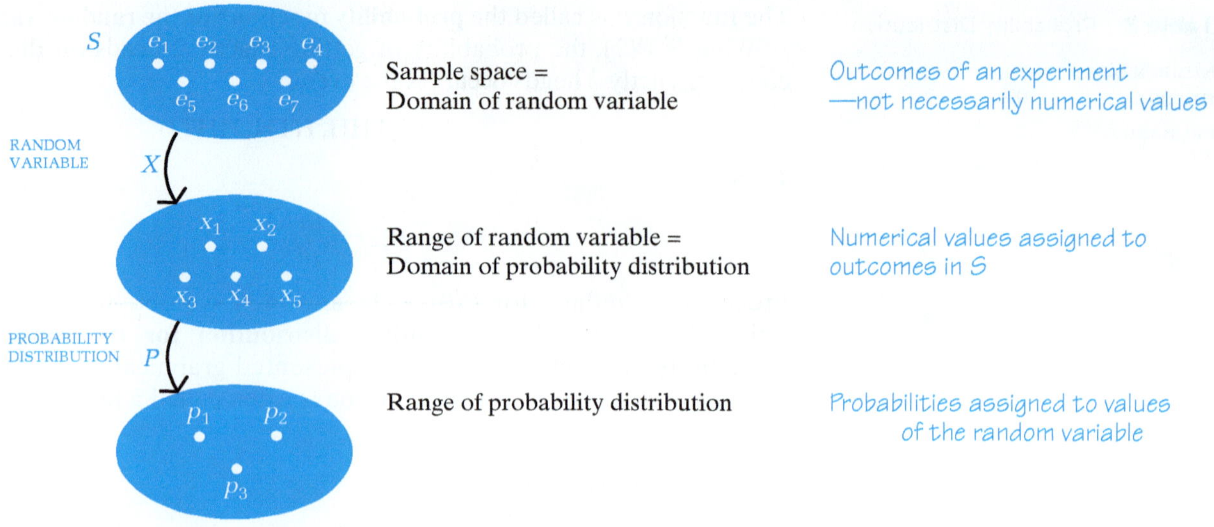

Figure 2 Probability distribution of a random variable for a finite sample space

EXAMPLE 1 A spinner is numbered from 1 to 4, and each number is as likely to turn up as any other. The random variable X represents the sum of the numbers obtained in two consecutive spins of the pointer. Find and graph the probability distribution of X.

SOLUTION The possible values of X are 2, 3, 4, 5, 6, 7, and 8. In order to determine the probability of each of these sums occurring, we consider the sample space S consisting of the sixteen possible outcomes listed in the following table:

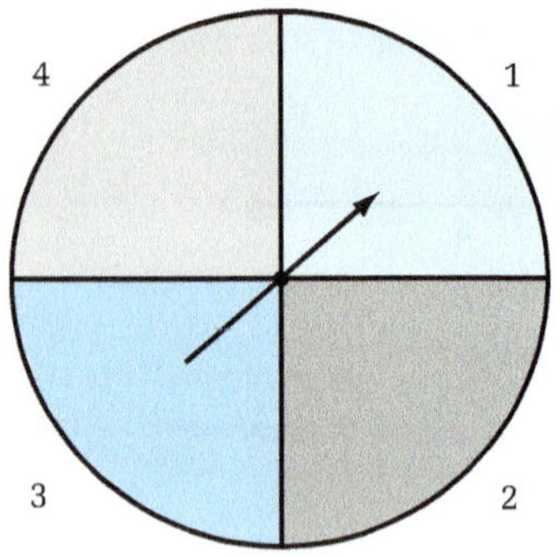

		Second Spin			
		1	2	3	4
	1	(1, 1)	(1, 2)	(1, 3)	(1, 4)
FIRST	2	(2, 1)	(2, 2)	(2, 3)	(2, 4)
SPIN	3	(3, 1)	(3, 2)	(3, 3)	(3, 4)
	4	(4, 1)	(4, 2)	(4, 3)	(4, 4)

In this sample space, the simple outcome (3, 4) is to be distinguished from the simple outcome (4, 3) even though both produce the same sum. The former indicates that the pointer stopped at 3 on the first spin and at 4 on the second spin, while the latter indicates that the pointer stopped at 4 on the first spin and at 3 on the second spin. Since each of the four possible outcomes of a single spin are equally likely, pairing all possible outcomes from the first spin with all possible outcomes from the second spin produces an equally likely sample space of combined outcomes. Thus, we can use Theorem 1 to compute the probability distribution of the random variable X, which assumes the values 2, 3, 4, 5, 6, 7, and 8:

$$P(2) = P(X = 2) = P(\{(1, 1)\}) = \tfrac{1}{16}$$
$$P(3) = P(X = 3) = P(\{(1, 2), (2, 1)\}) = \tfrac{2}{16}$$
$$P(4) = P(X = 4) = P(\{(1, 3), (2, 2), (3, 1)\}) = \tfrac{3}{16}$$

and so on. The probability distribution for X is

x	2	3	4	5	6	7	8
$P(x)$	$\frac{1}{16}$	$\frac{2}{16}$	$\frac{3}{16}$	$\frac{4}{16}$	$\frac{3}{16}$	$\frac{2}{16}$	$\frac{1}{16}$

And the graph is

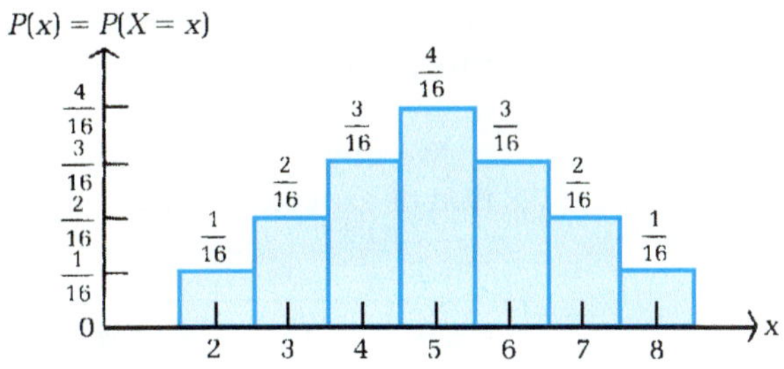

Matched Problem 1 A spinner is marked from 1 to 3, and each number is as likely to come up as any other. The random variable X represents the sum of the numbers obtained in two consecutive spins of the pointer. Find and graph the probability distribution of X.

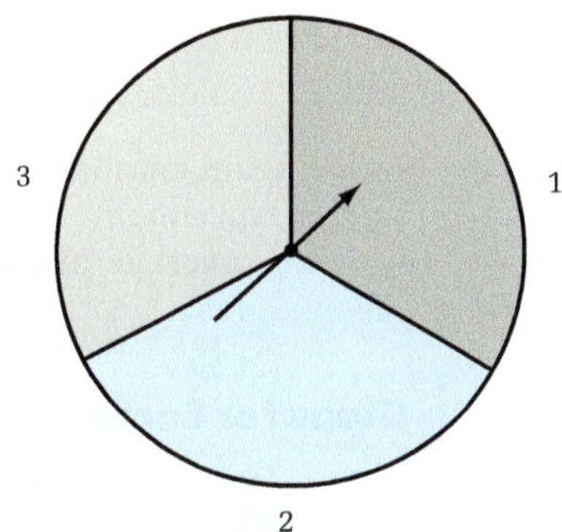

Expected Value of a Random Variable

If the experiment of tossing 3 coins were repeated 1,000 times, then based on the probability distribution in Table 2, we would expect to obtain the following results:

0 heads approximately	$\frac{1}{8}(1{,}000) = 125$ times
1 head approximately	$\frac{3}{8}(1{,}000) = 375$ times
2 heads approximately	$\frac{3}{8}(1{,}000) = 375$ times
3 heads approximately	$\frac{1}{8}(1{,}000) = 125$ times

Of course, it is not likely that we would obtain these exact values, but the actual results should be close to those predicted by the probabilities. Now suppose we were interested in the average number of heads per toss (the total number of heads in all tosses divided by the total number of tosses). Using the values listed above, we would expect the average number of heads per toss of the 3 coins, or the *expected value* $E(X)$, to be given by

$$E(X) = \frac{0 \cdot 125 + 1 \cdot 375 + 2 \cdot 375 + 3 \cdot 125}{1{,}000} \qquad \frac{\text{Total number of heads}}{\text{Total number of tosses}}$$

$$= 0\left(\frac{125}{1{,}000}\right) + 1\left(\frac{375}{1{,}000}\right) + 2\left(\frac{375}{1{,}000}\right) + 3\left(\frac{125}{1{,}000}\right)$$

$$= 0\left(\frac{1}{8}\right) + 1\left(\frac{3}{8}\right) + 2\left(\frac{3}{8}\right) + 3\left(\frac{1}{8}\right) + \frac{12}{8} = 1.5$$

It is important to note that the expected value is not a value that will necessarily occur in a single experiment (1.5 heads cannot occur in the toss of 3 coins), but it is an average of what occurs over a large number of experiments. Sometimes, we will toss more than 1.5 heads and sometimes less, but if the experiment were repeated many times, the average number of heads per experiment would approach 1.5.

We now make the above discussion more precise through the following definition of expected value:

Expected Value of a Random Variable X

Given the probability distribution for the random variable X,

x_i	x_1	x_2	$\cdots$	x_m
p_i	p_1	p_2	$\cdots$	p_m

where $p_i = P(x_i)$, we define the **expected value of X**, denoted **$E(X)$**, by the formula

$$E(X) = x_1p_1 + x_2p_2 + \cdots + x_mp_m$$

We again emphasize that the expected value is not to be expected to occur in a single experiment; it is a long-run average of repeated experiments—it is the weighted average of the possible outcomes, each weighted by its probability.

Steps for Computing the Expected Value of a Random Variable X

Step 1 Form the probability distribution of the random variable X.

Step 2 Multiply each image value of X, x_i, by its corresponding probability of occurrence p_i; then add the results.

EXAMPLE 2 What is the expected value (long-run average) of the number of dots facing up for the roll of a single die?

SOLUTION If we choose

$$S = \{1, 2, 3, 4, 5, 6\}$$

as our sample space, then each simple event is a numerical outcome reflecting our interest, and each is equally likely. The random variable X in this case is just the identity function (each number is associated with itself). Thus, the probability distribution for X is

x_i	1	2	3	4	5	6
p_i	$\frac{1}{6}$	$\frac{1}{6}$	$\frac{1}{6}$	$\frac{1}{6}$	$\frac{1}{6}$	$\frac{1}{6}$

Hence,

$$E(X) = 1(\tfrac{1}{6}) + 2(\tfrac{1}{6}) + 3(\tfrac{1}{6}) + 4(\tfrac{1}{6}) + 5(\tfrac{1}{6}) + 6(\tfrac{1}{6})$$
$$= \tfrac{21}{6} = 3.5$$

Matched Problem 2 Suppose the die in Example 2 is not fair and we obtain (empirically) the following probability distribution for X:

x_i	1	2	3	4	5	6
p_i	.14	.13	.18	.20	.11	.24

[*Note*: Sum = 1.]

What is the expected value of X?

EXAMPLE 3 A spinner device is numbered from 0 to 5, and each of the 6 numbers is as likely to come up as any other. A player who bets \$1 on any given number wins \$4 (and gets the bet back) if the pointer comes to rest on the chosen number; otherwise, the \$1 bet is lost. What is the expected value of the game (long-run average gain or loss per game)?

SOLUTION The sample space of equally likely events is

$$S = \{0, 1, 2, 3, 4, 5\}$$

Each simple outcome occurs with a probability of $\frac{1}{6}$. The random variable X assigns \$4 to the winning number and $-\$1$ to each of the remaining numbers. Thus, the probability distribution for X, called a **payoff table**, is as shown below:

Payoff Table (Probability Distribution for X)		
x_i	\$4	$-\$1$
p_i	$\frac{1}{6}$	$\frac{5}{6}$

The probability of winning \$4 is $\frac{1}{6}$ and of losing \$1 is $\frac{5}{6}$. We can now compute the expected value of the game:

$$E(X) = \$4(\tfrac{1}{6}) + (-\$1)(\tfrac{5}{6}) = -\$\tfrac{1}{6} \approx - ;\$0.1667 \approx -17¢ \text{ per game}$$

Thus, in the long run the player will lose an average of about 17¢ per game.

In general, a game is said to be **fair** if $E(X) = 0$. The game in Example 3 is not fair—the "house" has an advantage on the average, of about 17¢ per game.

Matched Problem 3 Repeat Example 3 with the player winning \$5 instead of \$4 if the chosen number turns up. The loss is still \$1 if any other number turns up. Is this now a fair game?

EXAMPLE 4 Suppose you are interested in insuring a car stereo system for \$500 against theft. An insurance company charges a premium of \$60 for coverage for 1 year, claiming an empirically determined probability of .1 that the stereo will be stolen some time during the year. What is your expected return from the insurance company if you take out this insurance?

SOLUTION This is actually a game of chance in which your stake is \$60. You have a .1 chance of receiving \$440 from the insurance company (\$500 minus your stake of \$60) and a .9 chance of losing your stake of \$60. What is the expected value of this "game"? We form a payoff table (the probability distribution for X) as shown below:

Payoff Table		
x_i	\$440	$-\$60$
p_i	.1	.9

Then we compute the expected value as follows:

$$E(X) = (\$440)(.1) + (-\$60)(.9) = -\$10$$

This means that if you insure with this company over many years and the circumstances remain the same, you would have an average net loss to the insurance company of \$10 per year.

Matched Problem 4 Find the expected value in Example 4 from the insurance company's point of view.

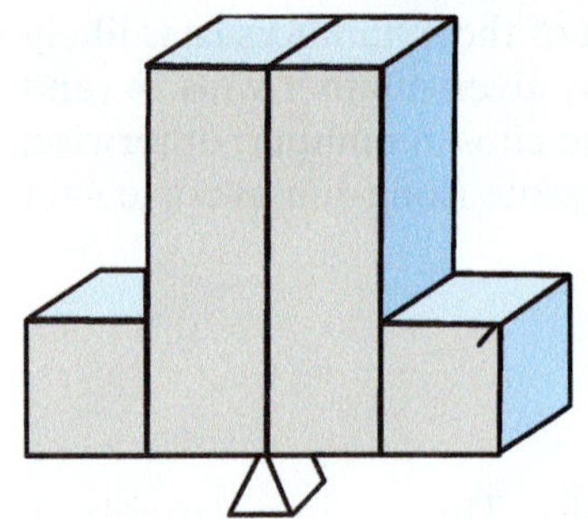

Figure 3 The balance point on the histogram is y = 1.5

Mean and Standard Deviation

Since the expected value of a random variable represents the long-run average of repeated experiments, it is often referred to as the **arithmetic average**, or **mean**. Traditionally, the Greek letter μ is used to denote the mean. Thus,

$$\mu = E(X) = x_1p_1 + x_2p_2 + \cdots + x_mp_m$$

is the mean of the random variable X. Geometrically, the mean is the center of the values of X and is often referred to as a *measure of central tendency*. For example, if the histogram in Figure 1 were drawn on a piece of wood of uniform thickness and the wood cut around the outside of the figure, then the resulting object would balance on a wedge placed at the mean $\mu = 1.5$ (see Figure 3).

Another numerical quantity that is used to describe the properties of a random variable is the *standard deviation*. This quantity gives a *measure of the dispersion*, or *spread*, of the random variable X about the mean μ.

Standard Deviation of a Random Variable X

Given the probability distribution for the random variable X,

x_i	x_1	x_2	$\cdots$	x_m
p_i	p_1	p_2	$\cdots$	p_m

and the mean

$$\mu = x_1p_1 + x_2p_2 + \cdots + x_mp_m$$

we define the **variance of *X***, denoted by ***V*(*X*)**, by the formula

$$V(X) = (x_1 - \mu)^2p_1 + (x_2 - \mu)^2 p_2 + \cdots + (x_m - \mu)^2 p_m$$

and the **standard deviation of *X***, denoted by $\boldsymbol{\sigma}$ (Greek letter sigma), by the formula

$$\sigma = \sqrt{V(X)}$$

In other words, the variance is the expected value of the squares of the distances from each value of X to the mean. Standard deviation is defined using the square root so that it will be expressed in the same units as the values of X.

Returning to the coin-tossing experiment, we have already shown that $\mu = E(X) = 1.5$. Thus,

$$\begin{aligned} V(X) &= (0 - 1.5)^2(\tfrac{1}{8}) + (1 - 1.5)^2(\tfrac{3}{8}) + (2 - 1.5)^2(\tfrac{3}{8}) + (3 - 1.5)^2(\tfrac{1}{8}) \\ &= .75 \end{aligned}$$

and

$$\sigma = \sqrt{V(X)} = \sqrt{.75} \approx .866$$

EXAMPLE 5 Find the variance and standard deviation for the random variable in example 2.

SOLUTION

$$\mu = E(X) = 3.5$$

$$\begin{aligned} V(X) &= (1 - 3.5)^2(\tfrac{1}{6}) + (2 - 3.5)^2(\tfrac{1}{6}) + (3 - 3.5)^2(\tfrac{1}{6}) \\ &\quad + (4 - 3.5)^2(\tfrac{1}{6}) + (5 - 3.5)^2(\tfrac{1}{6}) + (6 - 3.5)^2(\tfrac{1}{6}) \\ &= \tfrac{35}{12} \end{aligned}$$

$$\sigma = \sqrt{V(X)} = \sqrt{\tfrac{35}{12}} \approx 1.708$$

Matched Problem 5 Find the variance and standard deviation for the random variable in problem 2.

The standard deviation is often used to compare different probability distributions. Figure 4 shows four different probability distributions and their graphs. Notice the relationship between the standard deviation σ and the dispersion of the probability distribution about the mean. The tighter the cluster of the probability distribution about the mean, the smaller the standard deviation.

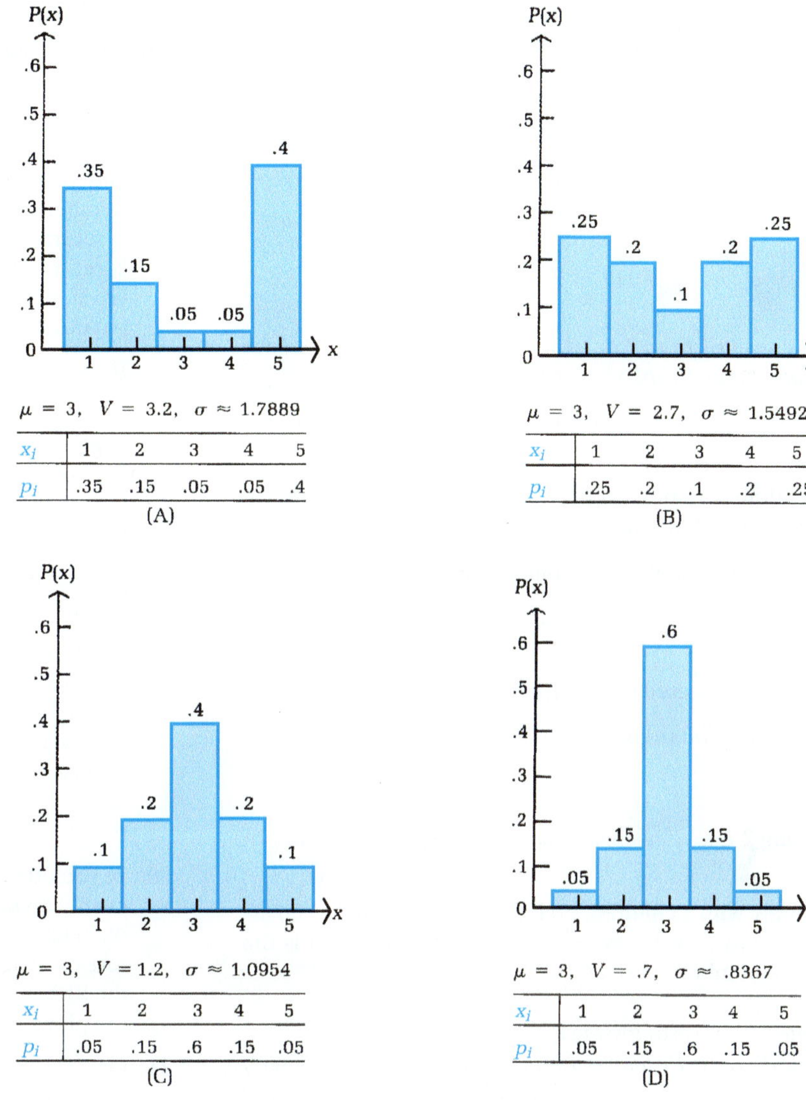

Figure 4

Answers to Matched Problems

1.

x	2	3	4	5	6
$P(x)$	$\frac{1}{9}$	$\frac{2}{9}$	$\frac{3}{9}$	$\frac{2}{9}$	$\frac{1}{9}$

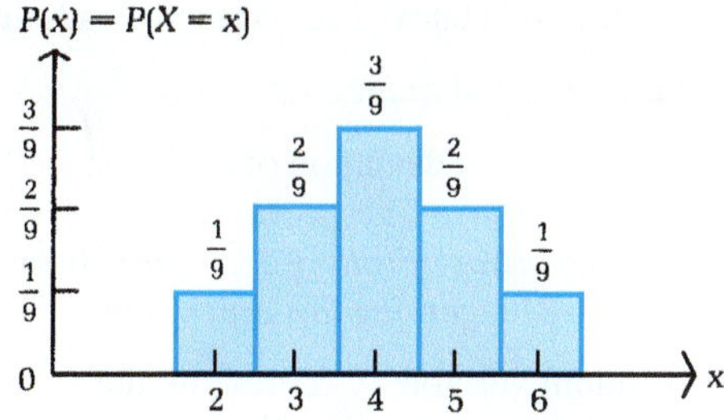

2. $E(X) = 3.73$ 3. $E(X) = \$0$; the game is fair
4. $E(X) = (-\$440)(.1) + (\$60)(.9) = \$10$; this amount, of course, is necessary to cover expenses and profit
5. $V(X) = 2.9571$; $\sigma \approx 1.720$

Exercises 10-1

A

In Problems 1–4, graph the probability distribution, and find the mean and standard deviation of the random variable X.

1.

x_i	−2	−1	0	1	2
p_i	.1	.2	.4	.2	.1

2.

x_i	−2	−1	0	1	2
p_i	.1	.1	.2	.2	.4

3.

x_i	−2	−1	0	1	2
p_i	.5	.2	.1	.1	.1

4.

x_i	−2	−1	0	1	2
p_i	.3	.1	.1	.1	.4

A spinner is marked from 1 to 10 and each number is as likely to turn up as any other. Problems 5–10 refer to this experiment.

5. Find a sample space S consisting of equally likely simple events.
6. The random variable X represents the number that turns up on the spinner. Find the probability distribution for X.
7. What is the probability of obtaining an even number?
8. What is the probability of obtaining a number that is exactly divisible by 3?
9. What is the expected value of X?
10. What is the standard deviation of X?

B

11. In tossing 2 fair coins once, what is the expected number of heads?
12. In a family with 2 children, excluding multiple births and assuming a boy is as likely as a girl at each birth, what is the expected number of boys?

An experiment consists of tossing a coin 4 times in succession. Answer the questions in Problems 13–18 regarding this experiment.

13. The random variable X represents the number of heads that occur in 4 tosses. Find and graph the probability distribution of X.
14. What is the probability of getting 2 or more heads?
15. What is the probability of getting an even number of heads?
16. What is the probability of getting more heads than tails?
17. Find the expected number of heads.
18. Find the standard deviation of X.

An experiment consists of rolling 2 fair dice. Answer the questions in Problems 19–24 regarding this experiment.

19. The random variable X represents the sum of the dots on the 2 up faces of the dice. Find and graph the probability distribution of X.
20. What is the probability that the sum is 7 or 11?
21. What is the probability that the sum is less than 6?
22. What is the probability that the sum is an even number?
23. Find the expected value of the sum of the dots.
24. Find the standard deviation of X.

C

25. After you pay $4 to play a game, a single fair die is rolled and you are paid back the number of dollars equal to the number of dots facing up. For example, if 5 dots turn up, $5 is returned to you for a net gain of $1. IF 1 dot turns up, $1 is returned to you for a net gain, or payoff, of −$3; and so on. If X is the random variable that represents net gain, or payoff, what is the expected value of X?
26. Repeat Problem 25 with the same game costing $3.50 for each play.
27. A player tosses 2 coins and wins $3 if 2 heads appear and $1 if 1 head appears, but loses $6 if 2 tails appear. If X is the random variable representing the player's net gain, what is the expected value of X?
28. Repeat Problem 27 if the player wins $5 if 2 heads appear and $2 if 1 head appears, but loses $7 if 2 tails appear.
29. Roulette wheels in the United States generally have 38 equally spaced slots numbered 00, 0, 1, 2, . . . , 36. A player who bets $1 on any given number wins $35 (and gets the bet back) if the ball comes to rest on the chosen number; otherwise, the $1 bet is lost. If X is the random variable that represents the player's net gain, what is the expected value of X?
30. In roulette (see Problem 29) the numbers from 1 to 36 are evenly divided between red and black. A player who bets $1 on black, wins $1 (and gets the bet back) if the ball comes to rest on black; otherwise (if the ball lands on red, 0 or 00), the $1 bet is lost. If X is the random variable that represents the player's net gain, what is the expected value of X?
31. One thousand raffle tickets are sold at $10 each. A ticket will be drawn at random, and the ticket holder will be paid $5,000. Suppose you buy 5 tickets.
    (A) Create a payoff table for the random variable X representing your net gain in this raffle.
    (B) What is the expected value of X?
32. Repeat Problem 31 with the purchase of 10 tickets.

Applications

Business & Economics

33. **Insurance.** The annual premium for a $5,000 insurance policy against the theft of a painting is $150. If the (empirical) probability that the painting will be stolen during the year is .01, what is your expected return from the insurance company if you take out this insurance?

34. Insurance. Repeat Problem 33 from the point of view of the insurance company.

35. Decision analysis. An oil company, after careful testing and analysis, is considering drilling in two different sites. It is estimated that site A will net \$30 million if successful (probability .2) and lose \$3 million if not (probability .8); site B will net \$70 million if successful (probability .1) and lose \$4 million if not (probability .9). Which site should the company choose according to the expected return from each site?

36. Decision analysis. Repeat Problem 35, assuming additional analysis caused the estimated probability of success in field B to be changed from .1 to .11.

Life Sciences

37. Genetics. Suppose that, at each birth, having a girl is not as likely as having a boy. The probability assignments for the number of boys in a 3-child family are approximated empirically from past records and are given in the table below. What is the expected number of boys in a 3-child family?

Number of Boys	
x_i	p_1
0	.12
1	.36
2	.38
3	.14

38. Genetics. A pink-flowering plant is of genotype RW. If two such plants are crossed, we obtain a red plant (RR) with probability .25, a pink plant (RW or WR) with probability .50, and a white plant (WW) with probability .25, as shown in the table at the top of the next page. What is the expected number of W genes present in a crossing of this type?

Number of W Genes Present	
x_i	p_i
0	.25
1	.50
2	.25

Social Sciences

39. Politics. A money drive is organized by a campaign committee for a candidate running for public office. Two approaches are considered:

A_1: A general mailing with a followup mailing

A_2: Door-to-door solicitation with followup telephone calls

From campaign records of previous committees, average donations and their corresponding probabilities are estimated to be:

A_1		A_2	
x_i **(return per person)**	p_i	x_i **(return per person)**	p_i
\$10	.3	\$15	.3
5	.2	3	.1
0	.5		.6
	1.0		1.0

What are the expected returns? Which course of action should be taken according to the expected returns?

10-2 Binomial Distributions

- Bernoulli Trials
- Binomial Formula (Brief Review)
- Binomial Distribution
- Application

Bernoulli Trials

If we toss a coin, either a head occurs or it does not. If we roll a die, either a 3 shows or it fails to show. If you are vaccinated for smallpox, either you contract smallpox or you do not. What do all these situations have in common? All can be classified as experiments with two possible outcomes, each the complement of the other. An experiment for which there are only two possible outcomes, E or E', is called a **Bernoulli experiment**, or **trial**, after Jacob Bernoulli (1654–1705), a Swiss scientist and mathematician who was one of the first people to study systematically the probability problems related to a two-outcome experiment.

In a Bernoulli experiment or trial, it is customary to refer to one of the two outcomes as a **success** S and to the other as a **failure** F. If we designate the probability of success by

$$P(S) = p$$

then the probability of failure will be

$$P(F) = 1 - p = q \qquad \textit{Note: } p + q = 1$$

EXAMPLE 6 We roll a fair die and ask for the probability of a 6 turning up. This can be viewed as a Bernoulli trial by identifying a success with a 6 turning up and a failure with any of the other numbers turning up. Thus,

$$p = \tfrac{1}{6} \quad \text{and} \quad q = 1 - \tfrac{1}{6} = \tfrac{5}{6}$$

Matched Problem 6 Identify p and q for a single roll of a fair die where a success is a number divisible by 3 turning up.

Now, suppose a Bernoulli experiment is repeated 5 times. How can we compute the probability of the outcome $SSFFS$? In order to answer this question, we must make two basic assumptions about the trials in a sequence of Bernoulli experiments. First, we will assume that the probability of a success remains the same from trial to trial. Second, we will assume that the trials are **independent**; that is, the outcome of one trial has no effect on the outcome of any of the other trials. With these two assumptions, it can be shown that *the probability of a sequence of events is equal to the product of the probability of each event in the sequence.* Thus,

$$\begin{aligned} P(SSFFS) &= P(S)P(S)P(F)P(F)P(S) \\ &= ppqqp \\ &= p^3q^2 \end{aligned}$$

In general, we define a sequence of Bernoulli trials as follows:

Bernoulli Trials

A sequence of experiments is called a **sequence of Bernoulli trials** if:

1. Only two outcomes are possible on each trial.
2. The probability of success p for each trial is a constant (probability of failure is then $q = 1 - p$).
3. All trials are independent.

In simple Bernoulli experiments, such as tossing a coin or rolling a die, it seems very reasonable to assume that the trials are independent. In more complicated situations, it can be very difficult to determine whether the trials are actually independent. We will assume that all the Bernoulli experiments we consider in this book have independent trials.

EXAMPLE 7 If we roll a fair die 5 times and identify a success in a single roll with a 1 turning up, what is the probability of the sequence $SFFSS$ occurring?

SOLUTION

$$p = \tfrac{1}{6} \qquad q = 1 - p = \tfrac{5}{6}$$

$$\begin{aligned} P(SFFSS) &= pqqpp = p^3q^2 \\ &= \left(\tfrac{1}{6}\right)^3\left(\tfrac{5}{6}\right)^2 \approx .003 \end{aligned}$$

Matched Problem 7 In Example 7, find the probability of the outcome $FSSSF$.

In most applications involving sequences of Bernoulli trials, we will be interested in the number of successes, rather than in a specific outcome of the form $SSFFS$. If X is the random variable associated with the number of successes in a sequence of Bernoulli trials, we would like to find the probability distribution for X. Since this probability distribution is closely related to the *binomial formula*, it will be helpful if we first review this important formula.

Binomial Formula (Brief Review)

To start, let us calculate directly the first five natural number powers of $(a + b)^x$:

$$(a + b)^1 = a + b$$
$$(a + b)^2 = a^2 + 2ab + b^2$$
$$(a + b)^3 = a^3 + 3a^2b + 3ab^2 + b^3$$
$$(a + b)^4 = a^4 + 4a^3b + 6a^2b^2 + 4ab^3 + b^4$$
$$(a + b)^5 = a^5 + 5a^4b + 10a^3b^2 + 10a^2b^3 + 5ab^4 + b^5$$

In general, it can be shown that a binomial expansion is given by the well-known **binomial formula**:

Binomial Formula

For n a natural number,

$$(a + b)^n = C_{n,0}a^n + C_{n,1}a^{n-1}b + C_{n,2}a^{n-2}b^2 + \cdots + C_{n,n}b^n$$

where

$$C_{n,r} = \frac{n!}{r!(n-r)!} \qquad n \geqslant r \geqslant 0$$

EXAMPLE 8 Use the binomial formula to expand $(q + p)^3$.

SOLUTION

$$(q + p)^3 = C_{3,0}q^3 + C_{3,1}q^2p + C_{3,2}qp^2 + C_{3,3}p^3$$
$$= q^3 + 3q^2p + 3qp^2 + p^3$$

Matched Problem 8 Use the binomial formula to expand $(q + p)^4$.

EXAMPLE 9 Use the binomial formula to find the fifth term in the expansion of $(q + p)^6$.

SOLUTION The fifth term is given by

$$C_{6,4}q^2p^4 = \frac{6!}{4!(6-4)!}q^2p^4 = 15\,q^2p^4$$

Matched Problem 9 Use the binomial formula to find the third term in the expansion of $(q + p)^7$.

Binomial Distribution

We now generalize the discussion of Bernoulli trials to *binomial distributions*. We start by considering a sequence of three Bernoulli trials. Let the random variable X_3 represent the number of successes in three trials, 0, 1, 2, or 3. We are interested in the probability distribution for this random variable.

Which outcomes of an experiment consisting of a sequence of three Bernoulli trials lead to the random variable values 0, 1, 2, and 3, and what are the probabilities associated with these values? Table 3 answers these questions completely.

Table 3

Simple Event	Probability Simple Event	X_3 x Successes in 3 Trials	$P(X_3 = x)$
FFF	$qqq = q^3$	0	q^3
FFS	$qqp = q^2p$		
FSF	$qpq = q^2p$	1	$3q^2p$
SFF	$pqq = q^2p$		
FSS	$qpp = qp^2$		
SFS	$pqp = qp^2$	2	$3qp^2$
SSF	$ppq = qp^2$		
SSS	$ppp = p^3$	3	p^3

The terms in the last column of Table 3 are the terms in the binomial expansion of $(q + p)^3$, as we saw in Example 8. The last two columns in Table 3 provide a probability distribution for the random variable X_3. Note that both conditions for a probability distribution (see Section 10-1) are met:

1. $0 \le P(X_3 = x) \le 1, \quad x \in \{0, 1, 2, 3\}$

2. $1 = 1^3 = (q + p)^3$ Recall that $q + p = 1$

$$= C_{3,0}q^3 + C_{3,1}q^2p + C_{3,2}qp^2 + C_{3,3}P^3$$
$$= q^3 + 3q^2p + 3pq^2 + p^3$$
$$= P(X_3 = 0) + P(X_3 = 1) + P(X_3 = 2) + P(X_3 = 3)$$

Reasoning in the same way for the general case, we see why the probability distribution of a random variable associated with the number of successes in a sequence of n Bernoulli trials is called a *binomial distribution*—the probability of each number is a term in the binomial expansion of $(q + p)^n$. For this reason, a sequence of Bernoulli trials is often referred to as a *binomial experiment*. In terms of a formula, we have Theorem 2:

THEOREM 2 Binomial distribution

$$P(X_n = x) = P(x \text{ success in } n \text{ trials})$$
$$= C_{n,x}p^xq^{n-x} \qquad x \in \{0, 1, 2, \dots, n)$$

where p is the probability of success and q is the probability of failure on each trial.

Informally, we will write $P(x)$ in place of $P(X_n = x)$.

EXAMPLE 10 If a fair coin is tossed 4 times, what is the probability of tossing:

(A) Exactly 2 heads? (B) At least 2 heads?

SOLUTIONS (A) Use Theorem 2 with $n = 4, x = 1, p = \frac{1}{2}$, and $q = \frac{1}{2}$;

$$P(2) = P(X_4 = 2)$$
$$= C_{4,2}\left(\frac{1}{2}\right)^2\left(\frac{1}{2}\right)^2$$
$$= \frac{4!}{2!2!}\left(\frac{1}{2}\right)^4 = .375$$

(B) Notice how this problem differs from part A. Here we have

$$P(X_4 \geq 2) = P(2) + P(3) + P(4)$$

$$= C_{4,2}\left(\frac{1}{2}\right)^2\left(\frac{1}{2}\right)^2 + C_{4,3}\left(\frac{1}{2}\right)^3\left(\frac{1}{2}\right)^1 + C_{4,4}\left(\frac{1}{2}\right)^4\left(\frac{1}{2}\right)^0$$

$$= \frac{4!}{2!2!}\left(\frac{1}{2}\right)^4 + \frac{4!}{3!1!}\left(\frac{1}{2}\right)^4 + \frac{4!}{4!0!}\left(\frac{1}{2}\right)^4$$

$$= .375 + .25 + .0625 = .6875$$

Matched Problem 10 If a fair coin is tossed 4 times, what is the probability of tossing:

(A) Exactly 1 head? (B) At most 1 head?

EXAMPLE 11 Suppose a fair die is rolled 3 times and a success on a single roll is considered to be rolling a number divisible by 3.

(A) Write the probability function for the binomial distribution.
(B) Construct a table for this binomial distribution.
(C) Draw a histogram for this binomial distribution.

SOLUTIONS (A) $p = \frac{1}{3}$ *Since two numbers out of six are divisible by 3*

$q = 1 - p = \frac{2}{3}$

$n = 3$

Hence,

$$P(x) = P(x \text{ successes in 3 trials}) = C_{3,x}(\tfrac{1}{3})^x(\tfrac{2}{3})^{3-x}$$

(B)

x	$P(x)$
0	$C_{3,0}(\frac{1}{3})^0(\frac{2}{3})^3 \approx .30$
1	$C_{3,3}(\frac{1}{3})^1(\frac{2}{3})^2 \approx .44$
2	$C_{3,2}(\frac{1}{3})^2(\frac{2}{3})^1 \approx .22$
3	$C_{3,3}(\frac{1}{3})^3(\frac{2}{3})^0 \approx .04$
	1.00

(C)

If we actually performed the binomial experiment described in Example 11 a large number of times with a fair die, we would find that we would roll no number divisible by 3 in 3 rolls of a die about 30% of the time, one number divisible by 3 in 3 rolls about 44% of the time, two numbers divisible by 3 in 3 rolls about 22% of the time, and three numbers divisible by 3 in 3 rolls only 4% of the time. Note that the sum of all the probabilities is 1, as it should be.

Matched Problem 11 Repeat Example 11 where the binomial experiment consists of 2 rolls of a die instead of 3 rolls.

We close our discussion of binomial distributions by stating (without proof) formulas for the mean and standard deviation of the random variable associated with the distribution.

Mean and Standard Deviation (Random Variable in a Binomial Distribution)

MEAN	$\mu = np$
STANDARD DEVIATION	$\sigma = \sqrt{npq}$

EXAMPLE 12 Compute the mean and standard deviation for the random variable in Example 11.

SOLUTION $n = 3 \qquad p = \frac{1}{3} \qquad q = 1 - \frac{1}{3} = \frac{2}{3}$

$$\mu = np = 3(\tfrac{1}{3}) = 1 \qquad \sigma = \sqrt{npq} = \sqrt{3(\tfrac{1}{3})(\tfrac{2}{3})} \approx .82$$

Matched Problem 12 Compute the mean and standard deviation for the random variable in Problem 11 above.

Application

Binomial experiments are associated with a wide variety of practical problems: industrial sampling, drug testing, genetics, epidemics, medical diagnosis, opinion polls, analysis of social phenomena, qualifying tests, and so on. Several types of applications are included in Exercise 10-2. We will now consider one application in detail.

EXAMPLE 13 **Medicine** The probability of recovering after a particular type of operation is .5. Let us investigate the binomial distribution involving 8 patients undergoing this operation.

(A) Write the function defining this distribution.

(B) Construct a table for the distribution.

(C) Construct a histogram for the distribution.

(D) Find the mean and standard deviation for the distribution.

SOLUTIONS (A) Letting a recovery be a success, we have

$$p = .5 \qquad q = 1 - p = .5 \qquad n = 8$$

Hence

$$P(x) = P(\text{Exactly } x \text{ successes in 8 trials}) = C_{8,x}(.5)^x(.5)^{8-x} = C_{8,x}(.5)^8$$

(B)

x	$P(x)$
0	$C_{8,0}(.5)^8 \approx .004$
1	$C_{8,1}(.5)^8 \approx .031$
2	$C_{8,2}(.5)^8 \approx .109$
3	$C_{8,3}(.5)^8 \approx .219$
4	$C_{8,4}(.5)^8 \approx .273$
5	$C_{8,5}(.5)^8 \approx .219$
6	$C_{8,6}(.5)^8 \approx .109$
7	$C_{8,7}(.5)^8 \approx .031$
8	$C_{8,8}(.5)^8 \approx .004$
	$.999 \approx 1$

(C)

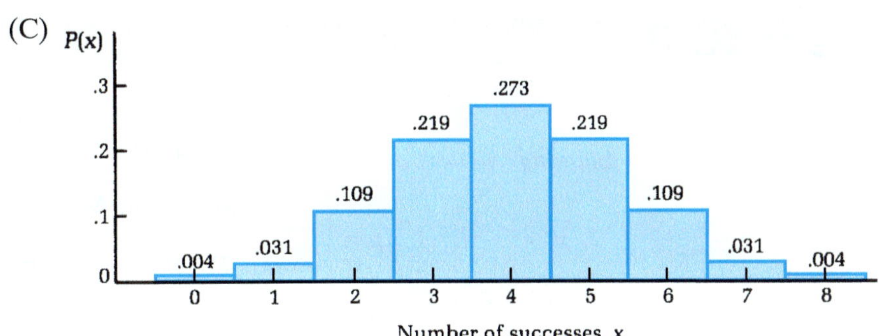

(D) $\mu = np = 8(.5) = 4 \qquad \sigma = \sqrt{npq} = \sqrt{8(.5)(.5)} \approx 1.41$

Matched Problem 13 Repeat Example 13 for 4 patients.

Answers to Matched Problems

6. $p = \frac{1}{3}, q = \frac{2}{3}$ **7.** $p^3q^2 = (\frac{1}{6})^3(\frac{5}{6})^2 \approx .003$

8. $C_{4,0}q^4 + C_{4,1}q^3p + C_{4,2}q^2p^2 + C_{4,3}qp^3 + C_{4,4}p^4 = q^4 + 4q^3p + 6q^2p^2 + 4qp^3 + p^4$

9. $C_{7,2}q^5p^2 = 21q^5p^2$

10. (A) $P(1) = C_{4,1}(\frac{1}{2})^1(\frac{1}{2})^3 = .25$ (B) $P(X_4 \leq 1) = P(0) + P(1)$

$= C_{4,0}(\frac{1}{2})^0(\frac{1}{2})^4 + C_{4,1}(\frac{1}{2})^1(\frac{1}{2})^3$

$= .3125$

11. (A) $P(x) = P(x \text{ successes in 2 trials}) = C_{2,x}(\frac{1}{3})^x(\frac{2}{3})^{2-x}, x \in \{0, 1, 2\}$

(B)

x	$P(x)$
0	$\frac{4}{9} \approx .44$
1	$\frac{4}{9} \approx .44$
2	$\frac{1}{9} \approx .11$

(C)

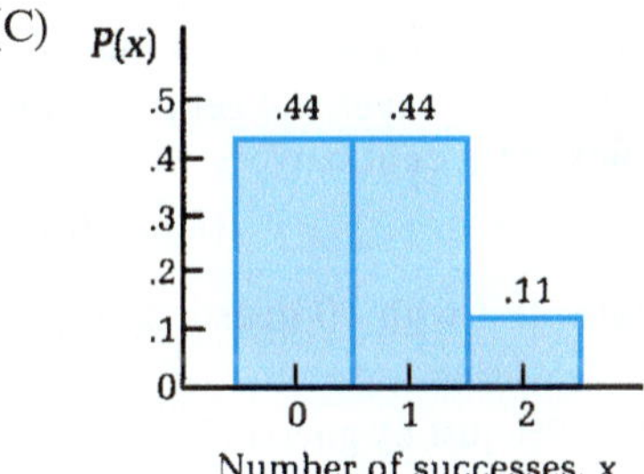

12. $\mu \approx .67; \sigma \approx .67$

13. (A) $P(x) = P(\text{Exactly } x \text{ successes in 4 trials}) = C_{4,x}(.5)^4$

(B)

x	$P(x)$
0	.06
1	.25
2	.38
3	.25
4	.06
	1.00

(C)

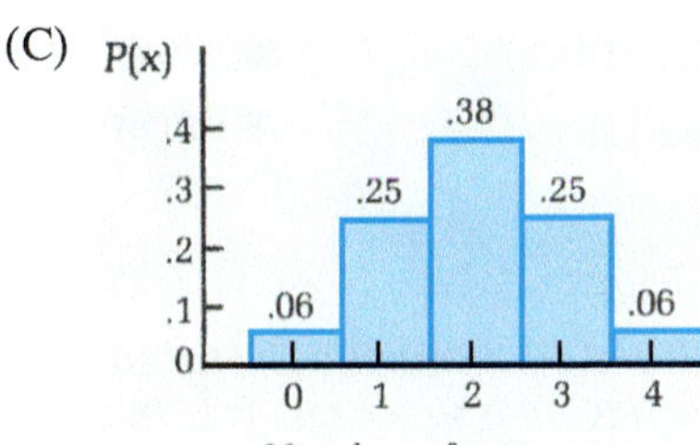

(D) $\mu = 2; \sigma = 1$

Exercises 10-2

A

Evaluate $C_{n,x}p^xq^{n-2}x$ for the following values of n, x, and p:

1. $n = 3, \quad x = 2, \quad p = \frac{1}{2}$
2. $n = 3, \quad x = 1, \quad p = \frac{1}{2}$
3. $n = 3, \quad x = 0, \quad p = \frac{1}{2}$
4. $n = 3, \quad x = 3, \quad p = \frac{1}{2}$
5. $n = 5, \quad x = 3, \quad p = .4$
6. $n = 5, \quad x = 0, \quad p = .4$

A fair coin is tossed 3 times. What is the probability of obtaining:

7. Exactly 2 heads?
8. Exactly 1 head?
9. 0 heads?
10. 3 heads?
11. At least 2 heads?
12. At least 1 head?

Construct a histogram for each of the binomial distributions in Problems 13–16. Compute the mean and standard deviation for each distribution.

13. $P(x) = C_{2,x}(.3)^x(.7)^{2-x}$
14. $P(x) = C_{2,x}(.7)^x(.3)^{2-x}$
15. $P(x) = C_{4,x}(.5)^x(.5)^{4-x}$
16. $P(x) = C_{6,x}(.5)^x(.5)^{6-x}$

B

A fair die is rolled 4 times. What is the probability of rolling:

17. Exactly three 2's?
18. Exactly two 3's?
19. No 1's?
20. All 4's?
21. At least one 6?
22. At least one 4?
23. If a baseball player has a batting average of .350, what is the probability that the player will get the following number of hits in the next 4 times at bat?
    (A) Exactly 2 hits
    (B) At least 2 hits
24. If a true–false test with 10 questions is given, what is the probability of scoring:
    (A) Exactly 70% just by guessing?
    (B) 70% or better just by guessing?

Construct a histogram for each of the binomial distributions in Problems 25–28. Compute the mean and standard deviation for each distribution.

25. $P(x) = C_{6,x}(.4)^x(.6)^{6-x}$
26. $P(x) = C_{6,x}(.6)^x(.4)^{6-x}$
27. $P(x) = C_{8,x}(.3)^x(.7)^{8-x}$
28. $P(x) = C_{8,x}(.7)^x(.3)^{8-x}$

C

In Problems 29 and 30, a coin is weighted so that the probability of a head occurring on a single toss is $\frac{3}{4}$. In 5 tosses of the coin, what is the probability of getting:

29. All heads or all tails?
30. Exactly 2 heads or exactly 2 tails?

Applications

Business & Economics

31. **Management training.** Each year a company selects a number of employees for a management training program given by a nearby university. On the average, 70% of those sent complete the program. Out of 7 people sent by the company, what is the probability that:
    (A) Exactly 5 complete the program?
    (B) 5 or more complete the program?
32. **Employee turnover.** If the probability of a new employee in a fast-food chain still being with the company at the end of 1 year is .6, what is the probability that out of 8 newly hired people:
    (A) 5 will still be with the company after 1 year?
    (B) 5 or more will still be with the company after 1 year?

33. **Quality control.** A manufacturing process produces, on the average, 6 defective items out of 100. To control quality, each day a sample of 10 completed items is selected at random and inspected. If the sample produces more than 2 defective items, then the whole day's output is inspected and the manufacturing process is reviewed. What is the probability of this happening, assuming that the process is still producing 6% defective items?
34. **Guarantees.** A manufacturing process produces, on the average, 3% defective items. The company ships 10 items in each box and wishes to guarantee no more than 1 defective item per box. If this guarantee accompanies each box, what is the probability that the box will fail to satisfy the guarantee?
35. **Quality control.** A manufacturing process produces, on the average, 5 defective items out of 100. To control quality, each day a random sample of 6 completed items is selected and inspected. If a success on a single trial (inspection of 1 item) is finding the item defective, then the inspection of each of the 6 items in the sample constitutes a binomial experiment, which has a binomial distribution.

(A) Write the function defining the distribution.

(B) Construct a table for the distribution.

(C) Draw a histogram.

(D) Compute the mean and standard deviation.

36. **Management training.** Each year a company selects 5 employees for a management training program given at a nearby university. On the average, 40% of those sent complete the course in the top 10% of their class. If we consider an employee finishing in the top 10% of the class a success in a binomial experiment, then for the 5 employees entering the program there exists a binomial distribution involving $P(x$ successes out of $5)$.

(A) Write the function defining the distribution.

(B) Construct a table for the distribution.

(C) Draw a histogram.

(D) Compute the mean and standard deviation.

Life Sciences

37. **Medical diagnosis.** A person with tuberculosis is given a chest x-ray. Four tuberculosis x-ray specialists examine each x-ray independently. If each specialist can detect tuberculosis 80% of the time when it is present, what is the probability that at least 1 of the specialists will detect tuberculosis in this person?

38. **Harmful side effects of drugs.** A pharmaceutical laboratory claims that a drug it produces causes serious side effects in 20 people out of 1,000, on the average. To check this claim, a hospital administers the drug to 10 randomly chosen patients and finds that 3 suffer from serious side effects. If the laboratory's claims are correct, what is the probability of the hospital obtaining these results?

39. **Genetics.** The probability that brown-eyed parents, both with the recessive gene for blue, will have a child with brown eyes is .75. If such parents have 5 children, what is the probability that they will have:

(A) All blue-eyed children?

(B) Exactly 3 children with brown eyes?

(C) At least 3 children with brown eyes?

40. **Gene mutations.** The probability of gene mutation under a given level of radiation is 3×10^{-5}. What is the probability of the occurrence of at least 1 gene mutation if 10^5 genes are exposed to this level of radiation?

41. **Epidemics.** If the probability of a person contracting influenza on exposure is .6, consider the binomial distribution for a family of 6 that has been exposed.

(A) Write the function defining the distribution.

(B) Construct a table for the distribution.

(C) Draw a histogram.

(D) Compute the mean and standard deviation.

42. **Side effects of drugs.** The probability that a given drug will produce a serious side effect in a person using the drug is .02. In the binomial distribution for 450 people using the drug, what are the mean and standard deviation?

Social Sciences

43. **Testing.** A multiple-choice test is given with 5 choices for each of 10 questions. What is the probability of passing the test with a grade of 70% or better just by guessing?

44. **Opinion polls.** An opinion poll based on a small sample can be unrepresentative of the population. To see why, let us assume that 40% of the electorate favours a certain candidate. If a random sample of 7 is asked their preference, what is the probability that a majority will favor the candidate?

45. **Testing.** A multiple-choice test is given with 5 choices for each of 5 questions. Answering each of the 5 questions by guessing constitutes a binomial experiment with an associated binomial distribution.

(A) Write the function defining the distribution.

(B) Construct a table for the distribution.

(C) Draw a histogram.

(D) Compute the mean and standard deviation.

46. **Sociology.** The probability that a marriage will end in divorce within 10 years is .4. What are the mean and standard deviation for the binomial distribution involving 1,000 marriages?

47. **Sociology.** If the probability is .60 that a marriage will end in divorce within 20 years after its start, what is the probability that out of 6 couples just married, in the next 20 years:

(A) None will be divorced?

(B) All will be divorced?

(C) Exactly 2 will be divorced?

(D) At least 2 will be divorced?

10-3 Poisson Distributions

- Poisson Approximation to the Binomial Distribution
- Poisson Distributions in General

Among the most important probability distributions encountered throughout probability studies are the binomial distribution, the *Poisson distribution*, and the *normal distribution*. The Poisson distribution will be investigated in this section and the normal distribution in Section 10-7. The Poisson distribution was first introduced by the French mathematician Siméon Denis Poisson (1781–1840) in a book on applications of probability theory to lawsuits and criminal trials published in 1837.

Poisson Approximation to the Binomial Distribution

Many important binomial applications involve very large values for n and very small values for p. As n increases and p decreases, computations of binomial values become increasingly tedious. The Poisson distribution provides an excellent approximation to the binomial distribution for large n and small p. To appreciate this point, consider the following example. If the probability of being dealt a royal flush (10, jack, queen, king and ace in one suit) in a 5-card poker hand is $\frac{1}{649,740}$, what is the probability of being dealt 2 royal flushes in 649,740 hands of 5 cards? Since each deal is a Bernoulli trial, we have a binomial experiment, and the problem can be solved using the binomial distribution:

$$\begin{aligned} P(2 \text{ royal flushes in } 649{,}740 \text{ deals}) &= C_{649,740,2}\left(\frac{1}{649{,}740}\right)^2\left(\frac{649{,}739}{649{,}740}\right)^{649{,}738} \\ &\approx .1839^* \end{aligned}$$

Compare this solution with the following Poisson approximation:

$$\begin{aligned} P(2 \text{ royal flushes in } 649{,}740 \text{ deals}) &\approx e^{-1}\left(\frac{1^2}{2!}\right) \\ &\approx .1839 \end{aligned}$$

This excellent approximation is not a coincidence. To understand why the approximation is so good, we start with the binomial distribution and let n increase without bound as we hold np constant at $np = \lambda$ (the Greek letter lambda is customarily used for this constant). To facilitate the process, we convert the formula for the binomial distribution into a more convenient form:

$$\begin{aligned} P(x_n = x) = C_{n,x}p^x q^{n-x} &= \frac{n!}{x!(n-x)!}p^x(1-p)^{n-x} \\ &= \frac{n!}{x!(n-x)!}\left(\frac{\lambda}{n}\right)^x\left(1-\frac{\lambda}{n}\right)^{n-x} \qquad \text{Since } p = \lambda/n \\ &= \frac{n(n-1)\cdot\cdots\cdot(n-x+1)}{n^x}\left(\frac{\lambda^x}{x!}\right)\frac{\left(1-\frac{\lambda}{n}\right)^n}{\left(1-\frac{\lambda}{n}\right)^x} \qquad (1) \end{aligned}$$

To find the limit of this last expression as $n \to \infty$, we proceed by finding the limit of each key part (remembering that x and λ are constant):

1. $$\begin{aligned} \lim_{n\to\infty}\frac{n(n-1)\cdot\cdots\cdot(n-x+1)}{n^x} &= \lim_{n\to\infty}\left(\frac{n}{n}\cdot\frac{n-1}{n}\cdot\cdots\cdot\frac{n-x+1}{n}\right) \\ &= \lim_{n\to\infty}\left[\frac{n}{n}\cdot\left(1-\frac{1}{n}\right)\cdot\cdots\cdot\left(1-\frac{x-1}{n}\right)\right] \\ &= 1 \end{aligned}$$

2. $$\lim_{n\to\infty}\left(1-\frac{\lambda}{n}\right)^x = 1^x = 1$$

---

*If your calculator cannot compute $C_{649,740,2}$ using the $C_{n,x}$ key because of the large volume of n, then use the formula

$$C_{649,740,2} = \frac{649{,}740!}{2!649{,}738!} = \frac{649{,}740 \cdot 649{,}739}{2}$$

3. $$\lim_{n\to\infty}\left(1-\frac{\lambda}{n}\right)^n = \lim_{n\to\infty}\left[\left(1-\frac{\lambda}{n}\right)^{n/\lambda}\right]^\lambda$$ Let $s = -\lambda/n$.

$$= \lim_{s\to 0}[(1+s)^{-1/s}]^\lambda$$ $n \to y$ implies $s = -\lambda/n \to 0$.

$$= \lim_{s\to 0}\left[\frac{1}{(1+s)^{1/s}}\right]^\lambda$$

$$= \left[\frac{1}{e}\right]^\lambda = e^{-\lambda}$$

Using limits 1–3 and expression (1), it follows that if $np = \lambda$, λ a positive constant, then

$$\lim_{n\to\infty} C_{n,x}p^x q^{n-x} = \frac{\lambda^x}{x!}e^{-\lambda} \qquad x = 0, 1, 2, \ldots$$

From the property of limits, we thus have the classic Poisson approximation to the binomial distribution:

Poisson Approximation to the Binomial Distribution

For n large and p small, and $\lambda = np$,

$$C_{n,x}p^x q^{n-x} \approx \frac{\lambda^x}{x!}e^{-\lambda} \qquad x \in \{0, 1, \ldots, n\}$$

RULE-OF-THUMB

Excellent approximations can be obtained when $n \geqslant 100$ and $np \leqslant 10$.

Table 4 and Figure 5 illustrate the Poisson approximation for the binomial distribution when $n = 100$ and $p = .02$. For the Poisson approximation we use $\lambda = np = 100(.02) = 2$.

Table 4 Poisson Approximation of the Binomial

x	Binomial $C_{100,x}(.02)^x(.98)^{100-x}$	Poisson $(2^x/x!)e^{-2}$
0	.1326	.1353
1	.2707	.2707
2	.2734	.2707
3	.1823	.1804
4	.0902	.0902
5	.0353	.0361
6	.0114	.0120
7	.0031	.0034
8	.0007	.0009
9	.0002	.0002
.	.	.
.	.	.
.	.	.
100		

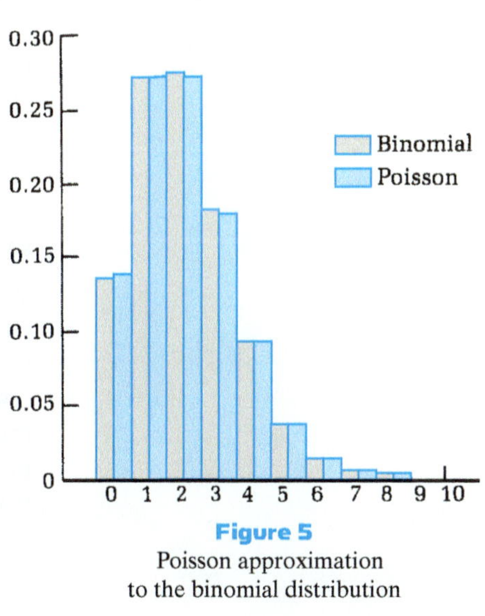

Figure 5
Poisson approximation to the binomial distribution

EXAMPLE 14 **Quality Control** A company manufactures inexpensive disposable flashlights. The management accepts a manufacturing process that yields, on the average, 2 defective flashlights per 100 produced. In an order for 200 flashlights, what is the probability that:

(A) The order will contain exactly 6 defective flashlights?
(B) The order will contain 6 or more defective flashlights?

SOLUTIONS This problem can be solved using a binomial distribution with $p = .02$ and $n = 200$. But since n is large (≥ 100) and $np = 4 \leq 10$, we can use the Poisson approximation and the calculations will be less tedious:

$$\frac{\lambda^x}{x!}e^{-\lambda} \qquad \text{with } \lambda = np = 4$$

(A) $P(\text{Exactly 6 defective flashlights}) = P(6) \approx \frac{4^6}{6!}e^{-4} \approx .1042$

[**Note:** The direct use of the binomial formula produces $P(6)\ C_{200,6}(.02)^6(.98)^{194} \approx .1047$.]

(B) $P(\text{6 or more defective flashlights})$

$$\begin{aligned}
&= P(6 \leq X \leq 200) \\
&= 1 - P(0 \leq X \leq 6) \qquad \text{Since } P(0) + P(1) + \cdots + P(200 = 1) \\
&= 1 - [P(0) + P(1) + P(2) + P(3) + P(4) + P(5)] \\
&\approx 1 - \left[\frac{4^0}{0!}e^{-4} + \frac{4^1}{1!}e^{-4} + \cdots + \frac{4^5}{5!}e^{-4}\right] \qquad \text{Poisson approximation} \\
&= 1 - e^{-4}\left[\frac{4^0}{0!} + \frac{4^1}{1!} + \frac{4^2}{2!} + \frac{4^3}{3!} + \frac{4^4}{4!} + \frac{4^5}{5!}\right] \\
&\approx 1 - .7851 = .2149
\end{aligned}$$

Matched Problem 14 A company manufactures inexpensive solar credit-card size calculators with company logos for business promotions. The management has accepted a manufacturing process that, on the average, produces 1 defective calculator per 100 produced. For an order of 300 calculators, what is the probability that:

(A) The order will contain no defective calculators?

(B) The order will contain at least 5 defective calculators?

(Compute answers to four decimal places.)

Poisson Distributions in General

We now turn to the Poisson distribution as a probability distribution in its own right. In fact, the Poisson distribution can be applied to many problems that have no direct connection to the binomial distribution. A **Poisson experiment** usually consists of counting the number of times a random event occurs in a given unit of measure such as time, distance, area, volume, weight, and so on. The possible values of the Poisson random variable are 0, 1, 2,. . . . Some applications that often involve Poisson distributions are the following:

Business & Economics

The number of items taken from inventory per hour

The number of telephone calls arriving at a switchboard per minute

The number of death claims arriving at an insurance company per day

The number of industrial accidents occurring at a company per month

The number of typesetting errors per page

The number of flaws per 1,000 feet of video tape

Life Sciences

The number of parts per million of a toxic substance found in 1 milliliter of water

The number of births or deaths per day in a city

The number of yeast cells suspended in a small drop of liquid

The number of diseased plants per acre

Social Sciences

The number of suicides per month from the Golden Gate Bridge

The number of accidents occurring on a certain stretch of highway per week

The number of automobiles arriving at a toll booth per hour

The number of unscheduled arrivals at a hospital emergency room per day

Three conditions (listed below) must be satisfied for ***X* to be a Poisson random variable**. (To simplify our statements of these conditions, we restrict these statements to time. Similar statements may be made for distance, area, volume, and other units of measure.)

1. The number of occurrences of an event over one unit of time is independent of the number of occurrences in any other nonoverlapping unit of time.
2. The expected or average number of occurrences of an event over any time interval is proportional to the length of the time interval. (For example, if on the average 5 calls arrive at a switchboard per minute, then 10 calls will arrive on the average over a 2 minute period.)
3. Two events cannot occur at exactly the same time.

A Poisson random variable X has the following probability distribution:

Poisson Distribution with Mean and Standard Deviation

$$P(X = x) = P(x \text{ occurrences per unit of measure})$$

$$= \frac{\lambda^x}{x!}e^{-\lambda} \qquad x \in \{0, 1, 2, \ldots\}$$

MEAN $\mu = \lambda$

STANDARD DEVIATION $\sigma = \sqrt{\lambda}$

Informally, we will write $P(x)$ in place of $P(X = x)$.

Most Poisson distributions of practical importance are fairly tightly clustered about the mean. In fact, most of the distribution is often found within 2–3 units of the mean (see Figure 5, where the mean is 2 and the standard deviation is $\sqrt{2}$).

Binomial and Poisson random variables are **discrete random variables.** A random variable is said to be discrete if its set of possible values (range) contains a finite number of elements or infinitely many elements that can be arranged into a sequence of the form $e_1, e_2, e_3, \ldots$ A binomial random variable has a finite discrete range, and the range for a Poisson random variable is discrete and infinite. In Sections 10-4–10-7, we will study *continuous* random variables. Probabilities of events involving discrete random variables are obtained by additions, whereas integrations are required for continuous random variables.

To show that the Poisson distribution is actually a probability distribution, we must show that

1. $0 \le P(X = x) \le 1, \quad x = 0, 1, 2, \ldots$
2. $P(0) + P(1) + P(2) + \ldots = 1$

Since λ^x, $x!$, and $e^{-\lambda}$ are all nonnegative, and $x!$ is never 0, we see that

$$P(X = x) = \frac{\lambda^x}{x!} e^{-\lambda} \geq 0$$

which establishes the left-hand part of the inequality in 1. (We will establish the right-hand part of the inequality in 1 after establishing 2.)

To establish 2, we use the following property of the exponential function:

$$\begin{aligned} e^{\lambda} &= \lim_{n \to \infty} \left(\frac{\lambda^0}{0!} + \frac{\lambda^1}{1!} + \frac{\lambda^2}{2!} + \cdots + \frac{\lambda^n}{n!} \right) \\ &= \frac{\lambda^0}{0!} + \frac{\lambda^1}{1!} + \frac{\lambda^2}{2!} + \cdots \end{aligned}$$

Using this fact, we have

$$\begin{aligned} P(0) + P(1) + P(2) + \cdots &= \frac{\lambda}{0!}e^{-\lambda} + \frac{\lambda^1}{1!}e^{-\lambda} + \frac{\lambda^2}{2!}e^{-\lambda} + \cdots \\ &= e^{-\lambda}\left(\frac{\lambda^0}{0!} + \frac{\lambda^1}{1!} + \frac{\lambda^2}{2!} + \cdots \right) = e^{-\lambda}e^{\lambda} = 1 \end{aligned}$$

To complete the task at hand, we need to establish the right-hand part of the inequality in 1. Since (from 2)

$$P(0) + P(1) + P(2) + \cdots = 1$$

and each term is nonnegative, it follows that $P(X = x) \leq 1$ for each x. Thus, the Poisson distribution is a probability distribution.

We now turn to some examples illustrating the use of the Poisson distribution in interesting applications.

EXAMPLE 15 **Arrival Rates** During a business day the switchboard in a large legal firm receives 180 calls per hour, on the average. Assuming that the calls are coming in randomly during the business day, compute the probability that during a particular minute:

(A) No calls will come in. (B) 10 calls will come in.

SOLUTIONS The three conditions of a Poisson random variable are met. Since, on the average, 180 calls arrive per hour (and the calls are coming in randomly), then the average number of calls per minute is $\frac{180}{60} = 3 = \lambda$. We use the Poisson distribution with $\lambda = 3$:

$$P(x) = \frac{\lambda^x}{x!} e^{-\lambda} = \frac{3^x}{x!} e^{-3} \qquad x = 0, 1, 2, \ldots$$

(A) $P(0) = \frac{3^0}{0!}e^{-3} = e^{-3} \approx .0498$

(B) $P(10) = \frac{3^{10}}{10!}e^{-3} \approx .0008$

Matched Problem 15 In Example 15, compute the probability of the switchboard receiving between 5 and 7 calls, inclusive, in a particular period of 2 minutes.

EXAMPLE 16 **Quality Control** Flaws occur randomly in the manufacturing of certain brand of blank audio cassette tapes, at the average rate of 2 flaws per 1,125 feet of tape produced (60 minutes of tape played at 3.75 inches per second). This flaw rate is acceptable to management. Random samples of 1,125 feet of tape are periodically inspected, and the management has agreed to shut down the manufacturing process for servicing if more than 4 flaws are found in a sample. Assuming the manufacturing process is operating normally:

(A) What is the probability that the plant will be shut down after an inspection?
(B) What is the probability that if you purchase a 60 minute tape, it will have no flaws?
(C) Compute the mean and standard deviation for the distribution.

SOLUTIONS The three conditions for a Poisson random variable are satisfied, so we use the Poisson distribution with $\lambda = 2$ (the average number of flaws per 1,125 feet of tape produced):

$$P(x) = \frac{\lambda^x}{x!}e^{-\lambda} = \frac{2^x}{x!}e^{-2} \qquad x - 0, 1, 2, \ldots$$

(A) $P(X > 4) = 1 - P(0 \leq X \leq 4)$ Since $P(0) + P(1) + \cdots + = 1$

$$= 1 - [P(0) + P(1) + P(2) + P(3) + P(4)]$$

$$= 1 - e^{-2}\left(\frac{2^0}{0!} + \frac{2^1}{1!} + \frac{2^2}{2!} + \frac{2^3}{3!} + \frac{2^4}{4!}\right)$$

$$\approx 1 - .9473 = .0527$$

Thus, if the manufacturing process is operating normally, then the probability that the plant will be (unnecessarily) shut down for service is .0527. [*Note:* The management's policy of shutting down the manufacturing process for service when more than 4 flaws occur in a sample means that they have chosen to reject the assumption that the manufacturing process is functioning normally when more than 4 flaws occur in a sample. Under this policy, they will make the wrong decision about 5% of the time when the manufacturing process is operating normally.]

(B) $P(0) = \frac{2^0}{0!}e^{-2} = e^{-2} \approx .1353$

(C) Mean: $\mu = \lambda = 2$ Standard deviation: $\sigma = \sqrt{\lambda} = \sqrt{2} \approx 1.4142$

Matched Problem 16 In Example 16, find the probability of a 60 minute tape chosen at random containing no more than 4 flaws.

Answers to Matched Problems

14. (A) $P(0) \approx \frac{3^0}{0!}e^{-3} \approx .0498$

(B) $P(X \leq 5) \approx 1 - e^{-3}\left(\frac{3^0}{0!} + \frac{3^1}{1!} + \frac{3^2}{2!} + \frac{3^3}{3!} + \frac{3^4}{4!}\right) \approx 1 - .8153 = .1847$

15. $P(5 \leq X \leq 7) = e^{-6}\left(\frac{6^5}{5!} + \frac{6^6}{6!} + \frac{6^7}{7!}\right) \approx .4589$

16. $e^{-2}\left(\frac{2^0}{0!} + \frac{2^1}{1!} + \frac{2^2}{2!} + \frac{2^3}{3!} + \frac{2^4}{4!}\right) \approx .9473$

(the accepted manufacturing process will produce 60 minute tapes containing no more than 4 flaws about 95% of the time.)

Exercises 10-3

Unless instructed to the contrary, compute all answers to four decimal places.

A

In Problems 1–8, compute each Poisson probability.

1. $P(0) = \frac{4^0}{0!}e^{-4}$

2. $P(0) = \frac{(0.5)^0}{0!}e^{-0.5}$

3. $P(2) = \frac{(0.25)^2}{2!}e^{-0.25}$

4. $P(3) = \frac{1^3}{3!}e^{-1}$

5. $P(X < 3) = e^{-2}\left(\frac{2^0}{0!} + \frac{2^1}{1!} + \frac{2^2}{2!}\right)$

6. $P(X \leq 2) = e^{-3}\left(\frac{3^0}{0!} + \frac{3^1}{1!} + \frac{3^2}{2!}\right)$

7. $P(X \geq 3) = 1 - e^{-2}\left(\frac{2^0}{0!} + \frac{2^1}{1!} + \frac{2^2}{2!}\right)$

8. $P(X > 2) = 1 - e^{-3}\left(\frac{3^0}{0!} + \frac{3^1}{1!} + \frac{3^2}{2!}\right)$

B

In Problems 9–12, compute P(3) for both the binomial distribution and the Poisson approximation to the binomial for each of the indicated values of n and p.

9. $n = 200, \quad p = .01$

10. $n = 1{,}000, \quad p = .003$

11. $n = 10{,}000, \quad p = .000\,05$

12. $n = 5{,}000, \quad p = .000\,02$

In Problems 13 and 14, use the Poisson approximation to the binomial distribution to determine the indicated probabilities.

13. The probability of rolling a pair of 1's (snake eyes) in a single roll of a pair of fair dice is $\frac{1}{36}$, what is the probability of:

(A) Rolling no snake eyes in 108 rolls of a pair of fair dice?

(B) Rolling 5 snake eyes in 108 rolls of a pair of fair dice?

14. The probability of drawing the ace of spades from a standard 52-card deck is $\frac{1}{52}$. What is the probability of:

(A) Not drawing an ace of spades in 104 single-card draws (with replacement) from a 52-card deck?

(B) Drawing an ace of spades 4 times in 104 single-card draws (with replacement) from a 52 card deck?

C

In Problems 15 and 16, find N to the nearest integer so that:

15. $P(0) = \frac{(N/2{,}500)^0}{0!}e^{-N/2{,}500} = .01$

16. $P(0) = \frac{(N/500)^0}{0!}e^{-N/500} = .05$

Applications

Business & Economics

17. Manufacturing. A small bottling company bottles a popular brand of mineral water using automated equipment. The process produces, on the average, 5 defective (improperly filled) bottles per 1,000 produced. In an order for 100 bottles, what is the probability of the order:

(A) Containing no defective bottles?

(B) Containing at most 2 defective bottles?

(C) Containing more than 2 defective bottles?

Approximate these binomial probabilities using Poisson approximations.

18. Manufacturing. A pencil manufacturing company produces, on the average, 2 defective pencils per 1,000 produced. In an order for 500 pencils, what is the probability that:

(A) There will be no defective pencils?

(B) There will be 4 defective pencils?

(C) There will be at least 4 defective pencils?

Approximate these binomial probabilities using Poisson approximations.

19. Hospital scheduling. A community hospital wishes to schedule its delivery room facilities to take care of new births effectively without an excess of staff. On the average, 21 births occur randomly per week. During a 8 hour work shift, what is the probability that:

(A) No birth will occur? (B) 4 births will occur?

(C) 4 or more births will occur?

20. Loading dock arrivals. Over a period of several months it was observed that, on the average, 16 trucks arrived randomly at a store's loading dock in an 8 hour business day. Compute the probability that during a given business hour:

(A) No trucks will arrive. (B) 5 trucks will arrive.

(C) 5 or more trucks will arrive.

21. Quality control. In manufacturing a certain brand of carpeting, flaws occur randomly at the acceptable average rate of 3 flaws per 10 square yards of carpet. Random samples of 1 square yard of carpet are inspected periodically, and the management has a policy of shutting down and servicing the equipment if an inspection reveals more than 1 flaw in a 1 square yard sample. Assuming the manufacturing process is operating normally:

(A) What is the probability that the plant will be shut down after an inspection?

(B) What is the probability that a given 1 square yard sample will have no flaws?

22. **Quality control.** Wire is produced with an acceptable manufacturing flaw rate of 2 per 1,000 feet of wire produced. Random samples of 100 feet of wire are inspected periodically, and the manufacturing process is shut down for servicing if an inspection reveals more than 1 flaw. Assuming the manufacturing process is operating normally:

    (A) What is the probability that the process will be shut down after an inspection?

    (B) What is the probability of a given sample having no flaws?

23. **Equipment failure.** A company's records show that their computer is down on the average of about twice per week (5 business days). Compute the probability that during a given business day:

    (A) The computer will not be down.

    (B) The computer will be down twice.

    (C) The computer will be down more than once.

24. **Equipment failure.** The elevator in a large old hotel breaks down, on the average, about 13 times a year. What is the probability that during a given week:

    (A) The elevator will not break down?

    (B) The elevator will break down once?

    (C) The elevator will break down more than once?

25. **Typesetting.** A typesetting company produces, on the average, 1 error for every 15 pages of typesetting. In 30 pages, what is the probability of having:

    (A) No typesetting errors?

    (B) At most 4 typesetting errors?

    (C) More than 4 typesetting errors?

26. **Industrial accidents.** The average number of industrial accidents per day in a manufacturing company is 1.2. If accidents occur randomly over the working time, what is the probability that in 1 work week (5 days), there will be:

    (A) 10 accidents?

    (B) At most 3 accidents?

    (C) At least 3 accidents?

Life Sciences

27. **Vaccine testing.** In a certain city, on the average, 4% of the adult population will contract a certain strain of influenza during any given year. To test a newly discovered vaccine, 200 healthy adults are randomly selected and are inoculated with the new vaccine. Assuming the vaccine has no effect at all, what is the probability that during the test year:

    (A) 2 out of the 200 inoculated adults contract influenza?

    (B) No more than 2 out of the 200 inoculated adults contract influenza?

    Approximate these binomial probabilities using Poisson approximations.

28. **Life Expectancy.** In a stable community where there is little change in mortality rates or size, it is found that at any given time (on the average) 0.4% of the population are over 100 years old. Assuming there has not been a change in size or mortality rates, what is the probability that in a random sample of 500 people from the community:

    (A) 5 people are over 100?

    (B) More than 4 people are over 100?

    Approximate these binomial probabilities using Poisson approximations.

29. **Organic pesticides.** An organic pesticide contains, on the average, 300,000 organisms per liter that are randomly distributed throughout the solution. The pesticide will be sprayed over a crop with a drop size of about $\frac{1}{100,000}$ liter.

    (A) What is the probability that a single drop will have no organisms in it?

    (B) How many organisms must be present per liter (on the average) so that the probability of a single drop not having any organisms will be .01? (Round your answer to the nearest thousand.)

30. **Plant nutrition.** Small slow-release nutrition pellets are randomly mixed throughout a dry planting mix. On the average, the mix contains 1,728 pellets per cubic foot.

    (A) What is the probability that 1 cubic inch of the mix will contain no pellets?

    (B) How many pellets must be present per cubic foot of mix (on the average) so that the probability of 1 cubic inch of the mix not containing any pellets is .1? (Round your answer to the nearest thousand.)

Social Sciences

31. **Traffic control.** A small river lock, accommodating one pleasure boat at a time, operates daily. It is found that, on the average, 5 boats arrive randomly per day. Compute the probabilities that on a given day:

    (A) No boat will arrive.

    (B) From 4 to 6 boats will arrive.

    (C) Fewer than 3 boats will arrive.

32. **Auto safety.** A major freeway interchange in Walnut Creek, California, handles about 280,000 vehicles per day, and, on the average, 1 car crashes every 30 hours. Compute the probability that:

    (A) No car will crash in a given 24 hour period.

    (B) 3 or more cars will crash in a 24 hour period.

33. **Emergency medical services.** A hospital wants to staff its emergency room from midnight to 6 AM so that adequate service is available most of the time. On the average, it is found that 18 emergency patients arrive randomly during these 6 hours. What is the probability that:

    (A) No more than 5 emergency patients arrive in 1 hour?

    (B) More than 5 emergency patients arrive in 1 hour?

    (C) No more than 2 emergency patients arrive over a 2 hour period?

34. **Suicides.** On the average, about 4 people per year jump from the Golden Gate Bridge to their deaths. The suicides occur randomly over time. What is the probability that:

    (A) No more than 2 suicides occur during a given year?

    (B) More than 6 suicides occur during a given year?

    (C) No more than 4 suicides occur over a 2 year period?

10-4 Continuous Random Variables

Continuous Random Variables

Suppose that during an 8 hour business day the switchboard of a large law firm receives an average of 3 calls per minute. As we saw in the preceding section, the number of calls received per minute is a Poisson random variable with the following discrete set of possible values (range):

$$S = \{0, 1, 2, \dots\}$$

Now we want to consider a random variable X representing the time interval (in minutes) between incoming calls. What are the values that this random variable can assume? It is reasonable to assume that all the values of X must lie in the interval 90, 480], since there are 480 minutes in an 8 hour business day. However, we cannot use the discrete set

$$S = \{0, 1, 2, \dots, 480\}$$

for the range of this random variable, because this set excludes noninteger values such as 2.5 minutes or $\sqrt{35}$ minutes. Although it may be very unlikely that X will be close to 0 or to 480, theoretically there is no reason to exclude any of the values in the interval [0, 480] as a possibility for the time between incoming calls. Thus, we assume that the range of X is the interval of real numbers [0, 480], and we call X *a continuous random variable.*

Continuous Random Variable

A **continuous random variable** is a random variable with a set of possible values (range) that is an interval of real numbers. This interval may be open or closed, and it may be bounded or unbounded.

The term *continuous* is not used in the same sense here as it was used in Section 1-1. In this case, it refers to the fact that the values of the random variable form a continuous set of real numbers, such as $[0, \infty]$, rather than a discrete set, such as $\{5, 9, 17\}$ or $\{1, 3, 5, \dots\}$.

Probability Density Function

If X is a discrete random variable, then the probability that X lies in a given interval can be computed by addition. For example, if X is a binomial or Poisson random variable, then

$$P(0 \leqslant X \leqslant 2) = P(X = 0) + P(X = 1) + P(X = 2)$$

The probability distribution function for X is used to evaluate each of the probabilities in the sum on the right and the results are added.

If X is a continuous random variable, the same approach will not work. Since X can now assume any real number in the interval [0, 2], it is impossible to write $P(0 \leqslant X \leqslant 2)$ as a finite or even an infinite sum. (What would be the second term in such a sum? That is, what is the "next" real number after 0? Think about this.) Instead, we introduce a new type of function, called a *probability density function*, and use integrals involving this function to compute the probability that a continuous random variable X lies in a given interval. For example, we might use a probability density function to find the probability that the actual amount of soda in a 12 ounce can is between 11.9 and 12.1 ounces, or that the speed of a car involved in an accident was between 60 and 65 miles per hour.

For convenience in stating definitions and formulas, we will assume that the value of a continuous random variable can be any real number; that is, the range is $(-\infty, \infty)$.

Probability Density Function

The function $f(x)$ is a **probability density function** for a continuous random variable X if:

1. $f(x) \geqslant 0$ for all $x \in (-\infty, \infty)$
2. $\int_{-\infty}^{\infty} f(x)\, dx = 1$
3. The probability that X lies in the interval $[c, d]$ is given by

$$P(c \leqslant X \leqslant d) = \int_{c}^{d} f(x)\, dx$$

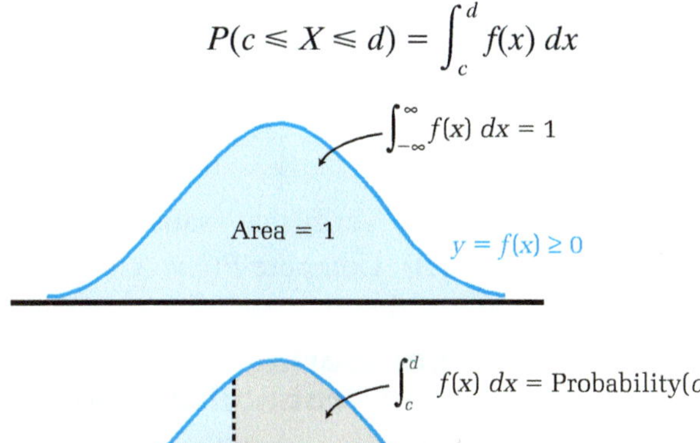

Range of $X = (-\infty, \infty) =$ Domain of f

EXAMPLE 17 Let: $f(x) = \begin{cases} 12x^2 - 12x^3 & \text{if } 0 \leqslant x \leqslant 1 \\ 0 & \text{otherwise} \end{cases}$

(A) Verify that f satisfies the first two conditions for a probability density function.
(B) Compute $P(\frac{1}{4} \leqslant X \leqslant \frac{3}{4})$, $P(X \leqslant \frac{1}{2})$, $P(X \geqslant \frac{2}{3})$, and $P(X = \frac{1}{3})$.

SOLUTIONS (A) For $0 \leqslant x \leqslant 1$, we have $f(x) = 12x^2 - 12x^3 = 12x^2(1 - x) \geqslant 0$. Since $f(x) = 0$ for all other values of x, it follows that $f(x) \geqslant 0$ for all x. Also,

$$\int_{-\infty}^{\infty} f(x)dx = \int_{0}^{1} (12x^2 - 12x^3)dx = (4x^3 - 3x^4)\Big|_0^1 = (4 - 3) - (0) = 1$$

(B)
$$\begin{aligned} P(\tfrac{1}{4} \leqslant X \leqslant \tfrac{3}{4}) &= \int_{1/4}^{3/4} f(x)\, dx \\ &= \int_{1/4}^{3/4} (12x^2 - 12x^3)dx \\ &= (4x^3 - 3x^4)\Big|_{1/4}^{3/4} \\ &= \tfrac{189}{256} - \tfrac{13}{256} = \tfrac{11}{16} \end{aligned}$$

$$\begin{aligned} P(X \leqslant \tfrac{1}{2}) &= \int_{-\infty}^{1/2} f(x)\, dx \\ &= \int_{0}^{1/2} (12x^2 - 12x^3)dx \\ &= (4x^3 - 3x^4)\Big|_0^{1/2} \\ &= \tfrac{5}{16} \end{aligned}$$

Note that $f(x) = 0$ for $x < 0$.

$$P(X \geq \tfrac{2}{3}) = \int_{2/3}^{\infty} f(x)\,dx$$ Note that $f(x) = 0$ for $x > 1$.

$$= \int_{2/3}^{1} (12x^2 - 12x^3)dx$$
$$= (4x^3 - 3x^4)\Big|_{2/3}^{1}$$
$$= 1 - \tfrac{16}{27} = \tfrac{11}{27}$$

$$P(X = \tfrac{1}{3}) = \int_{1/3}^{1/3} f(x)\,dx = 0$$

Matched Problem 17 Let: $f(x) = \begin{cases} 6x - 6x^2 & \text{if } 0 \leq x \leq 1 \\ 0 & \text{otherwise} \end{cases}$

(A) Verify that f satisfies the first two conditions for a probability density function.
(B) Compute $P(\frac{1}{3} \leq X \leq \frac{2}{3})$, $P(X \leq \frac{1}{5})$, $P(X \geq \frac{1}{2})$, $P(X = \frac{1}{4})$.

Comparing Probability Distribution Functions and Probability Density Functions

The last probability in Example 17 illustrates a fundamental difference between discrete and continuous random variables. In the discrete case, there is a *probability distribution* $P(x)$ that gives the probability of each possible value of the random variable. Thus, if c is one of the values of the random variable, then $P(X = c) = P(c)$. In the continuous case, the *integral of the probability density function* $f(x)$ gives the probability that the outcome lies in a certain interval. If c is any real number, then the probability that the outcome is *exactly* c is

$$P(X = c) = P(c \leq X \leq c) = \int_{c}^{c} f(x)\,dx = 0$$

Thus, $P(X = c) = 0$ for *any number* c and, since $f(c)$ is certainly not 0 for all values of c, we see that $f(x)$ does not play the same role for a continuous random variable as $P(x)$ does for a discrete random variable.

The fact that $P(X = c) = 0$ also implies that excluding either end point from an interval does not change the probability that the random variable lies in that interval; that is,

$$P(a < X < b) = P(a < X \leq b) = P(a \leq X < b) = P(a \leq X \leq b)$$
$$= \int_{a}^{b} f(x)\,dx$$

EXAMPLE 18 Use the probability density function in Example 17 to compute $P(.1 < X \leq .2)$ and $P(X > .9)$.

SOLUTION

$$P(.1 < X \leq .2) = \int_{.1}^{.2} f(x)\,dx \qquad f(x) = \begin{cases} 12x^2 - 12x^3 & \text{if } 0 \leq x \leq 1 \\ 0 & \text{otherwise} \end{cases}$$

$$= \int_{.1}^{.2} (12x^2 - 12x^3)dx$$

$$= (4x^3 - 3x^4)\Big|_{.1}^{.2}$$

$$= .0272 - .0037 = .0235$$

$$\begin{aligned} P(X > .9) &= \int_{.9}^{\infty} f(x)\,dx \\ &= \int_{.9}^{1} (12x^2 - 12x^3)dx \\ &= (4x^3 - 3x^4)\Big|_{.9}^{0} \\ &= 1 - .9477 = .0523 \end{aligned}$$

Matched Problem 18 Use the probability density function in Problem 17 to compute $P(.2 \leqslant X < .4)$ and $P(X < .8)$.

If $f(x)$ is the probability density function in Example 17, notice that $f(\frac{2}{3}) = \frac{16}{9} > 1$. Thus, a probability density function can assume values larger than 1. This illustrates another difference between probability density functions and probability distribution functions. In terms of inequalities, a probability distribution function must always satisfy $0 \leqslant P(x) \leqslant 1$, while a probability density function need only satisfy $f(x) \geqslant 0$. Despite these differences, we shall see that there are many similarities in the application of probability distribution functions and probability density functions.

EXAMPLE 19 **Shelf-Life** The shelf-life (in months) of a certain drug is a continuous random variable with probability density function (see the figure)

$$f(x) + \begin{cases} 50/(x + 50)^2 & \text{if } x \geqslant 0 \\ 0 & \text{otherwise} \end{cases}$$

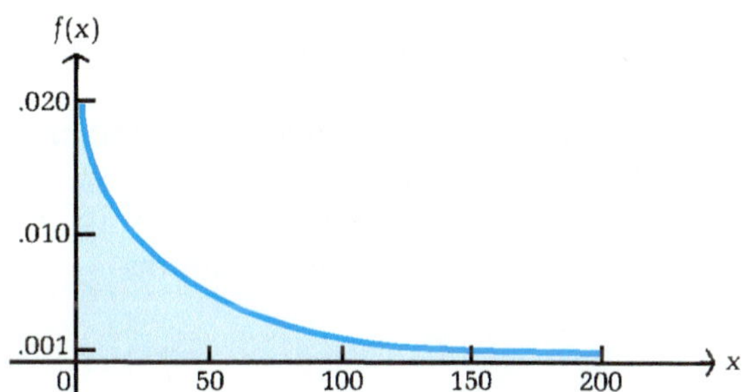

Find the probability that the drug has a shelf-life of:

(A) Between 10 and 20 months (B) At most 30 months
(C) Over 25 months

SOLUTIONS (A)

$$P(10 \leqslant X \leqslant 20) = \int_{10}^{20} f(x)\,dx = \int_{10}^{20} \frac{50}{(x + 50)^2}dx = \frac{-50}{x + 50}\Big|_{10}^{20}$$

$$= \left(-\frac{50}{70}\right) - \left(-\frac{50}{60}\right) = \frac{5}{42}$$

(B)

$$P(X \leqslant 30) = \int_{-\infty}^{30} f(x)\,dx = \int_{0}^{30} \frac{50}{(x + 50)^2}dx = \frac{-50}{x + 50}\Big|_{0}^{30}$$

$$= \left(-\frac{50}{80}\right) - (-1) = \frac{3}{8}$$

(C) $$P(X > 25) = \int_{25}^{\infty} f(x)\,dx = \int_{25}^{\infty} \frac{50}{(x + 50)^2}dx$$

This improper integral can be evaluated directly using the techniques discussed earlier. However, there is another method that does not involve evaluation of any improper integrals. Since f is a probability density function, we can write

$$1 = \int_{-\infty}^{\infty} f(x)\,dx$$

$$= \int_{0}^{\infty} \frac{50}{(x + 50)^2}\,dx \qquad \int_{a}^{b} f(x)\,dx = \int_{a}^{c} f(x)\,dx + \int_{c}^{b} f(x)\,dx$$

$$= \int_{0}^{25} \frac{50}{(x + 50)^2}\,dx + \int_{25}^{\infty} \frac{50}{(x + 50)^2}\,dx$$

Solving this last equation for $\int_{25}^{\infty} [50/(x + 50)^2]\,dx$, we have

$$\int_{25}^{\infty} \frac{50}{(x + 50)^2}\,dx = 1 - \int_{0}^{25} \frac{50}{(x + 50)^2}\,dx$$

$$= 1 - \frac{-50}{x + 50}\Bigg|_{0}^{25}$$

$$= 1 - \left(\tfrac{-50}{75}\right) - 1 = \tfrac{2}{3}$$

Thus, $P(X > 25) = \frac{2}{3}$.

Matched Problem 19 In Example 19 find the probability that the drug has a shelf-life of:

(A) Between 50 and 100 months (B) At most 20 months

(C) Over 10 months

Cumulative Probability Distribution Function

Each time we compute the probability for a continuous random variable, we must find the antiderivative of the probability density function. This antiderivative is used so often that it is convenient to give it a name.

Cumulative Probability Distribution Function

If f is a probability density function, then the associated **cumulative probability distribution function** $\boldsymbol{F}$ is defined by

$$F(x) = P(X \le x) = \int_{-\infty}^{x} f(t)\,dt$$

Furthermore,

$$P(c \le X \le d) = F(d) - F(c)$$

Figure 6 gives a geometric interpretation of these ideas.

Notice that $F(x) = \int_{-\infty}^{x} f(t)\,dt$ is a function of x, the upper limit of integration, not t, the variable in the integrand. We state some important properties of cumulative probability distribution functions in the next box. These properties follow directly from the fact that $F(x)$ can be interpreted geometrically as the area under the graph of $y = f(t)$ from $-\infty$ to x (see Figure 6B).

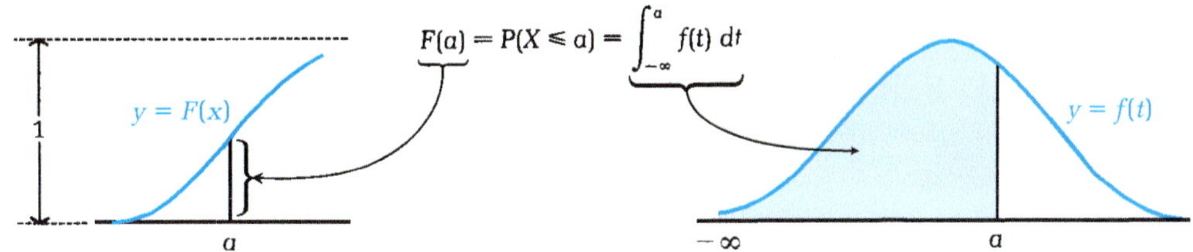

(A) Cumulative probability distribution function (B) Probability density function

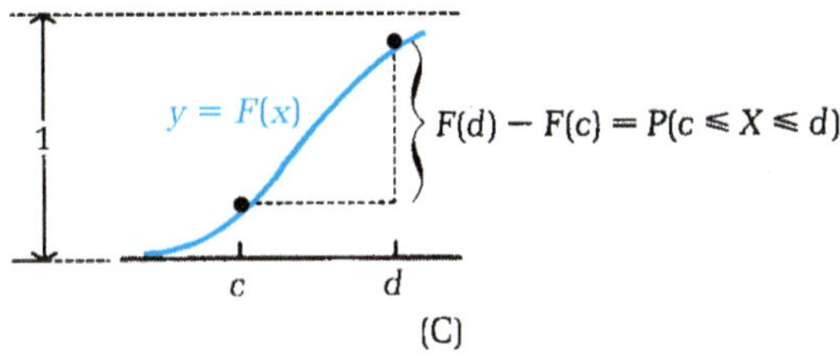

(C)

Figure 6

> **Properties of Cumulative Probability Distribution Functions**
> If f is a probability density function and
>
> $$If\,(x) = \int_{-\infty}^{x} f(t)\,dt$$
>
> is the associated cumulative probability distribution function, then:
>
> 1. $F'(x) = f(x)$ wherever f is continuous
> 2. $0 \le F(x) \le 1, -\infty < x < \infty$
> 3. $F(x)$ is nondecreasing on $(-\infty, \infty)$*

EXAMPLE 20 Find the cumulative probability distribution function for the probability density function in Example 17, and use it to compute $P(.1 \le X \le .9)$.

SOLUTION If $x < 0$, then

$$F(x) = \int_{-\infty}^{x} f(t)\,dt \qquad f(x) = \begin{cases} 12x^2 - 12x^3 & \text{if } 0 \le x \le 1 \\ 0 & \text{otherwise} \end{cases}$$

$$= \int_{-\infty}^{x} 0\,dt = 0$$

If $0 \le x \le 1$, then

$$F(x) = \int_{-\infty}^{x} f(t)\,dt = \int_{-\infty}^{0} f(t)\,dt + \int_{0}^{x} f(t)\,dt$$

$$= 0 + \int_{0}^{x} (12t^2 - 12t^3)\,dt = (4t^3 - 3t^4)\Big|_0^x$$

$$= 4x^3 - 3x^4$$

If $x > 1$, then

$$F(x) = \int_{-\infty}^{x} f(t)\,dt = \int_{-\infty}^{0} f(t)\,dt + \int_{0}^{1} f(t)\,dt + \int_{1}^{x} f(t)\,dt$$

$$= 0 + 1 + 0 = 1$$

* A function $F(x)$ is nondecreasing on (a, b) if $F(x_1) \le F(x_2)$ for $a < x_1 < x_2 < b$.

Thus,

$$F(x) = \begin{cases} 0 & \text{if } x < 0 \\ 4x^3 - 3x^4 & \text{if } 0 \leq x \leq 1 \\ 1 & \text{if } x > 1 \end{cases}$$

And

$$P(.1 \leq X \leq .9) = F(.9) - F(.1) = .9477 - .0037 = .944$$

See the figure.

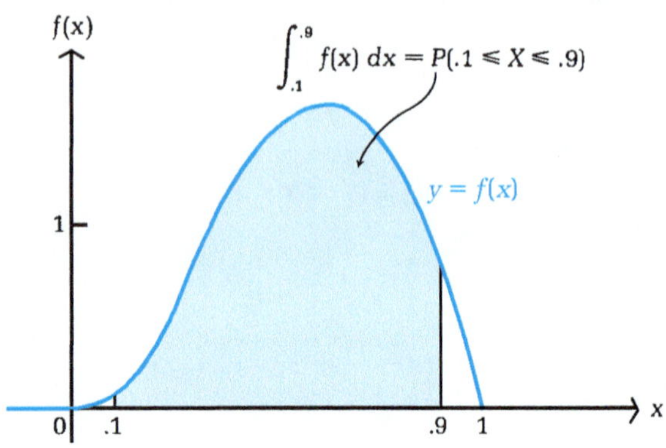

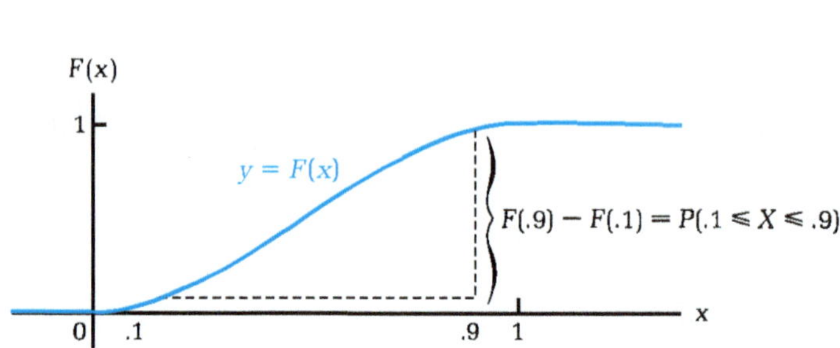

Matched Problem 20 Find the cumulative probability distribution function for the probability density function in Problem 17, and use it to compute $P(.3 \leq X \leq .7)$.

EXAMPLE 21 **Shelf-Life** Returning to the discussion of the shelf-life of a drug in Example 19, suppose a pharmacist wants to be 95% certain that the drug is still good when it is sold. How long is it safe to leave the drug on the shelf?

SOLUTION Let x be the number of months the drug has been on the shelf when it is sold. The probability that the shelf-life of the drug is less than the number of months it has been sitting on the shelf is $P(0 \leq X \leq x)$. The pharmacist wants this probability to be .05. Thus, we must solve the equation $P(0 \leq X \leq x) = .05$ for x. First, we will find the cumulative probability distribution function F. For $x < 0$, we see that $F(x) = 0$. For $x \geq 0$,

$$F(x) = \int_0^x \frac{50}{(50 + t)^2}\,dt = \left.\frac{-50}{50 + t}\right|_0^x = \frac{-50}{50 + x} - (-1) = 1 - \frac{50}{50 + x}$$

$$= \frac{x}{50 + x}$$

Thus,

$$F(x) = \begin{cases} 0 & \text{if } x < 0 \\ x/(50 + x) & \text{if } x \geq 0 \end{cases}$$

Now, to solve the equation $P(0 \leq X \leq x) = .05$, we solve

$$F(x) - F(0) = .05 \qquad F(0) = 0$$

$$\frac{x}{50 + x} = .05$$

$$x = 2.5 + .05x$$

$$.95x = 2.5$$
$$x \approx 2.6$$

If the drug is sold during the first 2.6 months it is on the shelf, then the probability that it is still good is .95.

Matched Problem 21 Repeat Example 21 if the pharmacist wants the probability that the drug is still good to be .99.

Answers to Matched Problems

17. (B) $\frac{13}{27}$; $\frac{13}{125}$; $\frac{1}{2}$; 0 **18.** .248; .896 **19.** (A) $\frac{1}{6}$ (B) $\frac{2}{7}$ (C) $\frac{5}{6}$

20. $F(x) = \begin{cases} 0 & \text{if } x - 0 \\ 3x^2 - 2x^3 & \text{if } 0 \le x \le 1 \\ 1 & \text{if } x \ge 1 \end{cases}$

$P(.3 \le X \le .7) = .568$

21. Approx. $\frac{1}{2}$ month, or 15 days

Exercise 10-4

A

In Problems 1 and 2, graph f, and show that f satisfies the first two conditions for a probability density function.

1. $f(x) = \begin{cases} \frac{1}{8}x & \text{if } 0 \le x \le 4 \\ 0 & \text{otherwise} \end{cases}$

2. $f(x) = \begin{cases} \frac{1}{9}x^2 & \text{if } 0 \le x \le 3 \\ 0 & \text{otherwise} \end{cases}$

3. Use the function in Problem 1 to find the indicated probabilities. Illustrate each probability with a graph.

(A) $P(1 \le X \le 3)$ (B) $P(X \le 2)$ (C) $P(X > 3)$

4. Use the function in Problem 2 to find the indicated probabilities. Illustrate each probability with a graph.

(A) $P(1 \le X \le 2)$ (B) $P(X \ge 1)$ (C) $P(X < 2)$

5. Use the function in Problem 1 to find the indicated probabilities.

(A) $P(X = 1)$ (B) $P(X > 5)$ (C) $P(X \le 5)$

6. Use the function in Problem 2 to find the indicated probabilities.

(A) $P(X) = 2)$ (B) $P(X > 4)$ (C) $P(X < 4)$

7. Find the graph the cumulative probability distribution function associated with the function in Problem 1.

8. Find and graph the cumulative probability distribution function associated with the function in Problem 2.

9. Use the cumulative probability distribution function from Problem 7 to find the indicated probabilities.

(A) $P(2 \le X \le 4)$ (B) $P(0 < X < 2)$

10. Use the cumulative probability distribution function from Problem 8 to find the indicated probabilities.

(A) $P(0 \le X \le 1)$ (B) $P(2 < X < 3)$

11. Use the cumulative probability distribution function from Problem 7 to find the value of x that satisfies each equation.

(A) $P(0 \le X \le x) = \frac{1}{4}$ (B) $P(0 \le X \le x) = \frac{1}{9}$

12. Use the cumulative probability distribution function from Problem 8 to find the value of x that satisfies each equation.

(A) $P(0 \le X \le x) + \frac{1}{8}$ (B) $P(0 \le X \le x) = \frac{1}{64}$

B

In Problems 13 and 14, graph f, and show that f satisfies the first two conditions for a probability density function.

13. $f(x) = \begin{cases} 2/(1+x)^3 & \text{if } x \ge 0 \\ 0 & \text{otherwise} \end{cases}$

14. $f(x) = \begin{cases} 2/(2+x)^2 & \text{if } x \ge 0 \\ 0 & \text{otherwise} \end{cases}$

15. Use the function in Problem 13 to find the indicated probabilities.

(A) $P(1 \le X \le 4)$ (B) $P(X > 3)$ (C) $P(X \le 2)$

16. Use the function in Problem 14 to find the indicated probabilities.

(A) $P(2 \le X \le 8)$ (B) $P(X \ge 3)$ (C) $P(X < 1)$

17. Find and graph the cumulative probability distribution function associated with the function in Problem 13.

18. Find and graph the cumulative probability distribution function associated with the function in Problem 14.

19. Use the cumulative probability distribution function from Problem 17 to find the value of x that satisfies each equation.

(A) $P(0 \le X \le x) + \frac{3}{4}$(B) $P(X \ge x) = \frac{1}{16}$

20. Use the cumulative probability distribution function from Problem 18 to find the value of x that satisfies each equation.

(A) $P(0 \le X \le x) = \frac{3}{4}$ (B) $P(X > x) = \frac{1}{3}$

In Problems 21–24, find the associated cumulative probability distribution function. Graph both functions (on separate sets of axes).

21. $f(x) = \begin{cases} \frac{3}{2}x - \frac{3}{4}x^2 & \text{if } 0 \le x \le 2 \\ 0 & \text{otherwise} \end{cases}$

22. $f(x) = \begin{cases} \frac{3}{4} - \frac{3}{4}x^2 & \text{if } -1 \le x \le 12 \\ 0 & \text{otherwise} \end{cases}$

23. $f(x) = \begin{cases} \frac{1}{2} + \frac{1}{2}x^3 & \text{if } -1 \le x \le 1 \\ 0 & \text{otherwise} \end{cases}$

24. $f(x) = \begin{cases} \frac{3}{4} - \frac{3}{8}\sqrt{x} & \text{if } 0 \le x \le 4 \\ 0 & \text{otherwise} \end{cases}$

In Problems 25–28, use a graphic calculator or a computer to approximate (to two decimal places) the value of x that satisfies the given equation for the indicated cumulative probability distribution function F(x).

25. $P(0 \le X \le x) = .2$ for $F(x)$ from Problem 21

26. $P(-1 \le X \le x) = .4$ for $F(x)$ from Problem 22

27. $P(-1 \le X \le x) = .6$ for $F(x)$ from Problem 23

28. $P(0 \le X \le x) = .7$ for $F(x)$ from Problem 24

In Problems 29–32, find the associated cumulative probability function, and use it to find the indicated probability.

29. Find $P(1 \le X \le 2)$ for

$$f(x) = \begin{cases} \ln x & \text{if } 1 \le x \le e \\ 0 & \text{otherwise} \end{cases}$$

30. Find $P(1 \le X \le 2)$ for

$$f(x) = \begin{cases} 3x/(8\sqrt{1+x}) & \text{if } 0 \le x \le 3 \\ 0 & \text{otherwise} \end{cases}$$

31. Find $P(X \ge 1)$ for

$$f(x) = \begin{cases} xe^{-x} & \text{if } x \ge 0 \\ 0 & \text{otherwise} \end{cases}$$

32. Find $P(X \ge e)$ for

$$f(x) = \begin{cases} (\ln x)/x^2 & \text{if } x \ge 1 \\ 0 & \text{otherwise} \end{cases}$$

C

In problems 33–36, F(x) is the cumulative probability distribution function for a continuous random variable X. Find the probability density function f(x) associated with each F(x).

33. $F(x) = \begin{cases} 0 & \text{if } x < 0 \\ x^2 & \text{if } 0 \le x \le 1 \\ 1 & \text{if } x > 1 \end{cases}$

34. $f(x) = \begin{cases} 0 & \text{if } x < 1 \\ \frac{1}{2}x - \frac{1}{2} & \text{if } 1 \le x \le 3 \\ 1 & \text{if } x > 3 \end{cases}$

35. $F(x) = \begin{cases} 0 & \text{if } x < 0 \\ 6x^2 - 8x^3 + 3x^4 & \text{if } 0 \le x \le 1 \\ 1 & \text{if } x > 1 \end{cases}$

36. $F(x) = \begin{cases} 1-(1/x^3) & \text{if } x \ge 1 \\ 0 & \text{otherwise} \end{cases}$

In Problems 37 and 38, find the associated cumulative distribution function.

37. $f(x) = \begin{cases} x & \text{if } 0 \le x \le 1 \\ 2 - x & \text{if } 1 < x \le 2 \\ 0 & \text{otherwise} \end{cases}$

38. $f(x) = \begin{cases} \frac{1}{4} & \text{if } 0 \le x \le 1 \\ \frac{1}{2} & \text{if } 1 < x \le 2 \\ \frac{1}{4} & \text{if } 2 < x \le 3 \\ 0 & \text{otherwise} \end{cases}$

Applications

Business & Economics

39. Electricity consumption. The daily demand for electricity (in millions of kilowatt-hours) in a large city is a continuous random variable with probability density function

$$f(x) = \begin{cases} .2 - .02x & \text{if } 0 \le x \le 10 \\ 0 & \text{otherwise} \end{cases}$$

(A) What is the probability that the daily demand for electricity is less than 8 million kilowatt-hours?

(B) What is the probability that 5 million kilowatt-hours will not be sufficient to meet the daily demand?

40. Gasoline consumption. The daily demand for gasoline (in millions of gallons) in a large city is a continuous random variable with probability density function

$$f(x) = \begin{cases} .4 - .08x & \text{if } 0 \le x \le 5 \\ 0 & \text{otherwise} \end{cases}$$

(A) What is the probability that the daily demand is less than 2 million gallons?

(B) What is the probability that 3 million gallons will not be sufficient to meet the daily demand?

41. Time-sharing. In a computer time-sharing network, the time it takes (in seconds) to respond to a user's request is a continuous random variable with probability density function given by

$$f(x) = \begin{cases} \frac{1}{10}e^{-x/10} & \text{if } x \ge 0 \\ 0 & \text{otherwise} \end{cases}$$

(A) What is the probability that the computer responds within 1 second?

(B) What is the probability that a user must wait over 4 seconds for a response?

42. Waiting time. The time (in minutes) a customer must wait in line at a bank is a continuous random variable with probability density function given by

$$f(x) = \begin{cases} \frac{1}{2}e^{-x/2} & \text{if } x \geq 0 \\ 0 & \text{otherwise} \end{cases}$$

(A) What is the probability that a customer waits less than 3 minutes?

(B) What is the probability that a customer waits more than 5 minutes?

43. Demand. The weekly demand for hamburger (in thousands of pounds) for a chain of supermarkets is a continuous random variable with probability density function given by

$$f(x) = \begin{cases} .003x\sqrt{100 - x^2} & \text{if } 0 \leq x \leq 10 \\ 0 & \text{otherwise} \end{cases}$$

(A) What is the probability that more than 4,000 pounds of hamburger are demanded?

(B) The manager of the meat department ordered 8,000 pounds of hamburger. What is the probability that the demand will not exceed this amount?

(C) The manager wants the probability that the demand does not exceed the amount ordered to be .9. How much hamburger meat should be ordered?

44. Demand. The demand for a weekly sports magazine (in thousands of copies) in a certain city is a continuous random variable with probability density function given by

$$f(x) = \begin{cases} \frac{3}{125}x\sqrt{25 - x^2} & \text{if } 0 \leq x \leq 5 \\ 0 & \text{otherwise} \end{cases}$$

(A) The magazine's distributor in this city orders 3,000 copies of the magazine. What is the probability that the demand exceeds this number?

(B) What is the probability that the demand does not exceed 4,000 copies?

(C) If the distributor wants to be 95% certain that the demand does not exceed the number ordered, how many copies should be ordered?

Life Sciences

45. Life expectancy. The life expectancy (in minutes) of a certain microscopic organism is a continuous random variable with probability density function given by

$$f(x) = \begin{cases} \frac{1}{5{,}000}(10x^3 - x^4) & \text{if } 0 \leq x \leq 10 \\ 0 & \text{otherwise} \end{cases}$$

(A) What is the probability that an organism lives for at least 7 minutes?

(B) What is the probability that an organism lives for at most 5 minutes?

46. Life expectancy. The life expectancy (in months) of plants of a certain species is a continuous random variable with probability density function given by

$$f(x) = \begin{cases} \frac{1}{36}(6x - x^2) & \text{if } 0 \leq x \leq 6 \\ 0 & \text{otherwise} \end{cases}$$

(A) What is the probability that one of these plants survives for at least 4 months?

(B) What is the probability that one of these plants survives for at most 5 months?

47. Shelf-life. The shelf-life (in days) of a perishable drug is a continuous random variable with probability density function given by

$$f(x) = \begin{cases} 800x/(400 + x^2)^2 & \text{if } x \geq 5 \\ 0 & \text{otherwise} \end{cases}$$

(A) What is the probability that the drug has a shelf-life of at most 20 days?

(B) What is the probability that the shelf-life exceeds 15 days?

(C) If the user wants the probability that the drug is still good to be .8, when is the last time it should be used?

48. Shelf-life. Repeat Problem 47 if

$$f(x) = \begin{cases} 200x/(100 + x^2)^2 & \text{if } x \geq 0 \\ 0 & \text{otherwise} \end{cases}$$

Social Sciences

49. Learning. The number of words per minute a beginner can type after 1 week of practice is a continuous random variable with probability density function given by

$$f(x) = \begin{cases} \frac{1}{20}e^{-x/20} & \text{if } x \geq 0 \\ 0 & \text{otherwise} \end{cases}$$

(A) What is the probability that a beginner can type at least 30 words per minute after 1 week of practice?

(B) What is the probability that a beginner can type at least 80 words per minute after 1 week of practice?

50. Learning. The number of hours it takes a chimpanzee to learn a new task is a continuous random variable with probability density function given by

$$f(x) = \begin{cases} \frac{4}{9}x^2 - \frac{4}{27}x^3 & \text{if } 0 \leq x \leq 3 \\ 0 & \text{otherwise} \end{cases}$$

(A) What is the probability that the chimpanzee learns the task in the first hour?

(B) What is the probability that the chimpanzee does not learn the task in the first 2 hours?

10-5 Expected Value, Standard Deviation, and Median of Continuous Random Variables

Expected Value and Standard Deviation

Earlier, we used finite sums and the probability distribution function to define the expected value or mean, the variance, and the standard deviation of a discrete random variable. In much the same way, we now use integration and the probability density function to define these quantities for a continuous random variable. Compare the formulas below with those in Section 10-1.

Expected Value and Standard Deviation for a Continuous Random Variable

Let $f(x)$ be the probability density function for a continuous random variable X. The **expected value, or mean, of X** is

$$\mu = E(X) = \int_{-\infty}^{\infty} xf(x)\,dx$$

The **variance** is

$$V(X) = \int_{-\infty}^{\infty} (x-\mu)^2 f(x)\,dx$$

and the **standard deviation** is

$$\sigma = \sqrt{V(X)}$$

Just as in the discrete case, the mean of a continuous random variable is a measure of central tendency for the variable, and the standard deviation is a measure of the dispersion of the variable about the mean. This is illustrated in Figure 7. The probability density function in Figure 7A has a standard deviation of 1. Most of the area under the curve is near the mean. In Figure 7B, the standard deviation is four times as large, and the area under the graph is much more spread out.

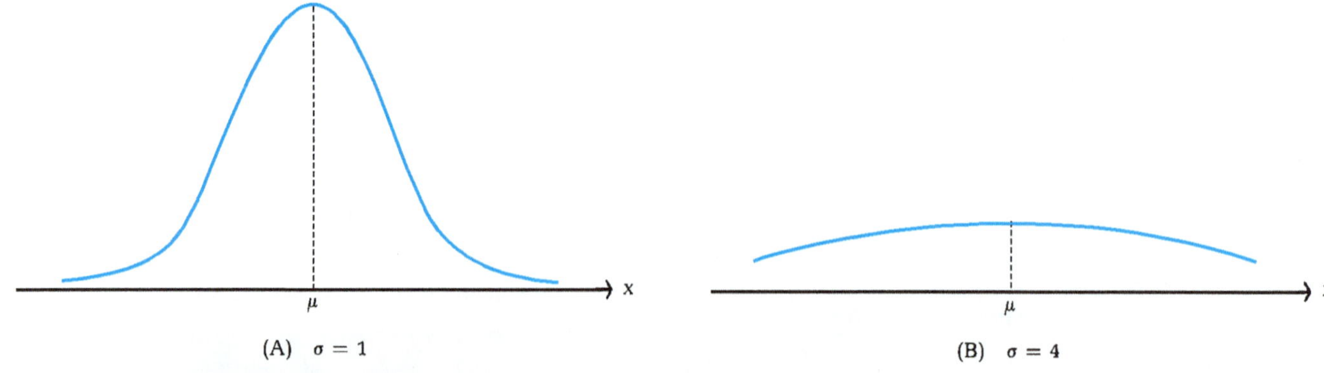

Figure 7

EXAMPLE 22 Find the mean, variance, and standard deviation for

$$f(x) = \begin{cases} 12x^2 - 12x^3 & \text{if } 0 \le x \le 1 \\ 0 & \text{otherwise} \end{cases}$$

SOLUTION

$$\mu = E(X) = \int_{-\infty}^{\infty} xf(x)\,dx = \int_0^1 x(12x^2 - 12x^3)dx = \int_0^1 (12x^3 - 12x^4)\,dx$$

$$= (3x^4 - \tfrac{12}{5}x^5)\Big|_0^1 = \tfrac{3}{5}$$

$$V(X) = \int_{-\infty}^{\infty} (x - \mu)^2 f(x)\, dx = \int_0^1 (x - \tfrac{3}{5})^2(12x^2 - 12x^3)\, dx$$

$$= \int_0^1 (x^2 - \tfrac{6}{5}x + \tfrac{9}{25})(12x^2 - 12x^3)\, dx$$

$$= \int_0^1 (\tfrac{108}{25}x^2 - \tfrac{468}{25}x^3 + \tfrac{132}{5}x^4 - 12x^5)\, dx$$

$$= (\tfrac{36}{25}x^3 - \tfrac{117}{25}x^4 + \tfrac{132}{25}x^5 - 2x^6)\Big|_0^1 = \tfrac{1}{25}$$

$$\sigma = \sqrt{V(X)} = \sqrt{\tfrac{1}{25}} = \tfrac{1}{5}$$

Matched Problem 22 Find the mean, variance, and standard deviation for

$$f(x) = \begin{cases} 6x - 6x^2 & \text{if } 0 \le x \le 1 \\ 0 & \text{otherwise} \end{cases}$$

The graph of the probability density function considered in Example 22 is shown in Figure 8, along with the indicated areas (computations omitted). Just as in the discrete case (see Figure 3), geometrically the mean represents the balance point for the region formed by the graph of f and the x axis. Notice that a vertical line through the mean does not divide the area under the graph of f into two regions with equal area, as you might expect. The value of x that does this is the *median*, which we will discuss later in this section.

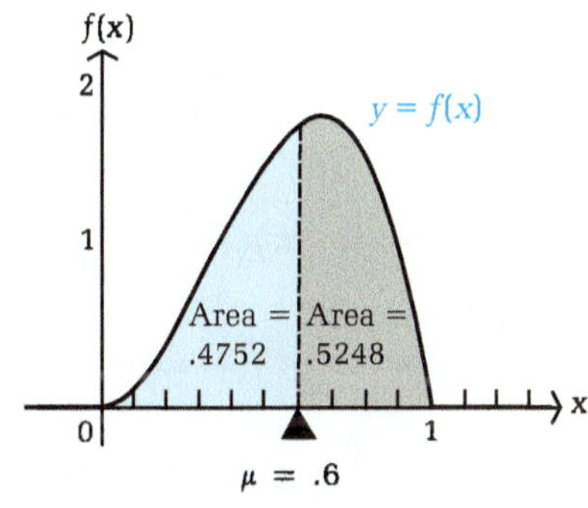

Figure 8 Mean of a continuous random variable

EXAMPLE 23 **Life Expectancy** The life expectancy (in hours) for a particular brand of light bulbs is a continuous random variable with probability density function

$$f(x) = \begin{cases} \frac{1}{100} - \frac{1}{20{,}000}x & \text{if } 0 \le x \le 200 \\ 0 & \text{otherwise} \end{cases}$$

(A) What is the average life expectancy of one of these light bulbs?

(B) What is the probability that a bulb will last longer than this average?

SOLUTIONS (A) Since the value of this random variable is the number of hours a bulb lasts, the average life expectancy is just the expected value of the random variable. Thus,

$$E(X) = \int_{-\infty}^{\infty} xf(x)\, dx = \int_0^{200} x(\tfrac{1}{100} - \tfrac{1}{20{,}000}x)\, dx$$

$$= \int_0^{200} \left(\tfrac{1}{100}x - \tfrac{1}{20,000}x^2\right) dx = \left(\tfrac{1}{200}x^2 - \tfrac{1}{60,000}x^3\right)\Big|_0^{200}$$

$$= \tfrac{200}{3} \text{ or } 66\tfrac{2}{3} \text{ hours}$$

(B) The probability that a bulb lasts longer than $66\frac{2}{3}$ hours is

$$P\left(X > \tfrac{200}{3}\right) = \int_{200/3}^{\infty} f(x)\, dx = \int_{200/3}^{200} \left(\tfrac{1}{100} - \tfrac{1}{20,000}x\right) dx$$

$$= \left(\tfrac{1}{100}x - \tfrac{1}{40,000}x^2\right)\Big|_{200/3}^{200}$$

$$= 1 - \tfrac{5}{9} = \tfrac{4}{9}$$

Matched Problem 23 Repeat Example 23 if the probability density function is

$$f(x) = \begin{cases} \frac{1}{200} - \frac{1}{90,000}x & \text{if } 0 \le x \le 300 \\ 0 & \text{otherwise} \end{cases}$$

Alternate Formula for Variance

The term $(x - \mu)^2$ in the formula for $V(X)$ introduces some complicated algebraic manipulations in the evaluation of the integral. We can use the properties of the definite integral to simplify this formula. Thus,

$$V(X) = \int_{-\infty}^{\infty} (x - \mu)^2 f(x)\, dx \qquad \text{Expand } (x - \mu)^2.$$

$$= \int_{-\infty}^{\infty} (x^2 - 2x\mu + \mu^2) f(x)\, dx \qquad \text{Multiply by } f(x)$$

$$= \int_{-\infty}^{\infty} [x^2 f(x) - 2x\mu f(x) + \mu^2 f(x)]\, dx$$

$$= \int_{-\infty}^{\infty} x^2 f(x)\, dx - \int_{-\infty}^{\infty} 2x\mu f(x)\, dx + \int_{-\infty}^{\infty} \mu^2 f(x)\, dx$$

$$= \int_{-\infty}^{\infty} x^2 f(x)\, dx - 2\mu \int_{-\infty}^{\infty} x f(x)\, dx + \mu^2 \int_{-\infty}^{\infty} f(x)\, dx \qquad \int_{-\infty}^{\infty} x f(x)\, dx = \mu, \quad \int_{-\infty}^{\infty} f(x)\, dx = 1$$

$$= \int_{-\infty}^{\infty} x^2 f(x)\, dx - 2\mu(\mu) + \mu^2(1)$$

$$= \int_{-\infty}^{\infty} x^2 f(x)\, dx - \mu^2$$

In general, it will be easier to evaluate $\int_{-\infty}^{\infty} x^2 f(x)\, dx$ than to evaluate $\int_{-\infty}^{\infty} (x - \mu)^2 f(x)\, dx$.

THEOREM 3 Alternate Formula for Variance

$$V(X) = \int_{-\infty}^{\infty} x^2 f(x)\, dx - \mu^2$$

EXAMPLE 24 Use the alternate formula for variance (Theorem 3) to compute the variance in Example 22.

SOLUTION From Example 22, we have $\mu = \int_{-\infty}^{\infty} x f(x)\, dx = \frac{3}{5}$. Thus,

$$\int_{-\infty}^{\infty} x^2 f(x)\, dx = \int_0^1 x^2(12x^2 - 12x^3)\, dx \qquad f(x) = \begin{cases} 12x^2 - 12x^3 & \text{if } 0 \le x \le 1 \\ 0 & \text{otherwise} \end{cases}$$

$$= \int_0^1 (12x^4 - 12x^5)\,dx$$

$$= (\tfrac{12}{5}x^5 - \tfrac{12}{6}x^6)\big|_0^1 = \tfrac{2}{5}$$

$$V(X) = \int_{-\infty}^{\infty} x^2 f(x)\,dx - \mu^2 = \tfrac{2}{5} - (\tfrac{3}{5})^2 = \tfrac{1}{25}$$

Matched Problem 24 Use the alternate formula for variance (Theorem 3) to compute the variance in Problem 22.

EXAMPLE 25 Find the mean, variance, and standard deviation for

$$f(x) = \begin{cases} 3/x^4 & \text{if } x \geq 1 \\ 0 & \text{otherwise} \end{cases}$$

SOLUTION

$$\mu = \int_{-\infty}^{\infty} xf(x)\,dx = \int_1^{\infty} x\frac{3}{x^4}\,dx = \lim_{R\to\infty} \int_1^R \frac{3}{x^3}\,dx$$

$$= \lim_{R\to\infty} \left[-\frac{3}{2}\left(\frac{1}{x^2}\right)\right]\bigg|_1^R = \lim_{R\to\infty}\left[-\frac{3}{2}\left(\frac{1}{R^2}\right) + \frac{3}{2}\right] = \frac{3}{2}$$

$$\int_{-\infty}^{\infty} x^2 f(x)\,dx = \int_1^{\infty} x^2\frac{3}{x^4}\,dx = \lim_{R\to\infty}\int_1^R \frac{3}{x^2}\,dx = \lim_{R\to\infty}\left(-\frac{3}{x}\right)\bigg|_1^R$$

$$= \lim_{R\to\infty}\left(-\frac{3}{R} + 3\right) = 3$$

$$V(X) = \int_{-\infty}^{\infty} x^2 f(x)\,dx - \mu^2 = 3 - \left(\frac{3}{2}\right)^2 = \frac{3}{4}$$

$$\sigma = \sqrt{V(X)} = \sqrt{\frac{3}{4}} = \frac{\sqrt{3}}{2} \approx .8660$$

Matched Problem 25 Find the mean, variance, and standard deviation for

$$f(x) = \begin{cases} 24/x^4 & \text{if } x \geq 2 \\ 0 & \text{otherwise} \end{cases}$$

Median

Another measurement often used to describe the properties of a random variable is the median. The **median** is the value of the random variable that divides the area under the graph of the probability density function into two equal parts (see Figure 9). If x_m is the median, then x_m must satisfy

$$P(X \leq x_m) = \tfrac{1}{2}$$

Generally, this equation is solved by first finding the cumulative probability distribution function.

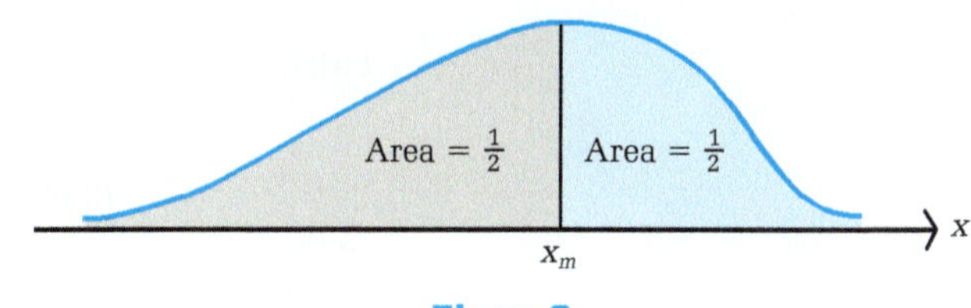

Figure 9

EXAMPLE 26 Find the median of the continuous random variable with probability density function

$$f(x) = \begin{cases} 3/x^4 & \text{if } x \geq 1 \\ 0 & \text{otherwise} \end{cases}$$

SOLUTION **Step 1** Find the cumulative probability distribution function. For $x < 1$, we have $F(x) = 0$. If $x \geq 1$, then

$$F(x) = \int_{-\infty}^{x} f(t)\, dt = \int_{1}^{x} \frac{3}{t^4}\, dt = -\frac{1}{t^3}\bigg|_1^x = -\frac{1}{x^3} + 1 = 1 - \frac{1}{x^3}$$

Step 2 Solve the equation $P(X \leq x_m) = \frac{1}{2}$ for x_m.

$$F(x_m) = P(X \leq x_m)$$

$$1 - \frac{1}{x_m^3} = \frac{1}{2}$$

$$\frac{1}{2} = \frac{1}{x_m^3}$$

$$x_m^3 = 2$$

$$x_m = \sqrt[3]{2}$$

Thus, the median is $\sqrt[3]{2} \approx 1.26$.

Matched Problem 26 Find the median of the continuous random variable with probability density function

$$f(x) = \begin{cases} 24/x^4 & \text{if } x \geq 2 \\ 0 & \text{otherwise} \end{cases}$$

EXAMPLE 27 **Life Expectancy** In Example 23, find the median life expectancy of a light bulb.

SOLUTION **Step 1** Find the cumulative probability distribution function. If $x < 0$, we have $F(x) = 0$. If $0 \leq x \leq 200$, then

$$F(x) = \int_{-\infty}^{x} f(t)\, dt \qquad f(x) = \begin{cases} \frac{1}{100} - \frac{1}{20{,}000}x \\ 0 \end{cases}$$

$$= \int_0^x \left(\frac{1}{100} - \frac{1}{20{,}000}t\right) dt$$

$$= \left(\frac{1}{100}t - \frac{1}{40{,}000}t^2\right)\bigg|_0^x$$

$$= \frac{1}{100}x - \frac{1}{40{,}000}x^2$$

If $x > 200$, then

$$F(x) = \int_{-\infty}^{x} f(t)\, dt = \int_{-\infty}^{0} f(t)\, dt + \int_{0}^{200} f(t)\, dt + \int_{200}^{x} f(t)\, dt$$

$$= 0 + 1 + 0 = 1$$

Thus,

$$F(x) = \begin{cases} 0 & \text{if } x < 0 \\ \frac{1}{100}x - \frac{1}{40{,}000}x^2 & \text{if } 0 \leq x \leq 200 \\ 1 & \text{if } x > 200 \end{cases}$$

Step 2 Solve the equation $P(X \le x_m) = \frac{1}{2}$ for x_m:

$$F(x_m) = P(X \le x_m) = \frac{1}{2}$$

$$\frac{1}{100}x_m - \frac{1}{40{,}000}x_m^2 = \frac{1}{2}$$

$$x_m^2 - 400x_m + 20{,}000 = 0$$ The solution must occur for $0 \le x_m \le 200$.

This quadratic equation has two solutions, $200 + 100\sqrt{2}$ and $200 - 100\sqrt{2}$. Since x_m must lie in the interval $[0, 200]$, the second root is the correct answer.

Thus, the median life expectancy is $200 - 100\sqrt{2} \approx 58.58$ hours.

Matched Problem 27 In Problem 23, find the median life expectancy of a light bulb.

If you compare Example 23 and 27, and Examples 25 and 26, you will see that the mean and the median generally are not equal (see Figure 10).

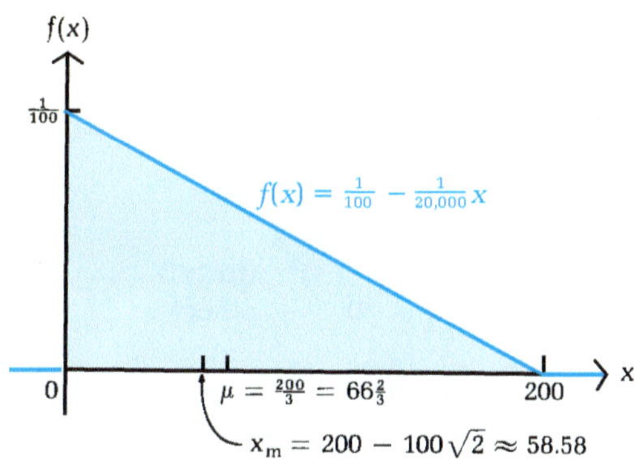

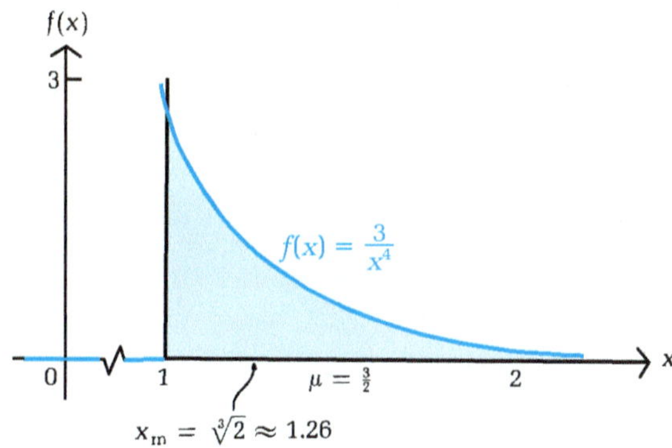

Figure 10

Answers to Matched Problems

22. $\mu = \frac{1}{2}$; $V(X) = \frac{1}{20}$; $\sigma \approx .2236$ **23.** (A) 125 hr (B) $\frac{133}{288}$

24. $\frac{1}{20}$ **25.** $\mu = 3$; $V(X) = 3$; $\sigma = \sqrt{3} \approx 1.732$

26. $x_m = \sqrt[3]{16} = 2\sqrt[3]{2} \approx 2.52$ **27.** $x_m = 450 - 150\sqrt{5} \approx 114.59$ hr

Exercise 10-5

A

In Problems 1–6, find the mean, variance, and standard deviation.

1. $f(x) = \begin{cases} \frac{1}{2}x & \text{if } 0 \le x \le 2 \\ 0 & \text{otherwise} \end{cases}$

2. $f(x) = \begin{cases} 3x^2 & \text{if } 0 \le x \le 1 \\ 0 & \text{otherwise} \end{cases}$

3. $f(x) = \begin{cases} \frac{1}{3} & \text{if } 0 \le x \le 2 \\ 0 & \text{otherwise} \end{cases}$

4. $f(x) = \begin{cases} \frac{1}{2} & \text{if } 1 \le x \le 3 \\ 0 & \text{otherwise} \end{cases}$

5. $f(x) = \begin{cases} 4 - 2x & \text{if } 1 \le x \le 2 \\ 0 & \text{otherwise} \end{cases}$

6. $f(x) = \begin{cases} 2 - \frac{1}{2}x & \text{if } 2 \le x \le 4 \\ 0 & \text{otherwise} \end{cases}$

In Problems 7–12, find the median.

7. $f(x) = \begin{cases} 2x & \text{if } 0 \le x \le 1 \\ 0 & \text{otherwise} \end{cases}$

8. $f(x) = \begin{cases} \frac{1}{8}x & \text{if } 0 \le x \le 4 \\ 0 & \text{otherwise} \end{cases}$

9. $f(x) = \begin{cases} \frac{1}{6}x & \text{if } 2 \le x \le 4 \\ 0 & \text{otherwise} \end{cases}$

10. $f(x) = \begin{cases} \frac{1}{4}x & \text{if } 1 \le x \le 3 \\ 0 & \text{otherwise} \end{cases}$

11. $f(x) = \begin{cases} \frac{1}{2} - \frac{1}{8}x & \text{if } 0 \le x \le 4 \\ 0 & \text{otherwise} \end{cases}$

12. $f(x) = \begin{cases} 2 - 2x & \text{if } 0 \le x \le 1 \\ 0 & \text{otherwise} \end{cases}$

B

In Problems 13–16, find the mean, variance, and standard deviation.

13. $f(x) = \begin{cases} 4/x^5 & \text{if } x \ge 1 \\ 0 & \text{otherwise} \end{cases}$

14. $f(x) = \begin{cases} 5/x^6 & \text{if } x \ge 1 \\ 0 & \text{otherwise} \end{cases}$

15. $f(x) = \begin{cases} 64/x^5 & \text{if } x \ge 2 \\ 0 & \text{otherwise} \end{cases}$

16. $f(x) = \begin{cases} 81/x^4 & \text{if } x \ge 3 \\ 0 & \text{otherwise} \end{cases}$

In Problems 17–24, find the median.

17. $f(x) = \begin{cases} 1/x & \text{if } 1 \le x \le e \\ 0 & \text{otherwise} \end{cases}$

18. $f(x) = \begin{cases} 1/(2x) & \text{if } 0 \le x \le e^2 \\ 0 & \text{otherwise} \end{cases}$

19. $f(x) = \begin{cases} 4/(2 + x)^2 & \text{if } 0 \le x \le 2 \\ 0 & \text{otherwise} \end{cases}$

20. $f(x) = \begin{cases} 2/(1 + x)^2 & \text{if } 0 \le x \le 1 \\ 0 & \text{otherwise} \end{cases}$

21. $f(x) = \begin{cases} 1/(1 + x)^2 & \text{if } x \ge 0 \\ 0 & \text{otherwise} \end{cases}$

22. $f(x) = \begin{cases} 3/(3 + x)^2 & \text{if } x \ge 0 \\ 0 & \text{otherwise} \end{cases}$

23. $f(x) = \begin{cases} 2e^{-2x} & \text{otherwise} \\ 0 & \end{cases}$

24. $f(x) = \begin{cases} e^{-x} & \text{if } x \ge 0 \\ 0 & \text{otherwise} \end{cases}$

C

In Problems 25 and 26, $f(x)$ is a continuous probability density function with mean μ and standard deviation σ; a and b are constants. Evaluate each integral, expressing the result in terms of a, b, μ, and σ.

25. $\int_{-\infty}^{\infty} (ax + b)f(x)\, dx$

26. $\int_{-\infty}^{\infty} (x - a)^2 f(x)\, dx$

*The **quartile points** for a probability density function are the values x_1, x_2, x_3 that divide the area under the graph of the function into four equal parts. Find the quartile points for the probability density functions in Problems 27–30.*

27. $f(x) = \begin{cases} \frac{1}{2}x & \text{if } 0 \le x \le 2 \\ 0 & \text{otherwise} \end{cases}$

28. $f(x) = \begin{cases} 3x^2 & \text{if } 0 \le x \le 1 \\ 0 & \text{otherwise} \end{cases}$

29. $f(x) = \begin{cases} 3/(3 + x)^2 & \text{if } x \ge 0 \\ 0 & \text{otherwise} \end{cases}$

30. $f(x) = \begin{cases} 1/(1 + x)^2 & \text{if } x \ge 0 \\ 0 & \text{otherwise} \end{cases}$

In Problems 31–34, use a graphic calculator or a computer to approximate the median of the indicated probability density function f to two decimal places.

31. $f(x) = \begin{cases} 4x - 4x^3 & \text{if } 0 \le x \le 1 \\ 0 & \text{otherwise} \end{cases}$

32. $f(x) = \begin{cases} 3x - 3x^5 & \text{if } 0 \le x \le 1 \\ 0 & \text{otherwise} \end{cases}$

33. $f(x) = \begin{cases} \ln x & \text{if } 1 \le x \le x \\ 0 & \text{otherwise} \end{cases}$

34. $f(x) = \begin{cases} xe^{-x} & \text{if } x \ge 0 \\ 0 & \text{otherwise} \end{cases}$

Applications

Business & Economics

35. **Profit.** A building contractor's profit (in thousands of dollars) on each unit in a subdivision is a continuous random variable with probability density function given by

$$f(x) = \begin{cases} \frac{1}{8}(10 - x) & \text{if } 6 \le x \le 10 \\ 0 & \text{otherwise} \end{cases}$$

(A) Find the contractor's expected profit.
(B) Find the median profit.

36. **Electricity consumption.** The daily consumption of electricity (in millions of kilowatt-hours) in a large city is a continuous random variable with probability density function

$$f(x) = \begin{cases} .2 - .02x & \text{if } 0 \le x \le 10 \\ 0 & \text{otherwise} \end{cases}$$

(A) Find the expected daily consumption of electricity.
(B) Find the median daily consumption of electricity.

37. **Waiting time.** The time (in minutes) a customer must wait in line at a bank is a continuous random variable with probability density function given by

$$f(x) = \begin{cases} \frac{1}{3}e^{-x/3} & \text{if } x \ge 0 \\ 0 & \text{otherwise} \end{cases}$$

Find the median waiting time.

38. **Product life.** The life expectancy (in years) of an automobile battery is a continuous random variable with probability density function given by

$$f(x) = \begin{cases} \frac{1}{2}e^{-x/2} & \text{if } x \ge 0 \\ 0 & \text{otherwise} \end{cases}$$

Find the median life expectancy.

39. **Water consumption.** The daily consumption of water (in millions of gallons) in a small city is a continuous random variable with probability density function given by

$$f(x) = \begin{cases} 1/(1 + x^2)^{3/2} & \text{if } x \ge 0 \\ 0 & \text{otherwise} \end{cases}$$

Find the expected daily consumption.

40. **Gasoline consumption.** The daily consumption of gasoline (in millions of gallons) in a large city is a continuous random variable with probability density function.

$$f(x) = \begin{cases} 4/(4 + x^2)^{3/2} & \text{if } x \ge 0 \\ 0 & \text{otherwise} \end{cases}$$

Find the expected daily consumption of gasoline.

Life Sciences

41. **Life expectancy.** The life expectancy of a certain microscopic organism (in minutes) is a continuous random variable with probability density function given by

$$f(x) = \begin{cases} \frac{1}{5{,}000}(10x^3 - x^4) & \text{if } 0 \le x \le 10 \\ 0 & \text{otherwise} \end{cases}$$

Find the mean life expectancy of one of these organisms.

42. **Life expectancy.** The life expectancy (in months) of plants of a certain species is a continuous random variable with probability density function given by

$$f(x) = \begin{cases} \frac{1}{36}(6x - x^2) & \text{if } 0 \le x \le 6 \\ 0 & \text{otherwise} \end{cases}$$

Find the mean life expectancy of one of these plants.

43. **Shelf-life.** The shelf-life (in days) of a perishable drug is a continuous random variable with probability density function given by

$$f(x) = \begin{cases} 800x/(400 + x^2)^2 & \text{if } x \ge 0 \\ 0 & \text{otherwise} \end{cases}$$

Find the median shelf-life.

44. **Shelf life.** Repeat Problem 43 if

$$f(x) = \begin{cases} 200x/(100 + x^2)^2 & \text{if } x \ge 0 \\ 0 & \text{otherwise} \end{cases}$$

Social Sciences

45. **Learning.** The number of hours it takes a chimpanzee to learn a new task is a continuous random variable with probability density function given by

$$f(x) = \begin{cases} \frac{4}{9}x^2 - \frac{4}{27}x^3 & \text{if } 0 \le x \le 3 \\ 0 & \text{otherwise} \end{cases}$$

What is the expected number of hours it will take a chimpanzee to learn the task?

46. **Voter turnout.** The number of registered voters (in thousands) who vote in an off-year election in a small town is a continuous random variable with probability density function given by

$$f(x) = \begin{cases} 1 - \frac{1}{8}x & \text{if } 4 \le x \le 8 \\ 0 & \text{otherwise} \end{cases}$$

(A) Find the expected voter turnout.
(B) Find the median voter turnout.

10-6 Uniform, Beta, and Exponential Distributions

- Uniform Distribution
- Beta Distribution
- Exponential Distribution

In this section we will examine several important probability density functions. In actual practice, we do not usually construct a probability density function for each experiment. Instead, we try to select a known probability density function that seems to give a reasonable description of the experiment. Thus, it is important to be familiar with the properties and applications of a variety of probability density functions.

Uniform Distribution

We begin with an example that will lead to some general observations.

EXAMPLE 28 **Waiting Time** A bus arrives every 30 minutes at a particular bus stop. If an individual arrives at the bus stop at a random time (that is, with no knowledge of the bus schedule), then the random variable X, representing the time this individual must spend waiting for the next bus, is said to be *uniformly distributed* on the interval [0, 30]. This means that the probability that X lies in a small interval of fixed length is independent of the location of that interval within [0, 30]. Thus, the probability of waiting between 0 and 5 minutes is the same as the probability of waiting between 5 and 10 minutes or the probability of waiting between 18 and 23 minutes. It can be shown that the probability density function of this uniformly distributed random variable is

$$f(x) = \begin{cases} \frac{1}{30} & \text{if } 0 \le x \le 30 \\ 0 & \text{otherwise} \end{cases}$$

Thus,

$$P(0 \le X \le 5) = \int_0^5 \frac{1}{30}\,dx = \frac{x}{30}\Big|_0^5 = \frac{5}{30} - \frac{0}{30} = \frac{1}{6}$$

$$P(5 \le X \le 10) = \int_5^{10} \frac{1}{30}\,dx = \frac{x}{30}\Big|_5^{10} = \frac{10}{30} - \frac{5}{30} = \frac{1}{6}$$

$$P(18 \le X \le 23) = \int_{18}^{23} \frac{1}{30}\,dx = \frac{x}{30}\Big|_{18}^{23} = \frac{23}{30} - \frac{18}{30} = \frac{1}{6}$$

Each of these probabilities is represented as an area under the graph of f in the figure.

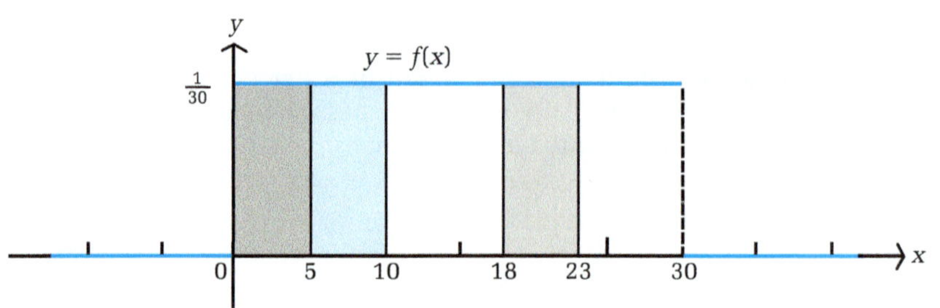

Matched Problem 28 Use the probability density function given in Example 28 to find the probability that the individual waits:

(A) Between 0 and 10 minutes (B) Between 10 and 20 minutes
(C) Between 17 and 27 minutes

In general, if the outcomes of an experiment lie in an interval $[a, b]$ and if the probability of the outcome lying in a small interval of fixed length is independent of the location of this small interval within $[a, b]$, then we say that the continuous random variable for this experiment is **uniformly distributed** on the interval $[a, b]$. The **uniform probability density function** is

$$f(x) = \begin{cases} \dfrac{1}{b-a} & \text{if } a \le x \le b \\ 0 & \text{otherwise} \end{cases}$$

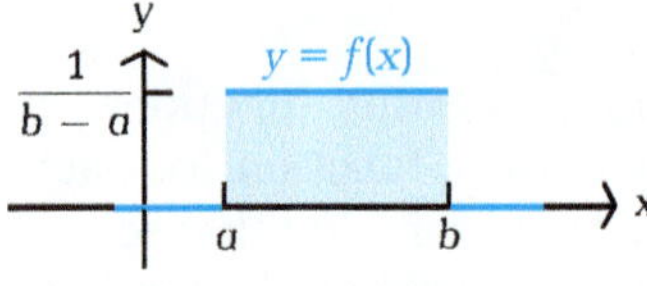

Figure 11

See Figure 11. Since $f(x) \ge 0$ and

$$\int_{-\infty}^{\infty} f(x)\, dx = \int_a^b \frac{1}{b-a}\, dx = \frac{x}{b-a}\Big|_a^b = \frac{b}{b-a} - \frac{a}{b-a} = 1$$

f satisfies the necessary conditions for a probability density function.

If F is the associated cumulative probability distribution function, then for $x < a$, $F(x) = 0$. For $a \leq x \leq b$, we have

$$F(x) = \int_{-\infty}^{x} f(t)\, dt = \int_a^x \frac{1}{b-a}\, dt = \frac{t}{b-a}\Big|_a^x$$

$$= \frac{x}{b-a} - \frac{a}{b-a} = \frac{x-a}{b-a}$$

For $x > b$, $F(x) = 1$.

Using the techniques discussed in the preceding section, it can be shown that (see Problems 25–28 in Exercise 10-6)

$$\mu = x_m = \frac{1}{2}(a+b) \qquad \text{and} \qquad \sigma = \frac{1}{\sqrt{12}}(b-a)$$

These properties are summarized in the box on the next page.

Uniform Probability Density Function

PROBABILITY DENSITY FUNCTION CUMULATIVE PROBABILITY DISTRIBUTION

$$f(x) = \begin{cases} \frac{1}{b-a} & \text{if } a \leq x \leq b \\ 0 & \text{otherwise} \end{cases} \qquad f(x) = \begin{cases} 0 & \text{if } x < b \\ \frac{x-a}{b-a} & \text{if } a \leq x \leq b \\ 1 & \text{if } x > b \end{cases}$$

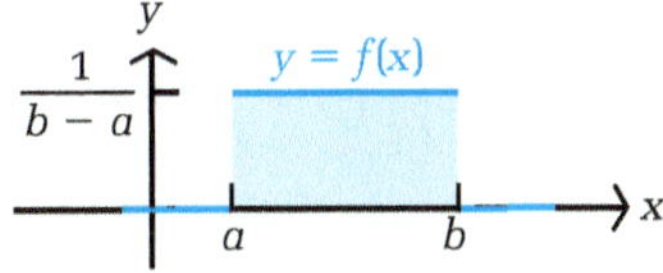

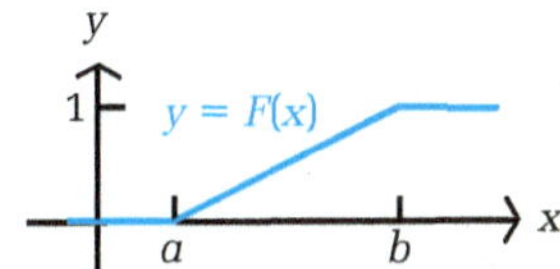

MEAN $\mu = \frac{1}{2}(a+b)$ MEDIAN $x_m = \frac{1}{2}(a+b)$

STANDARD DEVIATION $\sigma = \frac{1}{\sqrt{12}}(b-a)$

EXAMPLE 29 **Electrical Current** Standard electrical current is uniformly distributed between 110 and 120 volts. What is the probability that the current is between 113 and 118 volts?

SOLUTION Since we are told that the current is uniformly distributed on the interval [110, 120], we choose the uniform probability density function

$$f(x) = \begin{cases} \frac{1}{10} & \text{if } 110 \leq x \leq 120 \\ 0 & \text{otherwise} \end{cases}$$

Then

$$P(113 \leq X \leq 118) = \int_{113}^{118} \frac{1}{10}\, dx = \frac{x}{10}\Big|_{113}^{118} = \frac{118}{10} - \frac{113}{10} = \frac{1}{2}$$

Matched Problem 29 In Example 29, what is the probability that the current is at least 116 volts?

Beta Distribution

An area of applied statistics called *Bayesian inference,* named after Presbyterian minister Thomas Bayers (1702–1763), has received a great deal of attention in recent year. Effective use of this method requires experience with a variety of random variables with values that can be expressed as fractions or percentages. One particular random variable that has been used extensively in this area of statistics is the *beta random variable.* Typical applications of beta random variables include the percentage of fast-food restaurants that make a profit during their first year of operation, the percentage of time each year a manufacturing plant is shut down, the consumption of natural gas as a percentage of capacity, and the percentage of job application forms that contain errors.

A continuous random variable has a **beta distribution*** and is referred to as a **beta random variable** if its probability density function is the **beta probability density function**

$$f(x) = \begin{cases} (\beta + 1)(\beta + 2)x^\beta(1 - x) & \text{if } 0 \leqslant x \leqslant 1 \\ 0 & \text{otherwise} \end{cases}$$

where β is a constant, $\beta \geqslant 0$. The value of β is usually determined by examining the results of a particular experiment. The values of a beta random variable can be expressed as fractions or percentages; however, percentages should be converted to fractions before performing calculations involving a beta random variable.

First, we show that f satisfies the requirements for a probability density function:

$$f(x) = (\beta + 1)(\beta + 2)x^\beta(1 - x) \geqslant 0 \qquad 0 \leqslant x \leqslant 1$$

$$\begin{aligned} \int_{-\infty}^{\infty} f(x)\,dx &= \int_0^1 (\beta + 1)(\beta + 2)x^\beta(1 - x)\,dx \\ &= \int_0^1 (\beta + 1)(\beta + 2)(x^\beta - x^{\beta+1})\,dx \\ &= (\beta + 1)(\beta + 2)\left(\frac{x^{\beta+1}}{\beta + 1} - \frac{x^{\beta+2}}{\beta + 2}\right)\Bigg|_0^1 \\ &= (\beta + 1)(\beta + 2)\left(\frac{1}{\beta + 1} - \frac{1}{\beta + 2}\right) \\ &= (\beta + 2) - (\beta + 1) = 1 \end{aligned}$$

Thus, f is a probability density function.

If $F(x)$ is the associated cumulative probability distribution function, then for $x < 0$, $F(x) = 0$. For $0 \leqslant x \leqslant 1$, we have

$$\begin{aligned} F(x) &= \int_{-\infty}^{x} f(t)\,dt = \int_0^x (\beta + 1)(\beta + 2)t^\beta(1 - t)\,dt \\ &= (\beta + 1)(\beta + 2)\left(\frac{t^{\beta+1}}{\beta + 1} - \frac{t^{\beta+2}}{\beta + 2}\right)\Bigg|_0^x \\ &= (\beta + 1)(\beta + 2)\left(\frac{x^{\beta+1}}{\beta + 1} - \frac{x^{\beta+2}}{\beta + 2}\right) \\ &= (\beta + 2)x^{\beta+1} - (\beta + 1)x^{\beta+2} \end{aligned}$$

*There is a more general definition of a beta distribution, but we will not consider it here.

And for $x > 1$,

$$F(x) = 1$$

In general, it is not possible to solve the equation $F(x_m) = \frac{1}{2}$ for x_m. Thus, we will not discuss the median of a beta random variable. By straightforward (but tedious) integration we can show that

$$\mu = \frac{\beta + 1}{\beta + 3} \quad \text{and} \quad \sigma = \sqrt{\frac{2(\beta + 1)}{(\beta + 4)(\beta + 3)^2}}$$

The calculations are not included here. The above results are summarized in the box.

Beta Probability Density Function

$$f(x) = \begin{cases} (\beta + 1)(\beta + 2)x^{\beta}(1 - x) & \text{if } 0 \le x \le 1 \\ 0 & \text{otherwise} \end{cases} \quad \text{where } \beta \ge 0$$

$$F(x) = \begin{cases} 0 & \text{if } x < 0 \\ (\beta + 2)x^{\beta+1} - (\beta + 1)x^{\beta+2} & \text{if } 0 \le x \le 1 \\ 1 & \text{if } x > 1 \end{cases}$$

MEAN $\mu = \frac{\beta + 1}{\beta + 3}$ STANDARD DEVIATION $\sigma = \sqrt{\frac{2(\beta + 1)}{(\beta + 4)(\beta + 3)^2}}$

Figure 12 shows the graphs of $f(x)$ for some typical values of β.

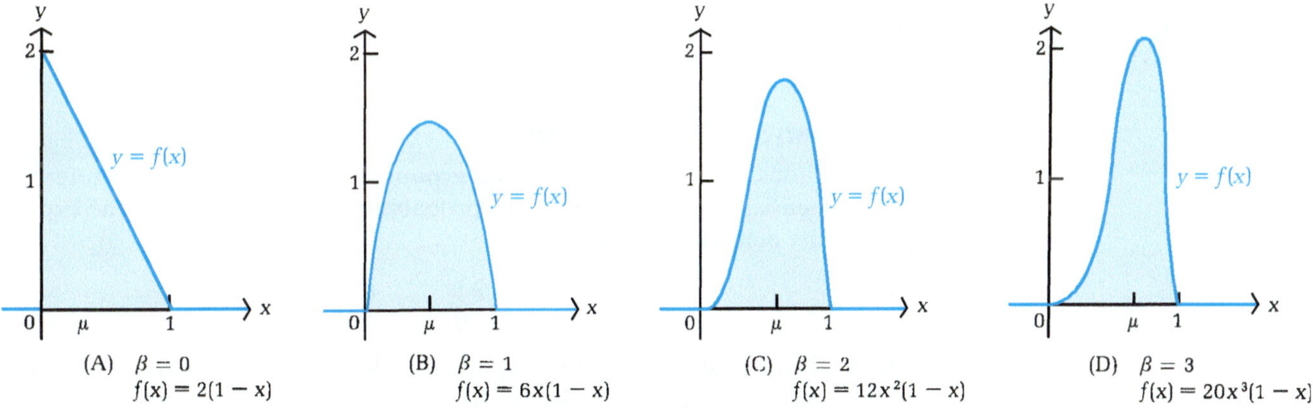

Figure 12
Beta probability density functions

EXAMPLE 30 **Income Tax** The annual percentage of correct income tax forms filed with the Internal Revenue Service is a beta random variable with $\beta = 8$.

(A) What is the probability that at least half the returns filed are correct?
(B) What is the expected percentage of correct returns?

SOLUTIONS Substituting $\beta = 8$ in the definition of the beta probability density function, we have

$$f(x) = \begin{cases} 90x^8(1 - x) & \text{if } 0 \le x \le 1 \\ 0 & \text{otherwise} \end{cases}$$

(A) $P(X \geq \frac{1}{2}) = \int_{1/2}^{1} 90x^8(1 - x)\,dx = \int_{1/2}^{1} (90x^8 - 90x^9)\,dx$

$$= (10x^9 - 9x^{10})\big|_{1/2}^{1} = 1 - \frac{11}{2^{10}} \approx .989$$

(B) $\mu = E(X) = \frac{\beta + 1}{\beta + 3} = \frac{8 + 1}{8 + 3} = \frac{9}{11} \approx .818$

Thus, we expect approximately 82% of the returns to be correct.

Matched Problem 30 In Example 30, what is the probability that at least 90% of the returns are correct?

EXAMPLE 31 **Learning** A psychologist is studying the learning abilities of children in a certain age group by giving them 5 minutes to learn to perform a particular task. Repeated trials of the experiment have led to the conclusions that the percentage of children who can learn to perform this task in 5 minutes is a beta random variable and that the average percentage of children who learn the task is 75%. What is the appropriate value of β for this experiment?

SOLUTIONS Since the average percentage of children who learn the task is 75% and the mean for any beta distribution is $(\beta + 1)/(\beta + 3)$, the value of β must satisfy

$$\frac{\beta + 1}{\beta + 3} = .75 \qquad \text{Convert 75\% to the decimal fraction .75.}$$

Solving this equation, we obtain $\beta = 5$.

Matched Problem 31 In Example 31, what is the probability that at least 75% of the children will learn the task in 5 minutes?

Exponential Distribution

A continuous random variable has an **exponential distribution** and is referred to as an **exponential random variable** if its probability density function is the **exponential probability density function**

$$f(x) = \begin{cases} (1/\lambda)e^{-x/\lambda} & \text{if } x \geq 0 \\ 0 & \text{otherwise} \end{cases}$$

where λ is the positive constant. Exponential random variables are used in a variety of applications, including studies of the length of telephone conversations, the time customers spend waiting in line at a bank, and the life expectancy of a machine part.

Since $f(x) \geq 0$ and

$$\int_{-\infty}^{\infty} f(x)\,dx = \int_{0}^{\infty} \frac{1}{\lambda} e^{-x/\lambda}\,dx$$

$$= \lim_{R\to\infty} \int_{0}^{R} \frac{1}{\lambda} e^{-x/\lambda}\,dx$$

$$= \lim_{R\to\infty} (-e^{-x/\lambda})\big|_{0}^{R}$$

$$= \lim_{R\to\infty} (-e^{-R/\lambda} + 1) = 1$$

f satisfies the conditions for a probability density function. If F is the cumulative distribution function, we see that $F(x) = 0$ for $x < 0$. For $x \geqslant 0$, we have

$$\begin{aligned} f(x) &= \int_{-\infty}^{x} f(t)\,dt = \int_{0}^{x} \frac{1}{\lambda} e^{-t/\lambda}\,dt \\ &= -e^{-t/\lambda}\Big|_0^x = 1 - e^{-x/\lambda} \end{aligned}$$

To find the median, we solve

$$\begin{aligned} F(x_{\mathrm{m}}) = P(X \leqslant x_{\mathrm{m}}) &= \frac{1}{2} \\ 1 - e^{-x_{\mathrm{m}}/\lambda} &= \frac{1}{2} \\ \frac{1}{2} &= e^{-x_{\mathrm{m}}/\lambda} \\ \ln\frac{1}{2} &= -\frac{x_{\mathrm{m}}}{\lambda} \\ x_{\mathrm{m}} &= -\lambda \ln\frac{1}{2} = \lambda \ln 2 \qquad \textit{Note}: \ln\tfrac{1}{2} = -\ln 2 \end{aligned}$$

Integration by parts can be used to show that $\mu = \lambda$ and $\sigma = \lambda$. The calculations are not included here. These results are summarized in the box at the top of the next page.

Exponential Probability Density Function

$$f(x) = \begin{cases} (1/\lambda)e^{-x/\lambda} & \text{if } x \geqslant 0 \\ 0 & \text{otherwise} \end{cases} \qquad F(x) = \begin{cases} 1 - e^{-x/\lambda} & \text{if } x \geqslant 0 \\ 0 & \text{otherwise} \end{cases}$$

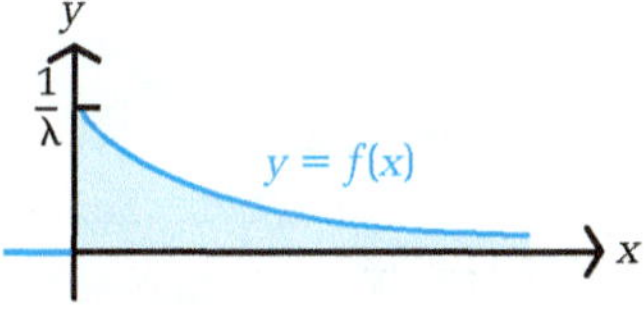

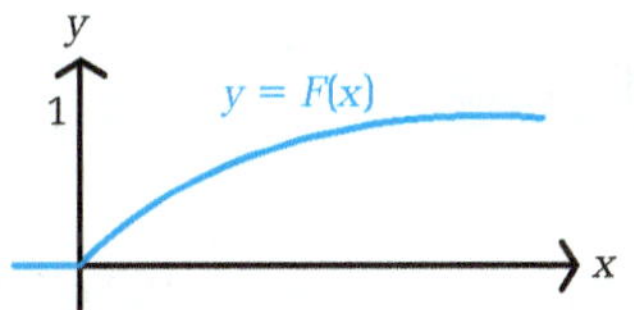

MEAN $\mu = \lambda$ MEDIAN $x_{\mathrm{m}} = \lambda \ln 2$

STANDARD DEVIATION $\sigma = \lambda$

EXAMPLE 32 **Arrival Rates** The length of time between calls received by the switchboard in a large legal firm is an exponential random variable. The average length of time between calls is 20 seconds. If a call has just been received, what is the probability that no calls are received in the next 30 seconds?

SOLUTION Let X be the random variable that represents the length of time between calls received (in seconds). Since the average length of time is 20 seconds, we have $\mu = \lambda = 20$. Thus, the probability density function for X is

$$f(x) = \begin{cases} \frac{1}{20}e^{-x/20} & \text{if } x \geqslant 0 \\ 0 & \text{otherwise} \end{cases}$$

and

$$P(X \geq 30) = \int_{30}^{\infty} \tfrac{1}{20}e^{-x/20}\,dx \qquad \int_0^{30} f(x)\,dx + \int_{30}^{\infty} f(x)\,dx = 1$$

$$= 1 - \int_0^{30} \tfrac{1}{20}e^{-x/20}\,dx = 1 - (-e^{-x/20})\Big|_0^{30} = e^{-1.5} \approx .223$$

Matched Problem 32 In Example 32, if a call has just been received, what is the probability that no calls are received in the next 10 seconds?

Referring to Example 32, if the average length of time between calls received by the switchboard is 20 seconds, then the switchboard must be receiving an average of 3 calls per minute [3(20 seconds) = 60 seconds, or 1 minute]. But in Section 10-3 we saw that the number of calls received each minute is a Poisson random variable. A moment's thought should convince you that this is not a coincidence. After all, the following two statements really say the same thing:

1. The average number of calls per minute is 3.
2. The average time between calls is $\frac{1}{3}$ minute, or 20 seconds.

In general, a Poisson random variable is concerned with the number of times an event occurs, which is a discrete concept. Associated with each Poisson random variable is an exponential random variable involving the length of time between occurrences of this event, which is a continuous concept. Thus, we see that there are applications that involve both discrete and continuous random variables.

Answers to Matched Problems

28. (A) $\frac{1}{3}$ (B) $\frac{1}{3}$ (C) $\frac{1}{3}$ **29.** $\frac{2}{5}$ **30.** .264

31. .555 **32.** $e^{-0.5} \approx .607$

Exercise 10-6

A

In Problems 1–6, find the probability density function f and the associated cumulative distribution function F for the continuous random variable X if:

1. X is uniformly distributed on [0, 2].
2. X is uniformly distributed on [3, 6].
3. X is a beta random variable with $\beta = 3$.
4. X is a beta random variable with $\beta = 5$.
5. X is an exponential random variable with $\lambda = \frac{1}{2}$.
6. X is an exponential random variable with $\lambda = \frac{1}{4}$.

In Problems 7–10, find the mean, median, and standard deviation of the continuous random variable X if:

7. X is uniformly distributed on [1, 5].
8. X is uniformly distributed on [2, 8].
9. X is an exponential random variable with $\lambda = 5$.
10. X is an exponential random variable with $\lambda = 3$.

In Problems 11 and 12, find the mean and standard deviation of the continuous random variable X if:

11. X is a beta random variable with $\beta = \frac{1}{2}$.
12. X is a beta random variable with $\beta = \frac{1}{3}$.

B

In Problems 13–18, X is a continuous random variable with mean μ. Find μ, and then find $P(X \leq \mu)$ if:

13. X is uniformly distributed on [0, 4].
14. X is uniformly distributed on [0, 10].
15. X is a beta random variable with $\beta = 0$.
16. X is a beta random variable with $\beta = 1$.
17. X is an exponential random variable with $\lambda = 1$.
18. X is an exponential random variable with $\lambda = 2$.

In Problems 19–22, X is a continuous random variable with mean μ and standard deviation σ. Find μ and σ, and then find $P(\mu - \sigma \leq X \leq \mu + \sigma)$ if:

19. X is uniformly distributed on [−5, 5].
20. X is uniformly distributed on [−2, 2].
21. X is an exponential random variable with $x_m = 6 \ln 2$.
22. X is an exponential random variable with $x_m = 4 \ln 2$.

In Problems 23 and 24, X is a beta random variable with mean μ. Find β, and then find $P(X \leq \mu)$ if:

23. $\mu = \frac{3}{5}$ **24.** $\mu = \frac{5}{7}$

Problems 25–28 refer to the uniformly distributed random variable X with probability density function

$$f(x) = \begin{cases} \frac{1}{b-a} & \text{if } a \leq x \leq b \\ 0 & \text{otherwise} \end{cases}$$

25. Show that $\mu = (a + b)/2$. **26.** Show that $x_m = (a + b)/2$.

27. Show that $\int_{-\infty}^{\infty} x^2 f(x)\,dx = (b^2 + ab + a^2)/3$.

28. Show that $V(X) = (b - a)^2/12$.

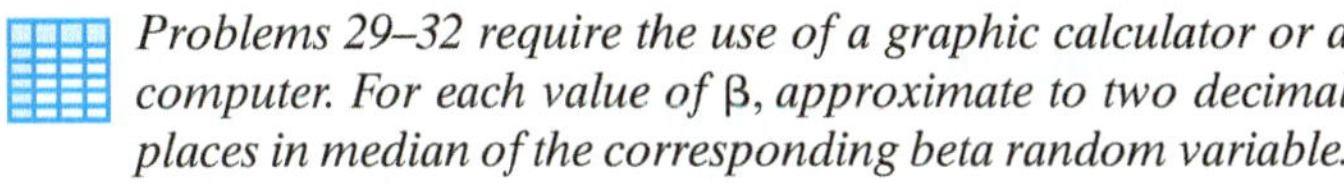

Problems 29–32 require the use of a graphic calculator or a computer. For each value of β, approximate to two decimal places in median of the corresponding beta random variable.

29. $\beta = 2$ **30.** $\beta = 3$ **31.** $\beta = 4$ **32.** $\beta = 5$

Applications

Business & Economics

33. Waiting time. The time (in minutes) applicants must wait for an officer to give them a driver's examination is uniformly distributed on the interval [0, 40]. What is the probability that an applicant must wait more than 25 minutes?

34. Waiting time. The time (in minutes) passengers must wait for a commuter plane in a large airport is uniformly distributed on the interval [0, 60]. What is the probability that a passenger waits less than 20 minutes?

35. Business failures. The percentage of restaurants that fail during the first year of operation is a beta random variable with $\beta = 2$.

(A) What is the expected percentage of failures?

(B) What is the probability that over 80% of the restaurants fail during the first year?

36. Business failures. The percentage of computer hobby stores that fail during the first year of operation is a beta random variable with $\beta = 4$.

(A) What is the expected percentage of failures?

(B) What is the probability that over 50% of the stores fail during the first year?

37. Absenteeism. The percentage of assembly line workers that are absent one Monday each month is a beta random variable. The mean percentage is 50%.

(A) What is the appropriate value of β?

(B) What is the probability that no more than 75% of the workers will be absent on one Monday each month?

38. Insurance. The percentage of insurance claims that contain errors is a beta random variable. The expected percentage of erroneous claims is 40%.

(A) What is the appropriate value of β?

(B) What is the probability that fewer than 25% of the claims are erroneous?

39. Communication. The length of time for telephone conversations (in minutes) is exponentially distributed. The average (mean) length of a conversation is 3 minutes. What is the probability that a conversation lasts less than 2 minutes?

Marty Haigney

40. Waiting time. The waiting time (in minutes) for customers at a drive-in bank is an exponential random variable. The average (mean) time a customer waits is 4 minutes. What is the probability that a customer waits more than 5 minutes?

41. Service time. The time between failures of a photo copier is an exponential random variable. Half the copiers require service during the first 2 years of operation. What is the probability that a copier requires service during the first year of operation?

42. Component failure. The life expectancy (in years) of a component in a microcomputer is an exponential random variable. Half the components fail in the first 3 years. The company that manufactures the component offers a 1 year warranty. What is the probability that a component will fail during the warranty period?

Life Sciences

43. Nutrition. The percentage of the daily requirement of vitamin *D* present in an 8 ounce serving of milk is a beta random variable with $\beta = .2$.

(A) What is the expected percentage of vitamin *D* per serving?

(B) What is the probability that a serving contains at least 50% of the daily requirement?

44. **Germination.** The percentage of a certain type of flower seeds that will germinate is a beta random variable with $\beta = 47$. What percentage of these seeds can be expected to germinate?

45. **Medical research.** A new test has been developed to detect a particular disease. The percentage of correct diagnoses obtained by using this test is a beta random variable with mean $\mu = .95$.

(A) What is the value of β?

(B) What is the probability that the percentage of correct diagnoses is greater than 90%?

46. **Medicine.** A scientist is measuring the percentage of a drug present in the bloodstream 10 minutes after an injection. The results indicate that the percentage of the drug present is a beta random variable with mean $\mu = .75$.

(A) What is the value of β?

(B) What is the probability that no more than 25% of the drug is present 10 minutes after an injection?

47. **Survival time.** The time of death (in years) after patients have contracted a certain disease is exponentially distributed. The probability that a patient dies within 1 year is .3.

(A) What is the expected time of death?

(B) What is the probability that a patient survives longer than the expected time of death?

48. **Survival time.** Repeat Problem 47 if the probability that a patient dies within 1 year is .5.

Social Sciences

49. **Education.** The percentage of entering students who complete the first year of college is a beta random variable with $\beta = 17$.

(A) What is the expected percentage of students who complete the first year?

(B) What is the probability that more than 95% of the students complete the first year?

50. **Voter turnout.** The percentage of registered voters who vote in a presidential election is a beta random variable with $\beta = 2.5$.

(A) What is the expected voter turnout?

(B) What is the probability that the voter turnout is greater than 50%?

51. **Learning.** The time (in minutes) it takes an adult to memorize a sequence of random digits is an exponential random variable. The average (mean) time is 2 minutes. What is the probability that it takes an adult over 5 minutes to memorize the digits?

52. **Psychology.** The time (in seconds) it takes rats to find their way through a maze is exponentially distributed. The average (mean) time is 30 seconds. What is the probability that it takes a rat over 1 minute to find a path through the maze?

10-7 Normal Distributions

- Normal Probability Density Functions
- The Standard Normal Curve
- Areas Under Arbitrary Normal Curves
- Approximating A Binomial Distribution With A Normal Distribution

We will now consider the most important of all the probability density functions, the *normal probability density function.* This function is at the heart of a great deal of statistical theory, and it is also a useful tool in its own right for solving problems. We will see that the normal probability density function also can be used to provide a good approximation to the binomial distribution.

Normal Probability Density Functions

A continuous random variable X has a **normal distribution** and is referred to as a **normal random variable** if its probability density function is the **normal probability density function**

$$f(x) = \frac{1}{\sigma\sqrt{2\pi}} e^{-(x-\mu)^2/2\sigma^2}$$

where μ is any constant and σ is any positive constant. It can be shown, but not easily, that

$$\int_{-\infty}^{\infty} f(x)\,dx = 1$$

$$E(X) = \int_{-\infty}^{\infty} xf(x)\,dx = \mu$$

and

$$V(X) = \int_{-\infty}^{\infty} (x-\mu)^2 f(x)\,dx = \sigma^2$$

Thus, μ is the mean of the normal probability density function and σ is the standard deviation. The graph of $f(x)$ is always a bell-shaped curve called a **normal curve**. Figure 13 illustrates three normal curves for different values of μ and σ.

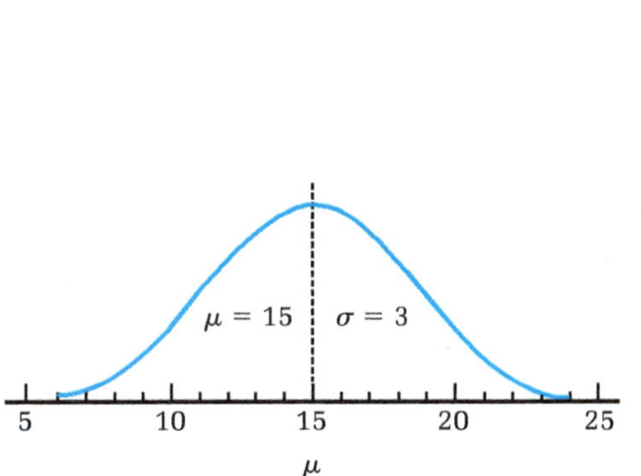

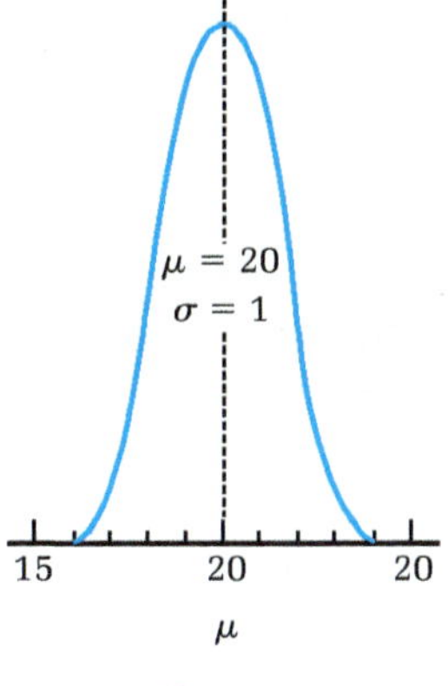

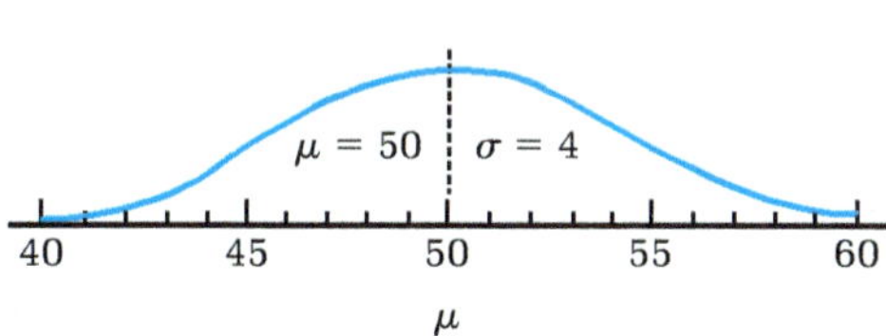

Figure 13
Normal probability distributions

The standard deviation measures the dispersion of the normal probability density function about the mean. A small standard deviation indicates a tight clustering about the mean and thus a tall, narrow curve; a large standard deviation indicates a large deviation from the mean and thus a broad, flat curve. Notice that each of the normal curves in Figure 13 is symmetric about a vertical line through the mean. This is true for any normal curve. Thus, the line $x = \mu$ divides the region under a normal curve into two regions with equal area. Since the total area under a normal curve is always 1, the area of each of these regions is .5. This implies that the median of a normal random variable is always equal to the mean (see Figure 14).

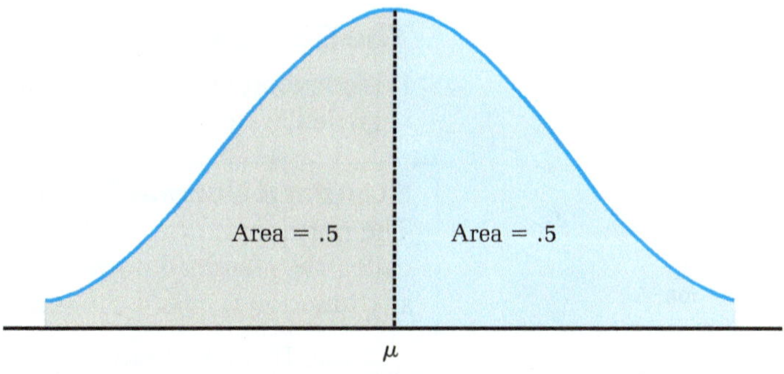

Figure 14
The mean and median of a normal random variable

The properties of the normal probability density function are summarized in the box on the next page for ease of reference.

Normal Probability Density Function

$$f(x) = \frac{1}{\sigma\sqrt{2\pi}} e^{-(x-\mu)^2/2\sigma^2} \qquad \sigma > 0$$

$y = f(x)$

μ $\quad x$

MEAN μ

MEDIAN μ

STANDARD DEVIATION σ

The graph of $F(x)$ is symmetric with respect to the line $x = \mu$.

The cumulative distribution function for a normal random variable is given formally by

$$F(x) = \frac{1}{\sigma\sqrt{2\pi}} \int_{-\infty}^{x} e^{-(t-\mu)^2/2\sigma^2}\, dt$$

It is not possible to express $F(x)$ as a finite combination of the functions we are familiar with. Furthermore, we cannot use antidifferentiation to evaluate probabilities such as

$$P(c \leq X \leq d) = \frac{1}{\sigma\sqrt{2\pi}} \int_{c}^{d} e^{-(x-\mu)^2/2\sigma^2}\, dx$$

Instead, we will use a table to approximate probabilities of this type. Fortunately, we can use the same table for all normal probability density functions, irrespective of the values of μ and σ. It is a remarkable fact that the area under a normal curve between the mean and a given number of standard deviations to the right (or left) of μ is the same regardless of the values of μ and σ (see Figure 15).

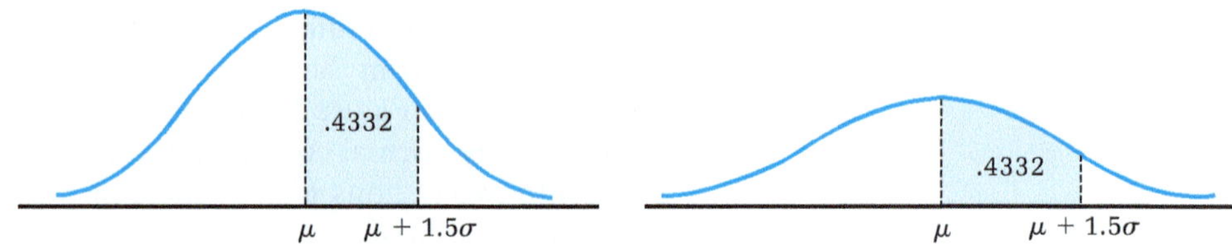

Figure 15

The Standard Normal Curve

It is convenient to relate the area under an arbitrary normal curve to the area under a particular normal curve called the *standard normal curve.*

> **Standard Normal Curve**
> The normal random variable Z with mean $\mu = 0$ and standard deviation $\sigma = 1$ is called the **standard normal random variable**, and the graph of its probability density function is called the **standard normal curve**.

Table III in Appendix B gives the area under the standard normal curve from 0 to z for values z in the range $0 \leq z \leq 3.99$. The values in this table, together with the familiar properties of area under a curve, can be used to compute probabilities involving the standard normal random variable.

EXAMPLE 33 Use Table III to compute the following probabilities for the standard normal random variable Z:

(A) $P(0 \leq Z \leq .88)$ (B) $P(Z \leq 1.45)$ (C) $P(.3 \leq Z \leq 2.73)$

SOLUTIONS

(A) From Table III, the area under the standard normal curve from $z = 0$ to $z = .88$ is .3106. Thus,

$$P(0 \leq Z \leq .88) = .3106$$

(B) $P(Z \leq 1.45)$ is the area under the standard normal curve over the interval $(-\infty, 1.45]$. Since Table III gives only the area over intervals of the form $[0, z_0]$, we must divide this region into two parts. Let A_1 be the area of the region over the interval $(-\infty, 0]$, and let A_2 be the area of the region over the interval $[0, 1.45]$, as shown in the figure. The median of the standard normal random variable is 0; thus, $A_1 = .5$. From Table III, $A_2 = .4265$. Adding these areas we have

$$P(Z \leq 1.45) = A_1 + A_2 = .5 + .4265 = .9265$$

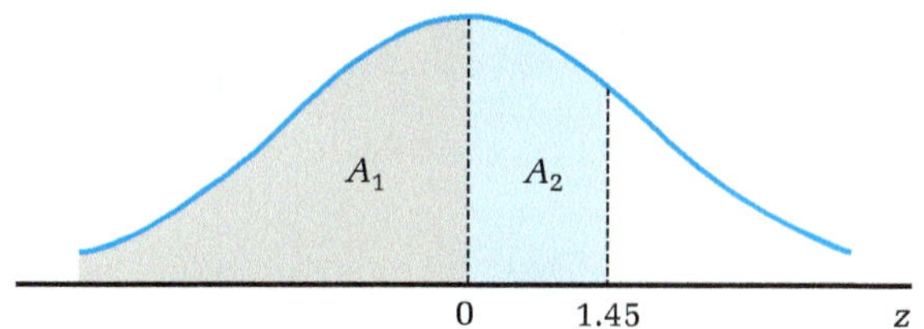

(C) This time we let A_1 be the area of the region from 0 to 2.73, and let A_2 be the area of the region from 0 to .3, as shown in figures below. Using the appropriate values from Table III, we have

$$P(.3 \leq Z \leq 2.73) = A_1 - A_2 = .4968 - .1179 = .3789$$

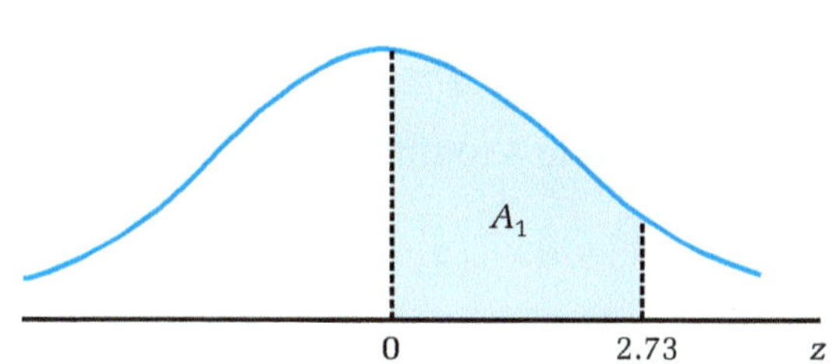

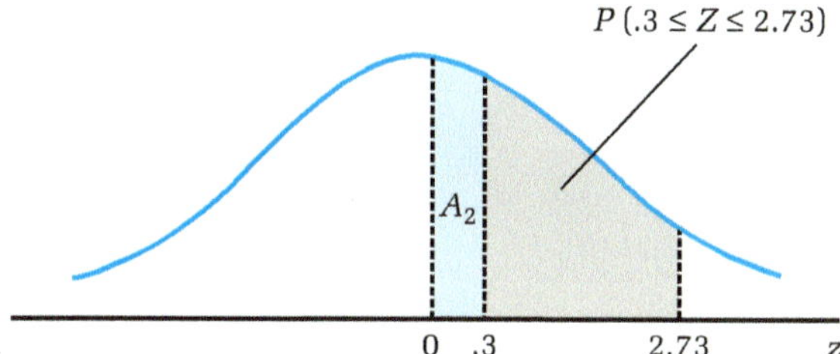

Matched Problem 33 Use Table III to find the following probabilities for the standard normal random variable Z:

(A) $P(0 \leq Z \leq 2.15)$ (B) $P(Z \leq .75)$ (C) $P(.7 \leq Z \leq 3.2)$

Areas Under Arbitrary Normal Curves

Now that we have seen how to use Table III to determine probabilities involving the standard normal random variable, we want to consider the more general case. Theorem 4 relates areas under any normal curve to corresponding areas under the standard normal curve. This will enable us to use Table III to find areas under any normal curve, regardless of the values of μ and σ.

THEOREM 4 If X is a normal random variable with mean μ and standard deviation σ, Z is the standard random variable, and

$$z_i = \frac{x_i - \mu}{\sigma} \qquad i = 1, 2 \tag{1}$$

then

$$P(x_1 \leq X \leq x_2) = P(z_1 \leq Z \leq z_2) \tag{2}$$

$$P(x_1 \leq X) = P(z_1 \leq Z) \tag{3}$$

$$P(X \leq x_2) = P(Z \leq z_2) \tag{4}$$

EXAMPLE 34 A manufacturing process produces light bulbs with life expectancies that are normally distributed with a mean of 500 hours and a standard deviation of 100 hours. What percentage of the light bulbs can be expected to last between 500 and 670 hours?

SOLUTION Since the total area under a normal curve is 1, the percentage of light bulbs that can be expected to last between 500 and 670 hours is the same as the area under the curve from 500 to 670 (see the figure).

Light bulb life expectancy

If X is the random variable associated with the life expectancy of a light bulb, then we must find

$$P(500 \leq X \leq 670)$$

First we use equation (1) in Theorem 4 to find the corresponding z values:

$$z_1 = \frac{x_1 - \mu}{\sigma} = \frac{500 - 500}{100} = 0 \qquad z_2 = \frac{x_2 - \mu}{\sigma} = \frac{670 - 500}{100} = 1.7$$

Next we use equation (2) in Theorem 4 to write

$$P(500 \leq X \leq 670) = P(0 \leq Z \leq 1.7)$$

Finally, we use Table III to find the area under the standard normal curve from $z = 0$ to $z = 1.7$. This area is .4554. Thus,

$$P(500 \leq X \leq 670) = P(0 \leq Z \leq 1.7)$$
$$= .4554$$

and we conclude that 45.54% of the light bulbs produced will last between 500 and 670 hours.

Matched Problem 34 Refer to Example 34. What percentage of the light bulbs can be expected to last between 500 and 750 hours?

EXAMPLE 35 Refer to Example 34. From all light bulbs produced, what is the probability that a light bulb chosen at random lasts between 380 and 500 hours?

SOLUTION The corresponding z values are

$$z_1 = \frac{x_1 - \mu}{\sigma} = \frac{380 - 500}{100} = -1.2 \qquad z_2 = \frac{x_2 - \mu}{\sigma} = \frac{500 - 500}{100} = 0$$

Thus,

$$P(380 \leq X \leq 500) = P(-1.2 \leq Z \leq 0)$$

Table III does not include negative values of z, but because normal curves are symmetric with respect to a vertical line through the mean, we simply use the absolute value of z in Table III (see the figure).

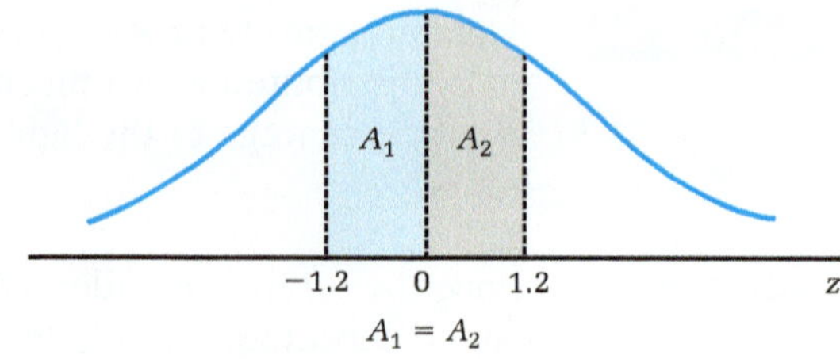

Area under the standard normal curve for negative z

The area under the standard normal curve from $z = -1.2$ to $z = 0$ is the same as the area from $z = 0$ to $z = 1.2$, which is .3849. Thus,

$$P(380 \le X \le 500) = P(-1.2 \le Z \le 0) \quad \text{Theorem 4}$$
$$= P(0 \le Z \le 1.2) \quad \text{Symmetry property of the normal curve}$$
$$= .3849 \quad \text{Table III}$$

Matched Problem 35 Refer to Example 35. What is the probability that a light bulb selected at random lasts between 400 and 500 hours?

Aproximating a Binomial Distribution With a Normal Distribution

In Section 10-3 we saw that a Poisson distribution can be used to approximate a binomial distribution when n is large:

Binomial distribution	Poisson approximation

$$C_{n,x}p^x q^{n-x} \approx \frac{\lambda^x}{x!}e^{-\lambda} \qquad \lambda = np, x = 0, 1, \ldots, n$$

Although this approximation is quite useful in some situations, it still requires a great deal of computation in problems involving the sum of a large number of terms. For example, the computation of $P(1 \le X \le 100)$ for a binomial random variable X with a large n value (say, $n > 200$) requires a sum of 100 terms, whether you use the binomial distribution or a Poisson approximation. Furthermore, since a table of values for Poisson approximations requires a separate section for each value of λ, no single table can be used for all possible Poisson distributions.

Fortunately, it turns out that probabilities involving the binomial distribution also can be approximated by an appropriately selected normal distribution and evaluated easily using Table III. To clarify ideas and relationships, let us consider an example of a binomial distribution with a relatively small n value. Then we will consider an example with a large n value.

EXAMPLE 36 **Market Research** A credit card company claims that their card is used by 40% of the people buying gasoline in a particular city. A random sample of 20 gasoline purchases is made. If the company's claim is correct, what is the probability that;

(A) From 6 to 12 people in the sample use the card?

(B) Fewer than 4 people in the sample use the card?

SOLUTIONS Before we start, it is useful to look at a histogram of the binomial distribution with a normal distribution fitted to it using the same mean and standard deviation. The mean and standard deviation of the binomial distribution are

$$\mu = np = (20)(.4) = 8 \qquad n = \text{Sample size}$$
$$\sigma = \sqrt{npq} = \sqrt{(20)(.4)(.6)} \approx 2.19 \qquad p = .4 \text{ (from the 40\% claim)}$$

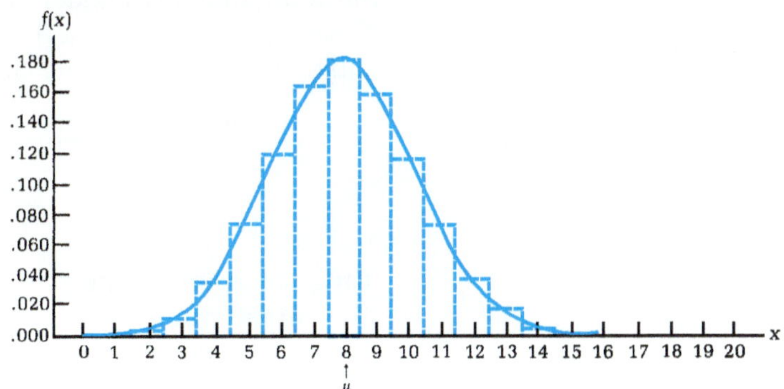

(A) To approximate the probability that 6 to 12 people in the sample use the credit card, we find the area under the normal curve from 5.5 to 12.5. We use 5.5 rather than 6, because the rectangle in the histogram corresponding to 6 extends from 5.5 to 6.5; and, reasoning in the same way, we use 12.5 instead of 12. To use Table III, we split the area into parts: A_1 to the left of the mean and A_2 to the right of the mean. A sketch is helpful:

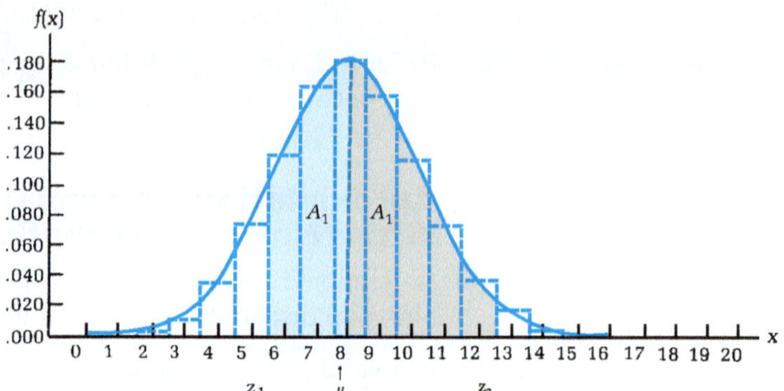

Areas A_1 and A_2 are found as follows:

$$z_1 = \frac{x - \mu}{\sigma} = \frac{5.5 - 8}{2.19} \approx -1.14 \qquad A_1 = .3729$$

$$z_2 = \frac{x - \mu}{\sigma} = \frac{12.5 - 8}{2.19} \approx 2.05 \qquad A_2 = .4798$$

$$\text{Total area} = A_1 + A_2 = .8527$$

Thus, the approximate probability that the sample will contain between 6 and 12 users of the credit card is .85 (assuming the firm's claim is correct).

(B) To use the normal curve to approximate the probability that the sample contains fewer than 4 users of the credit card, we must find the area A_1 under the normal curve to the left of 3.5. Again, a sketch is useful:

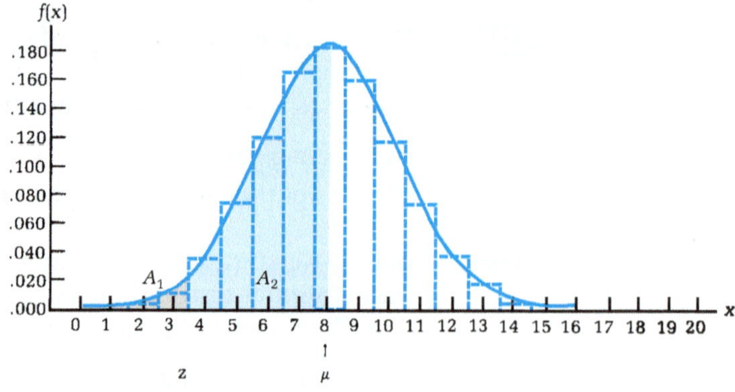

Since the total area under either half of the normal curve is .5, we first use Table III to find the area A_2 under the normal curve from 3.5 to the mean 8, and then subtract A_2 from .5:

$$z = \frac{x - \mu}{\sigma} = \frac{3.5 - 8}{2.19} \approx -2.05 \qquad A_2 = .4798$$

$$A_1 = .5 - A_2 = .5 - .4798 = .0202$$

Thus, the approximate probability that the sample contains fewer than 4 users of the credit card is .02 (assuming the company's claim is correct).

Matched Problem 36 In Example 36 use the normal curve to approximate the probability that in the sample there are:

(A) From 5 to 9 users of the credit card
(B) More than 10 users of the card

You no doubt are wondering how large n should be before a normal distribution provides an adequate approximation for a binomial distribution. Without getting too involved, the following rule-of-thumb provides a good test:

Rule-of-Thumb Test

Use a normal distribution to approximate a binomial distribution only if the interval $[\mu - 3\sigma, \mu + 3\sigma]$ lies entirely in the interval from 0 to n.

Note that in Example 36 the interval $[\mu - 3\sigma, \mu + 3\sigma] = [1.43, 14.57]$ lies entirely within the interval from 0 to 20; hence, the use of the normal distribution was justified.

EXAMPLE 37 **Quality Control** A company manufactures 50,000 ballpoint pens each day. The manufacturing process produces 50 defective pens per 1,000, on the average. A random sample of 400 pens is selected from each day's production and tested. What is the probability that the sample contains:

(A) At least 14 and no more than 25 defective pens?
(B) 33 or more defective pens?

SOLUTIONS Is it appropriate to use a normal distribution to approximate this binomial distribution? The answer is yes, since the rule-of-thumb test passes with ease:

$$\mu = np = 400(.05) = 20 \qquad p = \frac{50}{1{,}000} = .05$$

$$\sigma = \sqrt{npq} = \sqrt{400(.05)(.95)} \approx 4.36$$

$$[\mu - 3\sigma, \mu + 3\sigma] = [6.92, 33.08]$$

This interval is well within the interval from 0 to 400.

(A) To find the approximate probability of the number of defective pens in a sample being at least 14 and not more than 25, we find the area under the normal curve from 13.5 to 25.5. To use Table III, we split the area into an area to the left of the mean and an area to the right of the mean, as shown:

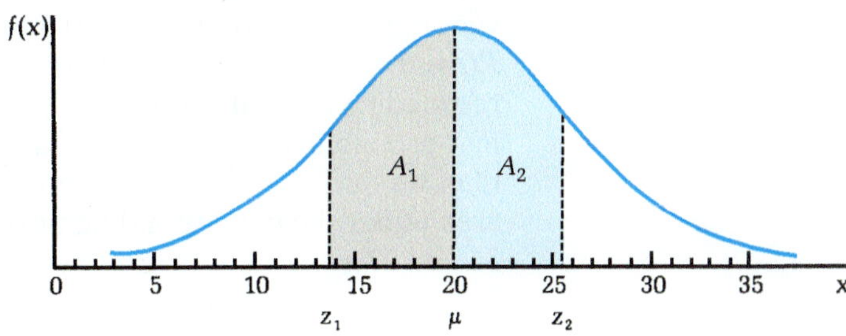

$$z_1 = \frac{x - \mu}{\sigma} = \frac{13.5 - 20}{4.36} \approx -1.49 \qquad A_1 = .4319$$

$$z_2 = \frac{x - \mu}{\sigma} = \frac{25.5 - 20}{4.36} \approx 1.26 \qquad A_2 = .3962$$

$$\text{Total area} = A_1 + A_2 = .8281$$

Thus, the approximate probability of the number of defective pens in the sample being at least 14 and not more than 25 is .83.

(B)

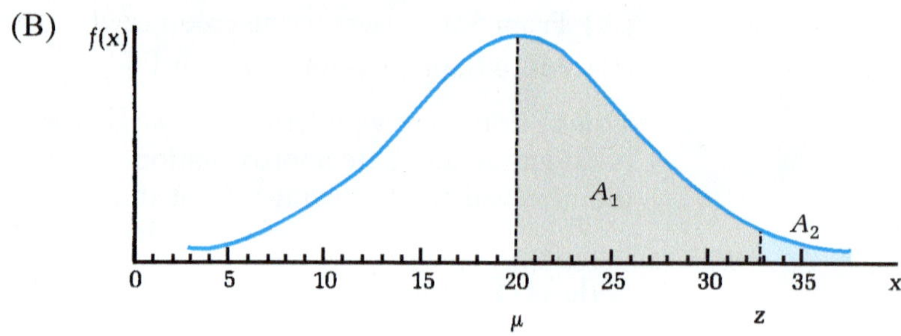

Since the total area under a normal curve from the mean on is .5, we find the area A_1 from Table III and subtract it from .5 to obtain A_2:

$$z = \frac{x - \mu}{\sigma} = \frac{32.5 - 20}{4.36} \approx 2.87 \quad A_1 = .4979$$

$$A_2 = .5 - A_1 = .5 - .4979 = .0021 \approx .002$$

Thus, the approximate probability of finding 33 or more defective pens in the sample is .002. If a random sample of 400 included more than 33 defective pens, then the management would conclude that either a rare event has happened and the manufacturing process is still producing only 50 defective pens per 1,000, on the average, or something is wrong with the manufacturing process and it is producing more than 50 defective pens per 1,000, on the average. The company might very well have a policy of checking the manufacturing process whenever 33 or more defective pens are found in a sample rather than believing a rare event has happened and that the manufacturing process is still running smoothly.

Matched Problem 37 Suppose in Example 37 that the manufacturing process produces 40 defective pens per 1,000, on the average. What is the approximate probability that in the sample of pens there are:

(A) At least 10 and no more than 20 defective pens?

(B) 27 or more defective pens?

When to Use the .5 Adjustment

If we are assuming a normal probability distribution for a continuous random variable (such as that associated with heights or weights of people), then we find $P(a \leq x \leq b)$, where a and b are real numbers, by finding the area under the corresponding normal curve from a to b (see Example 34). However, if we use a normal probability distribution to approximate a binomial probability distribution, then we find $P(a \leq x \leq b)$, where a and b are nonnegative integers, by finding the area under the corresponding normal curve from $a - .5$ to $b + .5$ (see Examples 36 and 37).

Answers to Matched Problems

33. (A) .4842 (B) .7734 (C) .2413 **34.** 49.38% **35.** .3413

36. (A) .80 (B) .13 **37.** (A) .83 (B) .004

Exercises 10-7

A

Use Table III to find the area under the standard normal curve from 0 to the indicated value of z:

1. 1 **2.** 2 **3.** − 3 **4.** −1

5. .9 **6.** −1.7 **7.** 2.47 **8.** −1.96

Given a normal distribution with mean 50 and standard deviation 10, use Theorem 4 and Table III to find the area under this normal curve from the mean to the indicated measurement.

9. 65 **10.** 75 **11.** 83 **12.** 79

13. 45 **14.** 38 **15.** 42 **16.** 26

B

In Problems 17–24, find the indicated probability for the standard normal random variable Z.

17. $P(-1.7 \leq Z \leq .6)$ **18.** $P(-.4 \leq Z \leq 2)$

19. $P(.45 \leq Z \leq 2.25)$ **20.** $P(1 \leq Z \leq 2.75)$

21. $P(Z \geq .75)$ **22.** $P(Z \leq -1.5)$

23. $P(Z \leq 1.88)$ **24.** $P(Z \geq -.66)$

Given a normal random variable X with mean 70 and standard deviation 8, find the indicated probabilities.

25. $P(60 \leq X \leq 80)$ **26.** $P(50 \leq X \leq 90)$

27. $P(62 \leq X \leq 74)$ **28.** $P(66 \leq X \leq 78)$

29. $P(X \geq 88)$ **30.** $P(X \geq 90)$

31. $P(X \leq 60)$ **32.** $P(X \leq 56)$

In Problems 33–40, use the rule-of-thumb test to check whether a normal distribution (with the same mean and standard deviation as the binomial distribution) is a suitable approximation for the binomial distribution with:

33. $n = 15, p = .7$ **34.** $n = 12, p = .6$

35. $n = 15, p = .4$ **36.** $n = 20, p = .6$

37. $n = 100, p = .05$ **38.** $n = 200, p = .03$

39. $n = 500, p = .05$ **40.** $n = 200, p = .08$

A binomial experiment consists of 500 trials with the probability of success for each trial .4. What is the probability of obtaining the number of successes indicated in Problems 41–48? Approximate these probabilities to two decimal places using a normal curve. (This binomial experiment easily passes the rule-of-thumb test, as you can check. When computing the probabilities, adjust the intervals as in Examples 36 and 37.).

41. 185–220 **42.** 190–205

43. 210–220 **44.** 175–185

45. 225 or more **46.** 212 or more

47. 175 or less **48.** 188 or less

To graph Problems 49–52, use a graphic calculator or a computer and refer to the normal probability distribution function with mean μ and standard deviation σ:

$$f(x) = \frac{1}{\sigma\sqrt{2\mu}} e^{-(x-\mu)^2/2\sigma^2}$$

49. Graph equation (1) with $\sigma = 5$ and:

(A) $\mu = 10$ (B) $\mu = 15$ (C) $\mu = 20$

Graph all three in the same viewing rectangle with the x range = $[-10, 50]$ and the y range = $[0, 0.1]$.

50. Graph equation (1) with $\sigma = 4$ and:

(A) $\mu = 8$ (B) $\sigma = 12$ (C) $\mu = 16$

Graph all three in the same viewing rectangle with the x range = $[-5, 30]$ and the y range = $[0, 0.1]$.

51. Graph equation (1) with $\mu = 20$ and:

(A) $\sigma = 2$ (B) $\sigma = 4$

Graph both in the same viewing rectangle with the x range = $[0, 40]$ and the y range = $[0, 0.2]$.

52. Graph equation (1) with $\mu = 18$ and:

(A) $\sigma = 3$ (B) $\sigma = 6$

Graph both in the same viewing rectangle with the x range = $[0, 40]$ and the y range = $[0, 0.2]$.

Applications

Business & Economics

53. Sales. Salespeople for a business machine company have average annual sales of \$200,000, with a standard deviation of \$20,000. What percentage of the salespeople would be expected to make annual sales of \$240,000 or more?

54. Guarantees. The average lifetime for a car battery of a certain brand is 170 weeks, with a standard deviation of 10 weeks. If the company guarantees the battery for 3 years, what percentage of the batteries sold would be expected to be returned before the end of the warranty period?

55. Quality control. A manufacturing process produces a critical part of average length 100 millimeters, with a standard deviation of 2 millimeters. All parts deviating by more than 5 millimeters from the mean must be rejected. What percentage of the parts must be rejected, on the average?

56. **Quality control.** An automated manufacturing process produces a component with an average width of 7.55 centimeters, with a standard deviation of 0.02 centimeter. All components deviating by more than 0.05 centimeter from the mean must be rejected. What percentage of the parts must be rejected, on the average?

57. **Marketing claims.** A company claims that 60% of the households in a given community use their product. A competitor surveys the community, using a random sample of 40 households, and finds only 15 households out of the 40 in the sample using the product. If the company's claim is correct, what is the probability of 15 or fewer households using the product in a sample of 40? What can you conclude?

58. **Labor relations.** A union representative claims 60% of the union membership will vote in favor of a particular settlement. A random sample of 100 members is polled, and out of these, 47 favor the settlement. What is the approximate probability of 47 or fewer in a sample of 100 favoring the settlement when 60% of all the membership favor the settlement? What can you conclude?

Social Sciences

59. **Medicine.** The average healing time of a certain type of incision is 240 hours, with a standard deviation of 20 hours. What percentage of the people having this incision would heal in 8 days or less?

60. **Agriculture.** The average height of a hay crop is 38 inches, with a standard deviation of 1.5 inches. What percentage of the crop will be 40 inches or more?

61. **Genetics.** In a family with 2 children, the probability that both children are girls is approximately .25. In a random sample of 1,000 families with 2 children, what is the approximate probability that 220 or fewer will have 2 girls?

62. **Genetics.** In Problem 61, what is the approximate probability of the number of families with 2 girls in the sample being at least 225 and not more than 275?

Social Sciences

63. **Testing.** Scholastic Aptitude Tests are scaled so that the mean score is 500 and the standard deviation is 100. What percentage of the students taking this test should score 700 or more?

64. **Politics.** Candidate Harkins claims that a private poll indicates she will receive 52% of the vote for governor. Her opponent, Mankey, secures the services of another pollster, who finds that 470 out of a random sample of 1,000 registered voters favor Harkins. If Harkins' claim is correct, what is the probability that only 470 or fewer will favor her in a random sample of 1,000? What can you conclude?

65. **Grading on a curve.** An instructor grades on a curve by assuming the grades on a test are normally distributed. If the average grade is 70 and the standard deviation is 8, find the test scores for each grade interval if the instructor wishes to assign grades as follows: 10% A's, 20% B's, 40% C's, 20% D's, and 10% F's.

66. **Psychology.** A test devised to measure aggressive-passive personalities was standardized on a large group of people. The scores were normally distributed, with a mean of 50 and a standard deviation of 10. If we want to designate the highest 10% as aggressive, the next 20% as moderately aggressive, the middle 40% as average, the next 20% as moderately passive, and the lowest 10% as passive, what ranges of scores will be covered by these five designations?

Chapter 10 Review

Important Terms, Symbols, and Concepts

10-0 Improper Integrals

If f is continuous over the indicated interval and the limit exists, then the improper integral on the left in each of the following equations is defined by

1. $\int_a^{\infty} f(x)\,dx = \lim_{b\to\infty} \int_a^b f(x)\,dx$

2. $\int_{-\infty}^{b} f(x)\,dx = \lim_{a\to-\infty} \int_a^b f(x)\,dx$

3. $\int_{-\infty}^{\infty} f(x)\,dx = \int_{-\infty}^{c} f(x)\,dx + \int_c^{\infty} f(x)\,dx$

where c is any point on $(-\infty, \infty)$, provided both integrals on the right exist.

The improper integral **converges** if the limit exists; otherwise it **diverges** (and no value is assigned to it).

A continuous income stream is **perpetual** if it never stops producing income. The **capital value,** CV, of a perpetual income stream $f(t)$, at a rate r compounded continuously, is defined by

$$CV = \int_0^{\infty} f(t)e^{-rt}\,dt$$

Capital value provides a method for expressing the worth (in terms of today's dollars) of an investment that will produce income indefinitely.

10-1 Finite Probability Models

Random experiment; sample space; simple outcome; event; simple event; compound event; event E occurs; equally likely; random variable; probability function; probability distribution of a random variable; histogram; expected value; payoff table; fair game; mean; variance; standard deviation

Probability distribution: $P(x) = P(X = x)$

Mean: $\mu = E(X) = x_1p_1 + x_2p_2 + \cdots + x_mp_m$

Variance: $V(X) = (x_1 - \mu)^2p_1 + (x_2 - \mu)^2p_2 + \cdots + (x_m - \mu)^2p_m$

Standard deviation: $\sigma = \sqrt{V(X)}$

10-2 Binomial Distributions

Bernoulli trial; success; failure; independent trials; sequence of Bernoulli trials; binomial formula; binomial distribution

$P(x) = P(x \text{ successes in } n \text{ trials}) = C_{n,x}p^xq^{n-x},$
$x \in \{0, 1, \ldots, n\}$

Mean: $\mu = np$; Standard deviation: $\sigma = \sqrt{npq}$

10-3 Poisson Distributions

Approximation to the binomial distribution; rule-of-thumb; Poisson experiment; Poisson random variable; Poisson probability distribution; discrete random variable

$C_{n,x}p^xq^{n-x} \approx \frac{\lambda^x}{x!}e^{-\lambda}, \pi = np, x \in \{0, 1, \ldots, n\}$

$P(x) = P(x \text{ occurrences per unit of measure}) = \frac{\lambda^x}{x!}e^{-\lambda},$

$x \in \{0, 1, 2, \ldots\}$

Mean: $\mu = \lambda$; Standard deviation: $\sigma = \sqrt{\lambda}$

10-4 Continuous Random Variables

Continuous random variable; probability density function; cumulative probability distribution function

$$P(c \le X \le d) = \int_c^d f(x)\,dx;$$

$$F(x) = P(X \le x) = \int_{-\infty}^{x} f(t)\,dt$$

10-5 Expected Value, Standard Deviation, and Median of Continuous Random Variables

Expected value; mean; variance; standard deviation; alternate formula for variance; median

$$\text{Mean: } \mu = E(X) = \int_{-\infty}^{\infty} xf(x)\,dx$$

$$\text{Variance: } V(x) = \int_{-\infty}^{\infty} (x - \mu)^2 f(x)\,dx = \int_{-\infty}^{\infty} x^2 f(x)\,dx - \mu^2$$

Standard deviation: $\sigma = \sqrt{V(X)}$;

Median: Solve $P(X \le x_m) = \frac{1}{2}$ for x_m

10-6 Uniform, Beta, and Exponential Distributions

Uniform probability density function:

$$f(x) = \begin{cases} \dfrac{1}{b - a} & \text{if } a \le x \le b \\ 0 & \text{otherwise} \end{cases}$$

Beta probability density function:

$$f(x) = \begin{cases} (\beta + 1)(\beta + 2)x^{\beta}(1 - x) & \text{if } 0 \leqslant x \leqslant 1 \\ 0 & \text{otherwise} \end{cases}$$

Exponential probability density function:

$$f(x) = \begin{cases} (1/\lambda)e^{-x/\lambda} & \text{if } x \geqslant 0 \\ 0 & \text{otherwise} \end{cases}$$

10-7 Normal Distributions

Normal distribution; normal random variable; normal probability density function; normal curve; standard normal random variable; standard normal curve; approximating binomial distributions with normal distributions; rule-of-thumb test

$x = (x - \mu)/\sigma$

Review Exercises

Work through all the problems in this chapter review and check your answers in the back of the book. (Answers to all review problems are there.) Where weaknesses show up, review appropriate sections in the text.

A

A spinner can land on any one of eight different sectors, and each sector is as likely to turn up as any other. The sectors are numbered in the figure in the margin. An experiment consists of spinning the dial once and recording the number in the sector that the spinner lands on. Problems 1–3 refer to this experiment.

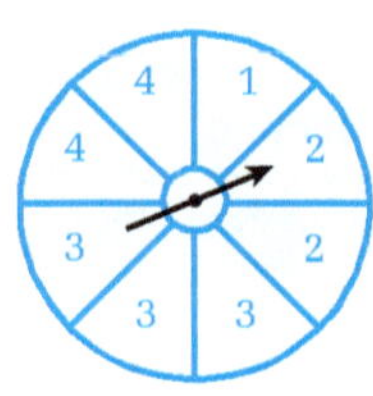

1. Find a sample space and probability distribution for this experiment.
2. What is the probability that the spinner stops on an even-numbered sector?
3. Find the expected value, variance, and standard deviation for the probability distribution found in Problem 1.
4. If a fair coin is tossed eight times, what is the probability of obtaining exactly 2 heads?
5. (A) Draw a histogram for the binomial distribution
$$P(x) = C_{3,x}(.4)^x(.6)^{3-x}$$
(B) What are the mean and standard deviation?
6. Find $P(3)$ for the binomial distribution with $n = 500$ and $p = .004$, and then find the Poisson approximation to this probability.
7. Find $P(X \leqslant 1)$ and $P(X > 2)$ for the Poisson random variable X with mean $\mu = 4$.

Problems 8–11 refer to the continuous random variable X with probability density function

$$f(x) = \begin{cases} 1 - \frac{1}{2}x & \text{if } 0 \leqslant x \leqslant 2 \\ 0 & \text{otherwise} \end{cases}$$

8. Find $P(0 \leqslant X \leqslant 1)$ and illustrate with the graph.
9. Find the mean, variance, and standard deviation.
10. Find and graph the associated cumulative probability distribution function.
11. Find the median.
12. If Z is the standard normal random variable, find $P(0 \leqslant Z \leqslant 2.5)$.
13. If X is a normal random variable with a mean of 100 and a standard deviation of 10, find $P(100 \leqslant X \leqslant 118)$.

B

14. (A) Construct a histogram for the binomial distribution
$$P(x) = C_{6,x}(.5)^x(.5)^{6-x}$$
(B) What are the mean and standard deviation?
15. The probability of drawing a king from a standard 52-card deck is $\frac{1}{13}$. Use the Poisson approximation to the binomial distribution to find the probability of drawing 5 kings in 117 single-card draws (with replacement) from a 52-card deck.

Problems 16–19 refer to the continuous random variable X with probability density function

$$f(x) = \begin{cases} \frac{5}{2}x^{-7/2} & \text{if } x \geqslant 1 \\ 0 & \text{otherwise} \end{cases}$$

16. Find $P(1 \leqslant X \leqslant 4)$ and illustrate with a graph.
17. Find the mean, variance, and standard deviation.

18. Find and graph the associated cumulative probability distribution function.

19. Find the median.

Problems 20–23 refer to a beta random variable X with $\beta = 5$.

20. Find and graph the probability density function.

21. Find $P(\frac{1}{4} \le X \le \frac{3}{4})$.

22. Find and graph the associated cumulative probability distribution function.

23. Find the mean and standard deviation.

Problems 24–27 refer to an exponentially distributed random variable X.

24. If $P(4 \le X) = e^{-2}$, find the probability density function.

25. Find $P(0 \le X \le 2)$.

26. Find the associated cumulative probability distribution function.

27. Find the mean, standard deviation, and median.

28. What are the mean and standard deviation for a binomial distribution with $p = .6$ and $n = 1{,}000$?

29. If the probability of success in a single trial of a binomial experiment with 1,000 trials is .6, what is the probability of obtaining at least 550 and no more than 650 successes in 1,000 trials? [**Hint:** Approximate with a normal distribution.]

30. Given a normal distribution with mean 50 and standard deviation 6, find the area under the normal curve:
(A) Between 41 and 62 (B) From 59 on'

C

31. If X is a beta random variable with mean $\mu = 8$, what is the value of β?

32. Find the mean and the median of the continuous random variable with probability density function

$$f(x) = \begin{cases} 50/(x+5)^3 & \text{if } x \ge 0 \\ 0 & \text{otherwise} \end{cases}$$

33. If $f(x)$ is a continuous probability density function with mean μ, and standard deviation σ, and a, b, and c are constants, evaluate the integral given below. Express the result in terms of μ, σ, and a, b, and c.

$$\int_{-\infty}^{\infty} (ax^2 + bx + c)f(x)\, dx$$

Applications

Business & Economics

34. **Quality control.** A manufacturing process produces, on the average, 6 defective items out of 100. To control quality, each day a sample of 10 completed items is selected at random and is inspected. If the sample produces more than 2 defective items, then the whole day's output is inspected and the manufacturing process is reviewed. What is the probability of this happening, assuming that the process is still producing 6% defective items and that the number of defective items is a binomial random variable?

35. **Equipment failure.** A company's records show that a printing press breaks down, on the average, 26 times a year. Assuming that the number of times the press breaks down per week is a Poisson random variable, find the probability that during a particular week:
(A) The printing press will not break down.
(B) The printing press will break down once.
(C) The printing press will break down more than 2 times.

36. **Demand.** The manager of a movie theater has determined that the weekly demand for popcorn (in pounds) is a continuous random variable with probability density function

$$f(x) = \begin{cases} \frac{1}{50}(1 - .01x) & \text{if } 0 \le x \le 100 \\ 0 & \text{otherwise} \end{cases}$$

(A) If the manager has 50 pounds of popcorn on hand at the beginning of the week, what is the probability that this will be enough to meet the weekly demand?
(B) If the manager wants the probability that the supply on hand exceeds the weekly demand to be .96, how much popcorn must be on hand at the beginning of the week?

37. **Credit applications.** The percentage of applications for a national credit card that are processed on the same day they are received is a beta random variable with $\beta = 1$.
(A) What is the probability that at least 20% of the applications received are processed the same day they arrive?
(B) What is the expected percentage of applications processed the same day they arrive?

38. **Computer failure.** A computer manufacturer has determined that the time between failures for its computers is an exponentially distributed random variable with a mean failure time of 4,000 hours. Suppose a particular computer has just been repaired.
(A) What is the probability that the computer operates for the next 4,000 hours without a failure?
(B) What is the probability that the computer fails in the next 1,000 hours?

39. **Waiting time.** Each business day the switchboard for a commercial airline receives an average of 120 calls per hour.
(A) Assume that the number of calls per minute is a Poisson random variable. What is the probability that 4 calls are received during a particular minute?

(B) Assume that the time between calls is an exponential random variable. If a call has just arrived, what is the probability that no other calls arrive during the next 45 seconds?

40. Radial tire failure. The life expectancy (in miles) of a certain brand of radial tire is a normal random variable with a mean of 35,000 and a standard deviation of 5,000. What is the probability that a tire fails during the first 25,000 miles of use?

41. Personnel screening. The scores on a screening test for new technicians are normally distributed with mean 100 and standard deviation 10. Find the approximate percent of applicants taking the test scoring:

(A) Between 92 and 108 (B) 115 or higher

42. Market research. A newspaper publisher claims that 70% of the people in a community read their newspaper. Doubting the assertion, a competitor randomly surveys 200 people in the community. Based on the publisher's claim (and assuming a binomial distribution):

(A) Compute the mean and standard deviation of the binomial distribution.

(B) Determine whether the rule-of-thumb test warrants the use of a normal distribution to approximate this binomial distribution.

(C) Calculate the approximate probability of finding at least 130 and no more than 155 readers in the sample.

(D) Determine the approximate probability of finding 125 or fewer readers in the sample.

Life Sciences

43. Medicine. The shelf-life (in months) of a certain drug is a continuous random variable with probability density function

$$f(x) = \begin{cases} 10/(x + 10)^2 & \text{if } x \geq 0 \\ 0 & \text{otherwise} \end{cases}$$

(A) What is the probability that the drug is still usable after 5 months?

(B) What is the median shelf-life?

44. Life expectancy. The life expectancy (in months) after dogs have contracted a certain disease is an exponentially distributed random variable. The probability of surviving more than 1 month is e^{-2}. After contracting this disease:

(A) What is the probability of the dog surviving more than 2 months?

(B) What is the mean life expectancy?

45. Harmful side effects of drugs. A drug causes harmful side effects in 25% of the patients treated with the drug. If the drug is administered to 100 patients, what is the probability that 30 or more of these patients will suffer from the side effects? (Use a normal distribution to approximate this binomial distribution.)

46. Measles epidemic. Measles are found to be spreading through a high-rise college dormitory. A public health official determines that the number of infected students per floor is a Poisson random variable with mean $\mu = 3$. Find the probability that a particular floor has:

(A) No infected students (B) 1 infected student

(C) More than 3 infected students

Social Sciences

47. Testing. The percentage of correct answers on a college entrance examination is a beta random variable. The mean score is 75%. What is the probability that a student answers over 50% of the questions correctly?

48. Testing. The IQ scores for 6-year-old children in a certain area are normally distributed with a mean of 108 and a standard deviation of 12. What percentage of children can be expected to have IQ scores of 135 or more?

49. Safety research. The intersection of 76th Street and Good Hope Road has the highest accident rate of all the intersections in a certain city. A study of the traffic flow at this intersection shows that, on an average weekday, 2,880 vehicles arrive at this intersection at random times during the afternoon rush hour (from 4:00 PM to 5:00 PM). If the number of vehicle arrivals per second during this rush hour is a Poisson random variable, find the probability that during a particular second:

(A) No vehicles arrive at the intersection.

(B) 1 vehicle arrives at the intersection.

(C) 5 or more vehicles arrive at the intersection

Linear Inequalities and Linear Programming

11

Chapter Problem

The revolutionary human transport machine invented by Dean Kamen in the late 1990s is now called a Segway and is manufactured by a company of the same name. Currently, Segway produces two versions, the i series Segway for personal use and the e series Segway for industrial use. Each machine is built on an assembly line and then is tested by a quality control department. The assembly line takes 1 hour to assemble one i series machine and 1.5 hours to assemble one e series machine. The quality control department can test one machine in either series in 30 minutes. The demand for these machines is far exceeding the production capabilities, and the company can sell all they can produce. Currently Segway is running two 10-hour shifts on the assembly line and one 9-hour shift in the quality control department. The company makes a profit of $500 on each i series machine and $650 on each e series machine. How many of each type of machine should they produce each day in order to maximize their daily profit?

Objectives

1. Graph the solution set for a system of linear inequalities.
2. Construct a mathematical model in the form of a linear programming problem and solve it graphically.
3. Understand the relationship between the graphical solution method and an algebraic method called the simplex method.
4. Use the simplex method to solve any linear programming problem.

Introduction

In this chapter we discuss linear inequalities in two and more variables. In addition, we introduce a relatively new and powerful mathematical tool called *linear programming*, which is used to solve a variety of interesting practical problems.

Taken from *College Mathematics for Business, Economics, Life Sciences, and Social Sciences*, Tenth Edition, by Raymond A. Barnett, Michael R. Ziegler, and Karl E. Byleen.

11-1 Systems of Linear Inequalities in Two Variables

- Graphing Linear Inequalities in Two Variables
- Solving Systems of Linear Inequalities Graphically
- Applications

Many applications of mathematics involve systems of inequalities rather than systems of equations. A graph is often the most convenient way to represent the solutions of a system of linear inequalities in two variables. In this section we discuss techniques for graphing both a single linear inequality in two variables and a system of linear inequalities in two variables.

Graphing Linear Inequalities in Two Variables

We know how to graph first-degree equations such as

$$y = 2x - 3 \quad \text{and} \quad 2x - 3y = 5$$

but how do we graph first-degree inequalities such as the following?

$$y \le 2x - 3 \quad \text{and} \quad 2x - 3y > 5$$

We will find that graphing these inequalities is almost as easy as graphing the equations, but first we must discuss some important subsets of a plane in a rectangular coordinate system.

A line divides the plane into two regions called **half-planes.** A vertical line divides it into **left** and **right half-planes;** a nonvertical line divides it into **upper** and **lower half-planes.** In either case, the dividing line is called the **boundary line** of each half-plane, as indicated in Figure 1.

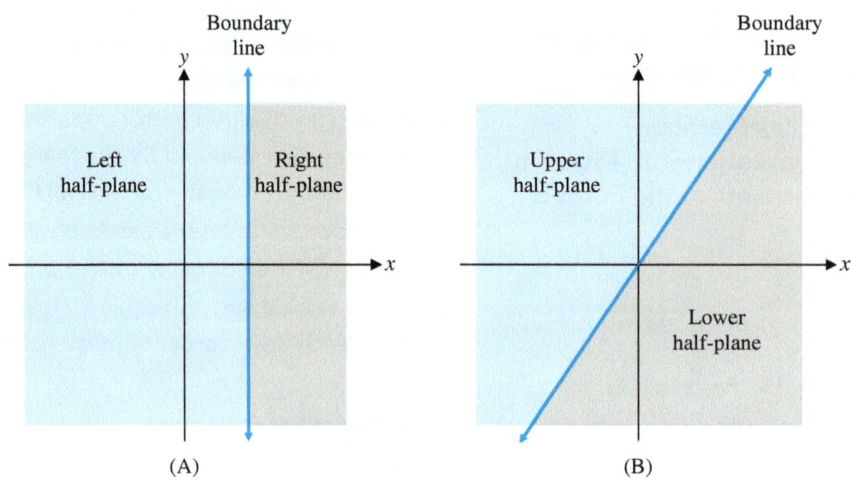

Figure 1

EXPLORE & DISCUSS 1

Consider the following linear equation and related linear inequalities:

(1) $3x - 4y = 24$ (2) $3x - 4y < 24$ (3) $3x - 4y > 24$

(A) Graph the line with equation (1).

(B) Find the point on this line with x coordinate 4 and draw a vertical line through this point. Discuss the relationship between the y coordinates of the points on this line and statements (1), (2), and (3).

(C) Repeat part (B) for $x = -4$. For $x = 12$.

(D) Based on your observations in parts (B) and (C), write a verbal description of all the points in the plane that satisfy equation (1), those that satisfy inequality (2), and those that satisfy inequality (3).

To investigate the half-planes determined by a linear equation such as $y - x = -2$, we rewrite the equation as $y = x - 2$. For any given value of x, there is exactly one value for y such that (x, y) lies on the line. For example, for $x = 4$, we have $y = 4 - 2 = 2$. For the same x and smaller values of y, the point (x, y) will lie below the line, since $y < x - 2$. Thus, the lower half-plane corresponds to the solution of the inequality $y < x - 2$. Similarly, the upper half-plane corresponds to $y > x - 2$, as shown in Figure 2.

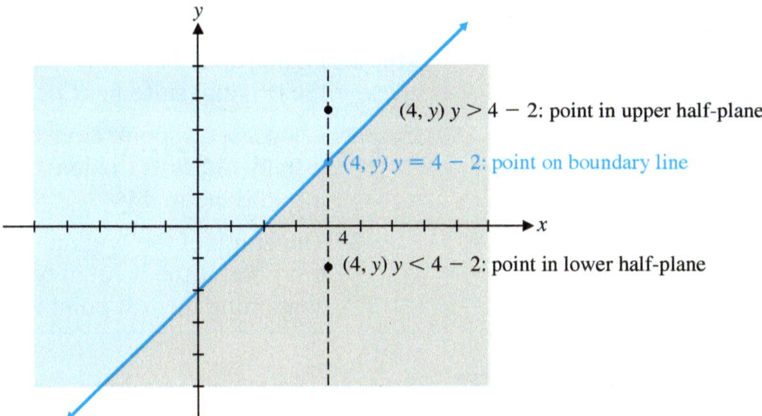

Figure 2

The four inequalities formed from $y = x - 2$ by replacing the = sign by $>$, $\geq$, $<$, and $\leq$, respectively, are

$$y > x - 2 \qquad y \geq x - 2 \qquad y < x - 2 \qquad y \leq x - 2$$

The graph of each is a half-plane, excluding the boundary line for $<$ and $>$, and including the boundary line for $\leq$ and $\geq$. In Figure 3, the half-planes are indicated with small arrows on the graph of $y = x - 2$ and then graphed as shaded regions. Excluded boundary lines are shown as dashed lines, and included boundary lines are shown as solid lines.

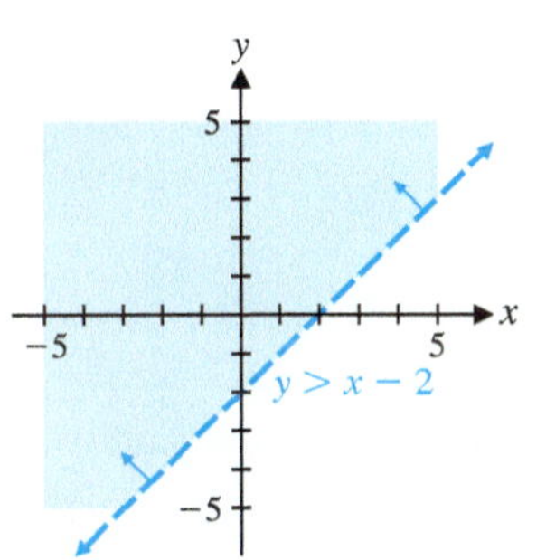

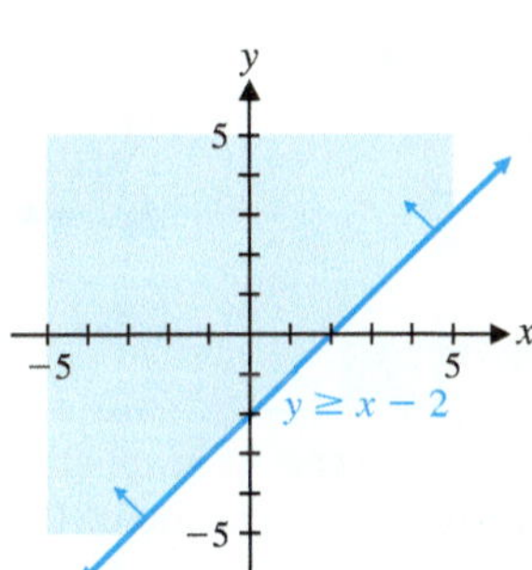

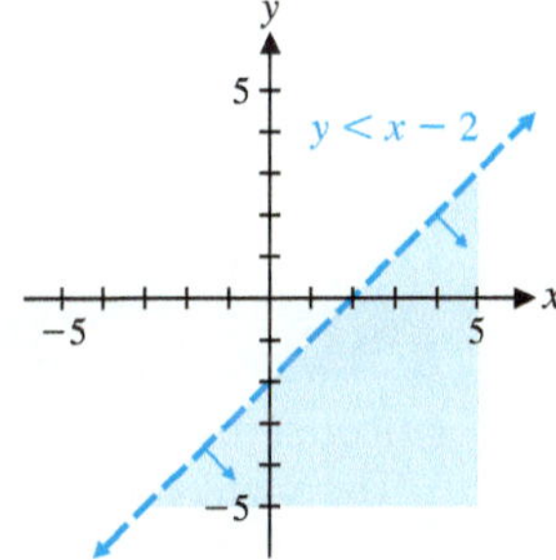

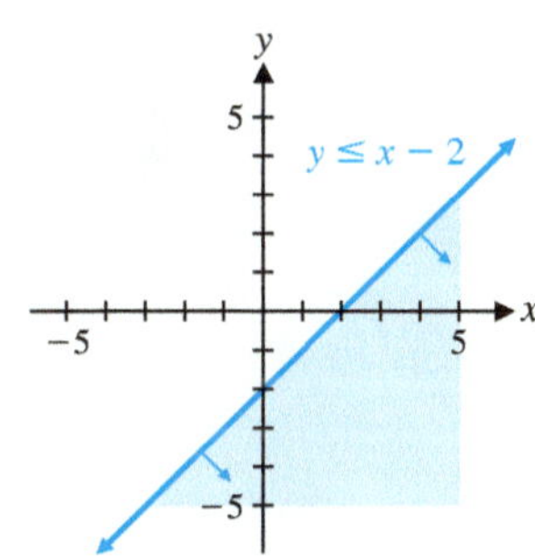

Figure 3

The preceding discussion suggests the following theorem, which is stated without proof:

THEOREM 1 Graphs of Linear Inequalities

The graph of the linear inequality

$$Ax + By < C \qquad \text{or} \qquad Ax + By > C$$

with $B \neq 0$, is either the upper half-plane or the lower half-plane (but not both) determined by the line $Ax + By = C$.

If $B = 0$ and $A \neq 0$, the graph of

$$Ax < C \qquad \text{or} \qquad Ax > C$$

is either the left half-plane or the right half-plane (but not both) determined by the line $Ax = C$.

As a consequence of this theorem, we state a simple and fast mechanical procedure for graphing linear inequalities.

PROCEDURE Graphing Linear Inequalities

Given $y = f(x)$, the **slope of the graph** at the point $(a, f(a))$ is given by

Step 1 First graph $Ax + By = C$ as a dashed line if equality is not included in the original statement or as a solid line if equality is included.

Step 2 Choose a test point anywhere in the plane not on the line [the origin $(0, 0)$ usually requires the least computation] and substitute the coordinates into the inequality.

Step 3 The graph of the original inequality includes the half-plane containing the test point if the inequality is satisfied by that point or the half-plane not containing the test point if the inequality is not satisfied by that point.

EXAMPLE 1 Graphing a Linear Inequality Graph $2x - 3y \le 6$.

SOLUTION **Step 1** Graph $2x - 3y = 6$ as a solid line, since equality is included in the original statement (Fig. 4)

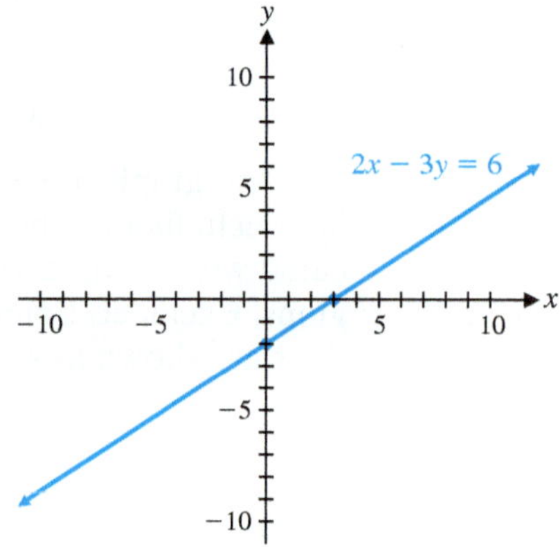

Figure 4

CONCEPTUAL INSIGHT

Recall that the line $2x - 3y = 6$ can be graphed by finding any two points on the line. The x and y intercepts are usually a good choice (see Fig. 4).

$2x - 3y = 6$

x	y
0	−2
3	0

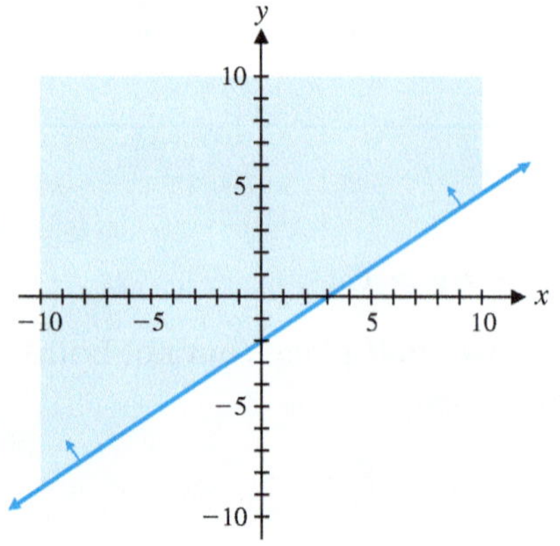

Figure 5

Step 2 Pick a convenient test point above or below the line. The origin (0, 0) requires the least computation, so substituting (0, 0) into the inequality, we get

$$2x - 3y \le 6$$
$$2(0) - 3(0) = 0 \le 6$$

This is a true statement; therefore, the point $(0, 0)$ is in the solution set.

Step 3 The line $2x - 3y = 6$ and the half-plane containing the origin form the graph of $2x - 3y \le 6$, as shown in Figure 5.

Matched Problem 1 Graph $6x - 3y > 18$.

EXAMPLE 2 **Graphing Inequalities** Graph:

(A) $y > -3$
(B) $2x \le 5$
(C) $x \le 3y$

SOLUTION (A)

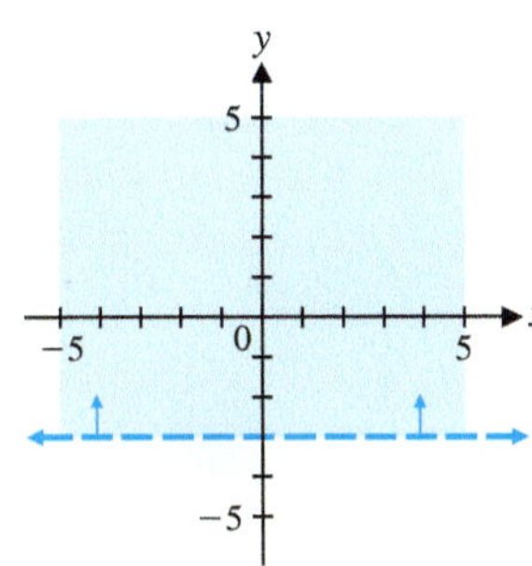

(B)

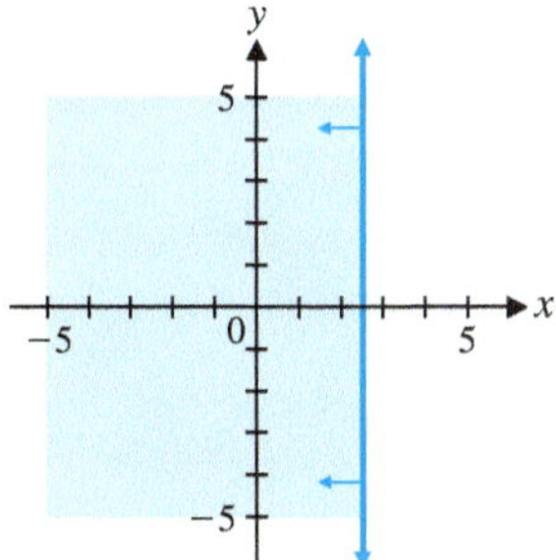

(C)

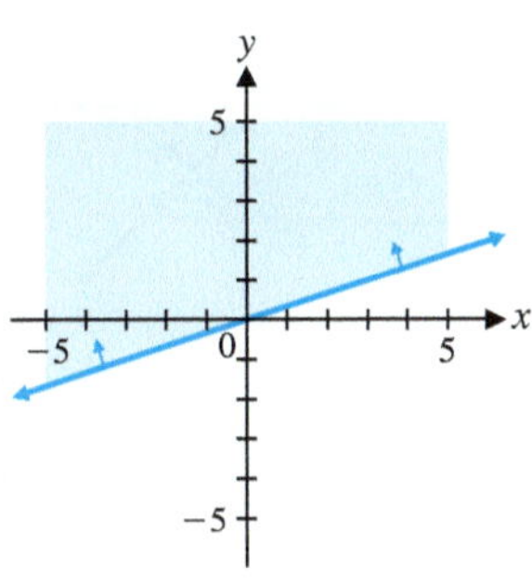

Matched Problem 2 Graph:

(A) $y < 4$ (B) $4x \ge -9$ (C) $3x \ge 2y$

Solving Systems of Linear Inequalities Graphically

We now consider systems of linear inequalities such as

$$\begin{aligned} x + y &\ge 6 \\ 2x - y &\ge 0 \end{aligned} \quad \text{and} \quad \begin{aligned} 2x + y &\le 22 \\ x + y &\le 13 \\ 2x + 5y &\le 50 \\ x &\ge 0 \\ y &\ge 0 \end{aligned}$$

We wish to **solve** such systems **graphically,** that is, to find the graph of all ordered pairs of real numbers (x, y) that simultaneously satisfy all the inequalities in the system. The graph is called the **solution region** for the system. (In many applications, the solution region is also called the **feasible region.**) To find the solution region, we graph each inequality in the system and then take the intersection of all the graphs. To simplify the discussion that follows, **we consider only systems of linear inequalities where equality is included in each statement in the system.**

EXAMPLE 3 **Solving a System of Linear Inequalities Graphically** Solve the following system of linear inequalities graphically:

$$\begin{aligned} x + y &\ge 6 \\ 2x - y &\ge 0 \end{aligned}$$

SOLUTION Graph the line $x + y = 6$ and shade the region that satisfies the linear inequality $x + y \ge 6$. This region is shaded with gray lines in Figure 6A. Next, graph the line $2x - y = 0$ and shade the region that satisfies the inequality $2x - y \ge 0$. This region is shaded with blue lines in Figure 6A. The solution region for the system of inequalities is the intersection of these two regions. This is the region shaded in both gray and blue in Figure 6A and redrawn in Figure 6B with only the solution region shaded. The coordinates of any point in the shaded region of Figure 6B specify a solution to the system. For example, the points (2, 4), (6, 3), and (7.43, 8.56) are three of infinitely many solutions, as can be easily checked. The intersection point (2, 4) is obtained by solving the equations $x + y = 6$ and $2x - y = 0$ simultaneously.

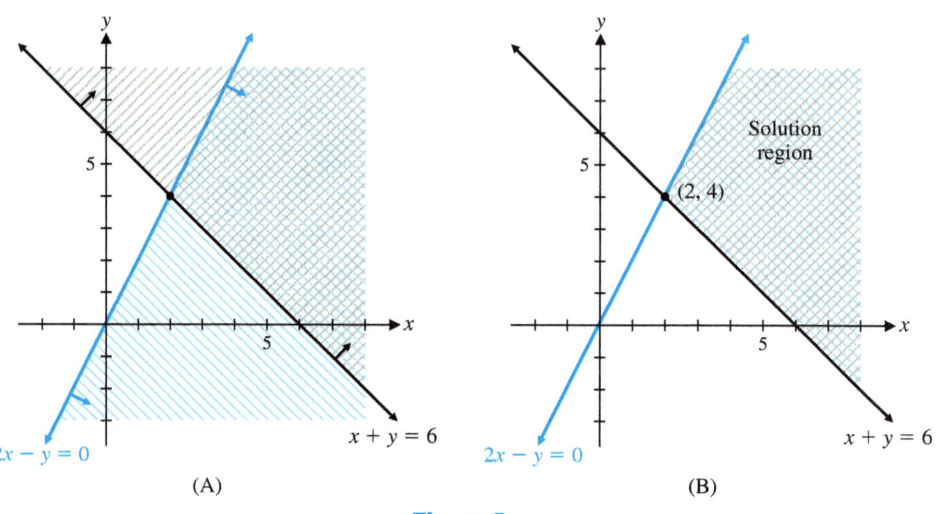

Figure 6

Matched Problem 3 Solve the following system of linear inequalities graphically:

$$3x + y \le 21$$
$$x - 2y \le 0$$

CONCEPTUAL INSIGHT

To check that you have shaded a solution region correctly, choose a test point in the region and check that it satisfies each inequality in the system. For example, choosing the point (5, 5) in the shaded region in Figure 6B, we have

$x + y \ge 6$	$2x - y \ge 0$
$5 + 5 \overset{?}{\ge} 6$	$10 - 5 \overset{?}{\ge} 0$
$10 \overset{\checkmark}{\ge} 6$	$5 \overset{\checkmark}{\ge} 0$

EXPLORE & DISCUSS 2 Refer to Example 3. Graph each boundary line and shade the region containing the points that do *not* satisfy each inequality. That is, shade the region of the plane that corresponds to the inequality $x + y < 6$ and then shade the region that corresponds to the inequality $2x - y < 0$. What portion of the plane is left unshaded? Compare this method with the one used in the solution to Example 3.

The points of intersection of the lines that form the boundary of a solution region will play a fundamental role in the solution of linear programming problems, which are discussed in the next section.

DEFINITION Corner Point

A **corner point** of a solution region is a point in the solution region that is the intersection of two boundary lines.

For example, the point (2, 4) is the only corner point of the solution region in Example 3 (Fig. 6).

EXAMPLE 4 **Solving a System of Linear Inequalities Graphically** Solve the following system of linear inequalities graphically, and find the corner points:

$$\begin{aligned} 2x + y &\le 22 \\ x + y &\le 13 \\ 2x + 5y &\le 50 \\ x &\ge 0 \\ y &\ge 0 \end{aligned}$$

SOLUTION The inequalities $x \ge 0$ and $y \ge 0$ indicate that the solution region will lie in the first quadrant.* Thus, we can restrict our attention to that portion of the plane. First, we graph the lines

$$\begin{aligned} 2x + y &= 22 \\ x + y &= 13 \\ 2x + 5y &= 50 \end{aligned}$$

Find the x and y intercepts of each line; then sketch the line through these points.

Next, choosing (0, 0) as a test point, we see that the graph of each of the first three inequalities in the system consists of its corresponding line and the half-plane lying below the line, as indicated by the small arrows in Figure 7. Thus, the solution region of the system consists of the points in the first quadrant that simultaneously lie on or below all three of these lines (see the shaded region in Fig. 7).

The corner points (0, 0), (0, 10), and (11, 0) can be determined from the graph. The other two corner points are determined as follows:

Solve the system	Solve the system
$2x + 5y = 50$	$2x + y = 22$
$x + y = 13$	$x + y = 13$
to obtain (5, 8).	to obtain (9, 4).

* The inequalities $x \ge 0$ and $y \ge 0$ occur frequently inv applications involving systems of inequalities, since x and y often represent quantities that cannot be negative (number of units produced, number of hours worked, and so on).

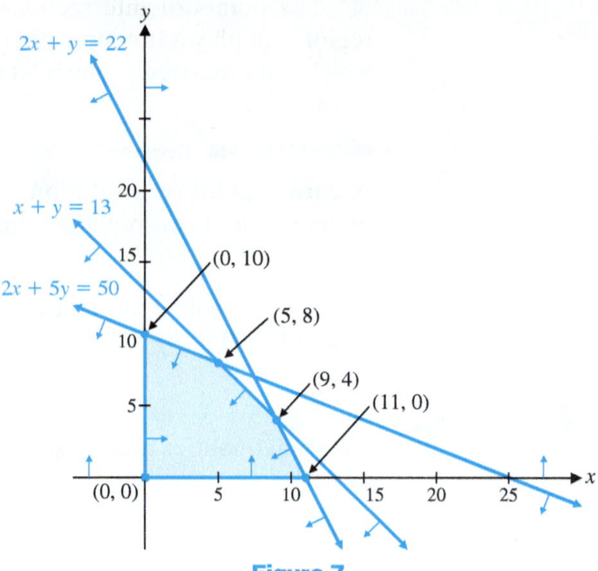

Figure 7

Note that the lines $2x + 5y = 50$ and $2x + y = 22$ also intersect, but the intersection point is not part of the solution region, and hence, is not a corner point.

Matched Problem 4 Solve the following system of linear inequalities graphically, and find the corner points:

$$\begin{aligned} 5x + y &\geq 20 \\ x + y &\geq 12 \\ x + 3y &\geq 18 \\ x &\geq 0 \\ y &\geq 0 \end{aligned}$$

If we compare the solution regions of Examples 3 and 4, we see that there is a fundamental difference between these two regions. We can draw a circle around the solution region in Example 4; however, it is impossible to include all the points in the solution region in Example 3 in any circle, no matter how large we draw it. This leads to the following definition:

DEFINITION Bounded and Unbounded Solution Regions

A solution region of a system of linear inequalities is **bounded** if it can be enclosed within a circle. If it cannot be enclosed within a circle, it is **unbounded.**

Thus, the solution region for Example 4 is bounded, and the solution region for Example 3 is unbounded. This definition will be important in the next section.

Applications

EXAMPLE 5 **Medicine** A patient in a hospital is required to have at least 84 units of drug A and 120 units of drug B each day (assume that an overdosage of either drug is harmless). Each gram of substance M contains 10 units of drug A and 8 units of drug B, and each gram of substance N contains 2 units of drug A and 4 units of drug B. How many grams of substances M and N can be mixed to meet the minimum daily requirements?

SOLUTION The question posed in the example indicates that the relevant variables are

$$x = \text{number of grams of substance } M \text{ used}$$
$$y = \text{number of grams of substance } N \text{ used}$$

Next, we arrange the information in the problem in a table with the variables related to columns of the table.

	Amount of Drug per Gram		Minimum Daily Requirement
	Substance M	Substance N	
Drug A	10 units	2 units	84 units
Drug B	8 units	4 units	120 units

Since there are 10 units of drug A in 1 gram of substance M and x number of grams of substance M in the daily dose of medication, there are $10x$ units of drug A in the daily dose. Using similar reasoning, we can list all the components of the daily dose.

$$10x = \text{number of units of drug } A \text{ in } x \text{ grams of substance } M$$
$$2y = \text{number of units of drug } A \text{ in } y \text{ grams of substance } N$$
$$8x = \text{number of units of drug } B \text{ in } x \text{ grams of substance } M$$
$$4y = \text{number of units of drug } B \text{ in } y \text{ grams of substance } N$$

The following conditions must be satisfied to meet daily requirements:

$$\begin{pmatrix}\text{number of units of}\\ \text{drug } A\\ \text{in } x \text{ grams of substance } M\end{pmatrix} + \begin{pmatrix}\text{number of units of}\\ \text{drug } A\\ \text{in } y \text{ grams of substance } N\end{pmatrix} \geq 84$$

$$\begin{pmatrix}\text{number of units of}\\ \text{drug } B\\ \text{in } x \text{ grams of substance } M\end{pmatrix} + \begin{pmatrix}\text{number of units of}\\ \text{drug } B\\ \text{in } y \text{ grams of substance } N\end{pmatrix} \geq 120$$

$$(\text{number of grams of substance } M \text{ used}) \geq 0$$
$$(\text{number of grams of substance } N \text{ used}) \geq 0$$

Converting these verbal statements into symbolic statements by using the variables x and y introduced above, we obtain the following model consisting of a system of linear inequalities:

$$10x + 2y \geq 84 \quad \text{Drug } A \text{ restriction}$$
$$8x + 4y \geq 120 \quad \text{Drug } B \text{ restriction}$$
$$x \geq 0 \quad \text{Cannot use a negative amount of } M$$
$$y \geq 0 \quad \text{Cannot use a negative amount of } N$$

Graphing this system of linear inequalities, we obtain the set of feasible solutions, or the feasible region (solution region), as shown in Figure 8. Thus, any point in the shaded area (including the straight-line boundaries) will meet the daily requirements; any point outside the shaded area will not. For example, 4 units of drug M and 23 units of drug N will meet the daily requirements, but 4 units of drug M and 21 units of drug N will not. (Note that the feasible region is unbounded.)

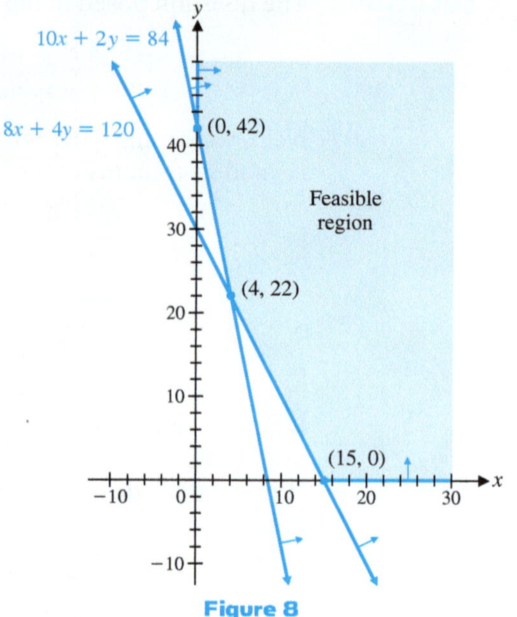

Figure 8

Matched Problem 5 **Resource Allocation** A manufacturing plant makes two types of inflatable boats, a two-person boat and a four-person boat. Each two-person boat requires 0.9 labor-hour in the cutting department and 0.8 labor-hour in the assembly department. Each four-person boat requires 1.8 labor-hours in the cutting department and 1.2 labor-hours in the assembly department. The maximum labor-hours available each month in the cutting and assembly departments are 864 and 672, respectively.

(A) Summarize this information in a table.

(B) If x two-person boats and y four-person boats are manufactured each month, write a system of linear inequalities that reflect the conditions indicated. Find the set of feasible solutions graphically.

Answers to Matched Problems

1. Graph $6x - 3y = 18$ as a dashed line (since equality is not included). Choosing the origin $(0, 0)$ as a test point, we see that $6(0) - 3(0) > 18$ is a false statement; thus, the lower half-plane determined by $6x - 3y = 18$ is the graph of $6x - 3y > 18$.

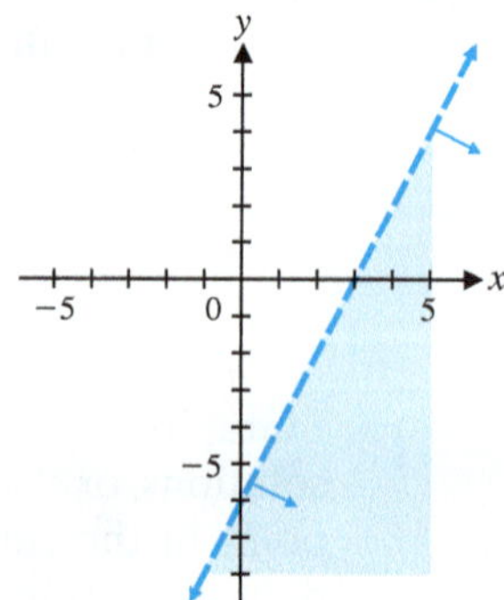

2. (A)

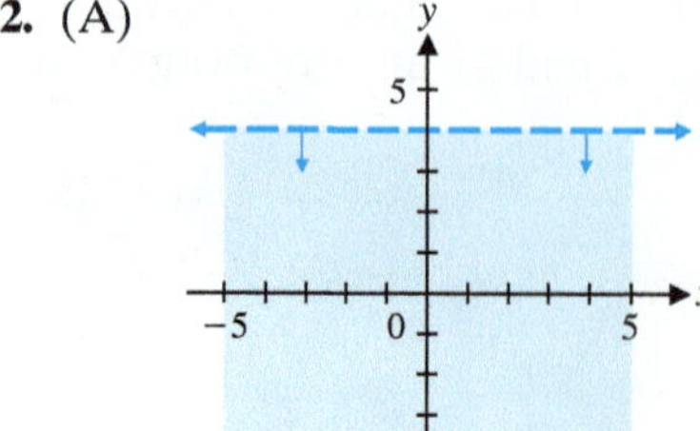

(B)

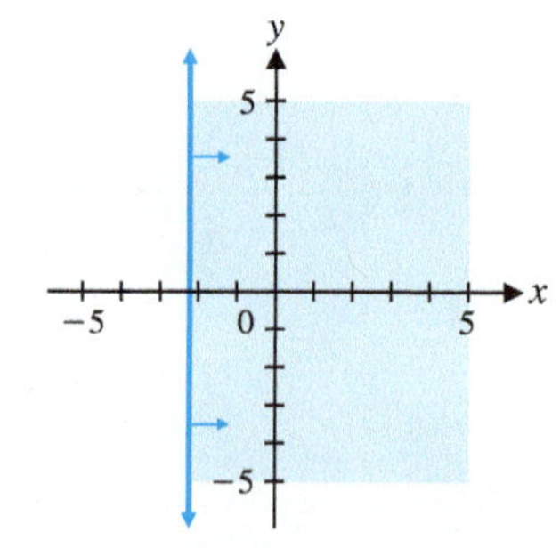

(C)

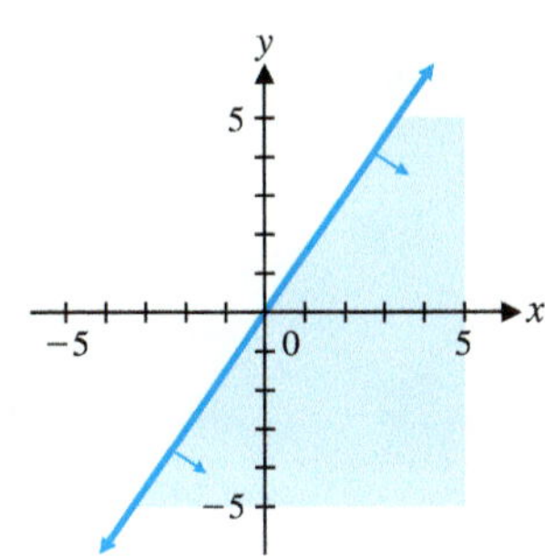

3.

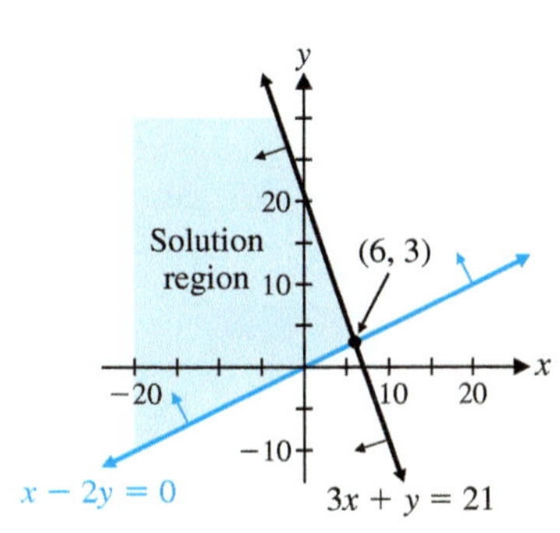

4.

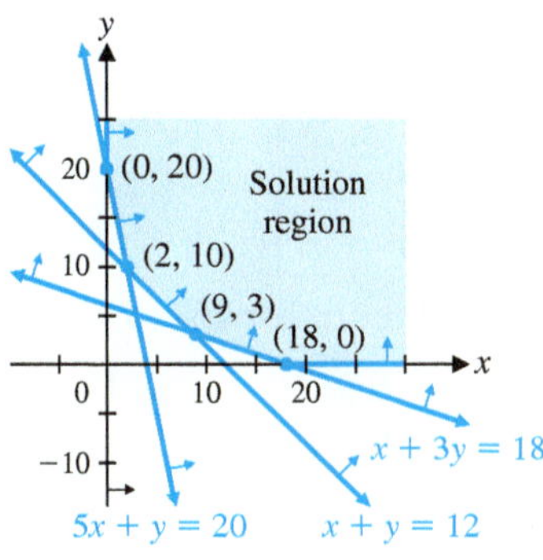

5. (A)

	Labor-Hours Required		Maximum Labor-Hours Available per Month
	Two-Person Boat	**Four-Person Boat**	
Cutting Department	0.9	1.8	864
Assembly Department	0.8	1.2	672

(B)
$$
\begin{aligned}
0.9x + 1.8y &\le 864\\
0.8x + 1.2y &\le 672\\
x &\ge 0\\
y &\ge 0
\end{aligned}
$$

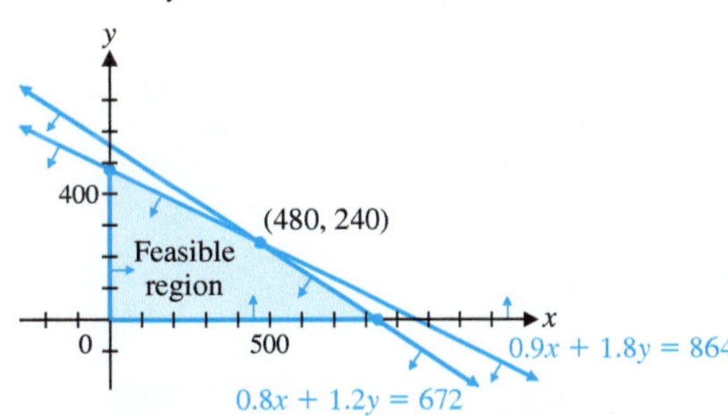

Exercise 11-1

A

Graph each inequality in Problems 1–10.

1. $y \le x - 1$ **2.** $y > x + 1$

3. $3x - 2y > 6$ **4.** $2x - 5y \le 10$

5. $x \ge -4$ **6.** $y < 5$

7. $6x + 4y \ge 24$ **8.** $4x + 8y \ge 32$

9. $5x \le -2y$ **10.** $6x \ge 4y$

In Problems 11–14,

(A) graph the set of points that satisfy the inequality.

(B) graph the set of points that do not satisfy the inequality.

11. $2x + 3y < 18$ **12.** $3x + 4y > 24$

13. $5x - 2y \ge 20$ **14.** $3x - 5y \le 30$

In Problems 15–18, match the solution region of each system of linear inequalities with one of the four regions shown in the figure.

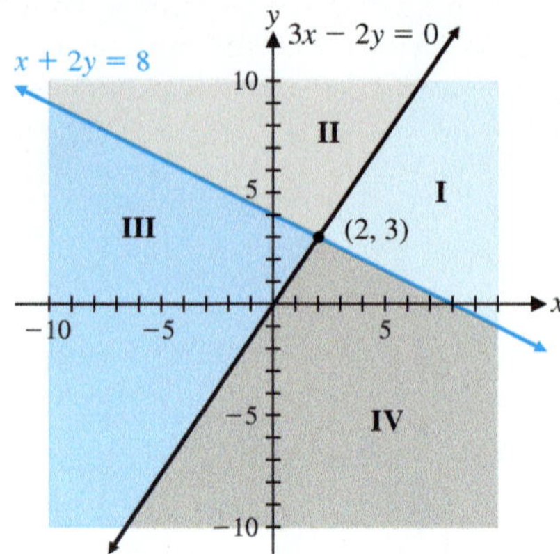

Figure for 15–18

15. $\begin{aligned} x + 2y &\le 8 \\ 3x - 2y &\ge 0 \end{aligned}$

16. $\begin{aligned} x + 2y &\ge 8 \\ 3x - 2y &\le 0 \end{aligned}$

17. $\begin{aligned} x + 2y &\ge 8 \\ 3x - 2y &\ge 0 \end{aligned}$

18. $\begin{aligned} x + 2y &\le 8 \\ 3x - 2y &\le 0 \end{aligned}$

In Problems 19–22, solve each system of linear inequalities graphically.

19. $\begin{aligned} 3x + y &\ge 6 \\ x &\le 4 \end{aligned}$

20. $\begin{aligned} 3x + 4y &\le 12 \\ y &\ge -3 \end{aligned}$

21. $\begin{aligned} x - 2y &\le 12 \\ 2x + y &\ge 4 \end{aligned}$

22. $\begin{aligned} 2x + 5y &\le 20 \\ x - 5y &\ge -5 \end{aligned}$

In Problems 23–26, solve each system two ways (see Explore–Discuss 2):

(A) by shading the points that satisfy each inequality in the system

(B) by shading the points that do not satisfy each inequality in the system

In each case, explain how you can recognize the solution region.

23. $\begin{aligned} x + y &\le 5 \\ 2x - y &\le 1 \end{aligned}$

24. $\begin{aligned} x - 2y &\le 1 \\ x + 3y &\ge 12 \end{aligned}$

25. $\begin{aligned} 2x + y &\ge 4 \\ 3x - y &\le 7 \end{aligned}$

26. $\begin{aligned} 3x + y &\ge -2 \\ x - 2y &\ge -6 \end{aligned}$

In Problems 27–30, match the solution region of each system of linear inequalities with one of the four regions shown in the figure. Identify the corner points of each solution region.

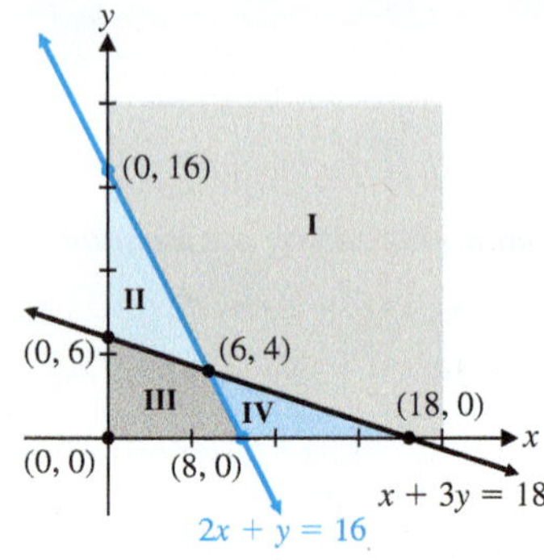

Figure for 15–18

27. $\begin{aligned} x + 3y &\le 18 \\ 2x + y &\ge 16 \\ x &\ge 0 \\ y &\ge 0 \end{aligned}$

28. $\begin{aligned} x + 3y &\le 18 \\ 2x + y &\le 16 \\ x &\ge 0 \\ y &\ge 0 \end{aligned}$

29. $\begin{aligned} x + 3y &\ge 18 \\ 2x + y &\ge 16 \\ x &\ge 0 \\ y &\ge 0 \end{aligned}$

30. $\begin{aligned} x + 3y &\ge 18 \\ 2x + y &\le 16 \\ x &\ge 0 \\ y &\ge 0 \end{aligned}$

Solve the systems in Problems 31–40 graphically, and indicate whether each solution region is bounded or unbounded. Find the coordinates of each corner point.

31. $\begin{aligned} 2x + 3y &\le 12 \\ x &\ge 0 \\ y &\ge 0 \end{aligned}$

32. $\begin{aligned} 3x + 4y &\le 24 \\ x &\ge 0 \\ y &\ge 0 \end{aligned}$

33. $\begin{aligned} 2x + y &\le 10 \\ x + 2y &\le 8 \\ x &\ge 0 \\ y &\ge 0 \end{aligned}$

34. $\begin{aligned} 6x + 3y &\le 24 \\ 3x + 6y &\le 30 \\ x &\ge 0 \\ y &\ge 0 \end{aligned}$

35. $\begin{aligned} 2x + y &\ge 10 \\ x + 2y &\ge 8 \\ x &\ge 0 \\ y &\ge 0 \end{aligned}$

36. $\begin{aligned} 4x + 3y &\ge 24 \\ 3x + 4y &\ge 8 \\ x &\ge 0 \\ y &\ge 0 \end{aligned}$

37. $\begin{aligned} 2x + y &\le 10 \\ x + y &\le 7 \\ x + 2y &\le 12 \\ x &\ge 0 \\ y &\ge 0 \end{aligned}$

38. $\begin{aligned} 3x + y &\le 21 \\ x + y &\le 9 \\ x + 3y &\le 21 \\ x &\ge 0 \\ y &\ge 0 \end{aligned}$

39. $\begin{aligned} 2x + y &\ge 16 \\ x + y &\ge 12 \\ x + 2y &\ge 14 \\ x &\ge 0 \\ y &\ge 0 \end{aligned}$

40. $\begin{aligned} 3x + y &\ge 24 \\ x + y &\ge 16 \\ x + 3y &\ge 30 \\ x &\ge 0 \\ y &\ge 0 \end{aligned}$

C

Solve the systems in Problems 41–50 graphically, and indicate whether each solution region is bounded or unbounded. Find the coordinates of each corner point.

41. $\begin{aligned} x + 4y &\le 32 \\ 3x + y &\le 30 \\ 4x + 5y &\ge 51 \end{aligned}$

42. $\begin{aligned} x + y &\le 11 \\ x + 5y &\ge 15 \\ 2x + y &\ge 12 \end{aligned}$

43. $\begin{aligned} 4x + 3y &\le 48 \\ 2x + y &\ge 24 \\ x &\le 9 \end{aligned}$

44. $\begin{aligned} 2x + 3y &\ge 24 \\ x + 3y &\le 15 \\ y &\ge 4 \end{aligned}$

45. $\begin{aligned} x - y &\le 0 \\ 2x - y &\le 4 \\ 0 \le x &\le 8 \end{aligned}$

46. $\begin{aligned} 2x + 3y &\ge 12 \\ -x + 3y &\le 3 \\ 0 \le y &\le 5 \end{aligned}$

47. $$\begin{aligned} -x + 3y &\geq 1 \\ 5x - y &\geq 9 \\ x + y &\leq 9 \\ x &\leq 5 \end{aligned}$$

48. $$\begin{aligned} x + y &\leq 10 \\ 5x + 3y &\geq 15 \\ -2x + 3y &\leq 15 \\ 2x - 5y &\leq 6 \end{aligned}$$

49. $$\begin{aligned} 16x + 13y &\leq 120 \\ 3x + 4y &\geq 25 \\ -4x + 3y &\leq 11 \end{aligned}$$

50. $$\begin{aligned} 2x + 2y &\leq 21 \\ -10x + 5y &\leq 24 \\ 3x + 5y &\geq 37 \end{aligned}$$

Problems 51 and 52 introduce an algebraic process for finding the corner points of a solution region without drawing a graph. We will have a great deal more to say about this process later in this chapter.

51. Consider the following system of inequalities and corresponding boundary lines:

$$\begin{aligned} 3x + 4y &\leq 36 & \qquad 3x + 4y &= 36 \\ 3x + 2y &\leq 30 & \qquad 3x + 2y &= 30 \\ x &\geq 0 & \qquad x &= 0 \\ y &\geq 0 & \qquad y &= 0 \end{aligned}$$

(A) Use algebraic methods to find the intersection points (if any exist) for each possible pair of boundary lines. (There are six different possible pairs.)

(B) Test each intersection point in all four inequalities to determine which are corner points.

52. Repeat Problem 51 for

$$\begin{aligned} 2x + y &\leq 16 & \qquad 2x + y &= 16 \\ 2x + 3y &\leq 36 & \qquad 2x + 3y &= 36 \\ x &\geq 0 & \qquad x &= 0 \\ y &\geq 0 & \qquad y &= 0 \end{aligned}$$

Applications

Business & Economics

53. Manufacturing: resource allocation. A manufacturing company makes two types of water skis, a trick ski and a slalom ski. The trick ski requires 6 labor-hours for fabricating and 1 labor-hour for finishing. The slalom ski requires 4 labor-hours for fabricating and 1 labor-hour for finishing. The maximum labor-hours available per day for fabricating and finishing are 108 and 24, respectively. If x is the number of trick skis and y is the number of slalom skis produced per day, write a system of linear inequalities that indicates appropriate restraints on x and y. Find the set of feasible solutions graphically for the number of each type of ski that can be produced.

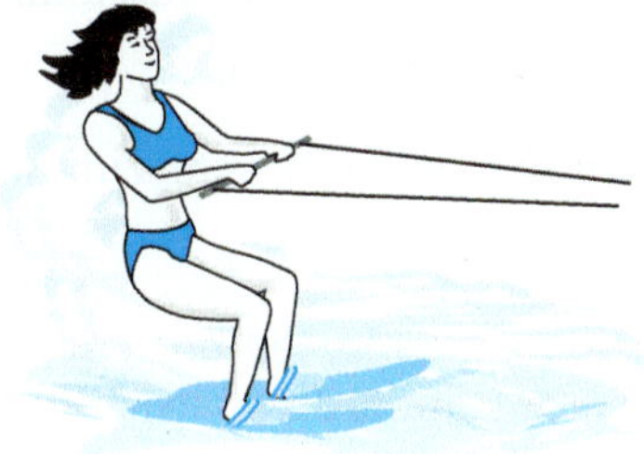

54. Manufacturing: resource allocation. A furniture manufacturing company manufactures dining room tables and chairs. A table requires 8 labor-hours for assembling and 2 labor-hours for finishing. A chair requires 2 labor-hours for assembling and 1 labor-hour for finishing. The maximum labor-hours available per day for assembly and finishing are 400 and 120, respectively. If x is the number of tables and y is the number of chairs produced per day, write a system of linear inequalities that indicates appropriate restraints on x and y. Find the set of feasible solutions graphically for the number of tables and chairs that can be produced.

55. Manufacturing: resource allocation. Refer to Problem 53. The company makes a profit of $50 on each trick ski and a profit of $60 on each slalom ski.

(A) If the company makes 10 trick skis and 10 slalom skis per day, the daily profit will be $1,100. Are there other production schedules that will result in a daily profit of $1,100? How are these schedules related to the graph of the line $50x + 60y = 1{,}100$?

(B) Find a production schedule that will produce a daily profit greater than $1,100 and repeat part (A) for this schedule.

(C) Discuss methods for using lines like those in parts (A) and (B) to find the largest possible daily profit.

56. Manufacturing: resource allocation. Refer to Problem 54. The company makes a profit of $50 on each table and a profit of $15 on each chair.

(A) If the company makes 20 tables and 20 chairs per day, the daily profit will be $1,300. Are there other production schedules that will result in a daily profit of $1,300? How are these schedules related to the graph of the line $50x + 15y = 1{,}300$?

(B) Find a production schedule that will produce a daily profit greater than $1,300 and repeat part (A) for this schedule.

(C) Discuss methods for using lines like those in parts (A) and (B) to find the largest possible daily profit.

Life Sciences

57. Nutrition: plants. A farmer can buy two types of plant food, mix A and mix B. Each cubic yard of mix A contains 20 pounds of phosphoric acid, 30 pounds of nitrogen, and 5 pounds of potash. Each cubic yard of mix B contains 10 pounds of phosphoric acid, 30 pounds of nitrogen, and 10 pounds of potash. The minimum monthly requirements are 460 pounds of phosphoric acid, 960 pounds of nitrogen, and 220 pounds of potash. If x is the number of cubic yards of mix A used and y is the number of cubic yards of mix B used, write a system of linear inequalities that indicates appropriate restraints on x and y. Find the set of feasible solutions graphically for the amounts of mix A and mix B that can be used.

58. Nutrition: people. A dietitian in a hospital is to arrange a special diet using two foods. Each ounce of food M contains 30 units of calcium, 10 units of iron, and 10 units of vitamin A. Each ounce of food N contains 10 units of calcium, 10 units of iron, and 30 units of vitamin A. The minimum requirements in the diet are 360 units of calcium, 160 units of iron, and 240 units of vitamin A. If x is the number of ounces of food M used and y is the number of ounces of food N used, write a system of linear inequalities that reflects the conditions indicated. Find the set of feasible solutions graphically for the amount of each kind of food that can be used.

Social Sciences

59. **Psychology.** In an experiment on conditioning, a psychologist uses two types of Skinner (conditioning) boxes with mice and rats. Each mouse spends 10 minutes per day in box A and 20 minutes per day in box B. Each rat spends 20 minutes per day in box A and 10 minutes per day in box B. The total maximum time available per day is 800 minutes for box A and 640 minutes for box B. We are interested in the various numbers of mice and rats that can be used in the experiment under the conditions stated. If we let x be the number of mice used and y the number of rats used, write a system of linear inequalities that indicates appropriate restrictions on x and y. Find the set of feasible solutions graphically.

11-2 Linear Programming in Two Dimensions: Geometric Approach

- Linear Programming Problem
- Linear Programming: General Description
- Geometric Solution of Linear Programming Problems
- Applications

Several problems discussed in the preceding section are related to a more general type of problem called a *linear programming problem*. Linear programming is a mathematical process that has been developed to help management in decision making, and it has become one of the most widely used and best-known tools of management science. We introduce this topic by considering an example in detail, using an intuitive geometric approach. Insight gained from this approach will prove invaluable when we later consider an algebraic approach that is less intuitive but necessary in solving most real-world problems.

AGREEMENT

For ease of generalization to the larger problems in later sections, we now change variable notation from letters such as x and y to subscript forms such as x_1 and x_2.

Linear Programming Problem

We begin our discussion with a concrete example. The geometric method of solution will suggest two important theorems and a simple general geometric procedure for solving linear programming problems in two variables.

EXAMPLE 1 **Production Scheduling** A manufacturer of lightweight mountain tents makes a standard model and an expedition model for national distribution. Each standard tent requires 1 labor-hour from the cutting department and 3 labor-hours from the assembly department. Each expedition tent requires 2 labor-hours from the cutting department and 4 labor-hours from the assembly department. The maximum labor-hours available per day in the cutting department and the assembly department are 32 and 84, respectively. If the company makes a profit of \$50 on each standard tent and \$80 on each expedition tent, how many tents of each type should be manufactured each day to maximize the total daily profit (assuming that all tents can be sold)?

SOLUTION This is an example of a linear programming problem. We begin by analyzing the question posed in this example.

DECISION VARIABLES According to the question in the last sentence of the example, the *objective* of management is to maximize profit. Since the profits for standard and expedition tents differ, management must *decide* how many of each type of tent to manufacture. Thus, it is reasonable to introduce the following **decision variables:**

$$\text{Let } x_1 = \text{number of standard tents produced per day}$$
$$x_2 = \text{number of expedition tents produced per day}$$

TABLE 1

	Labor-Hours per Tent		Maximum Labor-Hours Available per Day
	Standard Model	Expedition Model	
Cutting department	1	2	32
Assembly department	3	4	84
Profit per tent	$50	$80	

Now we summarize the manufacturing requirements, objectives, and restrictions in Table 1, with the decision variables related to the columns in the table.

OBJECTIVE FUNCTION Using the decision variables and the information in Table 1, we can form the **objective function** (we assume that all tents manufactured are actually sold):

$$P = 50x_1 + 80x_2 \quad \text{Objective function}$$

The **objective** is to find values of the decision variables that produce the **optimal value** (in this case, maximum value) of the objective function.

CONSTRAINTS The form of the objective function indicates that the profit can be made as large as we like simply by producing enough tents. But any manufacturing company has limits imposed by available resources, plant capacity, demand, and so on. These limits are referred to as **problem constraints.** Using the information in Table 1, we can determine two problem constraints.

CUTTING DEPARTMENT

$$\begin{pmatrix}\text{daily cutting}\\ \text{time for } x_1\\ \text{standard tents}\end{pmatrix} + \begin{pmatrix}\text{daily cutting}\\ \text{time for } x_2\\ \text{expedition tents}\end{pmatrix} \le \begin{pmatrix}\text{maximum labor-}\\ \text{hours available}\\ \text{per day}\end{pmatrix}$$
$$1x_1 \quad + \quad 2x_2 \quad \le \quad 32$$

ASSEMBLY DEPARTMENT CONSTRAINT

$$\begin{pmatrix}\text{daily assembly}\\ \text{time for } x_1\\ \text{standard tents}\end{pmatrix} + \begin{pmatrix}\text{daily assembly}\\ \text{time for } x_2\\ \text{expedition tents}\end{pmatrix} \le \begin{pmatrix}\text{maximum labor-}\\ \text{hours available}\\ \text{per day}\end{pmatrix}$$
$$3x_1 \quad + \quad 4x_2 \quad \le \quad 84$$

NONNEGATIVE CONSTRAINTS It is not possible to manufacture a negative number of tents; thus, we have the **nonnegative constraints**

$$x_1 \ge 0$$
$$x_2 \ge 0$$

which we usually write in the form

$$x_1, x_2 \ge 0$$

MATHEMATICAL MODEL We now have a **mathematical model** for the problem under consideration:

$$\text{Maximize} \quad P = 50x_1 + 80x_2 \quad \text{Objective function}$$

$$\text{subject to} \quad \left.\begin{aligned} x_1 + 2x_2 &\le 32 \\ 3x_1 + 4x_2 &\le 84 \end{aligned}\right\} \quad \text{Problem constraints}$$
$$x_1, x_2 \ge 0 \quad \text{Nonnegative constraints}$$

GRAPHIC SOLUTION **Solving** the set of linear inequality constraints **graphically** (see the preceding section), we obtain the feasible region for production schedules (Fig. 1).

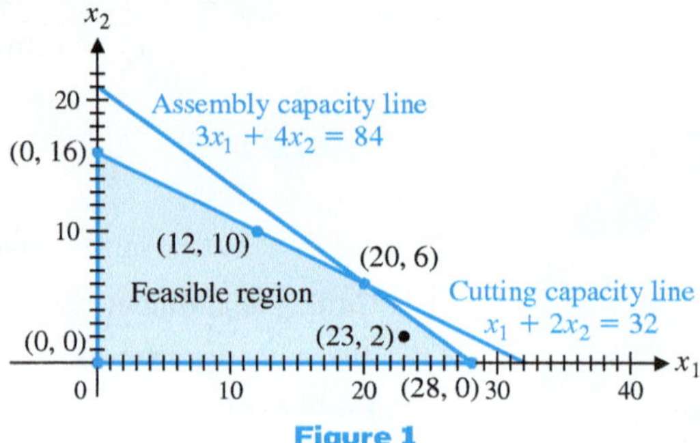

Figure 1

By choosing a production schedule (x_1, x_2) from the feasible region, a profit can be determined using the objective function

$$P = 50x_1 + 80x_2$$

For example, if $x_1 = 12$ and $x_2 = 10$, the profit for the day would be

$$\begin{aligned} P &= 50(12) + 80(10) \\ &= \$1{,}400 \end{aligned}$$

Or if $x_1 = 23$ and $x_2 = 2$, the profit for the day would be

$$\begin{aligned} P &= 50(23) + 80(2) \\ &= \$1{,}310 \end{aligned}$$

But the question is, out of all possible production schedules (x_1, x_2) from the feasible region, which schedule(s) produces the *maximum* profit? Thus, we have a **maximization problem.** Since point-by-point checking is impossible (there are infinitely many points to check), we must find another way.

By assigning P in $P = 50x_1 + 80x_2$ a particular value and plotting the resulting equation in the coordinate system shown in Figure 1, we obtain a **constant-profit line (isoprofit line).** Every point in the feasible region on this line represents a production schedule that will produce the same profit. By doing this for a number of values for P, we obtain a family of constant-profit lines (Fig. 2) that are parallel to each other, since they all have the same slope. To see this, we write $P = 50x_1 + 80x_2$ in the slope–intercept form

$$x_2 = -\frac{5}{8}x_1 + \frac{P}{80}$$

and note that for any profit P, the constant-profit line has slope. $-\frac{5}{8}$. We also observe that as the profit P increases, the x_2 intercept $(P/80)$ increases, and the line moves away from the origin.

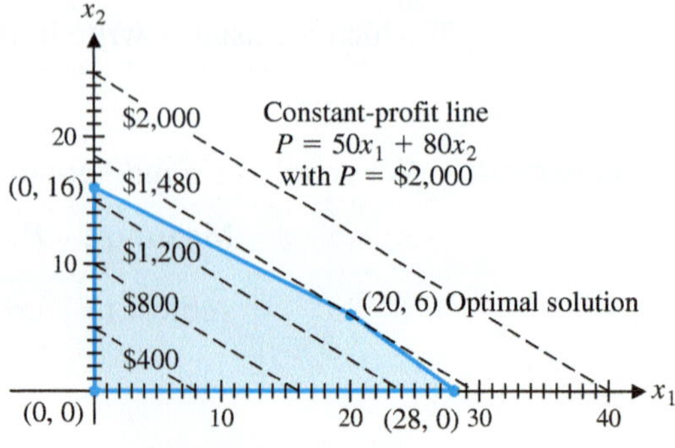

Figure 2 Constant-profit lines

Thus, the maximum profit occurs at a point where a constant-profit line is the farthest from the origin but still in contact with the feasible region. In this example, this occurs at (20, 6), as is seen in Figure 2. Thus, if the manufacturer makes 20 standard tents and 6 expedition tents per day, the profit will be maximized at

$$\begin{aligned} P &= 50(20) + 80(6) \\ &= \$1{,}480 \end{aligned}$$

The point (20, 6) is called an **optimal solution** to the problem, because it maximizes the objective (profit) function and is in the feasible region. In general, it appears that a maximum profit occurs at one of the corner points. We also note that the minimum profit ($P = 0$) occurs at the corner point (0, 0).

Matched Problem 1 A manufacturing plant makes two types of inflatable boats, a two-person boat and a four-person boat. Each two-person boat requires 0.9 labor-hour from the cutting department and 0.8 labor-hour from the assembly department. Each four-person boat requires 1.8 labor-hours from the cutting department and 1.2 labor-hours from the assembly department. The maximum labor-hours available per month in the cutting department and the assembly department are 864 and 672, respectively. The company makes a profit of \$25 on each two-person boat and \$40 on each four-person boat.

(A) Identify the decision variables.
(B) Summarize the relevant material in a table similar to Table 1 in Example 1.
(C) Write the objective function P.
(D) Write the problem constraints and the nonnegative constraints.
(E) Graph the feasible region. Include graphs of the objective function for $P = \$5{,}000$, $P = \$10{,}000$, $P = \$15{,}000$, and $P = \$21{,}600$.
(F) From the graph and constant-profit lines, determine how many boats should be manufactured each month to maximize the profit. What is the maximum profit?

Before proceeding further, let's summarize the steps we used to form the model in Example 1.

PROCEDURE Constructing the Model for an Applied Linear Programming Problem

Step 1 Introduce decision variables.

Step 2 Summarize relevant material in table form, relating the decision variables with the columns in the table, if possible (see Table 1).

Step 3 Determine the objective and write a linear objective function.

Step 4 Write problem constraints using linear equations and/or inequalities.

Step 5 Write nonnegative constraints.

EXPLORE & DISCUSS 1 Refer to the feasible region S shown in the figure on next page.

(A) Let $P = x_1 + x_2$. Graph the isoprofit lines through the points (5, 5) and (10, 10). Place a straightedge along the line with the smaller profit and slide it in the direction of increasing profit, without changing its slope. What is the maximum value of P? Where does this maximum value occur?

(B) Repeat part (A) for $P = x_1 + 10x_2$.

(C) Repeat part (A) for $P = 10x_1 + x_2$.

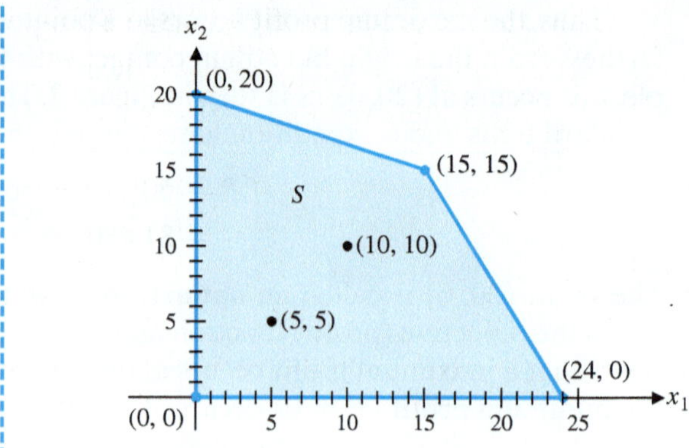

Linear Programming: General Description

In Example 1 and Matched Problem 1, the optimal solution occurs at a corner point of the feasible region. Is this always the case? The answer is a qualified yes, as will be seen in Theorem 1. First, we give a few general definitions.

A **linear programming problem** is one that is concerned with finding the **optimal value** (maximum or minimum value) of a linear **objective function** of the form

$$z = c_1x_1 + c_2x_2 + \cdots + c_nx_n$$

where the **decision variables** $x_1, x_2, \ldots, x_n$ are subject to **problem constraints** in the form of linear inequalities and equations. In addition, the decision variable must satisfy the **nonnegative constraints** $x_i \geq 0, i = 1, 2, \ldots, n$. The set of points satisfying both the problem constraints and the nonnegative constraints is called the **feasible region** for the problem. Any point in the feasible region that produces the optimal value of the objective function over the feasible region is called an **optimal solution.**

THEOREM 1 Fundamental Theorem of Linear Programming: Version 1

If the optimal value of the objective function in a linear programming problem exists, then that value must occur at one (or more) of the corner points of the feasible region.

Theorem 1 provides a simple procedure for solving a linear programming problem, *provided that the problem has an optimal solution—not all do.* In order to use Theorem 1, we must know that the problem under consideration has an optimal solution. Theorem 2 provides some conditions that will ensure that a linear programming problem has an optimal solution.

THEOREM 2 Existence of Optimal Solutions

(A) If the feasible region for a linear programming problem is bounded, then both the maximum value and the minimum value of the objective function always exist.

(B) If the feasible region is unbounded and the coefficients of the objective function are positive, then the minimum value of the objective function exists, but the maximum value does not.

(C) If the feasible region is empty (that is, there are no points that satisfy all the constraints), then both the maximum value and the minimum value of the objective function do not exist.

Theorem 2 does not cover all possibilities. For example, what happens if the feasible region is unbounded and one (or both) of the coefficients of the objective function are negative? Problems of this type must be solved by carefully examining the graph of the objective function for various feasible solutions, as we did in Example 1.

Geometric Solution of Linear Programming Problems

The discussion above leads to the following procedure for the geometric solution of linear programming problems with two decision variables:

PROCEDURE Geometric Solution of a Linear Programming Problem with Two Decision Variables

Step 1 Graph the feasible region. Then, if an optimal solution exists according to Theorem 2, find the coordinates of each corner point.

Step 2 Construct a **corner point table** listing the value of the objective function at each corner point.

Step 3 Determine the optimal solution(s) from the table in step 2.

Step 4 For an applied problem, interpret the optimal solution(s) in terms of the original problem.

Before we consider additional applications, let us use this procedure to solve some linear programming problems where the model has already been determined.

EXAMPLE 2 **Solving a Linear Programming Problem**

(A) Minimize and maximize

$$z = 3x_1 + x_2$$

subject to

$$\begin{aligned} 2x_1 + x_2 &\le 20 \\ 10x_1 + x_2 &\ge 36 \\ 2x_1 + 5x_2 &\ge 36 \\ x_1, x_2 &\ge 0 \end{aligned}$$

(B) Minimize and maximize

$$z = 10x_1 + 20x_2$$

subject to

$$\begin{aligned} 6x_1 + 2x_2 &\ge 36 \\ 2x_1 + 4x_2 &\ge 32 \\ x_2 &\le 20 \\ x_1, x_2 &\ge 0 \end{aligned}$$

SOLUTION (A) **Step 1** Graph the feasible region S. Then, after checking Theorem 2 to determine that an optimal solution exists, find the coordinates of each corner point. Since S is bounded, z will have both a maximum and a minimum on S (Theorem 2A) and these will both occur at corner points (Theorem 1).

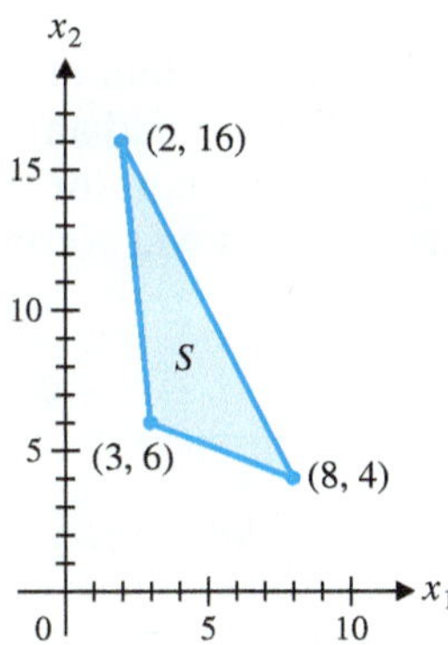

Step 2 Evaluate the objective function at each corner point, as shown in the table in the margin.

Step 3 Determine the optimal solutions from step 2. Examining the values in the table, we see that the minimum value of z is 15 at (3, 6) and the maximum value of z is 28 at (8, 4).

(B) **Step 1** Graph the feasible region S. Then, after checking Theorem 2 to determine that an optimal solution exists, find the coordinates of each corner point. Since S is unbounded and the coefficients of the objective function are positive, z has a minimum value on S but no maximum value (Theorem 2B).

Corner Point

(x_1, x_2)	$z = 3x_1 + x_2$
$(3, 6)$	15
$(2, 16)$	22
$(8, 4)$	28

Corner Point

(x_1, x_2)	$z = 10x_1 + 20x_2$
$(0, 20)$	400
$(0, 18)$	360
$(4, 6)$	160
$(16, 0)$	160

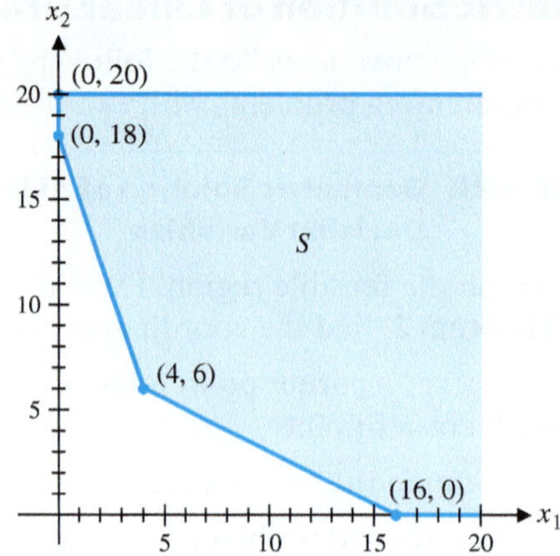

Step 2 Evaluate the objective function at each corner point, as shown in the table in the margin.

Step 3 Determine the optimal solution from step 2. The minimum value of z is 160 at $(4, 6)$ and at $(16, 0)$.

RESULT

The solution to Example 2 is a **multiple optimal solution.**

In general, if two corner points are both optimal solutions to a linear programming problem, then any point on the line segment joining them is also an optimal solution.

This is the only time that optimal solutions also occur at noncorner points.

Matched Problem 2 (A) Maximize and minimize $z = 4x_1 + 2x_2$ subject to the constraints given in Example 2A.

(B) Maximize and minimize $z = 20x_1 + 5x_2$ subject to the constraints given in Example 2B.

CONCEPTUAL INSIGHT

Determining that an optimal solution exists is a critical step in the solution of a linear programming problem. If you skip this step, you may examine a corner point table like the one in the solution of Example 2(B) and erroneously conclude that the maximum value of the objective function is 400.

Explore & Discuss 2

In Example 2B we saw that there was no optimal solution for the problem of maximizing the objective function P over the feasible region S. We want to add an additional constraint to modify the feasible region so that an optimal solution for the maximization problem does exist. Which of the following constraints will accomplish this objective?

(A) $x_1 \le 20$ (B) $x_2 \ge 4$ (C) $x_1 \le x_2$ (D) $x_2 \le x_1$

For an illustration of Theorem 2C, consider the following:

$$\begin{aligned} \text{Maximize} \quad & P = 2x_1 + 3x_2 \\ \text{subject to} \quad & x_1 + x_2 \ge 8 \\ & x_1 + 2x_2 \le 8 \\ & 2x_1 + x_2 \le 10 \\ & x_1, x_2 \ge 0 \end{aligned}$$

The intersection of the graphs of the constraint inequalities is the empty set (Fig. 3); hence, the *feasible region is empty*. If this happens, the problem should be reexamined to see if it has been formulated properly. If it has, the management may have to reconsider items such as labor-hours, overtime, budget, and supplies allocated to the project in order to obtain a nonempty feasible region and a solution to the original problem.

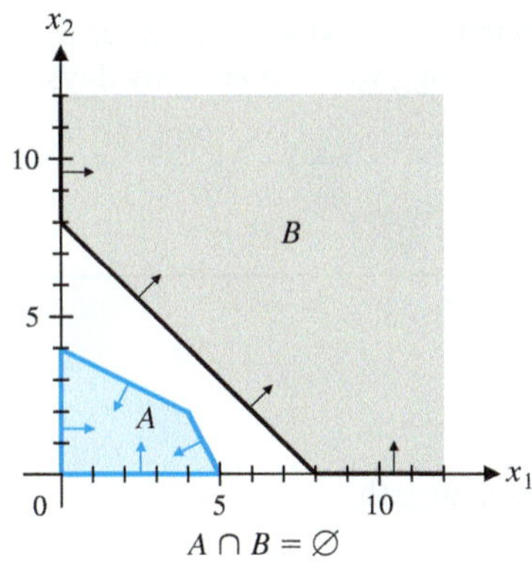

Figure 3

Applications

EXAMPLE 3 **Medicine** We now convert Example 5 from the preceding section into a linear programming problem. A patient in a hospital is required to have at least 84 units of drug *A* and 120 units of drug *B*, each day (assume that an overdosage of either drug is harmless). Each gram of substance *M* contains 10 units of drug *A* and 8 units of drug *B*, and each gram of substance *N* contains 2 units of drug *A* and 4 units of drug *B*. Now, suppose that both *M* and *N* contain an undesirable drug *D*, 3 units per gram in *M* and 1 unit per gram in *N*. How many grams of each of substances *M* and *N* should be mixed to meet the minimum daily requirements and at the same time minimize the intake of drug *D*? How many units of the undesirable drug *D* will be in this mixture?

SOLUTION First we construct the mathematical model.

Step 1 Introduce decision variables. According to the questions asked in this example, we must decide how many grams of substances *M* and *N* should be mixed to form the daily dose of medication. Thus, these two quantities are the decision variables:

$$x_1 = \text{number of grams of substance } M \text{ used}$$
$$x_2 = \text{number of grams of substance } N \text{ used}$$

Step 2 Summarize relevant material in a table, relating the columns to substances *M* and *N*.

	Amount of Drug per Gram		Minimum Daily Requirement
	Substance *M*	**Substance *N***	
Drug *A*	10 units	2 units	84 units
Drug *B*	8 units	4 units	120 units
Drug *D*	3 units	1 unit	

Step 3 Determine the objective and the objective function. The objective is to minimize the amount of drug *D* in the daily dose of medication. Using the decision variables and the information in the table, we form the linear objective function

$$C = 3x_1 + x_2$$

Step 4 Write the problem constraints. The constraints in this problem involve minimum requirements, so the inequalities will take a different form:

$$10x_1 + 2x_2 \geq 84 \quad \text{Drug A constraint}$$
$$8x_1 + 4x_2 \geq 120 \quad \text{Drug B constraint}$$

Step 5 Add the nonnegative constraints and summarize the model.

Minimize	$C = 3x_1 + x_2$	Objective function
subject to	$10x_1 + 2x_2 \geq 84$	Drug A constraint
	$8x_1 + 4x_2 \geq 120$	Drug B constraint
	$x_1, x_2 \geq 0$	Nonnegative constraints

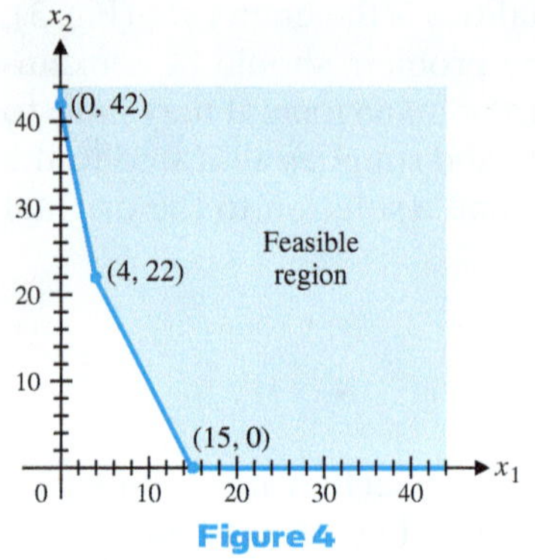

Figure 4

Corner Point (x_1, x_2)	$C = 3x_1 + x_2$
(0, 42)	42
(4, 22)	34
(15, 0)	45

Now we use the geometric method to solve the problem.

Step 1 Graph the feasible region. Then, after checking Theorem 2 to determine that an optimal solution exists, find the coordinates of each corner point. Solving the system of constraint inequalities graphically, we obtain the feasible region shown in Figure 4. Since the feasible region is unbounded and the coefficients of the objective function are positive, this minimization problem has a solution.

Step 2 Evaluate the objective function at each corner point, as shown in the table.

Step 3 Determine the optimal solution from step 2. The optimal solution is $C = 34$ at the corner point (4, 22).

Step 4 Interpret the optimal solution in terms of the original problem. If we use 4 grams of substance M and 22 grams of substance N, we will supply the minimum daily requirements for drugs A and B and minimize the intake of the undesirable drug D at 34 units. (Any other combination of M and N from the feasible region will result in a larger amount of the undesirable drug D.)

Matched Problem 3

Agriculture A chicken farmer can buy a special food mix A at 20¢ per pound and a special food mix B at 40¢ per pound. Each pound of mix A contains 3,000 units of nutrient N_1 and 1,000 units of nutrient N_2; each pound of mix B contains 4,000 units of nutrient N_1 and 4,000 units of nutrient N_2. If the minimum daily requirements for the chickens collectively are 36,000 units of nutrient N_1 and 20,000 units of nutrient N_2, how many pounds of each food mix should be used each day to minimize daily food costs while meeting (or exceeding) the minimum daily nutrient requirements? What is the minimum daily cost? Construct a mathematical model and solve using the geometric method.

CONCEPTUAL INSIGHT

Refer to Example 3. If we change the minimum requirement for drug A from 120 to 125, the optimal solution changes to 3.6 grams of substance M and 24.1 grams of substance N, correct to one decimal place.

Now refer to Example 1. If we change the maximum labor-hours available per day in the assembly department from 84 to 79, the solution changes to 15 standard tents and 8.5 expedition tents.

We can measure 3.6 grams of substance M and 24.1 grams of substance N, but how can we make 8.5 tents? Should we make 8 tents? Or 9 tents? If the solutions to a problem must be integers and the optimal solution found graphically involves decimals, rounding the decimal value to the nearest integer does not always produce the *optimal integer solution* (see Problem 36, Exercise 11-2). Finding optimal integer solutions to a linear programming problem is called *integer programming* and requires special techniques that are beyond the scope of this book. As mentioned earlier, if we encounter a solution like 8.5 tents per day, we will interpret this as an *average* value over many days of production.

Answers to Matched Problems

1. (A) x_1 = number of two-person boats produced each month

x_2 = number of four-person boats produced each month

(B)

	Labor-Hours Required		Maximum Labor-Hours Available per Month
	Two-Person Boat	**Four-Person Boat**	
Cutting department	0.9	1.8	864
Assembly department	0.8	1.2	672
Profit per boat	\$25	\$40	

(C) $P = 25x_1 + 40x_2$

(D) $0.9x_1 + 1.8x_2 \le 864$

$0.8x_1 + 1.2x_2 \le 672$

$x_1, x_2 \ge 0$

(E)

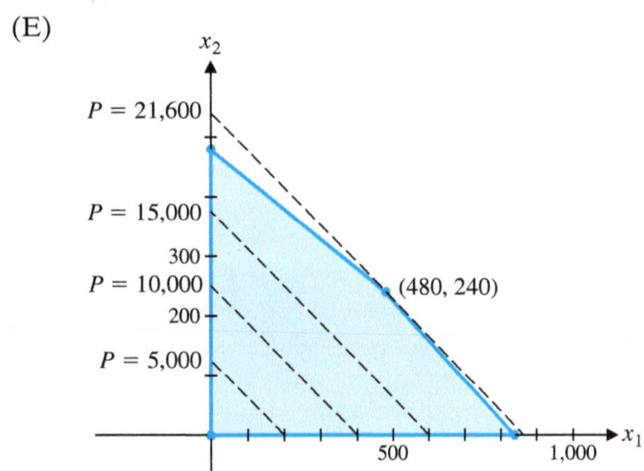

(F) 480 two-person boats, 240 four-person boats; max P = \$21,600 per month

2. (A) Min $z = 24$ at $(3, 6)$; max $z = 40$ at $(2, 16)$ and $(8, 4)$ (multiple optimal solution)

(B) Min $z = 90$ at $(0, 18)$; no maximum value

3. Min $C = 0.2x_1 + 0.4x_2$

subject to $3{,}000x_1 + 4{,}000x_2 \ge 36{,}000$

$1{,}000x_1 + 4{,}000x_2 \ge 20{,}000$

$x_1, x_2 \ge 0$

8 lb of mix A, 3 lb of mix B; min C = \$2.80 per day

Exercise 11-2

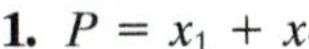

In Problems 1–4, graph the isoprofit lines through (3, 3) and also through (6, 6). Use a straight edge to identify the corner point where the maximum profit occurs (see Explore–Discuss 1). Confirm your answer by constructing a corner point table.

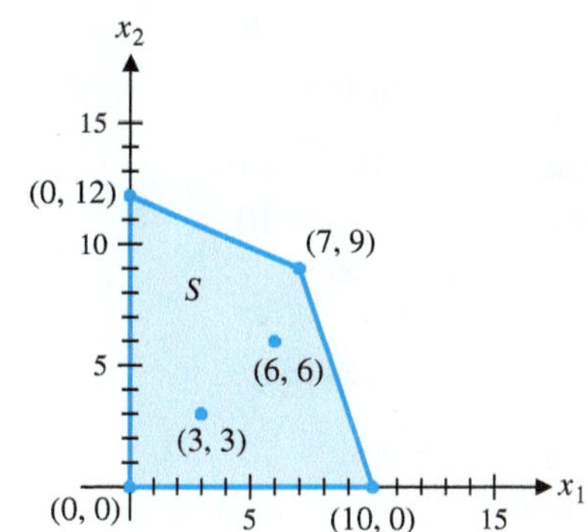

1. $P = x_1 + x_2$

2. $P = 4x_1 + x_2$

3. $P = 3x_1 + 7x_2$

4. $P = 9x_1 + 3x_2$

In Problems 5–8, graph the isocost lines through (9, 9) and also through (12, 12). Use a straight edge to identify the corner point where the minimum cost occurs. Confirm your answer by constructing a corner point table.

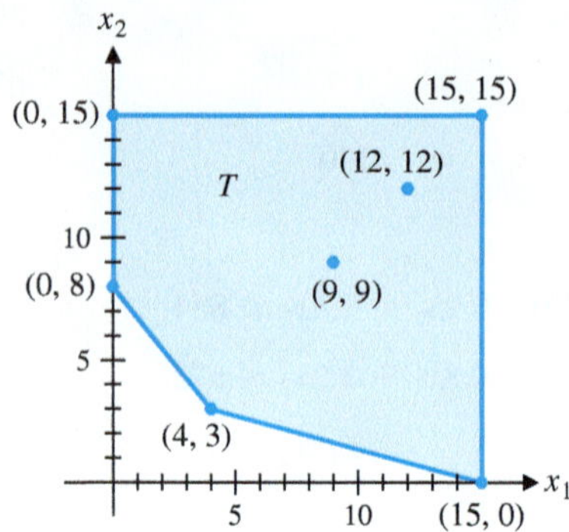

5. $C = 7x_1 + 4x_2$

6. $C = 7x_1 + 9x_2$

7. $C = 3x_1 + 8x_2$

8. $C = 2x_1 + 11x_2$

B

Solve the linear programming problems stated in Problems 9–26.

9. Maximize $P = 5x_1 + 5x_2$
subject to
$$\begin{aligned} 2x_1 + x_2 &\le 10 \\ x_1 + 2x_2 &\le 8 \\ x_1, x_2 &\ge 0 \end{aligned}$$

10. Maximize $P = 3x_1 + 2x_2$
subject to
$$\begin{aligned} 6x_1 + 3x_2 &\le 24 \\ 3x_1 + 6x_2 &\le 30 \\ x_1, x_2 &\ge 0 \end{aligned}$$

11. Minimize and maximize
$z = 2x_1 + 3x_2$
subject to
$$\begin{aligned} 2x_1 + x_2 &\ge 10 \\ x_1 + 2x_2 &\ge 8 \\ x_1, x_2 &\ge 0 \end{aligned}$$

12. Minimize and maximize
$z = 8x_1 + 7x_2$
subject to
$$\begin{aligned} 4x_1 + 3x_2 &\ge 24 \\ 3x_1 + 4x_2 &\ge 8 \\ x_1, x_2 &\ge 0 \end{aligned}$$

13. Maximize $P = 30x_1 + 40x_2$
subject to
$$\begin{aligned} 2x_1 + x_2 &\le 10 \\ x_1 + x_2 &\le 7 \\ x_1 + 2x_2 &\le 12 \\ x_1, x_2 &\ge 0 \end{aligned}$$

14. Maximize $P = 20x_1 + 10x_2$
subject to
$$\begin{aligned} 3x_1 + x_2 &\le 21 \\ x_1 + x_2 &\le 9 \\ x_1 + 3x_2 &\le 21 \\ x_1, x_2 &\ge 0 \end{aligned}$$

15. Minimize and maximize
$z = 10x_1 + 30x_2$
subject to
$$\begin{aligned} 2x_1 + x_2 &\ge 16 \\ x_1 + x_2 &\ge 12 \\ x_1 + 2x_2 &\ge 14 \\ x_1, x_2 &\ge 0 \end{aligned}$$

16. Minimize and maximize
$z = 400x_1 + 100x_2$
subject to
$$\begin{aligned} 3x_1 + x_2 &\ge 24 \\ x_1 + x_2 &\ge 16 \\ x_1 + 3x_2 &\ge 30 \\ x_1, x_2 &\ge 0 \end{aligned}$$

17. Minimize and maximize
$P = 30x_1 + 10x_2$
subject to
$$\begin{aligned} 2x_1 + 2x_2 &\ge 4 \\ 6x_1 + 4x_2 &\le 36 \\ 2x_1 + x_2 &\le 10 \\ x_1, x_2 &\ge 0 \end{aligned}$$

18. Minimize and maximize
$P = 2x_1 + x_2$
subject to
$$\begin{aligned} x_1 + x_2 &\ge 2 \\ 6x_1 + 4x_2 &\le 36 \\ 4x_1 + 2x_2 &\le 20 \\ x_1, x_2 &\ge 0 \end{aligned}$$

19. Minimize and maximize
$P = 3x_1 + 5x_2$
subject to
$$\begin{aligned} x_1 + 2x_2 &\le 6 \\ x_1 + x_2 &\le 4 \\ 2x_1 + 3x_2 &\ge 12 \\ x_1, x_2 &\ge 0 \end{aligned}$$

20. Minimize and maximize
$P = -x_1 + 3x_2$
subject to
$$\begin{aligned} 2x_1 - x_2 &\ge 4 \\ -x_1 + 2x_2 &\le 4 \\ x_2 &\le 6 \\ x_1, x_2 &\ge 0 \end{aligned}$$

21. Minimize and maximize
$P = 20x_1 + 10x_2$
subject to
$$\begin{aligned} 2x_1 + 3x_2 &\ge 30 \\ 2x_1 + x_2 &\le 26 \\ -2x_1 + 5x_2 &\le 34 \\ x_1, x_2 &\ge 0 \end{aligned}$$

22. Minimize and maximize
$P = 12x_1 + 14x_2$
subject to
$$\begin{aligned} -2x_1 + x_2 &\ge 6 \\ x_1 + x_2 &\le 15 \\ 3x_1 - x_2 &\ge 0 \\ x_1, x_2 &\ge 0 \end{aligned}$$

23. Maximize $P = 20x_1 + 30x_2$
subject to
$$\begin{aligned} 0.6x_1 + 1.2x_2 &\le 960 \\ 0.03x_1 + 0.04x_2 &\le 36 \\ 0.3x_1 + 0.2x_2 &\le 270 \\ x_1, x_2 &\ge 0 \end{aligned}$$

24. Minimize $C = 30x_1 + 10x_2$
subject to
$$\begin{aligned} 1.8x_1 + 0.9x_2 &\ge 270 \\ 0.3x_1 + 0.2x_2 &\ge 54 \\ 0.01x_1 + 0.03x_2 &\ge 3.9 \\ x_1, x_2 &\ge 0 \end{aligned}$$

25. Maximize $P = 525x_1 + 478x_2$
subject to
$$\begin{aligned} 275x_1 + 322x_2 &\le 3{,}381 \\ 350x_1 + 340x_2 &\le 3{,}762 \\ 425x_1 + 306x_2 &\le 4{,}114 \\ x_1, x_2 &\ge 0 \end{aligned}$$

26. Maximize $P = 300x_1 + 460x_2$
subject to
$$\begin{aligned} 245x_1 + 452x_2 &\le 4{,}181 \\ 290x_1 + 379x_2 &\le 3{,}888 \\ 390x_1 + 299x_2 &\le 4{,}407 \\ x_1, x_2 &\ge 0 \end{aligned}$$

In Problems 27 and 28, explain why Theorem 2 cannot be used to conclude that a maximum or minimum value exists. Graph the feasible regions and use graphs of the objective function $z = x_1 - x_2$ for various values of z to discuss the existence of a maximum value and a minimum value.

27. Minimize and maximize
$z = x_1 - x_2$
subject to
$$\begin{aligned} x_1 - 2x_2 &\le 0 \\ 2x_1 - x_2 &\le 6 \\ x_1, x_2 &\ge 0 \end{aligned}$$

28. Minimize and maximize
$z = x_1 - x_2$
subject to
$$\begin{aligned} x_1 - 2x_2 &\ge -6 \\ 2x_1 - x_2 &\ge 0 \\ x_1, x_2 &\ge 0 \end{aligned}$$

C

29. The corner points for the bounded feasible region determined by the system of linear inequalities
$$\begin{aligned} x_1 + 2x_2 &\le 10 \\ 3x_1 + x_2 &\le 15 \\ x_1, x_2 &\ge 0 \end{aligned}$$
are $O = (0, 0)$, $A = (0, 5)$, $B = (4, 3)$, and $C = (5, 0)$. If $P = ax_1 + bx_2$ and $a, b > 0$, determine conditions on a and b that will ensure that the maximum value of P occurs:

(A) Only at A

(B) Only at B

(C) Only at C

(D) At both A and B

(E) At both B and C

30. The corner points for the feasible region determined by the system of linear inequalities
$$\begin{aligned} x_1 + 4x_2 &\ge 30 \\ 3x_1 + x_2 &\ge 24 \\ x_1, x_2 &\ge 0 \end{aligned}$$
are $A = (0, 24)$, $B = (6, 6)$, and $D = (30, 0)$. If $C = ax_1 + bx_2$ and $a, b > 0$, determine conditions on a and b that will ensure that the minimum value of C occurs:

(A) Only at A

(B) Only at B

(C) Only at D

(D) At both A and B

(E) At both B and D

Applications

In Problems 31–46, construct a mathematical model in the form of a linear programming problem. (The answers in the back of the book for these application problems include the model.) Then solve by the geometric method.

Business & Economics

31. Manufacturing: resource allocation. A manufacturing company makes two types of water skis, a trick ski and a slalom ski. The relevant manufacturing data are given in the table.

	Labor-Hours per Ski		Maximum Labor-Hours
Department	**Trick Ski**	**Slalom Ski**	**Available per Day**
Fabricating	6	4	108
Finishing	1	1	24

(A) If the profit on a trick ski is \$40 and the profit on a slalom ski is \$30, how many of each type of ski should be manufactured each day to realize a maximum profit? What is the maximum profit?

(B) Discuss the effect on the production schedule and the maximum profit if the profit on a slalom ski decreases to \$25.

(C) Discuss the effect on the production schedule and the maximum profit if the profit on a slalom ski increases to \$45.

32. Manufacturing: resource allocation. A furniture manufacturing company manufactures dining room tables and chairs. The relevant manufacturing data are given in the table.

	Labor-Hours per Ski		Maximum Labor-Hours
Department	**Table**	**Chair**	**Available per Day**
Assembly	8	2	400
Finishing	2	1	120
Profit per unit	\$90	\$25	

(A) How many tables and chairs should be manufactured each day to realize a maximum profit? What is the maximum profit?

(B) Discuss the effect on the production schedule and the maximum profit if the marketing department of the company decides that the number of chairs produced should be at least four times the number of tables produced.

33. Manufacturing: production scheduling. A furniture company has two plants that produce the lumber used in manufacturing tables and chairs. In 1 day of operation, plant A can produce the lumber required to manufacture 20 tables and 60 chairs, and plant B can produce the lumber required to manufacture 25 tables and 50 chairs. The company needs enough lumber to manufacture at least 200 tables and 500 chairs.

(A) If it costs \$1,000 to operate plant A for 1 day and \$900 to operate plant B for 1 day, how many days should each plant be operated to produce a sufficient amount of lumber at a minimum cost? What is the minimum cost?

(B) Discuss the effect on the operating schedule and the minimum cost if the daily cost of operating plant A is reduced to \$600 and all other data in part (A) remain the same.

(C) Discuss the effect on the operating schedule and the minimum cost if the daily cost of operating plant B is reduced to \$800 and all other data in part (A) remain the same.

34. Manufacturing: resource allocation. An electronics firm manufactures two types of personal computers, a standard model and a portable model. The production of a standard computer requires a capital expenditure of \$400 and 40 hours of labor. The production of a portable computer requires a capital expenditure of \$250 and 30 hours of labor. The firm has \$20,000 capital and 2,160 labor-hours available for production of standard and portable computers.

(A) What is the maximum number of computers the company is capable of producing?

(B) If each standard computer contributes a profit of \$320 and each portable model contributes a profit of \$220, how much profit will the company make by producing the maximum number of computers determined in part (A)? Is this the maximum profit? If not, what is the maximum profit?

35. Transportation. The officers of a high school senior class are planning to rent buses and vans for a class trip. Each bus can transport 40 students, requires 3 chaperones, and costs \$1,200 to rent. Each van can transport 8 students, requires 1 chaperone, and costs \$100 to rent. Since there are 400 students in the senior class that may be eligible to go on the trip, the officers must plan to accommodate at least 400 students. Since only 36 parents have volunteered to serve as chaperones, the officers must plan to use at most 36 chaperones. How many vehicles of each type should the officers rent in order to minimize the transportation costs? What are the minimal transportation costs?

36. Transportation. Refer to Problem 35. If each van can transport 7 people and there are 35 available chaperones, show that the optimal solution found graphically involves decimals. Find all feasible solutions with integer coordinates and identify the one that minimizes the transportation costs. Can this optimal integer solution be obtained by rounding the optimal decimal solution? Explain.

37. Investment. An investor has \$60,000 to invest in a CD and a mutual fund. The CD yields 5% and the mutual fund yields on the average 9%. The mutual fund requires a minimum investment of \$10,000 and the investor requires that twice as much should be invested in CDs as in the mutual fund. How much should be invested in CDs and how much in the mutual fund to maximize the return? What is the maximum return?

38. Investment. An investor has \$24,000 to invest in bonds of AAA and B qualities. The AAA bonds yield on the average 6% and the B bonds yield 10%. The investor requires that at least three times as much money should be invested in AAA bonds as in B bonds. How much should be invested in each type of bond to maximize the return? What is the maximum return?

39. Pollution control. Because of new federal regulations on pollution, a chemical plant introduced a new, more expensive process to supplement or replace an older process used in the production of a particular chemical. The older process emitted 20 grams of sulfur dioxide and 40 grams of particulate matter into the atmosphere for each gallon of chemical produced. The new process emits 5 grams of sulfur dioxide and 20 grams of particulate matter for each gallon produced. The company makes a profit of 60¢ per gallon and 20¢ per gallon on the old and new processes, respectively.

(A) If the government allows the plant to emit no more than 16,000 grams of sulfur dioxide and 30,000 grams of particulate matter daily, how many gallons of the chemical should be produced by each process to maximize daily profit? What is the maximum daily profit?

(B) Discuss the effect on the production schedule and the maximum profit if the government decides to restrict emissions of sulfur dioxide to 11,500 grams daily and all other data remain unchanged.

(C) Discuss the effect on the production schedule and the maximum profit if the government decides to restrict emissions of sulfur dioxide to 7,200 grams daily and all other data remain unchanged.

40. Capital expansion. A fast-food chain plans to expand by opening several new restaurants. The chain operates two types of restaurants, drive-through and full-service. A drive-through restaurant costs \$100,000 to construct, requires 5 employees, and has an expected annual revenue of \$200,000. A full-service restaurant costs \$150,000 to construct, requires 15 employees, and has an expected annual revenue of \$500,000. The chain has \$2,400,000 in capital available for expansion. Labor contracts require that they hire no more than 210 employees, and licensing restrictions require that they open no more than 20 new restaurants. How many restaurants of each type should the chain open in order to maximize the expected revenue? What is the maximum expected revenue? How much of their capital will they use and how many employees will they hire?

Life Sciences

41. Nutrition: plants. A fruit grower can use two types of fertilizer in his orange grove, brand A and brand B. The amounts (in pounds) of nitrogen, phosphoric acid, and chloride in a

	Pounds per Bag	
	Brand *A*	**Brand *B***
Nitrogen	8	3
Phosphoric acid	4	4
Chloride	2	1

bag of each brand are given in the table. Tests indicate that the grove needs at least 1,000 pounds of phosphoric acid and at most 400 pounds of chloride.

(A) If the grower wants to maximize the amount of nitrogen added to the grove, how many bags of each mix should be used? How much nitrogen will be added?

(B) If the grower wants to minimize the amount of nitrogen added to the grove, how many bags of each mix should be used? How much nitrogen will be added?

42. Nutrition: people. A dietitian in a hospital is to arrange a special diet composed of two foods, M and N. Each ounce of food M contains 30 units of calcium, 10 units of iron, 10 units of vitamin A, and 8 units of cholesterol. Each ounce of food N contains 10 units of calcium, 10 units of iron, 30 units of vitamin A, and 4 units of cholesterol. If the minimum daily requirements are 360 units of calcium, 160 units of iron, and 240 units of vitamin A, how many ounces of each food should be used to meet the minimum requirements and at the same time minimize the cholesterol intake? What is the minimum cholesterol intake?

43. Nutrition: plants. A farmer can buy two types of plant food, mix A and mix B. Each cubic yard of mix A contains 20 pounds of phosphoric acid, 30 pounds of nitrogen, and 5 pounds of potash. Each cubic yard of mix B contains 10 pounds of phosphoric acid, 30 pounds of nitrogen, and 10 pounds of potash. The minimum monthly requirements are 460 pounds of phosphoric acid, 960 pounds of nitrogen, and 220 pounds of potash. If mix A costs \$30 per cubic yard and mix B costs \$35 per cubic yard, how many cubic yards of each mix should the farmer blend to meet the minimum monthly requirements at a minimal cost? What is this cost?

44. Nutrition: animals. A laboratory technician in a medical research center is asked to formulate a diet from two commercially packaged foods, food A and food B, for a group of animals. Each ounce of food A contains 8 units of fat, 16 units of carbohydrate, and 2 units of protein. Each ounce of food B contains 4 units of fat, 32 units of carbohydrate, and 8 units of protein. The minimum daily requirements are 176 units of fat, 1,024 units of carbohydrate, and 384 units of protein. If food A costs 5¢ per ounce and food B costs 5¢ per ounce, how many ounces of each food should be used to meet the minimum daily requirements at the least cost? What is the cost for this amount of food?

Social Sciences

45. Psychology. In an experiment on conditioning, a psychologist uses two types of Skinner boxes with mice and rats. The amount of time (in minutes) each mouse and each rat spends in each box per day is given in the table. What is the maximum number of mice and rats that can be used in this experiment? How many mice and how many rats produce this maximum?

	Time		Maximum Time
	Mice	**Rats**	**Available per Day**
Skinner box *A*	10 min	20 min	800 min
Skinner box *B*	20 min	10 min	640 min

46. Sociology. A city council voted to conduct a study on inner-city community problems. A nearby university was contacted to provide sociologists and research assistants. Allocation of time and costs per week are given in the table. How many sociologists and how many research assistants should be hired to minimize the cost and meet the weekly labor-hour requirements? What is the minimum weekly cost?

	Labor-Hours		Minimum Labor-Hours
	Sociologist	**Research Assistant**	**Needed per Week**
Fieldwork	10	30	180
Research center	30	10	140
Costs per week	\$500	\$300	

11-3 Geometric Introduction to the Simplex Method

- Standard Maximization Problems in Standard Form
- Slack Variables
- Basic and Nonbasic Variables: Basic Solutions and Basic Feasible Solutions
- Basic Feasible Solutions and the Simplex Method

The geometric method of solving linear programming problems provides us with an overview of the subject and some useful terminology. But, practically speaking, the method is useful only for problems involving two decision variables and relatively few problem constraints. What happens when we need more decision variables and more problem constraints? We use an algebraic method called the *simplex method,* which was developed by George B. Dantzig in 1947 while on assignment to the U.S. Department of the Air Force. Ideally suited to computer use, the method is used routinely on applied problems involving hundreds and even thousands of variables and problem constraints.

The algebraic procedures utilized in the simplex method require the problem constraints to be written as equations rather than inequalities. This new form of the linear programming problem also prompts the use of some new terminology. We introduce this new form and associated terminology through a simple example and an appropriate geometric interpretation. From this example we can illustrate what the simplex method does geometrically before we immerse ourselves in the algebraic details of the process.

Standard Maximization Problems in Standard Form

We now return to the tent production problem in Example 1 from the preceding section. Recall the mathematical model for the problem:

$$
\begin{aligned}
\text{Maximize} \quad & P = 50x_1 + 80x_2 && \text{Objective function} \\
\text{subject to} \quad & x_1 + 2x_2 \le 32 && \text{Cutting department constraint} \\
& 3x_1 + 4x_2 \le 84 && \text{Assembly department constraint} \\
& x_1, x_2 \ge 0 && \text{Nonnegative constraints}
\end{aligned}
\qquad (1)
$$

where the decision variables x_1 and x_2 are the number of standard and expedition tents, respectively, produced each day.

Notice that the problem constraints involve $\le$ inequalities with positive constants to the right of the inequality. Maximization problems that satisfy this condition are called *standard maximization problems.* In this and the next section we restrict our attention to standard maximization problems.

DEFINITION Standard Maximization Problem in Standard Form

A linear programming problem is said to be a **standard maximization problem in standard form** if its mathematical model is of the following form:

Maximize the objective function

$$P = c_1x_1 + c_2x_2 + \cdots + c_nx_n$$

subject to problem constraints of the form

$$a_1x_1 + a_2x_2 + \cdots + a_nx_n \le b \quad b \ge 0$$

with nonnegative constraints

$$x_1, x_2, \ldots, x_n \ge 0$$

Note: Mathematical model (1) is a standard maximization problem in standard form. Also note that the coefficients of the objective function can be any real numbers.

EXPLORE & DISCUSS 1

Find an example of a standard maximization problem in standard form involving two variables and one problem constraint such that

(A) The feasible region is bounded.

(B) The feasible region is unbounded.

Is it possible for a standard maximization problem to have no solution? Explain.

Slack Variables

To adapt a linear programming problem to the matrix methods used in the simplex process (as discussed in the next section), we convert the problem constraint inequalities into a system of linear equations by using a simple device called a *slack variable*. In particular, to convert the system of problem constraint inequalities from model (1),

$$\begin{aligned} x_1 + 2x_2 &\le 32 \quad \text{Cutting department constraint} \\ 3x_1 + 4x_2 &\le 84 \quad \text{Assembly department constraint} \end{aligned} \tag{2}$$

into a system of equations, we add variables s_1 and s_2 to the left sides of inequalities (2) to obtain

$$\begin{aligned} x_1 + 2x_2 + s_1 &= 32 \\ 3x_1 + 4x_2 + s_2 &= 84 \end{aligned} \tag{3}$$

The variables s_1 and s_2 are called **slack variables** because each makes up the difference (takes up the slack) between the left and right sides of an inequality in system (2). For example, if we produced 20 standard tents ($x_1 = 20$) and 5 expedition tents ($x_2 = 5$), then the number of labor-hours used in the cutting department would be $20 + 2(5) = 30$, leaving a slack of 2 unused labor-hours out of the 32 available. Thus, s_1 would have the value of 2.

Notice that if the decision variables x_1 and x_2 satisfy the system of constraint inequalities (2), then the slack variables s_1 and s_2 are nonnegative. We will have more to say about this later in this discussion.

Basic and Nonbasic Variables: Basic Solutions and Basic Feasible Solutions

Observe that system (3) has infinitely many solutions—just solve for s_1 and s_2 in terms of x_1 and x_2, and then assign x_1 and x_2 arbitrary values. Certain solutions of system (3), called *basic solutions,* are related to the intersection points of the (extended) boundary lines of the feasible region in Figure 1.

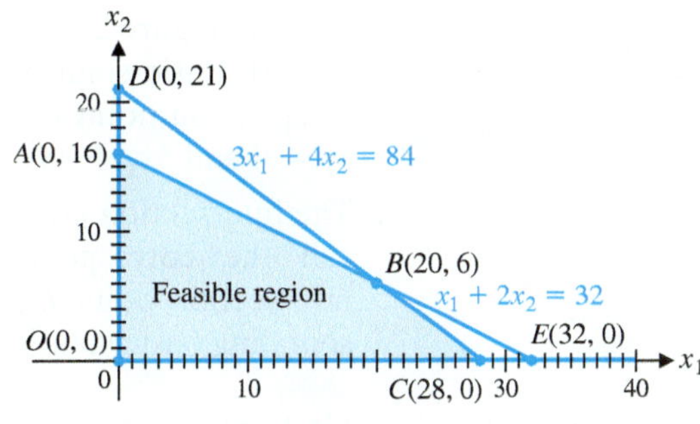

Figure 1

How are basic solutions to system (3) determined? System (3) involves four variables and two equations. We divide the four variables into two groups, called *basic variables* and *nonbasic variables*, as follows: Basic variables are selected arbitrarily with the restriction that there be as many basic variables as there are equations. The remaining variables are nonbasic variables.

Since system (3) has two equations, we can select any two of the four variables as basic variables. The remaining two variables are then nonbasic variables. A solution found by setting the two nonbasic variables equal to 0 and solving for the two basic variables is a *basic solution*. [Note that setting two variables equal to 0 in system (3) results in a system of two equations with two variables, which has exactly one solution, infinitely many solutions, or no solution.]

EXAMPLE 1 **Basic Solutions and the Feasible Region**

(A) Find two basic solutions for system (3) by first selecting s_1 and s_2 as basic variables, and then by selecting x_2 and s_1 as basic variables.

(B) Associate each basic solution found in part (A) with an intersection point of the (extended) boundary lines of the feasible region in Figure 1, and indicate which boundary lines produce each intersection point.

(C) Indicate which of the intersection points found in part (B) are in the feasible region.

SOLUTION (A) With s_1 and s_2 selected as basic variables, x_1 and x_2 are nonbasic variables. A basic solution is found by setting the nonbasic variables equal to 0 and solving for the basic variables. If $x_1 = 0$ and $x_2 = 0$, system (3) becomes

$$\begin{aligned} s_1 &= 32 \\ s_2 &= 84 \end{aligned} \qquad \begin{aligned} \overset{0}{x_1} + 2\overset{0}{x_2} + s_1 &= 32 \\ 3x_1 + 4x_2 + s_2 &= 84 \end{aligned}$$

and the basic solution is

$$x_1 = 0, \quad x_2 = 0, \quad s_1 = 32, \quad s_2 = 84 \tag{4}$$

If we select x_2 and s_1 as basic variables, x_1 and s_2 are nonbasic variables. Setting the nonbasic variables equal to 0, system (3) becomes

$$\begin{aligned} 2x_2 + s_1 &= 32 \\ 4x_2 &= 84 \end{aligned} \qquad \begin{aligned} \overset{0}{x_1} + 2x_2 + s_1 \phantom{+ \overset{0}{s_2}} &= 32 \\ 3x_1 + 4x_2 + \overset{0}{s_2} &= 84 \end{aligned}$$

Solving, we see that $x_2 = 21$ and $s_1 = -10$, and the basic solution is

$$x_1 = 0, \quad x_2 = 21, \quad s_1 = -10, \quad s_2 = 0 \tag{5}$$

(B) Basic solution (4)—since $x_1 = 0$ and $x_2 = 0$—corresponds to the origin O $(0, 0)$ in Figure 1, which is the intersection of the boundary lines $x_1 = 0$ and $x_2 = 0$. Basic solution (5)—since $x_1 = 0$ and $x_2 = 21$—corresponds to the intersection point $D(0, 21)$, which is the intersection of the boundary lines $x_1 = 0$ and $3x_1 + 4x_2 = 84$.

(C) The intersection point $O(0, 0)$ is in the feasible region; hence, it is natural to call the corresponding basic solution a *basic feasible solution*. The intersection point $D(0, 21)$ is not in the feasible region; hence, the corresponding basic solution is not feasible.

Matched Problem 1 (A) Find two basic solutions for system (3) by first selecting x_1 and s_1 as basic variables, and then by selecting x_1 and s_2 as basic variables.

(B) Associate each basic solution found in part (A) with an intersection point of the (extended) boundary lines of the feasible region in Figure 1, and indicate which boundary lines produce each intersection point.

(C) Indicate which of the intersection points found in part (B) are in the feasible region.

CONCEPTUAL INSIGHT

If we apply Gauss–Jordan elimination to system (3), we can express the solution set S as (verify this)

$$S = \{(x_1, x_2, s_1, s_2 \mid x_1 = 20 + 2s_1 - s_2, x_2 = 6 - 1.5s_1 + 0.5s_2\}$$

where s_1 and s_2 are any real numbers. This set contains an infinite number of points. We are interested in developing a systematic procedure that will find the finite subset of S consisting of the basic solutions of system (3). These basic solutions turn out to be the intersection points of any two of the four boundary lines of the feasible region.

Proceeding systematically as in Example 1 and Matched Problem 1, we can obtain all basic solutions to system (3). The results are summarized in Table 1, which also includes geometric interpretations of the basic solutions relative to Figure 1. Figure 2 summarizes these interpretations. A careful study of Table 1 and Figure 2 is very worthwhile. (Note that to be sure we have listed all possible basic solutions in Table 1, it is convenient to organize the table in terms of the zero values of the nonbasic variables.)

Table 1 Basic Solutions

Basic Solutions						
x_1	x_2	s_1	s_2	Intersection Point	Intersecting Boundary Lines	Feasible
0	0	32	84	$O(0,0)$	$x_1 = 0$ $x_2 = 0$	Yes
0	16	0	20	$A(0,16)$	$x_1 = 0$ $x_1 + 2x_2 = 32$	Yes
0	21	−10	0	$D(0,21)$	$x_1 = 0$ $3x_1 + 4x_2 = 84$	No
32	0	0	−12	$E(32,0)$	$x_2 = 0$ $x_1 + 2x_2 = 32$	No
28	0	4	0	$C(28,0)$	$x_2 = 0$ $3x_1 + 4x_2 = 84$	Yes
20	6	0	0	$B(20,6)$	$x_1 + 2x_2 = 32$ $3x_1 + 4x_2 = 84$	Yes

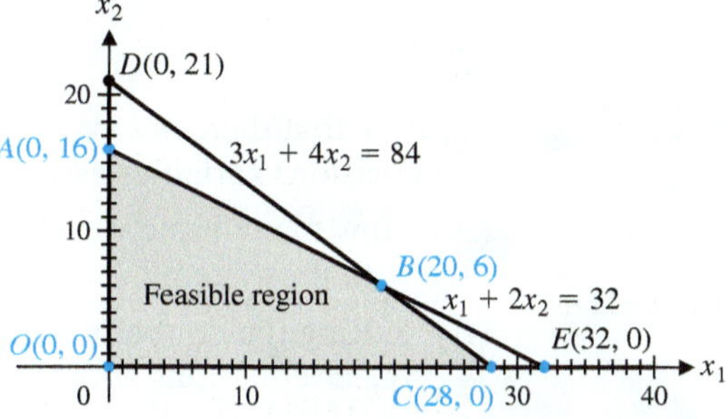

Figure 2 Basic feasible solutions: O, A, B, C
Basic solutions that are not feasible: D, E

CONCEPTUAL INSIGHT

In Table 1, observe that a basic solution that is not feasible includes at least one negative value and that a basic feasible solution does not include any negative values. That is, we can determine the feasibility of a basic solution simply by examining the signs of all the variables in the solution.

In Table 1 and Figure 2, observe that basic feasible solutions are associated with the corner points of the feasible region, which include the optimal solution to the original linear programming problem.

Before proceeding further, let us formalize the definitions alluded to in the discussion above so that they apply to standard maximization problems in general, without any reference to geometric forms.

DEFINITION Basic Variables and Nonbasic Variables; Basic Solutions and Basic Feasible Solutions

Given a system of linear equations associated with a linear programming problem (such a system will always have more variables than equations),

The variables are divided into two (mutually exclusive) groups, as follows: **Basic variables** are selected arbitrarily with the one restriction that there be as many basic variables as there are equations. The remaining variables are called **nonbasic variables.**

A solution found by setting the nonbasic variables equal to 0 and solving for the basic variables is called a **basic solution.** If a basic solution has no negative values, it is a **basic feasible solution.**

EXAMPLE 2 **Basic Variables and Basic Solutions** Suppose that there is a system of three problem constraint equations with eight (slack and decision) variables associated with a standard maximization problem.

(A) How many basic variables and how many nonbasic variables are associated with the system?

(B) Setting the nonbasic variables equal to 0 will result in a system of how many linear equations with how many variables?

(C) If a basic solution has all nonnegative elements, is it feasible or not feasible?

SOLUTION (A) Since there are three equations in the system, there should be three basic variables and five nonbasic variables.

(B) Three equations with three variables

(C) Feasible

Matched Problem 2 Suppose that there is a system of five problem constraint equations with 11 (slack and decision) variables associated with a standard maximization problem.

(A) How many basic variables and how many nonbasic variables are associated with the system?

(B) Setting the nonbasic variables equal to 0 will result in a system of how many linear equations with how many variables?

(C) If a basic solution has one or more negative elements, is it feasible or not feasible?

Basic Feasible Solutions and the Simplex Method

The following important theorem, which is equivalent to the fundamental theorem (Theorem 1 in the preceding section), is stated without proof:

> **THEOREM 1 Fundamental Theorem of Linear Programming: Version 2**
> If the optimal value of the objective function in a linear programming problem exists, then that value must occur at one (or more) of the basic feasible solutions.

Now you can understand why the concepts of basic and nonbasic variables and basic solutions and basic feasible solutions are so important—these concepts are central to the process of finding optimal solutions to linear programming problems.

EXPLORE & DISCUSS 2

If we know that a standard maximization problem has an optimal solution, how could we use a table of basic solutions like Table 1 to find the optimal solution?

We have taken the first step toward finding a general method of solving linear programming problems involving any number of variables and problem constraints. That is, **we have found a method of identifying all the corner points (basic feasible solutions) of a feasible region without drawing its graph.** This is a critical step if we want to consider problems with more than two decision variables. Unfortunately, the number of corner points increases dramatically as the number of variables and constraints increases. In real-world problems, it is not practical to find all the corner points in order to find the optimal solution. Thus, the next step is to find a method of locating the optimal solution without finding every corner point. The procedure for doing this is the simplex method mentioned at the beginning of this section.

The simplex method, using a special matrix and row operations, automatically moves from one basic feasible solution to another—that is, from one corner point of the feasible region to another—each time getting closer to an optimal solution (if one exists), until an optimal solution is reached. Then the process stops. A remarkable property of the simplex method is that in large linear programming problems it usually arrives at an optimal solution (if one exists) by testing relatively few of the large number of basic feasible solutions (corner points) available.

With this background, we are now ready to discuss the algebraic details of the simplex method in the next section.

Answers to Matched Problems

1. (A) Basic solution corresponding to basic variables x_1 and s_1: $x_1 = 28, x_2 = 0, s_1 = 4, s_2 = 0$. Basic solution corresponding to basic variables x_1 and s_2: $x_1 = 32, x_2 = 0, s_1 = 0, s_2 = -12$.
(B) The first basic solution corresponds to $C(28, 0)$, which is the intersection of the boundary lines $x_2 = 0$ and $3x_1 + 4x_2 = 84$. The second basic solution corresponds to $E(32, 0)$, which is the intersection of the boundary lines $x_2 = 0$ and $x_1 + 2x_2 = 32$.
(C) $C(28, 0)$ is in the feasible region (hence, the corresponding basic solution is a basic feasible solution); $E(32, 0)$ is not in the feasible region (hence, the corresponding basic solution is not feasible).

2. (A) Five basic variables and six nonbasic variables
(B) Five equations and five variables
(C) Not feasible

Exercise 11-3

A

1. Discuss the relationship between a standard maximization problem with two problem constraints and three decision variables, and the associated system of problem constraint equations. In particular, find each of the following quantities and explain how each was determined:
   (A) The number of slack variables that must be introduced to form the system of problem constraint equations
   (B) The number of basic variables and the number of nonbasic variables associated with the system
   (C) The number of linear equations and the number of variables in the system formed by setting the nonbasic variables equal to 0

2. Repeat Problem 1 for a standard maximization problem with three problem constraints and four decision variables.

3. Discuss the relationship between a standard maximization problem and the associated system of problem constraint equations, if the system of problem constraint equations has nine variables, including five slack variables. In particular, find each o f the following quantities and explain how each was determined:
   (A) The number of constraint equations in the system
   (B) The number of decision variables in the system
   (C) The number of basic variables and the number of nonbasic variables associated with the system
   (D) The number of linear equations and the number of variables in the system formed by setting the nonbasic variables equal to 0

4. Repeat Problem 3 if the system of problem constraint equations has 10 variables, including four slack variables.

5. Listed in the table below are all the basic solutions for the system

$$\begin{aligned} 2x_1 + 3x_2 + s_1 \quad &= 24 \\ 4x_1 + 3x_2 \quad + s_2 &= 36 \end{aligned}$$

For each basic solution, identify the nonbasic variables and the basic variables. Classify each basic solution as feasible or not feasible. Explain how you could use the basic feasible solutions to find the maximum value of $z = 2x_1 + 3x_2$.

	x_1	x_2	s_1	s_2
(A)	0	0	24	36
(B)	0	8	0	12
(C)	0	12	−12	0
(D)	12	0	0	−12
(E)	9	0	6	0
(F)	6	4	0	0

6. Repeat Problem 5 for the system

$$\begin{aligned} 2x_1 + x_2 + s_1 \quad &= 30 \\ x_1 + 5x_2 \quad + s_2 &= 60 \end{aligned}$$

whose basic solutions are given in the following table:

	x_1	x_2	s_1	s_2
(A)	0	0	30	60
(B)	0	30	0	−90
(C)	0	12	18	0
(D)	15	0	0	45
(E)	60	0	−90	0
(F)	10	10	0	0

7. Listed in the table below are all the possible choices of nonbasic variables for the system

$$\begin{aligned} 2x_1 + x_2 + s_1 \quad &= 50 \\ x_1 + 2x_2 \quad + s_2 &= 40 \end{aligned}$$

In each case, find the values of the basic variables and determine whether the basic solution is feasible.

	x_1	x_2	s_1	s_2
(A)	0	0	?	?
(B)	0	?	0	?
(C)	0	?	?	0
(D)	?	0	0	?
(E)	?	0	?	0
(F)	?	?	0	0

8. Repeat Problem 7 for the system

$$\begin{aligned} x_1 + 2x_2 + s_1 \quad &= 12 \\ 3x_1 + 2x_2 \quad + s_2 &= 24 \end{aligned}$$

B

Graph the systems of inequalities in Problems 9–12. Introduce slack variables to convert each system of inequalities to a system of equations, and find all the basic solutions of the system. Construct a table (similar to Table 1) listing each basic solution, the corresponding point on the graph, and whether the basic solution is feasible. (You do not need to list the intersecting lines.)

9. $x_1 + x_2 \le 16$
   $2x_1 + x_2 \le 20$
   $x_1, x_2 \ge 0$

10. $5x_1 + x_2 \le 35$
    $4x_1 + x_2 \le 32$
    $x_1, x_2 \ge 0$

11. $2x_1 + x_2 \le 22$
    $x_1 + x_2 \le 12$
    $x_1 + 2x_2 \le 20$
    $x_1, x_2 \ge 0$

12. $4x_1 + x_2 \le 28$
    $2x_1 + x_2 \le 16$
    $x_1 + x_2 \le 13$
    $x_1, x_2 \ge 0$

11-4 Simplex Method: Maximization with Problem Constraints of the Form ≤

- Initial System
- Simplex Tableau
- Pivot Operation
- Interpreting the Simplex Process Geometrically
- Simplex Method Summarized
- Application

We are now ready to develop the simplex method for a standard maximization problem. Specific details in the presentation of the method generally vary from one book to another, even though the underlying process is the same. The presentation developed here emphasizes concept development and understanding.

As pointed out in the preceding section, the simplex method is most useful when used with computers. Consequently, it is not intended that you become expert in manually solving linear programming problems using the simplex method. But it is important that you become proficient in constructing the models for linear programming problems so that they can be solved using a computer, and it is also important that you develop skill in interpreting the results. One way to gain this proficiency and interpretive skill is to set up and manually solve a number of fairly simple linear programming problems using the simplex method. This is the main goal here and in Sections 11-5 and 11-6. To assist you in learning to develop the models, the answer sections for Exercises 11-4, 11-5, and 11-6 contain both the model and its solution.

Initial System

We will introduce the concepts and procedures involved in the simplex method through an example—the tent production example we have discussed earlier. We restate the problem here in standard form for convenient reference:

$$\begin{aligned} \text{Maximize} \quad & P = 50x_1 + 80x_2 && \text{Objective function} \\ \text{subject to} \quad & x_1 + 2x_2 \le 32 && \\ & 3x_1 + 4x_2 \le 84 && \text{Problem constraints} \\ & x_1, x_2 \ge 0 && \text{Nonnegative constraints} \end{aligned} \tag{1}$$

Introducing slack variables s_1 and s_2, we convert the problem constraint inequalities in problem (1) into the following system of problem constraint equations:

$$\begin{aligned} x_1 + 2x_2 + s_1 &= 32 \\ 3x_1 + 4x_2 + s_2 &= 84 \\ x_1, x_2, s_1, s_2 &\ge 0 \end{aligned} \tag{2}$$

Since a basic solution of system (2) is not feasible if it contains any negative values, we have also included the nonnegative constraints for both the decision variables x_1 and x_2 and the slack variables s_1 and s_2. From our discussion in the last section, we know that out of the infinitely many solutions to system (2), an optimal solution is among the basic feasible solutions, which correspond to the corner points of the feasible region.

As part of the simplex method we add the objective function equation $P = 50x_1 + 80x_2$ in the form $-50x_1 - 80x_2 + P = 0$ to system (2) to create what is called the **initial system:**

$$\begin{aligned} x_1 + 2x_2 + s_1 &= 32 \\ 3x_1 + 4x_2 + s_2 &= 84 \\ -50x_1 - 80x_2 + P &= 0 \\ x_1, x_2, s_1, s_2 &\ge 0 \end{aligned} \tag{3}$$

When we add the objective function equation to system (2), we must slightly modify the earlier definitions of basic solution and basic feasible solution so that they apply to the initial system (3).

DEFINITION Basic Solutions and Basic Feasible Solutions for Initial Systems

1. The objective function variable P is always selected as a basic variable and is never selected as a nonbasic variable.*
2. Note that a basic solution of system (3) is also a basic solution of system (2) after P is deleted.
3. If a basic solution of system (3) is a basic feasible solution of system (2) after deleting P, then the basic solution of system (3) is called a **basic feasible solution** of system (3).
4. A basic feasible solution of system (3) can contain a negative number, but only if it is the value of P, the objective function variable.

These changes lead to a small change in the second version of the fundamental theorem (see Theorem 1, Section 11-3).

THEOREM 1 Fundamental Theorem of Linear Programming: Version 3
If the optimal value of the objective function in a linear programming problem exists, then that value must occur at one (or more) of the basic feasible solutions of the initial system.

With these adjustments understood, we start the simplex process with a basic feasible solution of the initial system (3), which we will refer to as an **initial basic feasible solution.** An initial basic feasible solution that is easy to find is the one associated with the origin.

Since system (3) has three equations and five variables, it has three basic variables and two nonbasic variables. Looking at the system, we see that x_1 and x_2 appear in all equations and s_1, s_2, and P each appears only once and each in a different equation. A basic solution can be found by inspection by selecting s_1, s_2, and P as the basic variables (remember, P is always selected as a basic variable) and x_1 and x_2 as the nonbasic variables to be set equal to 0. Setting x_1 and x_2 equal to 0 and solving for the basic variables, we obtain the basic solution:

$$x_1 = 0, \quad x_2 = 0, \quad s_1 = 32, \quad s_2 = 84, \quad P = 0$$

This basic solution is feasible since none of the variables (excluding P) are negative. Thus, this is the initial basic feasible solution we seek.

Now you can see why we wanted to add the objective function equation to system (2): A basic feasible solution of system (3) not only includes a basic feasible solution of system (2), but, in addition, it includes the value of P for that basic feasible solution of system (2).

The initial basic feasible solution we just found is associated with the origin. Of course, if we do not produce any tents, we do not expect a profit, so $P = \$0$. Starting with this easily obtained initial basic feasible solution, the simplex process moves through each iteration (each repetition) to another basic feasible solution, each time improving the profit, and the process continues until the maximum profit is reached. Then the process stops.

Simplex Tableau

To facilitate the search for the optimal solution, we now turn to matrix methods. Our first step is to write the augmented matrix for the initial system (3). This matrix is called the **initial simplex tableau,**† and it is simply a tabulation of the coefficients in system (3).

---

* The classification of P as a basic variable is not universally agreed upon. Some authors call it a nonbasic variable; others say it is neither basic nor nonbasic. The vast majority of beginning textbooks treat it as a basic variable, as we do here.

† The format of a simplex tableau can vary. Some authors place the objective function coefficients in the first row, rather than the last. Others do not include a column for P.

$$\begin{array}{c} \\ s_1 \\ s_2 \\ P \end{array}\begin{array}{c}\begin{array}{cccccc} x_1 & x_2 & s_1 & s_2 & P & \end{array}\\ \left[\begin{array}{ccccc|c} 1 & 2 & 1 & 0 & 0 & 32 \\ 3 & 4 & 0 & 1 & 0 & 84 \\ \hline -50 & -80 & 0 & 0 & 1 & 0 \end{array}\right]\end{array} \quad \text{Initial simplex tableau} \qquad (4)$$

In tableau (4), the row below the dashed line always corresponds to the objective function. Each of the basic variables we selected above, s_1, s_2, and P, is also placed on the left of the tableau so that the intersection element in its row and column is not 0. For example, we place the basic variable s_1 on the left so that the intersection element of the s_1 row and the s_1 column is 1 and not 0. The basic variable s_2 is similarly placed. The objective function variable P is always placed at the bottom. The reason for writing the basic variables on the left in this way is that this placement makes it possible to read certain basic feasible solutions directly from the tableau. If $x_1 = 0$ and $x_2 = 0$, the basic variables on the left of tableau (4) are lined up with their corresponding values, 32, 84, and 0, to the right of the vertical line.

Looking at tableau (4) relative to the choice of s_1, s_2, and P as basic variables, we see that each basic variable is above a column that has all 0 elements except for a single 1 and that no two such columns contain 1's in the same row. These observations lead to a formalization of the process of selecting basic and nonbasic variables that is an important part of the simplex method:

PROCEDURE Selecting Basic and Nonbasic Variables for the Simplex Method

Given a simplex tableau,

Step 1 Determine the number of basic variables and the number of nonbasic variables. These numbers do not change during the simplex process.

Step 2 *Selecting basic variables.* A variable can be selected as a basic variable only if it corresponds to a column in the tableau that has exactly one nonzero element (usually 1) and the nonzero element in the column is not in the same row as the nonzero element in the column of another basic variable. (This procedure always selects P as a basic variable, since the P column never changes during the simplex process.)

Step 3 *Selecting nonbasic variables.* After the basic variables are selected in step 2, the remaining variables are selected as the nonbasic variables. (The tableau columns under the nonbasic variables usually contain more than one nonzero element.)

The earlier selection of s_1, s_2, and P as basic variables and x_1 and x_2 as nonbasic variables conforms to this prescribed convention of selecting basic and nonbasic variables for the simplex process.

Pivot Operation

The simplex method will now switch one of the nonbasic variables, x_1 or x_2, for one of the basic variables, s_1 or s_2 (but not P), as a step toward improving the profit. For a nonbasic variable to be classified as a basic variable we need to perform appropriate row operations on the tableau so that the newly selected basic variable will end up with exactly one nonzero element in its column. In this process, the old basic variable will usually gain additional nonzero elements in its column as it becomes nonbasic.

Which nonbasic variable should we select to become basic? It makes sense to select the nonbasic variable that will increase the profit the most per unit change in that variable. Looking at the objective function

$$P = 50x_1 + 80x_2$$

we see that if x_1 stays a nonbasic variable (set equal to 0) and if x_2 becomes a new basic variable, then

$$P = 50(0) + 80x_2 = 80x_2$$

and for each unit increase in x_2, P will increase \$80. If x_2 stays a nonbasic variable and x_1 becomes a new basic variable, then (reasoning in the same way) for each unit increase in x_1, P will increase only \$50. So, we select the nonbasic variable x_2 to enter the set of basic variables, and call it the **entering variable.** (The basic variable leaving the set of basic variables to become a nonbasic variable is called the **exiting variable.** Exiting variables will be discussed shortly.)

We call the column corresponding to the entering variable the **pivot column.** Looking at the bottom row in tableau (4)—the objective function row below the dashed line—we see that the pivot column is associated with the column to the left of the P column that has the most negative bottom element. In general, the most negative element in the bottom row to the left of the P column *indicates* the variable above it that will produce the greatest increase in P for a unit increase in that variable. For this reason, we call the elements in the bottom row of the tableau, to the left of the P column, **indicators.**

We illustrate the indicators, the pivot column, the entering variable, and the initial basic feasible solution below:

Entering variable ↓ (above x_2)

$$\begin{array}{c} \\ s_1 \\ s_2 \\ P \end{array} \begin{array}{c} \begin{array}{cccccc} x_1 & x_2 & s_1 & s_2 & P & \end{array} \\ \left[\begin{array}{ccccc|c} 1 & 2 & 1 & 0 & 0 & 32 \\ 3 & 4 & 0 & 1 & 0 & 84 \\ \hline -50 & -80 & 0 & 0 & 1 & 0 \end{array}\right] \end{array} \qquad (5)$$

Initial simplex tableau

Indicators are shown in color.

↑ Pivot column (below x_2)

$$x_1 = 0, \quad x_2 = 0, \quad s_1 = 32, \quad s_2 = 84, \quad P = 0$$ Initial basic feasible solution

Now that we have chosen the nonbasic variable x_2 as the entering variable (the nonbasic variable to become basic), which of the two basic variables, s_1 or s_2, should we choose as the exiting variable (the basic variable to become nonbasic)? We saw above that for $x_1 = 0$, each unit increase in the entering variable x_2 results in an increase of \$80 for P. Can we increase x_2 without limit? No! A limit is imposed by the nonnegative requirements for s_1 and s_2. (Remember that if any of the basic variables except P become negative, we no longer have a feasible solution.) So we rephrase the question and ask: How much can x_2 be increased when $x_1 = 0$ without causing s_1 or s_2 to become negative? To see how much x_2 can be increased, we refer to tableau (5) or system (3) and write the two problem constraint equations with $x_1 = 0$:

$$\begin{aligned} 2x_2 + s_1 &= 32 \\ 4x_2 + s_2 &= 84 \end{aligned}$$

Solving for s_1 and s_2, we have

$$\begin{aligned} s_1 &= 32 - 2x_2 \\ s_2 &= 84 - 4x_2 \end{aligned}$$

For s_1 and s_2 to be nonnegative, x_2 must be chosen so that both $32 - 2x_2$ and $84 - 4x_2$ are nonnegative. That is, so that

$$\begin{aligned} 32 - 2x_2 &\ge 0 \\ -2x_2 &\ge -32 \\ x_2 &\le \tfrac{32}{2} = 16 \end{aligned} \qquad \text{and} \qquad \begin{aligned} 84 - 4x_2 &\ge 0 \\ -4x_2 &\ge -84 \\ x_2 &\le \tfrac{84}{4} = 21 \end{aligned}$$

For both inequalities to be satisfied, x_2 must be less than or equal to the smaller of the values, which is 16. Thus, x_2 can increase to 16 without either s_1 or s_2 becoming negative. Now, observe how each value (16 and 21) can be obtained directly from the following tableau:

Entering variable ↓ (above x_2)

$$\begin{array}{c} \\ s_1 \\ s_2 \\ P \end{array} \begin{array}{c} \begin{array}{cccccc} x_1 & x_2 & s_1 & s_2 & P & \end{array} \\ \left[\begin{array}{rrrrr|r} 1 & 2 & 1 & 0 & 0 & 32 \\ 3 & 4 & 0 & 1 & 0 & 84 \\ \hline -50 & -80 & 0 & 0 & 1 & 0 \end{array}\right] \end{array} \quad \begin{array}{l} \tfrac{32}{2} = 16 \text{ (smallest)} \\ \tfrac{84}{4} = 21 \\ \\ \end{array} \qquad (6)$$

↑ Pivot column (below x_2) column

From tableau (6) we can determine the amount the entering variable can increase by choosing the smallest of the quotients obtained by dividing each element in the last column above the dashed line by the corresponding *positive* element in the pivot column. The row with the smallest quotient is called the **pivot row,** and the variable to the left of the pivot row is the exiting variable. In this case, s_1 will be the exiting variable, and the roles of x_2 and s_1 will be interchanged. The element at the intersection of the pivot column and the pivot row is called the **pivot element,** and we circle this element for ease of recognition. Since a negative or 0 element in the pivot column places no restriction on the amount an entering variable can increase, it is not necessary to compute the quotient for negative or 0 values in the pivot column.

A negative or 0 element is never selected for the pivot element.

The following tableau illustrates this process, which is summarized in the next box.

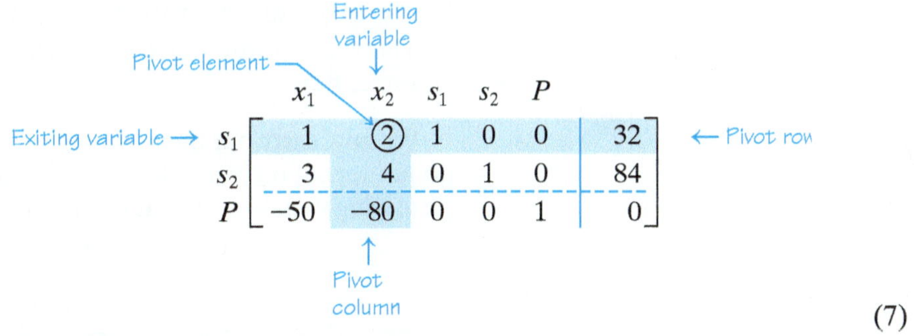

(7)

PROCEDURE Selecting the Pivot Element

Step 1 Locate the most negative indicator in the bottom row of the tableau to the left of the P column (the negative number with the largest absolute value). The column containing this element is the *pivot column.* If there is a tie for the most negative, choose either.

Step 2 Divide each *positive* element in the pivot column above the dashed line into the corresponding element in the last column. The *pivot row* is the row corresponding to the smallest quotient obtained. If there is a tie for the smallest quotient, choose either. If the pivot column above the dashed line has no positive elements, there is no solution, and we stop.

Step 3 The *pivot* (or *pivot element*) is the element in the intersection of the pivot column and pivot row.

Note: The pivot element is always positive and is never in the bottom row.

Remember: The entering variable is at the top of the pivot column, and the exiting variable is at the left of the pivot row.

In order for x_2 to be classified as a basic variable, we perform row operations on tableau (7) so that the pivot element is transformed into 1 and all other elements in the column into 0's. This procedure for transforming a nonbasic variable into a basic variable is called a *pivot operation,* or *pivoting,* and is summarized in the box.

PROCEDURE Performing a Pivot Operation

A **pivot operation,** or **pivoting,** consists of performing row operations as follows:

Step 1 Multiply the pivot row by the reciprocal of the pivot element to transform the pivot element into a 1. (If the pivot element is already a 1, omit this step.)

Step 2 Add multiples of the pivot row to other rows in the tableau to transform all other nonzero elements in the pivot column into 0's.

CONCEPTUAL INSIGHT

A pivot operation uses some of the same row operations as those used in Gauss–Jordan elimination, but there is one essential difference. **In a pivot operation, you can never interchange two rows.**

Performing a pivot operation has the following effects:

1. The (entering) nonbasic variable becomes a basic variable.
2. The (exiting) basic variable becomes a nonbasic variable.
3. The value of the objective function is increased, or, in some cases, remains the same.

We now carry out the pivot operation on tableau (7). (To facilitate the process, we do not repeat the variables after the first tableau, and we use "Enter" and "Exit" for "Entering variable" and "Exiting variable," respectively.)

$$
\begin{array}{r}
\\
\text{Exit} \rightarrow s_1 \\
s_2 \\
P
\end{array}
\begin{array}{c}
\begin{array}{ccccc} x_1 & \overset{\text{Enter}}{\overset{\downarrow}{x_2}} & s_1 & s_2 & P \end{array} \\
\left[\begin{array}{ccccc|c}
1 & \textcircled{2} & 1 & 0 & 0 & 32 \\
3 & 4 & 0 & 1 & 0 & 84 \\
\hdashline
-50 & -80 & 0 & 0 & 1 & 0
\end{array}\right]
\end{array}
\begin{array}{l}
\\
\tfrac{1}{2}R_1 \rightarrow R_1 \\
\\
\\
\end{array}
$$

$$
\sim \left[\begin{array}{ccccc|c}
\frac{1}{2} & \textcircled{1} & \frac{1}{2} & 0 & 0 & 16 \\
3 & 4 & 0 & 1 & 0 & 84 \\
\hdashline
-50 & -80 & 0 & 0 & 1 & 0
\end{array}\right]
\begin{array}{l}
\\
(-4)R_1 + R_2 \rightarrow R_2 \\
80R_1 + R_3 \rightarrow R_3
\end{array}
$$

$$
\sim \left[\begin{array}{ccccc|c}
\frac{1}{2} & 1 & \frac{1}{2} & 0 & 0 & 16 \\
1 & 0 & -2 & 1 & 0 & 20 \\
\hdashline
-10 & 0 & 40 & 0 & 1 & 1{,}280
\end{array}\right]
$$

We have completed the pivot operation, and now we must insert appropriate variables for this new tableau. Since x_2 replaced s_1, the basic variables are now x_2, s_2, and P, as indicated by the labels on the left side of the new tableau. Note that this selection of basic variables agrees with the procedure outlined on page 645 for selecting basic variables. We write the new basic feasible solution by setting the nonbasic variables x_1 and s_1 equal to 0 and solving for the basic variables by inspection. (Remember, the values of the basic variables listed on the left are the corresponding numbers to the right of the vertical line. To see this, substitute $x_1 = 0$ and $s_1 = 0$ in the corresponding system shown next to the simplex tableau.)

$$\begin{array}{c} \\ x_2 \\ s_2 \\ P \end{array}\begin{array}{c} \begin{array}{ccccc} x_1 & x_2 & s_1 & s_2 & P \end{array} \\ \left[\begin{array}{ccccc|c} \frac{1}{2} & 1 & \frac{1}{2} & 0 & 0 & 16 \\ 1 & 0 & -2 & 1 & 0 & 20 \\ \hline -10 & 0 & 40 & 0 & 1 & 1{,}280 \end{array}\right] \end{array} \quad \begin{array}{l} \frac{1}{2}x_1 + x_2 + \frac{1}{2}s_1 = 16 \\ x_1 - 2s_1 + s_2 = 20 \\ -10x_1 + 40s_1 + P = 1{,}280 \end{array}$$

$$x_1 = 0, \quad x_2 = 16, \quad s_1 = 0, \quad s_2 = 20, \quad P = \$1{,}280$$

A profit of \$1,280 is a marked improvement over the \$0 profit produced by the initial basic feasible solution. But we can improve P still further, since a negative indicator still remains in the bottom row. To see why, we write out the objective function:

$$-10x_1 + 40s_1 + P = 1{,}280$$

or

$$P = 10x_1 - 40s_1 + 1{,}280$$

If s_1 stays a nonbasic variable (set equal to 0) and x_1 becomes a new basic variable, then

$$P = 10x_1 - 40(0) + 1{,}280 = 10x_1 + 1{,}280$$

and for each unit increase in x_1, P will increase \$10.

We now go through another iteration of the simplex process (that is, we repeat the sequence of steps above) using another pivot element. The pivot element and the entering and exiting variables are shown in the following tableau:

$$\begin{array}{r} \\ x_2 \\ \text{Exit} \rightarrow s_2 \\ P \end{array}\begin{array}{c} \begin{array}{ccccc} \overset{\text{Enter}\downarrow}{x_1} & x_2 & s_1 & s_2 & P \end{array} \\ \left[\begin{array}{ccccc|c} \frac{1}{2} & 1 & \frac{1}{2} & 0 & 0 & 16 \\ \textcircled{1} & 0 & -2 & 1 & 0 & 20 \\ \hline -10 & 0 & 40 & 0 & 1 & 1{,}280 \end{array}\right] \end{array} \quad \begin{array}{l} \frac{16}{1/2} = 32 \\ \frac{20}{1} = 20 \\ \\ \end{array}$$

We now pivot on (the circled) 1. That is, we perform a pivot operation using this 1 as the pivot element. Since the pivot element is 1, we do not need to perform the first step in the pivot operation, so we proceed to the second step to get 0's above and below the pivot element 1. As before, to facilitate the process, we omit writing the variables, except for the first tableau.

$$\begin{array}{r} \\ x_2 \\ \text{Exit} \rightarrow s_2 \\ P \end{array}\begin{array}{c} \begin{array}{ccccc} \overset{\text{Enter}\downarrow}{x_1} & x_2 & s_1 & s_2 & P \end{array} \\ \left[\begin{array}{ccccc|c} \frac{1}{2} & 1 & \frac{1}{2} & 0 & 0 & 16 \\ \textcircled{1} & 0 & -2 & 1 & 0 & 20 \\ \hline -10 & 0 & 40 & 0 & 1 & 1{,}280 \end{array}\right] \end{array} \quad \begin{array}{l} (-\frac{1}{2})R_2 + R_1 \rightarrow R_1 \\ \\ 10R_2 + R_3 \rightarrow R_3 \end{array}$$

$$\sim \left[\begin{array}{ccccc|c} 0 & 1 & \frac{3}{2} & -\frac{1}{2} & 0 & 6 \\ 1 & 0 & -2 & 1 & 0 & 20 \\ \hline 0 & 0 & 20 & 10 & 1 & 1{,}480 \end{array}\right]$$

Since there are no more negative indicators in the bottom row, we are through. Let us insert the appropriate variables for this last tableau and write the corresponding basic feasible solution. The basic variables are now x_1, x_2, and P, so to get the corresponding basic feasible solution, we set the nonbasic variables s_1 and s_2 equal to 0 and solve for the basic variables by inspection.

$$\begin{array}{c} \\ x_2 \\ x_1 \\ P \end{array}\begin{array}{c} \begin{array}{ccccc|c} x_1 & x_2 & s_1 & s_2 & P & \end{array} \\ \left[\begin{array}{ccccc|c} 0 & 1 & \frac{3}{2} & -\frac{1}{2} & 0 & 6 \\ 1 & 0 & -2 & 1 & 0 & 20 \\ \hline 0 & 0 & 20 & 10 & 1 & 1{,}480 \end{array}\right] \end{array}$$

$$x_1 = 20, \quad x_2 = 6, \quad s_1 = 0, \quad s_2 = 0, \quad P = 1{,}480$$

To see why this is the maximum, we rewrite the objective function from the bottom row:

$$20s_1 + 10s_2 + P = 1{,}480$$
$$P = 1{,}480 - 20s_1 - 10s_2$$

Since s_1 and s_2 cannot be negative, any increase of either from 0 will make the profit smaller.

Finally, returning to our original problem, we conclude that a production schedule of 20 standard tents and 6 expedition tents will produce a maximum profit of \$1,480 per day, which is the same as the geometric solution obtained in Section 11-2. The fact that the slack variables are both 0 means that for this production schedule, the plant will operate at full capacity—there is no slack in either the cutting department or the assembly department.

Interpreting the Simplex Process Geometrically

We can now interpret the simplex process just completed geometrically in terms of the feasible region graphed in the preceding section. Table 1 lists the three basic feasible solutions we just found using the simplex method (in the order they were found). The table also includes the corresponding corner points of the feasible region illustrated in Figure 1.

TABLE 1

Basic Feasible Solution (Obtained Above)

x_1	x_2	s_1	s_2	$P(\$)$	Corner Point
0	0	32	84	0	$O(0, 0)$
0	16	0	20	1,280	$A(0, 16)$
20	6	0	0	1,480	$B(20, 6)$

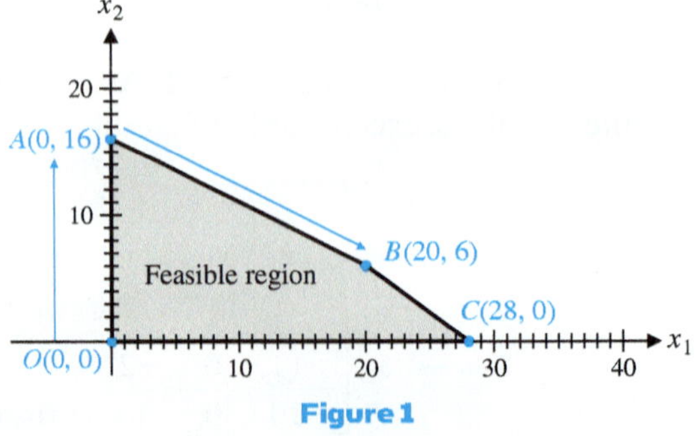

Figure 1

Looking at Table 1 and Figure 1, we see that the simplex process started at the origin, moved to the adjacent corner point $A(0, 16)$, and then to the optimal solution $B(20, 6)$ at the next adjacent corner point. This is typical of the simplex process.

Simplex Method Summarized

Before presenting additional examples, we summarize the important parts of the simplex method schematically in Figure 2.

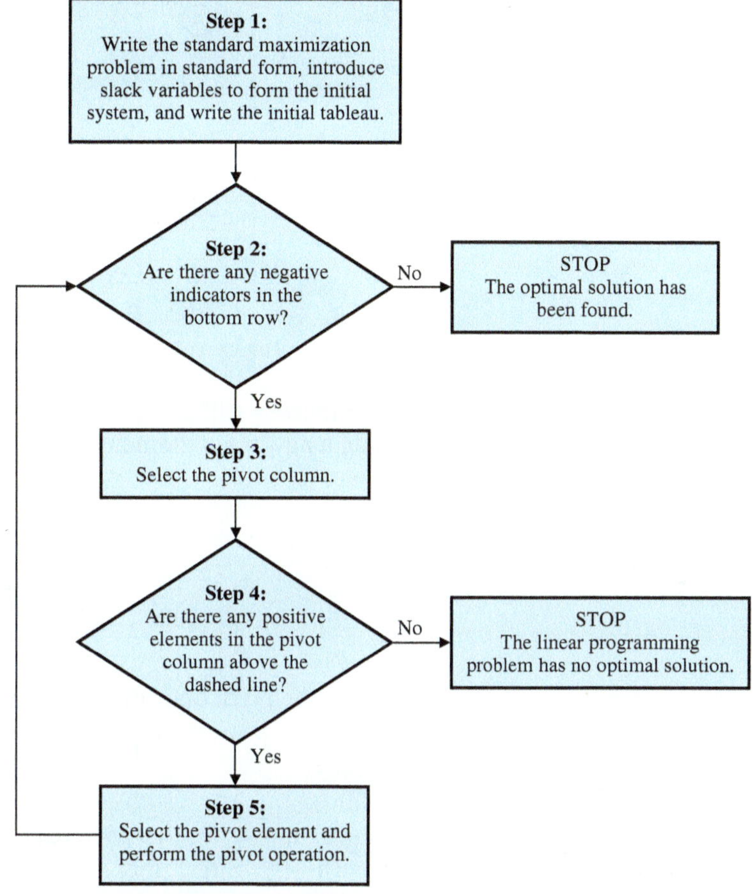

Figure 2 Simplex algorithm for standard maximization problems (Problem constraints are of the $\le$ form with nonnegative constants on the right. The coefficients of the objective function can be any real numbers.)

EXAMPLE 1 **Using the Simplex Method** Solve the following linear programming problem using the simplex method:

$$\begin{aligned} \text{Maximize} \quad & P = 10x_1 + 5x_2 \\ \text{subject to} \quad & 4x_1 + x_2 \le 28 \\ & 2x_1 + 3x_2 \le 24 \\ & x_1, x_2 \ge 0 \end{aligned}$$

SOLUTION Introduce slack variables s_1 and s_2, and write the initial system:

$$\begin{aligned} 4x_1 + x_2 + s_1 \qquad &= 28 \\ 2x_1 + 3x_2 \qquad + s_2 \qquad &= 24 \\ -10x_1 - 5x_2 \qquad\qquad + P &= 0 \\ x_1, x_2, s_1, s_2 &\ge 0 \end{aligned}$$

Write the simplex tableau, and identify the first pivot element and the entering and exiting variables:

$$\begin{array}{r} \\ \\ \text{Exit} \rightarrow s_1 \\ s_2 \\ P \end{array} \begin{array}{c} \text{Enter} \\ \downarrow \\ \end{array} \begin{matrix} & x_1 & x_2 & s_1 & s_2 & P & \\ \end{matrix}$$

$$\begin{matrix} \\ \text{Exit} \rightarrow s_1 \\ s_2 \\ P \end{matrix} \begin{matrix} \overset{\text{Enter}}{\downarrow} \\ x_1 \\ \end{matrix}$$

$$\begin{matrix} & x_1 & x_2 & s_1 & s_2 & P & \\ \text{Exit} \rightarrow s_1 & \textcircled{4} & 1 & 1 & 0 & 0 & 28 \\ s_2 & 2 & 3 & 0 & 1 & 0 & 24 \\ P & -10 & -5 & 0 & 0 & 1 & 0 \end{matrix} \quad \begin{matrix} \\ \frac{28}{4} = 7 \\ \frac{24}{2} = 12 \\ \\ \end{matrix}$$

Perform the pivot operation:

Enter ↓ (x_1 column)

$$\begin{array}{r} \\ \text{Exit}\rightarrow s_1 \\ s_2 \\ P \end{array} \begin{array}{c} \begin{array}{ccccc|c} x_1 & x_2 & s_1 & s_2 & P & \end{array} \\ \left[\begin{array}{ccccc|c} \textcircled{4} & 1 & 1 & 0 & 0 & 28 \\ 2 & 3 & 0 & 1 & 0 & 24 \\ \hdashline -10 & -5 & 0 & 0 & 1 & 0 \end{array}\right] \end{array} \quad \tfrac{1}{4}R_1 \rightarrow R_1$$

$$\sim \left[\begin{array}{ccccc|c} \textcircled{1} & 0.25 & 0.25 & 0 & 0 & 7 \\ 2 & 3 & 0 & 1 & 0 & 24 \\ \hdashline -10 & -5 & 0 & 0 & 1 & 0 \end{array}\right] \quad \begin{array}{l} \\ (-2)R_1 + R_2 \rightarrow R_2 \\ 10R_1 + R_3 \rightarrow R_3 \end{array}$$

$$\begin{array}{r} x_1 \\ \sim s_2 \\ P \end{array} \left[\begin{array}{ccccc|c} 1 & 0.25 & 0.25 & 0 & 0 & 7 \\ 0 & 2.5 & -0.5 & 1 & 0 & 10 \\ \hdashline 0 & -2.5 & 2.5 & 0 & 1 & 70 \end{array}\right]$$

Since there is still a negative indicator in the last row, we repeat the process by finding a new pivot element:

Enter ↓ (x_2 column)

$$\begin{array}{r} \\ x_1 \\ \text{Exit}\rightarrow s_2 \\ P \end{array} \begin{array}{c} \begin{array}{ccccc|c} x_1 & x_2 & s_1 & s_2 & P & \end{array} \\ \left[\begin{array}{ccccc|c} 1 & 0.25 & 0.25 & 0 & 0 & 7 \\ 0 & \textcircled{2.5} & -0.5 & 1 & 0 & 10 \\ \hdashline 0 & -2.5 & 2.5 & 0 & 1 & 70 \end{array}\right] \end{array} \quad \begin{array}{l} \\ \frac{7}{0.25} = 28 \\ \frac{10}{2.5} = 4 \\ \\ \end{array}$$

Performing the pivot operation, we obtain

Enter ↓ (x_2 column)

$$\begin{array}{r} \\ x_1 \\ \text{Exit}\rightarrow s_2 \\ P \end{array} \begin{array}{c} \begin{array}{ccccc|c} x_1 & x_2 & s_1 & s_2 & P & \end{array} \\ \left[\begin{array}{ccccc|c} 1 & 0.25 & 0.25 & 0 & 0 & 7 \\ 0 & \textcircled{2.5} & -0.5 & 1 & 0 & 10 \\ \hdashline 0 & -2.5 & 2.5 & 0 & 1 & 70 \end{array}\right] \end{array} \quad \begin{array}{l} \\ \\ \frac{1}{2.5}R_2 \rightarrow R_2 \\ \\ \end{array}$$

$$\sim \left[\begin{array}{ccccc|c} 1 & 0.25 & 0.25 & 0 & 0 & 7 \\ 0 & \textcircled{1} & -0.2 & 0.4 & 0 & 4 \\ \hdashline 0 & -2.5 & 2.5 & 0 & 1 & 70 \end{array}\right] \quad \begin{array}{l} (-0.25)R_2 + R_1 \rightarrow R_1 \\ \\ 2.5R_2 + R_3 \rightarrow R_3 \end{array}$$

$$\begin{array}{r} x_1 \\ \sim x_2 \\ P \end{array} \left[\begin{array}{ccccc|c} 1 & 0 & 0.3 & -0.1 & 0 & 6 \\ 0 & 1 & -0.2 & 0.4 & 0 & 4 \\ \hdashline 0 & 0 & 2 & 1 & 1 & 80 \end{array}\right]$$

Since all the indicators in the last row are nonnegative, we stop and read the optimal solution:

$$\text{Max } P = 80 \quad \text{at} \quad x_1 = 6, \quad x_2 = 4, \quad s_1 = 0, \quad s_2 = 0$$

(To see why this makes sense, write the objective function corresponding to the last row to see what happens to P when you try to increase s_1 or s_2.)

Matched Problem 1 Solve the following linear programming problem using the simplex method:

$$\begin{array}{ll} \text{Maximize} & P = 2x_1 + x_2 \\ \text{subject to} & 5x_1 + x_2 \le 9 \\ & x_1 + x_2 \le 5 \\ & x_1, x_2 \ge 0 \end{array}$$

EXPLORE & DISCUSS 1

Graph the feasible region for the linear programming problem in Example 1 and trace the path to the optimal solution determined by the simplex method.

EXAMPLE 2 **Using the Simplex Method** Solve using the simplex method:

$$\begin{aligned} \text{Maximize} \quad & P = 6x_1 + 3x_2 \\ \text{subject to} \quad & -2x_1 + 3x_2 \le 9 \\ & -x_1 + 3x_2 \le 12 \\ & x_1, x_2 \ge 0 \end{aligned}$$

SOLUTION Write the initial system using the slack variables s_1 and s_2:

$$\begin{aligned} -2x_1 + 3x_2 + s_1 \qquad\qquad &= 9 \\ -x_1 + 3x_2 \qquad + s_2 \qquad &= 12 \\ -6x_1 - 3x_2 \qquad\qquad + P &= 0 \end{aligned}$$

Write the simplex tableau and identify the first pivot element:

$$\begin{matrix} & x_1 & x_2 & s_1 & s_2 & P & \\ s_1 & -2 & 3 & 1 & 0 & 0 & 9 \\ s_2 & -1 & 3 & 0 & 1 & 0 & 12 \\ P & -6 & -3 & 0 & 0 & 1 & 0 \end{matrix}$$

↑ Pivot column

Since both elements in the pivot column above the dashed line are negative, we are unable to select a pivot row. We stop and conclude that there is no optimal solution.

Matched Problem 2 Solve using the simplex method:

$$\begin{aligned} \text{Maximize} \quad & P = 2x_1 + 3x_2 \\ \text{subject to} \quad & -3x_1 + 4x_2 \le 12 \\ & x_2 \le 2 \\ & x_1, x_2 \ge 0 \end{aligned}$$

EXPLORE & DISCUSS 2

In Example 2 we encountered a tableau with a pivot column and no pivot row, indicating a problem with no optimal solution.

(A) What happens if you violate the rule for selecting the pivot row, select the negative element -1 for the pivot, and then continue with the simplex method?

(B) There was another negative indicator in the tableau in Example 2. What happens if you violate the rule for selecting the pivot column, select the second column for a pivot column, and then continue with the simplex method?

(C) Graph the feasible region in Example 2, graph the lines corresponding to $P = 36$ and $P = 66$, and explain why this problem has no optimal solution.

Refer to Examples 1 and 2. In Example 1 we concluded that we had found the optimal solution because we could not select a pivot column. In Example 2 we concluded that the problem had no optimal solution because we selected a pivot column and then could not select a pivot row. Notice that we do not try to continue with the simplex method by selecting a negative pivot element or using a different column for the pivot column. Remember:

PROCEDURE Stopping the Simplex Method

If it is not possible to select a pivot column, the simplex method stops and we conclude that the optimal solution has been found. If the pivot column has been selected and it is not possible to select a pivot row, the simplex method stops and we conclude that there is no optimal solution.

Application

EXAMPLE 3 **Agriculture** A farmer owns a 100-acre farm and plans to plant at most three crops. The seed for crops A, B, and C costs \$40, \$20, and \$30 per acre, respectively. A maximum of \$3,200 can be spent on seed. Crops A, B, and C require 1, 2, and 1 workdays per acre, respectively, and there are a maximum of 160 workdays available. If the farmer can make a profit of \$100 per acre on crop A, \$300 per acre on crop B, and \$200 per acre on crop C, how many acres of each crop should be planted to maximize profit?

SOLUTION The farmer must decide on the number of acres of each crop that should be planted. Thus, the decision variables are

$$\begin{aligned} x_1 &= \text{number of acres of crop } A \\ x_2 &= \text{number of acres of crop } B \\ x_3 &= \text{number of acres of crop } C \end{aligned}$$

The farmer's objective is to maximize profit:

$$P = 100x_1 + 300x_2 + 200x_3$$

The farmer is constrained by the number of acres available for planting, the money available for seed, and the available work days. These lead to the following constraints:

$$\begin{aligned} x_1 + x_2 + x_3 &\le 100 && \text{Acreage constraint} \\ 40x_1 + 20x_2 + 30x_3 &\le 3{,}200 && \text{Monetary constraint} \\ x_1 + 2x_2 + x_3 &\le 160 && \text{Labor constraint} \end{aligned}$$

Adding the nonnegative constraints, we have the following model for a linear programming problem:

$$\begin{aligned} \text{Maximize } \quad & P = 100x_1 + 300x_2 + 200x_3 && \text{Objective function} \\ \text{subject to } \quad & \left.\begin{aligned} x_1 + x_2 + x_3 &\le 100 \\ 40x_1 + 20x_2 + 30x_3 &\le 3{,}200 \\ x_1 + 2x_2 + x_3 &\le 160 \end{aligned}\right\} && \text{Problem constraints} \\ & x_1, x_2, x_3 \ge 0 && \text{Nonnegative constraints} \end{aligned}$$

Next, we introduce slack variables and form the initial system:

$$\begin{aligned} x_1 + x_2 + x_3 + s_1 \qquad\qquad\qquad &= 100 \\ 40x_1 + 20x_2 + 30x_3 \qquad + s_2 \qquad\qquad &= 3{,}200 \\ x_1 + 2x_2 + x_3 \qquad\qquad + s_3 \qquad &= 160 \\ -100x_1 - 300x_2 - 200x_3 \qquad\qquad\qquad + P &= 0 \\ x_1, x_2, x_3, s_1, s_2, s_3 &\ge 0 \end{aligned}$$

Notice that the initial system has $7 - 4 = 3$ nonbasic variables and 4 basic variables. Now we form the simplex tableau and solve by the simplex method:

Enter ↓ (x_2)

$$\begin{array}{r} \\ s_1 \\ s_2 \\ \text{Exit} \rightarrow s_3 \\ P \end{array}
\begin{array}{ccccccc|c} x_1 & x_2 & x_3 & s_1 & s_2 & s_3 & P & \\ 1 & 1 & 1 & 1 & 0 & 0 & 0 & 100 \\ 40 & 20 & 30 & 0 & 1 & 0 & 0 & 3{,}200 \\ 1 & \textcircled{2} & 1 & 0 & 0 & 1 & 0 & 160 \\ \hline -100 & -300 & -200 & 0 & 0 & 0 & 1 & 0 \end{array}
\quad 0.5R_3 \rightarrow R_3$$

$$\sim \left[\begin{array}{ccccccc|c} 1 & 1 & 1 & 1 & 0 & 0 & 0 & 100 \\ 40 & 20 & 30 & 0 & 1 & 0 & 0 & 3{,}200 \\ 0.5 & \textcircled{1} & 0.5 & 0 & 0 & 0.5 & 0 & 80 \\ \hline -100 & -300 & -200 & 0 & 0 & 0 & 1 & 0 \end{array}\right]
\begin{array}{l} (-1)R_3 + R_1 \rightarrow R_1 \\ (-20)R_3 + R_2 \rightarrow R_2 \\ \\ 300R_3 + R_4 \rightarrow R_4 \end{array}$$

Enter ↓ (x_3)

$$\begin{array}{r} \\ \text{Exit} \rightarrow s_1 \\ s_2 \\ x_2 \\ P \end{array}
\begin{array}{ccccccc|c} x_1 & x_2 & x_3 & s_1 & s_2 & s_3 & P & \\ 0.5 & 0 & \textcircled{0.5} & 1 & 0 & -0.5 & 0 & 20 \\ 30 & 0 & 20 & 0 & 1 & -10 & 0 & 1{,}600 \\ 0.5 & 1 & 0.5 & 0 & 0 & 0.5 & 0 & 80 \\ \hline 50 & 0 & -50 & 0 & 0 & 150 & 1 & 24{,}000 \end{array}
\quad 2R_1 \rightarrow R_1$$

$$\sim \left[\begin{array}{ccccccc|c} 1 & 0 & \textcircled{1} & 2 & 0 & -1 & 0 & 40 \\ 30 & 0 & 20 & 0 & 1 & -10 & 0 & 1{,}600 \\ 0.5 & 1 & 0.5 & 0 & 0 & 0.5 & 0 & 80 \\ \hline 50 & 0 & -50 & 0 & 0 & 150 & 1 & 24{,}000 \end{array}\right]
\begin{array}{l} \\ (-20)R_1 + R_2 \rightarrow R_2 \\ (-0.5)R_1 + R_3 \rightarrow R_3 \\ 50R_1 + R_4 \rightarrow R_4 \end{array}$$

$$\sim \begin{array}{r} x_3 \\ s_2 \\ x_2 \\ P \end{array}
\left[\begin{array}{ccccccc|c} 1 & 0 & 1 & 2 & 0 & -1 & 0 & 40 \\ 10 & 0 & 0 & -40 & 1 & 10 & 0 & 800 \\ 0 & 1 & 0 & -1 & 0 & 1 & 0 & 60 \\ \hline 100 & 0 & 0 & 100 & 0 & 100 & 1 & 26{,}000 \end{array}\right]$$

All indicators in the bottom row are nonnegative, and we can now read the optimal solution:

$$x_1 = 0, \quad x_2 = 60, \quad x_3 = 40, \quad s_1 = 0, \quad s_2 = 800, \quad s_3 = 0, \quad P = \$26{,}000$$

Thus, if the farmer plants 60 acres in crop B, 40 acres in crop C, and no crop A, the maximum profit of \$26,000 will be realized. The fact that $s_2 = 800$ tells us (look at the second row in the equations at the start) that this maximum profit is reached by using only \$2,400 of the \$3,200 available for seed; that is, we have a slack of \$800 that can be used for some other purpose.

There are many types of software that can be used to solve linear programming problems by the simplex method. Figure 3 illustrates a solution to Example 3 in Excel, a popular spreadsheet for personal computers.

	A	B	C	D	E	F
1	Resources	Crop A	Crop B	Crop C	Available	Used
2	Acres	1	1	1	100	100
3	Seed($)	40	20	30	3,200	2,400
4	Workdays	1	2	1	160	160
5	Profit Per Acre	100	300	200	26,000	<-Total
6	Acres to plant	0	60	40		Profit

Figure 3

Matched Problem 3 Repeat Example 3 modified as follows:

	Investment per Acre			Maximum
	Crop *A*	**Crop *B***	**Crop *C***	**Available**
Seed cost	\$24	\$40	\$30	\$3,600
Workdays	1	2	2	160
Profit	\$140	\$200	\$160	

CONCEPTUAL INSIGHT

If you are solving a system of linear equations or inequalities, you know that you can always multiply both sides of an equation by any nonzero number and both sides of an inequality by any positive number without changing the solution set. This is still the case for the simplex method, however you must be careful when you interpret the results. For example, consider the second problem constraint in the model for Example 3:

$$40x_1 + 20x_2 + 30x_3 \le 3{,}200$$

Multiplying both sides by $\frac{1}{10}$ before introducing slack variables simplifies subsequent calculations. However, performing this operation has a side effect—it changes the units of the slack variable from dollars to tens of dollars. Compare the following two equations:

$$40x_1 + 20x_2 + 30x_3 + s_2 = 3{,}200 \quad s_2 \text{ represents dollars}$$
$$4x_1 + 2x_2 + 3x_3 + s'_2 = 320 \quad s'_2 \text{ represents tens of dollars}$$

In general, if you multiply a problem constraint by a positive number, remember to take this into account when you interpret the value of the slack variable for that constraint.

It can be shown that the feasible region for the linear programming problem in Example 3 has eight corner points, yet the simplex method found the solution in only two steps. Now you begin to see the power of the simplex method. In larger problems, the difference between the total number of corner points and the number of steps required by the simplex method is even more dramatic. A feasible region may have hundreds or even thousands of corner points, yet the simplex method will often find the optimal solution in 10 or 15 steps.

It is important to realize that in order to keep this introduction as simple as possible, we have purposely avoided certain degenerate cases that lead to difficulties. Discussion and resolution of these problems is left to a more advanced treatment of the subject.

Answers to Matched Problems

1. Max $P = 6$ when $x_1 = 1$ and $x_2 = 4$
2. No optimal solution
3. 40 acres of crop A, 60 acres of crop B, no crop C; max $P = \$17{,}600$ (since $s_2 = 240$, \$240 out of the \$3,600 will not be spent).

Exercises 11-4

A

For the simplex tableaux in Problems 1–4,

(A) *Identify the basic and nonbasic variables.*

(B) *Find the corresponding basic feasible solution.*

(C) *Determine whether the optimal solution has been found, an additional pivot is required, or the problem has no optimal solution.*

1.

$$\begin{array}{c} \begin{matrix} x_1 & x_2 & s_1 & s_2 & P \end{matrix} \\ \left[\begin{array}{rrrrr|r} 2 & 1 & 0 & 3 & 0 & 12 \\ 3 & 0 & 1 & -2 & 0 & 15 \\ \hdashline -4 & 0 & 0 & 4 & 1 & 50 \end{array}\right] \end{array}$$

2.

$$\begin{array}{c} \begin{matrix} x_1 & x_2 & s_1 & s_2 & P \end{matrix} \\ \left[\begin{array}{rrrrr|r} 1 & 4 & -2 & 0 & 0 & 10 \\ 0 & 2 & 3 & 1 & 0 & 25 \\ \hdashline 0 & 5 & 6 & 0 & 1 & 35 \end{array}\right] \end{array}$$

3.

$$\begin{array}{c} \begin{matrix} x_1 & x_2 & x_3 & s_1 & s_2 & s_3 & P \end{matrix} \\ \left[\begin{array}{rrrrrrr|r} -2 & 0 & 1 & 3 & 1 & 0 & 0 & 5 \\ 0 & 1 & 0 & -2 & 0 & 0 & 0 & 15 \\ -1 & 0 & 0 & 4 & 1 & 1 & 0 & 12 \\ \hdashline -4 & 0 & 0 & 2 & 4 & 0 & 1 & 45 \end{array}\right] \end{array}$$

4.

$$\begin{array}{c} \begin{matrix} x_1 & x_2 & x_3 & s_1 & s_2 & s_3 & P \end{matrix} \\ \left[\begin{array}{rrrrrrr|r} 0 & 2 & -1 & 1 & 4 & 0 & 0 & 5 \\ 0 & 1 & 2 & 0 & -2 & 1 & 0 & 2 \\ 1 & 3 & 0 & 0 & 5 & 0 & 0 & 11 \\ \hdashline 0 & -5 & 4 & 0 & -3 & 0 & 1 & 27 \end{array}\right] \end{array}$$

In Problems 5–8, find the pivot element, identify the entering and exiting variables, and perform one pivot operation.

5.

$$\begin{array}{c} \begin{matrix} x_1 & x_2 & s_1 & s_2 & P \end{matrix} \\ \left[\begin{array}{rrrrr|r} 1 & 4 & 1 & 0 & 0 & 4 \\ 3 & 5 & 0 & 1 & 0 & 24 \\ \hdashline -8 & -5 & 0 & 0 & 1 & 0 \end{array}\right] \end{array}$$

6.

$$\begin{array}{c} \begin{matrix} x_1 & x_2 & s_1 & s_2 & P \end{matrix} \\ \left[\begin{array}{rrrrr|r} 1 & 6 & 1 & 0 & 0 & 36 \\ 3 & 1 & 0 & 1 & 0 & 5 \\ \hdashline -1 & -2 & 0 & 0 & 1 & 0 \end{array}\right] \end{array}$$

7.

$$\begin{array}{c} \begin{matrix} x_1 & x_2 & s_1 & s_2 & s_3 & P \end{matrix} \\ \left[\begin{array}{rrrrrr|r} 2 & 1 & 1 & 0 & 0 & 0 & 4 \\ 3 & 0 & 1 & 1 & 0 & 0 & 8 \\ 0 & 0 & 2 & 0 & 1 & 0 & 2 \\ \hdashline -4 & 0 & -3 & 0 & 0 & 1 & 5 \end{array}\right] \end{array}$$

8.

$$\begin{array}{c} \begin{matrix} x_1 & x_2 & s_1 & s_2 & s_3 & P \end{matrix} \\ \left[\begin{array}{rrrrrr|r} 0 & 0 & 2 & 1 & 1 & 0 & 2 \\ 1 & 0 & -4 & 0 & 1 & 0 & 3 \\ 0 & 1 & 5 & 0 & 2 & 0 & 11 \\ \hdashline 0 & 0 & -6 & 0 & -5 & 1 & 18 \end{array}\right] \end{array}$$

In Problems 9–12,

(A) *Using slack variables, write the initial system for each linear programming problem.*

(B) *Write the simplex tableau, circle the first pivot, and identify the entering and exiting variables.*

(C) *Use the simplex method to solve the problem.*

9. Maximize $P = 15x_1 + 10x_2$
subject to $2x_1 + x_2 \le 10$
$x_1 + 3x_2 \le 10$
$x_1, x_2 \ge 0$

10. Maximize $P = 3x_1 + 2x_2$
subject to $5x_1 + 2x_2 \le 20$
$3x_1 + 2x_2 \le 16$
$x_1, x_2 \ge 0$

11. Repeat Problem 9 with the objective function changed to $P = 30x_1 + x_2$.

12. Repeat Problem 10 with the objective function changed to $P = x_1 + 3x_2$.

B

Solve the linear programming problems in Problems 13–28 using the simplex method.

13. Maximize $P = 30x_1 + 40x_2$
subject to $2x_1 + x_2 \le 10$
$x_1 + x_2 \le 7$
$x_1 + 2x_2 \le 12$
$x_1, x_2 \ge 0$

14. Maximize $P = 15x_1 + 20x_2$
subject to $2x_1 + x_2 \le 9$
$x_1 + x_2 \le 6$
$x_1 + 2x_2 \le 10$
$x_1, x_2 \ge 0$

15. Maximize $P = 2x_1 + 3x_2$
subject to $-2x_1 + x_2 \le 2$
$-x_1 + x_2 \le 5$
$x_2 \le 6$
$x_1, x_2 \ge 0$

16. Repeat Problem 15 with $P = -x_1 + 3x_2$.

17. Maximize $P = -x_1 + 2x_2$

subject to
$$\begin{aligned} -x_1 + x_2 &\le 2 \\ -x_1 + 3x_2 &\le 12 \\ x_1 - 4x_2 &\le 4 \\ x_1, x_2 &\ge 0 \end{aligned}$$

18. Repeat Problem 17 with $P = x_1 + 2x_2$.

19. Maximize $P = 5x_1 + 2x_2 - x_3$

subject to
$$\begin{aligned} x_1 + x_2 - x_3 &\le 10 \\ 2x_1 + 4x_2 + 3x_3 &\le 30 \\ x_1, x_2, x_3 &\ge 0 \end{aligned}$$

20. Maximize $P = 4x_1 - 3x_2 + 2x_3$

subject to
$$\begin{aligned} x_1 + 2x_2 - x_3 &\le 5 \\ 3x_1 + 2x_2 + 2x_3 &\le 22 \\ x_1, x_2, x_3 &\ge 0 \end{aligned}$$

21. Maximize $P = 2x_1 + 3x_2 + 4x_3$

subject to
$$\begin{aligned} x_1 + x_3 &\le 4 \\ x_2 + x_3 &\le 3 \\ x_1, x_2, x_3 &\ge 0 \end{aligned}$$

22. Maximize $P = x_1 + x_2 + 2x_3$

subject to
$$\begin{aligned} x_1 - 2x_2 + x_3 &\le 9 \\ 2x_1 + x_2 + 2x_3 &\le 28 \\ x_1, x_2, x_3 &\ge 0 \end{aligned}$$

23. Maximize $P = 4x_1 + 3x_2 + 2x_3$

subject to
$$\begin{aligned} 3x_1 + 2x_2 + 5x_3 &\le 23 \\ 2x_1 + x_2 + x_3 &\le 8 \\ x_1 + x_2 + 2x_3 &\le 7 \\ x_1, x_2, x_3 &\ge 0 \end{aligned}$$

24. Maximize $P = 4x_1 + 2x_2 + 3x_3$

subject to
$$\begin{aligned} x_1 + x_2 + x_3 &\le 11 \\ 2x_1 + 3x_2 + x_3 &\le 20 \\ x_1 + 3x_2 + 2x_3 &\le 20 \\ x_1, x_2, x_3 &\ge 0 \end{aligned}$$

C

25. Maximize $P = 20x_1 + 30x_2$

subject to
$$\begin{aligned} 0.6x_1 + 1.2x_2 &\le 960 \\ 0.03x_1 + 0.04x_2 &\le 36 \\ 0.3x_1 + 0.2x_2 &\le 270 \\ x_1, x_2 &\ge 0 \end{aligned}$$

26. Repeat Problem 25 with $P = 20x_1 + 20x_2$.

27. Maximize $P = x_1 + 2x_2 + 3x_3$

subject to
$$\begin{aligned} 2x_1 + 2x_2 + 8x_3 &\le 600 \\ x_1 + 3x_2 + 2x_3 &\le 600 \\ 3x_1 + 2x_2 + x_3 &\le 400 \\ x_1, x_2, x_3 &\ge 0 \end{aligned}$$

28. Maximize $P = 10x_1 + 50x_2 + 10x_3$

subject to
$$\begin{aligned} 3x_1 + 3x_2 + 3x_3 &\le 66 \\ 6x_1 - 2x_2 + 4x_3 &\le 48 \\ 3x_1 + 6x_2 + 9x_3 &\le 108 \\ x_1, x_2, x_3 &\ge 0 \end{aligned}$$

In Problems 29 and 30, first solve the linear programming problem by the simplex method, keeping track of the basic feasible solutions at each step. Then graph the feasible region and illustrate the path to the optimal solution determined by the simplex method.

29. Maximize $P = 2x_1 + 5x_2$

subject to
$$\begin{aligned} x_1 + 2x_2 &\le 40 \\ x_1 + 3x_2 &\le 48 \\ x_1 + 4x_2 &\le 60 \\ x_2 &\le 14 \\ x_1, x_2 &\ge 0 \end{aligned}$$

30. Maximize $P = 5x_1 + 3x_2$

subject to
$$\begin{aligned} 5x_1 + 4x_2 &\le 100 \\ 2x_1 + x_2 &\le 28 \\ 4x_1 + x_2 &\le 42 \\ x_1 &\le 10 \\ x_1, x_2 &\ge 0 \end{aligned}$$

Solve Problems 31 and 32 by the simplex method and also by the geometric method. Compare and contrast the results.

31. Maximize $P = 2x_1 + 3x_2$

subject to
$$\begin{aligned} -2x_1 + x_2 &\le 4 \\ x_2 &\le 10 \\ x_1, x_2 &\ge 0 \end{aligned}$$

32. Maximize $P = 2x_1 + 3x_2$

subject to
$$\begin{aligned} -x_1 + x_2 &\le 2 \\ x_2 &\le 4 \\ x_1, x_2 &\ge 0 \end{aligned}$$

In Problems 33–36, there is a tie for the choice of the first pivot column. Use the simplex method to solve each problem two different ways: first by choosing column 1 as the first pivot column, and then by choosing column 2 as the first pivot column. Discuss the relationship between these two solutions.

33. Maximize $P = x_1 + x_2$

subject to
$$2x_1 + x_2 \le 16$$
$$x_1 \le 6$$
$$x_2 \le 10$$
$$x_1, x_2 \ge 0$$

34. Maximize $P = x_1 + x_2$

subject to
$$x_1 + 2x_2 \le 10$$
$$x_1 \le 6$$
$$x_2 \le 4$$
$$x_1, x_2 \ge 0$$

35. Maximize $P = 3x_1 + 3x_2 + 2x_3$

subject to
$$x_1 + x_2 + 2x_3 \le 20$$
$$2x_1 + x_2 + 4x_3 \le 32$$
$$x_1, x_2, x_3 \ge 0$$

36. Maximize $P = 2x_1 + 2x_2 + x_3$

subject to
$$x_1 + x_2 + 3x_3 \le 10$$
$$2x_1 + 4x_2 + 5x_3 \le 24$$
$$x_1, x_2, x_3 \ge 0$$

Applications

In Problems 37–50, construct a mathematical model in the form of a linear programming problem. (The answers in the back of the book for these application problems include the model.) Then solve the problem using the simplex method. Include an interpretation of any nonzero slack variables in the optimal solution.

Business & Economics

37. Manufacturing: resource allocation. A small company manufactures three different electronic components for computers. Component A requires 2 hours of fabrication and 1 hour of assembly; component B requires 3 hours of fabrication and 1 hour of assembly; and component C requires 2 hours of fabrication and 2 hours of assembly. The company has up to 1,000 labor-hours of fabrication time and 800 labor-hours of assembly time available per week. The profit on each component, A, B, and C, is \$7, \$8, and \$10, respectively. How many components of each type should the company manufacture each week in order to maximize its profit (assuming that all components that it manufactures can be sold)? What is the maximum profit?

38. Manufacturing: resource allocation. Solve Problem 37 with the additional restriction that the combined total number of components produced each week cannot exceed 420. Discuss the effect of this restriction on the solution to Problem 37.

39. Investment. An investor has at most \$100,000 to invest in government bonds, mutual funds, and money market funds. The average yields for government bonds, mutual funds, and money market funds are 8%, 13%, and 15%, respectively. The investor's policy requires that the total amount invested in mutual and money market funds not exceed the amount invested in government bonds. How much should be invested in each type of investment in order to maximize the return? What is the maximum return?

40. Investment. Repeat Problem 39 under the additional assumption that no more than \$30,000 can be invested in money market funds.

41. Advertising. A department store chain has up to \$20,000 to spend on television advertising for a sale. All ads will be placed with one television station, where a 30-second ad costs \$1,000 on daytime TV and is viewed by 14,000 potential customers, \$2,000 on prime-time TV and is viewed by 24,000 potential customers, and \$1,500 on late-night TV and is viewed by 18,000 potential customers. The television station will not accept a total of more than 15 ads in all three time periods. How many ads should be placed in each time period in order to maximize the number of potential customers who will see the ads? How many potential customers will see the ads? (Ignore repeated viewings of the ad by the same potential customer.)

42. Advertising. Repeat Problem 41 if the department store increases its budget to \$24,000 and requires that at least half of the ads be placed in prime-time shows.

43. Construction: resource allocation. A contractor is planning to build a new housing development consisting of colonial, split-level, and ranch-style houses. A colonial house requires $\frac{1}{2}$ acre of land, \$60,000 capital, and 4,000 labor-hours to construct, and returns a profit of \$20,000. A split-level house requires $\frac{1}{2}$ acre of land, \$60,000 capital, and 3,000 labor-hours to construct, and returns a profit of \$18,000. A ranch house requires 1 acre of land, \$80,000 capital, and 4,000 labor-hours to construct, and returns a profit of \$24,000. The contractor has available 30 acres of land, \$3,200,000 capital, and 180,000 labor-hours.

(A) How many houses of each type should be constructed to maximize the contractor's profit? What is the maximum profit?

(B) A decrease in demand for colonial houses causes the profit on a colonial house to drop from \$20,000 to \$17,000. Discuss the effect of this change on the number of houses built and on the maximum profit.

(C) An increase in demand for colonial houses causes the profit on a colonial house to rise from \$20,000 to \$25,000. Discuss the effect of this change on the number of houses built and on the maximum profit.

44. Manufacturing: resource allocation. A company manufactures three-speed, five-speed, and ten-speed bicycles. Each bicycle passes through three departments, fabrication, painting & plating, and final assembly. The relevant manufacturing data are given in the table.

	Labor-Hours per Bicycle			Maximum Labor-Hours Available per Day
	Three-Speed	Five-Speed	Ten-Speed	
Fabrication	3	4	5	120
Painting & plating	5	3	5	130
Final assembly	4	3	5	120
Profit per bicycle ($)	80	70	100	

(A) How many bicycles of each type should the company manufacture per day in order to maximize its profit? What is the maximum profit?

(B) Discuss the effect on the solution to part (A) if the profit on a ten-speed bicycle increases to $110 and all other data in part (A) remains the same.

(C) Discuss the effect on the solution to part (A) if the profit on a five-speed bicycle increases to $110 and all other data in part (A) remains the same.

45. **Packaging: product mix.** A candy company makes three types of candy, solid-center, fruit-filled, and cream-filled, and packages these candies in three different assortments. A box of assortment I contains 4 solid-center, 4 fruit-filled, and 12 cream-filled candies, and sells for $9.40. A box of assortment II contains 12 solid-center, 4 fruit-filled, and 4 cream-filled candies, and sells for $7.60. A box of assortment III contains 8 solid-center, 8 fruit-filled, and 8 cream-filled candies, and sells for $11.00. The manufacturing costs per piece of candy are $0.20 for solid-center, $0.25 for fruit-filled, and $0.30 for cream-filled. The company can manufacture 4,800 solid-center, 4,000 fruit-filled, and 5,600 cream-filled candies weekly.

(A) How many boxes of each type should the company produce each week in order to maximize its profit? What is the maximum profit?

(B) Discuss the effect on the solution to part (A) if the number of fruit-filled candies manufactured weekly is increased to 5,000 and all other data in part (A) remain the same.

(C) Discuss the effect on the solution to part (A) if the number of fruit-filled candies and the number of solid-center candies manufactured weekly are each increased to 6,000 and all other data in part (A) remain the same.

46. **Scheduling: resource allocation.** A small accounting firm prepares tax returns for three types of customers: individual, commercial, and industrial. The tax preparation process begins with a 1-hour interview with the customer. The data collected during this interview is entered into a time-sharing computer system, which produces the customer's tax return. It takes 1 hour to enter the data for an individual customer, 2 hours for a commercial customer, and $1\frac{1}{2}$ hours for an industrial customer. It takes 10 minutes of computer time to process an individual return, 25 minutes to process a commercial return, and 20 minutes to process an industrial return. The firm has one employee who conducts the initial interview and two who enter the data into the computer. The interviewer can work a maximum of 50 hours a week, and each of the data-entry employees can work a maximum of 40 hours a week. The computer is available for a maximum of 1,025 minutes a week. The firm makes a profit of $50 on each individual customer, $65 on each commercial customer, and $60 on each industrial customer.

(A) How many customers of each type should the firm schedule each week in order to maximize its profit? What is the maximum profit?

(B) Discuss the effect on the solution to part (A) if the maximum number of hours the interviewer works is decreased to 35 hours per week and all other data in part (A) remains the same.

(C) Discuss the effect on the solution to part (A) if the maximum number of minutes of available computer time is decreased to 450 minutes per week and all other data in part (A) remains the same.

Life Sciences

47. **Nutrition: animals.** The natural diet of a certain animal consists of three foods, *A*, *B*, and *C*. The number of units of calcium, iron, and protein in 1 gram of each food and the average daily intake are given in the table. A scientist wants to investigate the effect of increasing the protein in the animal's diet while not allowing the units of calcium and iron to exceed their average daily intakes. How many grams of each food should be used to maximize the amount of protein in the diet? What is the maximum amount of protein?

	Units per Gram			Average Daily Intake (units)
	Food *A*	Food *B*	Food *C*	
Calcium	1	3	2	30
Iron	2	1	2	24
Protein	3	4	5	60

48. **Nutrition: animals.** Repeat Problem 47 if the scientist wants to maximize the daily calcium intake while not allowing the intake of iron or protein to exceed the average daily intake.

Social Sciences

49. **Opinion survey.** A political scientist has received a grant to fund a research project involving voting trends. The budget of the grant includes $3,200 for conducting door-to-door interviews the day before an election. Undergraduate students, graduate students, and faculty members will be hired

to conduct the interviews. Each undergraduate student will conduct 18 interviews and be paid \$100. Each graduate student will conduct 25 interviews and be paid \$150. Each faculty member will conduct 30 interviews and be paid \$200. Due to limited transportation facilities, no more than 20 interviewers can be hired. How many undergraduate students, graduate students, and faculty members should be hired in order to maximize the number of interviews that will be conducted? What is the maximum number of interviews?

50. **Opinion survey.** Repeat Problem 49 if one of the requirements of the grant is that at least 50% of the interviewers be undergraduate students.

APPENDIX A Basic Algebra Review

Appendix A reviews some important basic algebra concepts usually studied in earlier courses. The material may be studied systematically before beginning the rest of the book or reviewed as needed.

A-1 Real Numbers

The rules for manipulating and reasoning with symbols in algebra depend, in large measure, on properties of the real numbers. In this section we look at some of the important properties of this number system. To make our discussions here and elsewhere in the book clearer and more precise, we occasionally make use of simple *set* concepts and notation.

Set of Real Numbers

Informally, a **real number** is any number that has a decimal representation. Table 1 describes the set of real numbers and some of its important subsets. Figure 1 illustrates how these sets of numbers are related.

The set of integers contains all the natural numbers and something else—their negatives and 0. The set of rational numbers contains all the integers and something else—noninteger ratios of integers. And the set of real numbers contains all the rational numbers and something else—irrational numbers.

Table 1 Set of Real Numbers

Symbol	Name	Description	Examples
N	Natural numbers	Counting numbers (also called positive integers)	$1, 2, 3, \ldots$
Z	Integers	Natural numbers, their negatives, and 0	$\ldots, -2, -1, 0, 1, 2, \ldots$
Q	Rational numbers	Numbers that can be represented as a/b, where a and b are integers and $b \neq 0$; decimal representations are repeating or terminating	$-4, 0, 1, 25, \frac{-3}{5}, \frac{2}{3}, 3.67, -0.33\overline{3}, 5.272\,7\overline{27}$*
I	Irrational numbers	Numbers that can be represented as nonrepeating and nonterminating decimal numbers	$\sqrt{2}, \pi, \sqrt[3]{7}, 1.414\,213\ldots, 2.718\,281\,82\ldots$
R	Real numbers	Rational and irrational numbers	

*The overbar indicates that the number (or block of numbers) repeats indefinitely. The space after every third digit is used to help keep track of the number of decimal places.

Natural numbers (N)

Zero

Negatives of natural numbers

Integers (Z)

Noninteger ratios of integers

Rational numbers (Q)

Irrational numbers (I)

Real numbers (R)

Figure 1 Real numbers and important subsets

Real Number Line

A one-to-one correspondence exists between the set of real numbers and the set of points on a line. That is, each real number corresponds to exactly one point, and each point corresponds to exactly one real number. A line with a real number associated with each point, and vice versa, as shown in Figure 2, is called a **real number line**, or simply a **real line**. Each number associated with a point is called the coordinate of the point.

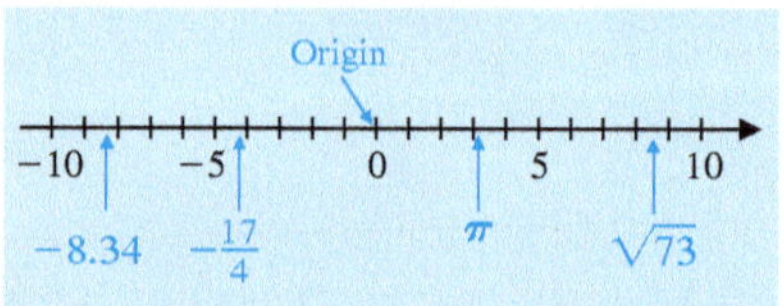

Figure 2 Real number line

The point with coordinate 0 is called the **origin**. The arrow on the right end of the line indicates a positive direction. The coordinates of all points to the right of the origin are called positive **real numbers**, and those to the left of the origin are called **negative real numbers**. The real number 0 is neither positive nor negative.

Basic Real Number Properties

We now take a look at some of the basic properties of the real number system that enable us to convert algebraic expressions into *equivalent forms*.

SUMMARY Basic Properties of the Set of Real Numbers

Let a, b, and c be arbitrary elements in the set of real numbers R.

Addition Properties

Associative: $(a + b) + c = a + (b + c)$

Commutative: $a + b = b + a$

Identity: 0 is the additive identity; that is, $0 + a = a + 0 = a$ for all a in R, and 0 is the only element in R with this property.

Inverse: For each a in R, $-a$, is its unique additive inverse; that is, $a + (-a) = (-a) + a = 0$ and $-a$ is the only element in R relative to a with this property.

Multiplication Properties

Associative: $(ab)c = a(bc)$

Commutative: $ab = ba$

Identity: 1 is the multiplicative identity; that is, $(1)a = a(1) = a$ for all a in R, and 1 is the only element in R with this property.

Inverse: For each a in R, $a \neq 0$, $1/a$ is its unique multiplicative inverse; that is, $a(1/a) = (1/a)a = 1$, and 1/a is the only element in R relative to a with this property.

Distributive Properties

$$a(b + c) = ab + ac \qquad (a + b)c = ac + bc$$

You are already familiar with the **commutative properties** for addition and multiplication. They indicate that the order in which the addition or multiplication of two numbers is performed does not matter. For example,

$$7 + 2 = 2 + 7 \quad \text{and} \quad 3 \cdot 5 = 5 \cdot 3$$

Is there a commutative property relative to subtraction or division? That is, does $a - b = b - a$ or does $a \div b = b \div a$ for all real numbers a and b (division by 0 excluded)? The answer is no, since, for example,

$$8 - 6 \neq 6 - 8 \quad \text{and} \quad 10 \div 5 \neq 5 \div 10$$

When computing

$$3 + 2 + 6 \quad \text{or} \quad 3 \cdot 2 \cdot 6$$

why don't we need parentheses to indicate which two numbers are to be added or multiplied first? The answer is to be found in the **associative properties**. These properties allow us to write

$$(3 + 2) + 6 = 3 + (2 + 6) \quad \text{and} \quad (3 \cdot 2) \cdot 6 = 3 \cdot (2 \cdot 6)$$

so it does not matter how we group numbers relative to either operation. Is there an associative property for subtraction or division? The answer is no, since, for example,

$$(12 - 6) - 2 \neq 12 - (6 - 2) \quad \text{and} \quad (12 \div 6) \div 2 \neq 12 \div (6 \div 2)$$

Evaluate each side of each equation to see why.

What number added to a given number will give that number back again? What number times a given number will give that number back again? The answers are 0 and 1, respectively. Because of this, 0 and 1 are called the **identity elements** for the real numbers. Hence, for any real numbers a and b,

$$0 + 5 = 5 \quad \text{and} \quad (a + b) + 0 = a + b$$
$$1 \cdot 4 = 4 \quad \text{and} \quad (a + b) \cdot 1 = a + b$$

We now consider **inverses**. For each real number a, there is a unique real number $-a$ such that $a + (-a) = 0$. The number $-a$ is called the **additive inverse** of a, or the **negative** of a. For example, the additive inverse (or negative) of 7 is -7, since $7 + (-7) = 0$. The additive inverse (or negative) of -7 is $-(-7) = 7$, since $-7 + [-(-7)] = 0$.

CONCEPTUAL INSIGHT

Do not confuse negation with the sign of a number. If a is a real number, $-a$ is the negative of a and may be positive or negative. Specifically, if a is negative, then $-a$ is positive and if a is positive, then $-a$ is negative.

For each nonzero real number a, there is a unique real number $1/a$ such that $a(1/a) = 1$. The number $1/a$ is called the **multiplicative inverse** of a, or the **reciprocal** of a. For example, the multiplicative inverse (or reciprocal) of 4 is $\frac{1}{4}$, since $4\left(\frac{1}{4}\right) = 1$. (Also note that 4 is the multiplicative inverse of $\frac{1}{4}$.) The number 0 has no multiplicative inverse.

We now turn to the **distributive properties**, which involve both multiplication and addition. Consider the following two computations:

$$5(3 + 4) = 5 \cdot 7 = 35 \qquad 5 \cdot 3 + 5 \cdot 4 = 15 + 20 = 35$$

Thus,

$$5(3 + 4) = 5 \cdot 3 + 5 \cdot 4$$

and we say that multiplication by 5 *distributes* over the sum $(3 + 4)$. In general, **multiplication distributes over addition** in the real number system. Two more illustrations are

$$9(m + n) = 9m + 9n \qquad (7 + 2)u = 7u + 2u$$

EXAMPLE 1 **Real Number Properties** State the real number property that justifies the indicated statement.

Statement	Property Illustrated
(A) $x(y + z) = (y + z)x$	Commutative $(\cdot)$
(B) $5(2y) = (5 \cdot 2)y$	Associative $(\cdot)$
(C) $2 + (y + 7) = 2 + (7 + y)$	Commutative $(+)$
(D) $4z + 6z = (4 + 6)z$	Distributive
(E) If $m + n = 0$, then $n = -m$.	Inverse $(+)$

Matched Problem 1 State the real number property that justifies the indicated statement.

(A) $8 + (3 + y) = (8 + 3) + y$

(B) $(x + y) + z = z + (x + y)$

(C) $(a + b)(x + y) = a(x + y) + b(x + y)$

(D) $5xy + 0 = 5xy$

(E) If $xy = 1$, $x \neq 0$, then $y = 1/x$.

Further Properties

Subtraction and *division* can be defined in terms of addition and multiplication, respectively:

DEFINITION Subtraction and Division

For all real numbers a and b,

Subtraction: $a - b = a + (-b)$ $\quad 7 - (-5) = 7 + [-(-5)] = 7 + 5 = 12$

Division: $a \div b = a\left(\frac{1}{b}\right), b \neq 0$ $\quad 9 \div 4 = 9\left(\frac{1}{4}\right) = \frac{9}{4}$

To subtract b from a, add the negative (the additive inverse) of b to a. To divide a by b, multiply a by the reciprocal (the multiplicative inverse) of b. Note that division by 0 is not defined, since 0 does not have a reciprocal. **0 can never be used as a divisor!**

The following properties of negatives can be proved using the preceding assumed properties and definitions.

THEOREM 1 Negative Properties

For all real numbers a and b,

1. $-(-a) = a$
2. $(-a)b = -(ab) = a(-b) = -ab$
3. $(-a)(-b) = ab$
4. $(-1)a = -a$
5. $\frac{-a}{b} = -\frac{a}{b} = \frac{a}{-b}, b \neq 0$
6. $\frac{-a}{-b} = -\frac{-a}{b} = -\frac{a}{-b} = \frac{a}{b}, b \neq 0$

We now state two important properties involving 0.

THEOREM 2 Zero Properties

For all real numbers a and b,

1. $a \cdot 0 = 0$ $\quad 0 \cdot 0 = 0 \quad (-35)(0) = 0$
2. $ab = 0$ if and only if $a = 0$ or $b = 0$

If $(3x + 2)(x - 7) = 0$, then either $3x + 2 = 0$ or $x - 7 = 0$.

Fraction Properties

Recall that the quotient $a \div b\ (b \neq 0)$ written in the form a/b is called a **fraction**. The quantity a is called the **numerator**, and the quantity b is called the **denominator**.

THEOREM 3 Fraction Properties

For all real numbers a, b, c, d, and k (division by 0 excluded):

1. $\frac{a}{b} = \frac{c}{d}$ if and only if $ad = bc$ $\quad \frac{4}{6} = \frac{6}{9}$ since $4 \cdot 9 = 6 \cdot 6$

2. $\frac{ka}{kb} = \frac{a}{b}$ $\quad \frac{7 \cdot 3}{7 \cdot 5} = \frac{3}{5}$

3. $\frac{a}{b} \cdot \frac{c}{d} = \frac{ac}{bd}$ $\quad \frac{3}{5} \cdot \frac{7}{8} = \frac{3 \cdot 7}{5 \cdot 8}$

4. $\frac{a}{b} \div \frac{c}{d} = \frac{a}{b} \cdot \frac{d}{c}$ $\quad \frac{2}{3} \div \frac{5}{7} = \frac{2}{3} \cdot \frac{7}{5}$

5. $\frac{a}{b} + \frac{c}{b} = \frac{a + c}{b}$ $\quad \frac{3}{6} + \frac{5}{6} = \frac{3 + 5}{6}$

6. $\frac{a}{b} - \frac{c}{b} = \frac{a - c}{b}$ $\quad \frac{7}{8} - \frac{3}{8} = \frac{7 - 3}{8}$

7. $\frac{a}{b} + \frac{c}{d} = \frac{ad + bc}{bd}$ $\quad \frac{2}{3} + \frac{3}{5} = \frac{2 \cdot 5 + 3 \cdot 3}{3 \cdot 5}$

Exercises A-1

All variables represent real numbers.

A

In Problems 1–6, replace each question mark with an appropriate expression that will illustrate the use of the indicated real number property.

1. Commutative property $(\cdot)$: $uv = ?$
2. Commutative property $(+)$: $x + 7 = ?$
3. Associative property $(+)$: $3 + (7 + y) = ?$
4. Associative property $(\cdot)$: $x(yz) = ?$
5. Identity property $(\cdot)$: $1(u + v) = ?$
6. Identity property $(+)$: $0 + 9m = ?$

In Problems 7–26, indicate true (T) or false (F).

7. $5(8m) = (5 \cdot 8)m$
8. $a + cb = a + bc$
9. $5x + 7x = (5 + 7)x$
10. $uv(w + x) = uvw + uvx$
11. $-2(-a)(2x - y) = 2a(-4x + y)$
12. $8 \div (-5) = 8\left(\frac{1}{-5}\right)$
13. $(x + 3) + 2x = 2x + (x + 3)$
14. $\frac{x}{3y} \div \frac{5y}{x} = \frac{15y^2}{x^2}$
15. $\frac{2x}{-(x + 3)} = -\frac{2x}{x + 3}$
16. $-\frac{2x}{-(x - 3)} = \frac{2x}{x - 3}$
17. $(-3)\left(\frac{1}{-3}\right) = 1$
18. $(-0.5) + (0.5) = 0$
19. $-x^2y^2 = (-1)x^2y^2$
20. $[-(x + 2)](-x) = (x + 2)x$
21. $\frac{a}{b} + \frac{c}{d} = \frac{a + c}{b + d}$
22. $\frac{k}{k + b} = \frac{1}{1 + b}$
23. $(x + 8)(x + 6) = (x + 8)x + (x + 8)6$
24. $u(u - 2v) + v(u - 2v) = (u + v)(u - 2v)$
25. If $(x - 2)(2x + 3) = 0$, then either $x - 2 = 0$ or $2x + 3 = 0$.
26. If either $x - 2 = 0$ or $2x + 3 = 0$, then $(x - 2)(2x + 3) = 0$.

B

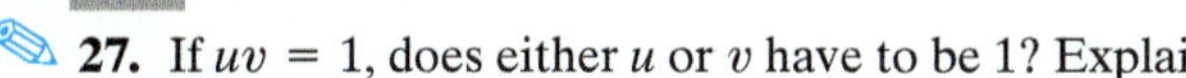

27. If $uv = 1$, does either u or v have to be 1? Explain.

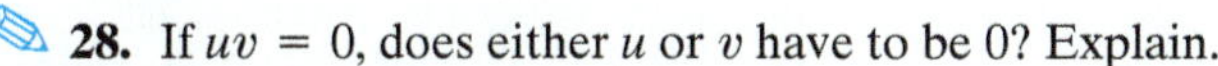

28. If $uv = 0$, does either u or v have to be 0? Explain.
29. Indicate whether the following are true (T) or false (F):
    (A) All integers are natural numbers.
    (B) All rational numbers are real numbers.
    (C) All natural numbers are rational numbers.
30. Indicate whether the following are true (T) or false (F):
    (A) All natural numbers are integers.
    (B) All real numbers are irrational.
    (C) All rational numbers are real numbers.

31. Give an example of a real number that is not a rational number.

32. Give an example of a rational number that is not an integer.

33. Given the sets of numbers N (natural numbers), Z (integers), Q (rational numbers), and R (real numbers), indicate to which set(s) each of the following numbers belongs:

(A) 8 (B) $\sqrt{2}$ (C) -1.414 (D) $\frac{-5}{2}$

34. Given the sets of numbers N, Z, Q, and R (see Problem 33), indicate to which set(s) each of the following numbers belongs:

(A) -3 (B) 3.14 (C) π (D) $\frac{2}{3}$

35. Indicate true (T) or false (F), and for each false statement find real number replacements for a, b, and c that will provide a counterexample. For all real numbers a, b, and c,

(A) $a(b - c) = ab - c$
(B) $(a - b) - c = a - (b - c)$
(C) $a(bc) = (ab)c$
(D) $(a \div b) \div c = a \div (b \div c)$

36. Indicate true (T) or false (F), and for each false statement find real number replacements for a and b that will provide a counterexample. For all real numbers a and b,

(A) $a + b = b + a$
(B) $a - b = b - a$
(C) $ab = ba$
(D) $a \div b = b \div a$

C

37. If $c = 0.151\,515\ldots$, then $100c = 15.151\,5\ldots$ and

$$100c - c = 15.151\,5\ldots - 0.151\,515\ldots$$
$$99c = 15$$
$$c = \tfrac{15}{99} = \tfrac{5}{33}$$

Proceeding similarly, convert the repeating decimal $0.090\,909\ldots$ into a fraction. (All repeating decimals are rational numbers, and all rational numbers have repeating decimal representations.)

38. Repeat Problem 37 for $0.181\,818\ldots$.

Use a calculator to express each number in Problems 39 and 40 as a decimal to the capacity of your calculator. Observe the repeating decimal representation of the rational numbers and the nonrepeating decimal representation of the irrational numbers.

39. (A) $\frac{13}{6}$ (B) $\sqrt{21}$ (C) $\frac{7}{16}$ (D) $\frac{29}{111}$

40. (A) $\frac{8}{9}$ (B) $\frac{3}{11}$ (C) $\sqrt{5}$ (D) $\frac{11}{8}$

Answers to Matched Problems

1. (A) Associative (+)
(B) Commutative (+)
(C) Distributive
(D) Identity (+)
(E) Inverse ($\cdot$)

A-2 Operations on Polynomials

- Natural Number Exponents
- Polynomials
- Combining Like Terms
- Addition and Subtraction
- Multiplication
- Combined Operations

This section covers basic operations on *polynomials*. Our discussion starts with a brief review of natural number exponents. Integer and rational exponents and their properties will be discussed in detail in subsequent sections. (Natural numbers, integers, and rational numbers are important parts of the real number system; see Table 1 and Figure 1 in Appendix A-1.)

Natural Number Exponents

We define a **natural number exponent** as follows:

DEFINITION Natural Number Exponent

For n a natural number and b any real number,

$$b^n = b \cdot b \cdot \cdots \cdot b \qquad n \text{ factors of } b$$
$$3^5 = 3 \cdot 3 \cdot 3 \cdot 3 \cdot 3 \qquad 5 \text{ factors of } 3$$

where n is called the exponent and b is called the **base**.

Along with this definition, we state the **first property of exponents**:

THEOREM 1 First Property of Exponents

For any natural numbers m and n, and any real number b:

$$b^m b^n = b^{m+n} \quad (2t^4)(5t^3) = 2 \cdot 5t^{4+3} = 10t^7$$

Polynomials

Algebraic expressions are formed by using constants and variables and the algebraic operations of addition, subtraction, multiplication, division, raising to powers, and taking roots. Special types of algebraic expressions are called *polynomials.* A **polynomial in one variable** x is constructed by adding or subtracting constants and terms of the form ax^n, where a is a real number and n is a natural number. A **polynomial in two variables** x and y is constructed by adding and subtracting constants and terms of the form $ax^m y^n$, where a is a real number and m and n are natural numbers. Polynomials in three and more variables are defined in a similar manner.

Polynomials		Not Polynomials	
8	0	$\frac{1}{x}$	$\frac{x - y}{x^2 + y^2}$
$3x^3 - 6x + 7$	$6x + 3$		
$2x^2 - 7xy - 8y^2$	$9y^3 + 4y^2 - y + 4$	$\sqrt{x^3 - 2x}$	$2x^{-2} - 3x^{-1}$
$2x - 3y + 2$	$u^5 - 3u^3v^2 + 2uv^4 - v^4$		

Polynomial forms are encountered frequently in mathematics. For the efficient study of polynomials, it is useful to classify them according to their *degree.* If a term in a polynomial has only one variable as a factor, then the **degree of the term** is the power of the variable. If two or more variables are present in a term as factors, then the **degree of the term** is the sum of the powers of the variables. The **degree of a polynomial** is the degree of the nonzero term with the highest degree in the polynomial. Any nonzero constant is defined to be a **polynomial of degree 0.** The number 0 is also a polynomial but is not assigned a degree.

EXAMPLE 1 **Degree**

(A) The degree of the first term in $5x^3 + \sqrt{3}x - \frac{1}{2}$ is 3, the degree of the second term is 1, the degree of the third term is 0, and the degree of the whole polynomial is 3 (the same as the degree of the term with the highest degree).

(B) The degree of the first term in $8u^3v^2 - \sqrt{7}uv^2$ is 5, the degree of the second term is 3, and the degree of the whole polynomial is 5.

Matched Problem 1

(A) Given the polynomial $6x^5 + 7x^3 - 2$, what is the degree of the first term? The second term? The third term? The whole polynomial?

(B) Given the polynomial $2u^4v^2 - 5uv^3$, what is the degree of the first term? The second term? The whole polynomial?

In addition to classifying polynomials by degree, we also call a single-term polynomial a **monomial**, a two-term polynomial a **binomial**, and a three-term polynomial a **trinomial**.

Combining Like Terms

The concept of *coefficient* plays a central role in the process of combining *like terms.* A constant in a term of a polynomial, including the sign that precedes it, is called the **numerical coefficient**, or simply, the **coefficient**, of the term. If a constant does not appear, or only a + sign appears, the coefficient is understood to be 1. If only a − sign appears, the coefficient is understood to be −1. Given the polynomial

$$5x^4 - x^3 - 3x^2 + x - 7 \quad = 5x^4 + (-1)x^3 + (-3)x^2 + 1x + (-7)$$

the coefficient of the first term is 5, the coefficient of the second term is −1, the coefficient of the third term is −3, the coefficient of the fourth term is 1, and the coefficient of the fifth term is −7.

The following distributive properties are fundamental to the process of combining *like terms.*

THEOREM 2 Distributive Properties of Real Numbers

1. $a(b + c) = (b + c)a = ab + ac$
2. $a(b - c) = (b - c)a = ab - ac$
3. $a(b + c + \cdots + f) = ab + ac + \cdots + af$

Two terms in a polynomial are called **like terms** if they have exactly the same variable factors to the same powers. The numerical coefficients may or may not be the same. Since constant terms involve no variables, all constant terms are like terms. If a polynomial contains two or more like terms, these terms can be combined into a single term by making use of distributive properties. The following example illustrates the reasoning behind the process:

$$\begin{aligned} 3x^2y - 5xy^2 + x^2y - 2x^2y &= 3x^2y + x^2y - 2x^2y - 5xy^2 \\ &= (3x^2y + 1x^2y - 2x^2y) - 5xy^2 \\ &= (3 + 1 - 2)x^2y - 5xy^2 \\ &= 2x^2y - 5xy^2 \end{aligned}$$

Note the use of distributive properties.

Free use is made of the real number properties discussed in Appendix A-1.

How can we simplify expressions such as $4(x - 2y) - 3(2x - 7y)$? We clear the expression of parentheses using distributive properties, and combine like terms:

$$\begin{aligned} 4(x - 2y) - 3(2x - 7y) &= 4x - 8y - 6x + 21y \\ &= -2x + 13y \end{aligned}$$

EXAMPLE 2 **Removing Parentheses** Remove parentheses and simplify:

(A) $$\begin{aligned} 2(3x^2 - 2x + 5) + (x^2 + 3x - 7) &= 2(3x^2 - 2x + 5) + 1(x^2 + 3x - 7) \\ &= 6x^2 - 4x + 10 + x^2 + 3x - 7 \\ &= 7x^2 - x + 3 \end{aligned}$$

(B) $(x^3 - 2x - 6) - (2x^3 - x^2 + 2x - 3)$

$$\begin{aligned} &= 1(x^3 - 2x - 6) + (-1)(2x^3 - x^2 + 2x - 3) \\ &= x^3 - 2x - 6 - 2x^3 + x^2 - 2x + 3 \\ &= -x^3 + x^2 - 4x - 3 \end{aligned}$$

Be careful with the sign here

(C) $$\begin{aligned} [3x^2 - (2x + 1)] - (x^2 - 1) &= [3x^2 - 2x - 1] - (x^2 - 1) \\ &= 3x^2 - 2x - 1 - x^2 + 1 \\ &= 2x^2 - 2x \end{aligned}$$

Remove inner parentheses first.

Matched Problem 2 Remove parentheses and simplify:

(A) $3(u^2 - 2v^2) + (u^2 + 5v^2)$

(B) $(m^3 - 3m^2 + m - 1) - (2m^3 - m + 3)$

(C) $(x^3 - 2) - [2x^3 - (3x + 4)]$

Addition and Subtraction

Addition and subtraction of polynomials can be thought of in terms of removing parentheses and combining like terms, as illustrated in Example 2. Horizontal and vertical arrangements are illustrated in the next two examples. You should be able to work either way, letting the situation dictate your choice.

EXAMPLE 3 **Adding Polynomials** Add horizontally and vertically:

$$x^4 - 3x^3 + x^2, \quad -x^3 - 2x^2 + 3x, \quad \text{and} \quad 3x^2 - 4x - 5$$

SOLUTION Add horizontally:

$$\begin{aligned}(x^4 - 3x^3 + x^2) &+ (-x^3 - 2x^2 + 3x) + (3x^2 - 4x - 5)\\ &= x^4 - 3x^3 + x^2 - x^3 - 2x^2 + 3x + 3x^2 - 4x - 5\\ &= x^4 - 4x^3 + 2x^2 - x - 5\end{aligned}$$

Or vertically, by lining up like terms and adding their coefficients:

$$\begin{array}{rrrrr} x^4 & -\ 3x^3 & +\ x^2 & & \\ & -\ x^3 & -\ 2x^2 & +\ 3x & \\ & & 3x^2 & -\ 4x & -\ 5 \\ \hline x^4 & -\ 4x^3 & +\ 2x^2 & -\ x & -\ 5 \end{array}$$

Matched Problem 3 Add horizontally and vertically:

$$3x^4 - 2x^3 - 4x^2, \quad x^3 - 2x^2 - 5x, \quad \text{and} \quad x^2 + 7x - 2$$

EXAMPLE 4 **Subtracting Polynomials** Subtract $4x^2 - 3x + 5$ from $x^2 - 8$, both horizontally and vertically.

SOLUTION

$$\begin{aligned}(x^2 - 8) &- (4x^2 - 3x + 5)\\ &= x^2 - 8 - 4x^2 + 3x - 5\\ &= -3x^2 + 3x - 13\end{aligned}$$

or

$$\begin{array}{rrr} x^2 & & -\ 8 \\ -4x^2 & +\ 3x & -\ 5 \\ \hline -3x^2 & +\ 3x & -\ 13 \end{array}$$

← Change signs and add.

Matched Problem 4 Subtract $2x^2 - 5x + 4$ from $5x^2 - 6$, both horizontally and vertically.

Multiplication

Multiplication of algebraic expressions involves the extensive use of distributive properties for real numbers, as well as other real number properties.

EXAMPLE 5 **Multiplying Polynomials** Multiply: $(2x - 3)(3x^2 - 2x + 3)$

SOLUTION

$$\begin{aligned}(2x - 3)(3x^2 - 2x + 3) &= 2x(3x^2 - 2x + 3) - 3(3x^2 - 2x + 3)\\ &= 6x^3 - 4x^2 + 6x - 9x^2 + 6x - 9\\ &= 6x^3 - 13x^2 + 12x - 9\end{aligned}$$

Or, using a vertical arrangement,

$$\begin{array}{llll} 3x^2 & -\ 2x & +\ 3 & \\ 2x & -\ 3 & & \\ \hline 6x^3 & -\ 4x^2 & +\ 6x & \\ & -\ 9x^2 & +\ 6x & -\ 9 \\ \hline 6x^3 & -\ 13x^2 & +\ 12x & -\ 9 \end{array}$$

Matched Problem 5 Multiply: $(2x - 3)(2x^2 + 3x - 2)$

Thus, to multiply two polynomials, multiply each term of one by each term of the other, and combine like terms.

Products of binomial factors occur frequently, so it is useful to develop procedures that will enable us to write down their products by inspection. To find the product $(2x - 1)(3x + 2)$, we proceed as follows:

$$(2x - 1)(3x + 2) = 6x^2 + 4x - 3x - 2$$
$$= 6x^2 + x - 2$$

The inner and outer products are like terms, so combine into a single term.

To speed the process, we do the step in the dashed box mentally.

Products of certain binomial factors occur so frequently that it is useful to learn formulas for their products. The following formulas are easily verified by multiplying the factors on the left.

THEOREM 3 Special Products

1. $(a - b)(a + b) = a^2 - b^2$
2. $(a + b)^2 = a^2 + 2ab + b^2$
3. $(a - b)^2 = a^2 - 2ab + b^2$

EXAMPLE 6 **Special Products** Multiply mentally, where possible.

(A) $(2x - 3y)(5x + 2y)$ (B) $(3a - 2b)(3a + 2b)$
(C) $(5x - 3)^2$ (D) $(m + 2n)^3$

SOLUTION

(A) $(2x - 3y)(5x + 2y) = 10x^2 + 4xy - 15xy - 6y^2$
$= 10x^2 - 11xy - 6y^2$

(B) $(3a - 2b)(3a + 2b) = (3a)^2 - (2b)^2$
$= 9a^2 - 4b^2$

(C) $(5x - 3)^2 = (5x)^2 - 2(5x)(3) + 3^2$
$= 25x^2 - 30x + 9$

(D) $(m + 2n)^3 = (m + 2n)^2(m + 2n)$
$= (m^2 + 4mn + 4n^2)(m + 2n)$
$= m^2(m + 2n) + 4mn(m + 2n) + 4n^2(m + 2n)$
$= m^3 + 2m^2n + 4m^2n + 8mn^2 + 4mn^2 + 8n^3$
$= m^3 + 6m^2n + 12mn^2 + 8n^3$

Matched Problem 6 Multiply mentally, where possible.

(A) $(4u - 3v)(2u + v)$
(B) $(2xy + 3)(2xy - 3)$
(C) $(m + 4n)(m - 4n)$
(D) $(2u - 3v)^2$
(E) $(2x - y)^3$

Combined Operations

We complete this section by considering several examples that use all the operations just discussed. Note that in simplifying, we usually remove grouping symbols starting from the inside. That is, we remove parentheses ( ) first, then brackets [ ], and finally braces { }, if present. Also,

DEFINITION Order of Operations

Multiplication and division precede addition and subtraction, and taking powers precedes multiplication and division.

$$2 \cdot 3 + 4 = 6 + 4 = 10, \quad \text{not} \quad 2 \cdot 7 = 14$$

$$\frac{10^2}{2} = \frac{100}{2} = 50, \quad \text{not} \quad 5^2 = 25$$

EXAMPLE 7 **Combined Operations** Perform the indicated operations and simplify:

(A) $3x - \{5 - 3[x - x(3 - x)]\} = 3x - \{5 - 3[x - 3x + x^2]\}$
$= 3x - \{5 - 3x + 9x - 3x^2\}$
$= 3x - 5 + 3x - 9x + 3x^2$
$= 3x^2 - 3x - 5$

(B) $(x - 2y)(2x + 3y) - (2x + y)^2 = 2x^2 - xy - 6y^2 - (4x^2 + 4xy + y^2)$
$= 2x^2 - xy - 6y^2 - 4x^2 - 4xy - y^2$
$= -2x^2 - 5xy - 7y^2$

Matched Problem 7 Perform the indicated operations and simplify:

(A) $2t - \{7 - 2[t - t(4 + t)]\}$

(B) $(u - 3v)^2 - (2u - v)(2u + v)$

Exercises A-2

A

Problems 1–8 refer to the following polynomials:

(A) $2x - 3$ (B) $2x^2 - x + 2$ (C) $x^3 + 2x^2 - x + 3$

1. What is the degree of (C)?
2. What is the degree of (A)?
3. Add (B) and (C).
4. Add (A) and (B).
5. Subtract (B) from (C).
6. Subtract (A) from (B).
7. Multiply (B) and (C).
8. Multiply (A) and (C).

In Problems 9–30, perform the indicated operations and simplify.

9. $2(u - 1) - (3u + 2) - 2(2u - 3)$
10. $2(x - 1) + 3(2x - 3) - (4x - 5)$
11. $4a - 2a[5 - 3(a + 2)]$
12. $2y - 3y[4 - 2(y - 1)]$
13. $(a + b)(a - b)$
14. $(m - n)(m + n)$
15. $(3x - 5)(2x + 1)$
16. $(4t - 3)(t - 2)$
17. $(2x - 3y)(x + 2y)$
18. $(3x + 2y)(x - 3y)$
19. $(3y + 2)(3y - 2)$
20. $(2m - 7)(2m + 7)$
21. $-(2x - 3)^2$
22. $-(5 - 3x)^2$
23. $(4m + 3n)(4m - 3n)$
24. $(3x - 2y)(3x + 2y)$
25. $(3u + 4v)^2$
26. $(4x - y)^2$
27. $(a - b)(a^2 + ab + b^2)$
28. $(a + b)(a^2 - ab + b^2)$
29. $[(x - y) + 3z][(x - y) - 3z]$
30. $[a - (2b - c)][a + (2b - c)]$

B

In Problems 31–44, perform the indicated operations and simplify.

31. $m - \{m - [m - (m - 1)]\}$
32. $2x - 3\{x + 2[x - (x + 5)] + 1\}$
33. $(x^2 - 2xy + y^2)(x^2 + 2xy + y^2)$
34. $(3x - 2y)^2(2x + 5y)$
35. $(5a - 2b)^2 - (2b + 5a)^2$
36. $(2x - 1)^2 - (3x + 2)(3x - 2)$

37. $(m - 2)^2 - (m - 2)(m + 2)$

38. $(x - 3)(x + 3) - (x - 3)^2$

39. $(x - 2y)(2x + y) - (x + 2y)(2x - y)$

40. $(3m + n)(m - 3n) - (m + 3n)(3m - n)$

41. $(u + v)^3$

42. $(x - y)^3$

43. $(x - 2y)^3$

44. $(2m - n)^3$

45. Subtract the sum of the last two polynomials from the sum of the first two: $2x^2 - 4xy + y^2$, $3xy - y^2$, $x^2 - 2xy - y^2$, $-x^2 + 3xy - 2y^2$

46. Subtract the sum of the first two polynomials from the sum of the last two: $3m^2 - 2m + 5$, $4m^2 - m$, $3m^2 - 3m - 2$, $m^3 + m^2 + 2$

C

In Problems 47–50, perform the indicated operations and simplify.

47. $[(2x - 1)^2 - x(3x + 1)]^2$

48. $[5x(3x + 1) - 5(2x - 1)^2]^2$

49. $2\{(x - 3)(x^2 - 2x + 1) - x[3 - x(x - 2)]\}$

50. $-3x\{x[x - x(2 - x)] - (x + 2)(x^2 - 3)\}$

51. If you are given two polynomials, one of degree m and the other of degree n, where m is greater than n, what is the degree of their product?

52. What is the degree of the sum of the two polynomials in Problem 51?

53. How does the answer to Problem 51 change if the two polynomials can have the same degree?

54. How does the answer to Problem 52 change if the two polynomials can have the same degree?

55. Show by example that, in general, $(a + b)^2 \neq a^2 + b^2$. Discuss possible conditions on a and b that would make this a valid equation.

56. Show by example that, in general, $(a - b)^2 \neq a^2 - b^2$. Discuss possible conditions on a and b that would make this a valid equation.

Applications

57. Investment. You have \$10,000 to invest, part at 9% and the rest at 12%. If x is the amount invested at 9%, write an algebraic expression that represents the total annual income from both investments. Simplify the expression.

58. Investment. A person has \$100,000 to invest. If \$$x$ are invested in a money market account yielding 7% and twice that amount in certificates of deposit yielding 9%, and if the rest is invested in high-grade bonds yielding 11%, write an algebraic expression that represents the total annual income from all three investments. Simplify the expression.

59. Gross receipts. Four thousand tickets are to be sold for a musical show. If x tickets are to be sold for \$20 each and three times that number for \$30 each, and if the rest are sold for \$50 each, write an algebraic expression that represents the gross receipts from ticket sales, assuming all tickets are sold. Simplify the expression.

60. Gross receipts. Six thousand tickets are to be sold for a concert, some for \$20 each and the rest for \$35 each. If x is the number of \$20 tickets sold, write an algebraic expression that represents the gross receipts from ticket sales, assuming all tickets are sold. Simplify the expression.

61. Nutrition. Food mix A contains 2% fat, and food mix B contains 6% fat. A 10-kilogram diet mix of foods A and B is formed. If x kilograms of food A are used, write an algebraic expression that represents the total number of kilograms of fat in the final food mix. Simplify the expression.

62. Nutrition. Each ounce of food M contains 8 units of calcium, and each ounce of food N contains 5 units of calcium. A 160-ounce diet mix is formed using foods M and N. If x is the number of ounces of food M used, write an algebraic expression that represents the total number of units of calcium in the diet mix. Simplify the expression.

Answers to Matched Problems

1. (A) 5, 3, 0, 5 (B) 6, 4, 6

2. (A) $4u^2 - v^2$
(B) $-m^3 - 3m^2 + 2m - 4$
(C) $-x^3 + 3x + 2$

3. $3x^4 - x^3 - 5x^2 + 2x - 2$

4. $3x^2 + 5x - 10$

5. $4x^3 - 13x + 6$

6. (A) $8u^2 - 2uv - 3v^2$
(B) $4x^2y^2 - 9$
(C) $m^2 - 16n^2$
(D) $4u^2 - 12uv + 9v^2$
(E) $8x^3 - 12x^2y + 6xy^2 - y^3$

7. (A) $-2t^2 - 4t - 7$ (B) $-3u^2 - 6uv + 10v^2$

A-3 Factoring Polynomials

- Common Factors
- Factoring by Grouping
- Factoring Second-Degree Polynomials
- Special Factoring Formulas
- Combined Factoring Techniques

A polynomial is written in factored form if it is written as the product of two or more polynomials. The following polynomials are written in factored form:

$$4x^2y - 6xy^2 = 2xy(2x - 3y) \qquad 2x^3 - 8x = 2x(x - 2)(x + 2)$$
$$x^2 - x - 6 = (x - 3)(x + 2) \qquad 5m^2 + 20 = 5(m^2 + 4)$$

Unless stated to the contrary, we will limit our discussion of factoring polynomials to polynomials with integer coefficients.

A polynomial with integer coefficients is said to be **factored completely** if each factor cannot be expressed as the product of two or more polynomials with integer coefficients, other than itself or 1. All the polynomials above, as we will see by the conclusion of this section, are factored completely.

Writing polynomials in completely factored form is often a difficult task. But accomplishing it can lead to the simplification of certain algebraic expressions and to the solution of certain types of equations and inequalities. The distributive properties for real numbers are central to the factoring process.

Common Factors

Generally, a first step in any factoring procedure is to factor out all factors common to all terms.

EXAMPLE 1 **Common Factors** Factor out all factors common to all terms.

(A) $3x^3y - 6x^2y^2 - 3xy^3$

(B) $3y(2y + 5) + 2(2y + 5)$

SOLUTION (A) $3x^3y - 6x^2y^2 - 3xy^3 = (3xy)x^2 - (3xy)2xy - (3xy)y^2$
$= 3xy(x^2 - 2xy - y^2)$

(B) $3y(2y + 5) + 2(2y + 5) = 3y(2y + 5) + 2(2y + 5)$
$= (3y + 2)(2y + 5)$

Matched Problem 1 Factor out all factors common to all terms.

(A) $2x^3y - 8x^2y^2 - 6xy^3$ (B) $2x(3x - 2) - 7(3x - 2)$

Factoring by Grouping

Occasionally, polynomials can be factored by grouping terms in such a way that we obtain results that look like Example 1B. We can then complete the factoring following the steps used in that example. This process will prove useful in the next subsection, where an efficient method is developed for factoring a second-degree polynomial as the product of two first-degree polynomials, if such factors exist.

EXAMPLE 2 **Factoring by Grouping** Factor by grouping.

(A) $3x^2 - 3x - x + 1$

(B) $4x^2 - 2xy - 6xy + 3y^2$

(C) $y^2 + xz + xy + yz$

SOLUTION (A) $3x^2 - 3x - x + 1$
$= (3x^2 - 3x) - (x - 1)$
$= 3x(x - 1) - (x - 1)$
$= (x - 1)(3x - 1)$

Group the first two and the last two terms. Factor out any common factors from each group. The common factor $(x - 1)$ can be taken out, and the factoring is complete.

(B) $4x^2 - 2xy - 6xy + 3y^2 = (4x^2 - 2xy) - (6xy - 3y^2)$
$= 2x(2x - y) - 3y(2x - y)$
$= (2x - y)(2x - 3y)$

(C) If, as in parts (A) and (B), we group the first two terms and the last two terms of $y^2 + xz + xy + yz$, no common factor can be taken out of each group to complete the factoring. However, if the two middle terms are reversed, we can proceed as before:

$$\begin{aligned} y^2 + xz + xy + yz &= y^2 + xy + xz + yz \\ &= (y^2 + xy) + (xz + yz) \\ &= y(y + x) + z(x + y) \\ &= y(x + y) + z(x + y) \\ &= (x + y)(y + z) \end{aligned}$$

Matched Problem 2 Factor by grouping.

(A) $6x^2 + 2x + 9x + 3$

(B) $2u^2 + 6uv - 3uv - 9v^2$

(C) $ac + bd + bc + ad$

Factoring Second-Degree Polynomials

We now turn our attention to factoring second-degree polynomials of the form

$$2x^2 - 5x - 3 \quad \text{and} \quad 2x^2 + 3xy - 2y^2$$

into the product of two first-degree polynomials with integer coefficients. Since many second-degree polynomials with integer coefficients cannot be factored in this way, it would be useful to know ahead of time that the factors we are seeking actually exist. The factoring approach we use, involving the *ac test*, determines at the beginning whether first-degree factors with integer coefficients do exist. Then, if they exist, the test provides a simple method for finding them.

THEOREM 1 ***ac* Test for Factorability**

If in polynomials of the form

$$ax^2 + bx + c \quad \text{or} \quad ax^2 + bxy + cy^2 \tag{1}$$

the product ac has two integer factors p and q whose sum is the coefficient b of the middle term; that is, if integers p and q exist so that

$$pq = ac \quad \text{and} \quad p + q = b \tag{2}$$

then the polynomials have first-degree factors with integer coefficients. If no integers p and q exist that satisfy equations (2), then the polynomials in equations (1) will not have first-degree factors with integer coefficients.

If integers p and q exist that satisfy equations (2) in the ac test, the factoring always can be completed as follows: Using $b = p + q$, split the middle terms in equations (1) to obtain

$$ax^2 + bx + c = ax^2 + px + qx + c$$
$$ax^2 + bxy + cy^2 = ax^2 + pxy + qxy + cy^2$$

Complete the factoring by grouping the first two terms and the last two terms as in Example 2. This process always works, and it does not matter if the two middle terms on the right are interchanged.

Several examples should make the process clear. After a little practice, you will perform many of the steps mentally and will find the process fast and efficient.

EXAMPLE 3 **Factoring Second-Degree Polynomials** Factor, if possible, using integer coefficients.

(A) $4x^2 - 4x - 3$

(B) $2x^2 - 3x - 4$

(C) $6x^2 - 25xy + 4y^2$

SOLUTION (A) $4x^2 - 4x - 3$

Step 1 Use the ac test to test for factorability. Comparing $4x^2 - 4x - 3$ with $ax^2 + bx + c$, we see that $a = 4, b = -4$, and $c = -3$. Multiply a and c to obtain

$$ac = (4)(-3) = -12$$

pq
(1)(−12)
(−1)(12)
(2)(−6)
(−2)(6)
(3)(−4)
(−3)(4)

All factor pairs of $-12 = ac$

List all pairs of integers whose product is −12, as shown in the margin. These are called **factor pairs** of −12. Then try to find a factor pair that sums to $b = -4$, the coefficient of the middle term in $4x^2 - 4x - 3$. (In practice, this part of Step 1 is often done mentally and can be done rather quickly.) Notice that the factor pair 2 and −6 sums to −4. By the ac test, $4x^2 - 4x - 3$ has first-degree factors with integer coefficients.

Step 2 Split the middle term, using $b = p + q$, and complete the factoring by grouping. Using $-4 = 2 + (-6)$, we split the middle term in $4x^2 - 4x - 3$ and complete the factoring by grouping:

$$\begin{aligned} 4x^2 - 4x - 3 &= 4x^2 + 2x - 6x - 3 \\ &= (4x^2 + 2x) - (6x + 3) \\ &= 2x(2x + 1) - 3(2x + 1) \\ &= (2x + 1)(2x - 3) \end{aligned}$$

The result can be checked by multiplying the two factors to obtain the original polynomial.

(B) $2x^2 - 3x - 4$

Step 1 Use the ac test to test for factorability:

$$ac = (2)(-4) = -8$$

pq
(−1)(8)
(1)(−8)
(−2)(4)
(2)(−4)

All factor pairs of $-8 = ac$

Does −8 have a factor pair whose sum is −3? None of the factor pairs listed in the margin sums to $-3 = b$, the coefficient of the middle term in $2x^2 - 3x - 4$. According to the ac test, we can conclude that $2x^2 - 3x - 4$ does not have first-degree factors with integer coefficients, and we say that the polynomial is **not factorable**.

(C) $6x^2 - 25xy + 4y^2$

Step 1 Use the ac test to test for factorability:

$$ac = (6)(4) = 24$$

Mentally checking through the factor pairs of 24, keeping in mind that their sum must be $-25 = b$, we see that if $p = -1$ and $q = -24$, then

$$pq = (-1)(-24) = 24 = ac$$

and

$$p + q = (-1) + (-24) = -25 = b$$

So the polynomial is factorable.

Step 2 Split the middle term, using $b = p + q$, and complete the factoring by grouping. Using $-25 = (-1) + (-24)$, we split the middle term in $6x^2 - 25xy + 4y^2$ and complete the factoring by grouping:

$$\begin{aligned} 6x^2 - 25xy + 4y^2 &= 6x^2 - xy - 24xy + 4y^2 \\ &= (6x^2 - xy) - (24xy - 4y^2) \\ &= x(6x - y) - 4y(6x - y) \\ &= (6x - y)(x - 4y) \end{aligned}$$

The check is left to the reader.

Matched Problem 3 Factor, if possible, using integer coefficients.

(A) $2x^2 + 11x - 6$ (B) $4x^2 + 11x - 6$ (C) $6x^2 + 5xy - 4y^2$

Special Factoring Formulas

The factoring formulas listed in the following box will enable us to factor certain polynomial forms that occur frequently. These formulas can be established by multiplying the factors on the right.

THEOREM 2 Special Factoring Formulas

Perfect square:	1. $u^2 + 2uv + v^2 = (u + v)^2$
Perfect square:	2. $u^2 - 2uv + v^2 = (u - v)^2$
Difference of squares:	3. $u^2 - v^2 = (u - v)(u + v)$
Difference of cubes:	4. $u^3 - v^3 = (u - v)(u^2 + uv + v^2)$
Sum of cubes:	5. $u^3 + v^3 = (u + v)(u^2 - uv + v^2)$

CAUTION Notice that $u^2 + v^2$ is not included in the list of special factoring formulas. In fact,

$$u^2 + v^2 \neq (au + bv)(cu + dv)$$

for any choice of real number coefficients a, b, c, and d.

EXAMPLE 4 **Factoring** Factor completely.

(A) $4m^2 - 12mn + 9n^2$ (B) $x^2 - 16y^2$ (C) $z^3 - 1$

(D) $m^3 + n^3$ (E) $a^2 - 4(b + 2)^2$

SOLUTION

(A) $4m^2 - 12mn + 9n^2 = (2m - 3n)^2$

(B) $x^2 - 16y^2 = x^2 - (4y)^2 = (x - 4y)(x + 4y)$

(C) $z^3 - 1 = (z - 1)(z^2 + z + 1)$ Use the ac test to verify that $z^2 + z + 1$ cannot be factored.

(D) $m^3 + n^3 = (m + n)(m^2 - mn + n^2)$ Use the ac test to verify that $m^2 - mn + n^2$ cannot be factored.

(E) $a^2 - 4(b + 2)^2 = [a - 2(b + 2)][a + 2(b + 2)]$

Matched Problem 4 Factor completely:

(A) $x^2 + 6xy + 9y^2$ (B) $9x^2 - 4y^2$ (C) $8m^3 - 1$

(D) $x^3 + y^3z^3$ (E) $9(m - 3)^2 - 4n^2$

Combined Factoring Techniques

We complete this section by considering several factoring problems that involve combinations of the preceding techniques.

PROCEDURE Factoring Polynomials

Step 1 Take out any factors common to all terms.

Step 2 Use any of the special formulas listed in Theorem 2 that are applicable.

Step 3 Apply the *ac* test to any remaining second-degree polynomial factors.

Note: It may be necessary to perform some of these steps more than once. Furthermore, the order of applying these steps can vary.

EXAMPLE 5 **Combined Factoring Techniques** Factor completely.

(A) $3x^3 - 48x$ (B) $3u^4 - 3u^3v - 9u^2v^2$

(C) $3m^4 - 24mn^3$ (D) $3x^4 - 5x^2 + 2$

SOLUTION

(A) $3x^3 - 48x = 3x(x^2 - 16) = 3x(x - 4)(x + 4)$

(B) $3u^4 - 3u^3v - 9u^2v^2 = 3u^2(u^2 - uv - 3v^2)$

(C) $3m^4 - 24mn^3 = 3m(m^3 - 8n^3) = 3m(m - 2n)(m^2 + 2mn + 4n^2)$

(D) $3x^4 - 5x^2 + 2 = (3x^2 - 2)(x^2 - 1) = (3x^2 - 2)(x - 1)(x + 1)$

Matched Problem 5 Factor completely.

(A) $18x^3 - 8x$ (B) $4m^3n - 2m^2n^2 + 2mn^3$

(C) $2t^4 - 16t$ (D) $2y^4 - 5y^2 - 12$

Exercises A-3

A

In Problems 1–8, factor out all factors common to all terms.

1. $6m^4 - 9m^3 - 3m^2$

2. $6x^4 - 8x^3 - 2x^2$

3. $8u^3v - 6u^2v^2 + 4uv^3$

4. $10x^3y + 20x^2y^2 - 15xy^3$

5. $7m(2m - 3) + 5(2m - 3)$

6. $5x(x + 1) - 3(x + 1)$

7. $4ab(2c + d) - (2c + d)$

8. $12a(b - 2c) - 15b(b - 2c)$

In Problems 9–18, factor by grouping.

9. $2x^2 - x + 4x - 2$

10. $x^2 - 3x + 2x - 6$

11. $3y^2 - 3y + 2y - 2$

12. $2x^2 - x + 6x - 3$

13. $2x^2 + 8x - x - 4$

14. $6x^2 + 9x - 2x - 3$

15. $wy - wz + xy - xz$

16. $ac + ad + bc + bd$

17. $am - 3bm + 2na - 6bn$

18. $ab + 6 + 2a + 3b$

B

In Problems 19–56, factor completely. If a polynomial cannot be factored, say so.

19. $3y^2 - y - 2$

20. $2x^2 + 5x - 3$

21. $u^2 - 2uv - 15v^2$

22. $x^2 - 4xy - 12y^2$

23. $m^2 - 6m - 3$

24. $x^2 + x - 4$

25. $w^2x^2 - y^2$

26. $25m^2 - 16n^2$

27. $9m^2 - 6mn + n^2$

28. $x^2 + 10xy + 25y^2$

29. $y^2 + 16$

30. $u^2 + 81$

31. $4z^2 - 28z + 48$

32. $6x^2 + 48x + 72$

33. $2x^4 - 24x^3 + 40x^2$

34. $2y^3 - 22y^2 + 48y$

35. $4xy^2 - 12xy + 9x$

36. $16x^2y - 8xy + y$

37. $6m^2 - mn - 12n^2$

38. $6s^2 + 7st - 3t^2$

39. $4u^3v - uv^3$

40. $x^3y - 9xy^3$

41. $2x^3 - 2x^2 + 8x$

42. $3m^3 - 6m^2 + 15m$

43. $8x^3 - 27y^3$

44. $5x^3 + 40y^3$

45. $x^4y + 8xy$

46. $8a^3 - 1$

C

47. $(x + 2)^2 - 9y^2$

48. $(a - b)^2 - 4(c - d)^2$

49. $5u^2 + 4uv - 2v^2$

50. $3x^2 - 2xy - 4y^2$

51. $6(x - y)^2 + 23(x - y) - 4$

52. $4(A + B)^2 - 5(A + B) - 6$

53. $y^4 - 3y^2 - 4$

54. $m^4 - n^4$

55. $15y(x - y)^3 + 12x(x - y)^2$

56. $15x^2(3x - 1)^4 + 60x^3(3x - 1)^3$

In Problems 57–60, discuss the validity of each statement. If the statement is true, explain why. If not, give a counterexample.

57. If n is a positive integer greater than 1, then $u^n - v^n$ can be factored.

58. If m and n are positive integers and $m \neq n$, then $u^m - v^n$ is not factorable.

59. If n is a positive integer greater than 1, then $u^n + v^n$ can be factored.

60. If k is a positive integer, then $u^{2k+1} + v^{2k+1}$ can be factored.

Answers to Matched Problems

1. (A) $2xy(x^2 - 4xy - 3y^2)$ (B) $(2x - 7)(3x - 2)$

2. (A) $(3x + 1)(2x + 3)$

(B) $(u + 3v)(2u - 3v)$

(C) $(a + b)(c + d)$

3. (A) $(2x - 1)(x + 6)$

(B) Not factorable

(C) $(3x + 4y)(2x - y)$

4. (A) $(x + 3y)^2$

(B) $(3x - 2y)(3x + 2y)$

(C) $(2m - 1)(4m^2 + 2m + 1)$

(D) $(x + yz)(x^2 - xyz + y^2z^2)$

(E) $[3(m - 3) - 2n][3(m - 3) + 2n]$

5. (A) $2x(3x - 2)(3x + 2)$

(B) $2mn(2m^2 - mn + n^2)$

(C) $2t(t - 2)(t^2 + 2t + 4)$

(D) $(2y^2 + 3)(y - 2)(y + 2)$

A-4 Operations on Rational Expressions

- Reducing to Lowest Terms
- Multiplication and Division
- Addition and Subtraction
- Compound Fractions

We now turn our attention to fractional forms. A quotient of two algebraic expressions (division by 0 excluded) is called a **fractional expression.** If both the numerator and the denominator are polynomials, the fractional expression is called a **rational expression**. Some examples of rational expressions are

$$\frac{1}{x^3 + 2x} \qquad \frac{5}{x} \qquad \frac{x + 7}{3x^2 - 5x + 1} \qquad \frac{x^2 - 2x + 4}{1}$$

In this section, we discuss basic operations on rational expressions. Since variables represent real numbers in the rational expressions we will consider, the properties of real number fractions summarized in Appendix A-1 will play a central role.

AGREEMENT Variable Restriction

Even though not always explicitly stated, we always assume that variables are restricted so that division by 0 is excluded.

For example, given the rational expression

$$\frac{2x + 5}{x(x + 2)(x - 3)}$$

the variable x is understood to be restricted from being 0, -2, or 3, since these values would cause the denominator to be 0.

Reducing to Lowest Terms

Central to the process of reducing rational expressions to *lowest terms* is the *fundamental property of fractions*, which we restate here for convenient reference:

THEOREM 1 Fundamental Property of Fractions

If a, b, and k are real numbers with $b, k \neq 0$, then

$$\frac{ka}{kb} = \frac{a}{b} \qquad \frac{5 \cdot 2}{5 \cdot 7} = \frac{2}{7} \qquad \frac{x(x + 4)}{2(x + 4)} = \frac{x}{2}, \quad x \neq -4$$

Using this property from left to right to eliminate all common factors from the numerator and the denominator of a given fraction is referred to as **reducing a fraction to lowest terms**. We are actually dividing the numerator and denominator by the same nonzero common factor.

Using the property from right to left—that is, multiplying the numerator and denominator by the same nonzero factor—is referred to as **raising a fraction to higher terms**. We will use the property in both directions in the material that follows.

EXAMPLE 1 **Reducing to Lowest Terms** Reduce each rational expression to lowest terms.

(A) $\dfrac{6x^2 + x - 1}{2x^2 - x - 1} = \dfrac{(2x + 1)(3x - 1)}{(2x + 1)(x - 1)}$ Factor numerator and denominator completely.

$= \dfrac{3x - 1}{x - 1}$ Divide numerator and denominator by the common factor $(2x + 1)$.

(B) $\dfrac{x^4 - 8x}{3x^3 - 2x^2 - 8x} = \dfrac{x(x - 2)(x^2 + 2x + 4)}{x(x - 2)(3x + 4)}$

$= \dfrac{x^2 + 2x + 4}{3x + 4}$

Matched Problem 1 Reduce each rational expression to lowest terms.

(A) $\dfrac{x^2 - 6x + 9}{x^2 - 9}$ (B) $\dfrac{x^3 - 1}{x^2 - 1}$

CONCEPTUAL INSIGHT

Using Theorem 1 to divide the numerator and denominator of a fraction by a common factor is often referred to as **canceling**. This operation can be denoted by drawing a slanted line through each common factor and writing any remaining factors above or below the common factor. Canceling is often incorrectly applied to individual terms in the numerator or denominator, instead of to common factors. For example,

$$\frac{14 - 5}{2} = \frac{9}{2}$$ Theorem 1 does not apply. There are no common factors in the numerator.

$$\frac{14 - 5}{2} \neq \frac{\overset{7}{\cancel{14}} - 5}{\underset{1}{\cancel{2}}} = 2$$ Incorrect use of Theorem 1. To cancel 2 in the denominator, 2 must be a factor of each term in the numerator.

Multiplication and Division

Since we are restricting variable replacements to real numbers, multiplication and division of rational expressions follow the rules for multiplying and dividing real number fractions summarized in Appendix A-1.

THEOREM 2 **Multiplication and Division**

If $a, b, c,$ and d are real numbers, then

1. $\dfrac{a}{b} \cdot \dfrac{c}{d} = \dfrac{ac}{bd}, \quad b, d \neq 0$ $\qquad \dfrac{3}{5} \cdot \dfrac{x}{x + 5} = \dfrac{3x}{5(x + 5)}$
2. $\dfrac{a}{b} \div \dfrac{c}{d} = \dfrac{a}{b} \cdot \dfrac{d}{c}, \quad b, c, d \neq 0$ $\qquad \dfrac{3}{5} \div \dfrac{x}{x + 5} = \dfrac{3}{5} \cdot \dfrac{x + 5}{x}$

EXAMPLE 2 **Multiplication and Division** Perform the indicated operations and reduce to lowest terms.

(A) $\dfrac{10x^3y}{3xy + 9y} \cdot \dfrac{x^2 - 9}{4x^2 - 12x}$

Factor numerators and denominators. Then divide any numerator and any denominator with a like common factor.

$$= \frac{\overset{5x^2}{\cancel{10x^3y}}}{\underset{3\cdot 1}{3\cancel{y}\cancel{(x+3)}}} \cdot \frac{\overset{1\cdot 1}{\cancel{(x-3)}\cancel{(x+3)}}}{\underset{2\cdot 1}{\cancel{4x}\cancel{(x-3)}}}$$

$$= \frac{5x^2}{6}$$

(B) $\dfrac{4 - 2x}{4} \div (x - 2) = \dfrac{\overset{1}{\cancel{2}}(2 - x)}{\underset{2}{\cancel{4}}} \cdot \dfrac{1}{x - 2}$ $\quad x - 2 = \dfrac{x - 2}{1}$

$$= \frac{2 - x}{2(x - 2)} = \frac{\overset{-1}{\cancel{-(x - 2)}}}{2\underset{1}{\cancel{(x - 2)}}}$$

$b - a = -(a - b)$, a useful change in some problems

$$= -\frac{1}{2}$$

Matched Problem 2 Perform the indicated operations and reduce to lowest terms.

(A) $\dfrac{12x^2y^3}{2xy^2 + 6xy} \cdot \dfrac{y^2 + 6y + 9}{3y^3 + 9y^2}$ $\qquad$ (B) $(4 - x) \div \dfrac{x^2 - 16}{5}$

Addition and Subtraction

Again, because we are restricting variable replacements to real numbers, addition and subtraction of rational expressions follow the rules for adding and subtracting real number fractions.

THEOREM 3 Addition and Subtraction

For a, b, and c real numbers,

1. $\dfrac{a}{b} + \dfrac{c}{b} = \dfrac{a + c}{b}, \quad b \neq 0$ $\qquad \dfrac{x}{x + 5} + \dfrac{8}{x + 5} = \dfrac{x + 8}{x + 5}$

2. $\dfrac{a}{b} - \dfrac{c}{b} = \dfrac{a - c}{b}, \quad b \neq 0$ $\qquad \dfrac{x}{3x^2y^2} - \dfrac{x + 7}{3x^2y^2} = \dfrac{x - (x + 7)}{3x^2y^2}$

We add rational expressions with the same denominators by adding or subtracting their numerators and placing the result over the common denominator. If the denominators are not the same, we raise the fractions to higher terms, using the fundamental property of fractions to obtain common denominators, and then proceed as described.

Even though any common denominator will do, our work will be simplified if the *least common denominator (LCD)* is used. Often, the LCD is obvious, but if it is not, the steps in the next box describe how to find it.

PROCEDURE Least Common Denominator

The least common denominator (LCD) of two or more rational expressions is found as follows:

1. Factor each denominator completely, including integer factors.
2. Identify each different factor from all the denominators.
3. Form a product using each different factor to the highest power that occurs in any one denominator. This product is the LCD.

EXAMPLE 3 **Addition and Subtraction** Combine into a single fraction and reduce to lowest terms.

(A) $\frac{3}{10} + \frac{5}{6} - \frac{11}{45}$ (B) $\frac{4}{9x} - \frac{5x}{6y^2} + 1$ (C) $\frac{1}{x-1} - \frac{1}{x} - \frac{2}{x^2-1}$

SOLUTION (A) To find the LCD, factor each denominator completely:

$$\left.\begin{aligned} 10 &= 2 \cdot 5 \\ 6 &= 2 \cdot 3 \\ 45 &= 3^2 \cdot 5 \end{aligned}\right\} \quad LCD = 2 \cdot 3^2 \cdot 5 = 90$$

Now use the fundamental property of fractions to make each denominator 90:

$$\begin{aligned} \frac{3}{10} + \frac{5}{6} - \frac{11}{45} &= \frac{9 \cdot 3}{9 \cdot 10} + \frac{15 \cdot 5}{15 \cdot 6} - \frac{2 \cdot 11}{2 \cdot 45} \\ &= \frac{27}{90} + \frac{75}{90} - \frac{22}{90} \\ &= \frac{27 + 75 - 22}{90} = \frac{80}{90} = \frac{8}{9} \end{aligned}$$

(B)
$$\left.\begin{aligned} 9x &= 3^2 x \\ 6y^2 &= 2 \cdot 3y^2 \end{aligned}\right\} \quad LCD = 2 \cdot 3^2 xy^2 = 18xy^2$$

$$\begin{aligned} \frac{4}{9x} - \frac{5x}{6y^2} + 1 &= \frac{2y^2 \cdot 4}{2y^2 \cdot 9x} - \frac{3x \cdot 5x}{3x \cdot 6y^2} + \frac{18xy^2}{18xy^2} \\ &= \frac{8y^2 - 15x^2 + 18xy^2}{18xy^2} \end{aligned}$$

(C)
$$\begin{aligned} &\frac{1}{x-1} - \frac{1}{x} - \frac{2}{x^2-1} \\ &= \frac{1}{x-1} - \frac{1}{x} - \frac{2}{(x-1)(x+1)} \qquad LCD = x(x-1)(x+1) \\ &= \frac{x(x+1) - (x-1)(x+1) - 2x}{x(x-1)(x+1)} \\ &= \frac{x^2 + x - x^2 + 1 - 2x}{x(x-1)(x+1)} \\ &= \frac{1 - x}{x(x-1)(x+1)} \\ &= \frac{\overset{-1}{\cancel{-(x-1)}}}{x\underset{1}{\cancel{(x-1)}}(x+1)} = \frac{-1}{x(x+1)} \end{aligned}$$

Matched Problem 3 Combine into a single fraction and reduce to lowest terms.

(A) $\frac{5}{28} - \frac{1}{10} + \frac{6}{35}$ (B) $\frac{1}{4x^2} - \frac{2x+1}{3x^3} + \frac{3}{12x}$

(C) $\frac{2}{x^2 - 4x + 4} + \frac{1}{x} - \frac{1}{x-2}$

Compound Fractions

A fractional expression with fractions in its numerator, denominator, or both is called a **compound fraction**. It is often necessary to represent a compound fraction as a **simple fraction**—that is (in all cases we will consider), as the quotient of two polynomials. The process does not involve any new concepts. It is a matter of applying old concepts and processes in the correct sequence.

EXAMPLE 4 **Simplifying Compound Fractions** Express as a simple fraction reduced to lowest terms:

(A) $\dfrac{\frac{1}{5+h} - \frac{1}{5}}{h}$ (B) $\dfrac{\frac{y}{x^2} - \frac{x}{y^2}}{\frac{y}{x} - \frac{x}{y}}$

SOLUTION We will simplify the expressions in parts (A) and (B) using two different methods—each is suited to the particular type of problem.

(A) We simplify this expression by combining the numerator into a single fraction and using division of rational forms.

$$\frac{\frac{1}{5+h} - \frac{1}{5}}{h} = \left[\frac{1}{5+h} - \frac{1}{5}\right] \div \frac{h}{1}$$
$$= \frac{5 - 5 - h}{5(5+h)} \cdot \frac{1}{h}$$
$$= \frac{-h}{5(5+h)h} = \frac{-1}{5(5+h)}$$

(B) The method used here makes effective use of the fundamental property of fractions in the form

$$\frac{a}{b} = \frac{ka}{kb} \qquad b, k \neq 0$$

Multiply the numerator and denominator by the LCD of all fractions in the numerator and denominator—in this case, x^2y^2:

$$\frac{x^2y^2\left(\frac{y}{x^2} - \frac{x}{y^2}\right)}{x^2y^2\left(\frac{y}{x} - \frac{x}{y}\right)} = \frac{x^2y^2\frac{y}{x^2} - x^2y^2\frac{x}{y^2}}{x^2y^2\frac{y}{x} - x^2y^2\frac{x}{y}} = \frac{y^3 - x^3}{xy^3 - x^3y}$$
$$= \frac{\overset{1}{\cancel{(y-x)}}(y^2 + xy + x^2)}{xy\underset{1}{\cancel{(y-x)}}(y+x)}$$
$$= \frac{y^2 + xy + x^2}{xy(y+x)} \quad \text{or} \quad \frac{x^2 + xy + y^2}{xy(x+y)}$$

Matched Problem 4 Express as a simple fraction reduced to lowest terms:

(A) $\dfrac{\dfrac{1}{2+h} - \dfrac{1}{2}}{h}$ (B) $\dfrac{\dfrac{a}{b} - \dfrac{b}{a}}{\dfrac{a}{b} + 2 + \dfrac{b}{a}}$

Exercises A-4

A

In Problems 1–18, perform the indicated operations and reduce answers to lowest terms.

1. $\dfrac{d^5}{3a} \div \left(\dfrac{d^2}{6a^2} \cdot \dfrac{a}{4d^3}\right)$

2. $\left(\dfrac{d^5}{3a} \div \dfrac{d^2}{6a^2}\right) \cdot \dfrac{a}{4d^3}$

3. $\dfrac{x^2}{12} + \dfrac{x}{18} - \dfrac{1}{30}$

4. $\dfrac{2y}{18} - \dfrac{-1}{28} - \dfrac{y}{42}$

5. $\dfrac{4m-3}{18m^3} + \dfrac{3}{4m} - \dfrac{2m-1}{6m^2}$

6. $\dfrac{3x+8}{4x^2} - \dfrac{2x-1}{x^3} - \dfrac{5}{8x}$

7. $\dfrac{x^2-9}{x^2-3x} \div (x^2 - x - 12)$

8. $\dfrac{2x^2+7x+3}{4x^2-1} \div (x+3)$

9. $\dfrac{2}{x} - \dfrac{1}{x-3}$

10. $\dfrac{5}{m-2} - \dfrac{3}{2m+1}$

11. $\dfrac{2}{(x+1)^2} - \dfrac{5}{x^2-x-2}$

12. $\dfrac{3}{x^2-5x+6} - \dfrac{5}{(x-2)^2}$

13. $\dfrac{x+1}{x-1} - 1$

14. $m - 3 - \dfrac{m-1}{m-2}$

15. $\dfrac{3}{a-1} - \dfrac{2}{1-a}$

16. $\dfrac{5}{x-3} - \dfrac{2}{3-x}$

17. $\dfrac{2x}{x^2-16} - \dfrac{x-4}{x^2+4x}$

18. $\dfrac{m+2}{m^2-2m} - \dfrac{m}{m^2-4}$

B

In Problems 19–30, perform the indicated operations and reduce answers to lowest terms. Represent any compound fractions as simple fractions reduced to lowest terms.

19. $\dfrac{x^2}{x^2+2x+1} + \dfrac{x-1}{3x+3} - \dfrac{1}{6}$

20. $\dfrac{y}{y^2-y-2} - \dfrac{1}{y^2+5y-14} - \dfrac{2}{y^2+8y+7}$

21. $\dfrac{1 - \dfrac{x}{y}}{2 - \dfrac{y}{x}}$

22. $\dfrac{2}{5 - \dfrac{3}{4x+1}}$

23. $\dfrac{c+2}{5c-5} - \dfrac{c-2}{3c-3} + \dfrac{c}{1-c}$

24. $\dfrac{x+7}{ax-bx} + \dfrac{y+9}{by-ay}$

25. $\dfrac{1 + \dfrac{3}{x}}{x - \dfrac{9}{x}}$

26. $\dfrac{1 - \dfrac{y^2}{x^2}}{1 - \dfrac{y}{x}}$

27. $\dfrac{\dfrac{1}{2(x+h)} - \dfrac{1}{2x}}{h}$

28. $\dfrac{\dfrac{1}{x+h} - \dfrac{1}{x}}{h}$

29. $\dfrac{\dfrac{x}{y} - 2 + \dfrac{y}{x}}{\dfrac{x}{y} - \dfrac{y}{x}}$

30. $\dfrac{1 + \dfrac{2}{x} - \dfrac{15}{x^2}}{1 + \dfrac{4}{x} - \dfrac{5}{x^2}}$

In Problems 31–38, imagine that the indicated "solutions" were given to you by a student whom you were tutoring in this class.

(A) Is the solution correct? If the solution is incorrect, explain what is wrong and how it can be corrected.

(B) Show a correct solution for each incorrect solution.

31. $\dfrac{x^2+4x+3}{x+3} = \dfrac{x^2+4x}{x} = x + 4$

32. $\dfrac{x^2-3x-4}{x-4} = \dfrac{x^2-3x}{x} = x - 3$

33. $\dfrac{(x+h)^2 - x^2}{h} = (x+1)^2 - x^2 = 2x + 1$

34. $\dfrac{(x+h)^3 - x^3}{h} = (x+1)^3 - x^3 = 3x^2 + 3x + 1$

35. $\dfrac{x^2-3x}{x^2-2x-3} + x - 3 = \dfrac{x^2-3x+x-3}{x^2-2x-3} = 1$

36. $\dfrac{2}{x-1} - \dfrac{x+3}{x^2-1} = \dfrac{2x+2-x-3}{x^2-1} = \dfrac{1}{x+1}$

37. $\dfrac{2x^2}{x^2-4} - \dfrac{x}{x-2} = \dfrac{2x^2-x^2-2x}{x^2-4} = \dfrac{x}{x+2}$

38. $x + \dfrac{x-2}{x^2-3x+2} = \dfrac{x+x-2}{x^2-3x+2} = \dfrac{2}{x-2}$

C

Represent the compound fractions in Problems 39–42 as simple fractions reduced to lowest terms.

39. $\dfrac{\dfrac{1}{3(x+h)^2} - \dfrac{1}{3x^2}}{h}$

40. $\dfrac{\dfrac{1}{(x+h)^2} - \dfrac{1}{x^2}}{h}$

41. $x - \dfrac{2}{1 - \dfrac{1}{x}}$

42. $2 - \dfrac{1}{1 - \dfrac{2}{a+2}}$

Answers to Matched Problems

1. (A) $\dfrac{x-3}{x+3}$ (B) $\dfrac{x^2+x+1}{x+1}$

2. (A) $2x$ (B) $\dfrac{-5}{x+4}$

3. (A) $\dfrac{1}{4}$ (B) $\dfrac{3x^2-5x-4}{12x^3}$ (C) $\dfrac{4}{x(x-2)^2}$

4. (A) $\dfrac{-1}{2(2+h)}$ (B) $\dfrac{a-b}{a+b}$

A-5 Integer Exponents and Scientific Notation

- Integer Exponents
- Scientific Notation

We now review basic operations on integer exponents and scientific notation.

Integer Exponents

DEFINITION Integer Exponents

For n an integer and a a real number:

1. For n a positive integer,

$$a^n = a \cdot a \cdot \cdots \cdot a \quad n \text{ factors of } a \qquad 5^4 = 5\cdot5\cdot5\cdot5$$

2. For $n = 0$,

$$a^0 = 1 \quad a \neq 0 \qquad 12^0 = 1$$

0^0 is not defined.

3. For n a negative integer,

$$a^n = \frac{1}{a^{-n}} \quad a \neq 0 \qquad a^{-3} = \frac{1}{a^{-(-3)}} = \frac{1}{a^3}$$

[If n is negative, then $(-n)$ is positive.]

Note: It can be shown that for *all* integers n,

$$a^{-n} = \frac{1}{a^n} \quad \text{and} \quad a^n = \frac{1}{a^{-n}} \quad a \neq 0 \qquad a^5 = \frac{1}{a^{-5}},\ a^{-5} = \frac{1}{a^5}$$

The following properties are very useful in working with integer exponents.

THEOREM 1 Exponent Properties

For n and m integers and a and b real numbers,

1. $a^m a^n = a^{m+n}$ $\qquad a^8a^{-3} = a^{8+(-3)} = a^5$

2. $(a^n)^m = a^{mn}$ $\qquad (a^{-2})^3 = a^{3(-2)} = a^{-6}$

3. $(ab)^m = a^m b^m$ $\qquad (ab)^{-2} = a^{-2}b^{-2}$

4. $\left(\dfrac{a}{b}\right)^m = \dfrac{a^m}{b^m} \quad b \neq 0$ $\qquad \left(\dfrac{a}{b}\right)^5 = \dfrac{a^5}{b^5}$

5. $\dfrac{a^m}{a^n} = a^{m-n} = \dfrac{1}{a^{n-m}} \quad a \neq 0$ $\qquad \dfrac{a^{-3}}{a^7} = \dfrac{1}{a^{7-(-3)}} = \dfrac{1}{a^{10}}$

Exponents are frequently encountered in algebraic applications. You should sharpen your skills in using exponents by reviewing the preceding basic definitions and properties and the examples that follow.

EXAMPLE 1 **Simplifying Exponent Forms** Simplify, and express the answers using positive exponents only.

(A) $(2x^3)(3x^5) = 2 \cdot 3x^{3+5} = 6x^8$

(B) $x^5x^{-9} = x^{-4} = \frac{1}{x^4}$

(C) $\frac{x^5}{x^7} = x^{5-7} = x^{-2} = \frac{1}{x^2}$ or $\frac{x^5}{x^7} = \frac{1}{x^{7-5}} = \frac{1}{x^2}$

(D) $\frac{x^{-3}}{y^{-4}} = \frac{y^4}{x^3}$

(E) $(u^{-3}v^2)^{-2} = (u^{-3})^{-2}(v^2)^{-2} = u^6v^{-4} = \frac{u^6}{v^4}$

(F) $\left(\frac{y^{-5}}{y^{-2}}\right)^{-2} = \frac{(y^{-5})^{-2}}{(y^{-2})^{-2}} = \frac{y^{10}}{y^4} = y^6$

(G) $\frac{4m^{-3}n^{-5}}{6m^{-4}n^3} = \frac{2m^{-3-(-4)}}{3n^{3-(-5)}} = \frac{2m}{3n^8}$

Matched Problem 1 Simplify, and express the answers using positive exponents only.

(A) $(3y^4)(2y^3)$ (B) m^2m^{-6} (C) $(u^3v^{-2})^{-2}$

(C) $\left(\frac{y^{-6}}{y^{-2}}\right)^{-1}$ (D) $\frac{8x^{-2}y^{-4}}{6x^{-5}y^2}$

EXAMPLE 2 **Converting to a Simple Fraction** Write $\frac{1-x}{x^{-1}-1}$ as a simple fraction with positive exponents.

SOLUTION First note that

$$\frac{1-x}{x^{-1}-1} \neq \frac{x(1-x)}{-1} \quad \text{A common error}$$

The original expression is a complex fraction, and we proceed to simplify it as follows:

$$\frac{1-x}{x^{-1}-1} = \frac{1-x}{\frac{1}{x}-1}$$

Multiply numerator and denominator by x to clear internal fractions.

$$= \frac{x(1-x)}{x\left(\frac{1}{x}-1\right)}$$

$$= \frac{x(1-x)}{1-x} = x$$

Matched Problem 2 Write $\frac{1+x^{-1}}{1-x^{-2}}$ as a simple fraction with positive exponents.

Scientific Notation

In the real world, one often encounters very large and very small numbers. For example,

- The public debt in the United States in 2008, to the nearest billion dollars, was

$10,025,000,000,000

- The world population in the year 2025, to the nearest million, is projected to be

7,947,000,000

- The sound intensity of a normal conversation is

0.000 000 000 316 watt per square centimeter*

It is generally troublesome to write and work with numbers of this type in standard decimal form. The first and last example cannot even be entered into many calculators as they are written. But with exponents defined for all integers, we can now express any finite decimal form as the product of a number between 1 and 10 and an integer power of 10, that is, in the form

$$a \times 10^n \qquad 1 \le a < 10, \quad a \text{ in decimal form}, \quad n \text{ an integer}$$

A number expressed in this form is said to be in **scientific notation.** The following are some examples of numbers in standard decimal notation and in scientific notation:

Decimal and Scientific Notation

$7 = 7 \times 10^0$	$0.5 = 5 \times 10^{-1}$
$67 = 6.7 \times 10$	$0.45 = 4.5 \times 10^{-1}$
$580 = 5.8 \times 10^2$	$0.0032 = 3.2 \times 10^{-3}$
$43{,}000 = 4.3 \times 10^4$	$0.000\,045 = 4.5 \times 10^{-5}$
$73{,}400{,}000 = 7.34 \times 10^7$	$0.000\,000\,391 = 3.91 \times 10^{-7}$

Note that the power of 10 used corresponds to the number of places we move the decimal to form a number between 1 and 10. The power is positive if the decimal is moved to the left and negative if it is moved to the right. Positive exponents are associated with numbers greater than or equal to 10; negative exponents are associated with positive numbers less than 1; and a zero exponent is associated with a number that is 1 or greater, but less than 10.

EXAMPLE 3 **Scientific Notation**

(A) Write each number in scientific notation:

7,320,000 and 0.000 000 54

(B) Write each number in standard decimal form:

4.32×10^6 and 4.32×10^{-5}

SOLUTION

(A) $7{,}320{,}000 = 7.320\,000. \times 10^6 = 7.32 \times 10^6$

6 places left

Positive exponent

$0.000\,000\,54 = 0.000\,000\,5.4 \times 10^{-7} = 5.4 \times 10^{-7}$

7 places right

Negative exponent

*We write 0.000 000 000 316 in place of 0.000000000316, because it is then easier to keep track of the number of decimal places. We follow this convention when there are more than five decimal places to the right of the decimal.

(B) $4.32 \times 10^6 = 4{,}320{,}000$ (6 places right; Positive exponent 6) $\quad 4.32 \times 10^{-5} = \frac{4.32}{10^5} = 0.000\ 043\ 2$ (5 places left; Negative exponent −5)

Matched Problem 3 (A) Write each number in scientific notation: 47,100; 2,443,000,000; 1.45
(B) Write each number in standard decimal form: 3.07×10^8; 5.98×10^{-6}

Exercises A-5

A

In Problems 1–14, simplify and express answers using positive exponents only. Variables are restricted to avoid division by 0.

1. $2x^{-9}$ **2.** $3y^{-5}$ **3.** $\frac{3}{2w^{-7}}$

4. $\frac{5}{4x^{-9}}$ **5.** $2x^{-8}x^5$ **6.** $3c^{-9}c^4$

7. $\frac{w^{-8}}{w^{-3}}$ **8.** $\frac{m^{-11}}{m^{-5}}$ **9.** $(2a^{-3})^2$

10. $7d^{-4}d^4$ **11.** $(a^{-3})^2$ **12.** $(5b^{-2})^2$

13. $(2x^4)^{-3}$ **14.** $(a^{-3}b^4)^{-3}$

In Problems 15–20, write each number in scientific notation.

15. 82,300,000,000 **16.** 5,380,000

17. 0.783 **18.** 0.019

19. 0.000 034 **20.** 0.000 000 007 832

In Problems 21–28, write each number in standard decimal notation.

21. 4×10^4 **22.** 9×10^6

23. 7×10^{-3} **24.** 2×10^{-5}

25. 6.171×10^7 **26.** 3.044×10^3

27. 8.08×10^{-4} **28.** 1.13×10^{-2}

B

In Problems 29–38, simplify and express answers using positive exponents only.

29. $(22 + 31)^0$ **30.** $(2x^3y^4)^0$

31. $\frac{10^{-3} \cdot 10^4}{10^{-11} \cdot 10^{-2}}$ **32.** $\frac{10^{-17} \cdot 10^{-5}}{10^{-3} \cdot 10^{-14}}$

33. $(5x^2y^{-3})^{-2}$ **34.** $(2m^{-3}n^2)^{-3}$

35. $\left(\frac{-5}{2x^3}\right)^{-2}$ **36.** $\left(\frac{2a}{3b^2}\right)^{-3}$

37. $\frac{8x^{-3}y^{-1}}{6x^2y^{-4}}$ **38.** $\frac{9m^{-4}n^3}{12m^{-1}n^{-1}}$

In Problems 39–42, write each expression in the form $ax^p + bx^q$ or $ax^p + bx^q + cx^r$, where a, b, and c are real numbers and p, q, and r are integers. For example,

$$\frac{2x^4 - 3x^2 + 1}{2x^3} = \frac{2x^4}{2x^3} - \frac{3x^2}{2x^3} + \frac{1}{2x^3} = x - \frac{3}{2}x^{-1} + \frac{1}{2}x^{-3}$$

39. $\frac{7x^5 - x^2}{4x^5}$ **40.** $\frac{5x^3 - 2}{3x^2}$

41. $\frac{5x^4 - 3x^2 + 8}{2x^2}$ **42.** $\frac{2x^3 - 3x^2 + x}{2x^2}$

Write each expression in Problems 43–46 with positive exponents only, and as a single fraction reduced to lowest terms.

43. $\frac{3x^2(x - 1)^2 - 2x^3(x - 1)}{(x - 1)^4}$

44. $\frac{5x^4(x + 3)^2 - 2x^5(x + 3)}{(x + 3)^4}$

45. $2x^{-2}(x - 1) - 2x^{-3}(x - 1)^2$

46. $2x(x + 3)^{-1} - x^2(x + 3)^{-2}$

In Problems 47–50, convert each number to scientific notation and simplify. Express the answer in both scientific notation and in standard decimal form.

47. $\frac{9{,}600{,}000{,}000}{(1{,}600{,}000)(0.000\,000\,25)}$

48. $\frac{(60{,}000)(0.000\,003)}{(0.0004)(1{,}500{,}000)}$

49. $\frac{(1{,}250{,}000)(0.000\,38)}{0.0152}$

50. $\frac{(0.000\,000\,82)(230{,}000)}{(625{,}000)(0.0082)}$

51. What is the result of entering 2^{3^2} on a calculator?

52. Refer to Problem 51. What is the difference between $2^{(3^2)}$ and $(2^3)^2$? Which agrees with the value of 2^{3^2} obtained with a calculator?

53. If $n = 0$, then property 1 in Theorem 1 implies that $a^ma^0 = a^{m+0} = a^m$. Explain how this helps motivate the definition of a^0.

54. If $m = -n$, then property 1 in Theorem 1 implies that $a^{-n}a^n = a^0 = 1$. Explain how this helps motivate the definition of a^{-n}.

C

Write the fractions in Problems 55–58 as simple fractions reduced to lowest terms.

55. $\dfrac{u+v}{u^{-1}+v^{-1}}$

56. $\dfrac{x^{-2}-y^{-2}}{x^{-1}+y^{-1}}$

57. $\dfrac{b^{-2}-c^{-2}}{b^{-3}-c^{-3}}$

58. $\dfrac{xy^{-2}-yx^{-2}}{y^{-1}-x^{-1}}$

Applications

Problems 59 and 60 refer to Table 1.

Table 1 U.S. Public Debt, Interest on Debt, and Population

Year	Public Debt ($)	Interest on Debt ($)	Population
2000	5,674,000,000,000	362,000,000,000	281,000,000
2008	10,025,000,000,000	451,000,000,000	304,000,000

59. **Public debt.** Carry out the following computations using scientific notation, and write final answers in standard decimal form.

(A) What was the per capita debt in 2008 (to the nearest dollar)?

(B) What was the per capita interest paid on the debt in 2008 (to the nearest dollar)?

(C) What was the percentage interest paid on the debt in 2008 (to two decimal places)?

60. **Public debt.** Carry out the following computations using scientific notation, and write final answers in standard decimal form.

(A) What was the per capita debt in 2000 (to the nearest dollar)?

(B) What was the per capita interest paid on the debt in 2000 (to the nearest dollar)?

(C) What was the percentage interest paid on the debt in 2000 (to two decimal places)?

Air pollution. *Air quality standards establish maximum amounts of pollutants considered acceptable in the air. The amounts are frequently given in parts per million (ppm). A standard of 30 ppm also can be expressed as follows:*

$$30 \text{ ppm} = \frac{30}{1{,}000{,}000} = \frac{3 \times 10}{10^6}$$
$$= 3 \times 10^{-5} = 0.000\,03 = 0.003\%$$

In Problems 61 and 62, express the given standard:

(A) In scientific notation

(B) In standard decimal notation

(C) As a percent

61. 9 ppm, the standard for carbon monoxide, when averaged over a period of 8 hours

62. 0.03 ppm, the standard for sulfur oxides, when averaged over a year

63. **Crime.** In 2008, the United States had a violent crime rate of 466 per 100,000 people and a population of 304 million people. How many violent crimes occurred that year? Compute the answer using scientific notation and convert the answer to standard decimal form (to the nearest thousand).

64. **Population density.** The United States had a 2008 population of 304 million people and a land area of 3,539,000 square miles. What was the population density? Compute the answer using scientific notation and convert the answer to standard decimal form (to one decimal place).

Answers to Matched Problems

1. (A) $6y^7$ (B) $\dfrac{1}{m^4}$ (C) $\dfrac{v^4}{u^6}$ (D) y^4 (E) $\dfrac{4x^3}{3y^6}$

2. $\dfrac{x}{x-1}$

3. (A) 4.7×10^4; 2.443×10^9; 1.45×10^0
(B) 307,000,000; 0.000 005 98

A-6 Rational Exponents and Radicals

- *n*th Roots of Real Numbers
- Rational Exponents and Radicals
- Properties of Radicals

Square roots may now be generalized to *nth roots*, and the meaning of exponent may be generalized to include all rational numbers.

*n*th Roots of Real Numbers

Consider a square of side r with area 36 square inches. We can write

$$r^2 = 36$$

and conclude that side r is a number whose square is 36. We say that r is a **square root** of b if $r^2 = b$. Similarly, we say that r is a **cube root** of b if $r^3 = b$. And, in general,

DEFINITION *n*th Root

For any natural number n,

$$r \text{ is an } \textbf{\textit{n}th root} \text{ of } b \text{ if } r^n = b$$

So 4 is a square root of 16, since $4^2 = 16$; -2 is a cube root of -8, since $(-2)^3 = -8$. Since $(-4)^2 = 16$, we see that -4 is also a square root of 16. It can be shown that any positive number has two real square roots, two real 4th roots, and, in general, two real nth roots if n is even. Negative numbers have no real square roots, no real 4th roots, and, in general, no real nth roots if n is even. The reason is that no real number raised to an even power can be negative. For odd roots, the situation is simpler. Every real number has exactly one real cube root, one real 5th root, and, in general, one real nth root if n is odd.

Additional roots can be considered in the *complex number system*. In this book, we restrict our interest to *real roots of real numbers*, and *root* will always be interpreted to mean "real root."

Rational Exponents and Radicals

We now turn to the question of what symbols to use to represent nth roots. For n a natural number greater than 1, we use

$$b^{1/n} \quad \text{or} \quad \sqrt[n]{b}$$

to represent a **real *n*th root of *b***. The exponent form is motivated by the fact that $(b^{1/n})^n = b$ if exponent laws are to continue to hold for rational exponents. The other form is called an ***n*th root radical**. In the expression below, the symbol $\sqrt{}$ is called a **radical**, n is the **index** of the radical, and b is the **radicand**:

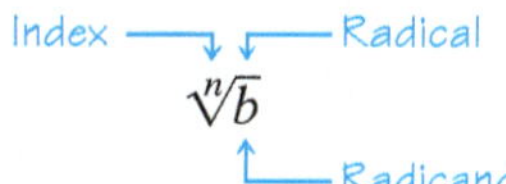

When the index is 2, it is usually omitted. That is, when dealing with square roots, we simply use $\sqrt{b}$ rather than $\sqrt[2]{b}$. If there are two real nth roots, both $b^{1/n}$ and $\sqrt[n]{b}$ denote the positive root, called the **principal *n*th root**.

EXAMPLE 1 **Finding *n*th Roots** Evaluate each of the following:

(A) $4^{1/2}$ and $\sqrt{4}$ (B) $-4^{1/2}$ and $-\sqrt{4}$ (C) $(-4)^{1/2}$ and $\sqrt{-4}$
(D) $8^{1/3}$ and $\sqrt[3]{8}$ (E) $(-8)^{1/3}$ and $\sqrt[3]{-8}$ (F) $-8^{1/3}$ and $-\sqrt[3]{8}$

SOLUTION (A) $4^{1/2} = \sqrt{4} = 2$ $(\sqrt{4} \neq \pm 2)$ (B) $-4^{1/2} = -\sqrt{4} = -2$
(C) $(-4)^{1/2}$ and $\sqrt{-4}$ are not real numbers
(D) $8^{1/3} = \sqrt[3]{8} = 2$ (E) $(-8)^{1/3} = \sqrt[3]{-8} = -2$
(F) $-8^{1/3} = -\sqrt[3]{8} = -2$

Matched Problem 1 Evaluate each of the following:

(A) $16^{1/2}$ (B) $-\sqrt{16}$ (C) $\sqrt[3]{-27}$ (D) $(-9)^{1/2}$ (E) $(\sqrt[4]{81})^3$

CAUTION The symbol $\sqrt{4}$ represents the single number 2, not ± 2. Do not confuse $\sqrt{4}$ with the solutions of the equation $x^2 = 4$, which are usually written in the form $x = \pm\sqrt{4} = \pm 2$.

We now define b^r for any rational number $r = m/n$.

DEFINITION Rational Exponents

If m and n are natural numbers without common prime factors, b is a real number, and b is nonnegative when n is even, then

$$b^{m/n} = \begin{cases} (b^{1/n})^m = (\sqrt[n]{b})^m \\ (b^m)^{1/n} = \sqrt[n]{b^m} \end{cases}$$

$8^{2/3} = (8^{1/3})^2 = (\sqrt[3]{8})^2 = 2^2 = 4$

$8^{2/3} = (8^2)^{1/3} = \sqrt[3]{8^2} = \sqrt[3]{64} = 4$

and

$$b^{-m/n} = \frac{1}{b^{m/n}} \quad b \neq 0 \qquad 8^{-2/3} = \frac{1}{8^{2/3}} = \frac{1}{4}$$

Note that the two definitions of $b^{m/n}$ are equivalent under the indicated restrictions on m, n, and b.

CONCEPTUAL INSIGHT

All the properties for integer exponents listed in Theorem 1 in Section A-5 also hold for rational exponents, provided that b is nonnegative when n is even. This restriction on b is necessary to avoid nonreal results. For example,

$$(-4)^{3/2} = \sqrt{(-4)^3} = \sqrt{-64} \qquad \text{Not a real number}$$

To avoid nonreal results, all variables in the remainder of this discussion represent positive real numbers.

EXAMPLE 2 **From Rational Exponent Form to Radical Form and Vice Versa** Change rational exponent form to radical form.

(A) $x^{1/7} = \sqrt[7]{x}$

(B) $(3u^2v^3)^{3/5} = \sqrt[5]{(3u^2v^3)^3}$ or $(\sqrt[5]{3u^2v^3})^3$ The first is usually preferred.

(C) $y^{-2/3} = \dfrac{1}{y^{2/3}} = \dfrac{1}{\sqrt[3]{y^2}}$ or $\sqrt[3]{y^{-2}}$ or $\sqrt[3]{\dfrac{1}{y^2}}$

Change radical form to rational exponent form.

(D) $\sqrt[5]{6} = 6^{1/5}$

(E) $-\sqrt[3]{x^2} = -x^{2/3}$

(F) $\sqrt{x^2 + y^2} = (x^2 + y^2)^{1/2}$ Note that $(x^2 + y^2)^{1/2} \neq x + y$. Why?

Matched Problem 2 Convert to radical form.

(A) $u^{1/5}$ (B) $(6x^2y^5)^{2/9}$ (C) $(3xy)^{-3/5}$

Convert to rational exponent form.

(D) $\sqrt[4]{9u}$ (E) $-\sqrt[7]{(2x)^4}$ (F) $\sqrt[3]{x^3 + y^3}$

EXAMPLE 3 **Working with Rational Exponents** Simplify each and express answers using positive exponents only. If rational exponents appear in final answers, convert to radical form.

(A) $(3x^{1/3})(2x^{1/2}) = 6x^{1/3+1/2} = 6x^{5/6} = 6\sqrt[6]{x^5}$

(B) $(-8)^{5/3} = [(-8)^{1/3}]^5 = (-2)^5 = -32$

(C) $(2x^{1/3}y^{-2/3})^3 = 8xy^{-2} = \dfrac{8x}{y^2}$

(D) $\left(\dfrac{4x^{1/3}}{x^{1/2}}\right)^{1/2} = \dfrac{4^{1/2}x^{1/6}}{x^{1/4}} = \dfrac{2}{x^{1/4-1/6}} = \dfrac{2}{x^{1/12}} = \dfrac{2}{\sqrt[12]{x}}$

Matched Problem 3 Simplify each and express answers using positive exponents only. If rational exponents appear in final answers, convert to radical form.

(A) $9^{3/2}$ (B) $(-27)^{4/3}$ (C) $(5y^{1/4})(2y^{1/3})$ (D) $(2x^{-3/4}y^{1/4})^4$

(E) $\left(\dfrac{8x^{1/2}}{x^{2/3}}\right)^{1/3}$

EXAMPLE 4 **Working with Rational Exponents** Multiply, and express answers using positive exponents only.

(A) $3y^{2/3}(2y^{1/3} - y^2)$ (B) $(2u^{1/2} + v^{1/2})(u^{1/2} - 3v^{1/2})$

SOLUTION

(A) $3y^{2/3}(2y^{1/3} - y^2) = 6y^{2/3+1/3} - 3y^{2/3+2}$

$= 6y - 3y^{8/3}$

(B) $(2u^{1/2} + v^{1/2})(u^{1/2} - 3v^{1/2}) = 2u - 5u^{1/2}v^{1/2} - 3v$

Matched Problem 4 Multiply, and express answers using positive exponents only.

(A) $2c^{1/4}(5c^3 - c^{3/4})$ (B) $(7x^{1/2} - y^{1/2})(2x^{1/2} + 3y^{1/2})$

EXAMPLE 5 **Working with Rational Exponents** Write the following expression in the form $ax^p + bx^q$, where a and b are real numbers and p and q are rational numbers:

$$\frac{2\sqrt{x} - 3\sqrt[3]{x^2}}{2\sqrt[3]{x}}$$

SOLUTION

$$\frac{2\sqrt{x} - 3\sqrt[3]{x^2}}{2\sqrt[3]{x}} = \frac{2x^{1/2} - 3x^{2/3}}{2x^{1/3}}$$ Change to rational exponent form.

$$= \frac{2x^{1/2}}{2x^{1/3}} - \frac{3x^{2/3}}{2x^{1/3}}$$ Separate into two fractions.

$$= x^{1/6} - 1.5x^{1/3}$$

Matched Problem 5 Write the following expression in the form $ax^p + bx^q$, where a and b are real numbers and p and q are rational numbers:

$$\frac{5\sqrt[3]{x} - 4\sqrt{x}}{2\sqrt{x^3}}$$

Properties of Radicals

Changing or simplifying radical expressions is aided by several properties of radicals that follow directly from the properties of exponents considered earlier.

THEOREM 1 **Properties of Radicals**

If c, n, and m are natural numbers greater than or equal to 2, and if x and y are positive real numbers, then

1. $\sqrt[n]{x^n} = x$ $\quad \sqrt[3]{x^3} = x$
2. $\sqrt[n]{xy} = \sqrt[n]{x}\sqrt[n]{y}$ $\quad \sqrt[5]{xy} = \sqrt[5]{x}\sqrt[5]{y}$
3. $\sqrt[n]{\dfrac{x}{y}} = \dfrac{\sqrt[n]{x}}{\sqrt[n]{y}}$ $\quad \sqrt[4]{\dfrac{x}{y}} = \dfrac{\sqrt[4]{x}}{\sqrt[4]{y}}$

EXAMPLE 6 **Applying Properties of Radicals** Simplify using properties of radicals.

(A) $\sqrt[4]{(3x^4y^3)^4}$ (B) $\sqrt[4]{8}\sqrt[4]{2}$ (C) $\sqrt[3]{\frac{xy}{27}}$

SOLUTION

(A) $\sqrt[4]{(3x^4y^3)^4} = 3x^4y^3$ — Property 1

(B) $\sqrt[4]{8}\sqrt[4]{2} = \sqrt[4]{16} = \sqrt[4]{2^4} = 2$ — Properties 2 and 1

(C) $\sqrt[3]{\frac{xy}{27}} = \frac{\sqrt[3]{xy}}{\sqrt[3]{27}} = \frac{\sqrt[3]{xy}}{3}$ or $\frac{1}{3}\sqrt[3]{xy}$ — Properties 3 and 1

Matched Problem 6 Simplify using properties of radicals.

(A) $\sqrt[7]{(x^3 + y^3)^7}$ (B) $\sqrt[3]{8y^3}$ (C) $\frac{\sqrt[3]{16x^4y}}{\sqrt[3]{2xy}}$

What is the best form for a radical expression? There are many answers, depending on what use we wish to make of the expression. In deriving certain formulas, it is sometimes useful to clear either a denominator or a numerator of radicals. The process is referred to as **rationalizing** the denominator or numerator. Examples 7 and 8 illustrate the rationalizing process.

EXAMPLE 7 **Rationalizing Denominators** Rationalize each denominator.

(A) $\frac{6x}{\sqrt{2x}}$ (B) $\frac{6}{\sqrt{7} - \sqrt{5}}$ (C) $\frac{x - 4}{\sqrt{x} + 2}$

SOLUTION

(A) $\frac{6x}{\sqrt{2x}} = \frac{6x}{\sqrt{2x}} \cdot \frac{\sqrt{2x}}{\sqrt{2x}} = \frac{6x\sqrt{2x}}{2x} = 3\sqrt{2x}$

(B) $\frac{6}{\sqrt{7} - \sqrt{5}} = \frac{6}{\sqrt{7} - \sqrt{5}} \cdot \frac{\sqrt{7} + \sqrt{5}}{\sqrt{7} + \sqrt{5}}$

$= \frac{6(\sqrt{7} + \sqrt{5})}{2} = 3(\sqrt{7} + \sqrt{5})$

(C) $\frac{x - 4}{\sqrt{x} + 2} = \frac{x - 4}{\sqrt{x} + 2} \cdot \frac{\sqrt{x} - 2}{\sqrt{x} - 2}$

$= \frac{(x - 4)(\sqrt{x} - 2)}{x - 4} = \sqrt{x} - 2$

Matched Problem 7 Rationalize each denominator.

(A) $\frac{12ab^2}{\sqrt{3ab}}$ (B) $\frac{9}{\sqrt{6} + \sqrt{3}}$ (C) $\frac{x^2 - y^2}{\sqrt{x} - \sqrt{y}}$

EXAMPLE 8 **Rationalizing Numerators** Rationalize each numerator.

(A) $\frac{\sqrt{2}}{2\sqrt{3}}$ (B) $\frac{3 + \sqrt{m}}{9 - m}$ (C) $\frac{\sqrt{2 + h} - \sqrt{2}}{h}$

SOLUTION

(A) $\frac{\sqrt{2}}{2\sqrt{3}} = \frac{\sqrt{2}}{2\sqrt{3}} \cdot \frac{\sqrt{2}}{\sqrt{2}} = \frac{2}{2\sqrt{6}} = \frac{1}{\sqrt{6}}$

(B) $\frac{3 + \sqrt{m}}{9 - m} = \frac{3 + \sqrt{m}}{9 - m} \cdot \frac{3 - \sqrt{m}}{3 - \sqrt{m}} = \frac{9 - m}{(9 - m)(3 - \sqrt{m})} = \frac{1}{3 - \sqrt{m}}$

(C) $\frac{\sqrt{2 + h} - \sqrt{2}}{h} = \frac{\sqrt{2 + h} - \sqrt{2}}{h} \cdot \frac{\sqrt{2 + h} + \sqrt{2}}{\sqrt{2 + h} + \sqrt{2}}$

$= \frac{h}{h(\sqrt{2 + h} + \sqrt{2})} = \frac{1}{\sqrt{2 + h} + \sqrt{2}}$

Matched Problem 8 Rationalize each numerator.

(A) $\frac{\sqrt{3}}{3\sqrt{2}}$ (B) $\frac{2 - \sqrt{n}}{4 - n}$ (C) $\frac{\sqrt{3 + h} - \sqrt{3}}{h}$

Exercises A-6

A

Change each expression in Problems 1–6 to radical form. Do not simplify.

1. $6x^{3/5}$ **2.** $7y^{2/5}$ **3.** $(32x^2y^3)^{3/5}$

4. $(7x^2y)^{5/7}$ **5.** $(x^2 + y^2)^{1/2}$ **6.** $x^{1/2} + y^{1/2}$

Change each expression in Problems 7–12 to rational exponent form. Do not simplify.

7. $5\sqrt[4]{x^3}$ **8.** $7m\sqrt[5]{n^2}$ **9.** $\sqrt[5]{(2x^2y)^3}$

10. $\sqrt{(8x^4y)^3}$ **11.** $\sqrt[3]{x} + \sqrt[3]{y}$ **12.** $\sqrt[3]{x^2 + y^3}$

In Problems 13–24, find rational number representations for each, if they exist.

13. $25^{1/2}$ **14.** $64^{1/3}$ **15.** $16^{3/2}$

16. $16^{3/4}$ **17.** $-49^{1/2}$ **18.** $(-49)^{1/2}$

19. $-64^{2/3}$ **20.** $(-64)^{2/3}$ **21.** $\left(\frac{4}{25}\right)^{3/2}$

22. $\left(\frac{8}{27}\right)^{2/3}$ **23.** $9^{-3/2}$ **24.** $8^{-2/3}$

In Problems 25–34, simplify each expression and write answers using positive exponents only. All variables represent positive real numbers.

25. $x^{4/5}x^{-2/5}$ **26.** $y^{-3/7}y^{4/7}$ **27.** $\frac{m^{2/3}}{m^{-1/3}}$

28. $\frac{x^{1/4}}{x^{3/4}}$ **29.** $(8x^3y^{-6})^{1/3}$ **30.** $(4u^{-2}v^4)^{1/2}$

31. $\left(\frac{4x^{-2}}{y^4}\right)^{-1/2}$ **32.** $\left(\frac{w^4}{9x^{-2}}\right)^{-1/2}$ **33.** $\frac{(8x)^{-1/3}}{12x^{1/4}}$

34. $\frac{6a^{3/4}}{15a^{-1/3}}$

Simplify each expression in Problems 35–40 using properties of radicals. All variables represent positive real numbers.

35. $\sqrt[5]{(2x + 3)^5}$ **36.** $\sqrt[3]{(7 + 2y)^3}$

37. $\sqrt{6x}\sqrt{15x^3}\sqrt{30x^7}$ **38.** $\sqrt[5]{16a^4}\sqrt[5]{4a^2}\sqrt[5]{8a^3}$

39. $\frac{\sqrt{6x}\sqrt{10}}{\sqrt{15x}}$ **40.** $\frac{\sqrt{8}\sqrt{12y}}{\sqrt{6y}}$

B

In Problems 41–48, multiply, and express answers using positive exponents only.

41. $3x^{3/4}(4x^{1/4} - 2x^8)$

42. $2m^{1/3}(3m^{2/3} - m^6)$

43. $(3u^{1/2} - v^{1/2})(u^{1/2} - 4v^{1/2})$

44. $(a^{1/2} + 2b^{1/2})(a^{1/2} - 3b^{1/2})$

45. $(6m^{1/2} + n^{-1/2})(6m - n^{-1/2})$

46. $(2x - 3y^{1/3})(2x^{1/3} + 1)$

47. $(3x^{1/2} - y^{1/2})^2$

48. $(x^{1/2} + 2y^{1/2})^2$

Write each expression in Problems 49–54 in the form $ax^p + bx^q$, where a and b are real numbers and p and q are rational numbers.

49. $\frac{\sqrt[3]{x^2} + 2}{2\sqrt[3]{x}}$ **50.** $\frac{12\sqrt{x} - 3}{4\sqrt{x}}$ **51.** $\frac{2\sqrt[4]{x^3} + \sqrt[3]{x}}{3x}$

52. $\frac{3\sqrt[3]{x^2} + \sqrt{x}}{5x}$ **53.** $\frac{2\sqrt[3]{x} - \sqrt{x}}{4\sqrt{x}}$ **54.** $\frac{x^2 - 4\sqrt{x}}{2\sqrt[3]{x}}$

Rationalize the denominators in Problems 55–60.

55. $\frac{12mn^2}{\sqrt{3mn}}$ **56.** $\frac{14x^2}{\sqrt{7x}}$ **57.** $\frac{2(x + 3)}{\sqrt{x - 2}}$

58. $\frac{3(x + 1)}{\sqrt{x + 4}}$ **59.** $\frac{7(x - y)^2}{\sqrt{x} - \sqrt{y}}$ **60.** $\frac{3a - 3b}{\sqrt{a} + \sqrt{b}}$

Rationalize the numerators in Problems 61–66.

61. $\frac{\sqrt{5xy}}{5x^2y^2}$ **62.** $\frac{\sqrt{3mn}}{3mn}$

63. $\frac{\sqrt{x + h} - \sqrt{x}}{h}$ **64.** $\frac{\sqrt{2(a + h)} - \sqrt{2a}}{h}$

65. $\frac{\sqrt{t} - \sqrt{x}}{t^2 - x^2}$ **66.** $\frac{\sqrt{x} - \sqrt{y}}{\sqrt{x} + \sqrt{y}}$

Problems 67–70 illustrate common errors involving rational exponents. In each case, find numerical examples that show that the left side is not always equal to the right side.

67. $(x + y)^{1/2} \neq x^{1/2} + y^{1/2}$ **68.** $(x^3 + y^3)^{1/3} \neq x + y$

69. $(x + y)^{1/3} \neq \frac{1}{(x + y)^3}$ **70.** $(x + y)^{-1/2} \neq \frac{1}{(x + y)^2}$

C

In Problems 71–82, discuss the validity of each statement. If the statement is true, explain why. If not, give a counterexample.

71. $\sqrt{x^2} = x$ for all real numbers x

72. $\sqrt{x^2} = |x|$ for all real numbers x

73. $\sqrt[3]{x^3} = |x|$ for all real numbers x

74. $\sqrt[3]{x^3} = x$ for all real numbers x

75. If $r < 0$, then r has no cube roots.

76. If $r < 0$, then r has no square roots.

77. If $r > 0$, then r has two square roots.

78. If $r > 0$, then r has three cube roots.

79. The fourth roots of 100 are $\sqrt{10}$ and $-\sqrt{10}$.

80. The square roots of $2\sqrt{6} - 5$ are $\sqrt{3} - \sqrt{2}$ and $\sqrt{2} - \sqrt{3}$.

81. $\sqrt{355 - 60\sqrt{35}} = 5\sqrt{7} - 6\sqrt{5}$

82. $\sqrt[3]{7 - 5\sqrt{2}} = 1 - \sqrt{2}$

In Problems 83–88, simplify by writing each expression as a simple or single fraction reduced to lowest terms and without negative exponents.

83. $-\frac{1}{2}(x - 2)(x + 3)^{-3/2} + (x + 3)^{-1/2}$

84. $2(x - 2)^{-1/2} - \frac{1}{2}(2x + 3)(x - 2)^{-3/2}$

85. $\dfrac{(x - 1)^{1/2} - x\left(\frac{1}{2}\right)(x - 1)^{-1/2}}{x - 1}$

86. $\dfrac{(2x - 1)^{1/2} - (x + 2)\left(\frac{1}{2}\right)(2x - 1)^{-1/2}(2)}{2x - 1}$

87. $\dfrac{(x + 2)^{2/3} - x\left(\frac{2}{3}\right)(x + 2)^{-1/3}}{(x + 2)^{4/3}}$

88. $\dfrac{2(3x - 1)^{1/3} - (2x + 1)\left(\frac{1}{3}\right)(3x - 1)^{-2/3}(3)}{(3x - 1)^{2/3}}$

In Problems 89–94, evaluate using a calculator. (Refer to the instruction book for your calculator to see how exponential forms are evaluated.)

89. $22^{3/2}$ **90.** $15^{5/4}$ **91.** $827^{-3/8}$

92. $103^{-3/4}$ **93.** $37.09^{7/3}$ **94.** $2.876^{8/5}$

In Problems 95 and 96, evaluate each expression on a calculator and determine which pairs have the same value. Verify these results algebraically.

95. (A) $\sqrt{3} + \sqrt{5}$ (B) $\sqrt{2 + \sqrt{3}} + \sqrt{2 - \sqrt{3}}$
(C) $1 + \sqrt{3}$ (D) $\sqrt[3]{10 + 6\sqrt{3}}$
(E) $\sqrt{8 + \sqrt{60}}$ (F) $\sqrt{6}$

96. (A) $2\sqrt[3]{2} + \sqrt{5}$ (B) $\sqrt{8}$
(C) $\sqrt{3} + \sqrt{7}$ (D) $\sqrt{3 + \sqrt{8}} + \sqrt{3 - \sqrt{8}}$
(E) $\sqrt{10 + \sqrt{84}}$ (F) $1 + \sqrt{5}$

Answers to Matched Problems

1. (A) 4 (B) -4
(C) -3 (D) Not a real number (E) 27

2. (A) $\sqrt[5]{u}$ (B) $\sqrt[9]{(6x^2y^5)^2}$ or $\left(\sqrt[9]{6x^2y^5}\right)^2$
(C) $1/\sqrt[5]{(3xy)^3}$ (D) $(9u)^{1/4}$ (E) $-(2x)^{4/7}$
(F) $(x^3 + y^3)^{1/3}$ (not $x + y$)

3. (A) 27 (B) 81
(C) $10y^{7/12} = 10\sqrt[12]{y^7}$ (D) $16y/x^3$
(E) $2/x^{1/18} = 2/\sqrt[18]{x}$

4. (A) $10c^{13/4} - 2c$ (B) $14x + 19x^{1/2}y^{1/2} - 3y$

5. $2.5x^{-7/6} - 2x^{-1}$

6. (A) $x^3 + y^3$ (B) $2y$ (C) $2x$

7. (A) $4b\sqrt{3ab}$
(B) $3(\sqrt{6} - \sqrt{3})$
(C) $(x + y)(\sqrt{x} + \sqrt{y})$

8. (A) $\dfrac{1}{\sqrt{6}}$ (B) $\dfrac{1}{2 + \sqrt{n}}$ (C) $\dfrac{1}{\sqrt{3 + h} + \sqrt{3}}$

A-7 Quadratic Equations

- Solution by Square Root
- Solution by Factoring
- Quadratic Formula
- Quadratic Formula and Factoring
- Application: Supply and Demand

In this section we consider equations involving second-degree polynomials.

DEFINITION Quadratic Equation

A **quadratic equation** in one variable is any equation that can be written in the form

$$ax^2 + bx + c = 0 \qquad a \neq 0 \qquad \text{Standard form}$$

where x is a variable and a, b, and c are constants.

The equations

$$5x^2 - 3x + 7 = 0 \qquad \text{and} \qquad 18 = 32t^2 - 12t$$

are both quadratic equations, since they are either in the standard form or can be transformed into this form.

We restrict our review to finding real solutions to quadratic equations.

Solution by Square Root

The easiest type of quadratic equation to solve is the special form where the first-degree term is missing:

$$ax^2 + c = 0 \qquad a \neq 0$$

The method of solution of this special form makes direct use of the square-root property:

THEOREM 1 Square-Root Property
If $a^2 = b$, then $a = \pm\sqrt{b}$.

EXAMPLE 1 **Square-Root Method** Use the square-root property to solve each equation.

(A) $x^2 - 7 = 0$ (B) $2x^2 - 10 = 0$

(C) $3x^2 + 27 = 0$ (D) $(x - 8)^2 = 9$

SOLUTION

(A) $x^2 - 7 = 0$

$x^2 = 7$ What real number squared is 7?

$x = \pm\sqrt{7}$ Short for $\sqrt{7}$ and $-\sqrt{7}$

(B) $2x^2 - 10 = 0$

$2x^2 = 10$

$x^2 = 5$ What real number squared is 5?

$x = \pm\sqrt{5}$

(C) $3x^2 + 27 = 0$

$3x^2 = -27$

$x^2 = -9$ What real number squared is -9?

No real solution, since no real number squared is negative.

(D) $(x - 8)^2 = 9$

$x - 8 = \pm\sqrt{9}$

$x - 8 = \pm 3$

$x = 8 \pm 3 = 5 \quad \text{or} \quad 11$

Matched Problem 1 Use the square-root property to solve each equation.

(A) $x^2 - 6 = 0$ (B) $3x^2 - 12 = 0$

(C) $x^2 + 4 = 0$ (D) $(x + 5)^2 = 1$

Solution by Factoring

If the left side of a quadratic equation when written in standard form can be factored, the equation can be solved very quickly. The method of solution by factoring rests on a basic property of real numbers, first mentioned in Section A-1.

CONCEPTUAL INSIGHT

Theorem 2 in Section A-1 states that if a and b are real numbers, then $ab = 0$ if and only if $a = 0$ or $b = 0$. To see that this property is useful for solving quadratic equations, consider the following:

$$x^2 - 4x + 3 = 0 \quad (1)$$
$$(x - 1)(x - 3) = 0$$
$$x - 1 = 0 \quad \text{or} \quad x - 3 = 0$$
$$x = 1 \quad \text{or} \quad x = 3$$

You should check these solutions in equation (1).

If one side of the equation is not 0, then this method cannot be used. For example, consider

$$x^2 - 4x + 3 = 8 \quad (2)$$
$$(x - 1)(x - 3) = 8$$
$$x - 1 \neq 8 \quad \text{or} \quad x - 3 \neq 8$$

$ab = 8$ does not imply that $a = 8$ or $b = 8$.

$$x = 9 \quad \text{or} \quad x = 11$$

Verify that neither $x = 9$ nor $x = 11$ is a solution for equation (2).

EXAMPLE 2 **Factoring Method** Solve by factoring using integer coefficients, if possible.

(A) $3x^2 - 6x - 24 = 0$ (B) $3y^2 = 2y$ (C) $x^2 - 2x - 1 = 0$

SOLUTION (A) $3x^2 - 6x - 24 = 0$ *Divide both sides by 3, since 3 is a factor of each coefficient.*

$x^2 - 2x - 8 = 0$ *Factor the left side, if possible.*

$(x - 4)(x + 2) = 0$

$x - 4 = 0$ or $x + 2 = 0$

$x = 4$ or $x = -2$

(B) $3y^2 = 2y$

$3y^2 - 2y = 0$ *We lose the solution $y = 0$ if both sides are divided by y ($3y^2 = 2y$ and $3y = 2$ are not equivalent).*

$y(3y - 2) = 0$

$y = 0$ or $3y - 2 = 0$

$3y = 2$

$y = \frac{2}{3}$

(C) $x^2 - 2x - 1 = 0$

This equation cannot be factored using integer coefficients. We will solve this type of equation by another method, considered below.

Matched Problem 2 Solve by factoring using integer coefficients, if possible.

(A) $2x^2 + 4x - 30 = 0$ (B) $2x^2 = 3x$ (C) $2x^2 - 8x + 3 = 0$

Note that an equation such as $x^2 = 25$ can be solved by either the square-root or the factoring method, and the results are the same (as they should be). Solve this equation both ways and compare.

Also, note that the factoring method can be extended to higher-degree polynomial equations. Consider the following:

$$x^3 - x = 0$$
$$x(x^2 - 1) = 0$$
$$x(x - 1)(x + 1) = 0$$
$$x = 0 \quad \text{or} \quad x - 1 = 0 \quad \text{or} \quad x + 1 = 0$$
$$\text{Solution: } x = 0, 1, -1$$

Check these solutions in the original equation.

The factoring and square-root methods are fast and easy to use when they apply. However, there are quadratic equations that look simple but cannot be solved by either method. For example, as was noted in Example 2C, the polynomial in

$$x^2 - 2x - 1 = 0$$

cannot be factored using integer coefficients. This brings us to the well-known and widely used *quadratic formula.*

Quadratic Formula

There is a method called *completing the square* that will work for all quadratic equations. After briefly reviewing this method, we will use it to develop the quadratic formula, which can be used to solve any quadratic equation.

The method of **completing the square** is based on the process of transforming a quadratic equation in standard form,

$$ax^2 + bx + c = 0$$

into the form

$$(x + A)^2 = B$$

where A and B are constants. Then, this last equation can be solved easily (if it has a real solution) by the square-root method discussed above.

Consider the equation from Example 2C:

$$x^2 - 2x - 1 = 0 \tag{3}$$

Since the left side does not factor using integer coefficients, we add 1 to each side to remove the constant term from the left side:

$$x^2 - 2x = 1 \tag{4}$$

Now we try to find a number that we can add to each side to make the left side a square of a first-degree polynomial. Note the following square of a binomial:

$$(x + m)^2 = x^2 + 2mx + m^2$$

We see that the third term on the right is the square of one-half the coefficient of x in the second term on the right. To complete the square in equation (4), we add the square of one-half the coefficient of x, $\left(-\frac{2}{2}\right)^2 = 1$, to each side. (This rule works only when the coefficient of x^2 is 1, that is, $a = 1$.) Thus,

$$x^2 - 2x + 1 = 1 + 1$$

The left side is the square of $x - 1$, and we write

$$(x - 1)^2 = 2$$

What number squared is 2?

$$x - 1 = \pm\sqrt{2}$$
$$x = 1 \pm \sqrt{2}$$

And equation (3) is solved!

Let us try the method on the general quadratic equation

$$ax^2 + bx + c = 0 \qquad a \neq 0 \tag{5}$$

and solve it once and for all for x in terms of the coefficients a, b, and c. We start by multiplying both sides of equation (5) by $1/a$ to obtain

$$x^2 + \frac{b}{a}x + \frac{c}{a} = 0$$

Add $-c/a$ to both sides:

$$x^2 + \frac{b}{a}x = -\frac{c}{a}$$

Now we complete the square on the left side by adding the square of one-half the coefficient of x, that is, $(b/2a)^2 = b^2/4a^2$, to each side:

$$x^2 + \frac{b}{a}x + \frac{b^2}{4a^2} = \frac{b^2}{4a^2} - \frac{c}{a}$$

Writing the left side as a square and combining the right side into a single fraction, we obtain

$$\left(x + \frac{b}{2a}\right)^2 = \frac{b^2 - 4ac}{4a^2}$$

Now we solve by the square-root method:

$$x + \frac{b}{2a} = \pm\sqrt{\frac{b^2 - 4ac}{4a^2}}$$

$$x = -\frac{b}{2a} \pm \frac{\sqrt{b^2 - 4ac}}{2a}$$ Since $\pm\sqrt{4a^2} = \pm 2a$ for any real number a

When this is written as a single fraction, it becomes the **quadratic formula**:

Quadratic Formula

If $ax^2 + bx + c = 0$, $a \neq 0$, then

$$x = \frac{-b \pm \sqrt{b^2 - 4ac}}{2a}$$

This formula is generally used to solve quadratic equations when the square-root or factoring methods do not work. The quantity $b^2 - 4ac$ under the radical is called the **discriminant**, and it gives us the useful information about solutions listed in Table 1.

Table 1

$b^2 - 4ac$	$ax^2 + bx + c = 0$
Positive	Two real solutions
Zero	One real solution
Negative	No real solutions

EXAMPLE 3 **Quadratic Formula Method** Solve $x^2 - 2x - 1 = 0$ using the quadratic formula.

SOLUTION $x^2 - 2x - 1 = 0$

$$x = \frac{-b \pm \sqrt{b^2 - 4ac}}{2a}$$ $a = 1, b = -2, c = -1$

$$= \frac{-(-2) \pm \sqrt{(-2)^2 - 4(1)(-1)}}{2(1)}$$

$$= \frac{2 \pm \sqrt{8}}{2} = \frac{2 \pm 2\sqrt{2}}{2} = 1 \pm \sqrt{2} \approx -0.414 \quad \text{or} \quad 2.414$$

CHECK $x^2 - 2x - 1 = 0$

When $x = 1 + \sqrt{2}$,

$$(1 + \sqrt{2})^2 - 2(1 + \sqrt{2}) - 1 = 1 + 2\sqrt{2} + 2 - 2 - 2\sqrt{2} - 1 = 0$$

When $x = 1 - \sqrt{2}$,

$$(1 - \sqrt{2})^2 - 2(1 - \sqrt{2}) - 1 = 1 - 2\sqrt{2} + 2 - 2 + 2\sqrt{2} - 1 = 0$$

Matched Problem 3 Solve $2x^2 - 4x - 3 = 0$ using the quadratic formula.

If we try to solve $x^2 - 6x + 11 = 0$ using the quadratic formula, we obtain

$$x = \frac{6 \pm \sqrt{-8}}{2}$$

which is not a real number. (Why?)

Quadratic Formula and Factoring

As in Section A-3, we restrict our interest in factoring to polynomials with integer coefficients. If a polynomial cannot be factored as a product of lower-degree polynomials with integer coefficients, we say that the polynomial is **not factorable in the integers**.

How can you factor the quadratic polynomial $x^2 - 13x - 2{,}310$? We start by solving the corresponding quadratic equation using the quadratic formula:

$$x^2 - 13x - 2{,}310 = 0$$

$$x = \frac{-(-13) \pm \sqrt{(-13)^2 - 4(1)(-2{,}310)}}{2}$$

$$x = \frac{-(-13) \pm \sqrt{9{,}409}}{2}$$

$$= \frac{13 \pm 97}{2} = 55 \quad \text{or} \quad -42$$

Now we write

$$x^2 - 13x - 2{,}310 = [x - 55][x - (-42)] = (x - 55)(x + 42)$$

Multiplying the two factors on the right produces the second-degree polynomial on the left.

What is behind this procedure? The following two theorems justify and generalize the process:

THEOREM 2 Factorability Theorem

A second-degree polynomial, $ax^2 + bx + c$, with integer coefficients can be expressed as the product of two first-degree polynomials with integer coefficients if and only if $\sqrt{b^2 - 4ac}$ is an integer.

THEOREM 3 Factor Theorem

If r_1 and r_2 are solutions to the second-degree equation $ax^2 + bx + c = 0$, then

$$ax^2 + bx + c = a(x - r_1)(x - r_2)$$

EXAMPLE 4 **Factoring with the Aid of the Discriminant** Factor, if possible, using integer coefficients.

(A) $4x^2 - 65x + 264$ (B) $2x^2 - 33x - 306$

SOLUTION (A) $4x^2 - 65x + 264$

Step 1 Test for factorability:

$$\sqrt{b^2 - 4ac} = \sqrt{(-65)^2 - 4(4)(264)} = 1$$

Since the result is an integer, the polynomial has first-degree factors with integer coefficients.

Step 2 Factor, using the factor theorem. Find the solutions to the corresponding quadratic equation using the quadratic formula:

$$4x^2 - 65x + 264 = 0 \quad \text{From step 1}$$

$$x = \frac{-(-65) \pm 1}{2 \cdot 4} = \frac{33}{4} \quad \text{or} \quad 8$$

Thus,

$$4x^2 - 65x + 264 = 4\left(x - \frac{33}{4}\right)(x - 8)$$
$$= (4x - 33)(x - 8)$$

(B) $2x^2 - 33x - 306$

Step 1 Test for factorability:

$$\sqrt{b^2 - 4ac} = \sqrt{(-33)^2 - 4(2)(-306)} = \sqrt{3{,}537}$$

Since $\sqrt{3{,}537}$ is not an integer, the polynomial is not factorable in the integers.

Matched Problem 4 Factor, if possible, using integer coefficients.

(A) $3x^2 - 28x - 464$ (B) $9x^2 + 320x - 144$

Application: Supply and Demand

Supply-and-demand analysis is a very important part of business and economics. In general, producers are willing to supply more of an item as the price of an item increases and less of an item as the price decreases. Similarly, buyers are willing to buy less of an item as the price increases, and more of an item as the price decreases. We have a dynamic situation where the price, supply, and demand fluctuate until a price is reached at which the supply is equal to the demand. In economic theory, this point is called the **equilibrium point**. If the price increases from this point, the supply will increase and the demand will decrease; if the price decreases from this point, the supply will decrease and the demand will increase.

EXAMPLE 5 **Supply and Demand** At a large summer beach resort, the weekly supply-and-demand equations for folding beach chairs are

$$p = \frac{x}{140} + \frac{3}{4} \quad \text{Supply equation}$$

$$p = \frac{5{,}670}{x} \quad \text{Demand equation}$$

The supply equation indicates that the supplier is willing to sell x units at a price of p dollars per unit. The demand equation indicates that consumers are willing to buy x units at a price of p dollars per unit. How many units are required for supply to equal demand? At what price will supply equal demand?

SOLUTION Set the right side of the supply equation equal to the right side of the demand equation and solve for x.

$$\frac{x}{140} + \frac{3}{4} = \frac{5{,}670}{x} \quad \text{Multiply by } 140x\text{, the LCD.}$$

$$x^2 + 105x = 793{,}800 \quad \text{Write in standard form.}$$

$$x^2 + 105x - 793{,}800 = 0 \quad \text{Use the quadratic formula.}$$

$$x = \frac{-105 \pm \sqrt{105^2 - 4(1)(-793{,}800)}}{2}$$

$$x = 840 \text{ units}$$

The negative root is discarded since a negative number of units cannot be produced or sold. Substitute $x = 840$ back into either the supply equation or the demand equation to find the equilibrium price (we use the demand equation).

$$p = \frac{5{,}670}{x} = \frac{5{,}670}{840} = \$6.75$$

At a price of \$6.75 the supplier is willing to supply 840 chairs and consumers are willing to buy 840 chairs during a week.

Matched Problem 5 Repeat Example 5 if near the end of summer, the supply-and-demand equations are

$$p = \frac{x}{80} - \frac{1}{20} \quad \text{Supply equation}$$

$$p = \frac{1{,}264}{x} \quad \text{Demand equation}$$

Exercises A-7

Find only real solutions in the problems below. If there are no real solutions, say so.

A

Solve Problems 1–4 by the square-root method.

1. $2x^2 - 22 = 0$

2. $3m^2 - 21 = 0$

3. $(3x - 1)^2 = 25$

4. $(2x + 1)^2 = 16$

Solve Problems 5–8 by factoring.

5. $2u^2 - 8u - 24 = 0$

6. $3x^2 - 18x + 15 = 0$

7. $x^2 = 2x$

8. $n^2 = 3n$

Solve Problems 9–12 by using the quadratic formula.

9. $x^2 - 6x - 3 = 0$

10. $m^2 + 8m + 3 = 0$

11. $3u^2 + 12u + 6 = 0$

12. $2x^2 - 20x - 6 = 0$

B

Solve Problems 13–30 by using any method.

13. $\frac{2x^2}{3} = 5x$

14. $x^2 = -\frac{3}{4}x$

15. $4u^2 - 9 = 0$

16. $9y^2 - 25 = 0$

17. $8x^2 + 20x = 12$

18. $9x^2 - 6 = 15x$

19. $x^2 = 1 - x$

20. $m^2 = 1 - 3m$

21. $2x^2 = 6x - 3$

22. $2x^2 = 4x - 1$

23. $y^2 - 4y = -8$

24. $x^2 - 2x = -3$

25. $(2x + 3)^2 = 11$

26. $(5x - 2)^2 = 7$

27. $\frac{3}{p} = p$

28. $x - \frac{7}{x} = 0$

29. $2 - \frac{2}{m^2} = \frac{3}{m}$

30. $2 + \frac{5}{u} = \frac{3}{u^2}$

In Problems 31–38, factor, if possible, as the product of two first-degree polynomials with integer coefficients. Use the quadratic formula and the factor theorem.

31. $x^2 + 40x - 84$

32. $x^2 - 28x - 128$

33. $x^2 - 32x + 144$

34. $x^2 + 52x + 208$

35. $2x^2 + 15x - 108$

36. $3x^2 - 32x - 140$

37. $4x^2 + 241x - 434$

38. $6x^2 - 427x - 360$

C

39. Solve $A = P(1 + r)^2$ for r in terms of A and P; that is, isolate r on the left side of the equation (with coefficient 1) and end up with an algebraic expression on the right side involving A and P but not r. Write the answer using positive square roots only.

40. Solve $x^2 + 3mx - 3n = 0$ for x in terms of m and n.

41. Consider the quadratic equation

$$x^2 + 4x + c = 0$$

where c is a real number. Discuss the relationship between the values of c and the three types of roots listed in Table 1 on page 578.

42. Consider the quadratic equation

$$x^2 - 2x + c = 0$$

where c is a real number. Discuss the relationship between the values of c and the three types of roots listed in Table 1 on page 578.

Applications

43. **Supply and demand.** A company wholesales shampoo in a particular city. Their marketing research department established the following weekly supply-and-demand equations:

$$p = \frac{x}{450} + \frac{1}{2} \quad \text{Supply equation}$$
$$p = \frac{6{,}300}{x} \quad \text{Demand equation}$$

How many units are required for supply to equal demand? At what price per bottle will supply equal demand?

44. **Supply and demand.** An importer sells an automatic camera to outlets in a large city. During the summer, the weekly supply-and-demand equations are

$$p = \frac{x}{6} + 9 \quad \text{Supply equation}$$
$$p = \frac{24{,}840}{x} \quad \text{Demand equation}$$

How many units are required for supply to equal demand? At what price will supply equal demand?

45. **Interest rate.** If P dollars are invested at $100r$ percent compounded annually, at the end of 2 years it will grow to $A = P(1 + r)^2$. At what interest rate will \$484 grow to \$625 in 2 years? (*Note:* If $A = 625$ and $P = 484$ find r.)

46. **Interest rate.** Using the formula in Problem 45, determine the interest rate that will make \$1,000 grow to \$1,210 in 2 years.

47. **Ecology.** To measure the velocity v (in feet per second) of a stream, we position a hollow L-shaped tube with one end under the water pointing upstream and the other end pointing straight up a couple of feet out of the water. The water will then be pushed up the tube a certain distance h (in feet) above the surface of the stream. Physicists have shown that $v^2 = 64h$. Approximately how fast is a stream flowing if $h = 1$ foot? If $h = 0.5$ foot?

48. **Safety research.** It is of considerable importance to know the least number of feet d in which a car can be stopped, including reaction time of the driver, at various speeds v (in miles per hour). Safety research has produced the formula $d = 0.044v^2 + 1.1v$. If it took a car 550 feet to stop, estimate the car's speed at the moment the stopping process was started.

Answers to Matched Problems

1. (A) $\pm\sqrt{6}$ (B) ± 2
   (C) No real solution (D) $-6, -4$
2. (A) $-5, 3$ (B) $0, \frac{3}{2}$
   (C) Cannot be factored using integer coefficients
3. $(2 \pm \sqrt{10})/2$
4. (A) Cannot be factored using integer coefficients
   (B) $(9x - 4)(x + 36)$
5. 320 chairs at \$3.95 each

APPENDIX B Special Topics

B-1 Sequences, Series, and Summation Notation

- Sequences
- Series and Summation Notation

If someone asked you to list all natural numbers that are perfect squares, you might begin by writing

$$1, 4, 9, 16, 25, 36$$

But you would soon realize that it is impossible to actually list all the perfect squares, since there are an infinite number of them. However, you could represent this collection of numbers in several different ways. One common method is to write

$$1, 4, 9, \ldots, n^2, \ldots \quad n \in N$$

where N is the set of natural numbers. A list of numbers such as this is generally called a *sequence*.

Sequences

Consider the function f given by

$$f(n) = 2n + 1 \tag{1}$$

where the domain of f is the set of natural numbers N. Note that

$$f(1) = 3, \quad f(2) = 5, \quad f(3) = 7, \quad \ldots$$

The function f is an example of a sequence. In general, a **sequence** is a function with domain a set of successive integers. Instead of the standard function notation used in equation (1), sequences are usually defined in terms of a special notation.

The range value $f(n)$ is usually symbolized more compactly with a symbol such as a_n. Thus, in place of equation (1), we write

$$a_n = 2n + 1$$

and the domain is understood to be the set of natural numbers unless something is said to the contrary or the context indicates otherwise. The elements in the range are called **terms of the sequence**; a_1 is the first term, a_2 is the second term, and a_n is the ***n*th term**, or **general term**.

$$\begin{aligned} a_1 &= 2(1) + 1 = 3 && \text{First term} \\ a_2 &= 2(2) + 1 = 5 && \text{Second term} \\ a_3 &= 2(3) + 1 = 7 && \text{Third term} \\ &\vdots \\ a_n &= 2n + 1 && \text{General term} \end{aligned}$$

The ordered list of elements

$$3, 5, 7, \ldots, 2n + 1, \ldots$$

obtained by writing the terms of the sequence in their natural order with respect to the domain values is often informally referred to as a sequence. A sequence also may be represented in the abbreviated form $\{a_n\}$, where a symbol for the nth term is written within braces. For example, we could refer to the sequence $3, 5, 7, \ldots, 2n + 1, \ldots$ as the sequence $\{2n + 1\}$.

If the domain of a sequence is a finite set of successive integers, then the sequence is called a **finite sequence**. If the domain is an infinite set of successive integers, then the sequence is called an **infinite sequence.** The sequence $\{2n + 1\}$ discussed above is an infinite sequence.

EXAMPLE 1 **Writing the Terms of a Sequence** Write the first four terms of each sequence:

(A) $a_n = 3n - 2$ (B) $\left\{\frac{(-1)^n}{n}\right\}$

SOLUTION (A) 1, 4, 7, 10 (B) $-1, \frac{1}{2}, \frac{-1}{3}, \frac{1}{4}$

Matched Problem 1 Write the first four terms of each sequence:

(A) $a_n = -n + 3$ (B) $\left\{\frac{(-1)^n}{2^n}\right\}$

Now that we have seen how to use the general term to find the first few terms in a sequence, we consider the reverse problem. That is, can a sequence be defined just by listing the first three or four terms of the sequence? And can we then use these initial terms to find a formula for the nth term? In general, without other information, the answer to the first question is no. Many different sequences may start off with the same terms. Simply listing the first three terms (or any other finite number of terms) does not specify a particular sequence.

What about the second question? That is, given a few terms, can we find the general formula for at least one sequence whose first few terms agree with the given terms? The answer to this question is a qualified yes. If we can observe a simple pattern in the given terms, we usually can construct a general term that will produce that pattern. The next example illustrates this approach.

EXAMPLE 2 **Finding the General Term of a Sequence** Find the general term of a sequence whose first four terms are

(A) 3, 4, 5, 6, . . . (B) 5, −25, 125, −625, . . .

SOLUTION (A) Since these terms are consecutive integers, one solution is $a_n = n, n \geq 3$. If we want the domain of the sequence to be all natural numbers, another solution is $b_n = n + 2$.

(B) Each of these terms can be written as the product of a power of 5 and a power of −1:

$$5 = (-1)^0 5^1 = a_1$$
$$-25 = (-1)^1 5^2 = a_2$$
$$125 = (-1)^2 5^3 = a_3$$
$$-625 = (-1)^3 5^4 = a_4$$

If we choose the domain to be all natural numbers, a solution is

$$a_n = (-1)^{n-1} 5^n$$

Matched Problem 2 Find the general term of a sequence whose first four terms are

(A) 3, 6, 9, 12, . . . (B) 1, −2, 4, −8, . . .

In general, there is usually more than one way of representing the nth term of a given sequence (see the solution of Example 2A). However, unless something is stated to the contrary, we assume that the domain of the sequence is the set of natural numbers N.

Series and Summation Notation

If $a_1, a_2, a_3, \ldots, a_n, \ldots$ is a sequence, the expression

$$a_1 + a_2 + a_3 + \cdots + a_n + \cdots$$

is called a **series**. If the sequence is finite, the corresponding series is a **finite series**. If the sequence is infinite, the corresponding series is an **infinite series**. We consider only finite series in this section. For example,

$$1, 3, 5, 7, 9 \qquad \text{Finite sequence}$$

$$1 + 3 + 5 + 7 + 9 \qquad \text{Finite series}$$

Notice that we can easily evaluate this series by adding the five terms:

$$1 + 3 + 5 + 7 + 9 = 25$$

Series are often represented in a compact form called **summation notation**. Consider the following examples:

$$\begin{aligned}\sum_{k=3}^{6} k^2 &= 3^2 + 4^2 + 5^2 + 6^2 \\ &= 9 + 16 + 25 + 36 = 86\end{aligned}$$

$$\begin{aligned}\sum_{k=0}^{2} (4k + 1) &= (4 \cdot 0 + 1) + (4 \cdot 1 + 1) + (4 \cdot 2 + 1) \\ &= 1 + 5 + 9 = 15\end{aligned}$$

In each case, the terms of the series on the right are obtained from the expression on the left by successively replacing the **summing index** $\boldsymbol{k}$ with integers, starting with the number indicated below the **summation sign** Σ and ending with the number that appears above Σ. The summing index may be represented by letters other than k and may start at any integer and end at any integer greater than or equal to the starting integer. If we are given the finite sequence

$$\frac{1}{2}, \frac{1}{4}, \frac{1}{8}, \ldots, \frac{1}{2^n}$$

the corresponding series is

$$\frac{1}{2} + \frac{1}{4} + \frac{1}{8} + \cdots + \frac{1}{2^n} = \sum_{j=1}^{n} \frac{1}{2^j}$$

where we have used j for the summing index.

EXAMPLE 3 **Summation Notation** Write

$$\sum_{k=1}^{5} \frac{k}{k^2 + 1}$$

without summation notation. Do not evaluate the sum.

SOLUTION

$$\begin{aligned}\sum_{k=1}^{5} \frac{k}{k^2 + 1} &= \frac{1}{1^2 + 1} + \frac{2}{2^2 + 1} + \frac{3}{3^2 + 1} + \frac{4}{4^2 + 1} + \frac{5}{5^2 + 1} \\ &= \frac{1}{2} + \frac{2}{5} + \frac{3}{10} + \frac{4}{17} + \frac{5}{26}\end{aligned}$$

Matched Problem 3 Write

$$\sum_{k=1}^{5} \frac{k + 1}{k}$$

without summation notation. Do not evaluate the sum.

If the terms of a series are alternately positive and negative, we call the series an **alternating series**. The next example deals with the representation of such a series.

EXAMPLE 4 **Summation Notation** Write the alternating series

$$\frac{1}{2} - \frac{1}{4} + \frac{1}{6} - \frac{1}{8} + \frac{1}{10} - \frac{1}{12}$$

using summation notation with

(A) The summing index k starting at 1

(B) The summing index j starting at 0

SOLUTION (A) $(-1)^{k+1}$ provides the alternation of sign, and $1/(2k)$ provides the other part of each term. So, we can write

$$\frac{1}{2} - \frac{1}{4} + \frac{1}{6} - \frac{1}{8} + \frac{1}{10} - \frac{1}{12} = \sum_{k=1}^{6} \frac{(-1)^{k+1}}{2k}$$

(B) $(-1)^j$ provides the alternation of sign, and $1/[2(j+1)]$ provides the other part of each term. So, we can write

$$\frac{1}{2} - \frac{1}{4} + \frac{1}{6} - \frac{1}{8} + \frac{1}{10} - \frac{1}{12} = \sum_{j=0}^{5} \frac{(-1)^j}{2(j+1)}$$

Matched Problem 4 Write the alternating series

$$1 - \frac{1}{3} + \frac{1}{9} - \frac{1}{27} + \frac{1}{81}$$

using summation notation with

(A) The summing index k starting at 1

(B) The summing index j starting at 0

Summation notation provides a compact notation for the sum of any list of numbers, even if the numbers are not generated by a formula. For example, suppose that the results of an examination taken by a class of 10 students are given in the following list:

87, 77, 95, 83, 86, 73, 95, 68, 75, 86

If we let $a_1, a_2, a_3, \ldots, a_{10}$ represent these 10 scores, then the average test score is given by

$$\frac{1}{10}\sum_{k=1}^{10} a_k = \frac{1}{10}(87 + 77 + 95 + 83 + 86 + 73 + 95 + 68 + 75 + 86)$$

$$= \frac{1}{10}(825) = 82.5$$

More generally, in statistics, the **arithmetic mean** $\bar{a}$ of a list of n numbers $a_1, a_2, \ldots, a_n$ is defined as

$$\bar{a} = \frac{1}{n}\sum_{k=1}^{n} a_k$$

EXAMPLE 5 **Arithmetic Mean** Find the arithmetic mean of 3, 5, 4, 7, 4, 2, 3, and 6.

SOLUTION $\bar{a} = \frac{1}{8}\sum_{k=1}^{8} a_k = \frac{1}{8}(3 + 5 + 4 + 7 + 4 + 2 + 3 + 6) = \frac{1}{8}(34) = 4.25$

Matched Problem 5 Find the arithmetic mean of 9, 3, 8, 4, 3, and 6.

Exercises B-1

A

Write the first four terms for each sequence in Problems 1–6.

1. $a_n = 2n + 3$

2. $a_n = 4n - 3$

3. $a_n = \dfrac{n+2}{n+1}$

4. $a_n = \dfrac{2n+1}{2n}$

5. $a_n = (-3)^{n+1}$

6. $a_n = \left(-\frac{1}{4}\right)^{n-1}$

7. Write the 10th term of the sequence in Problem 1.

8. Write the 15th term of the sequence in Problem 2.

9. Write the 99th term of the sequence in Problem 3.

10. Write the 200th term of the sequence in Problem 4.

In Problems 11–16, write each series in expanded form without summation notation, and evaluate.

11. $\sum_{k=1}^{6} k$

12. $\sum_{k=1}^{5} k^2$

13. $\sum_{k=4}^{7} (2k - 3)$

14. $\sum_{k=0}^{4} (-2)^k$

15. $\sum_{k=0}^{3} \dfrac{1}{10^k}$

16. $\sum_{k=1}^{4} \dfrac{1}{2^k}$

Find the arithmetic mean of each list of numbers in Problems 17–20.

17. 5, 4, 2, 1, and 6

18. 7, 9, 9, 2, and 4

19. 96, 65, 82, 74, 91, 88, 87, 91, 77, and 74

20. 100, 62, 95, 91, 82, 87, 70, 75, 87, and 82

B

Write the first five terms of each sequence in Problems 21–26.

21. $a_n = \dfrac{(-1)^{n+1}}{2^n}$

22. $a_n = (-1)^n(n - 1)^2$

23. $a_n = n[1 + (-1)^n]$

24. $a_n = \dfrac{1 - (-1)^n}{n}$

25. $a_n = \left(-\dfrac{3}{2}\right)^{n-1}$

26. $a_n = \left(-\dfrac{1}{2}\right)^{n+1}$

In Problems 27–42, find the general term of a sequence whose first four terms agree with the given terms.

27. $-2, -1, 0, 1, \ldots$

28. $4, 5, 6, 7, \ldots$

29. $4, 8, 12, 16, \ldots$

30. $-3, -6, -9, -12, \ldots$

31. $\frac{1}{2}, \frac{3}{4}, \frac{5}{6}, \frac{7}{8}, \ldots$

32. $\frac{1}{2}, \frac{2}{3}, \frac{3}{4}, \frac{4}{5}, \ldots$

33. $1, -2, 3, -4, \ldots$

34. $-2, 4, -8, 16, \ldots$

35. $1, -3, 5, -7, \ldots$

36. $3, -6, 9, -12, \ldots$

37. $1, \frac{2}{5}, \frac{4}{25}, \frac{8}{125}, \ldots$

38. $\frac{4}{3}, \frac{16}{9}, \frac{64}{27}, \frac{256}{81}, \ldots$

39. $x, x^2, x^3, x^4, \ldots$

40. $1, 2x, 3x^2, 4x^3, \ldots$

41. $x, -x^3, x^5, -x^7, \ldots$

42. $x, \dfrac{x^2}{2}, \dfrac{x^3}{3}, \dfrac{x^4}{4}, \ldots$

Write each series in Problems 43–50 in expanded form without summation notation. Do not evaluate.

43. $\sum_{k=1}^{5} (-1)^{k+1}(2k - 1)^2$

44. $\sum_{k=1}^{4} \dfrac{(-2)^{k+1}}{2k + 1}$

45. $\sum_{k=2}^{5} \dfrac{2^k}{2k + 3}$

46. $\sum_{k=3}^{7} \dfrac{(-1)^k}{k^2 - k}$

47. $\sum_{k=1}^{5} x^{k-1}$

48. $\sum_{k=1}^{3} \dfrac{1}{k}x^{k+1}$

49. $\sum_{k=0}^{4} \dfrac{(-1)^k x^{2k+1}}{2k + 1}$

50. $\sum_{k=0}^{4} \dfrac{(-1)^k x^{2k}}{2k + 2}$

Write each series in Problems 51–54 using summation notation with

(A) The summing index k starting at k = 1

(B) The summing index j starting at j = 0

51. $2 + 3 + 4 + 5 + 6$

52. $1^2 + 2^2 + 3^2 + 4^2$

53. $1 - \frac{1}{2} + \frac{1}{3} - \frac{1}{4}$

54. $1 - \frac{1}{3} + \frac{1}{5} - \frac{1}{7} + \frac{1}{9}$

Write each series in Problems 55–58 using summation notation with the summing index k starting at k = 1.

55. $2 + \dfrac{3}{2} + \dfrac{4}{3} + \cdots + \dfrac{n+1}{n}$

56. $1 + \dfrac{1}{2^2} + \dfrac{1}{3^2} + \cdots + \dfrac{1}{n^2}$

57. $\dfrac{1}{2} - \dfrac{1}{4} + \dfrac{1}{8} - \cdots + \dfrac{(-1)^{n+1}}{2^n}$

58. $1 - 4 + 9 - \cdots + (-1)^{n+1}n^2$

C

In Problems 59–62, discuss the validity of each statement. If the statement is true, explain why. If not, give a counterexample.

59. For each positive integer n, the sum of the series $1 + \dfrac{1}{2} + \dfrac{1}{3} + \cdots + \dfrac{1}{n}$ is less than 4.

60. For each positive integer n, the sum of the series $\dfrac{1}{2} + \dfrac{1}{4} + \dfrac{1}{8} + \cdots + \dfrac{1}{2^n}$ is less than 1.

61. For each positive integer n, the sum of the series $\dfrac{1}{2} - \dfrac{1}{4} + \dfrac{1}{8} - \cdots + \dfrac{(-1)^{n+1}}{2^n}$ is greater than or equal to $\dfrac{1}{4}$.

62. For each positive integer n, the sum of the series $1 - \dfrac{1}{2} + \dfrac{1}{3} - \dfrac{1}{4} + \cdots + \dfrac{(-1)^{n+1}}{n}$ is greater than or equal to $\dfrac{1}{2}$.

Some sequences are defined by a **recursion formula**—*that is, a formula that defines each term of the sequence in terms of one or more of the preceding terms. For example, if* $\{a_n\}$ *is defined by*

$$a_1 = 1 \quad \textit{and} \quad a_n = 2a_{n-1} + 1 \quad \textit{for } n \geq 2$$

then

$$a_2 = 2a_1 + 1 = 2 \cdot 1 + 1 = 3$$
$$a_3 = 2a_2 + 1 = 2 \cdot 3 + 1 = 7$$
$$a_4 = 2a_3 + 1 = 2 \cdot 7 + 1 = 15$$

and so on. In Problems 63–66, write the first five terms of each sequence.

63. $a_1 = 2$ and $a_n = 3a_{n-1} + 2$ for $n \geq 2$

64. $a_1 = 3$ and $a_n = 2a_{n-1} - 2$ for $n \geq 2$

65. $a_1 = 1$ and $a_n = 2a_{n-1}$ for $n \geq 2$

66. $a_1 = 1$ and $a_n = -\frac{1}{3}a_{n-1}$ for $n \geq 2$

If A is a positive real number, the terms of the sequence defined by

$$a_1 = \frac{A}{2} \quad \textit{and} \quad a_n = \frac{1}{2}\left(a_{n-1} + \frac{A}{a_{n-1}}\right) \quad \textit{for } n \geq 2$$

can be used to approximate $\sqrt{A}$ *to any decimal place accuracy desired. In Problems 67 and 68, compute the first four terms of this sequence for the indicated value of A, and compare the fourth term with the value of* $\sqrt{A}$ *obtained from a calculator.*

67. $A = 2$

68. $A = 6$

69. The sequence defined recursively by $a_1 = 1$, $a_2 = 1$, $a_n = a_{n-1} + a_{n-2}$ for $n \geq 3$ is called the *Fibonacci sequence*. Find the first ten terms of the Fibonacci sequence.

70. The sequence defined by $b_n = \frac{\sqrt{5}}{5}\left(\frac{1 + \sqrt{5}}{2}\right)^n$ is related to the Fibonacci sequence. Find the first ten terms (to three decimal places) of the sequence $\{b_n\}$ and describe the relationship.

Answers to Matched Problems

1. (A) 2, 1, 0, −1 (B) $\frac{-1}{2}, \frac{1}{4}, \frac{-1}{8}, \frac{1}{16}$
2. (A) $a_n = 3n$ (B) $a_n = (-2)^{n-1}$
3. $2 + \frac{3}{2} + \frac{4}{3} + \frac{5}{4} + \frac{6}{5}$
4. (A) $\sum_{k=1}^{5} \frac{(-1)^{k-1}}{3^{k-1}}$ (B) $\sum_{j=0}^{4} \frac{(-1)^j}{3^j}$
5. 5.5

B-2 Arithmetic and Geometric Sequences

- Arithmetic and Geometric Sequences
- *n*th-Term Formulas
- Sum Formulas for Finite Arithmetic Series
- Sum Formulas for Finite Geometric Series
- Sum Formula for Infinite Geometric Series
- Applications

For most sequences, it is difficult to sum an arbitrary number of terms of the sequence without adding term by term. But particular types of sequences—*arithmetic sequences* and *geometric sequences*—have certain properties that lead to convenient and useful formulas for the sums of the corresponding *arithmetic series* and *geometric series*.

Arithmetic and Geometric Sequences

The sequence $5, 7, 9, 11, 13, \ldots, 5 + 2(n - 1), \ldots$, where each term after the first is obtained by adding 2 to the preceding term, is an example of an arithmetic sequence. The sequence $5, 10, 20, 40, 80, \ldots, 5(2)^{n-1}, \ldots$, where each term after the first is obtained by multiplying the preceding term by 2, is an example of a geometric sequence.

DEFINITION Arithmetic Sequence

A sequence of numbers

$$a_1, a_2, a_3, \ldots, a_n, \ldots$$

is called an **arithmetic sequence** if there is a constant d, called the **common difference**, such that

$$a_n - a_{n-1} = d$$

That is,

$$a_n = a_{n-1} + d \qquad \text{for every } n > 1$$

DEFINITION Geometric Sequence

A sequence of numbers

$$a_1, a_2, a_3, \ldots, a_n, \ldots$$

is called a **geometric sequence** if there exists a nonzero constant r, called a **common ratio**, such that

$$\frac{a_n}{a_{n-1}} = r$$

That is,

$$a_n = ra_{n-1} \quad \text{for every } n > 1$$

EXAMPLE 1 **Recognizing Arithmetic and Geometric Sequences** Which of the following can be the first four terms of an arithmetic sequence? Of a geometric sequence?

(A) $1, 2, 3, 5, \ldots$ (B) $-1, 3, -9, 27, \ldots$

(C) $3, 3, 3, 3, \ldots$ (D) $10, 8.5, 7, 5.5, \ldots$

SOLUTION

(A) Since $2 - 1 \neq 5 - 3$, there is no common difference, so the sequence is not an arithmetic sequence. Since $2/1 \neq 3/2$, there is no common ratio, so the sequence is not geometric either.

(B) The sequence is geometric with common ratio -3. It is not arithmetic.

(C) The sequence is arithmetic with common difference 0, and is also geometric with common ratio 1.

(D) The sequence is arithmetic with common difference -1.5. It is not geometric.

Matched Problem 1 Which of the following can be the first four terms of an arithmetic sequence? Of a geometric sequence?

(A) $8, 2, 0.5, 0.125, \ldots$ (B) $-7, -2, 3, 8, \ldots$ (C) $1, 5, 25, 100, \ldots$

*n*th-Term Formulas

If $\{a_n\}$ is an arithmetic sequence with common difference d, then

$$\begin{aligned} a_2 &= a_1 + d \\ a_3 &= a_2 + d = a_1 + 2d \\ a_4 &= a_3 + d = a_1 + 3d \end{aligned}$$

This suggests that

THEOREM 1 ***n*th Term of an Arithmetic Sequence**

$$a_n = a_1 + (n - 1)d \qquad \text{for all } n > 1 \tag{1}$$

Similarly, if $\{a_n\}$ is a geometric sequence with common ratio r, then

$$\begin{aligned} a_2 &= a_1 r \\ a_3 &= a_2 r = a_1 r^2 \\ a_4 &= a_3 r = a_1 r^3 \end{aligned}$$

This suggests that

THEOREM 2 *n*th Term of a Geometric Sequence

$$a_n = a_1 r^{n-1} \quad \text{for all } n > 1 \tag{2}$$

EXAMPLE 2 Finding Terms in Arithmetic and Geometric Sequences

(A) If the 1st and 10th terms of an arithmetic sequence are 3 and 30, respectively, find the 40th term of the sequence.

(B) If the 1st and 10th terms of a geometric sequence are 3 and 30, find the 40th term to three decimal places.

SOLUTION (A) First use formula (1) with $a_1 = 3$ and $a_{10} = 30$ to find d:

$$\begin{aligned} a_n &= a_1 + (n-1)d \\ a_{10} &= a_1 + (10-1)d \\ 30 &= 3 + 9d \\ d &= 3 \end{aligned}$$

Now find a_{40}:

$$a_{40} = 3 + 39 \cdot 3 = 120$$

(B) First use formula (2) with $a_1 = 3$ and $a_{10} = 30$ to find r:

$$\begin{aligned} a_n &= a_1 r^{n-1} \\ a_{10} &= a_1 r^{10-1} \\ 30 &= 3r^9 \\ r^9 &= 10 \\ r &= 10^{1/9} \end{aligned}$$

Now find a_{40}:

$$a_{40} = 3(10^{1/9})^{39} = 3(10^{39/9}) = 64{,}633.041$$

Matched Problem 2 (A) If the 1st and 15th terms of an arithmetic sequence are -5 and 23, respectively, find the 73rd term of the sequence.

(B) Find the 8th term of the geometric sequence

$$\frac{1}{64}, \frac{-1}{32}, \frac{1}{16}, \ldots$$

Sum Formulas for Finite Arithmetic Series

If $a_1, a_2, a_3, \ldots, a_n$ is a finite arithmetic sequence, then the corresponding series $a_1 + a_2 + a_3 + \cdots + a_n$ is called a *finite arithmetic series*. We will derive two simple and very useful formulas for the sum of a finite arithmetic series. Let d be the common difference of the arithmetic sequence $a_1, a_2, a_3, \ldots, a_n$ and let S_n denote the sum of the series $a_1 + a_2 + a_3 + \cdots + a_n$. Then

$$S_n = a_1 + (a_1 + d) + \cdots + [a_1 + (n-2)d] + [a_1 + (n-1)d]$$

Reversing the order of the sum, we obtain

$$S_n = [a_1 + (n-1)d] + [a_1 + (n-2)d] + \cdots + (a_1 + d) + a_1$$

Something interesting happens if we combine these last two equations by addition (adding corresponding terms on the right sides):

$$2S_n = [2a_1 + (n-1)d] + [2a_1 + (n-1)d] + \cdots + [2a_1 + (n-1)d] + [2a_1 + (n-1)d]$$

All the terms on the right side are the same, and there are n of them. Thus,

$$2S_n = n[2a_1 + (n - 1)d]$$

and we have the following general formula:

THEOREM 3 Sum of a Finite Arithmetic Series: First Form

$$S_n = \frac{n}{2}[2a_1 + (n - 1)d] \qquad (3)$$

Replacing

$$[a_1 + (n - 1)d] \quad \text{in} \quad \frac{n}{2}[a_1 + a_1 + (n - 1)d]$$

by a_n from equation (1), we obtain a second useful formula for the sum:

THEOREM 4 Sum of a Finite Arithmetic Series: Second Form

$$S_n = \frac{n}{2}(a_1 + a_n) \qquad (4)$$

EXAMPLE 3 **Finding a Sum** Find the sum of the first 30 terms in the arithmetic sequence:

$$3, 8, 13, 18, \ldots$$

SOLUTION Use formula (3) with $n = 30$, $a_1 = 3$, and $d = 5$:

$$S_{30} = \frac{30}{2}[2 \cdot 3 + (30 - 1)5] = 2{,}265$$

Matched Problem 3 Find the sum of the first 40 terms in the arithmetic sequence:

$$15, 13, 11, 9, \ldots$$

EXAMPLE 4 **Finding a Sum** Find the sum of all the even numbers between 31 and 87.

SOLUTION First, find n using equation (1):

$$a_n = a_1 + (n - 1)d$$
$$86 = 32 + (n - 1)2$$
$$n = 28$$

Now find S_{28} using formula (4):

$$S_n = \frac{n}{2}(a_1 + a_n)$$
$$S_{28} = \frac{28}{2}(32 + 86) = 1{,}652$$

Matched Problem 4 Find the sum of all the odd numbers between 24 and 208.

Sum Formulas for Finite Geometric Series

If $a_1, a_2, a_3, \ldots, a_n$ is a finite geometric sequence, then the corresponding series $a_1 + a_2 + a_3 + \cdots + a_n$ is called a *finite geometric series.* As with arithmetic series, we can derive two simple and very useful formulas for the sum of a finite geometric series. Let r be the common ratio of the geometric sequence $a_1, a_2, a_3, \dots, a_n$ and let S_n denote the sum of the series $a_1 + a_2 + a_3 + \cdots + a_n$. Then

$$S_n = a_1 + a_1 r + a_1 r^2 + \cdots + a_1 r^{n-2} + a_1 r^{n-1}$$

If we multiply both sides by r, we obtain

$$rS_n = a_1 r + a_1 r^2 + a_1 r^3 + \cdots + a_1 r^{n-1} + a_1 r^n$$

Now combine these last two equations by subtraction to obtain

$$rS_n - S_n = (a_1 r + a_1 r^2 + a_1 r^3 + \cdots + a_1 r^{n-1} + a_1 r^n) - (a_1 + a_1 r + a_1 r^2 + \cdots + a_1 r^{n-2} + a_1 r^{n-1})$$
$$(r - 1)S_n = a_1 r^n - a_1$$

Notice how many terms drop out on the right side. Solving for S_n, we have

THEOREM 5 Sum of a Finite Geometric Series: First Form

$$S_n = \frac{a_1(r^n - 1)}{r - 1} \qquad r \neq 1 \qquad (5)$$

Since $a_n = a_1 r^{n-1}$, or $ra_n = a_1 r^n$, formula (5) also can be written in the form

THEOREM 6 Sum of a Finite Geometric Series: Second Form

$$S_n = \frac{ra_n - a_1}{r - 1} \qquad r \neq 1 \qquad (6)$$

EXAMPLE 5 **Finding a Sum** Find the sum (to 2 decimal places) of the first ten terms of the geometric sequence:

$$1, 1.05, 1.05^2, \dots$$

SOLUTION Use formula (5) with $a_1 = 1$, $r = 1.05$, and $n = 10$:

$$S_n = \frac{a_1(r^n - 1)}{r - 1}$$
$$S_{10} = \frac{1(1.05^{10} - 1)}{1.05 - 1}$$
$$\approx \frac{0.6289}{0.05} \approx 12.58$$

Matched Problem 5 Find the sum of the first eight terms of the geometric sequence:

$$100, 100(1.08), 100(1.08)^2, \dots$$

Sum Formula for Infinite Geometric Series

Given a geometric series, what happens to the sum S_n of the first n terms as n increases without stopping? To answer this question, let us write formula (5) in the form

$$S_n = \frac{a_1 r^n}{r - 1} - \frac{a_1}{r - 1}$$

It is possible to show that if $-1 < r < 1$, then r^n will approach 0 as n increases. The first term above will approach 0 and S_n can be made as close as we please to the second term, $-a_1/(r - 1)$ [which can be written as $a_1/(1 - r)$], by taking n sufficiently large. So, if the common ratio r is between -1 and 1, we conclude that the sum of an infinite geometric series is

THEOREM 7 Sum of an Infinite Geometric Series

$$S_\infty = \frac{a_1}{1 - r} \qquad -1 < r < 1 \tag{7}$$

If $r \le -1$ or $r \ge 1$, then an infinite geometric series has no sum.

Applications

EXAMPLE 6 **Loan Repayment** A person borrows \$3,600 and agrees to repay the loan in monthly installments over 3 years. The agreement is to pay 1% of the unpaid balance each month for using the money and \$100 each month to reduce the loan. What is the total cost of the loan over the 3 years?

SOLUTION Let us look at the problem relative to a time line:

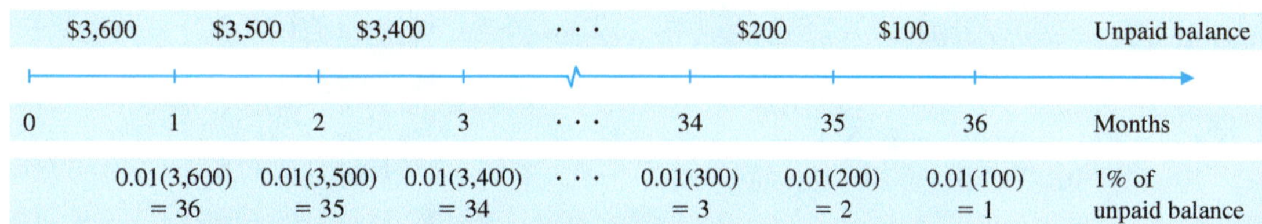

The total cost of the loan is

$$1 + 2 + \cdots + 34 + 35 + 36$$

The terms form a finite arithmetic series with $n = 36$, $a_1 = 1$, and $a_{36} = 36$, so we can use formula (4):

$$S_n = \frac{n}{2}(a_1 + a_n)$$

$$S_{36} = \frac{36}{2}(1 + 36) = \$666$$

We conclude that the total cost of the loan over 3 years is \$666.

Matched Problem 6 Repeat Example 6 with a loan of \$6,000 over 5 years.

EXAMPLE 7 **Economy Stimulation** The government has decided on a tax rebate program to stimulate the economy. Suppose that you receive \$1,200 and you spend 80% of this, and each of the people who receive what you spend also spend 80% of what they receive, and this process continues without end. According to the **multiplier principle** in economics, the effect of your \$1,200 tax rebate on the economy is multiplied many times. What is the total amount spent if the process continues as indicated?

SOLUTION We need to find the sum of an infinite geometric series with the first amount spent being $a_1 = (0.8)(\$1{,}200) = \$960$ and $r = 0.8$. Using formula (7), we obtain

$$S_\infty = \frac{a_1}{1 - r}$$

$$= \frac{\$960}{1 - 0.8} = \$4{,}800$$

Assuming the process continues as indicated, we would expect the \$1,200 tax rebate to result in about \$4,800 of spending.

Matched Problem 7 Repeat Example 7 with a tax rebate of \$2,000.

Exercises B-2

A

In Problems 1 and 2, determine whether the indicated sequence can be the first three terms of an arithmetic or geometric sequence, and, if so, find the common difference or common ratio and the next two terms of the sequence.

1. (A) $-11, -16, -21, \ldots$ (B) $2, -4, 8, \ldots$

(C) $1, 4, 9, \ldots$ (D) $\frac{1}{2}, \frac{1}{6}, \frac{1}{18}, \ldots$

2. (A) $5, 20, 100, \ldots$ (B) $-5, -5, -5, \ldots$

(C) $7, 6.5, 6, \ldots$ (D) $512, 256, 128, \ldots$

B

In Problems 3–8, determine whether the finite series is arithmetic, geometric, both, or neither. If the series is arithmetic or geometric, find its sum.

3. $\sum_{k=1}^{101}(-1)^{k+1}$ **4.** $\sum_{k=1}^{200} 3$

5. $1 + \frac{1}{2} + \frac{1}{3} + \cdots + \frac{1}{50}$

6. $3 - 9 + 27 - \cdots - 3^{20}$

7. $5 + 4.9 + 4.8 + \cdots + 0.1$

8. $1 - \frac{1}{4} + \frac{1}{9} - \cdots - \frac{1}{100^2}$

Let $a_1, a_2, a_3, \ldots, a_n, \ldots$ be an arithmetic sequence. In Problems 9–14, find the indicated quantities.

9. $a_1 = 7; d = 4; a_2 = ?; a_3 = ?$

10. $a_1 = -2; d = -3; a_2 = ?; a_3 = ?$

11. $a_1 = 2; d = 4; a_{21} = ?; S_{31} = ?$

12. $a_1 = 8; d = -10; a_{15} = ?; S_{23} = ?$

13. $a_1 = 18; a_{20} = 75; S_{20} = ?$

14. $a_1 = 203; a_{30} = 261; S_{30} = ?$

Let $a_1, a_2, a_3, \ldots, a_n, \ldots$ be a geometric sequence. In Problems 15–24, find the indicated quantities.

15. $a_1 = 3; r = -2; a_2 = ?; a_3 = ?; a_4 = ?$

16. $a_1 = 32; r = -\frac{1}{2}; a_2 = ?; a_3 = ?; a_4 = ?$

17. $a_1 = 1; a_7 = 729; r = -3; S_7 = ?$

18. $a_1 = 3; a_7 = 2{,}187; r = 3; S_7 = ?$

19. $a_1 = 100; r = 1.08; a_{10} = ?$

20. $a_1 = 240; r = 1.06; a_{12} = ?$

21. $a_1 = 100; a_9 = 200; r = ?$

22. $a_1 = 100; a_{10} = 300; r = ?$

23. $a_1 = 500; r = 0.6; S_{10} = ?; S_\infty = ?$

24. $a_1 = 8{,}000; r = 0.4; S_{10} = ?; S_\infty = ?$

25. $S_{41} = \sum_{k=1}^{41}(3k + 3) = ?$ **26.** $S_{50} = \sum_{k=1}^{50}(2k - 3) = ?$

27. $S_8 = \sum_{k=1}^{8}(-2)^{k-1} = ?$ **28.** $S_8 = \sum_{k=1}^{8} 2^k = ?$

29. Find the sum of all the odd integers between 12 and 68.

30. Find the sum of all the even integers between 23 and 97.

31. Find the sum of each infinite geometric sequence (if it exists).

(A) $2, 4, 8, \ldots$ (B) $2, -\frac{1}{2}, \frac{1}{8}, \ldots$

32. Repeat Problem 31 for:

(A) $16, 4, 1, \ldots$ (B) $1, -3, 9, \ldots$

C

33. Find $f(1) + f(2) + f(3) + \cdots + f(50)$ if $f(x) = 2x - 3$.

34. Find $g(1) + g(2) + g(3) + \cdots + g(100)$ if $g(t) = 18 - 3t$.

35. Find $f(1) + f(2) + \cdots + f(10)$ if $f(x) = \left(\frac{1}{2}\right)^x$.

36. Find $g(1) + g(2) + \cdots + g(10)$ if $g(x) = 2^x$.

37. Show that the sum of the first n odd positive integers is n^2, using appropriate formulas from this section.

38. Show that the sum of the first n even positive integers is $n + n^2$, using formulas in this section.

39. If $r = 1$, neither the first form nor the second form for the sum of a finite geometric series is valid. Find a formula for the sum of a finite geometric series if $r = 1$.

40. If all of the terms of an infinite geometric series are less than 1, could the sum be greater than 1,000? Explain.

41. Does there exist a finite arithmetic series with $a_1 = 1$ and $a_n = 1.1$ that has sum equal to 100? Explain.

42. Does there exist a finite arithmetic series with $a_1 = 1$ and $a_n = 1.1$ that has sum equal to 105? Explain.

43. Does there exist an infinite geometric series with $a_1 = 10$ that has sum equal to 6? Explain.

44. Does there exist an infinite geometric series with $a_1 = 10$ that has sum equal to 5? Explain.

Applications

45. **Loan repayment.** If you borrow \$4,800 and repay the loan by paying \$200 per month to reduce the loan and 1% of the unpaid balance each month for the use of the money, what is the total cost of the loan over 24 months?

46. **Loan repayment.** If you borrow \$5,400 and repay the loan by paying \$300 per month to reduce the loan and 1.5% of the unpaid balance each month for the use of the money, what is the total cost of the loan over 18 months?

47. **Economy stimulation.** The government, through a subsidy program, distributes \$5,000,000. If we assume that each person or agency spends 70% of what is received, and 70% of this is spent, and so on, how much total increase in spending results from this government action? (Let $a_1 = \$3{,}500{,}000$.)

48. **Economy stimulation.** Due to reduced taxes, a person has an extra $1,200 in spendable income. If we assume that the person spends 65% of this on consumer goods, and the producers of these goods in turn spend 65% on consumer goods, and that this process continues indefinitely, what is the total amount spent (to the nearest dollar) on consumer goods?

49. **Compound interest.** If $1,000 is invested at 5% compounded annually, the amount A present after n years forms a geometric sequence with common ratio $1 + 0.05 = 1.05$. Use a geometric sequence formula to find the amount A in the account (to the nearest cent) after 10 years. After 20 years. (*Hint*: Use a time line.)

50. **Compound interest.** If $\$P$ is invested at $100r\%$ compounded annually, the amount A present after n years forms a geometric sequence with common ratio $1 + r$. Write a formula for the amount present after n years. (*Hint*: Use a time line.)

Answers to Matched Problems

1. (A) The sequence is geometric with $r = \frac{1}{4}$. It is not arithmetic.

(B) The sequence is arithmetic with $d = 5$. It is not geometric.

(C) The sequence is neither arithmetic nor geometric.

2. (A) 139 (B) -2

3. -960

4. 10,672

5. 1,063.66

6. $1,830

7. $8,000

B-3 Binomial Theorem

- Factorial
- Development of the Binomial Theorem

The binomial form

$$(a + b)^n$$

where n is a natural number, appears more frequently than you might expect. The coefficients in the expansion play an important role in probability studies. The *binomial formula,* which we will derive informally, enables us to expand $(a + b)^n$ directly for n any natural number. Since the formula involves *factorials*, we digress for a moment here to introduce this important concept.

Factorial

For n a natural number, ***n* factorial**, denoted by $n!$, is the product of the first n natural numbers. **Zero factorial** is defined to be 1. That is,

DEFINITION *n* Factorial

$$n! = n \cdot (n - 1) \cdot \cdots \cdot 2 \cdot 1$$
$$1! = 1$$
$$0! = 1$$

It is also useful to note that $n!$ can be defined recursively.

DEFINITION *n* Factorial—Recursive Definition

$$n! = n \cdot (n - 1)! \qquad n \geq 1$$

EXAMPLE 1 **Factorial Forms** Evaluate.

(A) $5! = 5 \cdot 4 \cdot 3 \cdot 2 \cdot 1 = 120$

(B) $\dfrac{8!}{7!} = \dfrac{8 \cdot \cancel{7!}}{\cancel{7!}} = 8$

(C) $\dfrac{10!}{7!} = \dfrac{10 \cdot 9 \cdot 8 \cdot \cancel{7!}}{\cancel{7!}} = 720$

Matched Problem 1 Evaluate.

(A) $4!$ (B) $\frac{7!}{6!}$ (C) $\frac{8!}{5!}$

The following formula involving factorials has applications in many areas of mathematics and statistics. We will use this formula to provide a more concise form for the expressions encountered later in this discussion.

THEOREM 1 **For n and r integers satisfying $0 \le r \le n$,**

$$C_{n,r} = \frac{n!}{r!(n-r)!}$$

EXAMPLE 2 **Evaluating $C_{n,r}$**

(A) $C_{9,2} = \frac{9!}{2!(9-2)!} = \frac{9!}{2!7!} = \frac{9 \cdot 8 \cdot \cancel{7!}}{2 \cdot \cancel{7!}} = 36$

(B) $C_{5,5} = \frac{5!}{5!(5-5)!} = \frac{5!}{5!0!} = \frac{5!}{5!} = 1$

Matched Problem 2 Find

(A) $C_{5,2}$ (B) $C_{6,0}$

Development of the Binomial Theorem

Let us expand $(a + b)^n$ for several values of n to see if we can observe a pattern that leads to a general formula for the expansion for any natural number n:

$$(a+b)^1 = a + b$$
$$(a+b)^2 = a^2 + 2ab + b^2$$
$$(a+b)^3 = a^3 + 3a^2b + 3ab^2 + b^3$$
$$(a+b)^4 = a^4 + 4a^3b + 6a^2b^2 + 4ab^3 + b^4$$
$$(a+b)^5 = a^5 + 5a^4b + 10a^3b^2 + 10a^2b^3 + 5ab^4 + b^5$$

CONCEPTUAL INSIGHT

1. The expansion of $(a + b)^n$ has $(n + 1)$ terms.
2. The power of a decreases by 1 for each term as we move from left to right.
3. The power of b increases by 1 for each term as we move from left to right.
4. In each term, the sum of the powers of a and b always equals n.
5. Starting with a given term, we can get the coefficient of the next term by multiplying the coefficient of the given term by the exponent of a and dividing by the number that represents the position of the term in the series of terms. For example, in the expansion of $(a + b)^4$ above, the coefficient of the third term is found from the second term by multiplying 4 and 3, and then dividing by 2 [that is, the coefficient of the third term $= (4 \cdot 3)/2 = 6$].

We now postulate these same properties for the general case:

$$(a+b)^n = a^n + \frac{n}{1}a^{n-1}b + \frac{n(n-1)}{1 \cdot 2}a^{n-2}b^2 + \frac{n(n-1)(n-2)}{1 \cdot 2 \cdot 3}a^{n-3}b^3 + \cdots + b^n$$
$$= \frac{n!}{0!(n-0)!}a^n + \frac{n!}{1!(n-1)!}a^{n-1}b + \frac{n!}{2!(n-2)!}a^{n-2}b^2 + \frac{n!}{3!(n-3)!}a^{n-3}b^3 + \cdots + \frac{n!}{n!(n-n)!}b^n$$
$$= C_{n,0}a^n + C_{n,1}a^{n-1}b + C_{n,2}a^{n-2}b^2 + C_{n,3}a^{n-3}b^3 + \cdots + C_{n,n}b^n$$

And we are led to the formula in the binomial theorem:

THEOREM 2 Binomial Theorem

For all natural numbers n,

$$(a + b)^n = C_{n,0}a^n + C_{n,1}a^{n-1}b + C_{n,2}a^{n-2}b^2 + C_{n,3}a^{n-3}b^3 + \cdots + C_{n,n}b^n$$

EXAMPLE 3 **Using the Binomial Theorem** Use the binomial theorem to expand $(u + v)^6$.

SOLUTION

$$(u + v)^6 = C_{6,0}u^6 + C_{6,1}u^5v + C_{6,2}u^4v^2 + C_{6,3}u^3v^3 + C_{6,4}u^2v^4 + C_{6,5}uv^5 + C_{6,6}v^6$$
$$= u^6 + 6u^5v + 15u^4v^2 + 20u^3v^3 + 15u^2v^4 + 6uv^5 + v^6$$

Matched Problem 3 Use the binomial theorem to expand $(x + 2)^5$.

EXAMPLE 4 **Using the Binomial Theorem** Use the binomial theorem to find the sixth term in the expansion of $(x - 1)^{18}$.

SOLUTION

$$\text{Sixth term} = C_{18,5}x^{13}(-1)^5 = \frac{18!}{5!(18-5)!}x^{13}(-1)$$
$$= -8{,}568x^{13}$$

Matched Problem 4 Use the binomial theorem to find the fourth term in the expansion of $(x - 2)^{20}$.

Exercises B-3

A

In Problems 1–20, evaluate each expression.

1. $6!$
2. $7!$
3. $\frac{10!}{9!}$
4. $\frac{20!}{19!}$
5. $\frac{12!}{9!}$
6. $\frac{10!}{6!}$
7. $\frac{5!}{2!3!}$
8. $\frac{7!}{3!4!}$
9. $\frac{6!}{5!(6-5)!}$
10. $\frac{7!}{4!(7-4)!}$
11. $\frac{20!}{3!17!}$
12. $\frac{52!}{50!2!}$

B

13. $C_{5,3}$
14. $C_{7,3}$
15. $C_{6,5}$
16. $C_{7,4}$
17. $C_{5,0}$
18. $C_{5,5}$
19. $C_{18,15}$
20. $C_{18,3}$

Expand each expression in Problems 21–26 using the binomial theorem.

21. $(a + b)^4$
22. $(m + n)^5$
23. $(x - 1)^6$
24. $(u - 2)^5$
25. $(2a - b)^5$
26. $(x - 2y)^5$

Find the indicated term in each expansion in Problems 27–32.

27. $(x - 1)^{18}$; 5th term
28. $(x - 3)^{20}$; 3rd term
29. $(p + q)^{15}$; 7th term
30. $(p + q)^{15}$; 13th term
31. $(2x + y)^{12}$; 11th term
32. $(2x + y)^{12}$; 3rd term

C

33. Show that $C_{n,0} = C_{n,n}$ for $n \geq 0$.
34. Show that $C_{n,r} = C_{n,n-r}$ for $n \geq r \geq 0$.
35. The triangle next is called **Pascal's triangle**. Can you guess what the next two rows at the bottom are? Compare these numbers with the coefficients of binomial expansions.

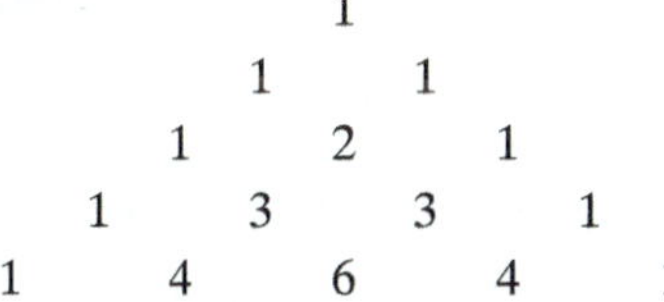

36. Explain why the sum of the entries in each row of Pascal's triangle is a power of 2. (*Hint:* Let $a = b = 1$ in the binomial theorem.)
37. Explain why the alternating sum of the entries in each row of Pascal's triangle (e.g., $1 - 4 + 6 - 4 + 1$) is equal to 0.
38. Show that $C_{n,r} = \frac{n - r + 1}{r}C_{n,r-1}$ for $n \geq r \geq 1$.
39. Show that $C_{n,r-1} + C_{n,r} = C_{n+1,r}$ for $n \geq r \geq 1$.

Answers to Matched Problems

1. (A) 24 (B) 7 (C) 336
2. (A) 10 (B) 1
3. $x^5 + 10x^4 + 40x^3 + 80x^2 + 80x + 32$
4. $-9{,}120x^{17}$

APPENDIX C Tables

Table I Basic Geometric Formulas

1. Similar Triangles

(A) Two triangles are similar if two angles of one triangle have the same measure as two angles of the other.

(B) If two triangles are similar, their corresponding sides are proportional:

$$\frac{a}{a'} = \frac{b}{b'} = \frac{c}{c'}$$

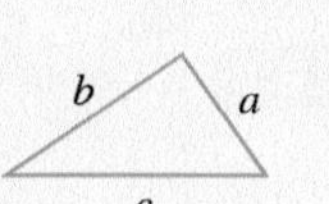

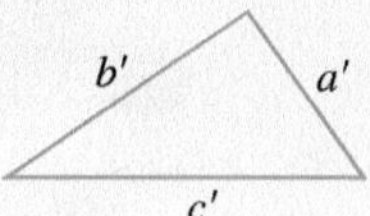

2. Pythagorean Theorem

$c^2 = a^2 + b^2$

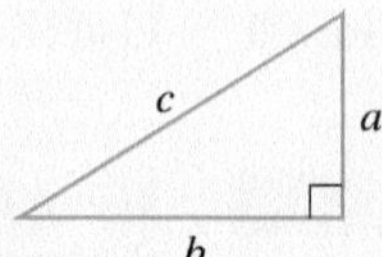

3. Rectangle

$A = ab$ Area

$P = 2a + 2b$ Perimeter

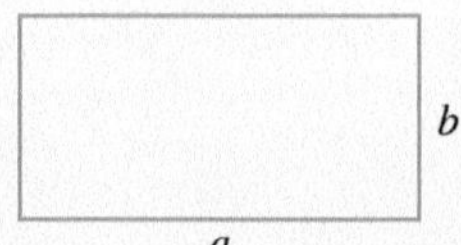

4. Parallelogram

$h =$ height

$A = ah = ab \sin \theta$ Area

$P = 2a + 2b$ Perimeter

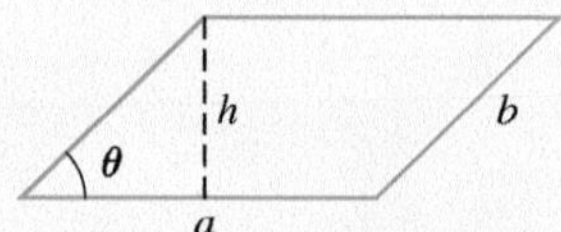

5. Triangle

$h =$ height

$A = \frac{1}{2}hc$ Area

$P = a + b + c$ Perimeter

$s = \frac{1}{2}(a + b + c)$ Semiperimeter

$A = \sqrt{s(s - a)(s - b)(s - c)}$ Area: Heron's formula

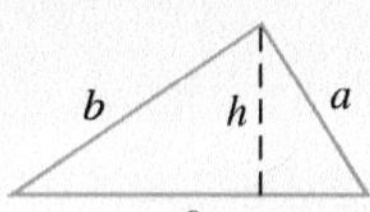

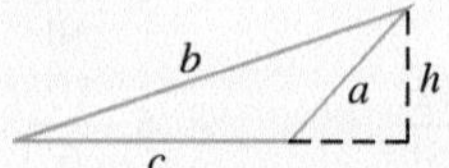

6. Trapezoid

Base a is parallel to base b.

$h =$ height

$A = \frac{1}{2}(a + b)h$ Area

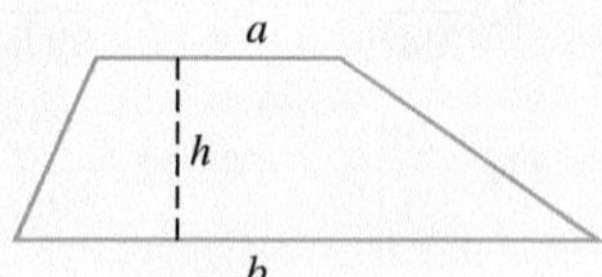

7. Circle

$R =$ radius

$D =$ diameter

$D = 2R$

$A = \pi R^2 = \frac{1}{4}\pi D^2$ Area

$C = 2\pi R = \pi D$ Circumference

$\frac{C}{D} = \pi$ For all circles

$\pi \approx 3.14159$

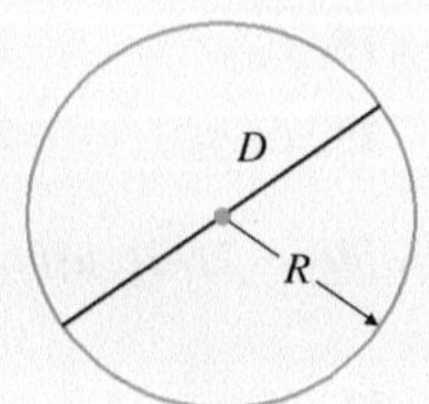

8. Rectangular Solid

$V = abc$ Volume

$T = 2ab + 2ac + 2bc$ Total surface area

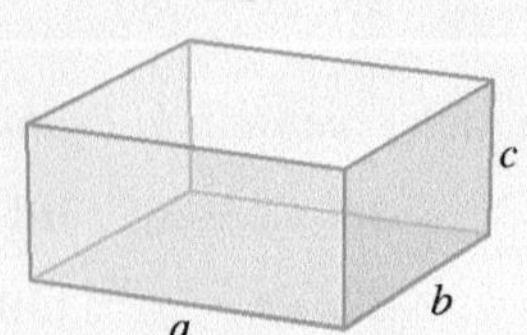

Table I Basic Geometric Formulas (continued)

9. Right Circular Cylinder

R = radius of base
h = height
$V = \pi R^2 h$ Volume
$S = 2\pi Rh$ Lateral surface area
$T = 2\pi R(R + h)$ Total surface area

10. Right Circular Cone

R = radius of base
h = height
s = slant height
$V = \frac{1}{3}\pi R^2 h$ Volume
$S = \pi Rs = \pi R\sqrt{R^2 + h^2}$ Lateral surface area
$T = \pi R(R + s) = \pi R(R + \sqrt{R^2 + h^2})$ Total surface area

11. Sphere

R = radius
D = diameter
$D = 2R$
$V = \frac{4}{3}\pi R^3 = \frac{1}{6}\pi D^3$ Volume
$S = 4\pi R^2 = \pi D^2$ Surface area

Table II Integration Formulas

[*Note:* The constant of integration is omitted for each integral, but must be included in any particular application of a formula. The variable u is the variable of integration; all other symbols represent constants.]

Integrals Involving u^n

1. $\int u^n\,du = \frac{u^{n+1}}{n+1}, \quad n \neq -1$

2. $\int u^{-1}\,du = \int \frac{1}{u}\,du = \ln|u|$

Integrals Involving $a + bu$, $a \neq 0$ and $b \neq 0$

3. $\int \frac{1}{a+bu}\,du = \frac{1}{b}\ln|a+bu|$

4. $\int \frac{u}{a+bu}\,du = \frac{u}{b} - \frac{a}{b^2}\ln|a+bu|$

5. $\int \frac{u^2}{a+bu}\,du = \frac{(a+bu)^2}{2b^3} - \frac{2a(a+bu)}{b^3} + \frac{a^2}{b^3}\ln|a+bu|$

6. $\int \frac{u}{(a+bu)^2}\,du = \frac{1}{b^2}\left(\ln|a+bu| + \frac{a}{a+bu}\right)$

7. $\int \frac{u^2}{(a+bu)^2}\,du = \frac{(a+bu)}{b^3} - \frac{a^2}{b^3(a+bu)} - \frac{2a}{b^3}\ln|a+bu|$

8. $\int u(a+bu)^n\,du = \frac{(a+bu)^{n+2}}{(n+2)b^2} - \frac{a(a+bu)^{n+1}}{(n+1)b^2}, \quad n \neq -1, -2$

9. $\int \frac{1}{u(a+bu)}\,du = \frac{1}{a}\ln\left|\frac{u}{a+bu}\right|$

10. $\int \frac{1}{u^2(a+bu)}\,du = -\frac{1}{au} + \frac{b}{a^2}\ln\left|\frac{a+bu}{u}\right|$

11. $\int \frac{1}{u(a+bu)^2}\,du = \frac{1}{a(a+bu)} + \frac{1}{a^2}\ln\left|\frac{u}{a+bu}\right|$

12. $\int \frac{1}{u^2(a+bu)^2}\,du = -\frac{a+2bu}{a^2u(a+bu)} + \frac{2b}{a^3}\ln\left|\frac{a+bu}{u}\right|$

Integrals Involving $a^2 - u^2$, $a > 0$

13. $\int \frac{1}{u^2-a^2}\,du = \frac{1}{2a}\ln\left|\frac{u-a}{u+a}\right|$

14. $\int \frac{1}{a^2-u^2}\,du = \frac{1}{2a}\ln\left|\frac{u+a}{u-a}\right|$

Integrals Involving $(a + bu)$ and $(c + du)$, $b \neq 0$, $d \neq 0$, and $ad - bc \neq 0$

15. $\int \frac{1}{(a+bu)(c+du)}\,du = \frac{1}{ad-bc}\ln\left|\frac{c+du}{a+bu}\right|$

16. $\int \frac{u}{(a+bu)(c+du)}\,du = \frac{1}{ad-bc}\left(\frac{a}{b}\ln|a+bu| - \frac{c}{d}\ln|c+du|\right)$

17. $\int \frac{u^2}{(a+bu)(c+du)}\,du = \frac{1}{bd}u - \frac{1}{ad-bc}\left(\frac{a^2}{b^2}\ln|a+bu| - \frac{c^2}{d^2}\ln|c+du|\right)$

18. $\int \frac{1}{(a+bu)^2(c+du)}\,du = \frac{1}{ad-bc}\frac{1}{a+bu} + \frac{d}{(ad-bc)^2}\ln\left|\frac{c+du}{a+bu}\right|$

19. $\int \frac{u}{(a+bu)^2(c+du)}\,du = -\frac{a}{b(ad-bc)}\frac{1}{a+bu} - \frac{c}{(ad-bc)^2}\ln\left|\frac{c+du}{a+bu}\right|$

20. $\int \frac{a+bu}{c+du}\,du = \frac{bu}{d} + \frac{ad-bc}{d^2}\ln|c+du|$

Table II (continued)

Integrals Involving $\sqrt{a + bu}$, $a \neq 0$ and $b \neq 0$

21. $\displaystyle\int \sqrt{a + bu}\, du = \frac{2\sqrt{(a + bu)^3}}{3b}$

22. $\displaystyle\int u\sqrt{a + bu}\, du = \frac{2(3bu - 2a)}{15b^2}\sqrt{(a + bu)^3}$

23. $\displaystyle\int u^2\sqrt{a + bu}\, du = \frac{2(15b^2u^2 - 12abu + 8a^2)}{105b^3}\sqrt{(a + bu)^3}$

24. $\displaystyle\int \frac{1}{\sqrt{a + bu}}\, du = \frac{2\sqrt{a + bu}}{b}$

25. $\displaystyle\int \frac{u}{\sqrt{a + bu}}\, du = \frac{2(bu - 2a)}{3b^2}\sqrt{a + bu}$

26. $\displaystyle\int \frac{u^2}{\sqrt{a + bu}}\, du = \frac{2(3b^2u^2 - 4abu + 8a^2)}{15b^3}\sqrt{a + bu}$

27. $\displaystyle\int \frac{1}{u\sqrt{a + bu}}\, du = \frac{1}{\sqrt{a}}\ln\left|\frac{\sqrt{a + bu} - \sqrt{a}}{\sqrt{a + bu} + \sqrt{a}}\right|, \quad a > 0$

28. $\displaystyle\int \frac{1}{u^2\sqrt{a + bu}}\, du = -\frac{\sqrt{a + bu}}{au} - \frac{b}{2a\sqrt{a}}\ln\left|\frac{\sqrt{a + bu} - \sqrt{a}}{\sqrt{a + bu} + \sqrt{a}}\right|, \quad a > 0$

Integrals Involving $\sqrt{a^2 - u^2}$, $a > 0$

29. $\displaystyle\int \frac{1}{u\sqrt{a^2 - u^2}}\, du = -\frac{1}{a}\ln\left|\frac{a + \sqrt{a^2 - u^2}}{u}\right|$

30. $\displaystyle\int \frac{1}{u^2\sqrt{a^2 - u^2}}\, du = -\frac{\sqrt{a^2 - u^2}}{a^2u}$

31. $\displaystyle\int \frac{\sqrt{a^2 - u^2}}{u}\, du = \sqrt{a^2 - u^2} - a\ln\left|\frac{a + \sqrt{a^2 - u^2}}{u}\right|$

Integrals Involving $\sqrt{u^2 + a^2}$, $a > 0$

32. $\displaystyle\int \sqrt{u^2 + a^2}\, du = \frac{1}{2}\left(u\sqrt{u^2 + a^2} + a^2\ln\left|u + \sqrt{u^2 + a^2}\right|\right)$

33. $\displaystyle\int u^2\sqrt{u^2 + a^2}\, du = \frac{1}{8}\left[u(2u^2 + a^2)\sqrt{u^2 + a^2} - a^4\ln\left|u + \sqrt{u^2 + a^2}\right|\right]$

34. $\displaystyle\int \frac{\sqrt{u^2 + a^2}}{u}\, du = \sqrt{u^2 + a^2} - a\ln\left|\frac{a + \sqrt{u^2 + a^2}}{u}\right|$

35. $\displaystyle\int \frac{\sqrt{u^2 + a^2}}{u^2}\, du = -\frac{\sqrt{u^2 + a^2}}{u} + \ln\left|u + \sqrt{u^2 + a^2}\right|$

36. $\displaystyle\int \frac{1}{\sqrt{u^2 + a^2}}\, du = \ln\left|u + \sqrt{u^2 + a^2}\right|$

37. $\displaystyle\int \frac{1}{u\sqrt{u^2 + a^2}}\, du = \frac{1}{a}\ln\left|\frac{u}{a + \sqrt{u^2 + a^2}}\right|$

38. $\displaystyle\int \frac{u^2}{\sqrt{u^2 + a^2}}\, du = \frac{1}{2}\left(u\sqrt{u^2 + a^2} - a^2\ln\left|u + \sqrt{u^2 + a^2}\right|\right)$

39. $\displaystyle\int \frac{1}{u^2\sqrt{u^2 + a^2}}\, du = -\frac{\sqrt{u^2 + a^2}}{a^2u}$

(continued)

Table II (continued)

Integrals Involving $\sqrt{u^2 - a^2}, a > 0$

40. $\int \sqrt{u^2 - a^2}\, du = \frac{1}{2}(u\sqrt{u^2 - a^2} - a^2 \ln|u + \sqrt{u^2 - a^2}|)$

41. $\int u^2\sqrt{u^2 - a^2}\, du = \frac{1}{8}[u(2u^2 - a^2)\sqrt{u^2 - a^2} - a^4 \ln|u + \sqrt{u^2 - a^2}|]$

42. $\int \frac{\sqrt{u^2 - a^2}}{u^2}\, du = -\frac{\sqrt{u^2 - a^2}}{u} + \ln|u + \sqrt{u^2 - a^2}|$

43. $\int \frac{1}{\sqrt{u^2 - a^2}}\, du = \ln|u + \sqrt{u^2 - a^2}|$

44. $\int \frac{u^2}{\sqrt{u^2 - a^2}}\, du = \frac{1}{2}(u\sqrt{u^2 - a^2} + a^2 \ln|u + \sqrt{u^2 - a^2}|)$

45. $\int \frac{1}{u^2\sqrt{u^2 - a^2}}\, du = \frac{\sqrt{u^2 - a^2}}{a^2 u}$

Integrals Involving $e^{au}, a \neq 0$

46. $\int e^{au}\, du = \frac{e^{au}}{a}$

47. $\int u^n e^{au}\, du = \frac{u^n e^{au}}{a} - \frac{n}{a}\int u^{n-1}e^{au}\, du$

48. $\int \frac{1}{c + de^{au}}\, du = \frac{u}{c} - \frac{1}{ac}\ln|c + de^{au}|, \quad c \neq 0$

Integrals Involving $\ln u$

49. $\int \ln u\, du = u \ln u - u$

50. $\int \frac{\ln u}{u}\, du = \frac{1}{2}(\ln u)^2$

51. $\int u^n \ln u\, du = \frac{u^{n+1}}{n+1}\ln u - \frac{u^{n+1}}{(n+1)^2}, \quad n \neq -1$

52. $\int (\ln u)^n\, du = u(\ln u)^n - n\int (\ln u)^{n-1}\, du$

Integrals Involving Trigonometric Functions of $au, a \neq 0$

53. $\int \sin au\, du = -\frac{1}{a}\cos au$

54. $\int \cos au\, du = \frac{1}{a}\sin au$

55. $\int \tan au\, du = -\frac{1}{a}\ln|\cos au|$

56. $\int \cot au\, du = \frac{1}{a}\ln|\sin au|$

57. $\int \sec au\, du = \frac{1}{a}\ln|\sec au + \tan au|$

58. $\int \csc au\, du = \frac{1}{a}\ln|\csc au - \cot au|$

59. $\int (\sin au)^2\, du = \frac{u}{2} - \frac{1}{4a}\sin 2au$

60. $\int (\cos au)^2\, du = \frac{u}{2} + \frac{1}{4a}\sin 2au$

61. $\int (\sin au)^n\, du = -\frac{1}{an}(\sin au)^{n-1}\cos au + \frac{n-1}{n}\int (\sin au)^{n-2}\, du, \quad n \neq 0$

62. $\int (\cos au)^n\, du = \frac{1}{an}\sin au(\cos au)^{n-1} + \frac{n-1}{n}\int (\cos au)^{n-2}\, du, \quad n \neq 0$

TABLE III Area Under the Standard Normal Curve

(Table Entries Represent the Area Under the Standard Normal Curve from 0 to z, $z \geq 0$)

z	.00	.01	.02	.03	.04	.05	.06	.07	.08	.09
0.0	0.0000	0.0040	0.0080	0.0120	0.0160	0.0199	0.0239	0.0279	0.0319	0.0359
0.1	0.0398	0.0438	0.0478	0.0517	0.0557	0.0596	0.0636	0.0675	0.0714	0.0753
0.2	0.0793	0.0832	0.0871	0.0910	0.0948	0.0987	0.1026	0.1064	0.1103	0.1141
0.3	0.1179	0.1217	0.1255	0.1293	0.1331	0.1368	0.1406	0.1443	0.1480	0.1517
0.4	0.1554	0.1591	0.1628	0.1664	0.1700	0.1736	0.1772	0.1808	0.1844	0.1879
0.5	0.1915	0.1950	0.1985	0.2019	0.2054	0.2088	0.2123	0.2157	0.2190	0.2224
0.6	0.2257	0.2291	0.2324	0.2357	0.2389	0.2422	0.2454	0.2486	0.2517	0.2549
0.7	0.2580	0.2611	0.2642	0.2673	0.2704	0.2734	0.2764	0.2794	0.2823	0.2852
0.8	0.2881	0.2910	0.2939	0.2967	0.2995	0.3023	0.3051	0.3078	0.3106	0.3133
0.9	0.3159	0.3186	0.3212	0.3238	0.3264	0.3289	0.3315	0.3340	0.3365	0.3389
1.0	0.3413	0.3438	0.3461	0.3485	0.3508	0.3531	0.3554	0.3577	0.3599	0.3621
1.1	0.3643	0.3665	0.3686	0.3708	0.3729	0.3749	0.3770	0.3790	0.3810	0.3830
1.2	0.3849	0.3869	0.3888	0.3907	0.3925	0.3944	0.3962	0.3980	0.3997	0.4015
1.3	0.4032	0.4049	0.4066	0.4082	0.4099	0.4115	0.4131	0.4147	0.4162	0.4177
1.4	0.4192	0.4207	0.4222	0.4236	0.4251	0.4265	0.4279	0.4292	0.4306	0.4319
1.5	0.4332	0.4345	0.4357	0.4370	0.4382	0.4394	0.4406	0.4418	0.4429	0.4441
1.6	0.4452	0.4463	0.4474	0.4484	0.4495	0.4505	0.4515	0.4525	0.4535	0.4545
1.7	0.4554	0.4564	0.4573	0.4582	0.4591	0.4599	0.4608	0.4616	0.4625	0.4633
1.8	0.4641	0.4649	0.4656	0.4664	0.4671	0.4678	0.4686	0.4693	0.4699	0.4706
1.9	0.4713	0.4719	0.4726	0.4732	0.4738	0.4744	0.4750	0.4756	0.4761	0.4767
2.0	0.4772	0.4778	0.4783	0.4788	0.4793	0.4798	0.4803	0.4808	0.4812	0.4817
2.1	0.4821	0.4826	0.4830	0.4834	0.4838	0.4842	0.4846	0.4850	0.4854	0.4857
2.2	0.4861	0.4864	0.4868	0.4871	0.4875	0.4878	0.4881	0.4884	0.4887	0.4890
2.3	0.4893	0.4896	0.4898	0.4901	0.4904	0.4906	0.4909	0.4911	0.4913	0.4916
2.4	0.4918	0.4920	0.4922	0.4925	0.4927	0.4929	0.4931	0.4932	0.4934	0.4936
2.5	0.4938	0.4940	0.4941	0.4943	0.4945	0.4946	0.4948	0.4949	0.4951	0.4952
2.6	0.4953	0.4955	0.4956	0.4957	0.4959	0.4960	0.4961	0.4962	0.4963	0.4964
2.7	0.4965	0.4966	0.4967	0.4968	0.4969	0.4970	0.4971	0.4972	0.4973	0.4974
2.8	0.4974	0.4975	0.4976	0.4977	0.4977	0.4978	0.4979	0.4979	0.4980	0.4981
2.9	0.4981	0.4982	0.4982	0.4983	0.4984	0.4984	0.4985	0.4985	0.4986	0.4986
3.0	0.4987	0.4987	0.4987	0.4988	0.4988	0.4989	0.4989	0.4989	0.4990	0.4990
3.1	0.4990	0.4991	0.4991	0.4991	0.4992	0.4992	0.4992	0.4992	0.4993	0.4993
3.2	0.4993	0.4993	0.4994	0.4994	0.4994	0.4994	0.4994	0.4995	0.4995	0.4995
3.3	0.4995	0.4995	0.4995	0.4996	0.4996	0.4996	0.4996	0.4996	0.4996	0.4997
3.4	0.4997	0.4997	0.4997	0.4997	0.4997	0.4997	0.4997	0.4997	0.4997	0.4998
3.5	0.4998	0.4998	0.4998	0.4998	0.4998	0.4998	0.4998	0.4998	0.4998	0.4998
3.6	0.4998	0.4998	0.4999	0.4999	0.4999	0.4999	0.4999	0.4999	0.4999	0.4999
3.7	0.4999	0.4999	0.4999	0.4999	0.4999	0.4999	0.4999	0.4999	0.4999	0.4999
3.8	0.4999	0.4999	0.4999	0.4999	0.4999	0.4999	0.4999	0.4999	0.4999	0.4999
3.9	0.5000	0.5000	0.5000	0.5000	0.5000	0.5000	0.5000	0.5000	0.5000	0.5000

Taken from *Additional Calculus Topics*, by Raymond A. Barnett, Michael R. Ziegler, and Karl E. Byleen.

APPENDIX D Exponential Functions

- Exponential Functions
- Base e Exponential Functions
- Growth and Decay Applications
- Compound Interest
- Continuous Compound Interest

This section introduces the important class of functions called *exponential functions.* These functions are used extensively in modeling and solving a wide variety of real-world problems, including growth of money at compound interest; growth of populations of people, animals, and bacteria; radioactive decay; and learning associated with the mastery of such devices as a new computer or an assembly process in a manufacturing plant.

Exponential Functions

We start by noting that

$$f(x) = 2^x \quad \text{and} \quad g(x) = x^2$$

are not the same function. Whether a variable appears as an exponent with a constant base or as a base with a constant exponent makes a big difference. The function g is a quadratic function, which we have already discussed. The function f is a new type of function called an *exponential function.* In general,

DEFINITION Exponential Function

The equation

$$f(x) = b^x \qquad b > 0, b \neq 1$$

defines an **exponential function** for each different constant b, called the **base.** The **domain** of f is the set of all real numbers, and the **range** of f is the set of all positive real numbers.

We require the base b to be positive to avoid imaginary numbers such as $(-2)^{1/2} = \sqrt{-2} = i\sqrt{2}$. We exclude $b = 1$ as a base, since $f(x) = 1^x = 1$ is a constant function, which we have already considered.

Asked to hand-sketch graphs of equations such as $y = 2^x$ or $y = 2^{-x}$, many students would not hesitate at all. [*Note:* $2^{-x} = 1/2^x = (1/2)^x$.] They would probably make up tables by assigning integers to x, plot the resulting points, and then join these points with a smooth curve as in Figure 1. The only catch is that we have not defined 2^x for all real numbers. We know what 2^5, 2^{-3}, $2^{2/3}$, $2^{-3/5}$, $2^{1.4}$, and $2^{-3.14}$ mean (that is, 2^p, where p is a rational number), but what does

$$2^{\sqrt{2}}$$

mean? The question is not easy to answer at this time. In fact, a precise definition of $2^{\sqrt{2}}$ must wait for more advanced courses, where it is shown that

$$2^x$$

names a positive real number for x any real number, and that the graph of $y = 2^x$ is indeed as indicated in Figure 1.

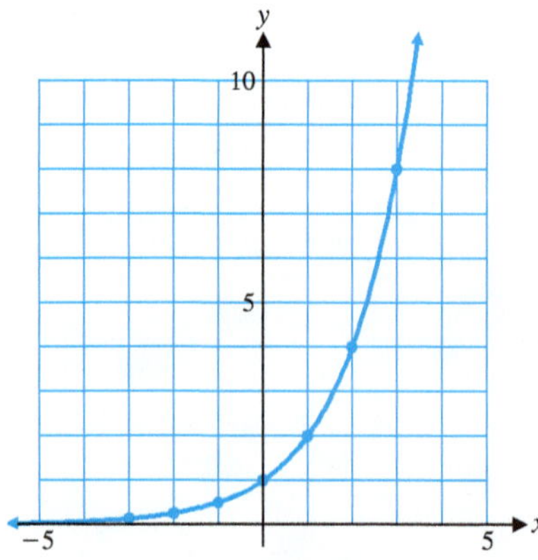

Figure 1 $y = 2^x$

It is useful to compare the graphs of $y = 2^x$ and $y = 2^{-x}$ by plotting both on the same set of coordinate axes, as shown in Figure 2A. The graph of

$$f(x) = b^x \quad b > 1 \text{ (Fig. 2B)}$$

looks very much like the graph of $y = 2^x$, and the graph of

$$f(x) = b^x \quad 0 < b < 1 \text{ (Fig. 2B)}$$

looks very much like the graph of $y = 2^{-x}$. Note that in both cases the x axis is a horizontal asymptote for the graphs.

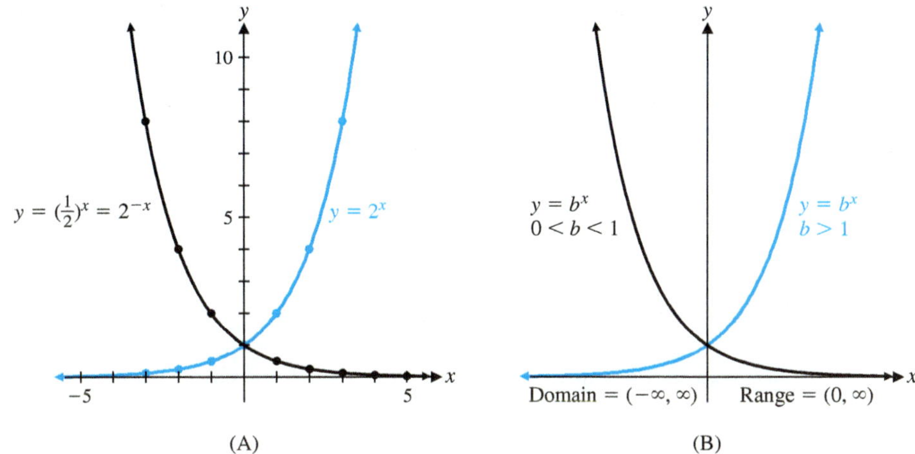

Figure 2 Exponential functions

The graphs in Figure 2 suggest the following important general properties of exponential functions, which we state without proof:

THEOREM 1 Basic Properties of the Graph of $f(x) = b^x, b > 0, b \neq 1$

1. All graphs will pass through the point (0, 1). *$b^0 = 1$ for any permissible base b.*
2. All graphs are continuous curves, with no holes or jumps.
3. The x axis is a horizontal asymptote.
4. If $b > 1$, then b^x increases as x increases.
5. If $0 < b < 1$, then b^x decreases as x increases.

CONCEPTUAL INSIGHT

Recall that the graph of a rational function has at most one horizontal asymptote and that it approaches the horizontal asymptote (if one exists) both as $x \to \infty$ *and* as $x \to -\infty$. The graph of an exponential function, on the other hand, approaches its horizontal asymptote as $x \to \infty$ *or* as $x \to -\infty$, but not both. In particular, there is no rational function that has the same graph as an exponential function.

The use of a calculator with the key $\boxed{y^x}$, or its equivalent, makes the graphing of exponential functions almost routine. Example 1 illustrates the process.

EXAMPLE 1 **Graphing Exponential Functions** Sketch a graph of $y = (\frac{1}{2})4^x, -2 \le x \le 2$.

SOLUTION Use a calculator to create the table of values shown. Plot these points, and then join them with a smooth curve as in Figure 3.

x	y
−2	0.031
−1	0.125
0	0.50
1	2.00
2	8.00

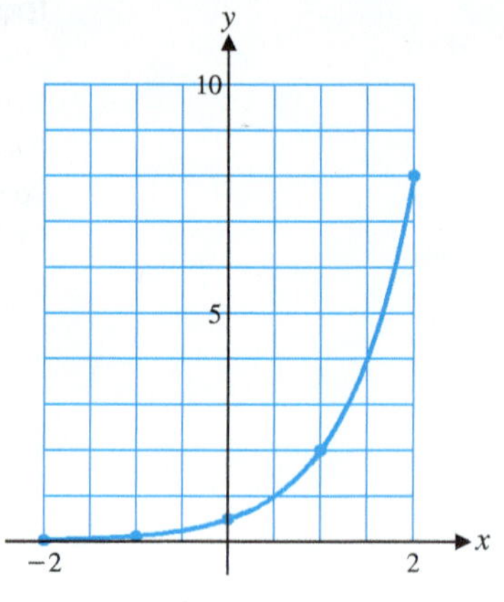

Figure 3 Graph of $y = (\frac{1}{2})4^x$

Matched Problem 1 Sketch a graph of $y = (\frac{1}{2})4^{-x}, -2 \le x \le 2$.

EXPLORE & DISCUSS 1 Graph the functions $f(x) = 2^x$ and $g(x) = 3^x$ on the same set of coordinate axes. At which values of x do the graphs intersect? For which values of x is the graph of f above the graph of g? Below the graph of g? Are the graphs close together as x increases without bound? Are the graphs close together as x decreases without bound? Discuss.

Exponential functions, whose domains include irrational numbers, obey the familiar laws of exponents. We summarize these exponent laws here and add two other important and useful properties.

THEOREM 2 Properties of Exponential Functions

For a and b positive, $a \ne 1$, $b \ne 1$, and x and y real,

1. Exponent laws:

$$a^x a^y = a^{x+y} \qquad \frac{a^x}{a^y} = a^{x-y} \qquad \frac{4^{2y}}{4^{5y}} = 4^{2y-5y} = 4^{-3y}$$

$$(a^x)^y = a^{xy} \qquad (ab)^x = a^x b^x \qquad \left(\frac{a}{b}\right)^x = \frac{a^x}{b^x}$$

2. $a^x = a^y$ if and only if $x = y$ *If* $7^{5t+1} = 7^{3t-3}$, *then* $5t + 1 = 3t - 3$, *and* $t = -2$.

3. For $x \ne 0$,

$a^x = b^x$ if and only if $a = b$ *If* $a^5 = 2^5$, *then* $a = 2$.

Base e Exponential Functions

Of all the possible bases b we can use for the exponential function $y = b^x$, which ones are the most useful? If you look at the keys on a calculator, you will probably see $\boxed{10^x}$ and $\boxed{e^x}$. It is clear why base 10 would be important, because our number system is a base 10 system. But what is e, and why is it included as a base? It turns out that base e is used more frequently than all other bases combined. The reason for this is that certain formulas and the results of certain processes found in calculus and more advanced mathematics take on their simplest form if this base is used. This is why you will see e used extensively in expressions and formulas that model real-world phenomena. In fact, its use is so prevalent that you will often hear people refer to $y = e^x$ as the exponential function.

The base e is an irrational number, and like π, it cannot be represented exactly by any finite decimal fraction. However, e can be approximated as closely as we like by evaluating the expression

$$\left(1 + \frac{1}{x}\right)^x \tag{1}$$

for sufficiently large x. What happens to the value of expression (1) as x increases without bound? Think about this for a moment before proceeding. Maybe you guessed that the value approaches 1, because

$$1 + \frac{1}{x}$$

approaches 1, and 1 raised to any power is 1. Let us see if this reasoning is correct by actually calculating the value of the expression for larger and larger values of x. Table 1 summarizes the results.

Table 1

x	$\left(1 + \frac{1}{x}\right)^x$
1	2
10	2.593 74 ...
100	2.704 81 ...
1,000	2.716 92 ...
10,000	2.718 14 ...
100,000	2.718 27 ...
1,000,000	2.718 28 ...
⋮	⋮

Interestingly, the value of expression (1) is never close to 1, but seems to be approaching a number close to 2.7183. In fact, as x increases without bound, the value of expression (1) approaches an irrational number that we call $\boldsymbol{e}$. The irrational number e to 12 decimal places is

$$e = \mathbf{2.718\ 281\ 828\ 459}$$

Compare this value of e with the value of e^1 from a calculator. Exactly who discovered the constant e is still being debated. It is named after the great Swiss mathematician Leonhard Euler (1707–1783).

DEFINITION Exponential Function with Base e

Exponential functions with base e and base $1/e$, respectively, are defined by

$$y = e^x \quad \text{and} \quad y = e^{-x}$$

Domain: $(-\infty, \infty)$

Range: $(0, \infty)$

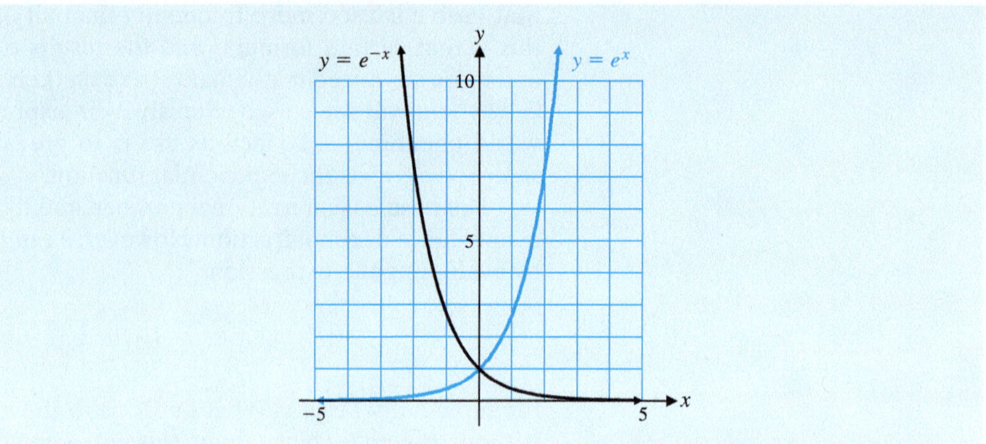

EXPLORE & DISCUSS 2

Graph the functions $f(x) = e^x$, $g(x) = 2^x$, and $h(x) = 3^x$ on the same set of coordinate axes. At which values of x do the graphs intersect? For positive values of x, which of the three graphs lies above the other two? Below the other two? How does your answer change for negative values of x?

Growth and Decay Applications

Most exponential growth and decay problems are modeled using base e exponential functions. We present two applications here and many more in Exercise Appendix D.

EXAMPLE 2 **Exponential Growth** Cholera, an intestinal disease, is caused by a cholera bacterium that multiplies exponentially by cell division as given approximately by

$$N = N_0 e^{1.386t}$$

where N is the number of bacteria present after t hours and N_0 is the number of bacteria present at the start ($t = 0$). If we start with 25 bacteria, how many bacteria (to the nearest unit) will be present:

(A) In 0.6 hour? (B) In 3.5 hours?

SOLUTION Substituting $N_0 = 25$ into the preceding equation, we obtain

$$N = 25e^{1.386t} \quad \text{The graph is shown in Figure 4.}$$

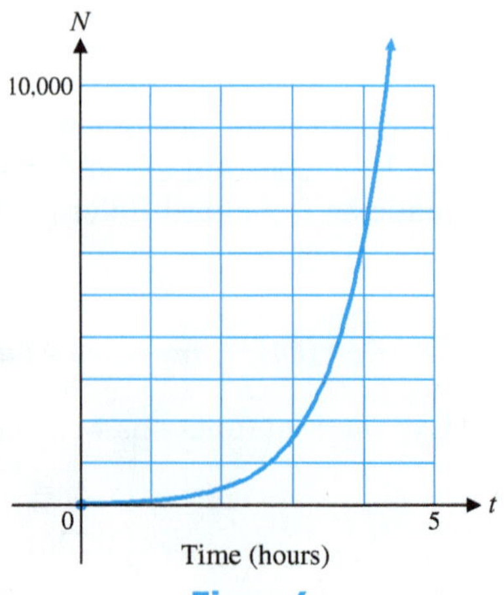

Figure 4

(A) Solve for N when $t = 0.6$:

$$N = 25e^{1.386(0.6)} \quad \text{Use a calculator.}$$
$$= 57 \text{ bacteria}$$

(B) Solve for N when $t = 3.5$:

$$N = 25e^{1.386(3.5)} \quad \text{Use a calculator.}$$
$$= 3{,}197 \text{ bacteria}$$

Matched Problem 2 Refer to the exponential growth model for cholera in Example 2. If we start with 55 bacteria, how many bacteria (to the nearest unit) will be present

(A) In 0.85 hour? (B) In 7.25 hours?

EXAMPLE 3 **Exponential Decay** Cosmic-ray bombardment of the atmosphere produces neutrons, which in turn react with nitrogen to produce radioactive carbon-14 (^{14}C). Radioactive ^{14}C enters all living tissues through carbon dioxide, which is first absorbed by plants. As long as a plant or animal is alive, ^{14}C is maintained in the living organism at a constant level. Once the organism dies, however, ^{14}C decays according to the equation

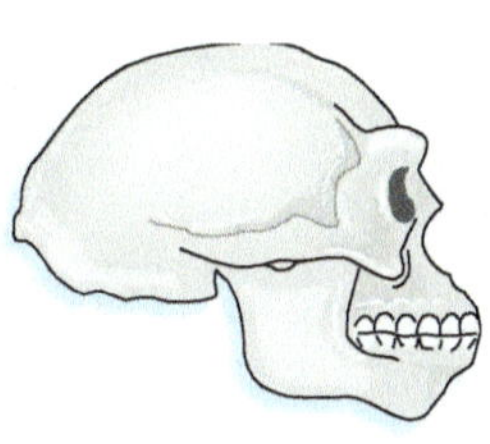

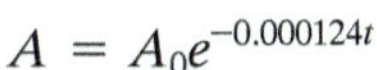

$$A = A_0e^{-0.000124t}$$

where A is the amount present after t years and A_0 is the amount present at time $t = 0$. If 500 milligrams of ^{14}C is present in a sample from a skull at the time of death, how many milligrams will be present in the sample in

(A) 15,000 years? (B) 45,000 years?

Compute answers to two decimal places.

SOLUTION Substituting $A_0 = 500$ in the decay equation, we have

$$A = 500e^{-0.000124t} \quad \text{See the graph in Figure 5.}$$

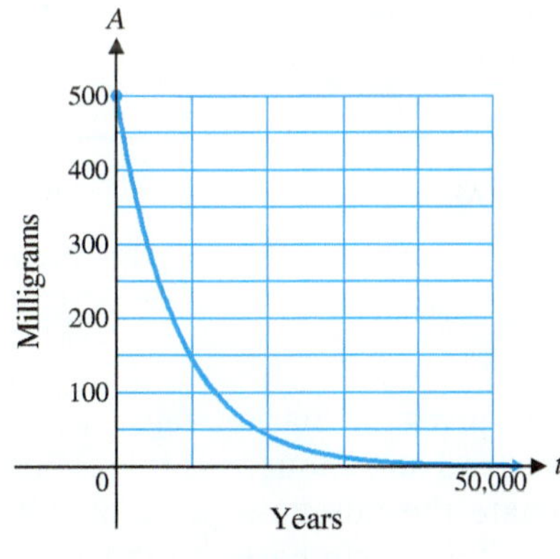

Figure 5

(A) Solve for A when $t = 15{,}000$:

$$A = 500e^{-0.000124(15{,}000)} \quad \text{Use a calculator.}$$
$$= 77.84 \text{ milligrams}$$

(B) Solve for A when $t = 45{,}000$:

$$A = 500e^{-0.000124(45{,}000)} \quad \text{Use a calculator.}$$
$$= 1.89 \text{ milligrams}$$

Matched Problem 3 Refer to the exponential decay model in Example 3. How many milligrams of ^{14}C would have to be present at the beginning in order to have 25 milligrams present after 18,000 years? Compute the answer to the nearest milligram.

EXPLORE & DISCUSS 3

(A) On the same set of coordinate axes, graph the three decay equations $A = A_0e^{-0.35t}$, $t \geq 0$, for $A_0 = 10, 20$, and 30.
(B) Identify any asymptotes for the three graphs in part (A).
(C) Discuss the long-term behavior for the equations in part (A).

EXAMPLE 4 **Depreciation** Table 2 gives the market value of a minivan (in dollars) x years after its purchase. Find an exponential regression model of the form $y = ab^x$ for this data set. Estimate the purchase price of the van. Estimate the value of the van 10 years after its purchase. Round answers to the nearest dollar.

Table 2

x	Value ($)
1	12,575
2	9,455
3	8,115
4	6,845
5	5,225
6	4,485

SOLUTION Enter the data into a graphing utility (Fig. 6A) and find the exponential regression equation (Fig. 6B). The estimated purchase price is $y_1(0) = \$14,910$. The data set and the regression equation are graphed in Figure 6C. Using the trace feature, we see that the estimated value after 10 years is \$1,959.

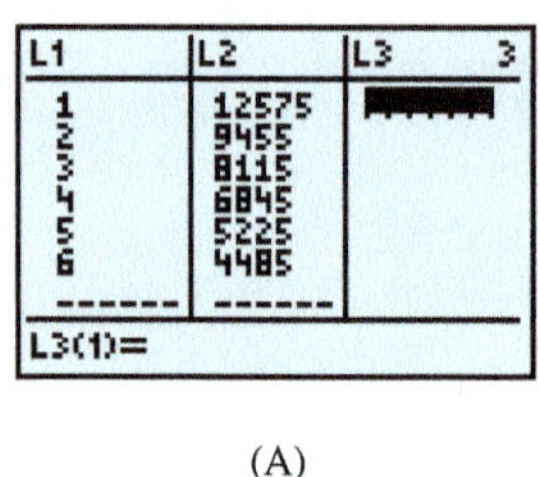

(A)

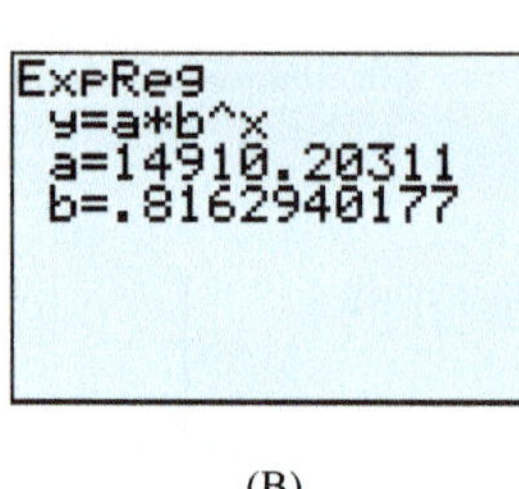

(B)

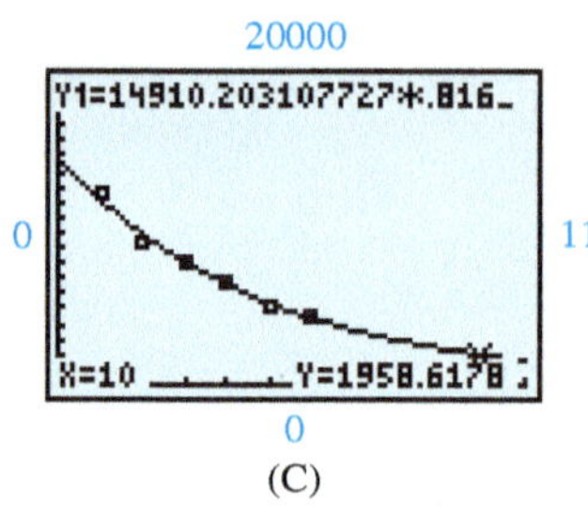

(C)

Figure 6

Matched Problem 4 Table 3 gives the market value of a luxury sedan (in dollars) x years after its purchase. Find an exponential regression model of the form $y = ab^x$ for this data set. Estimate the purchase price of the sedan. Estimate the value of the sedan 10 years after its purchase. Round answers to the nearest dollar.

Table 3

x	Value ($)
1	23,125
2	19,050
3	15,625
4	11,875
5	9,450
6	7,125

Compound Interest

We now turn to the growth of money at compound interest. The fee paid to use another's money is called **interest.** It is usually computed as a percent (called **interest rate**) of the principal over a given period of time. If, at the end of a payment period, the interest due is reinvested at the same rate, then the interest earned as well as the principal will earn interest during the next payment period. Interest paid on interest reinvested is called **compound interest,** and may be calculated using the following compound interest formula:

RESULT Compound Interest

If a **principal P (present value)** is invested at an annual **rate r** (expressed as a decimal) compounded m times a year, then the **amount A (future value)** in the account at the end of t years is given by

$$A = P\left(1 + \frac{r}{m}\right)^{mt}$$

Note: P could be replaced by A_0, but convention dictates otherwise.

EXAMPLE 5 **Compound Growth** If \$1,000 is invested in an account paying 10% compounded monthly, how much will be in the account at the end of 10 years? Compute the answer to the nearest cent.

SOLUTION We use the compound interest formula as follows:

$$\begin{aligned} A &= P\left(1 + \frac{r}{m}\right)^{mt} \\ &= 1{,}000\left(1 + \frac{0.10}{12}\right)^{(12)(10)} \quad \text{Use a calculator.} \\ &= \$2{,}707.04 \end{aligned}$$

The graph of

$$A = 1{,}000\left(1 + \frac{0.10}{12}\right)^{12t}$$

for $0 \le t \le 20$ is shown in Figure 7.

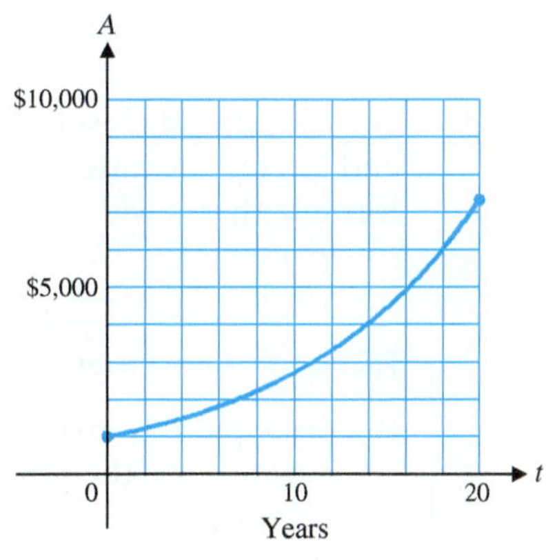

Figure 7

Matched Problem 5 If you deposit \$5,000 in an account paying 9% compounded daily, how much will you have in the account in 5 years? Compute the answer to the nearest cent.

EXPLORE & DISCUSS 4 Suppose that \$1,000 is deposited in a savings account at an annual rate of 5%. Guess the amount in the account at the end of 1 year if interest is compounded (1) quarterly, (2) monthly, (3) daily, (4) hourly. Use the compound interest formula to compute the amounts at the end of 1 year to the nearest cent. Discuss the accuracy of your initial guesses.

Continuous Compound Interest

Returning to the compound interest formula,

$$A = P\left(1 + \frac{r}{m}\right)^{mt}$$

suppose that the principal P, the annual rate r, and the time t are held fixed, and the number of compounding periods per year m is increased without bound. Will the amount A increase without bound, or will it tend to some limiting value?

Starting with $P = \$100$, $r = 0.08$, and $t = 2$ years, we construct Table 4 for several values of m with the aid of a calculator. Notice that the largest gain appears in going from annual to semiannual compounding. Then, the gains slow down as m increases. It appears that A gets closer and closer to \$117.35 as m gets larger and larger.

Table 4

Compounding Frequency	m	$A = 100\left(1 + \frac{0.08}{m}\right)^{2m}$
Annually	1	\$116.6400
Semiannually	2	116.9859
Quarterly	4	117.1659
Weekly	52	117.3367
Daily	365	117.3490
Hourly	8,760	117.3510

It can be shown that

$$P\left(1 + \frac{r}{m}\right)^{mt}$$

gets closer and closer to Pe^{rt} as the number of compounding periods m gets larger and larger. The latter is referred to as the **continuous compound interest formula,** a formula that is widely used in business, banking, and economics.

RESULT Continuous Compound Interest Formula

If a principal P is invested at an annual rate r (expressed as a decimal) compounded continuously, then the amount A in the account at the end of t years is given by

$$A = Pe^{rt}$$

EXAMPLE 6 **Compounding Daily and Continuously** What amount will an account have after 2 years if \$5,000 is invested at an annual rate of 8%

(A) Compounded daily? (B) Compounded continuously?

Compute answers to the nearest cent.

SOLUTION (A) Use the compound interest formula

$$A = P\left(1 + \frac{r}{m}\right)^{mt}$$

with $P = 5{,}000$, $r = 0.08$, $m = 365$, and $t = 2$:

$$A = 5{,}000\left(1 + \frac{0.08}{365}\right)^{(365)(2)} \quad \text{Use a calculator.}$$

$$= \$5{,}867.45$$

(B) Use the continuous compound interest formula

$$A = Pe^{rt}$$

with $P = 5{,}000$, $r = 0.08$, and $t = 2$:

$$A = 5{,}000e^{(0.08)(2)} \quad \text{Use a calculator.}$$

$$= \$5{,}867.55$$

Matched Problem 6 What amount will an account have after 1.5 years if \$8,000 is invested at an annual rate of 9%

(A) Compounded weekly? (B) Compounded continuously?

Compute answers to the nearest cent.

The formulas for simple interest, compound interest, and continuous compound interest are summarized in the box for convenient reference.

SUMMARY **Interest Formulas**

Simple interest $A = P(1 + rt)$

Compound interest $A = P\left(1 + \frac{r}{m}\right)^{mt}$

Continuous compound interest $A = Pe^{rt}$

Answers to Matched Problems

1.

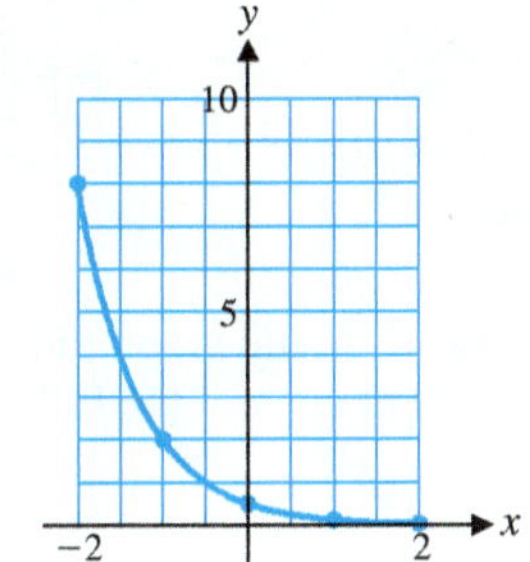

2. (A) 179 bacteria (B) 1,271,659 bacteria
3. 233 mg
4. Purchase price: \$30,363; value after 10 yr: \$2,864

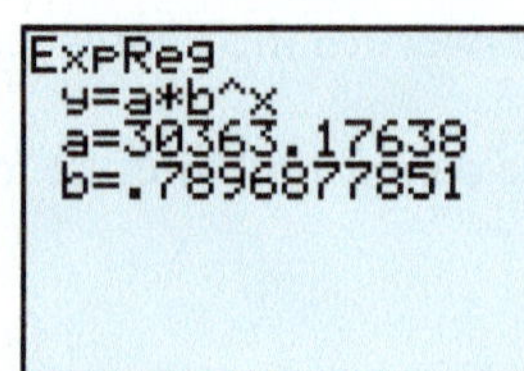

5. \$7,841.13 6. (A) \$9,155.23 (B) \$9,156.29

Exercise Appendix D

A

1. Match each equation with the graph of f, g, h, or k in the figure.
   (A) $y = 2^x$ (B) $y = (0.2)^x$
   (C) $y = 4^x$ (D) $y = (\frac{1}{3})^x$

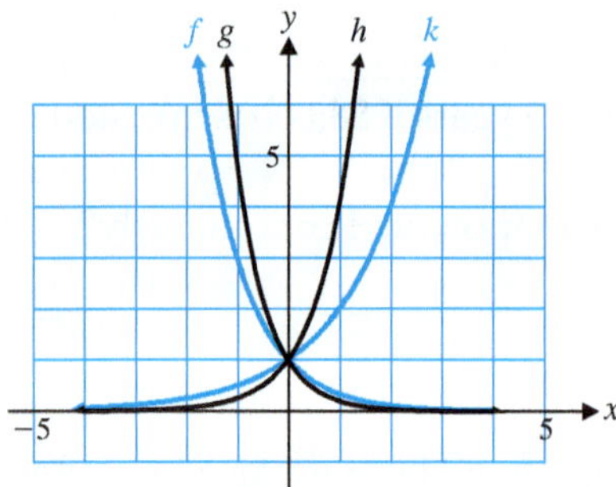

2. Match each equation with the graph of f, g, h, or k in the figure.
   (A) $y = (\frac{1}{4})^x$ (B) $y = (0.5)^x$
   (C) $y = 5^x$ (D) $y = 3^x$

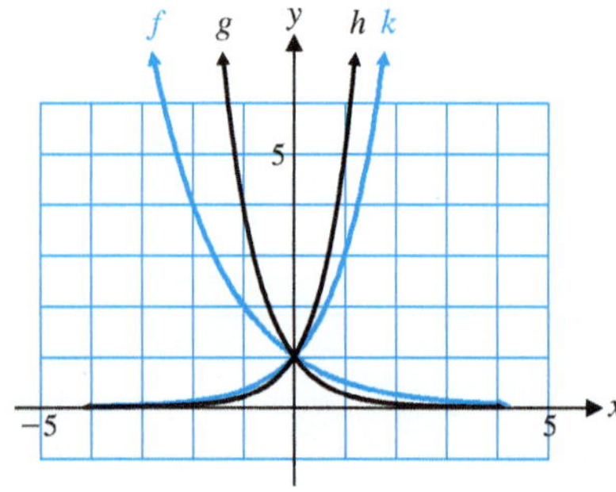

Graph each function in Problems 3–14 over the indicated interval.

3. $y = 5^x; [-2, 2]$
4. $y = 3^x; [-3, 3]$
5. $y = (\frac{1}{5})^x = 5^{-x}; [-2, 2]$
6. $y = (\frac{1}{3})^x = 3^{-x}; [-3, 3]$
7. $f(x) = -5^x; [-2, 2]$
8. $g(x) = -3^{-x}; [-3, 3]$
9. $y = -e^{-x}; [-3, 3]$
10. $y = -e^x; [-3, 3]$
11. $y = 100e^{0.1x}; [-5, 5]$
12. $y = 10e^{0.2x}; [-10, 10]$
13. $g(t) = 10e^{-0.2t}; [-5, 5]$
14. $f(t) = 100e^{-0.1t}; [-5, 5]$

Simplify each expression in Problems 15–20.

15. $(4^{3x})^{2y}$
16. $10^{3x-1}10^{4-x}$
17. $\dfrac{e^{x-3}}{e^{x-4}}$
18. $\dfrac{e^x}{e^{1-x}}$
19. $(2e^{1.2t})^3$
20. $(3e^{-1.4x})^2$

B

In Problems 21–28, describe verbally the transformations that can be used to obtain the graph of g from the graph of f.

21. $g(x) = -2^x; f(x) = 2^x$
22. $g(x) = 2^{x-2}; f(x) = 2^x$
23. $g(x) = 3^{x+1}; f(x) = 3^x$
24. $g(x) = -3^x; f(x) = 3^x$
25. $g(x) = e^x + 1; f(x) = e^x$
26. $g(x) = e^x - 2; f(x) = e^x$
27. $g(x) = 2e^{-(x+2)}; f(x) = e^{-x}$
28. $g(x) = 0.5e^{-(x-1)}; f(x) = e^{-x}$
29. Use the graph of f shown in the figure to sketch the graph of each of the following.
   (A) $y = f(x) - 1$ (B) $y = f(x + 2)$
   (C) $y = 3f(x) - 2$ (D) $y = 2 - f(x - 3)$

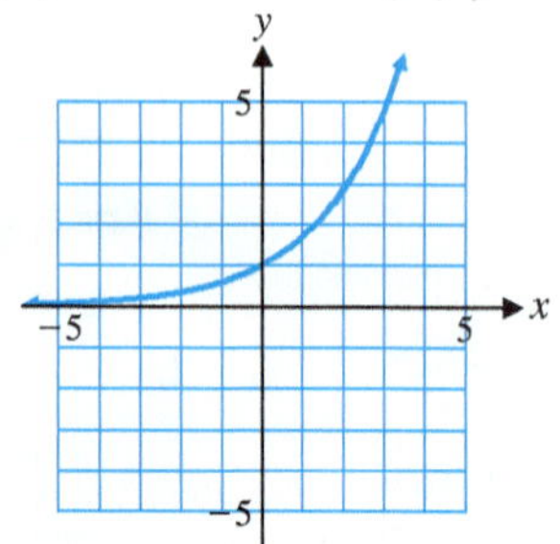

Figure for 29 and 30

30. Use the graph of f shown in the figure to sketch the graph of each of the following.
   (A) $y = f(x) + 2$ (B) $y = f(x - 3)$
   (C) $y = 2f(x) - 4$ (D) $y = 4 - f(x + 2)$

In Problems 31–40, graph each function over the indicated interval.

31. $f(t) = 2^{t/10}; [-30, 30]$

32. $G(t) = 3^{t/100}; [-200, 200]$

33. $y = -3 + e^{1+x}; [-4, 2]$

34. $y = 2 + e^{x-2}; [-1, 5]$

35. $y = e^{|x|}; [-3, 3]$

36. $y = e^{-|x|}; [-3, 3]$

37. $C(x) = \dfrac{e^x + e^{-x}}{2}; [-5, 5]$

38. $M(x) = e^{x/2} + e^{-x/2}; [-5, 5]$

39. $y = e^{-x^2}; [-3, 3]$

40. $y = 2^{-x^2}; [-3, 3]$

41. Find all real numbers a such that $a^2 = a^{-2}$. Explain why this does not violate the second exponential function property in Theorem 2.

42. Find real numbers a and b such that $a \neq b$ but $a^4 = b^4$. Explain why this does not violate the third exponential function property in Theorem 2 .

Solve each equation in Problems 43–48 for x.

43. $10^{2-3x} = 10^{5x-6}$

44. $5^{3x} = 5^{4x-2}$

45. $4^{5x-x^2} = 4^{-6}$

46. $7^{x^2} = 7^{2x+3}$

47. $5^3 = (x + 2)^3$

48. $(1 - x)^5 = (2x - 1)^5$

C

Solve each equation in Problems 49–52 for x. (Remember: $e^x \neq 0$ and $e^{-x} \neq 0$.)

49. $(x - 3)e^x = 0$

50. $2xe^{-x} = 0$

51. $3xe^{-x} + x^2e^{-x} = 0$

52. $x^2e^x - 5xe^x = 0$

Graph each function in Problems 53–56 over the indicated interval.

53. $h(x) = x(2^x); [-5, 0]$

54. $m(x) = x(3^{-x}); [0, 3]$

55. $N = \dfrac{100}{1 + e^{-t}}; [0, 5]$

56. $N = \dfrac{200}{1 + 3e^{-t}}; [0, 5]$

In Problems 57–60, approximate the real zeros of each function to two decimal places.

57. $f(x) = 4^x - 7$

58. $f(x) = 5 - 3^{-x}$

59. $f(x) = 2 + 3x + 10^x$

60. $f(x) = 7 - 2x^2 + 2^{-x}$

Applications

Business & Economics

61. Finance. Suppose that \$2,500 is invested at 7% compounded quarterly. How much money will be in the account in

(A) $\frac{3}{4}$ year? (B) 15 years?

Compute answers to the nearest cent.

62. Finance. Suppose that \$4,000 is invested at 6% compounded weekly. How much money will be in the account in

(A) $\frac{1}{2}$ year? (B) 10 years?

Compute answers to the nearest cent.

63. Money growth. If you invest \$7,500 in an account paying 8.35% compounded continuously, how much money will be in the account at the end of

(A) 5.5 years? (B) 12 years?

64. Money growth. If you invest \$5,250 in an account paying 7.45% compounded continuously, how much money will be in the account at the end of

(A) 6.25 years? (B) 17 years?

65. Finance. A person wishes to have \$15,000 cash for a new car 5 years from now. How much should be placed in an account now, if the account pays 6.75% compounded weekly? Compute the answer to the nearest dollar.

66. Finance. A couple just had a baby. How much should they invest now at 5.5% compounded daily in order to have \$40,000 for the child's education 17 years from now? Compute the answer to the nearest dollar.

67. Money growth. BanxQuote operates a network of Web sites providing real-time market data from leading financial providers. The following rates for 12-month certficates of deposit were taken from the Web sites:

(A) Stonebridge Bank, 2.15% compounded monthly

(B) DeepGreen Bank, 1.99% compounded daily

(C) Provident Bank, 1.95% compounded continuously

Compute the value of \$10,000 invested in each account at the end of 1 year.

68. Money growth. Refer to Problem 67. The following rates for 60-month certificates of deposit were also taken from BanxQuote Web sites:

(A) Oriental Bank & Trust, 3.25% compounded quarterly

(B) BMW Bank of North America, 3.16% compounded monthly

(C) BankFirst Corporation, 2.97% compounded daily

Compute the value of \$10,000 invested in each account at the end of 5 years.

69. Present value. A promissory note will pay \$50,000 at maturity $5\frac{1}{2}$ years from now. How much should you be willing to pay for the note now if money is worth 8% compounded continuously?

70. Present value. A promissory note will pay \$30,000 at maturity 10 years from now. How much should you be willing to pay for the note now if money is worth 7% compounded continuously?

71. Advertising. A company is trying to introduce a new product to as many people as possible through television advertising in a large metropolitan area with 2 million possible viewers. A model for the number of people N (in millions) who are aware of the product after t days of advertising was found to be

$$N = 2(1 - e^{-0.037t})$$

Graph this function for $0 \le t \le 50$. What value does N tend to as t increases without bound?

72. Learning curve. People assigned to assemble circuit boards for a computer manufacturing company undergo on-the-job training. From past experience it was found that the learning curve for the average employee is given by

$$N = 40(1 - e^{-0.12t})$$

where N is the number of boards assembled per day after t days of training. Graph this function for $0 \le t \le 30$. What is the maximum number of boards an average employee can be expected to produce in 1 day?

73. Sports salaries. Table 5 shows the average salaries for players in Major League Baseball (MLB) and the National Basketball Association (NBA) in selected years since 1990.

(A) Let x represent the number of years since 1990 and find an exponential regression model ($y = ab^x$) for the average salary in MLB. Use the model to estimate the average salary (to the nearest thousand dollars) in 2010.

(B) The average salary in MLB in 2000 was 1.984 million. How does this compare with the value given by the model of part (A)? How would the inclusion of the year 2000 data affect the estimated average salary in 2010? Explain.

Table 5 Average Salary (thousand $)

Year	MLB	NBA
1990	589	750
1993	1,062	1,300
1996	1,101	2,000
1999	1,724	2,400
2002	2,300	4,500

74. Sports salaries. Refer to Table 5.

(A) Let x represent the number of years since 1990 and find an exponential regression model ($y = ab^x$) for the average salary in the NBA. Use the model to estimate the average salary (to the nearest thousand dollars) in 2010.

(B) The average salary in the NBA in 1997 was $2.2 million. How does this compare with the value given by the model of part (A)? How would the inclusion of the year 1997 data affect the estimated average salary in 2010? Explain.

Life Sciences

75. Marine biology. Marine life is dependent upon the microscopic plant life that exists in the photic zone, a zone that goes to a depth where about 1% of the surface light remains. In some waters with a great deal of sediment, the photic zone may go down only 15 to 20 feet. In some murky harbors, the intensity of light d feet below the surface is given approximately by

$$I = I_0 e^{-0.23d}$$

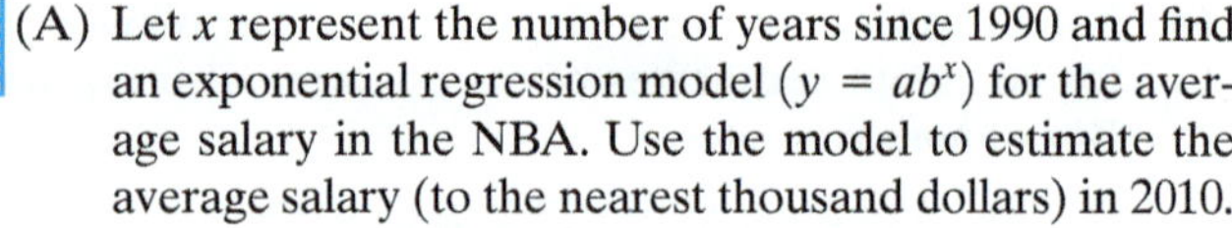

What percentage of the surface light will reach a depth of

(A) 10 feet? (B) 20 feet?

76. Marine biology. Refer to Problem 75. Light intensity I relative to depth d (in feet) for one of the clearest bodies of water in the world, the Sargasso Sea in the West Indies, can be approximated by

$$I = I_0 e^{-0.00942d}$$

where I_0 is the intensity of light at the surface. What percentage of the surface light will reach a depth of

(A) 50 feet? (B) 100 feet?

77. HIV/AIDS epidemic. The Joint United Nations Program on HIV/AIDS reported that HIV had infected 60 million people worldwide prior to 2002. Assume that number increases at an annual rate of 8% compounded continuously.

(A) Write an equation that models the worldwide spread of HIV, letting 2002 be year 0.

(B) Based on the model, how many people (to the nearest million) had been infected prior to 1999? How many would be infected prior to 2010?

(C) Sketch a graph of this growth equation from 2002 to 2010.

78. HIV/AIDS epidemic. The Joint United Nations Program on HIV/AIDS reported that 24.8 million people worldwide had died of AIDS prior to 2002. Assume that number increases at an annual rate of 19% compounded continuously.

(A) Write an equation that models total worldwide deaths from AIDS, letting 2002 be year 0.

(B) Based on the model, how many people (to the nearest million) had died from AIDS prior to 1999? How many would die from AIDS prior to 2010?

(C) Sketch a graph of this growth equation from 2002 to 2010.

Social Sciences

79. World population growth. It took from the dawn of humanity to 1830 for the population to grow to the first billion people, just 100 more years (by 1930) for the second billion, and 3 billion more were added in only 60 more years (by 1990). In 2002, the estimated world population was 6.2 billion with an annual growth rate of 1.25% compounded continuously.

(A) Write an equation that models the world population growth, letting 2002 be year 0.

(B) Based on the model, what is the expected world population (to the nearest hundred million) in 2010? In 2030?

(C) Sketch a graph of the equation found in part (A). Cover the years from 2002 through 2030.

80. Population growth in Ethiopia. In 2002, the estimated population in Ethiopia was 68 million people with an annual growth rate of 2.6% compounded continuously.

(A) Write an equation that models the population growth in Ethiopia, letting 2002 be year 0.

(B) Based on the model, what is the expected population in Ethiopia (to the nearest million) in 2010? In 2030?

(C) Sketch a graph of the equation found in part (A). Cover the years from 2002 through 2030.

81. Internet growth. The number of Internet hosts grew very rapidly from 1994 to 2003 (Table 6).

(A) Let x represent the number of years since 1994. Find an exponential regression model ($y = ab^x$) for this data set and estimate the number of hosts in 2010 (to the nearest million).

(B) Discuss the implications of this model if the number of Internet hosts continues to grow at this rate.

Table 6 Internet Hosts (Millions)

Year	Hosts
1994	2.4
1997	16.1
2000	72.4
2003	171.6

Source: Internet Software Consortium

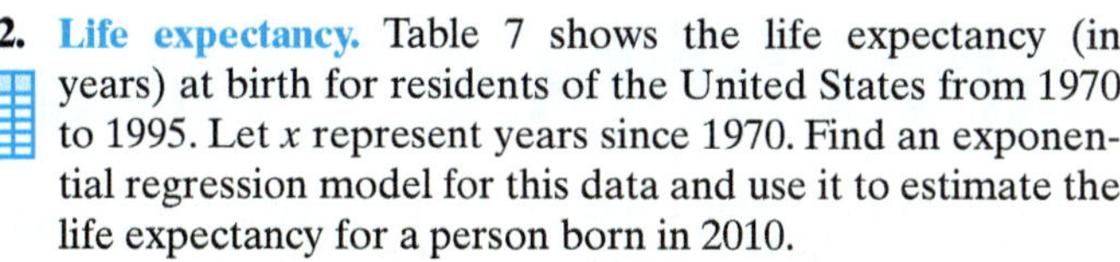

82. Life expectancy. Table 7 shows the life expectancy (in years) at birth for residents of the United States from 1970 to 1995. Let x represent years since 1970. Find an exponential regression model for this data and use it to estimate the life expectancy for a person born in 2010.

Table 7

Year of Birth	Life Expectancy
1970	70.8
1975	72.6
1980	73.7
1985	74.7
1990	75.4
1995	75.9
2000	76.9

APPENDIX E Logarithmic Functions

- Inverse Functions
- Logarithmic Functions
- Properties of Logarithmic Functions
- Calculator Evaluation of Logarithms
- Application

Find the exponential function keys $\boxed{10^x}$ and $\boxed{e^x}$ on your calculator. Close to these keys you will find $\boxed{\text{LOG}}$ and $\boxed{\text{LN}}$ keys. The latter represent *logarithmic functions,* and each is closely related to the exponential function it is near. In fact, the exponential function and the corresponding logarithmic function are said to be *inverses* of each other. In this section we will develop the concept of inverse functions and use it to define a logarithmic function as the inverse of an exponential function. We will then investigate basic properties of logarithmic functions, use a calculator to evaluate them for particular values of x, and apply them to real-world problems.

Logarithmic functions are used in modeling and solving many types of problems. For example, the decibel scale is a logarithmic scale used to measure sound intensity, and the Richter scale is a logarithmic scale used to measure the strength of the force of an earthquake. An important business application has to do with finding the time it takes money to double if it is invested at a certain rate compounded a given number of times a year or compounded continuously. This requires the solution of an exponential equation, and logarithms play a central role in the process.

Inverse Functions

Look at the graphs of $f(x) = \frac{x}{2}$ and $g(x) = \frac{|x|}{2}$ in Figure 1:

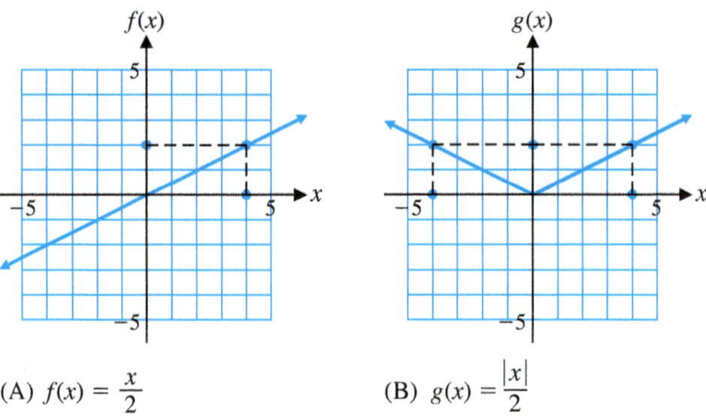

(A) $f(x) = \frac{x}{2}$ (B) $g(x) = \frac{|x|}{2}$

Figure 1

Because both f and g are functions, each domain value corresponds to exactly one range value. For which function does each range value correspond to exactly one domain value? This is the case only for function f. Note that for the range value 2, the corresponding domain value is 4. For function g the range value 2 corresponds to both -4 and 4. Function f is said to be *one-to-one*. In general,

DEFINITION One-to-One Functions

A function f is said to be **one-to-one** if each range value corresponds to exactly one domain value.

It can be shown that any continuous function that is either increasing or decreasing for all domain values is one-to-one. If a continuous function increases for some domain values and decreases for others, it cannot be one-to-one. Figure 1 shows an example of each case.

Explore & Discuss 1

Graph $f(x) = 2^x$ and $g(x) = x^2$. For a range value of 4, what are the corresponding domain values for each function? Which of the two functions is one-to-one? Explain why.

Starting with a one-to-one function f, we can obtain a new function called the *inverse* of f as follows:

DEFINITION Inverse of a Function

If f is a one-to-one function, then the **inverse** of f is the function formed by interchanging the independent and dependent variables for f. Thus, if (a, b) is a point on the graph of f, then (b, a) is a point on the graph of the inverse of f.

Note: If f is not one-to-one, then f **does not have an inverse.**

A number of important functions in any library of elementary functions are the inverses of other basic functions in the library. In this course, we are interested in the inverses of exponential functions, called *logarithmic functions.*

Logarithmic Functions

If we start with the exponential function f defined by

$$y = 2^x \tag{1}$$

and interchange the variables, we obtain the inverse of f:

$$x = 2^y \tag{2}$$

We call the inverse the **logarithmic function with base 2,** and write

$$y = \log_2 x \quad \text{if and only if} \quad x = 2^y$$

We can graph $y = \log_2 x$ by graphing $x = 2^y$, since they are equivalent. Any ordered pair of numbers on the graph of the exponential function will be on the graph of the logarithmic function if we interchange the order of the components. For example, $(3, 8)$ satisfies equation (1) and $(8, 3)$ satisfies equation (2). The graphs of $y = 2^x$ and $y = \log_2 x$ are shown in Figure 2. Note that if we fold the paper along the dashed line $y = x$ in Figure 2, the two graphs match exactly. The line $y = x$ is a line of symmetry for the two graphs.

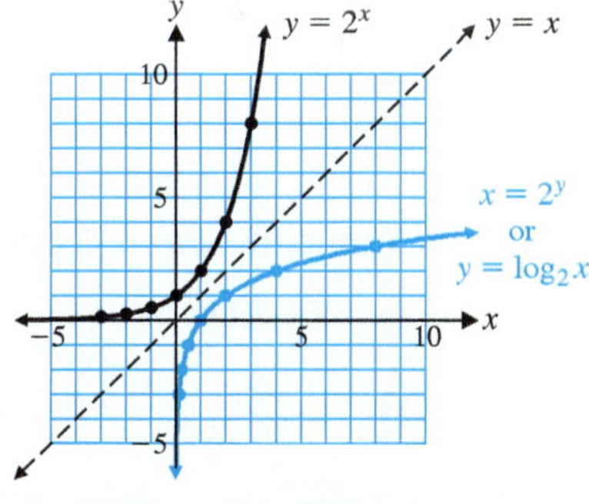

Figure 2

Exponential Function		Logarithmic Function	
x	$y = 2^x$	$x = 2^y$	y
−3	$\frac{1}{8}$	$\frac{1}{8}$	−3
−2	$\frac{1}{4}$	$\frac{1}{4}$	−2
−1	$\frac{1}{2}$	$\frac{1}{2}$	−1
0	1	1	0
1	2	2	1
2	4	4	2
3	8	8	3

Ordered pairs reversed

In general, since the graphs of all exponential functions of the form $f(x) = b^x$, $b \neq 1$, $b > 0$, are either increasing or decreasing, exponential functions have inverses.

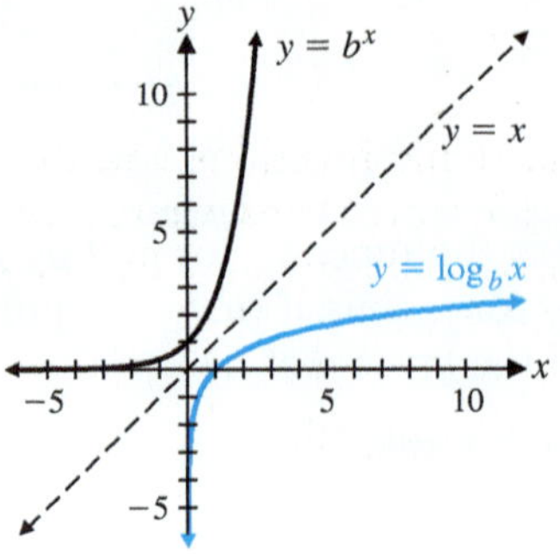

DEFINITION Logarithmic Functions

The inverse of an exponential function is called a **logarithmic function.** For $b > 0$ and $b \neq 1$,

Logarithmic form		*Exponential form*
$y = \log_b x$	is equivalent to	$x = b^y$

The **log to the base b of x** is the exponent to which b must be raised to obtain x. [**Remember**: A logarithm is an exponent.] The **domain** of the logarithmic function is the set of all positive real numbers, which is also the range of the corresponding exponential function; and the **range** of the logarithmic function is the set of all real numbers, which is also the domain of the corresponding exponential function. Typical graphs of an exponential function and its inverse, a logarithmic function, are shown in the figure in the margin.

CONCEPTUAL INSIGHT

Because the domain of a logarithmic function consists of the positive real numbers, the entire graph of a logarithmic function lies to the right of the y axis. In contrast, the graphs of polynomial and exponential functions intersect every vertical line, and the graphs of rational functions intersect all but a finite number of vertical lines.

The following examples involve converting logarithmic forms to equivalent exponential forms, and vice versa.

EXAMPLE 1 **Logarithmic–Exponential Conversions** Change each logarithmic form to an equivalent exponential form:

(A) $\log_5 25 = 2$ (B) $\log_9 3 = \frac{1}{2}$ (C) $\log_2(\frac{1}{4}) = -2$

SOLUTION

(A) $\log_5 25 = 2$ is equivalent to $25 = 5^2$

(B) $\log_9 3 = \frac{1}{2}$ is equivalent to $3 = 9^{1/2}$

(C) $\log_2(\frac{1}{4}) = -2$ is equivalent to $\frac{1}{4} = 2^{-2}$

Matched Problem 1 Change each logarithmic form to an equivalent exponential form:

(A) $\log_3 9 = 2$ (B) $\log_4 2 = \frac{1}{2}$ (C) $\log_3(\frac{1}{9}) = -2$

EXAMPLE 2 **Exponential–Logarithmic Conversions** Change each exponential form to an equivalent logarithmic form:

(A) $64 = 4^3$ (B) $6 = \sqrt{36}$ (C) $\frac{1}{8} = 2^{-3}$

SOLUTION

(A) $64 = 4^3$ is equivalent to $\log_4 64 = 3$

(B) $6 = \sqrt{36}$ is equivalent to $\log_{36} 6 = \frac{1}{2}$

(C) $\frac{1}{8} = 2^{-3}$ is equivalent to $\log_2(\frac{1}{8}) = -3$

Matched Problem 2 Change each exponential form to an equivalent logarithmic form:

(A) $49 = 7^2$ (B) $3 = \sqrt{9}$ (C) $\frac{1}{3} = 3^{-1}$

To gain a little deeper understanding of logarithmic functions and their relationship to the exponential functions, we consider a few problems where we want to find x, b, or y in $y = \log_b x$, given the other two values. All values are chosen so that the problems can be solved exactly without a calculator.

EXAMPLE 3 **Solutions of the Equation $y = \log_b x$** Find y, b, or x, as indicated.

(A) Find y: $y = \log_4 16$ (B) Find x: $\log_2 x = -3$

(C) Find y: $y = \log_8 4$ (D) Find b: $\log_b 100 = 2$

SOLUTION (A) $y = \log_4 16$ is equivalent to $16 = 4^y$. Thus,

$$y = 2$$

(B) $\log_2 x = -3$ is equivalent to $x = 2^{-3}$. Thus,

$$x = \frac{1}{2^3} = \frac{1}{8}$$

(C) $y = \log_8 4$ is equivalent to

$$4 = 8^y \quad \text{or} \quad 2^2 = 2^{3y}$$

Thus,

$$3y = 2$$

$$y = \tfrac{2}{3}$$

(D) $\log_b 100 = 2$ is equivalent to $100 = b^2$. Thus,

$b = 10$ *Recall that b cannot be negative.*

Matched Problem 3 Find y, b, or x, as indicated.

(A) Find y: $y = \log_9 27$ (B) Find x: $\log_3 x = -1$

(C) Find b: $\log_b 1{,}000 = 3$

Properties of Logarithmic Functions

Logarithmic functions have many powerful and useful properties. We list eight basic properties in Theorem 1.

THEOREM 1 Properties of Logarithmic Functions

If b, M, and N are positive real numbers, $b \neq 1$, and p and x are real numbers, then

1. $\log_b 1 = 0$
2. $\log_b b = 1$
3. $\log_b b^x = x$
4. $b^{\log_b x} = x, \quad x > 0$
5. $\log_b MN = \log_b M + \log_b N$
6. $\log_b \dfrac{M}{N} = \log_b M - \log_b N$
7. $\log_b M^p = p \log_b M$
8. $\log_b M = \log_b N$ if and only if $M = N$

The first four properties in Theorem 1 follow directly from the definition of a logarithmic function. Here we will sketch a proof of property 5. The other properties are established in a similar way. Let

$$u = \log_b M \quad \text{and} \quad v = \log_b N$$

Or, in equivalent exponential form,

$$M = b^u \quad \text{and} \quad N = b^v$$

Now, see if you can provide reasons for each of the following steps:

$$\log_b MN = \log_b b^u b^v = \log_b b^{u+v} = u + v = \log_b M + \log_b N$$

EXAMPLE 4 **Using Logarithmic Properties**

(A) $\log_b \dfrac{wx}{yz} = \log_b wx - \log_b yz$

$= \log_b w + \log_b x - (\log_b y + \log_b z)$

$= \log_b w + \log_b x - \log_b y - \log_b z$

(B) $\log_b(wx)^{3/5} = \frac{3}{5}\log_b wx = \frac{3}{5}(\log_b w + \log_b x)$

(C) $e^{x \log_e b} = e^{\log_e b^x} = b^x$

(D) $\dfrac{\log_e x}{\log_e b} = \dfrac{\log_e(b^{\log_b x})}{\log_e b} = \dfrac{(\log_b x)(\log_e b)}{\log_e b} = \log_b x$

Matched Problem 4 Write in simpler forms, as in Example 4.

(A) $\log_b \dfrac{R}{ST}$ (B) $\log_b\left(\dfrac{R}{S}\right)^{2/3}$ (C) $2^{u \log_2 b}$ (D) $\dfrac{\log_2 x}{\log_2 b}$

The following examples and problems, although somewhat artificial, will give you additional practice in using basic logarithmic properties.

EXAMPLE 5 **Solving Logarithmic Equations** Find x so that

$$\tfrac{3}{2}\log_b 4 - \tfrac{2}{3}\log_b 8 + \log_b 2 = \log_b x$$

SOLUTION

$\frac{3}{2}\log_b 4 - \frac{2}{3}\log_b 8 + \log_b 2 = \log_b x$

$\log_b 4^{3/2} - \log_b 8^{2/3} + \log_b 2 = \log_b x$ Property 7

$\log_b 8 - \log_b 4 + \log_b 2 = \log_b x$

$\log_b \dfrac{8 \cdot 2}{4} = \log_b x$ Properties 5 and 6

$\log_b 4 = \log_b x$

$x = 4$ Property 8

Matched Problem 5 Find x so that $3 \log_b 2 + \frac{1}{2}\log_b 25 - \log_b 20 = \log_b x$.

EXAMPLE 6 **Solving Logarithmic Equations** Solve: $\log_{10} x + \log_{10}(x + 1) = \log_{10} 6$.

SOLUTION

$\log_{10} x + \log_{10}(x + 1) = \log_{10} 6$

$\log_{10}[x(x + 1)] = \log_{10} 6$ Property 5

$x(x + 1) = 6$ Property 8

$x^2 + x - 6 = 0$ Solve by factoring.

$$(x+3)(x-2) = 0$$

$$x = -3, 2$$

We must exclude $x = -3$, since the domain of the function $\log_{10}(x+1)$ is $x > -1$ or $(-1, \infty)$; hence, $x = 2$ is the only solution.

Matched Problem 6 Solve: $\log_3 x + \log_3(x-3) = \log_3 10$.

EXPLORE & DISCUSS 2

Discuss the relationship between each of the following pairs of expressions. If the two expressions are equivalent, explain why. If they are not, give an example.

(A) $\log_b M - \log_b N$; $\dfrac{\log_b M}{\log_b N}$

(B) $\log_b M - \log_b N$; $\log_b \dfrac{M}{N}$

(C) $\log_b M + \log_b N$; $\log_b MN$

(D) $\log_b M + \log_b N$; $\log_b(M + N)$

Calculator Evaluation of Logarithms

Of all possible logarithmic bases, the base e and the base 10 are used almost exclusively. Before we can use logarithms in certain practical problems, we need to be able to approximate the logarithm of any positive number either to base 10 or to base e. And conversely, if we are given the logarithm of a number to base 10 or base e, we need to be able to approximate the number. Historically, tables were used for this purpose, but now calculators make computations faster and far more accurate.

Common logarithms (also called **Briggsian logarithms**) are logarithms with base 10. **Natural logarithms** (also called **Napierian logarithms**) are logarithms with base e. Most calculators have a key labeled "log" (or "LOG") and a key labeled "ln" (or "LN"). The former represents a common (base 10) logarithm and the latter a natural (base e) logarithm. In fact, "log" and "ln" are both used extensively in mathematical literature, and whenever you see either used in this book without a base indicated, they will be interpreted as follows:

NOTATION **Logarithmic Notation**

Common logarithm: $\log x = \log_{10} x$

Natural logarithm: $\ln x = \log_e x$

Finding the common or natural logarithm using a calculator is very easy. On some calculators, you simply enter a number from the domain of the function and press LOG or LN. On other calculators, you press either LOG or LN, enter a number from the domain, and then press ENTER. Check the user's manual for your calculator.

EXAMPLE 7 **Calculator Evaluation of Logarithms** Use a calculator to evaluate each to six decimal places:

(A) log 3,184 (B) ln 0.000 349 (C) log(−3.24)

SOLUTION (A) log 3,184 = 3.502 973 (B) ln 0.000 349 = −7.960 439

(C) log(−3.24) = Error* −3.24 *is not in the domain of the log function.*

Matched Problem 7 Use a calculator to evaluate each to six decimal places:

(A) log 0.013 529 (B) ln 28.693 28 (C) ln(−0.438)

We now turn to the second problem mentioned previously: Given the logarithm of a number, find the number. We make direct use of the logarithmic–exponential relationships, which follow from the definition of logarithmic function given at the beginning of this section.

RESULT Logarithmic–Exponential Relationships

$\log x = y$	is equivalent to	$x = 10^y$
$\ln x = y$	is equivalent to	$x = e^y$

EXAMPLE 8 **Solving $\log_b x = y$ for x** Find x to four decimal places, given the indicated logarithm:

(A) $\log x = -2.315$ (B) $\ln x = 2.386$

SOLUTION (A) $\log x = -2.315$ *Change to equivalent exponential form.*

$x = 10^{-2.315}$ *Evaluate with a calculator.*

$= 0.0048$

(B) $\ln x = 2.386$ *Change to equivalent exponential form.*

$x = e^{2.386}$ *Evaluate with a calculator.*

$= 10.8699$

Matched Problem 8 Find x to four decimal places, given the indicated logarithm:

(A) $\ln x = -5.062$ (B) $\log x = 2.0821$

EXAMPLE 9 **Solving Exponential Equations** Solve for x to four decimal places:

(A) $10^x = 2$ (B) $e^x = 3$ (C) $3^x = 4$

SOLUTION (A) $10^x = 2$ *Take common logarithms of both sides.*

$\log 10^x = \log 2$ *Property 3*

$x = \log 2$ *Use a calculator.*

$= 0.3010$

* Some calculators use a more advanced definition of logarithms involving complex numbers and will display an ordered pair of real numbers as the value of log(−3.24). You should interpret such a result as an indication that the number entered is not in the domain of the logarithm function as we have defined it.

(B)
$$\begin{aligned} e^x &= 3 && \textit{Take natural logarithms of both sides.} \\ \ln e^x &= \ln 3 && \textit{Property 3} \\ x &= \ln 3 && \textit{Use a calculator.} \\ &= 1.0986 \end{aligned}$$

(C)
$$\begin{aligned} 3^x &= 4 && \textit{Take either natural or common logarithms of both sides. (We choose commmon logarithms.)} \\ \log 3^x &= \log 4 && \textit{Property 7} \\ x \log 3 &= \log 4 && \textit{Solve for } x. \\ x &= \frac{\log 4}{\log 3} && \textit{Use a calculator.} \\ &= 1.2619 \end{aligned}$$

Exponential equations can also be solved graphically by graphing both sides of an equation and finding the points of intersection. Figure 3 illustrates this approach for the equations in Example 9.

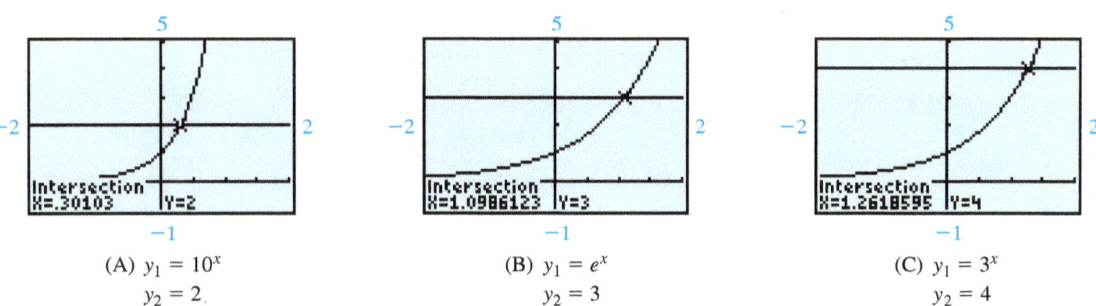

(A) $y_1 = 10^x$
$y_2 = 2$

(B) $y_1 = e^x$
$y_2 = 3$

(C) $y_1 = 3^x$
$y_2 = 4$

Figure 3 Graphical solution of exponential equations

Matched Problem 9 Solve for x to four decimal places:

(A) $10^x = 7$ (B) $e^x = 6$ (C) $4^x = 5$

Explore & Discuss 3 Discuss how you could find $y = \log_5 38.25$ using either natural or common logarithms on a calculator. [**Hint**: Start by rewriting the equation in exponential form.]

Remark—In the usual notation for natural logarithms, the simplifications of Example 4, parts (C) and (D) on page 758, become

$$e^{x \ln b} = b^x \quad \text{and} \quad \frac{\ln x}{\ln b} = \log_b x$$

With these formulas we can change an exponential function with base b, or a logarithmic function with base b, to expressions involving exponential or logarithmic functions, respectively, to the base e. Such **change-of-base formulas** are useful in calculus.

Application

A convenient and easily understood way of comparing different investments is to use their **doubling times**—the length of time it takes the value of an investment to double. Logarithm properties, as you will see in Example 10, provide us with just the right tool for solving some doubling-time problems.

EXAMPLE 10 **Doubling Time for an Investment** How long (to the next whole year) will it take money to double if it is invested at 20% compounded annually?

SOLUTION We use the compound interest formula:

$$A = P\left(1 + \frac{r}{m}\right)^{mt} \quad \text{Compound interest}$$

The problem is to find t, given $r = 0.20$, $m = 1$, and $A = 2p$; that is,

$$2P = P(1 + 0.2)^t$$

$$2 = 1.2^t \quad \text{Solve for } t \text{ by taking the natural or common}$$

$$1.2^t = 2 \quad \text{logarithm of both sides (we choose the}$$

$$\ln 1.2^t = \ln 2 \quad \text{natural logarithm).}$$

$$t \ln 1.2 = \ln 2 \quad \text{Property 7}$$

$$t = \frac{\ln 2}{\ln 1.2} \quad \text{Use a calculator.}$$

$$= 3.8 \text{ years} \quad [\text{Note: } (\ln 2)/(\ln 1.2) \neq \ln 2 - \ln 1.2]$$

$$\approx 4 \text{ years} \quad \text{To the next whole year}$$

When interest is paid at the end of 3 years, the money will not be doubled; when paid at the end of 4 years, the money will be slightly more than doubled.

Example 10 can also be solved graphically by graphing both sides of the equation $2 = 1.2^t$, and finding the intersection point (Fig. 4).

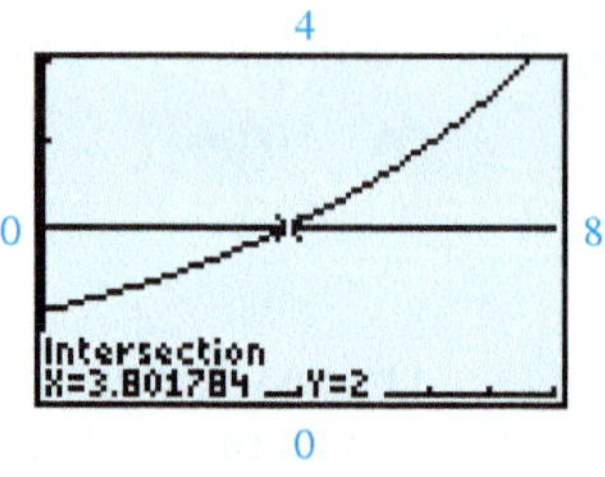

Figure 4 $y_1 = 1.2^x, y_2 = 2$

Matched Problem 10 How long (to the next whole year) will it take money to triple if it is invested at 13% compounded annually?

It is interesting and instructive to graph the doubling times for various rates compounded annually. We proceed as follows:

$$A = P(1 + r)^t$$
$$2P = P(1 + r)^t$$
$$2 = (1 + r)^t$$
$$(1 + r)^t = 2$$
$$\ln(1 + r)^t = \ln 2$$
$$t \ln(1 + r) = \ln 2$$
$$t = \frac{\ln 2}{\ln(1 + r)}$$

Figure 5

Figure 5 shows the graph of this equation (doubling time in years) for interest rates compounded annually from 1 to 70% (expressed as decimals). Note the dramatic change in doubling time as rates change from 1 to 20% (from 0.01 to 0.20).

Answers to Matched Problems

1. (A) $9 = 3^2$ (B) $2 = 4^{1/2}$ (C) $\frac{1}{9} = 3^{-2}$

2. (A) $\log_7 49 = 2$ (B) $\log_9 3 = \frac{1}{2}$ (C) $\log_3(\frac{1}{3}) = -1$

3. (A) $y = \frac{3}{2}$ (B) $x = \frac{1}{3}$ (C) $b = 10$

4. (A) $\log_b R - \log_b S - \log_b T$ (B) $\frac{2}{3}(\log_b R - \log_b S)$ (C) b^u (D) $\log_b x$

5. $x = 2$ **6.** $x = 5$

7. (A) $-1.868\,734$ (B) $3.356\,663$ (C) Not defined

8. (A) 0.0063 (B) 120.8092

9. (A) 0.8451 (B) 1.7918 (C) 1.1610

10. 9 yr

Exercises Appendix E

A

For Problems 1–6, rewrite in equivalent exponential form.

1. $\log_3 27 = 3$ **2.** $\log_2 32 = 5$ **3.** $\log_{10} 1 = 0$

4. $\log_e 1 = 0$ **5.** $\log_4 8 = \frac{3}{2}$ **6.** $\log_9 27 = \frac{3}{2}$

For Problems 7–12, rewrite in equivalent logarithmic form.

7. $49 = 7^2$ **8.** $36 = 6^2$ **9.** $8 = 4^{3/2}$

10. $9 = 27^{2/3}$ **11.** $A = b^u$ **12.** $M = b^x$

For Problems 13–24, evaluate without a calculator.

13. $\log_{10} 1$ **14.** $\log_e 1$ **15.** $\log_e e$

16. $\log_{10} 10$ **17.** $\log_{0.2} 0.2$ **18.** $\log_{13} 13$

19. $\log_{10} 10^3$ **20.** $\log_{10} 10^{-5}$ **21.** $\log_2 2^{-3}$

22. $\log_3 3^5$ **23.** $\log_{10} 1{,}000$ **24.** $\log_6 36$

For Problems 25–30, write in terms of simpler forms, as in Example 4.

25. $\log_b \frac{P}{Q}$ **26.** $\log_b FG$ **27.** $\log_b L^5$

28. $\log_b w^{15}$ **29.** $3^{p \log_3 q}$ **30.** $\frac{\log_3 P}{\log_3 R}$

B

For Problems 31–42, find x, y, or b without a calculator.

31. $\log_3 x = 2$ **32.** $\log_2 x = 2$

33. $\log_7 49 = y$ **34.** $\log_3 27 = y$

35. $\log_b 10^{-4} = -4$ **36.** $\log_b e^{-2} = -2$

37. $\log_4 x = \frac{1}{2}$ **38.** $\log_{25} x = \frac{1}{2}$

39. $\log_{1/3} 9 = y$ **40.** $\log_{49}(\frac{1}{7}) = y$

41. $\log_b 1{,}000 = \frac{3}{2}$ **42.** $\log_b 4 = \frac{2}{3}$

In Problems 43–52, discuss the validity of each statement. If the statement is always true, explain why. If not, give a counterexample.

43. Every polynomial function is one-to-one.

44. Every polynomial function of odd degree is one-to-one.

45. If g is the inverse of a function f, then g is one-to-one.

46. The graph of a one-to-one function intersects each vertical line exactly once.

47. The inverse of $f(x) = 2x$ is $g(x) = x/2$.

48. The inverse of $f(x) = x^2$ is $g(x) = \sqrt{x}$.

49. If $b > 0$ and $b \neq 1$, then the graph of $y = \log_b x$ is increasing.

50. If $b > 1$, then the graph of $y = \log_b x$ is increasing.

51. If f is one-to-one, then the domain of f is equal to the range of f.

52. If g is the inverse of a function f, then f is the inverse of g.

Find x in Problems 53–60.

53. $\log_b x = \frac{2}{3}\log_b 8 + \frac{1}{2}\log_b 9 - \log_b 6$

54. $\log_b x = \frac{2}{3}\log_b 27 + 2\log_b 2 - \log_b 3$

55. $\log_b x = \frac{3}{2}\log_b 4 - \frac{2}{3}\log_b 8 + 2\log_b 2$

56. $\log_b x = 3\log_b 2 + \frac{1}{2}\log_b 25 - \log_b 20$

57. $\log_b x + \log_b(x - 4) = \log_b 21$

58. $\log_b(x + 2) + \log_b x = \log_b 24$

59. $\log_{10}(x - 1) - \log_{10}(x + 1) = 1$

60. $\log_{10}(x + 6) - \log_{10}(x - 3) = 1$

Graph Problems 61 and 62 by converting to exponential form first.

61. $y = \log_2(x - 2)$

62. $y = \log_3(x + 2)$

63. Explain how the graph of the equation in Problem 61 can be obtained from the graph of $y = \log_2 x$ using a simple transformation.

64. Explain how the graph of the equation in Problem 62 can be obtained from the graph of $y = \log_3 x$ using a simple transformation.

65. What are the domain and range of the function defined by $y = 1 + \ln(x + 1)$?

66. What are the domain and range of the function defined by $y = \log(x - 1) - 1$?

For Problems 67 and 68, evaluate to five decimal places using a calculator.

67. (A) log 3,527.2 (B) log 0.006 913 2
(C) ln 277.63 (D) ln 0.040 883

68. (A) log 72.604 (B) log 0.033 041
(C) ln 40,257 (D) ln 0.005 926 3

For Problems 69 and 70, find x to four decimal places.

69. (A) $\log x = 1.1285$ (B) $\log x = -2.0497$
(C) $\ln x = 2.7763$ (D) $\ln x = -1.8879$

70. (A) $\log x = 2.0832$ (B) $\log x = -1.1577$
(C) $\ln x = 3.1336$ (D) $\ln x = -4.3281$

For Problems 71–78, solve each equation to four decimal places.

71. $10^x = 12$

72. $10^x = 153$

73. $e^x = 4.304$

74. $e^x = 0.3059$

75. $1.03^x = 2.475$

76. $1.075^x = 1.837$

77. $1.005^{12t} = 3$

78. $1.02^{4t} = 2$

Graph Problems 79–86 using a calculator and point-by-point plotting. Indicate increasing and decreasing intervals.

79. $y = \ln x$

80. $y = -\ln x$

81. $y = |\ln x|$

82. $y = \ln|x|$

83. $y = 2\ln(x + 2)$

84. $y = 2\ln x + 2$

85. $y = 4\ln x - 3$

86. $y = 4\ln(x - 3)$

C

87. Explain why the logarithm of 1 for any permissible base is 0.

88. Explain why 1 is not a suitable logarithmic base.

89. Write $\log_{10} y - \log_{10} c = 0.8x$ in an exponential form that is free of logarithms.

90. Write $\log_e x - \log_e 25 = 0.2t$ in an exponential form that is free of logarithms.

91. Let $p(x) = \ln x$, $q(x) = \sqrt{x}$, and $r(x) = x$. Use a graphing utility to draw graphs of all three functions in the same viewing window for $1 \le x \le 16$. Discuss what it means for one function to be larger than another on an interval, and then order the three functions from largest to smallest for $1 < x \le 16$.

92. Let $p(x) = \log x$, $q(x) = \sqrt[3]{x}$, and $r(x) = x$. Use a graphing utility to draw graphs of all three functions in the same viewing window for $1 \le x \le 16$. Discuss what it means for one function to be smaller than another on an interval, and then order the three functions from smallest to largest for $1 < x \le 16$.

Applications

Business & Economics

93. **Doubling time.** In its first 10 years the Gabelli Growth Fund produced an average annual return of 21.36%. Assume that money invested in this fund continues to earn 21.36% compounded annually. How long will it take money invested in this fund to double?

94. **Doubling time.** In its first 10 years the Janus Flexible Income Fund produced an average annual return of 9.58%. Assume that money invested in this fund continues to earn 9.58% compounded annually. How long will it take money invested in this fund to double?

95. **Investing.** How many years (to two decimal places) will it take \$1,000 to grow to \$1,800 if it is invested at 6% compounded quarterly? Compounded continuously?

96. **Investing.** How many years (to two decimal places) will it take \$5,000 to grow to \$7,500 if it is invested at 8% compounded semiannually? Compounded continuously?

97. **Investment.** A newly married couple wishes to have \$30,000 in 6 years for the down payment on a house. At what rate of interest compounded continuously (to three decimal places) must \$20,000 be invested now to accomplish this goal?

98. **Investment.** The parents of a newborn child want to have \$60,000 for the child's college education 17 years from now. At what rate of interest compounded continuously (to three decimal places) must a grandparent's gift of \$20,000 be invested now to achieve this goal?

99. **Supply and demand.** A cordless screwdriver is sold through a national chain of discount stores. A marketing company established price–demand and price–supply tables (Tables 1 and 2), where x is the number of screwdrivers people are willing to buy and the store is willing to sell each month at a price of p dollars per screwdriver.

Table 1 Price–Demand

x	$p = D(x)$ ($)
1,000	91
2,000	73
3,000	64
4,000	56
5,000	53

Table 2 Price–Supply

x	$p = S(x)$ ($)
1,000	9
2,000	26
3,000	34
4,000	38
5,000	41

(A) Find a logarithmic regression model ($y = a + b \ln x$) for the data in Table 1. Estimate the demand (to the nearest unit) at a price level of $50.

(B) Find a logarithmic regression model ($y = a + b \ln x$) for the data in Table 2. Estimate the supply (to the nearest unit) at a price level of $50.

(C) Does a price level of $50 represent a stable condition, or is the price likely to increase or decrease? Explain.

100. Equilibrium point. Use the models constructed in Problem 99 to find the equilibrium point. Write the equilibrium price to the nearest cent and the equilibrium quantity to the nearest unit.

Life Sciences

101. Sound intensity: decibels. Because of the extraordinary range of sensitivity of the human ear (a range of over 1,000 million millions to 1), it is helpful to use a logarithmic scale, rather than an absolute scale, to measure sound intensity over this range. The unit of measure is called the *decibel*, after the inventor of the telephone, Alexander Graham Bell. If we let N be the number of decibels, I the power of the sound in question (in watts per square centimeter), and I_0 the power of sound just below the threshold of hearing (approximately 10^{-16} watt per square centimeter), then

$$I = I_0 10^{N/10}$$

Show that this formula can be written in the form

$$N = 10 \log \frac{I}{I_0}$$

102. Sound intensity: decibels. Use the formula in Problem 101 (with $I_0 = 10^{-16}$ W/cm^2) to find the decibel ratings of the following sounds:

(A) Whisper: 10^{-13} W/cm^2

(B) Normal conversation: 3.16×10^{-10} W/cm^2

(C) Heavy traffic: 10^{-8} W/cm^2

(D) Jet plane with afterburner: 10^{-1} W/cm^2

103. Agriculture. Table 3 shows the yield (in bushels per acre) and the total production (in millions of bushels) for corn in the United States for selected years since 1950. Let x represent years since 1900. Find a logarithmic regression model ($y = a + b \ln x$) for the yield. Estimate (to one decimal place) the yield in 2010.

Table 3 United States Corn Production

Year	x	Yield (bushels per acre)	Total Production (million bushels)
1950	50	37.6	2,782
1960	60	55.6	3,479
1970	70	81.4	4,802
1980	80	97.7	6,867
1990	90	115.6	7,802
2000	100	139.6	10,192

104. Agriculture. Refer to Table 3. Find a logarithmic regression model ($y = a + b \ln x$) for the total production. Estimate (to the nearest million) the production in 2010.

Social Sciences

105. World population. If the world population is now 6.2 billion people and if it continues to grow at an annual rate of 1.25% compounded continuously, how long (to the nearest year) will it take before there is only 1 square yard of land per person? (The Earth contains approximately 1.68×10^{14} square yards of land.)

106. Archaeology: carbon-14 dating. The radioactive carbon-14 (^{14}C) in an organism at the time of its death decays according to the equation

$$A = A_0 e^{-0.000124t}$$

where t is time in years and A_0 is the amount of ^{14}C present at time $t = 0$. Estimate the age of a skull uncovered in an archaeological site if 10% of the original amount of ^{14}C is still present. [**Hint**: Find t such that $A = 0.1A_0$.]

ANSWERS

Chapter 1

Exercises 1-1

1. 2 **3.** 1.25 **5.** (A) 2 (B) 2 (C) 2 (D) 2 **7.** (A) 1 (B) 2 (C) Does not exist (D) 2 (E) No **9.** 2 **11.** 0.5
13. (A) 1 (B) 2 (C) Does not exist (D) Does not exist (E) No **15.** (A) 1 (B) 1 (C) 1 (D) 3 (E) Yes
17. (A) -2 (B) -2 (C) -2 (D) 1 (E) Yes **19.** (A) 2 (B) 2 (C) 2 (D) Does not exist (E) Yes
21. 12 **23.** 1 **25.** -4 **27.** -1.5 **29.** 3 **31.** 15 **33.** -6 **35.** $\frac{7}{5}$ **37.** 3

39. 

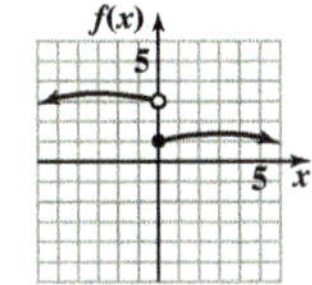

41.

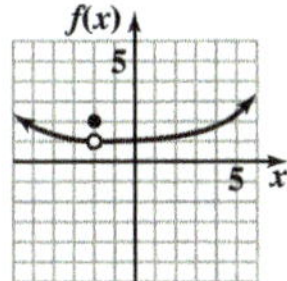

43. (A) 1 (B) 1 (C) 1 (D) 1
45. (A) 2 (B) 1 (C) Does not exist (D) Does not exist
47. (A) -6 (B) Does not exist (C) 6
49. (A) 1 (B) -1 (C) Does not exist (D) Does not exist
51. (A) Does not exist (B) $\frac{1}{2}$ (C) $\frac{1}{4}$ **53.** (A) -5 (B) -3 (C) 0
55. (A) 0 (B) -1 (C) Does not exist **57.** (A) 1 (B) $\frac{1}{3}$ (C) $\frac{3}{4}$ **59.** False **61.** True **63.** False **65.** 3 **67.** 4

69. (A) $\lim_{x\to1^-} f(x) = 2$, $\lim_{x\to1^+} f(x) = 3$

(B) $\lim_{x\to1^-} f(x) = 3$, $\lim_{x\to1^+} f(x) = 2$

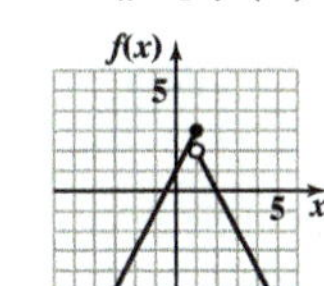

(C) $m = 1.5$

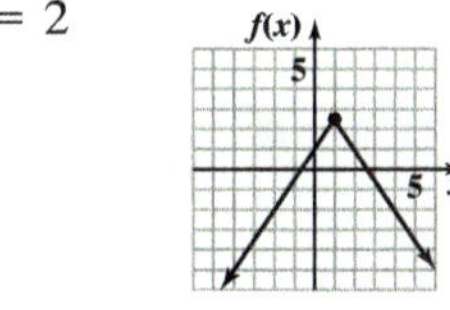

(D) The graph in (A) is broken when it jumps from (1, 2) up to (1, 3). The graph in (B) is also broken when it jumps down from (1, 3) to (1, 2). The graph in (C) is one continuous piece, with no breaks or jumps.

71. $2a$ **73.** $1/(2\sqrt{a})$

75. (A) $F(x) = \begin{cases} 0.99 & \text{if } 0 < x \le 20 \\ 0.07x - 0.41 & \text{if } x \ge 20 \end{cases}$

(B)

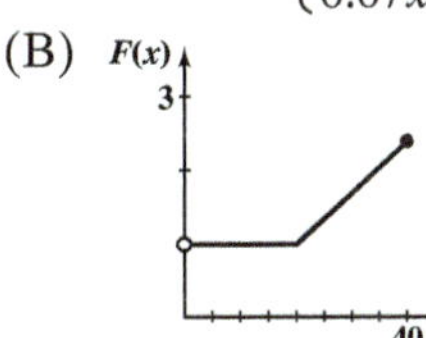

(C) All 3 limits are 0.99.

79. (A) $D(x) = \begin{cases} x & \text{if } 0 \le x < 300 \\ 0.97x & \text{if } 300 \le x < 1{,}000 \\ 0.95x & \text{if } 1{,}000 \le x < 3{,}000 \\ 0.93x & \text{if } 3{,}000 \le x < 5{,}000 \\ 0.9x & \text{if } x \ge 5{,}000 \end{cases}$

(B) $\lim_{x\to1{,}000} D(x)$ does not exist because
$\lim_{x\to1{,}000^-} D(x) = 970$ and $\lim_{x\to1{,}000^+} D(x) = 950$;
$\lim_{x\to3{,}000} D(x)$ does not exist because
$\lim_{x\to3{,}000^-} D(x) = 2{,}850$ and $\lim_{x\to3{,}000^+} D(x) = 2{,}790$

81. $F(x) = \begin{cases} 20x & \text{if } 0 < x \le 4{,}000 \\ 80{,}000 & \text{if } x \ge 4{,}000 \end{cases}$

$\lim_{x\to4{,}000} F(x) = 80{,}000$; $\lim_{x\to8{,}000} F(x) = 80{,}000$

83. $\lim_{x\to5} f(x)$ does not exist; $\lim_{x\to10} f(x) = 0$;
$\lim_{x\to5} g(x) = 0$; $\lim_{x\to10} g(x) = 1$

Exercises 1-2

1. -2 **3.** $-\infty$ **5.** Does not exist **7.** 0 **9.** (A) $-\infty$ (B) ∞ (C) Does not exist
11. (A) ∞ (B) ∞ (C) ∞ **13.** (A) 3 (B) 3 (C) 3 **15.** (A) $-\infty$ (B) ∞ (C) Does not exist
17. (A) ∞ (B) $-\infty$ **19.** (A) $-\infty$ (B) $-\infty$ **21.** $\lim_{x\to-3^-} f(x) = -\infty$; $\lim_{x\to-3^+} f(x) = \infty$; $x = -3$ is a vertical asymptote
23. $\lim_{x\to-2^-} h(x) = \infty$; $\lim_{x\to-2^+} h(x) = -\infty$; $\lim_{x\to2^-} h(x) = -\infty$; $\lim_{x\to2^+} h(x) = \infty$; $x = -2$ and $x = 2$ are vertical asymptotes
25. No zeros of denominator; no vertical asymptotes
27. $\lim_{x\to1^-} H(x) = -\infty$; $\lim_{x\to1^+} H(x) = \infty$; $\lim_{x\to3} H(x) = 2$; $x = 1$ is a vertical asymptote

29. $\lim_{x\to 0} T(x) = -\infty; \lim_{x\to 4} T(x) = \infty$; $x = 0$ and $x = 4$ are vertical asymptotes

31. (A) $\frac{47}{41} \approx 1.146$ (B) $\frac{407}{491} \approx 0.829$ (C) $\frac{4}{5} = 0.8$ **33.** (A) $\frac{2{,}011}{138} \approx 14.572$ (B) $\frac{12{,}511}{348} \approx 35.951$ (C) ∞

35. (A) $-\frac{8{,}568}{46{,}653} \approx -0.184$ (B) $-\frac{143{,}136}{1{,}492{,}989} \approx 0.096$ (C) 0

37. (A) $-\frac{7{,}010}{996} \approx -7.038$ (B) $-\frac{56{,}010}{7{,}996} \approx -7.005$ (C) -7

39. Horizontal asymptote: $y = 2$; vertical asymptote: $x = -2$

41. Horizontal asymptote: $y = 1$; vertical asymptotes: $x = -1$ and $x = 1$

43. No horizontal asymptotes; no vertical asymptotes **45.** Horizontal asymptote: $y = 0$; no vertical asymptotes

47. No horizontal asymptotes; vertical asymptote: $x = 3$

49. Horizontal asymptote: $y = 2$; vertical asymptotes: $x = -1$ and $x = 2$

51. Horizontal asymptote: $y = 2$; vertical asymptote: $x = -1$

53. False **55.** False **57.** True **59.** If $n \geq 1$ and $a_n > 0$, then the limit is ∞. If $n \geq 1$ and $a_n < 0$, then the limit is $-\infty$.

61. $\lim_{x\to\infty} f(x) = \infty; \lim_{x\to-\infty} f(x) = \infty$ **63.** $\lim_{x\to\infty} f(x) = \infty; \lim_{x\to-\infty} f(x) = -\infty$ **65.** $\lim_{x\to\infty} f(x) = \infty; \lim_{x\to-\infty} f(x) = -\infty$

67. $\lim_{x\to\infty} f(x) = -\infty; \lim_{x\to-\infty} f(x) = -\infty$

69. (A) $C(x) = 180x + 200$

(B) $\overline{C}(x) = \frac{180x + 200}{x}$

(C)

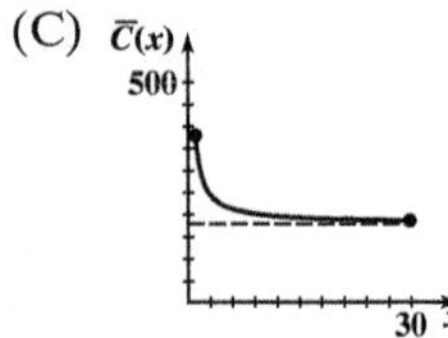

(D) \$180 per board

71. (A) $C_e(x) = 950 + 56x; \overline{C}_e(x) = \frac{950}{x} + 56$

(B) $C_c(x) = 900 + 66x; \overline{C}_c(x) = \frac{900}{x} + 66$

(C) At $x = 5$ years

(D) At $x = 5$ years

(E) $\lim_{x\to\infty} \overline{C}_e(x) = 56; \lim_{x\to\infty} \overline{C}_c(x) = 66$

73. The long-term drug concentration is 5 mg/ml.

75. (A) \$18 million

(B) \$38 million

(C) $\lim_{x\to 1^-} P(x) = \infty$

77. (C) $V_{\max} = 4, K_M = 20$

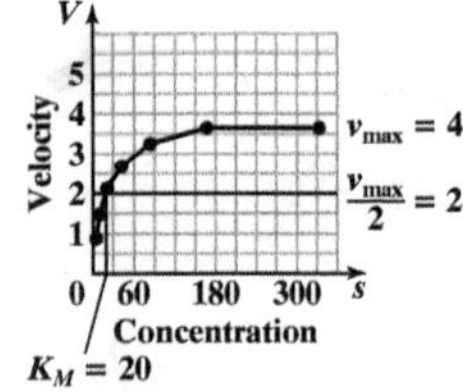

(D) $v(s) = \frac{4s}{20 + s}$

(E) $v = \frac{12}{7}$ when $s = 15$; $s = 60$ when $v = 3$

79. (A) $C_{\max} = 18, M = 150$

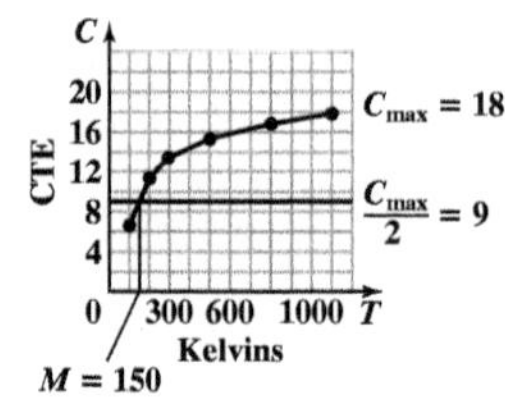

(B) $C(T) = \frac{18T}{150 + T}$

(C) $C = 14.4$ when $T = 600$ K; $T = 300$ K when $C = 12$

Exercises 1-3

1. f is continuous at $x = 1$, since $\lim_{x\to 1} f(x) = f(1)$.

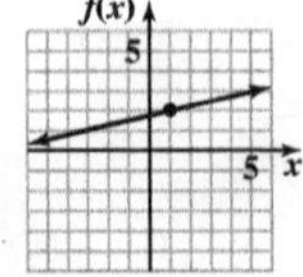

3. f is discontinuous at $x = 1$, since $\lim_{x\to 1} f(x) \neq f(1)$.

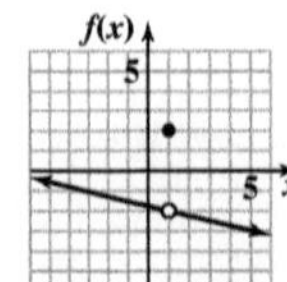

5. f is discontinuous at $x = 1$, since $\lim_{x\to 1} f(x)$ does not exist.

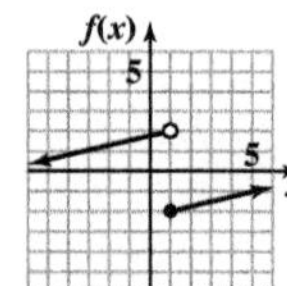

7. 1.9 **9.** 0.9 **11.** (A) 2 (B) 1 (C) Does not exist (D) 1 (E) No **13.** (A) 1 (B) 1 (C) 1 (D) 3 (E) No
15. 0.9 **17.** 2.05 **19.** (A) 1 (B) 1 (C) 1 (D) 3 (E) No **21.** (A) 2 (B) −1 (C) Does not exist (D) 2 (E) No

23. All x **25.** All x, except $x = -2$ **27.** All x, except $x = -4$ and $x = 1$ **29.** All x **31.** All x, except $x = \pm\frac{3}{2}$
33. (A) (B) 1 (C) 2 (D) No (E) All integers
35. $-3 < x < 4$; $(-3, 4)$ **37.** $x < 3$ or $x > 7$; $(-\infty, 3) \cup (7, \infty)$
39. $x < -2$ or $0 < x < 2$; $(-\infty, -2) \cup (0, 2)$ **41.** $-5 < x < 0$ or $x > 3$; $(-5, 0) \cup (3, \infty)$
43. (A) $(-4, -2) \cup (0, 2) \cup (4, \infty)$ (B) $(-\infty, -4) \cup (-2, 0) \cup (2, 4)$
45. (A) $(-\infty, -2.5308) \cup (-0.7198, \infty)$ (B) $(-2.5308, -0.7198)$
47. (A) $(-\infty, -2.1451) \cup (-1, -0.5240) \cup (1, 2.6691)$
(B) $(-2.1451, -1) \cup (-0.5240, 1) \cup (2.6691, \infty)$

49. $[6, \infty)$ **51.** $(-\infty, \infty)$ **53.** $(-\infty, -3] \cup [3, \infty)$ **55.** $(-\infty, \infty)$

57. Since $\lim_{x\to 1^-} f(x) = 2$ and $\lim_{x\to 1^+} f(x) = 4$, $\lim_{x\to 1} f(x)$ does not exist and f is not continuous at $x = 1$.

59. This function is continuous for all x.

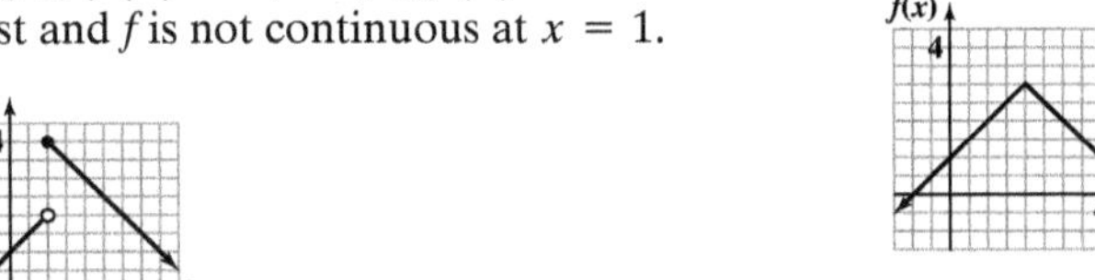

61. Since $\lim_{x\to 0} f(x) = 0$ and $f(0) = 1$, $\lim_{x\to 0} f(x) \neq f(0)$ and f is not continuous at $x = 0$.

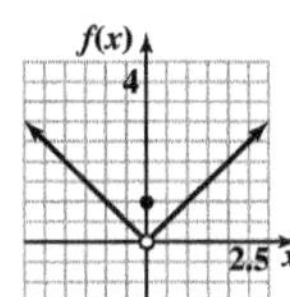

63. (A) Yes (B) No (C) Yes (D) No (E) Yes **65.** True **67.** False **69.** True
71. x intercepts: $x = -5, 2$

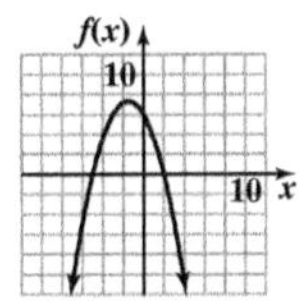

73. x intercepts: $x = -6, -1, 4$

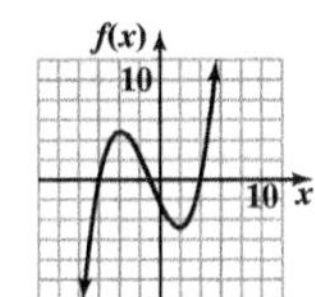

75. No, but this does not contradict Theorem 2, since f is discontinuous at $x = 1$.

77. (A) $P(x) = \begin{cases} 0.44 & \text{if } 0 < x \le 1 \\ 0.61 & \text{if } 1 < x \le 2 \\ 0.78 & \text{if } 2 < x \le 3 \\ 0.95 & \text{if } 3 < x \le 3.5 \end{cases}$ (B)
(C) Yes; no

81. (A) $S(x) = \begin{cases} 5 + 0.63x & \text{if } 0 \le x \le 50 \\ 14 + 0.45x & \text{if } 50 < x \end{cases}$
(B)

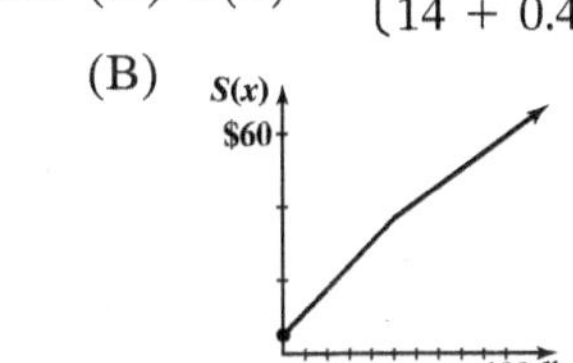

(C) Yes

83. (A)

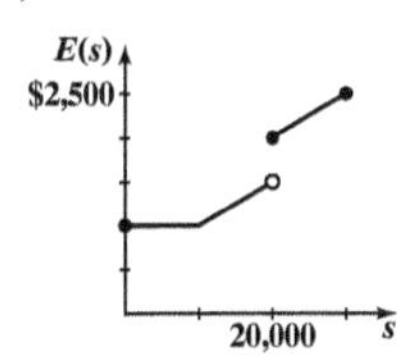

(B) $\lim_{s\to 10{,}000} E(s) = \$1{,}000$; $E(10{,}000) = \$1{,}000$
(C) $\lim_{s\to 20{,}000} E(s)$ does not exist; $E(20{,}000) = \$2{,}000$
(D) Yes; no

85. (A) t_2, t_3, t_4, t_6, t_7 (B) $\lim_{t\to t_5} N(t) = 7$; $N(t_5) = 7$ (C) $\lim_{t\to t_3} N(t)$ does not exist; $N(t_3) = 4$

Exercises 1-4

1. (A) −3; slope of the secant line through $(1, f(1))$ and $(2, f(2))$
(B) $-2 - h$; slope of the secant line through $(1, f(1))$ and $(1 + h, f(1 + h))$
(C) −2; slope of the tangent line at $(1, f(1))$
3. (A) 15 (B) 15 (C) $6 + 3h$ (D) 6 (E) 6 (F) 6 (G) $y = 6x - 3$
5. $f'(x) = 0$; $f'(1) = 0$, $f'(2) = 0$, $f'(3) = 0$ **7.** $f'(x) = 3$; $f'(1) = 3$, $f'(2) = 3$, $f'(3) = 3$
9. $f'(x) = -6x$; $f'(1) = -6$, $f'(2) = -12$, $f'(3) = -18$ **11.** $f'(x) = 2x + 6$; $f'(1) = 8$, $f'(2) = 10$, $f'(3) = 12$
13. $f'(x) = 4x - 7$; $f'(1) = -3$, $f'(2) = 1$, $f'(3) = 5$ **15.** $f'(x) = -2x + 4$; $f'(1) = 2$, $f'(2) = 0$, $f'(3) = -2$
17. $f'(x) = 6x^2$; $f'(1) = 6$, $f'(2) = 24$, $f'(3) = 54$ **19.** $f'(x) = -\frac{4}{x^2}$; $f'(1) = -4$, $f'(2) = -1$, $f'(3) = -\frac{4}{9}$
21. $f'(x) = \frac{3}{2\sqrt{x}}$; $f'(1) = \frac{3}{2}$, $f'(2) = \frac{3}{2\sqrt{2}}$ or $\frac{3\sqrt{2}}{4}$, $f'(3) = \frac{3}{2\sqrt{3}}$ or $\frac{\sqrt{3}}{2}$

23. $f'(x) = \dfrac{5}{\sqrt{x+5}}$; $f'(1) = \dfrac{5}{\sqrt{6}}$ or $\dfrac{5\sqrt{6}}{6}$, $f'(2) = \dfrac{5}{\sqrt{7}}$ or $\dfrac{5\sqrt{7}}{7}$, $f'(3) = \dfrac{5}{2\sqrt{2}}$ or $\dfrac{5\sqrt{2}}{4}$

25. $f'(x) = \dfrac{6}{(x+2)^2}$; $f'(1) = \dfrac{2}{3}$, $f'(2) = \dfrac{3}{8}$, $f'(3) = \dfrac{6}{25}$

27. (A) 5 (B) $3 + h$ (C) 3 (D) $y = 3x - 1$ **29.** (A) 5 m/s (B) $3 + h$ m/s (C) 3 m/s

31. Yes **33.** No **35.** Yes **37.** Yes

39. (A) $f'(x) = 2x - 4$ (B) $-4, 0, 4$
(C)

41. $v = f'(x) = 8x - 2$; 6 ft/s, 22 ft/s, 38 ft/s

43. (A) The graphs of g and h are vertical translations of the graph of f. All three functions should have the same derivative.
(B) $2x$

45. True

47. False

49. False

51. f is nondifferentiable at $x = 1$ **53.** f is differentiable for all real numbers

55. No **57.** No **59.** $f'(0) = 0$ **61.** 6 s; 192 ft/s

63. (A) \$8.75
(B) $R'(x) = 60 - 0.05x$
(C) $R(1{,}000) = 35{,}000$; $R'(1{,}000) = 10$; At a production level of 1,000 car seats, the revenue is \$35,000 and is increasing at the rate of \$10 per seat.

65. (A) $S'(t) = 1/\sqrt{t+10}$
(B) $S(15) = 10$; $S'(15) = 0.2$. After 15 months, the total sales are \$10 million and are increasing at the rate of \$0.2 million, or \$200,000, per month.
(C) The estimated total sales are \$10.2 million after 16 months and \$10.4 million after 17 months.

67. (A) $p'(t) = 328t + 161$
(B) $p(10) = 30{,}336$; $p'(10) = 413.4$; In 2015, 30,336 metric tons of tungsten are consumed and this quantity is increasing at the rate of 3,441 metric tons per year.

69. (A)

(B) $R(20) = 1{,}403.5$ billion kilowatts, $R'(20) = -10.7$ billion kilowatts. In 2020, 1,403.5 billion kilowatts will be sold and the amount sold is decreasing at the rate of 10.7 billion kilowatts per year.

71. (A) $P'(t) = 12 - 2t$
(B) $P(3) = 107$; $P'(3) = 6$. After 3 hours, the ozone level is 107 ppb and is increasing at the rate of 6 ppb per hour.

Exercises 1-5

1. 0 **3.** $9x^8$ **5.** $3x^2$ **7.** $-4x^{-5}$ **9.** $\dfrac{8}{3}x^{5/3}$ **11.** $-\dfrac{10}{x^{11}}$ **13.** $10x$ **15.** $2.8x^6$ **17.** $\dfrac{x^2}{6}$ **19.** 12 **21.** 2 **23.** 9 **25.** 2 **27.** $4t - 3$

29. $-10x^{-3} - 9x^{-2}$ **31.** $1.5u^{-0.7} - 8.8u^{1.2}$ **33.** $0.5 - 3.3t^2$ **35.** $-\dfrac{8}{5}x^{-5}$ **37.** $3x + \dfrac{14}{5}x^{-3}$ **39.** $-\dfrac{20}{9}w^{-5} + \dfrac{5}{3}w^{-2/3}$

41. $2u^{-1/3} - \dfrac{5}{3}u^{-2/3}$ **43.** $-\dfrac{9}{5}t^{-8/5} + 3t^{-3/2}$ **45.** $-\dfrac{1}{3}x^{-4/3}$ **47.** $-0.6x^{-3/2} + 6.4x^{-3} + 1$

49. (A) $f'(x) = 6 - 2x$ (B) $f'(2) = 2$; $f'(4) = -2$ (C) $y = 2x + 4$; $y = -2x + 16$ (D) $x = 3$

51. (A) $f'(x) = 12x^3 - 12x$ (B) $f'(2) = 72$; $f'(4) = 720$ (C) $y = 72x - 127$; $y = 720x - 2{,}215$ (D) $x = -1, 0, 1$

53. (A) $v = f'(x) = 176 - 32x$ (B) $f'(0) = 176$ ft/s; $f'(3) = 80$ ft/s (C) 5.5 s

55. (A) $v = f'(x) = 3x^2 - 18x + 15$ (B) $f'(0) = 15$ ft/s; $f'(3) = -12$ ft/s (C) $x = 1$ s, $x = 5$ s

57. $f'(x) = 2x - 3 - 2x^{-1/2} = 2x - 3 - \dfrac{2}{x^{1/2}}$; $x = 2.1777$ **59.** $f'(x) = 4\sqrt[3]{x} - 3x - 3$; $x = -2.9018$

61. $f'(x) = 0.2x^3 + 0.3x^2 - 3x - 1.6$; $x = -4.4607, -0.5159, 3.4765$

63. $f'(x) = 0.8x^3 - 9.36x^2 + 32.5x - 28.25$; $x = 1.3050$

69. $8x - 4$ **71.** $-20x^{-2}$ **73.** $-\dfrac{1}{4}x^{-2} + \dfrac{2}{3}x^{-3}$ **75.** True **77.** False **79.** True

81. (A) $S'(t) = 0.09t^2 + t + 2$
(B) $S(5) = 29.25$, $S'(5) = 9.25$. After 5 months, sales are \$29.25 million and are increasing at the rate of \$9.25 million per month.
(C) $S(10) = 103$, $S'(10) = 21$. After 10 months, sales are \$103 million and are increasing at the rate of \$21 million per month.

83. (A) $N'(x) = 3{,}780/x^2$
(B) $N'(10) = 37.8$. At the \$10,000 level of advertising, sales are increasing at the rate of 37.8 boats per \$1,000 spent on advertising.
$N'(20) = 9.45$. At the \$20,000 level of advertising, sales are increasing at the rate of 9.45 boats per \$1,000 spent on advertising.

85. (A)

```
CubicReg
 y=ax³+bx²+cx+d
 a=-.0016394337
 b=.1090342521
 c=-1.525096943
 d=54.7123987
```

(B) In 2016, 55.7% of male high-school graduates enroll in college and the percentage is decreasing at the rate of 1.9% per year.

87. (A) -1.37 beats/min
(B) -0.58 beat/min

89. (A) 25 items/h
(B) 8.33 items/h

Exercises 1-6

1. $\Delta x = 3$; $\Delta y = 45$; $\Delta y/\Delta x = 15$ **3.** 12 **5.** 12 **7.** $dy = (24x - 3x^2)\,dx$ **9.** $dy = \left(2x - \frac{x^2}{3}\right)dx$ **11.** $dy = -\frac{295}{x^{3/2}}\,dx$

13. (A) $12 + 3\,\Delta x$ (B) 12 **15.** $dy = (8x + 4)\,dx$ **17.** $dy = (1 - 9x^{-2})\,dx$ **19.** $dy = 1.4$; $\Delta y = 1.44$

21. $dy = -3$; $\Delta y = -3\frac{1}{3}$ **23.** 120 in.^{3}

25. (A) $\Delta y = \Delta x + (\Delta x)^2$; $dy = \Delta x$

(B)

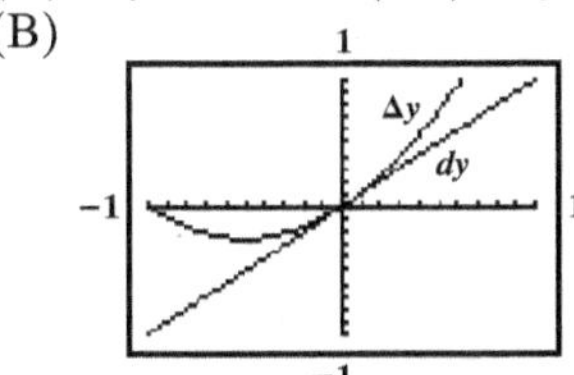

(C)

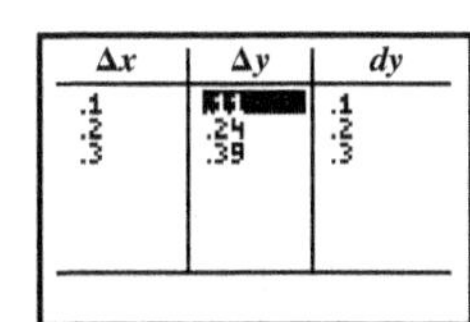

Δx	Δy	dy
.1	.11	.1
.2	.24	.2
.3	.39	.3

27. (A) $\Delta y = -\Delta x + (\Delta x)^2 + (\Delta x)^3$; $dy = -\Delta x$

(B)

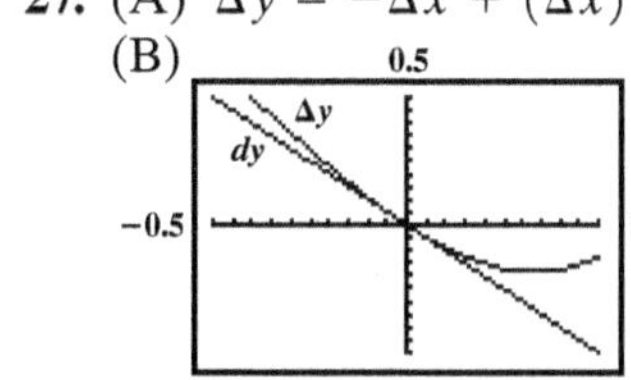

(C)

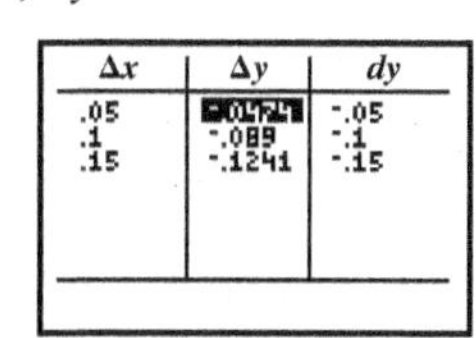

Δx	Δy	dy
.05	[illegible]	-.05
.1	-.089	-.1
.15	-.1241	-.15

29. True **31.** False **33.** $dy = \left(\frac{2}{3}x^{-1/3} - \frac{10}{3}x^{2/3}\right)dx$ **35.** $dy = 3.9$; $\Delta y = 3.83$ **37.** 40-unit increase; 20-unit increase

39. $-\$2.50$; \$1.25 **41.** -1.37/min; -0.58/min **43.** 1.26 mm^2 **45.** 3 wpm

47. (A) 2,100 increase (B) 4,800 increase (C) 2,100 increase

Exercises 1-7

1. $C'(x) = 0.8$ **3.** $C'(x) = 4.6 - 0.02x$ **5.** $R'(x) = 4 - 0.02x$ **7.** $R'(x) = 12 - 0.08x$ **9.** $P'(x) = 3.2 - 0.02x$

11. $P'(x) = 7.4 - 0.06x$ **13.** $\overline{C}(x) = 1.1 + \frac{145}{x}$ **15.** $\overline{C}'(x) = -\frac{145}{x^2}$ **17.** $P(x) = 3.9x - 0.02x^2 - 145$

19. $\overline{P}(x) = 3.9 - 0.02x - \frac{145}{x}$ **21.** True **23.** False **25.** (A) \$29.50 (B) \$30

27. (A) \$420
(B) $\overline{C}'(500) = -0.24$. At a production level of 500 frames, average cost is decreasing at the rate of 24¢ per frame.
(C) Approximately \$419.76

29. (A) \$14.70 (B) \$15

31. (A) $P'(450) = 0.5$. At a production level of 450 cassettes, profit is increasing at the rate of 50¢ per cassette.
(B) $P'(750) = -2.5$. At a production level of 750 cassettes, profit is decreasing at the rate of \$2.50 per cassette.

33. (A) \$13.50
(B) $\overline{P}'(50) = \$0.27$. At a production level of 50 mowers, the average profit per mower is increasing at the rate of \$0.27 per mower.
(C) Approximately \$13.77

35. (A) $p = 100 - 0.025x$, domain: $0 \le x \le 4{,}000$
(B) $R(x) = 100x - 0.025x^2$, domain: $0 \le x \le 4{,}000$
(C) $R'(1{,}600) = 20$. At a production level of 1,600 radios, revenue is increasing at the rate of \$20 per radio.
(D) $R'(2{,}500) = -25$. At a production level of 2,500 radios, revenue is decreasing at the rate of \$25 per radio.

37. (A) $p = 200 - \frac{1}{30}x$, domain: $0 \le x \le 6{,}000$
(B) $C'(x) = 60$
(C) $R(x) = 200x - (x^2/30)$, domain: $0 \le x \le 6{,}000$
(D) $R'(x) = 200 - (x/15)$
(E) $R'(1{,}500) = 100$. At a production level of 1,500 saws, revenue is increasing at the rate of \$100 per saw.
$R'(4{,}500) = -100$. At a production level of 4,500 saws, revenue is decreasing at the rate of \$100 per saw.

(F) Break-even points: (600, 108,000) and (3,600, 288,000)

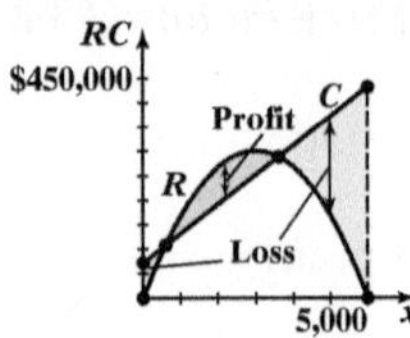

(G) $P(x) = -(x^2/30) + 140x - 72{,}000$
(H) $P'(x) = -(x/15) + 140$
(I) $P'(1{,}500) = 40$. At a production level of 1,500 saws, profit is increasing at the rate of \$40 per saw.
$P'(3{,}000) = -60$. At a production level of 3,000 saws, profit is decreasing at the rate of \$60 per saw.

39. (A) $p = 20 - 0.02x$, domain: $0 \le x \le 1{,}000$
(B) $R(x) = 20x - 0.02x^2$, domain: $0 \le x \le 1{,}000$
(C) $C(x) = 4x + 1{,}400$

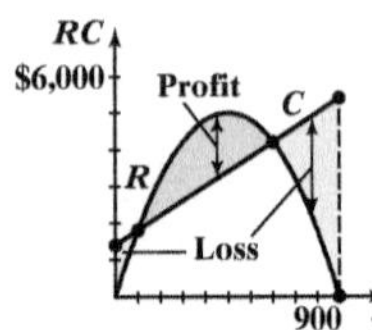

(D) Break-even points: (100, 1,800) and (700, 4,200)
(E) $P(x) = 16x - 0.02x^2 - 1{,}400$
(F) $P'(250) = 6$. At a production level of 250 toasters, profit is increasing at the rate of \$6 per toaster. $P'(475) = -3$. At a production level of 475 toasters, profit is decreasing at the rate of \$3 per toaster.

41. (A) $x = 500$
(B) $P(x) = 176x - 0.2x^2 - 21{,}900$
(C) $x = 440$
(D) Break-even points: (150, 25,500) and (730, 39,420); x intercepts for $P(x)$: $x = 150$ and $x = 730$

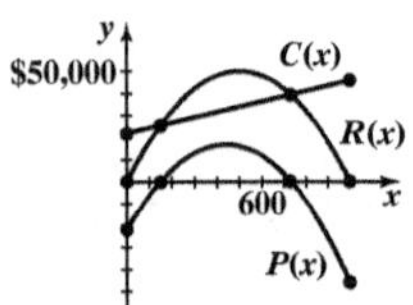

43. (A) $R(x) = 20x - x^{3/2}$
(B) Break-even points: (44, 588), (258, 1,016)

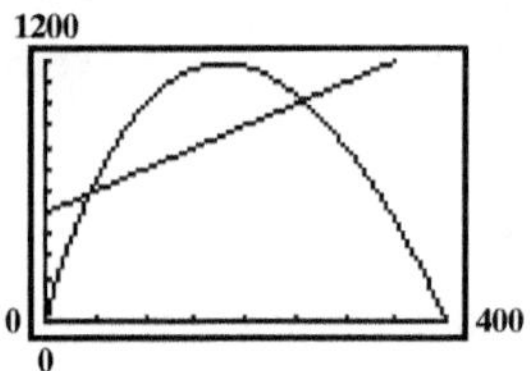

45. (A)

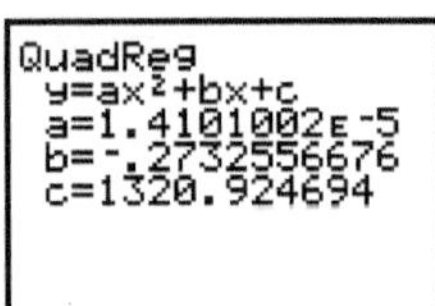

(B) Fixed costs ≈ \$721,680
Variable costs ≈ \$121

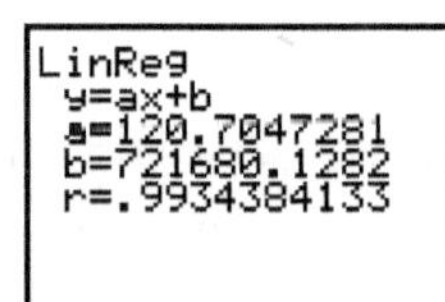

(C) (713, 807,703), (5,423, 1,376,227)
(D) $\$254 \le p \le \$1{,}133$

Chapter 1 Review Exercises

1. (A) 16 (B) 8 (C) 8 (D) 4 (E) 4 (F) 4 *(3-2)* **2.** $f'(x) = -3$ *(3-2)* **3.** (A) 22 (B) 8 (C) 2 (D) -5 *(3-1)* **4.** 1.5 *(3-1)* **5.** 3.5 *(3-1)* **6.** 3.75 *(3-1)* **7.** 3.75 *(3-1)* **8.** (A) 1 (B) 1 (C) 1 (D) 1 *(3-1)* **9.** (A) 2 (B) 3 (C) Does not exist (D) 3 *(3-1)* **10.** (A) 4 (B) 4 (C) 4 (D) Does not exist *(3-1)* **11.** (A) Does not exist (B) 3 (C) No *(3-3)* **12.** (A) 2 (B) Not defined (C) No *(3-3)* **13.** (A) 1 (B) 1 (C) Yes *(3-3)* **14.** 5 *(3-2)* **15.** 5 *(3-2)* **16.** ∞ *(3-2)* **17.** $-\infty$ *(3-2)* **18.** 0 *(3-1)* **19.** 0 *(3-1)* **20.** 0 *(3-1)* **21.** Vertical asymptote: $x = 2$ *(3-3)* **22.** Horizontal asymptote: $y = 5$ *(3-2)* **23.** $x = 2$ **24.** $f'(x) = 10x$ *(3-4)*

25. (A) -3 (B) 6 (C) -2 (D) 3 (E) -11 *(3-5)* **26.** $x^2 - 10x$ *(3-5)* **27.** $x^{-1/2} - 3 = \frac{1}{x^{1/2}} - 3$ *(3-5)* **28.** 0 *(3-5)*

29. $-\frac{3}{2}x^{-2} + \frac{15}{4}x^2 = \frac{-3}{2x^2} + \frac{15x^2}{4}$ *(3-5)* **30.** $-2x^{-5} + x^3 = \frac{-2}{x^5} + x^3$ *(3-5)* **31.** $f'(x) = 12x^3 + 9x^2 - 2$ *(3-5)*

32. $\Delta x = 2$, $\Delta y = 10$, $\Delta y/\Delta x = 5$ *(3-6)* **33.** 5 *(3-6)* **34.** 6 *(3-6)* **35.** $\Delta y = 0.64$; $dy = 0.6$ *(3-6)*

36. (A) 4 (B) 6 (C) Does not exist (D) 6 (E) No *(3-3)* **37.** (A) 3 (B) 3 (C) 3 (D) 3 (E) Yes *(3-3)* **38.** (A) $(8, \infty)$ (B) $[0, 8]$ *(3-3)* **39.** $(-3, 4)$ *(3-3)* **40.** $(-3, 0) \cup (5, \infty)$ *(3-3)* **41.** $(-2.3429, -0.4707) \cup (1.8136, \infty)$ *(3-3)*

42. (A) 3 (B) $2 + 0.5h$ (C) 2 *(3-4)* **43.** $-x^{-4} + 10x^{-3}$ *(3-4)* **44.** $\frac{3}{4}x^{-1/2} - \frac{5}{6}x^{-3/2} = \frac{3}{4\sqrt{x}} - \frac{5}{6\sqrt{x^3}}$ *(3-5)*

45. $0.6x^{-2/3} - 0.3x^{-4/3} = \frac{0.6}{x^{2/3}} - \frac{0.3}{x^{4/3}}$ *(3-4)* **46.** $-\frac{3}{5}(-3)x^{-4} = \frac{9}{5x^4}$ *(3-5)* **47.** (A) $m = f'(1) = 2$ (B) $y = 2x + 3$ *(3-4, 3-5)*
48. $x = 5$ *(3-4)* **49.** $x = -5, x = 3$ *(3-5)* **50.** $x = -1.3401, 0.5771, 2.2630$ *(3-4)* **51.** ± 2.4824 *(3-5)*
52. (A) $v = f'(x) = 16x - 4$ (B) 44 ft/sec *(3-5)* **53.** (A) $v = f'(x) = -10x + 16$ (B) $x = 1.6$ sec *(3-5)*

54. (A) The graph of g is the graph of f shifted 4 units to the right, and the graph of h is the graph of f shifted 3 units to the left:

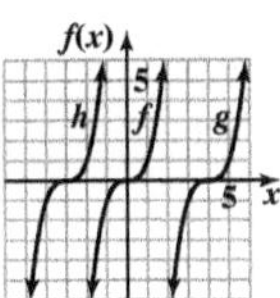

(B) The graph of g' is the graph of f' shifted 4 units to the right, and the graph of h' is the graph of f' shifted 3 units to the left:

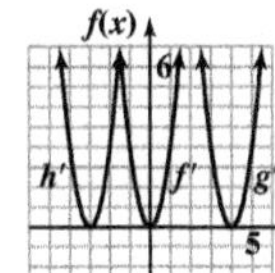

55. $(-\infty, \infty)$ *(3-3)* **56.** $(-\infty, 2) \cup (2, \infty)$ *(3-3)* **57.** $(-\infty, -4) \cup (-4, 1) \cup (1, \infty)$ *(3-3)* **58.** $(-\infty, \infty)$ *(3-3)*

59. $[-2, 2]$ *(3-3)* **60.** (A) -1 (B) Does not exist (C) $-\frac{2}{3}$ *(3-1)* **61.** (A) $\frac{1}{2}$ (B) 0 (C) Does not exist *(3-1)*

62. (A) -1 (B) 1 (C) Does not exist *(3-1)* **63.** (A) $-\frac{1}{6}$ (B) Does not exist (C) $-\frac{1}{3}$ *(3-1)*

64. (A) 0 (B) -1 (C) Does not exist *(3-1)* **65.** (A) $\frac{2}{3}$ (B) $\frac{2}{3}$ (C) Does not exist *(3-2)*

66. (A) ∞ (B) $-\infty$ (C) ∞ *(3-3)* **67.** (A) 0 (B) 0 (C) Does not exist *(3-2)* **68.** 4 *(3-1)* **69.** $\frac{-1}{(x+2)^2}$ *(3-1)*

70. $2x - 1$ *(3-4)* **71.** $1/(2\sqrt{x})$ *(3-4)* **72.** Yes *(3-4)* **73.** No *(3-4)* **74.** No *(3-4)* **75.** No *(3-4)* **76.** Yes *(3-4)* **77.** Yes *(3-4)*

78. Horizontal asymptote: $y = 5$; vertical asymptote: $x = 7$ *(3-2)*

79. Horizontal asymptote: $y = 0$; vertical asymptote: $x = 4$ *(3-2)*

80. No horizontal asymptotes; vertical asymptote: $x = 3$ *(3-2)*

81. Horizontal asymptotes: $y = 1$; vertical asymptotes: $x = -2, x = 1$ *(3-2)*

82. Horizontal asymptote: $y = 1$; vertical asymptotes: $x = -1, x = 1$ *(3-2)*

83. The domain of $f'(x)$ is all real numbers except $x = 0$. At $x = 0$, the graph of $y = f(x)$ is smooth, but it has a vertical tangent. *(3-4)*

84. (A) $\lim_{x \to 1^-} f(x) = 1$; $\lim_{x \to 1^+} f(x) = -1$ (B) $\lim_{x \to 1^-} f(x) = -1$; $\lim_{x \to 1^+} f(x) = 1$ (C) $m = 1$

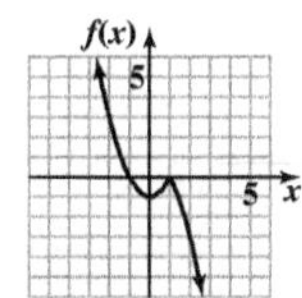

(D) The graphs in (A) and (B) have discontinuities at $x = 1$; the graph in (C) does not. *(3-2)*

85. (A) 1 (B) -1 (C) Does not exist (D) No *(3-4)*

86. (A) $S(x) = \begin{cases} 7.47 + 0.4x & \text{if } 0 \le x \le 90 \\ 24.786 + 0.2076x & \text{if } 90 < x \end{cases}$

(B)

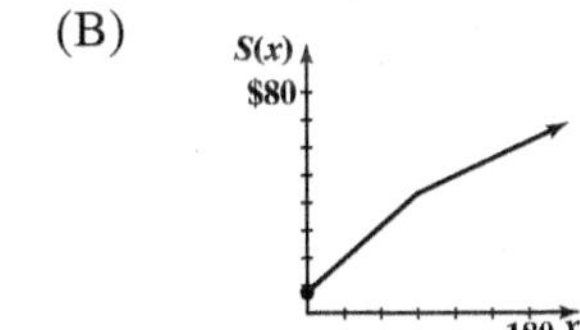

(C) Yes *(3-2)*

87. (A) \$179.90 (B) \$180 *(3-7)*

88. (A) $C(100) = 9{,}500$; $C'(100) = 50$. At a production level of 100 bicycles, the total cost is \$9,500, and cost is increasing at the rate of \$50 per bicycle.
(B) $\overline{C}(100) = 95$; $\overline{C}'(100) = -0.45$. At a production level of 100 bicycles, the average cost is \$95, and average cost is decreasing at a rate of \$0.45 per bicycle. *(3-7)*

89. The approximate cost of producing the 201st printer is greater than that of the 601st printer. Since these marginal costs are decreasing, the manufacturing process is becoming more efficient. *(3-7)*

90. (A) $C'(x) = 2$; $\overline{C}(x) = 2 + \frac{9{,}000}{x}$; $\overline{C}'(x) = \frac{-9{,}000}{x^2}$
(B) $R(x) = xp = 25x - 0.01x^2$; $R'(x) = 25 - 0.02x$; $\overline{R}(x) = 25 - 0.01x$; $\overline{R}'(x) = -0.01$
(C) $P(x) = R(x) - C(x) = 23x - 0.01x^2 - 9{,}000$; $P'(x) = 23 - 0.02x$;
$\overline{P}(x) = 23 - 0.01x - \frac{9{,}000}{x}$; $\overline{P}'(x) = -0.01 + \frac{9{,}000}{x^2}$
(D) (500, 10,000) and (1,800, 12,600)
(E) $P'(1{,}000) = 3$. Profit is increasing at the rate of \$3 per umbrella.
$P'(1{,}150) = 0$. Profit is flat.
$P'(1{,}400) = -5$. Profit is decreasing at the rate of \$5 per umbrella.
(F)

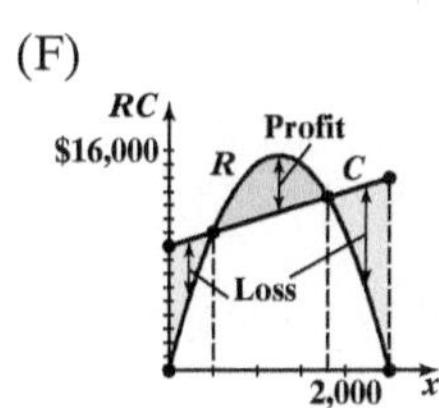

91. (A) 8 (B) 20 *(3-5)*

92. $N(9) = 27$; $N'(t) = 3.5$; After 9 months, 27,000 pools have been sold and the total sales are increasing at the rate of 3,500 pools per month. *(3-5)*

93. (A)

```
CubicReg
y=ax³+bx²+cx+d
a=.001225
b=-.0819285714
c=1.564642857
d=12.08428571
```

(B) $N(50) = 38.6$; $N'(50) = 2.6$. In 2010, natural gas consumption is 38.6 trillion cubic feet and is increasing at the rate of 2.6 trillion cubic feet per year *(3-4)*

94. (A)

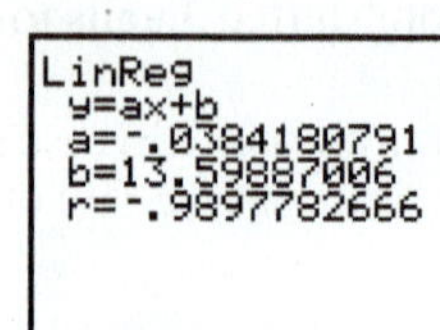

```
LinReg
y=ax+b
a=-.0384180791
b=13.59887006
r=-.9897782666
```

(B) Fixed costs: \$484.21; variable costs per kringle: \$2.11

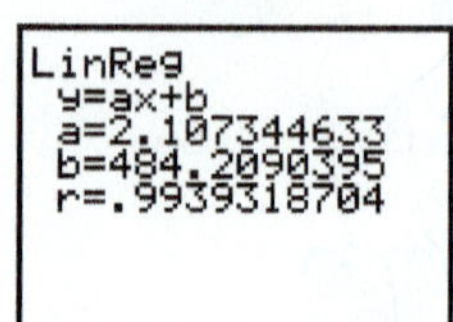

```
LinReg
y=ax+b
a=2.107344633
b=484.2090395
r=.9939318704
```

(C) (51, 591.15), (248, 1,007.62) (D) $\$4.07 < p < \$11.64$ *(3-7)*

95. $C'(10) = -1$; $C'(100) = -0.001$ *(3-5)*

96. $F(4) = 98.16$; $F'(4) = -0.32$; After 4 hours the patient's temperature is 98.16°F and is decreasing at the rate of 0.32°F per hour. *(3-5)*

97. (A) 10 items/h (B) 5 items/h *(3-5)*

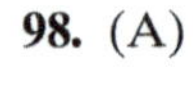

98. (A)

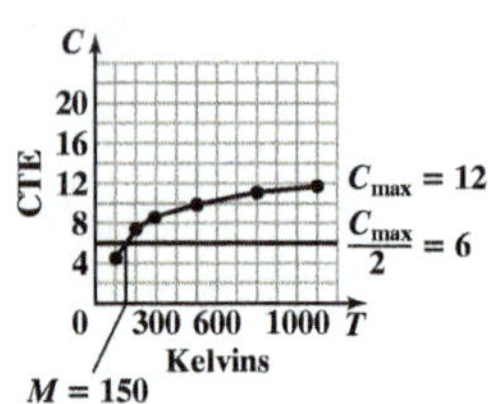

(B) $C(T) = \dfrac{12T}{150 + T}$

(C) $C = 9.6$ at $T = 600$ K, $T = 750$ K when $C = 10$ *(3-3)*

Chapter 2

Exercises 2-1

1. \$1,221.40; \$1,648.72; \$2,225.54

3.

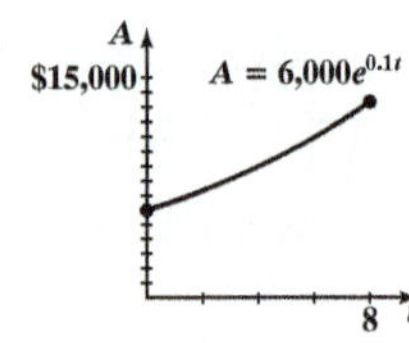

5. 11.55 **7.** 10.99 **9.** 0.14

11.

n	$[1 + (1/n)]^n$
10	2.593 74
100	2.704 81
1,000	2.716 92
10,000	2.718 15
100,000	2.718 27
1,000,000	2.718 28
10,000,000	2.718 28
↓	↓
∞	$e = 2.718\ 281\ 828\ 459\ldots$

13. $\lim_{n\to\infty}(1 + n)^{1/n} = 1$

15.

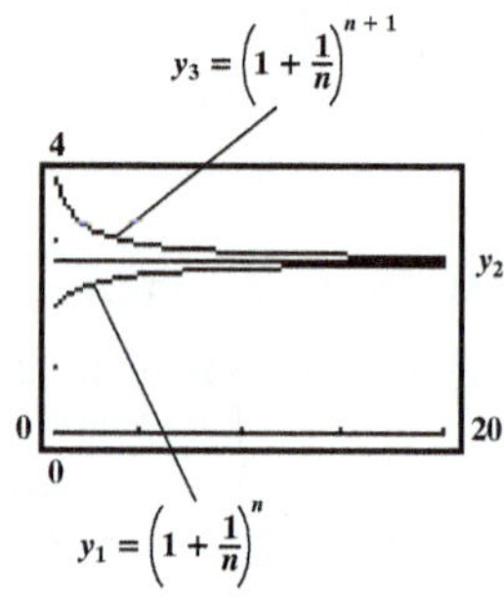

17. (A) \$15,143.71 (B) 14.16 yr

19. \$11,890.41

21. 8.11%

23. (A)

(B) $\lim_{t\to\infty} 10{,}000e^{-0.08t} = 0$

25. 17.33 yr **27.** 8.66% **29.** 7.3 yr

31. (A)

$$A = Pe^{rt}$$
$$2P = Pe^{rt}$$
$$2 = e^{rt}$$
$$rt = \ln 2$$
$$t = \frac{\ln 2}{r}$$

(B)

Although r could be any positive number, the restrictions on r are reasonable in the sense that most investments would be expected to earn a return of between 2% and 30%.

(C) The doubling times (in years) are 13.86, 6.93, 4.62, 3.47, 2.77, and 2.31, respectively.

33. $t = -(\ln 0.5)/0.000\,433\,2 \approx 1{,}600$ yr **35.** $r = (\ln 0.5)/30 \approx -0.0231$ **37.** 53.3 yr **39.** 1.39%

Exercises 2-2

1. $5e^x + 3$ **3.** $-\dfrac{2}{x} + 2x$ **5.** $3x^2 - 6e^x$ **7.** $e^x + 1 - \dfrac{1}{x}$ **9.** $\dfrac{3}{x}$ **11.** $5 - \dfrac{5}{x}$ **13.** $\dfrac{2}{x} + 4e^x$ **15.** $f'(x) = \dfrac{1}{x}$; $y = x + 2$

17. $f'(x) = 3e^x$; $y = 3x + 3$ **19.** $f'(x) = \dfrac{3}{x}$; $y = \dfrac{3x}{e}$ **21.** $f'(x) = e^x$; $y = ex + 2$ **23.** Yes; yes **25.** No; no

27. $f(x) = 10x + \ln 10 + \ln x$; $f'(x) = 10 + \dfrac{1}{x}$ **29.** $f(x) = \ln 4 - 3\ln x$; $f'(x) = -\dfrac{3}{x}$ **31.** $\dfrac{1}{x\ln 2}$ **33.** $3^x \ln 3$

35. $2 - \dfrac{1}{x\ln 10}$ **37.** $1 + 10^x \ln 10$ **39.** $\dfrac{3}{x} + \dfrac{2}{x\ln 3}$ **41.** $2^x \ln 2$ **43.** $(-0.82, 0.44)$, $(1.43, 4.18)$, $(8.61, 5503.66)$

45. (0.49, 0.49) **47.** (3.65, 1.30), (332, 105.11, 12.71) **51.** \$28,447/yr; \$18,664/yr; \$11,021/yr
53. $A'(t) = 5{,}000(\ln 4)4^t$; $A'(1) = 27{,}726$ bacteria/hr (rate of change at the end of the first hour); $A'(5) = 7{,}097{,}827$ bacteria/hr (rate of change at the end of the fifth hour)
55. At the 40-lb weight level, blood pressure would increase at the rate of 0.44 mm of mercury per pound of weight gain. At the 90-lb weight level, blood pressure would increase at the rate of 0.19 mm of mercury per pound of weight gain.
57. $dR/dS = k/S$

Exercises 2-3

1. $2x^3(2x) + (x^2 - 2)(6x^2) = 10x^4 - 12x^2$ **3.** $(x - 3)(2) + (2x - 1)(1) = 4x - 7$
5. $\frac{(x-3)(1) - x(1)}{(x-3)^2} = \frac{-3}{(x-3)^2}$ **7.** $\frac{(x-2)(2) - (2x+3)(1)}{(x-2)^2} = \frac{-7}{(x-2)^2}$
9. $3xe^x + 3e^x = 3(x+1)e^x$ **11.** $x^3\left(\frac{1}{x}\right) + 3x^2 \ln x = x^2(1 + 3\ln x)$
13. $(x^2 + 1)(2) + (2x - 3)(2x) = 6x^2 - 6x + 2$ **15.** $(0.4x + 2)(0.5) + (0.5x - 5)(0.4) = 0.4x - 1$
17. $\frac{(2x-3)(2x) - (x^2+1)(2)}{(2x-3)^2} = \frac{2x^2 - 6x - 2}{(2x-3)^2}$ **19.** $(x^2 + 2)2x + (x^2 - 3)2x = 4x^3 - 2x$
21. $\frac{(x^2-3)2x - (x^2+2)2x}{(x^2-3)^2} = \frac{-10x}{(x^2-3)^2}$ **23.** $\frac{(x^2+1)e^x - e^x(2x)}{(x^2+1)^2} = \frac{(x-1)^2e^x}{(x^2+1)^2}$
25. $\frac{(1+x)\left(\frac{1}{x}\right) - \ln x}{(1+x)^2} = \frac{1 + x - x\ln x}{x(1+x)^2}$ **27.** $xf'(x) + f(x)$ **29.** $x^3f'(x) + 3x^2f(x)$ **31.** $\frac{x^2f'(x) - 2xf(x)}{x^4}$
33. $\frac{f(x) - xf'(x)}{[f(x)]^2}$ **35.** $e^xf'(x) + f(x)e^x = e^x[f'(x) + f(x)]$ **37.** $\frac{f(x)\left(\frac{1}{x}\right) - (\ln x)f'(x)}{f(x)^2} = \frac{f(x) - (x\ln x)f'(x)}{x\,f(x)^2}$
39. $(2x + 1)(2x - 3) + (x^2 - 3x)(2) = 6x^2 - 10x - 3$ **41.** $(2.5t - t^2)(4) + (4t + 1.4)(2.5 - 2t) = -12t^2 + 17.2t + 3.5$
43. $\frac{(x^2+2x)(5) - (5x-3)(2x+2)}{(x^2+2x)^2} = \frac{-5x^2 + 6x + 6}{(x^2+2x)^2}$ **45.** $\frac{(w^2-1)(2w-3) - (w^2-3w+1)(2w)}{(w^2-1)^2} = \frac{3w^2 - 4w + 3}{(w^2-1)^2}$
47. $(1 + x - x^2)e^x + e^x(1 - 2x) = (2 - x - x^2)e^x$ **49.** $f'(x) = (1 + 3x)(-2) + (5 - 2x)(3)$; $y = -11x + 29$
51. $f'(x) = \frac{(3x-4)(1) - (x-8)(3)}{(3x-4)^2}$; $y = 5x - 13$ **53.** $f'(x) = \frac{2^x - x(2^x \ln 2)}{2^{2x}}$; $y = \left(\frac{1 - 2\ln 2}{4}\right)x + \ln 2$
55. $f'(x) = (2x - 15)(2x) + (x^2 + 18)(2) = 6(x - 2)(x - 3)$; $x = 2, x = 3$
57. $f'(x) = \frac{(x^2+1)(1) - x(2x)}{(x^2+1)^2} = \frac{1 - x^2}{(x^2+1)^2}$; $x = -1, x = 1$ **59.** $7x^6 - 3x^2$ **61.** $-27x^{-4} = -\frac{27}{x^4}$
63. $(w + 1)2^w \ln 2 + 2^w = [(w + 1)\ln 2 + 1]2^w$ **65.** $\frac{(4x^2+5x-1)(6x-2) - (3x^2-2x+3)(8x+5)}{(4x^2+5x-1)^2} = \frac{23x^2 - 30x - 13}{(4x^2+5x-1)^2}$
67. $9x^{1/3}(3x^2) + (x^3 + 5)(3x^{-2/3}) = \frac{30x^3 + 15}{x^{2/3}}$ **69.** $\frac{(1+x^2)\frac{1}{x\ln 2} - 2x\log_2 x}{(1+x^2)^2} = \frac{1 + x^2 - 2x^2\ln x}{x(1+x^2)^2 \ln 2}$
71. $\frac{(x^2-3)(2x^{-2/3}) - 6x^{1/3}(2x)}{(x^2-3)^2} = \frac{-10x^2 - 6}{(x^2-3)^2x^{2/3}}$ **73.** $g'(t) = \frac{(3t^2-1)(0.2) - (0.2t)(6t)}{(3t^2-1)^2} = \frac{-0.6t^2 - 0.2}{(3t^2-1)^2}$
75. $(20x)\frac{1}{x\ln 10} + 20\log x = \frac{20(1 + \ln x)}{\ln 10}$ **77.** $x^{-2/3}(3x^2 - 4x) + (x^3 - 2x^2)\left(-\frac{2}{3}x^{-5/3}\right) = -\frac{8}{3}x^{1/3} + \frac{7}{3}x^{4/3}$
79. $\frac{(x^2+1)[(2x^2-1)(2x) + (x^2+3)(4x)] - (2x^2-1)(x^2+3)(2x)}{(x^2+1)^2} = \frac{4x^5 + 8x^3 + 16x}{(x^2+1)^2}$
81. $\frac{e^t(1 + \ln t) - (t\ln t)e^t}{e^{2t}} = \frac{1 + \ln t - t\ln t}{e^t}$
83. (A) $S'(t) = \frac{(t^2+50)(180t) - 90t^2(2t)}{(t^2+50)^2} = \frac{9{,}000t}{(t^2+50)^2}$
(B) $S(10) = 60$; $S'(10) = 4$. After 10 months, the total sales are 60,000 DVDs and sales are increasing at the rate of 4,000 DVDs per month.
(C) Approximately 64,000 DVDs

85. (A) $\frac{dx}{dp} = \frac{(0.1p + 1)(0) - 4{,}000(0.1)}{(0.1p + 1)^2} = \frac{-400}{(0.1p + 1)^2}$
(B) $x = 800$; $dx/dp = -16$. At a price level of \$40, the demand is 800 DVD players per week and demand is decreasing at the rate of 16 players per dollar.
(C) Approximately 784 DVD players

87. (A) $C'(t) = \frac{(t^2 + 1)(0.14) - 0.14t(2t)}{(t^2 + 1)^2} = \frac{0.14 - 0.14t^2}{(t^2 + 1)^2}$
(B) $C'(0.5) = 0.0672$. After 0.5 h, concentration is increasing at the rate of 0.0672 mg/cm^3 per hour.
$C'(3) = -0.0112$. After 3 h, concentration is decreasing at the rate of 0.0112 mg/cm^3 per hour.

89. (A) $N'(x) = \frac{(x + 32)(100) - (100x + 200)}{(x + 32)^2} = \frac{3{,}000}{(x + 32)^2}$
(B) $N'(4) = 2.31$; $N'(68) = 0.30$

Exercises 2-4

1. $(3x^2 + 2)^3$ **3.** e^{-x^2} **5.** $y = u^4$; $u = 3x^2 - x + 5$ **7.** $y = e^u$; $u = 1 + x + x^2$ **9.** 3 **11.** $(-4x)$
13. $2x$ **15.** $4x^3$ **17.** $2(x + 3)$ **19.** $6(2x + 5)^2$ **21.** $-8(5 - 2x)^3$ **23.** $5(4 + 0.2x)^4(0.2) = (4 + 0.2x)^4$ **25.** $30x(3x^2 + 5)^4$
27. $5e^x$ **29.** $5e^{5x}$ **31.** $-18e^{-6x}$ **33.** $(2x - 5)^{-1/2} = \frac{1}{(2x - 5)^{1/2}}$ **35.** $-8x^3(x^4 + 1)^{-3} = \frac{-8x^3}{(x^4 + 1)^3}$ **37.** $-\frac{2}{x}$ **39.** $\frac{6x}{1 + x^2}$
41. $\frac{3(1 + \ln x)^2}{x}$ **43.** $f'(x) = 6(2x - 1)^2$; $y = 6x - 5$; $x = \frac{1}{2}$ **45.** $f'(x) = 2(4x - 3)^{-1/2} = \frac{2}{(4x - 3)^{1/2}}$; $y = \frac{2}{3}x + 1$; none
47. $f'(x) = 10(x - 2)e^{x^2 - 4x + 1}$; $y = -20ex + 5e$; $x = 2$ **49.** $12(x^2 - 2)^3(2x) = 24x(x^2 - 2)^3$
51. $-6(t^2 + 3t)^{-4}(2t + 3) = \frac{-6(2t + 3)}{(t^2 + 3t)^4}$ **53.** $\frac{1}{2}(w^2 + 8)^{-1/2}(2w) = \frac{w}{\sqrt{w^2 + 8}}$ **55.** $12xe^{3x} + 4e^{3x} = 4(3x + 1)e^{3x}$
57. $\frac{x^3\left(\frac{1}{1 + x}\right) - 3x^2 \ln(1 + x)}{x^6} = \frac{x - 3(1 + x)\ln(1 + x)}{x^4(1 + x)}$ **59.** $6te^{3(t^2+1)}$ **61.** $\frac{3x}{x^2 + 3}$
63. $-5(w^3 + 4)^{-6}(3w^2) = \frac{-15w^2}{(w^3 + 4)^6}$ **65.** $f'(x) = (4 - x)^3 - 3x(4 - x)^2 = 4(4 - x)^2(1 - x)$; $y = -16x + 48$
67. $f'(x) = \frac{(2x - 5)^3 - 6x(2x - 5)^2}{(2x - 5)^6} = \frac{-4x - 5}{(2x - 5)^4}$; $y = -17x + 54$ **69.** $f'(x) = \frac{1}{2x\sqrt{\ln x}}$; $y = \frac{x}{2e} + \frac{1}{2}$
71. $f'(x) = 2x(x - 5)^3 + 3x^2(x - 5)^2 = 5x(x - 5)^2(x - 2)$; $x = 0, 2, 5$
73. $f'(x) = \frac{(2x + 5)^2 - 4x(2x + 5)}{(2x + 5)^4} = \frac{5 - 2x}{(2x + 5)^3}$; $x = \frac{5}{2}$
75. $f'(x) = (x^2 - 8x + 20)^{-1/2}(x - 4) = \frac{x - 4}{(x^2 - 8x + 20)^{1/2}}$; $x = 4$ **77.** No; yes
79. $18x^2(x^2 + 1)^2 + 3(x^2 + 1)^3 = 3(x^2 + 1)^2(7x^2 + 1)$ **81.** $\frac{24x^5(x^3 - 7)^3 - (x^3 - 7)^4 6x^2}{4x^6} = \frac{3(x^3 - 7)^3(3x^3 + 7)}{2x^4}$
83. $\frac{1}{\ln 2}\left(\frac{6x}{3x^2 - 1}\right)$ **85.** $(2x + 1)(10^{x^2+x})(\ln 10)$ **87.** $\frac{12x^2 + 5}{(4x^3 + 5x + 7)\ln 3}$ **89.** $2^{x^3 - x^2 + 4x + 1}(3x^2 - 2x + 4)\ln 2$

91. (A) $C'(x) = (2x + 16)^{-1/2} = \frac{1}{(2x + 16)^{1/2}}$
(B) $C'(24) = \frac{1}{8}$, or \$12.50. At a production level of 24 cell phones, total cost is increasing at the rate of \$12.50 per cell phone and the cost of producing the 25th cell phone is approximately \$12.50.
$C'(42) = \frac{1}{10}$, or \$10.00. At a production level of 42 cell phones, total cost is increasing at the rate of \$10.00 per cell phone and the cost of producing the 43rd cell phone is approximately \$10.00.

93. (A) $\frac{dx}{dp} = 40(p + 25)^{-1/2} = \frac{40}{(p + 25)^{1/2}}$ (B) $x = 400$ and $dx/dp = 4$. At a price of \$75, the supply is 400 bicycle helmets per week and supply is increasing at the rate of 4 bicycle helmets per dollar.

95. (A) After 1 hr, the concentration is decreasing at the rate of 1.60 mg/mL per hour; after 4 hr, the concentration is decreasing at the rate of 0.08 mg/mL per hour. (B)

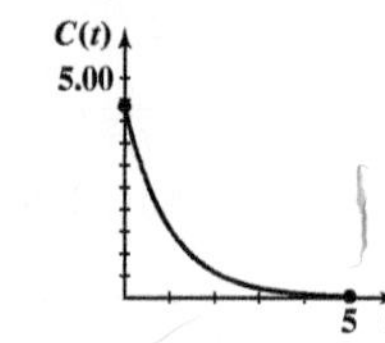

97. 2.27 mm of mercury/yr; 0.81 mm of mercury/yr; 0.41 mm of mercury/yr

99. (A) $f'(n) = n(n-2)^{-1/2} + 2(n-2)^{1/2}$

(B) $f'(11) = \frac{29}{3} \approx 9.67$. When the list contains 11 items, the learning time is increasing at the rate of 9.67 min per item. $f'(27) = \frac{77}{5} = 15.4$. When the list contains 27 items, the learning time is increasing at the rate of 15.4 min per item.

Exercises 2-5

1. $y' = -\frac{3}{5}$ **3.** $y' = \frac{3x}{2}$ **5.** $y' = 10x; 10$ **7.** $y' = \frac{2x}{3y^2}; \frac{4}{3}$ **9.** $y' = -\frac{3}{2y+2}; -\frac{3}{4}$ **11.** $y' = -\frac{y}{x}; -\frac{3}{2}$

13. $y' = -\frac{2y}{2x+1}; 4$ **15.** $y' = \frac{6-2y}{x}; -1$ **17.** $y' = \frac{2x}{e^y - 2y}; 2$ **19.** $y' = \frac{3x^2y}{y+1}; \frac{3}{2}$ **21.** $y' = \frac{6x^2y - y\ln y}{x+2y}; 2$

23. $x' = \frac{2tx - 3t^2}{2x - t^2}; 8$ **25.** $y'|_{(1.6,\,1.8)} = -\frac{3}{4}; y'|_{(1.6,\,0.2)} = \frac{3}{4}$ **27.** $y = -x + 5$ **29.** $y = \frac{2}{5}x - \frac{12}{5}; y = \frac{3}{5}x + \frac{12}{5}$

31. $y' = -\frac{1}{x}$ **33.** $y' = \frac{1}{3(1+y)^2 + 1}; \frac{1}{13}$ **35.** $y' = \frac{3(x-2y)^2}{6(x-2y)^2 + 4y}; \frac{3}{10}$ **37.** $y' = \frac{3x^2(7+y^2)^{1/2}}{y}; 16$

39. $y' = \frac{y}{2xy^2 - x}; 1$ **41.** $y = 0.63x + 1.04$ **43.** $p' = \frac{1}{2p-2}$ **45.** $p' = -\frac{\sqrt{10{,}000 - p^2}}{p}$

47. $\frac{dL}{dV} = \frac{-(L+m)}{V+n}$ **49.** $\frac{dT}{dv} = \frac{2}{k}\sqrt{T}$ **51.** $\frac{dv}{dT} = \frac{k}{2\sqrt{T}}$

Exercises 2-6

1. 30 **3.** $-\frac{16}{3}$ **5.** $-\frac{16}{7}$ **7.** Decreasing at 9 units/sec **9.** Approx. −3.03 ft/sec **11.** $dA/dt \approx 126$ ft²/sec
13. 3,768 cm³/min **15.** 6 lb/in.²/hr **17.** $-\frac{9}{4}$ ft/sec **19.** $\frac{20}{3}$ ft/sec **21.** 0.0214 ft/sec; 0.0135 ft/sec; yes, at $t = 0.000\,19$ sec
23. 3.835 units/sec **25.** (A) $dC/dt = \$15{,}000$/wk (B) $dR/dt = -\$50{,}000$/wk (C) $dP/dt = -\$65{,}000$/wk
27. $ds/dt = \$2{,}207$/wk **29.** (A) $dx/dt = -12.73$ units/month (B) $dp/dt = \$1.53$/month **31.** Approximately 100 ft³/min

Exercises 2-7

1. $\frac{35 - 0.8x}{35x - 0.4x^2}$ **3.** $-\frac{4e^{-x}}{7 + 4e^x}$ **5.** $\frac{5}{x(12 + 5\ln x)}$ **7.** 0 **9.** −0.017 **11.** −0.034 **13.** 1.013 **15.** 0.405 **17.** 11.8%

19. 5.4% **21.** −14.7% **23.** −431.6% **25.** $E(p) = \frac{450p}{25{,}000 - 450p}$ **27.** $E(p) = \frac{8p^2}{4{,}800 - 4p^2}$ **29.** $E(p) = \frac{0.6pe^p}{98 - 0.6e^p}$

31. (A) Inelastic (B) Unit elasticity (C) Elastic **33.** (A) Inelastic (B) Unit elasticity (C) Elastic

35. (A) $x = 6{,}000 - 200p \quad 0 \le p \le 30$
(B) $E(p) = \frac{p}{30 - p}$
(C) $E(10) = 0.5$; 5% decrease
(D) $E(25) = 5$; 50% decrease
(E) $E(15) = 1$; 10% decrease

37. (A) $x = 3{,}000 - 50p \quad 0 \le p \le 60$
(B) $R(p) = 3{,}000p - 50p^2$
(C) $E(p) = \frac{p}{60 - p}$
(D) Elastic on (30, 60); inelastic on (0, 30)
(E) Increasing on (0, 30); decreasing on (30, 60)
(F) Decrease
(G) Increase

39. Elastic on (3.5, 7); inelastic on (0, 3.5)
41. Elastic on $(25/\sqrt{3}, 25)$; inelastic on $(0, 25/\sqrt{3})$
43. Elastic on (48, 72); inelastic on (0, 48)
45. Elastic on $(25, 25\sqrt{2})$; inelastic on (0, 25)

47. $R(p) = 20p(10 - p)$

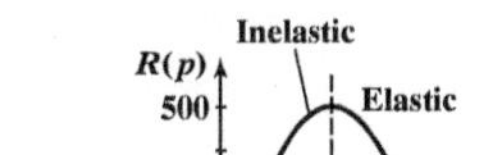
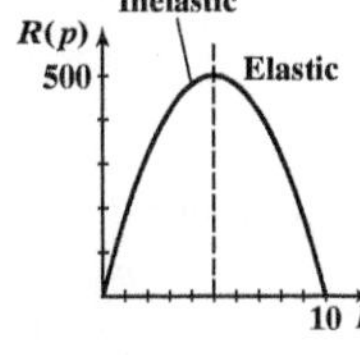

49. $R(p) = 40p(p - 15)^2$

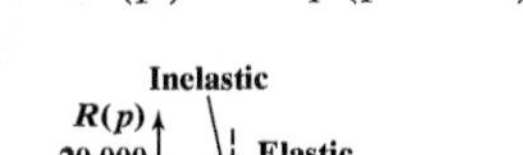
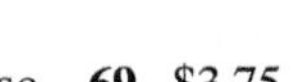

51. $R(p) = 30p - 10p\sqrt{p}$

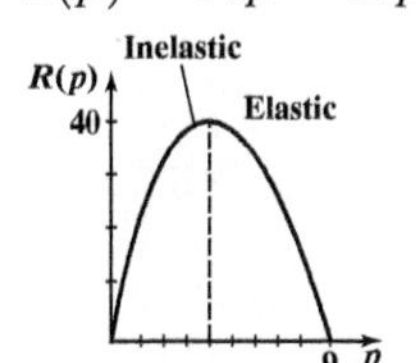

53. $\frac{3}{2}$ **55.** $\frac{1}{2}$

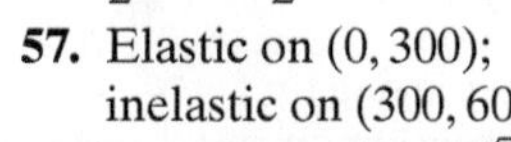

57. Elastic on (0, 300); inelastic on (300, 600)
59. Elastic on $(0, 10\sqrt{3})$; inelastic on $(10\sqrt{3}, 30)$
61. k
63. $75 per day

65. Increase **67.** Decrease **69.** $3.75 **71.** $p(t) = \frac{31}{0.31t + 18.5}$ **73.** −0.08

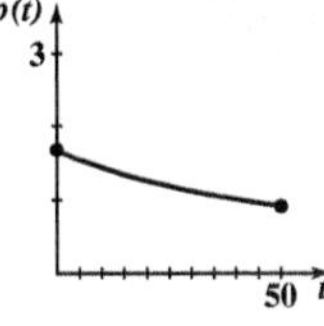

Chapter 2 Review Exercises

1. \$3,136.62; \$4,919.21; \$12,099.29 *(4-1)* **2.** $\frac{2}{x} + 3e^x$ *(4-2)* **3.** $2e^{2x-3}$ *(4-4)* **4.** $\frac{2}{2x+7}$ *(4-4)*

5. (A) $y = \ln(3 + e^x)$ (B) $\frac{dy}{dx} = \frac{e^x}{3 + e^x}$ *(4-4)* **6.** $y' = \frac{9x^2}{4y}; \frac{9}{8}$ *(4-5)* **7.** $dy/dt = 216$ *(4-6)*

8. (A) $x = 1{,}000 - 25p$ (B) $\frac{p}{40 - p}$ (C) 0.6; demand is inelastic and insensitive to small changes in price.

(D) $1{,}000p - 25p^2$ (E) Revenue increases *(4-7)*

9. -10 *(4-2)* **10.** $\lim_{n\to\infty}\left(1 + \frac{2}{n}\right)^n = e^2 \approx 7.389\ 06$ *(4-1)* **11.** $\frac{7[(\ln z)^6 + 1]}{z}$ *(4-4)* **12.** $x^5(1 + 6 \ln x)$ *(4-3)* **13.** $\frac{e^x(x - 6)}{x^7}$ *(4-3)*

14. $\frac{6x^2 - 3}{2x^3 - 3x}$ *(4-4)* **15.** $(3x^2 - 2x)e^{x^3 - x^2}$ *(4-4)* **16.** $\frac{1 - 2x \ln 5x}{xe^{2x}}$ *(4-4)* **17.** $y = -x + 2$; $y = -ex + 1$ *(4-4)*

18. $y' = \frac{3y - 2x}{8y - 3x}; \frac{8}{19}$ *(4-5)* **19.** $x' = \frac{4tx}{3x^2 - 2t^2}; -4$ *(4-5)* **20.** $y' = \frac{1}{e^y + 2y}; 1$ *(4-5)* **21.** $y' = \frac{2xy}{1 + 2y^2}; \frac{2}{3}$ *(4-5)*

22. $dy/dt = -2$ units/sec *(4-6)* **23.** 0.27 ft/sec *(4-6)* **24.** $dR/dt = 1/\pi \approx 0.318$ in./min *(4-6)*

25. Elastic for $5 < p < 15$; inelastic for $0 < p < 5$ *(4-7)*

26.

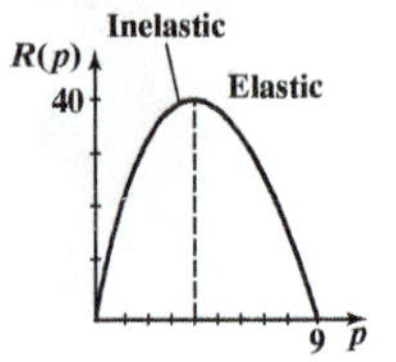

(4-7)

27. (A) $y = [\ln(4 - e^x)]^3$ (B) $\frac{dy}{dx} = \frac{-3e^x[\ln(4 - e^x)]^2}{4 - e^x}$ *(4-4)*

28. $2x(5^{x^2-1})(\ln 5)$ *(4-4)* **29.** $\left(\frac{1}{\ln 5}\right)\frac{2x - 1}{x^2 - x}$ *(4-4)*

30. $\frac{2x + 1}{2(x^2 + x)\sqrt{\ln(x^2 + x)}}$ *(4-4)* **31.** $y' = \frac{2x - e^{xy}y}{xe^{xy} - 1}; 0$ *(4-5)*

32. The rate of increase of area is proportional to the radius R, so the rate is smallest when $R = 0$, and has no largest value. *(4-6)*

33. Yes, for $-\sqrt{3}/3 < x < \sqrt{3}/3$ *(4-6)* **34.** (A) 15 yr (B) 13.9 yr *(4-1)*

35. $A'(t) = 10e^{0.1t}$; $A'(1) = \$11.05$/yr; $A'(10) = \$27.18$/yr *(4-1)* **36.** $R'(x) = (1{,}000 - 20x)e^{-0.02x}$ *(4-4)*

37. $p' = \frac{-(5{,}000 - 2p^3)^{1/2}}{3p^2}$ *(4-5)* **38.** $dR/dt = \$2{,}242$/day *(4-6)* **39.** Decrease price *(4-7)* **40.** 0.02125 *(4-7)*

41. -1.111 mg/mL per hour; -0.335 mg/mL per hour *(4-4)* **42.** $dR/dt = -3/(2\pi)$; approx. 0.477 mm/day *(4-6)*

43. (A) Increasing at the rate of 2.68 units/day at the end of 1 day of training; increasing at the rate of 0.54 unit/day after 5 days of training

(B) 7 days *(4-4)*

44. $dT/dt = -1/27 \approx -0.037$ min/operation hour *(4-6)*

Chapter 3

Exercises 3-1

1. (a, b); (d, f); (g, h) **3.** (b, c); (c, d); (f, g) **5.** c, d, f **7.** b, f

9. Local maximum at $x = a$; local minimum at $x = c$; no local extrema at $x = b$ and $x = d$ **11.** e **13.** d **15.** f

17. c **19.** (A) $f'(x) = 3x^2 - 12$ (B) $-2, 2$ (C) $-2, 2$ **21.** (A) $f'(x) = \frac{1}{3}(x + 5)^{-2/3}$ (B) -5 (C) -5

23. (A) $f'(x) = -\frac{6}{(x + 2)^2}$ (B) None (C) -2 **25.** (A) $f'(x) = \begin{cases} -1 \text{ if } x < 0 \\ 1 \text{ if } x > 0 \end{cases}$ (B) 0 (C) 0

27. Decreasing on $(-\infty, 1)$; increasing on $(1, \infty)$; $f(1) = -2$ is a local minimum

29. Increasing on $(-\infty, -4)$; decreasing on $(-4, \infty)$; $f(-4) = 7$ is a local maximum

31. Increasing for all x; no local extrema

33. Increasing on $(-\infty, -2)$ and $(3, \infty)$; decreasing on $(-2, 3)$; $f(-2) = 44$ is a local maximum, $f(3) = -81$ is a local minimum

35. Decreasing on $(-\infty, 1)$; increasing on $(1, \infty)$; $f(1) = 4$ is a local minimum

37. Increasing on $(-\infty, 2)$; decreasing on $(2, \infty)$; $f(2) = e^{-2} \approx 0.135$ is a local maximum

39. Increasing on $(-\infty, 8)$; decreasing on $(8, \infty)$; $f(8) = 4$ is a local maximum

41. Critical values: $x = -0.77, 1.08, 2.69$; decreasing on $(-\infty, -0.77)$ and $(1.08, 2.69)$; increasing on $(-0.77, 1.08)$ and $(2.69, \infty)$; local minima at $x = -0.77$ and $x = 2.69$; local maximum at $x = 1.08$

43. Critical values: $x = 1.34, 2.82$; decreasing on $(0, 1.34)$ and $(2.82, \infty)$; increasing on $(1.34, 2.82)$; local minimum at $x = 1.34$; and local maximum at $x = 2.82$

45. Critical values: 0.36, 2.15; increasing on $(-\infty, 0.36)$ and $(2.15, \infty)$; decreasing on $(0.36, 2.15)$; local maximum at $x = 0.36$; local minimum at $x = 2.15$

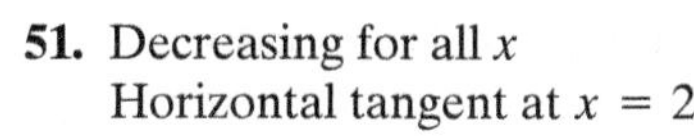

47. Increasing on $(-\infty, 4)$
Decreasing on $(4, \infty)$
Horizontal tangent at $x = 4$

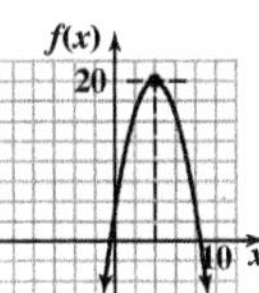

49. Increasing on $(-\infty, -1)$, $(1, \infty)$
Decreasing on $(-1, 1)$
Horizontal tangents at $x = -1, 1$

51. Decreasing for all x
Horizontal tangent at $x = 2$

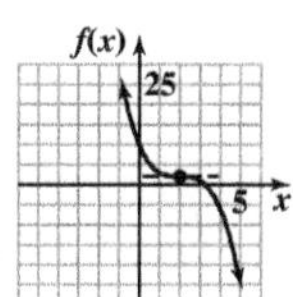

53. Decreasing on $(-\infty, -3)$ and $(0, 3)$; increasing on $(-3, 0)$ and $(3, \infty)$
Horizontal tangents at $x = -3, 0, 3$

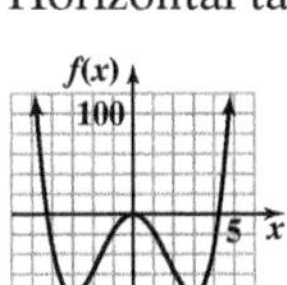

55.

57.

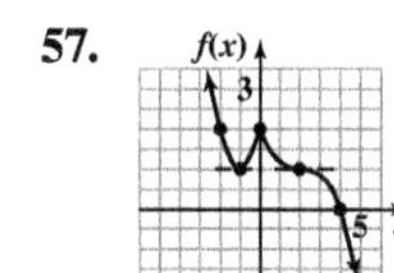

59.

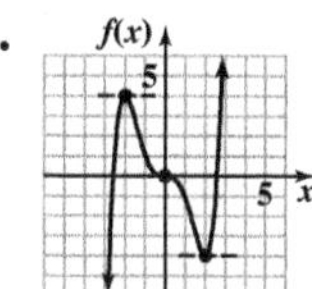

61.

63. g_4
65. g_6
67. g_2

69. Increasing on $(-1, 2)$; decreasing on $(-\infty, -1)$ and $(2, \infty)$; local minimum at $x = -1$; local maximum at $x = 2$

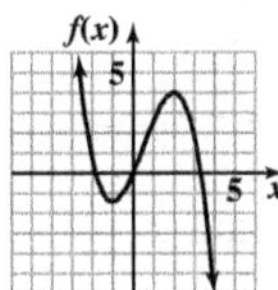

71. Increasing on $(-1, 2)$ and $(2, \infty)$; decreasing on $(-\infty, -1)$; local minimum at $x = -1$

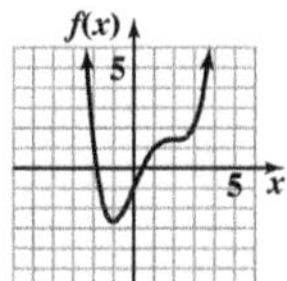

73. Increasing on $(-2, 0)$ and $(3, \infty)$; decreasing on $(-\infty, -2)$ and $(0, 3)$; local minima at $x = -2$ and $x = 3$; local maximum at $x = 0$

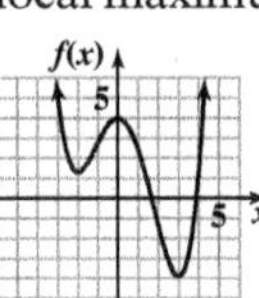

75. $f'(x) > 0$ on $(-\infty, -1)$ and $(3, \infty)$; $f'(x) < 0$ on $(-1, 3)$; $f'(x) = 0$ at $x = -1$ and $x = 3$

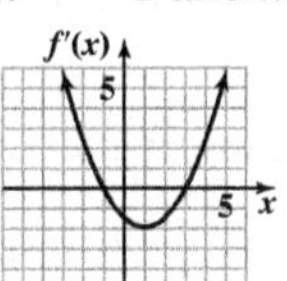

77. $f'(x) > 0$ on $(-2, 1)$ and $(3, \infty)$; $f'(x) < 0$ on $(-\infty, -2)$ and $(1, 3)$; $f'(x) = 0$ at $x = -2$, $x = 1$, and $x = 3$

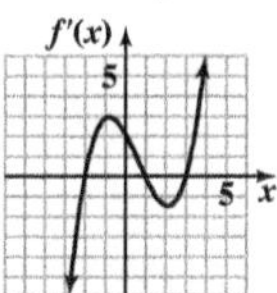

79. Critical values: $x = -2$, $x = 2$; increasing on $(-\infty, -2)$ and $(2, \infty)$; decreasing on $(-2, 0)$ and $(0, 2)$; local maximum at $x = -2$; local minimum at $x = 2$
81. Critical value: $x = -2$; increasing on $(-2, 0)$; decreasing on $(-\infty, -2)$ and $(0, \infty)$; local minimum at $x = -2$
83. Critical values: $x = 0$, $x = 4$; increasing on $(-\infty, 0)$ and $(4, \infty)$; decreasing on $(0, 2)$ and $(2, 4)$; local maximum at $x = 0$; local minimum at $x = 4$
85. Critical values: $x = 0$, $x = 4$, $x = 6$; increasing on $(0, 4)$ and $(6, \infty)$; decreasing on $(-\infty, 0)$ and $(4, 6)$; local maximum at $x = 4$; local minima at $x = 0$ and $x = 6$
87. (A) There are no critical values and no local extrema. The function is increasing for all x.
(B) There are two critical values, $x = \pm\sqrt{-k/3}$. The function increases on $(-\infty, -\sqrt{-k/3})$ to a local maximum at $x = -\sqrt{-k/3}$, decreases on $(-\sqrt{-k/3}, \sqrt{-k/3})$ to a local minimum at $x = \sqrt{-k/3}$, and increases on $(\sqrt{-k/3}, \infty)$.
(C) The only critical value is $x = 0$. There are no local extrema. The function is increasing for all x.
89. (A) The marginal profit is positive on $(0, 600)$, 0 at $x = 600$, and negative on $(600, 1{,}000)$.
(B)

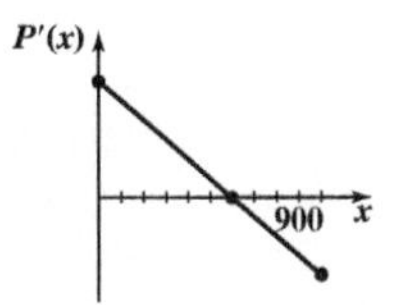

91. (A) The price decreases for the first 15 months to a local minimum, increases for the next 40 months to a local maximum, and then decreases for the remaining 15 months.
(B)

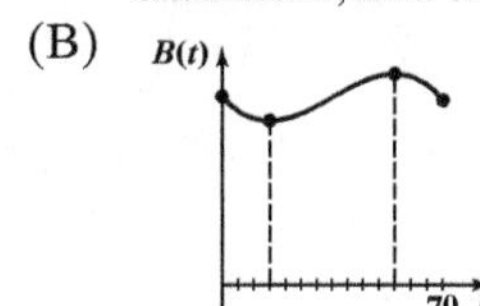

93. (A) $\overline{C}(x) = 0.05x + 20 + \dfrac{320}{x}$
(B) Critical value: $x = 80$; decreasing for $0 < x < 80$; increasing for $80 < x < 150$; local minimum at $x = 80$
95. $P(x)$ is increasing over (a, b) if $P'(x) = R'(x) - C'(x) > 0$ over (a, b); that is, if $R'(x) > C'(x)$ over (a, b).
97. Critical value: $t = 2$; increasing on $(0, 2)$; decreasing on $(2, 24)$; $C(2) = 0.07$ is a local maximum.

Exercises 3-2

1. (A) $(a, c), (c, d), (e, g)$ (B) $(d, e), (g, h)$ (C) $(d, e), (g, h)$ (D) $(a, c), (c, d), (e, g)$ (E) $(a, c), (c, d), (e, g)$ (F) $(d, e), (g, h)$ (G) d, e, g (H) d, e, g **3.** (C) **5.** (D) **7.** $12x - 8$ **9.** $4x^{-3} - 18x^{-4}$
11. $2 + \frac{9}{2}x^{-3/2}$ **13.** $8(x^2 + 9)^3 + 48x^2(x^2 + 9)^2 = 8(x^2 + 9)^2(7x^2 + 9)$

15. $-2e^{-x^2} + 4x^2e^{-x^2} = 2e^{-x^2}(2x^2 - 1)$ **17.** $\dfrac{x^3\left(-\frac{2}{x}\right) - (1 - 2\ln x)3x^2}{x^6} = \dfrac{6\ln x - 5}{x^4}$ **19.** $(-10, 2{,}000)$ **21.** $(0, 2)$ **23.** None

25. Concave upward for all x; no inflection points
27. Concave downward on $(-\infty, \frac{4}{3})$; concave upward on $(\frac{4}{3}, \infty)$; inflection point at $x = \frac{4}{3}$
29. Concave downward on $(-\infty, 0)$ and $(6, \infty)$; concave upward on $(0, 6)$; inflection points at $x = 0$ and $x = 6$
31. Concave upward on $(-2, 4)$; concave downward on $(-\infty, -2)$ and $(4, \infty)$; inflection points at $x = -2$ and $x = 4$
33. Concave upward on $(-\infty, \ln 2)$; concave downward on $(\ln 2, \infty)$; inflection point at $x = \ln 2$

35. **37.** **39.** **41.**

43. Domain: All real numbers
y intercept: 16; x intercepts: $2 - 2\sqrt{3}, 2, 2 + 2\sqrt{3}$
Increasing on $(-\infty, 0)$ and $(4, \infty)$
Decreasing on $(0, 4)$
Local maximum at $x = 0$, local minimum at $x = 4$
Concave downward on $(-\infty, 2)$
Concave upward on $(2, \infty)$
Inflection point at $x = 2$

45. Domain: All real numbers
y intercept: 2; x intercept: -1
Increasing on $(-\infty, \infty)$
Concave downward on $(-\infty, 0)$
Concave upward on $(0, \infty)$
Inflection point at $x = 0$

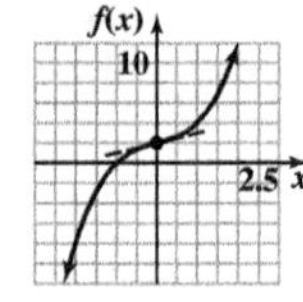

47. Domain: All real numbers
y intercept: 0; x intercepts: 0, 4
Increasing on $(-\infty, 3)$
Decreasing on $(3, \infty)$
Local maximum at $x = 3$
Concave upward on $(0, 2)$
Concave downward on $(-\infty, 0)$ and $(2, \infty)$
Inflection points at $x = 0$ and $x = 2$

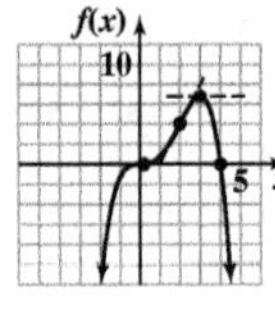

49. Domain: All real numbers
y intercept: 0; x intercepts: 0, 1
Increasing on $(0.25, \infty)$
Decreasing on $(-\infty, 0.25)$
Local minimum at $x = 0.25$
Concave upward on $(-\infty, 0.5)$ and $(1, \infty)$
Concave downward on $(0.5, 1)$
Inflection points at $x = 0.5$ and $x = 1$

51. Domain: All real numbers
y intercept: 27; x intercepts: $-3, 3$
Increasing on $(-\infty, -\sqrt{3})$ and $(0, \sqrt{3})$
Decreasing on $(-\sqrt{3}, 0)$ and $(\sqrt{3}, \infty)$
Local maxima at $x = -\sqrt{3}$ and $x = \sqrt{3}$
Local minimum at $x = 0$
Concave upward on $(-1, 1)$
Concave downward on $(-\infty, -1)$ and $(1, \infty)$
Inflection points at $x = -1$ and $x = 1$

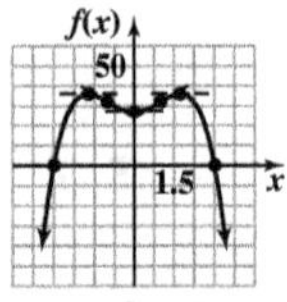

53. Domain: All real numbers
y intercept: 16; x intercepts: $-2, 2$
Decreasing on $(-\infty, -2)$ and $(0, 2)$
Increasing on $(-2, 0)$ and $(2, \infty)$
Local minima at $x = -2$ and $x = 2$
Local maximum at $x = 0$
Concave upward on $(-\infty, -2\sqrt{3}/3)$ and $(2\sqrt{3}/3, \infty)$
Concave downward on $(-2\sqrt{3}/3, 2\sqrt{3}/3)$
Inflection points at $x = -2\sqrt{3}/3$ and $x = 2\sqrt{3}/3$

55. Domain: All real numbers
y intercept: 0; x intercepts: 0, 1.5
Decreasing on $(-\infty, 0)$ and $(0, 1.25)$
Increasing on $(1.25, \infty)$
Local minimum at $x = 1.25$
Concave upward on $(-\infty, 0)$ and $(1, \infty)$
Concave downward on $(0, 1)$
Inflection points at $x = 0$ and $x = 1$

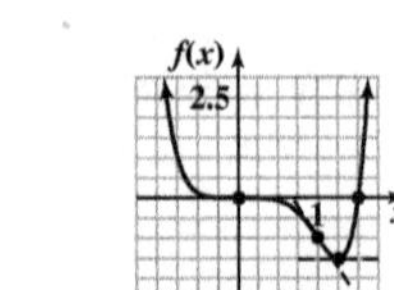

57. Domain: All real numbers
y intercept: 0; x intercept: 0
Increasing on $(-\infty, \infty)$
Concave downward on $(-\infty, \infty)$

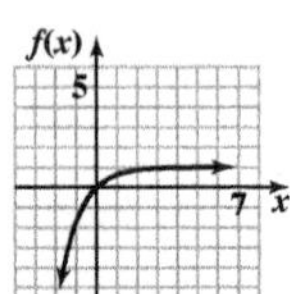

59. Domain: All real numbers
y intercept: 5
Decreasing on $(-\infty, \ln 4)$
Increasing on $(\ln 4, \infty)$
Local minimum at $x = \ln 4$
Concave upward on $(-\infty, \infty)$

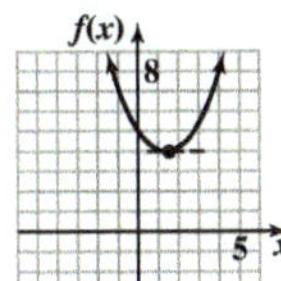

61. Domain: $(0, \infty)$
x intercept: e^2
Increasing on $(-\infty, \infty)$
Concave downward on $(-\infty, \infty)$

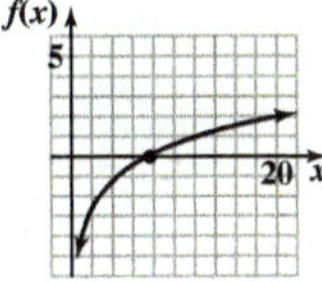

63. Domain: $(-4, \infty)$
x intercept: $e^2 - 4$
Increasing on $(-4, \infty)$
Concave downward on $(-4, \infty)$

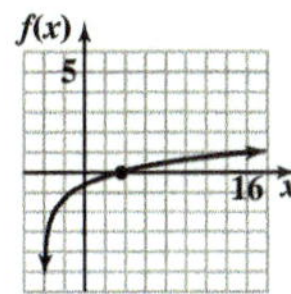

65.

x	$f'(x)$	$f(x)$
$-\infty < x < -1$	Positive and decreasing	Increasing and concave downward
$x = -1$	x intercept	Local maximum
$-1 < x < 0$	Negative and decreasing	Decreasing and concave downward
$x = 0$	Local minimum	Inflection point
$0 < x < 2$	Negative and increasing	Decreasing and concave upward
$x = 2$	Local maximum	Inflection point
$2 < x < \infty$	Negative and decreasing	Decreasing and concave downward

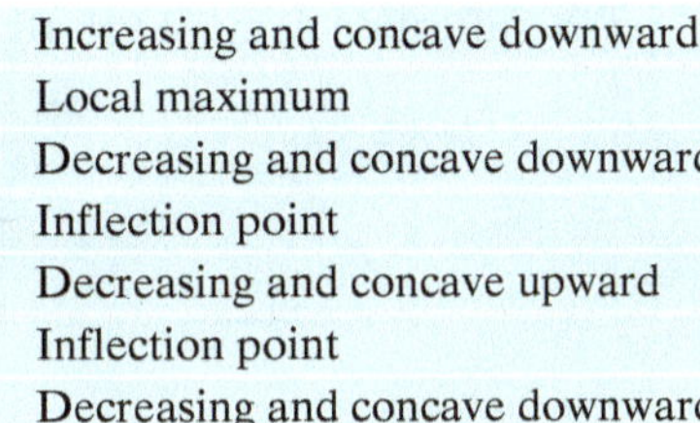

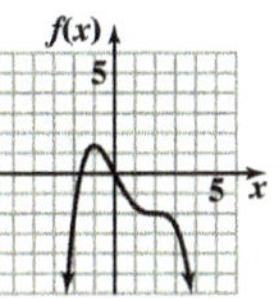

67.

x	$f'(x)$	$f(x)$
$-\infty < x < -2$	Negative and increasing	Decreasing and concave upward
$x = -2$	Local maximum	Inflection point
$-2 < x < 0$	Negative and decreasing	Decreasing and concave downward
$x = 0$	Local minimum	Inflection point
$0 < x < 2$	Negative and increasing	Decreasing and concave upward
$x = 2$	Local maximum	Inflection point
$2 < x < \infty$	Negative and decreasing	Decreasing and concave downward

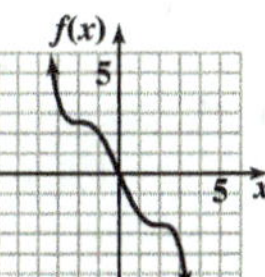

69. Domain: All real numbers
x intercepts: $-1.18, 0.61, 1.87, 3.71$
y intercept: -5
Decreasing on $(-\infty, -0.53)$ and $(1.24, 3.04)$
Increasing on $(-0.53, 1.24)$ and $(3.04, \infty)$
Local minima at $x = -0.53$ and $x = 3.04$
Local maximum at $x = 1.24$
Concave upward on $(-\infty, 0.22)$ and $(2.28, \infty)$
Concave downward on $(0.22, 2.28)$
Inflection points at $x = 0.22$ and $x = 2.28$

71. Domain: All real numbers
y intercept: 100; x intercepts: 8.01, 13.36
Increasing on $(-0.10, 4.57)$ and $(11.28, \infty)$
Decreasing on $(-\infty, -0.10)$ and $(4.57, 11.28)$
Local maximum at $x = 4.57$
Local minima at $x = -0.10$ and $x = 11.28$
Concave upward on $(-\infty, 1.95)$ and $(8.55, \infty)$
Concave downward on $(1.95, 8.55)$
Inflection points at $x = 1.95$ and $x = 8.55$

73. Domain: All real numbers
x intercepts: $-2.40, 1.16$; y intercept: 3
Increasing on $(-\infty, -1.58)$
Decreasing on $(-1.58, \infty)$
Local maximum at $x = -1.58$
Concave downward on $(-\infty, -0.88)$ and $(0.38, \infty)$
Concave upward on $(-0.88, 0.38)$
Inflection points at $x = -0.88$ and $x = 0.38$

75. Domain: All real numbers
x intercepts: $-6.68, -3.64, -0.72$; y intercept: 30
Decreasing on $(-5.59, -2.27)$ and $(1.65, 3.82)$
Increasing on $(-\infty, -5.59)$, $(-2.27, 1.65)$, and $(3.82, \infty)$
Local minima at $x = -2.27$ and $x = 3.82$
Local maxima at $x = -5.59$ and $x = 1.65$
Concave upward on $(-4.31, -0.40)$ and $(2.91, \infty)$
Concave downward on $(-\infty, -4.31)$ and $(-0.40, 2.91)$
Inflection points at $x = -4.31$, $x = -0.40$, and $x = 2.91$

77. If $f'(x)$ has a local extremum at $x = c$, then $f'(x)$ must change from increasing to decreasing or from decreasing to increasing at $x = c$. The graph of $y = f(x)$ must change concavity at $x = c$, and there must be an inflection point at $x = c$.

79. If there is an inflection point on the graph of $y = f(x)$ at $x = c$, then $f(x)$ must change concavity at $x = c$. Consequently, $f'(x)$ must change from increasing to decreasing or from decreasing to increasing at $x = c$, and $x = c$ is a local extremum for $f'(x)$.

81. The graph of the CPI is concave upward.

83. The graph of $y = C'(x)$ is positive and decreasing. Since marginal costs are decreasing, the production process is becoming more efficient as production increases.

85. (A) Local maximum at $x = 60$
(B) Concave downward on the whole interval $(0, 80)$

87. (A) Local maximum at $x = 1$
(B) Concave downward on $(-\infty, 2)$; concave upward on $(2, \infty)$

89. Increasing on $(0, 10)$; decreasing on $(10, 15)$; point of diminishing returns is $x = 10$, max $T'(x) = T'(10) = 500$

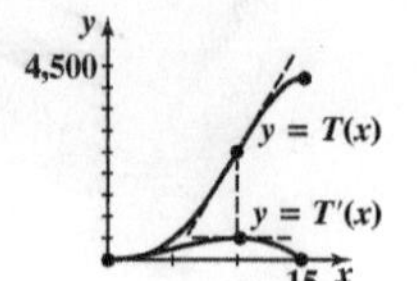

91. Increasing on $(24, 36)$; decreasing on $(36, 45)$; point of diminishing returns is $x = 36$, max $N'(x) = N'(36) = 3{,}888$

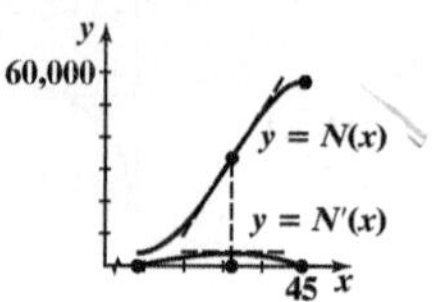

93. (A)

```
CubicReg
 y=ax³+bx²+cx+d
 a=-.005
 b=.485
 c=-1.85
 d=300
```

(B) 32 ads to sell 574 cars per month

95. (A) Increasing on $(0, 10)$; decreasing on $(10, 20)$
(B) Inflection point at $t = 10$
(C)

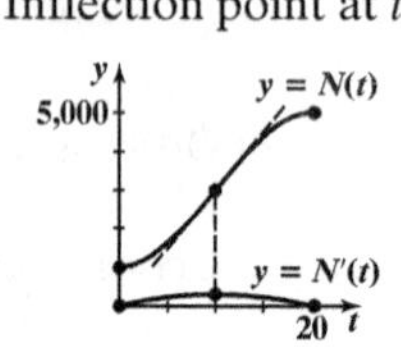

(D) $N'(10) = 300$

97. (A) Increasing on $(5, \infty)$; decreasing on $(0, 5)$
(B) Inflection point at $n = 5$
(C)

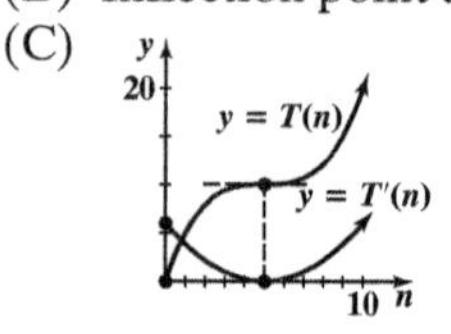

(D) $T'(5) = 0$

Exercises 3-3

1. $\frac{8}{3}$ **3.** $\frac{1}{2}$ **5.** 1 **7.** 4 **9.** 0 **11.** ∞ **13.** ∞ **15.** 5 **17.** ∞ **19.** $-\frac{1}{8}$ **21.** 17 **23.** 8 **25.** 0 **27.** ∞ **29.** ∞ **31.** $\frac{1}{3}$ **33.** -2 **35.** $-\infty$ **37.** 0 **39.** 0 **41.** 0 **43.** $\frac{1}{4}$ **45.** $\frac{1}{3}$ **47.** 0 **49.** 0 **51.** ∞ **53.** 1 **55.** 1

Exercises 3-4

1. (A) $(-\infty, b), (0, e), (e, g)$ (B) $(b, d), (d, 0), (g, \infty)$ (C) $(b, d), (d, 0), (g, \infty)$ (D) $(-\infty, b), (0, e), (e, g)$ (E) $x = 0$ (F) $x = b, x = g$ (G) $(-\infty, a), (d, e), (h, \infty)$ (H) $(a, d), (e, h)$ (I) $(a, d), (e, h)$ (J) $(-\infty, a), (d, e), (h, \infty)$ (K) $x = a, x = h$ (L) $y = L$ (M) $x = d, x = e$

3.

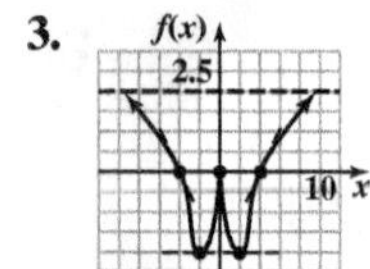

5.

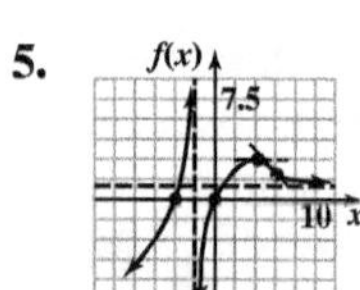

7.

9.

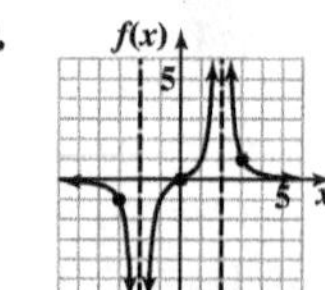

11. Domain: $[-4, \infty)$ y intercept: 2; x intercept: -4

13. Domain: All real numbers y intercept: 75; x intercept: 15

15. Domain: All real numbers, except 5 y intercept: -19

17. Domain: All real numbers, except 3
y intercept: -1; x intercept: -3
Horizontal asymptote: $y = 1$
Vertical asymptote: $x = 3$
Decreasing on $(-\infty, 3)$ and $(3, \infty)$
Concave upward on $(3, \infty)$
Concave downward on $(-\infty, 3)$

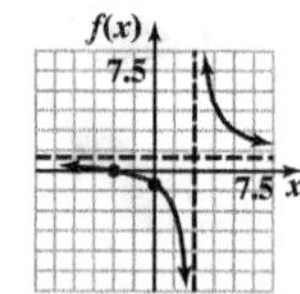

19. Domain: All real numbers, except 2
y intercept: 0; x intercept: 0
Horizontal asymptote: $y = 1$
Vertical asymptote: $x = 2$
Decreasing on $(-\infty, 2)$ and $(2, \infty)$
Concave downward on $(-\infty, 2)$
Concave upward on $(2, \infty)$

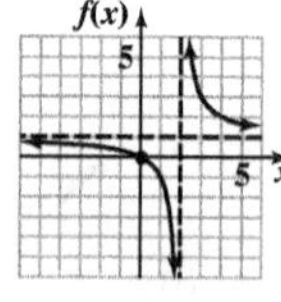

21. Domain: $(-\infty, \infty)$
y intercept: 10
Horizontal asymptote: $y = 5$
Decreasing on $(-\infty, \infty)$
Concave upward on $(-\infty, \infty)$

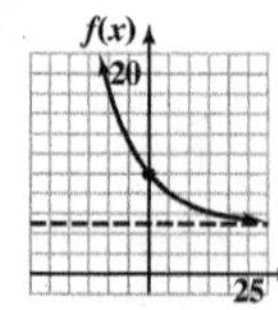

23. Domain: $(-\infty, \infty)$
y intercept: 0; x intercept: 0
Horizontal asymptote: $y = 0$
Increasing on $(-\infty, 5)$
Decreasing on $(5, \infty)$
Local maximum at $x = 5$
Concave upward on $(10, \infty)$
Concave downward on $(-\infty, 10)$
Inflection point at $x = 10$

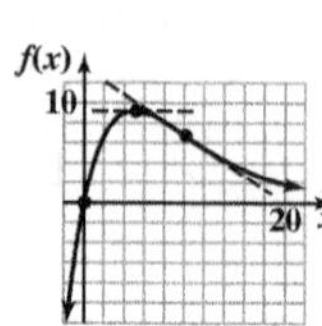

25. Domain: $(-\infty, 1)$
y intercept: 0; x intercept: 0
Vertical asymptote: $x = 1$
Decreasing on $(-\infty, 1)$
Concave downward on $(-\infty, 1)$

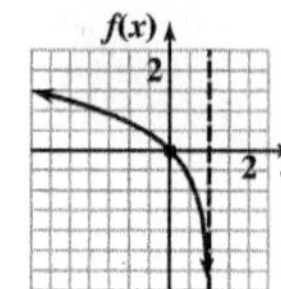

27. Domain: $(0, \infty)$
Vertical asymptote: $x = 0$
Increasing on $(1, \infty)$
Decreasing on $(0, 1)$
Local minimum at $x = 1$
Concave upward on $(0, \infty)$

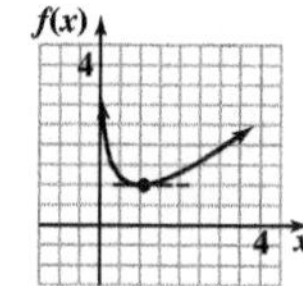

29. Domain: All real numbers, except ± 2
y intercept: 0; x intercept: 0
Horizontal asymptote: $y = 0$
Vertical asymptotes: $x = -2, x = 2$
Decreasing on $(-\infty, -2)$, $(-2, 2)$, and $(2, \infty)$
Concave upward on $(-2, 0)$ and $(2, \infty)$
Concave downward on $(-\infty, -2)$ and $(0, 2)$
Inflection point at $x = 0$

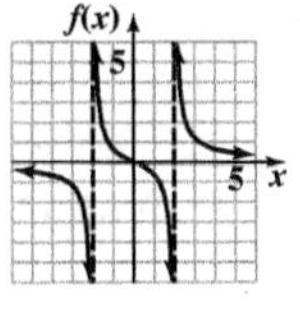

31. Domain: All real numbers
y intercept: 1
Horizontal asymptote: $y = 0$
Increasing on $(-\infty, 0)$
Decreasing on $(0, \infty)$
Local maximum at $x = 0$
Concave upward on $(-\infty, -\sqrt{3}/3)$ and $(\sqrt{3}/3, \infty)$
Concave downward on $(-\sqrt{3}/3, \sqrt{3}/3)$
Inflection points at $x = -\sqrt{3}/3$ and $x = \sqrt{3}/3$

33. Domain: All real numbers except -1 and 1
y intercept: 0; x intercept: 0
Horizontal asymptote: $y = 0$
Vertical asymptote: $x = -1$and $x = 1$
Increasing on $(-\infty, -1)$, $(-1, 1)$, and $(1, \infty)$
Concave upward on $(-\infty, -1)$ and $(0, 1)$
Concave downward on $(-1, 0)$ and $(1, \infty)$
Inflection point at $x = 0$

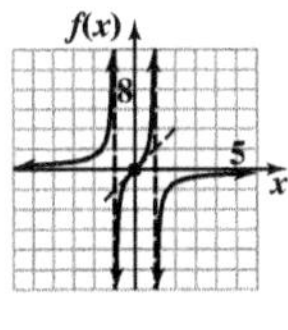

35. Domain: All real numbers except 1
y intercept: 0; x intercept: 0
Horizontal asymptote: $y = 0$
Vertical asymptote: $x = 1$
Increasing on $(-\infty, -1)$ and $(1, \infty)$
Decreasing on $(-1, 1)$
Local maximum at $x = -1$
Concave upward on $(-\infty, -2)$
Concave downward on $(-2, 1)$ and $(1, \infty)$
Inflection point at $x = -2$

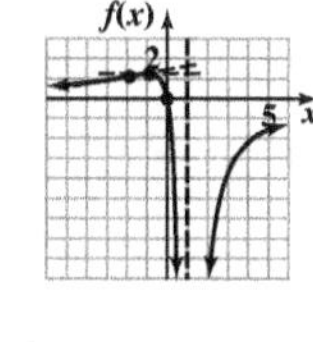

37. Domain: All real numbers except 0
Horizontal asymptote: $y = 1$
Vertical asymptote: $x = 0$
Increasing on $(0, 4)$
Decreasing on $(-\infty, 0)$ and $(4, \infty)$
Local maximum at $x = 4$
Concave upward on $(6, \infty)$
Concave downward on $(-\infty, 0)$ and $(0, 6)$
Inflection point at $x = 6$

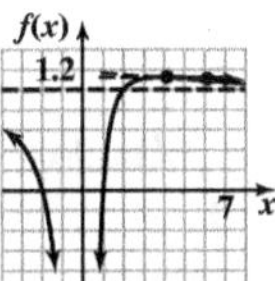

39. Domain: All real numbers except 1
y intercept: 0; x intercept: 0
Vertical asymptote: $x = 1$
Increasing on $(-\infty, 0)$ and $(2, \infty)$
Decreasing on $(0, 1)$ and $(1, 2)$
Local maximum at $x = 0$
Local minimum at $x = 2$
Concave upward on $(1, \infty)$
Concave downward on $(-\infty, 1)$

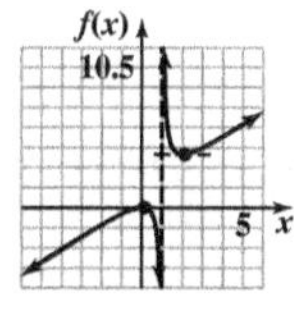

41. Domain: All real numbers except $-3, 3$
y intercept: $-\frac{2}{9}$
Horizontal asymptote: $y = 3$
Vertical asymptotes: $x = -3, x = 3$
Increasing on $(-\infty, -3)$ and $(-3, 0)$
Decreasing on $(0, 3)$ and $(3, \infty)$
Local maximum at $x = 0$
Concave upward on $(-\infty, -3)$ and $(3, \infty)$
Concave downward on $(-3, 3)$

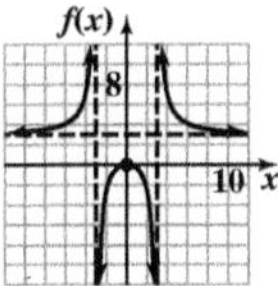

43. Domain: All real numbers except 2
y intercept: 0; x intercept: 0
Vertical asymptote: $x = 2$
Increasing on $(3, \infty)$
Decreasing on $(-\infty, 2)$ and $(2, 3)$
Local minimum at $x = 3$
Concave upward on $(-\infty, 0)$ and $(2, \infty)$
Concave downward on $(0, 2)$
Inflection point at $x = 0$

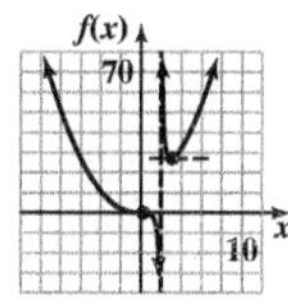

45. Domain: All real numbers
y intercept: 3; x intercept: 3
Horizontal asymptote: $y = 0$
Increasing on $(-\infty, 2)$
Decreasing on $(2, \infty)$
Local maximum at $x = 2$
Concave upward on $(-\infty, 1)$
Concave downward on $(1, \infty)$
Inflection point at $x = 1$

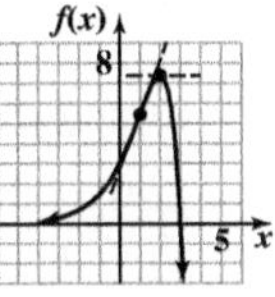

47. Domain: $(-\infty, \infty)$
y intercept: 1
Horizontal asymptote: $y = 0$
Increasing on $(-\infty, 0)$
Decreasing on $(0, \infty)$
Local maximum at $x = 0$
Concave upward on $(-\infty, -1)$ and $(1, \infty)$
Concave downward on $(-1, 1)$
Inflection points at $x = -1$ and $x = 1$

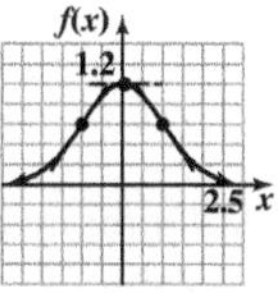

49. Domain: $(0, \infty)$
x intercept: 1
Increasing on $(e^{-1/2}, \infty)$
Decreasing on $(0, e^{-1/2})$
Local minimum at $x = e^{-1/2}$
Concave upward on $(e^{-3/2}, \infty)$
Concave downward on $(0, e^{-3/2})$
Inflection point at $x = e^{-3/2}$

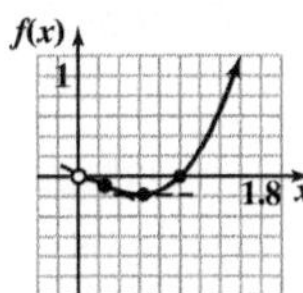

51. Domain: $(0, \infty)$
x intercept: 1
Vertical asymptote: $x = 0$
Increasing on $(1, \infty)$
Decreasing on $(0, 1)$
Local minimum at $x = 1$
Concave upward on $(0, e)$
Concave downward on (e, ∞)
Inflection point at $x = e$

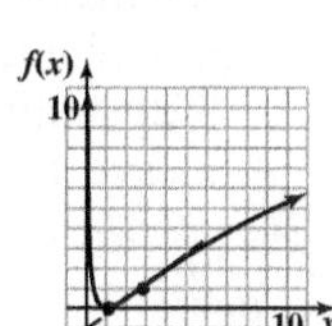

53. Domain: All real numbers except $-4, 2$
y intercept: $-1/8$
Horizontal asymptote: $y = 0$
Vertical asymptote: $x = -4$, $x = 2$
Increasing on $(-\infty, -4)$ and $(-4, -1)$
Decreasing on $(-1, 2)$ and $(2, \infty)$
Local maximum at $x = -1$
Concave upward on $(-\infty, -4)$ and $(2, \infty)$
Concave downward on $(-4, 2)$

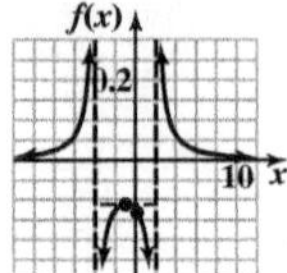

55. Domain: All real numbers except $-\sqrt{3}, \sqrt{3}$
y intercept: 0; x intercept: 0
Vertical asymptote: $x = -\sqrt{3}$, $x = \sqrt{3}$
Increasing on $(-3, -\sqrt{3})$, $(-\sqrt{3}, \sqrt{3})$, and $(\sqrt{3}, 3)$
Decreasing on $(-\infty, -3)$ and $(3, \infty)$
Local maximum at $x = 3$
Local minimum at $x = -3$
Concave upward on $(-\infty, -\sqrt{3})$ and $(0, \sqrt{3})$
Concave downward on $(-\sqrt{3}, 0)$ and $(\sqrt{3}, \infty)$
Inflection point at $x = 0$

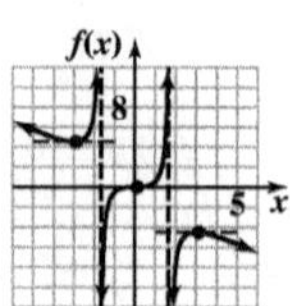

57. Domain: All real numbers except 0
Vertical asymptote: $x = 0$
Oblique asymptote: $y = x$
Increasing on $(-\infty, -2)$ and $(2, \infty)$
Decreasing on $(-2, 0)$ and $(0, 2)$
Local maximum at $x = -2$
Local minimum at $x = 2$
Concave upward on $(0, \infty)$
Concave downward on $(-\infty, 0)$

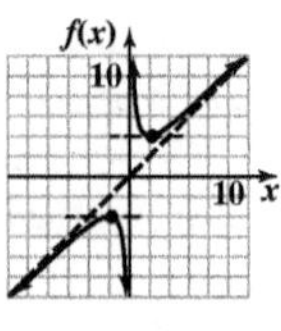

59. Domain: All real numbers except 0
x intercept: $\sqrt[3]{4}$
Vertical asymptote: $x = 0$
Oblique asymptote: $y = x$
Increasing on $(-\infty, -2)$ and $(0, \infty)$
Local maximum at $x = -2$
Decreasing on $(-2, 0)$
Concave downward on $(-\infty, 0)$ and $(0, \infty)$

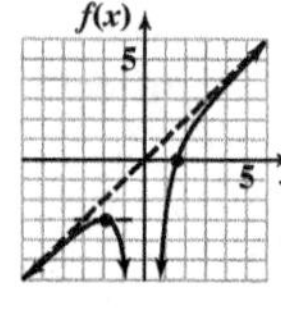

61. Domain: All real numbers except 0
x intercepts: $-\sqrt{3}, \sqrt{3}$
Vertical asymptote: $x = 0$
Oblique asymptote: $y = x$
Increasing on $(-\infty, 0)$ and $(0, \infty)$
Concave upward on $(-\infty, 0)$
Concave downward on $(0, \infty)$

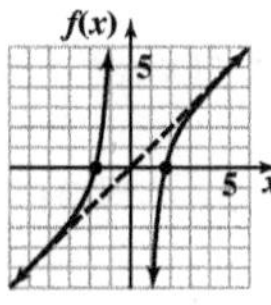

63. Domain: All real numbers except 0
Vertical asymptote: $x = 0$
Oblique asymptote: $y = x$
Increasing on $(-\infty, -2)$ and $(2, \infty)$
Decreasing on $(-2, 0)$ and $(0, 2)$
Local maximum at $x = -2$
Local minimum at $x = 2$
Concave upward on $(0, \infty)$
Concave downward on $(-\infty, 0)$

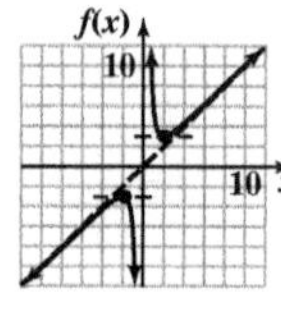

65. Domain: All real numbers except 2, 4
y intercept: $-3/4$; x intercept: -3
Vertical asymptote: $x = 4$
Horizontal asymptote: $y = 1$
Decreasing on $(-\infty, 2)$, $(2, 4)$, and $(4, \infty)$
Concave upward on $(4, \infty)$
Concave downward on $(-\infty, 2)$ and $(2, 4)$

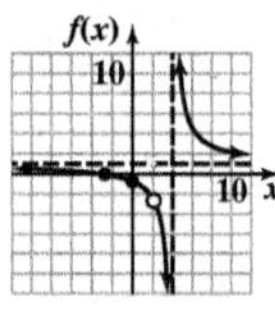

67. Domain: All real numbers except $-3, 3$
y intercept: 5/3; x intercept: 2.5
Vertical asymptote: $x = 3$
Horizontal asymptote: $y = 2$
Decreasing on $(-\infty, -3)$, $(-3, 3)$, and $(3, \infty)$
Concave upward on $(3, \infty)$
Concave downward on $(-\infty, -3)$ and $(-3, 3)$

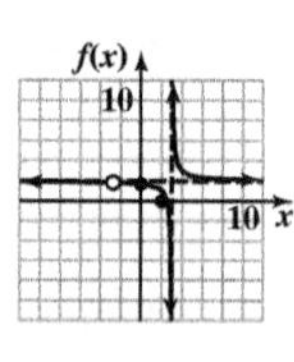

69. Domain: All real numbers except $-1, 2$
y intercept: 0; x intercepts: 0, 3
Vertical asymptote: $x = -1$
Increasing on $(-\infty, -3)$, $(1, 2)$, and $(2, \infty)$
Decreasing on $(-3, -1)$ and $(-1, 1)$
Local maximum at $x = -3$
Local minimum at $x = 1$
Concave upward on $(-1, 2)$ and $(2, \infty)$
Concave downward on $(-\infty, -1)$

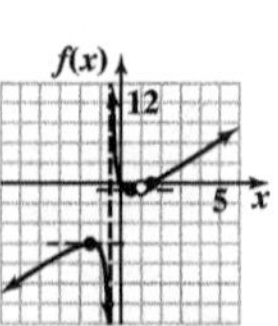

71. Domain: All real numbers except 1
y intercept: -2; x intercept: -2
Vertical asymptote: $x = 1$
Horizontal asymptote: $y = 1$
Decreasing on $(-\infty, 1)$ and $(1, \infty)$
Concave upward on $(1, \infty)$
Concave downward on $(-\infty, 1)$

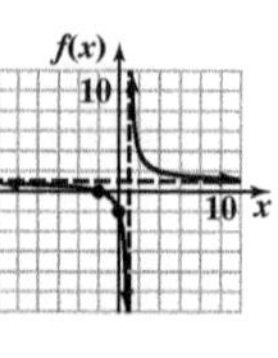

73.

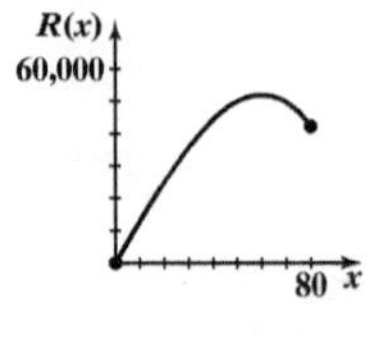

75. (A) Increasing on $(0, 1)$
(B) Concave upward on $(0, 1)$
(C) $x = 1$ is a vertical asymptote
(D) The origin is both an x and a y intercept
(E)

77. (A) $\overline{C}(n) = \dfrac{3{,}200}{n} + 250 + 50n$
(B)

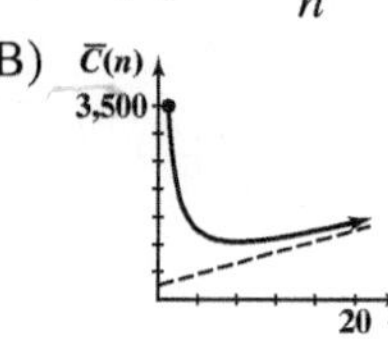

(C) 8 yr

79. (A) 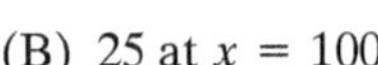

(B) 25 at $x = 100$

81. (A) 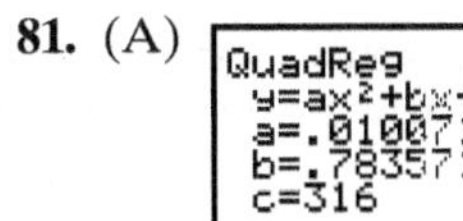

(B) Minimum average cost is \$4.35 when 177 pizzas are produced daily.

83. 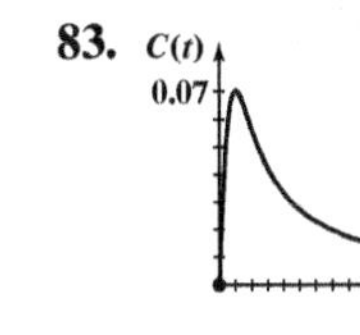

85. 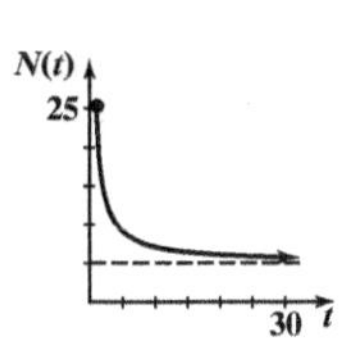

Exercises 3-5

1. Min $f(x) = f(0) = 0$; Max $f(x) = f(10) = 14$ **3.** Min $f(x) = f(0) = 0$; Max $f(x) = f(3) = 9$
5. Min $f(x) = f(1) = f(7) = 5$; Max $f(x) = f(10) = 14$ **7.** Min $f(x) = f(1) = f(7) = 5$; Max $f(x) = f(3) = f(9) = 9$
9. Min $f(x) = f(5) = 7$; Max $f(x) = f(3) = 9$ **11.** Max $f(x) = f(3) = 21$; Min $f(x) = f(2) = 17$
13. Max $f(x) = f(-1) = e \approx 2.718$; Min $f(x) = e^{-1} \approx 0.368$ **15.** Max $f(x) = f(0) = 9$; Min $f(x) = f(\pm 4) = -7$
17. Min $f(x) = f(1) = 2$; no maximum **19.** Max $f(x) = f(-3) = 18$; no minimum **21.** No absolute extrema
23. Max $f(x) = f(3) = 54$; no minimum **25.** No absolute extrema **27.** Min $f(x) = f(0) = 0$; no maximum
29. Max $f(x) = f(1) = 1$; Min $f(x) = f(-1) = -1$ **31.** Min $f(x) = f(0) = -1$, no maximum **33.** Min $f(x) = f(2) = -2$
35. Max $f(x) = f(2) = 4$ **37.** Min $f(x) = f(2) = 0$ **39.** No maximum **41.** Max $f(x) = f(2) = 8$
43. Min $f(x) = f(4) = 22$ **45.** Min $f(x) = f(\sqrt{10}) = 14/\sqrt{10}$ **47.** Min $f(x) = f(2) = \frac{e^2}{4} \approx 1.847$
49. Min $f(x) = f(3) = \frac{27}{e^3} \approx 1.344$ **51.** Max $f(x) = f(e^{1.5}) = 2e^{1.5} \approx 8.963$
53. Min $f(x) = f(e^{2.5}) = \frac{e^5}{2} \approx 74.207$ **55.** Max $f(x) = f(1) = -1$
57. (A) Max $f(x) = f(5) = 14$; Min $f(x) = f(-1) = -22$ (B) Max $f(x) = f(1) = -2$; Min $f(x) = f(-1) = -22$
(C) Max $f(x) = f(5) = 14$; Min $f(x) = f(3) = -6$
59. (A) Max $f(x) = f(0) = 126$; Min $f(x) = f(2) = -26$ (B) Max $f(x) = f(7) = 49$; Min $f(x) = f(2) = -26$
(C) Max $f(x) = f(6) = 6$; Min $f(x) = f(3) = -15$
61. (A) Max $f(x) = f(-1) = 10$; Min $f(x) = f(2) = -11$ (B) Max $f(x) = f(0) = f(4) = 5$; Min $f(x) = f(3) = -22$
(C) Max $f(x) = f(-1) = 10$; Min $f(x) = f(1) = 2$
63. Local minimum **65.** Unable to determine **67.** Neither **69.** Local maximum

Exercises 3-6

1. 7.5 and 7.5 **3.** 7.5 and −7.5 **5.** $\sqrt{15}$ and $\sqrt{15}$ **7.** $10\sqrt{2}$ ft by $10\sqrt{2}$ ft **9.** 37 ft by 37 ft
11. (A) Maximum revenue is \$125,000 when 500 phones are produced and sold for \$250 each.
(B) Maximum profit is \$46,612.50 when 365 phones are produced and sold for \$317.50 each.
13. (A) Max $R(x) = R(3{,}000) = \$300{,}000$
(B) Maximum profit is \$75,000 when 2,100 sets are manufactured and sold for \$130 each.
(C) Maximum profit is \$64,687.50 when 2,025 sets are manufactured and sold for \$132.50 each.
15. (A) 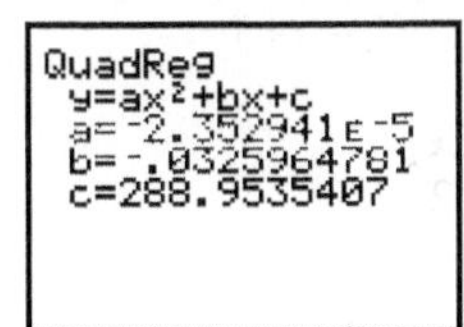

(B) 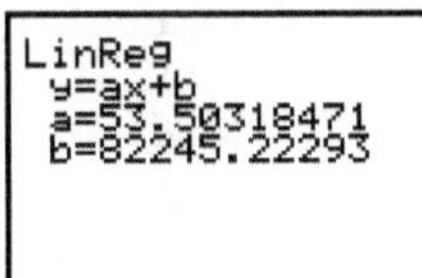

(C) The maximum profit is \$118,996 when the price per sleeping bag is \$195.

17. (A) \$4.80 (B) \$8 **19.** \$35; \$6,125 **21.** 40 trees; 1,600 lb **23.** $(10 - 2\sqrt{7})/3 = 1.57$ in. squares
25. 20 ft by 40 ft (with the expensive side being one of the short sides) **27.** 8 production runs per year
29. 10,000 books in 5 printings **31.** (A) $x = 5.1$ mi (B) $x = 10$ mi **33.** 4 days; 20 bacteria/cm^3
35. 1 month; 2 ft **37.** 4 yr from now

Chapter 3 Review Exercises

1. $(a, c_1), (c_3, c_6)$ *(5-1, 5-2)* **2.** $(c_1, c_3), (c_6, b)$ *(5-1, 5-2)* **3.** $(a, c_2), (c_4, c_5), (c_7, b)$ *(5-1, 5-2)* **4.** c_3 *(5-1)*
5. c_1, c_6 *(5-1)* **6.** c_1, c_3, c_5 *(5-1)* **7.** c_4, c_6 *(5-1)* **8.** c_2, c_4, c_5, c_7 *(5-2)*

9. *(5-2)* 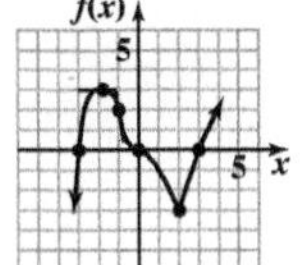

10. *(5-2)* 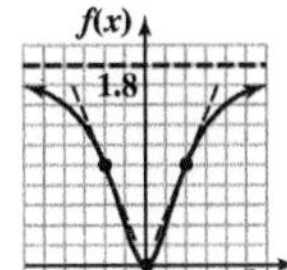

11. $f''(x) = 12x^2 + 30x$ *(5-2)*
12. $y'' = 8/x^3$ *(5-2)*
13. Domain: All real numbers, except 4
y intercept: $\frac{5}{4}$; x intercept: -5 *(5-2)*
14. Domain: $(-2, \infty)$
y intercept: $\ln 2$; x intercept: -1 *(5-2)*

15. Horizontal asymptote: $y = 0$;
Vertical asymptotes: $x = -2, x = 2$ *(5-4)*

16. Horizontal asymptote: $y = \frac{2}{3}$;
Vertical asymptote: $x = -\frac{10}{3}$ *(5-4)*

17. $(-\sqrt{2}, -20), (\sqrt{2}, -20)$ *(5-2)* **18.** $(-\frac{1}{2}, -6)$ *(5-2)* **19.** (A) $f'(x) = \frac{1}{5}x^{-1/5}$ (B) 0 (C) 0 *(5-1)*

20. (A) $f'(x) = -\frac{1}{5}x^{-6/5}$ (B) None (C) 0 *(5-1)*

21. Domain: All real numbers
y intercept: 0; x intercepts: 0, 9
Increasing on $(-\infty, 3)$ and $(9, \infty)$
Decreasing on $(3, 9)$
Local maximum at $x = 3$
Local minimum at $x = 9$
Concave upward on $(6, \infty)$
Concave downward on $(-\infty, 6)$
Inflection point at $x = 6$ *(5-4)*

22. Domain: All real numbers
y intercept: 16; x intercepts: $-4, 2$
Increasing on $(-\infty, -2)$ and $(2, \infty)$
Decreasing on $(-2, 2)$
Local maximum at $x = -2$
Local minimum at $x = 2$
Concave upward on $(0, \infty)$
Concave downward on $(-\infty, 0)$
Inflection point at $x = 0$ *(5-4)*

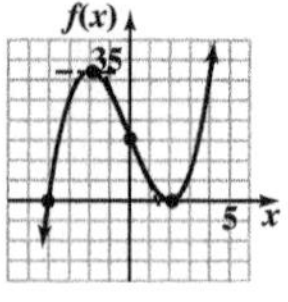

23. Domain: All real numbers
y intercept: 0; x intercepts: 0, 4
Increasing on $(-\infty, 3)$
Decreasing on $(3, \infty)$
Local maximum at $x = 3$
Concave upward on $(0, 2)$
Concave downward on $(-\infty, 0)$ and $(2, \infty)$
Inflection points at $x = 0$ and $x = 2$ *(5-4)*

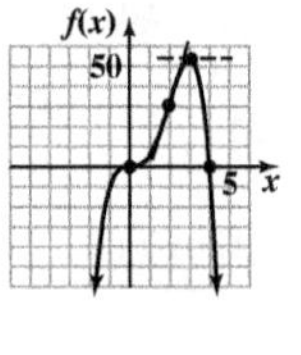

24. Domain: all real numbers
y intercept: -3; x intercepts: $-3, 1$
No vertical or horizontal asymptotes
Increasing on $(-2, \infty)$
Decreasing on $(-\infty, -2)$
Local minimum at $x = -2$
Concave upward on $(-\infty, -1)$ and $(1, \infty)$
Concave downward on $(-1, 1)$
Inflection points at $x = -1$ and $x = 1$ *(5-4)*

25. Domain: All real numbers, except -2
y intercept: 0; x intercept: 0
Horizontal asymptote: $y = 3$
Vertical asymptote: $x = -2$
Increasing on $(-\infty, -2)$ and $(-2, \infty)$
Concave upward on $(-\infty, -2)$
Concave downward on $(-2, \infty)$ *(5-4)*

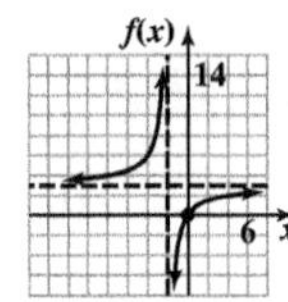

26. Domain: All real numbers
y intercept: 0; x intercept: 0
Horizontal asymptote: $y = 1$
Increasing on $(0, \infty)$
Decreasing on $(-\infty, 0)$
Local minimum at $x = 0$
Concave upward on $(-3, 3)$
Concave downward on $(-\infty, -3)$ and $(3, \infty)$
Inflection points at $x = -3, 3$ *(5-4)*

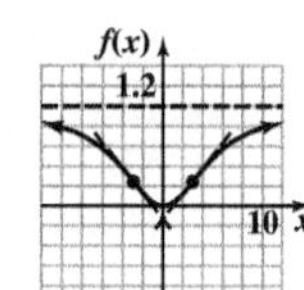

27. Domain: All real numbers except $x = -2$
y intercept: 0; x intercept: 0
Horizontal asymptote: $y = 0$
Vertical asymptote: $x = -2$
Increasing on $(-2, 2)$
Decreasing on $(-\infty, -2)$ and $(2, \infty)$
Local maximum at $x = 2$
Concave upward on $(4, \infty)$
Concave downward on $(-\infty, -2)$ and $(-2, 4)$
Inflection point at $x = 4$ *(5-4)*

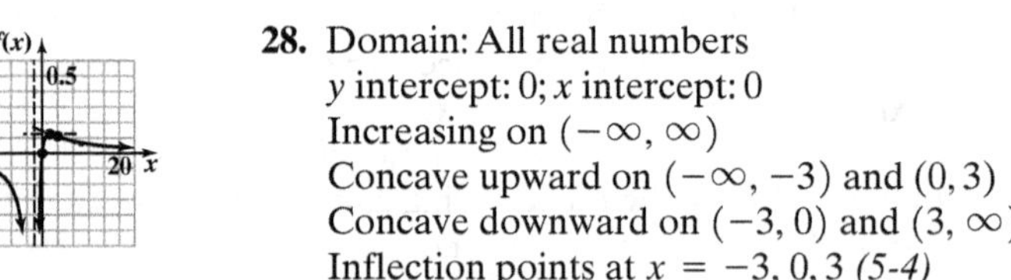

28. Domain: All real numbers
y intercept: 0; x intercept: 0
Increasing on $(-\infty, \infty)$
Concave upward on $(-\infty, -3)$ and $(0, 3)$
Concave downward on $(-3, 0)$ and $(3, \infty)$
Inflection points at $x = -3, 0, 3$ *(5-4)*

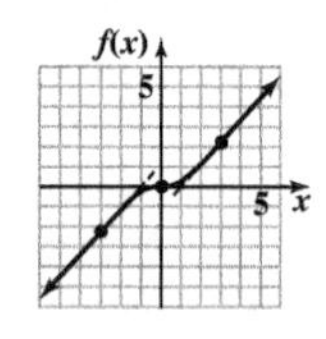

29. Domain: All real numbers
y intercept: 0; x intercept: 0
Horizontal asymptote: $y = 5$
Increasing on $(-\infty, \infty)$
Concave downward on $(-\infty, \infty)$

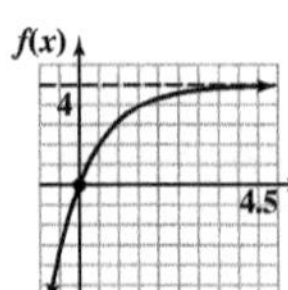

30. Domain: $(0, \infty)$
x intercept: 1
Increasing on $(e^{-1/3}, \infty)$
Decreasing on $(0, e^{-1/3})$
Local minimum at $x = e^{-1/3}$
Concave upward on $(e^{-5/6}, \infty)$
Concave downward on $(0, e^{-5/6})$
Inflection point at $x = e^{-5/6}$

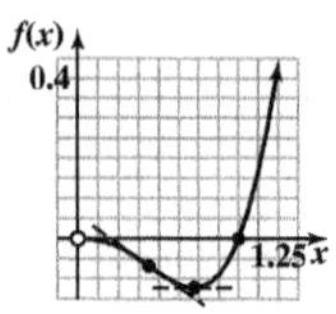

31. 3 *(5-3)* **32.** $-\frac{1}{5}$ *(5-3)* **33.** $-\infty$ *(5-3)* **34.** 0 *(5-3)* **35.** 0 *(5-3)*
36. 1 *(5-3)* **37.** 0 *(5-3)* **38.** 0 *(5-3)* **39.** 1 *(5-3)* **40.** 2 *(5-3)*

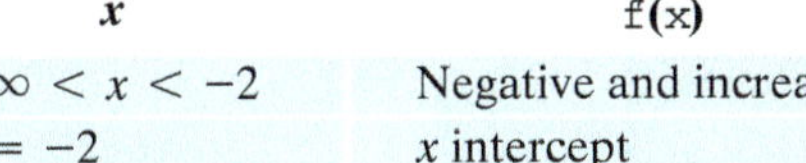
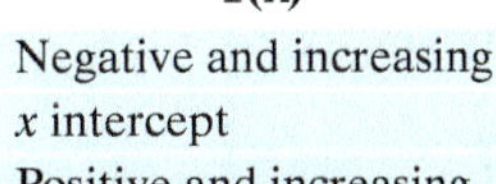
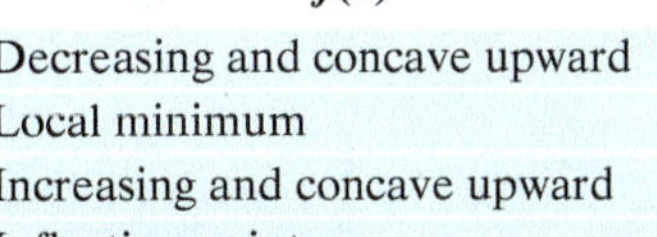
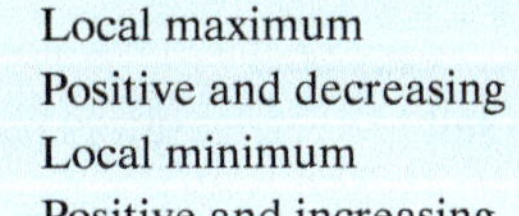

41.

x	$f'(x)$	$f(x)$
$-\infty < x < -2$	Negative and increasing	Decreasing and concave upward
$x = -2$	x intercept	Local minimum
$-2 < x < -1$	Positive and increasing	Increasing and concave upward
$x = -1$	Local maximum	Inflection point
$-1 < x < 1$	Positive and decreasing	Increasing and concave downward
$x = 1$	Local minimum	Inflection point
$1 < x < \infty$	Positive and increasing	Increasing and concave upward

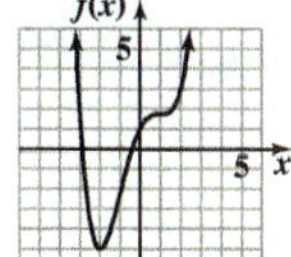

42. (C) *(5-2)* **43.** Local maximum at $x = -1$; local minimum at $x = 5$ *(5-5)*
44. Min $f(x) = f(2) = -4$; Max $f(x) = f(5) = 77$ *(5-5)* **45.** Min $f(x) = f(2) = 8$ *(5-5)*
46. Max $f(x) = f(e^{4.5}) = 2e^{4.5} \approx 180.03$ *(5-5)* **47.** Max $f(x) = f(0.5) = 5e^{-1} \approx 1.84$ *(5-5)*
48. Yes. Since f is continuous on $[a, b]$, f has an absolute maximum on $[a, b]$. But each endpoint is a local minimum; hence, the absolute maximum must occur between a and b. *(5-5)*
49. No, increasing/decreasing properties apply to intervals in the domain of f. It is correct to say that $f(x)$ is decreasing on $(-\infty, 0)$ and $(0, \infty)$. *(5-1)*
50. A critical value for $f(x)$ is a partition number for $f'(x)$ that is also in the domain of f. For example, if $f(x) = x^{-1}$, then 0 is a partition number for $f'(x) = -x^{-2}$, but 0 is not a critical value for $f(x)$ since 0 is not in the domain of f. *(5-1)*

51. Max $f'(x) = f'(2) = 12$ *(5-2, 5-5)*

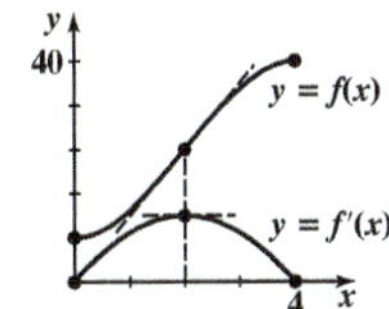

52. Each number is 20; minimum sum is 40 *(5-6)*
53. Domain: All real numbers
x intercepts: 0.79, 1.64; y intercept: 4
Increasing on $(-1.68, -0.35)$ and $(1.28, \infty)$
Decreasing on $(-\infty, -1.68)$ and $(-0.35, 1.28)$
Local minima at $x = -1.68$ and $x = 1.28$
Local maximum at $x = -0.35$
Concave downward on $(-1.10, 0.60)$
Concave upward on $(-\infty, -1.10)$ and $(0.60, \infty)$
Inflection points at $x = -1.10$ and $x = 0.60$ *(5-4)*

54. Domain: All real numbers
x intercepts; 0, 11.10; y intercept: 0
Increasing on $(1.87, 4.19)$ and $(8.94, \infty)$
Decreasing on $(-\infty, 1.87)$ and $(4.19, 8.94)$
Local maximum at $x = 4.19$
Local minima at $x = 1.87$ and $x = 8.94$
Concave upward on $(-\infty, 2.92)$ and $(7.08, \infty)$
Concave downward on $(2.92, 7.08)$
Inflection points at $x = 2.92$ and $x = 7.08$ *(5-4)*

55. Max $f(x) = f(1.373) = 2.487$ *(5-5)*
56. Max $f(x) = f(1.763) = 0.097$ *(5-5)*

57. (A) For the first 15 months, the graph of the price is increasing and concave downward, with a local maximum at $t = 15$. For the next 15 months, the graph of the price is decreasing and concave downward, with an inflection point at $t = 30$. For the next 15 months, the graph of the price is decreasing and concave upward, with a local minimum at $t = 45$. For the remaining 15 months, the graph of the price is increasing and concave upward.
(B)

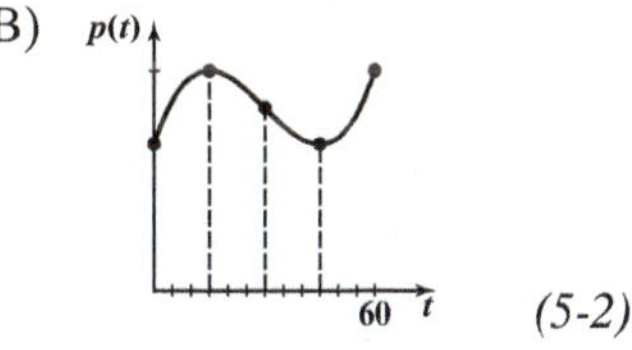

(5-2)

58. (A) Max $R(x) = R(10{,}000) = \$2{,}500{,}000$
(B) Maximum profit is \$175,000 when 3,000 readers are manufactured and sold for \$425 each.
(C) Maximum profit is \$119,000 when 2,600 readers are manufactured and sold for \$435 each. *(5-6)*
59. (A) The expensive side is 50 ft; the other side is 100 ft.
(B) The expensive side is 75 ft; the other side is 150 ft. *(5-6)*
60. \$49; \$6,724 *(5-6)* **61.** 12 orders/yr *(5-6)* **62.** Min $\overline{C}(x) = \overline{C}(200) = 50$
63. Min $\overline{C}(x) = \overline{C}(e^5) \approx \$49.66$ *(5-4)*

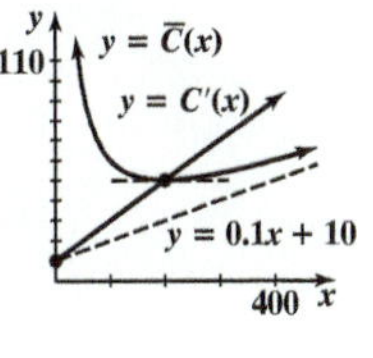

64. A maximum revenue of \$18,394 is realized at a production level of 50 units at \$367.88 each. *(5-6)*
65.

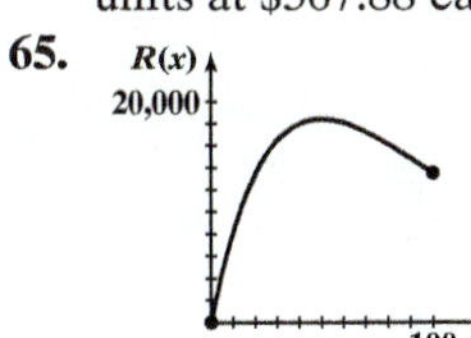

66. \$549.15; \$9,864 *(5-6)*
67. \$1.52 *(5-6)*
68. 20.39 feet *(5-6)*
69. (A)

```
QuadReg
 y=ax²+bx+c
 a=.0061285714
 b=.1224285714
 c=102.2
```

(B) Min $\overline{C}(x) = \overline{C}(129) = \$1.71$ *(5-4)*

70. Increasing on $(0, 18)$; decreasing on $(18, 24)$; point of diminishing returns is $x = 18$, max $N'(x) = N'(18) = 972$ *(5-2)*

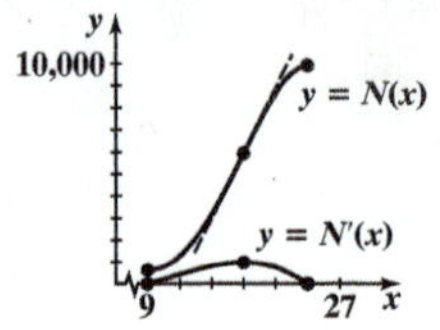

71. (A)

(B) 28 ads to sell 588 refrigerators per month *(5-2)*

72. 3 days *(5-1)*

73. 2 yr from now *(5-1)*

Chapter 4

Exercises 4-1

1. $7x + C$ **3.** $4x^2 + C$ **5.** $3x^3 + C$ **7.** $(x^6/6) + C$ **9.** $(-x^{-2}/2) + C$ **11.** $4x^{5/2} + C$ **13.** $3 \ln |z| + C$ **15.** $16e^u + C$ **17.** $y = 40x^5 + C$ **19.** $P = 24x - 3x^2 + C$ **21.** $y = e^x + 3x + C$ **23.** $x = 5 \ln|t| + t + C$ **25.** True **27.** False **29.** True

31. No, since one graph cannot be obtained from another by a vertical translation.

33. Yes, since one graph can be obtained from another by a vertical translation.

35. $(5x^2/2) - (5x^3/3) + C$ **37.** $2\sqrt{u} + C$ **39.** $-(x^{-2}/8) + C$ **41.** $4 \ln |u| + u + C$ **43.** $5e^z + 4z + C$ **45.** $x^3 + 2x^{-1} + C$ **47.** $2x^{3/2} + 4x^{1/2} + C$ **49.** $(e^x/4) - (3x^2/8) + C$ **51.** $C(x) = 2x^3 - 2x^2 + 3{,}000$ **53.** $x = 40\sqrt{t}$ **55.** $y = -2x^{-1} + 3 \ln|x| - x + 3$ **57.** $x = 4e^t - 2t - 3$ **59.** $y = 2x^2 - 3x + 1$ **61.** $x^2 + x^{-1} + C$ **63.** $\frac{1}{2}x^2 + x^{-2} + C$ **65.** $e^x - 2 \ln|x| + C$ **67.** $M = t + t^{-1} + \frac{3}{4}$ **69.** $y = 3x^{5/3} + 3x^{2/3} - 6$ **71.** $p(x) = 10x^{-1} + 10$ **73.** x^3 **75.** $x^4 + 3x^2 + C$ **83.** $\overline{C}(x) = 15 + \dfrac{1{,}000}{x}$; $C(x) = 15x + 1{,}000$; $C(0) = \$1{,}000$

85. (A) The cost function increases from 0 to 8, is concave downward from 0 to 4, and is concave upward from 4 to 8. There is an inflection point at $x = 4$.

(B) $C(x) = x^3 - 12x^2 + 53x + 30$; $C(4) = \$114{,}000$; $C(8) = \$198{,}000$

(C)

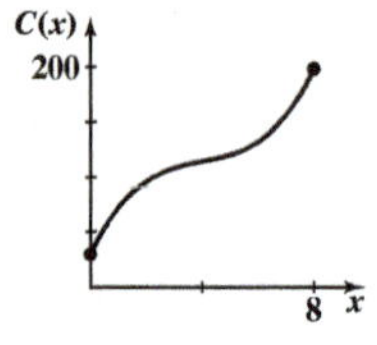

(D) Manufacturing plants are often inefficient at low and high levels of production.

87. $S(t) = 1{,}200 - 18t^{4/3}$; $50^{3/4} \approx 19$ mo

89. $S(t) = 1{,}200 - 18t^{4/3} - 70t$; $t \approx 4.05$ mo

91. $L(x) = 4{,}800x^{1/2}$; 24,000 labor-hours

93. $W(h) = 0.0005h^3$; 171.5 lb

95. 19,400

Exercises 4-2

1. $\frac{1}{3}(3x + 5)^3 + C$ **3.** $\frac{1}{6}(x^2 - 1)^6 + C$ **5.** $-\frac{1}{2}(5x^3 + 1)^{-2} + C$ **7.** $e^{5x} + C$ **9.** $\ln|1 + x^2| + C$ **11.** $\frac{2}{3}(1 + x^4)^{3/2} + C$ **13.** $\frac{1}{11}(x + 3)^{11} + C$ **15.** $-\frac{1}{6}(6t - 7)^{-1} + C$ **17.** $\frac{1}{12}(t^2 + 1)^6 + C$ **19.** $\frac{1}{2}e^{x^2} + C$ **21.** $\frac{1}{5}\ln|5x + 4| + C$ **23.** $-e^{1-t} + C$ **25.** $-\frac{1}{18}(3t^2 + 1)^{-3} + C$ **27.** $\frac{2}{5}(x + 4)^{5/2} - \frac{8}{3}(x + 4)^{3/2} + C$ **29.** $\frac{2}{3}(x - 3)^{3/2} + 6(x - 3)^{1/2} + C$ **31.** $\frac{1}{11}(x - 4)^{11} + \frac{2}{5}(x - 4)^{10} + C$ **33.** $\frac{1}{8}(1 + e^{2x})^4 + C$ **35.** $\frac{1}{2}\ln|4 + 2x + x^2| + C$ **37.** $\frac{1}{2}(5x + 3)^2 + C$ **39.** $\frac{1}{2}(x^2 - 1)^2 + C$ **41.** $\frac{1}{5}(x^5)^5 + C$ **49.** $\frac{1}{9}(3x^2 + 7)^{3/2} + C$ **51.** $\frac{1}{8}x^8 + \frac{4}{5}x^5 + 2x^2 + C$ **53.** $\frac{1}{9}(x^3 + 2)^3 + C$ **55.** $\frac{1}{4}(2x^4 + 3)^{1/2} + C$ **57.** $\frac{1}{4}(\ln x)^4 + C$ **59.** $e^{-1/x} + C$ **61.** $x = \frac{1}{3}(t^3 + 5)^7 + C$ **63.** $y = 3(t^2 - 4)^{1/2} + C$ **65.** $p = -(e^x - e^{-x})^{-1} + C$ **67.** $p(x) = \dfrac{2{,}000}{3x + 50} + 4$; 250 bottles **69.** $C(x) = 12x + 500 \ln(x + 1) + 2{,}000$; $\overline{C}(1{,}000) = \$17.45$

71. (A) $S(t) = 10t + 100e^{-0.1t} - 100$, $0 \le t \le 24$ (B) \$50 million (C) 18.41 mo

73. $Q(t) = 100 \ln(t + 1) + 5t$, $0 \le t \le 20$; 275 thousand barrels **75.** $W(t) = 2e^{0.1t}$; 4.45 g

77. (A) $-1{,}000$ bacteria/mL per day (B) $N(t) = 5{,}000 - 1{,}000 \ln(1 + t^2)$; 385 bacteria/mL (C) 7.32 days

79. $N(t) = 100 - 60e^{-0.1t}$, $0 \le t \le 15$; 87 words/min

81. $E(t) = 12{,}000 - 10{,}000\,(t + 1)^{-1/2}$; 9,500 students

Exercises 4-3

1. $y = 3x^2 + C$ **3.** $y = 7 \ln|x| + C$ **5.** $y = 50e^{0.02x} + C$ **7.** $y = \dfrac{x^3}{3} - \dfrac{x^2}{2}$ **9.** $y = e^{-x^2} + 2$

11. $y = 2 \ln|1 + x| + 5$

13. Figure B. When $x = 1$, the slope $dy/dx = 1 - 1 = 0$ for any y. When $x = 0$, the slope $dy/dx = 0 - 1 = -1$ for any y. Both are consistent with the slope field shown in Figure B.

15. $y = \dfrac{x^2}{2} - x + C$; $y = \dfrac{x^2}{2} - x - 2$

17. 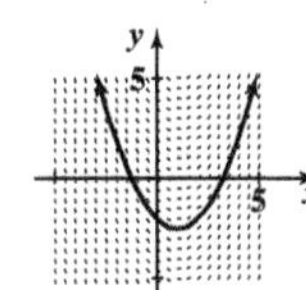

19. $y = Ce^{2t}$ **21.** $y = 100e^{-0.5x}$ **23.** $x = Ce^{-5t}$ **25.** $x = -(5t^2/2) + C$
27. Figure A. When $y = 1$, the slope $dy/dx = 1 - 1 = 0$ for any x. When $y = 2$, the slope $dy/dx = 1 - 2 = -1$ for any x. Both are consistent with the slope field shown in Figure A.
29. $y = 1 - e^{-x}$

31. 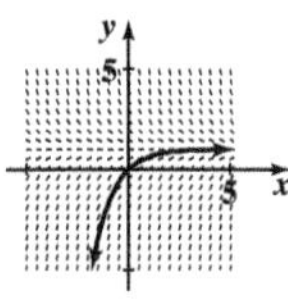

33. 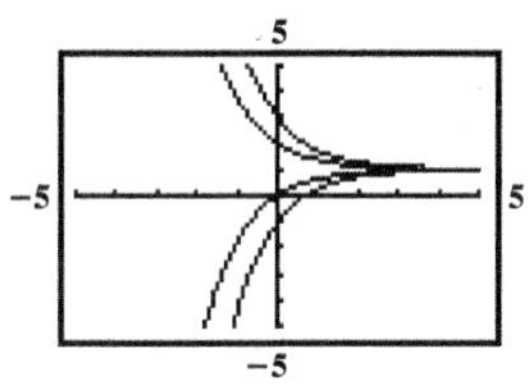

35. $y = \sqrt{25 - x^2}$ **37.** $y = -3x$ **39.** $y = 1/(1 - 2e^{-t})$

41. 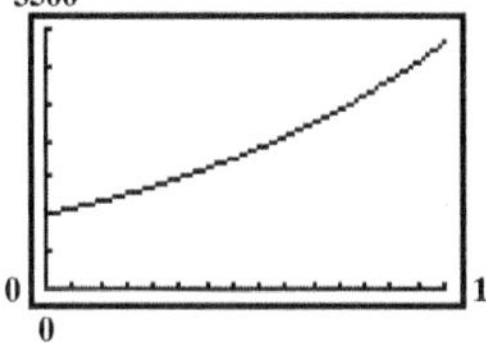

43. 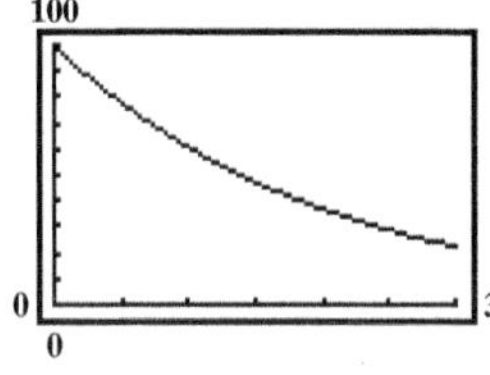

45. 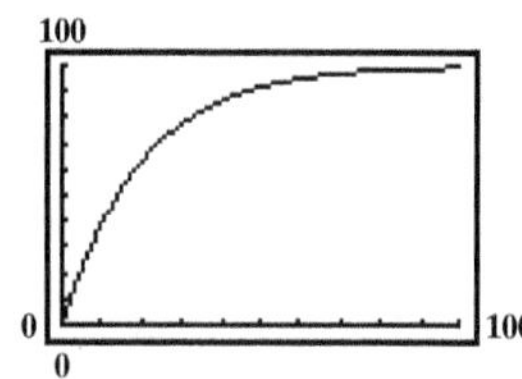

47. 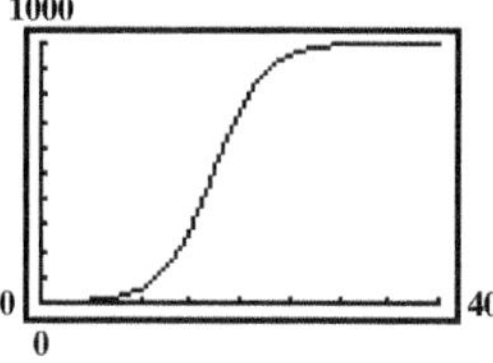

49. Apply the second-derivative test to $f(y) = ky(M - y)$. **51.** 2009 **53.** $A = 1{,}000e^{0.03t}$ **55.** $A = 8{,}000e^{0.016t}$

57. (A) $p(x) = 100e^{-0.05x}$
(B) \$60.65 per unit
(C) 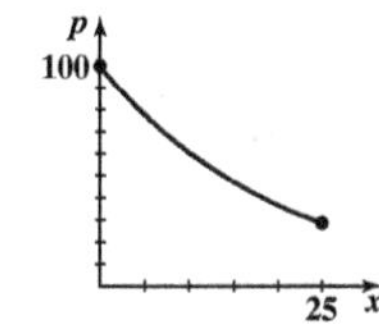

59. (A) $N = L(1 - e^{-0.051t})$
(B) 22.5%
(C) 32 days
(D) 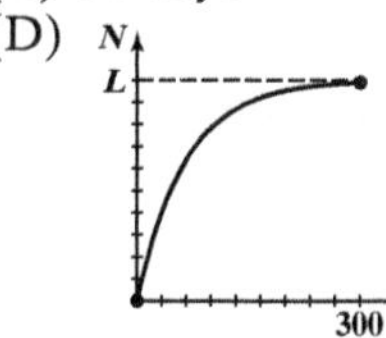

61. $I = I_0e^{-0.00942x}$; $x \approx 74$ ft
63. (A) $Q = 3e^{-0.04t}$ (B) $Q(10) = 2.01$ mL
(C) 27.47 hr
(D) 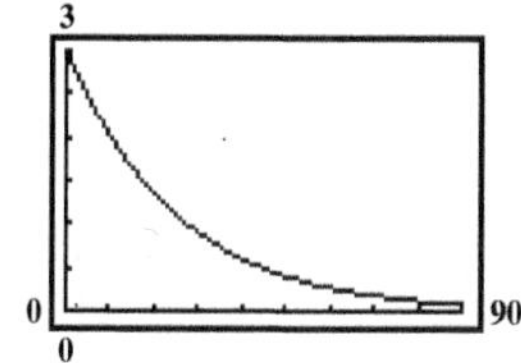

65. 0.023 117 **67.** Approx. 24,200 yr **69.** 104 times; 67 times
71. (A) 7 people; 353 people (B) 400
(C) 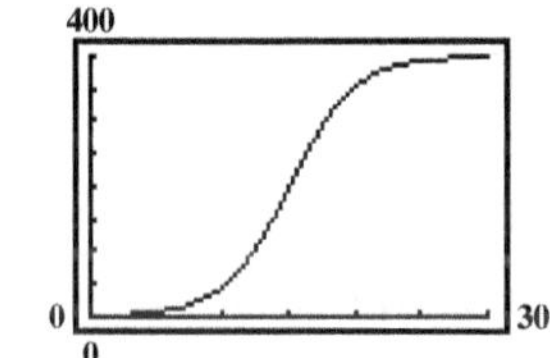

Exercises 4-4

1. C, E **3.** B **5.** H, I **7.** H
9. 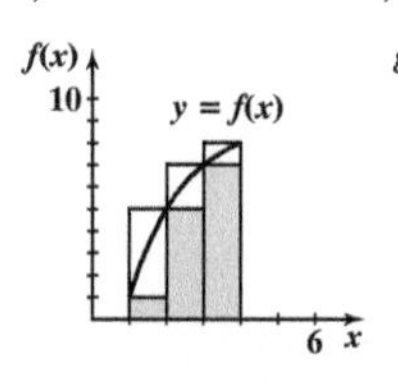

11. Figure A: $L_3 = 13$, $R_3 = 20$; Figure B: $L_3 = 14$, $R_3 = 7$
13. $L_3 \le \int_1^4 f(x)\,dx \le R_3$; $R_3 \le \int_1^4 g(x)\,dx \le L_3$; since $f(x)$ is increasing, L_3 underestimates the area and R_3 overestimates the area; since $g(x)$ is decreasing, the reverse is true.
15. In both figures, the error bound for L_3 and R_3 is 7.
17. $S_5 = -260$ **19.** $S_4 = -1{,}194$ **21.** $S_3 = -33.01$ **23.** $S_6 = -38$ **25.** -2.475
27. 4.266 **29.** 2.474 **31.** -5.333 **33.** 1.067 **35.** -1.066 **37.** 15 **39.** 58.5

41. -54 **43.** 248 **45.** 0 **47.** -183 **49.** False **51.** False **53.** False
55. $L_{10} = 286{,}100$ ft^2; error bound is 50,000 ft^2; $n \ge 200$
57. $L_6 = -3.53$, $R_6 = -0.91$; error bound for L_6 and R_6 is 2.63. Geometrically, the definite integral over the interval [2, 5] is the sum of the areas between the curve and the x axis from $x = 2$ to $x = 5$, with the areas below the x axis counted negatively and those above the x axis counted positively.
59. Increasing on $(-\infty, 0]$; decreasing on $[0, \infty)$
61. Increasing on $[-1, 0]$ and $[1, \infty)$; decreasing on $(-\infty, -1]$ and $[0, 1]$ **63.** $n \ge 22$ **65.** $n \ge 104$
67. $L_3 = 2{,}580$, $R_3 = 3{,}900$; error bound for L_3 and R_3 is 1,320
69. (A) $L_5 = 3.72$; $R_5 = 3.37$ (B) $R_5 = 3.37 \le \int_0^5 A'(t)\,dt \le 3.72 = L_5$
71. $L_3 = 114$, $R_3 = 102$; error bound for L_3 and R_3 is 12

Exercises 4-5

1. (A) $F(15) - F(10) = 375$
(B) $F'(x) = 6x$; Area = 375

3. (A) $F(15) - F(10) = 85$
(B) $F'(x) = -2x + 42$; Area = 85

5. 40 **7.** 72 **9.** 46.5 **11.** $e - 1 \approx 1.718$
13. $\ln 2 \approx 0.693$ **15.** 0 **17.** 48 **19.** -48
21. -10.25 **23.** 0 **25.** -2 **27.** 14
29. $5^6 = 15{,}625$ **31.** $\ln 4 \approx 1.386$
33. $20(e^{0.25} - e^{-0.5}) \approx 13.550$ **35.** $\frac{1}{2}$
37. $\frac{1}{2}(1 - e^{-1}) \approx 0.316$ **39.** 0

41. (A) Average $f(x) = 250$
(B)

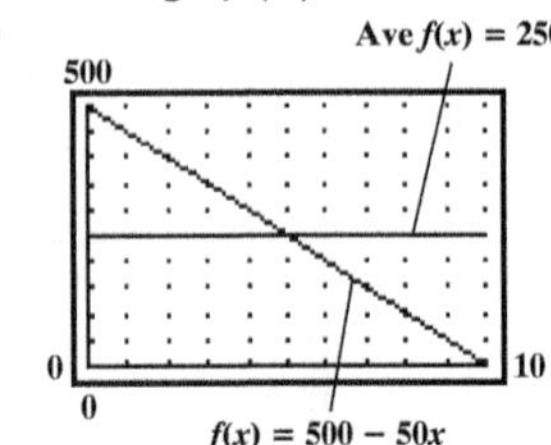

43. (A) Average $f(t) = 2$
(B)

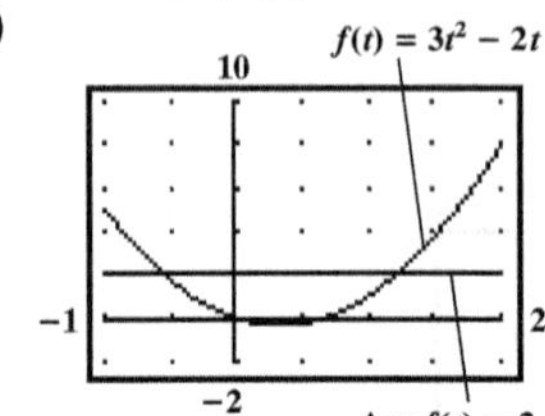

45. (A) Average $f(x) = \frac{45}{28} \approx 1.61$
(B)

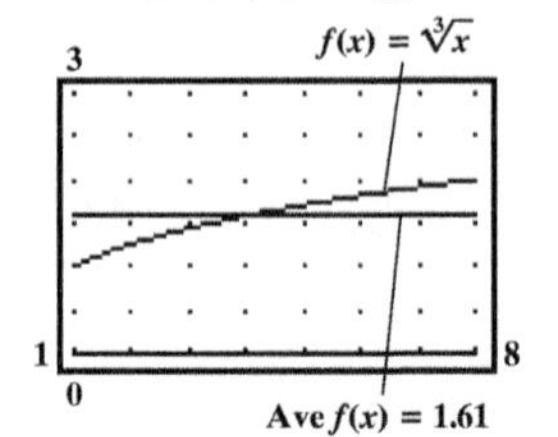

47. (A) Average $f(x) = 2(1 - e^{-2}) \approx 1.73$
(B)

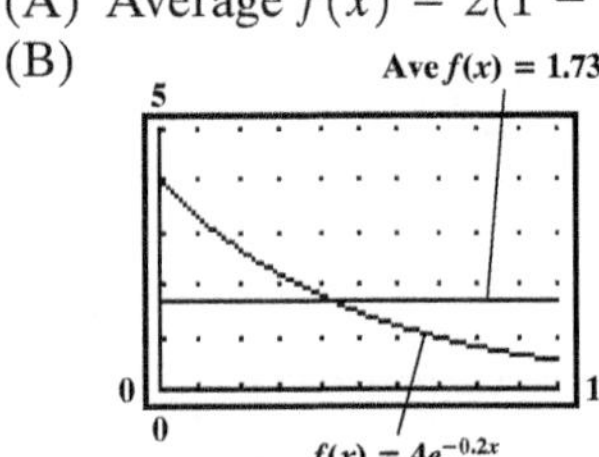

49. $\frac{1}{6}(15^{3/2} - 5^{3/2}) \approx 7.819$ **51.** $\frac{1}{2}(\ln 2 - \ln 3) \approx -0.203$ **53.** 0

55. 4.566 **57.** 2.214 **61.** $\int_{300}^{900} \left(500 - \frac{x}{3}\right) dx = \$180{,}000$

63. $\int_0^5 500(t - 12)\, dt = -\$23{,}750$; $\int_5^{10} 500(t - 12)\, dt = -\$11{,}250$

65. (A)

QuadReg
y=ax²+bx+c
a=-.0082142857
b=1.528571429
c=16

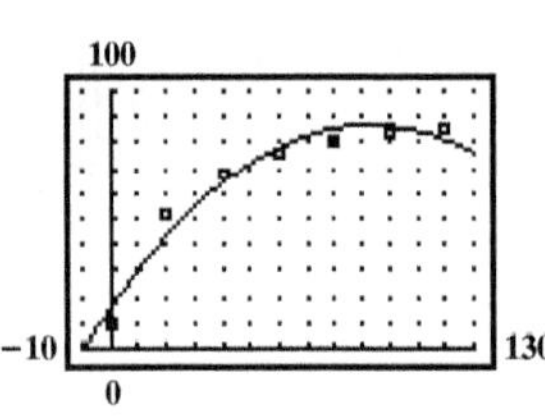

(B) 6,505

67. Useful life $= \sqrt{\ln 55} \approx 2$ yr; total profit $= \frac{51}{22} - \frac{5}{2}e^{-4} \approx 2.272$ or \$2,272
69. (A) \$420 (B) \$135,000

71. (A)

QuadReg
y=ax²+bx+c
a=3.074675325
b=-24.98073593
c=55.56363636

(B) \$100,505

73. $50e^{0.6} - 50e^{0.4} - 10 \approx \$6.51$
75. 4,800 labor-hours
77. (A) $I = -200t + 600$
(B) $\frac{1}{3}\int_0^3 (-200t + 600)\, dt = 300$
79. $100 \ln 11 + 50 \approx 290$ thousand barrels;
$100 \ln 21 - 100 \ln 11 + 50 \approx 115$ thousand barrels

81. $2e^{0.8} - 2 \approx 2.45$ g; $2e^{1.6} - 2e^{0.8} \approx 5.45$ g **83.** 10°C
85. $0.6 \ln 2 + 0.1 \approx 0.516$; $(4.2 \ln 625 + 2.4 - 4.2 \ln 49)/24 \approx 0.546$

Chapter 4 Review Exercises

1. $3x^2 + 3x + C$ *(6-1)* **2.** 50 *(6-5)* **3.** -207 *(6-5)* **4.** $-\frac{1}{8}(1 - t^2)^4 + C$ *(6-2)* **5.** $\ln|u| + \frac{1}{4}u^4 + C$ *(6-1)*
6. 0.216 *(6-5)* **7.** e^{-x^2} *(6-1)* **8.** $\sqrt{4 + 5x} + C$ *(6-1)* **9.** $y = f(x) = x^3 - 2x + 4$ *(6-3)*

10. (A) $2x^4 - 2x^2 - x + C$ (B) $e^t - 4\ln|t| + C$ *(6-1)* **11.** $R_2 = 72$; error bound for R_2 is 48 *(6-4)*
12. $\int_1^5 (x^2 + 1)\, dx = \frac{136}{3} \approx 45.33$; actual error is $\frac{80}{3} \approx 26.67$ *(6-5)* **13.** $L_4 = 30.8$ *(6-4)*
14. 7 *(6-5)* **15.** Width $= 2 - (-1) = 3$; height = average $f(x) = 7$ *(6-5)* **16.** $S_4 = 368$ *(6-4)* **17.** $S_5 = 906$ *(6-4)*
18. -10 *(6-4)* **19.** 0.4 *(6-4)* **20.** 1.4 *(6-4)* **21.** 0 *(6-4)* **22.** 0.4 *(6-4)* **23.** 2 *(6-4)* **24.** -2 *(6-4)*
25. -0.4 *(6-4)* **26.** (A) 1;1 (B) 4;4 *(6-3)*
27. $dy/dx = (2y)/x$; the slopes computed in Problem 26A are compatible with the slope field shown. *(6-3)*
29. $y = \frac{1}{4}x^2$; $y = -\frac{1}{4}x^2$ *(6-3)*

30. *(6-3)*

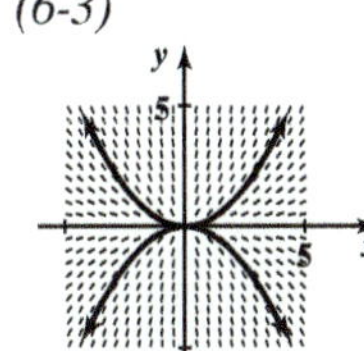

31. *(6-3)*

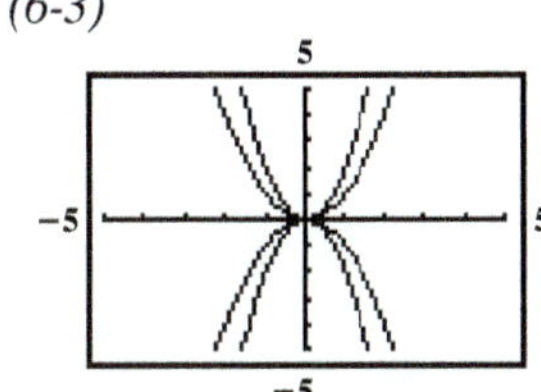

32. $\frac{2}{3}(2)^{3/2} \approx 1.886$ *(6-5)* **33.** $\frac{1}{6} \approx 0.167$ *(6-5)*
34. $-5e^{-t} + C$ *(6-1)* **35.** $\frac{1}{2}(1 + e^2)$ *(6-1)*
36. $\frac{1}{6}e^{3x^2} + C$ *(6-2)* **37.** $2(\sqrt{5} - 1) \approx 2.472$ *(6-5)*
38. $\frac{1}{2}\ln 10 \approx 1.151$ *(6-5)* **39.** 0.45 *(6-5)*
40. $\frac{1}{48}(2x^4 + 5)^6 + C$ *(6-2)* **41.** $-\ln(e^{-x} + 3) + C$ *(6-2)*
42. $-(e^x + 2)^{-1} + C$ *(6-2)*

43. $y = f(x) = 3\ln|x| + x^{-1} + 4$ *(6-2, 6-3)* **44.** $y = 3x^2 + x - 4$ *(6-3)*

45. (A) Average $f(x) = 6.5$
(B) *(6-5)*

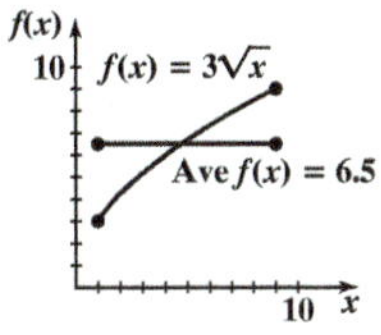

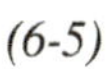

46. $\frac{1}{3}(\ln x)^3 + C$ *(6-2)*
47. $\frac{1}{8}x^8 - \frac{2}{5}x^5 + \frac{1}{2}x^2 + C$ *(6-2)*
48. $\frac{2}{3}(6 - x)^{3/2} - 12(6 - x)^{1/2} + C$ *(6-2)*
49. $\frac{1{,}234}{15} \approx 82.267$ *(6-5)*
50. 0 *(6-5)*
51. $y = 3e^{x^3} - 1$ *(6-3)*
52. $N = 800e^{0.06t}$ *(6-3)*

53. Limited growth

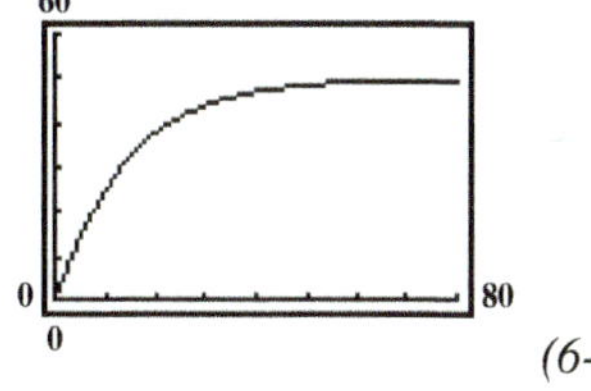

(6-3)

54. Exponential decay

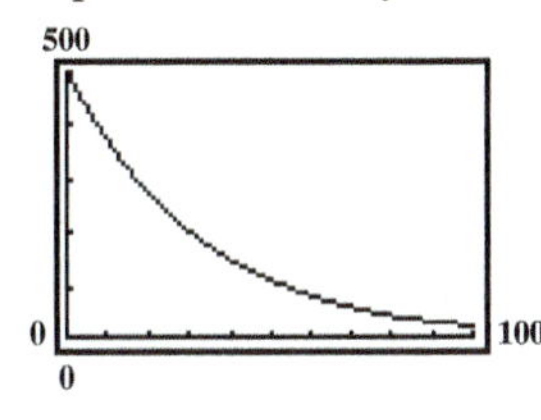

(6-3)

55. Unlimited growth

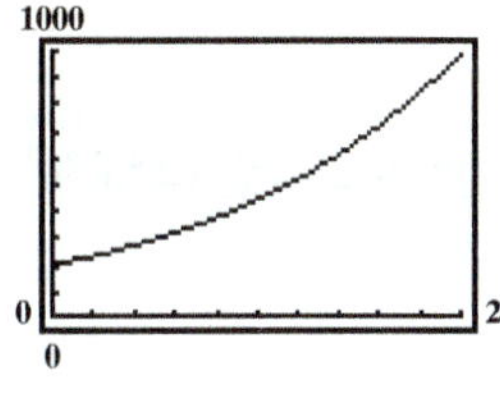

(6-3)

56. Logistic growth

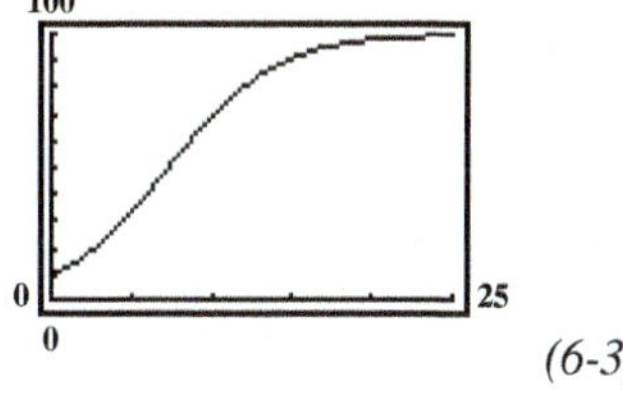

(6-3)

57. 1.167 *(6-5)* **58.** 99.074 *(6-5)* **59.** −0.153 *(6-5)* **60.** $L_2 = \$180{,}000$; $R_2 = \$140{,}000$; $\$140{,}000 \le \int_{200}^{600} C'(x)\,dx \le \$180{,}000$ *(6-4)*

61. $\int_{200}^{600}\left(600 - \frac{x}{2}\right)dx = \$160{,}000$ *(6-5)* **62.** $\int_{10}^{40}\left(150 - \frac{x}{10}\right)dx = \$4{,}425$ *(6-5)* **63.** $P(x) = 100x - 0.01x^2$; $P(10) = \$999$ *(6-3)*

64. $\int_0^{15}(60 - 4t)\,dt = 450$ thousand barrels *(6-5)* **65.** 109 items *(6-5)* **66.** $16e^{2.5} - 16e^2 - 8 \approx \$68.70$ *(6-5)*

67. Useful life $= 10\ln\frac{20}{3} \approx 19$ yr; total profit $= 143 - 200e^{-1.9} \approx 113.086$ or \$113,086 *(6-5)*

68. $S(t) = 50 - 50e^{-0.08t}$; $50 - 50e^{-0.96} \approx \$31$ million; $-(\ln 0.2)/0.08 \approx 20$ mo *(6-3)* **69.** 1 cm^2 *(6-3)* **70.** 800 gal *(6-5)*

71. (A) 133 million (B) About 61 years *(6-3)* **72.** $\frac{-\ln 0.04}{0.000\,123\,8} \approx 26{,}000$ yr *(6-3)*

73. $N(t) = 95 - 70e^{-0.1t}$; $N(15) \approx 79$ words/min *(6-3)*

Chapter 5

Exercises 5-1

1. $\int_a^b g(x)\,dx$ **3.** $\int_a^b[-h(x)]\,dx$

5. Since the shaded region in Figure C is below the x axis, $h(x) \le 0$; so, $\int_a^b h(x)\,dx$ represents the negative of the area.

7. 24 **9.** 51 **11.** 42 **13.** 6 **15.** 0.833 **17.** 2.350 **19.** 1 **21.** 10.5 **23.** 18 **25.** 44

27. $\int_a^b[-f(x)]\,dx$ **29.** $\int_b^c f(x)\,dx + \int_c^d[-f(x)]\,dx$ **31.** $\int_c^d[f(x) - g(x)]\,dx$

33. $\int_a^b[f(x) - g(x)]\,dx + \int_b^c[g(x) - f(x)]\,dx$

35. Find the intersection points by solving $f(x) = g(x)$ on the interval $[a, d]$ to determine b and c. Then observe that $f(x) \ge g(x)$ over $[a, b]$, $g(x) \ge f(x)$ over $[b, c]$, and $f(x) \ge g(x)$ over $[c, d]$.
Area $= \int_a^b[f(x) - g(x)]\,dx + \int_b^c[g(x) - f(x)]\,dx + \int_c^d[f(x) - g(x)]\,dx$.

37. 2.5 **39.** 7.667 **41.** 12 **43.** 15 **45.** 32 **47.** 36 **49.** 9 **51.** 2.832

53. $\int_{-3}^{3}\sqrt{9 - x^2}\,dx$; 14.137 **55.** $\int_0^4\sqrt{16 - x^2}\,dx$; 12.566 **57.** $\int_{-2}^{2}2\sqrt{4 - x^2}\,dx$; 12.566

59. 18 **61.** 1.858 **63.** 52.616 **65.** 8 **67.** 101.75 **69.** 17.979 **71.** 5.113 **73.** 8.290 **75.** 3.166 **77.** 1.385

79. Total production from the end of the fifth year to the end of the 10th year is $50 + 100\ln 20 - 100\ln 15 \approx 79$ thousand barrels.

81. Total profit over the 5-yr useful life of the game is $20 - 30e^{-1.5} \approx 13.306$, or \$13,306.

83. 1935: 0.412; 1947: 0.231; income was more equally distributed in 1947.
85. 1963: 0.818; 1983: 0.846; total assets were less equally distributed in 1983.
87. (A) $f(x) = 0.3125x^2 + 0.7175x - 0.015$ (B) 0.104
89. Total weight gain during the first 10 hr is $3e - 3 \approx 5.15$ g.
91. Average number of words learned from $t = 2$ hr to $t = 4$ hr is $15 \ln 4 - 15 \ln 2 \approx 10$.

Exercises 5-2

1. 0.43 **3.** 744.99 **5.** 10.27 **7.** 151.75 **9.** 93,268.66 **11.** (A) 10.72 (B) 3.28 (C) 10.72
13. (A) .75 (B) .11
(C) $f(x)$; 0.5; 12; x

15. 8 yr **17.** (A) .11 B) .10
19. $P(t \geq 12) = 1 - P(0 \leq t \leq 12) = .89$
21. $12,500
23. If $f(t)$ is the rate of flow of a continuous income stream, then the total income produced from 0 to 5 yr is the area under the graph of $y = f(t)$ from $t = 0$ to $t = 5$.

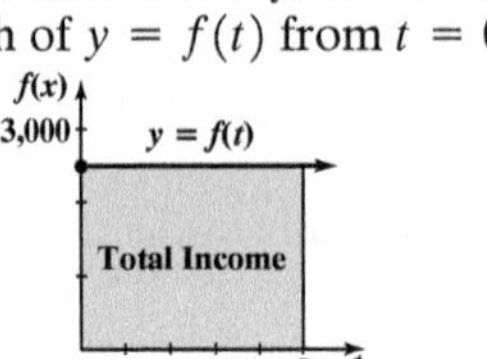

25. $8{,}000(e^{0.15} - 1) \approx \$1{,}295$
27. If $f(t)$ is the rate of flow of a continuous income stream, then the total income produced from 0 to 3 yr is the area under the graph of $y = f(t)$ from $t = 0$ to $t = 3$.

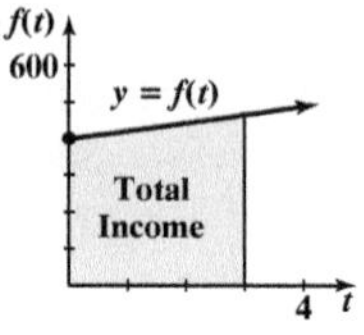

29. $255,562; $175,562 **31.** $6,779.52 **33.** $436.62
35. Clothing store: $66,420.83; computer store: $62,622.66; the clothing store is the better investment.
37. Bond: $12,062.30 business: $11,823.87; the bond is the better investment. **39.** $55,230
41. $\frac{k}{r}(e^{rT} - 1)$ **43.** $625,000
45. The shaded area is the consumers' surplus and represents the total savings to consumers who are willing to pay more than $150 for a product but are still able to buy the product for $150.

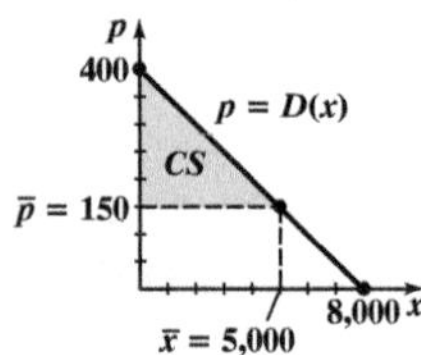

47. $9,900
49. The area of the region PS is the producers' surplus and represents the total gain to producers who are willing to supply units at a lower price than $67 but are still able to supply the product at $67.

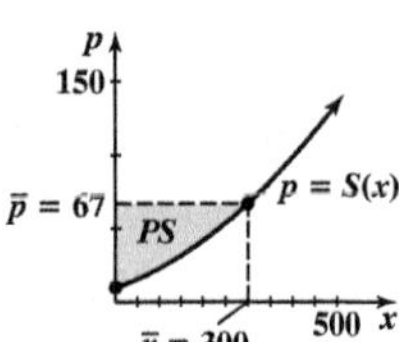

51. $CS = \$3{,}380$; $PS = \$1{,}690$

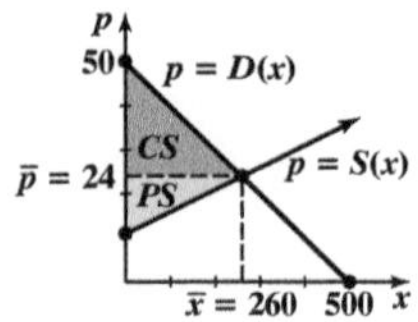

53. $CS = \$6{,}980$; $PS = \$5{,}041$

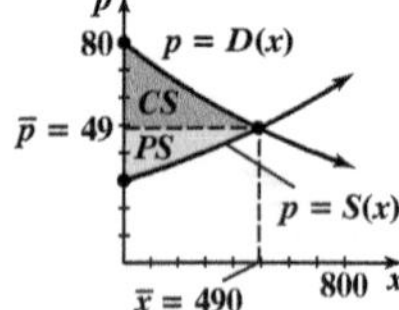

55. $CS = \$7{,}810$; $PS = \$8{,}336$

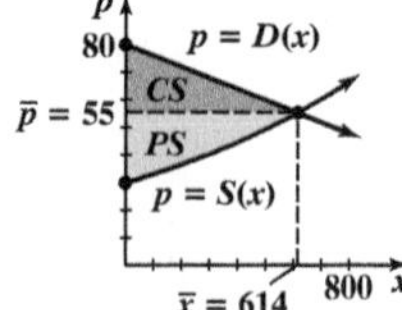

57. $CS = \$8{,}544$; $PS = \$11{,}507$

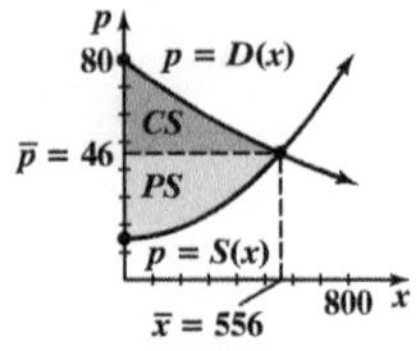

59. (A) $\bar{x} = 21.457$; $\bar{p} = \$6.51$
(B) $CS = 1.774$ or $1,774; $PS = 1.087$or $1,087

Exercises 5-3

1. $\frac{1}{3}xe^{3x} - \frac{1}{9}e^{3x} + C$ **3.** $\frac{x^3}{3}\ln x - \frac{x^3}{9} + C$ **5.** $u = x + 2$; $\frac{(x+2)(x+1)^6}{6} - \frac{(x+1)^7}{42} + C$ **7.** $-xe^{-x} - e^{-x} + C$
9. $\frac{1}{2}e^{x^2} + C$ **11.** $(xe^x - 4e^x)|_0^1 = -3e + 4 \approx -4.1548$ **13.** $(x \ln 2x - x)|_1^3 = (3 \ln 6 - 3) - (\ln 2 - 1) \approx 2.6821$
15. $\ln(x^2 + 1) + C$ **17.** $(\ln x)^2/2 + C$ **19.** $\frac{2}{3}x^{3/2}\ln x - \frac{4}{9}x^{3/2} + C$
21. $\frac{(x-3)(x+1)^3}{3} - \frac{(x+1)^4}{12} + C$ or $\frac{x^4}{4} - \frac{x^3}{3} - \frac{5x^2}{2} - 3x + C$

23. $\dfrac{(2x+1)(x-2)^3}{3} - \dfrac{(x-2)^4}{6} + C$ or $\dfrac{x^4}{2} - \dfrac{7x^3}{3} + 2x^2 + 4x + C$

25. The integral represents the negative of the area between the graph of $y = (x-3)e^x$ and the x axis from $x = 0$ to $x = 1$.

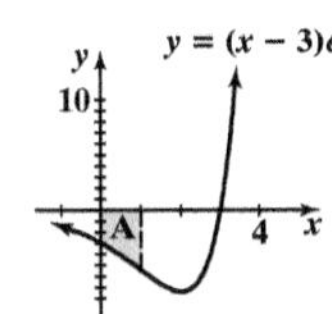

27. The integral represents the area between the graph of $y = \ln 2x$ and the x axis from $x = 1$ to $x = 3$.

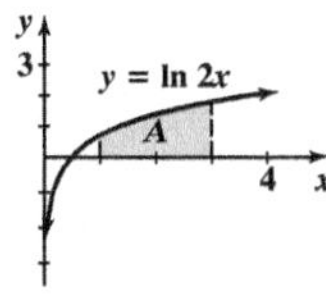

29. $(x^2 - 2x + 2)e^x + C$ **31.** $\dfrac{xe^{ax}}{a} - \dfrac{e^{ax}}{a^2} + C$ **33.** $\left(-\dfrac{\ln x}{x} - \dfrac{1}{x}\right)\Big|_1^e = -\dfrac{2}{e} + 1 \approx 0.2642$
35. $6\ln 6 - 4\ln 4 - 2 \approx 3.205$ **37.** $xe^{x-2} - e^{x-2} + C$ **39.** $\frac{1}{2}(1+x^2)\ln(1+x^2) - \frac{1}{2}(1+x^2) + C$
41. $(1+e^x)\ln(1+e^x) - (1+e^x) + C$ **43.** $x(\ln x)^2 - 2x\ln x + 2x + C$ **45.** $x(\ln x)^3 - 3x(\ln x)^2 + 6x\ln x - 6x + C$

47. 2 **49.** $\dfrac{1}{3}$ **51.** 1.56 **53.** 34.98 **55.** $\int_0^5 (2t - te^{-t})\,dt = \$24$ million

57. The total profit for the first 5 yr (in millions of dollars) is the same as the area under the marginal profit function, $P'(t) = 2t - te^{-t}$, from $t = 0$ to $t = 5$.

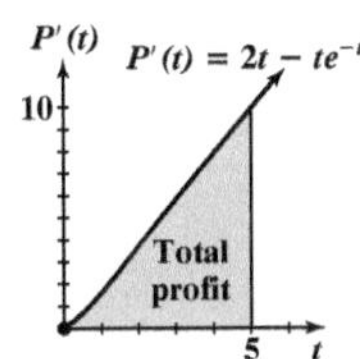

59. \$2,854.88 **61.** 0.264

63. The area bounded by $y = x$ and the Lorenz curve $y = xe^{x-1}$, divided by the area under the graph of $y = x$ from $x = 0$ to $x = 1$, is the Gini index of income concentration. The closer this index is to 0, the more equally distributed the income; the closer the index is to 1, the more concentrated the income in a few hands.

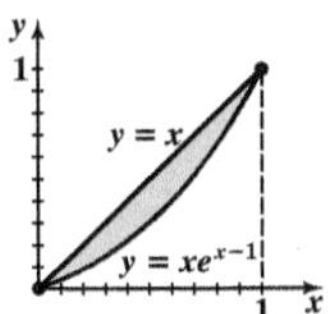

65. $S(t) = 1{,}600 + 400e^{0.1t} - 40te^{0.1t}$; 15 mo **67.** \$977

69. The area bounded by the price–demand equation, $p = 9 - \ln(x+4)$, and the price equation, $y = \overline{p} = 2.089$, from $x = 0$ to $x = \overline{x} = 1{,}000$, represents the consumers' surplus. This is the amount consumers who are willing to pay more than \$2.089 save.

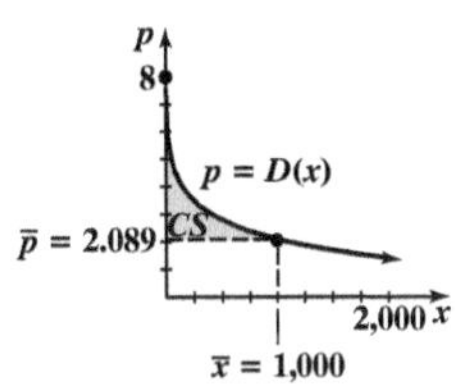

71. 2.1388 ppm
73. $N(t) = -4te^{-0.25t} - 40e^{-0.25t} + 80$; 8 wk; 78 words/min
75. 20,980

Exercises 5-4

1. $\ln\left|\dfrac{x}{1+x}\right| + C$ **3.** $\dfrac{1}{3+x} + 2\ln\left|\dfrac{5+2x}{3+x}\right| + C$ **5.** $\dfrac{2(x-32)}{3}\sqrt{16+x} + C$ **7.** $-\ln\left|\dfrac{1+\sqrt{1-x^2}}{x}\right| + C$
9. $\dfrac{1}{2}\ln\left|\dfrac{x}{2+\sqrt{x^2+4}}\right| + C$ **11.** $\frac{1}{3}x^3\ln x - \frac{1}{9}x^3 + C$ **13.** $x - \ln|1+e^x| + C$ **15.** $9\ln\frac{3}{2} - 2 \approx 1.6492$
17. $\frac{1}{2}\ln\frac{12}{5} \approx 0.4377$ **19.** $\ln 3 \approx 1.0986$ **21.** $-\dfrac{\sqrt{4x^2+1}}{x} + 2\ln|2x + \sqrt{4x^2+1}| + C$ **23.** $\frac{1}{2}\ln|x^2 + \sqrt{x^4-16}| + C$
25. $\frac{1}{6}(x^3\sqrt{x^6+4} + 4\ln|x^3 + \sqrt{x^6+4}|) + C$ **27.** $-\dfrac{\sqrt{4-x^4}}{8x^2} + C$ **29.** $\dfrac{1}{5}\ln\left|\dfrac{3+4e^x}{2+e^x}\right| + C$ **31.** $\frac{2}{3}(\ln x - 8)\sqrt{4+\ln x} + C$
33. $\frac{1}{5}x^2e^{5x} - \frac{2}{25}xe^{5x} + \frac{2}{125}e^{5x} + C$ **35.** $-x^3e^{-x} - 3x^2e^{-x} - 6xe^{-x} - 6e^{-x} + C$ **37.** $x(\ln x)^3 - 3x(\ln x)^2 + 6x\ln x - 6x + C$
39. $\frac{64}{3}$ **41.** $\frac{1}{2}\ln\frac{9}{5} \approx 0.2939$ **43.** $\dfrac{-1-\ln x}{x} + C$ **45.** $\sqrt{x^2-1} + C$ **47.** 31.38 **49.** 5.48 **51.** $3{,}000 + 1{,}500\ln\frac{1}{3} \approx \$1{,}352$

53.

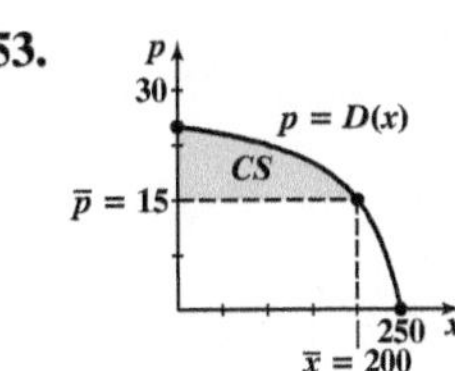

55. $C(x) = 200x + 1{,}000\ln(1 + 0.05x) + 25{,}000$; 608; \$198,773
57. \$18,673.95 **59.** 0.1407
61. As the area bounded by the two curves gets smaller, the Lorenz curve approaches $y = x$ and the distribution of income approaches perfect equality—all persons share equally in the income available.

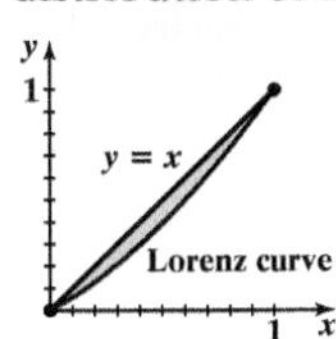

63. $S(t) = 1 + t - \frac{1}{1 + t} - 2\ln|1 + t|$; $24.96 - 2\ln 25 \approx \$18.5$ million

65. The total sales (in millions of dollars) over the first 2 yr (24 mo) is the area under the graph of $y = S'(t)$ from $t = 0$ to $t = 24$.

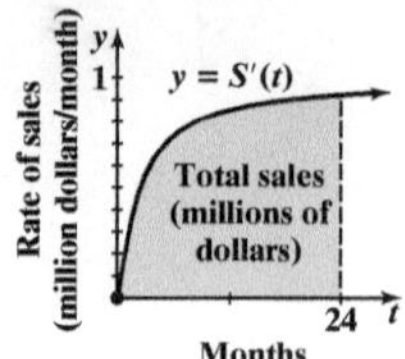

67. $P(x) = \frac{2(9x - 4)}{135}(2 + 3x)^{3/2} - 2{,}000.83$; 54; \$37,932

69. $100\ln 3 \approx 110$ ft

71. $60\ln 5 \approx 97$ items

73. The area under the graph of $y = N'(t)$ from $t = 0$ to $t = 12$ represents the total number of items learned in that time interval.

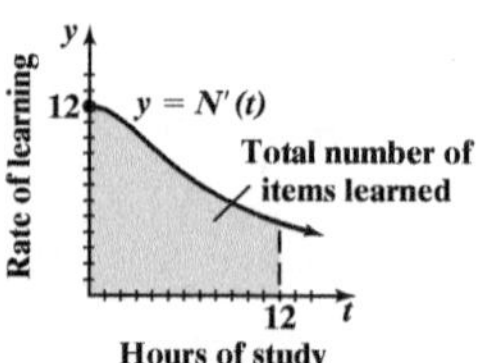

Chapter 5 Review Exercises

1. $\int_a^b f(x)\,dx$ *(7-1)* **2.** $\int_b^c [-f(x)]\,dx$ *(7-1)* **3.** $\int_a^b f(x)\,dx + \int_b^c [-f(x)]\,dx$ *(7-1)*

4. Area = 1.153 *(7-1)*

5. $\frac{1}{4}xe^{4x} - \frac{1}{16}e^{4x} + C$ *(7-3, 7-4)* **6.** $\frac{1}{2}x^2\ln x - \frac{1}{4}x^2 + C$ *(7-3, 7-4)*

7. $\frac{(\ln x)^2}{2} + C$ *(6-2)* **8.** $\frac{\ln(1 + x^2)}{2} + C$ *(7-2)* **9.** $\frac{1}{1 + x} + \ln\left|\frac{x}{1 + x}\right| + C$ *(7-4)*

10. $-\frac{\sqrt{1 + x}}{x} - \frac{1}{2}\ln\left|\frac{\sqrt{1 + x} - 1}{\sqrt{1 + x} + 1}\right| + C$ *(7-4)* **11.** 12 *(7-1)* **12.** 40 *(7-1)* **13.** 34.167 *(7-1)*

14. 0.926 *(7-1)* **15.** 12 *(7-1)* **16.** 18.133 *(7-1)* **17.** $\int_a^b [f(x) - g(x)]\,dx$ *(7-1)*

18. $\int_b^c [g(x) - f(x)]\,dx$ *(7-1)* **19.** $\int_b^c [g(x) - f(x)]\,dx + \int_c^d [f(x) - g(x)]\,dx$ *(7-1)*

20. $\int_a^b [f(x) - g(x)]\,dx + \int_b^c [g(x) - f(x)]\,dx + \int_c^d [f(x) - g(x)]\,dx$ *(7-1)*

21. Area = 20.833 *(7-1)*

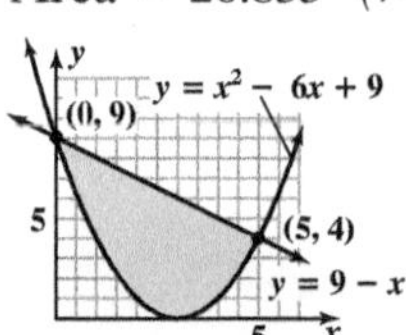

22. 1 *(7-3, 7-4)* **23.** $\frac{15}{2} - 8\ln 8 + 8\ln 4 \approx 1.955$ *(7-4)*

24. $\frac{1}{6}(3x\sqrt{9x^2 - 49} - 49\ln|3x + \sqrt{9x^2 - 49}|) + C$ *(7-4)*

25. $-2te^{-0.5t} - 4e^{-0.5t} + C$ *(7-3, 7-4)* **26.** $\frac{1}{3}x^3\ln x - \frac{1}{9}x^3 + C$ *(7-3, 7-4)*

27. $x - \ln|1 + 2e^x| + C$ *(7-4)*

28. (A) Area = 8

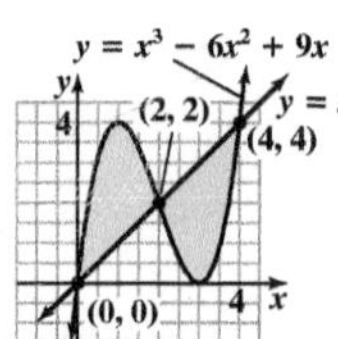

(B) Area = 8.38 *(7-1)*

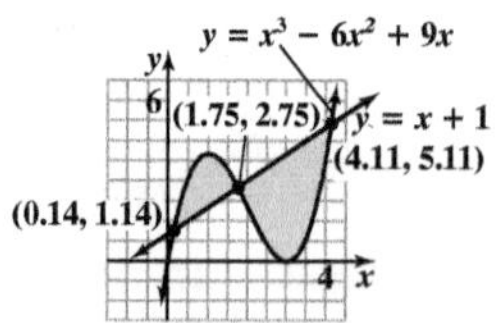

29. $\frac{1}{3}(\ln x)^3 + C$ *(6-2)*

30. $\frac{1}{2}x^2(\ln x)^2 - \frac{1}{2}x^2\ln x + \frac{1}{4}x^2 + C$ *(7-3, 7-4)*

31. $\sqrt{x^2 - 36} + C$ *(6-2)*

32. $\frac{1}{2}\ln|x^2 + \sqrt{x^4 - 36}| + C$ *(7-4)*

33. $50\ln 10 - 42\ln 6 - 24 \approx 15.875$ *(7-3, 7-4)*

34. $x(\ln x)^2 - 2x\ln x + 2x + C$ *(7-3, 7-4)*

35. $-\frac{1}{4}e^{-2x^2} + C$ *(6-2)*

36. $-\frac{1}{2}x^2e^{-2x} - \frac{1}{2}xe^{-2x} - \frac{1}{4}e^{-2x} + C$ *(7-3, 7-4)*

37. 1.703 *(7-1)* **38.** (A) .189 (B) .154 *(7-2)*

39. The probability that the product will fail during the second year of warranty is the area under the probability density function $y = f(t)$ from $t = 1$ to $t = 2$. *(7-2)*

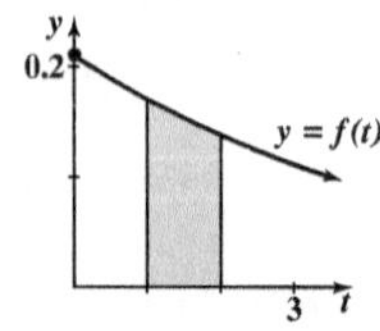

40. $R(x) = 65x - 6[(x + 1)\ln(x + 1) - x]$; 618/wk; \$29,506 *(7-3)*

41. (A) (B) \$8,507 *(7-2)*

42. (A) \$15,655.66 (B) \$1,454.39 *(7-2)*

43. (A)

(B) More equitably distributed, since the area bounded by the two curves will have decreased.

(C) Current = 0.3; projected = 0.2; income will be more equitably distributed 10 years from now. *(7-1)*

44. (A) $CS = \$2{,}250$; $PS = \$2{,}700$ (B) $CS = \$2{,}890$; $PS = \$2{,}278$ *(7-2)*

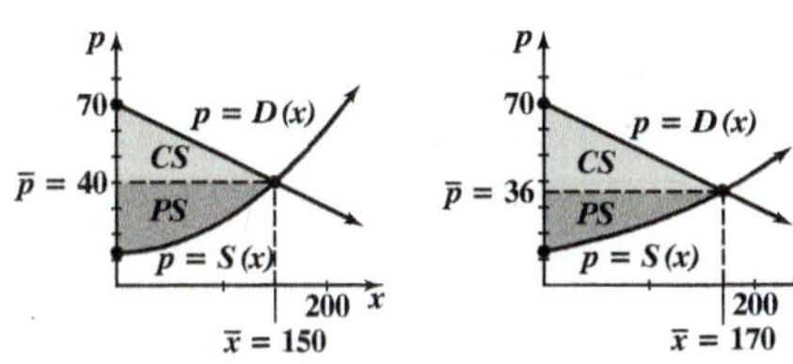

45. (A) 25.403 or 25,403 lb
(B) $PS = 121.6$ or \$1,216 *(7-2)*

46. 4.522 mL; 1.899 mL *(6-5, 7-4)*

47.

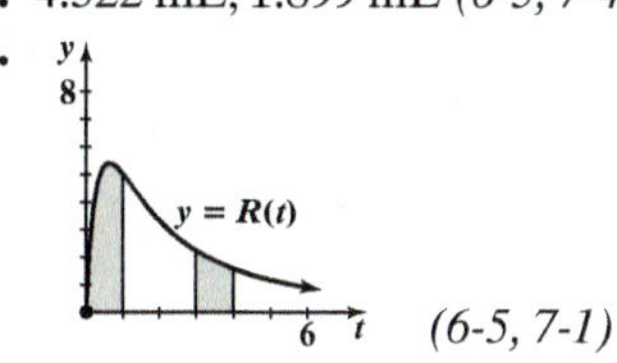

(6-5, 7-1)

48. .667; .333 *(7-2)*

49. The probability that the doctor will spend more than an hour with a randomly selected patient is the area under the probability density function $y = f(t)$ from $t = 1$ to $t = 3$. *(7-2)*

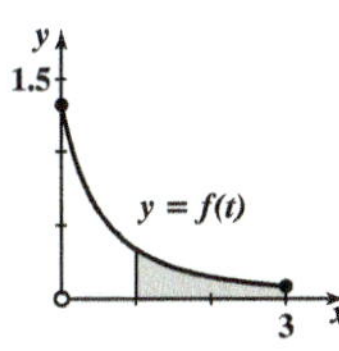

50. 45 thousand *(6-5, 7-1)* **51.** .368 *(7-2)*

Chapter 6

Exercises 6-1

1. −4 **3.** 11 **5.** 4 **7.** Not defined **9.** 59 **11.** −1 **13.** −64 **15.** 154,440 **17.** $192\pi \approx 603.2$ **19.** $3\pi\sqrt{109} \approx 98.4$

21. 118 **23.** 36,095.07 **25.** $f(x) = x^2 - 7$ **27.** $f(y) = 38y + 20$ **29.** $f(y) = -2y^3 + 5$ **31.** $y = 2$ **33.** $x = -\frac{1}{2}, 1$

35. −1.926, 0.599 **37.** $2x + h$ **39.** $2y^2$ **41.** $E(0, 0, 3)$; $F(2, 0, 3)$

43. (A) In the plane $y = c$, c any constant, $z = x^2$.
(B) The y axis; the line parallel to the y axis and passing through the point (1, 0, 1); the line parallel to the y axis and passing through the point (2, 0, 4)
(C) A parabolic "trough" lying on top of the y axis

45. (A) Upper semicircles whose centers lie on the y axis
(B) Upper semicircles whose centers lie on the x axis
(C) The upper hemisphere of radius 6 with center at the origin

47. (A) $a^2 + b^2$ and $c^2 + d^2$ both equal the square of the radius of the circle.
(B) Bell-shaped curves with maximum values of 1 at the origin
(C) A bell, with maximum value 1 at the origin, extending infinitely far in all directions.

49. \$13,200; \$18,000; \$21,300 **51.** $R(p, q) = -5p^2 + 6pq - 4q^2 + 200p + 300q$; $R(2, 3) = \$1{,}280$; $R(3, 2) = \$1{,}175$

53. 30,065 units **55.** (A) \$237,877.08 (B) 4.4% **57.** $T(70, 47) \approx 29$ min; $T(60, 27) = 33$ min

59. $C(6, 8) = 75$; $C(8.1, 9) = 90$ **61.** $Q(12, 10) = 120$; $Q(10, 12) \approx 83$

Exercises 6-2

1. $f_x(x, y) = 4$ **3.** $f_y(x, y) = -3x + 4y$ **5.** $\frac{\partial z}{\partial x} = 3x^2 + 8xy$ **7.** $\frac{\partial z}{\partial y} = 20(5x + 2y)^9$ **9.** 9 **11.** 3 **13.** −4 **15.** 0

17. 45.6 mpg; mileage is 45.6 mpg at a tire pressure of 32 psi and a speed of 40 mph

19. 37.6 mpg; mileage is 37.6 mpg at a tire pressure of 32 psi and a speed of 50 mph

21. 0.3 mpg per psi; mileage increases at a rate of 0.3 mpg per psi of tire pressure

23. $f_{xx}(x, y) = 0$ **25.** $f_{xy}(x, y) = 0$ **27.** $f_{xy}(x, y) = y^2e^{xy^2}(2xy) + e^{xy^2}(2y) = 2y(1 + xy^2)e^{xy^2}$ **29.** $f_{yy}(x, y) = \frac{2 \ln x}{y^3}$

31. $f_{xx}(x, y) = 80(2x + y)^3$ **33.** $f_{xy}(x, y) = 720xy^3(x^2 + y^4)^8$ **35.** $C_x(x, y) = 6x + 10y + 4$ **37.** 2 **39.** $C_{xx}(x, y) = 6$

41. $C_{xy}(x, y) = 10$ **43.** 6 **45.** $S_y(x, y) = \frac{x^3}{y} + 8ye^x$ **47.** $-1 + 8e^{-1} \approx 1.943$ **49.** $S_{yx}(x, y) = \frac{3x^2}{y} + 8ye^x$

51. $S_{yy}(x, y) = -\frac{x^3}{y^2} + 8e^x$ **53.** $3 + 8e^{-1} \approx 5.943$ **55.** $1 + 8e^{-1} \approx 3.943$

57. \$3,000; daily sales are \$3,000 when the temperature is 60° and the rainfall is 2 in.

59. −2,500 \$/in.; daily sales decrease at a rate of \$2,500 per inch of rain when the temperature is 90° and rainfall is 1 in.

61. −50 \$/in. per °F; S_r decreases at a rate of 50 \$/in. per degree of temperature

65. $f_{xx}(x, y) = 2y^2 + 6x$; $f_{xy}(x, y) = 4xy = f_{yx}(x, y)$; $f_{yy}(x, y) = 2x^2$

67. $f_{xx}(x, y) = -2y/x^3$; $f_{xy}(x, y) = (-1/y^2) + (1/x^2) = f_{yx}(x, y)$; $f_{yy}(x, y) = 2x/y^3$

69. $f_{xx}(x, y) = (2y + xy^2)e^{xy}$; $f_{xy}(x, y) = (2x + x^2y)e^{xy} = f_{yx}(x, y)$; $f_{yy}(x, y) = x^3e^{xy}$

71. $x = 2$ and $y = 4$

73. $x = 1.200$ and $y = -0.695$

75. (A) $-\frac{13}{3}$ (B) The function $f(0, y)$, for example, has values less than $-\frac{3}{13}$.
77. (A) $c = 1.145$ (B) $f_x(c, 2) = 0$; $f_y(c, 2) = 92.021$
79. $f_{xx}(x, y) + f_{yy}(x, y) = (2y^2 - 2x^2)/(x^2 + y^2)^2 + (2x^2 - 2y^2)/(x^2 + y^2)^2 = 0$
81. (A) $2x$ (B) $4y$
83. $P_x(1{,}200, 1{,}800) = 24$; profit will increase approx. \$24 per unit increase in production of type A calculators at the (1,200, 1,800) output level; $P_y(1{,}200, 1{,}800) = -48$; profit will decrease approx. \$48 per unit increase in production of type B calculators at the (1,200, 1,800) output level
85. $\partial x/\partial p = -5$: a \$1 increase in the price of brand A will decrease the demand for brand A by 5 lb at any price level (p, q); $\partial y/\partial p = 2$: a \$1 increase in the price of brand A will increase the demand for brand B by 2 lb at any price level (p, q)
87. (A) $f_x(x, y) = 7.5x^{-0.25}y^{0.25}$; $f_y(x, y) = 2.5x^{0.75}y^{-0.75}$
(B) Marginal productivity of labor $= f_x(600, 100) \approx 4.79$; Marginal productivity of capital $= f_y(600, 100) \approx 9.58$
(C) Capital
89. Competitive **91.** Complementary
93. (A) $f_w(w, h) = 6.65w^{-0.575}h^{0.725}$; $f_h(w, h) = 11.34w^{0.425}h^{-0.275}$
(B) $f_w(65, 57) = 11.31$: for a 65-lb child 57 in. tall, the rate of change in surface area is 11.31 in.^{2} for each pound gained in weight (height is held fixed); $f_h(65, 57) = 21.99$: for a child 57 in. tall, the rate of change in surface area is 21.99 in.^{2} for each inch gained in height (weight is held fixed)
95. $C_W(6, 8) = 12.5$: index increases approx. 12.5 units for a 1-in. increase in width of head (length held fixed) when $W = 6$ and $L = 8$; $C_L(6, 8) = -9.38$: index decreases approx. 9.38 units for a 1-in. increase in length (width held fixed) when $W = 6$ and $L = 8$.

Exercises 6-3

1. $f_x(x, y) = 4$; $f_y(x, y) = 5$; the functions $f_x(x, y)$ and $f_y(x, y)$ never have the value 0.
3. $f_x(x, y) = -1.2 + 4x^3$; $f_y(x, y) = 6.8 + 0.6y^2$; the function $f_y(x, y)$ never has the value 0.
5. $f(-2, 0) = 10$ is a local maximum. **7.** $f(-1, 3) = 4$ is a local minimum.
9. f has a saddle point at $(3, -2)$. **11.** $f(3, 2) = 33$ is a local maximum.
13. $f(2, 2) = 8$ is a local minimum. **15.** f has a saddle point at $(0, 0)$.
17. f has a saddle point at $(0, 0)$; $f(1, 1) = -1$ is a local minimum.
19. f has a saddle point at $(0, 0)$; $f(3, 18) = -162$ and $f(-3, -18) = -162$ are local minima.
21. The test fails at $(0, 0)$; f has saddle points at $(2, 2)$ and $(2, -2)$. **23.** f has a saddle point at $(0.614, -1.105)$.
25. $f(x, y)$ is nonnegative and equals 0 when $x = 0$, so f has a local minimum at each point of the y axis.
27. (B) Local minimum **29.** 2,000 type A and 4,000 type B; Max $P = P(2, 4) = \$15$ million
31. (A) When $p = \$110$ and $q = \$120$, $x = 80$ and $y = 40$; when $p = \$110$ and $q = \$110$, $x = 40$ and $y = 70$
(B) A maximum weekly profit of \$4,800 is realized for $p = \$100$ and $q = \$120$.
33. $P(x, y) = P(4, 2)$
35. 8 in. by 4 in. by 2 in.
37. 20 in. by 20 in. by 40 in.

Exercises 6-4

1. Max $f(x, y) = f(3, 3) = 18$ **3.** Min $f(x, y) = f(3, 4) = 25$
5. $F_x = -3 + 2\lambda = 0$ and $F_y = 4 + 5\lambda = 0$ have no simultaneous solution.
7. Max $f(x, y) = f(3, 3) = f(-3, -3) = 18$; Min $f(x, y) = f(3, -3) = f(-3, 3) = -18$
9. Maximum product is 25 when each number is 5. **11.** Min $f(x, y, z) = f(-4, 2, -6) = 56$
13. Max $f(x, y, z) = f(2, 2, 2) = 6$; Min $f(x, y, z) = f(-2, -2, -2) = -6$
15. Max $f(x, y) = f(0.217, 0.885) = 1.055$ **17.** $F_x = e^x + \lambda = 0$ and $F_y = 3e^y - 2\lambda = 0$ have no simultaneous solution.
19. Maximize $f(x, 5)$, a function of just one independent variable.
21. (A) Max $f(x, y) = f(0.707, 0.5) = f(-0.707, 0.5) = 0.47$
23. 60 of model A and 30 of model B will yield a minimum cost of \$32,400 per week.
25. (A) 8,000 units of labor and 1,000 units of capital; Max $N(x, y) = N(8{,}000, 1{,}000) \approx 263{,}902$ units
(B) Marginal productivity of money ≈ 0.6598; increase in production $\approx 32{,}990$ units
27. 8 in. by 8 in. by $\frac{8}{3}$ in. **29.** $x = 50$ ft and $y = 200$ ft; maximum area is 10,000 ft^2

Exercises 6-5

1. $y = 0.7x + 1$ **3.** $y = -2.5x + 10.5$ 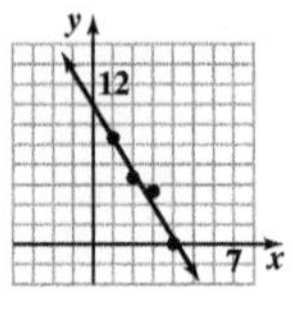 **5.** $y = x + 2$ 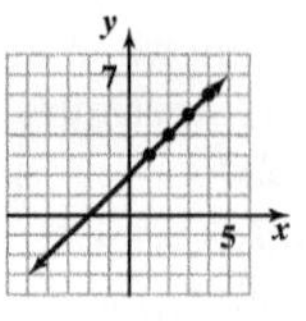
7. $y = -1.5x + 4.5$; $y = 0.75$ when $x = 2.5$
9. $y = 2.12x + 10.8$; $y = 63.8$ when $x = 25$
11. $y = -1.2x + 12.6$; $y = 10.2$ when $x = 2$
13. $y = -1.53x + 26.67$; $y = 14.4$ when $x = 8$

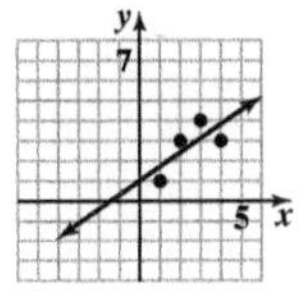

15. $y = 0.75x^2 - 3.45x + 4.75$

21. (A) $y = 1.52x - 0.16$; $y = 0.73x^2 - 1.39x + 1.30$ (B) The quadratic function

23. The normal equations form a system of 4 linear equations in the 4 variables a, b, c, and d, which can be solved by Gauss–Jordan elimination.

25. (A) $y = -65.4x + 3755.7$ (B) 2,709 crimes per 100,000 population

27. (A) $y = -0.48x + 4.38$ (B) \$6.56 per bottle

29. (A) $y = 0.0237x + 18.92$ (B) 19.87 ft

31. (A) $y = 0.0121x + 56.35$ (B) 58.77°F

Exercises 6-6

1. (A) $3x^2y^4 + C(x)$ (B) $3x^2$ **3.** (A) $2x^2 + 6xy + 5x + E(y)$ (B) $35 + 30y$

5. (A) $\sqrt{y + x^2} + E(y)$ (B) $\sqrt{y + 4} - \sqrt{y}$ **7.** (A) $\frac{\ln x \ln y}{x} + C(x)$ (B) $\frac{2 \ln x}{x}$

9. 9 **11.** 330 **13.** $(56 - 20\sqrt{5})/3$ **15.** 1 **17.** 16 **19.** 49 **21.** $\frac{1}{8}\int_1^5\int_{-1}^1 (x + y)^2\,dy\,dx = \frac{32}{3}$

23. $\frac{1}{15}\int_1^4\int_2^7 (x/y)\,dy\,dx = \frac{1}{2}\ln\frac{7}{2} \approx 0.6264$ **25.** $\frac{4}{3}$ cubic units **27.** $\frac{32}{3}$ cubic units **29.** $\int_0^1\int_1^2 xe^{xy}\,dy\,dx = \frac{1}{2} + \frac{1}{2}e^2 - e$

31. $\int_0^1\int_{-1}^1 \frac{2y + 3xy^2}{1 + x^2}\,dy\,dx = \ln 2$

35. (A) $\frac{1}{3} + \frac{1}{4}e^{-2} - \frac{1}{4}e^2$

(B)

(C) Points to the right of the graph in part (B) are greater than 0; points to the left of the graph are less than 0.

37. $\frac{1}{0.4}\int_{0.6}^{0.8}\int_5^7 \frac{y}{1 - x}\,dy\,dx = 30 \ln 2 \approx \$20.8$ billion

39. $\frac{1}{10}\int_{10}^{20}\int_1^2 x^{0.75}y^{0.25}\,dy\,dx = \frac{8}{175}(2^{1.25} - 1)(20^{1.75} - 10^{1.75}) \approx 8.375$ or 8,375 units

41. $\frac{1}{192}\int_{-8}^{8}\int_{-6}^{6}[10 - \frac{1}{10}(x^2 + y^2)]\,dy\,dx = \frac{20}{3}$ insects/ft^2

43. $\frac{1}{8}\int_{-2}^{2}\int_{-1}^{1}[100 - 15(x^2 + y^2)]\,dy\,dx = 75$ ppm

45. $\frac{1}{10,000}\int_{2,000}^{3,000}\int_{50}^{60} 0.000\,013\,3xy^2\,dy\,dx \approx 100.86$ ft

47. $\frac{1}{16}\int_8^{16}\int_{10}^{12} 100\frac{x}{y}\,dy\,dx = 600 \ln 1.2 \approx 109.4$

Exercises 6-7

1. $R = \{(x, y) | 0 \le y \le 4 - x^2, 0 \le x \le 2\}$
$R = \{(x, y) | 0 \le x \le \sqrt{4 - y}, 0 \le y \le 4\}$

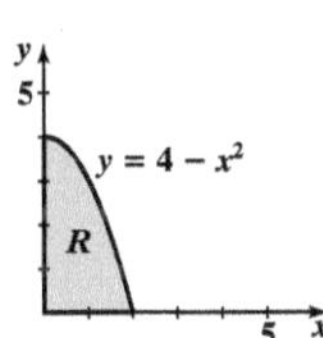

3. R is a regular x region:
$R = \{(x, y) | x^3 \le y \le 12 - 2x, 0 \le x \le 2\}$

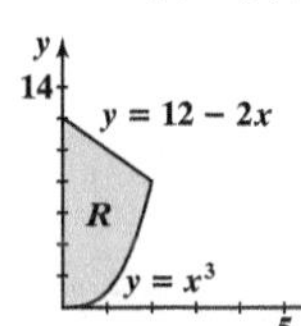

5. R is a regular y region:
$R = \{(x, y) | \frac{1}{2}y^2 \le x \le y + 4, -2 \le y \le 4\}$

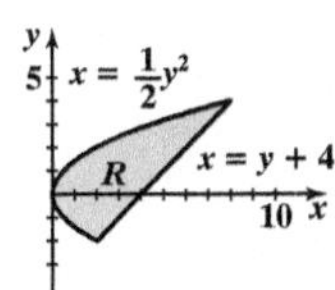

7. $\frac{1}{2}$ **9.** $\frac{39}{70}$

11. R consists of the points on or inside the rectangle with corners $(\pm 2, \pm 3)$; both

13. R is the arch-shaped region consisting of the points on or inside the rectangle with corners $(\pm 2, 0)$ and $(\pm 2, 2)$ that are not inside the circle of radius 1 centered at the origin; regular x region

15. $\frac{56}{3}$ **17.** $-\frac{3}{4}$ **19.** $\frac{1}{2}e^4 - \frac{5}{2}$

21. $R = \{(x, y) | 0 \le y \le x + 1, \ 0 \le x \le 1\}$
$\int_0^1\int_0^{x+1} \sqrt{1 + x + y}\,dy\,dx = (68 - 24\sqrt{2})/15$

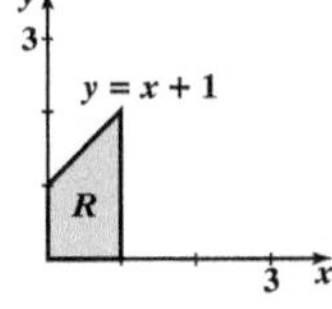

23. $R = \{(x, y) | 0 \le y \le 4x - x^2, 0 \le x \le 4\}$
$\int_0^4\int_0^{4x - x^2} \sqrt{y + x^2}\,dy\,dx = \frac{128}{5}$

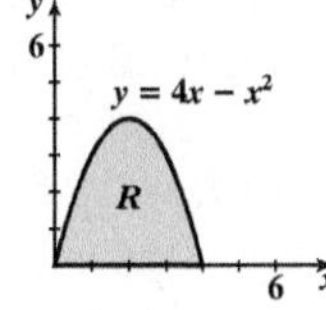

25. $R = \{(x, y) | 1 - \sqrt{x} \le y \le 1 + \sqrt{x}, 0 \le x \le 4\}$
$\int_0^4 \int_{1-\sqrt{x}}^{1+\sqrt{x}} x(y-1)^2 \, dy \, dx = \frac{512}{21}$

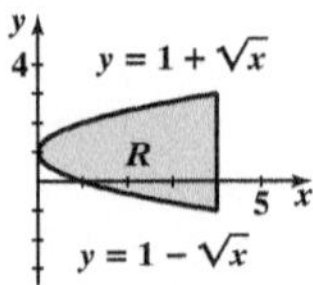

27. $\int_0^3 \int_0^{3-y} (x + 2y) \, dx \, dy = \frac{27}{2}$

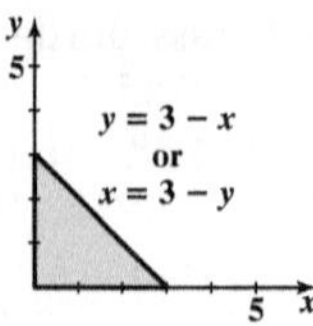

29. $\int_0^1 \int_0^{\sqrt{1-y}} x\sqrt{y} \, dx \, dy = \frac{2}{15}$

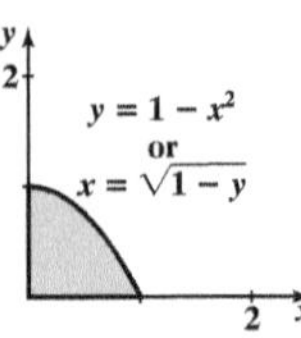

31. $\int_0^1 \int_{4y^2}^{4y} x \, dx \, dy = \frac{16}{15}$

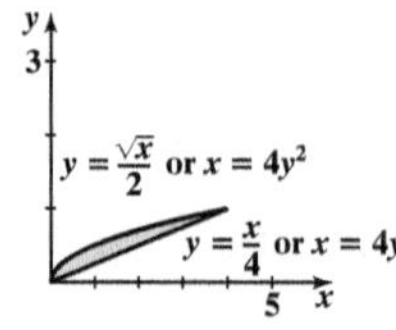

33. $\int_0^4 \int_0^{4-x} (4 - x - y) \, dy \, dx = \frac{32}{3}$

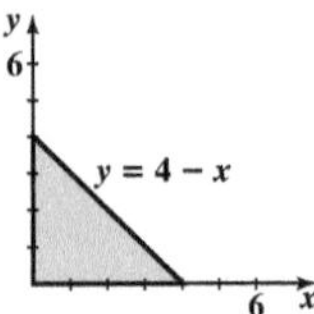

35. $\int_0^1 \int_0^{1-x^2} 4 \, dy \, dx = \frac{8}{3}$

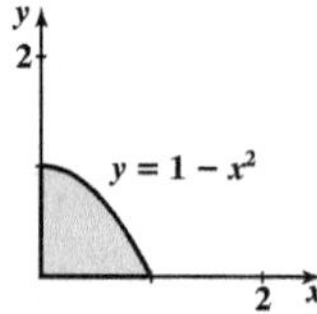

37. $\int_0^4 \int_0^{\sqrt{y}} \frac{4x}{1 + y^2} \, dx \, dy = \ln 17$

39. $\int_0^1 \int_0^{\sqrt{x}} 4ye^{x^2} \, dy \, dx = e - 1$

41. $R = \{(x, y) | x^2 \le y \le 1 + \sqrt{x}, 0 \le x \le 1.49\}$

$\int_0^{1.49} \int_{x^2}^{1+\sqrt{x}} x \, dy \, dx \approx 0.96$

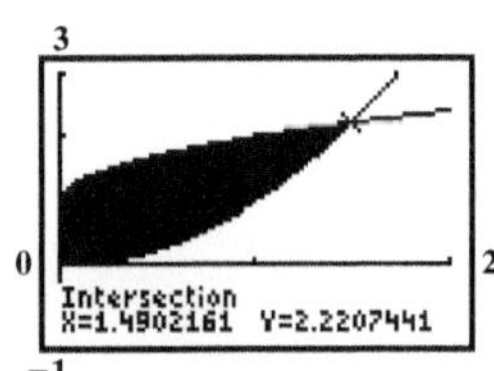

43. $R = \{(x, y) | y^3 \le x \le 1 - y, 0 \le y \le 0.68\}$

$\int_0^{0.68} \int_{y^3}^{1-y} 24xy \, dx \, dy \approx 0.83$

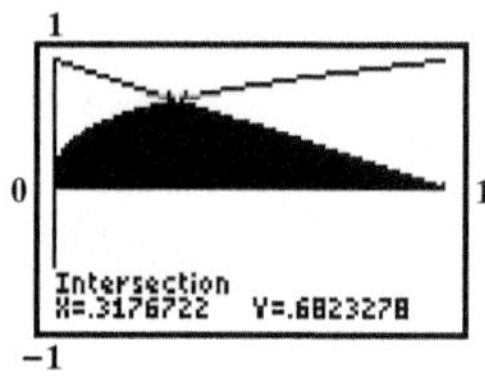

45. $R = \{(x, y) | e^{-x} \le y \le 3 - x, \quad -1.51 \le x \le 2.95\}$ Regular x region

$R = \{(x, y) | -\ln y \le x \le 3 - y, \quad 0.05 \le y \le 4.51\}$ Regular y region

$\int_{-1.51}^{2.95} \int_{e^{-x}}^{3-x} 4y \, dy \, dx = \int_{0.05}^{4.51} \int_{-\ln y}^{3-y} 4y \, dx \, dy \approx 40.67$

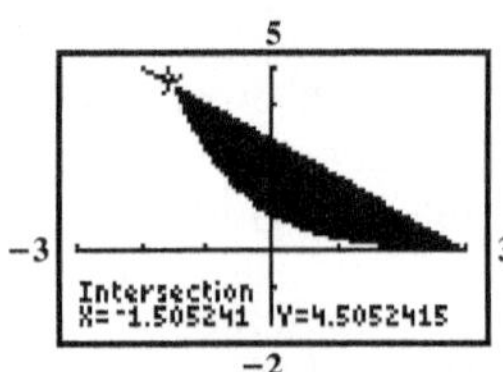

Chapter 6 Review Exercises

1. $f(5, 10) = 2{,}900$; $f_x(x, y) = 40$; $f_y(x, y) = 70$ *(8-1, 8-2)* **2.** $\partial^2 z/\partial x^2 = 6xy^2$; $\partial^2 z/\partial x \, \partial y = 6x^2 y$ *(8-2)*
3. $2xy^3 + 2y^2 + C(x)$ *(8-6)* **4.** $3x^2y^2 + 4xy + E(y)$ *(8-6)* **5.** 1 *(8-6)*
6. $f_x(x, y) = 5 + 6x + 3x^2$; $f_y(x, y) = -2$; the function $f_y(x, y)$ never has the value 0. *(8-3)*
7. $f(2, 3) = 7$; $f_y(x, y) = -2x + 2y + 3$; $f_y(2, 3) = 5$ *(8-1, 8-2)* **8.** $(-8)(-6) - (4)^2 = 32$ *(8-2)*
9. $(1, 3, -\frac{1}{2}), (-1, -3, \frac{1}{2})$ *(8-4)* **10.** $y = -1.5x + 15.5$; $y = 0.5$ when $x = 10$ *(8-5)* **11.** 18 *(8-6)* **12.** $\frac{8}{5}$ *(8-7)*
13. $f_x(x, y) = 2xe^{x^2+2y}$; $f_y(x, y) = 2e^{x^2+2y}$; $f_{xy}(x, y) = 4xe^{x^2+2y}$ *(8-2)*
14. $f_x(x, y) = 10x(x^2 + y^2)^4$; $f_{xy}(x, y) = 80xy(x^2 + y^2)^3$ *(8-2)*
15. $f(2, 3) = -25$ is a local minimum; f has a saddle point at $(-2, 3)$. *(8-3)* **16.** Max $f(x, y) = f(6, 4) = 24$ *(8-4)*
17. Min $f(x, y, z) = f(2, 1, 2) = 9$ *(8-4)* **18.** $y = \frac{116}{165}x + \frac{100}{3}$ *(8-5)* **19.** $\frac{27}{5}$ *(8-6)* **20.** 4 cubic units *(8-6)*

21. 0 *(8-6)* **22.** (A) 12.56 (B) No *(8-6)*
23. $F_x = 12x^2 + 3\lambda = 0$, $F_y = -15y^2 + 2\lambda = 0$, and $F_\lambda = 3x + 2y - 7 = 0$ have no simultaneous solution. *(8-4)* **24.** 1 *(8-7)*
25. (A) $P_x(1, 3) = 8$; profit will increase \$8,000 for a 100-unit increase in product *A* if the production of product *B* is held fixed at an output level of (1, 3).
(B) For 200 units of *A* and 300 units of *B*, $P(2, 3) = \$100$ thousand is a local maximum. *(8-2, 8-3)*
26. 8 in. by 6 in. by 2 in. *(8-3)*
27. $y = 0.63x + 1.33$; profit in sixth year is \$5.11 million *(8-4)*
28. (A) Marginal productivity of labor ≈ 8.37; marginal productivity of capital ≈ 1.67; management should encourage increased use of labor.
(B) 80 units of labor and 40 units of capital; Max $N(x, y) = N(80, 40) \approx 696$ units; marginal productivity of money ≈ 0.0696; increase in production ≈ 139 units
(C) $\frac{1}{1{,}000}\int_{50}^{100}\int_{20}^{40} 10x^{0.8}y^{0.2}\,dy\,dx = \frac{(40^{1.2} - 20^{1.2})(100^{1.8} - 50^{1.8})}{216} = 621$ items *(8-4)*
29. $T_x(70, 17) = -0.924$ min/ft increase in depth when $V = 70$ ft³ and $x = 17$ ft *(8-2)*
30. $\frac{1}{16}\int_{-2}^{2}\int_{-2}^{2}[100 - 24(x^2 + y^2)]\,dy\,dx = 36$ ppm *(8-6)* **31.** 50,000 *(8-1)*
32. $y = \frac{1}{2}x + 48$; $y = 68$ when $x = 40$ *(8-5)*
33. (A) $y = 0.4933x + 25.20$ (B) 84.40 people/mi² (C) 89.30 people/mi²; 97.70 people/mi² *(8-5)*
34. (A) $y = 1.069x + 0.522$ (B) 64.68 yr (C) 64.78 yr; 64.80 yr *(8-5)*

Chapter 7

Exercises 7-1

1.

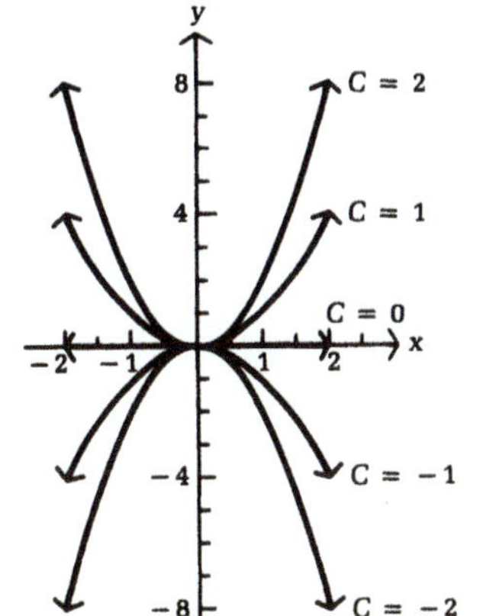

3.

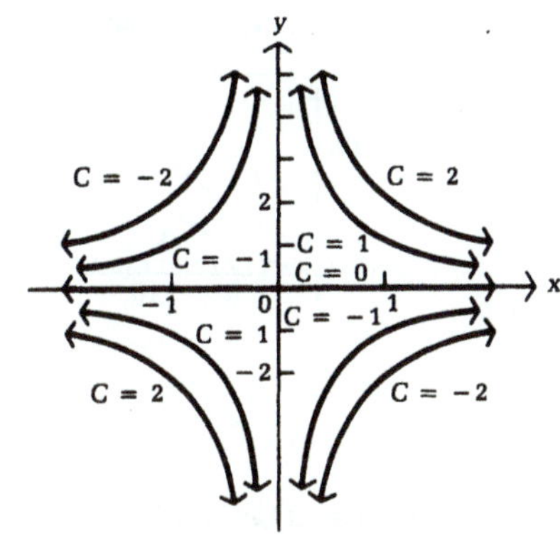

5. $y = 7e^x - 5x - 5$ **7.** $y = x^s - 2e^{2x}$ **9.** $y = x + \frac{2}{x}$ **15.** $y = \sqrt{9 - x^2}$ **17.** $y = 2 - e^x$

19. (A) $y = 2 - e^{-x}$
(B) $y = 2$
(C) $y = 2 + e^{-x}$

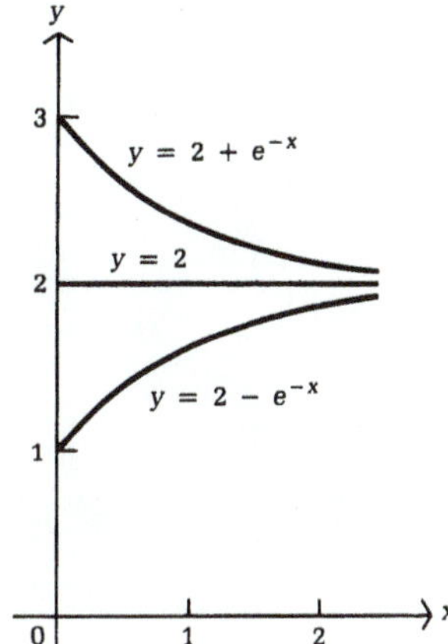

21. (A) $y = 2 - e^x$
(B) $y = 2$
(C) $y = 2 + e^x$

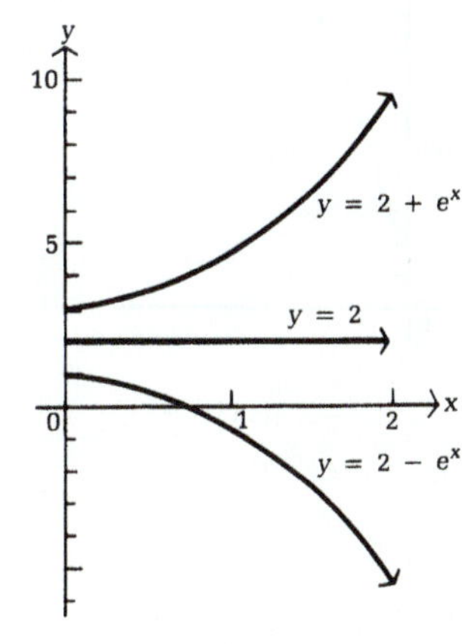

23. (A) $y = \frac{10}{1 + 9e^{-x}}$
(B) $y = 10$
(C) $y = \frac{10}{1 - 0.5e^{-x}}$

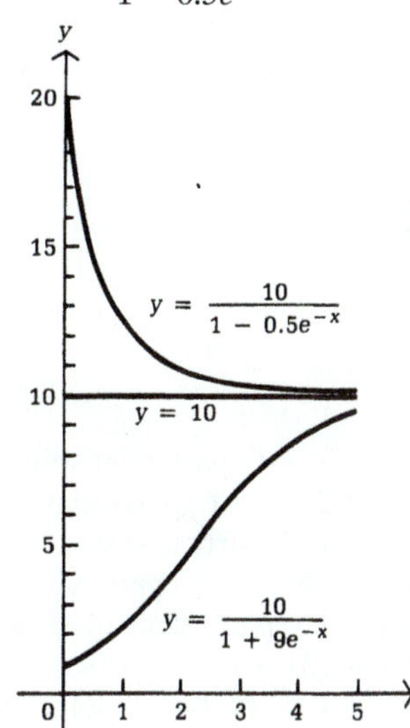

25. (A) $y = Cx^3 + 2$ for any C
(B) No particular solution exists
(C) $y = 2 - x^3$

27. (A)

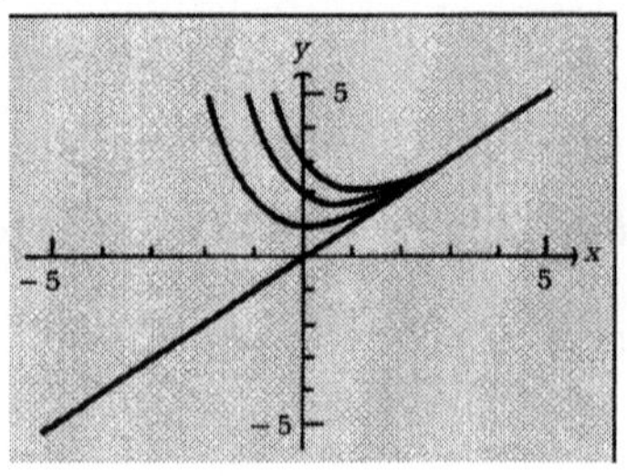

(B) Each graph decreases to a local minimum and then increases, approaching the line $y = x$ as x approaches ∞.

(C)

(D) Each graph is increasing for all x and approaches the line $y = x$ as x approaches ∞.

29. $\overline{p} = 5$
$p(t) = 5 - 4e^{-0.1t}$
$p(t) = 5 + 5\,e^{-0.1t}$

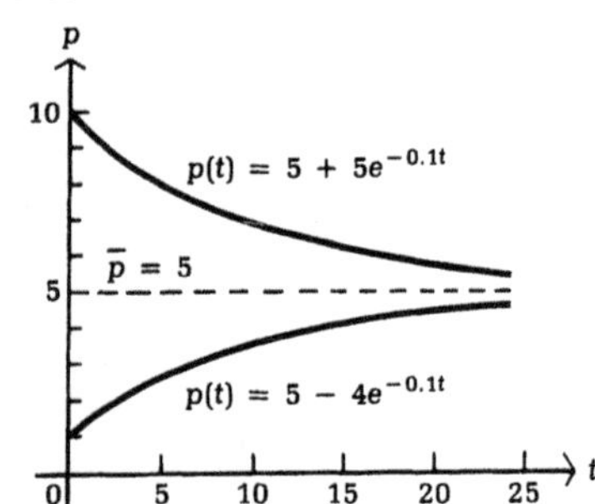

31. $A(t) = 2{,}500e^{0.08t} - 2{,}500$
$A(t) = 3{,}500e^{0.08t} - 2{,}500$

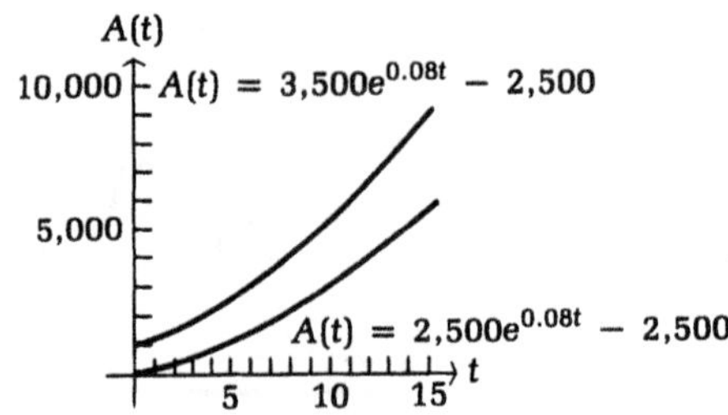

33. $\overline{N} = 200$
$N(t) = 200 - 150e^{-0.5t}$
$N(t) = 200 + 100e^{-0.5t}$

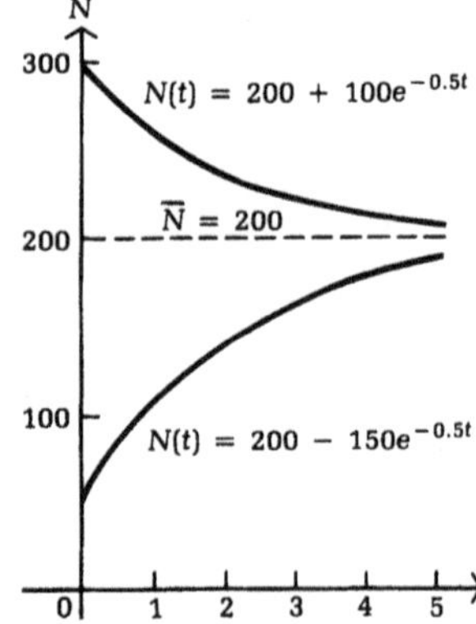

35. $N(t) = 200e^{2-2e^{-0.5t}}$
$\overline{N} = 200e^2 \approx 1{,}478$

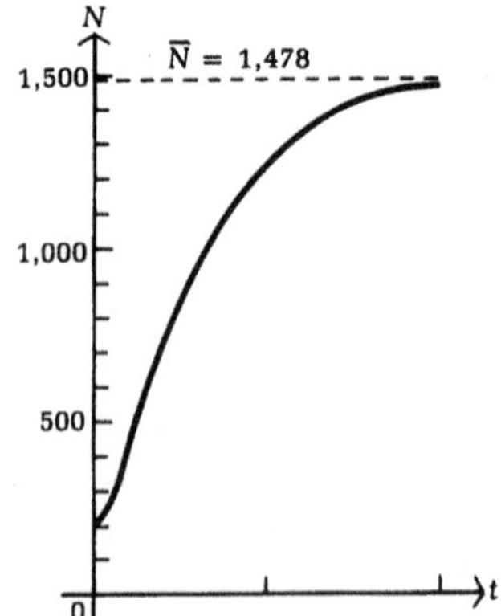

Exercises 7-2

1. General solution: $y = x + C$; particular solution: $y = x + 2$
3. General solution: $y = 2x^{1/2} + C$; particular solution: $y = 2x^{1/2} - 4$
5. General solution: $y = Ce^x$; particular solution: $y = 10e^x$
7. General solution: $y = 25 - Ce^{-x}$; particular solution: $y = 25 - 20e^{-x}$
9. General solution: $y = Cx$; particular solution: $y = 5x$
11. General solution: $y = (3x + C)^{1/3}$; particular solution: $y = (3x + 24)^{1/3}$
13. General solution: $y = Ce^{e^x}$; particular solution: $y = 3e^{e^x}$

15. General solution: $y = \ln(e^x + C)$; particular solution: $y = \ln(e^x + 1)$
17. General solution: $y = Ce^{x^2/2} - 1$; particular solution: $y = 3e^{x^2/2-1}$

19. General solution: $y = 2 - 1/(e^x + C)$; particular solution: $y = 2 - e^{-x}$ **21.** $y + \frac{1}{3}y^3 = x + \frac{1}{3}x^3 + C$

23. $\frac{1}{2}\ln|1 + y^2| = \ln|x| + (x^2/2) + C$ or $\ln(1 + y^2) = \ln(x^2) + x^2 + C$ **25.** $\ln(1 + e^y) = \frac{1}{2}x^2 + C$ **27.** $y = \sqrt{1 + (\ln x)^2}$

29. $y = \frac{1}{4}[x + \ln(x^2 + 3]^2$ **31.** $y = \sqrt{\ln(x^2 + e)}$ **33.** $5{,}000e^{1.2} \approx \$16{,}600$ **35.** $\dfrac{7 \ln 0.5}{\ln 0.8} \approx 22$ days **37.** $\dfrac{30 \ln 0.1}{\ln 0.5} \approx 100$ days

39. $\dfrac{4 \ln 0.2}{\ln 0.8} \approx 29$ yr **41.** $\dfrac{2 \ln \frac{5}{12}}{\ln \frac{5}{6}} \approx 9.6$ min **43.** $25 + 300e^{4 \ln(2/3)} = 84.26°\text{F}$ **45.** (A) $100e^{5 \ln 1.4} = 538$ bacteria (B) $\dfrac{\ln 10}{\ln 1.4} \approx 6.8$ hr

47. (A) $\dfrac{50{,}000}{1 + 499e^{2 \ln(99/499)}} \approx 2{,}422$ people (B) $\dfrac{10 \ln(1/499)}{\ln(99/499)} \approx 38.4$ days **49.** $I = As^k$

51. (A) $\dfrac{1{,}000}{1 + 199e^{7 \ln(99/199)}} \approx 400$ people (B) $\dfrac{\ln(3/3{,}383)}{\ln(99/199)} \approx 10$ days

53. 38 days **55.** $k = -0.2$; $M = \$11$ million **57.** 47 days

Exercises 7-3

1. $I(x) = e^{2x}$; $y = 2 + Ce^{-2x}$; $y = 2 - e^{-2x}$ **3.** $I(x) = e^x$; $y = -e^{-2x} + Ce^{-x}$; $y = -e^{-2x} + 4e^{-x}$
5. $I(x) = e^{-x}$; $y = 2xe^x + Ce^x$; $y = 2xe^x - 4e^x$ **7.** $I(x) = e^x$; $y = 3x^3e^{-x} + Ce^{-x}$; $y = 3x^3e^{-x} + 2e^{-x}$
9. $I(x) = x$; $y = x + (C/x)$; $y = x$ **11.** $I(x) = x^2$; $y = 2x^3 + (C/x^2)$; $y = 2x^3 - (32/x^2)$
13. $I(x) = e^{x^2/2}$; $y = 5 + Ce^{-x^2/2}$ **15.** $I(x) = e^{-2x}$; $y = -2x - 1 + Ce^{2x}$ **17.** $I(x) = x$; $y = e^x - (e^x/x) + (C/x)$
19. $I(x) = x$; $y = \frac{1}{2}x \ln x - \frac{1}{4}x + (C/x)$ **21.** $y = 1 + (C/x)$ **23.** $y = C(1 + x^2) - 1$ **25.** $y = Ce^{x^2} - 1$ **27.** $y = Ce^{kt}$

29. $A = 100{,}000 - 80{,}000e^{0.04t}$; $(\ln 1.25)/0.04 \approx 5.579$ yr; \$22,316 **31.** $30{,}000(1 - e^{-0.5}) \approx \$11{,}804$
33. $A = 32{,}000e^{0.08t} - 25{,}000$; \$22,738.39 **35.** $p(t) = 20 + 10e^{-3t}$; $\bar{p} = 20$ **37.** 622lb; $\frac{622}{250} \approx 2.5$ lb/gal

39. $600 - 200e^{-0.5} \approx 479$ lb; $3 - e^{-0.5} \approx 2.4$ lb/gal
41. $120 + 40e^{-0.15} \approx 154$ lb; $-(\ln \frac{3}{4})/0.005 \approx 58$ days; weight will approach 120 lb if the diet is maintained for a long period of time

43. $17.5\left(\dfrac{125 - 130e^{-0.15}}{1 - e^{-0.15}}\right) \approx 1{,}647$ cal

45. Student A: $0.9 - 0.8e^{-4.8} \approx 0.8934$ or 89.34%; student B: $0.7 - 0.3e^{-4.8} \approx 0.6975$ or 69.75%
47. 10.6 yr **49.** 6.2 hr

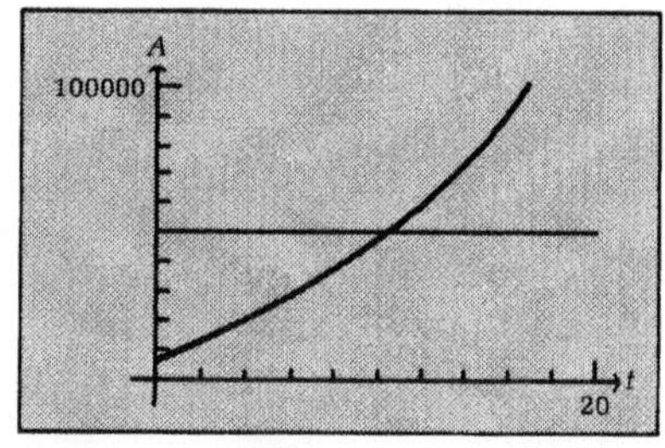

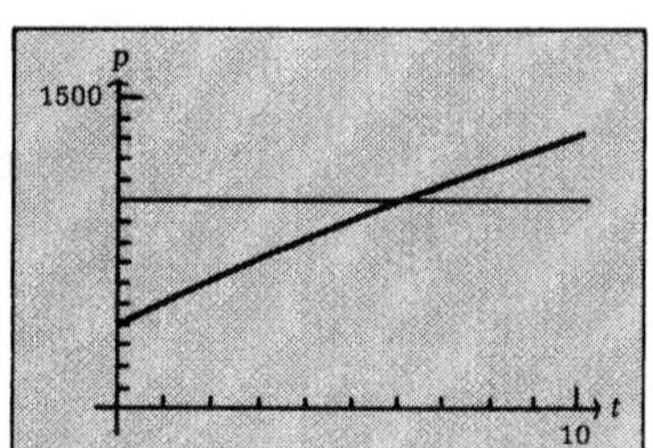

Exercises 7-4

1. $y = C_1e^{-x} + C_2e^{-2x}$ **3.** $y = C_1e^{-5x} + C_2e^{3x}$ **5.** $y = C_1 + C_2e^{-6x}$ **7.** $y = C_1e^{2x} + C_2xe^{2x}$ **9.** $y = 2e^x + e^{-x}$
11. $y = \frac{3}{2}e^{x/3} - \frac{1}{2}e^{3x}$ **13.** $y = 2e^{-x} + 6xe^{-x}$ **15.** $y = C_1e^{4x} + C_2e^{-x} - 3$ **17.** $y = 2e^x + e^{-2x} - 3$

19. $y = x + C_1e^{-0.5x} + C_2$ **21.** $y = -\frac{3}{2}x + C_1e^{2x} + C_2$ **23.** $y = e^{4x}$ **25.** $y = e^{2x} - e^{0.5x}$ **27.** $y = C_1 \cos x + C_2 \sin x$

29. $y = e^{2x}(C_1 \cos 3x + C_2 \sin 3x)$ **31.** $p(t) = -50e^{-0.1t} + 100e^{-0.2t} + 25$; $\bar{p} = 25$ **33.** $y = 2 - e^{-t}$; 2

Exercises 7-5

1. $x = C_1e^t + C_2e^{-2t}, y = 2C_1e^t - C_2e^{-2t}$ **3.** $x = C_1e^t + C_2e^{-t}, y = C_1e^t + 3C_2e^{-t}; x = 2e^t - e^{-t}, y = 2e^t - 3e^{-t}$
5. $x = C_1e^{3t} + C_2, y = C_1e^{3t} - 2C_2; x = e^{3t} + 1, y = e^{3t} - 2$ **7.** $x = C_1e^t + C_2e^{-t}, y = 3C_1e^t + C_2e^{-t}$
9. $x = C_1e^{2t} + C_2e^{-t} + 2, y = C_1e^{2t} + 2C_2e^{-t} + 4$
11. $p = 5e^{-2t} + 10e^{-3t} + 85, q = 10e^{-2t} + 10e^{-3t} + 80$; p decreases to the limiting value of 85; q decreases to the limiting value of 80
13. $x = 25e^{-4t} + 25e^{-6t}; 2y = 25e^{-4t} - 25e^{-6t}$; 30.5 units in compartment 1 and 3.0 units in compartment 2 after 6 min; 4.6 units in compartment 1 and 2.1 units in compartment 2 after 30 min; 0.5 unit in compartment 1 and 0.4 unit in compartment 2 after 1 hr

15.

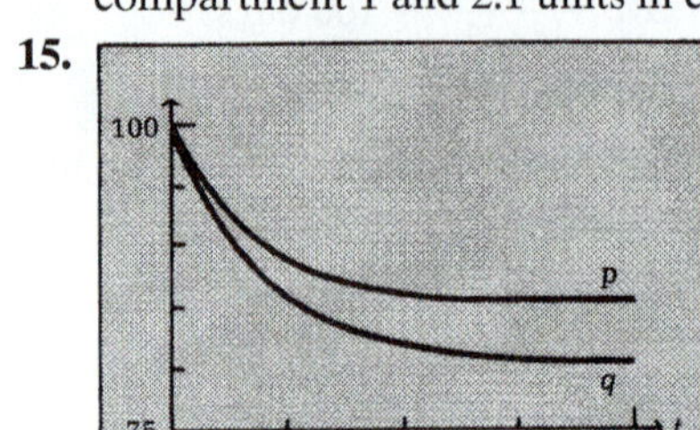

17.

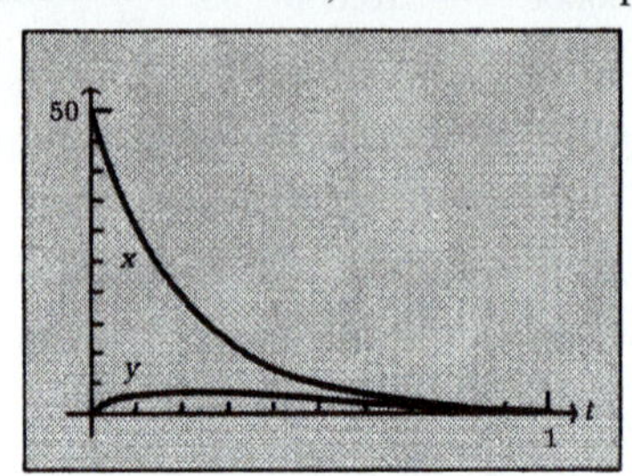

Chapter 7 Review Exercises

1.

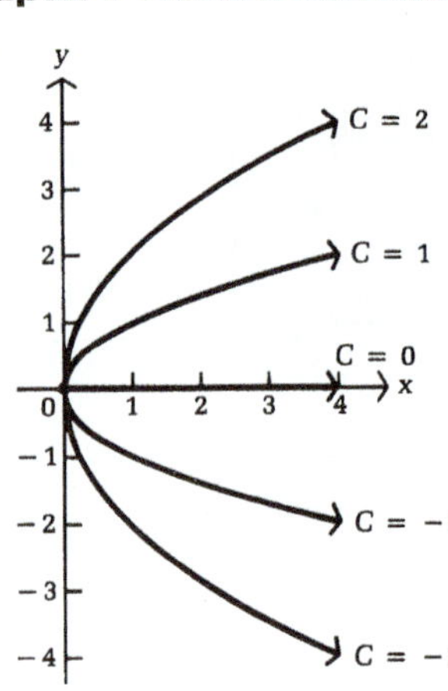

2.

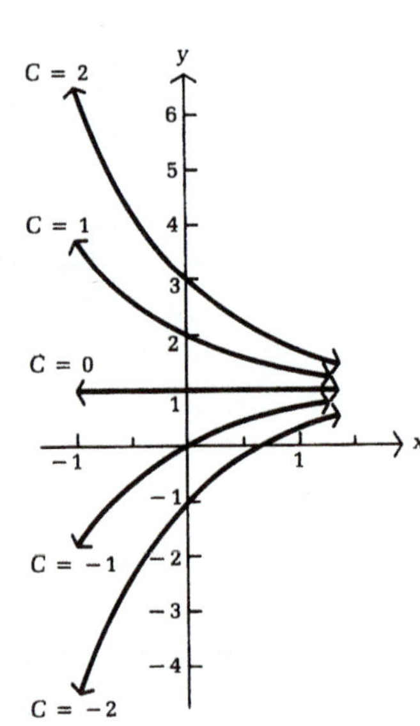

3. $y = C/x^4$
4. $y = \frac{1}{6}x^2 + (C/x^4)$
5. $y = -1/(x^3 + C)$
6. $y = e^x + Ce^{2x}$
7. $y = \frac{1}{2}x^7 + Cx^5$
8. $y = C(2 + x) - 3$
9. $y = C_1e^{7x} + C_2e^{-3x}$
10. $y = C_1e^{-6x} + C_2xe^{-6x}$
11. $y = 10 - 10e^{-x}$
12. $y = x - 1 + e^{-x}$
13. $y = e^2e^{-2e^{-x}}$

14. $y = (x^2 + 4)/(x + 4)$ **15.** $y = \sqrt{x^2 + 16} - 4$ **16.** $y = \frac{1}{3}x \ln x - \frac{1}{9}x + \frac{19}{9}(1/x^2)$ **17.** $y = \sqrt{1 + 2x^2}$ **18.** $y = (2x + 1)e^{-x^2}$
19. $y = \frac{1}{2}e^{-4x} + \frac{1}{2}$ **20.** $y = e^{4x} + e^{-4x} - 1$ **21.** $x = 1 - t; y = 1 - 2t$ **22.** $x = e^{2t} - e^{-4t} + 2; y = 2e^{2t} + e^{-4t} - 1$
23. $500e^{(-0.5)\ln 20} \approx \$236.44$ **24.** $(\ln \frac{1}{4})/(\ln \frac{3}{4}) \approx 5$ yr **25.** $p = 50 + 25e^{-t}; \overline{p} = 50$
26. $A = 100{,}000 - 40{,}000e^{0.05t}$; $(\ln 2.5)/0.5 \approx 18.326$ yr; \$91,630
27. $p = -25e^{-3t} - 25e^{-4t} + 100, q = 25e^{-3t} + 50e^{-4t} + 75$; p increases to a limiting value of 100; q decreases to a limiting value of 75
28. $y = 100 + te^{-t} - 100e^{-t}$ **29.** 300 lb
30. $x = 50e^{0.01t} + 25e^{0.04t}, y = 100e^{0.01t} - 25e^{0.04t}$; the first species increases without bound; the second species dies out after approx. 46.2 yr
31. (A) $200 - 199e^{(5/2)\ln(190/199)} \approx 23$ people (B) $\dfrac{2\ln(100/199)}{\ln(190/199)} \approx 30$ days

Chapter 8

Exercises 8-1

1. $-6/x^4$ **3.** $-e^{-x}$ **5.** $-486/(1 + 3x)^4$ **7.** $\dfrac{-15}{16}(1 + x)^{-7/2}$
9. $1 - x + \frac{1}{2}x^2 - \frac{1}{6}x^3 + \frac{1}{24}x^4$ **11.** $1 + 3x + 3x^2 + x^3$ **13.** $2x - 2x^2 + \frac{8}{3}x^3$ **15.** $1 + \frac{1}{3}x - \frac{1}{9}x^2 + \frac{5}{81}x^3$
17. (A) $p_3(x) = -1; -0.562 < x < 0.562$ (B) $p_4(x) = x^4 - 1$; for all x
19. $1 + (x - 1) + \frac{1}{2}(x - 1)^2 + \frac{1}{6}(x - 1)^3 + \frac{1}{24}(x - 1)^4$
21. $1 + 3(x - 1) + 3(x - 1)^2 + (x - 1)^3$ **23.** $3(x - \frac{1}{3}) - \frac{9}{2}(x - \frac{1}{3})^2 + 9(x - \frac{1}{3})^3$
25. $1 - 2x + 2x^2 - \frac{4}{3}x^3$; 0.604 166 67 **27.** $4 + \frac{1}{8}x - \frac{1}{512}x^2$; 4.123 046 9

29. $1 + \frac{1}{2}(x-1) - \frac{1}{8}(x-1)^2 + \frac{1}{16}(x-1)^3 - \frac{5}{128}(x-1)^4$; 1.095 437 5 **31.** $n!\,(4-x)^{-(n+1)}$

33. $3^n e^{3x}$ **35.** $-(n-1)!(6-x)^{-n}$

37. $\frac{1}{4} + \frac{1}{4^2}x + \frac{1}{4^3}x^2 + \frac{1}{4^4}x^3 + \cdots + \frac{1}{4^{n+1}}x^n$ **39.** $1 + 3x + \frac{3^2}{2!}x^2 + \frac{3^3}{3!}x^3 + \cdots + \frac{3^n}{n!}x^n$

41. $-\frac{1}{6}x - \frac{1}{6^2 2}x^2 - \frac{1}{6^3 3}x^3 - \cdots - \frac{1}{6^n n}x^n$ **43.** $-1 - (x+1) - (x+1)^2 - (x+1)^3 - \cdots - (x+1)^n$

45. $(x-1) - \frac{1}{2}(x-1)^2 + \frac{1}{3}(x-1)^3 - \frac{1}{4}(x-1)^4 + \cdots + \frac{(-1)^{n+1}}{n}(x-1)^n$

47. $\frac{1}{e^2} - \frac{1}{e^2}(x-2) + \frac{1}{2!e^2}(x-2)^2 - \frac{1}{3!e^2}(x-2)^3 + \cdots + \frac{(-1)^n}{n!e^2}(x-2)^n$

49. $3\ln 3 + (1 + \ln 3)(x-3) + \frac{1}{3 \cdot 2}(x-3)^2 - \frac{1}{3^2 2 \cdot 3}(x-3)^3$
$+ \frac{1}{3^3 3 \cdot 4}(x-3)^4 - \frac{1}{3^4 4 \cdot 5}(x-3)^5 + \cdots + \frac{(-1)^n}{3^{n-1}(n-1)\cdot n}(x-3)^n$

51. (A) $p_3(x) = 5x^2 + 1$ (B) 2 **53.** $n = 0, 3, 6$

55. $p_1(x) = x, p_2(x) = x - \frac{1}{2}x^2, p_3(x) = x - \frac{1}{2}x^2 + \frac{1}{3}x^3$

x	$p_1(x)$	$p_2(x)$	$p_3(x)$	$f(x)$
−0.2	−0.2	−0.22	−0.222 667	−0.223 144
−0.1	−0.1	−0.105	−0.105 333	−0.105 361
0	0	0	0	0
0.1	0.1	0.095	0.095 333	0.095 31
0.2	0.2	0.18	0.182 667	0.182 322

x	$\lvert p_1(x) - f(x)\rvert$	$\lvert p_2(x) - f(x)\rvert$	$\lvert p_3(x) - f(x)\rvert$
−0.2	0.023 144	0.003 144	0.000 477
−0.1	0.005 361	0.000 361	0.000 028
0	0	0	0
0.1	0.004 69	0.000 31	0.000 023

57.

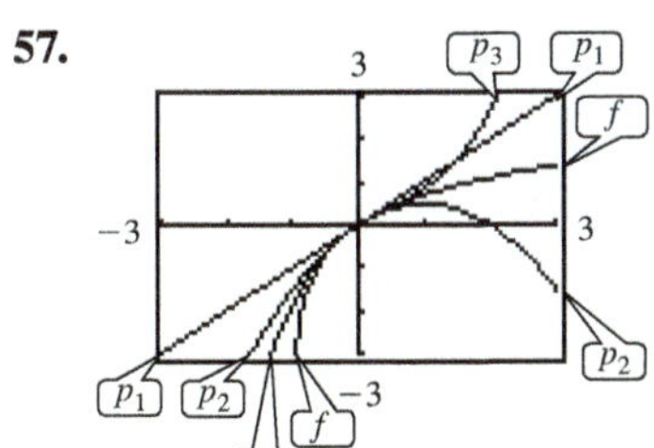

59. $-1.323 < x < 1.168$

61. $e^a\left[1 + (x-a) + \frac{1}{2!}(x-a)^2 + \frac{1}{3!}(x-a)^3 + \cdots + \frac{1}{n!}(x-a)^n\right] = e^a \sum_{k=0}^{n} \frac{1}{k!}(x-a)^k$

63. $\frac{1}{a} - \frac{1}{a^2}(x-a) + \frac{1}{a^3}(x-a)^2 - \frac{1}{a^4}(x-a)^3 + \cdots + \frac{(-1)^n}{a^{n+1}}(x-a)^n = \sum_{k=0}^{n} \frac{(-1)^k}{a^{k+1}}(x-a)^k$

65. $n \ge$ degree of f **67.** Yes **69.** $p_2(x) = 10 - 0.0005x^2$; \$9.85

71. $p_2(x) = 8 - \frac{3}{40}(x-60) - \frac{1}{1024}(x-60)^2$; \$7.12 **73.** $p_2(t) = 10 - 0.8t^2$; approx. \$18 million

75. $p_2(t) = -3 + \frac{3}{25}t^2$; 6.32 cm² **77.** $p_2(x) = 120 + 30(x-5) + \frac{3}{4}(x-5)^2$; 126.25 ppm **79.** $p_2(t) = 6 - 0.06t^2$; 27.5 wpm

81. **83.**

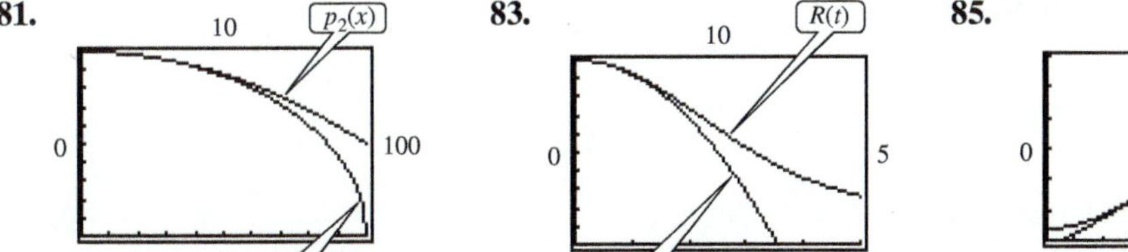

85.

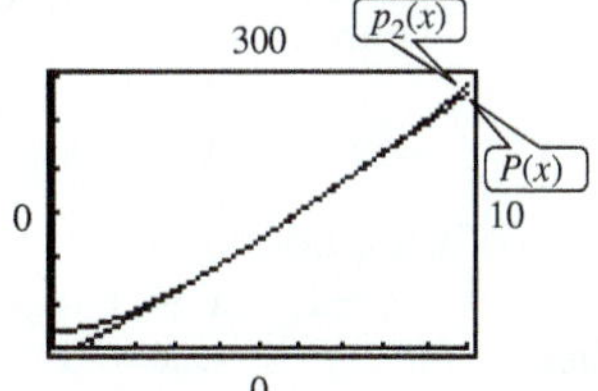

87.

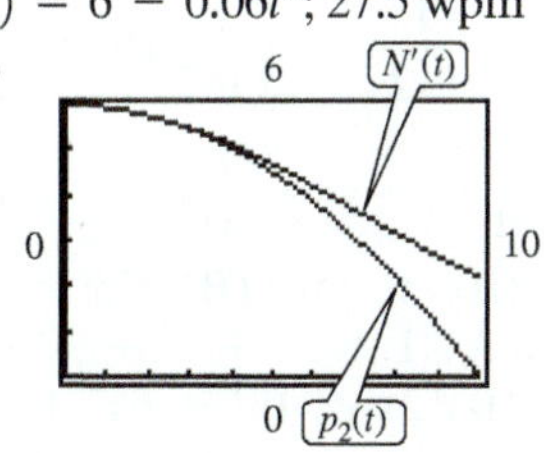

Exercises 8-2

1. $|x| < 1$ or $-1 < x < 1$ **3.** $|x| < \frac{1}{7}$ or $-\frac{1}{7} < x < \frac{1}{7}$ **5.** $|x| < 6$ or $-6 < x < 6$ **7.** $|x| < 1$ or $-1 < x < 1$

9. $|x| < 1$ or $-1 < x < 1$ **11.** $-\infty < x < \infty$ **13.** (A) 6 (B) Yes **15.** $|x + 3| < 1$ or $-4 < x < -2$

17. $-\infty < x < \infty$ **19.** $|x - 2| < 4$ or $-2 < x < 6$ **21.** $|x - 8| < 2$ or $6 < x < 10$

23. (A)

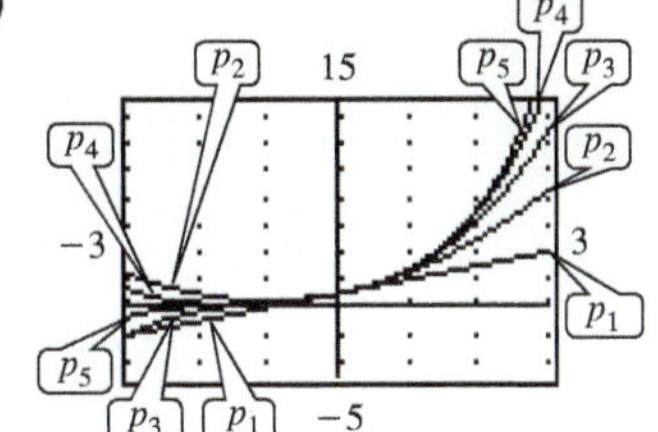

(B) $-\infty < x < \infty$

25. $p_n(x) = 1 + 4x + \frac{4^2}{2!}x^2 + \cdots + \frac{4^n}{n!}x^n; 1 + 4x + \frac{4^2}{2!}x^2 + \cdots + \frac{4^n}{n!}x^n + \cdots; -\infty < x < \infty$

27. $p_n(x) = 2x - \frac{2^2}{2}x^2 + \frac{2^3}{3}x^3 - \cdots + \frac{(-1)^{n-1}2^n}{n}x^n; 2x - \frac{2^2}{2}x^2 + \frac{2^3}{3}x^3 - \cdots + \frac{(-1)^{n-1}2^n}{n}x^n + \cdots; |x| < \frac{1}{2}$ or $-\frac{1}{2} < x < \frac{1}{2}$

29. $p_n(x) = 1 + \frac{1}{2}x + \frac{1}{2^2}x^2 + \cdots + \frac{1}{2^n}x^n; 1 + \frac{1}{2}x + \frac{1}{2^2}x^2 + \cdots + \frac{1}{2^n}x^n + \cdots; |x| < 2$ or $-2 < x < 2$

31. $p_n(x) = -(x-1) - \frac{1}{2}(x-1)^2 - \frac{1}{3}(x-1)^3 - \cdots - \frac{1}{n}(x-1)^n;$

$- (x-1) - \frac{1}{2}(x-1)^2 - \frac{1}{3}(x-1)^3 - \cdots - \frac{1}{n}(x-1)^n - \cdots; |x-1| < 1$ or $0 < x < 2$

33. $p_n(x) = -1 + (x-2) - (x-2)^2 + \cdots + (-1)^{n-1}(x-2)^n;$

$-1 + (x-2) - (x-2)^2 + \cdots + (-1)^{n-1}(x-2)^n + \cdots; |x-2| < 1$ or $1 < x < 3$

35. (A) $|x| < 4$ or $-4 < x < 4$
(B) The coefficients of the odd powers of x are 0.
(C) $|x| < 2$ or $-2 < x < 2$

37. (A) $\lim_{n\to\infty}\left|\frac{a_{n+1}}{a_n}\right|$ does not exist (B) $|x| < 1$ or $-1 < x < 1$

Exercises 8-3

1. $2 + 2x^2 + 2x^4 + \cdots + 2x^{2n} + \cdots; \ -1 < x < 1$
3. $2 + \frac{3}{2}x^2 + \frac{5}{6}x^3 + \cdots + \left[1 + \frac{(-1)^n}{n!}\right]x^n + \cdots; -1 < x < 1$
5. $x^3 - x^4 + x^5 - \cdots + (-1)^n x^{n+3} + \cdots; -1 < x < 1$
7. $-x^3 - \frac{1}{2}x^4 - \frac{1}{3}x^5 - \cdots - \frac{1}{n}x^{n+2} - \cdots; -1 < x < 1$
9. $1 + x^2 + \frac{1}{2}x^4 + \cdots + \frac{1}{n!}x^{2n} + \cdots; -\infty < x < \infty$
11. $3x - \frac{9}{2}x^2 + 9x^3 - \cdots + (-1)^{n-1}\frac{3^n}{n}x^n + \cdots; \ -\frac{1}{3} < x < \frac{1}{3}$
13. $\frac{1}{2} + \frac{1}{2^2}x + \frac{1}{2^3}x^2 + \cdots + \frac{1}{2^{n+1}}x^n + \cdots, -2 < x < 2$
15. $1 + 8x^3 + 8^2x^6 + \cdots + 8^nx^{3n} + \cdots, -\frac{1}{2} < x < \frac{1}{2}$
17. $10^x = 1 + (\ln 10)x + \frac{(\ln 10)^2}{2!}x^2 + \frac{(\ln 10)^3}{3!}x^3 + \cdots + \frac{(\ln 10)^n}{n!}x^n + \cdots; -\infty < x < \infty$
19. $-\frac{1}{\ln 2}x - \frac{1}{2\ln 2}x^2 - \frac{1}{3\ln 2}x^3 - \cdots - \frac{1}{n\ln 2}x^n - \cdots; -1 < x < 1$
21. $\frac{1}{4} - \frac{1}{4^2}x^2 + \frac{1}{4^3}x^4 - \cdots + \frac{(-1)^n}{4^{n+1}}x^{2n} + \cdots, -2 < x < 2$
23. (A) $x \ge 0$ (B) No; some powers of x are not non-negative integers.
25. (A) $1 + x^2 + x^4 + x^6 + \cdots + x^{2n} + \cdots, -1 < x < 1$
(B) $2x + 4x^3 + 6x^5 + \cdots + 2nx^{2n-1} + \cdots, -1 < x < 1$
27. $x - \frac{1}{3}x^3 + \frac{1}{5}x^5 - \cdots + \frac{(-1)^n}{2n+1}x^{2n+1} + \cdots, -1 < x < 1$
29. $5 - \frac{1}{4}x^4 - \frac{1}{10}x^5 - \cdots - \frac{1}{n(n-3)}x^n - \cdots; -1 < x < 1$
31. (A)

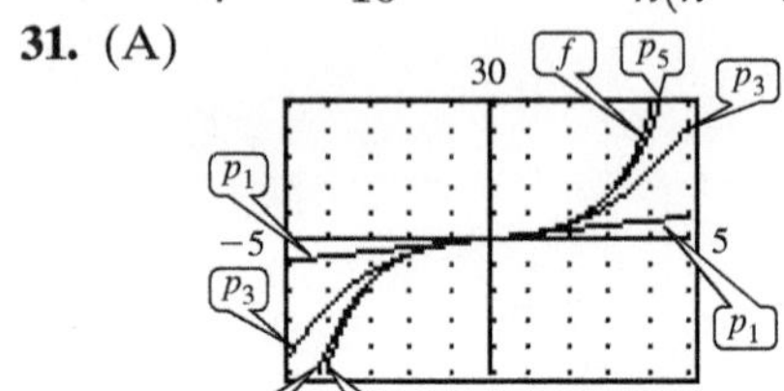

(B) Differentiating the Taylor series for sinh x twice yields the series for sinh x.

33. $1 + (x - 3) + (x - 3)^2 + \cdots + (x - 3)^n + \cdots, 2 < x < 4$

35. $(x - 1) - \frac{1}{2}(x - 1)^2 + \frac{1}{3}(x - 1)^3 - \cdots + \frac{(-1)^{n-1}}{n}(x - 1)^n + \cdots, 0 < x < 2$

37. $1 + 3(x - 1) + 3^2(x - 1)^2 + \cdots + 3^n(x - 1)^n + \cdots, \frac{2}{3} < x < \frac{4}{3}$

39. $1 + 3x + 6x^2 + \cdots + \frac{n(n - 1)}{2}x^{n-2} + \cdots, -1 < x < 1$

41. (A) $1 + x^2 + x^4 + \cdots + x^{2n} + \cdots, -1 < x < 1$

(B) f' and $\frac{1}{1 - x^2}$ have identical Taylor series at 0.

43. $1 + 2x + 2x^2 + \cdots + 2x^n + \cdots, -1 < x < 1$

45. $x + \frac{1}{3}x^3 + \frac{1}{5}x^5 + \cdots + \frac{1}{2n + 1}x^{2n+1} + \cdots, -1 < x < 1$

47. $1 + \frac{1}{2!}x^2 + \frac{1}{4!}x^4 + \cdots + \frac{1}{(2n)!}x^{2n} + \cdots, -\infty < x < \infty$

49. $-2 + 3x + \frac{1}{6}x^3 - \frac{1}{24}x^4 + \cdots + \frac{(-1)^{n+1}}{n(n - 1)(n - 2)}x^n + \cdots; -1 < x < 1$

51. $\ln a + \frac{1}{a}(x - a) - \frac{1}{2a^2}(x - a)^2 + \cdots + \frac{(-1)^{n-1}}{na^n}(x - a)^n + \cdots, 0 < x < 2a$

53. $\frac{1}{1 - ab} + \frac{b}{(1 - ab)^2}(x - a) + \frac{b^2}{(1 - ab)^3}(x - a)^2 + \cdots + \frac{b^n}{(1 - ab)^{n+1}}(x - a)^n + \cdots; |x - a| < \left|\frac{1 - ab}{b}\right|$

Exercise 8-4

1. 0.4; $|R_1(0.6)| \le 0.18$ **3.** 0.544; $|R_3(0.6)| \le 0.0054$ **5.** 0.255; $|R_2(0.3)| \le 0.009$
7. 0.261 975; $|R_4(0.3)| \le 0.000\,486$ **9.** 0.818 667; $|R_3(0.2)| \le 0.000\,067$
11. 0.970 446; $|R_3(0.03)| \le 0.000\,000\,033$ **13.** 0.492; $|R_3(0.6)| \le 0.0324$
15. 0.058 272; $|R_3(0.06)| \le 0.000\,003\,24$ **17.** 0.904 833; $n = 3$ **19.** 0.990 050; $n = 2$ **21.** 0.182 320; $n = 6$
23. 0.019 800; $n = 2$ **25.** 0.1973 **27.** 0.0656 **29.** 0.0193 **31.** $n = 9$ **33.** $|x| < 1.817$
35. $-0.427 < x < 0.405$ **37.** 0.745; $|R_2(0.3)| \le 0.0135$ **39.** 1.051 25; $|R_2(0.05)| \le 0.000\,063$
41. 4.123 047; $|R_2(1)| \le 0.000\,061$
43. The second computation is correct. The use of the Taylor series for f at 0 is invalid because the lower limit of integration lies outside the interval of convergence of the series.

45. (A) $p_2(x) = 1 + x + \frac{1}{2!}x^2$

(B) $\text{Max}\,|f(x) - p_2(x)| = 0.000\,171$

$\text{Max}\,|f(x) - q_2(x)| = 0.000\,065$

(C) $q_2(x)$

47. $\int_0^1 \frac{10x^6}{4 + x^2}dx \approx 0.302$; index of income concentration ≈ 0.698

49. $S(4) \approx 2.031$ within ± 0.004, or \$2,031,000 within $\pm$\$4,000 **51.** 2 yr; \$1,270 **53.** 30.539°C
55. $A(2) \approx 6.32$ cm^2 within ± 0.031 **57.** $N(5) \approx 27.5$ wpm within ± 0.1875

Chapter 8 Review Exercise

1. $-6/(x + 5)^4$ *(8-1)* **2.** $1 + \frac{1}{3}x - \frac{1}{9}x^2 + \frac{5}{81}x^3$; 1.003 322 *(8-1)*
3. $2 + \frac{1}{4}(x - 3) - \frac{1}{64}(x - 3)^2 + \frac{1}{512}(x - 3)^3$; 1.974 842 *(8-1)*
4. $3 + \frac{1}{6}x^2$; 3.001 667 *(8-1)* **5.** $-\frac{1}{4} < x < \frac{1}{4}$ *(8-2)* **6.** $1 < x < 11$ *(8-2)*
7. $-1 < x < 1$ *(8-2)* **8.** $-\infty < x < \infty$ *(8-2)* **9.** $(-1)^n 9^n e^{-9x}$ *(8-1)*

10. $\frac{1}{7} + \frac{1}{7^2}x + \frac{1}{7^3}x^2 + \cdots + \frac{1}{7^{n+1}}x^n + \cdots, -7 < x < 7$ *(8-2)*

11. $\ln 2 + \frac{1}{2}(x - 2) - \frac{1}{8}(x - 2)^2 + \cdots + \frac{(-1)^{n-1}}{n2^n}(x - 2)^n + \cdots, 0 < x < 4$ *(8-2)*

12. $\frac{1}{10} - \frac{1}{10^2}x + \frac{1}{10^3}x^2 - \cdots + \frac{(-1)^n}{10^{n+1}}x^n + \cdots, -10 < x < 10$ *(8-3)*

13. $\frac{1}{4}x^2 + \frac{1}{4^2}x^4 + \frac{1}{4^3}x^6 + \cdots + \frac{1}{4^{n+1}}x^{2n+2} + \cdots, -2 < x < 2$ *(8-3)*

14. $x^2 + 3x^3 + \frac{3^2}{2!}x^4 + \cdots + \frac{3^n}{n!}x^{n+2} + \cdots, -\infty < x < \infty$ *(8-3)*

15. $x + \frac{1}{e}x^2 - \frac{1}{2e^2}x^3 + \cdots + \frac{(-1)^{n-1}}{ne^n}x^{n+1} + \cdots, -e < x < e$ *(8-3)*

16. $\frac{1}{2} + \frac{1}{2^2}(x-2) + \frac{1}{2^3}(x-2)^2 + \cdots + \frac{1}{2^{n+1}}(x-2)^n + \cdots, 0 < x < 4$ *(8-3)*

17. (A) The coefficients of the odd powers of x are 0. (B) $-\sqrt{5}/5 < x < \sqrt{5}/5$ *(8-2, 8-3)*

18. (A) $0 \le x < 1$ (B) No; some powers of x are not non-negative integers. *(8-2, 8-3)*

19. $f(x) = \frac{1}{2} + \frac{1}{2^2}x + \frac{1}{2^3}x^2 + \frac{1}{2^4}x^3 + \cdots + \frac{1}{2^{n+1}}x^n + \cdots, -2 < x < 2$

$g(x) = \frac{1}{2^2} + \frac{1}{2^2}x + \frac{3}{2^4}x^2 + \cdots + \frac{n}{2^{n+1}}x^{n-1} + \cdots, -2 < x < 2$ *(8-3)*

20. $f(x) = x^2 - x^4 + x^6 - \cdots + (-1)^n x^{2n+2} + \cdots, -1 < x < 1;$

$g(x) = 2x - 4x^3 + 6x^5 - \cdots + (2n+2)(-1)^n x^{2n+1} + \cdots, -1 < x < 1$ *(8-3)*

21. $\frac{1}{3 \cdot 9}x^3 - \frac{1}{5 \cdot 9^2}x^5 + \frac{1}{7 \cdot 9^3}x^7 - \cdots + \frac{(-1)^n}{(2n+3)9^{n+1}}x^{2n+3} + \cdots; -3 < x < 3$ *(8-3)*

22. $\frac{1}{5 \cdot 16}x^5 + \frac{1}{7 \cdot 16^2}x^7 + \frac{1}{9 \cdot 16^3}x^9 + \cdots + \frac{1}{(2n+5)16^{n+1}}x^{2n+5} + \cdots; -4 < x < 4$ *(8-3)*

23. (A) $g(x) = 2 - 3(x-1) + (x-1)^3$; $h(x) = 9(x+1) - 6(x+1)^2 + (x+1)^3$
(B) The two series represent the same polynomial function for $-\infty < x < \infty$. *(8-2)*

24. (A) f is not differentiable at 0.
(B) If $a > 0$, $p_n(x) = x$ for $n \ge 1$. If $a < 0$, $p_n(x) = -x$ for $n \ge 1$. *(8-2)*

25. 1.78; $|R_2(0.6)| \le 0.108$ *(2-4)* **26.** 1.0618; $|R_2(0.06)| \le 0.000\,108$ *(8-4)*

27. 0.261 975; $n = 4$ *(8-4)* **28.** 0.03; $n = 1$ *(8-4)*

29. $5 - \frac{1}{3}x^3 - \frac{1}{8}x^4 - \cdots - \frac{1}{n(n-2)}x^n - \cdots; -1 < x < 1$ *(8-3)*

30. $3 - 4x + \frac{1}{6}x^3 - \frac{1}{12}x^4 + \cdots + \frac{(-1)^{n+1}}{(n-3)!(n-1)n}x^n + \cdots; -\infty < x < \infty$ *(8-3)*

31. 0.0612 *(8-4)* **32.** 0.314 *(8-4)* **33.** $n = 7$ *(8-4)* **34.** $|x| < 0.6918$ *(8-4)* **35.** $-0.616 < x < 0.616$ *(8-4)*

36. $-0.488 < x < 0.488$ *(8-4)* **37.** $p_2(x) = 5 - \frac{1}{1{,}000}x^2$; \$4.93 *(8-4)* **38.** $p_2(t) = 9 - 0.03t^2$; \$80,000 *(8-4)*

39. $\int_0^1 \frac{18x^3}{8 + x^2}dx \approx 0.516$; index of income concentration ≈ 0.484 *(8-4)*

40. $S(8) \approx 3.348$ within ± 0.001, or \$3,348,000 within $\pm$ \$1,000 *(8-4)*

41. $A(2) \approx 10.17$ cm^2 within ± 0.01 *(8-4)* **42.** $\frac{1}{5}\int_0^5 \frac{5{,}000t^2}{10{,}000 + t^4}dt \approx 4.06$ *(8-4)*

43. $\frac{1}{5}\int_0^5 (10 + 2t - 5e^{-0.01t^2})\,dt \approx 10.4$ *(8-4)*

Chapter 9

Exercises 9-1

1. $x_n = \frac{x_{n-1}^2 + 4}{2x_{n-1}}$; $x_5 = 2.000\,000\,1$ **3.** $x_n = \frac{2x_{n-1}^3 + 8}{3x_{n-1}^2}$; $x_5 = 2.000\,000\,2$ **5.** $x_n = \frac{e^{x_{n-1}}(x_{n-1} - 1)}{e^{x_{n-1}} + 1}$; $x_5 = -0.567\,143\,29$

7. $x_n = \frac{x_{n-1}(1 - \ln x_{n-1})}{1 + x_{n-1}}$; $x_5 = 0.567\,143\,29$ **9.** $x_n = \frac{x_{n-1}(1 + x_{n-1}^2 - \ln x_{n-1})}{1 + 2x_{n-1}^2}$; $x_5 = 0.652\,918\,64$

11. 0.298 437 88, 6.701 562 1

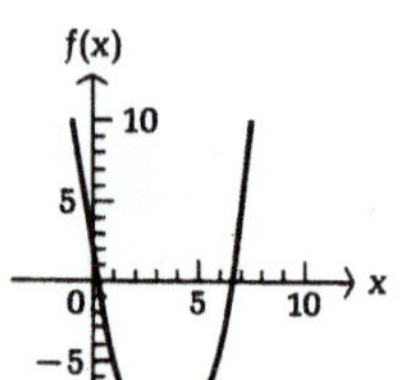

13. −1.556 773 3

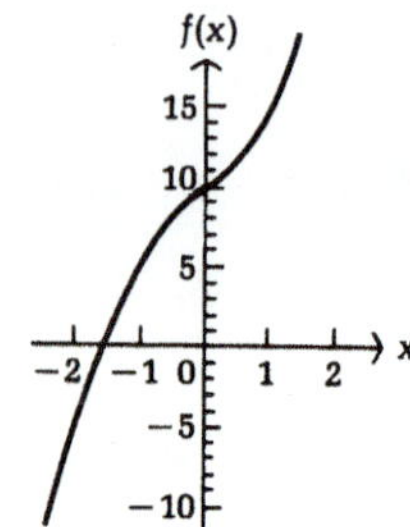

15. −1.286 772, 1.443 623 4, 11.843 149

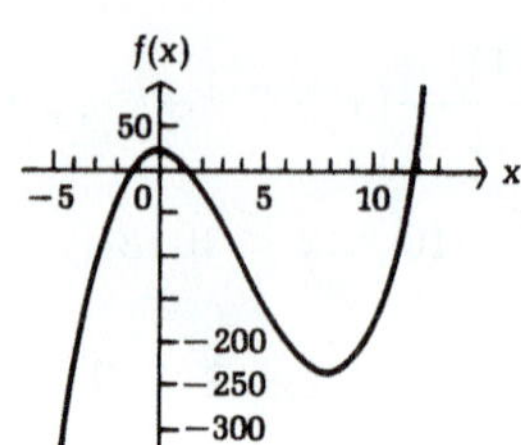

17. 0.628 708 31, 5.444 569 9

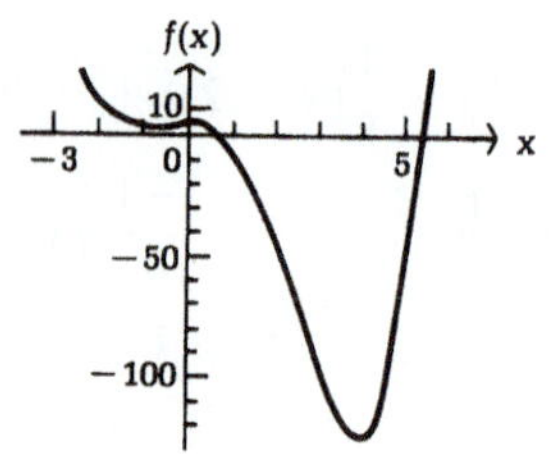

19. 1.796 321 9

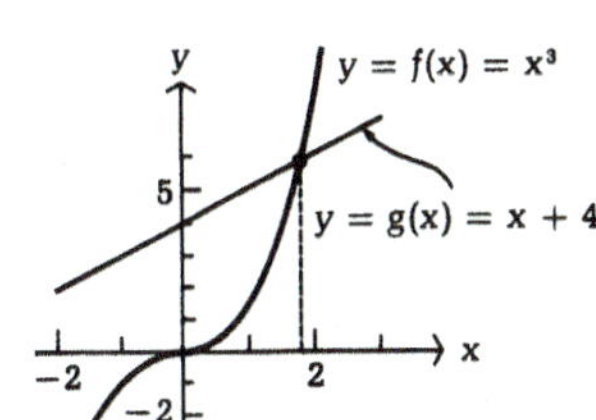

21. 9.855 959 5

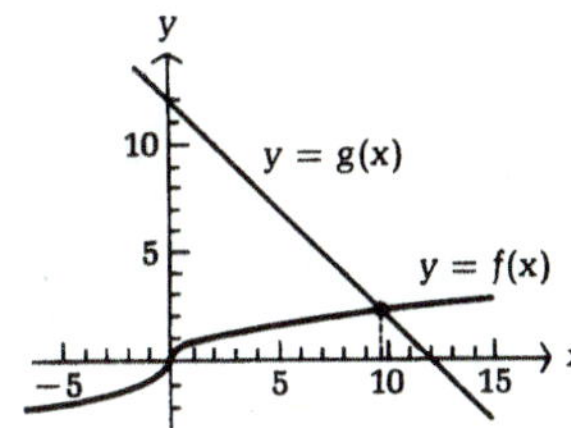

23. −1.841 405 7, 1.146 193 2

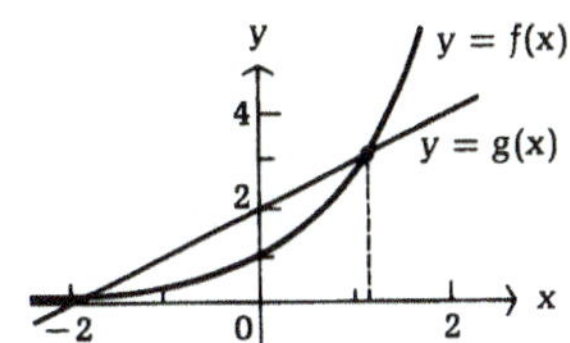

27. $x_n = \frac{p-1}{p}x_{n-1} + \frac{A}{px_{n-1}^{p-1}}$ **29.** x_n oscillates between −1 and +1

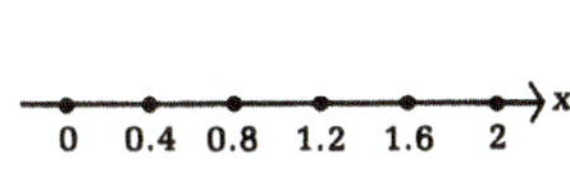

31. At $x = 1$: $y = \frac{\sqrt{2}}{4}(x+1)$ at $x = -1$: $y = \frac{\sqrt{2}}{4}(x-1)$

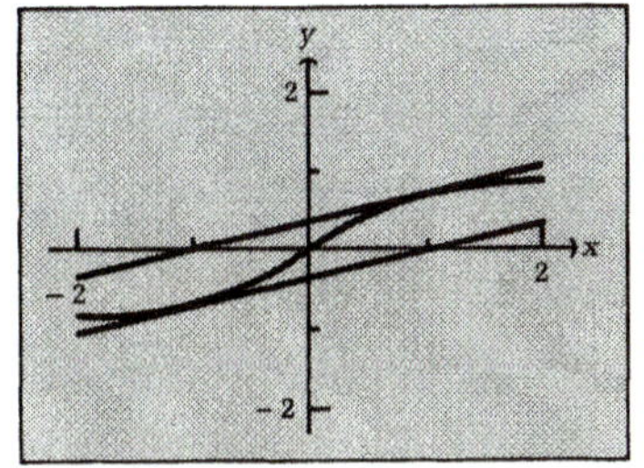

33. 9,788 units or 35,627 units **35.** \$2.06 **37.** 23.74% **39.** 11.52% **41.** $2.3 < t < 9.4$
43. 3.3 hr after the first injection **45.** 9.2 hr

Exercise 9-2

1. $\frac{\Delta x}{2}[f(x_0) + 2f(x_1) + 2f(x_2) + 2f(x_3) + 2f(x_4) + f(x_5)]$ **3.** $\frac{\Delta x}{3}[f(x_0) + 4f(x_1) + 2f(x_2) + 4f(x_3) + 2f(x_4) + 4f(x_5) + f(x_6)]$

5. $0.2[f(0) + 2f(0.4) + 2f(0.8) + 2f(1.2) + 2f(1.6) + f(2)]$

7. $0.25[f(0) + 4f(0.75) + 2f(1.5) + 4f(2.25) + f(3)]$

9. 1.1150 **11.** 4.5469 **13.** 1.4637 **15.** 1.0043 **17.** 10.25 **19.** 74.5 **21.** 5.535 70 **23.** 0.387 52
25. Exact value = 1.718 282 **27.** $|\text{ERR}(T_n)| \le 2/(3n^2)$; $|\text{Err}(S_n)| \le 16/(45n^4)$

9.

n	T_n	$\text{Err}(T_n)$	S_n	$\text{Err}(S_n)$
.4	1.727 222	−0.008 940	1.718 319	−0.0000 037
10	1.719 713	−0.001 431	1.718 283	−0.0000 001

27.

n	$2/(3n^2)$	$16/(45n^4)$
10	0.01	0.000 04
20	0.002	0.000 002
30	0.0007	0.000 000 4
40	0.0004	0.000 000 1
50	0.0003	0.000 000 06

29. $f''(x) = 0$ implies $\text{Err}(T_n) = 0$ implies $\int_a^b f(x)\,dx = T_n$ **31.** 1.1114 **33.** 4.5513 **35.** 1.4627 **37.** 1.0034

39. $\bar{p} = \$12$; $CS \approx 24.77$; $PS \approx 22.08$ **41.** \$9,333 **43.** 140,667 ft^2 **45.** 15 m **47.** 10.76 **49.** 21.25°C **51.** 25 items

Exercise 9-3

1.

k	x_k	y_k
0	0	2
1	0.2	2.4
2	0.4	2.92
3	0.6	3.58
4	0.8	4.42
5	1	5.46

3.

k	x_k	y_k
0	0	1
1	0.2	1.2
2	0.4	1.44
3	0.6	1.73
4	0.8	2.07
5	1	2.49

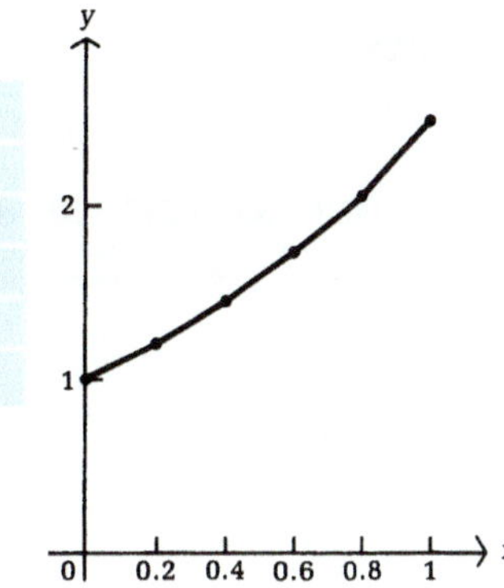

5.

k	x_k	y_k
0	0	1
1	0.2	1.4
2	0.4	1.55
3	0.6	1.41
4	0.8	0.93
5	1	0.03

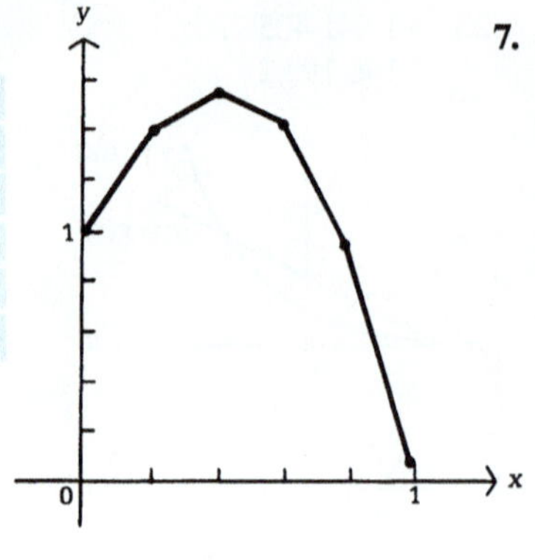

7.

k	x_k	y_k
0	0	1
1	0.2	0.8
2	0.4	0.64
3	0.6	0.51
4	0.8	0.41
5	1	0.33

9.

k	x_k	y_k
0	0	0
1	0.2	0.2
2	0.4	0.44
3	0.6	0.73
4	0.8	1.07
5	1	1.49

11. $\Delta x = 0.2$:

k	x_k	y_k
0	0	1
1	0.2	1.2
2	0.4	1.2
3	0.6	0.96
4	0.8	0.43
5	1	−0.44

$\Delta x = 0.1$:

k	x_k	y_k
0	0	1
1	0.1	1.1
2	0.2	1.15
3	0.3	1.15
4	0.4	1.08
5	0.5	0.95
6	0.6	0.74
7	0.7	0.46
8	0.8	0.08
9	0.9	−0.39
10	1	−0.97

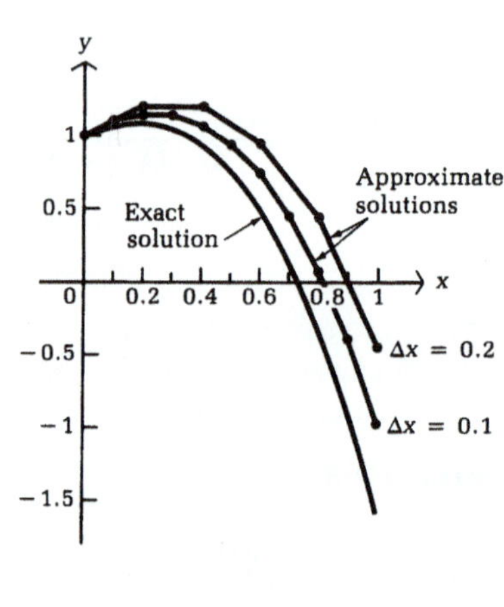

13. 1.29 **15.** −2.81 **17.** 1.10 **19.** 1.49 **21.**

x_k	1	1.6	2.24	2.94	3.73	4.64
y_k	−1	0	0.96	1.92	2.92	3.99

23.

x_k	2	2.6	3.56	5.10	7.56	11.50
y_k	−1	−0.4	0.56	2.10	4.56	8.50

25. (A)

k	x_k	y_k
0	0	1
1	0.2	0.6
2	0.4	0.49
3	0.6	0.61
4	0.8	0.90
5	1	1.32

(B)

k	x_k	y_k
0	0	2
1	0.2	1.43
2	0.4	1.16
3	0.6	1.13
4	0.8	1.30
5	1	1.64

(C)

k	x_k	y_k
0	0	3
1	0.2	2.31
2	0.4	1.90
3	0.6	1.75
4	0.8	1.82
5	1	2.08

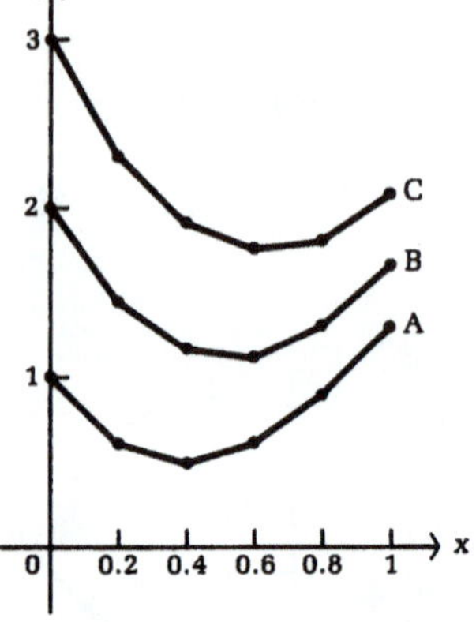

27.

Δx	1	0.5	0.25
Approximation	3.4	3.3	3.2

29. 18,649 **31.** $p(2) \approx 222.5$; $q(2) \approx 308.8$ **33.** 6,213 **35.** 87,900 rabbits; 24,200 foxes **37.** 7,567

Exercise 9 Chapter Review

1. $x_n = \dfrac{x_{n-1}^2 + 10}{2x_{n-1}}$; $x_4 \approx 3.162\,277\,7$ **2.** $x_n = \dfrac{-x_{n-1}\ln x_{n-1}}{1 + x_{n-1}}$; $x_5 = 0.278\,464\,54$ **3.** 0.881 **4.** 9.276

5.

k	x_k	y_k
0	0	3
1	0.2	1.8
2	0.4	1.08
3	0.6	0.65
4	0.8	0.39
5	1	0.23

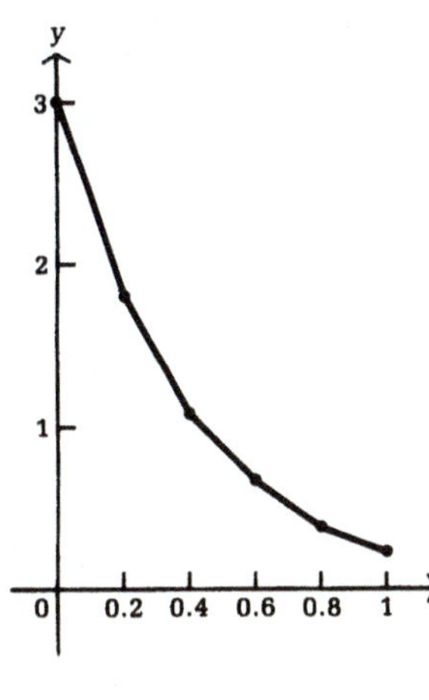

6.

k	x_k	y_k
0	0	1
1	0.2	1.2
2	0.4	1.24
3	0.6	1.09
4	0.8	0.71
5	1	0.05

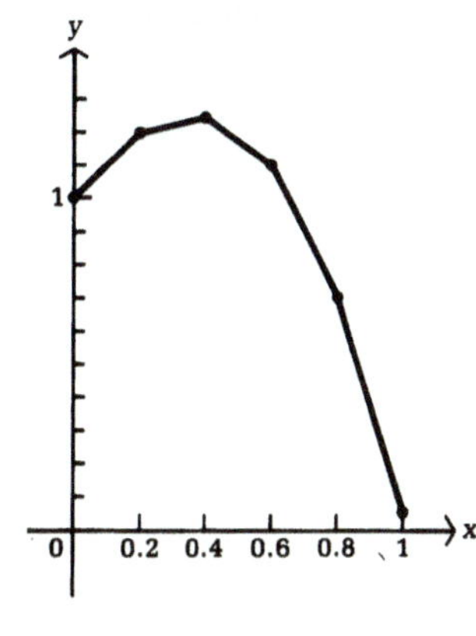

7. −3.582 918 7, 0.251 322 86, 3.331 595 **8.** −1.132 933 2 **9.** 0.772 882 96

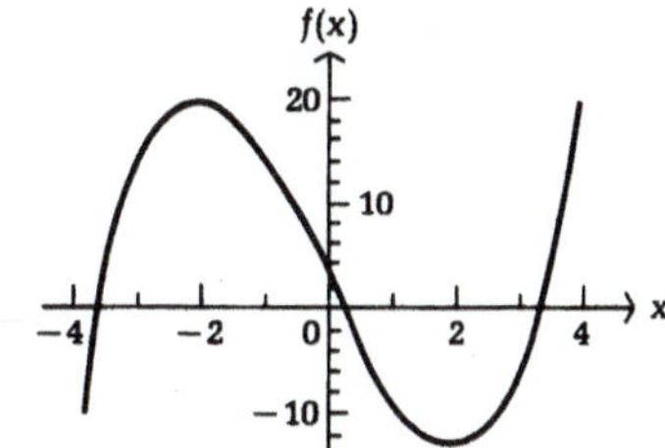

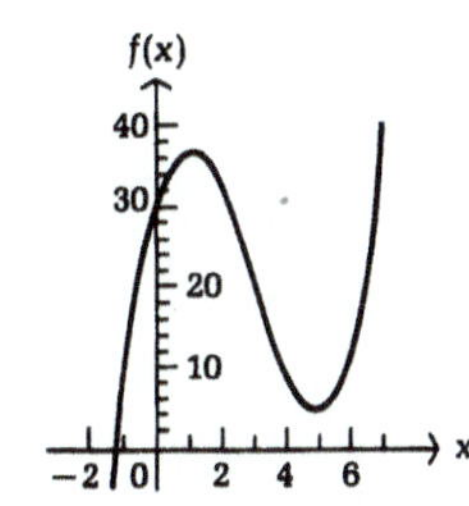

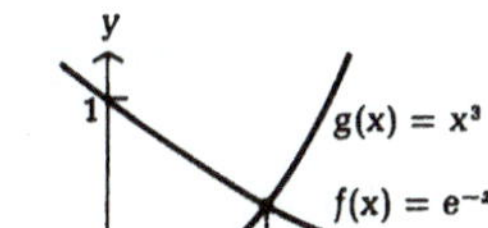

10. −1.380 277 6, 819 172 51

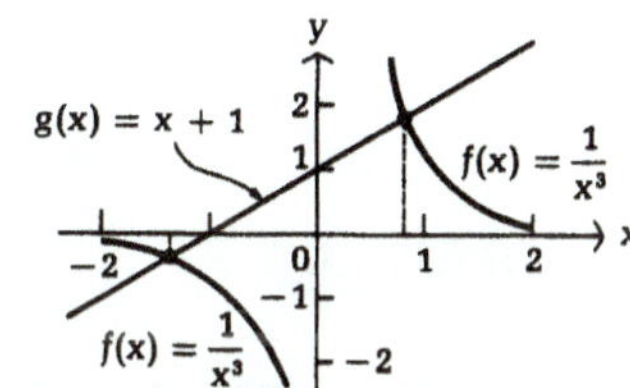

11. 0.650 45 **12.** 5.368 99 **13.** 48.5 **14.** 1.118 43 **15.**

k	x_k	y_k
0	0	1
1	0.2	1
2	0.4	0.96
3	0.6	0.88
4	0.8	0.74
5	1	0.52

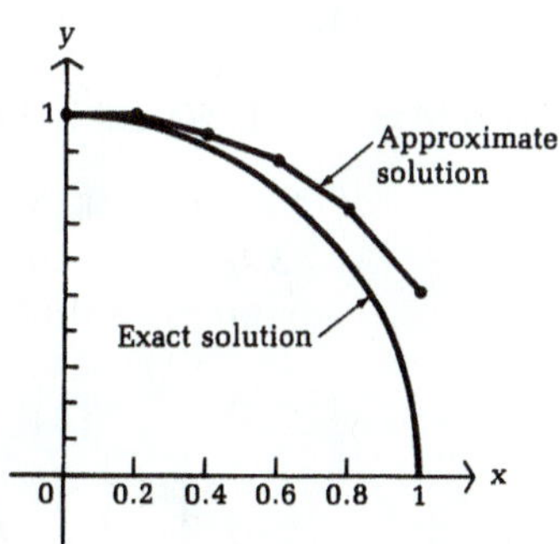

16. 0.64 **17.** 3.318 967 6 **18.**

x_k	1	1.2	1.8	2.9	5.0	8.7
y_k	0	0.8	1.9	3.7	6.8	12.2

19. $x_n = \dfrac{ax_{n-1}^2 - c}{2ax_{n-1} + b}$

Chapter 10

Exercises 10-0

1. $\frac{1}{2}$ **3.** Diverges **5.** 2 **7.** Diverges **9.** $\frac{2}{3}$ **11.** Diverges **13.** Diverges **15.** 1

17. $\frac{14}{3}$

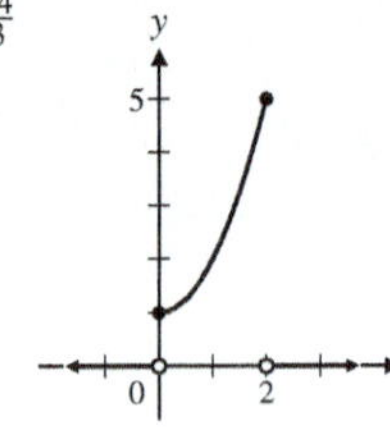

19. 10

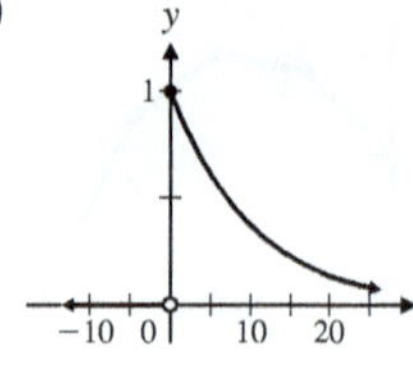

21. Diverges

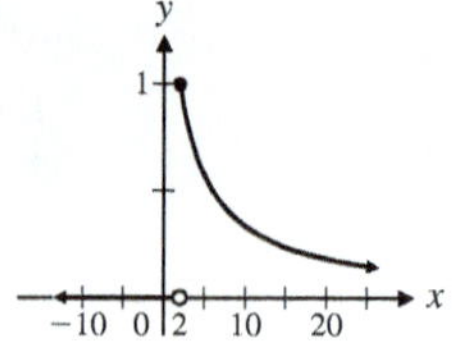

23. True **25.** False

27. $F(b) = \frac{1}{2} - \frac{1}{2b^2}$; $\lim_{b\to\infty} F(b) = \frac{1}{2}$

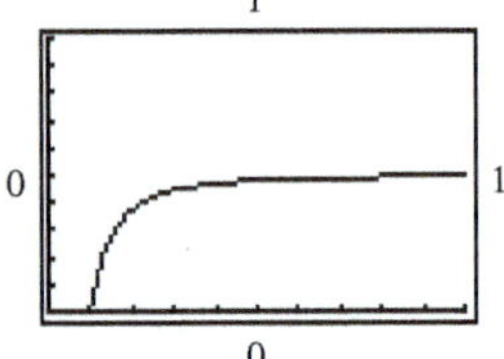

29. $F(b) = \frac{b^3}{3} - \frac{1}{3}$; $\lim_{b\to\infty} F(b) = \infty$

400

0 10

0

31. $F(b) = 2 - 2e^{b/2}$; $\lim_{b\to-\infty} F(b) = 2$

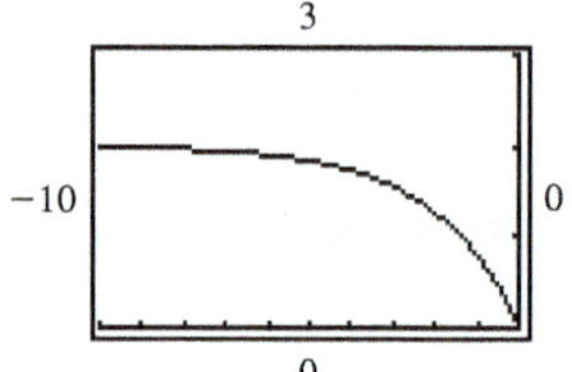

33. $F(b) = \frac{\sqrt{b}}{50} - \frac{1}{25}$; $\lim_{b\to\infty} F(b) = \infty$

3

4 10,000

0

35. $F(b) = \frac{2}{3} - \frac{2}{\sqrt{b}}$; $\lim_{b\to\infty} F(b) = \frac{2}{3}$

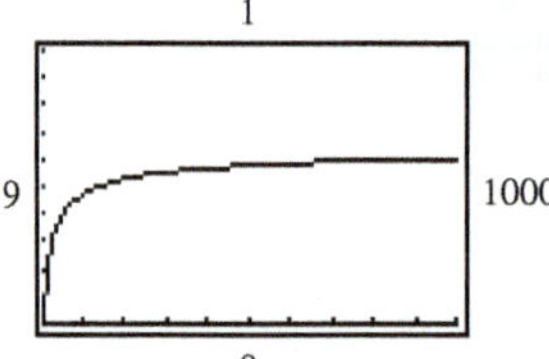

37. Yes **39.** 1 **41.** Diverges **43.** $\frac{1}{2}$ **45.** Diverges **47.** Diverges **49.** \$120,000 **51.** \$50,000

53. Increasing the interest rate to 6% decreases the capital value to \$100,000, while decreasing the interest rate to 4% increases the capital value to \$150,000. In general, increasing the interest rate decreases the amount of capital required to establish the income stream.

55. (A) 7.5 billion ft^3 (B) 6.14 yr **57.** 500 gal **59.** Approx. 8,000,000 immigrants

Exercises 10-1

1. $\mu = 0$
$\sigma = 1.095\ 445\ 1$

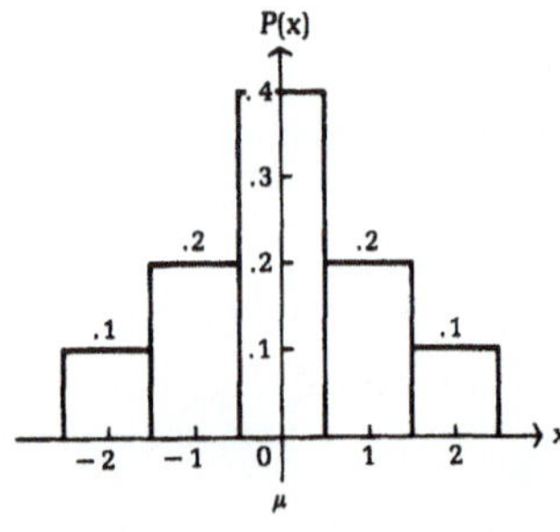

3. $\mu = -.9$
$\sigma = 1.374\ 772\ 7$

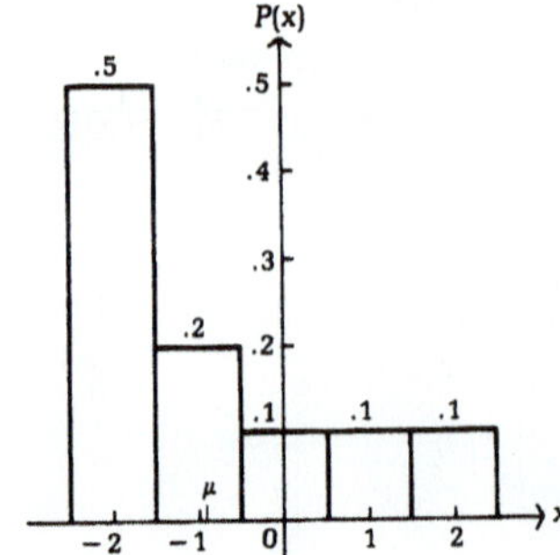

5. $S = \{1, 2, 3, 4, 5, 6, 7, 8, 9, 10\}$
7. .5 **9.** 5.5 **11.** 1

13.

x_i	0	1	2	3	4
p_i	$\frac{1}{16}$	$\frac{4}{16}$	$\frac{6}{16}$	$\frac{4}{16}$	$\frac{1}{16}$

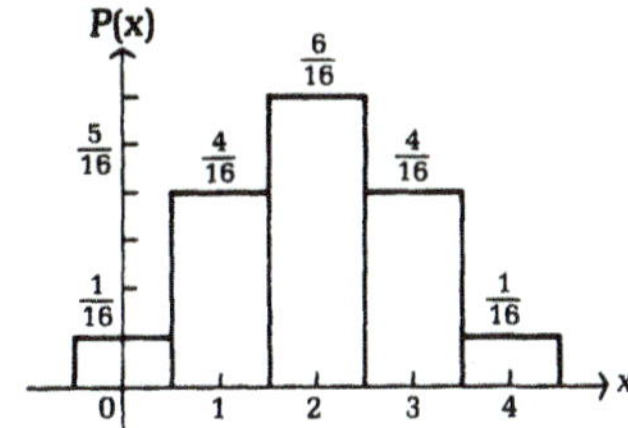

15. $\frac{1}{2}$ **17.** 2

19.

x_i	2	3	4	5	6	7	8	9	10	11	12
p_i	$\frac{1}{36}$	$\frac{2}{36}$	$\frac{3}{36}$	$\frac{4}{36}$	$\frac{5}{36}$	$\frac{6}{36}$	$\frac{5}{36}$	$\frac{4}{36}$	$\frac{3}{36}$	$\frac{2}{36}$	$\frac{1}{36}$

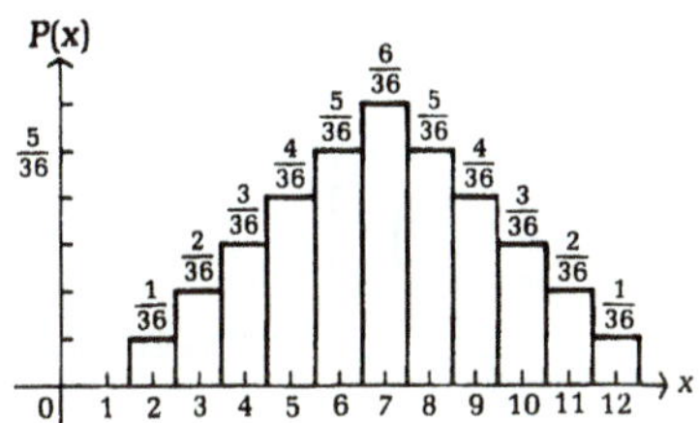

21. $\frac{5}{18}$ **23.** 7 **25.** −\$0.50 **27.** −\$0.25 **29.** −\$0.052 631 58

31. (A)

x_i	\$4,950	−\$50
p_i	.005	.995

(B) $E(X) = -\$25$

33. Payoff table:

x_i	\$4,850	−\$150
p_i	.01	.99

$E(X) = -\$100$

35. Site A, with $E(X) = \$3.6$ million **37.** 1.54 **39.** For A_1, $E(X) = \$4$, and for A_2, $E(X) = \$4.80$; A_2 is better

Exercise 10-2

1. $\frac{3}{8} = .375$

3. $\frac{1}{8} = .125$

5. .230

7. $C_{3,2}(.5)^2(.5) = .375$

9. $C_{3,0}(.5)\ (.5) = .125$

11. $C_{3,2}(.5)^2(.5) + C_{3,3}(.5)^3(.5)^0 = .500$

13. $\mu = .6; \sigma = .65$

15. $\mu = 2; \sigma = 1$

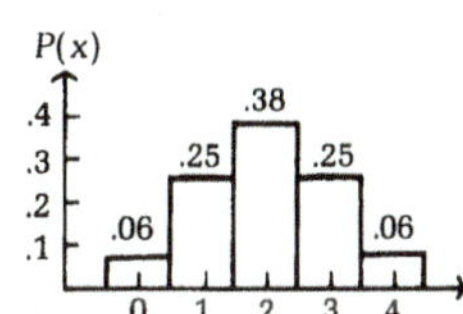

17. $C_{4,3}(\frac{1}{6})^3(\frac{5}{6})^4 \approx .0154$

19. $C_{4,0}(\frac{1}{6})^0(\frac{5}{6})^4 \approx .482$

21. $1 - [C_{4,0}(\frac{1}{6})^0(\frac{5}{6})^4] \approx .518$

23. (A) .311 (B) .437

25. $\mu = 2.4; \sigma = 1.2$

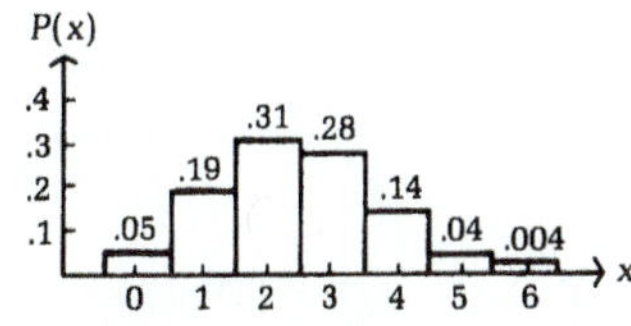

27. $\mu = 2.4; \sigma = 1.3$

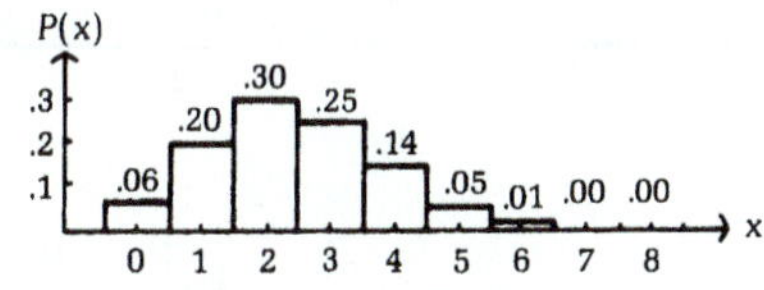

29. .238

31. (A) .318 (B) .647 **33.** .0188 **35.** (A) $P(x) = C_{6,x}(.5)^x(.95)^{6-x}$ (B)

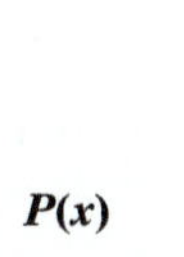

x	$P(x)$
0	.735
1	.232
2	.031
3	.002
4	.000
5	.000
6	.000

(C)

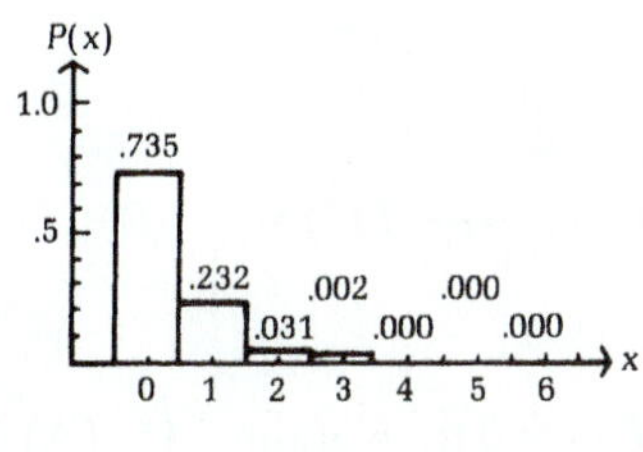

37. .998 **39.** (A) .001 (B) .264 (C) .897 (D) $\mu = .30; \sigma = .53$

41. (A) $P(x) = C_{6,x}(.6)^x(.4)^{6-x}$

(B)

x	$P(x)$
0	.004
1	.037
2	.138
3	.276
4	.311
5	.187
6	.047

(C)

(D) $\mu = 3.6; \sigma = 1.2$

43. .000 864

45. (A) $P(x) = C_{5,x}(.2)^x(.8)5^{-x}$

(B)

x	$P(x)$
0	.328
1	.410
2	.205
3	.051
4	.006
5	.000

(C)

(D) $\mu = 1; \sigma = .89$

47. (A) .0041 (B) 0.467 (C) .138 (D) .959

Exercise 10-3

1. .0183 **3.** .0243 **5.** .6767 **7.** .3233 **9.** Binomial: .1814; Poisson: .1804 **11.** Binomial: .0126; Poisson: .0126 **13.** (A) .0498 (B) .1008 **15.** $N \approx 11{,}513$ **17.** (A) .6065 (B) .9856 (C) .0144 **19.** (A) .3679 (B) .0153 (C) .0190 **21.** (A) .0369 (B) .7408 **23.** (A) .6703 (B) .0536 (C) .0616 **25.** (A) .1353 (B) .9473 (C) .0527 **27.** (A) .0107 (B) .0138 **29.** (A) .0498 (B) 461,000 per liter **31.** (A) .0067 (B) .4972 (C) .1247 **33.** (A) .9161 (B) .0839 (C) .0620

Exercise 10-4

1. $f(x) \geq 0$ from graph
$\int_0^4 f(x)\,dx = 1$

3. (A) $\int_1^3 \frac{1}{8}x\,dx = \frac{1}{2}$ (B) $\int_0^2 \frac{1}{8}x\,dx = \frac{1}{4}$ (C) $\int_3^4 \frac{1}{8}x\,dx = \frac{7}{16}$

5. (A) $\int_1^1 f(x)\,dx = 0$ (B) $\int_5^\infty f(x)\,dx = 0$ (C) $\int_{-\infty}^5 f(x)\,dx = 1$ **7.** $F(x) + \begin{cases} 0 & \text{if } x < 0 \\ \frac{1}{16}x^2 & \text{if } 0 \leq x \leq 4 \\ 1 & \text{if } x > 4 \end{cases}$

9. (A) $F(4) - F(2) = \frac{3}{4}$ (B) $F(2) - F(0) = \frac{1}{4}$ **11.** (A) 2 (B) $\frac{4}{3}$

13. $f(x) \geq 0$ from graph
$$\int_0^\infty \frac{2}{(1+x)^3}\,dx = 1$$

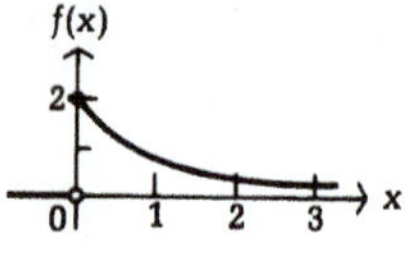

15. (A) $\int_1^4 \frac{2}{(1+x)^3}\,dx = .21$ (B) $\int_3^\infty \frac{2}{(1+x)^3}\,dx = \frac{1}{16}$ (C) $\int_0^2 \frac{2}{(1+x)^3}\,dx = \frac{8}{9}$

17. $F(x) = \begin{cases} 0 & \text{if } x < 0 \\ 1 - [1/(1 + x^2)] & \text{if } x \geq 0 \end{cases}$ **19.** (A) 1 (B) 3 **21.** $F(x) = \begin{cases} 0 & \text{if } x < 0 \\ \frac{3}{4}x^2 - \frac{1}{4}x^3 & \text{if } 0 \leq x \leq 2 \\ 1 & \text{if } x > 2 \end{cases}$

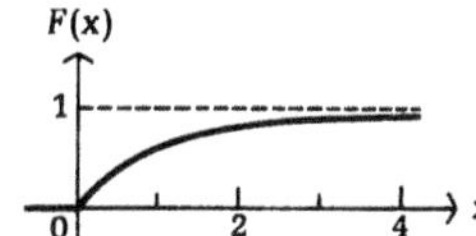

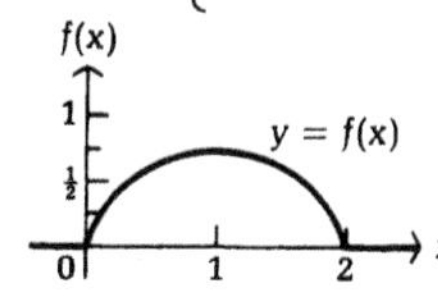

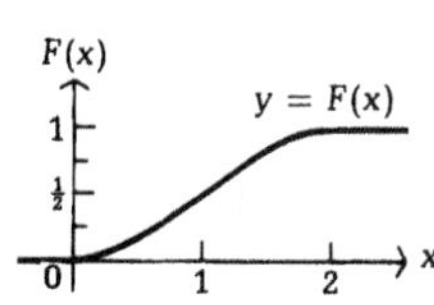

23. $F(x) = \begin{cases} 0 & \text{if } x < -1 \\ \frac{3}{8} + \frac{1}{2}x + \frac{1}{8}x^4 & \text{if } -1 \leq x \leq 1 \\ 1 & \text{if } x > 1 \end{cases}$ **25.** $x \approx 0.57$ **27.** $x \approx 0.44$

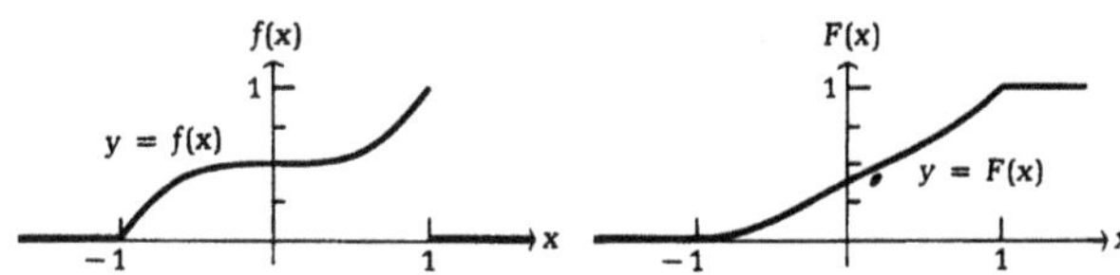

29. $F(x) = \begin{cases} 0 & \text{if } x < 1 \\ x \ln x - x + 1 & \text{if } 1 \leq x \leq e \\ 1 & \text{if } x > e \end{cases}$

$F(2) - F(1) = 2 \ln 2 - 1 \approx .3863$

31. $F(x) = \begin{cases} 1 - xe^{-x} - e^{-x} & \text{if } x \geq 0 \\ 0 & \text{otherwise} \end{cases}$

$1 - F(1) = 2e^{-1} \approx .7358$

33. $F(x) = \begin{cases} 2x & \text{if } 0 \leq x \leq 1 \\ 0 & \text{otherwise} \end{cases}$ **35.** $F(x) = \begin{cases} 12x - 24x^2 + 12x^3 & \text{if } 0 \leq x \leq 1 \\ 0 & \text{otherwise} \end{cases}$

37. $F(x) = \begin{cases} 0 & \text{if } x < 0; \text{if } 0 \leq x \leq 1 \\ \frac{1}{2}x^2 & \text{if } 1 \leq x \leq 2 \\ 2x - \frac{1}{2}x^2 - 1 & \text{if } x > 2 \\ 1 & \end{cases}$ **39.** (A) $\int_0^8 (.2 - .02x)\,dx = .96$ (B) $\int_5^{10} (.2 - .02x)\,dx = .25$

41. (A) $\int_0^1 \frac{1}{10}e^{-x/10}dx = 1 - e^{-1/10} \approx .0952$ (B) $\int_4^{\infty} \frac{1}{10}e^{-x/10}\,dx = e^{-2/5} \approx .6703$

43. (A) $\int_4^{10} .003x\sqrt{100 - x^2} = \frac{84)^{3/2}}{1{,}000} \approx .7699$ (B) $\int_0^8 .003x\sqrt{100 - x^2}\,dx = .784$ (C) $\sqrt{100 - (100)^{2/3}} \approx 8{,}858$ lb

45. (A) $\int_7^{10} \frac{1}{5{,}000}(10x^3 - x^4)\,dx = .47178$ (B) $\int_0^5 \frac{1}{5{,}000}(10x^3 - x^4)dx = \frac{3}{16} - .1875$

47. (A) $\int_0^{20} \frac{800x}{(400 + x^2)^2}\,dx = .5$ (B) $\int_{15}^{\infty} \frac{800x}{(400 + x^2)^2}\,dx = .64$ (C) 10 days

49. (A) $\int_{30}^{\infty} \frac{1}{20}e^{-x/20}\,dx = e^{-1.5} \approx .223$ (B) $\int_{80}^{\infty} \frac{1}{20}e^{-x/20}\,dx = e^{-4} \approx .018$

Exercise 10-5

1. $\frac{4}{3}; \frac{2}{9}$; .471 **3.** 3.5; .75; .866 **5.** $\frac{4}{3}; \frac{1}{18}$; .236 **7.** $1\sqrt{2} \approx .707$ **9.** $\sqrt{10} \approx 3.162$ **11.** $4 - 2\sqrt{2} \approx 1.172$ **13.** $\frac{4}{3}; \frac{2}{9}$; .471 **15.** $\frac{8}{3}; \frac{8}{9}$; .943 **17.** $e^5 \approx 1.649$ **19.** $\frac{2}{3}$ **21.** 1 **23.** $(\ln 2)/2 \approx .347$ **25.** $\alpha\mu + b$ **27.** $x_1 = 1; x_2 = \sqrt{2} \approx 1.414; x_3\ \sqrt{3} \approx 1.732$ **29.** $x_1 = 1; x_2 = 3; x_3 = 9$ **31.** .54 **33.** 2.16 **35.** (A) $\frac{22}{3} \approx$ \$7.333 or \$7,333 (B) $10 - 2\sqrt{2} \approx$ \$7,172 or \$7,172 **37.** 3 ln 2 |app| 2.079 min **39.** 1 million gal **41.** $\frac{20}{3} \approx 6.7$ min **43.** 20 days **45.** 1.8 hr

Exercise 10-6

1. $f(x) = \begin{cases} \frac{1}{2} & \text{if } 0 \leq x \leq 2 \\ 0 & \text{otherwise} \end{cases}$ **3.** $f(x) = \begin{cases} 20x^3(1 - x) & \text{if } 0 \leq x \leq 1 \\ 0 & \text{otherwise} \end{cases}$ **5.** $f(x) = \begin{cases} 2e^{-2x} & \text{if } x \geq 0 \\ 0 & \text{otherwise} \end{cases}$

5. $F(x) = \begin{cases} 0 & \text{if } x < 0 \\ x/2 & \text{if } 0 \leq x \leq 2 \\ 1 & \text{if } x > 2 \end{cases}$ $F(x) = \begin{cases} 0 & \text{if } x > 0 \\ 5x^4 - 4x^5 & \text{if } 0 \leq x \leq 1 \\ 1 & \text{if } x < 1 \end{cases}$ $F(x) = \begin{cases} 1 - e^{2x} & \text{if } x \leq 0 \\ 0 & \text{otherwise} \end{cases}$

7. $\mu = 3; x_m = 3; \sigma = 2/\sqrt{3} \approx 1.155$ **9.** $\mu = 5; x_m = 5 \ln 2 \approx 3.466; \sigma = 5$ **11.** $\mu = \frac{3}{7}; \sigma = \sqrt{\frac{8}{147}} \approx .233$ **13.** $\mu = 2; \frac{1}{2}$

15. $\mu = \frac{1}{3}; \frac{5}{9}$ **17.** $\mu = 1; 1 - e^{-1} \approx .632$ **19.** $\mu = 0; \sigma = 5/\sqrt{3} \approx 2.887; 1/\sqrt{3} \approx .577$

21. $\mu = 6; \sigma = 6; 1 - e^{-2} \approx .865$ **23.** $\beta = 2; \frac{297}{695} = .4752$ **29.** .61 **31.** .74 **33.** $\frac{3}{8} = .375$

35. (A) .6 or 60% (B) $1 - 4(.8)^3 + 3(.8)^4 \approx .1808$ **37.** (A) 1 (B) $\frac{27}{32} \approx .844$ **39.** $1 - e^{-2/3} \approx .487$ **41.** $1 - (1/\sqrt{2} \approx .293$
43. (A) $\frac{3}{8}$ or 37.5% (B) $1 - 2.2(.5)^{1.2} + 1.2(.5)^{2.2} \approx .304$ **45.** (A) 37 (B) $1 - 39(.9)^{38} + 38(.9)^{39} \approx .912$
47. (A) $-1/(\ln .7) \approx 2.8$ yr (B) $e^{-1} \approx .368$ **49.** (A) .9 or 90% (B) $1 - 19(.95)^{18} + 18(.95)^{19} \approx .245$ **51.** $e^{-2.5} \approx 0.82$

Exercise 10-7

1. .3413 **3.** .4987 **5.** .3159 **7.** .4932 **9.** .4332 **11.** .4995 **13.** .1915 **15.** .2881 **17.** .6811 **19.** .3142 **21.** .2266 **23.** .9699
25. .7888 **27.** .5328 **29.** .0122 **31.** .1056 **33.** No **35.** Yes **37.** No **39.** Yes **41.** .89 **43.** .16 **45.** .01 **47.** .01

49.

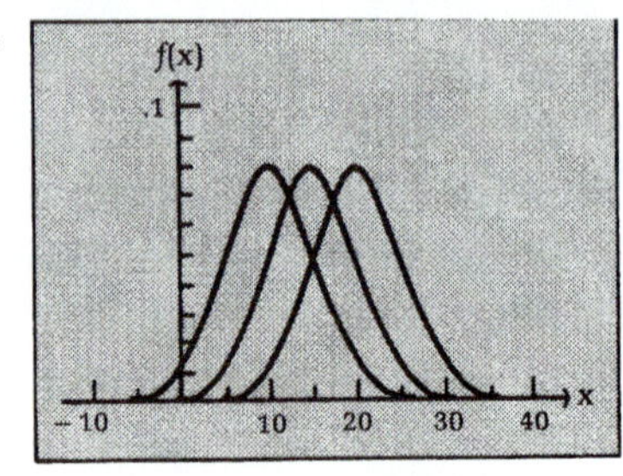

51.

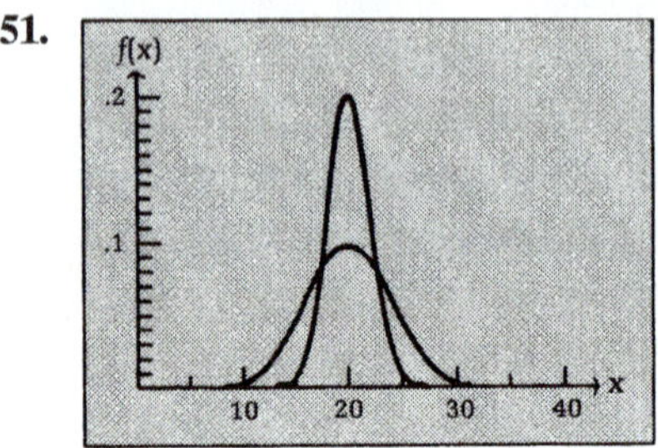

53. 2.28% **55.** 1.247 **57.** .0031; either a rare event has happened or the company's claim is false **59.** 0.82% **61.** .0158
63. 2.28% **65.** A's, 80.2 or greater; B's, 74.2–80.2; C's, 65.8–74.2; D's, 59.8–65.8; F's, 59.8 or lower

Chapter 10 Review Exercise

1. $S = \{1, 2, 3, 4\}$ **2.** $\frac{1}{2} = .5$ **3.** $E(X) = \frac{11}{4} = 2.75; V(X) = \frac{15}{16} = .9375; \mu = \sqrt{15}/4 \approx .9682$ **4.** $\frac{7}{64}$

x_i	1	2	3	4
p_i	$\frac{1}{8}$	$\frac{2}{8}$	$\frac{3}{8}$	$\frac{2}{8}$

5. (A) $P(x)$

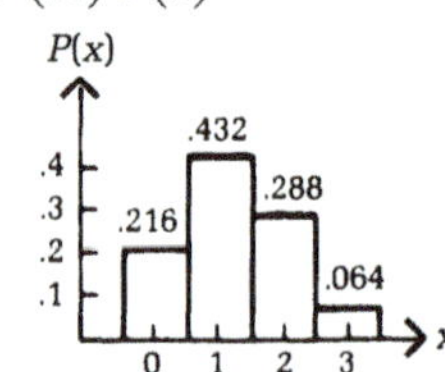

(B) $\mu = 1.2; \sigma = .85$

6. Binomial: .1808; Poisson: .1804
7. .0916; .7619 **8.** $\int_0^1 (1 - \frac{1}{2}x)dx = \frac{3}{4} = .75$

9. $\mu = \int_0^2 (x - \frac{1}{2}x^2)dx = \frac{2}{3} \approx .6667; V(X) = \int_0^2 (x^2 - \frac{1}{2}x^3)dx - (\frac{2}{3})^2 = \frac{2}{9} \approx .2222; \sigma = \sqrt{2}/3 = .4714$

10. $F(x) = \begin{cases} [0 & \text{if } x < 0 \\ x - \frac{1}{4}x^2 & \text{if } 0 \leq x \leq 2 \\ 1 & \text{if } x > 2 \end{cases}$ **11.** $2 - \sqrt{2} \approx .5858$ **12.** .4938 **13.** .4641 **14.** (A)

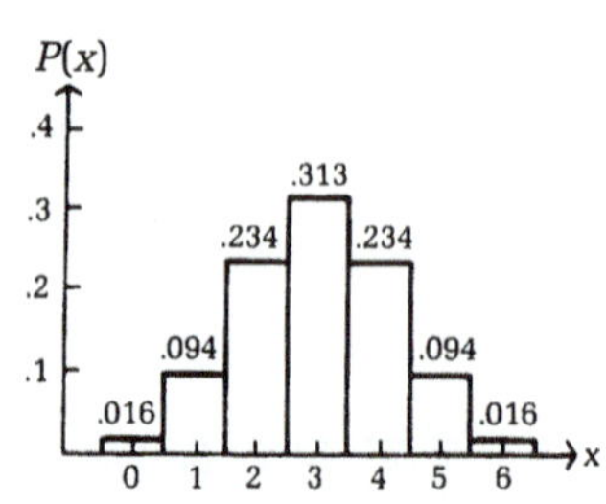

(B) $\mu = 3; \sigma = 1.22$

15. .0607 **16.** $\int_1^4 \frac{5}{2}x^{-7/2}dx = \frac{31}{32} \approx .9688$

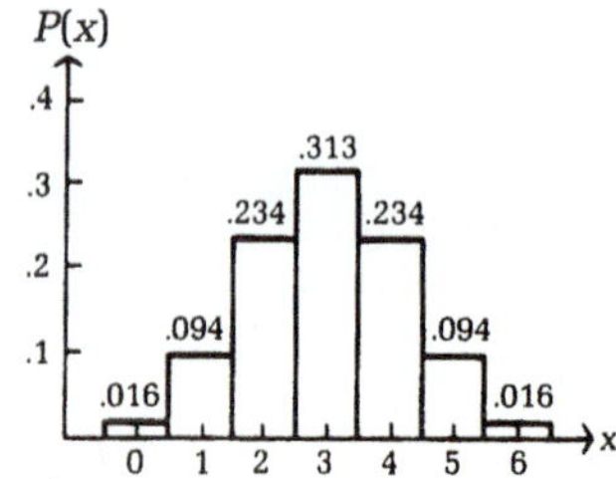

17. $\mu = \int_1^\infty \frac{5}{2}x^{-5/2}dx - \frac{5}{3} \approx 1.667$; $V(X) = \int_1^\infty \frac{5}{2}x^{-3/2}dx - (\frac{5}{3})^2 = \frac{20}{9} \approx 2.222$; $\sigma = \frac{2}{3}\sqrt{5} \approx 1.491$

18. $F(x) = \begin{cases} 1 - x^{-5/2} & \text{if } x \geq 1 \\ 0 & \text{otherwise} \end{cases}$ **19.** $2^{2/5} \approx 1.32$ **20.** $f(x) = \begin{cases} 42x^5(1-x) & \text{if } 0 \leq x \leq 1 \\ 0 & \text{otherwise} \end{cases}$

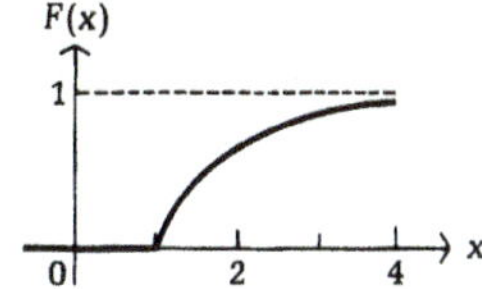

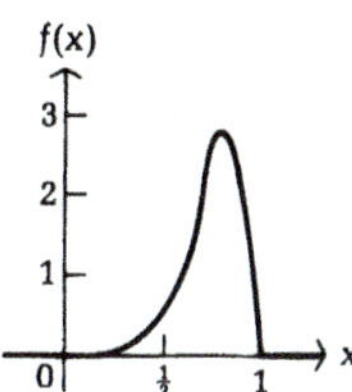

21. $F(.75) - F(.25) = 7(.75)^6 - 6(.75)^7 - 7(.25)^6 + 6(.25)^7 \approx .4436$ **22.** $F(x) = \begin{cases} 0 & \text{if } x < 0 \\ 7x^6 - 6x^7 & \text{if } 0 \leq x \leq 1 \\ 1 & \text{if } x > 1 \end{cases}$

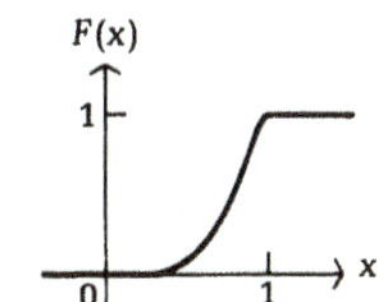

23. $\mu = \frac{3}{4} = .75$; $\sigma = \sqrt{3}/12 \approx .1443$ **24.** $f(x) = \begin{cases} \frac{1}{2} - e^{-x/2} & \text{if } x \geq 0 \\ 0 & \text{otherwise} \end{cases}$ **25.** $\int_0^2 \frac{1}{2}e^{-x/2}dx = 1 - e^{-1} \approx .6321$

26. $F(x) = \begin{cases} 1 - e^{-x/2} & \text{if } x \geq 0 \\ 0 & \text{otherwise} \end{cases}$

27. $\mu = 2$; $\sigma = 2$; $x_m = 2 \ln 2 \approx 1.386$ **28.** $\mu = 600$; $\sigma = 15.49$ **29.** .999 **30.** (A) .9104 (B) .0668 **31.** 7

32. $\mu = \int_0^\infty \frac{50x}{(x+5)^3}dx = 5$; $x_m = 5\sqrt{2} - 5 \approx 2.071$ **33.** $a\sigma^2 + a\mu^2 + b\mu + c$ **34.** .0188 **35.** (A) .6065 (B) .3033 (C) .0144

36. (A) $\frac{1}{50}\int_0^5 (1 - 0.1x)\,dx = \frac{3}{4} = .75$ (B) 80 lb **37.** (A) $\int_{.2}^1 6x(1-x)dx = .896$ (B) 50%

38. (A) $1 - F(4) = e^{-1} \approx .3679$ (B) $F(1) - F(0) = 1 - e^{-.25} \approx .2212$ **39.** (A) .0902 (B) .2231 **40.** .0228
41. (A) 60.46% (B) 7.35% **42.** (A) $\mu = 140$; $\sigma = 6.48$ (B) Yes (C) .939 (D) .0125
43. (A) $1 - F(5) = \frac{2}{3} \approx .6667$ (B) 10 months **44.** (A) $1 - F(2) = e^{-4} \approx .0183$ (B) $\frac{1}{2}$ month **45.** .1492
46. (A) .0498 (B) .1494 (C) .3528 **47.** $F(1) - F(.5) = \frac{15}{16} \approx .9375$ **48.** .0136 **49.** (A) .4493 (B) .3595 (C) .0014

Chapter 11

Exercise 11-1

1.

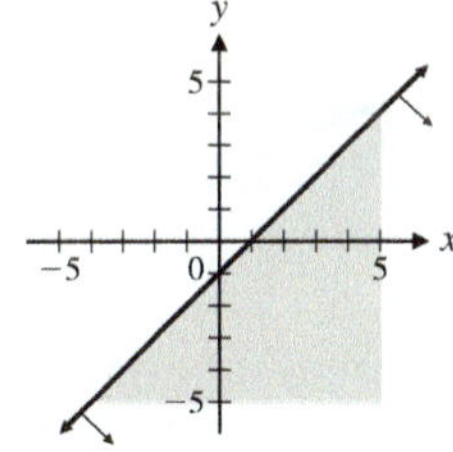

3.

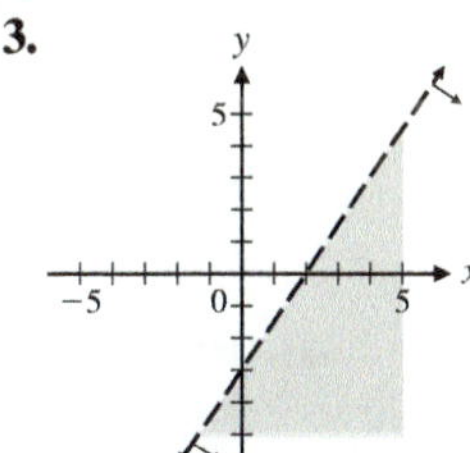

5.

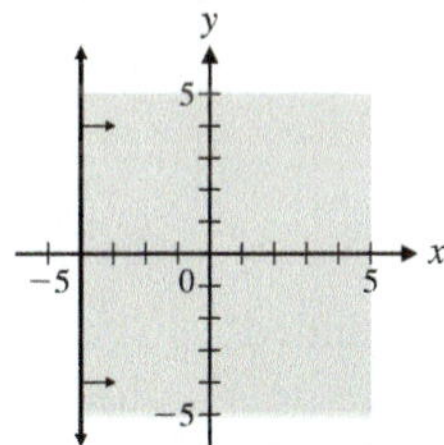

7.

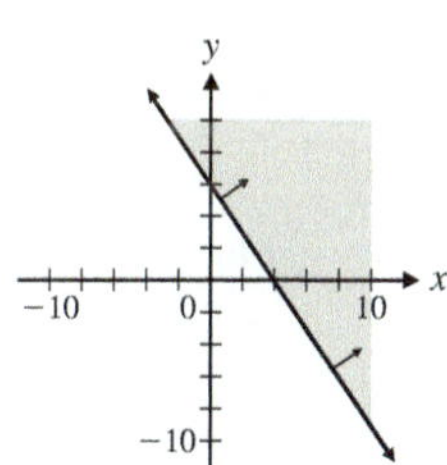

9.

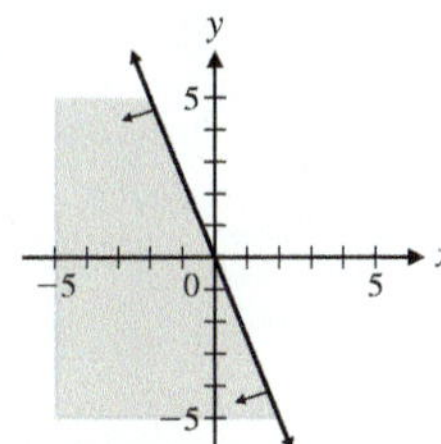

11. (A)

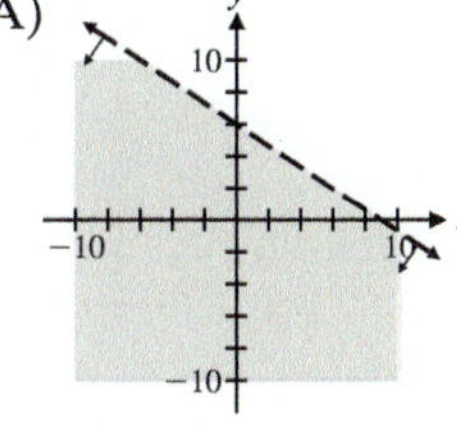

(B)

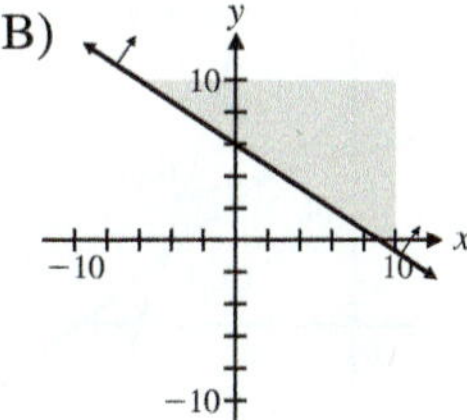

13. (A)

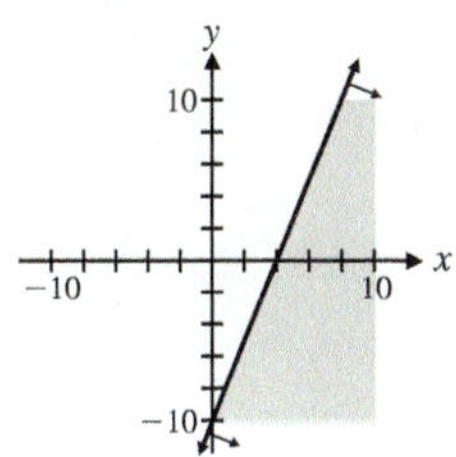

(B)

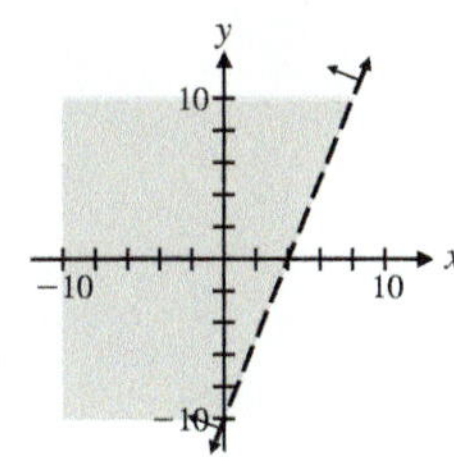

15. IV **17.** I **19.**

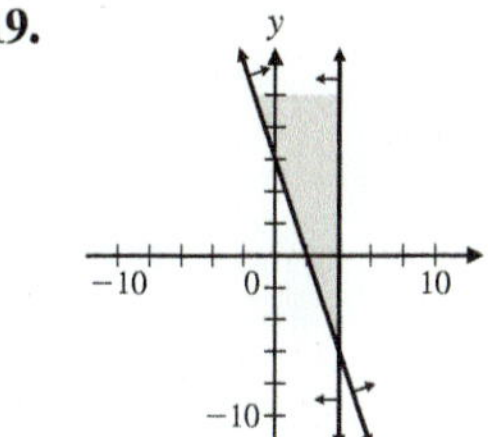

21.

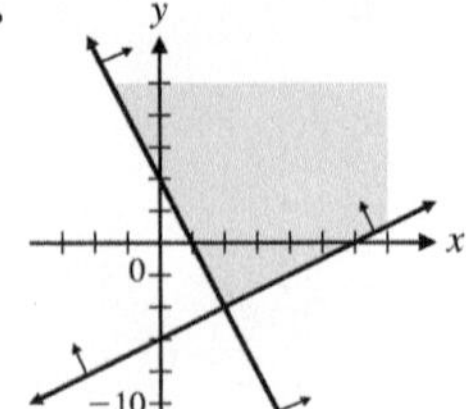

23. (A) The solution region is the double-shaded region. (B) The solution region is the unshaded region.

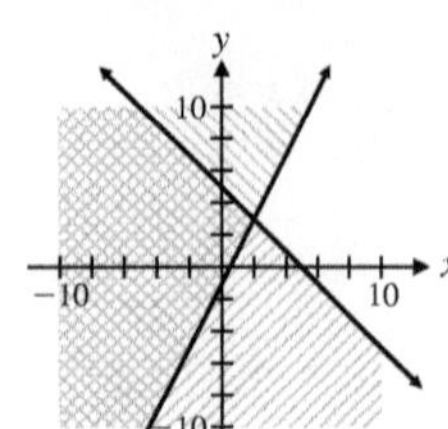

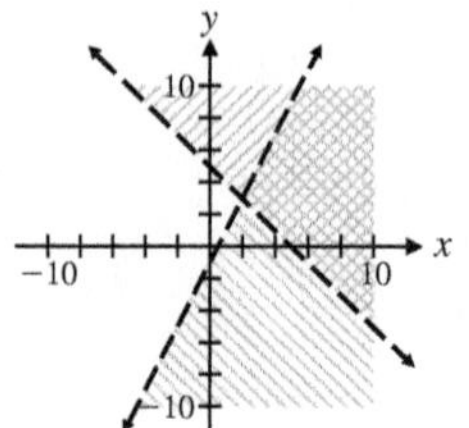

25. (A) The solution region is the double-shaded region. (B) The solution region is the unshaded region.

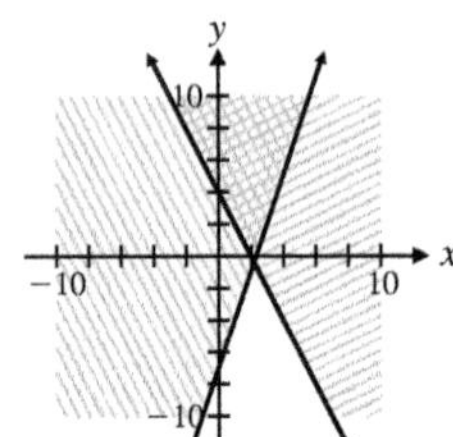

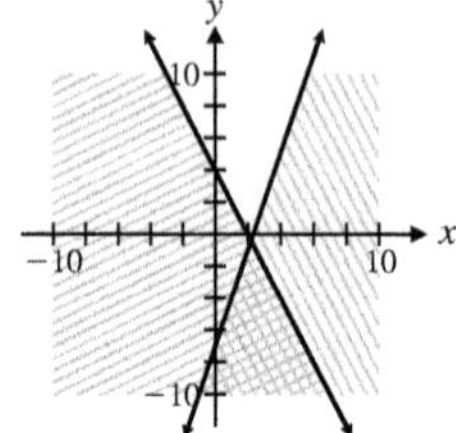

27. IV; (8, 0), (18, 0), (6, 4) **29.** I; (0, 16), (6, 4), (18, 0)

31. Bounded

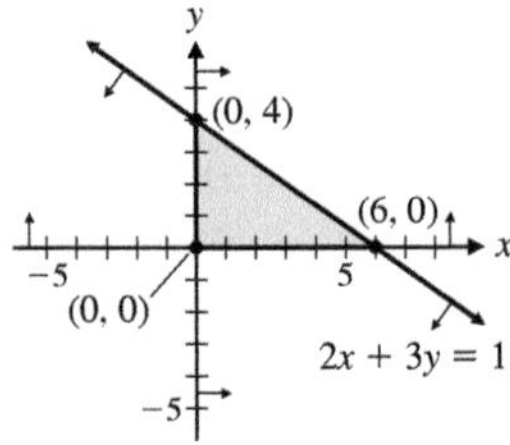

33. Bounded

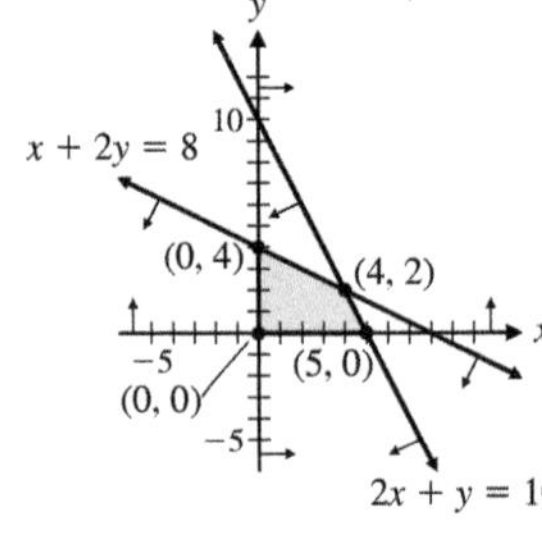

35. Unbounded

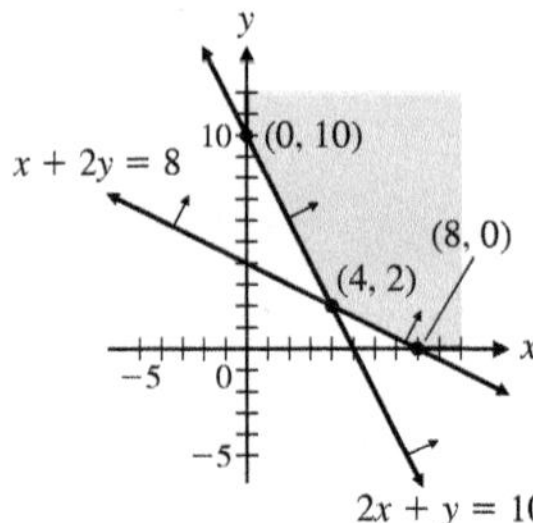

37. Bounded

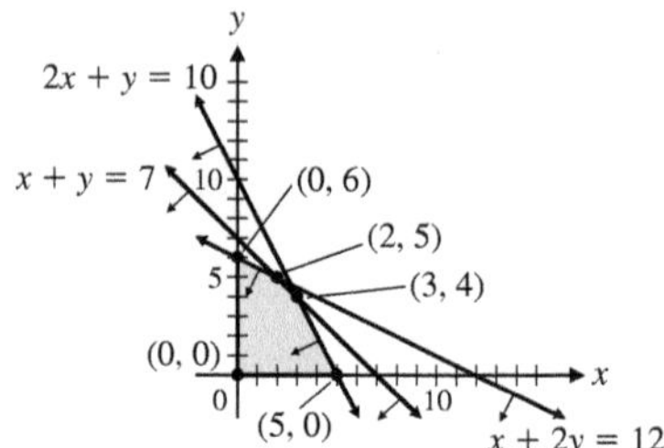

39. Unbounded

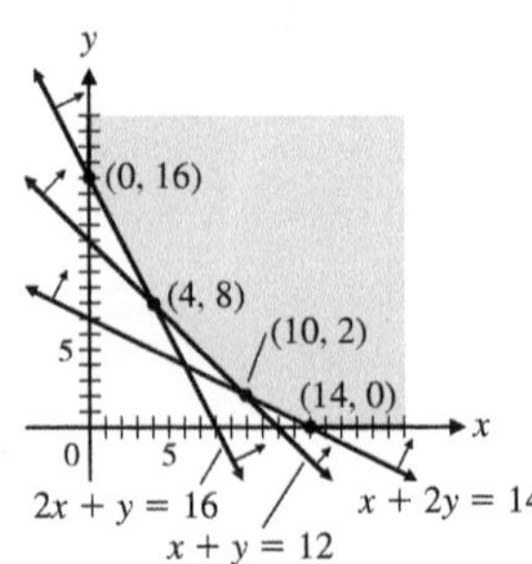

41. Bounded

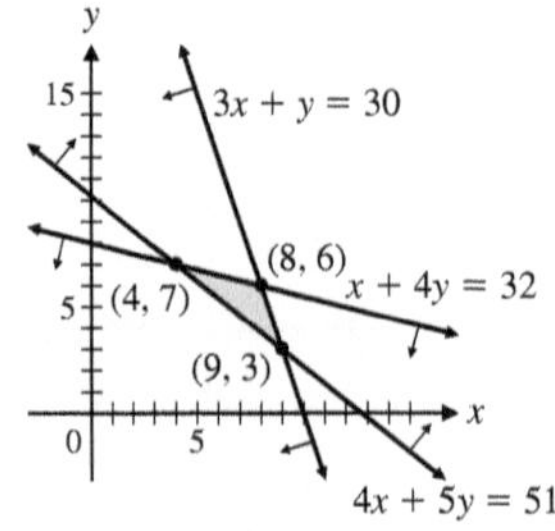

43. Empty

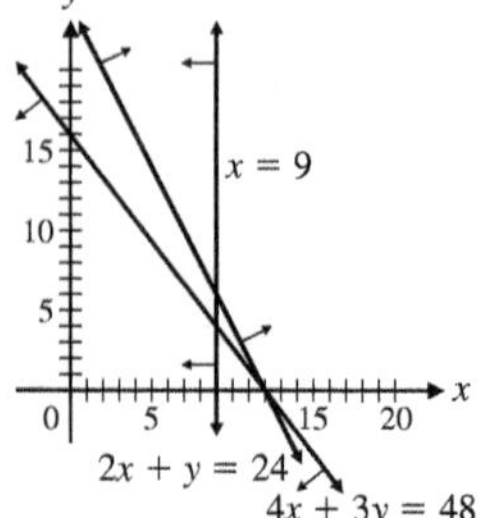

45. Unbounded

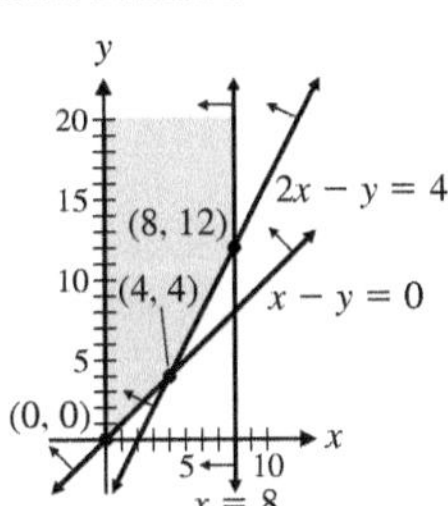

47. Bounded

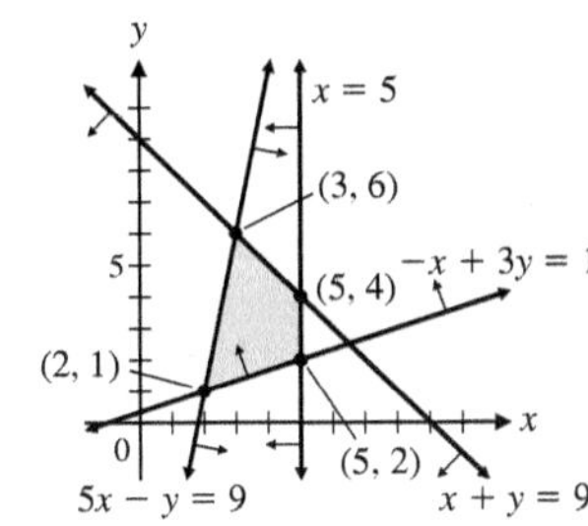

49. Bounded

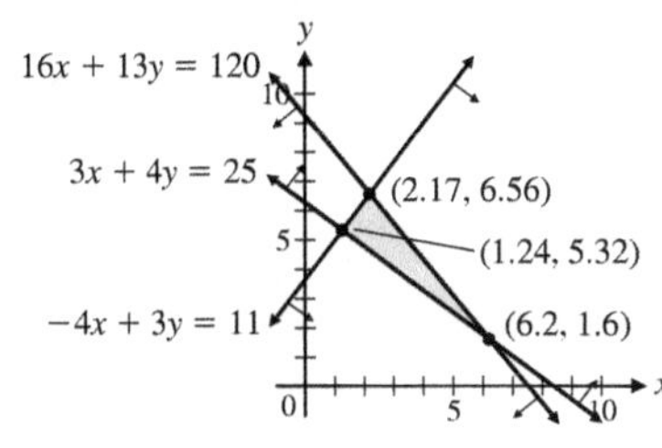

51. (A) $3x + 4y = 36$ and $3x + 2y = 30$ intersect at $(8, 3)$;
$3x + 4y = 36$ and $x = 0$ intersect at $(0, 9)$;
$3x + 4y = 36$ and $y = 0$ intersect at $(12, 0)$;
$3x + 2y = 30$ and $x = 0$ intersect at $(0, 15)$;
$3x + 2y = 30$ and $y = 0$ intersect at $(10, 0)$;
$x = 0$ and $y = 0$ intersect at $(0, 0)$
(B) $(8, 3), (0, 9), (10, 0), (0, 0)$

53.
$$\begin{aligned} 6x + 4y &\le 108 \\ x + y &\le 24 \\ x &\ge 0 \\ y &\ge 0 \end{aligned}$$

55. (A) All production schedules in the feasible region that are on the graph of $50x + 60y = 1{,}100$ will result in a profit of \$1,100.
(B) There are many possible choices. For example, producing 5 trick skis and 15 slalom skis will produce a profit of \$1,150. All the production schedules in the feasible region that are on the graph of $50x + 60y = 1{,}150$ will result in a profit of \$1,150.

57.
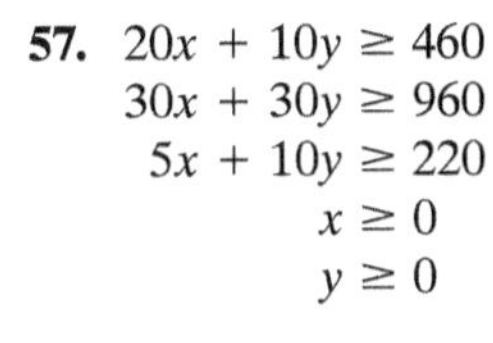
$$\begin{aligned} 20x + 10y &\ge 460 \\ 30x + 30y &\ge 960 \\ 5x + 10y &\ge 220 \\ x &\ge 0 \\ y &\ge 0 \end{aligned}$$

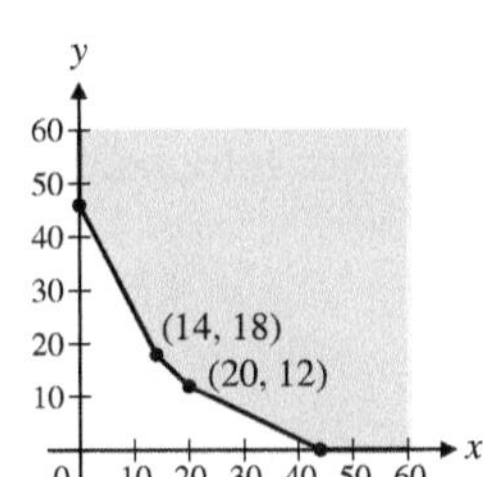

59.
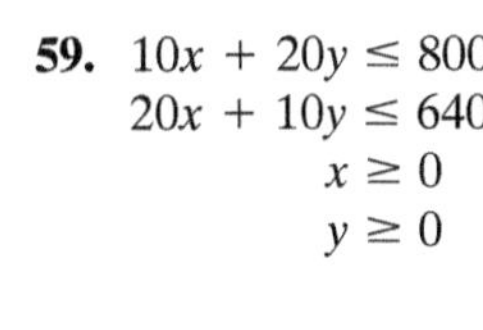
$$\begin{aligned} 10x + 20y &\le 800 \\ 20x + 10y &\le 640 \\ x &\ge 0 \\ y &\ge 0 \end{aligned}$$

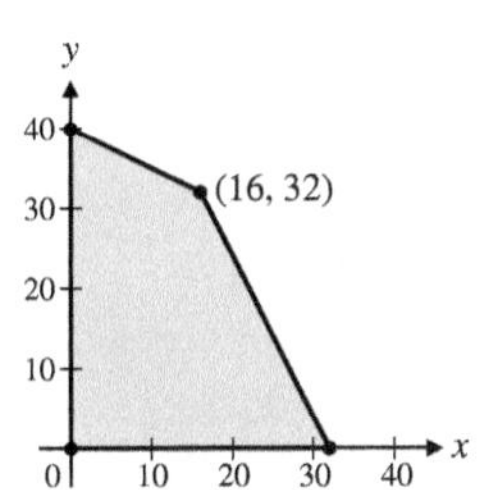

Exercise 11-2

1. Max $P = 16$ at $x_1 = 7$ and $x_2 = 9$
3. Max $P = 84$ at $x_1 = 7$ and $x_2 = 9$, at $x_1 = 0$ and $x_2 = 12$, and at every point on the line segment joining the preceding two points.
5. Min $C = 32$ at $x_1 = 0$ and $x_2 = 8$ **7.** Min $C = 36$ at $x_1 = 4$ and $x_2 = 3$ **9.** Max $P = 30$ at $x_1 = 4$ and $x_2 = 2$
11. Min $z = 14$ at $x_1 = 4$ and $x_2 = 2$; no max **13.** Max $P = 260$ at $x_1 = 2$ and $x_2 = 5$
15. Min $z = 140$ at $x_1 = 14$ and $x_2 = 0$; no max **17.** Min $P = 20$ at $x_1 = 0$ and $x_2 = 2$; Max $P = 150$ at $x_1 = 5$ and $x_2 = 0$
19. Feasible region empty; no optimal solutions
21. Min $P = 140$ at $x_1 = 3$ and $x_2 = 8$; Max $P = 260$ at $x_1 = 8$ and $x_2 = 10$, at $x_1 = 12$ and $x_2 = 2$, or at any point on the line segment from $(8, 10)$ to $(12, 2)$
23. Max $P = 26{,}000$ at $x_1 = 400$ and $x_2 = 600$ **25.** Max $P = 5{,}507$ at $x_1 = 6.62$ and $x_2 = 4.25$
27. Max $z = 2$ at $x_1 = 4$ and $x_2 = 2$; Min z does not exist
29. (A) $2a < b$ (B) $\frac{1}{3}a < b < 2a$ (C) $b < \frac{1}{3}a$ (D) $b = 2a$ (E) $b = \frac{1}{3}a$
31. (A) Let: x_1 = number of trick skis
x_2 = number of slalom skis produced per day.
Maximize $P = 40x_1 + 30x_2$
subject to
$$\begin{aligned} 6x_1 + 4x_2 &\le 108 \\ x_1 + x_2 &\le 24 \\ x_1 \ge 0, x_2 &\ge 0 \end{aligned}$$
Max profit = \$780 when 6 trick skis and 18 slalom skis are produced.
(B) Max profit decreases to \$720 when 18 trick skis and no slalom skis are produced.
(C) Max profit increases to \$1,080 when no trick skis and 24 slalom skis are produced.

33. (A) Let x_1 = number of days to operate plant A
x_2 = number of days to operate plant B
Minimize $C = 1000x_1 + 900x_2$
subject to $20x_1 + 25x_2 \geq 200$
$60x_1 + 50x_2 \geq 500$
$x_1 \geq 0, x_2 \geq 0$
Plant A: 5 days; Plant B: 4 days; min cost \$8,600
(B) Plant A: 10 days; Plant B: 0 days; min cost \$6,000
(C) Plant A: 0 days; Plant B: 10 days; min cost \$8,000

35. Let x_1 = number of buses
x_2 = number of vans
Minimize $C = 1200x_1 + 100x_2$
subject to $40x_1 + 8x_2 \geq 400$
$3x_1 + x_2 \leq 36$
$x_1 \geq 0, x_2 \geq 0$
7 buses, 15 vans; min cost \$9,900

37. Let x_1 = amount invested in the CD
x_2 = amount invested in the mutual fund
Maximize $P = 0.05x_1 + 0.09x_2$
subject to $x_1 + x_2 \leq 60{,}000$
$x_2 \geq 10{,}000$
$x_1 \geq 2x_2$
$x_1, x_2 \geq 0$
\$40,000 in the CD and \$20,000 in the mutual fund; max return is \$3,800

39. (A) Let
x_1 = number of gallons produced by the old process

x_2 = number of gallons produced by the new process
Maximize $P = 60x_1 + 20x_2$
subject to $20x_1 + 5x_2 \leq 16{,}000$
$40x_1 + 20x_2 \leq 30{,}000$
$x_1 \geq 0, x_2 \geq 0$
Max P = \$450 when 750 gal are produced using the old process exclusively.
(B) Max P = \$380 when 400 gal are produced using the old process and 700 gal are produced using the new process.
(C) Max P = \$288 when 1,440 gal are produced using the new process exclusively.

41. (A) Let x_1 = number of bags of brand A
x_2 = number of bags of brand B
Maximize $N = 8x_1 + 3x_2$
subject to $4x_1 + 4x_2 \geq 1000$
$2x_1 + x_2 \leq 400$
$x_1 \geq 0, x_2 \geq 0$
150 bags brand A, 100 bags brand B; max nitrogen 1,500 lb
(B) 0 bags brand A, 250 bags brand B; min nitrogen 750 lb

43. Let x_1 = number of cubic yards of mix A
x_2 = number of cubic yards of mix B
Minimize $C = 30x_1 + 35x_2$
subject to $20x_1 + 10x_2 \geq 460$
$30x_1 + 30x_2 \geq 960$
$5x_1 + 10x_2 \geq 220$
$x_1 \geq 0, x_2 \geq 0$
20 yd^3 A, 12 yd^3 B; \$1,020

45. Let x_1 = number of mice used
x_2 = number of rats used
Maximize $P = x_1 + x_2$
subject to $10x_1 + 20x_2 \leq 800$
$20x_1 + 10x_2 \leq 640$
$x_1 \geq 0, x_2 \geq 0$
48; 16 mice, 32 rats

Exercise 11-3

1. (A) 2 (B) 2 basic and 3 nonbasic variables (C) 2 linear equations with 2 variables
3. (A) 5
(B) 4
(C) 5 basic and 4 nonbasic variables
(D) 5 linear equations with 5 variables

5.

	Nonbasic	Basic	Feasible?
(A)	x_1, x_2	s_1, s_2	Yes
(B)	x_1, s_1	x_2, s_2	Yes
(C)	x_1, s_2	x_2, s_1	No
(D)	x_2, s_1	x_1, s_2	No
(E)	x_2, s_2	x_1, s_1	Yes
(F)	s_1, s_2	x_1, x_2	Yes

7.

	x_1	x_2	s_1	s_2	Feasible?
(A)	0	0	50	40	Yes
(B)	0	50	0	−60	No
(C)	0	20	30	0	Yes
(D)	25	0	0	15	Yes
(E)	40	0	−30	0	No
(F)	20	10	0	0	Yes

9.

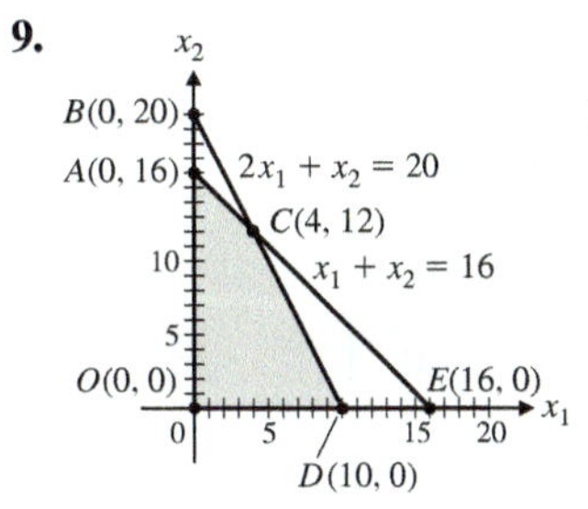

$$\begin{aligned} x_1 + x_2 + s_1 \qquad &= 16 \\ 2x_1 + x_2 \qquad + s_2 &= 20 \end{aligned}$$

x_1	x_2	s_1	s_2	Intersection Point	Feasible?
0	0	16	20	O	Yes
0	16	0	4	A	Yes
0	20	−4	0	B	No
16	0	0	−12	E	No
10	0	6	0	D	Yes
4	12	0	0	C	Yes

11.

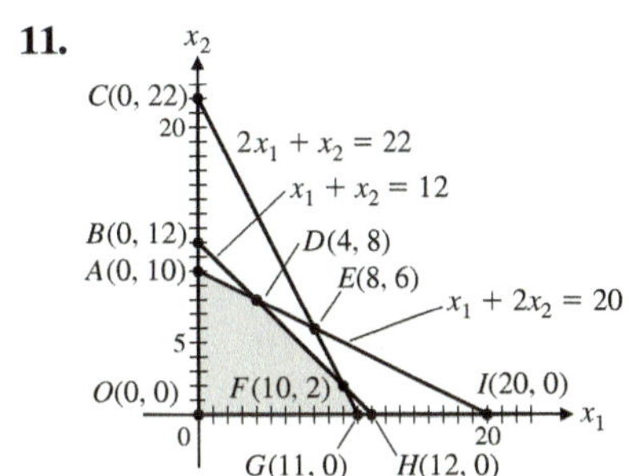

$$\begin{aligned} 2x_1 + x_2 + s_1 \qquad\qquad &= 22 \\ x_1 + x_2 \qquad + s_2 \qquad &= 12 \\ x_1 + 2x_2 \qquad\qquad + s_3 &= 20 \end{aligned}$$

x_1	x_2	s_1	s_2	s_3	Intersection Point	Feasible?
0	0	22	12	20	O	Yes
0	22	0	−10	−24	C	No
0	12	10	0	−4	B	No
0	10	12	2	0	A	Yes
11	0	0	1	9	G	Yes
12	0	−2	0	8	H	No
20	0	−18	−8	0	I	No
10	2	0	0	6	F	Yes
8	6	0	−2	0	E	No
4	8	6	0	0	D	Yes

Exercise 11-4

1. (A) Basic: x_2, s_1, P; nonbasic: x_1, s_2 (B) $x_1 = 0, x_2 = 12, s_1 = 15, s_2 = 0, P = 20$ (C) Additional pivot required

3. (A) Basic: x_2, x_3, s_3, P; nonbasic: x_1, s_1, s_2 (B) $x_1 = 0, x_2 = 15, x_3 = 5, s_1 = 0, s_2 = 0, s_3 = 12, P = 45$
(C) No optimal solution

5. Enter ↓ x_1; Exit → s_1

$$\begin{array}{c} \\ s_1 \\ s_2 \\ P \end{array} \left[\begin{array}{ccccc|c} x_1 & x_2 & s_1 & s_2 & P & \\ \textcircled{1} & 4 & 1 & 0 & 0 & 4 \\ 3 & 5 & 0 & 1 & 0 & 24 \\ \hline -8 & -5 & 0 & 0 & 1 & 0 \end{array}\right] \sim \begin{array}{c} x_1 \\ s_2 \\ P \end{array} \left[\begin{array}{ccccc|c} 1 & 4 & 1 & 0 & 0 & 4 \\ 0 & -7 & -3 & 1 & 0 & 12 \\ \hline 0 & 27 & 8 & 0 & 1 & 32 \end{array}\right]$$

7. Enter ↓ x_1; Exit → x_2

$$\begin{array}{c} \\ x_2 \\ s_2 \\ s_3 \\ P \end{array} \left[\begin{array}{cccccc|c} x_1 & x_2 & s_1 & s_2 & s_3 & P & \\ \textcircled{2} & 1 & 1 & 0 & 0 & 0 & 4 \\ 3 & 0 & 1 & 1 & 0 & 0 & 8 \\ 0 & 0 & 2 & 0 & 1 & 0 & 2 \\ \hline -4 & 0 & -3 & 0 & 0 & 1 & 5 \end{array}\right] \sim \begin{array}{c} x_1 \\ s_2 \\ s_3 \\ P \end{array} \left[\begin{array}{cccccc|c} 1 & \frac{1}{2} & \frac{1}{2} & 0 & 0 & 0 & 2 \\ 0 & -\frac{3}{2} & -\frac{1}{2} & 1 & 0 & 0 & 2 \\ 0 & 0 & 2 & 0 & 1 & 0 & 2 \\ \hline 0 & 2 & -1 & 0 & 0 & 1 & 13 \end{array}\right]$$

9. (A)

$$\begin{aligned} 2x_1 + x_2 + s_1 \qquad\qquad &= 10 \\ x_1 + 3x_2 \qquad + s_2 \qquad &= 10 \\ -15x_1 - 10x_2 \qquad\qquad + P &= 0 \end{aligned}$$

(B) Enter ↓ x_1; Exit → s_1

$$\begin{array}{c} \\ s_1 \\ s_2 \\ P \end{array} \left[\begin{array}{ccccc|c} x_1 & x_2 & s_1 & s_2 & P & \\ \textcircled{2} & 1 & 1 & 0 & 0 & 10 \\ 1 & 3 & 0 & 1 & 0 & 10 \\ \hline -15 & -10 & 0 & 0 & 1 & 0 \end{array}\right]$$

(C) Max $P = 80$ at $x_1 = 4$ and $x_2 = 2$

11. (A)

$$\begin{aligned} 2x_1 + x_2 + s_1 \qquad\qquad &= 10 \\ x_1 + 3x_2 \qquad + s_2 \qquad &= 10 \\ -30x_1 - x_2 \qquad\qquad + P &= 0 \end{aligned}$$

(B) Enter ↓ x_1; Exit → s_1

	x_1	x_2	s_1	s_2	P	
s_1	②	1	1	0	0	10
s_2	1	3	0	1	0	10
P	−30	−1	0	0	1	0

(C) Max $P = 150$ at $x_1 = 5$ and $x_2 = 0$

13. Max $P = 260$ at $x_1 = 2$ and $x_2 = 5$
15. No optimal solution exists.
17. Max $P = 7$ at $x_1 = 3$ and $x_2 = 5$
19. Max $P = 58$ at $x_1 = 12$, $x_2 = 0$, and $x_3 = 2$
21. Max $P = 17$ at $x_1 = 4$, $x_2 = 3$, and $x_3 = 0$
23. Max $P = 22$ at $x_1 = 1$, $x_2 = 6$, and $x_3 = 0$
25. Max $P = 26{,}000$ at $x_1 = 400$ and $x_2 = 600$
27. Max $P = 450$ at $x_1 = 0$, $x_2 = 180$, and $x_3 = 30$

29. Max $P = 88$ at $x_1 = 24$ and $x_2 = 8$

31. No solution **33.** Choosing either column produces the same optimal solution: max $P = 13$ at $x_1 = 3$ and $x_2 = 10$.
35. Choosing column 1: max $P = 60$ at $x_1 = 12$, $x_2 = 8$, and $x_3 = 0$. Choosing column 2: max $P = 60$ at $x_1 = 0$, $x_2 = 20$, and $x_3 = 0$.

37. Let x_1 = number of *A* components
x_2 = number of *B* components
x_3 = number of *C* components

Maximize $P = 7x_1 + 8x_2 + 10x_3$
subject to
$$\begin{aligned} 2x_1 + 3x_2 + 2x_3 &\le 1{,}000 \\ x_1 + x_2 + 2x_3 &\le 800 \\ x_1, x_2, x_3 &\ge 0 \end{aligned}$$

200 *A* components, 0 *B* components, and 300 *C* components; max profit is $4,400

39. Let x_1 = amount invested in government bonds
x_2 = amount invested in mutual funds
x_3 = amount invested in money market funds

Maximize $P = 0.08x_1 + 0.13x_2 + 0.15x_3$
subject to
$$\begin{aligned} x_1 + x_2 + x_3 &\le 100{,}000 \\ -x_1 + x_2 + x_3 &\le 0 \\ x_1, x_2, x_3 &\ge 0 \end{aligned}$$

$50,000 in government bonds, $0 in mutual funds, and $50,000 in money market funds; max return is $11,500

41. Let x_1 = number of ads placed in daytime shows
x_2 = number of ads placed in prime-time shows
x_3 = number of ads placed in late-night shows

Maximize $P = 14{,}000x_1 + 24{,}000x_2 + 18{,}000x_3$
subject to
$$\begin{aligned} x_1 + x_2 + x_3 &\le 15 \\ 1{,}000x_1 + 2{,}000x_2 + 1{,}500x_3 &\le 20{,}000 \\ x_1, x_2, x_3 &\ge 0 \end{aligned}$$

10 daytime ads, 5 prime-time ads, and 0 late-night ads; max number of potential customers is 260,000

43. (A) Let x_1 = number of colonial houses
x_2 = number of split-level houses
x_3 = number of ranch houses

Maximize $P = 20{,}000x_1 + 18{,}000x_2 + 24{,}000x_3$
subject to
$$\begin{aligned} \tfrac{1}{2}x_1 + \tfrac{1}{2}x_2 + x_3 &\le 30 \\ 60{,}000x_1 + 60{,}000x_2 + 80{,}000x_3 &\le 3{,}200{,}000 \\ 4{,}000x_1 + 3{,}000x_2 + 4{,}000x_3 &\le 180{,}000 \\ x_1, x_2, x_3 &\ge 0 \end{aligned}$$

20 colonial, 20 split-level, and 10 ranch houses; max profit is $1,000,000

(B) 0 colonial, 40 split-level, and 10 ranch houses; max profit is $960,000; 20,000 labor-hours are not used

(C) 45 colonial, 0 split-level, and 0 ranch houses; max profit is $1,125,000; 7.5 acres of land and $500,000 of capital are not used

45. (A) Let x_1 = number of boxes of assortment I
x_2 = number of boxes of assortment II
x_3 = number of boxes of assortment III

Maximize $P = 4x_1 + 3x_2 + 5x_3$
subject to
$$\begin{aligned} 4x_1 + 12x_2 + 8x_3 &\le 4{,}800 \\ 4x_1 + 4x_2 + 8x_3 &\le 4{,}000 \\ 12x_1 + 4x_2 + 8x_3 &\le 5{,}600 \\ x_1, x_2, x_3 &\ge 0 \end{aligned}$$

200 boxes of assortment I, 100 boxes of assortment II, and 350 boxes of assortment III; max profit is $2,850

(B) 100 boxes of assortment I, 0 boxes of assortment II, and 550 boxes of assortment III; max profit is $3,150; 200 fruit-filled candies are not used

(C) 0 boxes of assortment I, 50 boxes of assortment II, and 675 boxes of assortment III; max profit is $3,525; 400 fruit-filled candies are not used

47. Let x_1 = number of grams of food A
x_2 = number of grams of food B
x_3 = number of grams of food C

Maximize $P = 3x_1 + 4x_2 + 5x_3$
subject to
$$x_1 + 3x_2 + 2x_3 \le 30$$
$$2x_1 + x_2 + 2x_3 \le 24$$
$$x_1, x_2, x_3 \ge 0$$

0 g food A, 3 g food B, and 10.5 g food C; max protein is 64.5 units

49. Let x_1 = number of undergraduate students
x_2 = number of graduate students
x_3 = number of faculty members

Maximize $P = 18x_1 + 25x_2 + 30x_3$
subject to
$$x_1 + x_2 + x_3 \le 20$$
$$100x_1 + 150x_2 + 200x_3 \le 3{,}200$$
$$x_1, x_2, x_3 \ge 0$$

0 undergraduate students, 16 graduate students, and 4 faculty members; max number of interviews is 520

Appendix A

Exercises A-1

1. vu **3.** $(3 + 7) + y$ **5.** $u + v$ **7.** T **9.** T **11.** F **13.** T **15.** T **17.** T **19.** T **21.** F **23.** T **25.** T **27.** No **29.** (A) F (B) T (C) T **31.** $\sqrt{2}$ and π are two examples of infinitely many. **33.** (A) N, Z, Q, R (B) R (C) Q, R (D) Q, R **35.** (A) F, since, for example, $2(3 - 1) \ne 2 \cdot 3 - 1$ (B) F, since, for example, $(8 - 4) - 2 \ne 8 - (4 - 2)$ (C) T (D) F, since, for example, $(8 \div 4) \div 2 \ne 8 \div (4 \div 2)$. **37.** $\frac{1}{11}$ **39.** (A) 2.166 666 666... (B) 4.582 575 69... (C) 0.437 500 000... (D) 0.261 261 261...

Exercises A-2

1. 3 **3.** $x^3 + 4x^2 - 2x + 5$ **5.** $x^3 + 1$ **7.** $2x^5 + 3x^4 - 2x^3 + 11x^2 - 5x + 6$ **9.** $-5u + 2$ **11.** $6a^2 + 6a$ **13.** $a^2 - b^2$ **15.** $6x^2 - 7x - 5$ **17.** $2x^2 + xy - 6y^2$ **19.** $9y^2 - 4$ **21.** $-4x^2 + 12x - 9$ **23.** $16m^2 - 9n^2$ **25.** $9u^2 + 24uv + 16v^2$ **27.** $a^3 - b^3$ **29.** $x^2 - 2xy + y^2 - 9z^2$ **31.** 1 **33.** $x^4 - 2x^2y^2 + y^4$ **35.** $-40ab$ **37.** $-4m + 8$ **39.** $-6xy$ **41.** $u^3 + 3u^2v + 3uv^2 + v^3$ **43.** $x^3 - 6x^2y + 12xy^2 - 8y^3$ **45.** $2x^2 - 2xy + 3y^2$ **47.** $x^4 - 10x^3 + 27x^2 - 10x + 1$ **49.** $4x^3 - 14x^2 + 8x - 6$ **51.** $m + n$ **53.** No change **55.** $(1 + 1)^2 \ne 1^2 + 1^2$; either a or b must be 0 **57.** $0.09x + 0.12(10{,}000 - x) = 1{,}200 - 0.03x$ **59.** $20x + 30(3x) + 50(4{,}000 - x - 3x) = 200{,}000 - 90x$ **61.** $0.02x + 0.06(10 - x) = 0.6 - 0.04x$

Exercises A-3

1. $3m^2(2m^2 - 3m - 1)$ **3.** $2uv(4u^2 - 3uv + 2v^2)$ **5.** $(7m + 5)(2m - 3)$ **7.** $(4ab - 1)(2c + d)$ **9.** $(2x - 1)(x + 2)$ **11.** $(y - 1)(3y + 2)$ **13.** $(x + 4)(2x - 1)$ **15.** $(w + x)(y - z)$ **17.** $(a - 3b)(m + 2n)$ **19.** $(3y + 2)(y - 1)$ **21.** $(u - 5v)(u + 3v)$ **23.** Not factorable **25.** $(wx - y)(wx + y)$ **27.** $(3m - n)^2$ **29.** Not factorable **31.** $4(z - 3)(z - 4)$ **33.** $2x^2(x - 2)(x - 10)$ **35.** $x(2y - 3)^2$ **37.** $(2m - 3n)(3m + 4n)$ **39.** $uv(2u - v)(2u + v)$ **41.** $2x(x^2 - x + 4)$ **43.** $(2x - 3y)(4x^2 + 6xy + 9y^2)$ **45.** $xy(x + 2)(x^2 - 2x + 4)$ **47.** $[(x + 2) - 3y][(x + 2) + 3y]$ **49.** Not factorable **51.** $(6x - 6y - 1)(x - y + 4)$ **53.** $(y - 2)(y + 2)(y^2 + 1)$ **55.** $3(x - y)^2(5xy - 5y^2 + 4x)$ **57.** True **59.** False

Exercises A-4

1. $8d^6$ **3.** $\dfrac{15x^2 + 10x - 6}{180}$ **5.** $\dfrac{15m^2 + 14m - 6}{36m^3}$ **7.** $\dfrac{1}{x(x - 4)}$ **9.** $\dfrac{x - 6}{x(x - 3)}$ **11.** $\dfrac{-3x - 9}{(x - 2)(x + 1)^2}$ **13.** $\dfrac{2}{x - 1}$ **15.** $\dfrac{5}{a - 1}$ **17.** $\dfrac{x^2 + 8x - 16}{x(x - 4)(x + 4)}$ **19.** $\dfrac{7x^2 - 2x - 3}{6(x + 1)^2}$ **21.** $\dfrac{x(y - x)}{y(2x - y)}$ **23.** $\dfrac{-17c + 16}{15(c - 1)}$ **25.** $\dfrac{1}{x - 3}$ **27.** $\dfrac{-1}{2x(x + h)}$ **29.** $\dfrac{x - y}{x + y}$ **31.** (A) Incorrect (B) $x + 1$ **33.** (A) Incorrect (B) $2x + h$ **35.** (A) Incorrect (B) $\dfrac{x^2 - x - 3}{x + 1}$ **37.** (A) Correct **39.** $\dfrac{-2x - h}{3(x + h)^2x^2}$ **41.** $\dfrac{x(x - 3)}{x - 1}$

Exercises A-5

1. $2/x^9$ **3.** $3w^7/2$ **5.** $2/x^3$ **7.** $1/w^5$ **9.** $4/a^6$ **11.** $1/a^6$ **13.** $1/8x^{12}$ **15.** 8.23×10^{10} **17.** 7.83×10^{-1} **19.** 3.4×10^{-5} **21.** 40,000 **23.** 0.007 **25.** 61,710,000 **27.** 0.000 808 **29.** 1 **31.** 10^{14} **33.** $y^6/25x^4$ **35.** $4x^6/25$ **37.** $4y^3/3x^5$ **39.** $\frac{7}{4} - \frac{1}{4}x^{-3}$ **41.** $\frac{5}{2}x^2 - \frac{3}{2} + 4x^{-2}$ **43.** $\dfrac{x^2(x - 3)}{(x - 1)^3}$ **45.** $\dfrac{2(x - 1)}{x^3}$ **47.** 2.4×10^{10}; 24,000,000,000 **49.** 3.125×10^4; 31,250 **51.** 64 **55.** uv **57.** $\dfrac{bc(c + b)}{c^2 + bc + b^2}$ **59.** (A) \$32,977 (B) \$1,484 (C) 4.50% **61.** (A) 9×10^{-6} (B) 0.000 009 (C) 0.0009% **63.** 1,417,000

Exercises A-6

1. $6\sqrt[5]{x^3}$ **3.** $\sqrt[5]{(32x^2y^3)^3}$ **5.** $\sqrt{x^2 + y^2}$ (not $x + y$) **7.** $5x^{3/4}$ **9.** $(2x^2y)^{3/5}$ **11.** $x^{1/3} + y^{1/3}$ **13.** 5 **15.** 64 **17.** −7 **19.** −16
21. $\frac{8}{125}$ **23.** $\frac{1}{27}$ **25.** $x^{2/5}$ **27.** m **29.** $2x/y^2$ **31.** $xy^2/2$ **33.** $1/(24x^{7/12})$ **35.** $2x + 3$ **37.** $30x^5\sqrt{3x}$ **39.** 2 **41.** $12x - 6x^{35/4}$
43. $3u - 13u^{1/2}v^{1/2} + 4v$ **45.** $36m^{3/2} - \frac{6m^{1/2}}{n^{1/2}} + \frac{6m}{n^{1/2}} - \frac{1}{n}$ **47.** $9x - 6x^{1/2}y^{1/2} + y$ **49.** $\frac{1}{2}x^{1/3} + x^{-1/3}$ **51.** $\frac{2}{3}x^{-1/4} + \frac{1}{3}x^{-2/3}$
53. $\frac{1}{2}x^{-1/6} - \frac{1}{4}$ **55.** $4n\sqrt{3mn}$ **57.** $\frac{2(x + 3)\sqrt{x - 2}}{x - 2}$ **59.** $7(x - y)(\sqrt{x} + \sqrt{y})$ **61.** $\frac{1}{xy\sqrt{5xy}}$ **63.** $\frac{1}{\sqrt{x + h} + \sqrt{x}}$
65. $\frac{1}{(t + x)(\sqrt{t} + \sqrt{x})}$ **67.** $x = y = 1$ is one of many choices. **69.** $x = y = 1$ is one of many choices.
71. False **73.** False **75.** False **77.** True **79.** True **81.** False **83.** $\frac{x + 8}{2(x + 3)^{3/2}}$ **85.** $\frac{x - 2}{2(x - 1)^{3/2}}$ **87.** $\frac{x + 6}{3(x + 2)^{5/3}}$
89. 103.2 **91.** 0.0805 **93.** 4,588 **95.** (A) and (E); (B) and (F); (C) and (D)

Exercises A-7

1. $\pm\sqrt{11}$ **3.** $-\frac{4}{3}, 2$ **5.** −2, 6 **7.** 0, 2 **9.** $3 \pm 2\sqrt{3}$ **11.** $-2 \pm \sqrt{2}$ **13.** $0, \frac{15}{2}$ **15.** $\pm\frac{3}{2}$ **17.** $\frac{1}{2}, -3$ **19.** $(-1 \pm \sqrt{5})/2$
21. $(3 \pm \sqrt{3})/2$ **23.** No real solution **25.** $(-3 \pm \sqrt{11})/2$ **27.** $\pm\sqrt{3}$ **29.** $-\frac{1}{2}, 2$ **31.** $(x - 2)(x + 42)$
33. Not factorable in the integers **35.** $(2x - 9)(x + 12)$ **37.** $(4x - 7)(x + 62)$ **39.** $r = \sqrt{A/P} - 1$
41. If $c < 4$, there are two distinct real roots; if $c = 4$, there is one real double root; and if $c > 4$, there are no real roots.
43. 1,575 bottles at $4 each **45.** 13.64% **47.** 8 ft/sec; $4\sqrt{2}$ or 5.66 ft/sec

Appendix B

Exercises B-1

1. 5, 7, 9, 11 **3.** $\frac{3}{2}, \frac{4}{3}, \frac{5}{4}, \frac{6}{5}$ **5.** 9, −27, 81, −243 **7.** 23 **9.** $\frac{101}{100}$ **11.** $1 + 2 + 3 + 4 + 5 + 6 = 21$ **13.** $5 + 7 + 9 + 11 = 32$
15. $1 + \frac{1}{10} + \frac{1}{100} + \frac{1}{1{,}000} = \frac{1{,}111}{1{,}000}$ **17.** 3.6 **19.** 82.5 **21.** $\frac{1}{2}, -\frac{1}{4}, \frac{1}{8}, -\frac{1}{16}, \frac{1}{32}$ **23.** 0, 4, 0, 8, 0 **25.** $1, -\frac{3}{2}, \frac{9}{4}, -\frac{27}{8}, \frac{81}{16}$ **27.** $a_n = n - 3$
29. $a_n = 4n$ **31.** $a_n = (2n - 1)/2n$ **33.** $a_n = (-1)^{n+1}n$ **35.** $a_n = (-1)^{n+1}(2n - 1)$ **37.** $a_n = \left(\frac{2}{5}\right)^{n-1}$ **39.** $a_n = x^n$
41. $a_n = (-1)^{n+1}x^{2n-1}$ **43.** $1 - 9 + 25 - 49 + 81$ **45.** $\frac{4}{7} + \frac{8}{9} + \frac{16}{11} + \frac{32}{13}$ **47.** $1 + x + x^2 + x^3 + x^4$ **49.** $x - \frac{x^3}{3} + \frac{x^5}{5} - \frac{x^7}{7} + \frac{x^9}{9}$
51. (A) $\sum_{k=1}^{5}(k + 1)$ (B) $\sum_{j=0}^{4}(j + 2)$ **53.** (A) $\sum_{k=1}^{4}\frac{(-1)^{k+1}}{k}$ (B) $\sum_{j=0}^{3}\frac{(-1)^j}{j + 1}$ **55.** $\sum_{k=1}^{n}\frac{k + 1}{k}$ **57.** $\sum_{k=1}^{n}\frac{(-1)^{k+1}}{2^k}$ **59.** False **61.** True
63. 2, 8, 26, 80, 242 **65.** 1, 2, 4, 8, 16 **67.** $1, \frac{3}{2}, \frac{17}{12}, \frac{577}{408}$; $a_4 = \frac{577}{408} \approx 1.414\,216$, $\sqrt{2} \approx 1.414\,214$ **69.** 1, 1, 2, 3, 5, 8, 13, 21, 34, 55

Exercises B-2

1. (A) Arithmetic, with $d = -5$; −26, −31 (B) Geometric, with $r = -2$; −16, 32 (C) Neither (D) Geometric, with $r = \frac{1}{3}$; $\frac{1}{54}, \frac{1}{162}$
3. Geometric; 1 **5.** Neither **7.** Arithmetic; 127.5 **9.** $a_2 = 11, a_3 = 15$ **11.** $a_{21} = 82, S_{31} = 1{,}922$ **13.** $S_{20} = 930$
15. $a_2 = -6, a_3 = 12, a_4 = -24$ **17.** $S_7 = 547$ **19.** $a_{10} = 199.90$ **21.** $r = 1.09$ or -1.09 **23.** $S_{10} = 1{,}242, S_\infty = 1{,}250$
25. 2,706 **27.** −85 **29.** 1,120 **31.** (A) Does not exist (B) $S_\infty = \frac{8}{5} = 1.6$ **33.** 2,400 **35.** 0.999
37. Use $a_1 = 1$ and $d = 2$ in $S_n = (n/2)[2a_1 + (n - 1)d]$. **39.** $S_n = na_1$ **41.** No **43.** Yes
45. $48 + $46 + ⋯ + $4 + $2 = $600 **47.** About $11,670,000 **49.** $1,628.89; $2,653.30

Exercises B-3

1. 720 **3.** 10 **5.** 1,320 **7.** 10 **9.** 6 **11.** 1,140 **13.** 10 **15.** 6 **17.** 1 **19.** 816
21. $C_{4,0}a^4 + C_{4,1}a^3b + C_{4,2}a^2b^2 + C_{4,3}ab^3 + C_{4,4}b^4 = a^4 + 4a^3b + 6a^2b^2 + 4ab^3 + b^4$ **23.** $x^6 - 6x^5 + 15x^4 - 20x^3 + 15x^2 - 6x + 1$
25. $32a^5 - 80a^4b + 80a^3b^2 - 40a^2b^3 + 10ab^4 - b^5$ **27.** $3{,}060x^{14}$ **29.** $5{,}005p^9q^6$ **31.** $264x^2y^{10}$
33. $C_{n,0} = \frac{n!}{0!\,n!} = 1$; $C_{n,n} = \frac{n!}{n!\,0!} = 1$ **35.** 1 5 10 10 5 1; 1 6 15 20 15 6 1

SUBJECT INDEX

Note: Page numbers followed by fn refer to footnotes.

E

F

INDEX OF APPLICATIONS

Business & Economics

Life Sciences

Social Sciences